Lecture Notes in Artificial Intelligence 12278

Subseries of Lecture Notes in Computer Science

More information about this subseries at http://www.springer.com/series/1244

Meichun Liu · Chunyu Kit ·
Qi Su (Eds.)

Chinese Lexical Semantics

21st Workshop, CLSW 2020
Hong Kong, China, May 28–30, 2020
Revised Selected Papers

Springer

Editors
Meichun Liu
Department of Linguistics and Translation
City University of Hong Kong
Hong Kong SAR, China

Chunyu Kit
Department of Linguistics and Translation
City University of Hong Kong
Hong Kong SAR, China

Qi Su
School of Foreign Languages
Peking University
Beijing, China

ISSN 0302-9743 ISSN 1611-3349 (electronic)
Lecture Notes in Artificial Intelligence
ISBN 978-3-030-81196-9 ISBN 978-3-030-81197-6 (eBook)
https://doi.org/10.1007/978-3-030-81197-6

LNCS Sublibrary: SL7 – Artificial Intelligence

This Springer imprint is published by the registered company Springer Nature Switzerland AG
The registered company address is: Gewerbestrasse 11, 6330 Cham, Switzerland

Preface

This volume presents a collection of pioneering works in Chinese Lexical Semantics in the age of big data. The volume is born out of the 21st Chinese Lexical Semantics Workshop (CLSW2020), a 3-day conference organized and hosted by the Department of Linguistics and Translation at the City University of Hong Kong, in the face of COVID-19 pandemic, during May 28–30, 2020. The CLSW Steering Committee, the Organizing Committee, and the hosting department with its strong tech support team managed to host the conference via Zoom Webinar with realtime presentations of keynote speeches and panel discussions, as well as live broadcasting via YouTube and DouYu. It aimed to reach as wide an audience as possible to offer a scholarly feast during the unusual period of the pandemic. As members the Organizing Committee, we have to say, it was indeed a challenge! But thanks to the collective efforts and excellent teamwork of a wonderful cohort of scholars, colleagues, students, and volunteers, we successfully launched a rich conference with record-high participation rates:

6 Keynote speakers
24 Invited speakers in four pre-conference sessions
71 Paper presentations
51 Poster presentations
900 Registered participants
3000+ Online viewers

The theme of CLSW2020 was "Lexical Semantics, Artificial Intelligence, and Cognitive Science", which aimed to provide an opportunity for researchers, academics, and practitioners in these three frontline areas to meet, to interact, and hopefully to innovate. The papers were also clustered along the three themes:

Lexical semantics and general linguistics: 30
AI, big data, and natural language processing: 33
Cognitive science and experimental studies: 11

In total, there are 74 papers included in the volume, which were chosen from the 71 oral presentations and 51 poster presentations selected for the conference out of the original 233 submissions. This amounts to an acceptance rate of about 30% for the publication. A figure which attests to the competitiveness and the quality control of the selection process for the papers included in the volume.

The volume was made possible thanks to the kind help of the large number of contributing scholars, including the CLSW Publication Committee, reviewers, authors, and editorial assistants. We are sincerely grateful to all the contributing members listed in the organization pages.

It is our hope that the volume will help advance the studies of Chinese Lexical Semantics in the age of big data and eventually lead to more innovative cross-disciplinary collaborations.

April 2021

Meichun Liu
Chunyu Kit
Qi Su

Organization

Steering Committee Honorary Members

Shiwen Yu	Peking University, China
Chin-Chuan Cheng	Taiwan Normal University, Taiwan
Ka Yin Benjamin T'sou	The Education University of Hong Kong, Hong Kong SAR, China

Steering Committee Members

Yanbin Diao	Beijing Normal University, China
Chu-Ren Huang	The Hong Kong Polytechnic University, Hong Kong SAR, China
Jia-Fei Hong	Taiwan Normal University, Taiwan
Shu-Kai Hsieh	Taiwan University, Taiwan
Jing-Schmidt Zhuo	University of Oregon, USA
Donghong Ji	Wuhan University, China
Peng Jin	Leshan Normal University, China
Kim Teng Lua	China and Oriental Language Information Processing Society, Singapore
Meichun Liu	City University of Hong Kong, Hong Kong SAR, China
Qin Lu	The Hong Kong Polytechnic University, Hong Kong SAR, China
Xinchun Su	Xiamen University, China
Zhifang Sui	Peking University, China
Jie Xu	University of Macau, Macao SAR, China
Jiun-Shiung Wu	Taiwan Chung Cheng University, Taiwan
Hongying Zan	Zhengzhou University, China
Yangsen Zhang	Beijing Information Science and Technology University, China

Program Committee Chairs

Weiguang Qu	Nanjing Normal University, China
Chunyu Kit	City University of Hong Kong, Hong Kong SAR, China
Jia-Fei Hong	Taiwan Normal University, Taiwan

Program Committee

Xiaojing Bai	Tsinghua University, China
Siaw-Fong Chung	Taiwan Chengchi University, Taiwan
Minghui Dong	Institute for Infocomm Research, Singapore
Hong Gao	Nanyang Technological University, Singapore
Shu-Ping Gong	Taiwan Chiayi University, Taiwan
Shu-Kai Hsieh	Taiwan University, Taiwan
Donghong Ji	Wuhan University, China
Minghu Jiang	Tsinghua University, China
Peng Jin	Leshan Normal University, China
Ahrens Kathleen	The Hong Kong Polytechnic University, Hong Kong SAR, China
Huei-Ling Lai	Taiwan Cheng Chi University, Taiwan
Baoli Li	Henan University of Technology, China
Shih-Min Li	Taiwan Academy for Educational Research, Taiwan
Jingxia Lin	Nanyang Technological University, Singapore
Yao Liu	Institute of Scientific and Technical Information of China, China
Maofu Liu	Wuhan University of Science and Technology, China
Pengyuan Liu	Beijing Language and Culture University, China
Chia-Rung Lu	Taiwan University, Taiwan
Wei-Yun Ma	Academia Sinica, Taiwan
Yao Meng	Fujitsu Research and Development Center, China
Haihua Pan	Chinese University of Hong Kong, Hong Kong SAR, China
Likun Qiu	Ludong University, China
Weiguang Qu	Nanjing Normal University, China
Xiaodong Shi	Xiamen University, China
Zuoyan Song	Beijing Normal University, China
Xinchun Su	Xiamen University, China
Zhifang Sui	Peking University, China
Le Sun	Institute of Software, Chinese Academy of Sciences, China
Lei Wang	Peking University, China
Kam-Fai Wong	The Chinese University of Hong Kong, Hong Kong SAR, China
Yunfang Wu	Peking University, China
Jun Xia	Shenyang Normal University, China
Jiajuan Xiong	Southwestern University of Finance and Economics, China
Jie Xu	University of Macau, Macao SAR, China
Ruifeng Xu	Harbin Institute of Technology, China
Hongzhi Xu	University of Pennsylvania, USA
Yulin Yuan	Peking University, China
Weidong Zhan	Peking University, China

Lei Zhang	Northeast Normal University, China
Yangsen Zhang	Beijing Information Science and Technology University, China
Zezhi Zheng	Xiamen University, China

Publication Committee Chairs

Meichun Liu	City University of Hong Kong, Hong Kong SAR, China
Chunyu Kit	City University of Hong Kong, Hong Kong SAR, China
Qi Su	Peking University, China

Editorial Assistants

Yingying Ye	City University of Hong Kong, Hong Kong SAR, China
Jiaming Yang	City University of Hong Kong, Hong Kong SAR, China

Contents

Lexical Semantics and General Linguistics

The Quasi Causativity of Chinese Preposition "由(by)"

Enxu Wang(✉) and Chi Wenqian

College of Chinese Language and Literature, Jinan University,
No. 336 West Nanxinzhuang Road, Shizhong District, Jinan 250022, China

Abstract. Opinions regarding the functions of the agent-introducer "由(by)" fall into two types: that it is a passive marker and that it is a functional category. Neither of these two can capture the essence of "由(by)" constructions well. Based on on our corpus investigations, we propose that the syntactic status of "由(by)" as an agent-introducer should be analyzed as a quasi causative preposition, which can account for the source of the agent introduced by "由(by)" and the relationships between this usage and its other usages.

Keywords: "由(by)" · Passive marker · Quasi causativity (准使役性)

1 Introduction

Compared with other usages of the Chinese preposition "由(by)", the case where it introduces an agent into a sentence, such as "院长由国务院提名 (the president is nominated by the State Council)", has attracted more attention. This might be because "由(by)" can be translated into the English passive marker "by" like the Chinese passive marker " 被(by)". Therefore, earlier studies regarded it as a passive marker [1]. With the advance of relevant research, the passive marker status of "由(by)" is questioned by more and more scholars. A large number of studies have shown that there are huge differences between "由 (by)" and "被 (by)" in terms of syntactic functions, information focus, pragmatic tendency, tense and aspect characteristics, etc. [2, 3] Besides, most "由(by)" sentences are not passive sentences at all from the perspectives of agents, patients, predicates, pre- and post-components, adverbial components, grammatical functions, and pragmatic restrictions, even though there are some similarities between these two types of construction [4].

To avoid the above problems, reference [5] proposes that "由(by)" should be analyzed as a functional category. However, this proposal incurs some new problems. For instance, it breaks the relationship between the usage of "由(by)" that introduces an agent and other usages. It cannot capture the source of the agent that is introduced by "由(by)". Nor can it give a unified interpretation of this usage and other usages of "由(by)".

According to our investigation, the "由(by)" that introduces an agent is not a passive marker or a functional category, but a quasi causative preposition. This paper is organized as follows. Section 2 will discuss why the "由(by)" has causativity (使役性). Section 3 will analyze why it is quasi causativity (准使役性). The last section will focus on the advantages of the quasi causative preposition analysis.

M. Liu et al. (Eds.): CLSW 2020, LNAI 12278, pp. 3–13, 2021.
https://doi.org/10.1007/978-3-030-81197-6_1

2 Why “由(by)” Has Causativity(使役性)?

Previous studies have noted the causativity of “由(by)”. It is noted in [4:28] that a “由(by)” sentence can express a tone of advice, consultation, request, etc. It is pointed out in [3:43] that the agent behavior of “由(by)” sentences is not completely autonomous, but is usually aroused under someone‘s demands, persuasions, inducements, or triggered by specific occasions and situations. There are two types of “由(by)” sentences, namely, directive and declarative. The former type is used to give instructions, while the latter type describes the instructions or the completion of the instructions [6]. These studies do not solve the problems mentioned above, but they are consistent with our language intuition and can be a key to the problems in the study of “由(by)” sentences.

The main function of prepositions is to introduce optional arguments of predicates or clauses [7, 8]. The main purpose of the semantic structure of a preposition like “preposition + optional argument + predicate” is to find out what semantic components are included and what relationships exist among them [9]. Taking “从(from)” in “从北京出发 (departing from Beijing)” as an example, the semantic components of “从(from)” include “北京(Beijing)” and “出发(depart)”, and the semantic relationship between these components is “starting point – movement”. Unlike “从(from)”, the function of the preposition “由(by)” is more complicated. In addition to the basic function of “introducing optional arguments”, there is also a special function of “introduceing obligatory arguments”. Further details are as follows:

(1) The two functions of "由(by)" and its semantic structure.

a. **Basic function**: "由(by)" + optional argument + predicate, such as "由小路走 (moving by the pathway)"
b. **Causative function**: (causative force)...+ “由(by)”+ obligatory argument + predicate, such as “院长由国务院提名(the president is nominated by the State Council)”

There are many fundamental differences regarding syntactic structures and semantic interpretations between the case of “introducing optional arguments” and the case of “introducing obligatory arguments”. First of all, the former is an adverbial-center structure, where the “由(by) + optional argument” serves as the adverbial component of the predicate, while the latter is a causative structure, where the structure of “obligatory argument + predicate” is somehow dominated by a certain of causative force. In (1b), the action of “to nominate the president” is not performed autonomously by the State Council, but by some kind of causative force, such as regulations, rules, laws, etc., which bestows responsibilities upon the State Council and thus requires it to do so. Secondly, the semantic components of the former (optional arguments and predicates) are explicit, which are easily analyzed and identified. However, the semantic components of the latter, especially the “causative force”, are often implicit, which cannot be identified unless a specific semantic analysis is offered. Finally, there is only one single semantic relationship between semantic components of the former. The relationship in (1a), for example, is “path-movement”. In the latter, there are at least two semantic relationships, specifically, an "agent-action" relationship and a "cause-(be) caused" relationship (see (1b)). For the sake of space, this paper will not focus on the former, but on the latter.

Through our investigation, we find there is an inverse relationship between the causative force and the agent's autonomy of the latter usage: the stronger the causative force is, the weaker the agent's autonomy is, and vice versa. According to the degree of the agent's autonomy, this paper divides "由(by)" sentences into three categories: non-autonomous, partially autonomous, and autonomous sentences. Among the 255 randomly selected "由(by)" sentences, there are 188 non-autonomous sentences (accounting for 73.7%), 42 partially autonomous sentences (accounting for 16.5%), and 25 autonomous sentences (only 9.8%). In the following sub-sections, we will discuss them one by one.

2.1 Non-autonomous Agent: (Causative Force $_{strong}$)… "由(by)" + Agent $_{weak}$ + Predicate

"Non-autonomy" means that the agent can't dominate his behavior autonomously but only comply with the causative orders or rules. In such sentences, "由(by)" can be replaced by "让(let)", and the whole sentence means "让某人负责某事(to let someone be responsible for something)", where the causative force of "由(by)" is usually explicit, and appears before the "由(by)" sentence in the form of orders, instructions, regulations, decisions, requirements, obligations, requests, judgments, etc. For example, "制度(system)", "决定(decide)", "判决(judgment)" as in 2(a–c).

(2) Non-autonomy sentence with explicit causative force.

a. 全区坚持双日报制度,……对"双无单位"由开户行逐户登记审查， 对大户进行重点抽查,对确属"双无单位"由企业法人出具保证书。(The whole district adheres to the "double-day reporting" **system**, ... the banks shall register and review the "non-double-day organizations" one by one, conduct a key spot check on large accounts, and require the "non-double-day organizations" to provide a letter of guarantee issued by the enterprise legal person.)
b. 澳政府已决定……修建澳规模最大的电子信号拦截中心，人员编制120多名，由国防部单独管理。(The Australian government has **decided**… to build the largest electronic signal interception center in Australia, with more than 120 personnel, under the administration of the Ministry of National Defense alone.)
c. 琼北中级人民法院依法判决:……南方制药厂侵犯海口市制药厂名誉权而给该厂造成的直接经济损失472万余元，由南方制药厂偿付给海口市制药厂。(Qiongbei Intermediate People's Court passed a **judgment** according to law: … Nanfang Pharmaceutical Factory has infringed upon the reputation right of Haikou Pharmaceutical Factory and caused the Haikou Pharmaceutical Factory to suffer direct economic losses amounting to more than 4.72 million yuan, which shall be reimbursed by Nanfang Pharmaceutical Factory.)

In addition to the explicit causative forces, there are also some implicit forces that can be deduced from experience, common sense, or the context. For example, in (3a), according to the practices of the film industry (for example, the investor determines the director first, and then they shall jointly select the main actors), we know that the responsibilities of the director, "达式彪(Shibiao Da)", and the actor, "俞洛生(Luosheng Yu)", in the film are not determined by themselves, but by some kind of causative force. The same is true for example (3b). The duty of the deputy company commander to sign and approve the reimbursement is not inherent but is given to him by the rules and regulations of the army.

(3) Non-autonomy sentence with implicit causative force.

a. 此片由达式彪执导, 俞洛生饰杨少振。(This film is directed by Shibiao Da, and Luosheng Yu plays Shaozhen Yang.)
b. 事后, 由副连长签字, 在战士伙食费里报销了。(Afterwards, it was signed by the deputy company commander and reimbursed as the soldiers' food expenses.)
c. 战时, 意大利作战部队基本由北约盟军指挥。(During the war, the Italian combat forces were under the command of NATO allies.)

2.2 Partially-Autonomous Agent: (Causative Force $_{weak}$)... "由(by)" + Agent $_{strong}$ + Predicate

Although the agent in this type has a certain degree of autonomy, it is still affected by the causative force. In such sentences, "由(by)" can more often than not be replaced by "允许(allow)" or "听任(let)", and the whole sentence means "允许某人做某事(**to allow someone to do something**)", where the causative force of "由(by)" appears in the form of agreements, permissions, licenses, approvals, commitments, hopes, requests, etc. For example, the pharmaceutical factory needs the license of the State Drug Administration before producing the "Chaoyang pills" in (4a). Pawnshops need the permission of the owner before auctioning the pawn in (4b).

(4)a 中国医科院药用植物所药厂昨日召开新闻发布会宣布：治疗肝炎良药朝阳丸（原名朝阳丹）……，获国家药政局颁发的国家新药证和生产许可证，现已由该厂投入生产。(The pharmaceutical factory of the Institute of medicinal plants of the Chinese Medical College held a press conference yesterday to announce that the Chaoyang pills (formerly known as Chaoyang pellets), good medicine for hepatitis, has obtained the new national drug **license** and production **license** from the State Drug Administration, and has been put into production by their factory.)

b 终因赎出也难卖掉, 而宣布弃当, 由当铺进行了拍卖。(The owner finally gives up the pawns due to difficulties encountered and **agrees** to their auction by the pawnshop.)

c 《圣经》中有一句名言: "有人打你的右脸, 连左脸也转过来由他打。"(There is a famous saying in *the Bible*: " If anyone slaps you on the right cheek, turn to him the other also.")

Note: the "由(by)" that expresses the meaning of "allow" is often labeled as "verb" in dictionaries. However, the prepositional features of "由(by)" are more obvious, such as not being able to be used as a predicate alone, not being able to reduplicate, not being able to take temporal or aspectual components (see [5] for more details). This paper regards it as a preposition, in spite of a few exceptions, such as "身不由己 (to do something against one's own free will)".

2.3 Autonomous Agent: (Causative Force $_{default}$)… "由(by)" + Agent $_{very\ strong}$ + Predicate

Due to the default of the causative force, the behavior of this type is mainly determined by the agent himself. Based on our corpus survey, the cases where the agent can autonomously determine everything are often seen in the initiating act, where the agent acts as the initiator, originator, developer, organizer, founder, creator, record-holder, etc., as shown in (5a–d).

(5)a. 发起人:这是中国人民银行总行所属院校中的第一家实验性股份制银行，由湖南财经学院发起筹建……(**Initiator**: This is the first experimental joint-stock bank of the institutions under the People's Bank of China. This was initiated by Hunan University of Finance and Economics …)

b. 研制者：一种新型的儿童旅游鞋由上海胶鞋研究所试制成功并投入市场试销。(**Developer**: A new type of children's tourist shoes was successfully developed by Shanghai Rubber-soled Shoes Research Institute and put into the market for trial sale.)

c. 创办者：一家以陈列展销全国各地新书的"全国新书展销厅"昨日由《书讯报》和上海图书馆联合创立。(**Founder**: A national new book exhibition hall for displaying and selling new books from all over the country was founded yesterday by the *Book News* and Shanghai Library.)

d. 创造者:世界纪录198靶由葡萄牙选手保持。(**Record-holder**: The world record of 198 is held by a Portuguese athlete.)

Unlike Sects. 2.1 and 2.2, the "由(by)" sentences in this section tend to be "relativized". Statistic data show that nearly one-third of the "由(by)" sentences are not used independently but "relativized" into attributive components, as shown in example (6a–d).

(6)a. 近9000名观众，昨晚用鲜花、掌声和礼貌的态度参加了在广州天河体育馆由太平洋影音公司举办的音乐会。(Last night, nearly 9000 people attended the concert held by Pacific Video Company in the Tianhe Gymnasium in Guangzhou. They brought to the concert flowers, applause, and courtesy.)

b. 由北京电影制片厂与中国电影发行放映公司合资拍摄的这部影片，正在紧张筹备之中。(The film, jointly produced by Beijing Film Studio and China Film Distribution and Screening Company, is under preparation.)

c. 他们准备向我们这个世界输送由他们制造和控制的机器人。(They are ready to deliver robots that they make and control to our world.)

Upon careful observation, we will find that the "由(by)" structures in (6a–c) can be deleted without imposing any fundamental impact on producing and understanding the sentences. Since the structure of "由(by)" is not required by syntax and semantics, why does the speaker use it? We propose that there is a pragmatic purpose to highlight the initiator of the act.

The reason for this is that the initiator, founder, or creator is not only the first person to act but also the first person to be responsible for these things. When talking about these things or wanting to know more about them, the first person we will contact is the initiator. When something goes wrong, the first one we want to complain to is the initiator, too. It is exactly due to the important role of the initiator that speakers will give him special priority.

In actual discourse, there is more than one way to highlight the initiator. Reference [3:49] finds that the word "由(by)" often appears in the "是(be)…的" emphatic frame, which is reconfirmed in the investigation of this paper. In other words, most "由(by)" structures appearing in the "是(be)…的" frame is used to highlight the initiator, as shown in (7)

(7)a. 这一运动是由新西兰政府发起的。(The campaign was launched by the New Zealand government.)

b. 这一思想主要是由邓子恢同志提出来的。(This idea was mainly put forward by Comrade Zihui Deng.)

c. 这次为期七天的服务活动是由团秦皇岛市山海关区委组织的。(The seven-day service was organized by the Shanhaiguan District Committee of Qinhuangdao.)

.

As mentioned above, this use of "由(by)" mainly plays the role of "引出或凸显动作行为的发起者(to introduce or highlight the action's initiator)". In addition, we also observe that there are some "由(by)" sentences that highlight the self-consciousness of the agents, where the agents act as passive markers, as shown in example (8) below.

(8)a. 这时,女同胞已由一男青年护送进了旅店。(By this time, the female compatriot had been escorted into the hotel by a young man.)

b. 洋磁面盆里的蝌蚪, 由孩子们给迁居在墙角里新开的池塘里了。(Tadpoles in the ceramic basin have been moved to the new pond in the corner by the children.)

The ratio of "由(by)" as a passive marker is relatively small (only 5 sentences), which is in an absolutely inferior position as compared with the number of non-autonomous sentences, partially autonomous sentences, and autonomous sentences. Therefore, the usage of "由(by)" acting as a passive marker is not typical.

3 Why is It Quasi Causativity(准使役性)?

In the causative structure "causative force + causative verb + agent + predicate", the causative force and the autonomous force of agents often interact with each other [10].

It is the interaction of these two forces that makes the causative structure complicated. Taking "让(let)" as an example, if the causative force is strong, the autonomous force of the agent will be relatively weak, and the meaning of "让(let)" is roughly equivalent to "使(make)" in (9a), "派(dispatch)" in (9b), "叫(ask)" in (9c), "命令(order)", "要求(demand)" and so on.

(9) Causative force$_{strong}$ – autonomous force of agent $_{weak}$.

a. 我的父亲让我和母亲受了那么多苦，我恨他，我一心想报复。(My father **made** me and my mother suffer so much. I hate him and always want to get back at him)
b. 组织上决定让你去学习, 你准备一下, 明天去报到吧。(The organization has decided to **dispatch** you to study. You can make some preparations and go to register tomorrow.)
c. 爸爸拣来一桶沙土, 让妈妈做了一只沙袋。(Dad picked a bucket of sand and **asked** mom to make a sandbag.)

If the causative force is weak and the autonomous force of the agent is relatively strong, the meaning of "让(let)" will be roughly equivalent to "允许(allow)" in (10a) or "任由(let him)" in (10c). If the causative force is in default and the agent becomes the dominant force that influence the patient, "让(let)" will then act as a passive marker. For example, the "让(let)" in (11a–c) can be replaced by "被(by)".

(10) Causative force$_{weak}$ – autonomous force of agent $_{strong}$.

a. 只有发展民主制度, 让人民群众能充分发表意见,……领导者才可能在总结群众经验的基础上制定出符合实际的政策。(Only by developing the democratic system and **allowing** the masses to fully express their opinions,... can the leaders draw up practical policies based on summing up the experience of the masses.)
b. 请让我敬大家一杯!(Please **let** me propose a toast!)
c. 卫老塾师劝阻不住, 只好让他去了。(Wei, the old teacher, couldn't stop him, and had to **let him go**.)

(11) Causative force $_{default}$ – autonomous force of agent $_{strong}$

a. 谁知道哥哥耳朵软, 让媳妇在被窝里一哄,……连妹妹跟他说的真情话也不信了。(No one can figure out that the brother is so easily coaxed **by** his mate in bed … and doesn't believe what his sister told him to be true.)
b. 可惜连村子都让飞机炸平了，只剩下高高低低的烂泥巴塘。(Unfortunately, the village has been bombed **by** the plane, leaving only the rugged mire.)
c. 年轻人的裤管上沾满了稀泥, 胶鞋让泥浆灌满了, 脚脖子也让冰碴扎得通红。(The young man's trousers were covered with mud. His rubber shoes were filled with mud, and his ankles were reddened by hard ice bits.)

It is revealed in the above analysis that there are many semantic similarities between "由(by)" and "让(let)". To be specific, 1) the semantic structures are basically the same, in that both of them can be expressed as "causative force + 让(let)/由(by) + agent + predicate"; 2) the semantic components and the relationships among them are the same, which include three relationships: "strong causative force - weak agent force", "weak causative force - strong agent force" and "default causative force - strong agent force"; 3) following the common semantic extension path of "causation - permission – passive" in grammaticalization; 4) the conceptual meaning is basically the same, in that both of them have a causative meaning. Then, why do we not say that "由(by)" has causativity meaning? Why do we say that it is quasi causativity? The reasons are as follows:

(12)a From the point of view of syntactic functions, "由(by)" is closer to a preposition than a verb (see [5] for more details). However, causation is typically expressed by verbs, such as "使(make), 命令(order), 让(let), 叫(ask)", etc. Therefore, it is not suitable to regard the "由(by)" as a causative preposition.

b. From the perspective of predicate selection, "由(by)" sentences usually select verbs with weak transitivity, such as "执行(executing), 任命(appointing), 承担(undertake), 支撑(support), 看守(guard), 管理(manage), 代管(escrow)"[3]. Verbs with strong transitivity, such as "跳(jump), 洗(wash), 扔(throw), 写(write), 吃(eat), 喝(drink)", hardly enter a "由(by)" sentence. That is, we usually do not say any sentence like "由他跑 (to let him run)", "由他洗 (to let him wash)", "由他扔 (to let him throw)", etc. Comparatively speaking, restrictions on verbs in causative sentences are rare. Verbs with strong or weak transitivity can enter a causative sentence.

c. From the point of view of the effect of causative force, there is an important difference between "由(by)" sentences and causative sentences: the causative force of the latter is explicit, in that it appears with the agent in the same sentence and its influence on the agent is direct and is easy to be perceived (in 9a, 10a). The causative force of the former is implicit and does not appear with the agent in the same sentence but in the preceding sentence. Its influence on the agent is indirect and is not easy to be perceived (in 2a-c). In short, the causative influence of causative sentences is direct and explicit, but the causative influence of "由(by)" sentences is indirect and implicit.

As mentioned above, the Chinese word "由(by)" does not meet the syntactic, semantic requirements of typical causation. So, it should not be treated as a causative preposition. Given its similarities to causative constructions and the difference in its usage to introduce an agent from its other usages, this paper treats it, when in use to introduce an agent, as a quasi causative preposition. In other words, the so-called quasi causativity is a syntactic-semantic property similar to causative structures. For example, both of them can introduce the causer argument and form the structure of "causative force(causer) + cause + agent(causee) + predicate"[1]. The main difference between them is that

[1] Note: Not all causative structures have their results expressed explicitly. For example, the sentence让张三来决定 (let Zhang San make the decision) does not care about the causative result but cares only about the causative process.

the causer of quasi causative structures usually appears in the previous context and its influence on the causee is indirect and implicit.

4 Advantages of the Quasi Causative Preposition Analysis

Firstly, let‘s start with syntactic advantages. References [3, 5] noticed that treating “由(by)” as a passive marker will encounter some unexplained problems. For example, “by(被) + agent” can act as an adverbial component of the sentence. The agent of “by(被) + agent” can be omitted, etc., but such phenomena are not allowed by “由(by) + agent” sentences. Why is “由(by) + agent” so special? The reason is that "由(by)" is a quasi causative preposition, which makes the "由(by)" and the "被(by)" different.

“由(by)” is a quasi causative preposition that has properties similar to causative verbs. The “由(by)+ agent” is not an adverbial component but a pivotalt(兼语) whose agent acts as the receiver of causative forces and the initiator of actions simultaneously.

The agent of a typical causative sentence cannot be omitted (for example, we can‘t say “*张三, 我让 来决定(Zhang San, I let make decisions)”, see reference [5]). The same is true for “由(by) + agent” sentences.

Then, let’s look at the semantic advantages. The multiple meanings of polysemous words are not random. They are semantically interrelated. Once the semantic relation hidden behind the multiple meanings is revealed, it will be easier to grasp the multiple meanings of polysemous words [11, 12]. For example, “由(by)” has multiple meanings and usages. Introducing an agent is merely one of them, as shown in (13).

(13) Explanation of "由(by)" in reference [13:628-629]
VERB 听凭或听任(to allow or let): 信不信~你。((let you) believe it or not.)

PREP 1) 引出施事(to introduce agent): 院长~国务院提名。(the president is nominated by the State Council.)
2) 表示方式、原因或来源(to express the method, reason, or source): ~感冒引起了肺炎。(the pneumonia was caused by influenza.)
3) “从(from)”, including five conditions: a)表示处所起点或来源(to express the starting point or the source of place): ~南到北(from south to north); b)表示时间起点(to express the starting point of time): ~早上九点到晚上八点(from 9 am to 8 pm); c) 表示发展、变化、范围的起点(to express the starting point of development, change, and scope): ~蝌蚪变成青蛙(from tadpoles to frogs); d) 表示经过的路线,场所(to express the route or place passed by): ~这条路走近多了(it’s much closer via this road); e) 表示凭借、依据(to express the basis or evidence): ~地球应力场的变化探测地震发生的可能性(to detect the possibility of the earthquake by measuring changes of stress field of the earth).

At the beginning of this paper, we have noted that treating “由(by)” as a functional category will bring some new problems. For example, it will break the relationship between the usage of “由(by)” to introduce an agent and its other usages. It is not conducive to explaining where "由(by)" comes from. Nor is it possible to give a unified explanation of the multiple meanings of "由(by)" (see Part 1). Can the quasi causative preposition analysis solve these problems? The answer is definitely "yes".

Take "经过(pass by)" (the first meaning in [14]) as the starting point of our semantic analysis. We find that the semantic extension of "由(by)" follows two routes. One is from "经由路线(route passed by)" to "运动起点(starting point of movement)",carrying the meanings of "starting point of time", "starting point of place", "starting point of development and change" (in 13(3a-c)), etc. The other is from "经由路线(route passed by)" to "运动方式(method of movement)", which is further extended to the meanings of "方式或原因(method or reason)" (in 13(2)), "经由路线或场所(route or place passed by)" (in 13(3d)), "凭借或依据(basis or evidence)" (in 13(3e)) and so on.

Through the above two routes, we can connect most of the meanings of "由(by)" together, except for the meanings of "引出施事(to introduce an agent)" and "听凭或听任(to allow or let)". It seems that these meanings are not compatible with the others. Both of them are closely related to the meaning of "经由路线(route passed by)":

(14)a In an expression for "让某人负责某事 (to let someone be responsible for something)", the agent has weak autonomy and has to follow the instructions of the causative force. This meaning is related to the meaning in (13(3d)), but what the agent follows at this time is not a specific movement route but an abstract instruction.

b. In an expression for "允许或听任某人做某事 (to allow or let someone to do something)", the agent's autonomy is enhanced, but it is still influenced by the causative force, whose abstract directive meaning is also related to and especially extended from the meaning in (13(3d)).

c. When the meaning is "凸显动作行为的发起人 (to highlight the initiator of behavior)", the first person to do something is highlighted, which is related to the meaning of (13(3a–c)) and extended from the meaning of "starting point of movement".

In summary, compared with the functional category theory, treating "由(by)" as a quasi causative preposition has lots of advantages.This treatment not only reflects its actual usage but is also conducive to giving a uniform account for its multiple meanings and usages.

Acknowledgments. This paper is supported by the Youth Fund of the Ministry of Education (18YJC740095). The anonymous reviewers of CLSW2020 put forward many valuable comments. Here, please allow me to express my sincere thanks!

References

1. Wang, H.: Passive sentences in English and Chinese. Chinese language. Studies of the Chinese Language 6 (1983). In: Zhao Y. (eds.) Comparison of Chinese and foreign languages and teaching Chinese as a foreign language. Beijing Language and Culture University Press, Beijing, pp, 272–290 (1997). (in Chinese)

2. Bai, Q.: On the syntactic function of the “由(by)+施事(agent)” structure at the beginning of a sentence and its related problems. J. Beijing Normal Univ. **6**, 102–106 (1998). (in Chinese)
3. Zhang, Y.S.: On the passive sentence with the word “由(by)”—Concurrently discussing the difference between “由(by)” sentence and “被(by)” sentence. Lang. Sci. **3**, 38–53 (2004). (in Chinese)
4. Lü, W.H.: The “由(by)” sentence—Concurrently discussing the “被(by)” sentence. Langu. Teach. Linguist. Stud. **2**, 19–30 (1985). (in Chinese)
5. Xiong, Z.: A passive construction marked by “yóu” in mandarin Chinese. Mod. Foreign Lang. **1**, 12–22 (2010). (in Chinese)
6. Su, X.M.: Two types of directive-related “you”-construction in Chinese. J. Ocean Univ. China **5**, 117–122 (2014). (in Chinese)
7. Yuan, Y.L.: A study of Chinese Valence Grammar. The Commercial Press, Beijing (2010)
8. Liu, D.Q.: Handbook of grammar investigation and Research. 2nd edn. Shanghai Education House, Shanghai (2017). (in Chinese)
9. Guo, R.: Semantic structure and semantic analysis of Chinese form words. Chinese Teaching in the World **4**, 5–15 (2008). (in Chinese)
10. Talmy, L.: Towards a Cognitive Semantics(Vol1): Concept Structuring Systems. MIT Press, Cambridge & MA (2000)
11. Guo, Z.Y.: Exegetics (Revised). Zhonghua Book Company, Beijing (1986/2005). (in Chinese)
12. Wang, Y.L., Wang, C.: Research on the Core Meaning of Chinese Vocabulary. Peking University Press, Beijing (2014). (in Chinese)
13. Lü, Sh.X. (ed.): 800 Words in Modern Chinese (Revised). The Commercial Press, Beijing (1999). (in Chinese)
14. Li, X.J. (ed.): Dictionary of Modern Standard Chinese (3rd edition). Foreign Language Teaching and Research Press & Language and Culture Press, Beijing (2014). (in Chinese)

Event Structure and Logical Semantic Operations of the Chinese Character Shi (是) Used for Emphasis

Hongyu Nan[1(✉)] and Liming Yu[2(✉)]

[1] The School of International Education, Yancheng Teachers University, Yancheng 224000 Jiangsu, China
1643888947@qq.com
[2] The College of Literature and Journalism, Sichuan University, Chengdu, China
yingchuan80@sina.com

Abstract. The Chinese character shi (是) can indicate judgment or emphasis. When Shi (是) expresses judgment, it cannot be omitted from the sentence due to syntactic constraints. When shi (是) expresses emphasis, it can be omitted without affecting the integrity of the sentence. This paper claims that the bottom layer of Chinese language is a set of logical structures of event composed of semantic roles. Shi (是) expressing emphasis is the syntactic expression of a person's attitude. It lies between the syntactic structure layer and the event structure layer with the function of extracting semantic roles. Shi (是) expressing emphasis can: (1) extract all event semantic roles; (2) be placed before the syntactic position of the extracted semantic role (3) be realized as deshi (的是) when it extracts the semantic role of theme.

Keywords: Shi (是) · Emphasis · Judgment · Event structure · Syntactic projection

1 Introduction

Expressing an emphasis in a language is related to the expression of modality. The expression of a speaker's attitude involves a specific object expressed in the sentence. The expressions of modalities often have no rules to follow in terms of their syntax and semantics. They are often unpredictable. Therefore, modality has always been a confusing field in language research [1]. In Chinese language, modality is mainly manifested by lexical items. With a close scrutiny to the Chinese expressions of modality from the perspective of event structure, a well-structured system is observed. This article aims to i) study the extraction operation of the emphatic "shi" on the semantic role embedded in event structure, ii) systematically analyze the Chinese modalities, and iii) find out the characteristics of Chinese syntactic structure.

2 "*Shi*" Expressing Emphasis in Chinese

2.1 Two "*shi*"s in Chinese: Indicating Judgment or Emphasis

"*Shi*"s(是) in Chinese may express different meanings. Some researchers believe "*Shi*"

M. Liu et al. (Eds.): CLSW 2020, LNAI 12278, pp. 14–33, 2021.
https://doi.org/10.1007/978-3-030-81197-6_2

is used to express judgment [2–5]. Actually, it is not always the case, as can be seen from the comparison between the following two sentences:

(1) 你 是 高中生。
Ni shi gaozhongsheng。
you be high school student.
You are a high school student.
Ruined City by Jia Pingwa (贾平凹《废都》)

(2) 是 你 妨碍了 我！
Shi ni fang'aile wo！
be you hinder me
It is you who hinder me!
BigBreasts&WideHips by Mo Yan (莫言《丰乳肥臀》)

"*Shi*" in sentence (1) cannot be omitted. In sentence (2), it can be removed without affecting the event meaning, exemplified below,

(1b) 你是高中生。 → 你高中生。*
(2b) 是你妨碍了我！ → 你妨碍了我！

In sentence (1), there is no robust subject-predicate relationship between "you" and the identity as a high student. In order to establish a subject-predicate relationship, we must connect the two constituents with "*shi*". Therefore, "*shi*" in sentence (1) is required by the syntactic structure and cannot be omitted [6]. On the semantics of "*shi*" in sentence (2), there are three views: i) "*shi*" expresses an emphasis of attitude; (c.f. [7]); ii) "*shi*" is a focus marker (c.f. [8–11]); iii) "*shi*" expresses a modality confirmation (c.f. [12]). These views hold that "*shi*" in sentence (2) expresses the attitude of the speaker. While attitude must involve a target, the sentence expressing attitude must also involve an object. To be brief, "*shi*" in sentence (1) and (2) should be considered different, and in sentence (2), it expresses emphasis. Emphasis is a kind of modality, which is a method of information processing for the speaker to highlight a certain language segment with a certain linguistic means (such as morphology, function words, word order, prosody, etc.), so that the listener will pay attention to that segment [13].

When it comes to emphasis, not only "*shi*" in sentence (2), but also the structure "*shi……de*"(是……的) can express emphasis [14]. Regarding the function of "*shi……de*", previous studies have upheld different views: i) [15] has proposed three categories, namely judgment, affirmation and modal; ii) [16] has proposed three categories, namely structure, specific and emphasis; iii) [17] has proposed two categories: marking the past, and expressing a positive or confident tone. According to the above viewpoints, both affirmation and confident tone are related to emphasis. In other words, both "*shi*" in sentence (2) and the pattern "*shi…de*" convey the emphatic sense. Yet, their syntactic behaviors are different. The latter has a complementary relationship with the stressed constituent. Consider:

(3) 他 是 今天 看 的 电影。 (Self-constructed sentences)
Ta shi' jintian kan de dianying。
He be' today see de movie
It is today that he saw the movie!

(4) 他 是 今天′ 看 的 电影。
Ta shi jintian′ kan de dianying。
He be today′ see de movie
It is today that he saw the movie!

(5) 他 是′ 今天′ 看 的 电影。 *
Ta shi′ jintian′ kan de dianying *
He be′ today′ see de movie *
It is today that he saw the movie!

The stresses (marked with ′) in sentences (3) and (4) are on "*shi*" and "*Jintian*" (today) respectively. Note that the stress cannot be put on both *"shi"* and *"jintian"* at the same time, as shown in sentence (5). To elaborate, sentence (3) and (4) demonstrate the fact that stress can be put on different constituents within the pattern *"shi…de"*. However, for the emphatic "*shi*" (c.f. sentence (2)), stress can only be put on"*shi*". Therefore, the emphatic "*shi*" should have its own independent syntactic and semantic functions. This paper focuses on the emphatic "*shi*".

2.2 Syntactic Operation and Controversy Over "*shi*" Used for Emphasis

Compared with other viewpoints, linguistic research is now more inclined to consider the emphatic "*shi*" as a focus marker. The emphatic "*shi*" is syntactically flexible. It can appear not only before the subject in the sentence but also after the subject, and can be linked to different syntactic and semantic components:

(6) 他们 是 [追随]着 我 呢。
Tamen shi zhuisuizhe wo ne。
they be follow me SFP
They are following me!
Cat Country by Lao She (老舍《猫城记》)

(7) 这一次 日本兵 是 [用手] 拉开了 门。
Zheyici ribenbing shi yongshou lakaile men。
This time Japanese soldiers be by their hands open door
This time, the Japanese soldiers opened the door with their hands!
The Death of a Landlord by YuHua (余华《一个地主的死》)

(8) 我 是 [对你] 说 一句实心眼儿的话。
Wo shi duini shuo yijushixinyan'erdehua。
I be to you say a solid word
It is to you that I am saying a solid word!
Deep Night by Zhang Henshui (张恨水《夜深沉》)

(9) 父亲 是 [暗暗地] 为敢对自己挑战的好汉的儿子 自豪。
Fuqin shi anandi weiganduizijitiaozhandehaohandeerzi zihao。
my father be secretly for the son who dares to challenge his father pride
Secretly, my father is proud of the son who dares to challenge his father!
Unprecedented Confidence and Courage by Bing Xin (冰心《空前的信心和勇气》)

"*Shi*" appears before the verb "follow", the instrument adverbial "by hand", the object adverbial "to you", and the manner adverbial "secretly" in sentences (6), (7), (8), and (9) respectively. By analyzing the syntactic behavior of "*shi*" through comparing the focuses in each sentence, it is found that the emphatic "*shi*" can be marked on logical formula of the event structure, verb and degree [8], but cannot mark the theme that appears after the verb. "*Shi*" is often used after "*V* + *de*" (V的) to form the structure "V + *de* + *shi* + theme"(V的是 + theme), and the semantic of the whole sentence is the same as the situation where the syntactic component "*shi*" is the focus (often expressing new information). Examine:

(10) 楼下的 铺面房 开 的 是 烟草店。

Louxiade pumianfang kai de shi yancaodian。
downstairs shop open of be tobacco shop
The shop downstairs is a tobacco shop.

The Criminal Lu Yanshi by Yan Geling (严歌苓《陆犯焉识》)

The theme "tobacco shop" in sentence (10) serves as the new information. But whether "*de shi*" (的是) marks the theme as the focus needs to be analyzed. Also, whether the focus marked on the agent, time, place, instrument, verb, or degree is significant needs to be analyzed. It should be noted that the analysis of the syntactic distribution of the focus marker "*shi*" is a syntactic issue. [18] did not agree with this analysis. He believed that the emphatic "*shi*" was not a focus marker, nor could it involve in a syntactic operation, due to the distributional restriction that "*shi*" could only be located before the subject or the predicate. With such restriction, "*shi*" cannot appear as the focus marker. So far the analysis of the emphatic "*shi*" has mainly focused on the investigation of its syntactic and semantic functions, and other researchers have not paid attention to the semantic operation mechanism under "*shi*". Therefore, many semantic functions of the emphatic "*shi*" have actually been analyzed through its syntactic and pragmatic functions. This paper posits within the framework of event semantics an underlying layer of the Chinese language, and then analyzes the semantic extraction mechanism of the emphatic "*shi*" as well as the Chinese syntactic construction.

3 Event Logic Structure and Chinese Semantic Basis

This paper, from the perspective of event semantics, attempts to analyze this problem: the syntactic operations of the emphatic "*shi*" being applied to different semantic roles (e.g. agent, instrument, etc.) cannot be applied to the role of theme. At present, researchers have agreed that an event is conveyed by a verb and its arguments [19, 20]. Difficulties are encountered in the study of Chinese because of the fuzzy relationship between syntactic objects and the theme role in semantic structure. [21] proposed a hypothesis: the bottom layer of the Chinese language is a logical structure of event consisting of a set of semantic roles. The event structure is as follow:

(1) Core layer: including three components, namely agent, verb, theme;

(2) Additional layer: including four components, namely instrument, object, degree, manner;[1]
(3) Peripheral layer: including three components, namely time, location, direction.

The logical formula of the event structure is written as:

λxλyλzλjλkλlλmλnλo (∃e) [agent (e, x) &V (e) &theme (e, y) 【& instrument (e, z) & object (e, j) & degree (e, k) & manner (e, l) 】 【& time (e, m) & local (e, n) & direction (e, o) 】].[2]

The mapping between semantic roles and the syntactic positions are operated as: first, syntactic positions are available; and second, the relationships among semantic roles are ensured. Without considering the effect of pragmatics on the syntactic structure, the projection process is what is shown in Fig. 1:

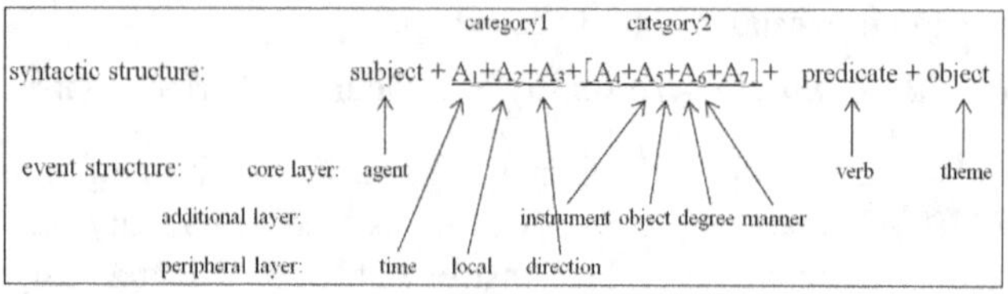

Fig. 1. Syntactic projection of event semantic roles[3]

The agent, the verb and the theme of the core layer of the event structure must be projected onto the subject, the predicate and the object positions in the syntactic structure respectively. There are two categories of items between the subject and the verb: category 1 includes A_1, A_2, and A_3, and category 2 includes A_4, A_5, A_6, and A_7. The semantic roles of the additional layer and the peripheral layer generally are not projected to the syntactic structure across categories. Semantic roles of the additional layer are projected to category 2 and semantic roles of the peripheral layer are projected to category 1. These are selective projections and do not necessarily appear in the syntactic structure. Figure 2 shows the model to express the projection of event semantic roles onto syntactic positions:

This model includes two levels, with "event tier" being the event structure layer, and "syntax tier" being the syntactic structure layer. While "Π" represents the semantic roles,

[1] In the event structure, the object refers to the person or thing that the action is performed for or to.

[2] Events include behavior and state. Because extraction of semantic role in states is similar to.that in behavioral events, it will not be analyzed for now. In the logical formula, 【】 marks additional semantic roles and peripheral semantic roles.

[3] Corpus analysis shows that the projections of instrument and manner can exchange positions in the syntactic structure. Within category 2, so can the projections of manner and object, and the projections of object and degree. But there is no exchange of syntactic position for projections of instrument and degree. Peripheral semantic roles include time, space, and direction, and because each semantic role itself is a category system, projection of peripheral semantic roles is more complicated. For the sake of simplicity, the usual projection order of the peripheral semantic roles is "time + location + direction".

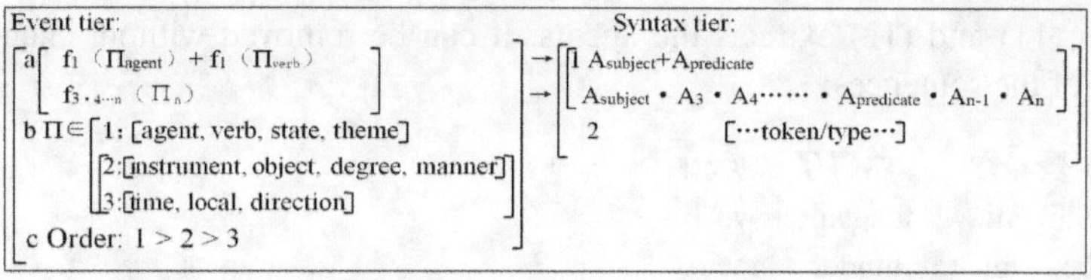

Fig. 2. The syntactic projection model of event semantic roles

the smallest component unit of the event structure is represented as "Π (e, x)", meaning that an event "e" requires the specific type of the semantic role "Π", which may be "agent, verb…", and the semantic role "Π" constrains the semantic status of the concept "x". The event structure layer (i.e. "Event tier" in Fig. 2) consists of three parts: (a) is a projection function, mapping to the grammatical positions of the syntactic structure; (b) is the whole structure of the event, including the core layer, additional layer, and peripheral layer; (c) is the order of projection (regardless of pragmatics, the core layer is projected first, then the additional layer, and finally the peripheral layer). The syntactic structure layer (i.e. "Syntax tier" in Fig. 2) includes two parts: the first is a grammatical position in the syntactic structure; the second is a word token or type, which can handle referrals and replacements. "f" represents the projection process, and subscripts "1, 2, 3…" are used if there are more than one projection functions. "A" represents the syntactic position. The whole formula reads as: every time "f" extracts a "Π", the event semantic role Π is projected to a syntactic position, and the semantic role projection function constrains the word order of the syntactic structure. The process of projection is: the core semantic role pairing "agent + verb" must be projected to the "subject + predicate" positions of the syntactic structure. When the 3rd, 4th… projections of other semantic roles of the event structure are performed, the semantic roles Π_3, Π_4… are sequentially projected to positions A_3, A_4… A_n after the syntactic position of A_1. Finally, prepositions are processed during the projection of additional semantic roles and peripheral semantic roles.

4 Extraction of Non-theme Roles by Emphasis "*shi*"

"*Shi*", which can extract event semantic roles, does not belong to the event structure, and its logical semantic functions traverse the event structure layer and the syntactic structure layer. Extraction of non-theme semantic roles and extraction of the theme role work differently. We will discuss them separately below.[4]

4.1 Extraction of the Agent by the Emphatic "*Shi*"

The semantic role agent of the event structure is generally the sender or bearer of the action, which is usually projected onto the subject position of a sentence. "*Shi*" in

4 "*ba*" (把) and "*bei*" (被) prepose the theme in the sentence and emphasizes the "cause/result", demonstrating marked structures. Thus, they will not be analyzed in this article.

sentences (11) and (12) extracts the agents. It can be removed without changing the meaning of the sentences:

(11) 是 你 妨碍了 我！
Shi ni fang'aile wo !
be you hinder me
You are the one who hinders me!
BigBreasts&WideHips by Mo Yan （莫言《丰乳肥臀》）

(12) 是 他 回来了。
Shi ta huilaile。
be he go back
He is the one who goes back!
Philosophy of Lao Zhang by Lao She (老舍《老张的哲学》)

If we delete "*shi*" from sentences (11) and (12), the sentences and the logical structures of the events will be as follows:

（11b）你妨碍我 （∃e）[agt（e，你）&妨碍（e）&Th（e，我）]
（12b）他回来了 （∃e）[agt（e，他）&回来了（e）]

The syntactic projection process of the extraction of the agent role by "*shi*" in sentences (11) and (12) is shown in Fig. 3:

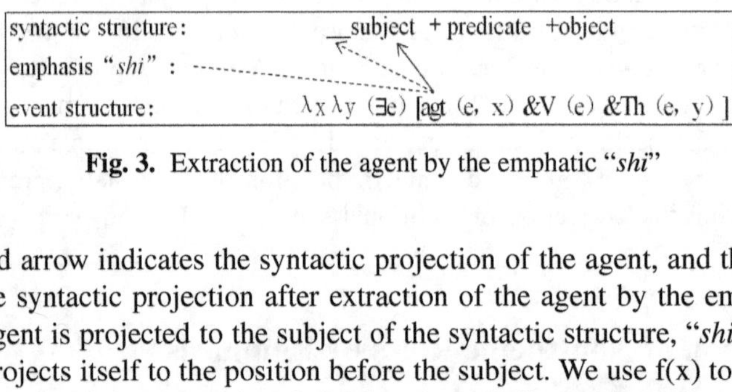

Fig. 3. Extraction of the agent by the emphatic "*shi*"

The solid arrow indicates the syntactic projection of the agent, and the dotted line indicates the syntactic projection after extraction of the agent by the emphatic "*shi*". When the agent is projected to the subject of the syntactic structure, "*shi*" extracts the agent and projects itself to the position before the subject. We use f(x) to represent the projection function of a semantic role of the event. "Π" represents the semantic role. "A" represents the syntactic position the semantic role is projected to. The projection function of the agent is:

$$\textbf{Projection of the agent}: [\mathbf{f}\ (\Pi_{\mathbf{agent}}) \rightarrow \mathbf{A}_{\mathbf{subject}}]$$

The function reads: if the agent is input to the function, it will be output as the syntactic subject. We use "↠" to indicate the extraction of the agent role by "*shi*":

$$\textbf{Extraction of the agent by "}\textit{shi}\textbf{"}: \mathbf{if}[\textit{shi} \twoheadrightarrow \mathbf{f}(\Pi_{\mathbf{agent}})] \rightarrow \left[\textit{shi} + \mathbf{A}_{\mathbf{subject}}\right]$$

The formula reads: if "*shi*" selects the agent as the item to emphasize from the event structure, and the agent is projected to the subject position of the syntactic structure, then "*shi*" is projected to the position before the subject.

4.2 Extraction of a Verb by the Emphatic "*Shi*"

The activity denoted by a verb is the core of the event structure, and the semantic role of an activity is usually projected to the predicate. "*Shi*" in sentences (13) and (14) extracts the verb, and can be removed without changing the meaning of the sentences:

(13) 他们 是 追随着 我呢。
Tamen shi zhuisuizhe wone。
they be follow me
They are following me!

Cat Country by Lao She (老舍《猫城记》)

(14) (我) 后来 仿佛 是 来到了 水边。
(wo) houlai fangfu shi laidaole shuibian。
we later seem be come to the river
It seems that I come to the river later!

To Live by Yu Hua (余华《活着》)

After deleting "*shi*" from sentences (13) and (14), we get:

(13b) 他们追随我 (∃e) [agt (e, 他们) &追随 (e) &Th (e, 我)]
(14b) (我) 来到水边 (∃e) [agt (e, 我) &来到 (e) &Th (e, 水边)]

The syntactic projection and the extraction of the verb by "*shi*" in sentences (13) and (14) are shown in Fig. 4:

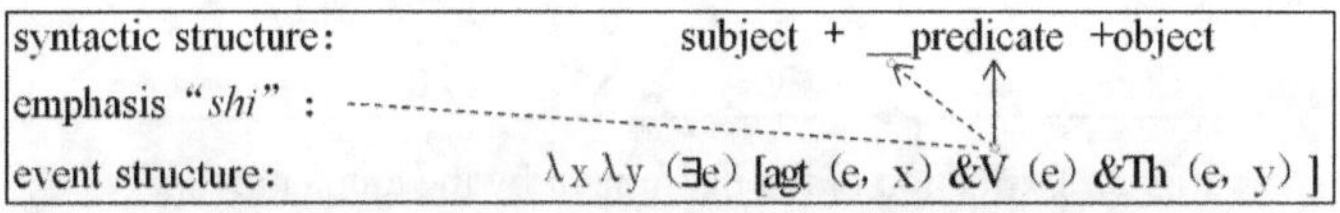

Fig. 4. The extraction of the verb by the emphatic "*shi*"

When the verb is projected to the predicate position of the syntactic structure, "*shi*" extracts the verb and projects itself to the position before the predicate. The projection function is:

Projection of the verb : $[\mathbf{f}\ (\Pi_{\mathbf{verb}}) \rightarrow \mathbf{A}_{\mathbf{predicate}}]$

The extraction of the verb by "*shi*" is as follows:

Extraction of the verb by "*shi*" : $\mathbf{if}[shi \twoheadrightarrow \mathbf{f}(\Pi_{\mathbf{verb}})] \rightarrow [shi + \mathbf{A}_{\mathbf{predicate}}]$

4.3 Extraction of the Instrument by the Emphatic "*Shi*"

The semantic role of instrument in the event structure is to extend the acting ability of an agent, and is usually projected to an instrumental adverbial position in the syntactic structure. "*Shi*" in sentences (15) and (16) extracts the instrument. "*Shi*" in the sentences can be removed without changing the meaning of the sentence:

(15) 这一次 日本兵 是 [用手] 拉开了 门。
Zheyici ribenbing shi yongshou lakaile men。
This time Japanese soldiers be by their hands open door
This time, the Japanese soldiers opened the door with their hands!
The death of a landlord by YuHua (余华《一个地主的死》)

(16) 仿佛 他 是 用两个拳头 在敲。
Fangfu ta shi yongliangequantou zaiqiao
as if he be with two fists knock
As if he was knocking with his two fists.
Events of April 3 by Yu Hua (余华《四月三日事件》)

If "*shi*" is deleted from sentences (15) and (16), the expression of the sentences and the logical structures will be as follows:

(15b) 日本兵用手拉门 (∃e) [agt (e, 日本兵) &拉 (e) &Th (e, 门) &instr (e, 手)]
(16b) 他用拳头敲(桌子) (∃e) [agt (e, 他) &敲 (e) &Th (e, 桌子) &instr (e, 拳头)]

The syntactic projection and the extraction of the instrument role by "*shi*" in sentences (15) and (16) are shown in Fig. 5:

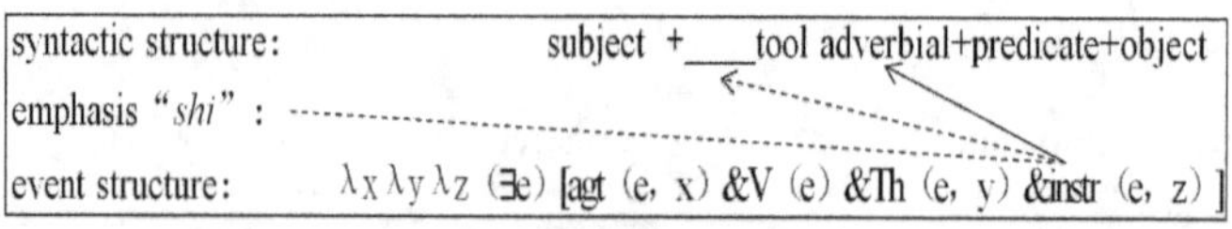

Fig. 5. Extraction of the instrument by the emphatic "*shi*"

The instrument role is projected to the instrument adverbial position of the syntactic structure. "*Shi*" extracts the instrument role and projects itself to the position before the instrument adverbial. The projection function is:

Projection of the instrument: $[\mathbf{f}\ (\Pi_{\mathbf{instrument}}) \rightarrow \mathbf{A}_{\mathbf{tool\ adverbial}}]$

The extraction of the instrument by "*shi*" is as follows:

Extraction of the instrument by "*shi*": $\mathbf{if}[shi \twoheadrightarrow \mathbf{f}(\Pi_{\mathbf{instrument}})] \rightarrow [shi + \mathbf{A}_{\mathbf{tool\ adverbial}}]$

4.4 Extraction of the Object by the Emphatic "*Shi*"

The object in the event structure expresses the direction of the action, and is usually projected to the position of a behavior-oriented adverbial in the syntactic structure.

"*Shi*" in sentences (17) and (18) extracts the object. "*Shi*" and can be deleted without changing the meaning of the sentence:

(17) 我 是 [对你] 说 一句实心眼儿的话。
Wo shi duini shuo yijushixinyan'erdehua。
I be to you say honest words
It is to you that I am saying honest words!
Deep Night by Zhang Henshui (张恨水《夜深沉》)

(18) 太太 是 对孩子 客气一点。
Taitai shi duihaizi keqiyidian。
Madam be to the children be kind
It is to the children that the madam is kind!
The Golden Family by Zhang Henshui (张恨水《金粉世家》)

If we delete "*shi*" from sentences (17) and (18), the expression of the sentences and the logical structures will be as follows:

(17b) 我对你说话 (∃e) [agt (e, 我) &说 (e) &Th (e, 话) &obj (e, 你)]
(18b) 太太对孩子客气 (∃e) [agt (e, 太太) &客气 (e) &obj (e, 孩子)]

The syntactic projection process and the extraction of the semantic role of object by "*shi*" in sentences (17) and (18) is shown in Fig. 6:

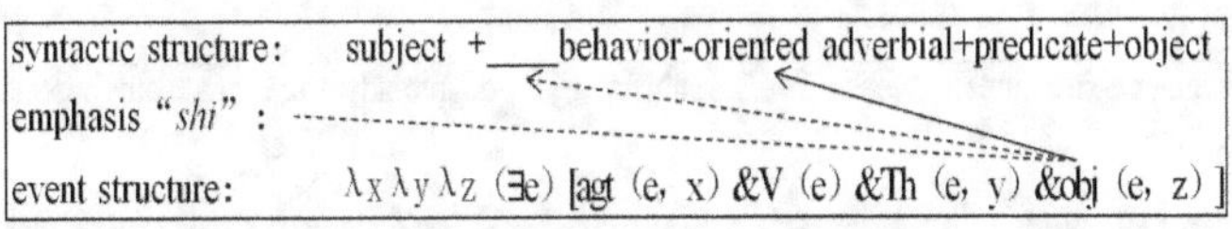

Fig. 6. Extraction of the object by the emphatic "*shi*"

When the object role is projected to a behavior-oriented adverbial position of the syntactic structure, "*shi*" extracts the object and projects itself to the position before the behavior-oriented adverbial. The projection function is:

Projection of the object : $[\mathbf{f}\ (\Pi_{\mathbf{object}}) \rightarrow \mathbf{A}_{\mathbf{behavior-oriented\ adverbial}}]$

The extraction of the object by "*shi*" is as follows:

Extraction of the object by "*shi*" : $\mathbf{if}[shi \twoheadrightarrow \mathbf{f}(\Pi_{\mathbf{object}})] \rightarrow [shi + \mathbf{A}_{\mathbf{behavior-oriented\ adverbial}}]$

4.5 Extraction of the Degree by the Emphatic "*Shi*"

The degree role in an event structure represents the measure on a continuous scale, which can be the measure of either the nature/state of entities or the measure of the speaker's mental state. It is projected to the position of a degree adverbial in the syntactic structure.

"*Shi*" in sentences (19) and (20) extracts the degree role. "*Shi*" again in these sentences can be deleted without changing the meaning of the sentence:

(19) 现在 我 是 十分 后悔呢。
Xianzai wo shi shifen houhuine。
now I be very regret
I am very regretful now.
The Golden Family by Zhang Henshui (张恨水《金粉世家》)

(20) 她 一直 是 非常 喜欢 写作。
Ta yizhi shi feichang xihuan xiezuo。
she always be very like writing
She has always liked writing very much.
My Writing at Shanghai by Chen Danyan (陈丹燕《我的写作与上海》)

After deleting "*shi*" from sentences (19) and (20), the sentences and the logical structures are as follows:

(19b) 我十分后悔 (∃e) [agt (e, 我) &后悔 (e) &Th (e, X) &degree (e, 十分)]
(20b) 她非常喜欢写作 (∃e) [agt (e, 她) &喜欢 (e) &Th (e, 写作) &degree (e, 非常)]

The syntactic projection process and the extraction of semantic role degree by "*shi*" in sentences (20) and (20) are shown in Fig. 7:

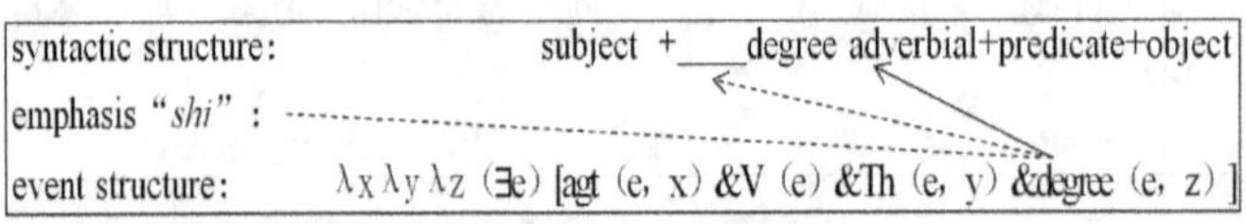

Fig. 7. Extraction of the degree by the emphatic "*shi*"

When the degree role is projected to the degree adverbial position of the syntactic structure, "*shi*" extracts the degree role and projects itself to the position before the degree adverbial. The projection function is:

$$\textbf{Projection of the degree}: [\mathbf{f}\ (\Pi_{\mathbf{degree}}) \rightarrow \mathbf{A}_{\mathbf{degree\ adverbial}}]$$

The extraction of the degree by "*shi*" is as follows:

$$\textbf{Extraction of the degree by "}\textit{shi}\textbf{"}: \mathbf{if}[\textit{shi} \twoheadrightarrow \mathbf{f}(\Pi_{\mathbf{degree}})] \rightarrow [\textit{shi} + \mathbf{A}_{\mathbf{degree\ adverbial}}]$$

4.6 Extraction of the Manner by the Emphatic "Shi"

The manner role in the event structure is the process of unfolding an action. It is projected to the position of a manner adverbial in the syntactic structure. "*Shi*" in sentences (21) and (22) extracts the manner role, and *"shi"* can be deleted without changing the meaning of the sentence:

（21）皇后　起先　是　亲自　出面　替李处温　解释。
Huanghou qixian shi qinzi chumian tilichuwen jieshi。
The queen at first be personally come to for Li Chuwen explain
At first the queen came to explain for Li Chuwen in person.

Incomplete Golden Goblet-National Territory By Xu Xingye （徐兴业《金瓯缺》）

（22）父亲 是暗暗地 为 敢 对自己 挑战的 好汉的 儿子 自豪。
Fuqin shi anandi wei gan duiziji tiaozhande haohande erzi zihao。
my father be secretly for the son who dares to challenge his father proud
My father is secretly proud of his son who dares to challenge him.

Unprecedented Confidence and Courage by Bing Xin（冰心《空前的信心和勇气》）

If we delete "*shi*" from the sentence (22) and (22), the expression of the whole sentence and the logical structure will be as follows:

（21b）皇后亲自出面　(∃e) [agt (e, 皇后) &出马 (e) &manner (e, 亲自)]
（22b）父亲暗暗地自豪　(∃e) [agt (e, 父亲) &自豪 (e) &manner (e, 暗暗地)]

The syntactic projection process and the extraction of the semantic role of manner for emphasis by "*shi*" in sentences (22) and (22) are shown in Fig. 8:

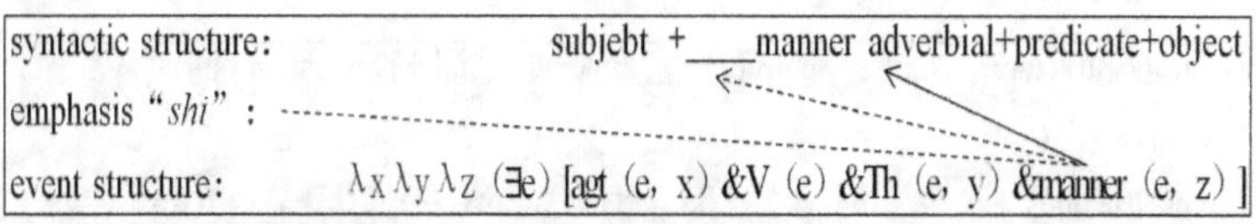

Fig. 8. Extraction of the manner by the emphatic "*shi*"

When the manner role is projected to the manner adverbial position of the syntactic structure, "*shi*" extracts the manner role and projects itself to the position before the manner adverbial. The representation for the projection is:

Projection of the manner: $[\mathbf{f}\ (\Pi_{\mathbf{manner}}) \rightarrow \mathbf{A}_{\mathbf{manner\ adverbial}}]$

The extraction of the manner by "*shi*" is as follows:

Extraction of the manner by "*shi*": $\mathbf{if}[shi \twoheadrightarrow \mathbf{f}(\Pi_{\mathbf{manner}})] \rightarrow [shi + \mathbf{A}_{\mathbf{manner\ adverbial}}]$

4.7 Extraction of the Time by the Emphatic "*Shi*"

The time role in the event structure gives the time of the event. It is projected to the position of a time adverbial of the syntactic structure. "*Shi*" in sentences (23) and (24) extracts the time, and it can be deleted without a change in the meanings of the sentences:

(23) 我 是 去年十月份 到 河北乡下。
Wo shi qunianshiyuefen dao hebeoxiangxia
I be last October go to the countryside in Hebei
I went to the countryside in Hebei last October.
It was last October that I went to the countryside in Hebei.
Dongcangji by Zong Pu（宗璞《东藏记》）

(24) 我 是 那时候 注意起了 苏宇。
Wo shi nashihou zhuyiqile suyu。
I be at that time note Su Yu
At that time, I noticed Su Yu.
It was at that moment that I began to notice Su Yu.
Yelling in the Drizzle by Yu Hua（余华《在细雨中呼喊》）

If we delete "*shi*" from sentences (23) and (24), the expression of the whole sentence and the logical structure will be as follows:

（23b）我去年到乡下 $(\exists e)$ [agt (e, 我) &到 (e) &Th (e, 乡下) &time (e, 去年)]
（24b）我那时候注意苏宇 $(\exists e)$ [agt (e, 我) &注意 (e) &Th (e, 苏宇) &time (e, 那时)]

The syntactic projection process and the extraction of time for emphasis by "*shi*" in sentences (23) and (24) are shown in Fig. 9:

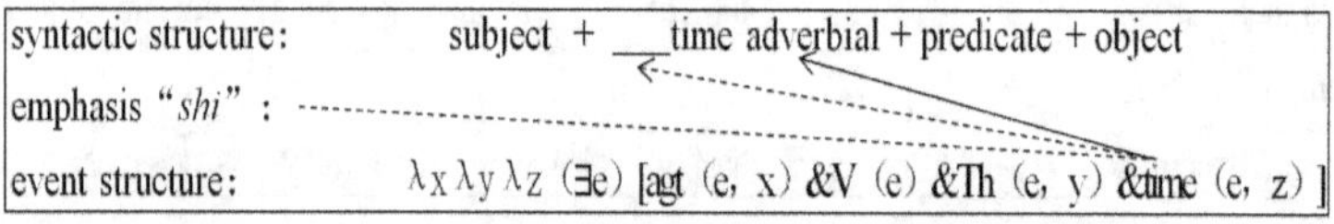

Fig. 9. Extraction of the time by the emphatic "*shi*"

When the time role is projected to the time adverbial position of the syntactic structure, "*shi*" extracts the time role and projects itself to the position before the time adverbial. The projection function is:

Projection of the time : $[\mathbf{f}\ (\Pi_{\mathbf{time}}) \rightarrow \mathbf{A}_{\mathbf{time\ adverbial}}]$

The extraction of the time by "*shi*" is as follows:

Extraction of the time by "*shi*" : $\mathbf{if}[shi \twoheadrightarrow \mathbf{f}(\Pi_{\mathbf{time}})] \rightarrow [shi + \mathbf{A}_{\mathbf{time\ adverbial}}]$

4.8 Extraction of the Space by the Emphatic "*Shi*"

The space in which an event occurs is usually manifested as location in the semantic structure, and is projected to an adverbial of place in the syntactic structure. It is uncommon to use the emphatic "*shi*" to extract the location directly in written Chinese, but such expressions often exist in spoken language. "*Shi*" in sentence (25) and (26) extracts the location, and it can be deleted without a change in the meaning of the sentences:

(25) 他们 是 在大街上 吵吵呢。
Tamen shi zaidajieshang chaochaone。 (Self-constructed sentence)
They be on the street quarrel
They are quarreling on the street.

(26) 他 是 在屋里 写 作业呢。
Ta shi zaiwuli xie zuoyene。 (Self-constructed sentence)
He be in the room write homework
He is doing his homework in the room!

If we delete "*shi*" from sentences (25) and (26), the expression of the sentence and the event structure are as follows:

(25b) 他们在大街上吵吵呢。 (∃e) [agt (e, 他们) &吵吵 (e) &local (e, 大街上)]
(26b) 他在屋里面写作业呢。 (∃e) [agt (e, 他) &写 (e) &Th (e, 作业) &local (e, 屋里)]

The syntactic projection process and the extraction of semantic role of location by "*shi*" in sentence (25) and (26) are shown in Fig. 10:

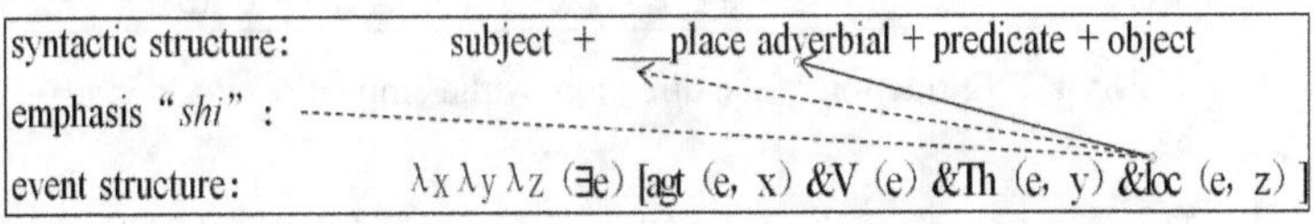

Fig. 10. Extraction of the location by the emphatic "*shi*"

When location is projected to the place adverbial position in the syntactic structure, "*shi*" extracts the location and projects itself to the position before the place of adverbial. Then the function of projection of the location role is:

$$\textbf{Projection of the location} : [\mathbf{f}\ (\Pi_{\mathbf{location}}) \rightarrow \mathbf{A}_{\mathbf{place\ adverbial}}]$$

The extraction of the location by "*shi*" is as follows:

$$\textbf{Extraction of the location by "}\textit{shi}\textbf{"}: \mathbf{if}[\textit{shi} \twoheadrightarrow \mathbf{f}(\Pi_{\mathbf{location}})] \rightarrow [\textit{shi} + \mathbf{A}_{\mathbf{place\ adverbial}}]$$

4.9 Extraction of the Direction by the Emphatic "*Shi*"

The direction role of an event structure, which is a kind of adverbial of place, implies the expansion process of the action in space. It is uncommon to use "*shi*" to emphasize the direction in written Chinese, but in the spoken language. "*Shi*" in sentences (25) and (26) extracts the direction role, and it can be deleted without changing the meaning of the sentence:

(27) 她 是 从上海 转车 到南京。
Ta shi congshanghai zhuanche daonanjing。
(Self-constructed sentence)
She be from shanghai transfer to Nan Jing
She transferred from Shanghai to Najing!

(28) 他们 是 朝另一个方向 去了。
Tamen shi chaolingyigefangxiang qule。

(Self-constructed sentence)

They be in the other direction go
They went in the other direction!

If "*shi*" is deleted, the expressions of the two sentences and the logical structures are as follows:

(27b) **她是从上海转车到南京。**
(∃e) [agt (e, 她) &转车到 (e) &Th (e, 南京) &direction$_{star}$ (e, 上海)]

(28b) **他们朝另一个方向去了**
(∃e) [agt (e, 他们) &去 (e) &direction$_{end}$ (e, 另一个方向)]

The syntactic projection process and the extraction of the semantic role of direction by "*shi*" in sentences (27) and (28) are shown in Fig. 11:

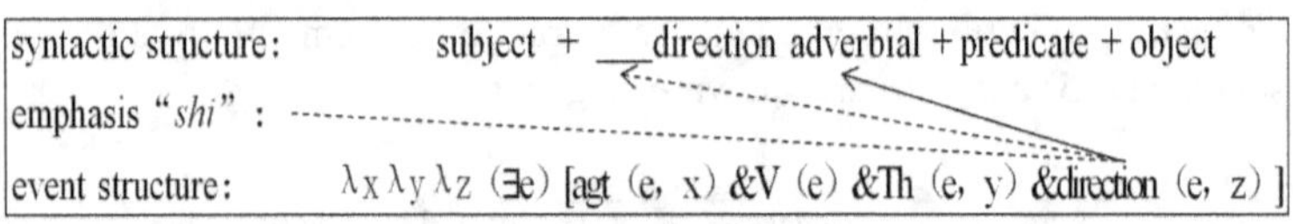

Fig. 11. Extraction of the direction by the emphatic *"shi"*

When the direction role is projected to the position of directional adverbials in the syntactic structure, "*shi*" extracts the direction and projects itself to the position before the direction adverbial. The projection function is:

$$\textbf{Projection of the direction} : [\mathbf{f}\ (\Pi_{\mathbf{direction}}) \rightarrow \mathbf{A}_{\mathbf{direction\ adverbial}}]$$

The extraction of the direction by "*shi*" is as follows:

$$\textbf{Extraction of the direction by "}\textit{shi}\textbf{"} : \mathbf{if}[\textit{shi} \twoheadrightarrow \mathbf{f}(\Pi_{\mathbf{direction}})] \rightarrow [\textit{shi} + \mathbf{A}_{\mathbf{direction\ adverbial}}]$$

We can see from the above analysis that the operation processes of the extraction of all the semantic roles (except for "theme") of the event structure by "*shi*" are similar. This is manifested in three aspects. First of all, "*shi*" belongs to the attitude of the speaker. There is no corresponding position in the event structure for it, and its semantic interpretation is speaker-oriented; secondly, "*shi*" selects the emphasized components from the event structure to operate the extraction of the semantic roles which include agent, instrument, object, degree, instrument, time, space and direction, but except for theme. Finally, the position of "*shi*" in the syntactic structure is determined by the syntactic position of the semantic role extracted by "*shi*". After a semantic role which has been extracted by "*shi*" is projected to a certain position in the syntactic structure, "*shi*" is projected in front of this position. While Π represents the semantic role selected by "*shi*" and A_n represents the syntactic position for the semantic role projection, the projection function of Π is:

$$\textbf{Projection of semantic roles} : [\mathbf{f}\ (\Pi) \rightarrow \mathbf{A_n}]$$

The formula reads: when extracting the semantic role Π from the event structure, Π is projected to the A_n position in the syntactic structure. The projection of emphasis "*shi*" is:

$$\textbf{Extraction of semantic roles by "}\textit{shi}\textbf{": If } [\textit{shi}\twoheadrightarrow f(\Pi)] \rightarrow [\underline{\textit{shi}}+A_n]$$

The formula reads: if "*shi*" selects the semantic role Π from the event structure for emphasis, and if Π is projected to the A_n position in the syntactic structure, then "*shi*" is projected to the position before A_n.

5 Extraction of the Theme by the Emphatic "*Shi*"

"*Shi*" can be projected directly to the syntactic structure after extracting a semantic role except for theme. Yet, it cannot enter the syntactic position after extracting the theme role. The form "*de* + *shi*" is used instead:[5]

(29) 楼下的 铺面房 开 的 是 烟草店。
Louxiade pumianfang kai de shi yancaodian。
Downstairs shop open of be tobacco shop
The shop downstairs is a tobacco shop.
The Criminal Lu Yanshi by Yan Geling (严歌苓《陆犯焉识》)

(30) 我 在工厂里 挣 的是 血汗钱。
Wo zaigongchangli zheng deshi xuehanqian。
I in the factory earn of be hard money
I earn hard money in the factory!
Chronicle of a Blood Merchant by Yu Hua (余华《许三观卖血记》)

"*de* + *shi*" in sentences (29) and (30) can extract the theme, and it can be deleted without changing the meaning of the sentence. If we delete "*de* + *shi*" from sentences (29) and (30), we can get:

(29b) 铺面房开烟草店 (∃e) [agt (e, 铺面房) &开 (e) &Th (e, 烟草店)]
(30b) 我挣汗钱 (∃e) [agt (e, 我) &挣 (e) &Th (e, 汗钱)]

The semantic operation of "*shi*" on the theme is similar to that of the non-theme roles. Firstly, "*shi*" appears between the event structure layer and the syntactic structure layer. Secondly, "*shi*" selects and extracts the semantic role of theme. The difference is that the semantic role of theme is projected to the object position and "*shi*" is projected before the object in the form of "*de* + *shi*". The syntactic projection process and the extraction of semantic role of theme by "*shi*" in sentences (29) and (30) are shown in Fig. 12:

[5] "*de* + *shi*" structure can accept not only the semantic role of the theme but also the semantic role of the agent. It is believed that "*shi*" followed by the agent indicates judgment and is not the emphatic "shi". Therefore it is excluded from this paper.

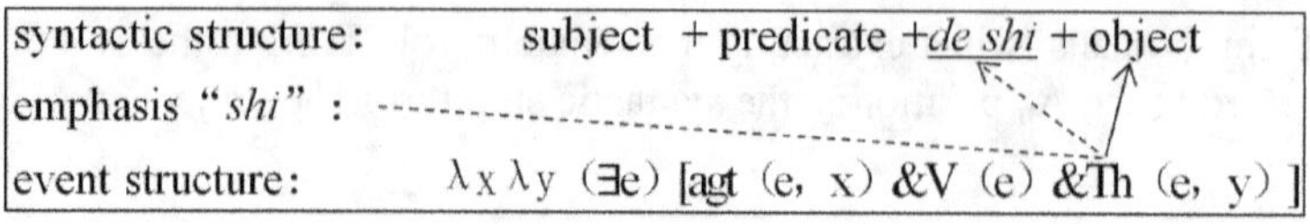

Fig. 12. Extraction of the theme by emphasis "*shi*"

When the theme is projected to the object position in the syntactic structure, "*shi*" extracts the theme and projects itself to the position before the object in the form of "*de* + *shi*". The projection function of the theme is:

$$\textbf{Projection of the theme} : [\mathbf{f}\ (\Pi_{\mathbf{theme}}) \rightarrow \mathbf{A}_{\mathbf{object}}]$$

The extraction of the theme by "*shi*" is as follows:

$$\textbf{Extraction of the theme by "}\textit{shi}\textbf{"} : \textbf{if}[\textit{shi} \twoheadrightarrow \mathbf{f}(\Pi_{\mathbf{theme}})\ ] \rightarrow [\textit{de shi} + \mathbf{A}_{\mathbf{object}}]$$

Although extraction of the theme role and extraction of non-theme semantic roles involve different syntactic forms, the process of extraction of theme by "*shi*" and the syntactic projection are the similar to the process of extraction and projection of non-theme semantic roles. "*Shi*" extracts the theme in the event structure and appears in the form of "*de* + *shi*" in the sentence, while the surface word order is not changed. "*Shi*" can extract all kinds of the semantic roles but generally it extracts only one role in a single event. The whole process is shown in Fig. 13:

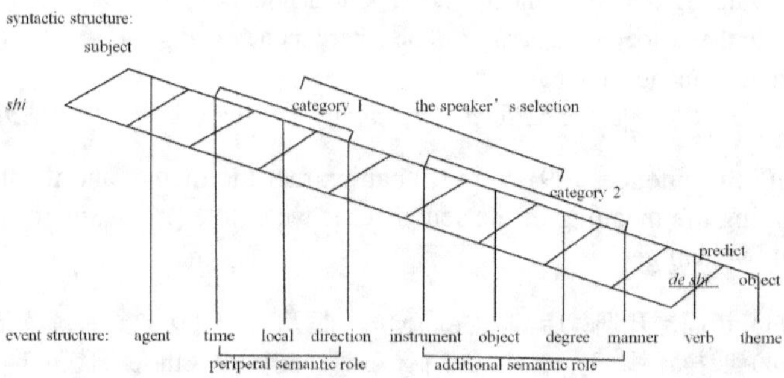

Fig. 13. Extraction of semantic roles by the emphatic "*shi*"

"*Shi*" can extract all the semantic roles, but when it extracts the theme role it is imposed of an additional constraint:

$$\textbf{Extraction of semantic roles by "}\textit{shi}\textbf{"} : \textbf{if}[\textit{shi} \twoheadrightarrow \mathbf{f}(\Pi)\ ] \rightarrow [\ \textit{shi} + \mathbf{A}]$$

(Restriction: if $\Pi \in$ theme , then "*shi*" is projected as "*de+shi*")

6 Other Words Showing Attitude and Their Extraction Functions

Similar to semantic roles extraction by "shi", words expressing the speaker's attitude like "*jiu*" (就), "*guoran*" (果然), "*pianpian*" (偏偏), "*yexu*" (也许), "*shenzhi*" (甚至)

and "*yinggai*" (应该) can also trigger extraction of semantic roles of the event structure. If we use "△" to represent these words, the syntactic form resulting from extracting non-theme semantic roles is "△ + Π". The form resulting from extracting the theme role is "*de* + △ + *shi* + theme", as shown in Fig. 14:

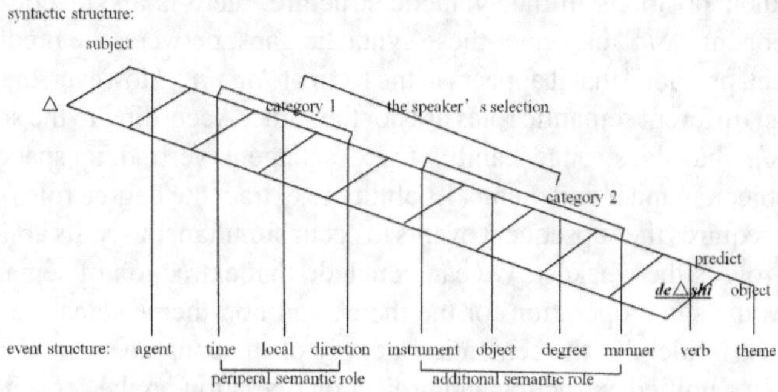

Fig. 14. Logical semantic operations and Chinese word order of "△"

we can further examine the Chinese syntactic structures. The bottom layer of the Chinese language is the event structure, and the surface layer is the syntactic structure. There is also a modal layer between the event layer and the syntactic layer. There are two discourse systems in Chinese corresponding to the event layer and the modal layer, namely:

1. A descriptive discourse system: corresponds to the event layer and deals with agent, verb, theme, instrument, object, degree, manner, time, space, and direction;
2. A narrative discourse system: corresponds to the modal layer and deals with "*shi*", "*jiu*" (就), "*guoran*" (果然), "*pianpian*" (偏偏), "*yexu*" (也许), "*shenzhi*" (甚至), "*yinggai*" (应该) and so on.

The syntactic system of Chinese reflects the mapping of narrative discourse system to the descriptive discourse system. The feature of Chinese is that a semantic role of the event layer is projected to a syntactic position, with a syntactic gap left before such syntactic position. This syntactic gap can be projected onto the modal layer. The operation process is shown in Fig. 15:

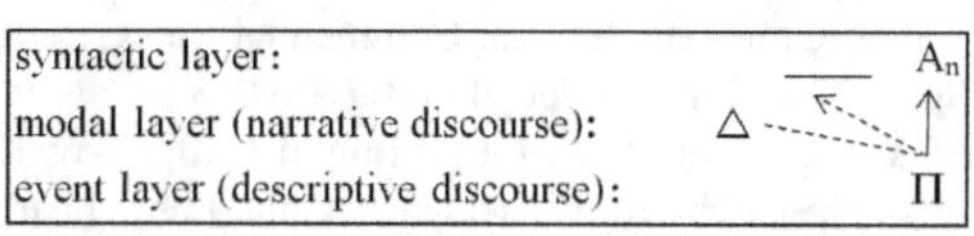

Fig. 15. Mechanism of extracting descriptive discourse to narrative discourse (Π is a semantic role; △ is a descriptive discourse element; A_n is a syntactic position).

7 Conclusion

"*Shi*" needs to select a semantic role in the event structure for emphasis. This appears as an operation on word order in the syntactic structure. The process of "*shi*" is projected before the emphasized semantic role. After the semantic roles of the event structure are projected to their positions in the syntactic structure, there is a syntactic gap in front of each component. "*Shi*" can enter these syntactic gaps, between the predicate and the syntactic object, provided that it appears in the form of "*de shi*". However, the functions of "*shi*" to extract different semantic roles are not the same. According to the statistics from the corpus, "*shi*" has the strongest ability to extract agent, verb, time, space (direction), instrument, object, manner and theme. Its ability to extract the degree role is quite weak, and usually it requires the subsequent events to occur simultaneously. Its ability to extract the direction role is the weakest. We can conclude that extraction of semantic roles by "*shi*" involves the same operation for the theme and non-theme semantic roles, which can: firstly, clearly identify the semantic category of the constituent to emphasize; and secondly, apply a unified extraction and projection operation on the semantic roles. The extraction of semantic roles may not be only for emphasis sense of "*shi*".

In a Chinese syntactic structure, not all event semantic roles are necessarily projected, and the syntactic gaps are not necessarily filled. Regardless of the syntactic structure, the semantic description of the event needs to be complete, and the expression of attitude/modality needs to be clear. This reflects the flexibility and diversity of the Chinese syntactic structures. Event structures can help us analyze Chinese in ideal cases and grasp the construction features of syntactic structures in Chinese. Sentences are the ultimate embodiment of multi-level interactive expressions. To be brief, in this paper, we have analyzed the multi-level structure in Chinese syntax, simplified the internal process of Chinese syntactic construction, and revealed the logical and semantic operation mechanism of Chinese modality expressions.

References

1. Lakoff, R.: The pragmatics of modality. The 8th Regional Meeting of Chicago Linguistic Society, p. 229. Chicago Linguistics Society, Chicago (1972)
2. Wang, L.: Modern Chinese Grammar, p. 115. The Commercial Press, Beijing (2011)
3. Lyu, S.: Chinese Grammar Analysis, p. 115. The Commercial Press, Beijing (1979)
4. Lyu, S.: Eight hundred Words in Modern Chinese. Beijing: The Commercial Press (first edition). 1980; The Complete Works of Lyu Shuxiang, vol. 5, p. 371. Liaoning Education Publishing House, Shenyang (2002)
5. Li, J.: The New Chinese Grammar, 1st edn. Commercial Press, Beijing. Changsha, Hunan (1924) Education Publishing House (reset edition) (2007)
6. Jin, L.: On "*shi...de*" and "*shi*" and "*de*". J. East Linguist. **13**, 52–65 (2013)
7. Zhou, Y.: Review on the research of "*shi*" sentence. Chin. Lang. Learn. **6**, 29–32 (1992)
8. Fang, M.: Syntactic Expression means of contrastive focus in Chinese. Stud. Chin. **4**, 279–288 (1995)
9. Liu, D., Xu, L.: Focus and background, topics and chinese "*Lian*" Sentences. Stud. Chin. Lang. **4**, 243–252 (1998)
10. Xu, J.: Universal Grammar Principles and Chinese Grammar Phenomena. Peking University Press, Beijing (2001)

11. Wang, P., Shi, F., Xiong, J., Shang, S.: Rhythm expression of the focus sentence in the Chinese Mandarin "*shi*". Appl. Linguist. **3**, 134–143 (2019)
12. Zhang, H.: Semantic function of "*shi*" in the "*shi*" configuration of modality confirmation. J. Peking Univ. (Philosophy and Social Sciences). **2**, 95–101 (2007)
13. Liu, D.: Handbook of Language Investigation and Research, 2nd edn, p. 219. Shanghai Educational Publishing House, Shanghai (2017)
14. Yang, K.: Looming Motivation of "*le*" from the perspective of expression function. Chin. Lang. Learn. **5**, 130–131 (2013)
15. Zhao, S.: About the Sentences of "is...". Lang. Teach. Linguist. Stud. **1**, 57–66 (1979)
16. Ma, S.: Conception of the Chinese Teaching Method of "*le, guo, shi...de, zhe, zai, ne*". In: The First International Chinese Teaching Seminar (1985)
17. Lyu, B.: Several Questions about the Structure of "*shi...de*". Lang. Teach. Linguist. Stud. **4**, 21–37 (1982)
18. Zhou, G.: The true role of "shi" - discussion mark - a comprehensive interpretation of the syntactic, semantic and pragmatic functions of "*shi*". Linguist. Res. **2**, 5–11 (2008)
19. Davidson: Essays on Actions and Events (edu). Oxord University Press, Oxford (2002)
20. Parsons, T.: Events in the Semantics of English. MIT Press, Cambridge, Mass (1990)
21. Nan, H.: Events in the Semantics of Chinese. Doctoral Dissertation of Sichuan University (2019)

Nominal Marker "de" in Modern Chinese and Its Role of Marking Presupposition

Liu jingyao(✉)

Institute For International Students, Nanjing University, Nanjing, China

Abstract. Based on the studies on the nominalization marker "de" in modern Chinese, the notion of the presupposition, and the theory of systematic functional grammar, this article attempts to demonstrate that the nominal structure "X + de" in modern Chinese can be listed as a presupposition trigger, where the "X" is presumed to be true, and the "de" is the marker of the presupposition.

Keywords: Nominalization marker "de" · "X + de" · Presupposition

1 Basic Usage of "de"

The function word "de" has been studied in the field of modern Chinese for quite a long time. Since Zhu (1961) published his essay, "About 'de'", Chinese linguists have been devoted to the studies of this word, and assigned it with the function of nominalization. Zhu was the first one who proposed that "de" is a nominal maker. Basically he claims that "de" has three main usages. The first one is to be an enclitic of an adverbial unit, i.e., "不住的" (in English, unstoppably), "十分的" (in English, thoroughly). The grammatical function of this kind of "de" structure is similar to that of the adverbials, and it was called "de_1". The second basic usage of "de" is to be an enclitic of a descriptive adjective unit, i.e., "红红的" (in English, red in color), "好好的" (in English, with good quality). The grammatical function of this kind of "de" structure is similar to that of the adjectives, and it was named "de_2". As for "de_3", it is the enclitic of a nominal unit. In addition to modifying adjectives (i.e., 坏的, in English, the bad one), verbs (i.e., 吃的, in English, food), and nouns (i.e., 木头的, in English, something made from wood), "de_3" also can modify personal pronouns, numerical structures, VO structures, SO structures and so on. All of these structures with "de_3" have a common grammatical function which is similar to that of nouns. In his essay, Zhu also used different methods to prove that a noun composition can modify another nouns. Zhu (1983) then went further in "de_3". He focused on the function of nominalization of this word, and pointed out that in the

The idea of this paper originated from a personal conversation between the author and Prof. Liu Meichun of the City University of Hong Kong. In 2015, Prof. Liu has published her paper "How to Make Sense of Chinese? -- Start with Chinese Function Words" in which she proposes the function of presupposition of Chinese function word "de". Here I want to express my thankfulness to Prof. Liu for her encouragement for me to investigate this grammar phenomenon further.

M. Liu et al. (Eds.): CLSW 2020, LNAI 12278, pp. 34–40, 2021.
https://doi.org/10.1007/978-3-030-81197-6_3

semantic of Modern Chinese, a main method of nominalization is to add "de_3" after predicative units. In the remainder of this article, we will focus on the "de_3".

The "X + de" structure, consisting of the nominalization marker "de_3", is usually used as the attributive, and sometimes as the appositive of a nominal component. In certain contexts, it can also exist independently, acting as the subject, predicative, or object, whose grammatical function is similar to the noun. There are several different viewpoints on the meaning and use rule of the independent "X + de" structure. The first one is the "ellipsis theory", supported by Ji (1965), Kong (1992), Zhang (1994). This theory argues that the independent "X + de" structure comes from the attributive structure whose head word is omitted, and whether the head noun can be omitted depends on the relationship between the attribute and the head noun. The second view claims that the "X + de" structure derives its nominal attribute from "de" itself, and it can function independently. In this theory, which is named "inherent theory", the key to distinguishing whether the structure can be used independently is the distinction between the "de_t" (which can be used to transfer meaning) and the "de_s" (which can only be used as self-reference). The representative linguists who support this theory are Zhu (1983), Liu et al. (2001), and Shao (2007). Shi (2009) pointed out the limitations of these two theory and proposed the third viewpoint. He explained the independent "X + de" structure with the conception of the zero-form indefinite pronoun. Though disagreements exist among all above explanations for the independent "X + de" structure, there is a consensus that this structure has the syntactic status of the noun composition. From the viewpoint of the coordination of modifier, He and Jiang (2011) also proved that the "X + de" structure is the nominalization marker rather than attributive marker.

Besides, "de" is also the marker of the modifier. When "X + de" acts as the attributive, undoubtedly, the "X" modifies the noun after the "de". When it exists independently, whether from the "ellipsis theory", "inherent theory", or "zero-form indefinite pronoun theory", the "X" is still the modifier who modifies the head noun (ellipsis theory), the main case of the predicative units which is extracted (inherent theory), or the zero-form indefinite pronoun after the "de" (indefinite pronoun theory). Liu (2015) elaborated this grammatical function, and demonstrated that "de" is a modifier marker before the noun. Given the above, we can summarize the basic usage of the "de". The first one is to be the marker of nominalization. That means, the "X + de" structure can always be the noun composition. The second one is to be the marker of the modifier. The "X" in the "X + de" is always used as the modifier.

2 Presupposition and Presupposition Triggers

The notion of presupposition was firstly applied in the field of philosophy, and then prevalent in the field of linguistic without a common definition. There are two main approaches to explain it. The first one is the semantic presupposition, and another one is the pragmatic presupposition.

He (1989: 181) explained the semantic presupposition as the referential condition that must be satisfied in order to verify a statement. For example:

(1) John is married.

The presupposition of this statement is:

(2) John exists.

Negation test is a popular way to identify whether the reference about a statement is the presupposition of it. After negating the statement, if the reference is still true, then it is the presupposition. For example, if we negate (1), and get a new statement:

(3) John is not married.

In this statement, (2) is still true. Thus we are sure that (2) is the presupposition of (1).

However, for the reason that the presuppositional relation is sensitive to contextual factors, more and more linguists adopt the pragmatic way to define this notion. Though there are controversies on how to elaborate the pragmatic presupposition, linguists agree that it is the common knowledge possessed by both sides of the conversation, which allows them to understand each other. Pragmatic presupposition is a thriving topic which has been studied by Levinson (1998), Goatly (2000), He (1989), Jiang (2000) and so on. In these studies, linguists found a series of words or structures that signify presuppositional relations, which are called presupposition triggers. Basically, to achieve a presupposition, speakers need to use the presupposition trigger. Levinson (1998) summarized thirteen presupposition triggers: 1) definite description; 2) factive verbs; 3) implicative verbs; 4) change of state verbs; 5) iteratives; 6) verbs of judging; 7) adverbial clauses of time; 8) cleft sentences; 9) structures and expressions indicating comparisons; 10) non-restrictive attributive clauses; 11) counterfactual conditionals; 12) questions; 13) moves of sentence stresses. Due to space limitations, we will not elaborate each kind of presupposition triggers here.

On the basis of Levinson's theory, Cheng (2003) pointed out that though pr2000esupposition triggers seem to be some simple words and structures, it is a complicated issue that what kind of words or forms can achieve the presupposition as it is closely related to the contextual factors and background information. Levinson did not exhaust all presupposition trigger types. Linguists such as Halliday (1994), Goatly (2000) and Cheng (2003) have noticed that nominalization is also a method to presuppose. Goatly () claimed that nominalization is a presupposition trigger, and to support this point, Cheng (2003) found the evidence from the theory of functional grammar. However, the existing studies about the relation between nominalization and presuppositions are all based on English. There is a lack of Chinese studies over this topic (except for Liu 2015). In Liu's paper, she proposes the high presuppositionality of modifier "de" in modern Chinese for the first time. On the basis of Liu's view, the remaining section of this article will pay further attention on elaborating how "de" and "X + de" structures achieve presuppositions.

3 "X + de" Structure and Its Pragmatic Presupposition

When elaborating the process of nominalization, Halliday (1994: 352) noted that it usually involves turning clausal patterns into nominal patterns. From the perspective of grammatical metaphor theory, Halliday thought that this process may lead to the loss of some information, but the "lost information" actually turns into presuppositions without explicit forms. Inspired by this explanation, we found that the noun composition "X + de" in modern Chinese are transferred from clausal patterns. Here are some examples:

(4) 书的颜色是白色——白色的书 (The book is white in color——the white book)
(5) 他们懂了——懂的人 (They have understood——people who have understood)
(6) 房子由木头制成——木头的房子 (The house is made from wood——wooden house)

In the following part, we will explain how the nominal structure "X + de" achieve the pragmatic presupposition.

(7a) 那个姑娘很漂亮。我很喜欢她。(The girl is pretty, and I like her.)
(7b) 我很喜欢那个漂亮的姑娘。(I like that pretty girl.)

Superficially, there is no obvious difference between these two sentences. Nevertheless, their real meaning are actually quite different. (7a) is a statement, expressing the opinion of the speaker. This statement may be correct (with truth value), or it could be wrong (without truth value). At the same time, the listener knows that this is only the speaker's opinion, he or she can agree with it or not. However, after transforming the clausal patter "那个姑娘很漂亮" into the nominal "X + de" structure "漂亮的姑娘", the opinion expressed by the clausal pattern turns in to presupposition. In (7b), the speaker thinks the girl is pretty and affirms that the listener will think so. In other words, the speaker impose his or her opinion on the listener, and in this way, "the girl is pretty" turns into an irrefutable fact accepted by both sides of the conversation.

When the "X + de" structure is used independently, it also marks presupposition. Here is the example:

(8a) 在今年招收的学生中, 有学生戴眼镜, 这部分学生数量很多。(Among the students enrolled this year, some of them wear glasses, and they take a quite large proportion.)
(8b) 在今年招收的学生中, 戴眼镜的很多。(Among the students enrolled this year, the students who wear glasses take a large proportion.)

In (8a), "有学生戴眼镜" (some of students were glasses) is an opinion without definite truth value, and the listener can refute it. However, in (8b), the nominal structure "戴眼镜的"(the students who wear glasses) turns this opinion into presupposition, and it becomes an actual fact.

Zhu (1961) claimed that the "X" in the "X + de" structure can be not only verbs, nouns and adjectives, but personal pronouns, numerical structures and so on. On the basis of his view, we will list how can different kinds of “X + de” structures achieve presuppositions.

	The “X+de”structures (contents in the squre brackets could be omitted)	The presuppositions
personal pronoun	我的 [书](my books)	我有书 (I have some books)
numerical structure	两公斤的[肉](the meat that weights 2 kilograms)	肉重两公斤 (The meat wights 2 kilograms)
verb-object structure	唱歌的[人](the one who sings)	有人唱歌(Someone sings)
verb-complement structure	洗好的[衣服] (the clothes that have been washed)	衣服洗好了(Someone has washed the clothes)
subject-object structure	我给他的[书] (the book that I gave him)	我给了他书(I gave him a book)
serial verb construction	送票给我的[人] (the one who gave me the ticket)	有人送票给我 (Someone gave me the ticket)
pivotal construction	请你看戏的[人](the one who invited you to watch the show)	有人请你看戏 (Someone invited you to watch the show)
combined structure	我和弟弟的[糖](candies of my brother and mine)	我和弟弟有糖(My brother and I have some candies)

From the list, we can tell that the presuppositions are achieved by transforming the clausal patterns into nominal patterns. The clausal pattern statements are opinions without definite truth value, while the “X” in the “X + de” structures are supposed to be with truth value, that is, accepted by both sides of the conversation. By applying this structure, the speaker imposes the presupposition on listeners, and “de” becomes the maker of the presupposition.

Our method to test whether the "X + de" structure can successfully realize the presupposition is to see whether it satisfies the two features of the general presupposition. The first one is that the truth value of the presupposition is the basis of the whole sentence, and the second one is that the presupposition is independent from the affirmative or negative form of the sentence. We firstly examine the former feature:

(9a) 昨天有报纸, 我看过了。(A newspaper was published yesterday, and I have read it.)
(9b) 昨天的报纸我看过了。(I have read the yesterday's newspaper.)

The structure "昨天的" in (9b) presupposes that "a newspaper was published yesterday", and this is also the precondition of (9b), because if no newspaper was published yesterday, then the whole sentence will make no sense. Thus we know that the "X + de" structure satisfies the first feature. How about the second feature? We negate (9b) and get:

(9c) 昨天的报纸我没看过。(I have not read the yesterday's newspaper.)

Obviously, the presupposition of (9c) is still that "a newspaper was published yesterday". Therefore, the "X + de" structure also satisfies the second feature of the general presupposition.

Actually, among the thirteen types of presupposition trigger listed by Levinson (1998), the definite description overlaps the "X + de" structure. The definite description includes the personal pronoun and the noun indicates possession, which in Chinese are the "X + de" structures that indicate possession (the X is personal pronoun or noun). However, what Levinson stresses are the circumstances in English, and do not involve other kind of "X + de" structures, thus we think that in the modern Chinese, the nominal structure "X + de" could be a specific kind of presupposition trigger. In this structure, the "X" is supposed to be true, and the "de" is the marker of the presupposition.

In his essay, Cheng (2003) elaborated the nominalized presupposition trigger in English with the notion of transitivity and thematic structure in the theory of systematic functional grammar. We also try to explain how the "X + de" structure in Chinese realize presupposition from these two perspectives. The first one is the transitivity. In the theory of systematic functional grammar, informative function means "expressing human's experiences in the real world (including the inner world) by language" (Hu et al. 1989: 71), and the transitivity is a semantic system to represent informative function. It divides those experiences into several processes, and stresses the factors that are related to these processes such as participates and contexts. The processes are always dynamic, however, the nominal structures "X + de" combine the dynamic processes and the factors that are related to them, and then form a new static item. Though in this combining process, some information loses, they can be inferred. The original dynamic processes may or may not occur, but people tend to believe the existence of static objects. (Goatly 2000: 77) And this is why the nominal structure "X + de" can realize presupposition.

Besides, according to the thematic structure, speakers usually put the "old" information at the thematic position. Meanwhile, the products of the nominalization are always put at the thematic position, and so does the nominal structure "X + de". Thus, this structure naturally carries the common knowledge and realize the presupposition. In other words, by nominalizing the clausal pattern, the "X + de" structure transfer the information of the discourse. It shifts the "new" information in the original clauses to a position which is for the "old" information. In this way, the structure completes the presupposition.

4 Conclusion

Based on the studies of the nominalization marker "de" in modern Chinese and the studies on the conception of the presupposition, we argue that in modern Chinese, the nominal structure "X + de" should be classified as a type of presupposition trigger, where the "X" is presumed to be true, and the "de" is the marker of the presupposition. We can elaborate how does this structure realize the presupposition from the notion of transitivity and thematic structure in the theory of systematic functional grammar.

References

1. Cheng, X.: Nominalization and presupposition. Foreign Lang. Res. **3**, 19–23 (2003)
2. Goatly, A.: Critical Reading and Writing - An Introductory Coursebook. Routledge, London (2000)
3. Halliday, M.A.K.: An Introduction to Functional Grammar. Edward Amold, London (1994)
4. He, C., Jiang, Y.: Nominative XP-de and semantics of -de. Contemp. Linguist. **1**, 53–66+97–98 (2011)
5. Zhaoxiong, H.: An Introduction to Pragmatics. Shanghai Foreign Language Education Press, Shanghai (1989)
6. Ziran, H.: Pragmatics and English Learning. Shanghai Foreign Language Education Press, Shanghai (1998)
7. Zhuanglin, H., Yongsheng, Z., Delu, Z.: A Survey of Systemic-Functional Grammar. Hunan Education Publishing House, Changsha (1989)
8. Yongxing, J.: On "About 'de.'" Stud. Chin. Lang. **5**, 363–364 (1965)
9. Wangqi, J.: Pragmatics: Theories and Applications. Beijing University Press, Beijing (2000)
10. Lingda, K.: Semantic rules on omitting head nouns in "n1+de+n2" structures. J. Anhui Normal Univ. **1**, 103–107 (1992)
11. Levinson, S.C.: Pragmatics. Cambridge University Press, Cambridge (1998)
12. Meichun, L.: How to understand Chinese grammar—from the perspective of the function words. J. Chin. Lang. Teach. **3**, 31–52 (2015)
13. Liu, Y., Pan, W., Gu, W.: Practical Modern Chinese Grammar. The Commercial Press, Beijing (2001)
14. Jingmin, S.: Introduction to Modern Chinese. Shanghai Education Press, Shanghai (2007)
15. Shi, D.: Indefinite pronoun and the independent "de" structure. Foreign Lang. Teach. Res. **2**, 5–13+82 (2009)
16. Jing, Z.: Language, Pragmatics and Grammar. Wenxin Press, Zhengzhou (1994)
17. Dexi, Z.: About "de." Stud. Chin. Lang. **12**, 1–15 (1961)
18. Dexi, Z.: Self-reference and transferred reference——grammatical and semantic function of Chinese nominalization markers "de, zhe, suo, zhi." Dialect **1**, 18–33 (1983)

An Analysis of the Word Formation Process for Negative Quasi-prefixes 非 "fei" and 无 "wu" in Modern Chinese

Yaoyao Ning(✉)

Department of Chinese Language and Literature, University of Macau, Macao, China
yb97711@umac.mo

Abstract. Though acting as very productive morphemes, the Chinese quasi-prefixes 非 "fei" and 无 "wu" have received inadequate attention in the studies of Chinese morphology. The current research, drawing on extensive corpus data, identifies the structural hierarchy of lexical units with 非 "fei" and 无 "wu" as quasi-prefixes through examination of corresponding syntactic, semantic and prosodic structures, analyzes their structural meanings within lexical units, and further explores their difference in word formation and the reasons for such difference. Our findings can provide a new model for analysing lexical units with 非 "fei" and 无 "wu" as quasi-prefixes, which may be applied to improve automatic word segmentation and part-of-speech tagging in Chinese information processing.

Keywords: Structural hierarchy · Structural meaning · Negation · Event coercion · Qualia structure

1 Introduction

In modern Chinese, 非 "fei" and 无 "wu" can be used as both negative adverbs and morphemes. When acting as morphemes, they are very productive, as reflected in many lexical units with them as quasi-prefixes in dictionary entries, new words, and colloquial expressions. Regarding their definition, some scholars consider them quasi-prefixes (e.g., [1, 2]) while others use the term prefixes (e.g., [3, 4]). The current research adopts the term quasi-prefix to distinguish them from the traditionally-defined prefixes that do not have a concrete meaning and are less productive [5, 6], such as 老 "lao", 阿 "a", and 第 "di".

Previous studies on 非 "fei" and 无 "wu" tend to explore their etymology, semantic meanings, and the types of morphemes they are attached to (e.g., [7, 8]), with little attention paid to the structural hierarchy of lexical units with them as quasi-prefixes. However, clarifying the structural hierarchy is essential for analyzing the word formation process involving 非 "fei" and 无 "wu". In this regard, existing research has mainly two views about the analysis of the internal structures (the first level of division) of the lexical units that contain them as quasi-prefixes. We may take 非企业法人*fei1 qi3ye4 fa3ren2* "non-corporate legal person" and 无脊椎动物*wu2 ji3zhui1 dong4wu4* "invertebrates"

M. Liu et al. (Eds.): CLSW 2020, LNAI 12278, pp. 41–52, 2021.
https://doi.org/10.1007/978-3-030-81197-6_4

as examples. These two lexical units are either divided into [非*fei1*- [企业法人*qi3ye4 fa3ren2*]] and [无*wu2*-[脊椎动物*ji3zhui1 dong4wu4*]] (e.g., [3]), or into [[非*fei1*-企业*qi3ye4*]法人*fa3ren2*]] and [[无*wu2*-脊椎 *ji3zhui1*]动物*dong4wu4*] (e.g., [4]). However, relevant studies tend to directly adopt one of the two without providing theoretical support for doing so. For example, Wang and Fu [6] analyze the prosodic structure of lexical units with suffix 式*shi4* "style" and summarize their lexical categories, including 字 "monomorphemic morphemes", 词 "words", 类词 "quasi-words", and 凝固短语 "solidified phrases". We can see that the researchers adopt the first analysis but do not provide any rationale for it. In addition, due to the asymmetry in word formation involving quasi-prefixes and quasi-suffixes [9], the patterns derived from the former cannot be generalized for the latter. To gain insight into the word formation process involving 非 "fei" and 无 "wu" as quasi-prefixes, this article, drawing on extensive corpus data, identifies the structural hierarchy of lexical units that contain them through examining their syntactic structure, semantic and prosodic structures, analyzes the structural meanings of the quasi-prefixes, and further investigates their difference in word formation and the reasons for such difference.

2 The Source and Classification of Corpus

The current research extensively searches and classifies lexical units with 非 "fei" and 无 "wu" as quasi-prefixes from the BCC multi-domain corpus, three Chinese dictionaries, including *Modern Chinese*[1], *Chinese Dictionary*[2] and *The New Dictionary of New Words*[3], and one reference book *Eight Hundred Words*[4]. A total of 420 lexical units are acquired from the BCC corpus through automatic word segmentation search. Among them, 155 cases, after manual verification, are finally sorted out, comprising 92 cases of 非 "fei" and 63 cases of 无 "wu". 71 out of the 155 cases are entries included in the aforementioned Chinese dictionaries or the reference book. Since most of these cases are composed of two or three components[5], the current study presents word examples in a form as provided in the following subsection[6], and detailed analysis of structural hierarchy will be illustrated later.

2.1 Types of Lexical Units with Quasi-prefix 非 "fei"

(1) [非-名词]$_{\text{名词}}$ "[fei-noun]n":

---

[1] *Modern Chinese Dictionary* (7th edition), Beijing: Commercial Press, 2016 edition, referred to as *Modern Chinese.*

[2] *Chinese Dictionary*, Shanghai: Shanghai Dictionary Publishing House, 2011 edition.

[3] *The New Dictionary of New Words* (1978-2018), Shanghai: Shanghai Dictionary Publishing House, 2018 edition.

[4] *Eight Hundred Words in Modern Mandarin Chinese*, Beijing: Commercial Press, 1999 edition, referred to as *Eight hundred words.*

[5] Refering to morphemes that can independently form words in the lexicon (e.g., [12, 13]). The current research abides by the entries in *Modern Chinese.*

[6] The examples in bold are entries in one of the dictionaries, including *Eight hundred words*, *Modern Chinese*, *Chinese Dictionary*, and *The New Dictionary of New Words.*

非-金属 *fei1- jin1shu3* "non-metallic"
非-党员 *fei1-dang3yuan3* "non-party member"
(2) [非-名词$_1$-名词$_2$]$_{名词}$ "[fei-noun1-noun2]n":
非-农-产业 *fei1- nong2- chan3ye4* "non-agriculture industry"
非-政府-组织 *fei1-zheng4fu3-zu3zhi1* "non-government organization"
(3) [非-动词-名词]$_{名词}$ "[fei-verb-noun]n":
非-卖-品 *fei1-mai4-pin3* "not for sale"
非-国有-企业 *fei1-guo2you3-qi3ye4* "non-state-owned enterprise"
(4) [非-形容词-名词]$_{名词}$ "[fei-adjective-noun]n":
非-发达-国家 *fei1- fa1da2- guo2jia1* "non-developed country"
非-正规-组织 *fei1- zheng4gui1- zu3zhi1* "informal organization"
(5) [非-属性词[7]-名]$_{名词}$ "[fei-attribute word-noun]n":
非-常规-能源 *fei1-chang2gui1-neng2yuan2* "non-conventional energy"
非-军用-机场 *fei1- jun1yong4 -ji1chang3* "non-military airport"

2.2 Types of Lexical Units with Quasi-prefix 无 "wu"

(6) [无-名词$_1$-名词$_2$]$_{名词}$ "[wu-noun1-noun2]n":
无-线-电 *wu2- xian4- dian4* "wireless"
无-脊椎-动物 *wu2- ji3zhui1-dong4wu4* "invertebrate"
(7) [无-名词-动词]$_{名词}$ "[wu-noun-verb]n":
无-氧-运动 *wu2-yang3 -yun4dong4* "anaerobic exercise"
无-条件-反射 *wu2 -tiao2jian4-fan3she4* "unconditional reflection"

We can see that only one case, i.e., (1), comprises two components, and its internal structure can be analysed into [fei-noun]n by a single dichotomy. However, this approach cannot be applied to the other cases composed of two components, e.g., (2)-(7). Their structural hierarchy will be examined in details in the following sections.

3 Structural Hierarchy of Lexical Units with 非 "fei" and 无 "wu" as Quasi-prefixes

In this section, the syntactic, semantic and prosodic structure of the examples (2)-(7) above are illustrated, and the compositional type of lexical units comprising three components are presented collectively in the form of [非"fei"/无"wu"-X-Y]n.

3.1 Syntactic Structure of [非"fei"/无"wu"-X-Y]n

The general structure for Chinese modifier-head constructions usually involves a subordinate category of the head word in question, that is, "AB is a kind of B". The basic form to identify modifier-head constructions is "Quantifier + AB→Quantifier + B", e.g., 一条鲤鱼*yi4 tiao2 li3yu2* "a carp"→一条鱼*yi4 tiao2 yu2* "a fish" and 一双拖鞋*yi4 shuang1 tuo1xie2* "a pair of slippers"→一双鞋*yi4 shuang1 xie2* "a pair of shoes". In

[7] The judgment of attribute words in this article follows *Modern Chinese.*

the current research this method is also applicable to [非"fei"/无"wu"-X-Y]n, and some specific examples are illustrated as follows.

(8) 非专业课: 一门非专业课 → 一门课

non-professional course: a non-professional course → a course

(9) 非油炸食品: 一种非油炸食品 → 一种食品

non-fried food: a kind of non-fried food → a kind of food

(10) 无声片: 一部无声片 → 一部片

silent film: a silent film → a film

(11) 无氧运动: 一次无氧运动 → 一次运动

anaerobic exercise: one anaerobic exercise → one exercise

Examples (8)–(11) are all modifier-head constructions in which [非"fei"/无"wu"-X]n as an entity serves as the attributive of [Y], modifying or qualifying the head noun, and thus, [非"fei"/无"wu"-X-Y]n can be analysed into [[非"fei"/无"wu"-X]-Y]n.

3.2 Semantic Structure of [非"fei"/无"wu"-X-Y]n

The semantic relationship within modifier-head constructions are diverse, including classification, position, and property. [X-Y] in [非"fei"-X-Y] is found to represent a classification relationship. Specifically, [X] is an attribute or feature of [Y] in the aspect of object, property, degree or function. Based on [X] (attribute feature), [Y] can be classified into different sub-categories, as in (12a) and (13a) below. Therefore, the negative quasi-prefix非"fei" negates [X] (attribute features), rather than [Y] (category). [非"fei"-X] specifies a new category of [Y] which is not altered in the word formation process. As such, the semantic relationship between [非"fei"-X] and [Y] is also a classification relationship, as in (12b) and (13b) below.

(12) a. 手术-疗法: 根据治疗方式"手术"对疗法分类。

surgical therapy: Classify the therapies according to the treatment method "surgery".

b. 非手术-疗法: 根据治疗方式"不涉及手术"对疗法分类。

non-surgical therapy: Classify treatments according to involvement of no "surgery".

(13) a. 国有-企业: 根据企业所属性质"国有"对企业分类。

state-owned enterprise: Classify enterprises according to the property of the "enterprise".

b. 非国有-企业: 根据企业所属性质"不属于国有"对企业分类。

non-state-owned enterprise: Classify enterprises according to the property of being not "state-owned".

Unlike in [非"fei"-X-Y]n, [X-Y] in [无"wu"-X-Y]n demonstrates an inclusive relationship. Specifically, [X] as a component or an attribute feature of [Y] loses its independent meaning and is semantically integrated into [Y]. As shown in cases (14) and (15) below, there is no such independent expression in the form of [X-Y] as 翼鸟 "winged bird" and 痛分娩 "painful delivery". Only for the purpose of emphasis is [X-Y] preceded by有*you3* "have/there be". Thus, the negation of the inclusive relationship between [X] and [Y] is achieved through using 无 "wu" to negate [X] (component or attribute feature), but not to negate [Y] (category). In sum, the analysis of the semantic relationships within the lexical units that contain非 "fei" and 无 "wu" as quasi-prefixes points to the internal structure of [非"fei"/无"wu"-X-Y]n as [[非"fei"/无"wu"-X]-Y]n.

(14) 无翼鸟: 翅膀是鸟的部件。
wingless bird: Wings are parts of a bird.
(15) 无痛分娩: 产生痛觉是分娩的属性特征。
painless childbirth: Pain sensation is an attribute of childbirth.

3.3 Prosodic Structure of [非"fei"/无"wu"-X-Y]n

The prosodic structure of [非"fei"/无"wu"-X-Y]n is mainly determined by its foot. According to Feng [14], the smallest, basic foot in Chinese is at least composed of two syllables and a prosodic word must contain at least one foot. As such, a two-syllable foot is called a "standard prosodic word" while a three-syllable foot, i.e., a super foot, is called a "super-prosodic word". In other words, a foot that contains more than three syllables must be a combination of standard prosodic words, of super-prosodic words, or of both. According to Feng [15], compounds and their word formation follow the left-to-right direction. Therefore, [非"fei"/无"wu"-X-Y]n as word formation structures should also follow the same foot operation rule, that is, taking syllables from left to right until the number of syllables legitimately satisfies the foot structural requirement. A number of examples of prosodic structure of [非"fei"/无"wu"-X-Y]n involving more syllables than two are given as follows in Table 1.

Table 1. Prosodic Structure of [非"fei"/无"wu"-X-Y]n

Number of syllables	Prosodic structure	Example
三音节 Three syllables	[2+1]	非卖品 *fei1 mai4 pin3* "not for sale": [[非卖*fei1 mai4*]品*pin3*] 无烟煤*wu2 yan1 mei2* "anthracite": [[无烟*wu2 yan1*]煤*mei2*]
四音节[8]Four syllables	[2+2]	非农产业*fei1nong2 chan3ye4* "anthracite": [[非农*fei1nong2*][产业*chan3ye4*]] 无轨电车*wu2gui3 dian4che1* "Trolleybus": [[无轨*wu2gui3*][电车*dian4che1*]]
五音节 Five syllables	[3+2]	非油炸食品*fei1 you2zha2 shi2pin3* "non-fried food": [[非油炸*fei1 you2zha2*][食品*shi2pin3*]] 无脊椎动物*wu2 ji3zhui1 dong4wu4* "invertebrate": [[无脊椎*wu2 ji3zhui1*][动物*dong4wu4*]]

[8] In the corpus there are four-syllable [非"fei"-X-Y]$_n$ lexical units, such as非专业课*fei1 zhuan1ye4ke4* "non-professional course". If following the left-to-right direction, its prosodic structure should be [2+2]. However, there are no such expressions as 非专*fei1 zhuan1* and 业课*ye4 ke4*. Duan [16] argues that the structure of lexical units can be adjusted to the best length, and thus, to meet the semantic requirement, the case can be processed as 非专业课程 *fei1 zhuan1ye4 ke4cheng2* "non-professional courses", following the [3+2] prosodic structure.

As shown in Table 1, a three-syllable lexical unit of [非"fei"/无"wu"-X-Y]n is a prosodic word; a four-syllable one is a combination of two standard prosodic words; and a five-syllable one is a combination of a super-prosodic word and a standard prosodic word. In all three types of prosodic structures listed in Table 1, the quasi-prefixes 非 "fei" and 无 "wu" combine with adjacent monosyllabic or disyllabic words [X] into a standard or a super foot. Therefore, the structural hierarchy of [非"fei"/无"wu"-X-Y]n based on prosodic structure is also [[非 "fei"/无 "wu"-X]-Y]n.

In sum, the analysis of the syntactic, semantic and prosodic structure gives a similar conclusion that the structural hierarchy of [非"fei"/无"wu"-X-Y]n is [[非"fei"/无"wu"-X]-Y]n instead of [非"fei"/无"wu"-[X-Y]]n. In other words, 非 "fei" and 无 "wu" are attached to the adjacent root [X], and the combination [非"fei"/无"wu"-X] functions as a attribute word[9], serving as an attributive to modify or qualify the head noun [Y].

4 Structural Meanings of the Quasi-prefixes 非 "fei" and 无 "wu"

According to Tan [18], the meaning of a word includes word meaning and structural meaning, and the latter is generated in the word formation processes, that is, through word combinations. For example, the interpretation of 玻璃瓶子*bo1li2 ping2zi3* "glass bottle" as "a bottle made of glass" is subject to both the word meaning and the semantic relationship within this modifier-head construction. However, the previous studies that examined the meaning of 非 "fei" and 无 "wu" seldom distinguish word meaning from structural meaning, simply regarding the meaning of 非 "fei" as 不是*bu2shi4* for "negative judgment" (e.g., [7, 19]) and 无 "wu" as 没有*mei2you3*, to indicate that something is not in anyone or someone's possession or it does not exist at all (e.g., [19]). The current research argues that such interpretation only presents word meaning, leaving out structural meaning. To fill this gap, the following section examines the structural meanings of 非 "fei" and 无 "wu" in lexical units.

4.1 Structural Meanings of the Quasi-prefix 非 "fei"

The basic word meaning of 非 "fei" is 不是*bu2shi4*, reflecting "negative judgment" [4]. When used as a quasi-prefix, 非 "fei" is interpreted in *Eight Hundred Words* and *Modern Chinese* as "not belonging to a certain range", which appears to be its structural meaning. However, this interpretation is get to find its theoretical support. To explore the features of 非"fei" in dictionary entries, the following section examines a number of the representative examples extracted from *Modern Chinese*.

(16) 非农: 不属于农村或农业的。

non-agricultural: not belonging to rural areas or agriculture.

(17) 非政府组织: 除政府和企业以外的其他非营利性的社会组织。

non-governmental organization: other non-profit social organizations except government and enterprises.

(18) 非婚生子女: 没有合法婚姻关系的男女所生的子女。

[9] Attribute words are mainly used as attributives for the function of classification or naming, to be followed by a noun or the particle 的 "de" [17].

children born out of wedlock: children born to men and women without a legal marriage relationship.

(19) 非卖品: 只用于展览、赠送等不出卖的物品。

not for sale: only used for exhibitions, gifts and other items that are not for sale.

(20) 非常规能源: 新能源。

unconventional energy: new energy.

We can see that only the interpretations in (16) and (17) reflect the structural meaning of 非 "fei", that is, "not belonging to a certain range or out of a certain range" which appears to be associated with the classification relation within the semantic structure [[非"fei"-X]-Y]n. As mentioned in Sect. 3.1 and 3.2, within the modifier-head construction of [[非"fei"-X]-Y]n, the semantic relationship between [非"fei"-X] and [Y] indicates a classification relationship. Specifically, [非"fei"-X] represents the semantic range opposite to [X] through using 非 "fei" to negate [X], and then modifies [Y] to form a new object category [[非"fei" -X]-Y]n that refers to "objects that are out of the range of [X]" or "all the objects that exclude [X]". Thus, the structural meaning of 非 "fei" as "not belonging to a certain range or out of a certain range" can be well justified.

4.2 Structural Meaning of the Quasi-prefix无 "wu"

The word meaning of 无 "wu" is 没有*mei2you3*, indicating something not possessed by anyone or something does not exist. However, its structural meaning in word formation cannot be inferred in this way. The following lexical units with 无 "wu" as a quasi-prefix show that a verb is added to reconstruct an event or a behavior in almost all the cases.

(21) 无产者: 不占有生产资料的人 (《现汉》)。

proletarian: A person who posseses no means of production (*Modern Chinese*).

(22) 无烟煤: 不冒烟或冒烟少的煤 (《汉语大词典》)。

anthracite: Coal that emits little or less smoke (*Chinese Dictionary*).

(23) 无影灯: 不形成阴影的灯 (《汉语大词典》》)。

shadowless lamp: A lamp that does not form a shadow (*Chinese Dictionary*).

(24) 无轨电车: 不用轨道的车 (《汉语大词典》》)。

trolley bus: A car that does not use rails (*Chinese Dictionary*).

(25) 无党派人士: 没有参加任何党派的人士 (《现汉》)。

non-partisan: Persons who have not joined any party (*Modern Chinese*).

This feature of semantic generation, according to the generative lexicon theory (GLT), embodies the event coercion which, as one of the generative mechanisms in semantics, refers to the process of reconstructing hidden events and revealing implicit meanings through the release of an invisible predicate. Similar phenomena exist extensively in predicate structures, subject-predicate structures, orientation structures and quasi-affix word formation in Chinese [20]. In addition, event coercion is associated with the qualia structure of nouns, which includes constitutive, formal, telic, and agentive roles, in addition to conventionalized attributes [21, 22]. Based on these concepts, Yuan [23] adds another six qualia roles to the qualia structure of Chinese nouns, including unit, evaluation, material, handle, action, and orientation. Our examination of lexical units in the pattern of [[无"wu"-X]-Y]n indicates that the verb semantically activated by 无 "wu" is associated with the qualia structure of the noun [X], which mainly plays the handle and agentive role in the following two types of situations.

(1.) [X] plays the constitutive role of [Y], i.e., [X] is a component or constituent of [Y]. The verb activated by 无 "wu" plays the handle role of [X], representing habitual actions, behaviors, or influences exerted by people or other objects on [X]. Two examples are given as follows.

(26) 无核枣: 去核的枣。核[处置角色: 去]

seedless date: jujube with the pip removed. pip [handle role: remove]

(27) 无声片: 没加声音的影片。声[处置角色: 加]

silent film: film without sound track added. sound [handle role: add]

(2.) [X] is a conventionalized attribute of [Y], and [X] represents a regular activity or attribute related to [Y] gained through previous access to or experience of [Y]. The verb semantically activated by 无 "wu" plays the agentive role of [X], referring to the way how [X] is formed or generated. Two examples are given as follows.

(28) 无翼鸟: 不长翅膀的鸟。翼[施成角色: 长]

wingless bird: A bird that does not grow wings. wing [agentive role: grow]

(29) 无尘粉笔: 不产生粉尘的粉笔。尘[施成角色: 产生]

dust-free chalk: chalk that does not produce dust. dust [agentive role: produce]

In sum, to understand and interpret lexical units of [[无"wu"-X]-Y]n, a verb needs to be added to reconstruct an event or a behavior and display its implicit semantic meaning. In this way, we can see that the structural meaning of 无 "wu" implies a hidden action or event not tangibly taking place.

5 Differences in Word Formation between the Quasi-prefixes 非 "fei" and 无 "wu"

The above analysis results indicate that although 非 "fei" and 无 "wu" are both negative quasi-prefixes, they vary in word formation: (1) 非 "fei" can be attached to nouns, verbs, adjectives and attribute words whereas无 "wu" usually is attached to nouns solely; (2) The structural meaning of 非 "fei" is "not belonging to a certain range or out of a certain range" while that of 无 "wu" indicates an action or an event not tangibly taking place. The cause of this difference is due to the different negation types and functions of the quasi-prefixes. Shen [24] syntactically distinguishes between the Chinese negative words that negate 是*shi4* "be" (e.g., 非 "fei") and those that negate 有*you3* "have/there be" (e.g., 无 "wu"), interprets非 "fei" as 不是*bu2shi4*, which reflects negative judgement, and considers 无 "wu" as 没有*mei2you3*, which indicates that something is not possessed by anyone or does not exist. In the following section, comparable word formation features and semantic relationship of these two quasi-prefixes for 是*shi4* "be" and 有*you3* "have/there be" are examined to discern whether morphologically 非 "fei" also negates 是*shi4* "be" and 无 "wu" also negates 有 *you3* "have/there be".

5.1 非 "fei" as the Negation of 是 *shi4* "be"

是*shi4* "be" as a judgment verb performs the function of affirmation and connection. Based on the semantic relations that是*shi4* "be" establishes between the adjacent items prior to and behind it, the semantic types of是*shi4* "be" include equivalence, classification, description, existence, possession, and so on [3, 25]. When是*shi4* "be" semantically performs the function of classification, it involves three syntactic combinations [3].

(30) 主语+是+名词: 鲸鱼是哺乳动物。

subject + be + noun: Whales are mammals.

(31) 主语+是+动词+的 (名词) : 这批货是新出厂的 (产品) 。

subject + be + verb + of (noun): This batch of goods are newly manufactured (product).

(32) 主语+是+形容词+的 (名词) : 衣服是旧的 (衣服) 。

subject + be + adjective + (noun): The clothes are old (clothes).

In the cases listed above, the item behind 是 *shi4* "be" can be a noun, a verb or an adjective. Semantically, 是 *shi4* "be" means that the preceding item belongs to the following category or type, and thus can be interpreted as "belonging". To negate 是*shi4* "be", 不 *bu4* "not" is added prior to 是 *shi4* "be", meaning "not belonging to" and indicating that the items prior to and behind 是 *shi4* "be" do not have a classification relationship. Similarly, the quasi-prefix 非 "fei" is attached to nouns, verbs and adjectives (attributive words) and semantically preforms the classification function, meaning "not belonging to". Therefore, the word formation features of 非 "fei" as quasi-prefixes are consistent with the negation features of 是 *shi4* "be" that represents a classification relationship.

5.2 无 "wu" as the Negation of 有 *you3* "have/there be"

As a verb, 有*you3* "have/there be" means "to possess" or "to exist", and refers to the semantic relationship between the items prior to and behind it instead of any specific action. Yuan [26] further classifies the possession relationship in 有*you3* "have/there be" constructions into the categories of vassal, containing and inclusion. To indicate an containing relationship, it takes the syntactic pattern "A+有*you3* "have/there be"+B", as shown in the following examples.

(33) 这辆自行车有四个轮子。

This bicycle has four wheels.

(34) 不少九年义务教育教材有明显特色。

There are distinct characteristics in many nine-year compulsory education textbooks.

The examples show that A and B are usually nominal, with A to indicate or contain B as its organic component or attribute. As shown in the cases above, "wheels" are integral parts of a "bicycle" and "characteristics" are connotative features of "education textbooks". Thus, an alteration or negation of the inclusion or containing relationship between A and B requires a new action, behavior, or prevention of certain event from happening. For example, the breakdown of the containing relationship between "wheels" and "bicycle" semantically implies an event or a behavior, that is, "remove or do not install the wheels".

The quasi-prefix 无 "wu" is mainly combined with nouns to form lexical units in the pattern of [[无"wu"-noun1]- noun2/verb]n, in which noun1 acts as a constituent or a conventional attribute of noun2, and 无 "wu" negates such containing relationship between the two nouns. The negating function of 无 "wu" cannot be achieved without reconstruction of the relationship of by an event or a behavior. For example, to negate the permanent containing relationship that between "fat" and "milk" in 无脂奶 *wu2zhi1nai3* "fat-free milk" that fat is an inherent component of milk, the action of "removing" is required, and accordingly, 无脂奶 *wu2zhi1nai3* "fat-free milk" should be interpreted as "milk with fat removed". Similarly, to negate the temporary containing relationship that between "sugar" and "coke" in 无糖可乐 *wu2 tang2 ke3le4* "sugar-free coke" that sugar is an artificially added ingredient, the behavior of "adding sugar" needs to be prevented, and thus, the semantic meaning of "sugar-free coke" is thus interpreted as "coke with no sugar added". In short, the word formation features of the quasi-prefix 无 "wu" is consistent with the negative usage of 有 *you3* "have/there be" that represents containing relationship, in that, 无 "wu" is often combined with nouns when forming lexical units, and semantically implies an event or a behavior.

To sum up, the essential difference in word formation between 非 "fei" and 无 "wu" lies in their different negation types and functions. Specifically, 非 "fei" negates 是 *shi4* "be" (indicating a classification relationship), and 无 "wu" negates 有*you3* "have/there be" (indicating a containing relationship), in consistence with their syntactic negative function as negative words.

6 Conclusion

The current research analyzes the word formation process involving the quasi-prefixes 非 "fei" and 无 "wu" through examining the structural hierarchy within the lexical units that contain these quasi-prefixes, and concludes that the internal structure of the lexical units is [[非"fei"/无"wu"-X]-Y]n instead of [非"fei"/无"wu"-[X-Y]]n. In other words, the root word that 非 "fei" and 无 "wu" are attached to can only be [X], rather than [X-Y]. Our findings provide evidence for that this new method may be applied to facilitate automatic word segmentation and part-of-speech tagging for similar types of lexical units in Chinese information processing.

Specifically, different Chinese information processing systems may follow different standards of automatic word segmentation and part-of-speech tagging to handle lexical units with quasi-prefixes 非 "fei" and 无 "wu" and produce different results. Take 非专业课 *fei1 zhuan1ye4 ke4* "non-professional course" for instance. ICTCLAS[10]segments it into "非 *fei1*/h专业课*zhuan1ye4ke4*/n", but LTP[11] segments it into "非专业 *fei1 zhuan1ye4*/b课*ke4*/n". The former treats 非 "fei" as a separate "preceding component"[12] and the "professional course" as a noun, whereas the latter regards "non-professional" as a distinguishing word (attribute word) and "course" as a separate noun. In addition, even the same system may produce different results for the same type of lexical units with 非 "fei" and 无 "wu" as quasi-prefixes. For example, the internal structure of 非政府组织 *fei1 zheng4fu3 zu3zhi1* "non-governmental organization" and 非官方组织*fei1 guan1fang1 zu3zhi1* "unofficial organization" are both [非-noun1-noun2]. However, ICTCLAS segments and tags them as "非政府*fei1 zheng4fu3*/b 组织*zu3zhi1*/n" and "非*fei1*/h 官方*guan1fang1*/n 组织*zu3zhi1*/n", respectively. The current research explicitly identifies the structural hierarchy of these lexical units as [[非"fei"/无"wu"-X]-Y]n,

10 ICTCLAS (Institute of Computing Technology, Chinese Lexical Analysis System) is a Chinese lexical analysis system developed by the Institute of Computing Technology, Chinese Academy of Sciences.

11 LTP (Language Technology platform) is a Chinese language processing system developed by The Social and Information Retrieval Research Center, Harbin Institute of Technology .

12 The term "preceding component" cited here is used to refer to a "quasi-prefix".

the root [X] as the unit for 非 "fei" and 无 "wu" to attach to, and a similar function for the part of [非"fei"/无"wu-X] as an attributive. In this way, similar types of lexical units can be consistently processed as [非"fei"/无"wu"-X]/b [Y]/n in automatic word segmentation and part-of-speech tagging.

References

1. Lü, S.X.: Problems of Chinese Grammar Analysis. The Commercial Press, Beijing (1999).(in Chinese)
2. Pan, W.G., Ye, B.Q., Han, Y.: Research on Chinese Word Formation. East China Normal University Press, Shanghai (2004). (in Chinese)
3. Lü, S.X.: Eight Hundred Words in Modern Mandarin Chinese. The Commercial Press, Beijing (1999). (in Chinese)
4. Edited by the Dictionary Editing Office, Institute of Linguistics, Chinese Academy of Social Sciences. Modern Chinese Dictionary, 7th edn. The Commercial Press, Beijing (2016). (in Chinese)
5. Ma, Q.Z.: The property, scope and classification of modern Chinese affixes. J. Chin. Linguist. **6**, 101–137 (1995). (in Chinese)
6. Wang, H.J., Fu, L.: Try to discuss quasi-affix in modern chinese. Lang. Sci. **5**, 3–17 (2005). (in Chinese)
7. Cui, T.: A Study on The 非 "fei"+X structure of modern Chinese. Master dissertation, Yanbian University (2009). (in Chinese)
8. Yin, H.L.: Research on Modern Chinese Quasi-affix. Hebei University Press, Baoding (2011). (in Chinese)
9. Jia, Z.L.: Analysis of asymmetry of Chinese Quasi-affix and Quasi-suffix. Acad. Res. Lang. **21**, 66–69 (2019). (in Chinese)
10. Edited by the Compilation Committee of Chinese Dictionary: Chinese Dictionary. Shanghai Dictionary Publishing House, Shanghai (2011). (in Chinese)
11. Kang, S.Y., Liu, H.R. (eds.): The New Dictionary of New Words (1978–2018). Shanghai Dictionary Publishing House, Shanghai (2018). (in Chinese)
12. Di Sciullo, A., Williams, E.: On the Definition of Word. The MIT Press, Cambridge (1987)
13. Dong, X.F.: Thesaurus and Morphology of Chinese, 2nd edn. Peking University Press, Beijing (2016). (in Chinese)
14. Lü, S.X., Rao, C.R.: On non-predicate adjectives Chinese. Language **2**, 81–85 (1981). (in Chinese)
15. Feng, S.L.: Discuss chinese "Prosodic Word." Chin. Soc. Sci. **1**, 161–176 (1996). (in Chinese)
16. Feng, S.L.: Research on Chinese Prosody Grammar. Peking University Press, Beijing (2005). (in Chinese)
17. Duan, M.S.: Rhythm in chinese. Contemp. Linguist. **4**, 203–209 (2000). (in Chinese)
18. Tan, J.C.: Word meaning, structure meaning and dictionary notation. Chin. Lang. **1**, 69–78 (2000). (in Chinese)
19. Sun, X.A.: Ideas on the issue of modern chinese affix. J. Yanshan Univ. **3**, 80–84 (2007). (in Chinese)
20. Song, Z.Y.: The Generative Lexicon Theory and The Study of Chinese Event Coercion. Peking University Press, Beijing (2015). (in Chinese)
21. Pustejovsky, J.: The Generative Lexicon. The MIT Press, Cambridge (1995)
22. Pustejovsky, J., Jezek, E.: Semantic coercion in language: beyond distributional analysis. Ital. J. Linguist. **20**(1), 181–214 (2008)

23. Yuan, Y.L.: On a descriptive system of qualia structure of chinese nouns and its application in parsing complex chinese grammatical phenomena Cases. Contemp. Linguist. **1**, 31–48 (2014). (in Chinese)
24. Shen, J.X.: Division of negatives and noun/verb division in English and Chinese. Chin. Lang. **5**, 387–399 (2010). (in Chinese)
25. Zhou, H.B.: The sentence pattern of "Shi" indicates the semantic type of judgment. J. Anhui Inst. Educ. **4**, 51–55 (1992). (in Chinese)
26. Yuan, Y.L.: A situational semantic analysis of yǒu constructions in chinese. World Chin. Teach. **3**, 291–307 (2009). (in Chinese)

A Semantic Study on Frame Constructions of Four Characters

Hu Wei(✉)

Department of Chinese Language and Literature of Jinan University, Guangzhou 510632, Guangdong, China

Abstract. The frame construction of four characters features structures both predictable and unpredictable, in terms of meanings. This study suggests that unpredictability shall not be taken as the only property of such structures. Components (A and B) included in the four-character frame construction should be semantically connected to a certain extent. Parallel relationship between the component A and component B in the construction is mainly divided into two groups, namely "similarity" and "opposite", which in turn forms the syntactic union structure. Semantic appreciation and overall semantic metaphor or metonymy are the shared characteristics of the four-character constructions. Frame markers are flexible in terms of their parts of speech, expressing different semantic meanings. This reflects the semantic diversity of the Chinese language. The overall framing effect of the four-character construction is incremental, and its conceptual blending is categorized at three levels, namely high blending level, medium blending level and low blending level.

Keywords: Frame construction · Semantic relation · Semantic type · Framing effect · Conceptual blending · Typology

1 Introduction

According to Goldberg (2006), "C is an independent construction if and only if C is a counterpart of form (Fi) and significance (Si), and any characteristic of its form or significance cannot be fully predicted from the components of C or other existing constructions". Seemingly, the four-character frame constructions in Chinese fall into such description. However, by studying the four-character frame constructions in Chinese, we have notice the meanings of the constructions are not totally unpredictable. Some constructions bear the predictable meanings. For instance, under the "一(*yi*, one) A 二(*er*, two) B" frame, there are predictable examples like "one woman and two men" and "one meat dish and two vegetable dishes", and unpredictable ones like "kill two birds with one stone". Therefore, unpredictability shall not be taken as the only property of the constructions.

Analyzing frame constructions in Chinese can thus enrich the theory construction grammar. In this paper, the constructions are suggested to be defined as follows:: constructions are a series of fixed expressions with set formats in language, where several

M. Liu et al. (Eds.): CLSW 2020, LNAI 12278, pp. 53–63, 2021.
https://doi.org/10.1007/978-3-030-81197-6_5

formats can be abstracted to an expression containing at least one X (or any other symbols) (the maxim of quality), X ≥ 10 (the maxim of quantity).

The author examine frame constructions of "XAYB" in which X can be the same as Y, for instance "连(*lian*, and) A带(*dai*, with) B", "又(*you*, both)A 又(*you*, and)B", etc.. The constructions consist of frame markers such as "连…带…(and; together with)" and "又…又…(both…and…)". "The person together with the car", "guns and cannons" and "running and jumping" are examples demonstrating concrete constructions.

Frame constructions are the most productive language constructions in Chinese and are also very important in research. The components in the frame constructions are important in their meanings as A and B are the foci in the utterance. Frame schemata are linguistic signs, among which aggregations like A and B are similar for the effect of metaphor. Frame markers, combining A and B can embody contiguity by means of metonymy.

In what follows, we examine four typical types of constructions with verb, numeral, adverb and conjunction as the frame markers. Examples cited below are mainly extracted from the CCL corpus.

2 Semantic Relationships Between the Components in the Constructions

When a four-character frame construction contains two components, say A and B, the two components must be semantically connected to a certain extent. The relations between A and B are mainly classified as similarity (i.e., the two elements are of the same category in meaning or logic) and opposite. In what follows, we will discuss the semantic relations between the two elements in different frame constructions in Chinese.

2.1 Frame Markers as Verbs

There are four types of frame constructions in Chinese in which the markers are verbs, namely i) A来 (*lai*, come) B去 (*qu*, go), ii) 有 (*you*, have) A有 (*you*, have) B; iii) 有 (*you*, present) A无 (*wu*, absent)B, iv) 无 (*wu*, without) A无(*wu*, without) B. In such constructions, it is observed that the two elements in the construction hold three kinds of relationships. The relationships are mainly parallel constructions (*binglie* 并列): i) A and B denote senses which are similar in meaning categories or logic (i.e., category labeled as 'Similarity'); ii) synonym; iii) A and B denote opposite meanings (i.e., category labeled as 'Opposite'). Similarity and opposite are shared in all four types, but synonymy is found in the constructions of "A来 (*lai*, come) B去 (*qu*, go)" and "无 (*wu*, without) A 无 (*wu*, without) B" only. Examples are presented below.

1) 来 (lai, come) B 去 (qu, go):

Similarity: 风来雨去(the wind comes and the rain goes), 你来他去(you come and he goes)

Opposite: 日来夜去 (the day comes and the night goes), 新来旧去 (the new comes and the old goes)

Synonymy: 推来搡去 (push forward and push back)

2) 有 (you, have) A有 (you, have) B:

Similarity: 有血有肉 (flesh and blood), 有谋有略 (astute and resourceful).
Opposite: 有来有回 (come and go), 有进有退 (advance and retreat).

3) 有 (you, present) A无 (wu, absent) B:

Similarity: 有气无力 (breath is present but vigor is absent), 有职无权 (hold a position without the authority), 有价无市 (having a price but no sales), 有价无货 (having a price but no goods available), 有党无派 (in a Party without grouping), 有人无岗 (having job seekers but no job), 有眼无珠 (to have eyes but not see), 有风无沙 (wind without sand), 有综合无系统 (integration without system).
Opposite: 有益无害 (be not harmful but helpful to), 有增无减 (having increase but no decrease), 有去无回 (leaving without returning), 有施无受 (giving without receiving), 有始无终 (having a beginning but no end), 有进无退 (forging ahead instead of retreating), 有花无实 (having flowers but no fruit), 有头无尾 (having a beginning but no end), 有名无实 (in name but not in deed).

4) 无 (wu, without) A无 (wu, without) B:

Similarity: 无怨无悔 (no complaint or regret), 无才无德 (no talent or virtue).
Opposite: 无好无恶 (no like or dislike), 无大无小 (no difference between the senior and the junior).
Synonymy: 无边无际 (no boundary or limit), 无牵无挂 (no care or concern).

2.2 Frame Markers as Numerals

Three types of frame constructions in Chinese include the markers which are numerals like 'one', 'two' and 'half', namely i) 一 (*yi*, one) A一 (*yi*, one) B, ii) 半 (*ban*, half) A 半 (*ban*, half) B, and iii) 一 (*yi*, one, first) A二 (*er*, two, second) B.

1) 一 (yi, one) A一 (yi, one) B:
According to the research of Shao (2013), there are three main types of semantic relations between A and B in "一A一B", namely i) parallel (including similarity, opposite and synonymy), ii) precedence and iii) arrangement, which constitute the syntactic union structure, serial verb structure and subject-predicate structure, respectively.

Parallel: 一菜一汤 (one dish and one soup), 一强一弱 (one stronger and one weaker) and 一言一语 (every word and sentence)
Precedence: 一高一低 (rise and fall)

Arrangement: (one for each)

2) 半 (ban, half) A半 (ban, half) B:
A and B included in "半A半B" only form parallel relations, including similarity (e.g., "half sitting half lying") and opposite (e.g., "half literary and half vernacular"), constituting the union structure.
3) 一 (yi, one, first) A二 (er, two, second) B:
A and B included in "一A二B" form three main types of semantic relations with each other. The first type of relations is parallel. The parallel relationship is dominated by similarity and no synonymy or opposite relationships are involved in the construction of "一 (*yi*, one/first) A二 (*er*, two/second)". The second type of relationships is precedence, which decides the sequence and continuity of movement and the components A and B therein are both verbs. Such relationship forms the serial verb structure, serving as predicate or adverbial mainly. The third type of relationships is declaration, which forms the structure serving as a predicate, adverbial or object. Semantic relations decide syntactic relations. These three different semantic relations of "一 A二B" result in the syntactic union structure, serial verb structure and subject-predicate structure, respectively.

Parallel:	一牛二虎(one ox and two tigers), 一女二男(one woman and two men)
Precedence:	(look and help), (swagger and swing)
Declaration:	(this (one) piece of wood can be used in two ways")

2.3 Frame Markers as Adverbs and Conjunctions

In "又 (*you*, both) A又 (*you*, and) B", the marker 又is an adverb and the elements A and B are adjectives or verbs. They form either the relationship of "similarity" or "opposite".

Similarity: 又高又壮 (both tall and strong), 又香又甜 (both fragrant and sweet).
Opposite: 又爱又恨 (love and hate), 又哭又笑 (cry and laugh).

Similarly, the two elements in the construction of "连 (*lian*, with) A带 (*dai*, with) B", in which the markers are conjunctions, also form either the relationships of "similarity" or "opposite".

Similarity: 连工带料 (work together with materials), 连蹦带跳 (hopping and skipping)
Opposite: 连新带旧 (both new and old), 连老带少 (both young and old)

3 Semantic Types of the Constructions

Several semantic types are found to be involved in the frame constructions, including semantic addition, semantic appreciation, overall semantic metaphor or metonymy and semantic split. The semantic types can be understood in the following ways:

i) Semantic addition: the meaning of the whole construction of AXBY can be interpreted as the meaning composited by AX + BY.
ii) Semantic appreciation: the meaning of the whole construction of AXBY can be interpreted as the meaning which more than the composition of the meaning of AX and BY (i.e., > AX + BY).
iii) Overall semantic metaphor or metonymy: In the frame construction of AXBY, the compositional meaning of AX and BY contributes a metaphorical meaning or metonymy (i.e., the meaning of the construction ≈ AX + BY, or ≠ A or B or AB).
iv) Semantic split: the meaning of the whole construction can be interpreted as A = 2B.
v) Semantic complex: A and B are originally morphemes attaching with each other to morph a word or phrase. The constructions can be interpreted as AB.

Among the above semantic types, semantic appreciation and overall semantic metaphor or metonymy are the shared characteristics of the frame constructions in discussion. Moreover, it will be shown in the following sub-sections that the frame markers are flexible in terms of their parts of speech. Be that as it may, the markers express different semantic meanings. Such features of the frame markers reflect the semantic diversity of the Chinese language. The following sub-sections present examples extracted from the corpus to illustrate the semantic types of the frame constructions.

3.1 Constructions with Verbal Frame Markers

1) A来 (*lai*, come) B去 (*qu*, go).

Semantic addition:	A来B去 = A来 + B去, e.g.: 你来他去 (you come, and he goes).
Semantic appreciation:	A来B去 > A来 + B去, e.g.: 风来雨去 (wind comes and rain goes), 南来北去 (comes from the south and goes to the north).
Overall semantic metaphor or metonymy:	A来B去 ≈ A来 + B去, e.g.: 明来暗去(的斗争) (lit.: overt come and covert go, i.e., "secretly contacting with someone").

2) 有 (*you*, have) A有 (*you*, have) B.

Semantic addition:	有A有B = 有A + 有B, e.g.: 有男有女 (have men and have women) = 有男人(have men) + 有女人(have women)
Semantic appreciation:	有A有B > 有A + 有B. E.g.: 有老有少 (have elderly and have youngster) = the elderly, the children, the middle-aged, etc.; 有滋有味 (tasty and savory) = 很有滋味 (very delicious).
Overall semantic metaphor or metonymy:	有A有B ≈ 有A + 有B, e.g., 有眉有眼 (with eyes and brows) describes the shape of the huge axe.

3) 有 (*you*, present) A无 (*wu*, absent) B.
Semantic addition: 有A无B = 有A + 无B, e.g.: 有音无字(having a pronunciation but no word), 有种无收(seeding but no harvest).
Semantic appreciation: 有A有B > 有A + 有B, e.g.: 有眼无珠(to have eyes but not see).
Overall semantic metaphor or metonymy: 有A有B ≈ 有A + 有B, e.g.: 有头有 (He always does things from the beginning to the end).

4) 无 (*wu*, without) A无 (*wu*, without) B.
Semantic addition: 无A无B = 无A + 无B, e.g.: no son or daughter.
Semantic appreciation: 无A无B > 无A + 无B, e.g.: 无情无义 (no gratitude or gratefulness, heartless).
Overall semantic metaphor or metonymy: 无A无B ≈ 无A + 无B, e.g.: 无法无天 (no laws human or divine).

3.2 Constructions with Numeral Frame Markers

1) 一 (*yi*, one) A一 (*yi*, one) B.
Semantic addition: 一A一B = A + B, e.g.: 一人一车 (one person and one car)。
Semantic appreciation: 一A一B > A + B, e.g.: 一砖一瓦 (a single brick and tile) in "do not move a single brick and tile (i.e., anything) here".
Overall semantic metaphor or metonymy: 一A一B ≈ A + B, e.g.:一板一眼 (one accented beat and one unaccented beat, every rhythm), 一朝一夕 (one morning and evening; a short time).

2) 半 (*ban*, half) A半 (*ban*, half) B
Semantic addition: 半A半B = 半A + 半B, e.g.: 半沙半草 (half sand half grassland).
Semantic appreciation: 半A半B>半A+半B, e.g.: 半掩半露 (half veiled half revealed, bashful).
Overall semantic metaphor or metonymy: 半A半B ≈ 半A + 半B, e.g.: 半鬼半人 (lit.: half man half ghost, i.e. very ugly and insidious).

3) 一 (*yi*, one, first) A二 (*er*, two, second) B
Semantic addition: 一A二B = A + 2B, e.g.: 一弓二箭 (one bow and two arrows), 一桥二路 (one bridge and two roads).
Semantic appreciation: 一A二B > A + 2B, e.g.: 一干二净 (neat and clean) meaning "all finished, without remainder", is similar to 一清二白 (perfectly clean), 一长二短 (unexpected misfortune), etc.

Overall semantic metaphor or metonymy:	一A二B ≈ A + 2B, e.g.: 一石二鸟 (kill two birds with one stone) meaning "getting two benefits out of one thing", is similar to 一穷二白 (poor and blank).
Semantic split:	一A二B = A = 2B, e.g.: 一州二城 (one state with two cities)

3.3 Constructions with Adverb Frame Markers

With "又A又B" as an example, semantic addition is not expressed by such type of constructions. Yet, semantic appreciation and overall semantic metaphor or metonymy are observed.

Semantic appreciation:	又A又B > A+B, e.g.:又锯又敲 (sawing and knocking)
Overall semantic metaphor or metonymy:	又A又B ≈ 又A + 又B, e.g.:又矮又小。

3.4 Constructions with Conjunction Frame Markers

Take "连(*lian*, and) A带(*dai*, with) B" as an example, four semantic types of the constructions which consist conjunctions as the frame markers are observed.

Semantic addition:	连A带B = A + B, e.g.: 连本带息 (the principal and interest) =本 (principal) + 息 (interest), also 连你带我 (you and me) and 连新带旧 (new and old).
Semantic appreciation:	连A带B > A + B, e.g.: 连哭带叫 (cry and shout) = 哭(cry) + 叫 (shout) + 闹 (make a noise), etc. Also 连头带脚 (head to feet), 连老带少 (young to old).
Overall semantic metaphor or metonymy:	连A带B≠ A or B or AB. In 连大带小 (the old and the young), the total number of entities involved is not two but can be larger three. "大(old) and 小(young)" are the metonymy of old people and young people.
Semantic complex:	连A带B = AB. AB here are united complex words, e.g.: 连亲带戚 (relatives within and outside the family) = 亲戚(relatives), also 连夫带妻 (husband and wife), 连编带造 (make up), 连敲带诈 (extort), and 连嘲带讽 (taunt).

4 Framing Effect

"Framing effect" refers to the semantic effect of the frame on the overall construction, which produces the constructional meaning. Conceptual blending theory sees semantic constructions as an on-line process. After integrating with frame and construction, the new framework meaning will emerge. The frame provides the constructional meaning and relational meaning to the whole construction, while variables express the specific meaning and practical meaning of the construction.

Some "A来B去" feature the framing effect of "recurrent and persistent", making A and B going on and on. Component A in "A来B去" format is mainly verbs, sometimes nouns, adjectives, pronouns, numerals, etc. The frame produces semantic appreciation and the framing effect of "recurrent and persistent" for A and B, generating the semantic feature of [+recurrent and persistent]. "走来跳去 (walking and jumping around)" means both "walking" and "jumping" and the repetition of the actions. Other related examples include "人来客去 (customers come and go), 蹦来跳去 (bouncing and jumping about), 翻来覆去 (toss and turn)".

Some "有A有B" amplifi the "framing effect" of "有A" and "有B", such as "有头有尾 (from the beginning to the end), 有血有肉 (flesh and blood), 有板有眼 (have both accented beats and unaccented beats, meaning orderly)", etc.

Some "有A无B" amplify the "framing effect" of "有A" or "无B", such as "有头无尾 (having a beginning but no end), 有眼无珠 (to have eyes but not see), 有恃无恐 (having no fear with backing)", etc.

Most "无A无B" phrases amplify the framing effect of negation, such as "无情无义(no gratitude or gratefulness)", criticizing the object for being heartless.

Some "又A又B" phrases amplify the framing effect of "又A" and "又B", focusing the expression on the peripheral significance of A and B together instead of the meanings of A and B as two independent components, i.e., the new pragmatic meaning produced after A and B enter the construction of "又A又B" format. For example, "又锯又敲 (sawing and knocking)" might also infer the sound of "sawing, knocking, grinding, chipping, drilling, striking", etc., highlighting the chaos of sounds.

Some "连A带B" phrases amplify the framing effect of "连A" and "带B", such as "连头带尾 (head to tail)" meaning head, body, limbs, tail and even the packaging and "连打带骂 (scolding and beating)" meaning "disciplining, teaching, instructing", etc.

Therefore, the overall framing effect of four-character frame constructions is amplification.

5 Conceptual Blending

Conceptual blending refers to the process of blending part of the meaning. two concepts from different fields are selectively extracted and then be put into a complex conceptual structure. Blending effect relies on the framework of blending and the linguistic elements involved. It generates emergent meaning. Conceptual blending consists of different levels. At the low-blending level, the overall meaning of constructions is based on the literal meaning of A and B, being the simple sum of "XA" and "YB". Such constructions, as temporary assemblies of words or morphemes in the context, are of poor coagulability. At the medium blending level, the constructions have constructional meaning, whose overall meaning cannot be achieved by simply adding up different parts of the construction, but by combining the literal meaning with the constructional meaning. At the high blending level, generally speaking, the constructions have extended meaning, which can be obtained by metaphor or metonymy.

In “A来B去”, the low blending level constructions include “我来你去 (I come and you go)” and etc.; the medium blending level constructions include “风来雨去 (the wind comes and the rain goes), 日来夜去 (the day comes and the night goes), 跑来走去 (running and walking about)” and etc.; and the high blending level constructions include “明来暗去 (overt come and covert go, i.e., “secretly contacting with someone”)” and others.

In “有A有B”, the low blending level constructions include “有你有我 (you and me)” and etc. At the medium level, there are “有滋有味 (tasty and savory), 有吃有喝 (having plenty to eat and drink)”, etc., and at the high level, there are “有眉有眼 (‘with eyes and brows’, describing the shape of the huge axe)” and others.

In “有A无B”, the low blending level constructions include “有音无字 (having a pronunciation but no word), 有种无收 (seeding but no harvest)”, etc. The medium blending level constructions include “有眼无珠 (to have eyes but not see)” and others, and the high blending level constructions include “有来无回 (come and never return), 有头无尾 (having a beginning but no end)”, etc.

In “无A无B”, the low blending level constructions include “无儿无女 (see above)” and others. At the medium level, the constructions include “无情无义 (no gratitude or gratefulness, heartless)” and others, and at the high level, the constructions include “无法无天 (no laws human or divine)” and others.

In “一A一B”, the low blending level constructions include “一远一近 (one far and one near), 一东一西 (one east and one west)”, etc. The medium blending level constructions include “一草一木 (every tree and bush), 一花一鸟 (every flower and bird)”, etc., and the high blending level constructions include “一唱一和 (one sings and one echoes, sing in chorus), 一手一足 (one hand and one foot, weak strength) and others.

In “半A半B”, the low blending level constructions include “半沙半草 (half sand half grassland)” and others. The medium blending level constructions include “半掩半露 (half veiled half revealed, bashful)” and others, and the high blending level constructions include “半鬼半人 (lit.: half man half ghost, i.e. very ugly and insidious)” and others.

In “一A二B”, the low blending level constructions include “一弓二箭 (one bow and two arrows), 一横二竖 (one horizontal stroke and two vertical strokes)” and etc. The medium blending level constructions include “一清二白 (perfectly clean), 一长二短 (unexpected misfortune)” and etc. The high blending level constructions include “一石二鸟 (kill two birds with one stone)” and others.

In “又A又B”, there are no low blending level construction. At the medium blending level, there are “又跑又跳(run and jump), 又长又黑(long and dark)”, etc. At the high blending level, the constructions include “又矮又小(short and small)” as in “the submarine is short and small”.

In “连A带B”, the low blending level constructions include “连本带息 (the principal and interest), 连你带我 (you and me)” and etc. The medium blending level constructions include “连老带少 (both young and old), 连吃带穿(food and clothing)” and etc. The high blending level constructions include “连大带小(see above)” and others.

According to the above, semantic appreciation and overall semantic metaphor or metonymy are the shared characteristics of these constructions. Frame markers can be of different or same parts of speech with differentiated semantic meanings. This reflects the semantic diversity of the Chinese language.

6 Typological Significance of Four-Character Frame Constructions

Our understanding of the reality includes the understandings of the whole real world, the material world and the spiritual world, which is the foundation of language use and the object of language expression and perception. The real world and the cognitive images are what human perceive about the reality. The reality can be mapped to the language, activating schematic constructions. With specific rules, instances of the constructions can be produced. The first two steps relate to people's perception of the real world, and the latter two language performance. This is also a process of "stimulation ↔ cognition ↔ feedback". To illustrate, back in the pre-Qin period, the Han people suffered from starvation and poor living in the real world. People then tended to search for expressions to describe their perception among their cognitive images, which then inspire the [X A Y B] schematic construction and the "无(*wu*, no) A无(*wu*, no) B" construction. Hence, they can express that "no fine dress or common clothes, how can I suffer through the year-end".

Four-character frame constructions are economical in language, rhythmic in pronunciation, symmetrical in grammar, and of typological value. The following part discusses the typological significance of four-character frame constructions, taking "连A带B" as an example.

According to Croft (2001), the formal schemata of constructions were not universal, because we could not find a series of structural characteristics to clearly define the so-called "equivalent" constructions in different languages. He also believed that constructions differed from language to language, i.e., different languages had different constructions, and constructions expressing the same function in different languages were significantly different in structure. For instance, in cases that an English speaker would like to express the meaning that A and B are together, they may make use of the construction of "both A and B", in case A and B are nouns or they may choose to use the conjunction "and" if A and B are verbs. A similar pattern in Chinese is "连A带B", which are a different structure compared to English. Apart from that, it is found that " 连A带B" constructions are used in Fujian dialect, Hakka, Cantonese and Beijing dialect while Cantonese also expresses "又A又B" to convey the similar meaning in different structures. "连A带B" or similar expressions in different dialects in China demonstrate the integrity of the Chinese language. Comparisons among different dialects thus indicates the vigor of "连A带B" or similar formats in Chinese language, which shows that the structure is of typological value.

7 Conclusion

Four-character frame constructions have both predictable and unpredictable constructions, in terms of their meanings. Unpredictability isn't the only property of constructions. This paper gives a new definition to the constructions: they are a series of fixed expressions with fixed formats in language, where several formats could be abstracted to an expression containing at least one X (or A or any other symbols) (the maxim of quality), X ≥ 10 (the maxim of quantity). Components included in the four-character frame constructions like A and B must be semantically connected to a certain extent. Parallel relations between A and B are mainly classified as similarity and opposite, which form the syntactical union structure. Semantic appreciation and overall semantic metaphor or metonymy are the shared characteristics of four-character constructions. The frame markers involved can be of different or same parts of speech with differentiated semantic meanings, which reflect the semantic diversity of the Chinese language. The overall framing effect of four-character frame constructions is amplification. Conceptual blending of frame constructions can be classified as high, medium and low. Four-character frame constructions are of typological significance.

References

1. Goldberg, A.: Constructions at Work. OUP, Oxford (2006)
2. Croft, W.: Radical Construction Grammar. Oxford University Press, Oxford (2001)
3. Changlai, C., et al.: Study on Fixed Phrases in Modern Chinese. Xuelin Press, Shanghai (2012)
4. Wei, H.: Diachronic and synchronic study on "A lai B qu." J. Yunnan Normal Univ. (Teach. Res. Chin. Foreign Lang.) **10**(6), 30–39 (2012)
5. Qiuzhen, Z., Wei, H.: Study on the "wu A wu B" frame construction. Guangxi Soc. Sci. **8**(11), 141–144 (2012)
6. Wei, H.: Comparative study on "ban A ban B, yi A yi B and yi A er B." J. Jinan Univ. **38**(5), 21–27 (2016)
7. Wei, H., Yuen, G.: Systematic study on the frame construction "lian A dai B." Acad. Res. **9**(7), 153–158 (2015)
8. Wei, H., Jia, W.: Comparative study on "you A you B" and "you A wu B." J. Ningxia Univ. **37**(1), 41–49 (2015)
9. Danqing, L.: Typology of Word Order and Theory of Prepositions. Commercial Press, Beijing (2003)
10. Jingmin, S.: On frame constructions in Chinese. Stud. Chin. Lang. **9**(3), 218–227 (2011)
11. Jingmin, S.: Dynamic Study of Chinese Grammar. Commercial Press, Beijing (2013)
12. Weishan, W.: The Rhythmic Framework of Chinese Language and Its Blending Effect. Xuelin Press, Shanghai (2011)

Angry Thunder and Vicious Frost: Remarks on the Unaccusativity of Chinese Weather Verbs

Sicong Dong[1], Jie Xu[2], and Chu-Ren Huang[3](✉)

[1] School of Humanities and Social Sciences, Harbin Institute of Technology, Shenzhen 518055, China
dongsicong@hit.edu.cn

[2] Department of Chinese Language and Literature, University of Macau, Macao, China
jiexu@um.edu.mo

[3] Department of Chinese and Bilingual Studies, The Hong Kong Polytechnic University, Hong Kong, China
churen.huang@polyu.edu.hk

Abstract. Thunder and frost are said in Sinitic languages to be controlled by higher powers, or to simply occur by themselves, or even to cast severe damage on human society as agents. Such diverse linguistic behaviours and meanings pose challenges and add complexity to the ongoing debate on the unaccusativity of weather verbs. We present in this paper an investigation into various weather verbs in Mandarin and other Sinitic languages based on dictionaries of different languages and corpus data. By a set of diagnostics, cases of unaccusative, unergative and transitive weather verbs have been attested in Sinitic languages. The majority of weather verbs are alternatively unaccusative or unergative, depending on which event structures they are associated with. Specifically, the unaccusative behaviour is linked to the view of weather events as happenstances, in the cognitive processing mode of sequential scanning; the unergative behaviour is linked to the view of weather events as activities, in the cognitive processing mode of summary scanning.

Keywords: Unaccusativity · Weather verb · Sinitic languages · Event structure · Thunder · Frost

1 Introduction

Since the Unaccusative Hypothesis [1] and Burzio's Generalisation [2], intransitive verbs have been widely known to consist of two subtypes: unaccusative verbs and unergative verbs. Unaccusatives refer to intransitive verbs whose subjects originate as objects, i.e., they take internal arguments as the subject, such as *die* in "the patient died". Unergatives, on the other hand, refer to intransitive verbs that take external arguments as the subject, such as *run* in "the man is running". Semantically, the two subtypes mainly differ in the way that unaccusative verbs normally describe actions involuntarily happened to the subjects, while unergative verbs typically show volitionality.

M. Liu et al. (Eds.): CLSW 2020, LNAI 12278, pp. 64–73, 2021.
https://doi.org/10.1007/978-3-030-81197-6_6

Such dichotomy of intransitive verbs can also be found in Chinese. For example, Chinese unaccusatives include 到 *dào* 'arrive', 死 *sǐ* 'die', 发生 *fāshēng* 'happen', 出现 *chūxiàn* 'appear', etc., and unergatives in Chinese include 睡 *shuì* 'sleep', 笑 *xiào* 'laugh', 哭 *kū* 'cry', 跳 *tiào* 'jump', etc. [3, 4]. Besides, Xu [5] validated one criterion for the identification of unaccusative or unergative verbs in Chinese: the argument of an unaccusative verb can only appear in postverbal position in attributive clauses, as shown in (1a), but on the contrary, the argument of an unergative verb can only be found before the verb in attributive clauses as shown in (1b). The glossing follows the Leipzig Rules.

(1) a. 死 了 一 条 狗 的 老头 / *一条狗死了的老头
sǐ ·le yī tiáo gǒu ·de lǎotóu
die PFV one CLF dog DE old-man
'An old man whose dog died'

b. 美国 人 旅游 的 季节 / *旅游美国人的季节
měiguó rén lǚyóu ·de jìjié
America people travel DE season
'American people's travelling season'

However, it can be highly debatable whether some verbs are unaccusative or unergative. A good example is weather verbs. As summarised by Levin and Krejci [6], at least four different conclusions have been reached regarding the unaccusative vs. unergative status of weather verbs: they were argued to be unaccusative, or not unaccusative, or both unaccusative and unergative, or neither, by various researchers. Most recently, Levin and Krejci [6] provided a new proposal for the ambiguous behaviours of weather verbs focusing on precipitation verbs, i.e., *rain*, *snow*, *hail*, etc. By answering the question that has long been asked whether it is the rain that rains or a higher force [7], they claimed that English weather *it* has semantic content, and precipitation verbs in English alternately pattern as unaccusatives or unergatives: they show unaccusative behaviour when linked to a directed motion event structure, a type of scalar change event structure, as in "a light rain rained on my head" where the subject is the substance; they show unergative behaviour when linked to a substance emission event structure, a type of activity event structure, as in "it rained (a light rain)" where the subject is the source.

The syntactic properties of Mandarin weather verbs, in contrast, seem not so controversial in previous studies. For example, Li [8] and Yang [9] treated weather verbs such as 下 *xià* 'fall' and 降 *jiàng* 'fall' as unaccusative verbs, claiming that the only argument of a weather verb appearing in the postverbal object position provided strong evidence.

Weather words in Sinitic languages (including Mandarin and traditionally termed *Chinese dialects*) have received much attention in recent years [10–18]. These studies covered a wide range of meteorological phenomena in more than 200 Sinitic languages or dialects and offered accounts for various morphosemantic and grammatical behaviours of weather expressions. They have demonstrated the great diversity of weather words in Mandarin and other Sinitic languages, as well as the potential benefits of the cross-linguistic data for research on language and weather. Following this line of research, this paper aims to revisit the unaccusativity of Chinese weather verbs based on data from Sinitic languages with a focus on thunder and frost expressions and argues that Chinese weather verbs are not uniformly unaccusative, but rather diverge in a similar way to the claim of Levin and Krejci [6], i.e., they could be unaccusative or unergative, depending on how the weather events are viewed.

2 Mandarin Weather Verbs

Standard Mandarin adopts the argument type to encode most weather events, i.e., the argument of a weather expression is primarily responsible for encoding the meteorological event [13, 19]. Hence, in weather expressions such as (2) and (3), a weather noun co-occurs with a verb to indicate the formation or appearance of a weather phenomenon. By *weather verbs*, therefore, we refer to such kind of verbs in Sinitic languages. Note that the preferred forms to describe the appearance of frost and fog in Mandarin are下霜 *xiàshuāng* 'to frost' and 起雾 *qǐwù* 'to fog' respectively, according to the data in Dong et al. [13].

(2) a.	下	雨	/ 下	雪	/ 下	雹子	/ 下	霜	/ 起	雾	了。	
	xià	yǔ	xià	xuě	xià	báo·zi	xià	shuāng	qí	wù	·le	
	fall	rain	fall	snow	fall	hail	fall	frost	rise	fog	LE	
	'It rained./		It snowed./		It hailed./		It frosted up./ It got foggy.'					
b.	雨	下	/ 雪	下	/ 雹子	下	/ 霜	下	/ 雾	起	了。	
	yǔ	xià	xuě	xià	báo·zi	xià	shuāng	xià	wù	qí	·le	
	rain	fall	snow	fall	hail	fall	frost	fall	fog	rise	LE	
	'It rained./		It snowed./		It hailed./		It frosted up./ It got foggy.'					
(3) a.	打	雷	/ 闪	电	/ 刮	风	了。					
	dǎ	léi	shǎn	diàn	guā	fēng	·le					
	hit	thunder	flash	lightning	blow	wind	LE					
	'It thundered./		Lightning flashed./ Wind blew.'									
b.	*雷	打	/*电	闪	/ *风	刮	了。					
	léi	dǎ	diàn	shǎn	fēng	guā	·le					
	thunder	hit	lightning	flash	wind	blow	LE					

It can be seen from (2) and (3) that these common Mandarin weather nouns can all appear as postverbal objects, which are the only arguments of the weather verbs. This serves as evidence for their unaccusativity according to the studies such as Yang [9]. Furthermore, such weather verbs behave differently in the way that one group can occur both before and after the weather nouns, including directional verbs co-occurring with atmospheric water such as precipitation, frost and fog, as in (2); the other group, such as the verbs for thunder, lightning and wind in (3), seem not to allow weather nouns to function as their subjects (at least when there are no additional elements). If this is the case, then the first group of verbs, i.e., 下 *xià* 'fall' and 起 *qǐ* 'rise', exhibits surface and deep unaccusativity when weather nouns are used after and before them respectively [20]; while the second group, i.e., 打 *dǎ* 'hit', 闪 *shǎn* 'flash' and 刮 *guā* 'blow', only shows surface unaccusativity. The major semantic difference between the two uses of the first group is that the preverbal arguments tend to be definite while the postverbal arguments tend to be indefinite.

Challenges, however, might be posed for such analysis. As mentioned earlier, typical unaccusatives are non-volitional verbs, but one obvious problem found here is that the verb 打 *dǎ* 'hit' describes a willed or volitional act, and is thus more likely to be unergative than unaccusative. Moreover, the usage of 打 *dǎ* 'hit' in Sinitic languages can paint a more comprehensive picture for us to rethink the grammatical categories of Chinese weather verbs.

3 Verbs for Thunder and Frost in Sinitic Languages

According to the large-scale data in Dong [12], Sinitic languages tend to use verbs with high activity and kinesis, e.g., 打 *dǎ* 'to hit', 扯 *chě* 'pull', etc., to state thundering and frosting. We thus consulted dictionaries of 44 Sinitic languages [21–23], which were compiled in the same format, to obtain detailed information of how Chinese people describe weather events involving thunder and frost. Our findings are shown below.

To state the event of thundering, Sinitic languages adopt two major types of verbs: verbs referring to striking activities, such as 打 *dǎ* 'hit' in Changsha (长沙) and 拍 *pāi* 'slap' in Haikou (海口); and verbs denoting sound making activities, such as 响 *xiǎng* 'make sound' in Lichuan (黎川), 吼 *hǒu* 'roar' in Wanrong (万荣), 叫 *jiào* 'shout' in Liuzhou (柳州) and 哀 *āi* 'whine' in Leizhou (雷州). Only postverbal thunder nouns co-occur with the first type of verbs, but pre- and postverbal nouns for thunder can both be found accompanied by the second type of verbs. As for the postverbal noun of frost, apart from打 *dǎ* 'hit', Sinitic languages mostly utilise directional verbs with it to describe the event, e.g., 打霜 *dǎshuāng* hit-frost in Danyang (丹阳), Nanchang (南昌) and Wuhan (武汉), 下霜 *xiàshuāng* fall-frost in Urumchi (乌鲁木齐), 落霜 *luòshuāng* fall-frost in Jinhua (金华), 降霜 *jiàngshuāng* fall-frost in Wenzhou (温州), 上霜 *shàngshuāng* rise-frost in Harbin (哈尔滨), etc.

So far, we have already found weather verbs having thunder nouns as subjects in Sinitic languages, but打 *dǎ* 'hit' seems to follow the same syntactic pattern as in Mandarin. More interesting, as discovered in our investigation, are the expressions of how thunder and frost affecting human beings and the environment in a harmful or even lethal way. Among the 44 dictionaries, 36 of them include entries denoting the event of thunder striking people, animals, or other objects on the ground. (Though technically speaking, it is the discharged electricity of lightning that strikes people or livestock but not the booming sound of thunder [24], which means that thunder and lightning may be intertwined in the conceptualisation of Chinese people.) Additionally, 8 of the dictionaries include entries of the event of frost severely damaging crops. Respective examples of lexical items indicating such events are: 雷打 *léidǎ* thunder-hit in Yangzhou (扬州) and Mouping (牟平), 雷劈 *léipī* thunder-chop in Wuhan (武汉) and Taiyuan (太原), 天打 *tiāndǎ* Tian-hit in Suzhou (苏州), 雷公劈 *léigōngpī* Lord_of_Thunder-chop in Nanning (南宁), 龙抓 *lóngzhuā* dragon-claw in Luoyang (洛阳), 佛拌 *fóbàn* Buddha-throw in Wenzhou (温州), etc.; 霜打 *shuāngdǎ* frost-hit in Suzhou (苏州) and Urumchi (乌鲁木齐), 霜杀 *shuāngshā* frost-kill in Lanzhou (兰州) and Yinchuan (银川), 霜冻 *shuāngdòng* frost-freeze in Luoyang (洛阳), etc.

Two characteristics of these expressions can be easily generalised. First, nouns related to thunder or frost all precede the verbs. That is, both打雷 *dǎléi* and 雷打 *léidǎ*, or 打霜 *dǎshuāng* and 霜打 *shuāngdǎ*, could coexist in one language, with one form denoting the appearance of the weather phenomenon, and the other denoting the damage it causes on human society. Even if the volition shown in 打雷 *dǎléi* 'to thunder' and 打霜 *dǎshuāng* 'to frost' might not be explicit enough, it is obvious and unquestionable in 雷打 *léidǎ* and 霜打 *shuāngdǎ* making the verb more unlikely to be unaccusative. Apart from this, further evidence comes from the fact that a second argument can appear as well in expressions of causing damage by thunder or frost. For example, in each of the sentences in Lanzhou (兰州) Chinese shown below in (4), quoted from Zhang and Mo [23], two arguments can be detected. One is the weather noun, and the other is the 'victim' of the weather event. In this sense, 打 *dǎ* 'hit', similar to other verbs with the same usage such as 劈 *pī* 'chop' and 杀 *shā* 'kill', is neither unaccusative nor unergative, but instead, a transitive verb. That is, thunder and frost are the agents, while the willow and poinsettia in (4) are the patients.

(4) a. 滨河路 上 的 一 棵 老 柳树 雷 打 掉 了。
bīnhélù shàng ·de yī kē lǎo liǔshù léi dǎ diào ·le
Binhe Road LOC DE one CLF old willow thunder hit drop LE
'One old willow on Binhe Road was struck down by lightning.'

b. 放 在 院子 里 的 一品红 霜 打 坏 了。
fàng zài yuàn·zi lǐ ·de yīpǐnhóng shuāng dǎ huài ·le
put be-at yard LOC DE poinsettia frost hit damaged LE
'The poinsettia put in the yard was damaged by frost.'

Second, except for 雷 *léi* 'thunder', several supernatural powers are said to harm people or other objects in case of thunder/lightning strike, e.g., 天 *tiān* 'heaven/god', 雷公 *léigōng* 'Lord of Thunder', 佛 *fó* 'Buddha', 龙 *lóng* 'dragon', etc. Those powers can all be considered as anthropomorphic, and as a punishment, may strike people who have done something evil, according to the traditional Chinese beliefs. Therefore, the participation of higher powers who are believed to uphold justice, strengthens the volition in these expressions, and thus the analysis of the verbs' being non-unaccusative is favourable. A question then arises: do these supernatural powers also play a role in 打雷 *dǎléi* 'to thunder' and 打霜 *dǎshuāng* 'to frost', or further, in 下雨 *xiàyǔ* 'to rain' and 下雪 *xiàxuě* 'to snow'?

4 Two Event Structures and Two Types of Verbs

Previous analysis of thunder and frost related verbs in Sinitic languages have provided insights on the unaccusative vs. unergative issue in several aspects: 1) the difference in showing surface and/or deep unaccusativity between certain weather verbs may not hold for all Sinitic languages, i.e., 雷 *léi* 'thunder' may act as subject; 2) unergatives, e.g., 吼 *hǒu* 'roar' and 叫 *jiào* 'shout', can be used as thunder verbs in Sinitic languages; 3) transitive verbs can also be attested in expressions of two extreme weather phenomena. In this section, we will re-examine the grammatical categories of weather verbs in Mandarin and other Sinitic languages, chiefly following the framework proposed by Levin and Krejci [6].

As introduced in Sect. 1, Levin and Krejci [6] claimed that English precipitation verbs could be unaccusative or unergative, depending on the event structure they are associated with. Specifically, the unaccusative precipitation verbs are linked to scalar change event structure, and the unergative are linked to activity event structure. Hence, event structures are crucial for the grammatical properties of verbs. By definition, an event of scalar change is characterised by a change in the value of a scalar-valued attribute of an entity, and the activity event schema encompasses any typical dynamic event of non-scalar change [6]. Therefore, in "a light rain rained on my head", the subject *rain* is a substance changing its value on a scale of path it moves along, i.e., from the sky to the speaker's head, thus the verb *rain* is used in a scalar change event and should be unaccusative. On the other hand, in "it rained (a light rain)", there is no such path scale, and what the sentence describes is simply an act of rain being emitted from the source *it*, thus the verb *rain* is used to denote an activity event and should be unergative. In addition, the two usages of *rain* have multiple parallel linguistic behaviours to typical unaccusative and unergative verbs [6]. We agree with such analysis in principle, and propose that the two event structures essentially reflect two views on weather events. When viewed as activities, weather events are presented as unergative; when viewed as happenstances [25], weather events are presented as unaccusative.

The two views on weather events, to some extent, can be adapted to Langacker's [26] two cognitive processing modes of summary scanning and sequential scanning. In summary scanning, all facets of events are conceived as coexistent and as a whole; in sequential scanning, each scene of an event is viewed successively rather than simultaneously, thus the change and details implicit in an event can be represented. Within such frameworks, we claim that the Mandarin weather verbs shown in (2) and (3) are not uniformly unaccusative. The verbs, when taking weather nouns as objects, are unergative verbs linked to the activity event structure, or they are viewed as activities and processed by summary scanning; when having weather nouns as subjects, they are unaccusative verbs linked to the scalar change event structure, or they are viewed as happenstances and processed by sequential scanning. By sentences in (2) and (3) alone, it would be difficult to make the distinction. Clearer evidence can be found when we consult a corpus.

We use raining event to conduct a case study. Expressions of 下雨 *xiàyǔ* 'to rain' and 雨下 *yǔ xià* 'rain falls' were searched in the CCL Corpus [27] (accessed 7 Jan 2020). The results show distinct behaviour between the two. By using 下雨 *xiàyǔ* 'to rain', people tend to regard the raining event as a whole, talking about whether or when it rains. Expressions similar to (5) are frequently used. However, people tend to describe the details of raining and talk about how it rains when using 雨下 *yǔ xià* 'rain falls', as typical examples shown in (6), suggesting that the raining event is viewed as a process with change and details. Therefore, expressions with 下雨 *xiàyǔ* 'to rain' indicate activity events, where 下 *xià* 'fall' functions as an unergative verb, and expressions using 雨下 *yǔ xià* 'rain falls' indicate scalar change events, where 下 *xià* 'fall' is unaccusative. In fact, the English sentence "it rained a light rain" can only be translated using 下+雨 as in 下了场小雨, and the Chinese translation of "a light rain rained on my head" should use 雨+下 as in 一场小雨下到我头上了.

(5) a. 一直 不 下雨
yīzhí bù xiàyǔ
continually NEG rain
'It has not rained for a long time.'

b. 要 下雨 了
yào xiàyǔ ·le
will rain LE
'It is going to rain.'

c. 不管 刮风 下雨
bùguǎn guāfēng xiàyǔ
regardless blow-wind rain
'No matter if it is windy or rainy.'

(6) a. 雨 下 得 很 密
yǔ xià ·de hěn mì
rain fall DE very intensive
'It is raining heavily.'

b. 雨 下 得 更 大 了
yǔ xià ·de gèng dà ·le
rain fall DE more big LE
'It is raining more heavily.'

c. 雨 下 个 不 停
yǔ xià ·ge bù tíng
rain fall GE NEG stop
'The rain does not stop.'

Furthermore, we can use the order of weather nouns and verbs in attributive clauses to identify unaccusatives and unergatives, as mentioned in Sect. 1. Examples in (7) show that both 下雨 *xiàyǔ* 'to rain' and 雨下 *yǔ xià* 'rain falls' can appear in attributive clauses, and are linked to two event structures, and two views on weather events or two cognitive processing modes respectively.

(7) a. 要 下雨 的 时候
yào xiàyǔ ·de shí·hou
will rain DE time
'When it is going to rain'

b. 雨 下 大 了 的 时候
yǔ xià dà ·le ·de shí·hou
rain fall big PFV DE time
'When it is raining more heavily'

Normally, a subject can be added to 下雨 *xiàyǔ* 'to rain', e.g., 天 *tiān* 'heaven/god' can appear in (5b), as 天要下雨了 'it is going to rain'. If English weather *it* has semantic content as argued by a number of linguists [6], there is no reason to treat 天 *tiān* 'heaven/god' as an expletive subject. Being a divine power in charge of everything, 天 *tiān* 'heaven/god' is deemed to be the aloft source and controller of various meteorological phenomena by ancient Chinese people, and weather substances such as rain, frost and fog are said to be originated from and sent down by it in Old Chinese [12].

This optional subject thus bolsters the unergative analysis of 下 *xià* 'fall' in下雨 *xiàyǔ* 'to rain' by bringing stronger volition. In addition, in imperatives, people only use下雨 *xiàyǔ* 'to rain' with the optional higher powers acting as addressees, e.g., 老天啊, 求求你别下雨了! 'Dear god, please stop raining!' This further confirms the volition and the connection of the higher powers with下雨 *xiàyǔ* 'to rain'.

The above distinctions hold not only for 下雪 *xiàxuě* 'to snow', 下霜 *xiàshuāng* 'to frost' and 起雾 *qíwù* 'to fog', but also for 打雷 *dǎléi* 'to thunder', 闪电 *shǎndiàn* 'lightning flashes' and 刮风 *guāfēng* 'wind blows'. These Mandarin weather verbs can be unaccusative or unergative according to how they are viewed and to which event structures they are linked. In fact, 打 *dǎ* 'hit', 闪 *shǎn* 'flash' and 刮 *guā* 'blow' do show unaccusative behaviour when used in more complex expressions, e.g., 雷打了半天 'it has been thundering for a long time', 电闪得更亮了 'the lightning is getting more dazzling', 风刮得很紧 'the wind is blowing quite hard'. In these expressions, the related events are treated as processes with details. Besides, the verbs can only appear after the weather nouns in attributive clauses when related to the view of weather events as happenstances. Optional subjects such as天 *tiān* 'heaven/god', 老天爷 *lǎotiānyé* 'god' and 雷公 *léigōng* 'Lord of Thunder' and the accompanying volition cannot be found here, but only in the unergative expressions, e.g., 打雷 *dǎléi* 'to thunder', 闪电 *shǎndiàn* 'lightning flashes' and 刮风 *guāfēng* 'wind blows'.

5 Conclusion

Weather verbs in Mandarin and other Sinitic languages are not uniformly unaccusative. Instead, cases of unaccusative, unergative, and even transitive verbs have been found in our investigation. The verb 打 *dǎ* 'hit', with an original meaning of striking, is demonstrated to be unergative in 打雷 *dǎléi* 'to thunder' and unaccusative in 雷打 *léi dǎ* 'to thunder' in Mandarin, and transitive in 雷打 *léidǎ* 'lightning strikes' and 霜打 *shuāngdǎ* 'frost damages' in Sinitic languages. The unaccusative vs. unergative dichotomy has several parallel syntactic and semantic behaviours, i.e., the way to describe weather events in terms of how they happen or whether they happen, the sequential order of weather verbs and nouns in attributive clauses, the (non-)existence of optional higher supernatural subjects and volition, etc. Such distinctions are related to two different views on the weather events. When viewed as activities, weather events are presented as unergative. On the other hand, when weather events are viewed as happenstances, they are encoded as unaccussative verbs with the weather substance as the topic. These may in turn correspond to two cognitive processing modes of summary scanning and sequential scanning.

Acknowledgement. This work was partially funded by University of Macau Research Committee project entitled "Grammatical Violations in Chinese Slogans and Titles and Their Linguistic Implications" (Project code: MYRG2018-00201-FAH).

References

1. Perlmutter, D.M.: Impersonal passives and the unaccusative hypothesis. In: Jeri, J.J., et al. (eds.) Proceedings of the Fourth Annual Meeting of the Berkeley Linguistics Society, pp. 157–189. Berkeley Linguistics Society, Berkely, CA (1978). https://doi.org/10.3765/bls.v4i0.2198
2. Burzio, L.: Italian Syntax: A Government-Binding Approach. Reidel, Dordrecht (1986)
3. He, Y.: A Generative Grammar of Mandarin Chinese. Peking University Press, Beijing (2011).(in Chinese)
4. Xu, J.: Formal Aspects of Chinese Grammar. World Scientific, Singapore (2017). https://doi.org/10.1142/10337
5. Xu, J.: Grammatical Principles and Grammatical Phenomena. Peking University Press, Beijing (2001).(in Chinese)
6. Levin, B., Krejci, B.: Talking about the weather: two construals of precipitation events in English. Glossa J. Gen. Linguist. **4**(1), 58 (2019). https://doi.org/10.5334/gjgl.794
7. Bleotu, A.C.: Why does IT always rain on me? On weather verbs. In: Proceedings of the First Central European Conference in Linguistics for Postgraduate Students, pp. 59–81 (2012)
8. Li, Y.-H. A.: Order and Constituency in Mandarin Chinese. Kluwer Academic Publishers, Dordrecht (1990). https://doi.org/10.1007/978-94-009-1898-6
9. Yang, S.: The unaccusative phenomenon - a study on the relationship between syntax and semantics. Contemp. Linguist. **1**(1), 30–43 (1999)
10. Dong, S.: Directions of fog. Yuyan Wenzi Zhoubao, 2, 1 August 2018. https://doi.org/10.13140/rg.2.2.15796.81289 (in Chinese)
11. Ren, H.: "Noun-Verb" Conversion in Archaic Chinese: from the Perspective of Lexical-Semantic Analysis. China Social Sciences Press, Beijing (2020) (in Chinese)
12. Dong, S.: A study on meteorological words in Chinese. Postdoctoral Report, Peking University (2019). https://doi.org/10.13140/rg.2.2.22648.29445 (in Chinese)
13. Dong, S., Yang, Y., Huang, C.-R., Ren, H.: Directionality and momentum of water in weather: a morphosemantic study of conceptualisation based on hantology. In: Hong, J.-F., Zhang, Y., Liu, P. (eds.) CLSW 2019. LNCS (LNAI), vol. 11831, pp. 575–584. Springer, Cham (2020). https://doi.org/10.1007/978-3-030-38189-9_59
14. Huang, C.-R., Dong, S.: From lexical semantics to traditional ecological knowledge: on precipitation, condensation and suspension expressions in Chinese. In: Hong, J.-F., Zhang, Y., Liu, P. (eds.) CLSW 2019. LNCS (LNAI), vol. 11831, pp. 255–264. Springer, Cham (2020). https://doi.org/10.1007/978-3-030-38189-9_27
15. Huang, C.-R., Dong, S., Yang, Y., Ren, H.: From language to meteorology: kinesis in weather events and weather verbs across Sinitic languages. Humanit. Soc. Sci. Commun. **8**, 4 (2021). https://doi.org/10.1057/s41599-020-00682-w
16. Dong, S., Yang, Y., Ren, H., Huang, C.-R.: Directionality of atmospheric water in Chinese: a lexical semantic study based on linguistic ontology. SAGE Open **11**(1), 1–13 (2021). https://doi.org/10.1177/2158244020988293
17. Dong, S., Huang, C.-R., Ren, H.: Towards a new typology of meteorological events: a study based on synchronic and diachronic data. Lingua **247**, 102894 (2020). https://doi.org/10.1016/j.lingua.2020.102894
18. Ren, H., Dong, S.: Ways to encode meteorological events in Classical Chinese: from a typological perspective. Bull. Linguist. Stud. **27**, 265–283 (2021)
19. Eriksen, P., Kittilä, S., Kolehmainen, L.: Weather and language. Lang. Linguist. Compass **6**(6), 383–402 (2012). https://doi.org/10.1002/lnc3.341
20. Levin, B., Rappaport Hovav, M.: Unaccusativity: At the Syntax-Lexical Semantics Interface. The MIT Press, Cambridge and London (1995)

21. Li, R. (ed.): Great Dictionary of Modern Chinese Dialects (42 Volumes). Jiangsu Education Publishing House, Nanjing (1993–2003) (in Chinese)
22. Tao, G. (ed.): Dictionary of Nantong Dialect. Jiangsu People's Publishing Ltd, Nanjing (2007).(in Chinese)
23. Zhang, W., Mo, C.: Dictionary of Lanzhou Dialect. China Social Sciences Press, Beijing (2009).(in Chinese)
24. Ahrens, D.: Essentials of Meteorology: An Invitation to the Atmosphere, 6th edn. Brooks/Cole, Cengage Learning, Belmont (2012)
25. McCawley, N.A.: From OE/ME 'impersonal' to 'personal' constructions: what is a 'subject-less' S. In: Steever, S.B., et al. (eds.) Papers from the Parasession on Diachronic Syntax: April 22, 1976, pp. 192–204. Chicago Linguistic Society, Chicago (1976)
26. Langacker, R.W.: Foundations of Cognitive Grammar, Vol. I: Theoretical Prerequisites. Stanford University Press, Stanford, CA (1987)
27. Zhan, W., Guo, R., Chang, B., Chen, Y., Chen, L.: The building of the CCL corpus: its design and implementation. Corpus Linguist. **6**(1), 71–86 (2019). http://ccl.pku.edu.cn:8080/ccl_corpus (in Chinese)

Degree Intensification and Sentential Functions in Chengdu Chinese

Jiajuan Xiong[1(✉)] and Feng-fan Hsieh[2]

[1] Southwestern University of Finance and Economics, Chengdu, China
[2] National Tsing Hua University, Hsinchu, Taiwan
siahonghoan@faculty.hss.nthu.edu.tw

Abstract. In Chengdu Chinese, degree intensification is inextricably associated with sentential functions, ranging from completing a sentence, encoding a focus to expressing exclamation. This correlation is exhibited in two ways: first, some intensifiers need to be paired with Sentence Final Particles (SFPs), such as the focus *ni* or the exclamatory *no*, and thus encode the relevant sentential functions. Second, some other intensifiers require no obligatory company of SFPs but still exhibit inalienable sentential functions, e.g., completing a sentence and expressing exclamation. Regardless of whether SFPs are present or not, the correlation between degree intensification and sentential functions is evident. Given this, we propose a double-layer DegP, with the lower $Deg_I P$ encoding intensification and the higher $Deg_S P$ fulfilling the sentential functions, to capture the phenomenon. This proposal can further explain the (non-)relativization of degree intensifiers in Chengdu as well as can shed new light on the semantically vacuous *hen* in Mandarin.

Keywords: Intensifiers · SFPs · Intensification · Sentential functions · DegP

1 Introduction

In Chengdu Chinese, degree intensifiers may vary in the following aspects: (i) some (at least one) degree intensifiers need to be accompanied by a particle *tɛ* (Chengdu counterpart of Mandarin DE), as shown in (1), whereas many others do not, as illustrated in (2)–(5); (ii) some degree intensifiers precede their modified AP or VP, as exemplified in (2)–(4), whereas some others follow the modified AP or VP, as in (1) and (5); (iii) some intensifiers require the obligatory presence of a sentence final particle (henceforth SFP), as shown in (3)–(5), but some others do not, as in (1) and (2).

M. Liu et al. (Eds.): CLSW 2020, LNAI 12278, pp. 74–86, 2021.
https://doi.org/10.1007/978-3-030-81197-6_7

(1) a. Tsɛ pɚ su iəutɕ'y **tɛ** **xən** **(o/io)**!
this CL book interesting DE very SFP/SFP
'This book is highly interesting.'
b. T'a ɕixuan k'an su **tɛ** **xən** **(o/io)**!
s/he like read book DE very SFP/SFP
'S/he likes reading very much.'

(2) a. Tsɛ pɚ su **tsɿ/xau** iəutɕ'y **(o/io)**!
this CL book very/very interesting SFP/SFP
'This is book is highly interesting.'
b. T'a **tsɿ/xau** ɕixuan k'an su **(o/io)**!
s/he very/very like read book SFP/SFP
'S/he likes reading very much.'

(3) a. Tsɛ pɚ su **to** iəutɕ'y ***(ni)**.
this CL book very interesting SFP
'This book is highly interesting.'
b. T'a **to** ɕixuan k'an su ***(ni)**.
s/he very like read book SFP
'S/he likes reading very much.'

(4) a. Tsɛ pɚ su **tsɿiəu** **namo** iəutɕ'y ***(no)**.
this CL book only so interesting SFP
'This book is highly interesting.'
b. T'a **tsɿiəu** **namo** ɕixuan k'an su ***(no)**.
s/he only so like read book SFP
'S/he likes reading very much.'

(5) a. Tsɛ pɚ su kuei **xən** ***(no)**.
this CL book expensive very SFP
'This book is too expensive.'
b. T'a k'an su k'an **xən** ***(no).**
s/he read book read very SFP
'S/he reads too much (e.g., for a long time at a stretch).'

The features of degree intensifiers, in terms of their distributional patterns and pairings with SFPs, are summarized in Table 1.

Table 1. The features of degree intensifiers in Chengdu Chinese

pre-AP/VP intensifiers		post-AP/VP intensifiers	
SFPs	[DEGREE + AP/VP]	[AP/VP + DEGREE]	[AP/VP + *tɛ* + DEGREE]
Ni	*To*	X	X
No	*tsɿiəu namo*	*xən, t'ɚ*	X
Ø/o/io	*tsɿ, xau*	X	*xən*

(X stands for incompatibility)

The pairings between degree intensifiers and SFPs will be the focus of this study. Specifically, we explore why degree intensifiers need to be paired with SFPs, and, in what way, the pairing is accomplished. This exploration starts with a study of SFPs in Chengdu Chinese in Sect. 2. In Sect. 3, the interactional patterns between degree intensifiers and SFPs are analyzed in detail. Then, in Sect. 4, we propose a double-layering DegP to capture both intensification meaning and sentential functions encoded by degree intensifiers, with or without an SFP. Finally, a conclusion is provided in Sect. 5.

2 A Sketch of SFPs in Chengdu Chinese

There is a rich repertoire of SFPs in Chengdu Chinese. Zhang et al. (2001: 372) have identified 25 sentence final particles. In fact, the exact number of SFPs is by no means easy to determine, for two reasons: first, some phonetically similar SFPs are considered as allophones, instantiating one and the same SFP; second, some SFPs take up multiple functions, with one item representing several SFPs. Most pertinent to our study are the SFPs *ni* and *no*, as they are required to be paired with degree intensifiers, as shown in Sect. 1. In Sects. 2.1 and 2.2, we analyze the two SFPs *ni* and *no*, respectively, in an attempt to determine their syntactic positions in the periphery (Rizzi 1997; Cinque 1999). Then, in Sect. 2.3, we adopt Tang's (2015, 2020) cartographic approach to analyze Chengdu SFPs, which include the focus, modality, interrogation and exclamation types, in order to delineate a cartographic sketch for Chengdu SFPs.

2.1 The SFP *ni*

The presence of the SFP *ni* in Chengdu Chinese brings about an assertive reading with regard to the whole proposition, as exemplified in (6a–d).

(6) a. T'a to ts'oŋmin *(**ni**).
s/he very intelligent SFP
'It is the case that s/he is highly intelligent.'
b. T'a ts'ɿ tɕ'i f'an *(**ni**).
s/he eat QI rice SFP
'It is the case that s/he is having a meal.'
c. T'a t'u ko taɕo **ni**.
s/he read EXP university SFP
'It is the case that s/he is a university graduate.'

The presence of *ni* turns a proposition into an assertion of the speaker. However, as shown in (6a–c), *ni* may or may not be obligatory. Specifically, the absence of *ni* in (6a) and (6b) results in ungrammaticality, whereas the removal of *ni* in (6c) only leads to the suspension of an assertive reading.[1] This assertive usage bears similarity with the

[1] The reason for the obligatory presence of *ni* in (6a) will be provided in Sect. 4. As for (6b), the SFP *ni* can be replaced by another SFP *tsai*. Unlike *ni*, *tsai* conveys the progressive meaning without any assertive sense. The relation between *ni* and *tsai* is out of the scope of this research. However, the semantic contrast between *ni* and *tsai*, in particular, the presence and absence of an assertive meaning, serves to add credence to the focus analysis of *ni*.

(shi) … de propositional assertion construction in Mandarin (Lee 2005; Xiong 2007; Paul and Whitman 2008; Mai 2013), in that (i) they bring about a focus reading of the proposition at issue; (ii) they are applicable to propositions anchored in the past, present or future time frame. In view of its semantic contribution, we analyze *ni* as a focus SFP.

2.2 The SFP *no*

The SFP *no* is versatile in Chengdu Chinese. As pointed out in Zhang et al. (2001), *no* can occur in declarative, interrogative, imperative and exclamatory sentences, as exemplified in (7a–d), in which (7b–d) are cited from Zhang et al. (2001).

(7)						
a. T‘a	ts‘ɿ	ko	fʻan	**no**.		(declarative)
s/he	eat	GUO	rice	SFP		
‘S/he had a meal already.’						
b. Tɕi	tian	**no**?				(interrogative)
which	time	SFP				
‘What’s the time?’						
c. Tsai	tsuaŋ	tian	**no**.			(imperative)
more	add	some	SFP			
‘Please add more.’						
d. Nan	tuei	su	t‘ɚ	**no**.		(exclamatory)
male	team	lose	very	SFP		
‘The men’s team suffered a crushing defeat!’						

Among the various functions of *no*, the exclamatory function in (7d) is of particular relevance to the current study, as it involves degree intensification. If *no* is taken as an exclamatory SFP, it is expected to behave on a par with other exclamatory SFPs, such as *o* or *io* as exemplified in (1) and (2). In fact, the exclamatory SFPs *no*, *o* and *io* are considered to be allophonic to one another (Zhang et al. 2001: 366). This proposal is rejected in this study, for the following reasons.[2]

First, it seems that the SFP *no* and the SFPs *o*/*io* are in a complementary distribution. Specifically, when *no* is required, it cannot be replaced by *o* or *io*, as shown in (8a, b). By contrast, when *o* and *io* are applicable, *no* is excluded, as exemplified in (9a, b).

Second, the occurrence of *no* is always obligatory, whereas the presence of the SFPs *o* and *io* are basically optional. This contrast is illustrated in (8) and (9).

---

[2] We thank one of the reviewers for proposing that the reason why *ni* and *no* cannot co-occur is probably due to a language-particular OCP constraint (c.f., Xu 2007; 2011). This phonological account is inspiring and interesting and it will be explored in a separate paper.

(8)	a. Nan	tuei	su	t'ɚ	***(no)/*o/*io/!**		
	male	team	lose	very	SFP/SFP/SFP		
	'The men's team suffered a crushing defeat!'						
	a. Tsɛ	pɚ	su	**tsʅiəu**	**namo**	iəutɕ'y	***(no)/*o/*io!**
	this	CL	book	only	so	interesting	SFP/SFP/SFP
	'This book is highly interesting.'						
(9)	a. Tsɛ	pɚ	su	iəutɕ'y	**tɛ** **xən**	**(o)/(io)/*no!**	
	this	CL	book	interesting	DE very	SFP/SFP/SFP	
	'This book is highly interesting.'						
	b. Tsɛ	pɚ	su	**tsʅ/xau**		iəutɕ'y	**(o)/(io)/*no!**
	this	CL	book	very/very		interesting	SFP/SFP/SFP
	'This is book is highly interesting.'						

Third, the exclamatory SFP *no* cannot be stacked on the focus SFP *ni*, whereas the exclamatory SFPs *o* and *io* can. This difference is shown in (10a, b).

(10)	a. Tsɛ	pɚ	su	to	iəutɕ'y	**ni**	**(*no).**
	This	CL	book	very	interesting	SFP	SFP
	Intended: 'It is indeed the case that this book is highly interesting!'						
	b. Tsɛ	pɚ	su	to	iəutɕ'y	**ni**	**o/io!**
	This	CL	book	very	interesting	SFP	SFP/SFP
	'This book is highly interesting!'						

To sum up, *no* should be distinguished from *o* and *io*, although all of them can express exclamation. We analyze the exclamatory *no* as a conflation between the finite *no* and the exclamatory *o*, giving rise to *no*. This insight is indeed expressed by Zhang et al. (2001: 366), even though they differ from us in taking *no*, *o* and *io* as allophones. In our analysis, the exclamatory SFP *no* is crucially different from the exclamatory SFPs *o* and *io*. Specifically, *no*, by virtue of combining two, carries double function, viz. rendering a clause finite and making it exclamatory.[3] By contrast, *o* and *io* are exclusively exclamatory. This analysis captures the above-mentioned differences between *no* and *o*/*io*: (i) the obligatory presence of *no* is due to its function to complete a sentence; (ii) the mutual exclusion between exclamatory *no* and focus *ni* follows from the fact that a conflation of two functional heads (FinP and CP) does not allow any insertion in between, e.g., by a focus *ni* (FocP) (see e.g., Rizzi 2018).

[3] It should be noted that *no* cannot occur in a non-finite context. For example, a non-finite complement allows a bare adjectival predicate (see Gu 2008), as shown in (i). However, the presence of [intensifier…*no*] renders it unacceptable, as illustrated in (ii).

(i)	ŋomən kuli		[t'a	zəntsən].		
	we encourage		s/he	hardworking		
	'We encourage her/him to be hardworking.'					
(ii)	*ŋomən kuli	[t'a	tsʅiəu	namo	zəntsən	**no**].
	we encourage	s/he	only	so	hardworking	SFP
	Intended: 'We encourage her/him to be very hardworking.'					

The above contrast proves that *no* marks finiteness.

Thus, the SFP *no*, in addition to initiating a question and encoding a directive, it can serve the purposes of completing a sentence and expressing exclamation. The contrast between the latter two (i.e., finiteness vs. exclamation) is in line with Erlewine's (2013) distinction between *v*P-domain SFPs and CP-domain SFPs as well as Zhang's (2019) differentiation between sentence final aspect particles (SFAPs) and sentence final particles (SFPs). The point is that these two functions can be merged to one and the same SFP *no* in Chengdu Chinese.

Given this, the exclamatory SFP *no* in Chengdu is taken as an amalgamative SFP combining both finiteness and exclamation. This stands in contrast to the exclamatory SFPs *o* and *io*, which are of the independent type in the sense that they are exclusively exclamatory.

2.3 More SFPs in Chengdu Chinese

Following Tang's (2015, 2020) classification of Cantonese SFPs, we identify various types of Chengdu SFPs, such as the focus, modality, interrogation, and exclamation types, by presenting and analyzing their typical examples. By means of studying the interactions of these SFPs, we work out a cartographic hierarchy, which provides a basis for our analysis of degree intensifiers and SFPs in Sect. 3.

Recall that, in Sects. 2.1 and 2.2, *ni* is identified as a focus SFP, whereas *no* can either be a finite SFP or an amalgamative exclamatory SFP, in addition to some other less-commonly-used functions, such as imperative and interrogative functions. Moreover, *o* and *io* are found to be independent exclamatory SFPs. In this section, we will further examine the SFPs *tɛ*, *tɛma*, *wa*, which will be exemplified in turn.

First, we provide two different contexts in which *tɛ* can be applied as an SFP, as below:

Context 1: The teacher is tutoring a student out of hours. This student thus gets impatient and absent-minded, saying that "I want to go home now". Then, the teacher replies:

(11) ŋo xai ɕiaŋ xuei tɕia **tɛ**.
I out_of_expectation want return home SFP
'As against your assumption, I want to go home as well.' (It functions as a refutation against the student's assumption that the teacher does not want to return home but enjoys the over-time work.)

Context 2: A cat broke a porcelain doll and hid it behind the sofa in order to avoid reproach from its owner. Upon finding the hidden doll, the pet owner says:

(12) tsɛ tsɿ maumi xai coŋmin **tɛ**.
this CL cat out_of_expectation clever SFP
'It is beyond my imagination that this cat is so clever.'

In these two contexts, the occurrence of *tɛ* yields an against-the-assumption reading. Specifically, the assumption could be either of the listener's or of the speaker's, as illustrated in context 2 and context 3, respectively. Evaluation against an assumption should be analyzed as a type of evaluative modality and thus *tɛ* is taken as a modality SFP.

Second, *tɛma* is a widely-used SFP, as exemplified below:

(13) T'a sɿ waŋ i-sən **tɛma**.
s/he be P.N. doctor SFP
'It is evident that s/he is Dr. Wang.'

The SFP *tɛma* mainly brings out an evidential reading, in the sense that the proposition should be evident either to the speaker or to the listener. For example, the above sentence may be uttered in the following contexts: (i) I did not recognize Dr. Wang, but I should have recognized her/him. In this context, this sentence can be paraphrased as "it just occurs to me that s/he is Dr. Wang, whom I should have recognized earlier". (ii) Since the listener did not recognize Dr. Wang, the speaker mentioned the fact, which is supposed to have been known to the listener. (iii) Since the listener failed to show due respect to Dr. Wang, the speaker uttered (13) in an attempt to remind the listener of proper behaviors in the presence of a respectable doctor. The semantic denominator of the above mentioned usages is evidentiality, which thus ascribes a modality membership to the SFP *tɛma.*

Third, the SFP *wa* is unambiguously interrogative, as illustrated in (14).

(14) T'a sɿ i-sən **wa**?
s/he be doctor SFP
'Is s/he a medical doctor?'

The SFPs discussed in this paper are listed and categorized in Table 2.

Table 2. Different types of Chengdu SFPs.

FinP	FocP	CP			
	Assurance	Modality		Interrogation	Exclamation
		Epistemic	Evidential		
no	*ni*	*tɛ*	*tɛma*	*wa*	*o, io; no*

The sequence of the SFPs in Table 2 can be exemplified in (15).

(15) a. T'a xai tu ko taɕo **no** **ni** **tɛ**.
s/he out_of-expectation study EXP university SFP SFP SFP
'It occurs to me that s/he is a university graduate.'
b. T'a tu ko taɕo **no** **ni** **tɛma**.
s/he study EXP university SFP SFP SFP
'It is evidential that s/he is a university graduate.'
c. T'a tu ko taɕo **ni** **wa**?
s/he study EXP university SFP SFP
'Is it the case that s/he is a university graduate?'
d. T'a tu ko taɕo **no** **ni** **o**!
s/he study EXP university SFP SFP SFP
'What a surprise fact that s/he is a university graduate!'

The SFPs in Chengdu take the linear order of [SFP_{FIN}+SFP_{FOC}+ SFP_{MOD}/SFP_{INTR}/SFP_{EXCL}], if any co-occurrence of them happen. In Sect. 3, we will examine how degree intensifiers interact with SFPs, so as to explore the nature of degree intensification in Chengdu Chinese.

3 The Interactions Between Degree Intensifiers and SFPs

In this section, we examine the interactional patterns between degree intensifiers and SFPs. In particular, we investigate (i) whether degree intensifiers require the obligatory presence of SFPs; (ii) if yes, what kinds of SFPs are required to be paired with degree intensifiers; (iii) in addition to the obligatorily paired SFPs, what other SFPs can be allowed to co-occur with degree intensifiers. The degree intensifiers will be examined in turn.

The degree intensifier *tɛ xən* does not require the obligatory presence of a SFP. When it interacts with SFPs, it defies the finite type of SFP but allows the presence of SFPs of the focus, modality, interrogation and exclamation types, as illustrated by the contrast between (16a) and (16b–f).

(16) a. T'a kauɕin **tɛ** **xən** (***no**).
s/he happy DE very SFP_{FIN}
'S/he is very happy.'
b. T'a kauɕin **tɛ** **xən** **ni**.
s/he happy DE very SFP_{FOC}
'It is the case that s/he is very happy.'
c. T'a xai kauɕin **tɛ** **xən** **tɛ**.
s/he out_of_expectation happy DE very SFP_{MOD}
'Out of my expectation, s/he is very happy.'
d. T'a kauɕin **tɛ** **xən** **tɛma**.
s/he happy DE very SFP_{MOD}
'It is evident that s/he is very happy.'
e. T'a kauɕin **tɛ** **xən** **wa?**
s/he happy DE very SFP_{INTR}
'Is s/he very happy?'
f. T'a kauɕin **tɛ** **xən** **o/io!**
s/he happy DE very SFP_{EXCL}
'S/he is very happy!'

The degree intensifiers *tsɿ* and *xau*, like *tɛ xən*, show no obligatory pairing with an SFP. However, they can co-occur with independent exclamatory SFPs, such as *o* and *io*. Apart from them, SFPs of other types, e.g., finiteness, focus, modality and interrogation, cannot be licensed. They are shown in (17).

(17) T'a **tsɿ/xau** kauɕin (***no**)/ (***ni**)/ (***tɛ**)/ (***tɛma**)/ (***wa**)/(**o/io**).
s/he very/very happy SFP$_{FIN}$/SFP$_{FOC}$/SFP$_{MOD}$/SFP$_{MOD}$/SFP$_{INTR}$/SFP$_{EXCL}$
'S/he is very happy!'

The degree intensifier *xən*, without the company of *tɛ*, is necessarily paired with the SFP *no*, as shown in (18a). This *no* should be analyzed as a finite SFP, as it can be stacked by other SFPs of the focus, modality, interrogation and exclamation types, as exemplified in (18b–f).

(18) a. T'a nuei **xən** **no.**
s/he tired very SFP$_{FIN}$
'S/he is too tired.'
b. T'a nuei **xən** **no** **ni.**
s/he tired very SFP$_{FIN}$ SFP$_{FOC}$
'It is the case that s/he is too tired.'
c. T'a xai nuei **xən** **no** **tɛ.**
s/he out_of_expectation tired very SFP$_{FIN}$ SFP$_{MOD}$
'(You may not be aware of the fact that) s/he is too tired.'
d. T'a nuei **xən** **no** **tɛma**.
s/he tired very SFP$_{FIN}$ SFP$_{MOD}$
'It is evident that s/he is too tired.'
e. T'a nuei **xən** **no** **wa**?
s/he tired very SFP$_{FIN}$ SFP$_{INTR}$
'Is s/he too tired?'
f. T'a nuei **xən** **no** **io**!
s/he tired very SFP$_{FIN}$ SFP$_{EXCL}$
'S/he is too tired! (a reminder)'

The degree intensifier *to* must be paired with the SFP$_{FOC}$ *ni* but excludes the SFPs lower than the focus layer, as shown in (19a). Moreover, it allows its co-occurrence with the SFPs higher than the focus type, as shown in (19b–e).

(19) a. T‘a **to** kauɕin (***no**) *(**ni**).
s/he very happy SFP$_{\text{FIN}}$ SFP$_{\text{FOC}}$
‘It is the case that s/he is very happy.’

b. T‘a xai **to** kauɕin **ni** **tɛ**.
s/he out_of_expectation very happy SFP$_{\text{FOC}}$ SFP$_{\text{MOD}}$
‘Out of my expectation, s/he is very happy.’

c. T‘a **to** kauɕin **ni** **tɛma**.
s/he very happy SFP$_{\text{FOC}}$ SFP$_{\text{MOD}}$
‘It is evident that s/he is very happy.’

d. T‘a **to** kauɕin **ni** **wa**?
s/he very happy SFP$_{\text{FOC}}$ SFP$_{\text{INTR}}$
‘Is it the case that s/he is very happy?’

e. T‘a **to** kauɕin **ni** **o/io**!
s/he very happy SFP$_{\text{FOC}}$ SFP$_{\text{EXCL}}$
‘It is indeed the case that s/he is very happy!’

The degree intensifier *tsʅiəu namo* requires the obligatory presence of the SFP *no*. This *no* cannot be of the finite type, as it bars the occurrence of SFPs higher than the finite type, such as the focus, modality and interrogation types. This is exemplified in (20a). Note that *no* can co-occur with *o* or *io*, with *no* necessarily preceding *o/io*, but not vice versa, as shown in (20b). In the latter case, *no* should be treated as a finite marker, as opposed to an exclamatory marker.

(20) a. T‘a **tsʅiəu namo** kauɕin (***ni**)/(***tɛ**)/ (***tɛma**)/ (***wa?**)/*(**no**)
s/he only so happy SFP$_{\text{FOC}}$/SFP$_{\text{MOD}}$/SFP$_{\text{MOD}}$/SFP$_{\text{INTR}}$/SFP$_{\text{EXCL}}$
‘S/he is very happy.’

b. T‘a **tsʅiəu namo** kauɕin *(**no**) (**o/io**).
s/he only so happy SFP$_{\text{FIN}}$ SFP$_{\text{EXCL}}$/SFP$_{\text{EXCL}}$
‘S/he is very happy!’

The above-illustrated interactional patterns between degree intensifiers and SFPs are summarized in Table 3.

Table 3. The Interactions between Degree Intensifiers and SFPs.

	Ø	FinP	FocP	CP				
			Assurance	Modality		Interrogation	Exclamation	
				Epistemic	Evidential			
		No	*ni*	*tɛ*	*tɛma*	*wa*	*no*	*o, io*
tɛxən	✓	X	✓	✓	✓	✓	✓	✓
tsʅ/xau	✓	X	X	X	X	X	X	✓
xən	X	✓	✓	✓	✓	✓	✓	✓
to	X	X	✓	✓	✓	✓	X	✓
tsʅiəu namo	X	X	X	X	X	X	✓	✓

(The symbols of ✓ and X stand for compatibility and incompatibility, respectively.)

As for each degree intensifier, the leftmost tick is most important, as SFPs are arranged from lower to higher positions, along the left-to-right ordering direction in the table. Since higher SFPs can, in principle, be stacked to lower SFPs, the leftmost collocation indicates the sentential type of degree intensifier. Along this line of analysis, *xən, to* and *tsɿiəu namo,* by virtue of being necessarily paired with focus finite SFP, focus SFP and exclamatory SFP, are analyzed as degree intensifiers of the finite, focus and exclamatory types, in that order. By contrast, *tɛ xən, tsɿ* and *xau* do not have to be paired with any SFPs. They differ in that *tɛ xən* allows the presence of SFPs higher than FinP, whereas *tsɿ* and *xau* only allow the independent exclamatory SFPs. Given this, *tɛ xən* is taken as a finite type degree intensifier, while *tsɿ* and *xau* are analyzed as the exclamatory type of degree intensifiers.

4 The Double-DegP Analysis

As shown in Sect. 3, degree intensification in Chengdu Chinese is closely related to SFPs of various types. We propose that degree intensifiers are of two functions, i.e., degree intensification and sentential marking. The latter one is versatile, ranging from marking the completion of a sentence, encoding focus to expressing exclamation. Syntactically, the double function is encoded by double layers of DegP, with the lower layer encoding the intensification ($Deg_I P$, with $_I$ indicating the meaning of intensification) while the higher one realizing the sentential function ($Deg_S P$, with s representing sentential function). The structure is shown in (21).

(21) <u>The DegP-shell analysis of degree intensification in Chengdu Chinese</u>:

$[_{DegSP}\ Deg_S{}^0\ [_{DegIP}\ [Deg_I{}^0\ [_{AP/VP} \ldots]]]]$

The lower layer $Deg_I P$ takes care of the function of intensification and thus host the degree words, whereas the higher layer $Deg_S P$ is connected with the sentential projections, such as FinP, FocP, CP etc. Thus, intensifiers are shown to co-occur with various types of SFPs. Moreover, it is $Deg_S P$ that renders the degree intensifiers some extra meanings and functions, e.g., sentence completion, focus-reading and exclamation-reading. This analysis can be supported by the following pieces of evidence.

First, the double function of one single construction is not a novel idea. For example, the *shi…de* cleft sentence in Mandarin Chinese is analyzed to involve two syntactic projections, i.e., a focus projection headed by *shi* and an aspect projection headed by *de*, even though the main function of the *shi…de* cleft is recognized as focus-encoding, but not aspect marking (Mai 2013). In a similar vein, although the degree intensifiers in Chengdu Chinese carry the main function of increasing the degree of a state, its function at the sentential level should also be taken into consideration while conducting syntactic analyses.

Second, this double DegP analysis can provide an explanation for the well-known Mandarin phenomenon that the degree word *hen* 'very' is required to encode positive polarity as well as to complete a sentence (see Sybesma 1999; Huang 2006; Gu 2008; Grano 2012, among others). These two features of *hen*, i.e. absence of intensification and completion of a sentence, follow naturally, if we analyze *hen* as the head of the higher

Deg$_S$P, with the lower DegIP being null. In other words, *hen* performs the sentential function only and it is thus described to be a semantically bleached degree adverb (Grano 2012). This analysis can be further verified, if we take into consideration the two different usages of the Mandarin *hen*'s Chengdu counterpart, i.e., *xən*, which can occur in the forms of [AP/VP *tɛ-xən*] (positive polarity) and [AP/VP *xən-no*] (degree intensification). Semantically, [AP/VP *tɛ-xən*] is similar to the positive usage of *hen*. In our analysis, both Mandarin *hen* and Chengdu *xən* are situated in the higher Deg$_S$P position, and compatible with the sentential function. The only difference is that the lower Deg$_I$P is null in Mandarin whereas it is realized by a neutral word *tɛ* in Chengdu Chinese. The word *tɛ* is degree-related but neutral in meaning. This is not far-fetched, as its Mandarin counterpart *de* is considered to head an extent phrase and thus degree-related (Sybesma 1999). Alternatively, *xən* can be positioned in the Deg$_I$P to express intensification. As a consequence, Deg$_S$P must be realized by other means, e.g., the presence of *no*, as in the case of [AP/VP *xən-no*].

Third, this analysis captures the contrastive syntactic behaviors of degree intensifiers. Notably, some intensified phrases, such as [AP/VP *tɛ-xən*], [*to* AP/VP *ni*], [AP/VP *xən-no*], can modify nominal phrases in the form of relative clauses. By contrast, some other intensified phrases, e.g., [*tsɿ/xau* AP/VP *Ø/(o)/(io)*] [*tsɿiəu namo* AP/VP *no*] cannot occur in relative clauses. This contrast follows from the fact that some intensifiers involve sentential functions at the CP layer, specifically, the exclamatory type of CP. By virtue of being at CP, they defy relativization.[4]

5 Concluding Remarks

This study is motivated by the observation that degree intensifiers in Chengdu Chinese generally require the presence of SFPs, of different types, such as finite, focus and exclamatory SFPs. There are two intensifiers that do not necessarily require the company of SFPs, but they can only be interpreted as exclamation. The correlation between degree intensification and sentential functions is captured by the proposal of double layer of DegP, with the lower Deg$_I$P taking care of intensification and the higher Deg$_S$P realizing the sentential functions.

References

Aoun, J., Li, Y.-H.A.: Essays on the Representational and Derivational Nature of Grammar: The Diversity of Wh-Constructions. MIT Press, Cambridge (2003)

Bianchi, V.: The raising analysis of relative clauses: a reply to Borsley. Linguist. Inq. **31**(1), 123–140 (2000)

Cinque, G.: Adverbs and Functional Heads. Oxford University Press, Oxford (1999)

Dikken, M.D., Singhapreecha, P.: Complex noun phrases and linkers. Syntax **7**(1), 1–54 (2004)

Erlewine, M.Y.: Sentence-final particles at the vP phase edge. In: Proceedings of the 25th North American Conference on Chinese Linguistics (NACCL-25) (2013)

Grano, T.: Mandarin hen and universal markedness in gradable adjectives. Nat. Lang. Linguist. Theory **30**(2), 513–565 (2012). https://doi.org/10.1007/s11049-011-9161-1

[4] The non-relativization of Chengdu intensifiers will be discussed in detail in a separate paper.

Gu, Y.: Studies of tense, aspect and Chinese time reference. In: Yang, S., Shengli, F. (eds.) Contemporary Linguistic Theories and Related Studies of Chinese, pp. 97–119. Commerce Press, Beijing (2008)

Hsieh, F.-F., Sybesma, R.P.E.: On the linearization of Chinese sentence-final particles: max spell out and why CP moves? Korea J. Chi. Lang. Lit. **1**(1), 53–90 (2011)

Huang, S.-Z.: Property theory, adjectives, and modification in Chinese. J. East Asian Linguis. **15**, 343–369 (2006)

Kayne, R.S.: The Antisymmetry of Syntax. MIT Press, Cambridge (1994)

Lee, H.-C.: On Chinese focus and cleft constructions. Ph.D. Dissertation, Taiwan National Tsing Hua University (2005)

Li, B.: Chinese final particles and the syntax of the periphery. Ph.D. Thesis, Universiteit Leiden (2006)

Lin, J.-W.: Aspectual selection and negation in Mandarin Chinese. Linguistics **41**(3), 425–459 (2003)

Lin, T.-H.J.: Finiteness of clauses and raising of arguments in Mandarin Chinese. Syntax **14**(1), 48–73 (2011)

Lin, T.-H.J.: Tense in Mandarin Chinese sentences. Syntax **18**(3), 320–342 (2015)

Liu, C.-S.L.: The positive morpheme in Chinese and the adjectival structure. Lingua **120**, 1010–1056 (2010)

Mai, Z.: Properties of the (Shi)…de focus construction in adult L2 acquisition and heritage language acquisition of Mandarin Chinese. Ph.D. Thesis, University of Cambridge (2013)

Rizzi, L.: The fine structure of the left periphery. In: Haegeman, L. (ed.) Elements of Grammar, pp. 281–337. Kluwer, Dordrecht (1997)

Rizzi, L.: Intervention effects in grammar and language acquisition. Probus **30**(2), 339–367 (2018). https://doi.org/10.1515/probus-2018-0006

Tang, S.-W.: Lectures on Cantonese Grammar. The Commercial Press, Hong Kong (2015)

Tang, S.-W.: Cartographic syntax of performative projections: evidence from Cantonese. J. East Asian Linguis. **29**, 1–30 (2020). https://doi.org/10.1007/s10831-019-09202-7

Xiong, Z.: Syntactic analysis of the construction 'Shi…de' (Shi…de Goujian Yanjiu). Stud. Chin. Lang. (Zhongguo yuwen) **4**, 321–330 (2007)

Xu, Z.: Inflectional morphology in optimality theory. Doctoral dissertation, Stony Brook University (2007)

Xu, Z.: Optimality theory and morphology. Lang. Linguist. Compass **5**(7), 466–484 (2011)

Zhang, N.N.: Sentence-final aspect particles as finite markers in Mandarin Chinese. Linguistics. **57**(5), 967–1023 (2019)

Zhang, Y., Zhang, Q., Deng, Y.: Grammar of Chengdu Chinese (Chengdu Fangyan Yufa Yanjiu). Bashu Publishing House (Bashu Shushe), Chengdu (2001)

Polysemy of *tuī* (推): An Image-Schema-Based Approach

Changle Yin(✉)

Department of Chinese Language and Literature, Peking University, Beijing 100871, China

Abstract. Under the guidance of cognitive semantics, this paper adopts an approach based on image schemas to analyze the meaning of the verb *tuī* (推) according to the research of Lakoff (1987) on the word *over*. The verb *tuī* can be divided into two sense branches: (1) To move the object or part of the object in the direction of force; (2) To cut or chop with a tool against the surface of the object. On this basis, we summarize 15 different image schemas. Next, this paper discusses the way the sense branches and how its variants are related (e.g.co-hyponymy analogue, specification and metonymy). Finally, the paper attains a radical-structured polysemous system of the verb *tuī*.

Keywords: *Tuī* (推) · Image schemas · Polysemy · Polysemous linking

1 Introduction

We have noticed that, in the past, most of the studies on Chinese *tuīlā* (推拉) verbs (i.e. push-pull verbs) mainly focus on the syntactic features. The polysemy of them has not been studied from the perspective of cognitive semantics.

Polysemy is a situation in which a form corresponds to several different but related meanings. Cognitive semantics regards polysemy as a category of sememes, which form a set of interrelated sememes through family similarity. Lakoff (1987) regarded polysemous words as the radical process of senses or sememe in the study of English preposition *over* and described the polysemous phenomenon of *over* through image schema. Following this line of methodology, Zhu (2006, 2010) analyzed the polysemy of the spatial motion verbs *shoū* (收, 'to receive') and *chuān* (穿, 'to across'). This paper attempts to extend this research model to the semantic analysis of Chinese spatial motion verb *tuī* (推), and explore the polysemy phenomenon by integrating the image schema of *tuī*. The verb *tuī* in Mandarin Chinese means "push" in English.

The verb *tuī* belongs to the spatial motion verb. The description of its motion situation involves the concepts of trajector, landmark and path. Trajector stands for the main part or the most prominent element in any relational structure, landmark refers to the reference point for positioning and path the trajectory of the main part moving from stage to stage.

We try to use the above concepts to analyze the image schemas of the verb *tuī* and sum up the polysemous system of it.

M. Liu et al. (Eds.): CLSW 2020, LNAI 12278, pp. 87–99, 2021.
https://doi.org/10.1007/978-3-030-81197-6_8

2 Image Schemas of the Verb *tuī*

2.1 Exerting Outward Force to Make an Object or a Part of the Object Move in the Direction of Force

We find that the most common sense branch of the verb *tuī* is exerting outward force to make an object or a part of the object move in the direction of force.

2.2 Agent Moving with the Trajector

In the image schemas of the verb *tuī*, the trajector (TR) is usually a three-dimensional entity, the landmark (LM) is usually far away from the position of the agent, and the agent and the trajector can move together. Such as[1]:

(1) 他 看到 一 个 推 婴儿车 的 女人。[1]
3sg see one CL push baby carriage DE women
"He saw a woman pushing a baby stroller."

(2) 有 时 候 我们 推 独 轮 车 搬运 地里 的 秫 秸。
sometimes 1pl push wheelbarrow carry field DE Kaoliang stalk
'Sometimes we push wheelbarrows to carry Kaoliang stalk in the field.'

It can be shown in Schema 1:

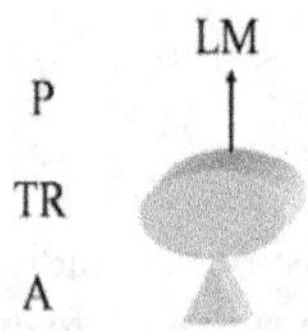

Schema 1. 3D AF ST

Schema 1 is formulated as 3D AF ST, that is, the trajector is three-dimensional (3D), the landmark is far away from the agent (AF), and the agent and the trajector move together (ST). We take (2) as an example. In sentence (2), 我们 ('we') is the agent and the 独轮车 ('wheelbarrow') is a three-dimensional trajectory. 我们 ('we') send out the action of *tuī*, moving away with the 独轮车 ('wheelbarrow').

We also find a kind of metonymic usage of *tuī*. Such as:

(3) 我 的 脚脖子 经常 抽筋，现在 做梦梦到 推 粪
1sg DE ankle often cramp now dream of push dung
上 山 还是要 抽 筋。
up mountain still cramp
"I often have cramps in my ankle. Now I dream of pushing dung up the mountain, but I still have cramps."

Such usage can be shown in Schema 2:

[1] The examples listed in this paper are mainly adapted from the BCC corpus of Beijing Language and Culture University. A few other of the examples are constructed by the author.

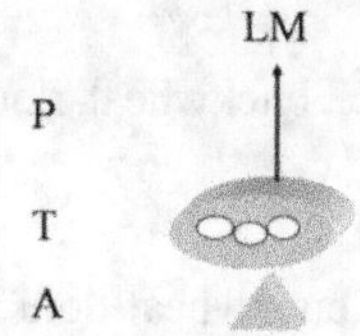

Schema 2. 3D AF ST T

Schema 2 is formulated as 3D AF ST T, that is, the trajector is three-dimensional (3D), the landmark is far away from the agent (AF), the agent and the trajector move together (same with trajector, ST), and the agent acts on the trajector with the aid of a tool (tool, T).

In (3), the trajector is 粪 ('dung'). However, the trajector is not the direct object of *tuī*. 粪 ('dung') is the contents of action *tuī* in the tool (container). It can be seen that the *tuī* and the trajectory 粪 ('dung') need to be mediated by certain tools, and the action *tuī* directly acts on the tools. This paper holds that the use of content instead of a container is obtained by metonymy in Schema 1.

The motion track of the agent and the trajector may not always be straight but may be circular. Examine the sentences below and Schema 3:

(4) 他 强迫 我 推 转盘。
3sg force 1sg push turntable
"He forced me to push the turntable."

(5) 他 的 任务 是 在 磨坊里 推 磨石。
3sg DE task COP in mill push millstone
"His task is to push the millstone in the mill."

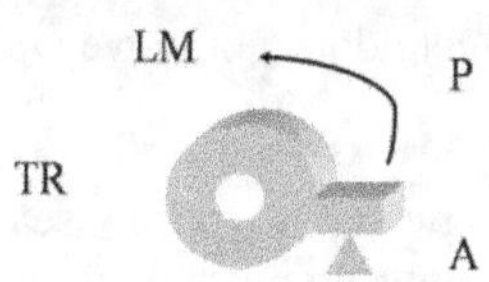

Schema 3. 3D ST R

Schema 3 is formulated as 3D ST R, that is, the trajectory is three-dimensional (3D), the agent and the trajector move together (ST), and the motion track is ring-shaped (R). Take (4) as an example. In (4), 他 ('he') is the agent, and 转盘 ('turntable') is a three-dimensional non-hollow trajector. 他 ('he') sent out a *tuī* action, outward force to make the trajectory 转盘 ('turntable') move together, forming a ring-shaped motion track. Schema 3 and Schema 1 form co-hyponymy analogue relations. In the two schemas, the motion track of the agent and the trajector are different. In Schema 1, the trajector moves away from the agent, and its motion track is not ring-shaped; in Schema 3, the motion track of the trajector and the agent is ring-shaped. The above two schemas belong to different subclasses of a relation.

We also find a kind of metonymic usage, exemplified:

(6) 嫂嫂 在 推 荞麦粉 预备
sister-in-law ZAI push buckwheat-flour prepare
做 荞 粑。
make buckwheat cake
'My sister-in-law is pushing buckwheat flour to prepare making a buckwheat cake.'

Such sense can be shown in Schema 4:

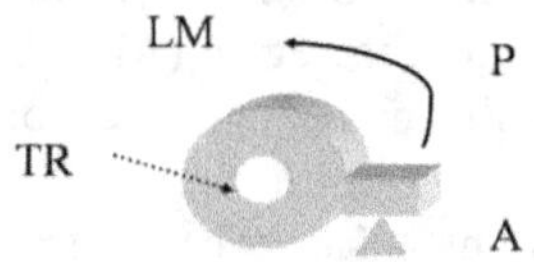

Schema 4. 3D ST R T[2]

Schema 4 is formulated as 3D ST R T, that is, the trajector is three-dimensional (3D), the agent and the trajector move together (ST), the motion track is ring-shaped (R), and the agent acts on the trajector with the aid of a tool (tool, T). In (6), the trajector is 荞麦粉 ('buckwheat flour'). However, the trajector is not the direct object of *tuī*. It is the action of *tuī* that makes 荞麦粉 ('buckwheat flour') to be 荞粑 ('a buckwheat cake') through a tool. It can be seen that some tools need to be used as a media between the action *tuī* and 荞麦粉 ('buckwheat flour'), and the action *tuī* directly acts on the tool. In our opinion, the use of products or materials instead of tools is obtained through metonymy in Schema 3.

2.2.1 Agent not Moving with the Trajector

We find that in most cases, the agent does not move with the trajector. For example,

(7) 黄 公 望 一推 座椅 站起来。
Huang Gongwang push seat stood up
"Huang Gongwang pushed his seat and stood up."
(8) 扣 扣 一推 棋 盘， 走 了 出去。
Koukou push chessboard go LE out
"Koukou pushed the chessboard and went out."

The moving situation can be shown in Schema 5:

Schema 5 is formulated as 3D AF NST, that is, the trajector is three-dimensional (3D), the landmark is far away from the agent (AF), and the agent does not move with the trajector (NST). In (7), for instance, Huang Gongwang (黄公望) is the agent, and the seat (座椅) is a three-dimensional trajectory. Huang Gongwang sent out the action *tuī* with outward force and makes the seat move in the direction of the force. The agent does not move with the trajector, i.e., the seat.

[2] In Schema 4, the position indicated by the dotted arrow is the position where the trajector (the material or the product) exists.

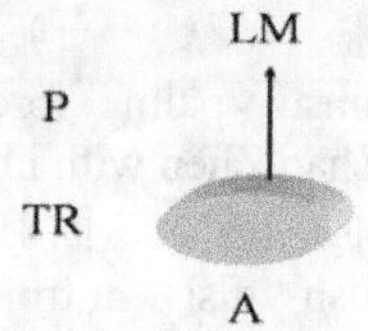

Schema 5. 3D AF NST[3]

Schema 5 and Schema 1 are in co-hyponymy analogue relations. The two schemas are different in whether the agent and the trajector move together. In Schema 1, the agent moves with the trajectory while in Schema 5, the agent does not. The above two schemas belong to different subclasses.

We find that the trajector can also be a two-dimensional entity. For example,

(9) 他 刚刚 伸 手 去 推 门帘，
3sg just stretch-out hand to push curtain
被 觉新 唤住 了。
by JueXin stop LE
"Just as he reached out to push the curtain, JueXin stopped him."

(10) 他 慌忙地 往回 推 钱。
3sg hurriedly back push money
"He hurriedly pushed back the money."

It can be shown in Schema 6:

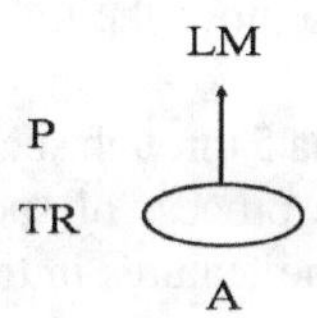

Schema 6. 2D AF NST

Schema 6 is formulated as 2D AF NST, that is, the trajector is two-dimensional (2D), the landmark is away from the agent (AF), and the agent does not move with the trajector (NST). In (9), for example, 他 ('he') is the agent, the门帘 ('curtain') is a two-dimensional non-hollow trajector. The person sent out the action of *tuī* to make the curtain move in the direction of force, and the agent did not move with the trajector.

Schema 5 and Schema 6 are in co-homology analogue relations. The two schemata are different in the dimension of the agent. In Schema 5, the agent is a three-dimensional entity; in Schema 6, the agent is a two-dimensional entity. The above two schemas belong to different subclasses.

In addition, the trajector can be part of the agent, as shown in Schema 7. Consider the following sentences:

---

[3] In Schema 5, since the agent does not move with the trajector, the letter A is used to mark the location of the agent. This way of marking is also used in other upcoming schemas.

(11) 柳再银 右手 推 掌 逼 退 赵真。
Liu Zaiyin right hand push palm force back Zhao Zhen
"Liu Zaiyin pushed back Zhao Zhen with his right hand."

(12) 表演者 猛地 推 拳， 面前 的 砖头
performer suddenly push fist in front of DE brick
变成 了 两 半。
turn-into LE two half
"The performer pushes his fist suddenly, and then the brick in front of him is broken in two."

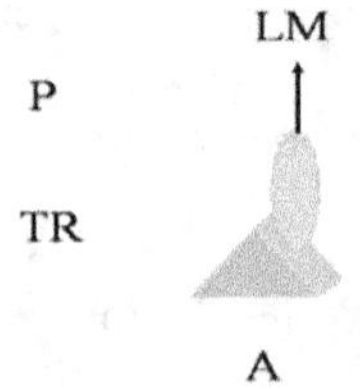

Schema 7. 3D PA AF NST

Schema 7 is formulated as 3D PA AF NST, that is, the trajector is three-dimensional (3D), the trajector is part of the agent (PA), the landmark is far away from the agent (AF), and the agent does not move with the trajector (NST). In (11) and (12), the trajectors 掌 (palm) and 拳 (fist) are three-dimensional entities, both of which are part of the agent. The agents, i.e. 柳再银 (Liu Zaiyin) and the 表演者 (performer) in sentence (11) and (12) respectively, pushed outward to move the trajectors 掌 (palm) and 拳 (fist) away from the agent.

Schema 7 is derived from Schema 5 through specialization. Zhu Yan (2010) pointed out that specialization refers to the process of specific meaning being derived from generalized meaning. There are some features in the generalized meaning that are not specified. In Schema 5, the trajector is a three-dimensional entity. In Schema 7, on top of the three-dimensional feature, the trajector is also part of the agent. Thus, Schema 7 is the specialization of Schema 5 in terms of the features of trajector.

There are also differences in the situation of landmarks. For example,

(13) 那 名 护士 在 推 注射器。
that CL nurse ZAI push syringe
"The nurse is pushing the syringe."

(14) 他 把 子弹 推 上 了 膛。
3sg BA bullet push into LE chamber
"He pushed the bullet into the chamber."

Such sense can be shown in Schema 8:

Schema 8 is formulated as 3D NST AF SU, that is, the trajector is three-dimensional (3D), the agent does not move with the trajector (NST), the landmark is away from the agent (away from, AF), and the landmark surrounds trajector (SU). Take (14) as an example. In (14), the syringe (注射器) serves as a landmark, and the plunger is the

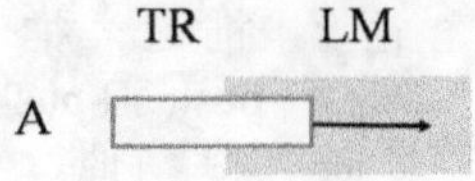

Schema. 8. 3D NST AF SU

trajector. The nurse moves the plunger toward the bottom of the syringe by the action of *tuī*. The landmark (i.e. the syringe) in this case surrounds the trajector (i.e. the plunger).

Schema 8 is derived from Schema 5 through specialization. In the description of Schema 8, the feature of "landmark surrounding trajector" is added. Thus, Schema 8 is the specialization of Schema 5 in terms of the features of the trajector.

The sense of *tuī* above can also be used in as a metonym. For example,

(15) 太阳 的 光线 像 注射器 推 药 一样，把 温暖
sun DE light like syringe push medicine same BA warmth
注 入 他们 的 体内。
infuse into 3pl DE body
"The light of the sun, like a syringe pushing medicine, infuses warmth into their bodies."

Such usage can be shown in Schema 9:

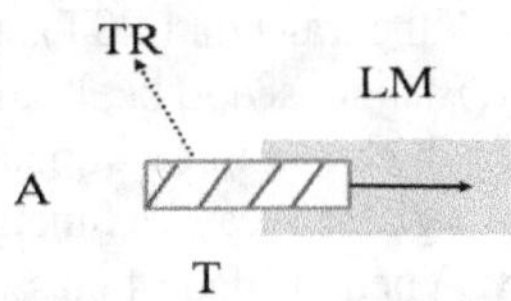

Schema 9. 3D NST AF SU T[4]

Schema 9 is formulated as 3D NST AF SU T, that is, the trajector is three-dimensional (3D), the agent does not move with the trajector (NST), the landmark is far away from the agent (AF), the landmark surrounds the trajector (SU), and the agent acts on the trajector with the aid of a tool (tool, T).

(15) uses metaphor and personification to personify the sun so that it can act in the same way as human beings do. The process of infusing warmth into the human body by the sun is compared to the process in which human operating syringes and pushing medicine into one's body. In the description of pushing medicine, a human act as the agent and the medicine is the trajector. However, the trajector is not the object directly affected by the action of *tuī*. Generally, the action of *tuī* directly acts on the plunger, and indirectly acts on the trajecto, i.e. medicine. We suggest that the Schema 9 is the result of Schema 8 through metonymy.

Differences in the moving direction of trajectory are also observed.

The first situation is that the trajector moves towards its original position, for example:

[4] In Schema 9, the position indicated by the dotted arrow is the position where the trajector exists.

(16) 他 得意地 用 手 往上 推 一 推 官 帽，
3sg jauntily use hand up push YI push official hat
以便 叫 行人 赏识 他 的 面貌。
so that let passer-by appreciate 3sg DE face
"He pushed his official hat up with his hand jauntily, so as to let the passers-by appreciate his face."

(17) 伊 牧师 推 了 推 眼镜， 看 着 伊 太太。
Yi pastor push LE push glasses look-at ZHE Mrs. Yi
"Pastor Yi pushed his glasses and looked at Mrs. Yi."

It can be shown in Schema 10:

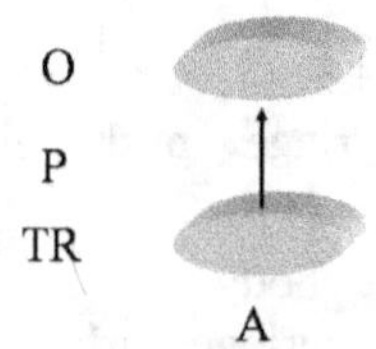

Schema 10. 3D NST O

Schema 10 is formulated as 3D NST O, that is, the trajector is three-dimensional (3D), the agent does not move with the trajector (NST), and the trajector moves towards the original position (O) where it is supposed to be. Example (17) for instance includes the trajector 'the pair of glasses' which is a three-dimensional non-hollow entity. Only in a certain position can the pair of glasses play its functions. In other words, there is an original position of the trajectory. When the pair of glasses is out of its original position, the agent, i.e., pastor Yi, moves it back to its original position by the action of *tuī*.

The relationship between Schema 10 and Schema 5 is co-hyponymy analogue. The two schemas are different in the direction of the motion of trajector. In Schema 5, the trajector moves away from the agent; in Schema 10, the trajector moves toward the original position. The above two schemas belong to different subclasses.

The second situation is that the trajector will return to its original position after the action *tuī* finishes. Such as:

(18) 陆 大麻 再次 用 手掌 向 上 推 鼻孔 。
Lu Dama again use palm towards up push nostrils
"Lu Dama pushed his nostrils up again with his palm."

(19) 小 珠 的 胖 手 用力 往前 推 耳朵。
Xiaozhu DE fat hand forcibly forward push ear
"Xiaozhu's fat hand pushed her ear forward forcibly."

It can be shown in Schema 11:

Schema 11 is formulated as 3D NSTR RE, that is, the trajector is three-dimensional (3D), the agent does not move with the trajector (NST), and the trajector returns to the original position after moving (RE).

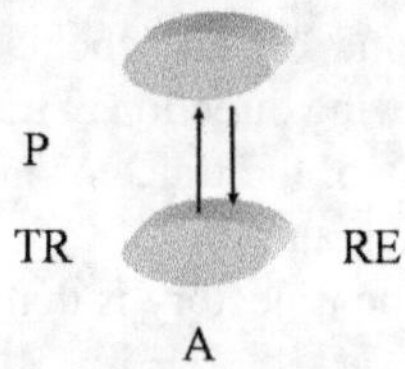

Schema 11. 3D NST RE

In (18) and (19), the trajectory 鼻孔 (nostrils) and 耳朵 (ear) are body parts, which are not easily displaced under normal circumstances. So, nostrils and ears will return to their original positions after the action *tuī* finishes.

The relationship between Schema 11 and Schema 5 is co-hyponymy analogue. The two schemas are different in the direction of the motion of trajector and belong to different subclasses. In Schema 5, the trajector moves away from the agent; in Schema 11, the trajector first leaves and then returns to its original position.

The third situation is that the trajector moves back and forth. For example,

(20) 网球 场 上 两 个 男人 在 来 回 推
tennis court on two CL man ZAI come return push
网球。
tennis ball
"Two men are pushing the tennis ball back and forth on the tennis court."

(21) 船夫 正 一 前 一 后 地
boatman ZHENG one forward one backward DE
推 着 船桨。
push ZHE paddle
"The boatmen were pushing the paddles forward and backward."

It can be shown in Schema 12:

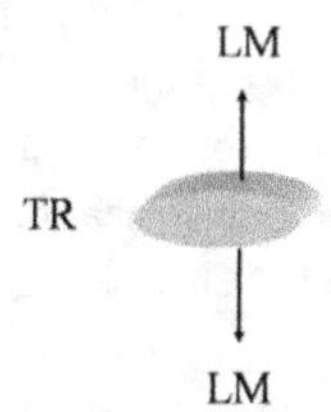

Schema 12. 3D NST BF

Schema 12 is formulated as 3D NST BF, that is, the trajector is three-dimensional (3D), the agent does not move with the trajector (NST), and the trajector moves back and forth (BF). In (20) and (21), the agents, i.e., the two men on the tennis court and the boatmen respectively, pushed the trajectory, i.e., the tennis ball and the paddles respectively, back and forth in the direction of the force.

The relationship between Schema 12 and Schema 5 is co-hyponymy analogue. The two schemas are different in the moving direction of trajector and thus belong to different subclasses. Specifically, in Schema 5, the trajector moves away from the agent and in Schema 12, the trajector moves back and forth.

The fourth way of moving of the trajectory is that it moves around. For example,

(22) 他 很 喜欢 推 牌九。
3sg very like push Pai Jiu
"He likes to push Pai Jiu (a kind of Chinese dominoes) very much."

It can be shown in Schema 13:

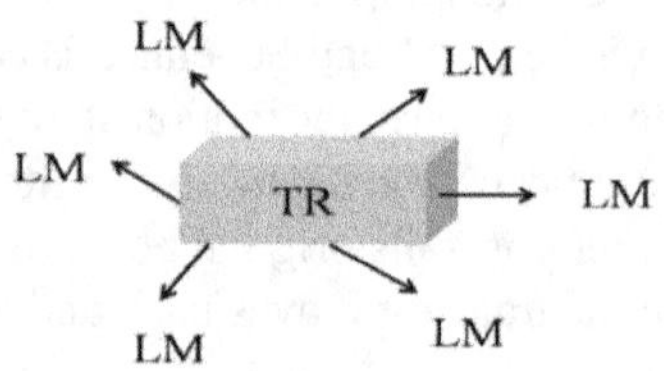

Schema 13. 3D NST AA

Schema 13 is formulated as 3D NST AA, that is, the trajector is three-dimensional (3D), the agent does not move with the trajector (NST), and the trajector moves all around (AA). In (22), the trajectory is Pai Jiu (牌九) which moves around by the action of *tuī*.

The relationship between Schema 13 and Schema 5 is co-hyponymy analogue. The two schemas are different in the moving direction of trajector. That is to say, in Schema 5, the trajector moves away from the agent while in Schema 13, the trajector moves all around. The above two schemas belong to different subclasses.

In the above four cases, there is a relation of co-hyponymy analogue among Schema 10, Schema 11, Schema 12, Schema 13, and Schema 5. Therefore, the relation of any two schemata between these five types of schemas is co-hyponymy analogue.

A special usage of *tuī* is also observed. For example,

(23) 年轻 的 掌 柜 为 我们 推 莜 面 窝 窝。
young DE shopkeeper for 1pl push buckwheat flour Wowo
"The young shopkeeper pushed buckwheat flour Wowo (i.e., a kind of Chinese food made of buckwheat flour) for us.

It can be shown in Schema 14:

Schema 14 is formulated as 3D NST AF EX, that is, the trajector is three-dimensional (3D), the agent does not move with the trajector (NST), the landmark is far away from the agent (AF), and the trajector is extended (EX). In (23), under the action of *tuī* by the agent, i.e., shopkeeper, the trajectory, i.e., the buckwheat flour is extended in shape and finally formed buckwheat flour Wowo.

Schema 14 is derived from Schema 5 through specialization. In Schema 5, the trajector is three-dimensional. In Schema 14, the trajector is not only three-dimensional but is also extended by the agent.

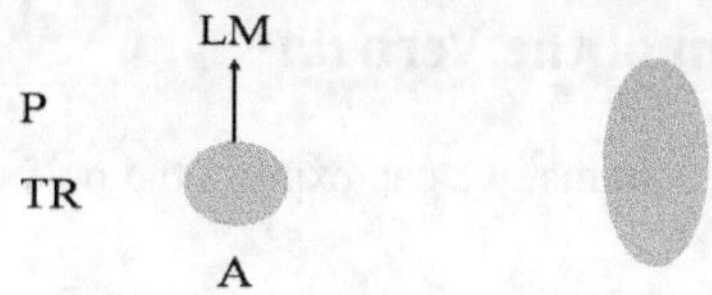

Schema 14. 3D NST AF EX

2.3 To Cut or Cut Forward with a Tool Against the Surface of an Object

The second sense branch of the verb *tuī* is "to cut or cut forward with a tool against the surface of an object". For example,

(24) 理发师 给 他 推 光 了 头。
barber for him shave bald LE head
"The barber shaved his head bald."

(25) 亨利 雇 邻居 的 孩子 替 他 推草。
Henry hire neighbor DE child for him weed
"Henry hired his neighbor's child to weed for him."

It can be shown in Schema 15:

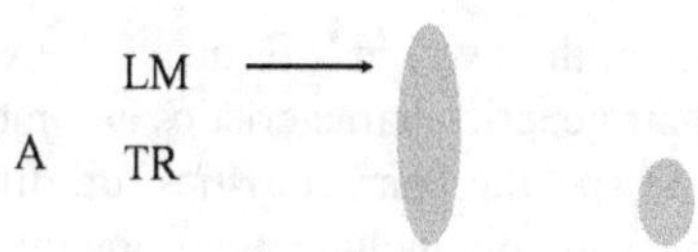

Schema 15. 3D STA REM

Schema 15 is formulated as 3D STA WF, that is, the trajector is three-dimensional (3D), and static (STA), a part of the trajector is removed by the agent (remove, REM). We take (25) as an example. In (25), the agent 亨利 (Henry) used tools such as a lawn mower to cut the surface of the lawn to remove part of it. In our analysis, the sense branch "cut or cut forward with the tool against the surface of the object" means to remove it.

Lakoff (1987: 430–433) found a kind of reflexive schema in his article. He used *out* as an example to illustrate the schema. The simplest use of *out* occurred in cases like *Harry ran out of the room.* In this case, the container (i.e., the room) is landmark, and the trajectory (i.e., Harry) moves from the interior to the exterior of the room. However, in the sentence *The syrup spread out*, the trajectory is the syrup which is not moving out with respect to any other landmark. The syrup, or at least part of it, is moving "out" relative to its own prior boundary. The syrup is its own landmark. Such a relation between a landmark and a trajectory is called reflexive. This paper holds that the Schema 15 is reflexive schema of Schema 1, 2, 3, 4, 5, 6, 7, 8, 9, 10, 11, 12, 13 and 14. In (24) and (25), the weed and hair moving away relative to its own prior boundary. Therefore, the relation between the two sense branches is reflexive.

3 Polysemous System of the Verb *tuī*

By summarizing the image schema, we can express the polysemous system of the verb *tuī* as follows:[5]

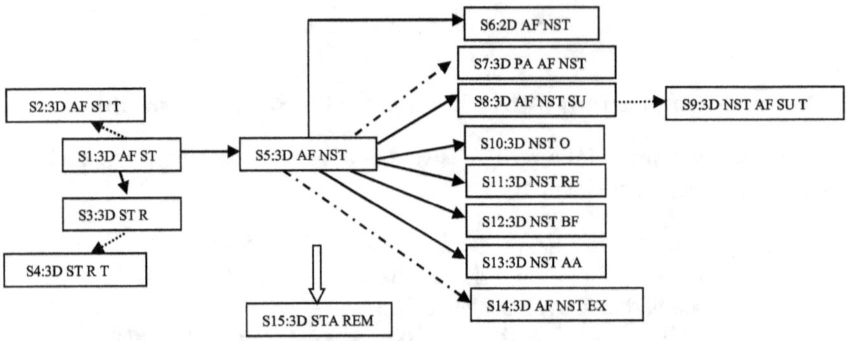

As can be seen from the above figure, the meanings of *tuī* are connected to form a radical-structured network. Among them, the meaning represented by S5 (Schema 5) has the most connection with other meanings, which can be regarded as the central meaning of the radical-structured network. The most important role of radical-structured network is co-hyponymy analogue.

Zhu (2010) pointed out that the two kinds of meanings with the relationship of co-hyponymy analogue, i.e., their generic characteristics and most specific characteristics, are the same, and only some specific characteristics are different. In the polysemous system of *tuī*, these differences mainly include the following aspects:

(1) Differences in TR spatial dimensions: There are two cases in which the TR is two-dimensional and the TR is three-dimensional, such as S5–S6.
(2) Differences in whether the agent and the TR move together: There are two situations in which the agent moves with TR and the agent does not move with the TR, such as S1–S5.
(3) Differences in the motion track of agent and TR: There are two situations in which the motion track is ring-shaped and the motion track is non-ring-shaped, such as S1–S3.
(4) Differences in whether the TR is part of the agent: There are two situations in which the TR is part of the agent and the TR is not part of the agent, such as S5–S7.
(5) Differences in whether LM surrounds TR: There are two cases in which the LM surrounds the TR and the LM does not surround the TR, such as S5–S8.
(6) Differences in the movement direction of TR: There are five situations in which the TR moves away from the agent, the TR moves toward its original position, the TR will return to its original position after the action *tuī* finishes, the TR moves back and forth and the TR moves around, e.g., S5–S10, S5–S11, S5–S12, S5–S13.
(7) Differences in whether the TR is extended: There are two cases in which the TR is extended and the TR is not extended, such as S5–S15.

[5] "S1" represents Schema 1; ⟶ represents analogy; - · - ·► represents specialization; ········► represents metonymy; ⇨ represents reflexive schema.

In the above differences, we can see that (6) forms the co-hyponymy analogue relationship with most of the other usage of *tuī*, and the movement direction of TR is the most important reason for the differences among the image schemas of *tuī*. In addition, although metonymy and specialization are also effective ways to connect different image schemas, their roles in polysemous system of *tuī* are obviously weaker than that of co-hyponymy analogue, in which specialization and metonymy appear three times respectively.

Acknowledgments. I am grateful to the anonymous reviewers of CLSW 2020 for helpful suggestions and comments. This paper is supported by the National Social Science Fund (19BYY030). All errors remain my own.

References

Lakoff, G.: Women, Fire, and Dangerous Things: What Catergories Reveal About the Mind, pp. 416–461. University of Chicago Press, Chicago (1987)

Wang, J.: An analysis of the image schema of "Wai". Sci. Educ. Art. Collects (1), 247–247 (2009). (in Chinese)

Zhang, M., Jiang, G.: An image schema-based analysis on polysemy of "Rao". J. Longyan Univ. (1), 63–69 (2014). (in Chinese)

Zhou, J.: A cognitive study of the basic image schema and metaphorical extension of the preposition "into", J. Anhui Univ. Technol. (Soc. Sci.) (2), 83–85 (2011). (in Chinese)

Zhu, Y.: Kernel/distinctive components and semantic analysis of Chinese action morphemes: a case study of Shou. Stud. Chin. Lang. (4), 27–34 (2006). (in Chinese)

Zhu, Y.: Polysemy of verb Chuan and the linking mechanisms among senses: an image-schema-based approach. Linguist. Sci. (3), 67–80 (2010). (in Chinese)

A Study on the Semantic Change of the Chinese Negative Adverb *Bùshèn* (不甚)

Xiao Han(✉) and Fangqiong Zhan(✉)

Department of Linguistics and Translation, City University of Hong Kong, 83, Tat Chee Avenue, Kowloon, Hong Kong SAR

xhan42-c@my.cityu.edu.hk, fangzhan@cityu.edu.hk

Abstract. The paper aims to account for the multiple interpretations of the Chinese negative degree adverb *bùshèn* (不甚). In *Chongbian Guoyu Cidian (Revised Chinese Dictionary)* (2015), it is defined as 'not very', however, in the CCL and BCC Modern Chinese corpora, we have found a few cases showing that *bùshèn* can only be interpreted as positive degree 'very'. Using the extensive classical and modern Chinese data, we argue that the semantic change has undergone a process of analogy (cf. e.g. Harris and Campbell 1995; Fischer 2007; Traugott and Trousdale 2013), i.e., the use of positive degree of *bùshèn* was analogized to the extant positive degree adverb *bùshèng* (不胜). The semantic change has occurred when *bùshèn* was followed by an adjective or a verb conveying emotions, such as *gănxiè* (感谢), which were the common collocations of the adverb *bùshèng*, e.g. *bùshèng gănxiè*.

Keywords: Semantics · Chinese · Degree adverb · Analogy · *Bùshèn* · 不甚 · *Bùshèng* · 不胜

1 Introduction

Bùshèn (不甚) in *Chongbian Guoyu Cidian (Revised Chinese Dictionary)* [1] is defined as 'not very' (不很), as in (1). However, a different interpretation could be found in some examples searched from the BCC Modern Chinese corpus [2], which expresses the meaning of 'very', see (2).

(1) 去美国前我对父亲的这句话不甚了然，三年后从美国回来才真正理解它的含义。

qù	mĕiguó	qián	wŏ	duì	fùqīn	de	zhè	jù	huà
go	USA	before	SG1	PREP	father	ASSOC	this	CL	word

bùshèn	liǎorán	sān	nián hòu	cóng	mĕiguó
NEG:very	understand	three	year later	PREP	USA

huílái	cái	zhēnzhèng	lĭjiĕ	tā	de
come:back	just	really	understand	SG3	ASSOC

hányì

M. Liu et al. (Eds.): CLSW 2020, LNAI 12278, pp. 100–107, 2021.
https://doi.org/10.1007/978-3-030-81197-6_9

meaning[1]

"Before I went to America, I didn't understand my father's words very well. It is until when I came back from America three years later that I understood the meaning."

Diwudai dianying dui Zhongguo dianying de zhuyao gongxian (2005)

(2) 从这次事件当中，我进一步地了解了上帝对我的恩惠，对于上帝对我的关怀我实在是不甚感激啊！

cóng zhè cì shìjiàn dāngzhōng wǒ jinyībù de
from this CL event PREP SG1 further DE
liǎojiě le shàngdì duì wǒ de ēnhuì
understand PFV God PREP SG1 ASSOC kindness
duìyú shàngdì duì wǒ de guānhuái
PREP God PERP SG1 ASSOC concern
wǒ shízài shì bùshèn gǎnjī a
SG1 really COP NEG:very appreciate PART

"From this event, I have further understood God's kindness to me, and I indeed really appreciate God's concern for me!"

Lubinxun piaoliu ji (n.d.)

The function of *bùshèn* in (2) is inconsistent with that in (1) and accordingly appeals to the need for further study. Yuan [3] discussed *bùshèng gǎnxiè* and *bùshèn gǎnxiè*, and considered the latter one as a misuse resulting from the fact that some dialect users cannot distinguish the pronunciation of the finals 'n' from 'ng'. Similarly, Pan [4] believed that two reasons could explain the confusing use of *bùshèn gǎnjī* (不甚感激) and *bùshèng gǎnjī* (不胜感激): a) tone change of '*sheng*', from the level tone to falling tone, caused by the change of its semantics from 'able to bear' to 'capacity, strength exceeds and wins, etc.' And b) the influence of the dialects from specific areas, such as Southwestern and Jianghuai regions. However, An [5] claimed that *bù* in *bùshèn* is used to supplement the syllable, which should be considered as a redundant rather than a negative morpheme in certain contexts.

We suggest the above accounts need to be reconsidered, and the reasons are as follows:

1) If it was the pronunciation issue, other words involving 'n' and 'ng' would also result in a similar confusion, which is not confined only to *bùshèn* and *bùshèng*.
2) If *bù* is a redundant morpheme, it does not explain why the meaning 'very' of *bùshèn* only occurs when *bù* collocates with adjectives or verbs which is also the common collocations of *bùshèng*.

Different from previous views, we argue that the semantics 'very' of *bùshèn* derives from the analogy to the usage of *bùshèng*, which can be justified by evidence based on

[1] The glosses used in this paper include: ASSOC: associative; CL: classifier; CONJ: conjunction; COP: copula; LOC: locative; NEG: negation; PART: particle; PFV: perfective aspect; PL1: 1st person plural pronoun; PREP: preposition; SG1: 1st person singular pronoun; SG3: 3rd person singular pronoun.

the extensive classical and modern data. Analogy, a process of pattern matching, has been recognized as a prime mechanism in language change and grammaticalization (cf. e.g. Harris and Campbell [6]; Fischer [7]; Traugott and Trousdale [8]). For analogy, the focus is on matching the original source with some extant structure that is considered to be similar in some respect and treated as an exemplar. In this case, language-users unconsciously match the usage of *bùshèn* with the pattern of *bùshèng*. We propose that *bùshèng* is the extant exemplar that enables the semantic change for *bùshèn*.

The rest of the paper is organized as: Sect. 2 and 3 state the semantics of *bùshèn* and *bùshèng*, respectively. Section 4 discusses the contexts in which the analogy occurred. Section 5 is the conclusion.

2 The Semantics of *Bùshèn*

Bù and *shèn* first co-appeared in Old Chinese[2], around the Spring and Autumn Period (770 BCE-476 BCE), where *bù* was a negative prefix attached to *shèn*. During this period, *shèn* following the negative morpheme *bù* was used as an adjective or a verb, see example (3) and (4).

(3) 上雨而不甚也。

shàngyǔ	ér	bù	shèn	yě
rain:in:time	but	NEG	torrential	PART

"It is a timely rain but not torrentially."

Gongyangzhuan (657 BCE)

(4) 虽古之所谓圣王明君者，其勤身而忧世不甚於此矣。

suī	gǔ	zhī	suǒwèi	shèng	wáng	míng
even:if	ancient	ASSOC	so:called	holy	emperor	wise

jūn	zhě	qí	qín	shēn	ér	yōu	shì
emperor	PART	SG3	diligent	body	CONJ	worry	country

bù	shèn	yú	cǐ	yǐ
NEG	more:than	PREP	SG3	PART

"Even the ancient holy wise emperors, their hard work and worry for the country cannot be compared with him."

Hanfeizi (280 BCE-233 BCE)

Bùshèn was first used as a negative degree adverb in *Hanshu* (105 CE) in the East Han Dynasty, see (5).

[2] The periodization for written Chinese adopted in the paper is as follows (Sun [9]): Old Chinese: 771 BCE to 220 CE. Middle Chinese: 220 CE to 960. Early Modern Chinese: 960 to 1900. Modern Chinese: 1900 to present.

(5) 吏卒起张掖，不甚远，然尚厮留甚众。

lìzú	qǐ	zhāngyè	bùshèn	yuǎn	rán	shàng
officers	depart	Zhangye	NEG:very	far	but	still
sīliú	shèn	zhòng				
separate	very	many:people				

"The officers departed from Zhangye. It is not a long way, but still a lot of people were left behind."

Hanshu (105 CE)

It was used consistently as 'not very' in the history of Chinese. However, limited cases from CCL corpus have shown [10] that *bùshèn* can be used as 'very' in Modern Chinese. In (6), *bùshèn* cannot be interpreted as 'not very', but can only be understood as 'very'.

(6) 八姑专爱钻戒，两弟则既爱财，又爱马，欲念不甚大

bāgū	zhuān	ài	zuànjiè	liǎngdì	
Bagu	special	love	diamond:ring	two:brothers	
zé	jì	ài	cái	yòu	ài
CONJ	CONJ	love	money	CONJ	love
mǎ	yùniàn	bùshèn	dà		
horse	desire	NEG:very	big		

"Bagu loves diamond rings especially, while the two brothers love both money and horses. They have much desire."

Xinhai geming qianhou (1979)

We argue that the semantic change of *bùshèn* from 'not very' to 'very' was analogized to the extant adverb *bùshèng*. We will discuss the process in detail in the following sections.

3 The Semantics of *Bùshèng*

The original meaning of *shèng* is 'able to bear' (禁得起). Diachronically, *bù* and *shèng* first co-appeared in Zhou Dynasty (1046 BCE-256 BCE), where *shèng* was used as a verb and *bù* was a negative adverb, as in (7), meaning 'not able to bear' which indicates a positive evaluation of the intended objects as either strong, heavy or hard to defeat.

Bù and *shèng* gradually condensed from a negative phrase to the degree adverb *bùshèng* through lexicalization. Han [11] believed the mechanisms of the process include semantic change, structure reanalysis, and rhythm evocation, accompanied by grammaticalization. At the beginning, *shèng* as a verb generated the semantics of 'to complete' (尽, 全) in negative and interrogative sentences when it co-occurred with an auxiliary such as 'can' (可, 能). Meanwhile, the frequency of the co-appearance of *bù* and *shèng* increased. The combination meant 'not complete'. Then the boundary shift happened and *bùshèng* reduced into a lexical item from a negative verbal phrase. In the processes of rhythm evocation and structure reanalysis, *bù* and *shèng* became more agglutinative and the combination of *bùshèng* generated the semantics of 'endless' and 'a great quantity'.

After that, the semantics of *bùshèng* was extended and abstracted to ‘very’. *Bùshèng* was gradually solidified to a degree adverb.

In (8), *bùshèng* was followed by an adjective, showing that it was no longer a phrase of negation plus verb, but already an adverb. Extended from its original meaning as a verb phrase, *bùshèng* as an adverb conveys a positive degree.

(7) 不胜而往，咎也。

bù	shèng	ér	wǎng	jiù	yě
NEG	competent	CONJ	go	calamity	PART

“If a person is unable to do something but insists on doing it, it will cause disaster.”

Zhouyi (1046 BCE-771 BCE)

(8) 猥蒙开示，深自庆幸，不胜欢喜，略附陈诚。

wěiméng	kāishì	shēn	zì	qìngxìng
be:granted:a:favor	enlighten	deeply	self	rejoice

bùshèng	huānxǐ	lüè	fù	chén	chéng
very	delighted	slightly	attach	expression	sincerity

“I am grateful to you for enlightening me. I feel deeply rejoice and very delighted. I would like to express my sincerity here.”

Guanghongmingji (664)

Bùshèng can be followed by an adjective or a verb when it is used as an adverb. The adjectives are generally those that express either positive or negative emotions, such as ‘delighted’ (欢喜) in (8) and ‘resentful’ (愤慨) in (9). Being attached to *bùshèng*, the emotional status of the adjectives is deepened or strengthened. When it modifies a verb, it again indicates the strengthened emotional status of the subject, see (10). In (10), *bùshèng*, followed by a verb ‘sigh with emotion’ (感慨), indicates that the subject has very complicated emotions while entering his hometown, because he had been away for three years. A few examples show that a noun can occur preceding *bùshèng*. In (11), ‘joy’ (喜悦) appears preceding *bùshèng*, and together the phrase means ‘full of joy (very happy)’.

(9) 说到最后一句，他猛把桌子拍了一下，露出不胜愤慨的神气。

shuō dào	zuìhòu	yī		jù	tā	měng	bǎ
speak arrive	last	one		CL	SG3	violently	BA
zhuōzi	pāi	le	yī	xià	lùchū	bùshèng fènkǎi	
table	slap	PFV	one	CL	expose	very resentful	
de	shénqì						
ASSOC	expression						

"Speaking of the last sentence, he slammed the table, revealing his indignation."

Shi (1930)

(10) 士会去秦三载，今日复进绛城，不胜感慨。

shìhuì	qù	qín	sān	zǎi	jīnrì	fù
Shihui	go	Qin:state	three	year	today	again
jìn jiàngchéng		bùshèng	gǎnkǎi			
enter Jiangcheng		very	sigh:with:emotion			

"Shihui went to the Qin state for three years. Today he reenters Jiangcheng with great emotion."

Dongzhou lieguo zhi (C.A. 1646)

(11) 心中却甜甜的喜悦不胜

xīn zhōng	què	tiántiánde	xǐyuè	bùshèng
heart LOC	but	sweet	joy	very

"Her heart is full of sweet joy."

Shendiao xialü (1959-1961)

The above examples show that *bùshèng* as a positive degree adverb usually modifies an adjective or verb to indicate the deepened emotional state of the subject, and that *bùshèng* occurs in the collocation where it is followed by an emotional adjective or verb.

4 Discussion

As shown in (8), in Middle Chinese around 664 CE, *bùshèng* was already found preceding an emotional adjective; and in Early Modern Chinese around 1646, it was found occurring with an emotional verb as in (10). The structure [*Bùshèng*+adjective/verb] meaning deepened emotional status, is the extant exemplar for [*Bùshèn*+adjective/verb] to be analogized to. If we take a closer look at the two adverbs:

[*Bùshèng*+ emotion adjective/verb]→ meaning deepened emotional status

↓ Analogy

[*Bùshèn*+ emotion adjective/verb]→ meaning deepened emotional status

We can see that they have partially identical forms, which by hypothesis enabled analogy to take place and resulted in the similar meaning. Earliest cases of *bùshèn* used as an adverb meaning 'very' were found in Modern Chinese corpus, much later than the occurrence of the [*Bùshèng*+adjective/verb] examples. We argue that the analogy occurred when *bùshèn* as an adverb modifying an adjective or a verb, functioned to indicate the positive degree of emotion, such as (6) and (12). Similar cases of the analogical use are not found when *shèn* is used as an adverb and *bù* functions as a negative prefix. The negative degree adverbial use of *bùshèn* co-exist with its positive expression. While the interpretation depends on specific contexts, the negative *bùshèn* does not collocate with emotional adjectives/verbs, as in (13).

(12) 心里早已是不甚感慨，说不出的唏嘘哀怨

xīn	lǐ	zǎoyǐ	shì	bùshèn		gǎnkǎi
heart	inside	already	COP	NEG:very		sigh:with:emotion
shuō bù	chù	de		xīxū	āiyuàn	
speak NEG	out	ASSOC		sigh	complain	

"He has already been not very emotional and has unspeakable sighs and complaints."

Wanquan fangdao shouce (2005)

(13) 大概在外国住久了，我们中国人的人情礼俗，他不甚了解。

dàgài	zài	wàiguó		zhù		jiǔ	le
probably	LOC	foreign:country		live		long	PFV
wǒmen	zhōngguórén		de		rénqíng		
PL1	Chinese:people		ASSOC		human:relationship		
lǐsú		tā	bùshèn		liǎojiě		
etiquette:and:custom		SG3	NEG:very		understand		

"It is probably because he has been lived in the foreign country for a long time, he doesn't know much about our Chinese customs."

Liangfuyin (1967)

In both (6) and (12), *bùshèn* indicates the strengthened emotion of the subject. In (6) it expresses the strong desire of the two brothers, and in (12) it signals the complicated emotions in the heart of the subject, which is implicit in the sentence. In (6), the two brothers not only love money, but also desire good horses. They are greedy and their desire is very strong. It is clear that *bùshèn* here expresses the meaning of 'very' rather than 'not very'. In (12), from the context, the implicit subject, being very emotional, again suggests *bùshèn* should be considered as 'very'.

As we have mentioned above, *bùshèng* followed by an adjective or verb indicating emotions are the common collocations, which set up an exemplar for *bùshèn* to match, resulting in the semantic change from 'not very' to 'very' in the cases of *bùshèn* conveying emotions.

5 Conclusion

Based on the extensive classical and modern Chinese data, it is found that *bùshèn* has the potential to express positive degree 'very' when used to indicate the strong degree of emotion. The semantic change was argued to have undergone an analogy to the usage of the existing negative adverb *bùshèng*.

References

1. Ministry of Education Mandarin Language Promotion Committee: Chongbian Guoyu Cidian (Revised Chinese Dictionary). Ministry of Education of the Republic of China, Taiwan (2015). http://dict.revised.moe.edu.tw/cgi-bin/cbdic/gsweb.cgi?o=dcbdic&searchid=Z00000021539
2. Xun, E.D., Rao, G.Q., Xiao, X.Y., Zang, J.J.: The construction of the BCC corpus in the age of big data. Corpus Linguist. **3**, 93–109, 118 (2016). Foreign Language Teaching and Research Press, Beijing
3. Yuan, L.G.: "Busheng Ganxie" yu "Bushen Ganxie" ["Very grateful" and "not very grateful"]. Chin. Teach. Stud. **11**, 38 (1983). Central China Normal University, Wuhan
4. Pan, Y.Y.: "Bushen Ganji" yu "Busheng Ganji" ["Not very grateful" and "very grateful"]. Mod. Chin. **14**, 32–34 (2017). Qufu Normal University, Qufu
5. An, H.M.: "Bu + Chengdu Fuci + X" Geshi de Ruohua Fouding Yanjiu [A study on the weakening negation of the format of "Bu + Degree Adverb + X"]. Master thesis, Shanghai Normal University, Shanghai (2019)
6. Harris, A., Lyle, C.: Historical Syntax in Cross-linguistic Perspective. Cambridge University Press, Cambridge (1995)
7. Fischer, O.: Morphosyntactic Change: Functional and Formal Perspectives. Oxford University Press, Oxford (2007)
8. Traugott, E.C., Trousdale, G.: Constructionalization and Constructional Changes. Oxford University Press, Oxford (2003)
9. Sun, C.F.: Word-Order Change and Grammaticalization in the History of Chinese. Stanford University Press, Stanford (1996)
10. Zhan, W.D., Guo, R., Chen, Y.R.: The building of the CCL corpus: its design and implementation. Corpus Linguist. **6**, 71–86 (2019). Foreign Language Teaching and Research Press, Beijing
11. Han, X.H.: The development of the degree adverb 不胜. J. Zhanjiang Normal Coll. **33**, 137–140 (2012). Zhanjiang Normal College, Zhanjiang

The Interplay Between Emojis and Linguistic Text

Helena Yan Ping Lau(✉) and Sophia Yat Mei Lee

Department of Chinese and Bilingual Studies, The Hong Kong Polytechnic University, Hong Kong, China
helena.lau@connect.polyu.hk, ym.lee@polyu.edu.hk

Abstract. This paper examines the interplay between emojis and the accompanying linguistic text when the two channels disagree with each other. In general, emojis are typically regarded as an explicit way to convey a specific emotion. However, it is observed that emojis, even of low ambiguity, may not explicitly express a corresponding emotion in a message. Based on the 831 comments extracted from Sina Weibo, we argue that the interplay between an emoji and its accompanying text may influence the expressed emotion. When the two channels mismatch, the overall emotion is often determined by the emotion expressed in the text regardless of the emotion type. It is observed that emojis expressing *happiness* occur most frequently in text expressing other emotions, as compared to emojis of other emotions. Moreover, emojis representing *happiness* are particularly common in text expressing *anger*. The overall emotion is mostly determined by the text (i.e. *anger*), meaning that the emojis representing *happiness* are frequently used ironically. We believe that this paper will lay the groundwork for the studies of emotion detection.

Keywords: Emotion analysis · Emoticon · Emoji · Social media

1 Introduction

Emoticons are ubiquitous in computer mediated communication (CMC). The term "emoticon" is a contraction of "emotion" and "icon" which is originally created to express one's emotional state by means of text characters, such as; -P. Traditional emoticons have been gradually replaced by emojis nowadays, that is, pictographs that represent a facial expression or gesture such as ☹. It is generally believed that emojis are an effective emotion indicator which can be used to detect emotions automatically. For example, the grinning face with smiling eyes emoji 😁 expresses a *happiness* emotion. When it appears in a comment, it is likely that the comment is expressing a *happiness* emotion. However, some studies indicated that emojis could serve as an amplifier or modifier when an emoji and its linguistic text were in disagreement. This poses a real challenge to the automatic emotion detection. In view of this, this paper investigates the interplay between emojis and linguistic texts in which the emojis are used, with a special attention being placed on the situations that the two channels conflict with each other. The paper aims to address the following questions:

M. Liu et al. (Eds.): CLSW 2020, LNAI 12278, pp. 108–117, 2021.
https://doi.org/10.1007/978-3-030-81197-6_10

1) Which channel determines the overall emotion when an emoji and its accompanying text disagree with each other?
2) If the overall emotion is determined by text, what is the function of emojis?

2 Related Work

Emotions can be classified into explicit emotions and implicit emotions. According to Lee (2015: 186), explicit emotions refer to "the presence of emotion-related information denoted by emotion keywords", and implicit emotions as "the emotion-related information that requires inference or connotation instead of being conveyed by emotion keywords". Explicit emotions can be directly denoted by emotion keywords; implicit emotions can be expressed by other linguistic cues such as emotion-related words and emotion-laden words (Pavlenko 2008).

Apart from textual information, previous work also proposed that emotions can be expressed by means of emoticons (Rezabek and Cochenour 1998; Wolf 2000; Crystal 2006; Dresner and Herring 2010). Therefore, the use of emoticons can be regarded as an explicit way of emotion expressions. The term emoticon is a contraction of "emotion" and "icon" which initially refers to a series of text characters such as punctuations and symbols to represent a facial expression or gesture, such as :) as a happy face and :(as a sad face. Some traditional emoticons can be used to input emojis nowadays. For example, when someone types in the happy face :), the emoticon is automatically transformed into its corresponding emoji 😊. The term "emoticon" is thereafter used in a broader sense to include both text characters representing facial expressions and gestures (i.e. emoticons) and colorful pictographs representing facial expressions and gestures (i.e. emojis). Given that traditional emoticons are rarely found in our corpus, we focus on the use of emojis provided by Sina Weibo.

Although emoticons have different communicative purposes, such as maintaining a conversational connection, expressing humor, strengthening a message and showing a close tie with others (Derks et al. 2008; Kelly and Watts 2015), they are primarily used to convey affect (Rezabek and Cochenour 1998; Wolf 2000; Crystal 2006; Dresner and Herring 2010). Therefore, most previous studies regarded emoticons (including emojis) as an explicit emotion indicator. Lo (2008) indicated that the presence of an emoticon in a message significantly improves the perception of writers' emotions. Hussien et al. (2016) showed that in the emotion detection task, the classifiers trained on an automatic annotated dataset using emojis even outperformed those classifiers trained on manually annotated dataset. Felbo et al. (2017) indicated that emojis can help classify emotions expressed in text accurately in many cases.

A great deal of research has proven that emoticons (including emojis) are an effective emotion indicator. However, some studies indicated that emoticons can serve as an amplifier or modifier when an emoticon and its linguistic text are in disagreement. Derks et al. (2008) explored whether the sentiment (i.e. positive and negative) of text can be altered by the presence of an emoticon. They indicated that when an emoticon and the accompanying text are at odds with each other, the sentiment becomes closer to neutral, but not enough to switch from one polarity to another. The claim is supported by Skovholt et al. (2014) who found that emoticons can soften negativity or strengthen positivity. Novak et al. (2015) proposed that text determines whether an emoji functions as an amplifier or a modifier. Riordan (2017) stated that it is the verbal content that determines the emotion of a message, and an emoji only alters the intensity of that emotion. Tian et al. (2017) argued that when emojis and the accompanying linguistic texts express different emotions, the overall communicated meaning is not a sum of the two channels. They observed a significant amount of positive emojis in text expressing *anger*, meaning that the emojis are used ironically.

3 Corpus Data

3.1 Dataset and Annotation.

Our data was taken from Sina Weibo, one of the most popular social media sites in China. We extracted both events (i.e. posts) and comments to construct an event-comment corpus (Lee and Lau 2020). The corpus was made up of 200 trending events and each of them includes 150 comments. It adds up to 30,000 comments for the entire corpus, of which 10,000 were manually annotated with emotion information.

Following Lee (2010), five basic emotions were annotated, namely *happiness*, *sadness*, *anger*, *fear* and *surprise*. We then extracted all the annotated comments containing emojis which are associated with a particular emotion (in appendix).

Of the 10,000 annotated comments, 831 comments were extracted. Annotators read through the comments and decided whether the emojis were used in a typical way. If the emotion expressed by emoji(s) agrees with the emotion expressed by the accompanying text, the use of emoji(s) is regarded as a typical one as in (1).

(1) 生活真的很难，想起我的爸爸
'Life is really hard, (I'm) thinking of my father '

The emojis in (1) are used in a typical way as both the text and emojis express a *sadness* emotion. However, if the two channels disagree with each other, the use of emoji(s) should be considered atypical. Annotators should then judge whether the overall emotion is determined by the emoji(s) or the accompanying text as in (2).

(2) 快开学了 要写完了 被撕了也是很绝望
'School is about to start. It is hopeless to see the homework being torn when it is almost done '

The emoji  in (2) expresses *happiness* whereas the text expresses *sadness* as indicated by the emotion word 绝望 'hopeless'. By looking at the context given in the post, it is not the writer's but a child's homework being torn. The emotion word 绝望 'hopeless' is not the emotion experienced by the writer but his/her prediction about the child's emotion state. Without the emoji , it seems that the writer feels a lot of sympathy for the child. However, when  is used, it indicates that the writer does not feel compassion for the child but finds the situation hilarious.

3.2 Data Analysis

To avoid situations in which an emoji is correlated with a specific emotion just by coincidence, we only study emojis of more than 5 occurrences. Besides, each selected emoji should be unambiguously expressing a particular emotion so that one can distinguish its typical and atypical use. Table 1 illustrates the distribution of the typical and atypical use of emojis of each emotion type.

Table 1. Typical and atypical use of emojis

	Typical (%)	Atypical (%)	Total
Happiness	319 (66.6%)	160 (33.4%)	479
Sadness	88 (92.6%)	7 (7.4%)	95
Anger	243 (94.6%)	14 (5.4%)	257
Fear	0 (0%)	0 (0%)	0
Surprise	0 (0%)	0 (0%)	0
Total	650 (78.2%)	181 (21.8%)	831

Of the 831 comments, 78.2% of emojis are used typically, and 21.8% atypically as shown in Table 1. As there are no emojis associated with *fear* and *surprise* have more than 5 occurrences, no statistics can be shown for the two emotions. Among the other three emotions, emojis of *happiness* (33.4%) are most likely used in an atypical way, followed by *sadness* (7.4%) and *anger* (5.4%). It shows that emojis expressing *happiness* often disagree with their linguistic text. In the next section, I examine the interplay between emojis of each emotion type and their accompanying text.

4 The Interplay Between Emoji and Linguistic Text

As emojis of *fear* and *surprise* are either ambiguous or of low occurrence (i.e. less than 5), the following three subsections discuss *happiness*, *sadness*, and *anger* only.

4.1 Emojis Expressing Happiness

We manually examined through all the extracted comments and summarized the distribution as in Table 2.

Table 2 shows that 66.6% of emojis expressing *happiness* occur in text that also expresses *happiness*. The remaining 43.4% are comments that contain an emoji indicating *happiness* but express another emotion in text. It is observed in Table 2 that the overall emotion is more often determined by text, accounting for 25.7%. Only 7.7% of comments are determined by emoji when the emoji(s) and text are at odds. In order to figure out which channel determines the overall emotion, the distribution of the atypical use of *happiness* emojis is shown in Table 3.

Table 2. The distribution of typical and atypical

Emojis	Typical Use	Atypical Use		Total
		Determined by Emoji	Determined by Text	
[微笑]	15	0	75	90
[good]	81	1	8	90
[赞]	67	0	5	72
[笑 cry]	37	18	13	68
[哈哈]	23	10	10	43
[嘻嘻]	20	4	4	28
[鼓掌]	19	0	2	21
[爱你]	13	0	1	14
[中国赞]	11	0	0	11
[太开心]	9	1	1	11
[偷笑]	6	3	2	11
[可爱]	8	0	2	10
[坏笑]	5	0	0	5
[赞啊]	5	0	0	5
Total	319 (66.6%)	37 (7.7%)	123 (25.7%)	479

Table 3. The distribution of the atypical use of happiness emojis

Possibilities of the atypical use of emojis: (Emoji of Happiness + Emotion in Text → Overall)	Total
(1) Happiness + Sadness → Sadness	12
(2) Happiness + Sadness → Happiness	16
(3) Happiness + Anger → Anger	93
(4) Happiness + Anger → Happiness	12
(5) Happiness + Fear → Fear	3
(6) Happiness + Fear → Happiness	4
(7) Happiness + Surprise → Surprise	16
(8) Happiness + Surprise → Happiness	4

Table 3 shows that when a *happiness* emoji occurs in text expressing *sadness* or *fear*, there is only a fifty-fifty chance that the overall emotion is determined by either channel. However, when a *happiness* emoji appears in text expressing *anger* or *surprise*, it is highly likely that the overall emotion is determined by text such as (3).

(2) 不该是刑事拘留嘛
'Shouldn't it be criminal detention '

The text in (3) expresses an *anger* emotion while the emoji expresses *happiness*. The rhetorical question in the text is raised to convey the underlying meaning that the punishment should be criminal detention. An *anger* emotion is connoted as the writer thinks the penalty is too light. Thus, the emoji in (3) is used ironically. It is a contempt smile which is composed of both disgust and resentment. In that case, the overall emotion is determined by the text and the emoji functions as an ironic means to strengthen the intensity of *anger* by showing disrespect for the someone with a sneer. The emoji is frequently used ironically. When appears in text that expresses *anger*, there is every likelihood that the overall emotion is an *anger* emotion.

As for the co-occurrence of a *happiness* emoji with a text expressing *surprise*, it is more likely that the overall emotion is determined by text. Consider (4).

(4) 居然是我们邯郸人，佩服
' Surprisingly, it's a person from Handan. Remarkable!'

In (4), the emoji expresses a *happiness* emotion and the text expresses a *surprise* emotion as indicated by 居然 'surprisingly'. (4) expresses a complex emotion 驚喜 '*delighted*' which is composed of *surprise* as the primary emotion and *happiness* as the secondary. Given that the complex emotion 驚喜 '*delighted*' belongs to the *surprise* emotion, the overall emotion follows the text. In that case, functions as a modifier to enrich the emotion complexity.

It is also possible yet less frequent that the overall emotion is determined by the *happiness* emoji when it co-occurs with text expressing *surprise*. Consider (5).

(5) 这小偷估计也是饿的不行了，十五个鸡蛋
'The thief was probably starving, (he/she had) 15 eggs '

In (5) is not expressing the complex emotion 驚喜 '*delighted*'. The text expresses *surprise* as it is surprising that the thief broken into the house and had 15 eggs. In that case, the overall emotion is determined by the emoji (i.e. *happiness*) as the writer finds it hilarious. In sum, when the emoji and the text interact to express 驚喜 '*delighted*', the overall emotion is determined by the text. However, if the interaction does not lead to the emotion 驚喜 '*delighted*', the overall emotion should follow the emoji.

4.2 Emojis Expressing Sadness

There are 5 types of emojis that are found to express *sadness*. The distribution of the typical and atypical use is demonstrated in Table 4.

Table 4 shows that most *sadness* emojis occurring in text also expresses *sadness*. There are only a few exceptional cases. All the possibilities are shown in Table 5.

Table illustrates that the atypical use of *sadness* emojis is mostly related to text expressing *happiness*. The main reason is because of the complex emotion, 感動 *moved*.

Table 4. The distribution of typical and atypical use of sadness emojis

Emojis	Typical Use	Atypical Use		Total
		Determined by Emoji	Determined by Text	
[悲伤]	25	0	4	29
[蜡烛]	27	0	2	29
[失望]	14	0	0	14
[伤心]	12	0	0	12
[可怜]	10	1	0	11
Total	61 (92.6%)	1 (1.1%)	4 (6.3%)	95

Table 5. The distribution of the atypical use of sadness emojis

Possibilities of the atypical use of emojis: (Emoji of Sadness + Emotion in Text → Overall)	Total
(1) Sadness + Happiness → Happiness	4
(2) Sadness + Happiness → Sadness	1
(3) Sadness + Anger → Anger	2

The *moved* emotion is composed of *happiness* (primary) + *sadness* (secondary). When text expresses *moved* (i.e. *happiness*), *sadness* emojis are not capable of overriding the *happiness* emotion, as *sadness* serves as a peripheral component of the *moved* emotion as in (6).

(6) 形式不重要，有朋友在 就很温暖
'It's not the form that matters. Having friends around you is heart-warming '

In (6), the text conveys *happiness* but expresses *sadness*. However, interacts with the text to represent tears of joy. Therefore, the emotion expressed should be *moved*, i.e. *happiness* + *sadness*, which belongs to the *happiness* emotion.

4.3 Emojis Expressing Anger

Table 6 illustrates the typical use and atypical use of these emojis.

From Table 6, we can see that not many *anger* emojis are used in an atypical way. When the emoji and text conflict with each other, the overall emotion is mostly determined by text. Table 7 demonstrates the distribution of the atypical use of *anger* emojis.

Table 7 shows that when an *anger* emoji co-occurs with text expressing other emotions, the overall emotion depends heavily on the text as in the cases of *happiness*, *sadness*, and *surprise*. Consider (7).

Table 6. The distribution of typical and atypical use of anger emojis

Emojis	Typical Use	Atypical Use		Total
		Determined by Emoji	Determined by Text	
[怒]	103	0	0	103
[挖鼻]	30	0	5	35
[费解]	28	0	2	30
[吃瓜]	24	0	2	26
[哼]	18	1	3	22
[鄙视]	14	0	0	14
[怒骂]	12	0	0	12
[吐]	8	0	1	9
[白眼]	6	0	0	6
Total	243 (94.6%)	1 (0.4%)	13 (5.1%)	257

Table 7. The distribution of the atypical use of anger emojis

Possibilities of the atypical use of emojis: (Emoji of Anger + Emotion in Text → Overall)	Total
(1) Anger + Happiness → Happiness	4
(2) Anger + Sadness → Sadness	4
(3) Anger + Sadness → Anger	1
(4) Anger + Fear → Fear	1
(5) Anger + Surprise → Surprise	3

(7) 小样，还收拾不了你了 耍无赖也不看看地方
'See how I get to you! How dare you play rouge before us!'

In (7), the text expresses *happiness* as the writer eventually got to the person who play rouge before him/her. As it is gratifying to see that person being punished, the writer uses to show an undisguised contempt which reflects that the writer is gloating over this. The writer is experiencing *appeased* which is a complex emotion consists of *happiness* and *anger*. Therefore, in (7) interacts with the text expressing *happiness* to expressed *appeased*. The emoji functions as a modifier to enrich the complexity of the *happiness* emotion.

5 Conclusion

This paper examines the interplay between emoji and the accompanying linguistic text when the two channels are at odds. Results show that *happiness* emojis are most often used in an atypical way, as compared to the other emotions. Furthermore, *happiness* emojis are often used ironically to express *anger*. When two channels conflict with each other, the overall emotion is generally determined by text regardless of the emotion type. When the overall emotion is determined by text, an emoji can function as an iconic means to strengthen the emotion expressed by text. It can also function as a modifier to enrich the complexity of a complex emotion. We believe that this linguistic account of the atypical use of emojis will enhance the existing automatic emotion classification models.

Acknowledgments. This work is supported by PolyU Research Grants (Project No. G-YBVK and 1-ZVMF).

Appendix: Emojis and Their Corresponding Emotions

Emotion	Emoji and its Corresponding Name						
Happiness	[中国赞]	[赞]	[赞啊]	[good]	[笑 cry]	[偷笑]	[嘻嘻]
	[太开心]	[鼓掌]	[哈哈]	[可爱]	[微笑]	[爱你]	[坏笑]
Sadness	[悲伤]	[可怜]	[失望]	[伤心]	[蜡烛]		
Anger	[怒]	[白眼]	[哼]	[挖鼻]	[费解]	[怒骂]	[鄙视]
	[吃瓜]	[吐]					
Fear	N/A						
Surprise	N/A						

References

Crystal, D.: Language and the Internet, 2nd edn. Cambridge University Press, Cambridge (2006)

Derks, D., Bos, A.E., Von Grumbkow, J.: Emoticons and online message interpretation. Soc. Sci. Comput. Rev. **26**(3), 379–388 (2008)

Dresner, E., Herring, S.C.: Functions of the nonverbal in CMC: emoticons and illocutionary force. Commun. Theor. **20**(3), 249–268 (2010)

Felbo, B., Mislove, A., Søgaard, A., Rahwan, I., Lehmann, S.: Using millions of emoji occurrences to learn any-domain representations for detecting sentiment, emotion and sarcasm (2017)

Hussien, W.A., Tashtoush, Y.M., Al-Ayyoub, M., Al-Kabi, M.N.: Are emoticons good enough to train emotion classifiers of arabic tweets? In: Proceedings of 2016 7th International Conference on Computer Science and Information Technology (CSIT), pp. 1–6. IEEE (2016)

Kelly, R., Watts, L.: Characterizing the inventive appropriation of emoji as relationally meaningful in mediated close personal relationships. In: Experiences of Technology Appropriation: Unanticipated Users, Usage, Circumstances, and Design (2015)

Lee, S.Y.M.: A Linguistic Approach to Emotion Detection and Classification. Doctoral Dissertation. The Hong Kong Polytechnic University, Hong Kong (2010)

Lee, S.Y.M.: A Linguistic Analysis of Implicit Emotions. In: Lu, Q., Gao, H. (eds.) Chinese Lexical Semantics: CLSW 2015. Lecture Notes in Computer Science, vol. 9332, pp. 185–194. Springer, Cham (2015). https://doi.org/10.1007/978-3-319-27194-1_19

Lee, S.Y.M., Lau, H.Y.P.: Annotating emotion-related information in an event-comment emotion corpus. In: Proceedings of the 12th Conference on Language Resources and Evaluation (LREC 2020), May 11–16, 2020, Marseille, France, pp. 1626–1635 (2020)

Lo, S.K.: The nonverbal communication functions of emoticons in computer-mediated communication. Cyberpsychol. Behav. **11**(5), 595–597 (2008)

Novak, P.K., Smailović, J., Sluban, B., Mozetič, I.: Sentiment of emojis. PLOS ONE **10**(12), e0144296 (2015). https://doi.org/10.1371/journal.pone.0144296

Pavlenko, A.: Emotion and emotion-laden words in the Bilingual Lexicon. Biling. Lang. Cogn. **11**(2), 147–164 (2008)

Rezabek, L., Cochenour, J.: Visual cues in computer-mediated communication: supplementing text with emoticons. J. Vis. Lit **18**(2), 201–215 (1998). https://doi.org/10.1080/23796529.1998.11674539

Riordan, M.A.: Emojis as tools for emotion work: communicating affect in text messages. J. Lang. Soc. Psychol. **36**(5), 549–567 (2017). https://doi.org/10.1177/0261927X17704238

Skovholt, K., Grønning, A., Kankaanranta, A.: The communicative functions of emoticons in workplace e-mails. J. Comput. Mediat. Commun. **19**(4), 780–797 (2014). https://doi.org/10.1111/jcc4.12063

Tian, Y., Galery, T., Dulcinati, G., Molimpakis, E., Sun, C.: Facebook sentiment: reactions and emojis. In: Proceedings of the Fifth International Workshop on Natural Language Processing for Social Media, pp. 11–16 (2017)

Wolf, A.: Emotional expression online: gender differences in emoticon use. Cyberpsychol. Behav. **3**(5), 827–833 (2000). https://doi.org/10.1089/10949310050191809

Subjective Negative Structure in Chinese: A Case Study of "*Da 'Big'* + NP *(De)*" Sentence with the Uncorrelated Relations

Xue Zhang(✉)

University International College, Macau University of Science and Technology, Avenida Wai Long, Taipa, Macau
xuzhang@must.edu.mo

Abstract. This paper focuses on "*Da 'big'* + NP *(de)*" structure in Chinese. Based on the predicate scale theory, the relations between the temporal NP and the event in a sentence have been classified into three categories: positively correlated, negatively correlated, and uncorrelated. By the analysis of Uncorrelated data, we argue that "*Da* + NP *(de)*" structure is an implicit subjective negation marker. The temporal NP in this structure refers to the speech time, rather than the semantic features and pragmatic information, which was traditionally considered critical for the structure.

Keywords: *Da* + NP*(de)* · Subjective negation · Correlated sentences · Speech time · Semantics

1 Introduction

"*Da 'big'* + temporal NP *(de)*" structure is a ubiquitous idiom construction in Chinese, for example:

(1) 大白天的, 你怎么就开始睡觉了?
Da bai-tian de, ni zen-me jiu kai-shi shui-jiao le.
big daytime de you why already start sleep SFP
'Why do you start sleeping in the daytime?'

(2) 大过年的, 别说不吉利的话。
Da guo nian de, bie shuo bu ji-li de hua.
big pass New-year de not say not lucky de utterance
'Don't say unlucky things at New Year.'

(3) 大夏天的, 他怎么还带皮帽子呢?
Da xia-tian de, ta zen-me hai dai pi mao-zi ne?
big summer de s/he how still wear leather hat SFP
'How can s/he still wear a leather hat in the summer?'

M. Liu et al. (Eds.): CLSW 2020, LNAI 12278, pp. 118–126, 2021.
https://doi.org/10.1007/978-3-030-81197-6_11

Since Song Yuzhu (1994) noticed the exclusive usage of "*Da* + temporal NP*(de)*" structure, scholars (e.g., Shen Yang 1996, Xiang Kaixi 1998, Zhao Jinguan 2006, Wu Changan 2007, Liu Yang 2009, Wang Yuan and Luo Qiongpeng 2017) have started discussing this structure from different perspective and conducted many studies on it. The focus of the previous study was the semantic characteristics of the structure. Considering the notion of "expectation", previous studies generally believed that "*Da* + temporal NP *(de)*" addresses a kind of "counter-expectation". Thus, when this structure is used, the speaker's subjective judgments and emotions, such as disappointment, doubt, indignation, etc., are conveyed. It has also been suggested in the previous studies that in the combination of "*Da* + temporal NP *(de)*" and its subsequent sentence, the pragmatic factors are necessarily significant. The usage of the structure relies on the mutual knowledge background, such as common sense, habitual knowledge, and culture. Furthermore, this background is decisive in the sense that the events referred to by the "temporal NP" and the subsequent sentences need to be unrelated according to the social consensus. Compare sentence (1), repeated below, with sentences (4) and (5):

(1) 大白天的, 你怎么就开始睡觉了?
Da bai-tian de, ni zen-me jiu kai-shi shui-jiao le?
big daytime de, you why just start sleep SFP
'Why do you start sleeping in the daytime?'

(4) *大白天的, 你怎么在上班。
*Da bai-tian de, ni zen-me zai shang-ban.
big daytime de, you why zai work.
Intended meaning: 'Why are you working during this daytime.'

(5) ? 大白天的, 你怎么在吃苹果。
?Da bai-tian de, ni zen-me zai chi ping-guo.
big daytime de, you why zai eat apple.
Intended meaning: 'Why are you eating an apple during this daytime.'

The event "sleeping" in the sentence (1) is typically opposed to what we generally believe to be a normal activity we should do in the daytime. Therefore, in the case of sentence (1), the "*Da* + temporal NP *(de)*" structure is used appropriately. In sentence (4), the event involved in the subsequent sentence, i.e., "working" and the temporal NP, i.e., "daytime" are compatible. Without a counter-expectation which violates our social consensus, the use of the structure in sentence (4) is confusing. As for sentence (5), the situation is more complicated. The event "eating an apple" is neither typically related to "daytime" nor typically unrelated. It is more like an "irrelevant" event. Such kind of "irrelevant" sentences like sentence (5) is the unexplored issue yet to be discussed and therefore becomes the focus of this paper.

2 From "Neither Tall nor Short" to "Eating an Apple"

Before discussing the irrelevance between the temporal NP and the event involved, we must first classify the relationship between "time" and "event". Moreover, this classification has a certain similarity with the classification of "degree" of the predicate in Chinese.

The study of semantic degree has developed rapidly in recent years. "Degree" is a critical semantic element and involves concepts such as "scale" and "polarity". Based on this finding, many adjectives and adverbs in Chinese have been well interpreted. However, within the framework of "degree" and "scale", it is often easier to quantify the extreme values, which are the polarities with apparent characteristics. Meanwhile, the position on a scale which are not at the two ends, and the process of quantitative change, are less noticed and interpreted.

We are taking the adjectives "tall" and "short" as examples. In terms of "degree", the two adjectives are of extreme values on the scale of "height" (Fig. 1). We can say that a 150 cm-height person is "short", and 190 cm is "tall". However, we cannot tell the exact numerical boundary between "tall" and "short". For 155 cm and 175 cm, it all can be called "neither tall nor short" according to our prevailing perception.

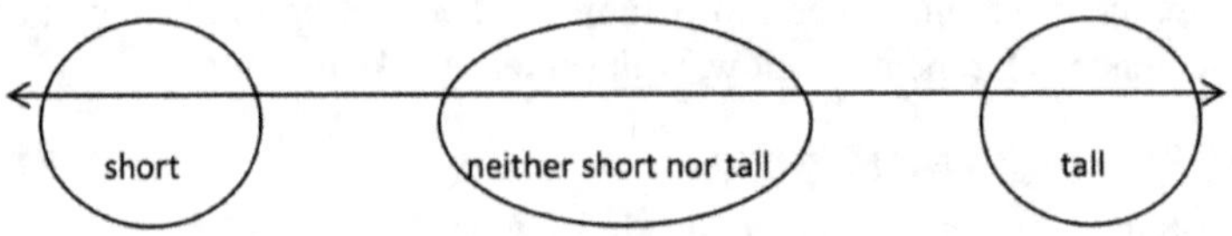

Fig. 1. Scale of height

As for the relationship between "time" and "event", we can consider the "scale/degree" concept of the predicate. With all events under the "daytime" aggregating as a "scale" (Fig. 2), "working" and "sleeping" should be events at the extreme ends on the scale, similar to how "tall" and "short" locate on the degree scale. Furthermore, events like "eating an apple" are like "neither tall nor short" degree that it does not occupy the extreme ends.

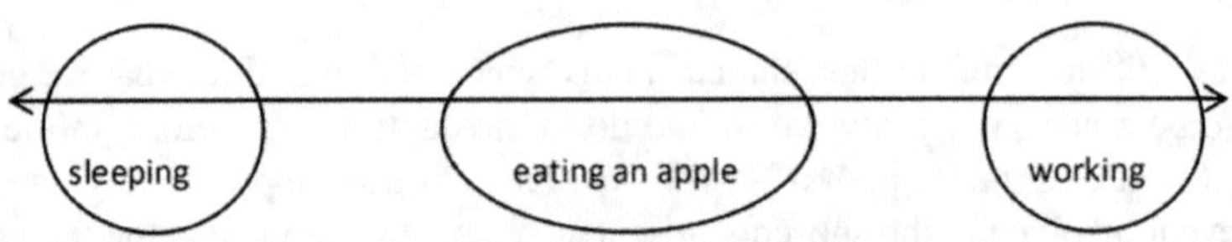

Fig. 2. Scale of daytime

Referring to the "degree" concept, the relationship between "time" and "event" can be classified into three categories:

Category A: Positive Correlation
The relationship between the "time" and the "event" conforms to human common sense and cultural background, such as "daytime" and "working".

Category B: Negative Correlation
The relationship between the "time" and the "event" runs counter to common sense or cultural background, such as "daytime" and "sleeping".

Category C: Uncorrelation
There is no conventional relationship between the "time" and the "event", such as "daytime" and "eating an apple".

In the previous research on "*Da* + NP *(de)*", the focus was on categories A and B. The following discussion in this article will focus on category C, the uncorrelated ones.

3 From "Unspeakable" to "Natural"

In the development of "*Da* + NP *(de)*", the category C of "uncorrelation" mentioned above has experienced the evolution from "unspeakable" to "speakable", which can be examined from two perspectives:

3.1 Uncorrelated Time Noun Phrase

Since the previous studies focused on the commonsense attribute of "temporal NP", the temporal NP without apparent social-cultural attribute was once considered unsuitable to be used in the structure. For example, Shen Yang (1996) pointed out that the temporal NP is "the days of its particularity":

① *Time words for festivals, traditional Chinese festivals, public holidays, etc. Such as New Year, New Year, weekends and so on. Besides, such as complex, the festival, Thanksgiving, Army Day, etc. cannot be used.*

② *The time word of the year, month, day, and week can be said. However, words of ordinal numbers cannot be said, such as this year, the day before yesterday, etc. Words of the month cannot be used, such as January and February, except Lunar January (Zheng Yue). The date cannot be said except for the first, fifteenth and thirtieth day on Lunar month.*

③ *Time words indicating seasons and climates can be used. However, spring, autumn, warm days, and cool days cannot be said.*

④ *Time of day cannot be said, such as "eight o'clock." Words like morning, afternoon, evening, dusk, etc. cannot be said.* (Shen Yang 1996)

However, we can find in the corpus that all the above temporal NP that is considered unsuitable actually can be used, as exemplified below:

(6) 啊!大圣诞的, 别这么没出息!想点好的: 回来也不是吃什么。大概是火鸡吧!
A! Da sheng-dan de, bie zhe-me mei chu-xi!
Interj big Christmas de, do-not such no accomplishment
Xiang dian hao de: hui-lai ye bu shi chi shen-me?

think some good de back also not be eat what
Da-gai shi huo-ji ba!
probably be turkey SFP
'Ah! Do not be a loser on such a Christmas day! Think about it: when you come back you are not eating something special. Probably turkey!'

(7) 这地球是要毁灭了么?怎么大八月的才十几度呢?
Zhe diqiu shi yao huimie le me?
this earth be want destroy le SFP
Zen-me da ba-yue de cai shi ji du ne?
why big August de just ten few degree SFP
'Is this earth going to be destroyed? Why is it only a dozen degrees in August?'

(8) 我们早来早回去, 大秋天的, 还都忙着下地哩。
Wo-men zao lai zao hui-qu, da qiu-tian de,
we early come early go-back, big Autumn de
hai dou mang zhe xia-di li.
still all busy zhe go-field SFP
'We shall come early and go back early. We are busy with planting fields in Autumn.'

(9) 大上午的, 路上就一个劲儿的堵。花了揪心的二十八块钱才算急火火的到了。
Da shang-wu de, lu shang jiu yi ge jin-er de du.
big morning de, road on just one CL strength de jam
Hua le jiu-xin de er-shi-ba kuai qian
spend le anxious de twenty-eight CL dollar
cai suan ji-huo-huo de dao le.
then count rush de arrive le
'The traffic is blocking in such a morning. I just rushed to arrive by reluctantly spending twenty-eight dollars.'

(10) 今天北京这路真是邪了门了, 大八点的竟然没堵车!
Jin-tian Beijing zhe lu zhen shi xie-le-men le,
today Beijing this road really be strange le,
da ba-dian de jing-ran mei du-che!
big eight-o'clock de unexpectedly no traffic-jam.
'Beijing's traffic is really a wicked one today. There is no jam at eight o'clock!'

3.2 Uncorrelated Event

The previous studies have pointed out that the events of the subsequent sentence of "*Da* + NP *(de)*" should correspond to the attributes of the "temporal NP", and hence "*da* (big)" is considered to mark a "significant interpretation" and "emphasizing a certain feature of things" (Xiang Kaixi 1998). Therefore, the events with significantly inapparent features cannot be used in the subsequent sentences after the "*Da* + NP *(de)*" structure. However, we find two groups of sentences as follows:

(11) 只见一位年轻的女士来到我的面前, 她气势汹汹地质问:"大白天的, 你敢偷车?"我一时懵了, 自己怎么变成偷自行车的人了呢?

Zhi jian yi wei nian-qing de nv-shi lai dao wo de
just see one CL young de lady come arrive I de
mian qian, ta qi-shi-xiong-xiong de zhi-men:
face front she angrily de inquire
"Da bai-tian de, ni gan tou che?"
big daytime de you dare steal bicycle
Wo yi shi meng le, zi-ji zen-me bian
I one moment confuse le self how turn
cheng tou zi-xing-che de ren le ne?
become steal bicycle de person le SFP
'I saw a young lady who came to me, and she asked aggressively: "How do you dare to steal my bicycle during the daytime like this?" I am in shock at that moment. How do I become a bike stealer?'

(12) 陆大可回头对田二掌柜道:"去吧去吧, 大白天的, 拿把剑舞持什么?上厨房给我们切盘羊头肉。我和成大掌柜好久不见了。"
Lu-Dake hui-tou dui Tian-Er zhang-gui dao:
Lu-Dake back-head to Tian-Er shopkeeper say
"qu ba qu ba, da bai-tian de, na ba jian
go SFP go SFP big daytime de, hold CL sword
wu-chi shen-me? shang chu-fang gei women qie pan
wave what go kitchen give we cut plate
yang tou rou. Wo he Cheng-Da zhang-gui hao jiu
sheep head meat I and Cheng-Da shopkeeper very long
bu jian le."
no see le

'Lu Dake turned to shopkeeper Tian Er and said, "Go, go! What do you do with the sword in such a daytime like this? Go to the kitchen and bring us some mutton. Shopkeeper Cheng Da and I have been away for a long time.'

(13) 我不聋, 你问问她, 大白天的, 不好好在屋里呆着, 满大街瞎唱什么?
Wo bu long, ni wen wen ta, da bai-tian de,
I not deaf you ask ask she, big daytime de
bu hao hao zai wu li dai zhe, man da jie
not good good at room inside stay zhe full big street
xia chang shen-me.
blindly sing what
'I am not deaf. You ask her. Why is she singing terribly in the daytime like this instead of staying in the house?'

(14) 从乡里到这儿有二十多里路, 这大清早的, 露水重得很, 舅舅的皮鞋上却是干干净净的。
Cong xiang li dao zhe-er you er-shi duo li lu,
from village inside to here have twenty some mile road
zhe da qing-zao de, lu-shui zhong de hen,
this big early-morning de dew heavy de very
jiu-jiu de pi-xie shang que shi gan-gan-jing-jing de.

uncle de leather-shoe on but be clean de
'There are more than twenty miles from the village to here. In this early morning, the dew is heavy, but uncle's leather shoes are clean.'

(15) 这个男人有五十多岁年纪, 一面撩开帐子, 一面打哈欠, 说:"你又狂什么?大清早的!"
Zhe ge nan-ren you wu-shi duo sui nian-ji,
this CL man have fifty some year-old age
yi mian liao kai zhang-zi, yi mian da-ha-qian, shuo:
one side lift open curtain one side yawn say
"ni you kuang shen-me? da qing-zao de!"
you again mad what big early-morning de
'This man is in his fifties, yawning while opening the curtain and says, "What are you crazy about early in the morning!"'

(16) 娴在里面生气地说, "大清早的你胡说些什么, 好好的怎么会寻死?要寻死的也是我, 不会是她。"
Xian zai li-mian sheng-qi de shuo,
Xian at inside angry de say
"da qing-zao de ni hu shuo xie shen-me,
big early-morning de you nonsense say some what
hao hao de zen-me hui xun si?
good good de how will look-for death
yao xun si de ye shi wo, bu hui shi ta."
want look-for death de also be me not will be she
'Xian said from inside angrily, "What nonsense do you say in the early morning? How can you attempt suicide? I am the one who wants to die, not her.'

Sentences (11) to (13) all use "*bai-tian* (daytime)", but the events in the subsequent sentences, i.e., stealing a bicycle, swaying a sword, singing at the street, respectively, are not related to "working" or "resting". Furthermore, the events in these three sentences are even difficult to generalize. Sentences (14) to (16) face the same situation. All three sentences are expressing "*qing-zao* (early morning)", and yet the commonality of the three subsequent sentence events are too rare to generalize. If there is any commonality, then the only one is that the sentence expresses a negative attitude of the speaker towards the occurrence of the events in the subsequent sentences.

4 When to "Quarrel"?

It has been demonstrated in this paper that in the construction of "*Da* + NP *(de)*", which belongs to the category C which involves uncorrelation relations between the temporal NP and the event in the subsequent sentence, has been used more frequently and naturally. It is noteworthy that the sentence with "*Da* + NP *(de)*" can always be understood, even if the temporal NP and the event involved are not related in some sense, the conversation participants can draw the meaning of "counter-expectation" from the sentence.

We have found these two sentences in the corpus:

(17) 大白天的, 你想吵架吗?
Da bai-tian de, ni xiang chao-jia ma?
big daytime de, you want quarrel SFP?
'Do you want to make a quarrel at the daytime like this?'
(18) 大晚上的, 你想吵架吗?
Da wan-shang de, ni xiang chao-jia ma?
big night de, you want quarrel SFP?
'Do you want to make a quarrel at the nighttime like this?'

Is there a generally-agreed and appropriate time for "quarrelling"? "*Day*" or "night", which one is more suitable for quarreling? These two sentences bring us a vital question. That is, what are the semantic expression and collocation requirements of "*Da* + NP *(de)*" structure? What is its theoretical support of the uncorrelation relation? For the questions above, we believe that the semantics of "*Da* + NP *(de)*" is a subjective negative marker. Furthermore, the semantic relationship and pragmatic background between the "temporal NP" and the "event" are not clear in the structure. Therefore, although there is a barrier for the listener to understanding the semantic meaning of the "uncorrelated sentence", the listener can still understand the "negation" expressed by the construction righteously.

This observation and finding make us necessarily re-examine the meaning and usage of the "*Da* + NP *(de)*" structure. Since it is learned that, in fact, there is no requirement of conventional connection between the "temporal NP" and the "event" in the structure, it is rigorous to modify the "temporal NP" as the speech time of the NP. The speech time of the NP can be a season time (大冬天的), a festival time (大过年的), a time slot (大白天的), or a time point (大八点的). In short, it refers to the time when the speaker speaks.

From this point of view, "*Da* + Speech Time of the NP *(de)*" is a construction that indicates the speaker's subjective negation. It expresses that the speaker is negating the events in subsequent sentences, regardless of whether negator words appearing in subsequent sentences or not. An uncorrelated relation between "Speech Time NP" and "Event" is the norm. The usage of negatively correlated relations (such as "daytime" and "sleeping") is a habitual tendency rather than a semantic requirement.

5 Subjective Negative Structure in Chinese

In the study of Chinese linguistics ontology and its application, constructional structures, or idioms have always been an inexplicable and unclear field that require exploration. Taking the "*Da* + NP *(de)*" as an example, the previous studies generally believed that the understanding and usage of this structure are subject to pragmatic factors and common sense, which makes this kind of construction difficult for second language learners to acquire and thus limited to advanced-level learners exclusively. The paper starts with "uncorrelated sentences" and discusses its core semantic features and usage constraints. We argue that "*Da* + NP *(de)*" is one of the subjective negative structures in Chinese. The expression has been thus amended to "*Da* + Speech Time of the NP *(de)*". This is a new perspective for Chinese construction/idiom study, which makes the formalization of an idiom possible.

The "Subjective Negative Structure" we are generalizing here, such as "*Da* + Speech Time of the NP *(de)*", can be analogous to Introductory Particle in ancient Chinese to some extent. It indicates that "I will deny the incident to be described below." Whether there are similar negative constructions in Chinese are yet to be studied. However, a semantically unspecific structure with specific components is undoubtedly worth an attention.

References

Liu, Y.: Re-investigation on time noun of da+NP(de) structure. J. Lang. Lit. Stud. **5A**, 97–98 (2009)

Shen, Y.: About da+time noun(de). Stud. Chin. Lang. **4**, 281–282 (1996)

Song, Y.-Z.: The Function of da as a distinguishing word. Stud. Chin. Lang. **6**, 447–447 (1994)

Wang, Y., Luo, Q.-P.: A semantic analysis of the structure "da (big) +temporal noun phrase" in a degree-based perspective. J. Yunnan Normal Univ. Teach. Stud. Chin. Foreign Lang. Ed. **15**(3), 36–45 (2017)

Wu, C.-A.: A brief introduction of da…de. Chin. Teach. World **2**, 62–66 (2007)

Xiang, K.-X.: The saliency and marked word da. Chin. Lang. Learn. **1**, 12–16 (1998)

Zhao, J.-G.: The matching mechanism of da+time-noun+de and its subsequent sentences. Mod. Chin. **9**, 30–31 (2006)

Diachronic Evolution of the Verb *Give*

Guoyan Lyu[1](✉), Haitao Chen[1], and Yanmei Gao[2]

[1] School of Foreign Studies, Beijing Information Science and Technology University, Beijing 100192, China

[2] School of Foreign Languages, Peking University, Beijing 100871, China

Abstract. The purpose of this study is to investigate the diachronic evolution of the most typical ditransitive verb, *give*, and its competition against other word forms with the similar meaning. Statistics from the historical chronicles between the 6th and 15th century show the competing trends. The results demonstrate that in OE and ME, the set of 'give' verbs with Germanic origin all have the meaning of taking and holding in their etymology. The prolixity of the verb's varied forms plays an important role in building its competitiveness.

Keywords: Diachronic evolution · Variants · Give · Dative alternation

1 Introduction

Since the early 20th century, the studies on predicates in languages have been the focus in many linguistic schools and perspectives. Among the predicates, ditransitive verbs have caught the attention of many influential figures in linguistics. Kittilä, for instance, stated that "if a language has only one ditransitive trivalent verb (on the basis of any feature of formal transitivity), then that verb is 'give' [1]." Much of the research on ditransitive constructions has also focused on the properties of the verb *give* which is often regarded as by far the most typical ditransitive verb. Bresnan and Nikitina found in their corpus study that *give* is the prototypical ditransitive verb which has the highest frequency, constituting more than 40% of all alternating dative verbs [2].

However, as suggested by Zhang based on his typological study of Chinese ditransitive constructions, the English verb *give,* in his term, V_{give}, might not be the underived general-purpose verb of giving but a substitution for it [3]. Similar hypothesis is also found in Chappell et al. [4] and Chappell and Peyraube [5]. Zhang made the statement that the scarcity of 'give' type Double Object Construction (DOC) in most modern Germanic languages may be attributed to the mono-transitive origin of their V_{give}, as what has been observed in a type of dialect in Southern China [1]. One piece of evidence is that *give* in English and *geben* ('give') in German as well as the V_{give} in nearly all modern Germanic languages are cognate to the Proto-Germanic **ʒebanan*, which is related to the Latin *habeō* 'hold, have' and is traceable to the Indo-European proto form: *ghabh-*, the cognates of which in other Indo-European languages function as a mono-transitive verb of handling, such as the Old Irish *gaibid* 'to take' [6].

In consideration of the above-mentioned possible correlation of the etymology of the verb *give* and its involvement in the dative alternation, the potential questions would

M. Liu et al. (Eds.): CLSW 2020, LNAI 12278, pp. 127–136, 2021.
https://doi.org/10.1007/978-3-030-81197-6_12

arise: When probably did the verb *give* become a verb of giving and involve in the typical dative structure? If there were other verbs expressing the similar sense of giving, how did *give* manage to become the typical verb of giving?

To find out the changes in the verb's lexical semantics and its structural trends, we incorporated the concept of frequency in our study. Bybee and Hopper documented the clear role of frequency in the emergence of grammatical structure [7]. Frequencies are also taken to represent degrees of entrenchment, with highly frequent constructions being central to the system, while constructions of lower frequency are regarded as less entrenched [8, 9]. The present study focuses more on token frequency; in other words, we pay special attention to the extent to which one particular verb is preferred over the other in discovering the usage differences for the verbs in question.

The study in this paper is intended to track the historical development[1] of the verb form *give*. Our focus is on eight historical chronicle-type of documents[2] in English with the purpose to find out the evolution of the verb *give* and its competitions with other verbs having the similar meaning.

2 Diachronic Development of the Verb *Give*

Rappaport Hovav and Levin argue that English verbs that denote transfer of possession such as *give*, *lend*, and *offer* only have a caused-possession sense [19]. The verb *give* has only one core meaning (i.e. the meaning components entailed in all uses of the verb independent of context) and this single meaning is expressed in two different structures. In other words, the verb itself lexicalizes caused-possession, hence the semantics of the verbs determine the caused-possession event. The similar result is found in the diachronic study (e.g. Lyu and Gao [20]) that the caused-possession verb *give* shows different structural trends in comparison with the caused-motion verb *send*. Goldberg also points out that 'give' represents the 'conceptual archetype' for ditransitive constructions which expresses the meaning 'CAUSE-RECEIVE' [21]. Therefore, it seems critical that the sense of caused-possession has to be in the core meaning of the verb, *give*. Yet if we look closely at the etymology and the early use of the verb, things were quite different.

2.1 Etymology of *Give*

As stated in Indo-European Language Revival Association [22], the verb *give* is a Germanic neologism, which is a replacement for *dō-* (Latin) "give". The form *gyfan* in OE

[1] The stages of English development are divided based on the *History of the English Language* (Burnley [10]): from Old English (700–1200, OE), Middle English (1200–1500, ME) to early Modern English (1500–1800, Early ModE).

[2] 1) *The Laws of the Earliest English Kings* (6–10th centuries) [11]; 2) *The Old English Version of Bede's Ecclesiastical History of the English People* (hereafter Bede, Early 10th) [12]; 3) *King Alfred's Anglo-Saxon Version of Boethius* (Boethius, Mid-10th) [13]; 4) *Beowulf* (Early 11th) [14]; 5) *The Anglo-Saxon Chronicle* (9–12th) including the Parker Manuscript (Manuscript A, 9–10th), Manuscript C&D (11th) and the Peterborough Manuscript (Manuscript E, 12th) [15]; 6) *Layamon's Brut* (13th) [16]; 7) *The Metrical Chronicle of Robert of Gloucester* (14th) [17]; 8) *The Brut or The Chronicles of England* (Brut, 15th) [18].

is a relatively new word, a substitution for the Latin word for 'give'. Wedgewood has a dictionary entry for *give* as related to Gaelic *gabh*, which means 'to take, lay hold of, seize' [23]. It states that "of this perhaps *give* is the causative, to cause another to take. In the same way to *take* was formerly used in the sense of 'deliver up to', or 'give'." The example given for *take* is used in the causative sense, meaning 'give': "*and privately taketh her the cloth*". As is shown here, the verb *give* is not only believed to have the original sense 'to take or to seize', which has a causative sense, but it is also similar to the verb *take*, which used to mean 'to deliver to or to give'.

Similarly, another explanation is provided by Kortlandt[3], in which it is assumed that "the verb can be derived from the Germanic prefix *ga-* plus the root of Hittie *epzi* 'seizes', Latin *apiscor* 'reach', *coepi* 'have begun'" [24]. And "the meaning 'reach' offers a suitable basis for the semantic remodeling of **ga-ef* < **ko(m)-hehp-*, thus establishing a direct connection with Latin *Coēpī* 'seized' < **ko(m)-hehp-*." Here, the verb *give* is reconstructed as having the original meaning of 'seizing'.

Based on the above investigation, though we have not found the actual instances of *give* carrying the meaning of taking and holding, occurring as a non-ditransitive verb, yet we did find that: First, the word *give* is relatively newer (no source found to be sure about when it came into being) than the giving verb *donner* in French, which originated in Latin. Second, also importantly, the word *give* has its root in Old High German, which originally meant taking and holding, instead of giving.

In the following sections, we will look into the competitions of *give* in different periods in history so as to find what makes it gradually become the most typical verb of giving in English.

2.2 Competition Between *Gyfan* and *Syllan*

Old English (OE) is an early form of the English language that was spoken and written in parts of what are now England and southern Scotland between the mid-5th century and the mid-12th century. It is a West Germanic language and is closely related to Old Frisian and experienced heavy influence from Old Norse, a member of the related North Germanic group of languages. In the development of English language, OE may be defined as the period of fully case-marked, Middle English (ME) as the period with semi-case-marked, and Modern English (ModE) as the period of case-lost language. As the case-system experienced gradual changes, the use of lexicon also witnessed competitions among different forms expressing relatively similar meanings.

In Henry Sweet's *The Oldest English Text*, there are two verbs mentioned to be used mainly to denote 'give'. They are *gyfan* ('give') and *syllan* ('sell'), among which, *gyfan* is a strong verb, and its variants include *a-gyfan* ('to give'), *for-gyfan* ('to grant', 'to give'), *of-gyfan* ('to desert') [25]. The word *syllan*, on the other hand, is a weak verb, and its variants include *ge-syllan* ('to give'), *a-syllan* ('to give away'). In OE, the form and the range of senses of *syllan* are actually varied widely. The verb *syllan*, corresponding

[3] When inquired in an email about actual instances of *give* when used as a transitive verb denoting seizing or reaching in OE, Prof. Kortlandt, responded that "the verb *reach* can be either transitive or intransitive, e.g. *he reached me an orange* for *he gave me an orange*, similarly German *jemandem Feuer reichen* for 'to give somebody a light'."

to the verb *sell* in PDE, is used interchangeably with *gyfan* in the similar constructions in OE. That means the verb includes the sense of giving or bestowing things gratuitously. Its prefixed forms, such as *asellan, gesyllan* or *ymbsyllan* denote similar meaning of delivering, with subtle distinctions in the manner of giving. In comparison with *syllan*, the verb *gyfan*, however, has the central meaning of giving as that in PDE and its varied forms including *agyfan, forgifan, ofgifan* and *gifu* have relatively diversified meanings. Yet, at this stage the major senses and use of *gyfan* and *syllan* are not much different in general.

If we compare the changing patterns of frequencies between the two forms of giving verbs, *syllan* and *gyfan*, we can see that their frequencies really affect their patterns of use. In other words, the token frequencies of *syllan* and *gyfan* alternate dramatically, in different periods in OE.

The distribution ratio of the two words are based on the token frequency of them in the documents. For example, in *The Laws of the Earliest English Kings*[4], the instances of *syllan* meaning 'giving gratuitously' or 'giving for free' are found along with the verb *gyfan*, which also mainly denotes 'giving' or 'granting'. The distribution of the two forms in all *The Laws of the Earliest English Kings* (late 6th–early 10th century) is illustrated as in the following table.

Table 1. Distribution of *syllan* and *gyfan* in The Laws of the Earliest English Kings (late 6th–early 10th century)

Forms	Tokens	Percentages (%)
syllan	107	73
gyfan	40	27
total	147	100

As illustrated in Table 1, the overall token frequency of the verb *syllan* is much higher than that of *gyfan*, with the ratio of 7:2 between them. It seems that the overall use of *syllan* enjoys a higher priority than the form *gyfan* in *The Laws of the Earliest English Kings* before the 10th century. And the result shown above includes all the forms of the two words, including their prefixed forms, such as *gesyllan* (similar to *syllan*, 'to give or to sell'), *foresyllan, forgyfan* and *agyfan*. The following Table 2 illustrates the specific distribution patterns of the basic forms *syllan* and *gyfan* along with their suffixed forms.

As shown in Table 2, in *The Laws of the Earliest English Kings*, among *syllan* and its varied forms, the more frequent form is actually *gesyllan* (accounting for 64% of all the forms of *syllan*), not the basic form, *syllan*. The form *gesyllan* has multiple senses, including 'to give, deliver, say, betray, sell, give up'. Similarly, instead of *gyfan*, the form

[4] *The Laws of the Earliest English Kings* contain seven law documents, among which we chose *The Kentish Laws* (including *The laws of Æthelberht* (issued around 6–7th century), *The Laws of Ine and of Alfred* (hereafter, *Ine and Alfred*); *Treaties with the Danes* (hereafter, *Treaties*); *The Laws of Edward and the Elder and of Aethelstan* (hereafter, Edward) (*The Laws of Edward* (I, II), the Ordinance relating to Charities, *The Laws of Aethelstan* (I, II, III, IV, V, VI)).

Table 2. Distribution of varied forms of *syllan* and *gyfan* in the Laws of the Earliest English Kings

SYLLAN (107 tokens in total)			GYFAN (73 tokens in total)		
Form	Token	Percentage (%)	Forms	Token	Percentage (%)
gesyllan	68	64	*agyfan*	27	68
syllan	36	34	*forgyfan*	8	20
foresyllan	2	2	*gyfan*	5	12

agyfan is found to be more frequent (68% of all forms), which also has varied meanings, including 'to restore, give back, give up, return, repay and reddere'. Comparing the core meanings of the form *gesyllan* and *agyfan*, we can see that in *The Laws of the Earliest English Kings*, the main verb denoting the meaning of giving is the verb form *gesyllan*, a prefixed form of *syllan*, instead of the verb *gyfan*. Then in less than a century, the word *syllan*, completely lost its original OE meaning by the year of 1300. The following Fig. 1 shows the comparative token frequencies of the two verbs against each other through seven centuries of time.

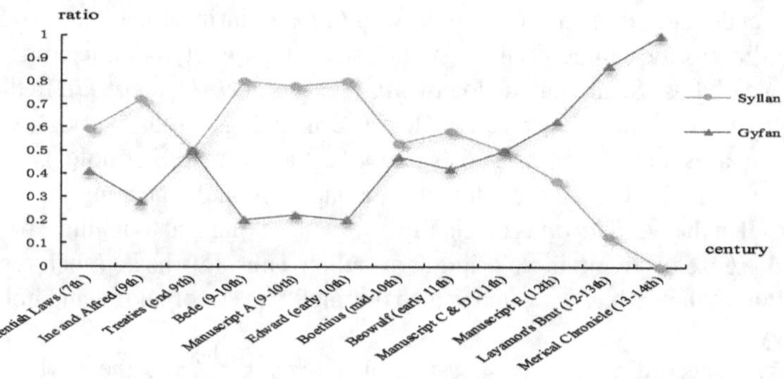

Fig. 1. Diachronic competition between *syllan* and *gyfan*

As shown in Fig. 1, the end of 9th-century and the mid-10th to 11th centuries witnessed a parallel distribution of *syllan* and *gyfan*. Starting from the 11–12th centuries as the English language began to make its transition from the period of OE to early ME, the form *gyfan* started to gain more and more strength in acquiring more space in the distribution. Until 13–14th the form *gyfan* completely won the competition over *syllan*, becoming the main verb denoting the meaning of giving gratuitously.

At the same time when the frequency of *syllan* is in decline with time, so are its suffixed variants (e.g., *gesyllan, asyllan*). On the other hand, the competition between the form *gyfan* and its competitors, such as *agyfan* ('give away'), *ongyfan* ('to give back'), *ofgyfan* ('to give up, quit'), *forgyfan* ('to grant, give') and *gegyfan* (expressing the meaning of giving in different manners) has been relatively severe. That is, almost all the suffixed forms of *gyfan* used to show strong competitiveness in trying to gain their status in use in OE and early ME. This may also explain why the form, such as *forgive*, managed to survive into the PDE but with a central meaning of pardoning. Similarly, the form *agyfan* is transformed into a phrasal verb, *give away*, and *ymbgyfan* to *give back*, partly due to the simplification of the inflection system in English.

In contrast, as for the suffixed forms of *syllan*, such as *gesyllan* ('to deliver'), *forsyllan* ('give out') and *ymbsyllan* ('give around'), their meanings were later identically focused on one sense, which was giving or delivering, just in different manners. This partly explains why as the basic form *syllan* lost its sense of giving or delivering gratuitously, all these varied forms went extinct without even being morphed into phrasal verbs. And the form *syllan* itself gradually lost its status of being polysemous, and instead, turned to be the major lexeme, *sell*, denoting the meaning of rendering things/people for a price.

Therefore, the prolific suffixed forms of *gyfan* might also help the stem-form to maintain its core power in sustaining its status of being the general-purpose verb of giving.

2.3 Competition Among *Gyfan*, *Niman* and *Tacan*

As discussed in the previous section, the verb *gyfan* seems to have won the competition against *syllan* in 12–13th centuries (frequency ratio of *syllan* vs. *gyfan* (including suffixed forms) is close to 1:7), yet it started another round of competition against another verb of 'taking and handling' *tacan* (the early form of *take*) during 12–14th centuries.

According to the online etymology dictionary (www. etymonline.com), the word *tacan* is one of those Scandinavian loanwords (such as *neck, leg, skin, kill* in the present-day English) [26]. The form *tacan* in late OE meaning 'to take, seize', came from a Scandinavian source (such as Old Norse *taka* 'take, grasp, lay hold'), from Proto-Germanic *takan-* (Gothic *tekan* 'to touch'), perhaps originally meaning 'to touch'. Here it suggests that the word *tacan* is originally a verb of 'taking and touching', which seems to carry no sense of giving in its original meaning. Thus, this bears much resemblance to the origin of the verb *give*, which is also originally a verb of taking and holding in its etymology.

However, according to the glossary of *Layamon's Brut*, the earlier form *bi-tæchen/bitechen* ('bitake') means 'to deliver, give' [16]. For example, in "*And Augel ich* ***take****þe an hond; al to-gadere Scotlond* ('And to Angel I **give** thee in hand Scotland altogether') [16]", the expression *take þe an hond* actually means 'give thee in hand', which shows that the form *tacan* in the 13th century means 'give' instead of 'take' in the modern sense, and it also occurs in the DOC, taking the IO 'Augel', and the DO 'Scotland'. Hence there is some overlap between the forms *tacan/bi-tacan* and *ȝeuen* ('give') in terms of the meaning and structure as well. And there are indeed cases found in *Layamon's Brut*, where the form *tacan* was used to replace *aȝeuen,* a variant of *ȝeuen* ('give'), in a later modified version. Then another question might be asked: what then was the form denoting the meaning of taking in the modern sense during that period of time?

In *Layamon's Brut* (12–13th century), we found the instances such as "***Nimeð*** eoure sexes; sele mine bernes ('Take your axes, my good warries.') [16]. Here it shows that the verb *niman* denotes the sense of taking, and is used only as a transitive verb, taking a

DO, as 'your axes'. It turns out to be always the case. According to the entries in etymology dictionaries ([22, 23, 27]), the verb *nimen/niman* originated from proto Germanic basically means 'to take, take away, seize, obtain, have hold, keep'. Here it shows that *niman* is also a verb of taking and holding and the structure it occurs has always been a sheer transitive structure.

In terms of the relative type-token frequencies, the following Table 3 illustrates the token frequency ratios of the four competitive forms: *ȝeuen, syllan, niman/biniman,* and *bitacan/tacan* in *Laymon's Brut*, all of which had the sense of giving and transferring.

Table 3. Distribution of four verbs in Layamon's brut (12–13th centuries)

Verbs of taking and holding (12–13th centuries)	Tokens	Percentages (%) (each against all four forms)
ȝeuen	83	34
niman/biniman	82	33
bitacan/tacan	70	28
syllan	13	**5**
Total	248	100

Table 3 indicates that both *ȝeuen and niman/biniman* have high frequencies in *Layamon's Brut*, which suggests that the verb *niman* still has rather strong competitiveness during this period. Yet, contrary to the high frequency of *ȝeuen,* the instances of *syllan* ('give') are lowered to only 5% among these verb forms. This means that *syllan* has lost its status as a verb of either giving or taking at least based on this text. Then in the beginning of the 14th century, in *The Metrical Chronicle of Robert of Gloucester* (hereafter, The Metrical Chronicle) which still maintained some aspects of OE, the status of *syllan* as a giving verb was completely in a loss. Therefore, the competition at this point was among *tacan, niman* and *give* as illustrated in Table 4.

Table 4. Distribution of four verbs in the metrical chronicle (14th century)

Taking and seizing verbs (14th century)	Tokens	Token percentages (%) (each against all five)
niman ('take')	194	53
Ʒiuen ('give')	81	22
tacan ('deliver/give')	***51***	15
bitacan ('give up')	***21***	5
biniman ('take away')	20	5
Total	367	100

As shown in Table 4, the verb *niman* meaning 'to take' is still shown to have competitiveness in the 14th-century chronicle. However, the form *tacan* indicates not enough competitiveness as a form meaning 'giving' or 'taking' against *niman* and *Ʒiuen*. When the verb *tacan* occurs in a ditransitive construction, the meaning is 'to give', for example, "*hii ne ssolde noʒt wurne* • Ac ***toke****him ostage god* • *at is owe wille* ('he should not wean but **gave** them good hostage at his own will.')" [17]. The form *tacan* takes both an IO and a DO. This means the verb *tacan* still occurs in the ditransitive construction and denotes the sense of giving. Besides the form *bitacan*, the word *tacan* has at least three other varied forms: *of take* (take way), *on take* (to behave) and *up take* (to deliver up). In contrast, *niman* has only one suffixed form: *biniman* (take way). And *niman* still occurs in the 15th-century manuscript, meaning 'to take'.

However, things changed significantly in the 15th century, which is shown in the text of *The Brut* (first published in 1480). The word *niman* still existed but in a quite limited way, while the verb *tacan* showed more flexibility in entering more types of constructions. In the 15th-century chronicle, the token frequency of *niman/biniman* is 19 and that of *tacan/bitacan* reaches 980. Aside from showing more token frequencies, the latter verb also narrowed its meaning down to 'take', losing the senses of 'to deliver' or 'to give' to the form *give.* It has become a purely transitive verb, participating in two-participant structures correspondingly. The meanings between *give* and *take* became completely distinct from each other, which means they each had their own semantic space from that period on.

In general, Fig. 2 illustrates the competing trend of the two forms between *tacan/bitacan* and *niman/geniman.*

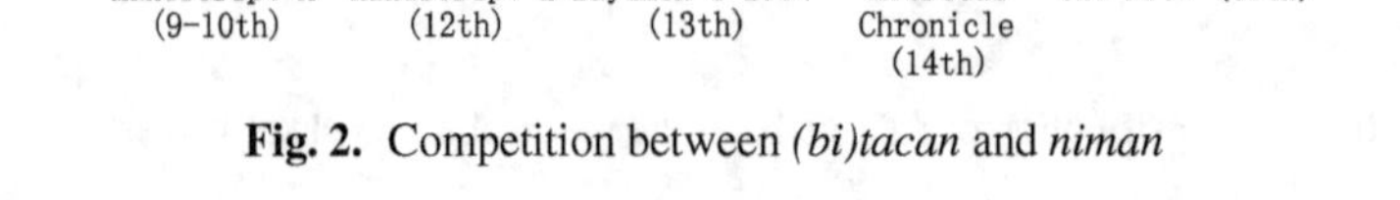

Fig. 2. Competition between *(bi)tacan* and *niman*

Figure 2 indicates that the competition between *tacan* and *niman* bears much resemblance to the one between *gyfan* and *syllan*; that is, the winning competitor at a point develops more varied suffixed forms than the losing competitor. Similar to *gyfan* and *syllan*, the verb *tacan* also has several prefixed forms carrying very different meanings, including *atacan* ('to overtake, catch'), *bitacan* ('to take'), *outtacan* ('to take out'), *ouertacan* ('to overtake') and *withtacan* ('to reprove') which help maintain the competitiveness of the verb stem. In the 9–13th centuries, with the emergence and the growing use of the verb *tacan/bitacan*, the older form *niman/geniman* gradually lost its status as the main verb denoting the sense of 'take'. In the 13–14th century, which is also the time when *gyfan* took the place of *syllan*, becoming the primary verb of giving; the meaning of the verb *tacan* gradually switched from an alternative verb of giving to a candidate verb for taking, hence competing against the more ancient verb *niman.* Then the verb *niman* almost went extinct in the 15th century.

What seems different from the competition between *tacan* against *niman* and that between *gyfan* against *syllan* is that the competition in the former case squeezed out the word *niman* forever; the latter left a space for *syllan.* As mentioned above, one of the reasons for *syllan* to survive in English lexicon has to do with the word's etymology, which includes the meaning 'sell' and which is not a space filled by *gyfan*; therefore, the word *syllan* filled that gap and managed to maintain its status. In contrast, the old form *niman* had one single sense, which was later taken away by the word *tacan*, and this may have caused the eventual extinction of the form *niman.*

3 Conclusion

In general, this paper has delved into the etymology and diachronic evolution of the forms of *give* in English. We have charted the competition patterns among the verbs of giving in chronicles of nearly nine centuries. The patterns reveal the different correlations between verb meanings and syntactic behaviors.

Our investigation of the chronicles and historical records indicates that the most representative examples of 'give' verbs were: *gyfan* (early form of *give*), *syllan* (early form of *sell*), *tacan* (early form of *take*), and *niman.* The verb *gyfan*, became the unique verb of giving starting around the 13–14th century. In the competition between *gyfan* and other verbs, its varied forms with diversified meanings play an important role in building the competitiveness of the verb. Also, it appears to be the case that in OE and ME, the set of 'give' verbs were almost all of purely Germanic origin. These four verbs all have the meaning of taking and holding in their etymologies, which is in line with what was suggested in previous studies [3–5]. In other words, the verb *give* is highly possible an originally two-participant verb of taking and holding, which later turns to a verb of giving participating in the dative alternation. Therefore, in our future study, one of the focuses will be on finding earlier evidence for the verb to occur as a sheer transitive verb meaning taking and holding.

References

1. Kittilä, S.: The anomaly of the verb 'give' explained by its high (formal and semantic) transitivity. Linguistics **44**, 569–612 (2006)
2. Bresnan, J., Nikitina, T.: On the gradience of the dative alternation. Stanford University (2003)
3. Zhang, M.: Revisiting the alignment typology of ditransitive constructions in Chinese dialects. Bull. Chin. Linguist. **4**(2), 87–270 (2011)
4. Chappell, H., Li, M., Peyraube, A.: Chinese linguistics and typology: the state of the art. Linguist. Typology **11**, 187–211 (2007)
5. Chappell, H., Peyraube, A.: The diachronic syntax of ditransitive constructions from Archaic Chinese to early Southern Min (Sinitic). Paper presented at the Conference on Ditransitive Constructions, Max Planck Institute for Evolutionary Anthropology, Leipzig (Germany), 23–25 November (2007)

6. Orel, V.: A Handbook of Germanic Etymology. Brill, Leiden (2003). https://archive.org/search.php?query=Germanic%20Etymology
7. Bybee, J., Hopper, P. (eds.): Frequency and the Emergence of Linguistic Structure. John Benjamins, Amsterdam (2001)
8. Langacker, R.: A dynamic usage-based model. In: Barlow, M., Kemmer, S. (eds.) Usage-Based Models of Language, pp. 1–63. CSLI Publications, Stanford (2000)
9. Bybee, J.: Frequency of Use and the Organization of Language. Oxford University Press, New York (2007)
10. Burnley, D.: The History of the English Language: A Sourcebook. Longman, London (1992)
11. Attenborough, L. (trans. and ed.): The Laws of the Earliest English Kings. Cambridge University Press (1922)
12. Miller, T. (ed.): The Old English Version of Bede's Ecclesiastical History of the English People with a Translation, vol. 3. Early English Text Society, London (1890)
13. Fox, S.: King Alfred's Anglo-Saxon Version of Boethius with a Literal English Translation. Covent Garden, London (1864)
14. Pintzuk, S., Kroch, A.: The rightward movement of complements and adjuncts in the old English of beowulf. Lang. Var. Chang. **1**, 115–143 (1989)
15. Gomme, E.E.C.: The Anglos-Saxon Chronicle: Newly Translated. George Bell and Sons, London (1909)
16. Madden, F. (ed.): Layamon's Brut or Chronicle of Britain, a Poetical Semi-Saxon Paraphrase of the Brut of Wace, vol. 3. The Society of Antiquaries of London, London (1847)
17. Wright, W.A.: The Metrical Chronicle of Robert of Gloucester, vol. 1, vol. 2. Eyre and Spottiswoode, London (1887)
18. Brie, F.W.D. (ed.): The Brut or The Chronicles of England, vol. 2. The Early English Text Society, London (1906)
19. Rappaport Hovav, M., Levin, B.: The English dative alternation: the case for verb sensitivity. J. Linguist. **44**(01), 129–167 (2008)
20. Lyu, G., Gao, Y.: A diachronic study of structure patterns of ditransitive verbs. In: Hong, J.-F., Zhang, Y., Liu, P. (eds.) CLSW 2019. LNCS (LNAI), vol. 11831, pp. 265–274. Springer, Cham (2020). https://doi.org/10.1007/978-3-030-38189-9_28
21. Goldberg, A.E.: Constructions: A Construction Grammar Approach to Argument Structure. Chicago University Press, Chicago (1995)
22. Indo-European Language Revival Association: An Etymological Dictionary of the Proto-Indo-European Language: A Revised Edition of Julius Pokorny's Indogermanisches Etymologisches Wörterbuch. Dnghu Association (2007) http://dnghu.org
23. Wedgewood, H.: A Dictionary of English Etymology, 2nd edn. Trübner & Co., Paternoster Row (1872)
24. Kortlandt, F.: The Germanic fifth class of strong verbs. North-Western European Language Evolution **19**, 101–107 (1992). Reprinted in Kortlandt, F.: Studies in Germanic, Indo-European and Indo-Uralic, pp. 211–214. Rodopi, Amsterdam (2010)
25. Sweet, H. (ed.): The Oldest English Texts. EETS, London (1885)
26. Poussa, P.: The evolution of early standard English: the creolization hypothesis. Studia Anglica Posnaniensia **14**, 69–85 (1982)
27. Bosworth, J.: A Dictionary of the Anglo-Saxon Language. Longman, London (1838)

The Memetic Transmission of Chinese *yi*-item Idioms

Mengya Shang[1], Liangbin Tian[2], Junling Wu[3], and Huibin Zhuang[4](✉)

[1] Shangqiu University, Shangqiu, China
[2] Henan University, Kaifeng, China
[3] Shandong University, Jinan, China
[4] School of Culture and Communication, Shandong University, Weihai 264209, P. R. China
huibinzhuang@aliyun.com

Abstract. Simple as it is, the numeral *yi* (one), an indispensable constituent element in Chinese idioms, demonstrates many meanings. Taking memetics into consideration, this paper studies the transmission mechanism of Chinese idioms whose initial character are *yi* from linguistic memetic perspective. It is found that Chinese *yi*-item idioms can be classified into strong linguistic memes and weak linguistic memes. Furthermore, Chinese *yi*-item idioms have their transmission mechanism. Specifically speaking, their meanings can be divided into two groups, that is, one group with the same content being inherited in various forms and the other group with the same content being inherited directly, while their forms pertain to the mode of the identical set pattern being followed by different contents. The findings are expected to help people explore forms and meanings of Chinese *yi*-item idioms from different perspectives.

Keywords: Chinese *yi*-item idioms · Memes · Linguistic memetics · Transmission

1 Introduction

In the numerals, *yi* (one) is the smallest positive integer but the biggest number in philosophy [1]. Chinese idioms whose initial character is *yi*, among which *yi* have various meanings, mainly include the literal meaning: "one", like "一呼百应" (*yī hū bǎi yìng* "hundreds respond to one call") and the functional meanings: "a few, small, low", for instance, "一鳞半爪" (*yī lín bàn zhǎo* "fragments"); "same, concentrated", for example, "一心一意" (*yī xīn yī yì* "undivided attention"), "full, completeness", for example, "一反常态" (*yī fǎn cháng tài* "be not one's usual self") and "no change", like "一去不返" (*yī qù bù fǎn* "be gone forever"). With regard to the collocations, *yi* can exist before nouns, verbs and adjectives. For instance, "一心一意" ((lit. one heart one mind), "一去不返" (lit. one leave not return) and "一寒如此" (*yī hán rú cǐ*, lit. one cold thus "be extremely impoverished and frustrated"). Consequently, *yi* is an indispensable constituent element in Chinese idioms. So far, however, there has been little discussion about Chinese idioms with *yi* as the initial character.

M. Liu et al. (Eds.): CLSW 2020, LNAI 12278, pp. 137–147, 2021.
https://doi.org/10.1007/978-3-030-81197-6_13

In 1976, Richard Dawkins, the zoologist of the University of Oxford, published *The Selfish Gene.* Dawkins regarded a meme as "a noun that conveys the idea of a unit of cultural transmission, or a unit of imitation" [2]. In 1999, *The Meme Machine* published by Dr. Susan Blackmore established the disciplinary position of memes. In her masterpiece, Blackmore [3] emphasized that the core of memes is imitation, and she believed that imitation includes "any kind of copying of ideas and behaviour from one person to another". Therefore, in a broad sense, a meme is described as anything that is passed on by imitation. For surviving, memes need to be replicated and transmitted continuously. There are two ways to do so: "the same core content being inherited in various forms and the identical set pattern being followed by different contents" [4]. The former is of "memetic genotype" [5], and the latter is of "memetic phenotype" [5]. Meanwhile, when memes are copied and transmitted, they will experience four stages. They are "assimilation, retention, expression and transmission" [6]. Specifically, the existent memes must be first noticed, understood, and accepted by the host. Next, in order to be spread widely, memes must be transformed from abstract patterns into concrete entities that the host can perceive. Then, the expression of memes requires tangible vehicles or media as carriers, which can be books, photographs, CDs, etc. Finally, memes are usually spread from one host to one or more potential hosts. Beyond this, Dawkins [2] proposed three qualities that made for high survival value among memes: "longevity, fecundity, and copying-fidelity". Longevity refers to the length of time that memes last in human brains or material carriers. Fecundity means the replication speed of memes, which is the most important factor. The more rapidly the memes replicate, the broader the spread range is and the more the number of transmissions are. Copying-fidelity indicates that the more faithful the meme is, the more likely the original version will be retained. Furthermore, according to three qualities of memes: copying-fidelity, fecundity and longevity, memes can be divided into strong memes and weak memes due to their transmission strength [5]. Strong memes refer to those that are replicated with high faithfulness, copied with more chances, spread in a wide range, and live much longer. Nevertheless, weak memes do the opposite. Moreover, strong memes and weak memes do not have absolutely clear but relatively distinct delimitation [7].

As is discussed above, memes need various carriers to represent determinate information. Language is a kind of carriers of memes, and in the memetic view, the language itself is a kind of memes which can be embodied in letters, words, phrases, clauses, paragraphs and texts. "In memetical view, language memes reveal the law of speech spread and language replication" [5] from a new angle. Since 2012, He Ziran led his team to establish a series of conceptions on linguistic memes. They proposed that "a linguistic meme is a representation of information which is widely spread with the intention of the meme's host, making use of linguistic structure and by means of repeating and analogizing" [8]. Thus, this paper attempts to explore the transmission mechanism of Chinese *yi*-item idioms with *yi* as an initial character through linguistic memetics.

2 Previous Studies on Chinese *yi*-item Idioms

In the literature, Chinese idioms with *yi* 'one' have drawn much attention from scholars, such as Liu Shu'e [9], Yang Lijun [10], Zhang Jinxia [11], and Wei Zaijiang [12] who have discussed their meanings, functions and collocations respectively.

Liu Shu'e [9] took the use of *yi* in Chinese idioms as numeral (cardinal numbers and ordinal numbers), adverbial and adjective and analyzed their grammatical meanings and functions. She put forwards that *yi* has plentiful senses as a constituent element in Chinese idioms. After researching the Chinese idioms of *yi* X *bu* Y, Yang Lijun [10] found that "*yi* X *bu* Y" can express the grammatical meaning of certain quantity"or express emphasis. The significance of idioms depends on the lexical categories and the meanings of "X" and "Y", and it is also relative to the meaning of a small quantity of *yi* and the negative meaning of *bu*. Zhang Jinxia [11] discussed the numeral collocations of *yi* in Chinese idioms. She found that *yi* can be matched with *ban* (half), *yi* (one), *er* (two), *san* (three), *jiu* (nine), *shi* (ten), *bai* (hundred), *qian* (thousand) and *wan* (ten thousand) to form some fixed patterns. The first property of those collocations is that they cannot be understood separately. The second one is that most of them hold extended meanings or exaggerated meanings. Wei Zaijiang [12] analyzed cognitive attributes of the Chinese polar languages — the *yi* polar idiom constructions from the perspective of cognitive linguistics. He argued that the production and use of polar language constructions are influenced by a variety of internal and external factors, such as human experience, cognition, semantics, pragmatics and so on. Obviously, the previous studies have explored Chinese *yi*-item idioms from various points of view but few studies can be found to explore *yi*-item idioms from linguistic memetic perspective, which is the aim of the current study.

3 Research Objects

The data used in this paper mainly involves the four-character Chinese idioms with *yi* (one) as the initial character, which can be abbreviated to "Chinese *yi*-item idioms". *Chinese Idiom Dictionary* [13] includes 357 entries of Chinese *yi*-item idioms with four characters, and *A Dictionary on Philological Studies of Chinese Idioms* [14] includes totally 251 entries of Chinese *yi*-item idioms, plus 36 entries which are not four-character idioms. After excluding 174 entries of identical idioms and 36 entries that are not four-character ones, the present study takes 398 entries of Chinese *yi*-item idioms as the research objects. The English interpretations of idioms in this paper mainly come from *A Chinese-English Dictionary of Chinese Idioms and Phrases* [15] and *A Modern Chinese-English Dictionary* [16] published by Foreign Language Teaching and Research Press. With regard to methodology, the qualitative method is mainly employed in this paper. Meanwhile, the study also adopts the method of observation to analyze the forms and meanings of Chinese *yi*-item idioms.

4 Linguistic Memetic Studies on Chinese *yi*-item Idioms

This part analyzes linguistic memes on Chinese idioms with *yi* as the initial character in depth.

4.1 Language Memes on Chinese *yi*-item Idioms

In *The Meme Machine*, Blackmore illustrated a meme with an example "when you hear a story and pass on the gist to someone else, you have copied a meme."[3] Here, the core part of becoming a meme is imitation. Imitation is not simply a kind of repetition but a kind of replication fused with the hosts' ideas, perception, experience and background. However, not everything can be memes. If representations are to become memes, "they must meet three demands: firstly, they should have determinate content; secondly, they must be replicable; thirdly, they must be able to exert their effects if they wish to be selected" [17].

Chinese idioms are the crystallization of experience and wisdom of Chinese people, and are believed to be extracted from their daily life since the ancient times. As a result, they contain profound culture and have spread for a long time. Undoubtedly, Chinese *yi*-item idioms, as one of the typical representations, must convey certain cultural information. For instance, "一清二楚" (*yī qīng èr chǔ* "be perfectly clear") means that somebody is perfectly clear with something. What is more, replication equals to imitation. To a certain extent, imitation can be embodied by the fecundity "whose a rough measure could be obtained by counting the number of times it is referred to in successive years" [2]. Thus, by searching 398 Chinese *yi*-item idioms in Modern Chinese Corpus of CCL, the author obtains the frequency of every *yi*-item idiom. The frequency of "一清二楚" is 811, which shows that the idiom can be replicable. Let us look at an example.

(1) 我不能和她闹个一清二楚, 我怕对妈有什么不好, 尽管妈已经不在了。(当代, 张洁《世界上最疼我的那个人去了》)

This sentence is abstracted from the novel written by the contemporary famous writer Zhang Jie. The sentence means that "I cannot argue with her perfectly clearly, because I am afraid that it would be somewhat not good for my mother even though my mother has been gone". In the novel, the idiom "一清二楚" is adopted to remind the host not to think back a lot, for fear of disturbing her mother's rest, which expresses Zhang Jie's deeply missing for her mother. When memes are copied and transmitted, they will pass four stages: assimilation, retention, expression and transmission. When Zhang Jie used "一清二楚" to convey her emotion, she had accepted the word and included it in her cognitive system. Otherwise, she could not transmit the idiom in a written form. Thus, the idiom has had an influence on the writer so that it could be spread. As a consequence, the idiom "一清二楚" has determinate content. It has been replicated and it could exert an influence on the user. Therefore, "一清二楚" is a meme.

In a word, linguistic meme is referred to as "a representation of information which is widely spread with the intention of the meme's host, making use of linguistic structure and by means of repeating and analogizing" [7]. Thus, the host can be everyone that passes on an idiom. In this way, when 811 "一清二楚" are spread from one person to another in different situations, they must carry with their hosts' feelings and ideas. Therefore, "一清二楚" is absolutely a linguistic meme.

Aside from the afore-mentioned, and according to the frequency searched out in CCL and the analysis made by the author, Chinese *yi*-item idioms can also be divided into strong linguistic memes and weak linguistic memes. Strong linguistic memes refer to representations that have high copying-fidelity, high rate of replication and long survival-time in which fecundity is a crucial element, while weak linguistic memes are the reverse. Through analyzing the data, the author finds out Chinese *yi*-item idioms with higher frequency have a powerful transmission ability. Some characteristics of this kind of idioms are as follows. Firstly, their structures are stable, that is, their forms and meanings do not usually vary in different contexts; secondly, they are often used in written and spoken languages; thirdly, their meanings are simple, which can be deduced from their constituent elements. However, Chinese *yi*-item idioms with lower frequency carry much more attributes of ancient Chinese so that they lack practicability in modern times. Their forms are classic and they have double meanings. In most cases, people cannot understand their meanings from their literal expressions. They are elegant, complicated and inaccessible, thus they cannot be copied easily. As a result, they are weak linguistic memes. Let us take "一如既往" (*yī rú jì wǎng* "just as in the past") and "一副急泪" (*yī fù jí lèi* "a string of tears for urgency") as examples.

(2) 人类在解决环境与发展问题上面临的困难还很多,道路还很漫长。中国将一如既往地与世界各国同舟共济、携手合作、积极行动, 为保护人类生存的地球环境和共同繁荣而奋斗。(《中国的环境保护》, 选自《中国政府白皮书》)

The frequency of "一如既往" is 3146 while it is zero for "一副急泪" in CCL. As a matter of fact, "一如既往" has been widely used in works, newspapers, and documents in modern times. In this case, "一如既往" is a strong linguistic meme, while "一副急泪" is a weak linguistic meme undoubtedly.

Beyond this, what we need to notice is that the strength and weakness of linguistic memes are a kind of relative concept without a strict standard of quantity. [7] Making a distinction is to "let people make clear that linguistic memes have clear differences in fidelity, fecundity and longevity, which are the basis of linguistic use and variation" [7].

4.2 Transmission Modes of Memes on Chinese *yi*-item Idioms

The transmission modes of memes on Chinese *yi*-item idioms include two facets, namely meaning and form. We will talk about them one by one in the following sub-sections.

4.2.1 Transmission Modes of Meanings

The genotypic mode of linguistic memes reveals that the essential information of memes does not vary but is adopted in different situations or in different ways. According to linguistic memes, they can be further classified into two types, namely the same content being inherited in various forms and the same content being inherited directly.

Here are some examples taken to elaborate. For instance, the same content "being defeated" is expressed by two idioms "一败如水" (*yī bài rú shuǐ* literally "the army overwhelmed likes water on the ground") and "一败涂地" (*yī bài tú dì* literally "be ready to dash one's brains out against the ground once defeated"). Because the same core information is represented by different idioms in every group, they belong to the same content being inherited in various forms of memetic genotype. The second is that the same content is inherited directly, such as, "一般见识" (*yī bān jiàn shi* "lower oneself to the same level as sb."), "一悲一喜" (*yī bēi yī xǐ* "be sad and happy"), "一本万利" (*yī běn wàn lì* "make big profits with a small capital"), "一本正经" (*yī běn zhèng jīng* "in all seriousness") and so on. Because their meanings are spread by themselves directly, they pertain to the type of the same content being inherited directly of the memetic genotype.

Therefore, as for the transmission of meanings of Chinese *yi*-item idioms with *yi* as the initial character, they can be divided into two types, namely the same content being inherited in various forms and the same content being inherited directly.

4.2.2 Transmission Mode of Forms

Another transmission mode of linguistic memes is the identical set pattern followed by different contents. That is memetic phenotype [7]. In other words, some information maintains the general forms, but the specific contents are different. With regard to Chinese *yi*-item idioms, when some of them are transmitted, they constitute different constructions so that the new *yi*-item idioms can imitate and enrich the structures of the original ones. Chinese *yi*-item idioms can be classified into three types and different idioms satisfy the demand of different structures.

The first type indicates that the third character is same, There are five kinds of such idioms: numbers that include *ban* (half), *yi* (one), *er* (two), *san* (three), *si* (four), *jiu* (none), *shi* (ten), *bai* (hundred), *qian* (thousand) and *wan* (ten thousand), negative words that cover *bu* (no), *wu* (not have), *nan* (cannot) and *mo* (do not) as well as *zhi* (a structural auxiliary word, equaling to "of" in English), *er* (a functional word expressing a kind of associated relations) and *ru* (like). The second type shows that the fixed part appears in the second character, like *wu* (not have), *yan* (discourse), *pian* (piece), *dai* (generation), *bi* (pen) and *biao* (look). The third type involves some specific cases. That is to say, in the process of transmitting, the *yi*-item idioms do not change their structures or forms, yet their Chinese characters or contents are replaced by others, and the new characters produce new or similar meanings. More specific description is as follows.

The first type includes five kinds of idioms. First, the third character is a number. The structure of "*yi* X *yi* Y" has a larger number, for example, "一草一木" (*yī cǎo yī mù* literally "every tree and grass", figuratively "every small thing") and "一步一趋" (*yī bù yī qū* literally "follow step by step", figuratively "imitate and follow others"). Second, the third character includes three patterns, namely "*yi* X *er* Y", "*yi* X *liang* Y" and "*yi* X *shuang* Y", like "一箭双雕" (*yī jiàn shuāng diāo* "kill two birds with one stone"). What's more, other seven patterns include "*yi* X *san* Y", for example, "一日三秋" (*yī rì sān qiū* literally "one day seems three years", deeply "miss sb. very much"), "*yi* X *ban* Y", for example, "一官半职" (*yī guān bàn zhí* "an unimportant official post"), "*yi* X *jiu* Y", for instance, "一言九鼎" (*yī yán jiǔ dǐng* literally "one word is as heavy as nine tripods", figuratively "one's words carry great weight"), "*yi* X *shi* Y", for example "一目十行" (*yī mù shí háng* literally "learn ten lines at a glance", figuratively "read rapidly"), "*yi* X *bai* Y", for instance, "一呼百应" (*yī hū bǎi yìng* "hundreds respond to a single call"), "*yi* X *qian* Y" taking an example with "一刻千金" (*yī kè qiān jīn* literally "one half-hour is worth a thousand pieces of gold", figuratively "every minute is precious"), "*yi* X *wan* Y" taking "一碧万顷" (*yī bì wàn qǐng* "the dark green reaches far beyond the horizon") as an example. Another idiom is "一谦四益" (*yī qiān sì yì* "modesty benefits people a lot"). In modern times, as a weak meme, "一谦四益" has gradually disappeared. Nevertheless, in respect to its form, it imitates the structure of the first character being *yi* and the third character being *si*, so "一谦四益" belongs to the first type. To sum up, all patterns mentioned above constitute the type that the third character is a numeral. Therefore, they are the phenotype of linguistic memes. Second, some idioms form the pattern of "*yi* X *zhi* Y", but they do not necessarily have different meanings because some of them have the same information representation. Third, the third character can be a negative word, and the structures contain "*yi* X *bu* Y", "*yi* X *wu* Y", "*yi* X *nan* Y" and "*yi* X *mo* Y", for example, "一尘不染" (*yī chén bù rǎn* literally "be not stained with a particle of dust", figuratively "maintain one's original pure character"). The last two kinds include the one with the third characters being *er* and the other with the third characters being *ru*. The structures are "*yi* X *er* Y" and "*yi* X *ru* Y" respectively. The first structure means a kind of associated relations, for instance, "一蹴而就" (*yī cù ér jiù* literally "accomplish in one move", deeply "things are easy to do"). The structure of "*yi* X *ru* Y" commonly expresses the rhetoric of metaphor, for example, "一廉如水" (*yī lián rú shuǐ* literally "an official being like water", figuratively "be a white handed official").

The second type is related to the second character, which contains six kinds of *yi*-item idioms. That is to say, the second characters are *wu*, *yan*, *pian*, *dai*, *bi* and *biao* and the patterns are "*yi wu* X Y", "*yi yan* X Y", "*yi pian* X Y", "*yi dai* X Y", "*yi bi* X Y" and "*yi biao* X Y" respectively. For instance, "一言难尽" (*yī yán nán jìn*) literally depicts that "it's a long story not to be told in one sentence", but it deeply says that "the thing is tortuous and complicated"; "一片至诚" (*yī piàn zhì chéng*) means to "be earnest and sincere", in which *yi pian* literally represents the "quantity", while it actually expresses "the degree of modifying abstract or concrete things". What is more, the literal meaning of "一代风流" (*yī dài fēng liú*) is the people talented in letters and unconventional in life style, and later, its meaning is that the people opening up the prevailing custom are admired by others in an era. Besides, "一笔勾销" and "一表人才" belong to the structures of "*yi bi* X Y" and "*yi biao* X Y" respectively. Therefore, this kind of *yi*-item idioms can be classified as a phenotype of linguistic memes.

The third type involves some specific cases, for example, "一脉相承" (*yī mài xiāng chéng*), "一脉相传" (*yī mài xiāng chuán*) and "一脉相通" (*yī mài xiāng tōng*). Their meanings are "in the same strain", but their first two characters are the same, that is, forming the structure of "*yi mai* X Y". Although the senses of the first two characters are the same, the other two characters are different. Hence, they should be regarded as the memetic phenotype. In fact, in this type, most of them have the same content, but their literal Chinese characters show the diverse ones. Thus, they belong to the mode of different contents in same structures. Other examples for such kind of idioms include "一枕黄粱" (*yī zhěn huáng liáng*) and "一枕黑甜" (*yī zhěn hēi tián*). "一枕" depicts "sleep". "一枕黄粱" is used to "have a vain dream", and "一枕黑甜" portrays "sleeping soundly". The two idioms apply the structure of "*yi zhen* X Y" to form a superficial pattern, but they have distinctive contents, which can be thought of as the mode of memetic phenotype.

Therefore, as for the transmission of forms of Chinese *yi*-item idioms with *yi* as its first character, they belong to the mode of the identical set pattern followed by different contents.

5 Motivations of the Transmission of Chinese *yi*-item Idioms

Until now, we have presented a detailed description of Chinese *yi*-item Idioms, concerning its nature of memes, transmission mechanism, as well as the classification according to forms and meanings. However, another highly-relevant question occurs to us, that is, if Chinese *yi*-item idioms are linguistic memes in nature, then what motivate the transmission so that they can be widely spread and lead to their memetic nature ultimately?

For this question, Chen & He's research on "*Chi* + NP_{-f}" could provide us with some beneficial guidance [18]. More specifically, Chen and He [18] believed that the reasons why "*Chi* + NP_{-f}" has the potential to become a strong linguistic meme are complex, but generally speaking, the main reasons are possibly from social practice, linguistic features, and pragmatic requirements. We believe that the transmission of Chinese *yi*-item idioms also has the similar motivations and in the next section we will discuss the question mentioned above from those three facets.

5.1 Social Motivation

Xie and He [19] pointed out that linguistic meme is the result of social practice and it usually has close relation with certain social event. For example, owing to the popularity of western buffet, on some occasions in China, people are inclined to adopt such way of treating guests, such as the dinner of a meeting or a conference. Moreover, it is such social practices (or new fashion) that have led to the production and transmission of the phrase "吃自助餐" (*chī zì zhù cān* "have a buffet"). In the same way, Chinese *yi*-item idioms have the similar feature, for example, "一针一线" (*yì zhēn yī xiàn* "one needle and one piece of thread"), "一纸空文" (*yī zhǐ kōng wén* "one piece of paper without words"), etc. As we can see, it is the invention and popularization of "needle, thread" and "paper" that make "一针一线" and "一纸空文" be created and spread. In this respect, social development should be the foundation of their transmission and social practices finally motivate the creation and transmission of these idioms. Distin [17] once put forward that the success of a meme depends on the brain structure, the meme's stability as well as its penetrability. Because of the high frequency and popularization of those Chinese *yi*-item idioms, they obviously have acquired such penetrability. For example:

(3) 美国的目的是"一石二鸟": 一是稳住局势, 并为巴以重开谈判提供依据; 二是要扶持一个"负责任"的巴新政权, 在中东树立"民主典范"。(BCC)

(4) 过分追求出口贸易顺差来实现陶瓷出口大国和强国的梦想, 其实是一枕黄粱, 它只会让我们离真正的陶瓷出口大国和强国的终极目标越走越远。(BCC)

In the above examples, "一石二鸟" (*yì shí èr niǎo* "one stone shot two birds") and "一枕黄粱" (*yì zhěn huáng liáng* "a daydream") are all produced in ancient times, but now they are still used in daily life and are employed in some new discourses of modern times. According to Li and He [8], people in society require contextualization in language use and language understanding, and the employment of "一石二鸟" and "一枕黄粱" in (3) and (4) respectively support this point. Specifically speaking, the change of social context can contribute to the variation of linguistic meme in the process of copying and transmission, for example, "一石二鸟" may possibly mean "two birds are shot by one stone at a time" originally, but in (3), the original meaning is evidently diserted and there is a new meaning in the new context, namely "America's policy has two goals: one is to control the situation, and the other is to help build a new government that is responsible". "一枕黄粱" in (4) also has the similar feature.

5.2 Linguistic Motivation

Distin [10] believed that people usually choose those memes that can take advantage of the contemporary cultural environment, and when in a new social practice (or some new things appear), the corresponding linguistic forms are required to represent those new social practices or new things [18]. In this case, Distin [10] further stated that if the copying process (or transmission process) starts from the preexisting linguistic forms, the process will be more effective than those processes that create ones and then transmit. After all, expressions depending on preexisting linguistic forms are easier to be recognized, memorized and employed [20]. Chen and He [18] found that the production and transmission of "*Chi* + NP_{-f}" fully embody this feature mentioned above. Obviously, "*Chi* + NP_{-f}" imitate the regular syntactic order, namely "V + NP", which will save people's energy to create a new pattern to represent a new form and master the new pattern, as people have the preexisting knowledge of "V + NP".

With regard to Chinese *yi*-item idioms, we find that they also have such characteristic. As we have discussed in 4.2.2, Chinese *yi*-item idioms can be divided into different types according to their transmission mode of forms. These types mainly include Chinese *yi*-item idioms whose third character is the same, such as "*yi…ban…*" "*yi…yi…*" "*yi…er…*" "*yi…si…*", etc. Specifically, take "*yi…yi…*" as an example, we have "一草一木" "一步一趋" "一心一意", etc. Another type is related to the second character, that is, the second character may be the same or different and the latter two characters are usually different. For example, we have "一代风流", "一代天骄" (*yí dài tiān jiāo* "a person who is quite outstanding"), etc., which belong to the structure "*yi dai* X Y". These examples show that many Chinese *yi*-item idioms have the preexisting structure to imitate. Moreover, it is these preexisting linguistic forms that linguistic memes could be recognized, memorized and copied easily [18].

5.3 Pragmatic Motivation

Yuan [21] proposed that pursuing economy is the tendency of language development diachronically and people also observe the "principle of economy" when they choose language synchronically, as economy principle accords with people's pursuit of energy saving. Chen and He [18] also mentioned that the production and transmission of linguistic memes may have two pragmatic motivations. One is to satisfy communicative and representational needs and the other is due to the goal of energy saving. In a word, language is an information transmission system, and it is natural for people to adopt economic principle in communication. More specifically, people would like to spend the least encoding effort to obtain the richest and most effective information in a communicative process.

Chinese idioms are the language essence of China, and they can, to a great extent, strengthen the expressive power of Chinese in daily life [22]. To be more specific, Chinese idioms are usually concise in structure but rich in meaning. For example, there are a lot of Chinese idioms, such as "望梅止渴" (*wàng méi zhǐ kě* "people won't feel thirsty when they think about the smoked plum"), which is the condensation of a historical story. As for the Chinese *yi*-item idioms in this paper, though most of them may not come from the condensation of a story or historical events, all of them apparently carry more information with succinct structure than those non-idiomatic expressions. For instance:

(5) a. 他冷冷一笑道: 说他跟我<u>一丘之貉</u>, 是大大地抬举了他。(BCC)
b.他冷冷一笑道: 说他跟我<u>一路人</u>, 是大大地抬举了他。
c.他冷冷一笑道: 说他跟我<u>一样是同类的坏人</u>, 是大大地抬举了他。

As we can see, the only difference among these three sentences is the employment of "一丘之貉" (*yì qiū zhī hé* "raccoon dogs of the same mountain"). However, although examples in (5) are all grammatical sentences with almost similar meaning, but it is obvious that (5a) is the most suitable one. In terms of (5b), "一路人" can be considered as the synonym of "一丘之貉", but according to our knowledge of this Chinese idiom, "一丘之貉" carries more meaning than "一路人", that is, "一丘之貉" has the extra (negative) meaning in modifying people, while "一路人" is usually used as a neutral expression without context. As for the (5c), it almost has the same meaning as (5a), because "同类的坏人" is actually the further explanation of "一丘之貉". However, in terms of the simplicity, (5c) is obviously more complex than (5a). He [4] pointed out that strong linguistic memes usually have rhetoric effect, while "一丘之貉" apparently has such effect, which is often the pursuit of expressive effect for speakers. Therefore, "一丘之貉" in the sentence is not only more economic, but also more life-like and has more expressive power.

6 Conclusion

This paper probes into the transmission mechanism of Chinese *yi*-item idioms whose initial character is *yi* from the perspective of linguistic memetics. After a careful analysis, the paper finds out that Chinese *yi*-item idioms are linguistic memes, and they can be divided into strong linguistic memes and weak linguistic memes. More importantly, the transmission modes of *yi*-item idioms concern about the transmission of form and the transmission of meaning.

In terms of the transmission of meaning, it includes two modes which are the same content being inherited in various forms, like "一败如水" and "一败涂地" and the same content being inherited directly, like "一般见识". As for the forms of *yi*-item idioms, they mainly pertain to the mode of the identical set pattern followed by different contents. There are mainly three types. The first type indicates the third character of the *yi*-idioms is similar, including numbers, negative words as well as *zhi*, *er* and *ru*. The fixed structures of numbers manly contain "*yi* X *yi* Y", "*yi* X *er* Y", "*yi* X *san* Y", "*yi* X *ban* Y", "*yi* X *jiu* Y", "*yi* X *shi* Y", "*yi* X *bai* Y", "*yi* X *qian* Y", "*yi* X *wan* Y". Moreover, the third character is a negative word, and the structures contain "*yi* X *bu* Y", "*yi* X *wu* Y", "*yi* X *nan* Y" and "*yi* X *mo* Y". Finally, the third characters are *er* and *ru*, and the structures are "*yi* X *er* Y" and "*yi* X *ru* Y" respectively. The second type shows that the fixed patterns appear in the second character. The fixed patterns include "*yi wu* X Y", "*yi yan* X Y", "*yi pian* X Y", "*yi dai* X Y", "*yi bi* X Y" and "*yi biao* X Y" respectively. The third type involves specific cases, for example, "一脉相承" (*yī mài xiāng chéng*), "一脉相传" (*yī mài xiāng chuán*) and "一脉相通" (*yī mài xiāng tōng*) have the same structure of "*yi mai* X Y". In a word, in the process of transmitting, the *yi*-item idioms do not change their structures or forms, and yet their Chinese characters or contents are replaced by others, and the new words produce new or similar meanings.

Finally, we probe into the motivations of Chinese *yi*-item idioms. Generally speaking, the motivations mainly include social motivations, linguistic motivations and pragmatic motivations, which actually are common factors for linguistic memes and thus can be applied to other linguistic memes as well.

Acknowledgments. We would like to express our sincere gratitude to the anonymous CLSW reviewers for their inspiration, guidance, and comments of the arguments in this paper. The usual disclaimers apply. The study is supported by the Social Science Foundation of Henan Province (No. 2020BYY002).

References

1. Zhang, K.: On the abstract meanings of numeral idioms. Thinking Wisdom **02**, 44–45 (1993)
2. Dawkins, R.: The Selfish Gene: The 4th, Anniversary Oxford University Press, New York (2016)
3. Blackmore, S.: The Meme Machine. Oxford University Press, Oxford (1999)
4. He, Z.: Memes in languages. Linguist. Sci. **06**, 54–64 (2005)
5. He, Z., Xie, C., Chen, X.: New Developments in Pragmatics: Relevance, Adaption and Memetics. Shanghai Educational Publishing House, Shanghai (2007)
6. Heylighen, F.: What makes a meme successful? Selection criteria for cultural evolution. In: Proceedings of the 15th International Congress on Cybemetics, pp. 423–418. de Cybernetique, Namur (1998)
7. He, Z., Xinren, C., et al.: Linguistic Memes: Theory and Application. Jinan University Press, Guangzhou (2014)
8. Li, J., Ziran, H.: Subjectivity and contextualization of linguistic memes. Foreign Lang. Res. **2**, 59–64 (2014)
9. Liu, S.: On *yi* in idioms. Chin. Teach. World **04**, 283–284 (1992)
10. Yang, L.: On the Chinese "*yi* x *bu* y" idioms. J. Hubei Univ. Philos. Soc. Sci. **06**, 75–77 (2001)
11. Zhang, J.: A study on the collocations of *yi* with other numerals in the four-character set phrases. J. Shandong Normal Univ. Humanit. Soc. Sci. **56**(02), 27–31 (2011)
12. Wei, Z.: The cognitive motivation of Chinese "one" quantity scalar polar construction. Foreign Languages in China **12**(03), 50–56 (2015)
13. Wang, T.: Chinese Idiom Dictionary (the New Edition). Shanghai Lexicographical Publishing House, Shanghai (2007)
14. Liu, J.: A Dictionary on Philological Studies of Chinese Idioms. The Commercial Press, Beijing (1989)
15. Hu, Z. (Ed.): A Chinese-English Dictionary of Chinese Idioms and Phrases. Shanghai Jiao Tong University Press, Shanghai (2010)
16. A Modern Chinese-English Dictionary. Foreign Language Teaching and Research Press, Beijing (1997)
17. Distin, K.: The Selfish Meme: A Critical Reassessment. Cambridge University Press, Cambridge (2005)
18. Chen, X., Ziran, H.: A memetic-syntactic account of "Chi+NP-f". J. Zhejiang Int. Stud. Univ. **1**, 9–16 (2018)
19. Xie, C., Ziran, H.: Some notes on language memes. Mod. Foreign Lang. **1**, 30–39+108–109 (2007)
20. Chen, X.: Extensions of the Chinese passive construction: a memetic account. East Asian Pragmatics **2**(1), 59–74 (2017)
21. Yuan, H.: A pragmatic approach to the principle of language economy. J. Guangdong Ocean Univ. **5**, 86–90 (2014)
22. Tian, L.: A cognitive grammar approach to parodic use of Chinese "V1O1+V2O2" idioms: exemplified by "*wang* X *xing tan*". J. Zhejiang Int. Stud. Univ. **06**, 41–48 (2019)

Sequencing of Grammatical Items in Textbooks Oriented Towards International Chinese Language Education

Jinghua Li[1] and Zhimin Wang[2](✉)

[1] College of Chinese Studies, Faculty of Linguistic Sciences, Beijing Language and Culture University, Beijing, China
lijinghua@blcu.edu.cn

[2] Research Institute of International Chinese Language Education, Beijing Language and Culture University, Beijing, China

Abstract. The selection and arrangement of grammar is important in International Chinese Language Education. This paper collects and sorts out some grammatical items in some authoritative Chinese textbooks. Through quantitative investigation, it is found that although different textbooks have different writing styles and reference standards for grammar, there are some commonalities despite the complexity involved. In this paper, the core parameter model and the position parameter model are designed, and the grammatical items are sequentially studied in the two dimensions of commonness and difficulty. After the adjustment of weight coefficients and interval statistics, the two models have achieved good results in determining the two-dimension sequential values.

Keywords: Core grammar · Grammar sequencing · Teaching grammar · International Chinese language education

1 Introduction

Grammar teaching is the main thread in international Chinese language education. The establishment and sequencing of grammatical items have been a core and difficult issue in Chinese language teaching. Scholars like Yang (2000), Yang (2001), Lv (2002) and Lu (2003) summarized and discussed from a theoretical perspective the principles for selecting and sequencing grammatical items in textbooks, reaching a consensus that the treatment of grammatical items in Chinese language textbooks is a science requiring a holistic consideration of the degree of difficulty of grammatical items themselves and a discrimination and sequencing of learners' acquisition patterns. At present, a variety of existing grammar teaching syllabi and textbooks have their own selection and sequencing of grammatical items, which, however, are largely based on empirical approaches without being supported by a quantitative analysis of large-sized corpora.

Extracting and sequencing commonly used grammatical items is reliant upon large-scale corpora, but the level of digitalization in existing grammatical resources is still

M. Liu et al. (Eds.): CLSW 2020, LNAI 12278, pp. 148–160, 2021.
https://doi.org/10.1007/978-3-030-81197-6_14

low and the body of the relevant research is still small. Drawing on "The Syllabus on Grammar for Teaching Chinese as a Second Language" Xing and Zhang (2005) labeled 20,000 sentences in their preliminary statistics of the frequencies of certain grammatical items in verbal and written texts. Peng, Song and Zhao (2014) established a "grammatical corpus for international Chinese language teaching" by combining the international Chinese language textbook corpus with the basic data resources for Chinese information processing. Tang, Yang and Su (2015) built a knowledge library consisting of 121 commonly used grammatical points based on the three-plane theory. These studies represent preliminary achievements made in Chinese linguistic research oriented towards practical projects, namely, the construction of fundamental resources in the field of grammar teaching. Despite their massive scales, these projects have yet to involve such specific tasks as model design for extraction and sequencing of commonly used grammatical items.

There have been scholarly achievements in the extraction of commonly used words and model design. For instance, the *Frequency Dictionary of Contemporary Chinese* utilized, for the first time, word frequencies and their distributions as a criterion for measuring commonly used words. Sun (2000) proposed that word frequency and dispersion coefficient should also be incorporated into the consideration of the degree of word usage, and the most crucial issue is to design an appropriate dispersion coefficient. Su (2007) proposed a frequency grade calculation method applicable to multiple corpora, so that the generalities of different sizes of corpora can be compared after transforming frequency into frequency grade. Wang and Yu (2019) realized an update function from old to new words in large-scale corpora by establishing a criterion of continuous distribution of statistical times. The above studies and extraction approaches provide a good path for the study of this paper.

As of today, substantial achievements have been made in developing international Chinese language textbooks. The growing number of an already large body of new textbooks represents a consensual thinking of the front-line teachers and experts on selection of grammatical items, as well as a convergence of empirical achievements. Developing a model to objectively presenting expert experience in sequencing by drawing on reasonable factors pertaining to selection and arrangement of grammatical items in a considerable number of selected Chinese language textbooks is a direction well worth exploring for research dealing with grammatical item sequencing.

Therefore, this paper proposes two models, namely, the extraction model and the sequencing model, for core grammatical items oriented for international Chinese language education by collecting and sorting out grammatical items from multiple textbooks with a view of examining grammatical item sequencing in the dimensions of commonness and difficulty.

2 Collection and Sorting-Out of Grammatical Information from Chinese Language Textbooks

A total of 10 comprehensive Chinese language textbooks are selected in this paper. Comprehensive textbooks are selected because grammar constitutes one of the most important priorities in textbooks of this nature, for which the systematic, rigorous and scientific selection and arrangement are highly emphasized. The 10 comprehensive Chinese language textbooks selected in this paper are all widely used authoritative textbooks over the past 20 years, which are presented as follows (Table 1):

Table 1. The comprehensive textbooks selected in this paper

Textbook name	Volume name	Quantity
New practical Chinese reader	1–4	4
Boya Chinese	Elementary (I, II), pre-intermediate (I, II)	4
Road to success	Lower elementary (1, 2), elementary (1, 2), upper elementary (1, 2, 3)	7
Erya Chinese	Basic Chinese: comprehensive course (Parts 1, 2), elementary Chinese: comprehensive course (Parts 1, 2)	4
Developing Chinese	I, II	2
Hanyu Jiaocheng	Volume 1 (Part 2), volume 2 (Parts 1, 2)	3
Experiencing Chinese	Parts 1, 2	2
Level Chinese	4	4
Contemporary Chinese	1–3	3
Jump high	0–3	4
Total		37

All of the above comprehensive textbooks are large-scale series textbooks oriented to long-term Chinese learners covering the elementary, intermediate and advanced levels. To ensure the completeness of the selected grammatical items, this paper follows the criterion of the "full teaching phase of a grammatical item" when intercepting grammatical items, leading to a variation in the number of volumes across selected textbooks. Be that as it may, these textbooks all include the basic grammatical items in Chinese language teaching.

The 2106 grammatical items are labeled, classified and summarized in a one-by-one fashion, and a descriptive framework of grammatical items consisting of three categories, namely, the basic information, label information and remark information, with a total of 11 properties, is established. The specific fields are as shown in Table 2.

Table 2. The descriptive framework of grammar information

	Field
Basic information	Serial number, name, explanation
Label information	Category, sub-category
Remark information	Textbook title, lesson title, volume, lesson, page, example sentence

There are 550 non-repetitive grammatical points in this framework. Specifically, 267 items only appear once, and 283 items appear twice or more times. The most frequently appearing grammatical item is "asking questions with interrogative pronouns", which appears 55 times. The frequency distribution of these 550 grammatical items is shown in the following Fig. 1.

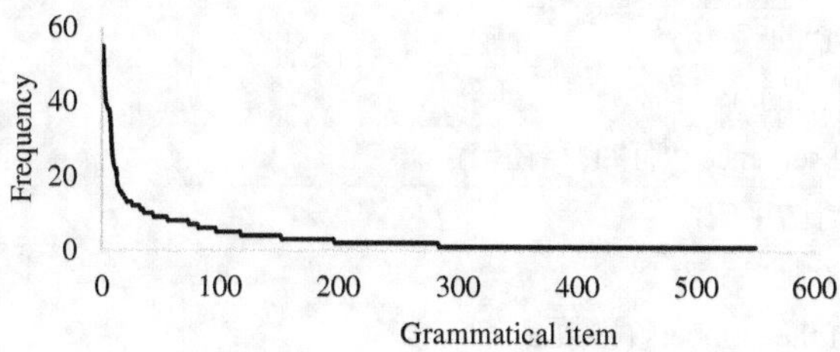

Fig. 1. The frequency distribution of grammatical items

Apart from the frequency of appearance, denoted by *F*, in all textbooks, each grammatical item also has a position value, denoted by *L*, in textbooks. Due to the differences in the volumes and chapters in each textbook, the position value *L* cannot be compared. For example, even if a certain grammatical item appears in chapter 10 in both textbooks A and B, it does not mean such a grammatical item is arranged at the same position in terms of its difficulty level as textbook A may have 30 chapters for its grammar teaching stage in comparison with the 50 chapters that textbook B has. To enable calculation and comparison, it is imperative to convert the position value of each grammatical item in a standardized way by assuming that all textbooks have a uniform length and then calculating the relative position of a grammatical item on this uniform scale. The relative position is thus deemed as the sequence value of the grammatical item, which is denoted by *S*. The calculation equation of *S* is as follows:

$$S = \frac{\mathrm{L} - \bar{L}}{\sigma} \quad (1)$$

Where *S* denotes the sequence value of the grammatical item; *L* denotes the position value of the grammatical item in textbook; $\bar{L}$ denotes the mean value of position values; and σ denotes the standard deviation of position values.

As the standardized sequence value S may be negative, the following conversion is used:

$$\mathrm{S}' = 50 + 28\mathrm{S} \quad (2)$$

The linear converted *S*', ranging from 1 to 100, does not change the original S relations.

Through such a conversion, the position of a grammatical item across different textbooks is changed from a sequential variable to an interval variable, leading to a comparability between different positions. The smaller the S', the nearer of the grammatical item to the front in the textbook, and vice versa.

The statistics carried out for this study shows that there are 23 grammatical items that are commonly used in the 10 textbooks (see Table 3).

Table 3. The Common grammatical items appearing in all 10 textbooks

1. "Bei"-structure ("被"字句)
2. "De"-structure ("的"字句)
3. "Shì…de" ("是…的")
4. "You" extential sentence ("有"的存在句)
5. "Bi"-structur ("比"字句)
6. "Ba"-structure ("把"字句)
7. Enumeration of the numbers (称数法)
8. Degree complement (程度补语)
9. Verbal reduplication (动词的重叠)
10. Rhetorical questions (发问句)
11. Expression of approximate numbers (概数的表达)
12. "Guo" ("过")
13. Simple directional complement (简单趋向补语)
14. Resultative complement (结果补语)
15. Potential complement (可能补语)
16. Sentence with verbal constructions in series (连动句)
17. "Le" ("了")
18. Extended meanings of directional complements (趋向补语的引申义)
19. Expressions of time (时间的表达)
20. Adverbial of time (时间状语)
21. Adjective-predicated sentences (形容词谓语句)
22. "Yao/kuai…le" ("要/快….了")
23. Interrogative denotation of interrogative pronouns (疑问代词表疑问)

These 23 common items are important nodes running through the Chinese language grammar teaching. Linearly sequencing these nodes by their position and difficulty will produce the main thread of the grammar teaching phase of the entire textbook. To examine the degree of consistency in grammatical structure across the 10 selected textbooks and ensure that there is no outliers pertaining to the collected grammatical items, this paper analyzes the correlation of the 23 common items in the 10 textbooks and their internal consistency.

The result shows that the Pearson correlation values of S' pairs among the 23 common items fall into the range of (0.620–0.909) across different textbooks, indicating that they are all significantly correlated. Additionally, the internal consistency (denoted by α coefficient, see Eq. 3) of the common grammatical items across the 10 textbooks is 0.969. This indicates a relatively high consistency in terms of the treatment of common grammatical items across the 10 textbooks, and it also justifies that the grammatical items are arranged in a scientific, systematic and stable manner in comprehensive Chinese language textbooks, which can thus serve as credible materials for grammatical item sequencing research.

$$\alpha = \left(\frac{n}{n-1}\right)\left(\frac{\sigma^2 - \sum \sigma'^2}{\sigma^2}\right) \tag{3}$$

where n is the number of grammatical items; σ^2 is total variance; and σ'^2 is the variance of each grammatical item.

3 Core and Positional Parameters of Grammatical Items

Commonly used and important grammatical items should be taught in Chinese language education. Undoubtedly, the 23 common grammatical items extracted in this research fall into this category. However, they cannot be sequenced by the degree of commonness in teaching, and the commonly used items should not necessarily be taught first. For example, the auxiliary "了" (le) as a common grammatical item should not be arranged in the early stage of teaching. Commonness does not mean easiness. Thus, sequencing of grammatical items must be examined on two dimensions of commonness and difficulty. Being selected into Chinese language textbooks points to commonness of a grammatical item, while how a grammatical item is arranged points to its difficulty. Following this thinking, this paper establishes two parameters to describe the degrees of commonness and difficulty of a grammatical item, namely the core parameter and the positional parameter.

3.1 Core Parameter

Core grammatical items refer to the most commonly used and most important grammatical items. The total frequency (F) of the appearance of a grammatical item in textbooks provides an important reference for its commonness, but it is also necessary to consider the distributional information of such an item across different textbooks. For grammatical items with an equal frequency, the commonness of those that are widely distributed in textbooks should be higher than those that are less widely distributed, that is, the frequency and distribution should be inversely proportional. Therefore, a core value model for grammatical items is designed in this paper using frequency and relative position information as factors, which is presented as follows:

$$U = \frac{F}{Stdev(S')} \tag{4}$$

where U is the core value of a grammatical item; F is the frequency of a grammatical item; S' is the relative position of the grammatical item; and $Stdev\ (S')$ is the standard deviation of S'.

Stdev (S') reflects the changes of a grammatical item across textbooks. The larger the *Stdev (S')*, the less stable of the distribution of the grammatical item across textbooks; the smaller the *Stdev (S')*, the higher the consistency in the recognition of the grammatical item across textbooks.

Through the calculation and sequencing of the model, the top 10 grammatical items in terms of their core value *U* are obtained and shown in Table 4.

Table 4. Extraction results of the core value model

Grammatical item	F	F (textbooks)	Stdev (S')	*U*
Chinese word order (汉语的语序)	6	6	0.638151	9.402167
Numeral-measure word phrase (数量词组)	2	2	0.260411	7.680167
Verb "jingguo" (动词"经过")	2	2	0.404584	4.943347
Interrogative denotation of interrogative pronoun (疑问代词表疑问)	55	10	11.97551	4.592708
Adverb "ye" (副词"也")	8	8	1.864626	4.290404
Verb "wei" (动词"为")	2	2	0.636434	3.142509
Extended meanings of directional complements (趋向补语的引申义)	46	10	16.67881	2.75799
"Le" ("了")	40	10	15.81312	2.529545
"Er" and "Liang" ("二"和"两")	9	7	3.732855	2.411023
Expressions of time (时间的表达)	23	10	10.3971	2.212154

As can be seen from the above table, the *U* values of grammatical items having a relatively high frequency, including "interrogative denotation of interrogative pronouns" (疑问代词表疑问), "extended meanings of directional complements" (趋向补语的引申义) and "le" ("了"), are among the top 10 items. Actually, these three items are also common grammatical items in the 10 textbooks, and so the result is completely reasonable. However, other grammatical items with relatively lower frequency and standard deviation, such as the verb "jingguo" (动词"经过") and "numeral-measure word phrase" (数量词组), are also among the top 10, which is apparently contradictory to common sense. The cause of such problem lies in overlooking the weights of frequency and standard deviation. The proportion of frequency in extraction of core grammatical items should be larger than that of the standard deviation; otherwise, the above circumstance where some low-frequency grammatical items that appear only in 1 or 2 textbooks in a rather concentrated distributions may enter the core grammar category. Thus, the following modifications are made to the model:

$$U' = \frac{F^2}{Stdev(S')} \tag{5}$$

The top 10 *U* values following the modification are as follows (Table 5):

Table 5. Extraction results of the modified core value model

Grammatical item	F	F (textbooks)	Stdev (S')	*U'*
Interrogative denotation of interrogative pronouns	55	10	11.97551	252.5989
Extended meanings of directional complements	46	10	16.67881	126.8676
"Le"	40	10	15.81312	101.1818
Resultative complement	40	10	19.52051	81.96507
"Bi"-structure	39	10	20.57070	73.94012041
Potential complement	35	10	17.00961	72.0181
Chinese word order	6	6	0.638151	56.413
Expressions of time	23	10	10.3971	50.87955
Enumeration of the numbers	38	10	28.94641	49.88529
"Ba"-structure	29	10	18.11052	46.43709

The table above presents the sequence after the model is modified. As can be seen, the "verb 'jingguo'" (动词"经过") and "verb 'wei'" (动词"为") that actually have a relatively low frequency are no longer sequenced ahead of the order, which is consistent with our expectation. After the adjustment, the sequential order of common grammatical items such as the "resultative complement" (结果补语), "potential complement" (可能补语) and "ba-structure" ("把"字句), has been significantly changed, indicating that the modified model is able to extract the common core grammatical items. In addition, a non-common grammatical item, the "Chinese word order" (汉语的语序), also enters the top-10 list, which is due to the fact that the item appears in 6 textbooks with extremely high consistency, which significantly boosts its core value. The item "Chinese word order" (汉语的语序) does not have an advantage in total frequency because its relatively low complexity makes it sufficient for the item to appear only once in each textbook. Grammatical items with a higher total frequency tend to be those entailing larger content in teaching. These grammatical items are relatively complicated in terms of structure, semantics and usage and often involve key grammatical points or difficulties in teaching. Thus, they are usually separated into smaller parts in teaching, which explains their high total frequency. The item "Chinese word order" does not have a high frequency in total. However, it appears in the majority of textbooks and has extremely high degrees of recognition and consistency, making it natural to incorporate them into the common grammatical items. This also confirms that it is reasonable to take the distributional value into the model.

3.2 Positional Parameter

The positional information of grammatical items in textbooks actually also represents the difficulty level of the items. A smaller relative positional value *S'* indicates that the position of a grammatical item is at the earlier part of a textbook to accommodate with learners' acquisition; to the contrary, the larger the *S'*, the later of the position of the grammatical item and certain basic or cumulative knowledge is needed in order to master the item. The major references for the positional parameter model are the average score of *S'*, denoted by *Avg (S')*, that is, the average value of the position of a grammatical item across different textbooks, and the distributional value *Stdev (S')*. *Stdev (S')* reflects the degree of ramification of different textbooks in the order of item position. Larger ramification causes a broader distribution and thus a larger *Stdev (S')*, and vice versa.

Grammatical items ordered in front should be those that are arranged in front consistently by multiple textbooks. The smaller the *Avg (S')*, the smaller the *Stdev (S')*, and vice versa. Assume that there are two grammatical items with an identical Avg (S') denoted by G1 and G2, if G2 has a relatively smaller distributional value, then the degrees of recognition assigned by different texts on them are different, and the recognition of G2 is higher than that of G1. Thus, the positional value of G2 should be larger than that of G1, and G2 should not be assigned with a same earlier position as with G1. Therefore, the positional value of an item is positively proportional to its recognition in the positional parameter model. In the meantime, similar to the core parameter model, the weight of the average positional value is also considered, and the model is written as:

$$D = \text{Avg(S')}^2 \times \text{Stdev(S')} \tag{6}$$

where D denotes the positional value of a grammatical item. The following table presents the first and last 5 grammatical items extracted by the model.

Table 6. Extraction results of the positional parameter model

	Grammatical item	F	Avg (S')	Stdev (S')	D
First 5 items	Chinese word order (汉语的语序)	6	2.628147052	0.638150768	4.407807
	Adverb "ye" (副词"也")	8	4.386028125	1.864626415	35.87027
	Punctuation mark (标点符号)	3	4.876844473	2.425575083	57.68894
	Numeral-measure word phrase (数量词组)	2	17.15167984	0.260411001	76.60774
	Attributives and "de" expressing possession (表领属的定语及"的")	4	6.096116377	3.051990876	113.42
Last 5 items	Shouxian…ranhou… ("首先…然后…")	2	69.77040381	36.00647196	175276.2
	V "lai" V "qu" ("V来V去")	2	65.79721151	41.7673556	180822.3
	"Zai…dehua, jiu…" ("再…的话,就…")	2	72.74738206	35.01259156	185293
	"Haoji/duo" + Measure word ("好几/多 + 量词")	2	63.1242208	47.93564751	191007.6
	Preposition "zai" (介词"在")	9	74.41242041	36.66025782	202995.5

The first 5 items are basically in line with the order of appearance in Chinese language textbooks. For example, the "Chinese word order" (汉语的语序), which is the basic "S + V + O" structure, is a typical typological characteristics of the Chinese language, a commonality shared by many languages, making it easily acceptable for the majority of learners. In addition, the teaching content of this grammatical item is also relatively monotonous, which is often incorporated into the self-introduction function in the first lesson. Thus, it is often the first grammatical item appearing in a textbook. The appearance of the "numeral-measure word phrase" (数量词组) in the first 5 items is not consistent with our expectation because it should appear after the "numerals" (数词) and "measure words" (量词). A further examination reveals that the average order value of the item is 17.15, an excessively small value, leading to a smaller D value and thus a relatively front position. Similar problem also occurs to the last 5 items in Table 6: despite the relatively small mean values of the last 5 items, they are nevertheless ordered in front due to their excessively large distributional value. Thus, the distributional value has an excessive effect on D value.

The distributional value Stdev (S') is calculated based on the S' value linear converted from the relatively sequential value. S' ranges from 1 to 100, while Stdev (S') ranges from 0 to 60. To reduce the effect of Stdev (S') on the D value, this paper equally divides the value of Stdev (S') into 5 intervals with each interval assigned with a value equally distributed between 1 and 2. The assigned distribution range value of Stdev (S') in 0–9 is 1, and that in 10–19 is 1.2, and so forth. While taking the effect of the distribution value Stdev (S') on D value into account, this treatment also effectively controls such an effect so that the D value is more approximate to the mean value.

The extraction results after the modification are as follows (Table 7):

Table 7. Extraction results of the modified positional parameter model

	Grammatical item	F	Avg (S')	Std-R	D'
First 5 items	Chinese word order (汉语的语序)	6	2.628147052	1	6.907156929
	Adverb "ye" (副词"也")	8	4.386028125	1	19.23724272
	Punctuation mark (标点符号)	3	4.876844473	1	23.78361202
	Attributives and "de" expressing possession) (表领属的定语及"的")	4	6.096116377	1	37.16263489
	Questions with "ma" (用"吗"的疑问句)	10	7.396197845	1.2	45.58645214
Last 5 items	"Yu" ("于")	2	91.69294667	1	8407.596469
	Verb "wei" (动词"为")	2	93.10558171	1	8668.649346
	Condensed compound sentences (紧缩复句)	3	93.16576667	1	8679.860079
	Compound sentences (复句)	4	94.06461092	1	8848.151028
	"Ze" ("则")	2	94.19431265	1	8872.568535

The sequencing of grammatical items is more reasonable following the modification. For example, in the first 5 items, the rank of the "numeral-measure word phrase"

(数量词组) falls to the 26th place, with a D' value of 294.18, next to positions of the "numeral" (数词) (9th place, D' = 145.3356516) and "measure words" (量词) (11th place, D' = 159.2077829), which appears to be more rational. The last 5 items all have smaller mean positional values. They are able to retain a position corresponding to their mean positional values because of a relatively higher consistency in their distribution. This also confirms that portioning positional distribution values into intervals can effectively control the effect of significant changes in distribution value on positional value, and thus obtain a more stable positional value.

To further examine the reasonableness of the order of the positional values, this paper selects a few grammatical items whose sequences are consistently agreed in the area of international Chinese language teaching for verification, and all their results prove the reliability of the model. Two examples are provided here.

There have been a scholarly consensus in terms of the sequences for "resultative complement" (结果补语), the "ba-structure" ("把"字句) and auxiliary "le" ("了"). Yang (2000) argued that "there is no ground for the appearance of the 'ba-structure' ("把"字句) if 'resultative complement' (结果补语) and 'directional complement' (趋向补语) have not been taught". "It is impossible for 'resultative complement' (结果补语) to appear without the appearance of 'le' ("了")". The extraction results of this paper also justifies such a sequence, as shown in the following Table 8:

Table 8. Verification example of extraction results of the positional parameter model (1)

Grammatical item	Order	F	Avg (S')	Stdev (S')	Std-R	D'
"Le" ("了")	89	40	38.02439	15.813123	1.2	1204.879
Resultative complement (结果补语)	158	40	56.29664	19.520511	1.2	2641.093
"Ba"-structure ("把"字句)	165	29	57.70353	18.110524	1.2	2774.747

Additionally, the sequence of grammatical items should follow a structural order from basic to derivative and a semantic order from basic meaning to extended meaning. Lv (2002) and Lu (2003) both illustrated this point using the sequence of directional complements. They argued that a reasonable order should be: "simple directional complements" (简单趋向补语) → "compound directional complements" (复合趋向补语) → "extended meaning of directional complements" (复合趋向补语的引申用法), which is consistent with the order of D' values extracted in this paper, which are presented as follows (Table 9):

Other items with relatively definitive sequences include "interrogative denotation of interrogative pronouns" (疑问代词表疑问), "interrogative pronouns to mean 'any'" (疑问代词表任指) and "interrogative pronoun virtual usage" (疑问代词表虚指), as well as the order of the "yes-no questions" (是非问), "alternative questions" (选择问) and "rhetorical questions" (反问句). The order of the D' values extracted in this paper are highly consistent with that demonstrated by experts. Thus, the positional parameter model proposed by this paper is highly effective in sequencing grammatical items in an appropriate and reasonable way.

Table 9. Verification example of extraction results of the positional parameter model (2)

Grammatical item	Order	F	Avg (S')	Stdev (S')	Std-R	D'
Simple directional complement (简单趋向补语)	119	15	46.05559	14.567173	1.2	1767.597
Extended meanings of directional complements (复合趋向补语)	171	10	59.36209	17.946773	1.2	2936.548
Extended meanings of directional complements (复合趋向补语的引申用法)	217	46	69.83032	16.678811	1.2	4063.561

4 Conclusion

By examining 10 comprehensive series of textbooks and collecting and sorting out grammatical items in international Chinese language teaching, this paper proposes that sequencing description of grammatical items should be carried out on two dimensions of commonness and difficulty. On this basis, the core and positional parameter models are designed. After modifying these models through approaches like adjusting the parametric weights and assigning value intervals, the two models render fairly good effectiveness in determining the sequential value on the two dimensions.

This paper has achieved to provide a scientific and quantitative support for course establishment, syllabus development and textbook compilation in international Chinese language education. In the future, the scope of examination can be extended in order to build a resource library on grammatical items for textbooks used in international Chinese language education, so that the fundamental work in the area of teaching resources construction can be strengthened, the informationization of such resources can be advanced and better services can be provided.

Acknowledgments. This research project is supported by Major Program of National Social Science Foundation of China (18ZDA295). The work is also supported by "BLCU Youth Talent Development Program".

References

Lu, F.: Dividing of the hierarchy and arranging of the items in grammar of Chinese teaching as a foreign language. Chin. Lang. Learn. **2**, 54–60 (2003)

Lv, Wenhua: Principles and strategy for arranging grammar items in Chinese textbooks. Chin. Teach. World **4**, 5+87–96 (2002)

Peng, W., Song, J., Zhao, M.: Construction of grammar resource base for International Chinese teaching. Distance Educ. China **8**, 92–96 (2014)

Su, X.: The role of measurement methods in vocabulary research and frequency statistics. Yangtze River Acad. **2**, 118–124 (2007)

Sun, M.: A preliminary study on lexical usage. Appl. Linguist. **1**, 6–10 (2000)

Tan, X., Yang, L., Su, J.: A study on the construction of grammar knowledge-base and the annotation of grammar points for TCSL. J. Chin. Inf. Process. **6**, 58–65 (2015)
Wang, Z., Yu, S.: New updates on teaching Chinese vocabulary based on large-scale corpus - a case of"nouns"in graded Chinese syllables characters and words (GCSC). Lexicographical Stud. **5**, 64–74+122 (2019)
Yang, D.: Some problems in the determination and arrangement of grammar points in primary Chinese textbooks. Chin. Teach. World **2**, 81–88 (2001)
Xing, H., Zhang, W.: Annotation and statistical study of modern Chinese grammar items. In: Zhao, J. (eds.) Duiwai Hanyu Jiaoxue de Quanfangwei Tansuo. The Commercial Press, Beijing (2005) 147–163
Yang, J.: The sequencing of grammar items in the primary stage of teaching Chinese as a foreign language. Lang. Teach. Linguist. Stud. **3**, 11–16 (2000)

A Study on the Construction Features of "Pan X"

Haifeng Yang[1(✉)] and Hongjie Yue[2(✉)]

[1] School of Law, Humanities and Sociology, Wuhan University of Technology, Wuhan, China
yanghaifengtl@whut.edu.cn

[2] College of Chinese Language and Culture, Nankai University, Tianjin, China

Abstract. From the diachronic and synchronic perspectives, this paper conducts a comprehensive study on the structural features of the online buzzword "pan (盘) X" which has become popular since 2019. At the diachronic level, the "pan (盘) X" construction, in the process of evolving from a grammatical construction to a rhetorical construction, on the one hand, retains the original semantic categories of agentive meaning and causative meaning; on the other hand, it has a high degree of dependency on the specific social environment characteristics of the prescriptive scenes. At the synchronic level, the construction of "pan (盘) X" is flexible in terms of syntactic structure with rich metaphorical meanings and has the semantic feature of controlling. The study on the construction features of "pan (盘) X" shows that there are some similarities between network language and other unconventional languages in essence, and the study of network language represented by "pan (盘) X" construction can deepen our understandings of the nature of network language.

Keywords: "Pan (盘) X" · Network buzzword · Grammatical construction · Rhetorical construction

1 Introduction

In the review of network buzzwords in 2019, "pan X" has been listed on the rankings of both *Yuyan Wenzi Zhoubao* and Baidu Boiling Point, showing its wide popularity and great influence.

"Pan X" originated from the crosstalk program *Antique Collectables* performed by crosstalk comedians Hetang Meng and Jiuliang Zhou in September 2018. Originally, it referred to rubbing and squeezing of walnut. Specifically, by repeatedly rubbing and squeezing with hands, walnuts are covered with patina and become shiny and more valuable. Later, the performers expanded the objects of "pan" from walnuts to lions, meatballs, tree-root carvings, etc., and won warm applause from the audience by using these unconventional language collocations and exaggerated performance at stage. After the program had been broadcast, people followed suit one after another and imitated the performers to "pan" all kinds of objects, and some even proposed that "万物皆可盘 (everything can be pan-ed)".

M. Liu et al. (Eds.): CLSW 2020, LNAI 12278, pp. 161–172, 2021.
https://doi.org/10.1007/978-3-030-81197-6_15

Before the new usage of "pan X" became popular, "pan" could match with objects and form the verb-object structure, such as "盘发 (*pan fa*, have hair rolled up)", "盘粮 (*pan liang*, make an inventory of food)", "盘炕 (*pan kang*, build a brick bed)" and "盘店 (*pan dian*, transfer all the goods in a shop to others)". However, in the new usage, the meaning of "pan" has transferred [1], and the object "X" is no longer limited to specific objects such as hair, food, bed, shop, etc. The function of "pan X" cannot be "strictly predictable from its component parts" [2]. For example:

(1) 外地男子抖音哗众!阜阳交警隔空喊话, 皖黔交警联手"盘"他!
Wàidì nánzǐ Dǒuyīn huázhòng! Fùyáng jiāojǐng gé kōng hǎn huà, Wǎn Qián jiāojǐng liánshǒu "pán" tā!
'A non-local man acted ostentatiously to impress others on Tik Tok! The traffic police of Fuyang warned him from a distance, and the traffic police of Anhui province and Guizhou province worked together to "take" him down!'

(2) 中国民航大学招生智能问答机器人上线了?盘它!
Zhōngguó Mínháng Dàxué zhāoshēng zhìnéng wèn dá jīqìrén shàngxiàn le? Pán tā!
'The intelligent Q&A robot for enrollment in Civil Aviation University of China launched? Chat with it!'

To sum up, the new usage of "pan X" structure can be classified into the construction category. It consists of a constant "pan" and a variable "X", and belongs to the "partially filled idiom" construction among the eight construction types classified by Goldberg according to size and complexity [2].

Some scholars have carried out relevant studies on the catchword "pan X". At the synchronic level, scholars mainly focused on the semantic and pragmatic features of the catchword. Semantically, some classified the meaning of "pan X" into liking and disgusting according to emotional coloring [3]. Some summarized the meanings according to specific situations [1, 4]. Pragmatically, "pan X" has a novel and unexpected expression effect and it can reflect people's subjective initiative. Its entertainment function is greater than the education function with weak pragmatic foundation [5, 6]. In terms of time and cultural property, Li [3] suggested that "pan X", as a grass-root vocabulary, is the embodiment of social entertainment culture and spiritual culture of senses of accomplishment, release and gain, while Su [7] believed that it reflects people's anxiety and blind pursuit of speed.

At the diachronic level, Ma [8] proposed that the semantic evolution of "pan X" includes the change of reference range and the change of style. Cai [9], on the basis of the meanings in modern and contemporary Chinese language, claimed that the meaning of "pan" has extended from the meaning of rotation to rubbing and squeezing, fondness and other meanings. For the motivation of the evolution, the explanations in the previous studies were given based on the promotion by Internet and multimedia, metaphorical mechanism, language meme theory, and economic principle [8, 10].

In this paper, the authors try to explore the diachronic evolution mechanism of "pan X" construction by combining the concepts of grammatical construction and rhetorical construction, the interactive view of construction coercion, and the Generative Lexicon Theory. It also tests, supplements and improve the existing synchronic research results by combining the diachronic features, metaphorical mechanism and related corpora of "pan X", hoping to provide a reference for the construction research of network buzzwords. The corpora cited in this paper are mainly People's Daily APP, Sina Weibo APP, BLCU Corpus Center (BCC), WeChat official accounts and other channels.

2 Evolution Mechanism of "Pan X" Constructions

Liu first proposed the concept of "rhetorical construction". He divided constructions into grammatical constructions and rhetorical constructions. Moreover, he also elaborated the evolution path of "existing grammatical construction → rhetorical construction → new grammatical construction". In his opinion, grammatical construction and rhetorical construction are located at "two ends of a continuum", and "rhetorical construction is formed by a certain rhetorical motivation added to a basic grammatical construction" [11, 12]. We propose that the evolution mechanism of "pan X" construction has experienced the transformation from grammatical construction to rhetorical construction.

2.1 Stage I: The Expansion of the Range of the Variable "X"

With the popularity of the crosstalk showing *Antique Collectables*, there has been an upsurge of the frequency of use of "pan X" on the Internet. People have imitated the performers to record videos of playing with lipstick, cactus and children with their hands. The more nonsense they are, the better they perform to attract attention and make people laugh. Many nouns that are not supposed to be compatible with "pan" have then entered the structure of "pan X".

At this stage, the form and function of "pan X" construction remains roughly the same as that of the original form, i.e. "盘核桃 (*pan hetao*)", which is "predictable". Thus, the structure of "pan X" at this stage is still a grammatical construction.

Considering the basic meaning of "pan" (i.e., rubbing and squeezing something by hand, making it be covered with patina and become shiny and more valuable), it is believed that the meaning of the structure involves agentive and causative meanings. The former gives rise to the action feature of "rubbing and squeezing by hand", while the latter the causative feature of "making something covered with patina, becoming shiny and more valuable". The causative meaning and the agentive meaning are highly related as an agent can be the causer of the event.

2.2 Stage II: The Generalization of the Meaning of the Constant "Pan"

With more and more unconventional "X" entering the construction "pan X", the meanings of "pan" have gradually changed. Here we take an adaption of one of Jay Chou's songs named *Simple Love* as an example:

(3) 我想大声宣布对你依依不舍
连隔壁邻居都盘到我现在的感受
河边的风在盘着头发飘动
盘着你的手一阵莫名感动

我想带你回我的外婆家
一起盘着日落一直到我们都睡着
Wǒ xiǎng dà shēng xuānbù duì nǐ yīyī-bùshě
Lián gébì línjū dōu pán dào wǒ xiànzài de gǎnshòu
Hé biān de fēng zài pánzhe tóufà piāodòng
Pánzhe nǐ de shǒu yī zhèn mòmíng gǎndòng
Wǒ xiǎng dài nǐ huí wǒ de wàipó jiā
Yīqǐ pánzhe rìluò yīzhí dào wǒmen dōu shuì zháo

'I want to loudly announce my reluctance to part with you
Even the next-door neighbors can feel my feelings now
The wind by the river is swirling around the hair
Holding your hand, I am touched inexplicably
I want to take you back to my grandmother's house
We are going to watch the sunset together until we fall asleep'

"Pan" seems to be a generalized verb, whose meaning changes with the change of context. We can explain the phenomenon by the views on qualia structure and qualia role contained of Generative Lexicon Theory. Generative Lexicon Theory has proposed four semantic representations of lexicon: argument structure, event structure, qualia structure and lexical inheritance structure, among which qualia structure is the core of various generative attributes in lexicon. Yuan, through the description system of qualia role and syntactic format, established ten qualia roles for the qualia structure of nouns in Chinese, such as form, composition, unit and handling, and others, hoping to find the syntactic-semantic interface of nouns [13]. Due to the conventional relationship between qualia roles and the corresponding nouns, when the nouns with specific qualia roles are triggered, the verbs expressing these qualia roles will be naturally activated by the listener. For example:

(4) a. 河边的风在盘着头发飘动。
Hé biān de fēng zài pánzhe tóufa piāodòng.
'The wind by the river is pan-ing hair.'
b. 河边的风在吹着头发飘动。
Hé biān de fēng*zài chuīzhe tóufa piāodòng.*
'The wind by the river is swirling around the hair.'
(5) a. 连隔壁邻居都盘到我现在的感受。
Lián gébì línjū dōu pán dào wǒ xiànzài de gǎnshòu.
'Even the next-door neighbors can pan my feelings now.'
b. 连隔壁邻居都猜到我现在的感受。
Lián gébì *línjū dōu cāi dào wǒ xiànzài de gǎnshòu.*
'Even the next-door neighbor can feel my feelings now.'

In example (4), the verb meaning of "swirl around" is activated by "wind" and "hair" and is provided by the action role of "wind" and the handling role of "hair". Such meaning is temporarily assigned to "pan". From the hearer's perspective, the sentence is interpreted as (4b). Similarly, in example (5), the action role of "neighbor" and the handling role of "my feeling" activate the verb meaning of "feel", and "pan" thus carries this meaning temporarily.

From the perspective of coercion, the meaning change of "pan" is actually the result of the joint coercion of variable "X" and the context. In addition, the variable "X" is also suppressed by "pan". Nouns such as "feeling" and "sunset" cannot have formed verb-object collocation with "pan". However, through the "centripetal coercion effect" [14], "X" adjusts "some aspects in the formal structure" [15] and temporarily serves as the object of "pan", thus becoming fit for "pan" and realizing the coercion. The above bidirectional coercion process reflects the interaction mechanism between different components of coercion.

"Pan X" construction at this stage turns less "predictable" and begins to evolve from a grammatical construction to a rhetorical construction. Liu pointed out that one of the important attributes of rhetorical constructions is its dependency on specific discourse scenes, which become the prescriptive scenes of rhetorical construction. Some scenes strongly depend on the details of specific social environments. If the constructions cannot enter these special environments, they can only stay in the state of non- deducible [11]. The "pan X" construction in this stage belongs to this kind of typical rhetorical construction.

2.3 Stage III: The Weakening of Action Features and Enhancement of Emotional Coloring

Compared with the previous stages, although the generalized meaning of the "pan X" construction still remains, the action meaning of the "pan X" construction in this stage is weakened, and it has obvious emotional coloring. For example:

(6) 年度最硬核摄影大赛火热开启, 拍医护, 拿大奖, 盘他!
Niándù zuì yìnghé shèyǐng dàsài huǒrè kāiqǐ, pāi yīhù, ná dà jiǎng, pán tā!
'The biggest photography competition of the year is launched. Photograph medical care staff and win awards. Participate in it!'

(7) 有越来越多的人喜欢上了这个女裁判, 最近更是有网友直呼"兄弟们盘她"。
Yǒu yuè lái yuè duō de rén xǐhuan shàngle zhègè nǚ cáipàn, zuìjìn gèng shì yǒu wǎngyǒu zhí hū "xiōngdimen pán tā".
'More and more people fall in love with this female referee. Recently, even some netizens have called "Brothers, express love to her".'

(8) 司机不打表、拒载、拼客? 盘他!
Sījī bù dǎbiǎo, *jùzài, pīnkè? Pán tā!*
'The taxi driver did not initiate the meter, refuse to take passengers or refuse carpooling? Teach him a lesson!'

Examples (6) and (7) express the hospitality of the organizers and people's love for the female referee respectively, while example (8) reflects the speaker's aversion to some the driver's infringement acts.

For the causes of such semantic bias, on the one hand, Tang explained them from the perspective of metaphorical mechanism [1]; on the other hand, their solidification is inseparable from the cause of "language itself and language use" [16]. Taking aversion as an example, according to the collected corpora, a large part of the network news utilizes the "pan X" construction to denounce and criticize some illegal behaviors. For example:

(9) 暴力抗拒执法?盘他!
Bàolì kàngjù zhífǎ? Pán tā!
'Violent resistance to law enforcement? Teach him a lesson!'

(10) 男子坐行驶汽车引擎盖, 网友: 警方快盘他。
Nánzǐ zuò*xíngshǐ qìchē yǐnqíng gài, wǎngyǒu: jǐngfāng kuài pán tā.*
'A man sits on the hood of the driving car. Netizens: Catch him, police.'

Of course, "pan X" construction may also express neutrality, which depends on the context. Similar to the second stage, the prescriptive scenes of "pan X" construction in this stage also "strongly depends on the details of specific social environments". So, it is still a typical rhetorical construction.

To sum up, "pan X" construction has only completed the first half of the path of "existing grammatical construction → rhetoric construction → new grammatical construction" but has not realized the transformation to the new grammatical construction in the end. Instead, it becomes a typical rhetorical construction.

3 Component Features of "Pan X" Construction

The construction of "pan X" mainly includes two components: the constant "pan" and the variable "X". Their characteristics are discussed separately below.

3.1 Component Features of "Pan"

According to *Modern Chinese Dictionary (7th Edition)*, "pan" has three parts of speech: noun, verb and quantifier. "Pan" in the construction of "pan X" is a bivalent verb, and "X" often assumes the role of patient forming a verb-object structure with "pan", and the agent usually precedes "pan".

The basic meaning of "pan" (i.e., rubbing and squeezing something by hand, making it be covered with patina and become shiny and more valuable) is the closest meaning to the verb meaning of "whirling around" in *Modern Chinese Dictionary*. In order to confirm the relationship between the two, the authors have consulted *Chinese Characters Form and Meaning Analysis Dictionary*. There are seven definitions of "pan" in *Chinese Characters Form and Meaning Analysis Dictionary*:

(1) A kind of utensils which is flat and shallow in shape, e.g., 茶盘 (tea plate), 托盘 (tray)

(2) A thing shaped like a plate, e.g., 磨盘 (millstone), 方向盘 (steering wheel)

(3) To swirl and coil around something, e.g., 盘山公路 (winding mountain road)

(4) To build something by piling up bricks, stones, earth, etc., e.g., 盘炕 (build a brick bed), 盘灶 (build a stove)

(5) To hold an inquiry or make an inventory, e.g., 盘查 (interrogate), 盘问 (cross-examinate)

(6) The commodity price in old terms, e.g., 平盘 (flat plate), 开盘 (opening quotation)

(7) A quantifier, e.g., 一盘磨 (a millstone), 一盘棋 (a game of chess)

According to the parts of speech of "pan" in the construction "pan X", this paper mainly studies the verb meaning of "pan" which is related to the meaning items (3), (4) and (5) above. According to *Chinese Characters Form and Meaning Analysis Dictionary*, the shape of dish is round, and so, when it is used as a verb, it means "swivel and whirl around", such as "盘曲 (*panqu*, whirl)". At the same time, it also extends the meaning of "build bricks in a circle", such as "盘灶 (*pan zao*, build a brick bed)". Then, from "swirl around", "repeatedly (check)" is derived.

According to Taylor's "semantic chain" theory, the relationship between the meanings of polysemous words is chained. For example, meaning A can extend to meaning B based on some common attributes, while meaning B can extend again to meaning C based on some other common attributes, and so on. Finally, the numerous meanings are formed [17]. Leech believed that the extension of word meanings is mainly realized through metaphor or metonymy [18]. Both metaphor and metonymy play an important role in human cognition. The former is mapping across conceptual domains based on similarity, while the latter is mapping in the same domain based on connection.

The extension of the meanings in (3), (4) and (5) above can be regarded as the extension of metaphor. The first two are based on the similarity of shape, while the third one is derived from the psychological similarity. The abstract concept of "repeatedly (check)" is metaphorical which is derived from the shape feature of "pan". The above extended path is shown in Fig. 1:

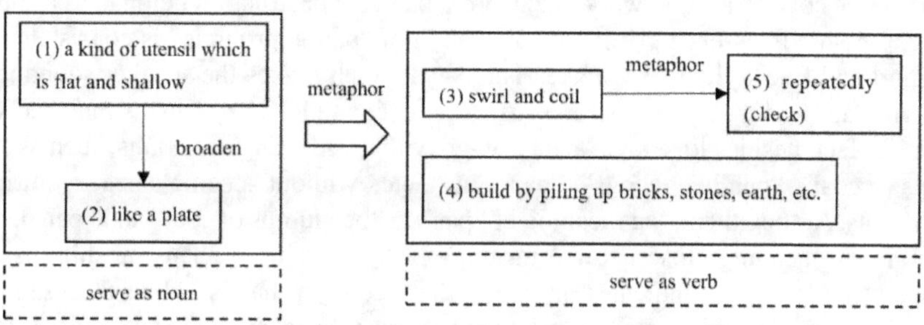

Fig. 1. The schematic diagram of semantic extension process of "pan" (part)

According to the process of semantic extension and the theory of "semantic chain", it can be concluded that the basic meaning of "pan" in the construction of "pan X" is derived from the verb meaning of "whirl around". After the word meaning has been widely used in the field of antique collectables, it has become a fixed term. The meaning related to purpose, i.e., "making it be covered with patina and become shiny and more valuable", has then been added.

3.2 Component Features of "X"

As can be seen from the above, the meaning scope of "X" that can enter the "pan X" structure is very wide, ranging from concrete to abstract (such as "feel (my present feeling)"), from objects to human beings (such as "盘爱豆 (*pan aidou*, dote on idols)"), and even some people put forward that "万物皆可盘 (everything can be *pan*-ed)".

In addition, in terms of the part of speech, "X" can be pronouns, nouns and quantitative phrases, and the probability of their occurrence decreases in turn. For example,

(11) 训练场竟有两大坑, 激起全队斗志, 盘他!
Xùnliàn chǎng jìng yǒu liǎng dà kēng, jī qǐ quán duì dòuzhì, pán tā!
'There are two big pits in the training ground, which arouses the morale of the whole team, defeat them!'

(12) 美女帮闺蜜盘渣男, 闺蜜却这样做, 结果优秀了。
Měinǚ bāng guīmì pán zhānán, guīmì què zhèyàng zuò, jiéguǒ yōuxiù le.
'A beautiful woman helped her best female friend to punish the love rat, but her best friend did this, and the result was "excellent".'

(13) 又盘一个!13年命案逃犯落网。
Yòu pán yī gè! 13 Nián mìng'àn táofàn luòwǎng.
'Caught another one! A 13-year outlaw of a homicide case has been arrested.'

The structure of "pan + pronoun" in example (11) has the widest application range, which can be explained by the following two reasons. Firstly, the hearer can infer the object referred by the pronoun together with other information such as context and paralanguage. In this way, misunderstanding will not exist and it conforms to the economy principle. Secondly, as a network buzzword, "pan X" construction belongs to a special oral expression in essence [19]. The collocation of "pan + pronoun" is consistent with the short and free style of the spoken language, which makes the speaker sound more powerful, and so it becomes a strong structure. In example (12), "pan" is matched with nouns. In fact, besides love rats, "pan" can go with a wide range of nouns, such as "pan lumps of coal", "pan hedgehogs", "pan idols", etc., without specific category attribute constraints. Among them, the meaning of "pan" in "pan lumps of coal" and "pan hedgehogs" is the meaning closest to the meaning of "pan walnuts". This meaning is also relatively basic, which indicates an action that occurs in hands, while the usages of "pan idols" and "pan love rats" are relatively abstract. In example (13), "pan" is seldom matched with quantitative phrases.

4 Structural Features of "Pan X"

4.1 Syntactic Features

In terms of syntactic function, "pan" and "X" constitute a verb-object relationship, and often serve as the predicate of a sentence. The subject may be an agent or a patient. Sometimes the subject can be omitted, leaving "pan X" alone constituting a verbal non-subject-predicate sentence, such as:

(14) 外地男子抖音哗众!阜阳交警隔空喊话, 皖黔交警联手 "盘"他!
Wàidì nánzǐ Dǒuyīn huázhòng! Fùyáng jiāojǐng gé kōng hǎn huà, Wǎn Qián jiāojǐng liánshǒu "pán" tā!

'A non-local man acted ostentatiously to impress others on Tik Tok! Fuyang traffic police warned him from a distance, and the traffic police of Anhui province and Guizhou province worked together to "take" him down!'

(15) 出轨男就该盘他, 为什么原谅?
Chūguǐ nán jiù gāi pán tā, wèishénme yuánliàng?
'An unfaithful man should be punished. Why should we forgive him?'

(16) 藏在古人诗词里的"三春"有多迷人?盘它!
Cáng zàigǔrén shīcí lǐ de "Sānchūn" yǒu duō mírén? Pán tā!
'How charming is the "Sanchun" hidden in ancient poems? Study it!'

In terms of the syntactic structure, "pan X" construction is flexible. In practical application, if the object "X" is known, it can sometimes be omitted. "Pan" and "X" can also contain "死 (*si*)" in between to indicate the degree of "pan", i.e. "pan-si X". In addition, "pan", like many other monosyllabic verbs, is suitable for "VV", "V 不(No) V" and other structures. For example:

(17) 网友举报违法车友会并要求"盘他", 亳州警方: 已盘。
Wǎngyǒu jǔbào wéifǎ chēyǒu huì bìng yāoqiú "pán tā", Bózhōu jǐngfāng: yǐ pán.
'Netizens reported an illegal car club and requested the police to "punish it". Bozhou Police said it was already punished.'

(18) 甭管他苏什么拉底, 万物皆可盘, 盘死他。
Béng guǎn tā Sū shénme lādǐ, wànwù jiē kě pán, pán sǐ tā.
'No matter who Socrates is, everything can be studied. Study him in depth.'

(19) 我是没时间陪他玩了, 有兴趣的去盘盘他。
Wǒ shì méi shíjiān péi tā wán le, yǒu xìngqù de qù pánpán tā.
'I do not have time to play with him. Anyone interested can do that.'

(20) 用灵魂诠释舞蹈, 用活力展现热血, 你就说这样的他你盘不盘!
Yòng línghún quánshì wǔdǎo, yòng huólì zhǎnxiàn rèxuè, nǐ jiù shuō zhèyàng de tā nǐ pán bù pán!
'He interprets dances with soul and shows enthusiasm with vitality. Tell me whether you will fall in love with him or not!'

4.2 Semantic Features

The semantics of the construction "pan X" has varied from "rubbing and squeezing something by hand, making it be covered with patina and become shiny and more valuable" to a generalized meaning to a grammaticalized one with strong emotional coloring. Some scholars have noticed these changes and analyzed the causes and paths of semantic changes from the perspectives of metaphorical mechanism and Event-domain Cognitive Model. For example, Wang found that the metaphorical meaning of "pan X" construction includes the process of metaphorical meaning and the result of metaphorical meaning [4]; Tang analyzed the semantic shift of "pan" and the semantic broadening of "X" in "pan X" by using the Event-domain Cognitive Model and explained the possibility of people understanding the generalized usage of "pan X" by using the metaphorical mechanism [1].

Although they all used metaphorical mechanism for analysis, the metaphorical meanings summarized by these two scholars are different. On the one hand, we can attribute them to different inductive angles. On the other hand, we can find that the metaphorical meaning of process and metaphorical meaning of result summarized by Wang are the preservation of the agentive meaning and causative meaning of the "pan X" construction originally used as a grammatical construction (rather than new semantic categories caused by metaphorical mechanism). Besides, the metaphorical meanings such as "to caress pets or lovers", "to pursue and worship idols or beautiful things" and "to cope with difficulties or negative emotions" summarized by Tang are essentially descriptions of detailed features of the specific social environments on which the prescriptive scenes of the rhetoric construction "pan X" depends.

In addition, in the process of semantic change, there is a semantic feature running throughout the whole process, that is, the semantic connotation of controlling. Li noticed the desire for control embodying in the "pan X" construction [3] but did not make an in-depth analysis of the causes.

In fact, this meaning of control comes from the original action meaning of "pan". The basic meaning of "pan" in the construction of "pan X" is "rubbing and squeezing something by hand, making it be covered with patina and become shiny and more valuable". "Hand" itself is an important part of body. One of its metaphorical mapping domains is "the power, control and influence of someone or an organization" because when we hold something in our hands that we can often control [20].

(21) 蒋先生既然这样做，他手底下就分成了两派，互相攻击，互相诋毁。(《战争和人》)
Jiǎng xiānsheng jìrán zhèyàng zuò, tā shǒudǐxià jiù fēnchéngle liǎng pài, hùxiāng gōngjī, hùxiāng dǐhuǐ. (Zhànzhēng Hé Rén)
'Since Mr. Jiang has done this, his subordinates are divided into two factions, attacking each other and slandering each other.' (*War and Man*)

(22) 凤举道: "打草惊蛇也不要紧,我料他们跑不出我的手掌心。" (《金粉世家》)
Fèngjǔ dào: "dǎcǎo-jīngshé yě bù yàojǐn, wǒ liào tāmen pǎo bù chū wǒ de shǒuzhǎngxīn."(Jīn Fěn Shìjiā)
'"It does not matter if we act rashly and alert the enemy. I believe that they can never hide from us," Fengju said.' (*The Story of A Noble Family*)

Therefore, although the construction of "pan X" has different semantic features in diverse contexts, it has a semantic connotation of controlling, that is, other things in the world can be included in the category of speaker's judgment or influence, which is full of strong subjective coloring. Especially, when the "pan X" construction enters stage III and the emotional coloring of speakers becomes more prominent, the semantic connotation of control becomes more prominent, too.

The semantic feature of control, as well as the above agentive and causative categories, is also consistent with Wang's viewpoint on how the construction meaning comes into being, that is, "the construction meanings come from the usages of main verbs … and are extracted from the prototype usages of verbs" [21].

4.3 Pragmatic Features

Scholars such as Che and Xu [5] have described and studied the pragmatic features of the "pan X" construction to varying degrees. Two points that need to be made here are as follows: Firstly, as a typical rhetorical construction, the prescriptive scenes of "pan X" construction have a strong dependency on specific social environments and thus, its meaning can be fully deduced only when it is put into a complete context. In this case, it is more difficult to understand the construction. However, due to its wide use and high frequency in use (at least during the time when it is still popular), the level of difficulty in understanding it is relatively reduced. Secondly, it should be noted that the "rhetoric" in the "rhetoric construction" mentioned here "generally refers to the emergence of "a large number of atypical, abnormal sentences whose use are limited by certain situations" caused by various factors in the communication process" rather than a specific rhetorical device in the traditional sense [22].

5 Conclusion

This paper has studied the construction features of the network buzzword "pan X" in 2019 from the diachronic and synchronic perspectives. The component and construction involved in the "pan X" have also been examined. First of all, by introducing the concepts of grammatical construction and rhetorical construction, we find that the development of "pan X" construction has undergone a transformation from an existing grammatical construction to a typical rhetorical construction, and the internal mechanism of this change has been explained by using the theories of construction coercion and Generative Lexicon. Secondly, through the analysis of components and constructions in grammar, semantics and pragmatics, we have a comprehensive understanding of the current use of "pan X" construction. Besides, we have also reviewed and supplemented the results from the previous studies.

By studying the examples of "pan X" construction, we find that the combination and complementation of grammatical construction and rhetorical construction can solve the contradiction between deducible constructions and non-deducible constructions. At the same time, we also find that the network language represented by the "pan X" construction is similar to other atypical and abnormal expressions in essence. To be specific, they are all products of different communicative needs calling for the language level. In addition, the diachronic research approach of "existing grammatical construction → rhetorical construction → new grammatical construction" can play an important role in testing and improving the existing research from the perspectives of metaphor and metonymy mechanism. Thus, this study can contribute to the construction system research of cognitive linguistics.

Acknowledgements. This paper is a partial outcome of the program "Research on Chinese Grammaticalization Mode" supported by "the Fundamental Research Funds for the Central Universities WUT (2020VI055)". This paper is a partial outcome of the Cooperative Education Program "Online Class Construction of 'Introduction to Teaching Chinese as the Second Language', Based on Numerous Terminals" of the Chinese Education Ministry (The Program Code: 201801049043) and also a partial outcome of the Program of OEB teaching: An Investigation of Constructing and

Reforming the Class and Teaching Mode of 'Introduction to Teaching Chinese as the Second Language', Based on the SPOC platform, (The Program Code: w2020049), which is a key program of WUT's research of teaching method reform. The program is supported by "The Fundamental Research Funds for the Central Universities" (WUT: 2021VI033 A Study on the Semantic Evolution of Willing Verbs from the Perspective of Typology.).

References

1. Tang, Y.T.: Analysis of internet buzzwords in 2018 from the perspective of event-domain cognitive model. Overseas English **18**, 68–70 (2019). (in Chinese)
2. Goldberg, A.E.: Constructions: a new theoretical approach to language. J. Foreign Lang. **7**(3), 219–224 (2003)
3. Li, J.: Analysis of "pan ta" and its cultural connotation. Culture J. **6**, 210–212 (2019). (in Chinese)
4. Wang J.: Study of metaphor in languages from the perspective of "pan ta". Res. Transm. Competence **21**, 241 (2019). (in Chinese)
5. Che, F., Xu, L.L.: Multi-faceted view of "pan ta". Yuyan Wenzi Zhoubao. **2** (2019). (in Chinese)
6. Jin, H.: Analysis of the buzzword "pan ta." Lang. Plan. **8**, 76–78 (2019). (in Chinese)
7. Su, X.: Why is "pan ta" so popular. Yuyan Wenzi Bao. **3** (2019). (in Chinese)
8. Ma, H.: Study of the internet buzzword "pan ta." Mod. Chin. **4**, 154–158 (2019). (in Chinese)
9. Cai, C.: The diachronic development of "pan." Sinogram Culture **21**, 147–148 (2019). (in Chinese)
10. Yang, J.: Analysis of the buzzword "pan ta" from the perspective of the theory of memetics. English Square **9**, 35–36 (2019). (in Chinese)
11. Liu, D.W.: From grammatical construction to rhetorical construction (part 1). Contemp. Rhetoric **3**, 7–17 (2010). (in Chinese)
12. Liu, D.W.: From grammatical construction to rhetorical construction (part 2). Contemp. Rhetoric **4**, 14–23 (2010). (in Chinese)
13. Yuan, Y.L.: On a descriptive system of qualia structure of Chinese nouns and its application in parsing complex Chinese grammatical phenomena. Contemp. Linguist. **1**, 31–48 (2014). (in Chinese)
14. Michaelis, L.A.: Type shifting in construction grammar: an integrated approach to aspectual coercion. Cogn. Linguist. **15**(1), 1–67 (2004)
15. Shi, C.: The linguistic value of construction coercion analysis. Contemp. Rhetoric **2**, 12–28 (2015). (in Chinese)
16. Shao, J.M.: On frame constructions in Chinese. Stud. Chin. Lang. **3**, 218–227 (2011). (in Chinese)
17. Taylor, J.R.: Linguistic Categorization: Prototypes in Linguistic Theory, pp.106–115. Foreign Language Teaching and Research Press, Beijing (2001)
18. Leech, G.: Semantics: The Study of Meaning. Penguin, Harmondsworth (1981)
19. Tang, M.Y.: New Exploration of Network Language. Henan People's Publishing House, Zhengzhou (2010). (in Chinese)
20. Liu, W.J.: Study of Nominal Body Metaphor from the Perspective of Cognitive Linguistics, p. 28. Xidian University, Xi'an (2011). (in Chinese)
21. Wang, Y.: Revision on construction coercion: lexical coercion and inertia coercion. Foreign Lang. Teach. **12**, 5–9 (2009). (in Chinese)
22. Lu, J.M.: From grammatical construction to rhetorical construction to grammatical construction. Contemp. Rhetoric **1**, 1–9 (2016). (in Chinese)

The Grammaticalization of Mandarin *Diao*

Gaowa Jing[1,2](✉)

[1] School of Chinese Studies, Zhejiang International Studies University, Hangzhou 310023, China
[2] Institute of Linguistics, Chinese Academy of Social Sciences, Beijing 100732, China

Abstract. This paper examines the grammaticalization of *diao* from a main verb to a resultative marker in Mandarin. *Diao* is a multifunctional word that has undergone a process of semantic evolution in its verbal meaning, from 'to swing', 'to throw', 'to fall down', to 'to disappear'. At the time when it contracted the verbal sense of falling, it came to be used in the Predicate-Complement structure and the sequence V-*diao* emerged. As the scope of the collocating verb preceding *diao* extends, the senses of *diao* also become extended from 'removal' to 'completion' or 'state change'. Diverse mechanisms such as metonymy, analogy, metaphor, generalization and the influence of Chinese southern dialects all contribute to the grammaticalization of *diao*. The development of *diao* is found to be in accordance with the development of aspectual markers in some other languages.

Keywords: *Diao* · Semantic evolution · Grammaticalization · Completive

1 Introduction

Diao is a common verb complement in contemporary Mandarin with a high frequency of use. The verb *diao* has multiple meanings, which are derived from the basic and original sense "to drop/fall down". It can be used as the main predicate, e.g., *diao yanlei* 掉眼泪 'to drop tears' or *diao jia* 掉价 'to lower the price'. It can also be used after a displacement or removal verb, which indicates that the object is detached from the source point, such as *peng diao* 碰掉 'to touch off' and *reng diao* 扔掉 'to throw away', or disappear like *shan diao* 删掉 'to delete off' and *ca diao* 擦掉 'to wipe off'. In recent years, the usage of *diao* has expanded to a phase marker, occurring after a creation verb, such as *xie diao* 写掉 'to write up (a paper)', or after some stative verbs such as *feng diao* 疯掉 'to be crazy'. This paper aims to explore the grammaticalization of *diao* in Mandarin from a cross-linguistic perspective.

Liu [1] put the grammatical meaning of V-*diao* into different categories: the object disengages, object disappear function, and the completion of events as well as state changes. *Diao* indicating object separation and object disappearance are more common in Mandarin, while the functional *diao* which indicates event completion and status changes, is used less often. Strictly speaking, *xie diao* 写掉 and *feng diao* 疯掉 should be *xie le* 写了 'to have written' and *feng le* 疯了 'to become crazy'. With the semantic evolution of *diao*, in recent years, the structure V-*diao* that represents event completion or state change has gradually increased in Mandarin. Jing and Xia [2] described the

M. Liu et al. (Eds.): CLSW 2020, LNAI 12278, pp. 173–186, 2021.
https://doi.org/10.1007/978-3-030-81197-6_16

multifunctionality of the removal markers which correspond with Mandarin *diao* in Chinese southern dialects and discussed how they are used as completive or perfect markers. There are cases of evolution from throwing verbs to completive and perfect markers in world languages, like *pelita* in Korean [3] and *shimau* in Japanese [4]. The development of *diao* corresponds with the general development of aspectual markers in Chinese dialects and some other languages.

This paper is organized in four subsequent sections. Section 2 analyzes the core sense and its semantic extension of Mandarin *diao*, and compares *diao* with *le* to discuss the nature of *diao*. Section 3 describes the path and mechanism of the grammaticalization of *diao*; Sect. 4 further discusses removal markers similar to *diao* from a typological perspective; Sect. 5 offers a summary of the paper's findings and a conclusion.

2 *Diao* in Mandarin Chinese

2.1 Multifunctionality of *Diao*

Diao is a multifunctional word in Mandarin Chinese. It can express 'to fall down', 'to disappear', 'removal', 'completion' and 'change of state' as in the following examples.

Fall Down. As a verb, the core sense of *diao* is 'to fall down', expressing 'to leave the source point downwards'. It can be used as a predicate alone, as in the following examples.

(1) a. 树叶掉了下来。

shuye	*diao*	le	xia lai.
leaves	*diao*	PERF	Down

'The leaves fell down.'

b. 男子汉不能轻易掉眼泪。

nanzihan	bu	neng	qingyi	*diao*	yanlei.
man	NEG	can	easy	*diao*	tear

'A man cannot shed tears easily.'

c. 房价什么时候能往下掉?

fangjia	shenme	shihou	neng	wang	xia	*diao*?
house price	what	time	can	to	down	*diao*

'When will house prices fall?'

Diao in (1a-c) expresses leaving the source point from the top to down. *The Modern Chinese Dictionary (Seventh Edition)* lists *'to lower'* as in (1c) as a meaning item of *diao*, which has a direct relationship with the direction downwards.

The meaning of *diao* varies in different uses, as summarized below:

Disappear. *Diao* can express disappearance when it is used with non-movable subjects without directed motion, as in the following examples.

(2) a. 钱包掉了。
qianbao *diao* le.
purse *diao* PERF
'The purse is gone.'

b. 这件衣服掉色。
zhe jian yifu *diao* se.
this CL cloth *diao* color
'The color of this piece of clothing will come out.'

Diao in example (2) loses its downward direction during the semantic extension, and it extends from 'to leave the source point' to 'out of sight' or 'away from a certain nature'.

Removal. When *diao* is used as a complement after a verb expressing the sense of motional contact, it can denote 'removal' as in the examples below.

(3) a. 赶紧拿掉。
ganjin na *diao*.
quickly get *diao*
'Get it off quickly.'

b. 你把我的手表碰掉了。
ni ba wode shoubiao peng *diao* le.
you ACC my watch touch *diao* PERF
'You knocked my watch off.'

Completion. *Diao* can also express the completion of an event when used with non-spatial verbs. The verbs which collocate with it range from removal verbs to common active verbs. *Diao* is often located at the complement of *Ba*-sentences and *Bei*-sentences.

(4) a. 赶紧把婚结掉。
ganjin ba hun jie *diao*.
quickly ACC married finish *diao*
'Get married quickly！'

b. 他这么快把事情办掉了。
ta zheme kuai ba shiqing ban *diao* le.
he so quickly ACC things do *diao* PERF
'He has got things done so quickly.'

c. 这部新上映的美剧一共 13 集，得赶紧看掉。
zhe bu xin shangying de meiju
this CL newly release REL American drama

yigong 13 ji, dei ganjin kan *diao*.
total 13 CL have to quickly watch *diao*

'This newly released American drama has a total of 13 episodes, we have to hurry to watch (all of it).'

Change of State. *Diao* can express the change of a state when used with stative predicates. Influenced by its lexical meaning, it often expresses unpleasant emotions (such as regret or complaints). Thus, seldom does *diao* collocate with positive adjectives. For example, *feng diao* 疯掉 'to be crazy' is correct while **gaoxing diao* *高兴掉 'to be happy' is odd. Furthermore, *diao* prefers decremental rather than incremental predicates. For example, *leng diao* 冷掉 'to become cold' makes sense, while *re diao* 热掉 'to become hot' sounds odd.

To sum up, *diao* is a multifunctional word in Mandarin Chinese, which develops over time from its core sense 'to leave the source point'. The functional range of *diao* is extended from a verb to a complement and then a phase marker. In a way, its use as a phase marker is similar to the aspectual marker *le* 了. This paper will focus on the functional extension of *diao* and compare it with *le* in the next section.

2.2 The Nature of *Diao* in the V-*Diao* Structure

Diao can express completion, but it is different from *le*. The following section analyzes *diao* in terms of grammatical functions, negative form, verb selection and sentence-completing.

Grammatical Functions. Termination and completion are different notions in Chinese [5, 6]. Completive *diao* provides an event perfectivity, while *le* signals boundness or termination, rather than completion.

(5) a. *这部电视剧他昨天看掉了，但是没看完。

zhe	bu	dianshiju	ta	zuotian	kan	*diao*	le
this	CL	TV drama	he	yesterday	watch	*diao*	PERF
danshi	mei	kan	wan.				
but	NEG	watch	CPL				

*'He watched (all the episodes of) the TV drama yesterday, but he didn't finish all of them.'

b. 这部电视剧他昨天看了，但是没看完。

zhe	bu	dianshiju	ta	zuotian	kan	le,
this	CL	TV drama	he	yesterday	watch	PERF
danshi	mei	kan	wan.			
but	NEG	watch	CPL			

'He watched the TV drama yesterday, but he didn't finish all of it.'

The different acceptability between (5a) and (5b) is due to the different nature of the event endpoint provided by *diao* and *le*. *Diao* provides an event with a natural endpoint and conveys completion. Thus, the event marked with *diao* is finished. (5a) does not make sense because there are semantic contradictions between the *kan diao* 看掉 and the following clause *mei kan wan* 没看完 ('not to finish reading'). In contrast, *le* in (5b) indicates termination, but not necessarily completion, and the event marked with *le* has not necessarily reached its natural endpoint. *Le* just signals the event happened [6, 7]. Thus, (5b) makes sense.

Negative Form. An event marked by *le* or *diao* can be negated with *mei*. The negative form of V-*diao* is *mei* V-*diao,* while the negative form of *V-le* is *mei V,* in which *le* is not needed, as shown in examples (6) and (7).

(6) a. 张老板的货卖掉了。

Zhang	laoban	de	huo	mai	*diao*	le.
Zhang	boss	GEN	goods	sell	*diao*	PERF

'Boss Zhang's goods sold out.'

b. 张老板的货没卖掉。

Zhang	laoban	de	huo	mei	mai	*diao.*
Zhang	boss	GEN	goods	NEG	sell	*diao*

' Boss Zhang's goods did not sell out.'

(7) a. 张老板的货卖了。

Zhang	laoban	de	huo	mai	*le.*
Zhang	boss	GEN	goods	sell	PERF

'Boss Zhang's goods sold.' (The goods were sold out, or only parts of the goods were sold.)

b. 张老板的货没卖。

Zhang	laoban	de	huo	mei	mai.
Zhang	boss	GEN	goods	NEG	sell

'Boss Zhang's goods did not sell.'

Verb Selection. Verbs that collocate with *diao* are subject to both situation types and semantics.

Situation Types. Situation aspects are traditionally called Aktionsart, a classification which was originally proposed by Vendler [8]. In Smith's framework [5], these situation types can be distinguished with three binary features: [±static], [±durative] and [±telic].

Not all situation types can co-occur with *diao*, as shown in example (8). (8 a-e) are examples of *diao* with various situation types. V-*diao* structures are not compatible with states and semelfactives.

Table 1. The situation types (Smith [5])

	Static	Durative	Telic	Examples
State	+	+	NA	xiang 想, 'to think'; love 爱, 'to love';
Activity	–	+	–	chi 吃, 'to eat';he 喝, 'to drink';
Accomplishment	–	+	+	pao yi gongli 跑一公里, 'to run a mile'; kan sanshi ye 看30页, 'to read 30 pages';
Semelfactive	–	–	–	kesou 咳嗽, 'to cough'
Achievement	–	–	+	si死, 'to die'; lan烂, 'to be rotten'

(8) a. state — *xiang *diao*; *ai *diao*
think *diao*; *love *diao*

b. activity — *gai *diao*; chi *diao*;
build *diao*; eat *diao*;

c. accomplishment — pao *diao* yi gongli; kan *diao* sanshi ye
run *diao* one mile; read *diao* thirty page

d. semelfactive — *kesou *diao*
cough *diao*

e. achievement — si *diao*; lan *diao*
die *diao*; rot *diao*

Verbal Semantics. *Diao* tends to match with verbs indicating reduction or removal and seldom with creative verbs. Thus, *chi* 吃 'to eat' in example (8b) can match with *diao*, whereas the creation verb *gai* 盖 'to build' does not hold.

In example (9 a-d), *chi diao* 吃掉 'to eat up' and *chai diao* 拆掉 'to tear down' make sense, but *zuo diao* 做掉 'to finish making' and *jian diao* 建掉 'to finish building' sound odd. Conversely, the range of the verbs preceding *le* is much greater.

(9) a. 吃掉这碗饭。

chi	*diao*	zhe	wan	fan.
eat	*diao*	this	bowl	meal

‘Eat up this bowl of rice.’

b. *做掉这碗饭。

*zuo	*diao*	zhe	wan	fan
make	*diao*	this	bowl	meal

c. 拆掉一座楼。

chai	*diao*	yi	zuo	lou.
demolish	*diao*	one	CL	building

‘Tear down a building.’

d. *建掉一座楼。

*jian	*diao*	yi	zuo	lou
build	*diao*	one	CL	building

Sentence-Completing. *Diao* often cannot complete a sentence alone. It is necessary to have *le* to present realis context and complete a sentence after *diao* as in (10a), or follow-up clauses to complete the sentence as in (10c).

(10) a. 他吃掉了这么大一碗饭。

ta	chi	*diao*	le	zheme	da	yi	wan	fan
he	eat	*diao*	PERF	so	big	one	bowl	rice

‘He ate up such a big bowl of rice.’

b. ?他吃掉这么大一碗饭。

?ta	chi	*diao*	zheme	da	yi	wan	fan
he	eat	*diao*	so	big	one	bowl	rice

c. 他吃掉这么大一碗饭，又喝了一大碗汤。

ta	chi	*diao*	zheme	da	yi	wan	fan
he	eat	*diao*	so	big	one	bowl	rice
you	he	le	yi	da	wan	tang	
again	drink	PERF	one	big	bowl	soup	

‘He ate up such a big bowl of rice and drank another bowl of soup.’

Le can complete a sentence. So in (10a), *chi diao* followed by *le* is correct. The oddity in (10b) is due to *chi diao* being unable to conclude a sentence alone in a realis context. The following clause after the *chi diao* event in (10c) successfully concludes the sentence.

To conclude, *diao* has a variety of meanings. The fundamental meaning of *diao* is ‘to fall down’ and extends to ‘to disappear’, ‘removal’, ‘completion’ and ‘a change of state’. When used as a complement after a verb, *diao* is essentially a completive, which provides the natural endpoint of the event, while *le* provides an arbitrary endpoint and signals the occurrence of the event. The different endpoints make the difference between *diao* and *le* in the form of negation, verbal-selection and sentence-completing. It also reflects the different degree of grammaticalization between *diao* and *le*. This paper will discuss the grammaticalization of *diao* in the next section.

3 The Grammaticalization of *Diao*

The diverse functions of *diao* in contemporary Mandarin is the result of grammaticalization along with semantic change and extension. This section describes the grammaticalization process of *diao* and analyzes the mechanisms involved.

3.1 The Grammaticalization Process

Along with the semantic change, *diao* was gradually grammaticalized. This paper divides the grammaticalization process of *diao* into 4 stages.

Stage 1. The original meaning of *diao* was 'to swing'. In the pre-Qin period, it was a transitive verb with a patient object, as shown in example below.

(11) …不能掉其尾…（《国语》）

bu neng *diao* qi wei (pre-Qin period)

NEG ability swing DEM tail

'…can not wag its tail'

In the Han Dynasty and the Wei and Jin Dynasty, *diao* kept the usage as a verb, indicating 'to swing', such as *diao bi* 掉臂 'to swing the arms'.

Stage 2. In the Tang Dynasty, *diao* keeps the usage as a main predicate, with a semantic extension from 'to wag/swing' to 'to throw/fling'. It collocates with a post-verbal object 'rope or whip'.

(12) 登山不拄杖, 上马能掉鞭（白居易《题裴晋公女几山刻石诗后》）

dengshan bu zhu zhang, shangma neng *diao* bian

climb mountains NEG lean crutches, ride.a.horse ability throw whip

(Tang Dynasty)

'climbing without sticks, whip while riding a horse'

In the Song Dynasty, there are cases of *diao* denoting a volitional act 'to drop/throw down'.

(13) . . . 解下一个刺绣香囊，从那窗窟龙子掉出. . . （《万秀娘仇报山亭儿》）

jiexia	yi	ge	cixiu	xiangnang,
take off	one	CL	embroidery	sachet
cong	na	chuangkulongzi	*diao*	chu......
from	tha	window	throw	out

(Southern Song Dynasty)

'...took off an embroidered sachet and threw it out from that window...'

The acquisition of 'to throw' is the key to the semantic evolution of *diao*. There has been less study of this in previous scholarship. The sense of 'to throw' provides *diao* with the characteristics 'leave the source point' and makes semantic preparations for 'fall down'. Completives in some southern Chinese dialects and other languages are derived from the throwing verbs. Besides, verbs representing 'discard' collocated with *diao* and triggered the sense of *diao* from 'to swing' to 'to drop/abandon', as is shown in (14).

(14) 掉弃兵革 （韩愈《元和圣德诗》）

*diao*qi bingge (Tang Dynasty)

abandon wars

'abandoned (the idea of) wars'

At the latest in the early Southern Song Dynasty, *diao* got the meaning of 'to drop', which is the potential results and a further development of 'to throw'.

(15) …如掉在水中一般(《朱子语类》卷一百六)

ru	*diao*	zai	shui	zhong	yiban	(Southern Song Dynasty)
resemble	drop	LOC	water	LOC	resemble	

'like falling into water'

Stage 3. *Diao* and the removal verbs constitute the V-*diao* Predicate-Complement structure (PC structure for short) in the Southern Song Dynasty. For example, *geng diao* 更掉 'to change' and *chu diao* 除掉 'to remove'.

(16) …更掉了这个…(《朱子语类》卷一百三十九)

geng *diao* le zhe ge (Southern Song Dynasty)

change *diao* PERF this CL

'changed this'

Stage 4. Further along the extension, *diao* in VC structure may indicate the completion of event by profiling the 'completive change' as a motional-path change, as shown in (17a-b).

(17) a.用掉了六十多两银子（《二刻拍案惊奇》）

yong *diao* le liushi duo liang yinzi (Ming Dynasty)

spend COMP PERF 60 more tael silver

'spent more than sixty taels of silver'

b.替人家办掉两桩事情（《官场现形记》）

ti renjia ban *diao* liang zhuang shiqing (Qing Dynasty)

BEN others do COMP two CL things

'finish two things for others'

When the first verb in VC gets generalized from an event to a state, *diao* naturally denotes a 'change of state'. It has to be noted that the use of *diao* preserves the original path schema in a downward movement in the extended sense of stative change, and typically collocates with a V implying a downward or derogative sense, such as *lan* 烂 in example (18).

(18) 把绳都烂掉在山上…（《醒世恒言》）

ba sheng dou lan *diao* zai shan shang (Ming Dynasty)

ACC rop all rot COMP LOC montain LOC

'…Rot the rope on the mountain…'

The grammaticalization stages of *diao* are illustrated in Table 2.

Table 2. Grammaticalization stages of *Diao*

Gramm. stages	Descriptions
Stage 1	*Diao* as main predicate denoting 'wag, swing' (Tr. or Intr)
Stage 2	*Diao* as main predicate denoting 'throw down'
Stage 3	*Diao* as a V_2 in VV sequence denoting 'detached, removal'
Stage 4	*Diao* as V_2 in VV sequence denoting 'event completion' or 'change of state' due to space-to-time transfer

3.2 Mechanisms for Grammaticalization

This section analyzes the Grammaticalization mechanisms of *diao*. Diverse mechanisms such as metonymy, analogy, metaphor, generalization and the influence of Chinese southern dialects contributed to the grammaticalization of *diao*.

Metonymy. The grammaticalization of *diao* is related to a larger event and its end phase, which is based on metonymy. The original meaning of *diao* was 'to swing' and then 'to throw'. When *diao* focuses on the end phase of the motion, it acquires the sense of 'falling down'. When aligned with spatial motion verbs such as pao 跑 'to run', it profiles the result and expresses 'away from the source point'. When collocated with active verbs such as *mai* 买 'to buy', it highlights the endpoint of the event and expresses completion.

Analogy. Liu [1] and Cao [9] clarified that the formation of PC structure V-*diao* is due to the fact that *diao* was often in the position of the second verb of the compound predicates and gradually became a complement and then further grammaticalized. However, this paper puts forward another possibility, that is, the formation of PC structure V-*diao* is due to the analogy mechanism.

When *diao* acquired the sense of 'falling down', PC structures had been widely used. Wang [10] claimed that the PC structure was formed in the Han Dynasty. However, Jiang [11] believed that only a limited number of verbs formed the PC structure in the Han Dynasty. It was during the Period of Wei Jin and the Northern and Southern Dynasties that PC structures became widely used. Although there is some controversy as to when PC structures formed, they had been widely used since the period when *diao* acquired the sense of 'falling down'. Therefore, we can infer that, affected by the analogy mechanism, when *diao* acquired the meaning of 'falling down' it was able to be located behind verbs and form the V-*diao* PC structure.

Metaphor. *Diao* extended from a removal verb in the spatial domain to a phase marker in the temporal domain, which was due to the metaphor mechanism. Removal *diao* highlights the endpoint of the spatial movement in the spatial domain. Completive *diao* signals completion in the temporal domain, focusing on the endpoint of the event. In addition to the spatial and temporal domains, *diao* can also express 'away' in cognitive domains, which is a more abstract concept, such as *wang diao* 忘掉 'to forget' and *mo diao jiyi* 抹掉记忆 'to erase memory'.

Generalization. The grammaticalization of *diao* is related to generalization. The removal verbs combined with *diao* indicate 'moving' or 'disappearing', such as *pao* 抛 'to throw', *fu* 拂 'to wipe gently' and so on. Gradually, some active verbs, such as *shuo* 说 'to say' and even achievement verbs like *si* 死 'to die' or *wang* 忘 'to forget' and some adjectives with passive meanings like *lan* 烂 'to be rotten' can collocate with *diao.* In present usage, the verbs that go with *diao* are more extensive, and even some creation verbs can match it, such as *mai* 买 'to buy'.

The Influence of Chinese Southern Dialects. The evolution of *diao* is also influenced by southern dialects. Jing's survey [12] on the frequency of V-*diao* in the literary works of southern dialect speaking writers and Beijing dialect speaking writers found that the frequency of V-*diao* in the works by southern dialect writers is much higher than those by Beijing writers. The frequency of V-*diao* in Mandarin and the Beijing dialect has gradually increased since the Ming and Qing Dynasties. The range of verbs in the V-*diao* structure has become more and more extensive. Removal markers (Jing & Xia [2]) exist widely in Chinese southern dialects, of which the syntactic distributions and functions are similar to Mandarin *diao*. Thus, we infer that the grammaticalization of *diao* in Mandarin is due to the influence of the removal markers in Chinese southern dialects.

4 Further Discussion

The morphemes which collocate with verbs and express 'loss' or 'depletion' are collectively referred to as removal markers in Jing & Xia [2]. The most distinctive characteristic of Chinese removal markers is that they can only make collocations with verbs which have the semantic feature of 'reduction' or 'loss' in irrealis context. In the process of evolution, the functions of the removal markers came into being sequentially: removal, completion, realis deviation, anterior continuing, and even perfective.

The occurrence of V-*diao* in Mandarin is less frequent than in Southern dialects, which is related to the removal marks of the Southern dialects. The multifunctionality of removal markers reflects the degree of semantic expansion. Although the Mandarin *diao* has been further expanded in recent years, it has not yet developed the perfect function.

When we broaden our horizons, we find that there are examples of perfect or completive markers in some languages that come from removal verbs. *Pelita* develops semantically and functionally on the basis of the sense 'throwing away' and can be used as a completive mark after a verb (Baik [3]). *Shimau* in Japanese is also a multifunctional word. When used as a verb, it means 'to put aside', or 'to complete'. It can also be used as a perfect marker (Ono [4]).

In addition to the Japanese *shimau* and Korean *pelita*, some languages have developed the use of the completive or perfective, like *pet* in Palaung and the *kai* in Fore. The *wara* in Diyari also further develops the usage of the past tense. (Heine & Kuteva [13]).

5 Conclusion

This paper has examined the multifunctionality of *diao* in contemporary Mandarin and its grammaticalization. *Diao* is essentially a completive in the V-*diao* structure, which provides a natural endpoint to the event. *Diao* can express 'falling down', 'disappearing', 'removal', 'completion' and 'change of state'. The original meaning of *diao* was 'to swing'. Then, it expanded to 'to throw', 'to fall down', and 'to disappear'. As the scope of the collocating verbs expanded, the senses of *diao* extended from 'displacement' in the spatial domain to 'completion' and 'change of state' in the temporal domain. This paper has also taken the view that various mechanisms of grammaticalization, such as metonymy, analogy, metaphor, generalization and influence by Chinese southern dialects have operated in the course of grammaticalization.

Finally, this paper suggests two issues which are worth studying further. Firstly, semantic change is based on cognitive transfer. Thus, it will be necessary to explain the semantic expansion of *diao* in detail. Secondly, since removal markers exist widely in southern Chinese dialects and are found in other languages, it would be intriguing to investigate the functions of the removal markers in different dialects or languages and analyze the grammaticalization process from the typological perspective.

Acknowledgments. The findings presented in this paper draw on wider research conducted as part of projects supported by the Scientific Research Fund of Zhejiang Provincial Education Department (Grant no. Y202045508), the China Postdoctoral Science Foundation (Grant no. 2019M660921), and the National Social Science Foundation of China (Grant no. 19AYY004) whose support I would like to acknowledge. Special thanks to Prof. Danqing Liu who read a previous version of this paper in Chinese and made many valuable suggestions, and Prof. Meichun Liu who provided a number of instructive ideas for further study. I would also like to thank my colleague Dr James O'Donnell who did greatly help in editing the final version of this paper.

Abbreviations

ACC Accusative
CPL Completive
CL Noun class
NEG Negation
PERF Perfect
REL Relative

References

1. Liu, Y.: "V diao" de Yuyi Leixing yu "Diao" de Xuhua "V 掉"的语义类型与"掉"的虚化 [Semantic Categories of V Diao and Grammaticalization of Diao]. Zhongguo Yuwen [Stud. Chin. Lang.] **2**, 133–143+191–192 (2007)
2. Jing, G., Xia, L.: Hanyu Fangyan Quchuyi Biaoji de Duogongnengxing yanjiu 汉语方言去除义标记的多功能性研究 [A study of the multifunctionality of removal markers in Chinese dialect]. Yuyan Yanjiu [Stud. Lang. Linguist.] **3**, 46–54 (2020)
3. Baik, J.: On grammatical evolution of Korean displacement verb pelita 'throw away' from a grammaticalization perspective. Int. J. Lang. Linguist. **1**, 49–56 (2016)
4. Ono, T.: The Grammaticalization of the Japanese verbs Oku and Shimau. Cogn. Linguist. **4**, 367–390 (1992)
5. Smith, C.: The Parameter of Aspect. Kluwer Acadamic Press, Boston (1991)
6. Liu, M.: Tense and aspect in Mandarin Chinese. In: Wang, W.S.-Y., Sun, C. (eds.) The Oxford Handbook of Chinese Linguistics, pp. 274–289. Oxford University Press, Oxford (2015)
7. Chu, C.: "Wancheng" "Youjie" he "Le_1" de Yinxian: Yuyi Jufa yu Pianzhang de Fen'gong Hezuo "完成""有界"和"了$_1$"的隐现: 语意, 句法与篇章的分工合作 [A study of "Perfectivity" "Boundedness" and "the Looming of Le_1" from the Interface of Semantics, Syntax and Discourse]. In: Chin, C., Kwok, B., Tsou, B. (eds.) Xiandai Hanyu Yuyanxue zhi Fu: Zhao Yuanren Xiansheng Jinian Lunwenji 现代汉语语言学之父: 赵元任先生纪念论文集 [Commemorative Essays for Yuen Ren Chao, Father of Modern Chinese Linguistics], pp. 162–172. Crane Publishing, Taipei (2016)
8. Vendler, Z.: Linguistics in Philosophy. Cornell University Press, Ithaca (1967)
9. Cao, J.: "V Diao" De Yufahua "V掉"的语法化 [the Grammaticalization of "V Diao"], Qiannan Minzu Shifan Xueyuan Xuebao. [J. Qiannan Normal Univ. Nationalities] **2**, 16–21+67 (2009)
10. Wang, L.: Hanyushi Gao 汉语史稿 [The History of Chinese]. Science Press, Beijing (1958)

11. Jiang, S.: Hanyu Dongjieshi Chansheng de Shidai 汉语动结式产生的时代 [the times when Chinese verb-resultative complement emerged]. In: Jiang, S.: Hanyu Cihui Yufa Shi Lunwenji 汉语词汇语法史论文集 [Collection of Papers on the History of Chinese Vocabulary and Grammar], pp. 240–262. The Commercial Press, Beijing (2001)
12. Jing, G.: Hanyu Wancheng Fanchou Biaoji de Yufahua Yanjiu——yi Jieshuyi Dongci he Quchuyi Dongci Weili 汉语完成范畴标记的语法化研究——以结束义动词和去除义动词为例 [Grammaticalization of Chinese Completion Markers——Taking 'Finish' and 'Removal' Verbs as Examples]. Beijing Normal University, Beijing (2017)
13. Heine, B., Kuteva, T.: World Lexicon of Grammaticalization. Cambridge University Press, Cambridge (2002)

Mandarin *Physical Contact* Verbs: A Frame-Based Constructional Approach

Meichun Liu[1], Tianqi He[3(✉)], Hongfeng He[2], and Yifan Cao[2]

[1] Department of Linguistics and Translation, City University of Hong Kong, 83 Tat Chee Avenue, Kowloon Tong, Kowloon, Hong Kong
meichliu@cityu.edu.hk

[2] Department of Chinese Language and Literature, Hua Zhong University of Science and Technology, 1037 Luo Yu Avenue, Hong Shan District, Wuhan, China
caoyifan@hust.edu.cn

[3] Department of Chinese Language and Literature, Central South University, 605 Lu Shan South Avenue, Yue Lu District, Changsha, China
458047682@qq.com

Abstract. This study aims to provide a preliminary analysis and classification of Mandarin *Physical Contact* (MPC) verbs, such as *dǎ/qiāo/pāi/jī* 打/敲/拍/擊 'hit/ knock/ pat/ strike', *cā/shì* 擦/拭 'wipe', *wò/zhuā* 握/抓 'hold', under the lexical-constructional approach. The conceptual schema of physical contact with salient semantic components is proposed as the base for various lexicalized profiles in different semantic frames. Constructional patterns and features are also identified as morpho-syntactic criteria for distinguishing semantic frames of MPC verbs. Based on semantic-to-syntactic evidence, MPC verbs are distinguished into three primary frames: verbs of *Contact by Impact* (CI verbs), verbs of *Contact by Surface Motion* (CS verbs) and verbs of *Contact by Holding* (CH verbs). CI verbs involves force and impact, which tend to collocate with force-related adverbials and impactive resultatives. CS verbs involve the coding of locus and locatum and thus can take either the Substance-Figure or Substance-Ground as direct object. CH verbs indicate (intentional) prolonged contact without motion, most of which can collocate with stative or resultative markers. This study provides a comprehensive scheme with lexical-constructional criteria for analyzing the semantic distinctions of MPC verbs, which constitutes a major task in the construction of Mandarin VerbNet and plays a significant role in the study of verbal semantics.

Keyword: Mandarin *physical contact* verbs · Lexical-constructional approach · *Contact by impact* verbs · *Contact by surface motion* verbs · *Contact by holding* verbs

1 Introduction

CONTACT is generally considered as one of the most basic semantic components of human languages as it can be observed from various verb categories of different languages.

M. Liu et al. (Eds.): CLSW 2020, LNAI 12278, pp. 187–205, 2021.
https://doi.org/10.1007/978-3-030-81197-6_17

Numerous studies concerning English verbal semantics have noticed the ubiquity of CONTACT as a semantic characteristic. [1] proposed that English *Spray, Poke, Hug* verbs all imply the act of contact, although they encode very different event categories. [2] attempted to describe the semantic properties of English verbs from the perspective of dynamicity and divided physical interactions between the agent and the real world into three stages: before, during and after contact. On the other hand, some scholars considered CONTACT to be a primitive concept and used it as a distinguishing semantic feature to classify English verb classes. [3] distinguished English *Running, Speaking, Hitting, Cutting* and *Change of State* verbs with neuroscientific evidence by contrasting five distinctive semantic components: ACTION, MOTION, CONTACT, CHANGE OF STATE and TOOL USE. [4] even argued that English verbs of *Instrument of Communication* such as *'Telephone'* and *'Skype'* also behave as *Contact* verbs in which a communicator tries to get in touch with an addressee by means of the instrument. This can be viewed as a metaphorical transfer from physical to communicational contact.

Moreover, it is observed that verbs of other languages also lexicalize CONTACT as an essential lexical feature. [5] compared the language-specific lexicalization patterns of English and Korean *Motion* verbs and found that Korean *Putting* verbs also entail CONTACT as a semantic feature. [6] further specified the core semantics of Korean *Putting* verb *'kki-ta'* as profiling the surface contact between Figure and Ground. [7] provided a thorough classification of Jaminjung verbs of *Contact/Force*. [8] analyzed the phonological characteristics of Japanese mimetic verbs and found that certain consonants of mimetic verbs can convey the meaning of CONTACT. [9] pointed out that the semantic properties of Japanese *'tuku'* verbs are distinct from *'suru'* verbs in that they inherit the meaning of *'surface contact'* from the source verb *'suku'*.

In sum, previous studies tend to consider CONTACT as an essential semantic component that is lexically encoded and distinguished in verb meanings, which has been testified with cross-linguistic evidence.

1.1 Previous Studies on *English Physical Contact* (EPC) Verbs

As mentioned above, CONTACT is a universal lexical semantic feature in action verbs. Previous works on *English Physical Contact* (EPC) verbs suggested that they encompass a diverse range of event types and further distinguished the verbs into different subsets based on the manner and degree of contact. For example, [10] differentiated verbs of *Contact by Impact* and verbs of *Contact* on the basis of whether directed motion can be syntactically encoded. It was proposed that *Contact by Impact* verbs may participate in conative alternation that implies directed motion, but pure verbs of contact may not. Examples from Levin are cited here:

(1) *Contact by Impact* verbs:
a. Paula hit the fence (with the stick).
b. Paula hit at the fence (with the stick). (Levin 1993: 149)

(2) *Contact* verbs:
a. Carrie touched the cat.
*b. *Carrie touched at the cat.* (Levin 1993: 155)

It was proposed that *Contact* verbs simply describe surface contact without encoding a directed motion through impact. There are also works concentrating on the specific group of *Contact by Impact* verbs and the two prototypical members *hit* and *break* were further contrasted with an analysis of their varied morpho-syntactic behaviors [11]. In [12], English verbs *Strike, Hit* and *Beat* were taken as *Physical Contact* verbs and examined in terms of specific lexical patterns, as opposed to one Swedish verb slå, which involved a wide range of semantic fields such as Physical Contact, Open, Postural, Fighting and even Disconnection.

On the other hand, some studies focused on the syntactic and semantic properties of verbs of *Surface Contact*. [13] treated *Hit* and *Slap* as verbs of *Surface Contact* and examined their semantic-syntactic differences under a frame-based approach. [14] listed *Comb, Rub, Scratch, Shovel, Sweep, Wipe* as *Surface Contact* verbs on the basis that they 'do not require their normal direct object, even when quantized, to be analyzed as an incremental theme'.

Some works went a step further to explore the force dynamic relation entailed in the act of contact and its semantic implication and syntactic patterns. [15] proposed that English *Restrain* verbs such as *Restrain, Prohibit, Bar* 'seem to imply some physical contact, and may therefore have an accompaniment phrase in the subject', as in *'John restrained me from watching television, but I managed to get to do it now'*. In addition, it is commonly believed that verbs involving 'Contact' indicate a transitive relation and display most of the morpho-syntactic correlates of high transitivity [3, 16–19].

In sum, previous literature pointed to two major sub-classes of *Contact* verbs that are most extensively studied, i.e., verbs of *Contact by Impact* and verbs of *Surface Contact*.

1.2 Previous Studies on Mandarin *Physical Contact* (MPC) Verbs

With regard to Chinese verbs of *Physical Contact*, such as *dǎ/pāi/qiāo/mō* 打/敲/拍/摸 'hit/knock/pat/touch', there are a number of studies that took different approaches and covered different ranges of verbs. Some major works took a comparative approach and analyzed Mandarin *Physical Contact* verbs from the perspective of cross-linguistic comparison. An extensive investigation of the contrastive lexicalization patterns of English and Chinese *Physical Contact* verbs has been conducted by [20], which argued that both MPC and EPC verbs are naturally combined with the *Motion* frame. It was suggested that the motion of Contact and Path are more closely combined in Chinese Physical Contact verbs due to the usage of co-verb *bǎ*, which requires resultative complements. [21] further explored the different semantic-syntactic behaviors of verbs of *Contact by Impact* in English and Chinese, and classified Chinese *Contact by Impact* verbs according to the specific body part involved.

Other studies focused mainly on the classifying criteria of MPC verbs. [22] attempted to distinguish the potential sub-types of MPC verbs according to [10]'s approach of diathesis alternation. Six subcategories have been proposed on the basis of their syntactic and semantic patterns. Moreover, [23] considered all MPC verbs as bivalent verbs and categorized them into three subcategories on the basis of semantic characteristics such as Contact, Attachment, Continuity, Motion and Impact.

Although both semantic and syntactic behaviors have been considered in previous literature, there still lacks a consensus on the scope and criteria in defining *Contact* verbs. This study attempts to offer a consistent, principled and comprehensive framework for analyzing Mandarin verbs of *Contact*. It aims to tease out the detailed lexical semantic distinctions encoded in MPC verbs, especially the language-specific lexicalization patterns unique to the Chinese verbal lexicon. By doing so, this paper presents a preliminary effort in providing a systematic categorization of MPC verbs on the basis of corpus distributions of the semantic-to-syntactic features from a frame-based constructional approach [24, 25].

1.3 Frame-Based Constructional Approach

The varied syntactic behaviors of contact verbs call for a systematic and fine-tuned investigation of MPC verbs that can be incorporated in a more comprehensive framework of lexical information representation. As an attempt in constructing such a framework, the present study looks closely into the lexicalized distinctions in MPC verbs with a frame-based constructional approach, based on theoretical premises in Frame Semantics [26, 27] and Construction Grammar [28–30]. The approach has been adopted to build a linguistically valid and computationally feasible lexical semantic database called the **Mandarin VerbNet** (http://verbnet.lt.cityu.edu.hk/#/), see [31–34], with the aim of providing *'lexical semantic information based on grammatical descriptions and anchored in linguistic theories'* [24]. The approach incorporates both semantic and syntactic analyses in defining event frames that verbs are anchored in. Each semantic frame is defined with frame-specific elements, i.e., characteristics of participant role of the frame [35] and collo-constructional patterns associated with the event frame as a form-meaning mapping construct in the sense of Constructional Grammar [28]. Frames are further structured into a hierarchical scheme: Archi-frame (major semantic domains with a well-defined conceptual schema) to primary frame (subparts of the schema with varied sets of core frame elements) to basic frame (basic-level categories with distinct semantic profiles). For illustrations of the framework, please see [24, 36]. The analyses of MPC verbs to be presented in this paper are mainly based on distributional data found in Sinica corpus and Gigaword.

In the next section, the lexical semantic components underlying the different frames of MPC verbs will be identified, which serve as the foundation to distinguish possible sub-types as a first step. We will then discuss how the semantic components are realized in varied collo-constructions in Sect. 3, which provides a corpus-based analysis of the defining patterns in different sub-types and a general classification of MPC verbs based on distinctive semantic features lexicalized in the verbal roots as well as the associated constructional patterns. Section four then summarizes the study with a conclusion.

2 Lexicalization Patterns of MPV Verbs

[37] discussed lexical semantic properties by analyzing the relations between meaning components and surface expressions. It was proposed that the semantic components are correlated with surface expressions and the correspondence between the two is not one-to-one; that is, a set of semantic features can be conveyed by a single surface form and vice versa. This approach is helpful to the exploration of lexicalized semantic components and corresponding surface patterns of Mandarin *Physical Contact* verbs, which serve as the starting point to further specify possible sub-types of MPC verbs, according to the observed semantic-to-syntactic distinctions in different groups of verbs.

2.1 Major Observations of Syntactic Variations

[20, 21] provided a pilot study of the semantic properties of *Mandarin Physical Contact* (MPC) verbs, which may essentially involve motional contact. However, MPC verbs include a wide and diverse range of verbs that display quite varied syntactic behaviors or collo-constructional patterns. For example, impactive contact verbs such as *dǎ/pāi/qiāo/jī* 打/拍/敲/擊 'hit/pat/knock/strike' behave very differently from motion verbs (e.g., *gǔn*滚 'roll') in several respects: 1) while motional Path with Goal (marked by *dào* 到 'to') or Direction (marked by *jìn/chū/shàng/xià* 進/出/上/下 'into/out of/up to/down to) is the characteristic of Motion verbs, it does not typically occur with contact verbs; 2) manner adverbials of force (e.g., *měng liè* 猛烈 'violently') is frequently found with impact verbs but not with motion verbs; 3) resultatives due to strong impact (e.g., *pò* 破 'broken', *suì* 碎 'smashed') are frequently used with contact verbs but not motion verbs. The differences are illustrated below with corpus examples and distributions:

(3) Different PPs: Path-PP (*dào*+Loc. Gaol) vs. Result-PP (*dào*+Affected Patient)

a. 球 一路 **滚到 外野**
Qiú yīlù ***gǔn-dào wàiyě***
Ball all-the-way roll-DAO outfield
'The ball rolled all the way to the outfield.'

b. 李 老師 可能 有 **踢到** **學生**
Lǐ lǎoshī kěnéng yǒu tī-dào xué shēng
Li teacher maybe have kick-DAO student
'Mr. Chen might have kicked the student.'

(4) Different Manner Adverbials: Forceful adverbials with MPC

a. 球 **慢慢** 滚 向 底線
Qiú mànmàn gǔn xiàng dǐ xiàn.
Ball slowly roll toward baseline
'The ball rolled slowly toward the baseline.'

b. 吳國棟 **用力** 打 了 三下 銅鑼
Wú Guódòng yònglì dǎ le sānxià tóngluó.
WuGuodong violently hit LE three-times gong
'Wu Guodong banged the gong three times.'

(5) Different Resultative in V-R sequences: Impactive vs. non-impactive result

a. 籃球 歪歪斜斜 地 **滚遠** 了
Lán qiú wāiwāixiéxié de gǔn-yuǎn le
Basketball askew DE roll-far LE
'The basketball rolled away askew.'

b. 民眾 曾 攔車 **打破** 車輛玻璃
Mínzhòng céng lánchē dǎ-pò chēliàngbōlí.
People once stop-cars hit broken car-window
'People have stopped cars to break the windows.'

The above examples demonstrate the distinctive syntactic behaviors of *Motion* (*gŭn* 滚 'roll') in contrast to *Contact* verbs (*dă/tī* 打/踢 'hit/kick'). While Motion-oriented verbs require a Path-complement and collocate most frequently with adverbs of motional manner or speed, Contact-based verbs do not require a directional Path and prefer forceful adverbs and impactive results. As shown in Table 1, a comparison between the verbs *gŭn* 滚 'roll' and *tī* 踢 'kick' in Gigaword Corpus may further illustrate the differences. It is clear that motion verb *gŭn* 滚 'roll' is more likely to take a motional path-PP:

Table 1. Path-PP in motion vs. Result-PP in contact in Gigaword.

	gŭn 滚 'roll'	*tī* 踢 'kick'
Path-PP (*dào* + Locative Goal)	**93/94 (98.94%)**	94/420 (22.38%)
Result-PP (*dào* + Affected Patient)	1/94 (1.06%	**324/420 (77.62%)**

Further observations of refined grammatical features are found with MPC verbs. As also mentioned in [20], a postverbal Path-complement may be triggered by the use of disposal co-verb *bă* in a *BA*-construction, such as *tā bă huāpíng cóng zhuōshàng zhuàng xià lái le* 他把花瓶从桌上撞下来了 'he bumped the vase down from the table'. However, even with a postverbal PP-complement, most MPC verbs prefer the stative Endpoint-PP with the locative marker *zài* 在 'at/on', instead of the Goal-PP marked by the motional marker *dào* 到 'to/unto'. In other words, MPC verbs may lexically profile contact with an affected figure as an end result, rather than profiling a motional path[1]. This semantic preference can be seen from the frequent co-occurrence with the locative marker *zài* 在 'at/on' in different types of contact verbs, as shown blow:

[1] Please note that the group of surface contact verbs such as *cā/săo/tú/mò* 擦/掃/塗/抹 'rub/sweep/paint/smear', which encode contact by surface motion may occasionally encode a motional path as in *bă huīchén săo dào dìshàng* 把灰尘掃到地上 'Sweep the dust to the floor'.

(6) Verbs of *Contact by Impact* (CI verbs): *dǎ* 打 'hit'
李文生 則 用 鐵桿 **打在** 球道 中央
Lǐ Wénshēng zé yòng tiěgǎn dǎ-zài qiúdào zhōngyāng.
Li Wensheng ZE use iron stick hit-ZAI fairway middle
'Li Wensheng hit the fairway in the middle with an iron stick.'

(7) Verbs of *Contact by Surface Motion* (CS verbs): *cā* 擦 'rub'
研究小組 把 含有 Lunacin的 乳液 **擦在** 實驗鼠 身上
Yánjiū xiǎozǔ bǎ hányǒu Lunacin de rǔyè cā-zài shíyànshǔ shēn-shàng.
Research team BA contain Lunacin DE lotion rub-ZAI lab-rat body-on
'The research team rubbed the lotion containing Lunacin on the rats.'

(8) Verbs of *Contact by Holding* (CH verbs): *wò* 握 'hold'
大陸民眾 都 將 美元 **握在** 手中
Dàlù mínzhòng dōu jiāng měiyuán wò-zài shǒu-zhōng.
People in Mainland all BA dollars hold-ZAI hand-in
'People in the Mainland are holding dollars in their hands.'

As shown above, *Physical Contact* verbs tend to collocate with *zài* 在 'at/on' when coding a path-Endpoint as the Ground for contact, which is a common characteristic shared by all contact verbs. Nevertheless, finer distinctions on the Ground component can still be found among different types of MPC verbs. While the contact Ground may include human or non-human locations for verbs of *Impactive Contact* and *Surface Contact* ((6) and (7)), it is quite restricted for verbs of *Holding,* which only allow the body part 'hand' to be coded as a contact Ground. This finer distinction indicates that verbs of *Holding* are lexically specified with a designated body part 'hand' in their meanings.

It is clear that different types of MPC verbs display various ranges of semantic-to-syntactic specificities and hence can be further distinguished into potential subcategories. If we compare the three semantically distinct groups of MPC verbs mentioned above, namely *Contact by Impact* (CI verbs), *Contact by Surface Motion* (CS verbs) and *Contact by Holding* (CH verbs), it will be obvious that they differ in the following four ways:

1) In terms of direct objects, only CS verbs such as *cā/shì* 擦/拭 'wipe', but not CI or CH verbs, can 'take either the Substance-Figure or the Surface-Ground as the direct object' [25].

(9) Objects for CS verbs: either Figure (Substance) or Ground (Surface)

a. 如 仍 繼續 在 陽光下 活動， 仍 應 擦 **防曬油**
Rú réng jìxù zài yángguāng-xià huódòng, réng yīng cā fángshàiyóu.
If still continue ZAI sun-XIA activity still should rub sunscreen
'Rub the sunscreen if you continue your activities in the sun.'

b. 大會 工作人員 拚命 用 拖把 和 抹布 擦 **地板**
Dàhuì gōngzuòrényuán pīnmìng yòng tuōbǎ hé mòbù cā dìbǎn.
Conference staff desperately use mop and cloth wipe floor
'Conference staff wiped the floor desperately with mops and wiping cloths.'

As shown above, the prototypical CS verb *cā* 擦 'rub' can allow the alternation of encoding either the Figure (Substance) or the Ground (Surface) as the direct Object, which is not attested with CI and CH verbs.

2) In terms of adverbial adjuncts, only CI and CH verbs tend to collocate with forceful manner adverbs, while CS verbs are more likely to co-occur with instrumental adjuncts marked by *yòng* 用 'use'. The distributional skewing in taking manner vs. instrument among three different verbs are given in Table 2; the difference in manner adverbs is provided in Table 3; and corpus examples are given in (10).

Table 2. Manner vs. instrumental adjunct distribution in Gigaword.

	dǎ 打 'hit'	wò 握 'hold'	cā 擦 'rub'
Manner	1195/89698 (1.33%)	249/3404 (7.32%)	32/2363 (1.35%)
Instrumental	512/89698 (0.51%)	18/3404 (0.53%)	114/2363 (4.82%)

Table 3. Force-related vs. non-force-related manner distribution in Gigaword.

	dǎ 打 'hit'	wò 握 'hold'	cā 擦 'rub'
Force-related	810/1195 (67.78%)	239/249 (95.98%)	13/32 (40.63%)
Non-force-related	385/1195 (32.22%)	10/249 (4.02%)	19/32 (59.38%)

(10) Different adverbials: Manner vs. Instrumental adverbials

a. 他 曾 **猛** 敲 樓頂 違建 的 門
Tā céng měng qiāo lóudǐng wéijiàn de mén.
He once violent knock roof built-illegally DE door
'He knocked on the illegally built door on the roof violently.'

b. 她 的 生母 顏刊 一直 **緊緊** 握著 她 的手
Tā de shēngmǔ Yán Kān yīzhí jǐnjǐn wò-zhe tā de shǒu.
She DE mother Yan Kan always tight hold-ZHE she DE hand
'Her mother Yan Kan kept holding her hand tightly.'

c. 她 **用** 毛巾 擦 一把 汗
Tā yòng máojīn cā yībǎ hàn.
She use towel wipe handful sweat
'She wiped off the sweat with a towel.'

As shown above, CI and CH verbs frequently occur with manner adjuncts while CS verbs tend to choose instrumental adjuncts. In terms of manner adjuncts, CI and CH verbs show a clear preference in collocating with force-related adjuncts.

3) Regarding VR constructions, CI verbs frequently take resultatives indicating change of state by impact, such as *suì/pò* 碎/破 'broken'; CH verbs tend to collocate with resultatives denoting tightly gripping, like *jǐn* 緊 'tightly'; and CS verbs mostly enter VR constructions implying appearance or disappearance, such as *diào/xià* 掉/下 'off'. Distributional evidence is given in Table 4 and detailed examples are listed in (11).

Table 4. Distributions of VR constructions in Gigaword.

Change of State	dǎ 打 'hit'	wò 握 'hold'	cā 擦 'rub'
Impactive initiated	36991/89698 (44.22%)	0	38/2363 (1.61%)
Gripping related	1713/89698 (1.91%)	161/3404 (4.73%)	0
Appearance related	0	0	149/2363 (6.31%)

(11) Different VR constructions: impactive vs. gripping vs. appearance

a. 嫌犯 **打碎** 玻璃
Xiánfàn dǎ-suì bōlí.
Suspect hit-broken glass
'The suspect broke the glass.'

b. 鄧穎超 同志 **握緊** 他 的 手
Dèng Yǐngchāo tóngzhì wò-jǐn tā de shǒu.
Deng Yingchao comrade hold-tight he DE hand
'Comrade Deng Yingchao held his hands tightly.'

c. 她 趕緊 **擦掉** 眼淚
Tā gǎn jǐn cā-diào yǎnlèi.
She quickly wipe-off tears
'He wiped off his tears quickly.'

As shown above, the three representative verbs show different preferences in V-R combinations: CI verb *dǎ*打 'hit' prefers impactive change of state as in (11a); CH verb *wò* 握 'hold' tends to co-occur with gripping-related resultatives as in (11b); CS verb *cā* 擦 'wipe' frequently collocates with appearance-pertaining resultatives as in (11c).

4) With respect to aspectual marking, only CH verbs frequently co-occur with the durative state marker *zhe* 著, coding prolonged contact, as shown below.

Table 5. Distributional skewing with durative *zhe* 著in Gigaword.

	***dǎ*打 'hit'**	***wò*握 'hold'**	***cā*擦 'rub'**
zhe 著 'ZHE'	3454/89698 (3.85%)	**999/3404 (29.35%)**	189/2363 (8.00%)

(12) Durative state markers:
青年 **握著** 吐信 的 蛇
Qīngnián wò-zhe tǔxìn de shé.
Youngman hold-ZHE spit DE snake
'The young man is holding the spitting snake.'

In sum, MPC verbs display quite various patterns in the corpus and therefore, an extensive study on the semantic components conflated in the verb roots and the corresponding syntactic behaviors of MPC verbs is significant in further distinguishing their potential subcategories.

2.2 Semantic Components Conflated by MPC Verbs

Taking *Physical Contact* as an overarching semantic domain (an Archi-frame), we can clearly define its conceptual schema as encoding 'an Agent engaged in a motional Act involving a specific Body Part or by use of an Instrument, in a specific Manner or Direction that brings the Agent in Contact with a patient-like Figure which may be affected in some way as a Result', as represented below.

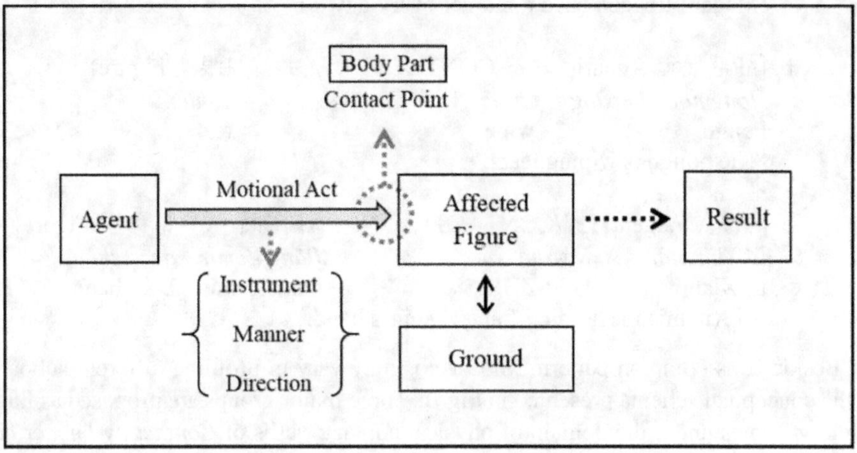

Fig. 1. A conceptual schema of Mandarin *Physical Contact* verbs.

As shown in the schema, four core semantic components (Agent/Affected Figure/Ground/Body Part) are proposed as the essential frame elements in MPC verbs, and four additional elements (Result/Instrument/Manner/Direction) may be found in expressing contact events. Based on semantic-to-syntactic distinctions, three major frames (Primary Frames) can be further distinguished: *Contact by Impact* (forceful contact), *Contact by Surface Motion* (surface contact), *Contact by Holding* (grasping contact). The following sections will present corpus-based evidence for the semantic classifications of MPC verbs.

3 Syntactic Patterns of Mandarin *Physical Contact* Verbs

As MPC verbs are highly transitive, they can easily enter the prototypical transitive pattern: 'Agent + CONTACT V + Figure' [3, 16–19].

(13) Transitive patterns:

a. [雷根夫人/Agent] 以前 每天 [打/CONTACT] [她/Figure]
Léigēn fūrén yǐqián měitiān dǎ tā.
Madam Reagen before everyday hit she
'Madam Reagen used to hit her every day.'

b. [紅毛猩猩/Agent] [擦/CONTACT] [眼淚/Figure]
Hóngmáo xīngxīng cā yǎnlèi.
Pongo wipe tears
'The pongo is wiping tears.'

c. [李先念/Agent] [握/CONTACT]著 [江澤民 的 手/Figure]
Lǐ Xiānniàn wò-zhe Jiāng Zémín de shǒu.
Li Xiannian hold-ZHE Jiang Zemin DE hand
'Li Xiannian is holding Jiang Zemin's hand.'

Besides this common pattern, MPC verbs may vary in profiling different subparts of the conceptual schema presented in Fig. 1. Three major groups are proposed as three primary frames under the domain of physical contact: verbs of *Contact by Impact* (CI verbs), verbs of *Contact by Surface Motion* (CS verbs) and verbs of *Contact by Holding* (CH verbs). Distinctive features of core participants and argument expressions will be provided below with corpus-based examples to demonstrate how the three groups differ from one another.

3.1 Verbs of *Contact by Impact* (CI Verbs)

As mentioned, CI verbs encode impactive contact initiated by an Agent who comes into contact with an Affected Figure. Given the nature of contact by impact, CI verbs frequently occur with forceful Manner or impactive Result. What is more important as a characteristic feature of CI verbs is that the Affected Figure, as a salient core argument under impact, can be expressed with three distinct syntactic forms: 1) as the direct object as in (14a), 2) as the conative target marked by the goal marker *cháo*朝 'toward' as in (14b), and 3) as the post-verbal locative complement marked by *zài* 'at/on' as in (14c). This three-way alternation in coding the Affected Figure is taken to be the crucial collo-constructional criteria for CI verbs.

(14) Three-way alternation in coding the Affected Figure:

a. 陳建汶 曾 用槍 打 **他**
Chén Jiànwèn céng yòng-qiāng dǎ tā.
Chen Jianwen once use-gun hit him
'Chen Jianwen has once shot at him with a gun.'

b. 男子 揮 拳 **朝 她的 臉** 打 兩下
Nánzǐ huī quán cháo tā-de liǎn dǎ liǎngxià.
Man swing fist toward she DE face hit twice
'The man swung his fit and hit at her face twice.'

c. 馬賢生 一拳 打在 **呂憲強 臉上**
Mǎ Xiánshēng yīquán dǎ-zài Lǚ Xiànqiáng liǎn-shàng.
Ma Xiansheng one-fist hit-ZAI Lǚ Xianqiang face-on
'Ma Xiansheng hit Lv Xianqiang on his face.'

The three-way alternation exemplified above demonstrates the language-specific, and type-specific constructional patterns for CI verbs. It shows that the salient object under contact by impact can be viewed as an affected patient or impactee, such as *tā* 他 'him' in (14a), or as an object aimed at, as *tā de liǎn* 她的脸 'her face' in (14b), or alternatively as a locative Ground, as *zài Lǚ Xiànqiáng liǎnshàng* 在吕憲強臉上 'at Lǚ Xianqiang's face' in (14c). The syntactic patterns also reveal that impactive contact can be perceived as a highly transitive event with Agent and Patient roles as well as a spatial event with a locative argument.

3.2 Verbs of *Contact* by *Surface Motion* (CS Verbs)

CS verbs encode physical contact by surface motion, such as *cā* 擦 'wipe', *tú* 塗 'paint', *mò* 抹 'rub'. [38] proposed that all CS verbs can enter the Locus-Locatum alternation in taking either the Substance or the Surface as direct object (*cā hūichén* 擦灰塵 'wipe the dust' vs. *cā zhuōzi* 擦桌子 'wipe the table'). The Locus-Locatum alternation (Substance vs. Surface, or Figure vs. Ground) is indeed a characteristic of Mandarin CS verbs as shown below (cited from [38]):

(15) Two-way alternation in coding the semantic role of the Object

a. Locus as Object:

故意 拿 瓶 清潔劑 擦 **桌椅**

Gùyì ná píng qīngjiéjì cā zhuō-yǐ.

Purposefully take bottle detergent wipe table-chair

'(S/he) Purposefully got a bottle of detergent to wipe the table(s) and chair(s).'

b. Locatum as Object:

在 毫無準備 下 難以 招架，而 頻頻 擦 **汗**

Zài háowú zhǔnbèi xià nányǐ zhāojià, er pínpín cā hàn.

In little-preparation under hard against so frequently wipe sweat

'...without any preparation, (s/he) was unable to answer the question, and could only stop wiping the sweat off.'

The above alternations are also found in Placing verbs [33] and termed as 'Substance-Figure vs. Surface-Ground Alternation'. CS verbs are considered to be MPC verbs since they also display the transitive pattern observed in MPC verbs (as illustrated in (13) above). However, unlike the three-way alternation of CI verbs (as shown in (14) above), CS verbs involve surface contact by Motion. Thus, they may encode a motional Path more frequently than other contact verbs. CS verbs can readily occur in the 'V-*lái*-V-*qù* V-來-V-去 V-come-V-go' construction to denote motional path, in either literal or metaphorical uses:

(16) V-lái-V-qù Construction:

a. 隊醫 趕緊 拿 冰塊 在 他 右腿 **擦-來-擦-去**

Duìyī gǎnjǐn ná bīngkuài zài tā yòutuǐ cā -lái -cā -qù.

Doctor hurry use ice-cube ZAI he right-leg rub-come-rub-go

'The doctor quickly rubbed his right leg back and forth with an ice cube.'

b. 好奇 的 眼光 從 我們 身上 **掃-來-掃-去**

Hǎoqí de yǎnguāng cóng wǒmen shēn-shàng sǎo-lái-sǎo-qù.

Curious DE sight from we body-on sweep-come-sweep-go

'Curious eyes swept back and forth on our bodies.'

The same 'V-*lái*-V-*qù* V-來-V-去 'V-come-V-go' construction would render a reciprocal reading with prototypical CI verbs (e.g., '*dǎ-lái-dǎ-qù* 打來打去 'hit each other'), but not a path reading. Thus, CS verbs differ from CI verbs in highlighting the Manner of placing Substance onto or off a Surface-Ground without a strong Impact.

3.3 Verbs of *Contact by Holding* (CH Verbs)

CH verbs profile the act of holding onto something by hand. [39] conducted a pilot study of Mandarin *Hold* and *Carry* verbs and proposed that *wò* 握 hold' is a typical *Hold* verb denoting prolonged contact with a gripping hand. Other members include: *zuàn* 攥 'grab', *zhuā* 抓 'seize', *bào* 抱 'hug', *niē* 捏 'squeeze', *bǐng* 秉 'grasp', *jiā* 挾 'clip', *zhí* 執 'hold', etc. Unlike the other two groups, CH verbs typically occur with a durative marker *zhe*著 to express a prolonged state of 'an agent holding onto an Affected Figure by hand'. In general, CH verbs display the prototypical transitive pattern in taking an overt object, but may collocate with very different manner adjuncts:

(17) 王子 緊緊地 握 著 她 的 手
Wángzǐ jǐnjǐn-de wò zhe tā de shǒu.
Prince tightly hold ZHE she DE hand
'The prince holds her hand tightly.'

In (17), the hand movement of gripping, which necessarily involves the application of force, may allow a manner adverb denoting tightness, such as *jǐnjǐn de* 緊緊地 'tightly'. What is unique in CH verbs is that the designated Body Part 'hand' is lexically so salient that it may be expressed syntactically as a core argument. There is a four-way alternation in coding the essential Body Part 'hand' with various grammatical roles: 1) as the Subject; 2) as the Object; 3) as the locative Ground 4) as the Instrument, exemplified below:

(18) Four-way alternation in expressing the salient component 'hand':

a. as the Subject (controlling entity):
你 的 雙手 老 握著 東西 不 放開
Nǐ de shuāngshǒu lǎo wò-zhe dōngxī bú fàngkāi.
You DE hand always hold-ZHE something not release
'Your hands always hold on to something.'

b. as the Object (affected Figure):
老人 平和 而 有力地 握著 厚實 的 **雙手**
Lǎorén pínghé ér yǒulì-de wò-zhe hòushí de shuāngshǒu.
Old-man peacefully but powerfully hold-ZHE thick DE hands
'The old man is holding his thick hands peacefully but powerfully.'

c. as the locative Ground (in *zài*-PP):
媽媽 總是 把 小鳥 握在 **手上**
Māma zǒngshì bǎ xiǎoniǎo wò-zài shǒu-shàng.
Mother always BA bird hold-ZAI hand-on
'My mother always holds the bird in her hands.'

d. as the Instrument:
小男孩 已 可以 **用手** 握 杯子
Xiǎo-nánhái yǐ kěyǐ yòng-shǒu wò bēizǐ.
Little-boy already able use-hand hold cup
'The little boy is able to hold the cup with his hands.'

Giving the saliency of 'hand' as the inherent and defining feature of CH verbs, it may undergo morphological processes to form N-V compounds, such as *shǒu-wò* 手握 'hold by hand', and *shǒu-zhuā* 手抓 'grasp by hand':

(19) a. 黃鴻松 手握 方向盤
Huáng Hóngsōng shǒu-wò fāngxiàngpán.
Huang Hongsong hand-hold steering-wheel
'Huang Hongsong holds the steering wheel in his hands.'

b. 他 手抓 鐐銬
Tā shǒu-zhuā liàokào.
He hand-grasp shackle
'He grasped the shackles in his hand.'

Mandarin CH verbs profile an event in which 'an Agent use hand as the Body Part to get Contact with the Figure in the Manner of Impact and the Figure is located on the Ground by the Contact action'.

As shown above, the sub-classes of MPC verbs are distinguished based on their semantic profiles and morpho-syntactic variations. To sum up, the hierarchical taxonomy of MPC verbs can be represented as follows:

Table 6. Preliminary classification of MPC verbs.

Archi-frame	Primary frame
Physical contact	Contact by impact (CI verbs)
	Contact by surface motion (CS verbs)
	Contact by holding (CH verbs)

4 Conclusion

This study provides a frame-based constructional characterization and classification of Mandarin *Physical Contact* (MPC) verbs according to their semantic-to-syntactic peculiarities. MPC verbs are extensively analyzed and distinguished into different semantic frames defined with frame semantic elements and collo-constructional patterns. It is proposed that MPC verbs behave differently from *Motion* verbs in conceptual schemata, argument expressions and semantic profiles, which leads to the distinction of three major classes (Primary Frames).

In terms of conceptual schemata, MPC verbs involve semantic components such as Body Part, Affected Figure, and locative Ground marked by *zài* 'at/on', which are more relevant to Contact than to Motion (self-motion or caused-motion). Thus, MPC verbs are divided into three primary frames: verbs of *Contact by Impact* (CI verbs), verbs of *Contact by Surface Motion* (CS verbs), and verbs of *Contact by Holding* (CH verbs). The first major class, *Contact by Impact* (CI) verbs, features impactive contact and displays a three-way alternation in expressing the Affected Figure, as the direct object, the conative goal, or the locative endpoint. The second major class, *Contact by Surface Motion* (CS) verbs, profiles surface contact with motion and renders quite distinct syntactic behavior in allowing both the Figure and the Ground to be coded as the direct object (Locus-Locatum alternation). They may also profile (with Goal-PP) the motional path for surface contact. Lastly, the unique class of holding verbs, i.e., *Contact by holding* (CH verbs), is differentiated based on their typical morpho-syntactic behavior in co-occurring with durative *-zhe* to express a prolonged state of contact and in coding the lexically designated body part 'hand' with syntactic and semantic saliency. Given that most MPC verbs encode hand motions, they are typically written with the semantic radical for 'hand' (手 'hand'). This shows that Chinese written characters are meaning-carrying, and most MPC verbs illustrate an interesting meaning-to-form association in their verb roots.

In sum, this study is significant in presenting a comprehensive scheme in distinguishing the lexical semantic categories of MPC verbs from a frame-based constructional approach. It identifies unique semantic and syntactic criteria for characterizing the subclasses of physical contact verbs, as lexically distinct from motion verbs. The proposed categorization of semantic frames with constructional patterns provides language-specific and domain-specific analyses of MPC verbs that are linguistically revealing and computationally applicable. It ultimately contributes to the advance of Chinese lexical semantic studies and the construction of lexical databases such as Mandarin VerbNet (http://verbnet.lt.cityu.edu.hk/#/).

Acknowledgments. We gratefully acknowledge the grant support (Project #9360163) of the Hong Kong Institute of Data Science, City University of Hong Kong.

References

1. Webster, M., Marcus, M.: Automatic acquisition of the lexical semantics of verbs from sentence frames. In: Proceedings of the 27th Annual Meeting on Association for Computational Linguistics, pp. 177–184 (1989)
2. Cohen, P.R., Oates, T.: A dynamical basis for the semantic content of verbs. In: Working Notes of the AAAI 1998 Workshop on The Grounding of Word Meaning: Data and Models, pp. 5–8 (1998)
3. Kemmerer, D., Castillo, J.G., Talavage, T., Patterson, S., Wiley, C.: Neuroanatomical distribution of five semantic components of verbs: evidence from fMRI. Brain Lang. **107**, 16–43 (2008)
4. De Clerck, B., Verroens, F., Willems, D., Colleman, T.: The syntactic flexibility of (new) verbs of instrument of communication: a corpus-based study. Funct. Lang. **18**, 57–86 (2011)
5. Choi, S., Bowerman, M.: Learning to express motion events in English and Korean: the influence of language-specific lexicalization patterns. Cognition **41**, 83–121 (1991)
6. Kawachi, K.: Korean putting verbs do not categorize space contrastively in terms of "tightness of fit." Lingua **117**, 1801–1820 (2007)
7. Schultze-Berndt, E.F.: Simple and complex verbs in Jaminjung: A study of event categorisation in an Australian language. [Sl: sn] (2000)
8. Tsujimura, N., Fried, M., Boas, H.C.: A constructional approach to mimetic verbs. In: Grammatical Constructions: Back to the Roots, pp. 137–154 (2005)
9. Sugahara, T., Hamano, S.: A corpus-based semantic analysis of Japanese mimetic verbs. In: Iconicity: East Meets West, vol. 14, p. 143 (2015)

10. Levin, B.: English Verb Classes and Alternations: A Preliminary Investigation. University of Chicago Press, Chicago (1993)
11. Kemmerer, D.: Why can you hit someone on the arm but not break someone on the arm?—a neuropsychological investigation of the English body-part possessor ascension construction. J. Neurolinguistics **16**, 13–36 (2003)
12. Viberg, Å.: Physical contact verbs in English and Swedish from the perspective of crosslinguistic lexicology. In: Advances in Corpus Linguistics, pp. 327–352 (2004)
13. Fillmore, C.J.: The grammar of hitting and breaking. In: Readings in English Transformational Grammar, pp. 120–133 (1970)
14. Hovav, M.R., Levin, B.: Change of state verbs: implications for theories of argument projection. In: Annual Meeting of the Berkeley Linguistics Society, vol. 28, no. 1, pp. 269–280 (2002)
15. Gruber, J.S.: Studies in lexical relations. Doctoral dissertation, Massachusetts Institute of Technology (1965)
16. Hopper, P.J., Thompson, S.A.: Transitivity in grammar and discourse. Language **56**, 251–299 (1980)
17. Gropen, J.H.: Learning locative verbs: how universal linking rules constrain productivity. Doctoral dissertation, Massachusetts Institute of Technology (1989)
18. Dixon, R.M.W.: A typology of causatives: form, syntax and meaning. In: Dixon, R.M.W., Aikhenvald, A.Y. (eds.) Changing Valency: Case Studies in Transitivity. Cambridge University Press, Cambridge (2000)
19. Ackerman, F., Moore, J.: Proto-Properties and Grammatical Encoding: A Correspondence Theory of Argument Selection. CSLI Publications, Stanford (2001)
20. Gao, H.: Notions of motion and contact for physical contact verbs. In: Proceedings of the 18th Scandinavian Conference of Linguistics, vol. 2, pp. 193–209. Lund University (2001)
21. Gao, H., Cheng, C.C.: Verbs of contact by impact in English and their equivalents in Mandarin Chinese. Lang. Linguist. **4**, 485–508 (2003)
22. Wang, B.: A lexical semantics approach: contact verbs of Chinese and the argument realization (汉语接触动词的语义与论元表达——从词汇语义学的角度探讨). Stud. Lang. Linguist. **3**, 66–73 (2008)
23. Zhao, M.: The valence study of Mandarin contact verbs (现代汉语接触动词的配价研究). J. Hefei Normal Univ. **26**, 107–110 (2008)
24. Liu, M.C., Chiang, T.Y.: The construction of Mandarin VerbNet: a frame-based study of statement verbs. Lang. Linguist. **9**, 239–270 (2008)
25. Liu, M., Chang, J.: From caused-motion to spatial configuration: placement verbs in Mandarin. Lang. Linguist. **20**, 180–224 (2019)
26. Fillmore, C.J.: Frame semantics. In: The Linguistic Society of Korea (ed.) Linguistics in the Morning Calm, pp. 111–137. Hanshin Publishing Co., Seoul (1982)
27. Fillmore, C.J.: Frames and the semantics of understanding. Quaderni di semantica **6**, 222–254 (1985)
28. Goldberg, A.E.: Constructions: A Construction Grammar Approach to Argument Structure. University of Chicago Press, Chicago (1995)
29. Goldberg, A.E.: Constructions: a new theoretical approach to language. Trends Cogn. Sci. **7**, 219–224 (2003)
30. Goldberg, A.E.: Constructions at Work: The Nature of Generalization in Language. Oxford University Press on Demand (2006)
31. Liu, M.C., Chang, J.C.: Semantic annotation for Mandarin verbal lexicon. In: 2016 International Conference on Asian Language Processing (IALP), pp. 30–36 (2016)

32. Liu, M.: A frame-based morpho-constructional approach to verbal semantics (框架为本, 构式为用-基于语料库实证的汉语动词语义分析与分类). In: Kit, C., Liu, M. (eds.) Frontiers of Empirical and Corpus Linguistics (《实证和语料库语言学前沿》). China Social Sciences Press, Beijing (2018)
33. Liu, M.: The construction and annotation of a semantically enriched database: the Mandarin VerbNet and its NLP applications. In: Su, Q., Zhan, W. (eds.) From Minimal Contrast to Meaning Construct. FCL, vol. 9, pp. 257–272. Springer, Singapore (2020). https://doi.org/10.1007/978-981-32-9240-6_18
34. Liu, M., Wan, M.: Chinese verbs and their categorization: construction and application of the semantic network of Chinese verbs (中文动词及分类研究: 中文动词词汇语义网的构建及应用). Lexicographical Stud. (辞书研究) **2**, 42–60 (2019)
35. Fillmore, C.J., Atkins, B.T.: Toward a frame-based lexicon: the semantics of RISK and its neighbors. In: Frames, Fields, and Contrasts: New Essays in Semantic and Lexical Organization, vol. 103, pp. 75–102 (1992)
36. Liu, M.: Emotion in lexicon and grammar: lexical-constructional interface of Mandarin emotional predicates. Lingua Sinica **2**(1), 1–47 (2016). https://doi.org/10.1186/s40655-016-0013-0
37. Talmy, L.: Lexicalization patterns: semantic structure in lexical forms. Lang. Typology Syntactic Description **3**, 36–149 (1985)
38. Liu, M.C.: Mandarin Verbal Semantics: A Corpus-Based Approach. Crane Publishing Co., Taipei (2002)
39. He, T., Liu, M., He, H.: Lexical semantics of Mandarin Carry and hold verbs: a frame-based constructional analysis of ná 拿 and wò 握. In: Su, Q., Zhan, W. (eds.) From Minimal Contrast to Meaning Construct. FCL, vol. 9, pp. 39–50. Springer, Singapore (2020). https://doi.org/10.1007/978-981-32-9240-6_3

Functional Diffusion and Decategorization of *Bai*: From a Colour Term to an Implicit Negative Adverb

Jinghan Zeng[1(✉)] and Yulin Yuan[2]

[1] School of Chinese Language and Literature of Beijing Normal University, Beijing, China
jhzeng@bnu.edu.cn

[2] Department of Chinese Language and Literature, Peking University, Beijing, China

Abstract. This paper focuses on the process of decategorization of *bai* from synchronic and diachronic perspectives by studying the nominal *bai*, the adjectival *bai*, and the adverbial *bai* in Mandarin. The central assertion is that the adverbial *bai*, with an implicit negative meaning in Mandarin, is the result of the evolutional process of decategorization applied to the nominal *bai*, i.e., a colour term for white. First, the basic evolution path of *bai* is traced by studying the semantic evolution of its different parts of speech. Second, it is proved that the evolution path of *bai* is a process of decategorization resulting from changes in the functions of *bai* caused by semantic overgeneralization and the lack of independence in a syntactic category.

Keywords: *Bai* · Semantic evolution · Decategorization · Metonymy

1 Introduction: Polysemy of the Adverbial *Bai* in Mandarin

The adverbial *bai* has two senses in Mandarin: 'doing something in vain' as in (1) and 'getting something for free' as shown in (2):[1]

(1) 许多人抗战八年的苦是白吃的了。
Xu-duo ren kang-zhan ba-nian de ku shi
many people Anti-Japanese War eight-years Stru effort be
bai chi le
bai have Asp[1]
'Many people have suffered for nothing in the eight years of the Anti-Japanese war.'

(2) 昨天白吃了您几块白薯。
Zuo-tian bai chi le nin ji-kuai bai-shu
yesterday *bai* eat Asp you several-pieces sweet-potato
'Yesterday I ate several of your sweet-potatoes but didn't pay for them.'

[1] The following abbreviations are used in this paper: Adv = adverb; Asp = aspect marker; Aux = auxiliary verb; Cl = classifier; Conj = conjunction; Mod = modal particle; Neg = negation; Prep = preposition; Pron = pronoun; Quan = quantifier; Stru = structural auxiliary.

M. Liu et al. (Eds.): CLSW 2020, LNAI 12278, pp. 206–214, 2021.
https://doi.org/10.1007/978-3-030-81197-6_18

The senses of the adverbial *bai* in (1–2) are just opposite to each other. *Bai-chi* in (1) means having paid the cost without getting the expected result. *Bai-chi* in (2) means having taken something without paying the due cost. Yuan (2014) generalized the two senses as *paying out without the wanted result* and *getting without the deserved cost*. Zeng and Yuan (2015a) studied the semantic features of *bai* and found that it had implicit negative meanings. Other similar adverbs like *gan, xia, kong* (干, 瞎, 空) were further studied to delve deeper into research on such implicit negative adverbs. (Zeng and Yuan 2018).

Yet, some questions are yet to be answered: Since the adverbial *bai* is not a negative adverb like *bu/mei* in Mandarin, why does it have negative meanings? Where does the negative meaning come from? To answer these questions, the semantic evolution of *bai* needs to be studied.

2 The Semantic Evolution and Generalization of *Bai*

2.1 The Referential Function of *Bai*

In Archaic Chinese, *bai* was a noun denoting a kind of colour. Such use of *bai* was attested from different sources. Consider the two quotations below (also listed in Zeng and Yuan (2015b)):

《说文解字》: 白, 西方色也, 阴用事, 物色白。从入合二。二, 阴数。

'*Bai*, a color to represent the west, has something to do with *Yin*.' (*Shuo Wen Jie Zi*)

《说文通训定声》: (白) 字从日, 上象日未出, 初生微光。按日未出地平时, 先露其光恒白, 今苏俗语'昧爽'曰东方发白, 是也, 字当从日丿, 指事, 训太阳之明也。

'*Bai* is closely connected with the meaning of the sun, and it represents the white light of the sun before it rises above the sea level. Today, the expression *mei-shuang* in the Suzhou dialect means the white light in the east. The character pattern of *bai* consists of the parts 日 and 丿.' (*Shuo Wen Tong Xun Ding*).

Bai is different from colours like red and green, which have specific colour features. Zeng and Yuan (2015b) pointed out that *bai* did *not have a specific colour value* and that the colour of *bai* was colourless. Thus, there was a [+negation] in the semantic feature set of *bai*. The Modern Chinese Dictionary (the 6th edition) describes *bai* as having the opposite meaning of black and being like the colour of frost or snow (像霜或雪的颜色, 跟黑相对). Physically, frost and snow has no colour, but their structures can reflect different light waves to form a composite light ray. When reflected onto a human's eyes, it is of a specific hue called 'white'. This supports the assertion by Zeng and Yuan (2015b).

Bai as a noun can also refer to the prominent attribute of white colour, that is, *whiteness*. In this case, the sense denoted by the nominal *bai* is generalized. Consider sentence (3), in which *bai* refers to the attribute of whiteness, rather than colourlessness as mentioned above.

(3) 石之白，石之堅，見與不見……（公孫龍子・堅白論）
Shi zhi bai, shi zhi jian, xian yu bu xian
rock Stru white rock Stru hard appear Conj Neg appear
'The white ness of rocks and the hardness of rocks appeared or not appeared…(*Gong Sun Long Zi*)'

The generalization of the meaning of *bai* can further develop in the sense that the sense denoted by *bai* can be more abstract. It can be further generalized to the sense of plainness in 空白 (blank) and 留白 (leaving spaces) in Modern Chinese for instance. Such generalization witnesses the evolution of *bai* from referring to colour attribute (no specific colour value) to referring to a state of plainness.

2.2 Generalization and Declarative Function of *Bai*

Apart from being a noun, *bai* can also be an adjective, describing properties pertaining to whiteness or plainness. In Archaic Chinese, the adjectival *bai* has four senses, namely 'white', 'light', 'pure' and 'plain', among which the first three are mostly attested in Archaic Chinese. Examples (4) to (7) demonstrates the four senses, respectively.

(4) 載之白旗，以爲天下諸侯僇。（墨子・明鬼下）
Zai zhi bai qi, yi wei tian -xia zhu -hou lu
carry Stru white flag Prep Prep world governors kill
'Carrying the white flag makes his life sought after by local rulers in the nation. (*Mozi*)'

(5) 命則處幽，吾將罷兮，願及白日之未暮。（楚辭・九章・思美人）
Ming ze chu you, wu jiang pi xi, yuan-ji bai ri
destiny Conj live dark -place I Aux tired Mod wish light day
zhi wei mu
Stru Neg dark
'Staying in dark places and feel ing tired, I wish I could seize time and make a difference. (*The Songs of Chu*)'

(6) 是故……身死而名彌白。（战国・荀子・榮辱篇）
Shi gu ... shen si er ming mi bai
is because body die but reputation more pure
'This is the reason why… the body is d ead but th e reputation becomes purer. (*Warring States*)'

(7) 白民之南，建木之下。（吕氏春秋・有始）
Bai-min zhi nan, jian-mu zhi xia
Common-people Stru south holy-tree Stru under
'(It's) In the south of where the common people live, under the holy tree. (*Lv's Spring and Autumn Annals*)'

When *bai* is used as an adjective, we still can find that the feature of [+negation] in the adjective, which is also the feature observed in nominal *bai* in Archaic Chinese. Thus, the feature is inherited during the change of *bai* from a noun to an adjective. The [+negation] feature is observed in the examples above in the following ways: *Bai-qi* in

(4) means *white flag*. In other words, *bai* denotes the hue of the flag as a kind of color with no specific color attributes. *Bai-ri* in (5) means the bright sunlight and daytime. That is to say, *bai* refers to the light that has no specific wavelength, and the colour of this light has no specific hue. *Bai* in (6) means pure. In other words, it describes things that contain no impurity, e.g., a reputed person without a black mark. *Bai-min* in (7) means the common people while *bai* means 'plain' here. In this case, the common people are not government officials and do not have any other specific social attributes. To be brief, from (4) to (7), it can be observed that the semantic feature of [+negation] of *bai*; and the attributes of *bai* become generalized, from a specific color to certain attributes, making *bai* more abstract.

In the Southern and Northern Dynasties, the frequency of use of the adjectival *bai* to mean 'plain' increased. Most of them modified food-words as shown in (8):

(8) 都督曰："……作白粥以投之。（南北朝・世说新语・汰侈）
Du-du yue: zuo bai-zhou yi tou zhi
Military-governor say cook plain-porridge to give him
'The military governor said, 'cook some plain porridge to serve the guest. (*Southern and Northern Dynasties·A New Account of the Tales of the World*)'

Bai-zhou in (8) means plain porridge. In such usage, *bai* is an adjective modifying the noun attributing the noun phrase with the meaning of 'without any other specific attributes', e.g., without other ingredients in (8).

Meanwhile, in the Southern and Northern Dynasties, *bai* was also used to modify verbs, as shown in (9):

(9) 斫取臆肉，去骨，奠如白煮之者。（南北朝・齐民要术・炙法第八十）
Zhuo qu yi-rou, qu gu, dian ru bai zhu zhi zhe
cut take breast cut-off bone put as *bai* boil it Pron.
'Remove the bones and cut off the flesh on the chest, and cook it in boiling water. (*Qi Min Yao Shu*)'

Bai-zhu in (9) means to boil, i.e., using only water to cook the food without any other special skills. In other words, the adverbial *bai* was also used to express the sense of 'without any other specific attributes'. Until now, expressions containing *bai* modifying a verb are still used to denote the way of cooking in Modern Chinese, e.g., 白切鸡/白切羊肉 (plain-cook-chicken/plain-cook-lamb).

During the Ming Dynasty and the Qing Dynasty, *bai* not only modified *zhu*, but also other kinds of verbs. At this stage, the verbs that could be modified by *bai* gradually increased. The newly emerged senses include 'always', 'in vain', 'without a reason', 'without paying' and 'without specific purpose', as demonstrated in examples (10–14) respectively. It is believed that these newly emerged senses are the result of further generalization from the senses at earlier stages of the development of *bai*.

(10) 也要往宅里看看姑娘，白不得个闲。（金瓶梅）

Ye yao wang zhai-li kan-kan gu-niang,
also want towards house-inside look-look lady
bai bu de ge xian
bai Neg get Cl availability
'I will go to the house to visit the lady, to always keep myself busy. (*The Golden Lotus*)'

(11) 姑娘却不可叫我白花钱。（儿女英雄传）

Gu-niang que bu ke jiao wo bai hua qian
lady yet Neg can make me *bai* spend money
'Yet you, Miss, please don't make my money wasted. (*Legend of Heroes*)'

(12) 谁知湘云有择席之病，虽在枕上，只白睡不着。（红楼梦）

Shui zhi Xiang-yun you ze xi zhi bing,
who know Xiangyun have choose bed Stru disease
sui zai zhen-shang, zhi bai shui-bu-zhao
although at pillow-on but *bai* cannot-sleep
'Yet Xiangyun has a habit of choosing her bed. Although lying in bed, she cannot fall asleep with no reason. (*A Dream in Red Mansions*)'

(13) 他们白听戏，白吃，已经便宜了……（红楼梦）

Ta-men bai ting xi, bai chi, yi-jing pian-yi le
They *bai* listen opera *bai* eat already benefit Mod
'They watched the opera performance and ate food for free. They had already taken so many advantages. (*A Dream in Red Mansions*)'

(14) 若是白来逛逛呢，便罢。（红楼梦）

Ruo shi bai lai guang-guang ne, bian ba
if be *bai* come stroll Mod Conj end
'If you just come for a stroll, then it's ok. (*A Dream in Red Mansions*)'

In (10), the meaning of *bai* is 'always'. Put it in another way, it means to do something without any change all along. In (11), the meaning of *bai* is 'doing something in vain'. That is to say, the expected result is not obtained after the event, e.g., the event of paying the money. In (12), the meaning of *bai* is 'doing something without a rational reason', i.e., the girl cannot fall asleep without a reason. In (13), the meaning of *bai* is 'doing something without a cost'. In (14), when *bai* is used, the meaning of 'doing something casually' is expressed. In other words, no specific purpose is involved in the action. It can be observed that the core semantics of *bai* in sentences (10–14) became more abstract.

Moreover, *bai* in these sentences demonstrates a common feature which is the negative feature. Yang (2007) studied the adverbial *bai* and defined it as *doing something with no reasons and under no limitation*, emphasizing the negative feature (i.e., [+negation]) in its semantic interpretation. Zeng and Yuan (2015a, 2015b) named this negative feature implicit negation, which is not a negative marker in its surface structure but implies negative meanings in its deep structure.

It is suggested that the adverbial *bai* appears because of the transformation of *bai*'s functions in an utterance. The nominal *bai* refers to a colour with an unspecified colour attribute, while the adjectival *bai* denotes the whiteness and plainness of white things. The adverbial *bai* reflects the plain ways of completing an action, implying that the action is not done in any specific ways. The path of evolution of *bai* can be described as follows showing the change of functions of *bai*: *Referring to the colour (white) → referring to its characteristic attributes (whiteness) → profiling the salient feature (plain, adj) → extending to the manner of null effort or null result (plainly).*

3 Variations in Syntactic Features of *Bai*

3.1 From Noun to Adjective: Non-self-Sufficiency in Grammatical Function

The semantic generalization of *bai* leads to the fuzziness in syntactic categories and non-self-sufficiency in grammatical function during its evolution process. Before decategorization, the nominal *bai* can act as an independent syntactic component in a sentence. Thus, the syntactic function of the nominal *bai* is self-sufficient. For example, in sentence (15) below, *bai* is a noun, referring to the colour white. The nominal *bai* is an independent object in the first clause and an independent subject in the second clause.

(15) 青以白非黃，白以青非碧。（公孫龍子・通變論）
Qing yi bai fei huang, bai yi qing fei bi
cyan Prep white Neg yellow white Prep cyan Neg viridity
'Classified with white, cyan is different from yellow. Classified with cyan, *bai* is different from viridity. (*Gongsun Long Zi*)'

Bai can also be an independent predicate in a sentence and be modified by an adverb of degree. For example:

(16) 是故窮則不隱，通則大明，身死而名彌白。（荀子・榮辱篇）
Shi gu qiong ze bu yin, tong ze da
be reason destitution Conj Neg conceal succeed Conj much
ming, shen si er ming mi bai
illustrious body die Conj reputation more pure
'Therefore, great people are not concealed under destitution and get greatly illustrious when they succeed. Even though they will die, their reputations will become much cleaner and purer. (*Xuncius*)'

In (16), *bai* serves as the predicate and is modified by the adverb of degree *mi* 'more'. At this stage, *bai* was strengthening its adjectival characteristics and thus gradually becoming an adjective. For example, the adjectival *bai* with the self-sufficient features in its syntactic function could act as the predicate independently, which is a necessary component in a sentence and cannot be omitted from the sentence, as shown in (17):

(17) 白馬為非馬者：言白，所以名色；言馬，所以名形也。（公孫龍子・跡府）
Bai ma wei fei ma zhe: yan bai, suo yi
white horse as Neg horse Aux talk white Aux Prep
ming se yan ma, suo yi ming xing ye
name color talk horse Aux Prep name shape Aux
'People saying '*a white horse is not a horse*' means that when saying *white*, one uses the word to describe the color; and when saying *horse*, one uses the word to describe the shape. (*Gongsun Long Zi*)'

The development of *bai* from a noun to an adjective is not a qualitative change. In ancient times, *bai*, as an essential syntactic element, could not be omitted. After the emergence of the adverbial *bai*, it changed qualitatively from a necessary component to a non-essential component in a sentence.

3.2 From Adjective to Adverb: The Syntactic Function Fading Out

The adjectival *bai* is also self-sufficient in terms of its syntactic functions. Yet, during the Southern and Northern Dynasties, the syntactic self-sufficiency of the adjectival *bai* gradually vanished, which had an impact on its semantic meaning. *Bai zhou* in (8), for instance, refers to plain porridge cooked in boiling water, without any other ingredients added or using any other cooking methods. The semantic meaning of *zhou* (referring to porridge) and *bai zhou* (referring to plain porridge), are similar especially when the modifier *bai* is omitted. A speaker intending to refer to the porridge containing other ingredients will attach other attributes before *zhou* (porridge), such as *hong-dou zhou* (red bean porridge) or *xiao-mi zhou* (millet porridge). In other words, *bai,* as an adjective, becomes unnecessary and can be omitted at this stage of development.

Moreover, *bai*, as an adverb in the expression of *bai zhu* (boil) (see (9)), also highlights the non-self-sufficiency. Whether *bai* is present in the sentence shows almost no difference in the meaning. *Bai zhu* refers to the cooking of food in water, where *bai* means that no other ingredient or seasoning is added, while *zhu* means boiling. Therefore, the meaning of *bai zhu* is the same as *zhu*. Meanwhile, as the core semantics of 'plain' is non-addition, the meaning is also associated with null effort or null result, given that the expected result is not achieved. Whichever meaning the adverbial *bai* expresses, it is similar to the adjectival *bai*, in the sense that it is optional in the sentence without affecting the grammaticality of the sentence.

In some cases, the adverbial *bai* is not describing the way an event happens, but only expressing the modality or mood of a sentence, which makes *bai* syntactically optional, with no effect on the VP, as illustrated in (18):

(18) 天色不晴，街上泥泞也深，白没个人儿来耍耍。（歧路灯）

Tian-se bu qing, jie-shang ni-ning ye shen, bai mei ge
weather Neg sunny street-on muddy Adv deep *bai* Neg Cl
ren-er lai shua-shua
man come play

'The weather is not sunny, and the street is very muddy. Unexpectedly nobody comes to play with me. (*Qilu Deng*)'

In (18), *bai* means *unexpectedly*, and the sentence means that no one comes to play is wholly unexpected. Here, *bai* is optional, and the propositional meaning of the clause after its omission remains the same. Instead of the manner of an action or behavior, *bai* refers to the speaker's attitude of unexpectedness which is a subjective mood toward the situation.

At this stage of development, *bai* has been highly grammaticalized, and it modifies the whole clause rather than the VP only and it occupies a higher adverbial position in the syntactic structure. Rizzi (1997) has divided a sentence structure into three layers: the lexical layer, the inflectional layer, and the complementizer layer. The lexical layer represents the propositional meaning of a sentence, which is the lowest one among the three layers. The inflectional layer represents the tense and aspect of a sentence, which is the middle layer. The complementizer layer represents the modal meaning of a sentence, reflecting the world knowledge of the speaker, which is the top layer. The verb phrase (VP) modified by the adverbial *bai* is located at the lexical level, and the auxiliary words expressing tense and aspect are located in the inflectional layer, while *bai* is located in the position of the complementizer layer, indicating the speaker's subjective attitude and cognitive world.

This kind of subjective evaluation function is more obvious in the modern Chinese adverb *bai,* as exemplified in (19):

(19) 转了个圈子，啥作用也没有起，白浪费了时间。（烈火金刚）
Zhuan le ge quan-zi, sha zuo-yong ye mei-you qi,
turn Asp Cl circle what effect Adv Neg have
bai lang-fei shi-jian.
bai waste time
'After going around, it didn't work at all. Time was wasted. (*Lie Huo Jin Gang*)'

In (19), *bai* means that the VP modified by it is *doing things with no expected result or in vain.* Yet, the VP *lang-fei le shi-jian* 'having wasted time' already indicates that it is a waste of time, without the expected result. Thus, *bai* is a redundant component in this sentence. With closer scrutiny, the constituent modified by *bai* here is not the verb but the whole clause. That is to say, *bai* is used here to express the speaker's subjective attitude. Such attitude is related to a deontic modal meaning, which means time should not be wasted.

At this stage of the development of *bai*, the adverbial *bai* is a high-level adverbial locating at the high-level position of the syntactic structure. The structure of the whole clause should be described as follows:

COMP[*bai* VP [*lang-fei le* NP [*shi-jian*]]]

In this light, the adverbial *bai* has become a non-self-sufficient and non-essential component in syntax.

4 Discussion and Conclusion

In Mandarin, *bai* has three parts of speech: noun, adjective, and adverb. It is intriguing why a linguistic form expresses various meanings and why the syntactic categories change during the same historical period. We claim that the reason for this phenomenon is that *bai* has experienced decategorization during its evolution. Categorization is a cognitive activity of classifying things. Parts of speech is the categorization of the grammatical functions of words in a language. While decategorization, which is inverse to the process of categorization, proposed by Hopper and Thompson (1984), refers to how the original category features gradually disappear, and the new category features gradually emerge. Decategorization is a significant step in language evolution. The new category of parts of speech which is derived from the language system is one of the results of decategorization.

This paper studies the semantic evolution and syntactic transformation process of *bai* from diachronic and synchronic perspectives. By investigating the meanings and usages of *bai* as a noun, an adjective, and an adverb from Archaic Chinese to Modern Chinese, we suggest that the transformation of *bai* from a noun to an adjective and from an adjective to an adverb is a decategorization process. To be specific, *bai* was initially a noun, referring to the color with no specific color value; and the adjectival *bai* emerged after decategorization and denoted the color attribute with no specific color value. During the Southern and Northern Dynasties, *bai* began to appear before verbs of cooking. *Bai* gradually occupied the position and category of an adverb. This adverbial use signifies the action of carrying out a certain activity without any additional materials, tools, or methods. The decategorization process of *bai* causes semantic generalization and the vanishing of syntactic category features. The track of the decategorization is simplified as this: Denoting things having certain attributes (noun category) → Denoting these attributes (adjective category) → Denoting manners in which actions are completed (adverb category).

Figuring out the process of decategorization of *bai* will improve the semantic processing of *bai* used as different parts of speech and language teaching. Moreover, the study of *bai* is only a beginning. The study of implicit negative adverbs (such as, *gan* 干, *xia* 瞎, *kui* 亏, *and cha-dian-er* 差点儿, etc.) in Mandarin would benefit the semantic construal and Chinese information processing, like machine translation.

Acknowledgments. We thank the anonymous reviewer for the helpful comments and suggestions. This work was supported by funds from the China Postdoctoral Science Foundation (2020M670191) and the Youth Teacher Fund Project of Beijing Normal University (310422133).

References

Guo, L.: The adverb *bai* and *baibai* in Proto-Mandarin. J. Chin. Linguist. **3**, 230–244 (1988). (in Chinese)

Hopper, P.J., Thompson, S.A.: The discourse basis for lexical categories in universal grammar. Language **60**(4), 703–752 (1984)

Ma, S., Shen, P.: The adverb *bai* in a dream of red mansions and the heroine. Stud. Chin. Lang. **6**, 461–463 (1981). (in Chinese)

Rizzi, L.: The fine structure of the left periphery. In: Haegeman, L. (ed.) Elements of Grammar, pp. 281–377. Dordrecht Kluwer (1997)

Yang, R.: The interpretation and origin of the adverbial *bai* in modern Chinese. In: Grammaticalization and Grammar Study, vol. 3. Commercial Press, Beijing (2007)

Yuan, Y.: Conception-driven and syntax-directed constructions and construal of sentences: a case study of the interpretation of sentences with the adverb *bai*. Stud. Chin. Lang. **362**, 402–417 (2014). (in Chinese)

Zeng, J., Yuan, Y.: The study on the implicit negative meaning of *bai* in Chinese. Macao J. Linguist. **46**(2), 4–14 (2015a). (in Chinese)

Zeng, J., Yuan, Y.: The implicit negation and Counter-Expectation of *bai* in mandarin Chinese. In: Lu, Q., Gao, H. (eds.) Chinese Lexical Semantics. CLSW 2015. LNCS, vol. 9332, pp. 47–61. Springer, Cham (2015b). https://doi.org/10.1007/978-3-319-27194-1_6

Zeng, J., Yuan, Y.: A Study on the Counter-expectation and Semantic Construal. In: Wu, Y., Hong, J.F., Su, Q. (eds.) Chinese Lexical Semantics. CLSW 2017. LNCS, vol. 10709, pp. 17–26. Springer, Cham (2018). https://doi.org/10.1007/978-3-319-73573-3_2

On the Behavior and Event Structure of 'Eating Steamed Buns Dipped in Human Blood'

Ren-feng Duann(✉)

Center for General Education, National Taitung University,
369, Section 2, Daxue Road, Taitung 950, Taiwan
rduann@nttu.edu.tw

Abstract. This paper, employing Conceptual Metaphor Theory [1, 2] and adjusting Extended Event Structure [3], aims to disclose the behavior and event structure of a Chinese multi-word expression (MWE) 'eating steamed buns dipped in human blood'. We have found that this expression, coined by Xun Lu in 1919, has undergone metaphorical extension. Regarding the event structure, we think that it presents an event that consists of two separate subevents, which are temporally ordered but not necessarily in a cause and effect relation. Aside from the temporal ordering of the subevents, we have found that one of the subevents is so prominent that it is foregrounded and represents the whole event, which indicates 'event headedness' [3: 72]. We thus argue that this MWE as a whole is used metonymically in texts from news reports or social media.

Keywords: Conceptual Metaphor Theory · Extended Event Structure · Multi-word expression in Chinese

1 Introduction

Multi-word expressions (MWEs), also known as 'multi-word items', 'fixed expressions', 'lexical phrases', etc., are defined as 'lexical items which consist of more than one "word" and have some kind of unitary meaning or pragmatic function' [4: 120], and their interpretations are 'idiosyncratic' and 'cross word boundaries (or spaces)' [5: 2]. Multi-word expressions in Chinese have attracted scholars' attention in recent decades. Previous research has focused on the metaphorical meanings and compositionality of Chinese idioms, e.g., [6, 7], the identification, classification, distribution, use and deep knowledge of Chinese MWEs, e.g., [8], the function of Chinese MWE in Teaching Chinese as a Second Language, e.g., [9], the automatic extraction of Chinese MWEs, e.g., [10], machine translation, e.g., [11], etc. The event structure of Chinese MWEs is yet to be investigated. This paper, incorporating Conceptual Metaphor Theory [1, 2], aims to disclose the event of 'eating steamed buns dipped in human blood (吃人血饅頭)'. Moreover, this paper, applying Extended Event Structure [3], which has been employed mainly to address single verbs, extends the discussion to MWE, which is a contribution to this theory.[1]

[1] I am indebted to the reviewer for pointing this out. Moreover, I thank the reviewer for referring me to the conceptual blending theory to analyze the data presented in this paper. However, the

M. Liu et al. (Eds.): CLSW 2020, LNAI 12278, pp. 215–222, 2021.
https://doi.org/10.1007/978-3-030-81197-6_19

We have found that this expression, coined by Xun Lu (魯迅) in 1919, has undergone metaphorical extension. Regarding its event structure, our analysis shows that two event structures can be retrieved from this MWE based on its contexts. One event structure presents an event composed of two separate subevents, which are temporally ordered but not necessarily in a cause and effect relation. The other event structure presents a causation, in which the first subevent is likely to cause the second subevent. Aside from the temporal ordering of the subevents, we have found that one of the subevents is so prominent that it is foregrounded and represents the whole event, which indicates 'event headedness' [3: 72]. We thus argue this MWE as a whole is used metonymically in news reports or social media.

The rest of this paper is organized as follows. Section 2 briefly introduces Conceptual Metaphor Theory and Extended Event Structure. Section 3 describes the methodology, and Sect. 4 the results, followed by the concluding remarks in Sect. 5.

2 Theoretical Backgrounds

Conceptual Metaphor Theory (CMT, [1, 2]) contends that people make sense of their experience in the world through the interactions between their bodies and things around them, and people tend to use concrete ideas to understand abstract ideas. A conceptual metaphor is formulated as TARGET DOMAIN IS SOURCE DOMAIN, in which the TARGET DOMAIN refers to a comparatively abstract notion and the SOURCE DOMAIN a comparatively concrete concept. This theory highlights two aspects of metaphor: (a) it is pervasive in everyday life, and (b) it is deeply rooted in language, thought and action. CMT has changed the way metaphor is defined: instead of merely an embellishment in language, it is a constitutive part of thought, which helps people understand and experience the world. Quasi-ubiquitous in language, thought and action, conceptual metaphors can pass unnoticed. Conceptual metaphors thus function as powerful tools to define what is taken as 'reality', which provides text producers of all types of sources (e.g., news reports or social media) with a good way to guide the reader to accept their value judgments.

Pustejovsky [3], noting the inadequacy of verbal semantics, proposes the Extended Event Structure, which represents the relations between an event and its proper subevents so as to explicate the internal aspects of events. For events composed of two subevents, Pustejovsky [3: 69–70] proposes three temporal ordering relations:

a. 'exhaustive ordered part of' relation, in which a larger event, *e3*, is composed of two subevents, *e1* and *e2*, and *e1* precedes *e2*.
b. 'exhaustive overlap part of' relation, in which a larger event *e3* contains two subevents *e1 and e2*, which occur simultaneously.
c. 'exhaustive ordered overlap' relation, in which a larger event *e3* consists of two subevents *e1* and *e2*, and *e1* starts before *e2*.

adoption of this theory would lead to a change of the theoretical framework and thus a de facto new paper, which will be my next research project.

Aside from the temporal ordering information of the subevents of a larger event, Pustejovsky [3: 72] addresses 'event headedness', which deals with the relative prominence of the subevents. That is, event headedness functions to reveal the 'foregrounding and backgrounding of event arguments'.

The utilization of CMT and the 'exhaustive ordered part of' relation of Extended Event Structure will enable us to better analyze different uses of the expression 'eating steamed buns dipped in human blood'. Moreover, we will argue that, with one subevent more prominent than the other, this MWE as a whole is used metonymically in texts nowadays.

3 Methodology

3.1 Data

The data at issue were taken from the short fiction [12] and Google results by searching the expression 'steamed buns dipped in human blood (人血饅頭)' in Chinese from 11 to 13 February 2020. We manually checked these texts so as to see how this expression behaves in the context. We aim to provide a qualitative analysis of the different uses of 'eating steamed buns dipped in human blood (吃人血饅頭)'.

3.2 Method

We employed CMT to investigate the metaphoricity of this expression. That is, we first examined its original meaning in *Medicine* by Xun Lu, and checked whether there was an incongruity between its original meaning and the meaning emerging from its uses in texts nowadays. Identifying the subevents of this MWE, we then utilized Extended Event Structure to analyze the temporal ordering and relative prominence of the subevents so as to explicate the possible event structures of this expression.

4 Results

4.1 The Metaphoricity of 'Steamed Buns Dipped in Human Blood'

'Steamed buns dipped in human blood', or 人血饅頭 in Chinese, was first coined by Xun Lu in 1919 in his short fiction, *Medicine*, in which the writer criticized the general public's follies and ignorance in the late Ching dynasty, when the revolutionary martyrs were trying to overthrow this imperial dynasty due to its decline. In this story, the general public did not understand or accept the revolutionary martyrs' goal. Meanwhile, they believed that these martyrs' blood, when ingested, would cure tuberculosis. The executioners dipped steamed buns in the martyrs' blood as soon as the martyrs were beheaded so as to make profit by selling these steamed buns to those who were eager to recover from tuberculosis.

The original meaning of this MWE is literal, as shown in the following excerpt from [12]:

1.

包好，	包好！	這樣	的	趁熱
bāohǎo	bāohǎo	zhèyàng	de	chènrè
guarantee-cure	gurantee-cure	this kind	DE	while-it-is-warm

吃下。	這樣	的	人血	饅頭，
chīxià	zhèyàng	de	rénxiě	mántou
eat	this kind	DE	human-blood	steamed bun

什麼	癆病	都	包好！ [12]
shénme	láobìng	dū	bāohǎo
whatever	tuberculosis	all	guarantee-cure

'You'll be cured! You'll be cured! Eat this steamed bun while it is warm. Steamed buns dipped in human blood can definitely cure all kinds of tuberculosis!' (Author's translation. Same for all the other translations.)

It is thus obvious that in this short fiction, the steamed buns dipped in human blood are considered a medicine, which is to be taken by the patient with tuberculosis. Eating steamed buns dipped in human blood is thus meant literally in this fiction.

Consider the excerpt below:

2.

誰	吃	香港	抗爭	者
shéi	chī	Xiānggǎng	kàngzhēng	zhě
who	eat	Hong Kong	protest	people

的	人血	饅頭？ [13]
de	rénxiě	mántou
DE	human blood	steamed bun

'Who is eating the steamed buns dipped in Hong Kong protesters' blood?'

3.

華女	掃	日本	口罩	大吃
huánǚ	sǎo	Rìběn	kǒuzhào	dàchī
Chinese woman	buy	Japan	masks	pig out on

「人血	饅頭」	出	po
'rénxiě	mántou'	chū	po
'human blood	steamed bun'	show	post

曬	$141 萬 [14]
shài	141 wàn
show off	1.41 million

'A Chinese woman who bought a large number of masks in Japan pigged out on "steamed buns dipped in human blood". She showed off the HK1.41 million she had made on a social media.'

4.

吃	高以翔	人血	饅頭！
chī	GāoYǐXiáng	rénxiě	mántou
eat	Godfrey Gao	human blood	steamed bun

歐陽娜娜	人氣	慘跌 [15]
ŌuyángNànà	rénqì	cǎndié
Nana Ouyang	popularity	tumble

'Eating steamed buns dipped in Godfrey Gao's blood! Nana Ouyang's popularity tumbled.'

An examination of the full texts of the three excerpts shows that the use of 'eating steamed buns dipped in human blood' is no longer literal. Instead, it has undergone semantic extension. The backdrop of Excerpt (2) is the anti-extradition protest in Hong

Kong in 2019, in which most of the protesters were high school or university students, and quite a number of students were hurt or even (allegedly) killed by the police. In the form of a question, this excerpt asks who has made (political) profits out of the blood shed by the protesters. The context of (3) is the outbreak of COVID-19 in China, in which facemasks were considered a protection to fend off the virus but they were out of stock almost everywhere in Hong Kong and China. A Chinese woman, buying 10 boxes of masks in Japan and selling them to people in China, made a fortune of 1.41 million HK dollars and showed off in her social media account. In Excerpt (4), what happened was the sudden death of Godfrey Gao, a celebrity. Afterwards, Nana Ouyang, another celebrity, posted condolence in her social media account, which was then considered by netizens as making use of Gao's death to boost her popularity. Excerpts (2), (3) and (4) demonstrate a semantic incongruity between their uses and the original meaning; that is, this expression is used metaphorically in these excerpts: it is no longer that someone literally eats steamed buns dipped in human blood, as described in [12], but that someone makes profits out of other people's sacrifice, suffering or even death. The conceptual metaphor thus formulated is MAKING PROFITS OUT OF OTHER PEOPLE'S SACRIFICE/SUFFERING/DEATH IS EATING STEAMED BUNS DIPPED IN HUMAN BLOOD.

4.2 The Event Structures of 'Eating Steamed Buns Dipped in Human Blood'

The original meaning and the metaphorical extension show that there are two parties involved in the consumption of this type of steamed buns: those who sacrifice themselves, suffer or die, and those who make profits out of the blood shed by the former. Using the 'exhaustive ordered part of' relation, we explicate the event structure of this MWE as follows:

Event Structure 1

A larger event, *e3*, eating steamed buns dipped in human blood, is composed of two subevents, *e1*, an event in which a group of people sacrifice themselves, suffer or die, and *e2*, an event in which another group of people make profits by exploiting *e1*; *e1* precedes *e2*, but *e1* is not necessarily the cause of *e2*.

Unlike the 'exhaustive ordered part of' relation described in [3], which indicates causation or inchoativeness between *e1* and *e2*, the subevents of 'eating steamed buns dipped in human blood' do not show a clear causation or inchoativeness. Instead, *e1* is more like a means or strategy utilized by the agent to achieve the purpose in *e2*.

Take Excerpt (2), which is the headline of a piece of opinion found on the online forum of ETtoday, a news agency in Taiwan. With the violent anti-extradition protests in Hong Kong, which started from early June 2019 and continued till the end of 2019 (*e1*), the Taiwan government led by the Democratic Progressive Party (DPP) kept on publicizing and even overstating the protests so as to elicit Taiwan people's hostility/phobia against China. The DPP's goal was to win the election (*e2*). We thus argue that, despite the temporal order of the two subevents, the connection between *e1* and *e2* may not necessarily be causative or inchoative. Our data have shown that *e1* functions as a means by which the agent of *e2* achieves the purpose in *e2*.

On the other hand, a close look at Excerpt (5) shows that the above event structure does not apply:

5.

武漢	病疫		期間，	太子		Banny	
Wǔhàn	bìngyì		qíjiān	Tàizǐ		Banny	
Wuhan	epidemic		period	Prince Edward		Banny	
賣	假	豬咀		食	人血		饅頭 [16]
mài	jiǎ	zhūjǔ		shí	rénxiě		mántou
sell	fake	gas mask		eat	human blood		steamed bun

'In the outbreak of COVID-19, Banny eats steamed buns dipped in human blood by selling fake 3M gas masks at Prince Edward.'

Despite the identical value embedded in Excerpts (2)–(5), i.e., to condemn the agents' unwarranted purposes, it is found that the order of the subevents of (5) is not the same as that of (2)–(4). In Excerpt (5), what happened first is that someone sold fake 3M gas masks in order to make profits. This behavior put the buyers at the risk of getting infected, which would probably lead to their suffering or even death. This interpretation is evidenced in the text subsequent to the headline (5-1):

5-1.

我	保證		你	屋企人，		你	朋友，	
wǒ	bǎozhèng		nǐ	wūqìrén		nǐ	péngyǒu	
I	guarantee		you	family		you	friend	
你	身邊		所有人		都	會	見到	你，
nǐ	shēnbiān		suǒyǒurén		dū	huì	jiàndào	nǐ
you	around		all people		all	will	see	you
知	你	賣	假野		害死	人，		
zhī	nǐ	mài	jiǎyě		hàisǐ	rén		
know	you	sell	faked mask		kill	people		
知	你	食	人血		饅頭 [16]			
zhī	nǐ	shí	rénxiě		mántou			
know	you	eat	human blood		steamed bun			

'I guarantee that your family, your friends, and people around you will see you and know that you cause people's death by selling fake gas masks. They'll know you are eating steamed buns dipped in human blood.'

We thus formulate the event structure of Excerpt (5) as follows:

Event Structure 2

A larger event, *e3*, eating steamed buns dipped in human blood, is composed of two subevents, *e1*, an event in which a group of people make profits, and *e2*, an event in which another group of people probably suffer or die; *e1* precedes *e2*, and *e1* has the potential to cause *e2*.

4.3 Event Headedness and the Metonymic Nature of This MWE

Extended Event Structure deals not only with the temporal ordering but also with the relative prominence of the subevents. The above analysis indicates that, whether it occurs after (*e2* in Event Structure 1) or before (*e1* in Event Structure 2) the other subevent, the event in which people make profits is foregrounded. That is, this subevent is the head event in both event structures.

The event headedness reveals the metonymic nature of this MWE 'eating steamed buns dipped in human blood'. A metaphorical event composed of two temporally ordered events, this expression highlights the agent's atrocity while underplaying the other event, be it the goal the agent wants to achieve or the cause which leads to the cruelty. One of the subevents is so prominent that it is foregrounded and represents the whole event. We thus argue this MWE as a whole is used metonymically in texts from news reports or social media.

5 Concluding Remarks

We have analyzed the Chinese MWE 'eating steamed buns dipped in human blood' by employing Conceptual Metaphor Theory and Extended Event Structure. This term, originally being used literally, has undergone metaphorical extension and is commonly used in texts of different sources, ranging from news reports to netizens' comments on social media. Investigating the event structure of this expression, we have found that, despite the transmission of the same value, there are two internal structures of the subevents. In Event Structure 1, the event which leads to a group of people's sacrifice/suffering/death precedes the event in which another group makes profits by exploiting the preceding subevent, and the preceding event functions as a means or strategy of the agent in the subsequent event. Unlike what is argued in [3], our data analysis shows that there is no necessary causative or inchoative relation between the two subevents. Event Structure 2, on the other hand, involves causation, in which the first subevent, i.e. someone making profits, is likely to cause the suffering or loss of life in the subsequent subevent. We have also disclosed the relatively prominent subevent in both event structures; that is, the event in which a group of people make profits. We argue that the head subevent reveals the metonymicity of this MWE when it is used in a text, as the prominent subevent is foregrounded and represents the whole event. In future study, we will adopt the conceptual blending theory in the analysis of this MWE, and examine other Chinese multi-word expressions so as to reveal their behavior and event structures.

Acknowledgments. I would like to thank Dr. Jiajuan Xiong and the anonymous reviewers for their constructive comments and suggestions. All remaining errors are mine.

References

1. Lakoff, G., Johnson, M.: Metaphors We Live By. The University of Chicago Press, Chicago, London (1980)
2. Lakoff, G., Johnson, M.: Metaphors We Live By, 2nd edn. The University of Chicago Press, Chicago, London (2003)
3. Pustejovsky, J.: The Generative Lexicon. The MIT Press, Cambridge, London (1995)
4. Moon, R.: Multi-word items. In: Taylor, J.R. (ed.) The Oxford Handbook of the Word, pp. 120–140. Oxford University Press, New York (2015)

5. Sag, I.A., Baldwin, T., Bond, F., Copestake, A., Flickinger, D.: Multiword expressions: a pain in the neck for NLP. In: Gelbukh, A. (ed.) CICLing 2002. LNCS, vol. 2276, pp. 1–15. Springer, Heidelberg (2002). https://doi.org/10.1007/3-540-45715-1_1
6. Hsieh, S.C.-y.: Compositionality in plant fixed expressions. In: Machery, E., Werning, M., Schurz, G. (eds.) The Compositionality of Meaning and Content, vol. 2: Applications to Linguistics, Psychology, and Neuroscience, pp. 107–122. Ontos Verlag, Frankfurt Paris Ebikon Lancaster, New Brunswick (2005)
7. Yu, N.: From Body to Meaning in Culture: Papers on Cognitive Semantics Studies of Chinese. John Benjamins, Amsterdam, Philadelphia (2009)
8. Wang, S.: Chinese Multiword Expressions. Theoretical and Practical Perspectives. Springer, Singapore (2020). https://doi.org/10.1007/978-981-13-8510-0
9. Gao, S.: Processing advantage of formulaic sequences in Chinese reading by native and second language speakers: an eye-tracking study 母語者和第二語言學習者漢語閱讀中語塊加工優勢的眼動研究. Chin. Teach. World **31**, 560–575 (2017)
10. Piao, S., Sun, G., Rayson, P., Yuan, Q.: Automatic extraction of Chinese multiword expressions with a statistical tool. In: Proceedings of the Workshop on Multi-word-expressions in Multilingual Context, pp. 17–24. Association for Computational Linguistics, Stroudsburg (2006)
11. Ren, Z., Liu, Y., Cao, J., Liu, Q., Huang, Y.: Improving statistical machine translation using domain bilingual multiword expression. In: Anastasiou, D., Hashimoto, C., Nakov, P., Kim, S.N. (eds.) Proceedings of the Workshop on Multiword Expressions: Identification, Interpretation, Disambiguation and Applications, pp. 47–54. Association for Computational Linguistics, Stroudsburg (2009)
12. Lu, X.: Medicine (1919). http://ftp.pcsh.ntpc.edu.tw/~selenesue/%E9%AD%AF%E8%BF%85%20%E8%97%A5.pdf. Accessed 11 Feb 2020
13. Shan, H.-Z.: Who is eating the steamed buns dipped in Hong Kong protesters' blood? 誰吃香港抗爭者的人血饅頭? (2020). https://forum.ettoday.net/news/1598899#ixzz6SihzxcVN. Accessed 12 Feb 2020
14. Apple Daily: A Chinese woman who bought a large number of masks in Japan pigged out on 'steamed buns dipped in human blood'. She showed off the HK1.41 million she had made on a social media 華女掃日本口罩大吃「人血饅頭」出po曬$141萬 (2020). https://hk.appledaily.com/china/20200213/TQETYTHRLRIORHVWKAFHZYXQG4/. Accessed 13 Feb 2020
15. SETN: Eating the steamed bun dipped in Godfrey Gao's blood! Nana Ouyang's popularity tumbled 吃高以翔人血饅頭!歐陽娜娜人氣慘跌 (2019). https://tw.news.yahoo.com/%E5%90%83%E9%AB%98%E4%BB%A5%E7%BF%94%E4%BA%BA%E8%A1%80%E9%A5%85%E9%A0%AD-%E6%AD%90%E9%99%BD%E5%A8%9C%E5%A8%9C%E4%BA%BA%E6%B0%A3%E6%85%98%E8%B7%8C-095502681.html. Accessed 12 Feb 2020
16. Jchann: In the outbreak of COVID-19, Banny eats steamed buns dipped in human blood by selling faked 3M gas masks at Prince Edward 武漢病疫期間, 太子Banny賣假豬咀食人血饅頭. http://jchann.net/56654/. Accessed 13 Feb 2020

The Lexical Semantics and Grammatical Distinctions of Sui-X: Taking the Example of *Suídì* 随地 and *Suíchù* 随处

Zhao Xu(✉)

Institute for International Students, Nanjing University, Nanjing 210093, Jiangsu, China

Abstract. Both *suídì* 随地 and *suíchù* 随处 mean 'everywhere/randomly' in Mandarin Chinese, but they often cause confusions to non-native speakers. For instance, why should one say *suídì-tǔtán* 随地吐痰 'spit randomly on the ground' but not *suídì-pǎobù* 随地跑步 'jog randomly on the ground'? Also, why is it forbidden to say *suídì-kàn* -随地看 'look randomly'? To answer these questions, this study examines the historical development of *suídì* 随地 and *suíchù* 随处 across different stages of Chinese and proposes a functional account for their semantic features and their collo-constructional preferences. And this study shows that semantic and grammatical constraints of contemporary usages of *Suí* 随-X can be traced back to their grammatical and functional changes in historical developments. It also demonstrates how form and meaning are correlated in rendering different collo-constructional preferences and semantic extensions.

Keywords: *Suídì* 随地 · *Suíchù* 随处 · Lexicalization · Functional change · Constructional semantics · Lexical adhesion

1 Introduction

This analysis of the distinction of *suídì* 随地 and *suíchù* 随处 was motivated by the author's observation of the grammatical errors made by international students in the Chinese language class, demonstrated as follows:

(1) *我喜欢随地跑步。

Wǒ xǐhuan suídì pǎobù.

I like everywhere run.

'I like jogging randomly on the ground.'

(2) *地上摆着很多书，你可以随地看。

Dìshàng bǎizhe hěnduō shū, nǐ kěyǐ suídì kàn.

Ground-on put many books, you can everywhere look.

'There are many books on the ground. You can read them as you like.'

These sentences sound weird to Chinese native speakers. No instance of *suídì-kàn* 随地看 in (2) is recorded in the Corpus of Beijing Language and Culture University,

M. Liu et al. (Eds.): CLSW 2020, LNAI 12278, pp. 223–234, 2021.
https://doi.org/10.1007/978-3-030-81197-6_20

but 100 cases of *suíchù-kàn* 随处看 'Look around (everywhere)' are located. As both *suídì* 随地 and *suíchù* 随处 can be translated into "everywhere/anywhere/wherever", it is significant to figure out why 随地 is not compatible with "*kàn* 看/see" and "*pǎobù* 跑步/run". Furthermore, if we substitute the VP with a different one as in (3), the sentence sounds natural.

(3) 落叶随地可见。

Luòyè suídì kějiàn.

Fallen leaves everywhere can see.

'Fallen leaves can be seen everywhere.'

How can one explicate that the expression *sui-di-ke-jian* 随地可见 'can be seen everywhere' is appropriate, while *sui-di-kan* 随地看 'look everywhere' is not? What are the semantic and grammatical distinctions in the two expressions? What are the collocational constraints in the form of 随-X? To address these questions, this study provides a detailed analysis of the lexicalization processes of *suídì* 随地 and *suíchù* 随处 and specifies their semantic characteristics and grammatical functions.

2 The Lexicalization Process of *Suídì* 随地

The diachronic language data in an Ancient Chinese Corpus indicate that *suídì* 随地 experienced a lexicalization process from a V-O structure to an adverb in ancient Chinese. In addition, by analyzing the meaning of *suídì* 随地 in specific contexts, this paper argues that *suídì* 随地 remains its spatial meaning with an emphasis on geographical surface, and thus often modifies events involving the ground surface.

As recorded in the Ancient Chinese Corpus of Center for Chinese Linguistics PKU, *suídì* 随地 first appeared in 'Taiping Jing'太平经 during the Han Dynasty.

(4) 随天为意，随地为理。

Suí tiān wéi yì, suí dì wéi lǐ.

Follow the heaven as willingness, follow the earth as principle.

'Submit to Heaven and Earth. Follow their will and principles.'

In the Han Dynasty, *suídì* 随地 was a verb-object phrase *suí* 随+*dì* 地 (Wang, 2010; Wu, 2012; Jiang & Guo, 2016; Fan, 2018), meaning 'follow or submit to Earth', as *dì* 地 'Earth' was treated as a god.

The semantics of the verb-object phrase *suí* 随+*dì* 地 changed from 'submit to Earth' into 'based on the geographical situation' in the Wei and Jin Dynasties. For example,

(5) 分精兵数十为一处，

F n j ngb ng shùshí wéi y chù,

Distribute elite troops tens of into one place,

随地形势置之。（魏晋《北史》）

Suídì xíngshì zhì zh .

according to terrain resettle them.

'Order tens of the elite troops in one place, choosing the place according to the terrain.'

In the above sentence, *suí* 随 occurs with the locational object *dì* 地 to refer to the precondition or geographical condition for placing the troops. The verbal feature of 随 is weakened and the adverbial feature of *suí* 随+*dì* 地 is enhanced. In this preverbal use, "*suí* 随" is partly like a verb and partly like a preposition (Li&Thompson, 1981). It makes sense to take *suí* 随 as a co-verb, with the meaning of "according to/based on".

During the Tang and Song Dynasties, *suídì* 随地 was still in the form of a verb-object phrase, i.e. *suí* 随+*dì* 地. As *dì* 地 has meant both 'ground' and 'region' since the early Qin Period, the phrase *suí* 随+*dì* 地 could be translated into 'based on the geographical situations' and 'based on the different regions' according to context. However, there was a clear semantic emphasis when using the phrase *suí* 随+*dì* 地 during this period of time. By searching the Ancient Chinese Corpus of Center for Chinese Linguistics PKU and the Ancient Chinese Corpus of State Language Work Committee, we can find 34 sentences with the verb-object phrase *suí* 随+*dì* 地, 23 instances of which means 'based on the geographical situations'. For instance,

(6) 去声随地急，

Qùsh ng suídì jí

The thunder based on the geographical situations rumble,

残势傍楼斜。（唐《全唐诗》）

Cánshì bànglóu xié.

rest sound surround buildings leant.

'The thunder rumbles differently according to the geographical terrain, and its sounds linger around the leant buildings.'

The phrase in the other 11 sentences has the meaning of 'based on the different regions'. For example,

(7) 方言俚语，随地随时各自不同。（宋《朱子语录》）

Fāngyán lǐyǔ, suídìsuíshí gèzì bùtóng.

Dialects slangs, according to different regions and stage itself distinct.

'Dialects and slangs are distinct according to different regions and times.'

In the above sentence, the object *dì* 地 occurs with *shí* 时. The which refer to human cognition of temporality and spatiality.

The adverb *suídì* 随地 first appeared in the Yuan Dynasty with the meaning of 'everywhere'.

(8) 若不肯与，不日南侵，江山难保。

Ruò bùkěn yǔ, búrì nánqīn, jiāngshān nán bǎo.

If don't agree, immediately invade to south, the nation difficult defend.

'If you do not agree, we will invade the south immediately. Your nation will be difficult to defend.'

随地打猎，延入塞内。（元《汉宫秋》）

Suídì dǎliè, yánrù sāinèi

Randomly hunting and killing, reach the inside of the wall.

'Hunting and killing randomly, until I reach the inside of the wall.'

Context and semantic orientation are the important bases for judging the syntactic function of words (Heine, 2002). In this sentence, *Suídì-dǎliè* 随地打猎 is the precise description of *búrì-nánqīn* 不日南侵 in the former sentence, emphasizing the severe destruction of the war. Therefore, according to 'form-function mapping construction' (Liu, 2017), *suídì* 随地 is used to modify the manner and scope of the verb *dǎliè* 打猎, which can be seen as the beginning of its use as an adverb.

In the Qing Dynasty, there were an increasing number of *suídì* 随地 functioning as adverbs with the meaning of 'everywhere' or 'randomly'. Regardlessly, the adverb *suídì* 随地 still emphasized that the actions and events happened on the ground. For example,

(9) 自从他老人家到处察访，随地擒拿。（《施公案》）

Zìcóng tā lǎorénjia dàochù cháfǎng, suídì qínná.

Since he old man everywhere investigate, everywhere arrest.

'Since the old man himself started to investigate everywhere and arrest (criminals) everywhere.'

(10) 男女单身独行，仍须随处随地倍加留神。

Nánnǚ dānshēn dúxíng, réngxū suíchùsuídì bèijiā liúshén.

Men and women single walk alone, still need everywhere especially careful.

'When men and women walk unaccompanied, they should still be more careful everywhere.'

In example (9), according to the information provided by *dàochù* 到处 'everywhere' in the former sentence, we understand that *suídì* 随地 functions as an adverb of scope and indicates that the old man arrests people everywhere. However, the verb *qínná* 擒拿 in Chinese is not simply expressing the meaning of 'to arrest', but also 'to pin someone down on the ground'. Therefore, even having been lexicalized as an adverb, *suídì* 随地 still maintains the meaning of 'geographical surface/ground' and required the verb to have the orientation 'down to the ground' in general. In example (10), *suíchù* 随处 and *suídì* 随地co-occur. It does not make sense if an adverb with the same meaning is repeated in a sentence. there must be a semantic distinction between *suíchù* 随处 and *suídì* 随地, and I will provide my explanation in the next section.

In the Qing Dynasty, *suídì* 随地 began to collocate with *suíshí* 随时 frequently expressing the meaning of 'whenever and wherever' (It first appeared in the Song Dynasty but with only one instance spotted in the data of the Song Dynasty and that of the Yuan Dynasty respectively.) Under this circumstance, *suídì* 随地 does not express the apparent semantic orientation to 'geographical surface' but denotes the special concept.

In summary, *suídì* 随地in Mandarin Chinese was derived from the verb-object phrase *suí* 随+*dì* 地 'submit to Earth' in the Han Dynasty. The phrase expressed the meaning of 'according to geographical surface' from the Wei and the Jin Dynasties and was then lexicalized into an adverb meaning 'everywhere' or 'randomly' in the Yuan Dynasty. Being used as a verb-object phrase or an adverb, *suídì* 随地 semantically emphasizes 'the spatial meaning with an emphasis on the geographical surface'.

3 The Lexicalization Process of *Suíchù* 随处

Unlike *suídì* 随地, which has undergone a process of lexicalization, I would argue that *suíchù* 随处 has not been lexicalized.

According to the Ancient Chinese Corpus of CCL, *suíchù* 随处 first appeared in the Wei and the Jin Dynasties,

(11) 明月必背上著箭，

Míngyuè bì bèishàng zhù jiàn,

Bright moon definitely on the back carry bow and arrow,

'They will definitely carry bows and arrows on the back,'

丰乐随处即下手，

Fēnglè suíchù jí xiàshǒu,

Name everywhere immediately shoot,

'Fengle shoot randomly and immediately at everywhere,'

数虽多，去兄远矣。（魏晋《北史》）

Shù suī duō, qù xiōng yuǎnyǐ.

Quantity although much, compare brother much less.

'Although the quantity of preys is large, it is much less when comparing with that belongs to his brother.'

The meaning and the grammatical function of *suíchù* 随处 can be deduced by referring to the meanings of *jí* 即 and *shù-suī-duō* 数虽多, with *jí* 即indicating the quick actions and *shù-suī-duō* 数虽多 the fruitful results. *Suíchù* 随处 is used to indicate many actions and a large scope of the shooting rather than expressing the meaning of 'according to different features of different regions'. Therefore, according to the meaning of 'everywhere/not limited in a certain place' it can be clearly seen that *suíchù* 随处 functions as an adverb.

Since the Sui Dynasty and the Tang Dynasty onwards, the meaning of 'everywhere/not limited in a certain region' of *suíchù* 随处 became stabilized. The CCL Ancient Chinese Corpus has recorded 132 sentences that contain *suíchù* 随处, and *suíchù* 随处 in 129 sentences functions as an adverb with the meaning of 'not confined to one place'. For example,

(12) 佛出世立教，

Fó chū shì lì jiào,

Buddha out society establish religion,

'The buddha went into the society and established Buddhism,'

与师随处度人。（隋唐《禅源诠序》）

Yǔ shī suíchù dù rén.

With master everywhere moralize people.

'Enlightening people everywhere with the master.'

However, in the remaining three sentences, the meaning and grammatical functions of *suíchù* 随处 are ambiguous, as they are all in the form of '随处-而-V'. One of these sentences is analyzed below as an example.

(13) 如月在天，只一而已；

Rú yuè zài tiān, zhī yī éryǐ;

Like moon in sky, only one only;

'Just like the moon in the sky, only one;'

及散在江湖，则随处而见。（宋《朱子语录》）

Jí sàn zài jiānghú, zé suíchù ér jiàn.

When scatter in world, then everywhere see.

'When it is floating around the world, it can be seen everywhere.'

The reason why the meaning and the grammatical function of *suíchù* 随处 are ambiguous is that the conjunction *ér* 而 indicates the continuity of the actions, e.g. the action of *suí* 随 (Xue,1991). The syntactic structure of *suíchù-ér-jiàn* 随处而见 should thus be understood as 'V-O-Conj-V' In this context where the verb *sàn* 散'to scatter' emphasizes the wide area the moon can appear and *yuè* 月 emphasizes the similarity of what people would see, it is inappropriate to understand *suíchù-ér-jiàn* 随处而见 as 'see the moon according to different places'. Instead, *suíchù* 随处 means 'everywhere/ not confined to one place' which is the meaning when *suíchù* 随处 is an adverb. This is how the ambiguity is observed in the phrase. The ambiguity can be justified in books such as '*Bǎizhànqífǎ*' 百战奇法 compiled in the Ming Dynasty which changed *suíchù-ér-jiàn*

随处而见 into *suíchù-kě-jiàn* 随处可见 'can be seen everywhere' in order to avoid the ambiguity mentioned above.

Suíchù 随处 has maintained its semantics and grammatical function over that period of time until now. There is only one exception recorded in CCL Ancient Chinese Corpus, attesting *suíchù* 随处 used as a verb-object phrase expressing the meaning 'according to different places' in the Ming Dynasty,

(14) 灾患缠缚，随处而安。（明《贤弈编》）

Zāi huàn chán fù, suí chù ér ān.

Disaster worry interwine contrain, obedient to places and keep calm.

In this sentence, *suí* 随 can be understood as 'obedient to' and matches with the verb *ān* 安 to constitute the syntactic structure 'V-O-Conj-V'.

Despite this exception, we can still draw the conclusion that *suíchù* 随处 generally functioned as an adverb with the meaning of 'everywhere/not confined to one place' during the whole period.

But it is intriguing that while both of them are in the form of '随-X', *suídì* 随地 has undergone a lexicalization process changing from a verb-object phrase to an adverb and *suíchù* 随处 was an adverb at the very beginning without undergoing the process of lexicalization. By considering the basic human need to express time and space concept, we have studied the lexicalization of *suíshí* (随时) 'whenever/anytime' and finally found the reason.

Suíshí 随时 was also used as a verb-object phrase when it was used in the early Qin Period to express the meaning of 'follow/based on the time'. For example,

(15) 夫圣人随时以行，是谓守时。（春秋《国语》）

Fū shèngrén suí shí yǐ xíng, shì wèi shǒushí.

Modal particle sage follow time to act, it is called obedient time.

'The sage acts based on the time, which is called obedience to time.'

Suíshí 随时 maintained its use as a phrase from the Spring-Autumn Period to the Han Dynasty, and then extended its meaning to 'anytime/not confined a certain time' in both the Wei and Jin Dynasties.

(16) 得随时念佛三昧门：

Dé suíshí niànfó sānmèimén:

Should anytime pray to Buddha Buddhism concept:

'You should pray to the Buddhas whenever you can:'

于一切时常见诸佛。（魏晋《东晋译经》）

yú yíqiè shícháng jiàn zhūfó.

at all moments see every Buddha.

'then you will see all the gods whenever you want to.'

In the above sentence, according to the context provided by *yú-yíqiè-shícháng* 于一切时常 'at any moment', it is clear that *suíshí* 随时 does not mean 'to follow the time' but 'not to be confined to any time' and to emphasize the frequency of the verb *niànfó* 念佛 'pray to buddha'. It thus shows that *suíshí* 随时 has begun to function as an adverb in the Wei and Jin Dynasties. Jiang&Guo (2016) has argued that it was due to the influence of the introduction of Buddhist Literature in the Wei and Jin Dynasties that the phrase 随-X began to lexicalize. It can also be seen that example (16) is extracted from a Buddhist book. It is worth noting that *suíchù* 随处 appeared in the same period of time and could be used with *suíshí* 随时 to express the meaning of 'whenever and wherever'. For example,

(17) 但随时随处，息业养神。（隋唐《禅源诠序》）

Dàn suíshí suíchù, x yè yang shén.

But whenever wherever, stop business recover energy.

'But we should stop the business and recover energy whenever and wherever we need.'

Therefore, we deduce that in order to meet the need of expressing both concepts of time and space, *suíchù* 随处 was created to match with the adverb *suíshí* 随时, expressing the meaning of 'whenever and wherever', while *suí-dì* 随-地 maintained the grammatical function of V-O phrase to match with the V-O phrase *suí-shí* 随-时, expressing the meaning of 'according to different geographical situation/places and according to different times.'

Suíchù 随处 has kept co-occurring with *suíshí* 随时 frequently since the Sui Dynasty and the Tang Dynasty and, as mentioned in Sect. 1, it was not until the Qing Dynasty that *suídì* 随地 began to collocate with *suíshí* 随时 in a larger number of cases.

In conclusion, we have studied the semantics and grammatical functions of *suídì* 随地 and *suíchù* 随处 and found that they have different functions. *Suídì* 随地 was derived from a V-O phrase *suí-dì* 随-地 in the Han Dynasty and underwent a change of meaning from 'follow the earth' to 'based on the geographical situations/places'. It emphasized that the actions and events involve movement towards the ground. *Suídì* 随地 began to be used as an adverb in the Yuan Dynasty with the meaning of 'everywhere', but still kept in its meaning the emphasis on movement towards the ground. On the other hand, *suíchù* 随处 was an adverb once it appeared in the Wei and the Jin Dynasties in order to match with the adverb *suíshí* 随时, expressing the meaning of 'everywhere/not confined to one place'. It emphasizes the concept of three-dimensional space rather than the geographical surface of the ground, which provides us with an insight to answer questions on grammar asked by international students in Chinese language classes.

4 Answers to Questions in CSL Class

4.1 Grammatical Rules of "*Suídì* 随地/*Suíchù* 随处+VP"

Committing the grammatical errors in the CSL class mentioned in Sect. 1, students matched the wrong verbs with *suídì* 随地. To explain the specific syntax of *suídì* 随地/*suíchù* 随处+VP, let us compare the following sentences,

(18) 私人场所禁止随处乱走。

Sīrén chǎngsuǒ jìnzhǐ suíchù luànzǒu.

Private place forbidden everywhere randomly walk.

'It is forbidden to walk randomly in private places.'

(19) *私人产所禁止随地乱走。

Sīrén chǎngsuǒ jìnzhǐ suídì luànzǒu.

Private place forbidden everywhere randomly walk.

'It is forbidden to walk everywhere on the ground in private places.'

The different meanings of *suíchù* 随处 and *suídì* 随地 involve certain attributes of the verbs. In the examples (18–19), the verb *zǒu* 走 is explained as 'using the feet to move the body', which involves the direction of horizontal movement. As I mentioned in Sect. 2, *suíchù* 随处 emphasizes actions happening in three-dimensional space, which allows *zǒu* 走 to take place. The phrase *jìnzhǐ-suíchù-luànzǒu* 禁止随处乱走 is to constrain the horizontal direction of the verb *zǒu* 走. However, *suídì* 随地 only emphasizes actions involving movement toward the ground, which requires the verb to have a clear upward/downward orientation, such as *zòu* 坐 'sit', *tùtán* 吐痰 'spit', *rēng-lājī* 扔垃圾'throw garbage' etc. (Zhou, 2019). Therefore, the 'three-dimensional displacement' cannot be simplified as 'the movement down to the ground'.

Suíchù 随处 is applied more broadly to phrasal verbs than *suídì* 随地, because the semantics 'three-dimensional space' expressed by *chù* 处is apparently broader than *dì* 地. Therefore, verbs with no specific directions are more likely to match with *suíchù* 随处, such as *suíchù-zhāngwàng* 随处张望 'look around everywhere', *suíchù-yóudàng* 随处游荡 'wandering everywhere' etc. However, verbs with a clear meaning of upward/downward direction, are more likely to match with *suídì* 随地, such as *jiǎn* 捡 'pick up', *tang* 躺 'lie down', *dǎo* 倒 'fall down' etc.

Based on the analysis above, it is easy to understand why *suídì-kàn* 随地看 'look randomly' is wrong to say in Mandarin Chinese, as *kàn* 看 has no specific direction involved. However, why is the phrase *suídì-kějiàn* 随地可见 'can be seen everywhere' frequently acceptable in Mandarin Chinese? The reason is that *suídì* 随地 in the phrase *suídì-kějiàn* 随地可见 is an adverb of scope rather than a manner adverb.

(20) 他正在随地捡落叶。

Tā zhèngzài suídì jiǎn luòyè.

He is randomly pick up fallen leaves.

'He is randomly picking up the fallen leaves everywhere.'

(21) 落叶随地可见。

Luòyè suídì kějiàn.

Fallen leaves everywhere can be seen.

'Fallen leaves can be seen everywhere.'

The dynamic semantics of *jiǎn* 捡 'pick up' should collocate with a manner adverb. *Suídì* 随地 can take up the grammatical function of a manner adverb to modify the manner in this case. This is the reason why *suídì* 随地 is compatible with *jiǎn* 捡. Furthermore, the syntactic structure 'manner adverb-verb' can be used with an aspect marker *zhe* 着 or *le* 了 which modifies the state of the verb (Li&Thompson, 1981). For example,

(22) 他随地扔了一袋垃圾。

Tā suídì rēngle yīdài lājī.

He randomly threw a bag of garbage.

'He randomly threw away a bag of garbage.'

However, used with the static visual results expressed by *jiàn* 见, *suídì* 随地 does not illustrate the manner or process of the verb, but indicates the scope of the visual result. *Suídì* 随地 functions as an adverb of scope to show the broad range of the leaves falling on the ground. Therefore, an adverb of scope is allowed to match with the resultative verb *jiàn* 见.

To summarize, the meaning expressed by *dì* 地 requires the verb to involve an upwards/downwards direction while 'three-dimensional space' expressed by *chù* 处 suits a broader range of verbs. In addition, there is a mutual influence between the adverb and the verb. Whether *suídì* 随地 functions as a manner adverb or an adverb of scope largely depends on the dynamic attributes of the verb.

4.2 The Lexical Adhesion of 随地 and 随处 in Mandarin Chinese

Influenced by their meanings and grammatical functions in ancient Chinese, *suídì* 随地 and *suíchù* 随处 show distinct lexical adhesion in Mandarin Chinese. As I mentioned in Sect. 2, *suíchù* 随处 was used as an adverb when it appeared in the Wei and Jin Dynasties. It has maintained its adverb function since the Wei and Jin Dynasties till now. By searching the CCL Modern Chinese Corpus, we found that *suíchù* 随处 in all the 2300 sentences expresses the meaning of 'not confined to one place/everywhere'. And in BCC (Beijing Language and Culture University Corpus Center) Chinese Corpus, there are only 2 exceptions:

(23) 走在海边的沙子上面，

Zǒuzài hǎibiān de shāzi shàngmian,
Walk seaside sand on,

感觉随处而异。

gǎnjué suíchù ér yì.
feel according to places different

'The feeling of walking on the beach is different in different places.'

有的地方干软，有的地方潮润。

Yǒude dìfāng gān ruǎn, yǒude dìfāng cháo rùn.
Some places dry soft, some places damp wet.

'Some places are dry and soft. Some places are damp and wet.'

According to the context, we can see that the subject's experiences or feelings vary according to different places. *suíchù* 随处 expresses the meaning of 'according to different places' rather than 'everywhere'. Therefore, *suí* 随 functions as a coverb in the clause and collocates with the noun *chù* 处 to constitute a Coverb-N phrase that serves to denote a locational adverbial modifying the following predicate. The other examples are found to share the similar structure and function.

Suídì 随地 is relatively weaker in lexical adhesion and denotes a different meaning. It is not rare to see its use as a V-O phrase to denote different ground surfaces. Phrases such as *suídì-ér-jìan* 随地而建 'constructed according to different geographical surfaces', *suídì-ér-yì* 随地而异 'being distinct according to different regions' are all occasionally used in Chinese literature.

Therefore, in a class of Chinese as Second Language, a Chinese teacher can answer students' questions by elucidating the semantic differences of *dì* 地 and *chù* 处, as they denote different attributes in directionality, dynamic and static features. Finally, the V-O structure of *suídì* 随地, which is frequently used in Modern Chinese, can be best explained with an insight from historical data that shows clearly its preference to have a meaning referring to surface-related actions.

5 Conclusion

This study aims to provide a clear and precise answer to the grammatical confusions concerning *suídì* 随地 and *suíchù* 随处 in CSL classes by sorting out the lexicalization process of the two adverbs in Ancient and Modern Chinese. It suggests that the two seemingly similar adverbs may have undergone distinct diachronic processes with different functional changes. By explicating the meanings and grammatical functions of *suídì* 随地 and *suíchù* 随处, TCSL teachers would have a better perception of how form and meaning are correlated in rendering varied collo-constructional preferences and range of semantic extensions.

References

1. Heine, B., Kuteva, T.K.: World Lexicon of Grammaticalization. Cambridge University Press, Cambridge (2002)
2. Li, C.N., Thompson, S.A.: Mandarin Chinese: A Functional Reference Grammar. University of California Press, Berkeley (1981)
3. Liu, M.: The Form-Function Mapping Principles in Chinese Grammar: A Morpho-Constructional Approach. International Chinese Language Education (2017)
4. Givon, T.: English Grammar: A Function-Based Introduction. John Benjamins Publishing Company, Amsterdam (1993)
5. Fan, S.: The Comparative Study on "Suidi" and "Suichu". Huazhong Normal University (2018)
6. Jiang, L., Guo, X.: Commonness and individuality of the lexical process of verbal-object disyllabic words: taking the example of "suiyi" and "suibian". Research Ancient Chin. Lang. (03), 81–89 (2016)
7. Jiao, H.: A Study of "sui-x" Model Adverbs in Modern Chinese Language. Huaibei Normal University (2015)
8. Kang, J., Fang, X.: The lexicalization and generation mechanism of the Chinese time adverbs "Suishi". J. China West Normal Univ. (04), 46–50 (2016)
9. Wang, L.: Comparative Study on General Adverb "Daochu (到处)", "Chuchu (处处)" and "Suichu (随处)". Nanjing Forestry University (2014)
10. Wang, X.: The lexicalization and grammaticalization of "Suibian" (随便)——also on the general law of the evolution of Chinese verb—object phrase. J. Yunnan Normal Univ. **08**(5), 28–32 (2010)
11. Wu, B.: The lexical process of the verb "sui." J. Wenzhou Univ. **25**(04), 42–47 (2012)
12. Xue, F.: The analysis on the semantics and grammatical functions of conjunction er. Lang. Stud. **01**, 55–62 (1991)
13. Zhou, H.: The spatial directional feature of the verb observed from verb-direction combination. J. Xinjiang Univ. **1.47**(01), 120–129 (2019)

Grammatical Development and Semantic Change of the *Qià*-Based Lexical Cluster: From Objective Match to Subjective Evaluation

Ying Lu[1](✉) and Meichun Liu[2]

[1] Teacher's College, Beijing Union University, Beijing 100101, China
[2] Department of Linguistics and Translation, City University of Hong Kong, Kowloon, Hong Kong
meichliu@cityu.edu.hk

Abstract. This study aims to analyze the semantic change of *qià* (恰), which has experienced a grammatical shift from a monosyllabic adverb to a set of related disyllabic words. Q*ià* first appeared in the Six Dynasties as a pre-verbal adverb to express a coincidental situation in time, space, or quantity. It then began to collocate with a predictive element to form a cluster of *qià*-based disyllabic words with partially shared meanings. Due to semantic weakening of *qià* and strengthening of a collocating predicate in the lexicalization processes, this lexical cluster acquired semantically varied meanings with different pragmatic implications. The collocation of *qià* with verbal elements, such as *qià-rú* (恰如) and *qià-sì* (恰似), appeared as verbal phrases, and the adjectival *qiàhǎo* (恰好) and the adverbial *qiàqià* (恰恰) were also derived in the Tang and Song periods. Later in the Ming Dynasty, the originally adjectival *qiàqiǎo* (恰巧) was used more frequently as an adverb, and a new adjective *qiàdàng* (恰当) was formed to take up the adjectival function of *qiàhǎo*. The three adverbials *qiàhǎo*, *qiàqià* and *qiàqiǎo* were then fully lexicalized. Although these disyllabic words are all derived from *qià*, they differ in semantic and pragmatic functions in profiling various facets and evaluative polarities in terms of expectation or presupposition and eventually acquired more subjectivized meanings. This study shows that the grammaticalization processes of the lexical cluster were triggered by usage-based collocational patterns and have undergone semantic changes in line with the pragmatic force of subjectification.

Keywords: *Qià* (恰)-based lexical cluster · Semantic and grammatical change · Subjective expectation · Objective match · Grammaticalization path

1 Introduction: Multiple Meanings of *Qià*-Based Lexical Cluster

In modern Chinese, there is a set of words containing the morpheme *qià* (恰), such as the verbs *qiàrú* (恰如) and *qiàsì* (恰似), the adjective *qiàdàng* (恰当), and the adverbs *qiàqiǎo* (恰巧), *qiàqià* (恰恰) and *qiàhǎo* (恰好). Although they differ in lexical category and meaning, they all share a common semantic core, expressing an exact match of a situation in time, space, quantity, condition, etc. For example,[1]

[1] Examples in (1) and (3) are cited from pages 446 and 445 of [8], and (2) is from page 717 of [9].

M. Liu et al. (Eds.): CLSW 2020, LNAI 12278, pp. 235–252, 2021.
https://doi.org/10.1007/978-3-030-81197-6_21

(1) a.他的处境[恰如]笼中之鸟。
Tā de chǔjìng [*qiàrú*] lóngzhōng zhī niǎo.
He is in a situation [just like] a caged bird.
b.这消息[恰似]晴天霹雳，令人十分震惊。
Zhè xiāoxi [*qiàsì*] qíngtiānpīlì, lìng rén shífēn zhènjīng.
The news came [just as] a bolt from the blue.
(2) 有些字眼用得不太[恰当]。
Yǒu xiē zìyǎn yòng de bú tài [*qiàdàng*].
Some words are not [very proper].
(3)a.左边锋疾射入网，[恰]在此时鸣笛终场。
Zuǒbiānfēng jíshè rù wǎng, [*qià*] zài cǐshí míngdí zhōngchǎng.
The outside-left sent the ball into the net with a quick shot [just] before the whistle blew for full time.
b.前面一棵大树，[恰恰]挡住了视线。
Qiánmiàn yì kē dà shù, [*qiàqià*] dǎngzhù le shìxiàn.
There is a big tree in front of us, [just] blocking our view.
c.八个人一桌，十六个人[恰好]坐两桌。
Bā gè rén yì zhuō, shíliù gè rén [*qiàhǎo*] zuò liǎng zhuō.
Eight persons per table, sixteen people are [just] two tables.
d.我毕业后，[恰巧]他们厂需要一个技术员，就把我调来了。
Wǒ bìyè hòu, [*qiàqiǎo*] tāmēn chǎng xūyào yí gè jìshùyuán, jiù bǎ wǒ diào lái le.
Their factory [coincidentally] needed a technician. So I was transferred here after my graduation.

These seven words are related *qià*-words, sharing a common element *qià*. They were often viewed as lexemes of ‘congruence’ or ‘coincidence’ and analyzed as near-synonyms sharing the semantic features [+ coincidence, + subjectivity] [1] and a tone of congruence [2], or as referring to the coincidence of events and situations relevant to speaker’s expectation [3]. However, there may be a divergence in the polarity of expectation: *qiàqià* is often used to emphasize counter-expectation with an adversative result; *qiàhǎo* is more likely to be used to express expected and positive information; *qiàqiǎo* is relatively neutral in expressing the coincidence and contingency of objective facts [4–6]. In general, *qià* refers to things or situations that are logically consistent with one another, and *qià*-based modal adverbs all express an expected mood [3, 7]. So the subtle differences lie in the speaker’s expectation and subjective judgment. This study attempts to answer the following questions: Is the lexical set under study able to cover all possibilities in terms of expectation? Are they consistent in coding the same dimension of expectation in varied contexts? In what ways are the different categories of *qià*-words (adverbs, verbs, and adjectives) interrelated?

In many dictionaries, these *qià*-words are mutually definable. For example, *qiàqià* is interpreted in [8] as follows: "1) the same as *qiàhǎo*, mostly used in written language; 2) *Qiàqià* [+*jiù*] +*shì* is used in positive and negative contrast sentences to strengthen an affirmative tone". In [10], it is interpreted as 'exactly; precisely.' But these accounts seem unable to explain a number of instances of *qiàqià* in the CCL corpus.[2] For example, the following sentences may not be replaced by *qiàhǎo*:

(4)a. 他觉得只要把事情做到尽善尽美，就一定会万无一失。可这种脱离现实的有为，[恰恰]可以理解为过分的欲望，是不合理的，它不是真正的积极。
Ta juéde zhǐyào bǎ shìqing zuòdào jìnshànjìnměi, jiù yídìng huì wànwúyìshī. Kě zhè zhǒng tuōlí xiànshí de yǒuwéi, [*qiàqià*] kěyǐ lǐjiě wéi guòfèn de yùwàng, shì bù hélǐ de, tā bú shì zhēnzhèng de jījí.
He felt that he would make everything foolproof, providing he could do it perfectly. But this kind of achievement, divorced from reality, can [only] be understood as an excessive desire, which is unreasonable and not really positive.

b. 在"男女平等"的大旗下，中国学校以统一的内容、统一的方式、统一的标准来要求男孩和女孩，而这[恰恰]不是真正意义上的平等。
Zài nán nǚ píngděng de dàqí xià, zhōngguó xuéxiào yǐ tǒngyī de nèiróng, tǒngyī de fāngshì, tǒngyī de biāozhǔn lái yāoqiú nánhái hé nǚhái, ér zhè [*qiàqià*] bú shì zhēnzhèng yìyì shàng de píngděng.
Under the banner of "gender equality," Chinese schools require boys and girls to do the same things in the same way and meet the same standard, which is not [exactly] real equality.

c. 不在其位，难谋其政啊！而且你我现在所没有的[恰恰]是权力！懂吗？
Bú zài qíwèi, nán móu qí zhèng ā! érqiě nǐ wǒ xiànzài suǒ méiyǒu de [*qiàqià*] shì quánlì! Dǒng mā?
It is hard to do anything if one is not in the position to do so! And what you and I do not have now is [precisely] power! Understood?

The uses of *qiàqià* in (4) are pretty different from *qiàhǎo*. If exchanged, *qiàhǎo* will be semantically incompatible with these negatively - expressed contexts. In (4a), *qiàqià* may be replaced by *qíshí* (其实), since the sentence expresses an implied adversative relationship (A → M1 ~ B),[3] which is usually used to convey a counter-expectation. In comparison, *qiàqià* in (4b) is used before the negator *búshì* (不是), signaling an expectational contrast with higher semantic intensity. Lastly, *qiàqià* in (4c) is not used to strengthen a positive mood but a negative mood.

Based on the above observations, we can see that *qià-words*, represented by *qiàqià*, can convey a subjective evaluation of a situation against the speaker's expectation. In terms of expectational polarity, *qià-words* may express three scenarios of related but different semantic features:

[2] The Center for Chinese Linguistics Corpus (CCL): ccl.pku.edu.cn:8080/ccl_corpus/.

[3] Implied adversative relation is a kind of presupposition in adversative relationship, which means "if A, then more likely (M1) not B, but the result is B". See [11] and [12] for more details.

i. The neutral, objective observation of 'coincidence,' i.e., a factual situation objectively described by the speaker as an exact match with another situation in space, time, quantity, condition, etc., generally marked by the adverbs *qià* and *qiàqiǎo*.
ii. The positive evaluation of a subjectively interpreted coincidence or 'appropriateness' whereby the speaker subjectively projects a match with a situation. This coincidence satisfies the speaker's wishes or intentions, making the speaker evaluate it as desirable and appropriate. This is a subjectively - evaluated positive matching, typically expressed by the verbs *qiàsì* and *qiàrú*, the adjective *qiàdàng* and the adverb *qiàhǎo*.
iii. The negative evaluation of a subjectively interpreted as 'counter-expectation.' This refers to a situation in opposition to the speaker's expectation or an expectation shared by the hearer and the society. Whether the sentence signals an adversative or contrastive relation, the objective coincidence is interpreted as a reverse match with the subjective expectation of some sort, which is typically expressed by the adverb *qiàqià*.

Among the three possibilities, i) is more in line with the original meaning of *qià*-based words, while ii) and iii) are further developed with pragmatic implications, which have not been sufficiently explored yet. The sections below will provide more details on the historical path of their grammatical and semantic changes.

2 Occurrences of *Qià* in Middle Ancient Chinese

The character *qià*, a late addition in *Shuowen Jiezi*, was not included by Xu Shen in the first edition. By examining an ancient Chinese corpus, we found five earliest appearances of *qià* in the texts from the Six Dynasties.[4] *Qià* is used before a predicate indicates a coincidental/situation as in (5) or a timely encounter as in (6). It refers to an objectively described situation without subjective evaluation of personal expectation or preference.

[4] This paper use + the ancient Chinese corpus of the Chinese Linguistics Research Center of Peking University. The corpus has more than 170 million words. Besides 5 examples, another example from Tao Yuanming's poem of "Guiniao • qiyi" is 和风弗恰, 翻翮求心, which is 和风不洽 in some editions. *Qià* 洽 is explained in Guangyun as 'harmony'. There are doubts about the character *qià* 恰 here, which is more likely to be *qià*洽.

(5)a. 遇太守郑瑕出猎，见田草深，遣人纵火爇之。信纯卧处，[恰]当顺风。犬见火来，乃以口拽纯衣，纯亦不动。
Yù tàishǒu zhèngxiá chūliè, jiàn tián cǎo shēn, qiǎn rén zòng huǒ ruò zhī. Xìnchún wòchù, [*qià*] dāng shùnfēng. Quǎn jiàn huǒ lái, nǎi yǐ kǒu zhuài chún yī, chún yì bú dòng.
The satrap Zheng Xia went hunting and sent people to set fire to the deep grass in the field. The place where Xinchun was sleeping was [just] in the downwind direction. The dog dragged his clothes by mouth when it saw the fire, but Xinchun was in a deep sleep.
b. 寻至秦国，以枕于市货之，[恰]遇秦妃东游，亲见度卖金枕，疑而索看。
Xúnzhì Qínguó, yǐ zhěn yú shì huò zhī, [*qià*] yù qín fēi dōng yóu, qīn jiàn dù mài jīnzhěn, yí ér suǒkàn.
(When Xindaodu) went to the Qin state, he tried to sell the golden pillow in the market. At that time, a concubine of the Prince of Qin [just] happened to see him on tour to the east. She suspected that the pillow was fake and asked for checking.

(6)a. 李冰當時随地理形便造桥，[恰]有形似北斗之排列。
Lǐbīng dāngshí suí dìlǐ xíngbiàn zào qiáo, [*qià*] yǒu xíng sì běidǒu zhī páiliè.
Li Bing exploited the geographic resources to build bridges, which were arranged [precisely] like the shape of the Big Dipper.
b. [恰]对妆台，诸窗并开。遥看已识，试唤便回。
[*Qià*] duì zhuāngtái, zhù chuāng bìng kāi. Yáo kàn yǐ shí, shì huàn biàn huí.
She was [just] facing the dressing table, and the windows were all opened. I recognized her when I saw her from afar. She replied when I called her.

Some previous research suggested that in the Six dynasties, *qià* could indicate an overlap or exact tally of things. This suggestion is reasonable, but it may not be feasible to take *qià* in the above examples as denoting only a positive view 'just right, exactly right.' The instances of *qià* in (5) and (6) already vary in coding different kinds of coincidences.

In the Tang and Song dynasties, *qià* was used much more frequently. In the examples below, *qià* is used syntactically to modify a verbal predicate as in (7a), an adjectival predicate as in (7b), and a quantifier as in (7c). Due to frequent collocation, the adjectival phrase *qià-hǎo* began to be formed, with *qià* modifying *hǎo*, as in (8a) and continued to evolve to an adverb, as in (8b). Besides, *qià* was further combined with verbs of resemblance to form a predicate of equative assertion, such as *qiàsì* and *qiàrú*, as in (9).

(7)a.秋水才深四五尺，野航[恰]受两三人。
Qiū shuǐ cái shēn sì wǔ chǐ, yě háng [*qià*] shòu liǎng sān rén.
Water (in the river) is only four or five feet deep in autumn, and the wild sailing boat can [just] carry two or three people.

b.毕卓醉狂潘氏少，倾来掷去[恰]相宜。
Bìzhuó zuì kuáng Pānshì shào, qīng lái zhì qù [*qià*] xiāngyí.
Bi Zhuo is drunk, and Pan is young. It is [just] not surprising that they are throwing things around.

c.不须更向沧溟望，惆怅欢情[恰]一年。
Bù xū gèng xiàng cāngmíng wàng, chóuchàng huānqíng [*qià*] yì nián.
It is not necessary to gaze at the sea. Our forlorn love has [just] been a year.

(8)a.只是难得到[恰好]处，不着意又失了，才着意又过了，所以难。
Zhǐshì nándé dào [*qià hǎo*] chù, bù zhuóyì yòu shī le, cái zhuóyì yòu guò le, suǒyǐ nán.
It is hard to get to the [precisely] right point. You will miss it if you do not care, but you will also get too far if you do. That is why it is hard.

b.府尹大怒，喝道："胡说！世间不信有这等巧事！他家失去了十五贯钱，你却卖的丝[恰好]也是十五贯钱，这他分明是支吾的说话了。"
Fǔyǐn dà nù, hèdào: "Húshuō! Shìjiān bú xìn yǒu zhè děng qiǎo shì! Tā jiā shīqù le shíwǔ guàn qián, nǐ què mài de sī [*qiàhǎo*] yě shì shíwǔ guàn qián, zhè tā fēnmíng shì zhīwǔ de shuōhuà le."
The magistrate was very angry and shouted: "Nonsense! How can there be such a coincidence on earth! His family lost fifteen sets of coins, and the silk you sold [just happened] to worth also fifteen sets of coins. This is clearly a lie."

(9)a.愁颜[恰似]烧残烛，珠泪阑干。
Chóu yán [*qiàsì*] shāo cán zhú, zhū lèi lángān.
A sad face is [just like] a candle burning to the end, leaving pearl-shaped tears on the guardrail.

b.贪人好聚财，[恰如]枭爱子。
Tān rén hào jù cái, [*qiàrú*] xiāo ài zǐ.
A greedy man loves accumulating wealth, [just as] an owl loves its young.

c.林云："丹霞患哑庞公患聋。"士云："[恰]是。"
Línyún: "Dānxiá huàn yǎ Páng gōng huàn lóng." Shì yún: "[*qià*] shì."
Lin said: "Danxia suffers from mutism while Pang gong suffers from deafness." Shi replied: "[Exactly] yes."

With regard to syntactic functions, the verbs *qiàsì* and *qiàrú* may take an eventive complement, as in (9a), or connect two clauses as a high-level predicate to make a subjective comment on the whole proposition, as in (9b), and it can be used alone to answer a question independently, as in (9c).

The instances of *qià* in (7)–(9) above show a process of morphological change with semantic weakening. That is, it changes from modifying a predicate as an independent adverb as in (7) to forming *qià*-V compounds like *qiàsì*, *qiàrú,* and *qiàhǎo*, with the semantic head on the V-element (*sì*, *rú* and *hǎo*). Along with disyllabification, the newly derived *qiàhǎo* with an adjectival V (*hǎo* 'good') further developed into a modal adverb to indicate the speaker's evaluation of something good. In the Song dynasty, *qiàhǎo* was mainly used as an adjective 'proper' to modify a noun (see 8a). At the end of the Song dynasty and the beginning of the Yuan dynasty, it was used more often as an adverb to express a coincidental situation consistent with the speaker's subjective view. At the same time, it underwent a semantic change to refer to various scenarios of coincidence, not necessarily a good or a proper one (see 8b).

In the Tang dynasty, *qiàqià* was formed through reduplication, which first appeared in Tang poetry in ancient Chinese. There are many controversies on the interpretation of *qiàqià* in Tang Poetry, including 'full heart, warbler singing, everywhere, dense, frequent, fusion,' or 'appropriate' as shown in the following examples.

(10)a. 年光[恰恰]来，满瓮营春酒。
Niánguāng [*qiàqià*] lái, mǎn wèng yíng chūn jiǔ.
New spring is [just] coming. I have prepared a full bottle of wine.
b. 留连戏蝶时时舞，自在娇莺[恰恰]啼。
Liúlián xìdié shíshí wǔ, zìzài jiāo yīng [*qiàqià*] tí.
The butterflies lingered and danced while the warblers were [just] singing.
c. 栾栌与户牖，[恰恰]金碧繁。
Luánlú yú hùyǒu, [*qiàqià*] jīnbì fán.
The beams and the windows of the house are [just] all shining gold and jade.
d. [恰恰]用心时，[恰恰]无心用。无心[恰恰]用，常用[恰恰]无。
[*Qiàqià*] yòng xīn shí, [*qiàqià*] wú xīn yòng. Wú xīn [*qiàqià*] yòng, cháng yòng [*qiàqià*] wú.
Whenever you [just] want to concentrate on something, you might [just] fail; it [just] comes if you [just] concentrate without deliberately doing it.

Examples in (10) are from Tang poetry and the transformation text, showing that from which the controversial interpretations of *qiàqià* are derived and evolved into varied readings. However, the diverse interpretations may stem from the collocational associations of *qiàqià* with various contextual elements in the poems. In view of the general lexical-semantic principles, it is necessary to offer a unified analysis of the lexical semantics of *qià* as a form-meaning mapping unit so as to integrate its contextually different usages. Considering the intertextuality of poetic clauses, *qiàqià* can be appropriately interpreted as conveying basically the concept of 'precisely matching and appropriate,' which is continued in the Song dynasty, as shown below.

(11)a. 最怜[恰恰]新眠起，云雨初收。斜倚琼楼。叶叶眉心一样愁。
Zuì lián [*qiàqià*] xīn mián qǐ, yún yǔ chū shōu. Xié yǐ qíong lóu. Yèyè méixīn yíyàng chóu.
The most pitiful scene was she had [just] woken up, leaning against the high building. The clouds and rains had just gathered and were as sad as her eyebrows.

b. 问答证明。[恰恰]相应。照中失默。便见侵凌。证明问答。相应[恰恰]。
Wèndá zhèngmíng. [*Qiàqià*] xiāngyìng. Zhào zhōng shī mò. Biàn jiàn qīnlíng. Zhèngmíng wèndá. Xiāngyìng [*qiàqià*].
The answer [precisely] proved the question clearly, and vice versa.

c. 自今夏来，觉见得才是圣人说话，也不少一个字，也不多一个字，[恰恰]地好，都不用一些穿凿。
Zì jīnxià lái, jué jiàn dé cáishì shèngrén shuōhuà, yě bù shǎo yí gè zì, yě bù duō yí gè zì, [*qiàqià*] de hǎo, dōu bú yòng yìxiē chuānzáo.
Since this summer, I have started to understand that a sage's speech has not a single word less or more than necessary, but is [just] right and in need of no farfetched interpretation.

Examples in (11) come from the Song poetry, Buddhist scriptures, and quotations. We can see that *qiàqià* was still used to describe a favored coincidence or a precise match in the Song period. Respectively, *qiàqià* describes temporal coincidence in (11a), precise scope in (11b), an appropriate degree in (11c). At this time, *qiàhǎo* was still used as an adjective indicating a proper match and not yet used as an adverb. In contrast, the adverb *qiàqià* was used for the positive evaluation of precise match of some sort of expectation, which is different from the negative evaluation of 'counter-expectation' in modern Chinese. In a word, the semantic functions of *qiàqià* and *qiàhǎo* were not fully developed in the Tang and Song periods.

3 Occurrences of *Qià* in Early-Modern Chinese

In the CCL corpus of Yuan Dynasty texts, the two characters *qià* 恰 and *qià* 洽 are interchangeable in writing, as often seen in Yuan drama, Sanqu, and novels, referring to the same meaning 'appropriately and precisely' as exemplified in (12). This appears to be one of the reasons for the various interpretations of *qià* 恰. At the end of the Song Dynasty and the beginning of the Yuan Dynasty, *qiàqià* still expressed the positive evaluation of appropriate or precise match as in (13a), but was gradually used to refer to counter-expectation as in (13b). This is due to the functional change of the adjectival *qiàhǎo* being turned into an adverb to indicate a positive evaluation of 'appropriate match.' Thus, there appeared a functional differentiation between the existent adverbial *qiàqià* and the newly derived adverbial *qiàhǎo*, as shown in (14).

(12)a. 想他每富家，杀羊也那宰马，每日里笑[恰]。

Xiǎng tā měi fù jiā, shā yáng yě nà zǎi mǎ, měi rì lǐ xiào[*qià*].

I am thinking about wealthy families. They slaughter sheep and horses for food, and laugh [well] every day.

b. 端端正正美人姿，月里嫦娥还喜[恰]。

Duānduānzhèngzhèng měirén zī, yuè lǐ cháng'é hái xǐ[*qià*].

She is a mannered and beautiful woman and more [appropriately] adored than the goddess on the moon.

(13)a. 须尽道、人间罕见。更[恰恰]占得，美景良辰，小春天暖。

Xū jìn dào, rénjiān hǎnjiàn. Gèng [*qiàqià*] zhàndé, měijǐng liángchén, xiǎo chūn tiān nuǎn.

Your young appearance is rare in the world, while we [precisely] enjoyed the wonderful time with beautiful scenery on those spring-like days in a winter month.

b. 残妆凤枕流清泪，景中情谁唤起？听西园[恰恰]莺啼。万里书难到，三春人未归：此恨谁知？

Cán zhuāng fèng zhěn liú qīng lèi, jǐng zhōng qíng shuí huàn qǐ? Tīng xī yuán [*qiàqià*] yīng tí. Wàn lǐ shū nán dào, sān chūn rén wèi guī: cǐ hèn shuí zhī?

My tears run down the phoenix pillow. Whose love does the scenery make me recall? I hear the warbler, [just] singing in the west garden. Ten thousand miles away, it must be hard to receive my letters. My lover is still not back in the third month of spring. Who knows my regret?

(14)a. 这客人被贼一箭射的昏了。苏醒回来，[恰好]有捕盗的官，到那里巡哨。那客人就把这缘故告了。

Zhè kèrén bèi zéi yí jiàn shè de hūn le. Sūxǐng huílái, [*qiàhǎo*] yǒu bǔdào de guān, dào nàlǐ xúnshào. Nà kèrén jiù bǎ zhè yuángù gào le.

The guest was struck by a thief's arrow. When he regained consciousness, an officer who was on patrol [just] came. The guest reported the incident to the officer.

b. 大人不知，[恰才]一个人，把这个妇人，[恰待]要勒死他。[恰好]撞着小人，救活他性命。

Dàrén bù zhī, [*qiàcái*] yí gè rén, bǎ zhè gè fùrén, [*qiàdài*] yào lēi sǐ tā. [*Qiàhǎo*] zhuàng zhe xiǎorén, jiùhuó tā xìngmìng.

Your Highness does not know that [just] at that moment a man [just] wanted to strangle this woman. I [just] happened to see them and saved her life.

The examples in (14) demonstrate that *qiàhǎo* (with the adjectival morpheme *hǎo* 'good') has undergone the functional shift to an adverb in the Yuan period, expressing the speaker's positive attitude towards a 'good' coincidence of events. Such coincidences are taken to be good and proper. But as shown in (8) above, in the late Song and early Yuan periods, the semantic function of *qiàhǎo* was not limited to only expressing positively - viewed coincidences. To remain functionally differentiated from the other adverb *qiàqià* and semantically distinct in its own lexical meaning, *qiàhǎo* is narrowed in its semantic scope and more exclusively used for a positive evaluation. It is also noted that there are occurrences of *qià* co-occurring with other morphemes, such as *cái* (才) 'just now' and *dài* (待) 'await' to form disyllabic adverbs, *qiàcái* (恰才) and *qiàdài* (恰待), as shown in (14b). Both of them were semantically similar to *qiàqià*, but became obsolete in later stages of Chinese.

In the Ming Dynasty, the adverb *qiàqià* did not always appear in contexts of a favorable or 'proper' coincidence, but often in contexts of less desirable coincidence, as in (15). In contrast, the other adverb *qiàhǎo* was used more often to express the positively expected coincidence by the speaker, as in (16). This contrastive development began in the Yuan Dynasty and was accomplished in the Ming dynasty. The subjective evaluation functions of the two adverbs, *qiàqià* vs. *qiàhǎo*, have thus been formally differentiated and entrenched.

(15)a. 今上神明威断，动法皇祖，而罹灾之岁，亦属壬寅，[恰恰]六十年。
Jīn shàng shénmíng wǔduàn, dòng fǎ huáng zǔ, ér lí zāi zhī suì, yì shǔ rén yín, [*qiàqià*] liùshí nián.
The emperor was wise and decisive, and he had ruled like the imperial ancestors. The year of the disaster was also a *Renyin* year, [precisely] sixty years afterward.

b. "如何不前不后，[恰恰]里到家便死，不信有恁般凑巧！"
"Rúhé bù qián bú hòu, [*qiàqià*] lǐ dào jiā biàn sǐ, bú xìn yǒu nènbān còuqiǎo!"
He said: "How could he die no sooner or later, but [just] after arriving at home? I do not believe this is just a coincidence!"

(16)a. 玄德欲弃了古城去守汝南，[恰好]刘辟、龚都差人来请。
Xuándé yù qì le gǔchéng qù shǒu rǔnán, [*qiàhǎo*] Liúpì, Gōngdū chāi rén lái qǐng.
Xuande wanted to abandon Gucheng and go to defend Runan. [Just] then, Liu Bi and Gong Du sent people to invite him.

b. 尊者心想："我正要见他见儿，他[恰好]就来请我，却不是有些夙缘？"
Zūnzhě xīn xiǎng: "wǒ zhèng yào jiàn tā jiàner, tā [*qiàhǎo*] jiù lái qǐng wǒ, què bú shì yǒu xiē fèngyuán?"
The Venerable said to himself: "I just want to see him, and [just] at this time he comes to invite me. Is it predestined?"

The cluster of *qià*-based lexical items was lexicalized disyllabic compounds in the Ming dynasty. This was manifested in three respects. First, the existing disyllabic compounds were further consolidated and frequently used, such as *qiàrú* and *qiàsì*, which became common verbs in the Tang and Song Dynasties, as in (17). Secondly, *qià* began to co-occur with *dàng* 'proper,' and form a new adjective, *qiàdàng* 'proper and suitable,' in the Ming Dynasty as in (18a). Thirdly, there was a new combination, *qiàqiǎo* 'precisely coincidental' used as an adverb, as in (18b).

(17)a. 当夜月黑，军士皆执灯火，明耀天地，[恰如]白昼。
Dāngyè yuè hēi, jūnshì jiē zhí dēnghuǒ, míng yào tiāndì, [*qiàrú*] báizhòu.
In the night, the moon was dark. All the soldiers held lamps and torches in their hands, lighting up the whole world and making it [just like] daytime.
b. "怎的两人写来[恰似]出于一手？真是才子佳人，可称双绝！"
"Zěn de liǎngrén xiě lái [*qiàsì*] chūyú yìshǒu? Zhēnshì cáizǐ jiārén, kě chēng shuāng jué!"
Why do the two of them write [as if] they were the same person? What a talented man and what a talented beauty! What an adorable couple!

(18)a. 江老见说话蹊跷，反怀着一肚子鬼胎道："敢怕有甚不[恰当]处。
Jiānglǎo jiàn shuōhuà qīqiāo, fǎn huái zhe yí dù zi guǐtāi dào: "gǎn pà yǒu shèn bú [*qiàdàng*] chù."
Seeing that he was speaking in a strange way, Jiang said with an ulterior motive: "I am worrying whether there is something [inappropriate]."
b. 王殷士在西宁城上，窥探朱兵，[恰巧]杨璟驾着飞天炮，直打过来，把头顶打得粉碎。
Wáng Yīnshì zài Xīníng chéng shàng, kuītàn Zhū bīng, [*qiàqiǎo*] Yǎng Jǐng jià zhe fēi tiān pào, zhí dǎ guò lái, bǎ tóudǐng dǎ dé fěnsuì.
Wang Yinshi was spying on Zhu's troops on the wall of Xining City when Yang Jing [coincidentally] drove a flying cannon there. The cannonball hit him and smashed his head.

During the Qing Dynasty and the early years of the Republic of China, the use of the adjective *qiàdàng* increased significantly, and its lexical status and semantic function were fully established. It was used for describing 'appropriate and proper', which has continued to modern times.

(19)a. 宝钗笑道："你的号早有了：'无事忙'。三字[恰当]得很！"
Bǎochāi xiào dào: "nǐ de hào zǎo yǒu le: 'Wúshìmáng.' Sān zì [*qiàdàng*] de hěn!"
Baochai said with a smile, "Your title is already this: 'Busy for Nothing'. These words are very [appropriate]!"
b. 此乃驾驭悍将之术，用在英布身上，真是[恰当]。
Cǐ nǎi jiàyù hànjiàng zhī shù, yòng zài Yīngbù shēnshàng, zhēn shì [*qiàdàng*].
This is the way to control fierce generals, and it is really [appropriate] to use it on Ying Bu.

The adverbial use of *qiàqià* was further extended in the Qing Dynasty, to indicate precise matching of expectations as well as counter-expectations. This can be evidenced in the contrastive pair of examples below. In (20a), *qiàqià* can be replaced with adverbials denoting 'unexpectedly' such as *méi xiǎngdào* (没想到), *jìngrán* (竟然), *shuíliàodào* (谁料到) 'surprisingly, unexpectedly,' while in (20b) *qiàqià* can be replaced with the factual markers such as *zhēngshì* (正是), *zhēnde* (真的) 'truly, exactly' to denote a precisely expected match.

(20)a.雪婆自己伸手，在小姐香云上拔一只紫金朱松簪，[恰恰]与江潮的挖耳一般精细，一般长短。
Xuěpó zìjǐ shēn shǒu, zài xiǎojiě xiāngyún shàng bǎ yì zhī zǐjīn zhūsōng zān, [*qiàqià*] yú Jiāngcháo de wāěr yìbān jīngxì, yìbān chángduǎn.
Xuepo reached out and pulled out a purple gold pine hairpin from the young lady's hair. It was [exactly] as fine and of the same length as Jiang Chao's earpick.
b.当即拆开纸包，大家一瞧，个个吓得伸舌，复将昨日玩过的纸牌，仔细一检，[恰恰]少一张。
Dāngjì chāikāi zhǐbāo, dàjiā yì qiáo, gègè xià de shēn shé, fù jiāng zuórì wán guò de zhǐpái, zǐxì yì jiǎn, [*qiàqià*] shǎo yì zhāng.
They immediately opened the paper bag for a look. Everyone was so scared. Then they carefully checked the cards that they played with the day before. [Exactly] one card was missing.

It can be seen from (20) that both an expected and an unexpected outcome may be a 'precise match' but differ in expectational polarities. In (20a), *qiàqià* is used with an unexpected match, while in (20b) it refers to an expected match.

What is more, *qiàqiǎo* has a more diverse range of uses in this period. It could be used as an adjective or adverb, referring to good or bad coincidences from a neutral and objective perspective.

(21)a.欧阳德说道："师弟不用夸口，你乃事逢[恰巧]。"
Ōuyáng Dé shuōdào: "shīdì bú yòng kuā kōu, nǐ nǎi shì féng [*qiàqiǎo*]."
Ōuyáng Dé said: "Don't boast, my younger martial brother. You just [happen to be] in the right place."
b.船已将到济南，[恰巧]有些微微的小雨，衬着空间轻烟，风景实在可爱。
Chuán yǐ jiāng dào jǐnán, [*qiàqiǎo*] yǒu xiē wēiwēi de xiǎo yǔ, chèn zhe kōngjiān qīng yān, fēngjǐng shízài kě'ài.
The ship is arriving at Jinan. There [happen to be] light rains and mist. The scenery is really lovely.

4 Occurrences of *Qià* in Modern Chinese

In modern Chinese, *qià* is predominantly used in a disyllabic word, and rarely used alone. It may be combined with a number of adverbial elements to form a compound, such as *qiàyě* (恰也) 'precisely also' and *qiàwéi* (恰为) 'precisely is.'

(22)a.玉成夫妇，[恰也]是不曾经过这种事的，跟着也就叫嚷起来。
Yùchéng fūfù, [*qiàyě*] shì bù céng jīng guò zhè zhǒng shì de, gēn zhe yě jiù jiàorǎng qǐlái.
Yucheng and his wife [just] had never been through such a thing. They began to shout.

b.此时[恰为]金风送爽之晨，梧叶新黄，柿叶初红，松柏旧绿，乌桕乍紫。
Cǐshí [*qiàwéi*] jīn fēng song shuǎng zhī chén, wū yè xīn huáng, shì yè chū hóng, sōngbǎi jiù lǜ, wūbǎi zhà zǐ.
Right now is [just] a golden autumn morning, phoenix tree leaves are turning yellow, persimmon leaves are turning red, pines and cypresses are getting dark green, and Chinese sapiums are turning purple.

It was also combined with verbs of resemblance in *qiàrú* and *qiàsì* 'precisely similar, to express a comparative meaning. They are followed by a noun or verb phrase, often collocating with *yìbān* (一般) or *yíyàng* (一样) 'same kind' in the sentence and indicating the various facets of a matching situation.

(23)a.所有人都不知道他在心里正与一个人作着热烈交谈，情话[恰如]潮涌。
Suǒyǒu rén dōu bù zhīdào tā zài xīn lǐ zhèng yú yí gè rén zuò zhe rèliè jiāotán, qínghuà [*qiàrú*] cháo yǒng.
No one knew that he was talking animatedly with a person in his heart, and his words of love were [just like] the waving tides.

b.金军恶狠狠地进山搜杀，[恰似]进了迷魂阵一般。
Jīn jūn è'hěnhěn de jìnshān sōushá, [*qiàsì*] jìn le míhúnzhèn yìbān.
The fierce Jin troops went up into the mountain to search and kill, but [just like] entering an enchanting maze.

As for the adjectival *qiàdàng*, its grammatical and semantic function has been relatively fixed since its birth in the Ming dynasty. It is used to describe appropriate situations that are positively evaluated, as it retains the central meaning of *qià* in denoting appropriateness.

(24)a.历史可以歌颂，但是歌颂要安于本分，要用得[恰当]，不要使读者讨厌。
Lìshǐ kěyǐ gēsòng, dànshì gēsòng yào ān yú běnfèn, yào yòng dé [*qiàdàng*], bú yào shǐ dúzhě tǎoyàn.
History can be eulogized, but eulogizing should be based on facts and used [correctly], and should not irritate the readers.

b.因此，食物的选择、搭配要[恰当]，使之充分满足少年对各种营养素的需要量，防止偏食。
Yìncǐ, shíwù de xuǎnzě, dāpèi yào [*qiàdàng*], shǐzhī chòngfèn mǎnzú shàonián duì gèzhǒng yíngyǎngsù de xūyàoliàng, fángzhǐ piānshí.
Therefore, the choice and matching of foods should be [appropriate], so as to fully meet teenagers' needs for various nutrients and prevent unbalanced diets.

As for the three adverbs, *qiàqiǎo*, *qiàhǎo,* and *qiàqià,* they are commonly used in modern Chinese and retain the semantics of the morpheme *qià*. They may all describe an exact match between an expected outcome and an actual situation. As they are morphologically and semantically *qià*-based, these three adverbs can be used interchangeably in most contexts. However, they display some subtle differences in semantic and pragmatic functions in marking presupposed expectation and subjectivity.

Specifically, *qiàqiǎo* is used to signal the current situation as a precise coincidence, no matter if it is viewed as good or bad by the speaker. It indicates an unexpected status against an expected one.

(25)a. 两人选了一条宁静的巷道走着，从黑暗中出来一个讨钱的乞丐。他摸摸口袋，[恰巧]口袋里没有零钱，他想对曾说，又不好意思开口。
Liǎng rén xuǎn le yì tiáo níngjìng de xiàngdào zǒu zhe, cóng hēi'àn zhōng chūlái yí gè tǎo qián de qǐgài. Tā mōmō kǒudài, [*qiàqiǎo*] kǒudài lǐ měiyǒu língqián, tā xiǎng duì zēng shuō, yòu bùhǎoyìsī kāikǒu.
They walked in a quiet lane, and a beggar came out of the darkness begging for money. He touched his pocket and found that there [happened to be] no small change. He wanted to tell Zeng, but found it embarrassing.

b. 大家一怔，回头望去，原来阮大铖不迟不早，[恰巧]在这当儿回来了。
Dàjiā yí zhèng, huí tóu wàng qù, yuánlái Ruǎn Dàchéng bù chí bù zǎo, [*qiàqiǎo*] zài zhè dàng'er huílái le.
They were stunned. When they looked back, they saw that Ruan Dacheng had just come back. It was [coincidental], not late or early.

Next, *qiàqiǎo* is used more often when the condition and meets the speaker's subjective intention or desire, which is considered a reasonable and appropriate match. Both the semantic and pragmatic implications are positive.

(26)a. 柳生便在近旁的一家茶亭落坐，要了一碗茶，喝毕，觉得腹中饥饿难忍，正思量着，[恰好]一个乡里人捧着许多薄饼来卖。
Liǔshēng biàn zài jìnpáng de yìjiā chátíng luòzuò, yào le yì wǎn chá, hē bì, juéde fùzhōng jī' è nánrěn, zhèng sīliàng zhē, [*qiàhǎo*] yí gè xiānglǐ rén pěng zhē xǔduō báobǐng lái mài.
Mr. Liu sat down in a nearby tea pavilion and asked for a bowl of tea. After drinking it, he felt so hungry and was thinking about finding some food. A country-folk [happened to] come to sell pancakes.

b. 远远盯梢的人果然像个尾巴似的跟着。快走近公共汽车站时，[恰好]一辆公共汽车开来停站。家霆对童霜威说："爸爸，快！上车！"
Yuǎnyuǎn dīngshāo de rén guǒrán xiàng gè wěiba sìde gēn zhē. Kuài zǒujìn gōnggòng qìchē zhàn shí, [*qiàhǎo*] yí liàng gōnggòng qìchē kāilái tíngzhàn. Jiātíng duì Tóng Shuāngwēi shuō: "Bàbà, kuài! Shàng chē!"
The man who was watching them from afar followed like a tail. As they approached the bus stop, a bus [coincidentally] came and stopped. Jiating said to Tong Shuangwei, "Dad, hurry up! Get on the bus!"

Lastly, *qiàqià* may be used if the speaker describes the current situation as an exact match of the unexpected from either the speaker's or the hearer's perspective. It is pragmatically associated more with counter-expectation. Specifically, it may show two kinds of logical relations of a counter-expectation for an adversative or scalar contrast, as exemplified below. The common ground of the two examples is that they both involve a reverse or opposite situation against the expected.

(27) a. 这样理解产权虽然行得通，但有一个缺陷，就是有些行为对某甲来说是被禁止的，但对某乙来说却可能[恰恰]是合法的。
Zhèyàng lǐjiě chǎnquán suīrán xíng de tōng, dàn yǒu yí gè quēxiàn, jiù shì yǒu xiē xíngwéi duì mǒu jiǎ lái shuō shì bèi jìnzhǐ de, dàn duì mǒu yǐ lái shuō què kěnéng [*qiàqià*] shì héfǎ de.
Although it is feasible to understand property rights in this way, there is a defect that some behaviors are prohibited for A, but maybe legal for B.

b. 国家与政治的产生，不仅与道德无关，而且[恰恰]是人类非道德的产物。
Guójiā yú zhèngzhì de chǎnshēng, bùjǐn yú dàodé wúguān, érqiě [*qiàqià*] shì rénlèi fēi dàodé de chǎnwù.
The emergence of the state and politics is not only irrelevant to morality, but also [exactly] the product of human immorality.

In the above examples, (27a) involves an adversative or opposite relation, while (27b) involves a scalar-value contrast. In both instances, *qiàqià* signals a reverse situation or counter-expectation.

Finally, the frequencies of these three adverbs are also related to their semantic and pragmatic functions. The number of tokens of *qiàqià* is the highest in the CCL corpus, as there are 1147 examples of *qiàqiǎo*, 3102 examples of *qiàhǎo*, and 6460 examples of *qiàqià* in the corpus. This shows that *qiàqià* is semantically more flexible and pragmatically more feasible in most contexts. Its use helps convey more expressive information in which the speaker describes the situation with subjectively projected evaluation or expectation. It is more explicit in the sense that with *qiàqià*, the speaker also expresses his own interests, opinions, views, and values at the interpersonal level. In contrast, there are very few purely neutral situations [13], which can explain why the other adverb *qiàqiǎo* 'precisely' used with objective statements has the lowest frequency of occurrence. Compared with the expected situation, counter-expected coincidence is much more common as it conveys more expressive information in a given context [14]. That accounts for why *qiàqià* is twice more frequently used than *qiàhǎo*.

5 Semantic Change of *Qià*-Based Words in Lexicalization

Now we can conclude about the evolutional development of *qià*-based words. In general, the grammatical evolution of *qià*-based disyllabics is mainly due to its frequent collocation with predicates and the semantic shift from marking objective coincidence to subjective views, in line with the general tendency of semantic change towards subjectivization [15]. The lexicalization of derived new words is clearly associated with the trend of disyllabization of *qià* due to usage-based entrenchment [22]. Specifically, the process of disyllabization changes from a free monosyllabic *qià* to a cohesively formed cluster of lexemes conveying distinct but related meanings. The grammatical process is also accompanied by the semantic and pragmatic change from expressing objective coincidence to subjective evaluation, which is consistent with the tendency of changing from objectively situated meaning to subjectively evaluated meaning [15]. Once subjectivization is established, it will further develop to intersubjectivity [16], which involves a more interactive manner for two-way interaction and stance-taking.

During the Six Dynasties, *qià* started as a monosyllabic, independent adverb, indicating a coincidence or precise match in an objective situation. It was used mainly to describe objective observations and the relation between events. At this time, *qià* often occurred alone before a predicate and was not lexically combined with other components to form its disyllable.

By the Tang and Song periods, *qià* was extended with more diverse usages in a richer scope of contexts. It could modify verbal, adjectival, and quantifier predicates, and started to exhibit trends of disyllabic change. It may be combined with verbs of resemblance *sì/rú* (似/如) 'resemble' and identification verb *shì* (是) 'be.' The newly derived *qiàsì* and *qiàrú* also extended from verb phrases to adverbial use for the whole sentence. *Qià* was also combined with the adjective *hǎo* 'good' to form the adjectival compound *qiàhǎo* 'precisely good,' which changed into an adverbial in later stages. At the same time, the disyllabic adverb *qiàqià* was formed through the reduplication of *qià*, frequently seen in Tang poetry. In the Tang and Song Dynasties, *qiàqià* was primarily used in relatively neutral or positive contexts, similar to the use of *qiàhǎo* in modern Chinese.

As a result of functional division, we see that in the Yuan Dynasty *qiàhǎo* 'precisely' underwent a semantic-pragmatic change to encode counter-expectations so as to be functionally distinct from *qiàhǎo* 'appropriately good.' On the other hand, with the categorical shift of *qiàhǎo* 'appropriately good' into an adverb, a new lexical coinage, *qiàdàng* 'appropriate' was formed later in the Ming period. In the Yuan period, *qià* was flexible to form disyllabic words with other morphemes, such as *qiàcai* 'just now' and *qiàdài* 'precisely await for.' That is why Li [17] believes that some time adverbs containing *qià* as a morpheme appeared after the Yuan period. In a word, it shows that the basic meaning of *qià* is weakened and extended in the process of disyllabization.

Later in the Ming Dynasty, the functional differentiation of the adverbs *qiàhǎo* 'appropriately' and *qiàqià* 'precisely' was completed. The most commonly used *qiàqià* often appeared to describe an unexpected coincidence, while *qiàhǎo* was used more to describe an appropriate or good coincidence. Through disyllabization, *qià* was further integrated into disyllabic compounds, such as *qiàrú* and *qiàsì* 'exactly similar to'. And the new compound *qiàdàng* 'appropriate' appeared to take over the adjectival function from the formerly used adjective *qiàhǎo*.

In the Qing Dynasty and the early stage of Modern Chinese, the use of *qiàdàng* 'appropriate' increased significantly and the pragmatic implication of *qiàqià* was also established. It described a proper match that was out of the speaker's expectation, or profiled a situation violating a common expectation. On the other hand, the use of *qiàqiǎo* 'appropriately' is also extended to a more neutral range of contexts, without necessarily expressing a positively valued coincidence.

In conclusion, the paths of grammatical and semantic changes of the *qià*-based lexical cluster well demonstrated how a single morpheme gives rise to a group of lexemes that are morphologically related, grammatically diversified, functionally distinct, and yet complementary. This study is significant in showing that the process of lexical development clearly manifests the general tendencies of semantic change towards a range of subjectively evaluated, interpersonal meanings, which are entrenched through frequent usage-based patterns. After all, lexical and grammatical changes are the mind's responses to repetitions from usage to grammar.

Acknowledgments. We are grateful to Cheng Lei of Wuhan University, Yang Dan of Beijing Union University, and He Tianqi of the City University of Hong Kong for their kind help with the historical data. This research was supported by the Beijing Philosophy and Social Sciences Youth Fund Project under the grant No. 19YYC019 (A Study of Emphatic Markers Based on Beijing Natural Spoken Language Corpus), the National Social Science Fund of China under the grant No. 16BYY114 (A Thematic Study of Chinese Conjunction Based on the Semantic Map), and the Academic Research Project of Beijing Union University (No. JS10202003). All errors remain our own.

References

1. Ding, Y.: Studies on the coincidental modal adverbs. Master thesis, Shanghai Foreign Languages University, Shanghai (2010)
2. Liu, X.L.: Diachronic investigation of the fit modal adverbs. Master thesis, Jiangxi Normal University, Nanchang (2017)
3. Luo, Z.B.: A survey of the modal adverb in the Ming and the Qing Dynasty. Doctor thesis, Hunan Normal University, Changsha (2013)
4. Chao, D.J.: Studies on Chinese modality adverbs denoting coincidence. Master ,thesis. Guangxi Normal University, Guilin (2005)
5. Xie, Y.: Research on the origin, evolution and formation of the reduplicative adverb. Master thesis, Shanghai Normal University, Shanghai (2017)
6. Liu, F.M.: Research on the teaching of Chinese as a foreign language coincidence of modal adverbs. Master thesis, Guangxi University for Nationalities, Nanning (2019)
7. Li, S.Y.: The Study of Chinese Expression Adverbs in the Middle Ancient Times. Shandong University, Jinan (2010)
8. Lv, S.X.: Xiandai Hanyu Babai Ci. The Commercial Press, Beijing (2001)
9. Yuan, Y.L.: The Adjective Dictionary of Sentence Making. The Commercial Press, Beijing (2018)
10. Modern Chinese Dictionary, 7th edn. The Commercial Press, Beijing (2018)
11. Wang, W.X.: New Interpretation of Chinese Compound Sentence. The Press of Huadong Normal University, Shanghai (1997)
12. Xing, F.Y.: The Research on Compound Sentence of Chinese. The Commercial Press, Beijing (2001)
13. Hyland, K.: Metadiscourse. Foreign Language Teaching and Research Press, Beijing (2008)
14. Dahl, O.: Grammaticalization and the Lift Cycles of Construction. Ms., Stockholm University, Stockholm (2000)
15. Traugott, E.C.: Subjectification in grammaticalization. In: Stein, D., Wright, S. (eds.) Subjectivity and Subjectivisation: Linguistic Perspectives, pp. 31–54. Cambridge University Press, Cambridge (1995)
16. Traugott, E.C.: Intersubjectification and intersubjectification: a reassessment. In: Brems, L. (eds.) Intersubjectivity and Intersubjectification in Grammar and Discourse, pp. 29–71. John Benjamins Publishing Company, Amsterdam (2010)
17. Li, Z.J.: The Grammaticalization of Chinese Empty Words. Academia Press, Shanghai (2017)

18. Heine, B., Claudi, U., Hünnemeyer, F.: Grammaticalization: A Conceptual Framework. University of Chicago Press, Chicago (1991)
19. Liu, M.C.: From motion verb to the linking element: discourse explanations for the grammaticalization of JIU in Mandarin Chinese. J. Chin. Linguist. **25**, 258–289 (1997)
20. Wang, E.X., Yuan, Y.L.: The qualia role distribution in word meaning and its influence on word interpretation-taking 'color(yán)+Noun(míng)' compound words as an example. J. Foreign Lang. **41**, 31–41 (2018)
21. Yuan, Y.L.: From politeness expression to conditional adverb: the origin and evolution of words with the morpheme *kui* (亏)(the *kui*-words). Essays Linguist. **60**, 61–87 (2019)
22. Bybee, J.: From usage to grammar: the mind's response to repetition. Language **82**, 711–733 (2006)

Analysis of the Characteristics of Metaphors in News Reviews on the China-US Trade War

Xi Wang and Zhimin Wang(✉)

Research Institute of International Chinese Language Education, Beijing Language and Culture University, Beijing, China
wanx08biu@sina.com

Abstract. How do we talk about the China-US trade war that has been the focus of China-US diplomacy? To study it, we collect news reviews regarding the trade war from the time when US President Trump took office in 2016 to December 2019, analyze the metaphors appearing in these news reviews, summarize their distributional characteristics and label them to build a metaphorical economic and trade metaphor resources. Through analysis, we find interesting differences in metaphors in terms of their positions in a concept dictionary and their numbers in these economic and trade articles. The first sentences in news reviews are found most likely to be metaphorical ones, and metaphors with concepts of water, human, disease, music, road, war and the like as source domains are outstanding. We also find that many metaphors have formed a fixed mapping of concepts and that metaphorical expressions often use source domain words or expressions with contrasting meanings and strong emotional tendencies. In addition, a conceptual metaphor may have multiple metaphorical expressions, but a more systematic view is that their source domain words belong to the same semantic category consistently.

Keywords: Metaphor · Wordlist · Conceptual metaphor · Source domain · Target domain · China-US economic and trade

1 Introduction

Metaphors are a cognitive phenomenon, a comparison without marker words. Regarding metaphor research, various scholars have already achieved many results. Feng [1] believed that conceptual metaphors can ensure the establishment of textual coherence and are not affected by the type and style of the article. Some scholars have also studied the construction of metaphorical knowledge bases. Wang [2] proposed a definition of the metaphor of noun phrases. At the same time, she designed and built a Chinese metaphor knowledge base, using machine learning and rule-assisted methods to identify Chinese noun phrases. The field of metaphor research is relatively broad, involving economic, political, and cultural aspects. Chen [3] used the MIPVU program, with the support of the Wmatrix corpus tool, to extract conceptual metaphors to conduct metaphorical research on English economic discourse. He found that economic metaphors are related to associated concepts such as animals, plants, diseases, water flow and war. Liang [4] used

M. Liu et al. (Eds.): CLSW 2020, LNAI 12278, pp. 253–267, 2021.
https://doi.org/10.1007/978-3-030-81197-6_22

architectural theory combined with conceptual metaphors to analyze China's national image and found various metaphors such as diseases and slogans.

The study of metaphors by various scholars provides inspiration for our research. So how can a corpus of economic and trade metaphors be constructed? How do we compile an economic and trade metaphorical vocabulary? Are there certain rules in the distribution and semantics of the source and target domains of economic and trade metaphors? These need to be studied in depth.

This study is based on a corpus of "China-US trade war" that we collected and annotated for constructing a metaphoric corpus and obtaining a metaphorical wordlist for economic and trade. In order to make best use of the corpus, we first located sentences containing metaphorical phenomena in order to retrieve their original texts, counted the occurrences of the words in source and target domains, and then combined them using the Synonym Dictionary of Harbin Institute of Technology Information Retrieval Laboratory (extended version; hereinafter referred to as the Synonym Dictionary). Next we located the source domain words in this synonym dictionary and produced a metaphorical wordlist. In the following sections of this paper, some high-frequency, typical source domains are analyzed and conclusions are drawn from the analysis.

2 Selection and Overview of Corpus

2.1 Source of Corpus

The source of the corpus for this study is a set of news reviews from *People's Daily*, the most important official media in China. These news reviews are considered to have proper authority. More specifically, we conducted a site search on the official website of *People's Daily* (http://www.people.com.cn/) using "China-US Trade" as keyword to retrieve news reviews from the time when US President Trump took office in 2016 to December 2019. A total of 175 news reviews, of 355,908 words, related to the China-US trade and economic theme were selected through screening. The maximum number of words in a single review is 8239, the minimum is 818, and the average is 2034. At the same time, we also counted the number of reviews on http://www.people.com.cn/ each year.

Figure 1 shows that the number of news reviews on China-US economic and trade relations increased continuously from 2016 to 2019, and the growth rate from November 2016 when Trump took office to 2018 when he launched the China-US trade war indicates a major impact of the trade war on China that has drawn more and more attention in the Chinese society.

We carry out Chinese word segmentation, count words, calculate word frequency (the proportion of a word count to the total of all words), and then sort word in the descending order of frequency, so as to obtain a ranked frequency distribution table.

In Table 1, we can see that among the top 15 words, there are 6 function words, accounting for 40%, and 9 content words, accounting for 60%, which are basically the core words for news reviews on China-US economic and trade relations that reflect the main topic of these news reviews. There are 7 monosyllable words, accounting for 47%, and 8 disyllable words, accounting for 53%. The corpus for this research consists of

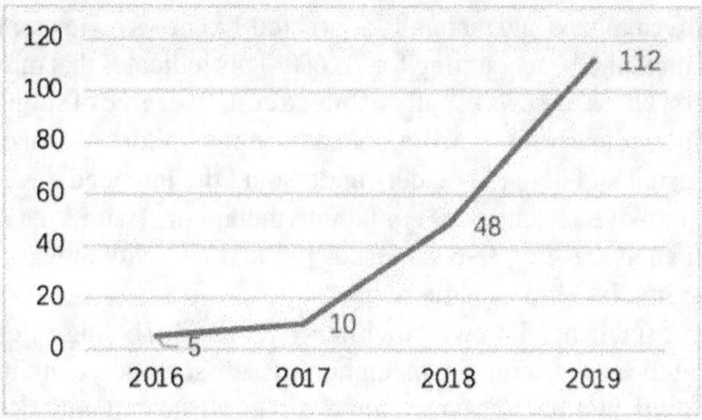

Fig. 1. Changes in the number of China-US economic and trade news reviews from 2016 to 2019

Table 1. The top 15 most frequently used words in China-US economic and trade news

Serial number	Word	Frequency	Proportion (%)
1	of (的)	8647	4.4696
2	China (中国)	2975	1.5378
3	United States (美国)	2312	1.1951
4	in/at (在)	1770	0.9149
5	and (和)	1759	0.9092
6	Economic (经济)	1548	0.8002
7	is (是)	1548	0.8002
8	Development (发展)	1224	0.6327
9	China and the US (中美)	1220	0.6306
10	le (了)	1194	0.6172
11	dui (对)	1170	0.6048
12	Trade (贸易)	1164	0.6017
13	World (世界)	870	0.4497
14	Economy and trade (经贸)	865	0.4471
15	Year (年)	783	0.4047

texts from a small field of economy and trade, where disyllabic words account for a high proportion.

2.2 Basic Situation of Metaphors in News Reviews

There are many metaphors in news reviews. We take these metaphors as our research data, covering as many metaphorical expressions as we can collect about economic and trade, and build a small corpus of metaphors. Hopefully, it can provide support to future studies on identification and evaluation of metaphors.

We statistically analyzed and manually annotated 175 news reviews, and found 134 of them containing metaphors, accounting for 76.6%. This indicates that most news reviews contain metaphors. News reviews usually exhibit a certain degree of subjectivity. In order to convey their ideas clearly, commentators can use metaphors to make their language more vivid than usual and help the readers understand the intended meaning. In the 134 news reviews, a total of 531 sentences containing metaphors (with a period as delimiter) are marked, with an average of 3.96 sentences per text and with at least 1 metaphorical sentence and maximally 33 in a single review.

This research establishes its own labeling system and labeling rules. It counts and organizes all sentences that contain metaphors. Each sentence containing metaphors is assigned a 7-digit number. The first three digits are text number, the middle two are paragraph number in a text, and the last two are position number of a sentence in a paragraph. For example, a sentence number 0010101 means that it is the first sentence of paragraph 1 in the first text; another one 0041203 means the third sentence of paragraph 12 in the fourth text. In this way, our data set is established.

We found certain rules for the locations of metaphors. The paragraph numbers can reflect the approximate distribution of metaphors. Our data show that the top three paragraph numbers in the descending order of occurrences are 59 (03), 58 (05), 06 (05), indicating that the sentences containing metaphors are mostly distributed in paragraphs 3, 5 and 6. Due to different writing habits and writing requirements of different reporters, the number of paragraphs in a news review is different. The minimum number of paragraphs for a single news review is 5 and the maximum 67. However, according to the style characteristics of news reviews, paragraph 3 usually enters formal discussion, for in most news reviews, paragraphs 3–6 are the main body of an article. So, the above distribution is quite in line with expectation.

In addition, a sentence number represents the exact position of a metaphorical sentence in a paragraph. Table 2 shows that the first sentence of a paragraph has the highest frequency of metaphors, with a total of 193 times. Despite sentences in the other places

Table 2. Distribution of metaphorical sentences in different positions in a paragraph

Sentence position	Number of metaphors	Percentage
01	193	36.3%
02	145	27.3%
03	89	16.8%
04	58	10.9
05	25	4.7%
06	9	1.7%
07	6	1.1%
08	5	0.9%
09	1	0.2%

also contain a certain number of metaphors, it can be observed that using a metaphor in the first sentence of a paragraph is a relatively common writing strategy in news reviews. The use of metaphorical expressions at the beginning of a paragraph can attract readers' attention and set the tone for the entire paragraph and even the whole text, so as to make the next texts easy to understand. As an official media of the government, *People's Daily* often uses metaphors in news reviews, which can help get close to people and avoid stereotyped preaching to a certain extent while conveying political guidance.

3 Exploration of Economic and Trade Metaphoric Vocabulary

In order to explore the economic and trade metaphor wordlist, we calculate the frequencies of these words in source and target domains and integrate them with the Synonym Dictionary. At the same time, we also have to delimit the economic and trade metaphors because not only the economic and trade terms in source or target domains are used as metaphors. We assume that both common and typical metaphor expressions in economic and trade articles are metaphors of our interest, and hence need to be included into our corpus. For this study we only consider cases where the source domain is a noun.

3.1 Distribution of Words in Source Domains

In this study, all sentences in the corpus are examined and every sentence containing a metaphor is marked. In the original text, a source domain is marked with angle brackets " < >". Then, source and target domains are listed separately, and the statistics are derived from the latter. There are 328 source domain word types in total, some of which are given below for illustration.

Source domain words of count 1 are the largest in number, of 218 words. The highest occurrence count is 31, for one expression only. Because in news reviews, commentators use expressions different from the plain and objective style of news reporting, some

Table 3. Frequencies of exemplar source domain words

Source domain	Frequency	Source domain	Frequency
Big stick	31	Large ship	6
War	30	Stabilizer	6
Ballast stone	20	Small river	6
Gate	17	Small lake	6
Sea	16	Storm	5
Wind and rain	16	Orbit	5
Winner	16	Golden key	5
Road	12	Crossroad	4
Thruster	12	Transcript	3
Trap	12	Cake	3
Card	9	Spare tire	2

Table 4. Metaphorical expressions with "big stick" as the source domain

Target domain	Frequency	Conceptual metaphor	Examples
Tariff	24	Tariff is the big stick	Undoubtedly, some people in the United States wantonly wield the <big stick> of tariffs, and the impact of their reactionary force on the United States itself has reached its limit
Trade protection	5	Trade protection is the big stick	For a time, the United States showed an aggressive posture of vowing to swing the <big stick> of trade protection at China
Trade sanctions	1	Trade sanctions is the big stick	If the US continues to raise the <big stick> of trade sanctions, including tariff increases, and stubbornly provoke a trade war, China will have no choice but to fight back
"301 Investigation"	1	"301 Investigation" is the big stick	After all, the < big stick > of "301 Investigation" that the United States wields against China cannot hurt China's bones

source domain words appear more frequently than others. Table 3 shows that the top three of the most frequent ones are "big stick", "war" and "ballast stone". One may observe an interesting qualitative phenomenon in this government official media that the commentator has formed a fixed expression and a stable mapping between some concept pairs. We can take the most frequent one "big stick" as an example to show this (Table 4)

When the "big stick" is used as the source domain, the most frequently used target domain is "tariff" and the most common phrase is "wielding the big stick of tariff". Obviously, this metaphor gives a negative impression because "big stick" is a clumsy, savage weapon that symbolizes the use of force, and "wielding the big stick of tariff" means repressing in a savage and unjust way, giving the public a clear imagery.

3.2 Distribution of Target Domain

This study also carries out a statistical analysis of target domains. 278 target domains appear in the news reviews, some of which are given below as examples.

Table 5. Exemplar target domains with frequencies

Target domain	Frequency	Target domain	Frequency
Difficulty	47	Economic and trade cooperation	9
Trade	36	The best interests of China and the US	8
World economy	27	Chinese economic development	6
Tariff	26	Huawei chip	3
China	23	China's overall strength	2

The target domain with the highest occurrences is "difficulty", in a total of 47 times, while the lowest occurrence count is only 1, for 185 target domains. When the target domain is "difficulty", the source domain is mostly bad weather, such as "storm and rain", "strong winds and big waves", "wind, frost, rain and snow", "floating clouds", etc., or an unpleasant geographical object and obstacle, such as "cliff", "vortex", "ditch", "ridge", "stumbling block", and so on. The "trade war" has indeed had a big impact, receiving many expressions about "difficulty" in the news reviews.

According to our analysis, the source and the target domains are quite different in terms of their types. Using the Chinese Concept Dictionary (hereinafter referred to as CCD) [5] as the standard classification scheme, the source domains are divided into three types: living body, non-living body and natural phenomenon. Living bodies include animals, plants and human beings while non-living bodies include artificial objects, natural objects, materials, etc., which are universal and concrete. Some target domains belong to the behavior or event category, such as "China-US summit" and "money supply". Target domains are usually more elaborate and abstract, and more dependent on original text contexts, such as "the number of non-resident residents' intellectual property applications", "the spirit of the contract that some American politicians are saying", "the largest joint point of China-US interests", etc. This is consistent with the basic characteristics of metaphors.

3.3 Construction of Economic and Trade Metaphoric Vocabulary

We compile a wordlist centered on the source domains of metaphors appearing in news reviews about the China-US trade war, aiming to examine the characteristics of the source domains and formulate some rules. We use the Synonym Dictionary to count source and target domains, and identify the emotional tendencies of the source domains as well as their locations in the Synonym Dictionary.

After locating a word in the Synonym Dictionary, we find its location and make a new label for it, which is convenient for machine to handle later. Note that a word may appear in multiple locations in the dictionary. For example, "hat" appears many times, with "Bq05A01 = 帽(hat) 帽子(hat) 冠(crown) 笠(hat) 盔(helmet) 冕(crown) 冠冕(crown) 帽盔(helmet) 头盔(helmet)", "Dd15A14 = 罪名(convict) 帽子(hat)", etc. Referring to its original texts, we find 8 occurrences of "hats" in related collocations as "put a hat on one's head", all with a derogatory sense to mean "crime". So, we decide that its final location is Dd15A14. Listed below are more examples of three synonyms for illustration.

Table 6. Examples of economic and trade metaphor words

Source domain	Target domain	Appraisal	Location
Trick	US tariff increase	Derogatory	Df09C03 = 花招(trick) 把戏(trick) 噱头(gimmick)
Disease	Problems in US economic development	Derogatory	Dl01A01 = 疾病(disease) 病症(illness) 病痛(illness)
Big stick	Tariff	Derogatory	Bo13A01 = 棒(stick) 棒子(stick) 棍子(stick)
Sea	World economy	Commendatory	Be05A04 = 海洋(ocean) 大海(sea) 海洋(ocean)
Mansion	China-American community of interest	Commendatory	Bn01B06 = 大厦(building) 高楼(skyscraper) 摩天楼(skyscraper)
Cake	Interest	Commendatory	Br09A17# 糕(cake) 蛋糕(cake) 发糕(cake)
Fence	Trade barriers	Derogatory	Bn07B01 = 篱笆(fence) 篱(fence) 藩篱(fence)
Wind and rain	Difficulties encountered in China-US relations	Derogatory	Da08A02 = 风浪(Winds and waves) 风霜(winds and frost) 风雨(winds and rains)
Common divisor	Common interests	Neutral	Dn03A41# 序数(Ordinal) 平均数(Mean) 公约数(Common Divisor)
Golden key	Comprehensively deepen reform	Commendatory	Bo03A38 = 钥匙(key) 钥(key)
War	Trade	Derogatory	Di11B01 = 战争(war) 战事(war)

We can see from Table 6 that metaphorical expressions are mostly sentimental. In addition, in the process of labeling synonyms, some source domain words not available in the Synonym Dictionary get no replacement objects. There are a total of 32 word types of them, as listed here: 牌局 (poker game), 玩法 (playing strategy), 健忘症 (amnesia), 失忆症 (amnesia), 棋局 (chess game), 换挡 (shift gears), 糊涂账 (complex difficult accounts), 红酒 (red wine), 混水/浑水 (muddy water), 偏架 (break up the fight with partiality), 脏水 (dirty water), 微信朋友圈 (WeChat moments), 黑屋子 (dark room), 牌 (cards), 游戏规则 (game rules), 地球村 (global village), 赢者 (winner), 硬骨头 (unyielding person), 备胎 (spare tire), 闲棋 (free play), 减压阀 (pressure reducing valve), 家门口 (house gate), 先手棋 (first to move in chess game), 蜜月期 (honeymoon),

白脸 (bad cop), 二级跳 (twice hop), 沉渣 (sediment), 回头浪 (undertow), 狂风骤雨 (squally showers), 疾风骤雨 (stormy showers), 顽瘴 (stubborn disease) and 癣疥之疾 (some slight ailment).

4 Semantic Classification of Source Domain Words

In this section we follow the initial concept nodes of the CCD dictionary to classify source domain words by semantic criteria and select a few representative types of them with relatively high frequency to demonstrate our analysis.

4.1 The Source Domain is a Natural Object

In our corpus, when source domains are natural objects, most of them are related to water. So, we use water metaphors as examples.

(1) *China is holding high the banner of peace, development, cooperation and win-win situation, injecting a "<warm current>" into the fledging world economy, and winning more and more supporters and peers worldwide.*
(2) *The Chinese economy is the <sea>, not a <small pond>. <Strong winds and heavy rains> can overwhelm a <small pond >, but cannot overwhelm the <sea>.*
(3) *It is impossible and unhistorical for the world economy's <sea> to retreat into isolated <small lakes> and <small rivers>.*

Many metaphorical expressions in our corpus use the method of contrast, listing a set of antonymous or contrastive source domain words, such as "sea" and "small pond" (or "small lake") that oppose each other in terms of size. Using them to praise or derogate gives a strong contrast and increases the richness of language. Listing all water metaphors and their positions in the Synonym Dictionary, we get the following table.

Table 7 shows that these source domain words can activate their synonyms in the Synonym Dictionary at the same locations for a replacement. For example, "sea", once located, can be replaced with another word of the same group, such as "ocean". These locations facilitates the use of machine to perform metaphor identification and further reasoning within a comprehensive system. Some words share the same location, such as "water flow" and "river", "big water" and "flood". Each of them can activate the entire set of words at the same location. Some words are in two different locations, such as "winds and waves", indicating that there are more words to be triggered for addition to the collection of words for a source domain. The more words in a location, the more choices are available for a word in the original text.

However, there are some words that cannot be replaced in this way. For example, "tributary" and "port" may not be synonymous. In a text, "tributary" may represent "different voices in the international community", but carries no meaning as "port". In addition, some words are not included in the Synonym Dictionary. According to our statistics, there are six unlisted words related to water metaphors, which are 脏水 (dirty water), 疾风骤雨 (stormy showers), 狂风骤雨 (squally showers), 浑水 (muddy water), 混水

Table 7. Locations of source domain words for water metaphors in the synonym dictionary

Source domain	Location
Sea	Be05A01 = 海洋(ocean) 大海(ocean) 海洋(ocean)
Tributary	Be05B07 = 支流(tributary) 港(port)
Torrent	Be05B12 = 急流(rapids) 激流(rapids) 奔流(torrents)
Warm current	Be05B14@ 暖流(warm current)
Inland river	Be05B16# 内河(inland river) 内陆河(inland river) 运河(canal)
Small lake	Be06A01 = 湖泊(lake) 湖(lake) 湖水(lake water)
Small pond	Be06B01 = 池塘(pond) 池沼(pond) 池子(pond)
Quagmire	Be06C01 = 泥坑(mud pit) 泥潭(quagmire) 泥塘(mud pond)
Flood	Bg01A03 = 洪水(flood) 洪流(torrent) 山洪(torrent)
Wave	Bg02B01 = 波浪(wave) 浪(wave) 浪花(wave spray)
Swirl	Bg02B04 = 漩涡(vortex) 涡流(vortex) 涡旋(vortex)
Wind and rain	Da08A02 = 风浪(wind and waves) 风霜(wind and frost) 风雨(wind and rain)

(muddy water), and 回头浪 (undertow). It is worth noting that "疾风骤雨 (stormy showers)" and "狂风骤雨 (squally showers)" have only one character different from "暴风骤雨 (wild winds and sudden downpour)" in Da02C02, but are not included in the Synonym Dictionary, suggesting that there is still room for its improvement. If more words could be included to expand the resources for our use, a greater support would be available to our research.

4.2 The Source Domain is an Organism

An organism here include one of human, animal or plant. In the annotated corpus, the cases of "source domain is human" has the largest number within "source domain is an organism". Therefore, let us take human organism metaphors as instances for our analysis.

(4) *An Italian political commentator said in the Evening Post that the United States is using tariffs as a means to show off its "economic <muscle>" to countries around the world and believes that China's traditional export-oriented economy development model cannot bear the impact of US tariffs.*

(5) *After all, the United States' wielding the big stick of the "301 Investigation" against China cannot hurt China's <muscles and bones>.*

(6) *The United States has released a lot of inconsistent and "face-smacking" information through social media and other unconventional methods. Its various kinds of arguments have changed drastically, making people call them "<mood> changeable" and "elusive".*

A conceptual metaphor can be drawn from this exemplar sentence: "The country is a person." Organizing all words in a source domain that are related to people in a given text, we can get the following classification:

Human organ: head, arms, shirtless, face, bones, neck, muscles, human body.

Name of the person: Ah Q, Uncle Sam.

Human attribute: strength, signs, volume, corporeity, physique, posture, spirit, energy, physical fitness, step, pace, footsteps.

Human spirit: mind, emotion.

Our statistics show that it is very rare for a source domain to be a "whole person", except some person names such as "Ah Q" and "Uncle Sam". Most cases are human organs, inherent attributes and temperament of a person. Therefore, when the machine retrieves and recognizes metaphors in economic and trade texts and encounters words for human organs, human spirit, personal attributes and the like, they have a high probability to be metaphorical expressions.

4.3 The Source Domain is an Abstract Object

When a source domain is abstract, its word can be an abstract object like disease, drama, and music. When a source domain is a disease or a disease-related object, its conceptual metaphor is similar to this one that "difficulty encountered in the course of economic development is a disease". A few examples of disease metaphor are given as follows.

(7) *If you measure the <pulse> wrong, naturally you can't prescribe the right <prescription>.*

(8) *When you are <sick>, you force others to take <medicine>.*

(9) *It should be said that many of the propositions during the Trump campaign showed that he was aware of the above two root causes that caused the <illness> of "twin double deficit" of the US economic and trade deficit and fiscal deficit.*

Sorting source domain words of disease metaphor and their locations in the Synonym Dictionary, we get the following table.

Source domain words of disease metaphor can be roughly divided into two categories. The first category is "disease", which includes the general term "disease" and various specific symptoms and the second is related to "disease treatment", including "prescription", "pill" and "pulse". Table 8 shows that the frequently-used general terms for "disease" can be unified at its location in the Synonym Dictionary. Some words in written and classical registers are not available, such as "癣疥之疾 (some slight ailment)" and "顽瘴 (stubborn disease)". Certain common diseases are also not included, such as "健忘症 (amnesia)" and "失忆症 (amnesia)". Again, the dictionary needs to be further expanded properly

In addition, when a source domain is music, its conceptual metaphor is like "China-US economic and trade cooperation is music". A variety of metaphorical expressions can be derived from a given text and the source domain words for music metaphors include: main theme, reverb, sound, tone, old tune, chords, tunes, symphonies, etc.

When a source domain is drama, its conceptual metaphor is like "China-US economic and trade cooperation is a drama". The metaphorical expressions appearing in a text for

Table 8. Positions of source domain words of disease metaphor in the synonym dictionary

Source domain	Location
Disease	Dl01A01 = 疾病(disease) 病症(illness) 病痛(illness)
Intestinal obstruction	Ib17D03 = 肠梗阻(Intestinal obstruction) 肠阻塞(Intestinal obstruction)
Reassurance pill	Br13A47# 胶丸(capsule) 定心丸(Reassurance pill)
Chronic illness	Dl01A02 = 沉疴(Chronic illness) 沉痼(Chronic illness) 痼疾(Chronic illness)
Blindness	Ib19C01 = 瞎(blind) 失明(blind) 盲(blind)
medicine	Br13A01 = 药品(medicine) 药物(medicine) 药(medicine)
Prescription	Dk15B01 = 处方(prescription) 方子(prescription) 药方(prescription)
Vein	Bk16A02 = 脉(pulse) 脉络(pulse)

this type of metaphors include: clown-style performance, protagonist, curtain, front stage, behind the scenes, the role of "disruptor", performance, etc.

4.4 Source Domain is an Artifact

When a source domain is an artifact, its conceptual metaphors include roads, ships, tools, cars, buildings, etc. Here we use road metaphors as examples to illustrate our analysis.

From Table 9, we can see that the conceptual metaphor "China-US cooperation is a road" subsumes two types of metaphorical expression: "Good China-US cooperation is a good road" and "China-US confrontation and conflict are going astray". The expressions "broad road" and "dangerous erroneous route" in a case also exhibit a contrast. On the "misguided road", there are words for difficulties, such as "ditch", "ridge", "blocker" and "stumbling block". Only when China and the United States move forward in the same "direction" can mutual benefits and win-win results be achieved.

4.5 Source Domain is an Event

When a source domain is an event, the source domain word "war" has the highest frequency. Exemplar sentences are given as follows and more exemplar metaphorical expressions appearing in a text of war metaphor are listed in Table 10 below.

(10) *There is no <winner> in a <Trade War>.*

(11) *Under this kind of "public opinion <offensive>", some domestic audiences have generated negative mentalities of "fear of America" and "respect for America, which is worth noticing.*

(12) *The United States regards China as an <enemy> and attempts to decouple China from the global economy.*

Table 9. Distribution of source words of road metaphor

Metaphorical emotion	Frequency	Source domain	Example
Commendatory	7	Sunshine boulevard, right track, track, fast track, broad road, the new road, good road	Promoting China-US confrontation and conflict is dangerous straying, and promoting China-US cooperation and win-win is a < broad road >
			This is not a private road, but we go hand in hand on < Sunshine Boulevard >
Derogatory	11	Kaner, ditch, fork road, dangerous astray, dead end, private path, old road, obstacle, one-way road, stumbling block	Promoting China-US confrontation and conflict is a < dangerous stray >, and promoting China-US cooperation and win-win is the broad road
			Win-win cooperation is the general trend and the will of the people. The closed and rigid can only enter a <dead end>
Neutral	2	Direction, crossroads	Human beings are standing at the <crossroads> of history
			Under the current global economic background, the two major powers of China and the United States are going in the <same direction> to resolve disputes through consultations, which is of great benefit to both countries and the world

Among all these metaphorical expressions, "trade war" has the highest frequency and is the closest one to the conceptual metaphor "trade is war". Many types of warfare metaphor are also found in our corpus, such as "public opinion warfare", "information warfare". Although their types are different, they share the same essence of war metaphors. There are subcategories of words about war, like combating party, weapon, strategy, etc. Therefore, the metaphorical expressions of "war" can be expanded into warfare elements in order to improve the accuracy of machine search and recognition of war metaphors.

Table 10. Metaphorical expressions of "the source domain is war"

War type	Trade war, public opinion war, information war, psychological war, hot wars, cold war, tough war, technology war
Warring parties	Winner, loser, opponent, strategic opponent, enemy, ally
Combat elements	Combat experience, tactic, spearhead, drum, trigger, war, strategy, heavy artillery shell, cold weapon
Other	Wound, alliance under the city wall, offensive

5 Conclusion

This study takes 175 news reviews of *People's Daily* on the "China-US Trade War" as its research object and summarizes the distributional characteristics of the metaphors involved after detailed annotations for constructing economic and trade metaphor knowledge resources. Through analysis, we find interesting differences in metaphors in terms of their locations in the Synonym Dictionary and their numbers in these economic and trade articles. The first sentence of a paragraph is found most probable to be a metaphorical sentence, and metaphors are found most likely to have concepts of water, human organism, disease, music, war, and others of the like as source domains. In addition, many economic and trade metaphors have formed a fixed mapping of concepts, in that many metaphorical expressions often use source domain words (or phrases) in highly collocational pairs for commendation and criticism, conveying a strong emotional tendency. Furthermore, a conceptual metaphor usually has a variety of metaphorical expressions, and their source domain words can be classified into semantic categories in a systematic way. This study also confirms the importance of the Synonym Dictionary in lending support to investigate metaphors, but some words from our corpus are not included in it yet, indicating the need to have it further expanded in the future. We are looking forward to conducting a more finer-grained analysis of metaphor resources, in hopes of providing more useful input to machine recognition of metaphors, metaphor evaluation, and the use of metaphors in machine writing of news reviews.

Acknowledgments. This work was supported by the Major Program of National Social Science Foundation of China (18ZDA295), the Fundamental Research Funds for the Central Universities (19PT03), Beijing Language and Culture University (BLCU) First-Class Team Support Plan (JC201902), and the BLCU supported project for young researchers program (supported by the Fundamental Research Funds for the Central Universities) (20YCX093).

References

1. Feng, X.-H: Metaphor – The Basis of Thinking – The Frame of the Text. University of International Business and Economics Press, Beijing (2004). (in Chinese)
2. Wang, Z.-M.: Research on Recognition of Chinese Noun Phrase Metaphors. Beijing Language and Culture University Press, Beijing (2010).(in Chinese)

3. Chen, L.: A corpus-based study of English economic discourse metaphor. J. PLA Univ. Foreign Lang. **41**(04), 34–42 (2018). (in Chinese)
4. Liang, J.-Y: An analysis of the structural metaphor of China's national image—taking the 2016 China column of the economist as an example. Foreign Lang. Res. **35**(06), 23–29 (2018). (in Chinese)
5. Institute of Computational Linguistics, Beijing University: Chinese Concept Dictionary (2000). (in Chinese)

A Study on the Evolution of Post-nominal Plural Expressions in Chinese Based on Tagged Corpora

Bing Qiu(✉)

Department of Chinese Language and Literature, School of Humanities and Social Sciences, Tsinghua University, Beijing 100084, China
qiubing@mail.tsinghua.edu.cn

Abstract. This article, basing its research on tagged corpora, obtains an exhaustive collection of post-nominal plural expressions that appeared in Old Chinese, Middle Chinese, and Pre-modern Chinese in different historical periods. According to statistics on word frequency, "deng" (等), "bei" (辈), and "men" (们) are the dominant plural expressions in each period of history. This article hence conducts a quantitative survey and provides a systematic description of these three plural expressions in regard to their collocation with proper nouns and common nouns as well as their distributions. The following observations are made on the basis of statistical evidence: In Old Chinese, the one and only plural expression was "deng" (等), which could only be used after proper nouns as a collective marker. From Middle Chinese on, especially in Chinese translations of Buddhist scriptures, "deng" (等) began to be used substantially after common nouns and took on a new function of plural marking. It was also in this period that "bei" (辈) emerged as a new plural expression, mainly used after common nouns as a plural marker. When it came to Pre-modern Chinese, "deng" (等) inherited its usage in Old and Middle Chinese, being a collective marker, while the function of plural marking was taken up by the newly emerged plural expression "men" (们).

Keywords: Tagged corpus · Proper noun · Common noun · Plural · Collective marker · Plural marker

1 Introduction

"Number" is one of the most important grammatical categories and its manifestation is usually associated with the type of the language in discussion. Languages use either classifiers or number markers to indicate the category of number. Generally, in classifier languages, the extensions of nouns are collective and there are no plural markers. However, in Chinese, a classifier language that uses quantifiers, there is also a plural marker "men" (们) [1, 2], which is used after common nouns and pronouns to indicate plural number. At present, the academia has believed that "men" (们), when used after nouns, can be divided into two types. One is to serve as a plural marker to indicate plurality (and only indefinite plurals), when used after common nouns or common noun phrases;

M. Liu et al. (Eds.): CLSW 2020, LNAI 12278, pp. 268–275, 2021.
https://doi.org/10.1007/978-3-030-81197-6_23

the other is to serve as a collective marker to indicate collections, when used after proper nouns or proper noun phrases [3]. Plural expressions after proper nouns and those after common nouns are of two distinct properties [4].

The plural marker "men" (们) did not emerge until Pre-modern Chinese. So, besides "men" (们), what were the plural expressions in Old and Middle Chinese? Were they also of two distinct properties? How did plural expressions of different properties come into being? Did these two types of plural expressions emerge at the same time or not? If one type came before the other, which one came first and were they related with each other in some way? These issues have not been systematically studied or fully discussed in previous literature.

This article, basing its research on tagged corpora, aims to present the trends and characteristics of the historical evolution of post-nominal plural expressions of different types. To achieve this, we will i) obtain an exhaustive collection of plural expressions that appeared in Old Chinese, Middle Chinese, and Pre-modern Chinese; ii) according to the statistics on word frequency, take the dominant plural expressions in various historical periods as the research object and iii) conduct a quantitative surveyproviding providing a systematic description of those expressions in regard to their collocation with proper nouns and common nouns as well as their distributions.

2 Tagged Corpora and Research Methods

2.1 Tagged Corpora

The Academia Sinica Ancient Chinese Corpus (hereafter "Sinica Corpus") in Taiwan, China, is adopted in this research, which has three sub-corpora, namely, the Old Chinese Tagged Corpus, the Middle Chinese Tagged Corpus, and the Pre-modern Chinese Tagged Corpus. So far the Sinica Corpus covers the majority of the most important linguistic data in Old Chinese (from Pre-Qin to Western Han), Middle Chinese (Eastern Han, Wei, Jin, Southern and Northern Dynasties), and Pre-modern Chinese (from Tang and Five Dynasties on), with some of which are tagged and are available for online retrieval.

The Old Chinese Tagged Corpus has been completed and offers 36 pieces of literature open for public access, including *Shangshu* (尚书), *Shijing* (诗经), *Zhouyi* (周易) and so on, which add up to around 2.5 million characters. This sub-corpus is currently the largest tagged corpus of Old Chinese and all linguistic data in it have been tagged with word segmentation and parts-of-speech.

The Middle Chinese Tagged Corpus incorporates not only works of native Chinese literature in this period but a large quantity of Chinese translations of Buddhist scriptures. The former include *Baopuzi Neipian* (抱朴子内篇), *Shishuo Xinyu* (世说新语), *Luoyang Qielan Ji* (洛阳伽蓝记), etc., adding up to over 0.3 million characters, and the latter include *Daoxing Banruo Jing* (道行般若经), *Fo Shuo Dousha Jing* (佛说兜沙经), *Zhongben Qijing* (中本起经), etc., adding up to over 1.17 million characters.

The Pre-modern Chinese Tagged Corpus now provides language data from 20 pieces of literature for online retrieval, including *Dunhuang Bianwen Ji Xinshu* (敦煌变文集新书), *Zutang Ji* (祖堂集), *Zhuzi Yulei* (朱子语类) and so on, adding up to 5.2 million characters. They are also tagged with word segmentation as well as parts-of-speech.

2.2 Research Methods

In the tagged corpora, common nouns are tagged as "Na" and proper nouns as "Nb"; post-nominal plural expressions are tagged as "T" (for auxiliary words) in the Old Chinese Tagged Corpus and "T6" (for dependent words indicating plurals) in the Middle and Pre-modern Chinese Tagged Corpora. This article conducts a statistical survey of all the plural expressions that occur in the three corpora, which are listed in Table 1 along with their frequencies.

Table 1. Post-nominal plural expressions and their frequencies in the tagged corpora.

Tagged corpus	Post-nominal plural expressions and their frequencies /times of occurrence
Old Chinese tagged corpus	"deng" (等)/404
Middle Chinese tagged corpus	"deng" (等)/1658, "bei" (辈)/263, "zhilei" (之类)/158, "lei" (类)/86, "zhishu" (之属)/58, "zhideng" (之等)/11, "zhiliu" (之流)/3, "zhibei" (之辈)/3, "dengbei" (等辈)/3, "cao" (曹)/2
Pre-modern Chinese tagged corpus	"men" (们)/2907, "deng" (等)/2898, "mei" (每)/392, "zhilei" (之类)/269, "hang" (行)/89, "bei" (辈)/44, "dengjian" (等件)/21, "zhishu" (之属)/15, "denglei" (等类)/12, "men" (门)/5, "lei" (类)/5, "chai" (侪)/1, "cao" (曹)/1, "deng" (等等)/1

It can be seen from Table 1 that "deng" (等) had always been a major post-nominal plural expression since Old Chinese until Pre-modern Chinese; and "bei" (辈) and "men" (们), the two other dominant plural expressions, emerged in Middle Chinese and Pre-modern Chinese, respectively. Through the history of the Chinese language, "men" (们) was also written as "弭", "伟", "满", "懑", "瞒", "门", "每", etc., and accordingly, in the tagged corpora, two other written forms of "men", besides "men" (们), are "门" and " 每". Nowadays "men" (们) is the main plural expression in Modern Chinese, which has the highest frequency of use. Taking the above facts into consideration, this article selects up to two post-nominal plural expressions with the highest frequencies in each period as the object of our statistical survey and research, i.e. "deng" (等) in Old Chinese, "deng" (等) and "bei" (辈) in Middle Chinese, and "deng" (等) and "men" (们) in Pre-modern Chinese.

In the following section, basing our research on the tagged corpora, we are going to provide an exhaustive description of "deng" (等), "bei" (辈), and "men" (们), the three major post-nominal plural expressions in different periods of the history of the Chinese language. We also conduct a statistical survey of their collocation with proper nouns and common nouns as well as their distributions in various periods. The trends of their historical evolution will also be presented in a quantitative manner.

3 Trends of Evolution of the Post-nominal Plural Expressions Based on Tagged Corpora

In the Old Chinese Tagged Corpus, only "deng" (等) is tagged as "T", with a total frequency of 404. Among them, 339 are used after proper nouns, and all the rest are after pronouns. A good example is found: "贯高、赵午等十馀人皆相谓曰: '乃吾等非也。……' "(GuanGao, ZhaoWu and so on said to each other: " This is our fault.") from "Zhang Er Chen Yu Liezhuan" in *Shiji* (《史记·张耳陈馀列传》), where the first "deng" (等) is used after the proper nouns for names "贯高" and "赵午", while the second "deng" (等) is after a first-person pronoun "吾". Statistics show that "deng" (等), as a post-nominal plural expression, was only seen after proper names in Old Chinese, and never after common pronouns.

In Middle Chinese, "deng" (等) was still the dominant plural expression, but its usage and distribution changed greatly compared with that in Old Chinese. In the Middle Chinese Tagged Corpus, "deng" (等) appears for 1,446 times, with 1,165 of which are used after common nouns and only 281 of which are used after proper nouns. Such significant change of the plural expression "deng" (等) is to some extent related to the large-scale language contact that occurred in the Middle Chinese period. At that time, Buddhism was introduced into central China from its country of origin, India, by way of the Western Region. which brought about a large number of Buddhist scriptures mainly in Sanskrit being translated into Chinese, forming the first large-scale language contact in the history of Chinese. It is believed that plural markers in Chinese are not a result of the development of the Chinese language itself, but a product of language contact between Chinese and Indo-European languages [4] (through translations of Indian Buddhist scriptures). Taking this historical context into account, in the following paragraphs, we analyze the differences between the native Chinese literature and the Chinese translations of Buddhist scriptures in terms of the plural expressions. In the nativeChinese literature of the Middle Chinese period, a total of 70 cases have been identified where "deng" (等) is used after nouns as a plural expression, 26 of which are after common nouns and 44 of which after proper nouns. It can be seen that in Middle Chinese, "deng" (等), when used in native literature, inherited its usage in Old Chinese, namely mainly after proper nouns. In the Chinese translations of Buddhist scriptures, a total of 1,376 cases of the plural expression "deng" (等) being used after nouns have been identified. Among them, 1,139 are after common nouns, including "天子" in *Daming dujing* (《大明度经》), "佛", "王子", "龙王", "大臣" in *Bei Hua Jing* (《悲华经》), "菩萨" in *Fo Shuo Dousha Jing*(《佛说兜沙经》), "弟子", "乞儿", "大王", "兄弟", "国王" in *Dazhuangyan Lun Jing* (《大庄严论经》), "比丘" in *Liudu Jijing* (《六度集经》), "女", "导师", "亲族", "眷属" in *Fo Ben Xing Jijing* (《佛本行集经》), "童子", "女人" in *Chuyao Jing* (《出曜经》), etc. Only the other 237 are used after proper nouns, including names such as "须菩提" in *Daoxing Banruo Jing* (《道行般若经》), "迦叶" in *Zhongben Xing Jing* (《中本行经》). Such distribution of "deng" (等) displays a marked difference from that in Old Chinese as well as that in the native Chinese literature of the same period.

In Middle Chinese, a new plural expression "bei" (輩) emerged, which was not in frequent use back in Old Chinese. Only 16 cases have been found in the corpus, all of which are used as nouns and mostly after a numeral word, meaning a hundred chariots or chariots in rows, such as "一辈", "三辈", "五辈", "八辈", "五六辈", "十辈", "十余

辈", etc. In Middle Chinese, however, the sense of "bei" (辈) changed and a new usage emerged, namely serving as a plural expression after nouns. A total of 205 cases have been identified in the corpus. Among them, 196 cases appear in the Chinese translations of Buddhist scriptures, 195 of which are used after common nouns, such as "菩萨" in *Fo Shuo Dousha Jing* (《佛说兜沙经》), "女人" in *Daoxing Banruo Jing* (《道行般若经》), "妓女", "母人" in *Zhongben Qijing* (《中本起经》) "众生", "人", "童子", "仙人", "小儿", "婇女", "王", "眷属", "仙", "天女", "女", "男子", "商人", "师", "大臣", "乞儿", "贫儿" in *Fo Ben Xing Jijing* (《佛本行集经》), etc., leaving only 1 after a proper noun. In the native Chinese literature, 9 cases of "bei"(辈) have been found, with 7 after common nouns such as in "道士辈" in *Baopuzi Neipian* (《抱朴子内篇》) and "门生辈" in *Shishuo Xinyu* (《世说新语》) and 2 after proper nouns. It is hence observed that "bei"(辈), was a newly emerged plural expression in Middle Chinese, and is characterized as having mainly been used after common nouns ever since it came into being.

In Pre-modern Chinese, "deng"(等) was still a dominant plural expression and in addition to it, "men"(们) emerged as a new plural expression. In the tagged corpus, a total of 2,489 cases of "deng"(等) are found to have been used after nouns, of which, 1,944 are used after proper nouns and 635 after common nouns. It can be said that such usage and distribution of "deng"(等) inherited that in the native Chinese literature of the Middle Chinese period. In contrast, 2,765 cases of "men"(们) are found to have been used after nouns, with 2,748 after common nouns and merely 17 after proper nouns. Based on the above statistics, it can easily be seen that "deng"(等) and "men"(们), as the two dominant plural expressions in Pre-modern Chinese, had a complementary distribution, with the former mainly after proper nouns and the latter mostly after common nouns.

As is noted in Sect. 2.1, the length of the incorporated texts, which is measured by the total number of characters, varies across the three corpora: 2.5 million characters in the Old Chinese Tagged Corpus, 1.4 million in the Middle Chinese Tagged Corpus, and 5.2 million in the Pre-modern Tagged Corpus. In the Middle Chinese Tagged Corpus, native Chinese literature and Chinese translations of Buddhist scriptures also differ in their quantities; the former has merely 0.3 million characters while the latter has over 1.1 million. The length of the incorporated texts in a corpus, i.e. the total number of characters, is directly related to the frequencies of the plural expressions under research. It is beyond doubt that the larger the text corpus is, the higher the frequencies the expressions appear in. Out of this concern, we calculate the average frequencies per ten thousand characters and the results are shown in Table 2.

In order to present the distribution of the post-nominal plural expressions across different historical periods in a clearer manner, we have transformed the data in Table 2 and into graphs in Fig. 1. Several observations can be made from Fig. 1.

Firstly, in Old Chinese, "deng"(等) was the only plural expression and could only be used after proper nouns. From Middle Chinese on, "deng"(等) could be used after common nouns, especially in Chinese translations of Buddhist scriptures, where such usage of "deng"(等) had a relatively high frequency, occurring for 7.39 times per ten thousand characters. In the native Chinese literature of the same period, however, "deng"(等) was still mainly used after proper nouns and such usage continued in Pre-modern Chinese.

Table 2. Frequencies of collocations of plural expressions with proper nouns and common nouns in different periods, measured by times of occurrence per ten thousand characters.

Period	Plural expression	After proper nouns		After common nouns	
		Total number of occurrences/times	Frequency/ times per 10000 characters	Total number of occurrences/times	Frequency /times per 10000 characters
Old Chinese	"deng"(等)	339	1.22	0	0
Middle Chinese translations of Buddhist scriptures	"deng"(等)	237	1.54	1129	7.39
	"bei"(辈)	1	0.01	195	1.27
Middle Chinese native literature	"deng"(等)	44	1.22	26	0.72
	"bei"(辈)	2	0.05	7	0.19
Pre-modern Chinese	"deng"(等)	1944	1.21	635	3.72
	"men"(们)	17	0.03	2748	5.26

Secondly, the two plural expressions "bei"(辈) and "men"(们), which newly emerged in Middle Chinese and Pre-modern Chinese respectively, were characterized by having mostly been used after common nouns ever since they came into being. Thus, "bei"(辈) and "men"(们) seldom occurred after proper nouns, with frequencies of merely 0.01 times and 0.03 times per ten thousand characters.

Thirdly, when the first large-scale language contact in the history of Chinese came in the Middle Chinese period, Chinese translations of Buddhist scriptures as a direct product of such contact, accordingly, displayed distinct characteristics that were different from native Chinese literature in terms of the usage and distribution of plural expressions. In native Chinese literature, "deng"(等) was mainly used after proper nouns in different periods. It is generally believed that this plural expression used after a proper noun indicates a collective sense, referring to a collection of entities represented by that proper noun, rather than a plural sense of the proper noun itself (Yang, 2015). Take a sentence in "Qin benji" of *Shiji* (《史记·秦本纪》) for example: "缪公于是复使孟明视等将兵伐晋,战于彭衙。" Here the collocation "孟明视等" refers to the person "孟明视" and his fellows, while "孟明视" remains a singular proper noun. Therefore we conclude that the plural expression "deng"(等) was mostly used as a collective marker in the native Chinese literature in different historical periods.

In the Chinese translations of Buddhist scriptures, plural expressions that occurred in large numbers after common nouns are plural markers in a strict sense, such as in "是比丘辈闻是三昧四事助欢喜,入高明之智。" in *Banzhou Sanmei Jing* (《般舟三昧经》) translated by Zhilou Jiachen (支娄迦谶) in the Eastern Han Dynasty. Here "bei"(

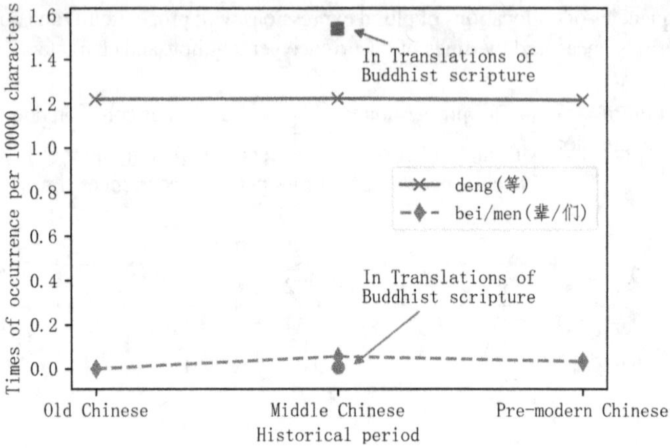

(a) After proper Nouns.

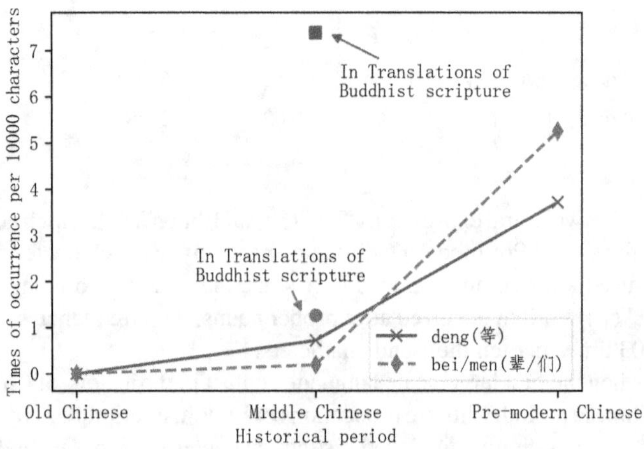

(b) After common Nouns.

Fig. 1. Trends of evolution of the post-nominal plural markers in different historical periods.

輩) denotes an indefinite plural, used in the same way as the plural marker "men"(们) in Modern Chinese.

It can be seen that the plural expressions after proper nouns and common nouns are of different properties. Statistics show that both "deng"(等) and "bei"(輩), which were used after common nouns and occurred substantially in the Chinese translations of Buddhist scriptures, having distinct distributions and frequencies from those in the native Chinese literature. This shows to some degree that plural markers in Chinese are not a result of the development of the Chinese language itself, but a result of language contact (translation of Indian Buddhist scriptures).

4 Conclusion

The category of plurals has always been a key and hot issue in Chinese studies. At present, the academia believes that plural expressions in Chinese are of two different properties, namely collective markers used after proper nouns and plural markers used after common nouns. This article, based on tagged corpora, conducts an exhaustive survey and analysis of plural expressions that have occurred in the history of the Chinese language. Statistics show that the only plural expression in Old Chinese was the collective marker "deng"(等), which was only used after proper nouns. In Middle Chinese, especially in translations of Buddhist scriptures, "deng"(等) began to occur in large numbers after common nouns, acquiring the usage of plural marking. In this period, a new plural expression "bei"(辈) emerged, with a sole function of plural marking after common nouns. When it comes to Pre-modern Chinese, "deng"(等), inheriting its usage in Old and Middle Chinese, was mainly used as a collective marker, while the function of plural marking was mostly taken by the newly emerged plural marker "men"(们).

Plural markers, which in the strict sense, refer to plural expressions used after common nouns, emerged in large quantities in the translations of Buddhist scriptures during Middle Chinese, which witnessed distinct usage of plural expressions from that in the native literature. Since there existed a characteristic and comprehensive plural marking system in Sanskrit, which employed inflections to indicate singularity, duality and plurality, we can further infer that the formation of Chinese plural marking is to some extent related to language contact. Chinese translations of Buddhist scriptures, as a typical product of unnatural language contact based on textural translation, mixed with foreign language components mainly from Sanskrit in their counterparts in the Chinese language.

Acknowledgments. The work was supported by the National Social Science Foundation of China (Grant No. 16BYY143). The author would also like to thank the national youth talent support plan in the Ten Thousand Talent Program.

References

1. Shuxiang, L.: About "men"(们). Chinese Language Monthly, 1949 (79, 80); included in Chinese Grammar Essays, pp. 145–168. Science Press, Beijing (1955)
2. Dan, X.: The expression of the concept of plural in the early period of the Western Han Dynasty from The Book of the Warring States Strategists (《战国纵横家书》). Res. His. Linguis. **2**, 83–94 (2009)
3. Yang, Y.: Plural marker "men"(们) and collective marker "men"(们). Lang. Teach. Res. **6**, 78–88 (2015)
4. Zhu, Q.: The relationship between the generation of plural marks of Chinese nouns and personal pronouns and the translation of Buddhist scriptures. Chinese Lang. J. **16**, 10–43 (2014)

Deriving Various Affected Subjects in *Bei*-Passives

Yu-Yin Hsu(✉) and Jun Chen

Department of Chinese and Bilingual Studies, The Hong Kong Polytechnic University, Kowloon, Hong Kong
yyhsu@polyu.edu.hk, jun061013.chen@connect.polyu.hk

Abstract. This paper investigates the thematic types and derivations of subjects in Mandarin *bei*-passives. Based on an examination of the categorial status of *bei* discussed in the literature, we try a new approach by adopting Liu and Huang's view that the mechanism of deriving *bei*-passives is argument raising (in which *bei* is a semi-lexical verb heading the Voice projection), and propose a simplified and unified account of *bei*-passives' Experiencer and Possessor subjects, as well as their typical Patient subjects. Specifically, we propose that the high applicative *gei* proposed in Hsu and Qu can account structurally for the source of the Experiencer in *bei*-passives. We then present how the structure proposed in this paper accounts for Possessor related subjects in *bei*-passives, and explain the interpretations and the possible derivation of an extra implicit affected subject in *bei*-passives.

Keywords: *Bei* passives · Raising · High applicative · Implicit subject · *Gei* 'give' · Mandarin

1 Introduction

This paper examines the interpretations and derivations of an interesting but understudied aspect of the Mandarin *bei*-passive construction – an atypical thematic role expressed by the subject. Typically, a Mandarin *bei*-passive allows the optional use of semantic Agent. Accordingly, researchers have often referred to those passive sentences with an overt Agent as the long *bei*-passive, as in (1), and those without an overt Agent as the short *bei*-passive, as in (2) [1–3].

(1) Long *bei*-passive:
Zhangsan bei Lisi da-le.
Zhangsan BEI Lisi hit-LE
'Zhangsan was hit by Lisi.' ([3]: 112)

(2) Short *bei*-passive:
Zhangsan bei da-le.
Zhangsan BEI hit-LE
'Zhangsan was hit.' ([3]: 112)

M. Liu et al. (Eds.): CLSW 2020, LNAI 12278, pp. 276–287, 2021.
https://doi.org/10.1007/978-3-030-81197-6_24

While the subject in a canonical passive sentence is often the semantic Patient of the event, some researchers have noted that the Mandarin *bei*-passive allows the subject position to be occupied by a non-Patient argument. In (3), for example, the subject *Zhangsan* is the possessor of the direct object *baba* 'father'. However, in that sentence, *Zhangsan* is only indirectly associated with the bandits' killing of his father. In a more common passive form of the same sentence, as in (4), the subject *Zhangsan's father* is the direct sufferer of the event.

(3) Zhangsan bei tufei da-si-LE baba.
Zhangsan BEI bandits hit-die-LE father
'Zhangsan was father-killed by the bandits.' ([3]: 142)

(4) Zhangsan(-de) baba bei tufei da-si-le.
Zhangsan's father BEI bandits hit-die-LE
'Zhangsan's father was killed by the bandits.'

The subject in (3) can be understood as the Possessor of the Patient of the event. This is one of the special subjects found in Mandarin *bei*-passives. Some scholars have therefore suggested that the subjects of *bei*-passives can be vaguely associated with the event described [2], and others suggested that they may or may not be overtly expressed [4]. For example, (5) is clearly about the event 'Wangwu hit a home run'. However, when this sentence is presented in isolation, it is unclear what role *Lisi* plays in that event.

(5) Lisi bei Wangwu jichu-le yi-zhi quan-lei-da.
Lisi BEI Wangwu hit-LE one-CL home-run
'Lisi had Wangwu hit a home run on him.' ([3]: 462-463)

Unlike (5), the subject position in (6a) is empty, and the semantic Patient (i.e., *zhe ye* 'this page') is retained in its original object position together with the Agent (i.e., *wo* 'I') following *bei*. In other words, the segment after *bei* in (6a) is a complete sentence by itself, and yet, readers of the full sentence may still perceive that some person or entity external to the event being described was unwillingly, adversatively, or unfortunately affected (see [5] for further discussion). Example (6b) represents a similar type of sentence, in which the implicit subject seems to be associated with the speaker's comment about the event 'he succeeded' – as implied by *zui-ou-de shi* 'what's the most upsetting is', and *juran* 'surprisingly'.

(6) a. Juran bei wo zai Taiwan-Yahoo shang zhaodao zhe ye.
surprisinglyBEI I on Taiwan-Yahoo on find this page
'Surprisingly, this page was found by me on Taiwan Yahoo.'
b. Zui-ou-de shi juran hai bei ta chenggong-le.
most-upsetting be surprisingly even BEI he succeed-LE.
'What's the most upsetting is that, surprisingly, he even succeeded.' ([4]: 430)

Although previous studies have mentioned similar examples of such atypical subjects in *bei*-passives (cf. [6]), to the best of our knowledge, there have been no systematic investigations of either how such subjects are derived, or the sources of the interpretations they are given. In this paper, therefore, we entertain the proposal that Mandarin *bei* is a semi-lexical verb that takes a raising structure to form long passives, and show that adopting the high applicative projection proposed by [7], in combination with a revised version of the raising structure discussed in [5], facilitates a unified account of different thematic types of subjects in Mandarin *bei*-passives.

2 Issues Around Canonical *Bei*-Passives

2.1 The Categorial Status of *Bei*

Mandarin *bei*-passives have been a topic with extensive discussion among linguists, involving both the surface realization of participants' roles in an event, and the categorial status of *bei*. Regarding the latter, four main proposals have been made, i.e., that *bei* – is a verb [1–4, 8, 9]; is a preposition [10, 11]; occupies a dual-status category [12–14]; and is a semi-lexical verb [5, 15].

Earlier studies have suggested that, much like the *by*-phrase in English passives, *bei* can be understood as a preposition introducing an Agent argument [10, 11]. However, some challenges remain if one adopts this understanding. First, if *bei* were a preposition, many short *bei* passives would exhibit preposition stranding (e.g., *Zhangsan* ***bei*** __ *da-le*), which is generally not allowed for Prepositional Phrases (PPs) cross-linguistically [3, 4]. Second, one might wonder why *bei* and its following NP (e.g., *bei Lisi* in (7)) cannot occur at the beginning of a sentence, unlike typical PPs (e.g., *dui Lisi* 'to Lisi' in (7)).

(7) a. * Bei Lisi Zhangsan zuotian da-le.
BEI Lisi Zhangsan yesterday hit-LE.
'By Lisi, Zhangsan was hit yesterday.' ([3]: 116)
b. Dui Lisi Zhangsan hen keqi.
to Lisi Zhangsan very polite.
'To Lisi, Zhangsan is very polite.' ([3]: 116)

Therefore, most of those who argue that *bei* is a verb [1–3] see it as the main verb which selects a clause (IP) as its complement to form the long *bei* passive (e.g., (8)). This involves an A'-movement of a null operator (NOP) that is later co-indexed with the subject through predication. When the main verb *bei* selects a VP to form a short passive (e.g., (9)), the process involves an A-movement of a PRO object to be controlled by the subject.

(8) Zhangsan$_i$ bei [$_{IP}$ NOP$_i$ [$_{IP}$ Lisi da-le t$_i$]].
Zhangsan BEI Lisi hit-LE
'Zhangsan was hit by Lisi.' ([3]: 138)

(9) Zhangsan$_i$ bei [$_{VP}$ PRO$_i$ da-le t$_i$].
Zhangsan BEI hit-LE
'Zhangsan was hit.' ([3]: 138)

To support the analysis of A'-movement in long passives, some studies assume that the trace can be replaced by a resumptive pronoun [3, 8]. However, [14] point out that this is impossible, as evidenced by the contrast between (10) and (11). Instead, they propose two categories of *bei* in Mandarin: *bei*$_1$, a functional category taking a VP as its complement, and *bei*$_2$ which is understood as a preposition [13, 14].

(10)	Zhangsan$_i$	bei	tufei	t$_i$	daduan-le	Pro$_i$	tui.
	Zhangsan	bei	bandits		hit-broken-LE		leg

'Zhangsan had his leg "hit to broken" by the bandits.' ([14]: 219)

(11)	*Zhangsan	bei	tufei	**ta**	daduan-le	Pro$_i$	tui.
	Zhangsan	bei	bandits	him	hit-broken-LE		leg

'Zhangsan had his leg "hit to broken" by the bandits.' ([14]: 219)

[14] noted that, when both *bei*$_1$ and *bei*$_2$ co-occur and form a long passive sentence with the structure shown in (12a), only one *bei* is pronounced, and ascribed this to haplology. When the *bei*$_2$ PP is not expressed, a short passive is formed (i.e., (12b)). This dual-status analysis of *bei*, which assumes the A-movement of the Patient object – raising from inside of the VP to the subject position for the nominative case [2, 3] – has the advantage of presenting a unified account of long and short *bei*-passive sentences.

(12) a. [$_{\text{PassiveP}}$ *bei*$_1$ [$_{\text{VP}}$ [$_{\text{PP}}$ *bei*$_2$ NP] VP]] ([14]: 220)
b. [$_{\text{PassiveP}}$ *bei*$_1$ VP] ([14]: 220)

Arguing in favor of the above dual-status analysis, Shi [13] shows that a subject-oriented adverb (e.g., *guyi* 'intentionally') can refer either to an implicit Agent pro, as in (13), or to the overt Agent: e.g., *duifang duiyuan* 'players of the other team' in (14). This suggests that the short *bei*-passive (e.g., (13)) can be derived by omitting the Agent from a long passive (e.g., (14)).

(13)	Dazhi	bei	pro	guyi	zhuangshang-le.
	Dazhi	BEI		intentionally	crash-injury-LE.

'Dazhi was intentionally given a "crash-injury".' ([13]: 115)

(14)	Dazhi bei	duifang-duiyuan	guyi	zhuangshang-le.
	Dazhi BEI	the.other.team-players	intentionally	crash-injury-LE.

'Dazhi was intentionally given a "crash-injury" from players of the other team.' ([13]: 116)

However, examples like (15) show that haplology [14] may not be a tenable solution in Chinese, because even the same type of *bei* sometimes co-occurs within a sentence to signal different structures. Thus, it is unclear why haplology has to apply to the two different categories of *bei* exemplified in (12a).

(15)	Zhangsan bei	bei	shitou	jizhong-de	ren	sha-le.
	Zhangsan BEI	BEI	the stone	hit-DE	the person	kill-LE.

'Zhangsan was killed by the person who was hit by the stone.'

Moreover, while the dual-status analysis assumes a derived pure Patient subject, sentences like (16) indicate that the matrix subject position can also be an Agentive role rather than a pure Patient, because *guyi* 'intentionally' can modify it. Such sentences suggest that the subject of *bei*-passives is in the matrix subject position, and one possible explanation is that the subject may be directly merged at that position. Yet, this has not been discussed in any version of the dual-status account. Nonetheless, if external-merge of the subject is allowed, how such an Agentive subject obtains an additional Patient interpretation still calls for explanation.

(16) Dazhi guyi bei duifang-duiyuan zhuangshang-le.
Dazhi intentionally BEI the.other.team-players crash-injury-LE.
'Dazhi intentionally got a "crash-injury" from players of the other team.' ([13]: 116)

In light of examples like (16), [5] propose that *bei* is a semi-lexical verb that can be decomposed to Experience (Exp) and Become (Bec), forming a control structure. As shown in (17), *bei* merges with the Voice head which takes *v*P as its complement, and undergoes a cyclic head movement to Bec, and then to Exp. In this derivation, the Patient PRO raises from VP object position to Spec,VoiceP and is licensed at Spec,BecP. The Experiencer *Zhangsan* is merged with the Spec,ExpP and co-indexed with the PRO in that position, and later moves to Spec,TP where it serves as the subject of the sentence. This derivation accounts for the fact that the subject of a *bei*-passive may have an Experiencer sense (e.g., *Zhangsan* in (17)) in addition to a typical Patient role.

(17) Control structure of *bei*
$[_{TP}$ Zhangsan$_i$ $[_{ExpP}$ t_i bei$_j$ $[_{BecP}$ PRO$_i$ t_j $[_{VoiceP}$ t_i t_j $[_{vP}$ Lisi $[_{vP}$ t_i da-le$_k$ $[_{VP}$ t_k $t_i]]]]]]]$.
Zhangsan BEI Lisi hit-LE
'Zhangsan got hit by Lisi.' ([5]: 396)

However, as highlighted by Liu and Huang [5], both long and short *bei*-passive sentences allow idiom-chunk subjects. As the structure in (18) shows, this kind of sentence can be derived by raising the NP *pianyi* 'advantage' cyclically to the matrix subject position. When this happens, *bei* only projects up to BecP, not beyond it to ExpP.

(18) Raising structure of *bei*
$[_{TP}$ Pianyi$_i$ dou $[_{BecP}$ t_i bei$_j$ $[_{VoiceP}$ t_i t_j $[_{vP}$ (ta yi-ge-ren) $[_{vP}$ t_i zhan-guang-le$_k$ $[_{VP}$ t_k $t_i]]]]]]$!
advantage all BEI (he alone) take-empty-LE
'All the advantage was taken (by him single-handedly).' ([5]: 389)

According to the same proposal [5], when there is no ExpP, the Experiencer can either be introduced by the applicative *gei* as an applicative argument (e.g., *gei wo* in (19)), or be implied from the discourse without being uttered.

(19) Pianyi dou bei ta (gei wo) zhan-guang-le!
advantage all BEI he on me take-empty-LE
'All the advantage was taken by him on me!' ([5]: 390)

Liu and Huang [5] then conclude that when the subject of *bei*-passives is animate, it can be ambiguous as to whether that subject is an Experiencer-Patient or a pure Patient; and that, while passives with an Experiencer-Patient subject such as those in (16) and (17) are accounted for by the control structure, those with a pure Patient subject are accounted for by the raising structure. In the latter, the raising *bei* is derived from the control *bei* after argument suppression, in which an Experiencer may either be introduced as an applicative argument, as in (19), or remain implicit. Crucially, according to this proposal, the Experiencer subject introduced by the control *bei* has its own source, which is independent of the lower applicative Experiencer introduced by *gei*.

While this control-and/or-raising account of *bei* nicely captures several patterns of passives in Mandarin, we would like to bring up some data that suggest that animacy may not be the reason of an Experiencer flavored subject as suggested by [8].

First, animate subjects do not always show such Experiencer-Patient ambiguity in *bei*-passives and some data show that Experiencer subjects of *bei*-passives can be associated with the applicative position. In this case, such an Experiencer subject does not express the type of Experiencer-Patient subject [5] related to (16–17), suggesting different derivation sources of the two types of thematic roles. As (20) shows, the Experiencer subject *Lisi* can be related to the applicative Experiencer position (*e*) after the applicative *gei*, while the Patient ('one home run') remains in its object position. One may wonder how the control-and/or-raising account [5] can explain whether *bei* derives a control or a raising structure in cases like this, which contains a pure Experiencer subject.

(20) Lisi bei Wangwu (gei) *e* jichu-le yi-zhi quan-lei-da.
Lisi BEI Wangwu Appl hit-LE one-CL home-run
'Lisi had been affected by Wangwu's hitting of a home run.

In light of examples like (20), we suggest that the Experiencer reading is determined by the association of the subject and the experiencer role introduced by the applicative *gei*, but not by an accidental side-effect of animacy. This account can also explain the affected Experiencer role for the matrix subject of *bei*-passive in the event described, but crucially not the Patient, when *gei* in sentences like (20) is covert. Concerning the sentences like (16–17) whose subjects are Patient with an experiencer sense, we thank one of the anonymous reviewers for this comment and will show in Sect. 3 that this is determined by the interaction of the structures of applicative *gei* and *bei*, rather than the *bei* structure alone as assumed in past analyses.

Second, it is not clear why the choices of the types of *bei* structure should be contingent on the occurrence of the (implicit) Experiencer introduced by *gei*. While *gei* can introduce an (extra) Experiencer to an event, such a beneficiary light-verb structure emerged in Chinese syntax around 221 B.C. [16]. However, typical Patient-as-subject *bei*-passives appeared between 220 and 589 A.D., and Experiencer/empty subject *bei*-passives between 618 and 907 A.D. [17]. Diachronically speaking, it therefore seems odd to propose that *bei* should be conceptualized as two sub-components – Experience and Become. Moreover, the occurrence of an Experiencer is independent of the passive structure. For example, in the active sentence in (21), depending on context (e.g., *Lisi* bet a lot of money as the result of a baseball game) and with respect to the event described, *Lisi* can be interpreted as having either a benefactive or a malefactive role in Wangwu's home-run event.

(21) Wangwu gei Lisi jichu-le yi-zhi quan-lei-da.
Wangwu Appl Lisi hit-LE one-CL home-run
'Lisi got Wangwu hit a home run for/on him.'

Third, the subject of *bei*-passives can have a Possessor role, which typically is not introduced by the applicative head *gei* (as in (3) repeated below), and is not the (direct) Patient of the event.

(3) Zhangsan bei tufei da-si-le baba
Zhangsan BEI the bandits hit-die-LE the father
'Zhangsan was "father-killed" by the bandits.'

Therefore, while we accept the prior proposal that *bei* is a semi-lexical verb [5], we depart from their proposal of "*bei* as Become and Experience". The remainder of this paper will present a revised and simplified version of that proposal's raising *bei* structure. In combination with the high applicative phrase [7], our new proposal can derive both atypical and typical subjects in *bei*-passives, while avoiding the inconsistency in the previous proposals.

3 New Raising Structure of *Bei*-Passives

Given that both short and long passives allow idiom-chunk subjects, and that the Experiencer realization introduced by *gei* is independent of passives, we propose a simplified raising structure of *bei*-passives, as shown in (22). That is, we propose that *bei* is a semi-lexical verb, whose status is like a light verb accounting for Voice, but is different from the Chomskyan light verb *v* (a view also held by [5, 13]). The structure is shown in (22b).[1]

(22) Revised raising structure of *bei*-passives
a. Pianyi dou bei ta-yi-ge-ren zhan-guang le.
advantage all BEI he-one-CL-person take-empty LE
'All the advantage was taken by him alone.'

b.

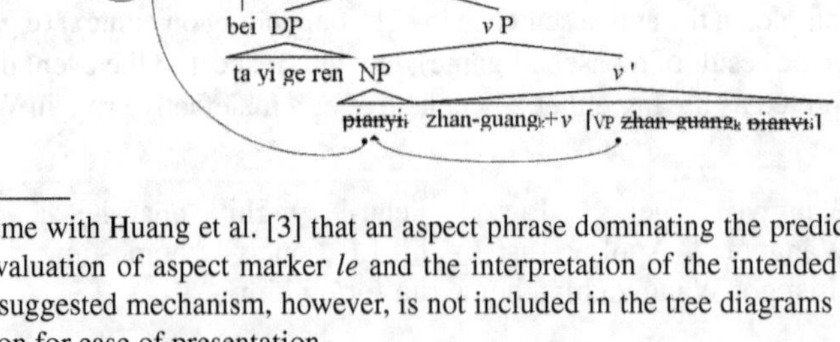

[1] We assume with Huang et al. [3] that an aspect phrase dominating the predicate (*v*P) accounts for the valuation of aspect marker *le* and the interpretation of the intended event. The detail of their suggested mechanism, however, is not included in the tree diagrams and the following discussion for ease of presentation.

We assume that *bei* is a light-verb introducing an affected sense (cf. [18]) to its specifier, and its status may be similar to the Voice head and makes the (tucked-in) internal argument (cf. [19]) available for movement (cf. [5]: 395). However, we depart from previous suggestions that *bei* invokes BecP-ExpP [5], and propose that the internal argument (pure Patient and non-Patient) can move to Spec,TP directly. Pure Patient subjects of *bei*-passives will be derived as proposed by Liu and Huang [5].

When a *bei*-passive contains a pure Experiencer subject, the Experiencer is initially merged with the specifier of the high applicative head (Appl_HP) dominated by the *v*P [7], deriving sentences like (21). Assuming with [7] that the directly merged high-applicative specifier obtains the Experiencer role and can undergo further raising, we propose that the passivization process of such sentences should be like that shown in (23).

(23) Raising structure for Experiencer subject

a. Zhangsan bei Lisi (gei) zhanguang-le pianyi.
Zhangsan BEI Lisi on/for take-empty-LE advantage
'Zhangsan was affected by Lisi's taking all the advantage.'

b.

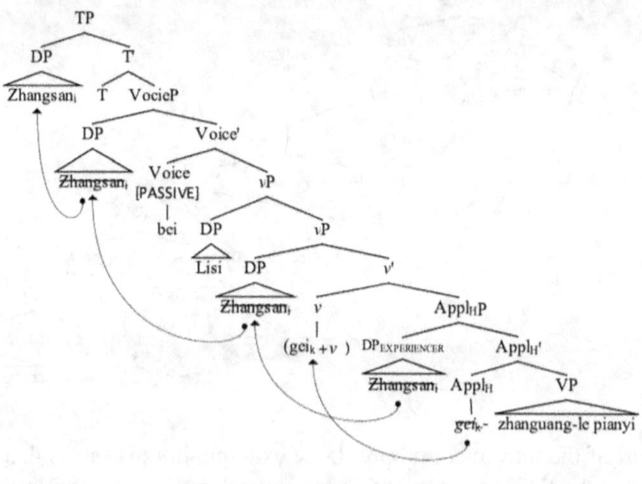

That is, the Experiencer introduced by the high applicative head, *Zhangsan*, first gets tucked into Spec,*v*P and then cyclically moves to Spec,TP. The applicative head *gei* that stay in the original positions may optionally be pronounced. This proposal accounts for the fact that an Experiencer subject of *bei*-passive does not always carry a Patient flavor (contrary to [5]), and that sometimes the *bei*-passive with a pure Experiencer subject may not contain an overt *gei* (cf. [7]). While assuming with previous studies that *bei* and applicative *gei* both show light-verb properties, in the structure (23) we reserve the notion, small *v*, to the light-verb that introduces the thematic Agent of an event, and distinguish it from *gei* and *bei* in the structure for ease of discussion.

4 Possessor Subject and Implicit Subject in *Bei*-Passives

Before we conclude this paper, we would like to discuss some other types of thematic subjects of *bei*-passives. We will first show how the current proposal may derive pure

Possessor subjects in *bei*-passives and discuss an alternative account for the implicit subjects of Mandarin *bei*-passives.

It was suggested by Hsu and Qu [7] that only phrases at Spec,Appl$_H$P that are not derived from Spec,Appl$_L$P may undergo further raising, and that the possessor of a direct object may follow a similar mechanism by first moving to Spec,Appl$_H$P from its nominal-internal position (e.g., the specifier of the Possessor Phrase cf. [20–22]) and then undergoing further raising to Spec,TP to form a passive sentence. This idea can be directly incorporated into the current proposal. A passive sentence with a Possessor subject, such as (3), would have a structure like (24).

(24) a. Zhangsan bei tufei gei da-si-le baba.
Zhangsan BEI bandits Appl$_H$ hit-die-LE father
'Zhangsan got affected by [the] bandits' killing [of his] father.

b.

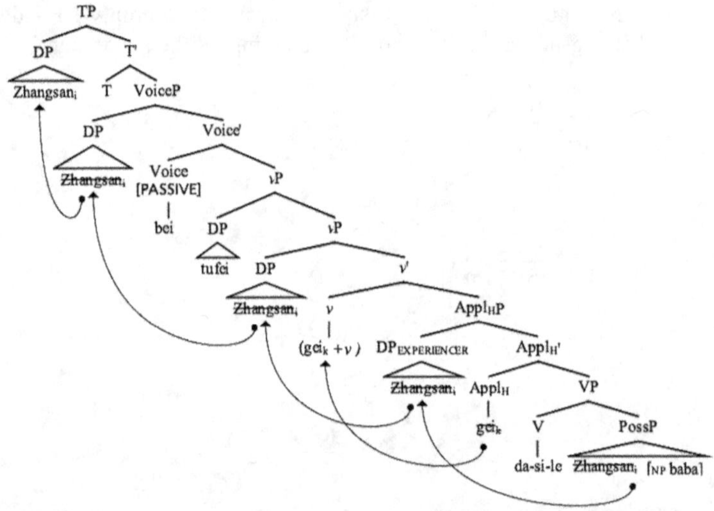

In support of the movement proposal, the example in (25) shows that a pronoun *ta* cannot occur in the Experiencer position introduced by the high applicative *gei* and be co-indexed with the subject *Zhangsan* (cf. [14]).

(25) *Zhangsan$_i$ bei tufei gei ta$_i$ da-si-le baba.
Zhangsan BEI bandits Appl$_H$ him hit-die-LE father
'Zhangsan got affected by [the] bandits' killing [of his] father.

Alternatively, a typical passive with a Patient subject as in (26a) may be derived by tucking in the whole direct object (*Zhangsan's father*) after Voice *bei* merges with *v*P (as per the revised raising structure illustrated in (22)), thus yielding the sentence, 'Zhangsan's father was killed by the bandits'. As predicted by the movement approach, the pronoun *ta* is not allowed in the original Patient position, as shown in (26b).

(26) a. Zhangsan baba bei tufei da-si-le .
Zhangsan father BEI bandits hit-die-LE
'Zhangsan's father was killed by [the] bandits.
b.*Zhangsan baba$_i$ bei tufei da-si-le ta$_i$.
Zhangsan father BEI bandits hit-die-LE him
'Zhangsan's father was killed by [the] bandits.

Last but not least, it is worth-noting that when the canonical thematic roles associated with the main predicate of an event all stay in their original positions, an extra-affectee interpretation somehow may still be available to the position before *bei*. When this occurs, we find that such an affectee is most naturally interpreted either as the speaker/narrator of a sentence like (27), or a topic mentioned in previous discourse – usually an unwilling, unfortunate affectee [6], and this implied subject (the narrator) tends to express the event as a surprise.

(27) Bei tufei gei Qi.Guo da-pao-LE da.jiangjun
BEI bandits on/for Qi.nation hit-run-LE general
'[Someone] got affected by the bandits' defeating the General of Qi'

The availability of such an extra affectee interpretation in sentences like (27) suggests that Chinese syntax does not employ subject expletives, and that Chinese may not have impersonal passives in the sense of [23]. We agree with previous studies that this kind of affectee interpretation is not identical to thematic roles [6]. We suggest that it is instead part of *bei*'s light-verb lexical semantics [5] to specify phrases' unwillingness, unfortunateness, or surprisal readings by merging with the associated phrases as *bei*'s specifier. The specifier phrase may later raise to the Sepc,TP and serve as the matrix subject of a passive sentence. Such an implied subject is initially merged at the matrix subject position.

In addition, the position before *bei* in sentences like (27) may be an overt affected nominal external to the whole event, who is only indirectly affected, but for whom the whole sentence represents a subjective comment delivered by the narrator. For example, considering a situation where a witness or a reporter wishes to comment on a battle between two countries (Qi and Zheng), Zheng played a trick so that they did not defeat the general of Qi themselves, and instead let the general of Qi be defeated by the bandits. The narrator could express such an unusual, unexpected event with a sentence like (28), where the extra-affectee can be made implicit as the context allows.

(28) (Zheng.Guo) guyi bei tufei gei Qi.Guo da-pao-le dajiangjun
Zheng.nation intentionally BEI bandits on/for Qi.nation hit-run-LE the.general
'The Country of Zheng intentionally got affected by the bandits' defeating the General of Qi.'

While the subject of a *bei*-passive sentence may later undergo topicalization to a topic position representing a discourse topic (like most of the subjects in Chinese sentences), we would like to point out that such a subject can surface as the matrix subject. Example (28) shows that whether the subject is overt or not, a subject-oriented adverb like *guyi* 'intentionally' modifies the referent of the subject located before *bei*.

5 Concluding Remarks

In this paper, we have proposed a simplified raising structure to account for various thematic types of subjects in Mandarin *bei*-passives, based on a modification of the raising structure proposed by [5]. This modification involves accommodating an optional realization of the high applicative *gei*-construction [7]. We have showed how the revised raising structure of *bei* can straightforwardly account for the pure Experiencer and Possessor subjects as well as the typical Patient subjects of *bei*-passive sentences.

In addition, based on our proposed derivation, we have elaborated possible interpretations of the empty subject in *bei*-passives, and showed how such empty subjects might be structurally licensed. In other words, assuming that *bei* is by nature a semi-lexical verb, we agreed with [5] that the control analysis of *bei*-passive is not the only way to derive different thematic types of subjects in it; and that, in fact, an argument-raising account could straightforwardly derive various thematic subjects in a consistent and systematic way, while avoiding the inconsistencies introduced by previous proposals.

If the raising structure proposed in this paper is on the right track, we expect the same or similar mechanisms to be available in Chinese languages other than Mandarin. Due to space limitations, however, we will leave that cross-dialectal and cross-linguistic investigation to future studies.

References

1. Ting, J.: Deriving the bei-construction in Mandarin Chinese. J. East Asian Linguis. **7**(4), 319–354 (1998)
2. Huang, C.T.: Chinese passive in comparative perspective. Tsing Hua J. Chin. Stud. **29**(4), 423–509 (1999)
3. Huang, C.T., Li, Y.H., Li, Y.F.: The Syntax of Chinese. Cambridge University Press, Cambridge (2009)
4. Her, O.S.: Unifying the long passive and the short passive: on the bei construction in Taiwan Mandarin. Lang. Linguist. **10**(3), 421–470 (2009)
5. Liu, N., Huang, C.-T.: Control and raising passives, and why Mandarin does not smuggle. J. East Asian Linguis. **25**(4), 385–404 (2016). https://doi.org/10.1007/s10831-016-9148-3
6. Kuo, W.J.: Null-subject passives in southern Chinese dialects. Tamkang J. Chin. Lit. **34**, 257–281 (2016)
7. Hsu, Y.Y., Qu, T.: V-gei double object construction and extra argument in Mandarin. In: The 33rd Pacific Asia Conference on Language, Information and Computation, PACLIC-33, pp. 76–84 (2019)
8. Feng, S.L.: Interactions between Morphology, Syntax and Prosody in Chinese. Peking University Presses (1997)
9. Tang, S.W.: Syntactic analysis of Chinese passives. Contemp. Linguist. **10**(4), 308–319 (2008)
10. Wang, H.: The *Ba*-constructions and the *Bei*-constructions. Education Press, Shanghai (1957)
11. Lü, S.X.: The Eight Hundred Words in Modern Chinese. The Commercial Press, Beijing (1980)
12. Shi, D.X.: On "*bei*" passive in Chinese. Chin. Linguist. **1**, 38–48 (2005)
13. Shi, D.X.: Probing into the long and short "passive" words and expressions. J. Qinghai Natl. Inst. **34**(3), 111–117 (2008)
14. Shi, D.X., Hu, J.H.: The syntactic status of "*bei*". Contemp. Linguist. **7**(3), 213–224 (2005)

15. Liu, N.: The structures of Chinese long and short Bei passives revisited. Lang. Linguist. **17**(6), 857–889 (2016)
16. Hsu, Y.Y., Wang, T.: From modern to ancient Chinese: a corpus approach to beneficiary structure. In: Hong, J.F., Zhang, Y., Liu, P. (eds.) Chinese Lexical Semantics, CLSW 2019. Lecture Notes in Computer Science, vol. 11831. Springer, Cham (2019).https://doi.org/10.1007/978-3-030-38189-9_73
17. Kuo, W.J.: The null-subject bei-constructions in the "Zhuzi Yulei". Bull. Depart. Chin. Lit. N.T.U. **37**(6), 175–220 (2012)
18. Li, Y.H.A.: Passive, BA, and topic constructions. In: Order and Constituency in Mandarin Chinese. Studies in Natural Language and Linguistic Theory, vol. 19, pp. 153–203. Springer, Dordrecht (1990). https://doi.org/10.1007/978-94-009-1898-6_6
19. Richards, N.: What Moves Where When in Which Languages? Doctoral Dissertation, Massachusetts Institute of Technology (1997)
20. Hsu, Y.Y., Ting, J.: On the possessive multiple nominative construction in Mandarin Chinese. In: 14th Annual Conference of the International Association of Chinese Linguistics (2006)
21. Hsu, Y.Y.: Possessor extraction in Mandarin Chinese. University of Pennsylvania Working papers in Linguistics, vol. 15, no. 1 (2008). Article 12
22. Hsu, Y.Y.: Possessor, appositive, and beyond: a study of nominal-initial pronouns and proper names in Chinese. In: George, F., James, L., Ronald, F. (eds.) Festschrift for Steve Franks. Slavica Publishers, Bloomington, IN (2019)
23. Perlmutter, D.M.: Impersonal passives and the unaccusative hypothesis. Proc. Ann. Meet. Berkeley Linguist. Soc. **38**, 157–189 (1978)

Affective Semantics and Regulatory Modes of the Word "可" in Text-Based Sentiment Analysis

Tuya Naren[1] and Xiaoyin Xu[2(✉)]

[1] Capital Normal University, Beijing 100089, China
nrty0910@cnu.edu.cn

[2] Harvard Medical School, Brigham and Women's Hospital, Boston, MA 02115, USA
xxu@bwh.harvard.edu

Abstract. A sentimental sentence usually contains three important components: first, the core expression component; second, the modifier component; and third, the regulatory component. The three components do not exist independently; rather, they interactively form a sentimental sentence. Therefore, in text-based sentiment analysis, while the core expression components are considered the essence of a sentence, all three components need to be considered in a layered structure. The semantics of "可[KE]" in text exhibits a high degree of variation, increasing the difficulty in text-based sentimental analysis. This paper focuses on the screening, classification, and analysis of sentimental sentences involving "可" within the framework of text-based affective analysis. We extracted models and patterns in affective expressions in sentences involving "可". The purpose of this work is to extract the common usage and patterns of "可" in affective expressions, which, in turn, will be helpful for sentimental analysis.

Keywords: Text-based affective analysis · "可" · Sentiment semantics · Mode of regulatory component

1 Introduction

In recent years, text-based sentiment analysis has attracted a great deal of attention from academics and been widely applied [1, 2]. Related research centers around are solving the following three problems: (1) how to identify subjective sentences in text; (2) how to determine the sentimental polarity of text; and (3) how to measure the sentimental intensity of text [3, 4]. Overall, the main mission of text-based sentiment analysis is to find subjective sentences in texts and determine the sentiment polarity, type and strength they convey, a task that involves modeling the patterns of sentiment expressions and identifying words and phrases that convey sentiments. As of today, the common approach is to rely on a dictionary-based collection of sentimental words to classify words and phrases into different categories of sentiment expressions [5]. For almost every language, including Chinese [6], such dictionaries have been created and used to develop the computerized analysis of sentiments [7–11]. More recently, a domain-based

M. Liu et al. (Eds.): CLSW 2020, LNAI 12278, pp. 288–302, 2021.
https://doi.org/10.1007/978-3-030-81197-6_25

sentimental word dictionary has been proposed to address the problem that the same words may have different sentimental meanings in various domains [12]. Then, given a sentimental word dictionary, deep learning-based analysis is widely adopted to train on dictionary-based models to identify sentiments [12–14]. However, such approaches cannot comprehensively capture sentimental expressions in texts as a language may convey sentiments in a versatile way, sometimes without the use of sentimental words but through special linguistic patterns [15] which are often of dynamic nature [16]. This phenomenon poses a challenge to the computerized analysis of text-based sentiments, and, thus, researchers have been investigating linguistic rules or statistical methods to capture such sentimental expression [17, 18].

The problem we aim to address in this work is the sentimental expressions consisting of Chinese word "可[pronounced "KE"]" in texts. The word "可" generally means "can", "may", or "be able to". As a modal, "可" finds its wide application in Chinese and can be used together with verbs, adverbs, adjectives, interrogative adverbs, and interjections. This versatility, on the one hand, translates into rich linguistic expressions as the word can be used in many scenarios, while, on the other hand, it also creates a challenge for text-based sentimental analysis as the word conveys wide-ranging sentiments and can modulate a speaker's intention and preference in both strong and subtle ways. Accurate modeling and analysis of sentiments conveyed by "可" can improve the comprehensiveness of text-based analysis as the word frequently appears in writing. According to the analysis of Wang [19], there are 20 modal verbs, including "可", among the 1500 high-frequency Chinese words. "可" has a vocabulary grade of 290; word order of 99; frequency of 0.02642%; and a degree of usage of 69.3, all of which convey its importance.

In this work, we set out to analyze patterns and modes of expressing sentiment which involves the use of "可". In Sect. 2, we analyze the affective semantics of "可" in terms of its usage to express permission as well as the concepts of 'can', 'can be used to', 'worth', and recommendation. In Sect. 3, we elaborate on the types of modal constructions of "可" in sentimental expression. In Sect. 4, we analyze the patterns of sentimental expressions given by "可" from the perspective of assigning a sentiment to a sentence or strengthening or weakening sentiments. We summarize our contribution and conclusions in Sect. 5.

2 The Affective Semantics of "可"

Modal words have the characteristics of polysemy, and some modal words have cross polysemy among dynamic modality, moral modality and cognitive modality. When polysemous modal words express different modal meanings, different features will appear in form and function. "可" is an admissible modal verb, meaning it belongs to the lower classification of cognitive modality that has multiple semantics, such as "permission", "can", "can be used to", "worth", "agreement", and "desire".

2.1 Permission

"可" may mean permission or authorization, including general pre-authorization, authorization given after some types of applications and authorization under specific conditions or. The meaning of "可" is the same as "可以 (can, may)" as the two are interchangeable in a sentence without affecting the expression in this type of usage. The negative form of "可" is "不可 (cannot, may not)". For example:

(1) 会议资料可公开。
Meeting materials may be made public.
(2) 答完试题可退场。
After finishing the examination, one can leave the room.
(3) 不可信口雌黄!
One cannot make irresponsible remarks!

The first two sentences have two layers of meanings, whereby the first meaning conveys "permission". These two sentences mean, respectively that i) making the meeting materials public is permitted and ii) one is permitted to leave the room after finishing the examination and does not have to wait until the time is up. The second layer of meaning conveys the intention that the actions following 可 are optional and not mandatory. In the first example, it therefore means that it is optional for the meeting materials to be made public. In the second sentence, it means that people taking the examination can choose to stay in the room. Example (3) has the meaning of the first layer and is in a negative form. It does not include the second layer of meaning. When used to express "permission", "可" also expresses agreement or consent of the speaker or writer toward an object, a person, or an event, and it can be used to mean agreement, consent, and other similar sentiments.

2.2 "Can"

"可" also has the meaning "can", which means that a certain goal, or a certain purpose and an expectation that is desired by the speaker can be achieved in a certain way. Under this meaning, "可" can be exchanged with "可以 (can, may)" with the meaning of the sentence unchanged. Examples follow.

(1) 她在比赛现场，可鼓舞士气。
If she is at the scene of the match, she will inspire everyone's morale.
(2) 他来带队，可提高工作效率。
If he leads the team, the working efficiency can be improved.
(3) 积极干预，可平稳房价。
Proactive intervention can stabilize housing prices.

In examples (1) and (2), both "at the scene of the match" and "leading the team" are "means" through which the goals of "inspiring everyone's morale" and "improving work efficiency" can be achieved. And, in example (3), to "stabilize housing price" through "proactive intervention" conveys a sentiment of expectation or anticipation. What "可" expresses is usually positive sentiment related to a sentiment such as expectation.

2.3 "Can Be Used to"

"可" can indicate or introduce something that has a certain function and can be replaced with "可以 (can, may), 能 (can, may)", without changing the meaning of a sentence. Examples follow.

(1) 棉花籽可炸油。
Cotton seeds can be used for oil extraction.
(2) 甘蔗和甜菜可用来制糖。
Sugar cane and sugar beet can be used to make sugar.
(3) 橘皮可入药，称之为"陈皮"
Orange peel can be used in medicine; it is called "CHENPI".

In these examples, oil extraction, sugar making and the making of medicine are the actual uses of "cotton", "sugar cane" and "orange peel". The uses are based on the experience and knowledge of the speaker. At the same time, the examples also express the objective modal meaning of "可". So, the polarity of sentiment is neutral.

2.4 "Worth"

"可" may have the meaning of "worth" when it is used in the form of "可 + Verb + de", indicating that the verb refers a certain act or manner recognized by the speaker or something the speaker is willing to do subjectively. It expresses feelings and attitudes that are approved or disapproved, worthwhile or not. Examples follow.

(1) 我和他没什么可说的。
I have nothing to say to him.
(2) 这样的好货可再来一打儿。
We can have a dozen more of such good things.
(3) 发生了这样的事没什么可同情的。
There is nothing to sympathize with when such a thing happens.

Examples (1) and (3) show the disapproving attitude and loathing sentiment of the speaker. The word "可说的" in example (1) and "可再来一打儿" in example (2) refer to something that the speaker considers worth doing. "可" expresses the positive sentiment of praise and the acknowledgement to "good things". Therefore, "可" can express the speaker's sentiment and attitude related to the attitude of approval or disapproval.

2.5 Suggestion

"可" also has the meaning of recommendation and suggestion. It has similar meaning to "可以 (can, may)" and "应当 (should)". Yet, when the mood is low, the sentiment and attitude only reflect a kind of constructive opinion of the speaker. The speaker does not guarantee that the proposed method can fully meet the conditions required to solve the problem, meaning it may not be the only way to solve the problem. Examples follow.

(1) 极端天气可在家办公。
One may work at home in extreme weather.
(2) 这项任务可由小李接替。
This task can be taken over by Xiao Li.
(3) 春节期间工作可放一放，敞开了玩儿几天。
During the Spring Festival, work can be set aside, and one can relax for a few days.

The above examples reflect suggestions of the speaker which are based on certain conditions, rather than mandatory requirements imposed by the speaker. In addition, "可" also means that the advice provided by the speaker is merely one of the ways or means to solve the problem. Other approaches to the problem are not excluded. Whether the listener accepts such advice and puts it into practice is not strongly required by the speaker. Therefore, what is expressed is a positive sentiment that is related to the types of sentiment, such as a suggestion, request, hope, etc.

2.6 Desire

"可" can also be used to reflect the speaker's strong wishes. It is commonly used in exclamatory sentences. It can be replaced with "终于", "可算", "总算", all meaning "finally". Examples follow.

(1) 这回可称心如意了。
You finally get it as you wish this time.
(2) 出差这么久，你可回来了！
After a long business trip, you are finally back!

Such sentences usually express positive sentiments, such as joy and approval, when the expectation of the speaker is achieved. They have high sentimental intensity and a positive sentimental polarity. However, it may also be used to express negative sentiments as shown in the following examples.

(3) 现在可真不妙啊！
The current situation is not very good!
(4) 你这回可摊上大麻烦了！
This time you are in big trouble!

In example (3), the use of "可" highlights the negative assessment of the current situation the speaker involves. In particular, it emphasizes the fact that the current situation is opposite to the desire of the speaker. In example (4), the use of "可" reflects the worries of the speaker, expressing a negative sentiment.

From the analysis of this section, we can observe that "可" may express sentiment when used in sentences expressing the meaning of "permission", "can", "can be used to", "worth", "agreement", and "desire". Within the framework of sentiment analysis, identifying "可" in a phrase allows recognition of implicit sentiments in the sentences.

3 Patterns of Phrases Containing "可"

Phrases containing "可" have some particular patterns. from the patterns, one can recognize what sentimental sense is expressed by the whole sentence, and, further on, classify the underlying sentiments based on analysis of the context. next, we elaborate on two common patterns.

3.1 Nothing to Do with ("没什么可 + (动词) + 的"[Nothing KE + Verb + DE])

In the phrase of "没什么可 + (动词) + 的", the word "可" contributes the subjective opinion to the phrase it appears whereby "可" conveys a sentiment of "no need" or "unnecessary", negating the meaning of the phrase. This is in contrast to the phrase "没什么 + (动词) + 的", which retains a neutral statement. Examples follow.

(1) (a) 该说的我都说了，现在没什么说的了。
I finished what I have to say. There is nothing left to talk more.
(b) 该说的我都说了，现在没什么可说的了。
I finished what I have to say. There is nothing worth saying.

(2) (a) 我负责的部分已经完成，没什么做的了。
I finished my part of the work. There is nothing left to do.
(b) 我负责的部分已经完成，没什么可做的了。
I finished my part of the work. There is nothing worth doing.

In example (1), the sentiment of the speaker is neutral in sentence (a) as it merely states a fact, while in sentence (b) the speaker expresses an opinion that all that is important has been said already and nothing else is worth mentioning. Therefore it conveys a negative sentiment. Similarly, in example (2), sentence (a) just states the fact that all the work has been finished and there is no remaining job to do, while sentence (b) emphasizes the completion of the work, such that there is nothing else worth doing, expressing a feeling of exhaustion.

From the analysis of the above examples, we can note that while the two phrases "没什么 + (动词) + 的" and "没什么可 + (动词) + 的" have similar structures and express the similar meanings, the latter, because of the use of "可", strengthens the sentimental meaning. Therefore, we can recognize that "可" can function to introduce and modulate sentiments. Sentiment is conveyed through the linguistic structure, rather than sentimental words.

As a linguistic structure, "没什么可 + (动词) + 的" can be used in two types of structure. In the first type, it functions as a short phrase to complement a long sentence and produces an additional explanation or a judgment. In this case, the short phrase containing "可" can be integrated into a long sentence. In the second type, phrases containing "可" expresses sentiments on its own. This case can be considered as a condensation of the first usage. In both cases, the phrase of "没什么可 + (动词) + 的" expresses a negative or declining feeling of the speaker as the subjects of the talk. Be it a person, a thing, or a behavior, all are deemed as not worthy, unnecessary, or insufficient. Examples follow.

(1) (a) 这个博物馆，没什么可参观的。
There is *really* nothing to see in this museum.
(b) 这个博物馆没什么可参观的。
There is nothing to see in this museum.
(2) (a) 他写的书，没什么可看的。
There is *indeed* nothing to read in his book.
(b) 他写的书没什么可看的。
There is nothing to read in his book.

In both (a) sentences of the above examples, the comma separates a long sentence into two parts. The separate phrases containing "可" convey a strong negative evaluation for the museum and the book. Implicitly, the phrases indicate criticism in both sentences. In the (b) sentences of the two examples, the evaluation, though negative, is of less personal preference. Comparing the (a) sentences versus the (b) sentences, we can observe that, when the phrase containing "可" is presented as a separate short sentence, it becomes the focus of the statement, i.e., the emphasis is more on the feeling of the speaker, instead of the subjects of the sentences, i.e. the museum and the book; while in the long sentence expression as in (b), the focus is on the subjects, i.e., the museum and the book. Distinguishing the two types of structures is of benefit to accurate sentiment analysis because different strengths of sentiments are involved. In the first type of structure, a comma separates the phrase containing "可" from the first part of the sentence. The first part of the sentence describes the subject while the phrase containing "可" is the evaluation of the subject. In the second type of structure, that is, one long sentence containing "可", the effect of "可" is on the whole sentence.

3.2 Nothing Worth of ("没什么可 + (情感词) + 的" [Nothing KE + sentimental word + de])

When "可" is used in phrases of "没什么可 + (情感词) + 的" where "可" is followed by a sentimental word, the phrases imbue a meaning opposite to the sentimental word. In this kind of phrases, the effect of "可" is to negate the polarity of the sentimental word following it. For example, if the sentimental word means "happy", then the whole phrase has the meaning of "nothing worth to be happy about". Examples follow.

(1) 这件事我没什么可欣慰的。
About this thing, there is nothing that I feel happy.
(2) 这样的结果没什么可高兴的。
This result has nothing to feel glad about.
(3) 考不上那所大学也没什么可遗憾的。
There is nothing to feel pity about that one did not get into that college.

In example (1), the phrase of "可" expresses the disappointment of the speaker by emphasizing the negation of feeling happy about. In example (2), similarly, though the phrase contains the word "glad", it expresses the speaker's disappointment at the result. Example (3) illustrates the reversion of a negative sentiment. Though usually it is a pity that one did not get into college, example (3) emphasizes speaker's notion that it is not a pity that if one did not get into the particular college. In all these examples, the use of "没什么可 + (情感词) + 的" highlights the negation of the sentiment, effectively giving a stronger sentiment to the sentences. All these examples demonstrate the power of negation of the phrases containing "可". When it comes to the identification of sentiment, if one relies on the polarity of sentimental words only, one will mistakenly identify positive sentiments for examples (1) and (2). To identify sentiments accurately, one should consider not only to identify the sentimental words but also consider the linguistic pattern of "没什么可 + (情感词) + 的" whereas "没什么" plays the role of negating the sentimental word. When sentimental words are involved in "没什么可 + (情感词) + 的", the polarity of the whole sentence is changed to the opposite polarity. Therefore, we can consider "没什么可 + (情感词) + 的" as a marker for sentiment. Only when we also take the linguistic pattern into consideration can we identify sentimental sense of the text correctly. The advantage of identifying such phrases is that the phrases are highly distinguishable in computerized analysis of text. The task is then to accurately parse the phrase to determine the correct sentiment.

3.3 May A or B ("可A可B" [KE + A + KE + B])

The pattern "may A or (may) B" is a fixed expression. This pattern is primarily used to express "synonymous", "antonymous", "quasi-synonymous" and "related synonymous" relationships between item A and item B. The constituent components can be verbs, pronouns, locational words, and adjectives, etc. Examples include: "可来可往 (may come or go)", "可走可留 (may go or stay)", "可你可我 (maybe you or me)", "可好可坏 (may be good or bad)", "可南可北 (may be south or north)", "可长可短 (may be long or short)", "可多可少 (may be more or less)" and so on.

This pattern may contain paired opposition or unpaired opposition. Opposition in a strict sense is the result of a logical operation. This kind of opposition is understood as paired opposition. Opposition in a broad sense involve two objects which do not predict each other but are incidentally paired. This kind of opposition is understood as unpaired opposition. Detailed illustration is given in the following subsections.

3.3.1 Paired Opposition

Paired opposition is opposition in a strict sense that, if described in relation between A and B, has the following characteristics: (1) A and B can foretell each other; (2) B can be inferred from A and vice versa; and (3) the semantic relationship between A and B is clear. For example, in the phrase "可高可矮 (may be tall or short)", the words "tall" and "short" are in a typical antonymous relationship, and paired opposition occurs in such phrases.

The combination form of "may A or B" has the characteristic of paired opposition. When non-sentimental words appear in paired opposition in the structure "may A or B", they express sentiments. The sentiments are neither positive nor negative, but neutral, meaning one can choose either A or B. Examples follow.

(1) 工钱可多可少。
Wage can be either high or low.
(2) 房子可大可小，重要的是交通便利。
A house can be either big or small. What is important is that traffic is convenient.
(3) 现在自行车可有可无，街上随时可以租赁到。
Nowadays it does not matter whether one owns a bicycle or not since rental bikes are everywhere on streets.

In the above examples, though word A and B are antonyms, they do not constitute opposite meanings in the speaker's mind. In other words, the antonymy between A and B can actually be cancelled out, while the "may A or B" pattern emphasizes the fact that both A and B are acceptable. In example (1), the "may A or B" phrase indicates that the speaker accepts either a high wage or a low wage. In example (2), it emphasizes that the speaker does not consider the size of the house important. In example (3), it expresses the opinion of the speaker that it is not necessary for one to own a bicycle. These examples demonstrate the attitude of the speaker that both A and B are acceptable choices, and the speaker does not prefer one to the other. Hence, such usage of "可" conveys a neutral sentiment and a low strength.

3.3.2 Unpaired Opposition

Unpaired opposition is opposition in a broad sense. If using the relationship between A and B to describe opposition in a broad sense, it has the following characteristics: (1) A and B may not be able to foretell each other; (2) A and B are incidental pairs, and A and B cannot infer each other; and (3) semantically, A and B have similar meanings and the relationship between them is fuzzy. The pattern "may A or B" has the characteristic of unpaired opposition. For example, in the phrase "可悲可叹 (lamentable and pitiful)", the words involved "lamentable" and "pitiful" appear as a unpaired opposition.

The requirement for "A" and "B" in unpaired opposition appearing in the pattern "may A or B" is that they have similar rather than opposite meanings; and therefore, speakers have broad choices of words to be used in the phrase. For instance, replacing "lamentable and pitiful with "pitiful and compassionate" is allowed. The words "lamentable" and "pitiful" in the structure convey a sentiment evolving from lamentable to pitiful, representing a negative sentiment. When expressing sentiment, "may A or B" pattern allows one to understand the speaker is expressing the meanings of both A and B. Examples follow.

(1) 当个铁面无私，可丁可卯的检查员。
You should try to become a qualified inspector. Be strict and fair.
(2) 到今日这番田地，真是可悲可叹。
Today's situation is indeed lamentable and pitiful.
(3) 两名运动员在赛场上的表现可圈可点。
The performance of the two athletes on the field is remarkable and noticeable.

In example (1), both "strict" and "fair" are not sentimental words. However, when used in the "may A or B" pattern, it represents an attitude towards responsibility, thus giving the sentence a positive sentiment. In example (2), "lamentable" and "pitiful" are both negative words, expressing a sentiment of regret by the speaker. Example (3) is a relatively special because the two Chinese words are not sentimental. In fact, when used separately, the two words mean to note or highlight noticeable texts in books. Here, as they are used in the "may A or B", they contribute positive meaning to the sentence.

3.3.3 Strengthening and Weakening Sentiment

The "may A or B" pattern can either strengthen sentiments or weaken them, depending on the interaction of the words A and B. When A and B are synonyms, the phrase "may A or B" actually conveys an emphasis in "may A and B", in which A and B re-enforce each other in expressing sentiment. Examples follow.

(1) 考上了清华，真是可喜可贺！
It is worth feeling happy and be congratulated for being admitted to the Tsinghua University.

(2) 英雄的事迹可歌可泣。
Heroic deed is laudable and memorable.

(3) 这样的老师是可亲可敬的。
Such teachers are amiable and respectable.

In example (1), because Tsinghua University is a well-recognized university, the speaker feels that being admitted to it is worth both congratulations and best wishes. Therefore, "may A or B" effectively combines the two good feelings to make the sentiment more intense than it would be if only "may A" or "may B" is used. Similar observation can be made about example (2). In example (3), the very positive evaluation about amiable and respectable teachers is strengthened by the "may A or B" expression, which is stronger than the expression in which either "amiable" or "respectable" is used to convey the positive sentiment.

The "may A or B" phrase may also reduce the strength of expressing sentiment, especially when A and B are of opposite polarity. This kind of usage usually weaken the intensity of the sentiment in the phrase. Examples follow.

(1) (a) 你想找个什么样的女朋友？
What kind of girlfriend do you want to find?
(b) 可美可丑。
Either pretty or average.

(2) (a) 你喜欢读什么样的书啊？
What kind of books do you like to read?
(b) 可阳春白雪可下里巴人。
Can be in either eloquent and cliché language or layman's language.

In example (1), "pretty" and "average" are the opposite descriptions about a person. The use of "may A or B" reduces the emphasis on prettiness and can be considered as to weaken the sentiment of the whole sentence. In example (2), eloquent and cliché language and layman's language are opposite in styles. By acknowledging that either style is fine with the speaker, the answer gives rise to ambiguity in sentiment, effectively wearing down the strength in the sentence. This can happen when one tries to avoid giving a definitive answer. Thus, "may A or B" in these cases have the effect of reducing the strength of sentiments or making the sentimental polarity ambiguous.

To summarize, in this section, we have elaborated on two patterns of phrases involving "可" from the perspective of sentiment analysis. The reason that these patterns can express sentiment is due to the linguistic patterns in using the word "可" and has little or nothing to do with whether sentimental words appear in the sentence. This points to the acknowledgment that, in Chinese, expressing sentiment does not rely only on sentimental words.

4 Sentiment Modulation of "可"

"可" has the ability of modulating sentiment. It has three modes of modulating a sentence, i.e. assigning, weakening, and strengthening sentiments, depending on whether sentimental words appear with "可" or not. When "可" is used without sentimental words, it can assign sentiments to a sentence. When it appears with sentimental words, it can either strengthen or weaken the sentimental sense in a sentence.

4.1 Assigning

When "可" and sentiment words are not present at same time, "可" can assign sentiment to sentences. In particular, when "可" means "许可 (allow, permit)", "能够 (can)", "建议 (recommend, suggest)", and "用途 (be used on, be used at)", and there are no sentiment words in the sentence, "可" can assign sentiment to sentences. Take the meaning of permission as an example. When one needs a permission for an act or event, e.g., a permission to put up a poster, the act or event of putting up a poster is a usual and normal act or event that can happen in the world. In other words, one can understand the act or event as an unmarked information. In this case, the act or event that needs a permission is of "normal value". On the contrary, unusual and special things are a kind of marked information to one's understanding, e.g., putting up the poster upside down. In this case, the act or event is of "abnormal value". Abnormal values are more often related to the expression of subjective opinions. Therefore, "可" when conveying the meaning of permission may express speaker's "normal value" and "abnormal value". So even if there are no sentimental words in a sentence, it still expresses sentiment.

Examples follow.

(1) 玫瑰可蓝色的。
A rose may be blue.

(2) "福"字可倒贴。
The character "fortune" can be posted upside down.

(3) 有些国家可安乐死。
Euthanasia is allowed in some countries.

In example (1), though most roses we see in our daily life are red, yellow, white, or pink, it is possible that some roses are of blue color, though they are rare. Then in example (1), it means one can choose blue roses if necessary, conveying the approval or permission of the speaker. In example (2), in a normal case, the Chinese character "福" should be posted in the correct way, but, in some cases, it can be posted upside down to convey a special meaning. In this example, the character is *allowed* to appear upside down. In example (3), "可" emphasizes that euthanasia is legally allowed in some countries. In the above examples, the words "blue", "upside down" and "euthanasia" are not sentimental words, but, with the use of "可", they express subjective approval. Therefore, "可" imbues sentences with sentiments.

4.2 Weakening

"可" also has the function of weakening the sentimental intensity of a sentence. When "可" and the sentiment word co-exist in a sentence, the sentiment polarity of the sentence does not change because of "可", no matter whether the sentiment is positive or negative, but the intensity of the sentiment is weakened. This happens in sentences in which "可" is used to mean "suggesting" and "permitting" or in confirmatory questions. Examples follow.

(1) 大家有意见可畅述。
Feel free to express your opinions.
(2) 对不文明行为可斥责。
You may criticize uncivilized behavior.
(3) 他的话可言过其实?
Is his words exaggerating?

Examples (1) and (2) embody the meaning of permission expressed by "可" and contain the meaning of "non-mandatory". "Non-mandatory" means that there is a choice. So, "可" in the sentence weakens the intensity of the sentimental word "feel free" and "criticize", thus, resulting in a decrease in the sentiment intensity of the entire sentence. "言过其实 (exaggerating)" in example (3) is a negative sentiment word. The example (3) is questioning whether the statement is factual, casting a doubtful sentiment. Since "可" is used in an interrogative question, the intensity of the question itself is weakened by "可". Hence, we believe that a sentence containing no sentiment word but "可" can have the intensity of its sentiment affected. Yet this does not change the polarity of the sentiment, regardless of whether it is a declarative sentence or an interrogative sentence.

4.3 Enhancing

"可" may have the function of enhancing the sentimental intensity of sentences. The sentiment intensity of a sentence will be strengthened when its meaning is desirability or worth, especially when a sentiment word is present. It commonly happens in exclamation sentences. Examples follow.

(1) 可把你盼来了!
You are here finally!
(2) 这次脸可丢大啦!
This time it is too embarrassing!
(3) 忙了好几天了，可完工了!
We have been busy several days. It is finally done.

In example (1), "可" emphasizes the sentiment of "expectation" as the speaker's hope becomes reality. It expresses sentiments such as: hope, joy, and surprise. In example (2), the word "可" emphasizes the feeling of embarrassment. In example (3), though the sentence has no sentimental word, the use of "可" expresses a strong sentiment towards the completion of the work, emphasizing the desire for early completion.

In terms of desirability and worth, the commonality of "可" is the subjective recognition of someone or something by the speaker, and a deeper evaluation based on the recognition. Givón stated that modality refers to the speaker's attitude towards the proposition, including epistemic attitudes and valuative attitudes [20]. Cognitive attitude includes possibility, certainty, justification, etc. and evaluative attitude refers to ability, obligation, manipulation, intention, and desirability, etc. But "可" has functions such as expressing desirability and possibility, so it is not only an expression of attitude, but also an expression of sentiment.

5 Summary

In this work, we have systematically analyzed the use of "可" for conveying and modulating sentiments in texts. We conclude that, because of its versatility, on the one hand, "可" may exert strong or subtle adjustment to the underlying sentiment of sentences that involve sentimental words; and on the other hand, "可" has the ability of conveying sentiments on its own. Therefore, from the perspective of sentiment analysis, in addition to identifying sentimental words in a corpus, it is necessary to screen texts for the appearance of "可" as it may affect the strength and polarity of emotion given by the sentimental words. In sentences without sentimental words, the appearance of "可" requires correct identification of the starting and end points of the "可" phrases, such as "可喜可贺", which is a strong positive sentiment, or "没什么可惜的", which refers to something not worthy or pitiable. In the latter case, we can see that it is critical to parse the phrase because, if we only extract "可惜的", which means "pity", then the polarity of the sentiment is inverted, leading to incorrect analysis.

It should be pointed out that not all the cases of "可" are related to sentiment or sentiment modulation as some intact phrases like "可行性"(feasibility), "可爱"(lovely), "可笑" (laughable) should not be parsed into individual characters, but rather considered as a whole word or phrase. Therefore, it is important to design computational algorithms to first exclude cases of "可" that are not sentiment-related, like "可行性" (feasibility), and then to identify cases that should not be further parsed, like "可爱" (lovely), and also to utilize the patterns and modes described in this paper to recognize the underlying sentiment.

As a modal word, "可" manifests versatility in expressing sentiment, similar to the modal verbs CAN, COULD, MAY, MIGHT, SHALL, SHOULD, WILL, WOULD, and MUST in English [21]. Contrary to the modal verbs in English, which have different forms for strong epistemic or deontic modality, such as MUST and SHALL, and for weak epistemic or deontic modality, such as MAY and SHOULD, "可" does not vary in form and its strength in sentiment expression solely depends on the linguistic pattern in which it appears, i.e. whether and how sentimental words are present in the phrases containing "可". This adds unique characteristics to sentiment analysis involving "可"

and requires judicious modeling of "可" for accurate understanding of the underlying sentiments.

The contribution of this work is to elucidate the roles and functions of "可", a widely used modal word, in sentiment expression and modulation. To the best of our knowledge, this is the first work focusing on "可" for its effect on text-based sentiment analysis. We have identified different linguistic patterns that "可" is embodied in sentiment expression. The work fills a gap in our existing knowledge about comprehensive modeling of sentiment which should not be limited to words in sentimental dictionaries. The impact of the work is that it provides linguistic insight into sentiment expression conveyed by non-sentimental words. Based on these findings, one can conceive of computational techniques not only to extract sentimental words but also to consider additional linguistic rules for accurate assessment of sentiments. Due the targeted scope of this paper, we have not proposed any particular machine learning or deep learning techniques for identifying and classifying sentiments enabled by "可". In our future work, we plan to develop computational models based on the understanding of "可" usage for more robust and accurate sentimental analysis.

Acknowledgments. The work of T. Naren was supported by Beijing Municipal Fellowship Award 2018PC03.

References

1. Taboada, M., Brooke, J., Tofiloski, M., Voll, K., Stede, M.: Lexicon-based methods for sentiment analysis. Comput. Linguist. **37**(2), 267–307 (2011)
2. Cambria, E.: Affective computing and sentiment analysis. IEEE Intell. Syst. **31**(2), 102–107 (2016)
3. Zhang, S., Wei, Z., Wang, Y., Liao, T.: Sentiment analysis of Chinese micro-blog text based on extended sentiment dictionary. Future Gener. Comput. Syst. **81**, 395–403 (2018)
4. Abdi, A., Shamsuddin, S.M., Hasan, S., Piran, J.: Deep learning-based sentiment classification of evaluative text based on multi-feature fusion. Inf. Process. Manage. **56**(4), 1245–1259 (2019)
5. Tsai, A.C.-R., Wu, C.-E., Tsai, R.T.-H., Hsu, J.Y.-J.: Building a concept-level sentiment dictionary based on commonsense knowledge. IEEE Intell. Syst. **28**(2), 22–30 (2013)
6. Xu, G., Yu, Z., Yao, H., Li, F., Meng, Y., Wu, X.: Chinese text sentiment analysis based on extended sentiment dictionary. IEEE Access **7**, 43749–43762 (2019)
7. Peng, H., Cambria, E., Hussain, A.: A review of sentiment analysis research in Chinese language. Cogn. Comput. **9**(4), 423–435 (2017)
8. Kaur, H., Mangat, V., Krail, N.: Dictionary based sentiment analysis of hinglish text. Int. J. Adv. Res. Comput. Sci. **8**(5) (2017)
9. Miyakawa, S., Saitoh, F., Ishizu, S.: A quality table-based method for sentiment expression word identification in Japanese. In: International Conference of Design, User Experience, and Usability. Springer (2017). https://doi.org/10.1007/978-3-319-58637-3_4
10. Park, S.-M., Na, C.-W., Choi, M.-S., Lee, D.-H., On, B.-W.: KNU Korean sentiment lexicon: Bi-LSTM-based method for building a Korean sentiment lexicon. J. Intell. Inform. Syst. **24**(4), 219–240 (2018)
11. Moreno-Sandoval, L.G., et al.: CSL: a combined Spanish Lexicon-resource for polarity classification and sentiment analysis. In: International Conference on Enterprise Information Systems. SCITEPRESS (2017)

12. Rosa, R.L., Schwartz, G.M., Ruggiero, W.V., Rodríguez, D.Z.: A knowledge-based recommendation system that includes sentiment analysis and deep learning. IEEE Trans. Industr. Inf. **15**(4), 2124–2135 (2018)
13. Chen, L.-C., Lee, C.-M., Chen, M.-Y.: Exploration of social media for sentiment analysis using deep learning. Soft. Comput. **24**(11), 8187–8197 (2019). https://doi.org/10.1007/s00500-019-04402-8
14. Yang, P., Chen, Y.: A survey on sentiment analysis by using machine learning methods. In: 2017 IEEE 2nd Information Technology, Networking, Electronic and Automation Control Conference (ITNEC). IEEE (2017)
15. Song, C., Wang, X.-K., Cheng, P.-F., Wang, J.-Q., Li, L.: SACPC: a framework based on probabilistic linguistic terms for short text sentiment analysis. Knowl.-Based Syst. 105572 (2020)
16. Poria, S., Cambria, E., Gelbukh, A., Bisio, F., Hussain, A.: Sentiment data flow analysis by means of dynamic linguistic patterns. IEEE Comput. Intell. Mag. **10**(4), 26–36 (2015)
17. Ding, X., Liu, B.: The utility of linguistic rules in opinion mining. In: Proceedings of the 30th annual international ACM SIGIR conference on Research and development in information retrieval (2007)
18. Li, F., Han, C., Huang, M., Zhu, X., Xia, Y., Zhang, S., Yu, H.: Structure-aware review mining and summarization. In Proceedings of the 23rd International Conference on Computational Linguistics (Coling 2010) (2010)
19. Wang, Z.: Expressional functions of modal verb in sentence. J. Liaoning Inst. Technol. **1** (2002)
20. Givón, T.: Functionalism and grammar. John Benjamins Publishing, Amsterdam (1995)
21. Gisborne, N.: Dynamic modality. SKASE J. Theor. Linguist. **4**(2), 44–61 (2007)

Accomplishment Predicates in Mandarin and Their Lexical Semantics

Xiaoqian Zhang, Yingyi Luo(✉), and Jianhua Hu

Institute of Linguistics, Chinese Academy of Social Sciences, Beijing, China

Abstract. This paper discusses the lexical semantics of Mandarin accomplishment predicates, which are known as allowing both culminating reading and non-culminating reading. We will firstly point out the limits of previous studies on the semantics of Mandarin accomplishments. Then we will categorise accomplishments into three groups on the basis of the endpoints that they encode and provide a modal base analysis for their semantics. Next, we will propose one pragmatic constraint to which accomplishments are subject when they allow non-culminating reading. Lastly, we will argue for the necessity of distinguishing different types of endpoints that contribute to the telicity of accomplishment predicates.

Keywords: Accomplishment predicates · Lexical semantics · Modal base analysis · Non-culminating reading and culminating reading

1 Accomplishment Predicates in Mandarin

This paper focuses on the lexical semantics of accomplishment predicates[1] in Mandarin. Accomplishment predicates are known as denoting complex events that are composed of a process and a natural endpoint ensuing from the process. As perfective aspect requires that the event time be included in the topic time, as shown in (1), generally, when an accomplishment predicate occurs in the context of perfective aspect, the whole event (i.e., process + natural endpoint) should complete within the temporal interval introduced by the topic time.

(1) Perfective aspect:
$\lambda P \lambda t_T \exists e[\tau(e) \subseteq t_T \ \& \ P(e)]$[2]

Take (2) for instance. The temporal adverbial *zuotian* 'yesterday' introduces the topic time. When the perfective marker *-le* co-occurs with the accomplishment predicate *write one letter*, the event of *writing one letter* reached its endpoint within the temporal interval that *yesterday* refers to, which induces a culminating reading.

---

[1] There is consensus in the literature that the temporal properties of a situation type are determined by the combination of verbs, NP complements, PPs, etc. Therefore, the accomplishments in this paper refer to accomplishment predicates or verb phrases rather than verbs *per se*.

[2] We adopt the Neo-Davidsonian framework of event semantics and assume that verbs and their projections denote sets of eventualities (states and events). Verbs are associated with their arguments by thematic roles like initiator, theme and so on. τ is temporal trace function ([16]), which applies to eventualities and yields their run time. t_T is variable over topic time.

M. Liu et al. (Eds.): CLSW 2020, LNAI 12278, pp. 303–319, 2021.
https://doi.org/10.1007/978-3-030-81197-6_26

(2) 他昨天写了一封信，然后寄出去了。
Tā zuótiān xiě-le yì fēng xìn, ránhòu jìchūqù-le.
PRO3SG[3] yesterday write-PFV one CLF letter then send-PFV
'He wrote one letter yesterday and sent it.'

In addition to the culminating reading, it has long been noted in the literature that Mandarin accomplishment predicates also allow non-culminating reading in the perfective context[4]. Specifically, *write-PFV one letter* can be continued by a clause suggesting that the event of *writing one letter* was not completed, as illustrated in (3).

(3) 他昨天写了一封信，没写完。
Tā zuótiān xiě-le yì fēng xìn, méi xiě-wán.
PRO3SG yesterday write-PFV one CLF letter NEG write-finish
'He wrote one letter yesterday and he didn't finish it.'

Since the classical analyses of the semantics of accomplishments and perfective markers cannot account for the sources of the non-culminating reading of accomplishment predicates[5], this cross-linguistic phenomenon arouses great interest among scholars and there are mainly three hypotheses in the literature.

1.1 Lexical Aspect Hypothesis

Some linguists accounted for the non-culminating reading of accomplishments in terms of lexical aspect, c.f. [7, 13, 29–32] and [36], etc. [31] claims that the accomplishment predicates in Mandarin are not equivalent to those in English. They do not imply the culmination of endpoint. That is why they can express non-culminating reading in the perfective context.

The lexical representations of Mandarin and English accomplishment predicates are provided respectively in (4)–(5).

(4) Lexical representation of Mandarin accomplishment predicates:
[x ACT]

(5) Lexical representation of English accomplishment predicates:
[[x ACT] CAUSE [BECOME [y <STATE>]]]

[3] Abbreviations used throughout in the glosses: 3=third person; CLF=classifier; EXP=experiential; NEG=negation; PFV=perfective; PRO=pronoun; PTCL=particle; SG=singular.

[4] Most linguists seem to agree that accomplishments in Mandarin can express a non-culminating reading. However, native speakers do not agree unanimously on it. In this paper, we only follow the intuition of the native speakers who accepted the non-culminating reading of accomplishments. We will discuss in another paper the reasons concerning the native speakers' variation in their judgement.

[5] We only deal with accomplishment predicates that comprise verbs of simple form but not resultative verbal compounds.

As shown by the lexical representation in (4), Mandarin accomplishment predicates describe a process, which suggests that they are more like activity predicates. On the contrary, English accomplishments not only comprise a process but also a natural endpoint that marks a change of state.

Although the lexical aspect account provides an explanation for the non-culminating reading of Mandarin accomplishment predicates, it is problematic under two aspects. On the one hand, the lexical representation in (4) can only express a process reading but fails to imply an intentionality meaning to reach a certain endpoint, which is required by the semantics of Mandarin accomplishment predicates. Take (6) for instance.

(6) 李四昨天差点儿吃了一个苹果。
Lǐsì zuótiān chàdiǎn'er chī-le yí gè píngguǒ.
Lisi yesterday almost eat-PFV one CLF apple
'Lisi almost ate one apple yesterday.'

In the scenario described by (6), the event of *eating one apple* only began but was not completed. Despite the non-culminating reading of this sentence, it implies that Lisi had the intention to reach the endpoint of *eating one apple* provided by the quantized object *one apple*, and *chadian'er* 'almost' targets this endpoint. The lexical representation in (4) cannot capture such an intentionality meaning.

In fact, (4) can only capture the process reading triggered by typical activity predicates, e.g. (7). The object *apple(s)* is non-quantized and thus provides no endpoint that *chadian'er* 'almost' can target. It can only target the initial point of the event of *eating apple(s)*. Sentence (7) means that Lisi did not even begin to eat apple(s).

(7) 李四昨天差点儿吃了苹果。
Lǐsì zuótiān chàdiǎn'er chī-le píngguǒ.
Lisi yesterday almost eat-PFV apple
'Lisi almost ate apple(s) yesterday.'

In a word, the comparison between (6) and (7) suggests that it would be an oversimplification to treat Mandarin accomplishment predicates like *chi yi ge pingguo* 'eat one apple' as typical activity predicates like *chi pingguo* 'eat apple(s)'.

On the other hand, we find that it is not always the case that all accomplishment predicates in Mandarin can express a non-culminating reading. For instance, example (8) indicates that the natural endpoint of the accomplishment predicate *xie yi ge zi* 'write one character' cannot be cancelled, and only the culminating reading is available.

(8) 他昨天写了一个字，#没写完。
Tā zuótiān xiě-le yí gè zì, #méi xiě-wán.
PRO3SG yesterday write-PFV one CLF character NEG write-finish
'He wrote one character yesterday, #and he didn't finish it.'

If Mandarin accomplishment predicates denote a process, they should always give rise to a non-culminating reading. The infelicity of example (8) poses a challenge to the lexical aspect account.

1.2 Grammatical Aspect Hypothesis

[8, 19, 20, 27, 28, 34] and [22] adopted an approach of grammatical aspect. Unlike the hypothesis of lexical aspect, the grammatical aspect account assumes that Mandarin accomplishment predicates are like their English counterparts and also encode a natural endpoint. The non-culminating reading arises from the presence of the perfective marker *-le*. More precisely, [27] and [28] propose that verbal *-le* can express either termination or completion. When it expresses termination, a non-culminating reading is triggered; when it expresses completion, a culminating reading is available.

Furthermore, [1, 14, 22] and [23] propose that there exist typological differences concerning the semantics of perfective operators in different languages. The perfective operators in Russian, English and French are "strong" operators, while those in Mandarin, Hindi and Thai are "weak" operators. The Mandarin perfective operator (as a weak operator) is defined in (9).

(9) $[[\mathrm{PFV_M}]] =$
$\lambda P \exists e[\tau(e) \subseteq t_T \ \& \ \mathrm{MAX}(e, P)]$ ([23])

As with the classical definition of perfective operator in (1), the weak perfective operator also requires that the topic time include the event time. But unlike the classical definition, the weak operator does not require event completion, but only requires event maximality, as defined in (10).

(10) MAX(e, P) if e is a part of a possible P-event and it is not a proper part of any actual event that is part of a possible P-event. ([22], cited from [1])

The definition shows that the maximal event e is either a complete P-event or a P-event that ceases to develop further towards a P-event in the actual world. When the maximal event is a complete P-event, it expresses a culminating reading; when the maximal event is a P-event that ceases to develop, it expresses a non-culminating reading.

The approach of grammatical aspect also has limits. Firstly, we find that the non-culminating reading is also available when verbal *-le* is absent, cf. (11) which does not comprise verbal *-le*, but experiential *-guo*.

(11) 他写过一封信，没写完。
Tā xiě-guo yì fēng xìn, méi xiě-wán.
PRO3SG write-EXP one CLF letter NEG write-finish
'He had the experience of writing one letter, and he didn't finish it.'

Experiential *-guo* is also a perfective marker in Mandarin. As with verbal *-le*, it can trigger both a terminative reading and a completive reading. If we adopt the proposal of grammatical aspect, we have to admit that experiential *-guo* is also ambiguous between a terminative reading and a completive reading. However, this solution is rather *ad hoc* for lack of other independent evidence.

Secondly, we will argue that verbal *-le* is not a maximal operator, contra [22]'s proposal. Examine sentence (12).

(12) 李四昨天吃了一包薯片，没吃光，现在还在吃。
Lǐsì zuótiān chī-le yì bāo shǔpiàn, méi chī-guāng,
Lisi yesterday eat-PFV one bag potato chips NEG eat-complete
xiànzài hái zài chī.
now still PROG eat
'Lisi ate one bag of potato chips yesterday, and he didn't finish them. He is still eating them.'

The example suggests that the event of *eating one bag of potato chips* can continue after having terminated. This is not predicted by the definition of maximal operator requiring that the event of *eating one bag of potato chips* should not develop further after its termination.

1.3 Conversational Implicature Hypothesis

Building on the data from languages such as St'át'mcets, Skwxwu7mesh, French and German, [2, 21] and [24] assume that accomplishment predicates are vague as to whether their endpoints are reached or not and the culminating reading arises from conversational implicature.

This hypothesis raises two questions in explaining the Mandarin data. Firstly, according to [6], if the accomplishment predicate is vague as to the attainment of the endpoint, it is possible to add a continuation clause emphasizing that the endpoint is reached. By contrast, if the accomplishment predicate entails the attainment of the endpoint, the continuation clause will cause redundancy.

Example (13) suggests that the presence of the continuation clause specifying the attainment of the endpoint is redundant in English.

(13) ?He opened the window. So the window was opened ([6]).

Likewise, example (14) is also very odd to many native speakers of Mandarin Chinese. The implausible continuation indicates that the telicity of the accomplishment predicate in Mandarin is not due to the conversational implicature.

(14) 李四吃了一个苹果，？吃光了。
Lǐsì chī-le yí gè píngguǒ, ?chī-guāng-le.
Lisi eat-PFV one CLF apple eat-complete-PFV
'Lisi ate one apple, ?and he finished it.'

Secondly, according to [18], in the VP-ellipsis construction, there is some formal identity between the antecedent and the elided VP. Consider (15).

(15) 李四吃了一个苹果，张三也是。
Lǐsì chī-le yí gè píngguǒ, Zhāngsān yě shì .
Lisi eat-PFV one CLF apple Zhangsan also be
Interpretation 1: 'Lisi finished eating one apple; Zhangsan finished eating one apple.'
Interpretation 2: 'Lisi ate one apple and didn't finish it; Zhangsan ate one apple and didn't finish it.'
#Interpretation 3: 'Lisi finished eating one apple; Zhangsan ate one apple and didn't finish it.'
#Interpretation 4: 'Lisi ate one apple and didn't finish it; Zhangsan finished eating one apple.'

Suppose that whether the endpoint was reached or not is vague in the meaning of *chi yi ge pingguo* 'eat one apple', this sentence may have four possible combinations of interpretations, i.e., in both situations described by the two clauses, the apple can be either completely eaten or partially eaten. However, only two readings rather than four readings are available for sentence (15), i.e., in both situations, the apple must be either completely eaten or partially eaten. The interpretations of the VP-ellipsis construction challenge the proposal, according to which the meaning of accomplishments is vague as to the attainment of an endpoint.

1.4 Organisation of the Paper

We have pointed out the limits of the previous proposals in explaining the non-culminating reading of Mandarin accomplishment predicates. We will then propose an alternative account, namely a modal base analysis, for the lexical representation of Mandarin accomplishments. The rest of the paper is organised as follows. In Sect. 2, we will firstly classify the accomplishment predicates into three groups according to the number of endpoints that an accomplishment predicate encodes and then propose that the natural endpoint of accomplishments is in the scope of a modal operator, which warrants that the natural endpoint is not necessarily reached in the actual world, and thus makes the non-culminating interpretation available. Section 3 presents one constraint to which accomplishments must be subject when allowing non-culminating reading. In Sect. 4, on the basis of two pieces of empirical evidence, we will argue for the necessity of distinguishing different sources of endpoints.

2 A Modal Base Account

2.1 Three Sources of Natural Endpoint

Sentences (16a)–(16c) all comprise an accomplishment predicate and the natural endpoint of each accomplishment can be cancelled in the presence of a negation clause.

(16) a. 他吃了一个苹果，没吃光。
Tā chī-le yí gè píngguǒ, méi chī-guāng.
PRO3SG eat-PFV one CLF apple NEG eat-complete
'He ate one apple. He didn't finish it.'

b. 他杀了一只猪，没杀死。
Tā shā-le yì zhī zhū, méi shā-sǐ.
PRO3SG kill-PFV one CLF pig NEG kill-die
'He killed one pig. The pig was not dead.'

c. 他做了一个实验，没做完。
Tā zuò-le yí gè shíyàn, méi zuò-wán.
PRO3SG conduct-PFV one CLF experiment NEG conduct-finish
'He conducted one experiment. He didn't finish conducting the experiment.'

Note that the secondary predicates of the resultative verbal compounds in the negation clauses of (16a)–(16c) are not the same, which are *guang* 'complete', *si* 'die' and *wan* 'finish' respectively. We assume that the use of the different lexical words reveals the difference in terms of the nature of endpoints that they target. Specifically, *guang* 'complete' of (16a) targets the incremental theme endpoint[6]; *si* 'die' of (16b) targets the property endpoint; *wan* 'finish' of (16c) targets the script endpoint. This subsection is devoted to the different sources of endpoint that contribute to the culminating reading of accomplishment predicates.

Firstly, examine *eat one apple*. The verb *chi* 'eat' is an incremental verb and the object *yi ge pingguo* 'one apple' is thus an incremental theme that provides an endpoint. It is extensively discussed in the literature that there is homomorphism between the event described by the incremental verb and the "abstract" scale that measures the change undergone by the incremental theme, cf. [3, 4, 15–17, 26, 33, 35] and [25]. More precisely, as the event of *eating* progresses, more and more parts of the apple are consumed. When the last part of the apple is eaten, the event of *eating one apple* is completed. The default reading of *eat one apple* in the context of perfective aspect is a culminating reading. The continuation clause appears to cancel the incremental theme endpoint that *guang* 'complete' targets.

Secondly, the endpoint associated with the predicate *sha yi zhi zhu* 'kill one pig' of (16b) is a property endpoint. Specifically, the object *one pig* is not an incremental theme and does not provide an incremental theme endpoint. It is the pig's property that supplies a natural endpoint: when the pig enters into the state of *being dead*, the event of *killing one pig* reaches its culmination point. The negation clause suggests that the secondary predicate *si* 'die' can be used to target the property endpoint of *kill one pig*.

Thirdly, consider the predicate *zuo yi ge shiyan* 'conduct one experiment' of (16c). What the secondary predicate *wan* 'finish' targets is neither an incremental theme endpoint nor a property endpoint, but a "script" endpoint. Specifically, the event structure of *conduct one experiment* is based on an "action script", which is "prototypical courses

[6] We can also replace *guang* by *wan* to target the incremental theme endpoint.

of actions performed by the agent" [12, p. 43][7]. We assume that the completion of the last action composing the event of *conducting one experiment* marks a script endpoint. The event of *conducting one experiment* unfolds as more and more actions are realised: when the last action is realised, the event in question culminated.

2.2 Categorisation of Accomplishments in Mandarin

We have shown that there exist three different sources of endpoint that contribute to the culminating reading of accomplishments. They are incremental theme endpoint, property endpoint and script endpoint. Notice that it is not always the case that one accomplishment predicate only encodes one source of endpoint. It is often the case that one predicate encodes more than one source of endpoint. This subsection classifies the accomplishment predicates on the basis of the number of sources of endpoint that one accomplishment predicate incorporates.

As suggested in Table 1, accomplishment predicates can be categorised into three groups based on the number of endpoints that they encode[8]. The first group comprises predicates that have one source of endpoint, e.g., *eat one apple* (incremental theme endpoint). The second group comprises predicates that have two sources of endpoint like *repair one radio* (property endpoint + script endpoint)[9]. The third group comprises predicates that have three sources of endpoint such as *make a tableful of dumplings* (incremental theme endpoint + property endpoint + script endpoint)[10]. We will discuss their semantics in detail in Subsect. 2.4.

2.3 The Initial Point of Accomplishments Cannot Be Cancelled

Although Mandarin accomplishments allow the cancellation of the natural endpoint in the perfective context (16a–16c), we find that the subtraction of the initial point is impossible, e.g. (17a)–(17c).

[7] Rothstein holds a similar viewpoint and proposes that an accomplishment is composed of stages that have "a natural and inherent order" determined by the real-world knowledge [26, p. 107].

[8] Although all accomplishments could have a culminating reading derived from the culmination of the script endpoint, given that the event structure of all accomplishments can be mapped onto an action script, we find that this culminating reading is not always an informative one. Take *eat one apple* for instance. The script of *eat one apple* is composed of a number of sub-events of *taking one bite of one apple*. The completion of the last sub-event marks the completion of the event of *eating one apple*. However, this culminating reading is not an informative one. The most salient reading is associated with the culmination of the incremental theme endpoint. Therefore, we will exclude the non-informative culminating reading from our classification of accomplishment predicates.

[9] The most salient endpoint associated with *repair one radio* is the property endpoint. Moreover, as the event of *repairing one radio* is composed of some prototypical operations such as *testing the components that do not work*, *tightening the bolts* and so on, *repair one radio* also incorporates a script endpoint, corresponding to the completion of the last stage of *repairing one radio*.

[10] As more and more dumplings are made with the unfolding of the event of *making dumplings*, the predicate *make a tableful of dumplings* encodes an incremental theme endpoint. The same predicate is also associated with a property endpoint concerning the cookedness of the dumplings. Moreover, the completion of the last stage of *turning off the fire* also provides a script endpoint.

Table 1. Categorisation of accomplishments according to number of sources of endpoint

One source	icremental theme 吃一个苹果 eat one apple	property 劝李四 persuade Lisi	script 做体检 have a medical check-up
Two sources	incremental theme + property 擦一张桌子 wipe one table	incremental theme + script 盖一幢房子 build one house	property + script 修一个收音机 repair one radio
Three sources	incremental theme + property + script 做一桌饺子 make a tableful of dumplings		

(17) a. 他吃了一个苹果，没吃光。#因为没开始吃。
Tā chī-le yí gè píngguǒ, méi chī-guāng. #Yīnwéi méi kaīshǐ chī.
PRO3SG eat-PFV one CLF apple NEG eat-complete because NEG begin eat
'He ate one apple. He didn't finish it. #Because he didn't begin.'

b. 他杀了一只猪，没杀死。#因为没开始杀。
Tā shā-le yì zhī zhū, méi shā-sǐ. #Yīnwéi méi kaīshǐ shā.
PRO3SG kill-PFV one CLF pig NEG kill-die because NEG begin kill
'He tried to kill one pig. The pig was not killed. #Because he didn't begin.'

c. 他做了一个实验，没做完。#因为没开始做。
Tā zuò-le yí gè shíyàn, méi zuò-wán. #Yīnwéi méi kaīshǐ zuò.
PRO3SG conduct-PFV one CLF experiment NEG conduct-finish because NEG begin conduct
'He conducted one experiment. He didn't finish conducting the experiment. #Because he didn't begin.'

The first sentences in examples (17a)–(17c) are all perfective sentences comprising accomplishments. They cannot be continued by the sentences specifying that the initial sub-events of *eating one apple*, *killing one pig* and *conducting one experiment* did not take place. This suggests that in the perfective context, the initial point of accomplishments cannot be cancelled, unlike the endpoint. Our formalised analysis in Subsect. 2.4 must take into consideration the difference between the initial point and the endpoint in terms of the possibility of their cancellability.

2.4 Formal Analysis

We have proposed that Mandarin accomplishment predicates in the perfective context can express both a culminating reading and a non-culminating reading. No matter which reading is involved, it is always true that the event initiator has the intention that the accomplishment culminates, see [9] and [6]. To capture the intentionality meaning, we propose to introduce a modal base in the lexical representation of Mandarin accomplishment predicates. Following [6], we assume that this modal base contains the set of possible worlds compatible with the initiator x's intentions, i.e., x's intention set I_X. According to [11], "an individual x intends P iff P is true in all worlds in x's intention set I_X". Therefore, if P is an accomplishment in the intention set of x, then it is true that the event initiator x has an intention that P culminates. A universal modal operator $\Box$ is introduced to quantify over the modal base I_X.

Now let us focus on how the culminating reading and the non-culminating reading of Mandarin accomplishment predicates are derived. Following [4], we propose that an accomplishment, as a durative event, is composed of at least three sub-events, i.e., an initial sub-event, an intermediary sub-event and a final sub-event. We assume that the different sub-events may occur in different possible worlds. We have argued in Subsect. 2.3 that the initial sub-event of an accomplishment must occur in the actual world. Therefore, it must be placed outside the scope of the modal base I_X. On the contrary, both the final sub-event of the accomplishment and the result state ensuing from it are placed within the scope of the modal base I_X to capture the fact that the event initiator intends that the accomplishment culminates. When the world of evaluation is not contained in the modal base, i.e., the endpoint did not culminate in the actual world, we obtain a non-culminating reading. By contrast, when the world of evaluation is contained in the modal base, we obtain a culminating reading.

We have proposed in Subsect. 2.2 that accomplishment predicates can be classified into three groups according to the number of sources of endpoint that they encode. We select one accomplishment predicate from each of the three groups and provide their lexical semantics in (18)–(20).

(18) [Lisi eat one apple] $\rightarrow$
$\lambda e\ \exists e_i\ \exists e{\ominus}e_i{\ominus}e_f[\text{eat}(e{\ominus}e_i{\ominus}e_f) \wedge \text{theme}(e{\ominus}e_i{\ominus}e_f)=\text{one apple} \wedge \text{initiator}(e) = \text{Lisi} \wedge \text{INI}_E(e_i, e) \wedge \text{eat}(e_i) \wedge \text{theme}(e_i) = \text{one apple} \wedge \Box_{I\text{Lisi}} \exists e_f\ \exists s\ [\text{FIN}_E(e_f, e) \wedge \text{eat}(e_f) \wedge \text{theme}(e_f) = \text{one apple} \wedge \text{cause}(e_f, s) \wedge \text{disappear}(s) \wedge \text{theme}(s) = \text{one apple}]]$

For any e, $e' \in U_E$, $\text{INI}_E(e', e)$ is true iff e' is initial in e, i.e., there exists no e'' < e that precedes e' in e, and analogously for $\text{FIN}_E(e', e)$. ([5])

(19) [Lisi kill one pig] $\rightarrow$
$\lambda e\ \exists e_i\ \exists e{\ominus}e_i{\ominus}e_f[\text{kill}(e{\ominus}e_i{\ominus}e_f) \wedge \text{theme}(e{\ominus}e_i{\ominus}e_f)=\text{one pig} \wedge \text{initiator}(e) = \text{Lisi} \wedge \text{INI}_E(e_i, e) \wedge \text{kill}(e_i) \wedge \text{theme}(e_i) = \text{one pig} \wedge \Box_{I\text{Lisi}} \exists e_f\ \exists s\ [\text{FIN}_E(e_f, e) \wedge \text{kill}(e_f) \wedge \text{theme}(e_f) = \text{one pig} \wedge \text{cause}(e_f, s) \wedge \text{be dead}(s) \wedge \text{theme}(s) = \text{one pig}]]$

(20) [Lisi make a whole table of dumplings] $\rightarrow$
$\lambda e\ \exists e_i\ \exists e{\ominus}e_i{\ominus}e_f[\text{make}(e{\ominus}e_i{\ominus}e_f) \wedge \text{theme}(e{\ominus}e_i{\ominus}e_f)=\text{a whole table of dumplings} \wedge \text{initiator}(e) = \text{Lisi} \wedge \text{INI}_E(e_i, e) \wedge \text{make}(e_i) \wedge \text{theme}(e_i) = \text{a whole table of dumplings} \wedge \Box_{I\text{Lisi}} \exists e_f\ \exists s_1\ \exists s_2\ [\text{FIN}_E(e_f, e) \wedge \text{make}(e_f) \wedge \text{theme}(e_f) = \text{a whole table of dumplings} \wedge \text{cause}(e_f, s_1) \wedge \text{be affected}(s_1) \wedge \text{theme}(s_1)= \text{a whole table of dumplings} \wedge \text{cause}(e_f, s_2) \wedge \text{be cooked}(s_2) \wedge \text{theme}(s_2) = \text{a whole table of dumplings}]]$

Among the three accomplishment predicates, *eat one apple* incorporates an incremental theme endpoint, *kill one pig* encodes both a property endpoint and a script endpoint and *make a tableful of dumplings* encodes all three endpoints. The final sub-event and the result states ensuing from the completion of the final sub-event are within the scope of the modal operator to derive the interpretation that the event initiator intends that the accomplishment culminates but the accomplishment does not necessarily reach its endpoint in the actual world.

3 Restrictions on the Non-culminating Reading of Accomplishments

As indicated in Subsect. 1.1, it is not true that all accomplishment predicates in Mandarin allow non-culminating reading. Two analyses exist in the literature that investigate the conditions accomplishments must satisfy so as to allow non-culminating reading.

3.1 An Account of the Property of Objects

[29] propose that Mandarin accomplishments can have their telos subtracted only when (i) the verbs are creation verbs and (ii) the object of the creation verb, namely the created object must be considered as the relevant object prior to the culmination of event. Specifically, [29] assume that there are two classes of created objects, i.e. No Partial Object (NPO) and Allows Partial Object (APO). No Partial Objects are those that "cannot be considered the relevant object until the process of creation has reached its inherent end point", while Allows Partial Objects are those that "can be considered the relevant object before culmination" [29, p. 202].

Consider (21)–(22), which contain the creation verb *xie* 'write'. According to [29], the object *yi feng xin* 'one letter' is an Allows Partial Object and thus the telos of the predicate *write one letter* can be cancelled. On the contrary, the object *yi ge zi* 'one character' belongs to the class of NPOs. It follows that the predicate *write one character* cannot have its telos subtracted.

(21) 他昨天写了一封信，没写完。
Tā zuótiān xiě-le yì fēng xìn, méi xiě-wán.
PRO3SG yesterday write-PFV one CLF letter NEG write-finish
'He wrote one letter yesterday, and he didn't finish it.' (=3)

(22) 他昨天写了一个字，#没写完。
Tā zuótiān xiě-le yí gè zì, #méi xiě-wán.
PRO3SG yesterday write-PFV one CLF character NEG write-finish
'He wrote one character yesterday, #and he didn't finish it.' (=8)

Although it seems that the hypothesis of the properties of objects can account for the difference between (21) and (22), there exist counter-examples that challenge this hypothesis.

Firstly, we find that the endpoint of the accomplishment can be cancelled even if the verb is not a creation verb. For instance, the verb *du* 'read' in (23) is not a verb of creation, but *du yi ben shu* 'read one book' can still give rise to a non-culminating reading.

(23) 他读了一本书，没读完。
Tā dú-le yì běn shū, méi dú-wán.
PRO3SG read-PFV one CLF book NEG read-finish
'He read one book, and he didn't finish it.'

Secondly, we notice that there exist cases where accomplishments comprising NPOs can have their telos subtracted. Consider (24). According to [29], *na ge zi* 'that character' is a No Partial Object. But the predicate *write that character* can nevertheless trigger a non-culminating reading in the scenario presented in (24), contra [29]'s prediction.

(24) 美术老师让张三在操场墙上写一个四乘四米的"静"字。张三刚刚写了那个字，可是由于字太大，没写完，他准备明天继续写。

Měishù lǎoshī ràng Zhāngsān zài cāochǎng qiáng shàng xiě yí
art teacher ask Zhangsan at playground wall on write one
gè sì chéng sì mǐ de "jìng" zì. Zhāngsān gānggāng
CLF four multiply four metre LNK jing character Zhangsan just
xiě-le nà gè zì, kěshì yóuyú zì tài dà, méi
write-PFV DEM CLF character but because character very big NEG
xiě-wán, tā zhǔnbèi míngtiān jìxù xiě.
write-finish PRO3SG plan tomorrow continue write

'The art teacher asked Zhangsan to write one character "jing" (in the size of 4 metres by 4 metres) on the wall of the playground. Zhangsan has written that character just now. But because the character was too big, he didn't finish writing it. He plans to continue tomorrow.'

One way to rescue [29]'s proposal is to assume that whether a created object can be considered as a relevant object or not is determined by the context, i.e., *yi ge zi* 'one character' is an NPO in the scenario presented in (22) but an APO in (24). However, we still need to explain how the contexts affect the interpretation of the created object.

3.2 An Account of the Scalarity of Predicates

[13] propose the Scalar Hypothesis specifying the condition under which accomplishment predicates can trigger a non-culminating reading.

The Scalar Hypothesis:

Only those stems that denote induced normative gradable changes can lead to the IE.[11] [13, p. 252].

The definition of normative gradable change is given as follows.

A predicate P describes an induced normative gradable change if and only if whenever it holds of an event e, (i) an activity A holds during an initial subinterval of the temporal trace of e ($\tau(e)$) and a normative gradable property C holds during a final subinterval of $\tau(e)$, (ii) the occurrence of A causes C to hold, and (iii) d > d' (d' is the degree at the initial subinterval of $\tau(e)$ and d the degree at the final subinterval of τ (e)). [13, pp. 252–253].

Let us re-examine the event described by the predicate *xie yi ge zi* 'write one character', cf. (22). As the object *one character* is an incremental theme, the number of strokes which are written at the final subinterval of $\tau(e)$ is greater than those at the initial subinterval of $\tau(e)$, i.e. d > d'. According to the definition of normative gradable change, we can conclude that the predicate *write one character* describes an induced normative gradable change. However, the predicate *write one character* cannot give rise to a non-culminating reading, which is not predicted by the proposal of [13].

[11] IE refers to Incompleteness Effect.

3.3 A New Account - Duration Constraint

We have pointed out that neither the property of the created object nor the scalarity of the predicate determine whether an accomplishment can trigger a non-culminating reading or not. We will argue below that the event's duration is an important factor in determining whether it is easy or not for an accomplishment predicate to give rise to a non-culminating reading. Our proposal is presented as follows.

(25) **Duration Constraint**
The longer duration an accomplishment requires to reach its natural endpoint, the easier it is for it to trigger a non-culminating reading arising from the cancellation of the same endpoint.

With this pragmatic constraint in mind, let us come back to the pair of sentences in (21)–(22). Although an event of *writing one character* is composed of at least three sub-events, i.e., a beginning, an end and a medial subevent, e.g., [4], the interval between the initial point and the endpoint is often very short under normal circumstances. Our proposal thus correctly predicts the difficulty of achieving a non-culminating reading. By contrast, in the case of the event of *writing one letter*, the interval between its initial point and endpoint is much longer than the interval over which the event of *writing one character* occurs. It follows that it is much easier for it to trigger a non-culminating interpretation arising from the cancellation of the incremental theme endpoint. Sentence (24) corroborates our hypothesis, i.e., once the interval between the initial point and the endpoint of the event of *writing one character* is extended due to the enrichment of contextual information, the non-culminating reading becomes available.

4 The Necessity of Distinguishing Different Sources of Endpoint

We have proposed in Subsect. 2.1 that there exist three sources that contribute to the culminating reading of accomplishments. Previous studies often neglected the distinction among them. As a result, when researchers studied the non-culminating reading of accomplishments, they were either not precise about which source of endpoint is involved ([13, 31]) or they only dealt with one source of endpoint ([10, 29]). In what follows, building on some empirical evidence, we will argue for the necessity of distinguishing different sources of endpoint. Specifically, we will examine examples that comprise both a perfective sentence containing an accomplishment predicate and a continuation sentence negating the culmination of one of the endpoints of the event described by the accomplishment predicate. We will show that some linguistic facts cannot be properly explained if we intermingle different sources of endpoint.

Firstly, consider (26a). The negation clause is infelicitous in this sentence. [13] assume that this is due to the fact that the predicate *cast one ballot* cannot have its natural endpoint subtracted. However, we find that the endpoint of *casting one ballot* can be cancelled in (26b), contrary to [13]'s assumption.

(26) a. 他投了一张票。#没投完。
Tā tóu-le yì zhāng piào. #Méi tóu-wán.
PRO3SG cast-PFV one CLF ballot NEG cast-finish
'He cast one ballot. #He didn't finish voting.' (adapted from [13])

b. 他投了一张票。没投上。
Tā tóu-le yì zhāng piào. Méi tóu-shàng.
PRO3SG cast-PFV one CLF ballot NEG cast-on
'He cast one ballot. He didn't succeed.'

[13] make a wrong prediction because the authors were not aware of the existence of different sources of endpoint and drew their conclusion only from the fact that the script endpoint of *cast one ballot* cannot be cancelled. Once we consider another type of endpoint, namely the property endpoint, we can obtain an opposite result, as shown in (26b). The contrast between (26a) and (26b) suggests that we need to consider the existence of different types of endpoint when studying the non-culminating reading of accomplishments. The cancellability of endpoints may vary depending on their types: it seems that a propernty endpoint can be cancelled more easily than a script endpoint.

Secondly, compare (27a) with (27b).

(27) a. 李四写了一封信。没写完，只写了一半信。
Lǐsì xiě-le yì fēng xìn. Méi xiě-wán, zhǐ xiě-le yí bàn xìn.
Lisi write-PFV one CLF letter NEG write-finish only write-PFV one half letter
'Lisi wrote one letter. He didn't finish and only wrote half of it.'

b. 李四烤了一只鸡。#没烤完，只烤了一半鸡。
Lǐsì kǎo-le yì zhī jī. #Méi kǎo-wán, zhǐ kǎo-le yí bàn jī.
Lisi bake-PFV one CLF chicken NEG bake-finish only bake-PFV one half chicken
'Lisi baked one chicken. #He didn't finish and baked only half of it.'

The natural endpoint of *xie yi feng xin* 'write one letter' can be cancelled by the continuation sentence asserting that half of the letter was not finished, but the same strategy does not apply to the predicate *kao yi zhi ji* 'bake one chicken'. The difference between (27a) and (27b) can only be accounted for if we seriously take into consideration the possibility of the existence of different types of endpoints. Specifically, although *write one letter* and *bake one chicken* are both accomplishments, only *write one letter* incorporates an incremental theme endpoint, but not *bake one chicken*. Since the continuation sentence "only + verb-*le* + one half + object" can only apply to a scenario where an incremental theme endpoint is targeted, it makes sense that only *write one letter* is compatible with the continuation sentence. On the contrary, we can use another continuation sentence that specifically targets the script endpoint to negate the endpoint of *bake one chicken*, cf. (28).

(28) 李四烤了一只鸡。没烤完，鸡刚放到烤箱里就停电了。
Lǐsì kǎo-le yì zhī jī. Méi kǎo-wán. Jī gāng fàngdào kǎoxiāng
Lisi bake-PFV one CLF chicken NEG bake-finish chicken just put oven
lǐ jiù tíng diàn le.
in once cut electricity PTCL
'Lisi baked one chicken. He didn't finish baking. The electricity was cut off as soon as the chicken was put into the oven.'

In a word, the use of different lexical words targeting the endpoints and the way an endpoint can be subtracted vary according to the types of endpoints that are involved, which justifies the necessity of distinguishing different sources of endpoint.

5 Concluding Remarks

This paper discusses the semantics of Mandarin accomplishment predicates that allow both a non-culminating reading and a culminating reading. After reviewing the problems raised by the previous studies accounting for the sources of non-culminating reading of accomplishments, we have presented a different analysis, which is the modal base approach. We assume that Mandarin accomplishments incorporate a modal base containing the set of possible worlds compatible with the initiator's intentions. The final sub-event and the result state ensuing from it are placed in the scope of the modal operator to guarantee that the accomplishment does not necessarily culminate in the actual world, so that a non-culminating reading can arise. We hope that our work will contribute to typological studies on the semantics of accomplishments across different languages. An experimental study concerning the differences and similarities among different types of accomplishments in different languages will also be one of our main research focuses in the near future.

Acknowledgments. This research was supported by the Research Project of Institute of Linguistics, CASS (Grant No. YB21–08), the CASS Innovative Research Project "Psycholinguistic Studies on Language Acquisition and Development" and the National Social Science Fund for Young Scholars of China (Grant No. 18CYY024).

References

1. Altshuler, D.: A typology of partitive aspectual operators. Nat. Lang. Linguist. Theor. **32**(3), 735–775 (2014). https://doi.org/10.1007/s11049-014-9232-1
2. Bar-el, L., Davis, H., Matthewson, L.: On non-culminating accomplishments. Proc. NELS **35**, 79–92 (2005)
3. Beavers, D.: On affectedness. Nat. Lang. Linguist. Theor. **29**, 335–370 (2011)
4. Beavers, J.: Scalar complexity and the structure of events. In: Dölling, J., Heyde-Zybatow, T., Schäfer, M. (eds.) Event Structures in Linguistic Form and Interpretation, vol. 5, pp. 245–265. Walter de Gruyter, Berlin (2007)
5. Beavers, J.: Lexical aspect and multiple incremental themes. In: Demonte, V., McNally, L. (eds.) Telicity, Change, and State. A Cross-Categorial View of Event Structure, pp. 23–59. Oxford University Press, Oxford (2012)
6. Beavers, J., Lee, J.: Intentionality, scalar change, and non-culmination in Korean caused change-of-state predicates. Linguistics (2020)
7. Chu, C.C.: Some semantic aspects of action verbs. Lingua **40**, 43–54 (1976)
8. Demirdache, H., Liu, J., Martin, F., Sun, H.: Licensing non-culminating accomplishments in Mandarin. Experimental & theoretical evidence. Paper presented at Workshop on Non-culminating, Irresultative and Atelic Readings of Telic Predicates (TELIC 2017), Universität Stuggart (2017)

9. Guéron, J., Vogeleer, S.: Non-culminating accomplishments: Subject, speaker and syntactic structure. Draft (2019)
10. van Hout, A., et al.: Agent control and the acquisition of event culmination in Basque, Dutch, English, Spanish, and Mandarin. In: LaMendola, M., Scott, J. (eds.) Proceedings of the 41st annual Boston University Conference on Language Development. Somerville, MA, pp. 323–332. Cascadilla Press (2017)
11. Inman, M.V.: Semantics and Pragmatics of Colloquial Sinhala Involitive Verbs. Ph.D. thesis, Stanford University (1993)
12. Kearns, K.: Telic senses of deadjectival verbs. Lingua **117**, 26–66 (2007)
13. Koenig, J.P., Chief, L.C.: Scalarity and state-changes in Mandarin (and other languages). In: Bonami, O., Cabredo Hofherr, P. (eds.) Empirical Issues in Syntax and Semantics, vol. 7, pp. 241–262 (2008)
14. Koenig, J.P., Muansuwan, N.: How to end without ever finishing: Thai semi-perfectivity. J. Semant. **17**, 147–182 (2000)
15. Krifka, M.: Thematic relations as links between nominal reference and temporal constitution. In: Sag, I., Szabolcsi, A. (eds.) Lexical Matters, pp. 29–53. CSLI Publications, Chicago University Press, Chicago (1992)
16. Krifka, M.: Nominal reference, temporal constitution and quantification in event semantics. In: Bartsch, R., van Benthem, J.F.A.K., van Emde Boas, P. (eds.) Semantics and Contextual Expression, vol. 11, pp. 75–115. Foris Publications, Dordrecht (1989)
17. Krifka, M.: The origins of telicity. In: Rothstein, S. (ed.) Events and Grammar, vol. 70, pp. 197–235. Kluwer Academic Publishers, Dordrecht (1998)
18. Lakoff, G.: Irregularity in Syntax. Holt, Rinehart, and Winston, New York, NY (1970)
19. Lin, J.W.: Time in a language without tense: The case of Chinese. J. Semant. **23**, 1–53 (2006)
20. Liu, X.: Xiandai Hanyu Yanjiu, chap. Xiandai hanyu ciwei "le" de yufa yiyi [The grammar of suffix "-le" in Modern Chinese]. pp. 1–20. Language and Culture University Press, Beijing (1998)
21. Martin, F.: Explaining the link between agentivity and non-culminating causation. Proc. SALT **25**, 246–266 (2015)
22. Martin, F.: Non-culminating accomplishments. Manuscript under review (2018)
23. Martin, F., Gyarmathy, Z.: A finer-grained typology of perfective operators. In: Piñon, C. (ed.) Empirical Issues in Syntax and Semantics, vol. 12, pp. 187–216 (2019)
24. Martin, F., Schäfer, F.: Sublexical modality in defeasible causative verbs. In: Arregui, A., Rivero, M.L., Salanova, A. (eds.) Modality across Syntactic Categories. Oxford University Press, pp. 87–102 (2017)
25. Ramchand, G.C.: Verb Meaning and the Lexicon: A First Phase Syntax. Cambridge University Press, Cambridge (2008)
26. Rothstein, S.: Structuring Events: A Study in the Semantics of Lexical Aspect. Blackwell Publishing, Oxford (2004)
27. Smith, C.S.: The Parameter of Aspect. Kluwer Academic Publishers, Dordrecht (1991)
28. Smith, C.S.: Aspectual viewpoint and situation type in Mandarin Chinese. J. East Asian Linguis. **3**, 107–146 (1994)
29. Soh, H.L., Kuo, J.Y.C.: Perfective aspect and accomplishment situations in Mandarin Chinese. In: Verkuyl, H.J., De Swart, H., Van Hout, A. (eds.) Perspectives on Aspect, vol. 32, pp. 199–216. Springer, Netherlands (2005)
30. Sybesma, R.: Why Chinese verb-*le* is a resultative predicate. J. East Asian Linguis. **6**, 215–261 (1997)
31. Tai, J.H.Y.: Verbs and times in Chinese: Vendler's four categories. In: Testen, D., Mishra, V., Drogo, J. (eds.) Papers from the Parasession on Lexical Semantics, Chicago Linguistic Society, vol. 20, pp. 289–296 (1984)

32. Tai, J.H.Y., Chou, J.Y.: On the equivalent of 'kill' in Mandarin Chinese. J. Chin. Lang. Teach. Assoc. **10**, 48–52 (1975)
33. Wechsler, S.: Resultatives under the 'event-argument homomorphism' model of telicity. In: Erteschik-Shir, N., Rapoport, T. (eds.) The Syntax of Aspect: Deriving Thematic and Aspectual Interpretation, pp. 255–273. Oxford University Press (2005)
34. Wu, J.S.: The semantics of the perfective *le* and its context-dependency: An SDRT approach. J. East Asian Linguis. **14**, 299–336 (2005)
35. Wyngaerd, G.V.: Measuring events. Language **77**, 61–90 (2001)
36. Zhang, A.: On non-culminating accomplishments in Mandarin. Ph.D. thesis, The University of Chicago (2018)

A Survey on Chinese as Second Language Learners' Acquisition of the Structural Auxiliary Word "de" (的)

Haifeng Yang[1(✉)], Junhua Chen[2], and Zhuoda Yang[1(✉)]

[1] Department of Chinese Linguistics, Wuhan University of Technology, Wuhan, Hubei Province, People's Republic of China

[2] School of Foreign Languages, Wuhan University of Technology, Wuhan, Hubei Province, People's Republic of China

Abstract. It is comparatively complicated for foreign students to grasp the usage of "de" (的), as a structural auxiliary word. With the second language acquisition theory, interlanguage theory, and related theories of error analysis, as well as natural linguistic data collected from South African students' Chinese composition and homework and the data from HSK dynamic composition corpus, this study takes 83 international students from National education school in Wuhan University of Technology as our research subjects. In the study, we have designed some questionnaires, aiming to investigate and analyze Chinese as Second Language learner's acquisition of the structural auxiliary word "de" (的), and to try to understand the mechanism and motivation from a micro perspective. It is found that (1) the average error rate of the use of "de" (的) for various types of attributions is 48.5%, with different error rates as to specific type of attributives; (2) the ability of the learners to distinguish between "de" (的) and "de" (地) in the object position is lower than that in the predicate position; and (3) the learners can relatively easily distinguish between "de" (的) and "de" (得), compared to their performance in differentiating "de" (的) from "de" (地). Finally, from the perspective of L2 learners, this study provides some suggestions are proposed as to the teaching of structural auxiliary word "de" (的), including textbook compilation, exercise design, and classroom teaching, in the hope of offering some help and enlightenment in teaching Chinese as a second language and in Chinese ontology.

Keywords: Structural auxiliary word "de" (的) · Second language acquisition · Comparison research · Error analysis

1 Introduction

Among function words in modern Chinese, the word "de" (的) is the most frequently used and thoroughly studied (Shi Dingxu 2008). The in-depth studies on modern Chinese grammar focusing on "de" (的) began with the paper "On 'de (的)'" by Zhu Dexi (1961). He adopted a markedly different method for analysis by dividing "de" (的) into three

M. Liu et al. (Eds.): CLSW 2020, LNAI 12278, pp. 320–332, 2021.
https://doi.org/10.1007/978-3-030-81197-6_27

types: the first type is used to attach to an adverbial unit, the second type to adjectives, and the third type to the nominals. This view aroused a great controversy, with some scholars supporting it while others opposing to it (Lv Shuxiang 1962; Lu Jianming 1965; Ji Yongxin 1965; Shi Yuzhi 2000; Lu Bingfu 2003; Shi Dingxu 2008).

This paper takes the line of analysis that "de" (的) is an attributive marker with syntactic, pragmatic and prosodic functions. The differences in structural auxiliary words between Chinese and other languages lie not only in syntax, but also in semantics and pragmatics (Liu Danqing 2008).

In linguistics, there is a clear functional division of the structural auxiliary words "de" (的), "de" (地) and "de" (得), as an attributive marker, an adverbial marker, and a complement marker respectively (Zhang Bin and Yisheng 2002). However, when it comes to the actual use of these markers, speakers may not clearly distinguish among these three structural auxiliary words in their utterances, especially the use of "de" (的) and "de" (地). It is, thus, not difficult to see the interchangeable use of these three structural auxiliary words undoubtedly increases the difficulty for L2 learners to acquire the structural auxiliary words.

In the field of teaching Chinese as a second language, the errors of using structural auxiliary word "de" (的) made by L2 learners have been analyzed and studied (Tong Huijun 1988; Yang Qinghui 1996). But these analyses and research, from the pedagogical perspective, have only pointed out the structural and semantic errors by syntactic and semantic comparison between error sentences and grammatical sentences, while the actual acquisition and the competency of distinguishing "de" (的) from other structural auxiliary words are into consideration.

On such basis of previous precious studies (Lu Shuxiang 1982; Peking University Chinese Department 1996; Liu Yuehua 1983; Zhang Guoxian 1994; Cui Xiliang 2001), we aim to conduct an investigation and review of the L2 learners' acquisition of structural auxiliary word "de" (的).

The current common practice of investigating the acquisition of a grammar rule by second language learners is to observe the errors from second language learners. The error data are mainly obtained by recording learners' daily talk or speech and by collecting second language learners' compositions and homework. This method focuses on observing the second language learners' acquisition of the second language in an unconscious state. However, due to the application of avoidance strategy in the learning process of second language learners, learners tend to use the language in a relatively "safer" manner by avoiding breaking grammatical rules. In such a way, the learners may not show their competency in using the language and thus we may not be able to examine their actual acquisition. So in order to observe their actual acquisition of a target grammatical rule, this study has deliberately placed the learners in a conscious environment, by designing some well-targeted questionnaires for them. In some cases, the auxiliary "de" (的) is optional. In this way, the situation is more complicated. In order to avoid the learners applying the avoidance strategy, so as to better understand their acquisition of the structural auxiliary words, this paper adopts the Second Language Acquisition (SLA) theory, interlanguage theory and related error analysis theory. We obtain the natural data from South African students' compositions and homework and the data from HSK Dynamic Composition Corpus in Beijing Language University. 83

L2 learners from National teaching school of Wuhan University of Technology are our research subjects. We will conduct questionnaires to survey, observe and analyze the L2 learners' acquisition of the structural auxiliary word "de" (的). Finally from the perspective of learners, we put forward some pedagogical suggestions of the teaching of structural auxiliary words in the field of teaching Chinese as a second language, including textbook compilation, exercise design, and classroom teaching, in the hope of giving some enlightenment and help for the teaching of Chinese as a second language and Chinese ontology research.

2 L2 Learners' Acquisition of "de" (的) with Various Types of Attributives

In this article, a questionnaire is designed to survey the second language learners' acquisition of the structural auxiliary word "de" (的). The questionnaire mainly involves the following two themes of structural auxiliary word "de": (1) The acquisition situation of "de" (的) with various types of attributives; (2) The ability of the learners to distinguish the structural auxiliary words "de" (的) and "de" (地), and "de" (得).

2.1 Basic Information of the L2 Learners

The research subjects of this study are 83 international students of the National Education School of Wuhan University of Technology. The were divided into two groups in this study according to their Chinese proficiency level, namely (i) elementary and intermediate level learners and (ii) advanced level learners. Learners in the first group account for 65% of the total number of participants. The duration of their learning of Chinese varies from two months to six months. They are at Level 3 of HSK, with a vocabulary of 1000 Chinese words. Learners in this group are mainly exposed to some simple usages of "de" with various types of attributives, such as personal pronouns as attributives. Learners in the second group account for 35% of the total number of participants. Their duration of Chinese learning is six months to one year. They are at HSK level 4 or 5 and obtain a vocabulary of 2000–2500 words. Learners at such levels have already learned some relatively complex and rare grammatical knowledge of "de", such as the use of "de" with localizers as an attributive, and the optional omission of "de" (的) when it is connected to unnecessarily related components.

Due to the discrepancies between the two groups of students, e.g., level of proficiency, learning time and accumulation of Chinese culture, we have designed two different questionnaires for them. On the one hand, it is believed that the questionnaires are convenient for subjects to answer questions better. On the other hand, it is hoped to use the questionnaires to improve the reliability and validity of this present study. Moreover, we have consulted the data from HSK Dynamic Composition Corpus in Beijing Language University, making this research more reliable and comprehensive.

2.2 The Design of the Questionnaire

We have designed a questionnaire by referring to the following resources: i) some exercises on structural auxiliary words in the learning materials including "Modern Chinese Lecture Notes on Chinese Functional Words", "Nine hundred Cases of Sentence Analysis of Chinese Sentences" and "Chinese Course", ii) the studies of Chinese ontology, iii) the corpus of errors in compositions and homework of South African students collected in previous teaching practice, iv) and the HSK Dynamic Composition Corpus in Beijing Language University. The questionnaire contains four question types including blank filling, error correction, multiple choice, and ordering, with a total of 50 questions based on the principles of comprehensive involvement, emphasis highlighted, and diverse question types.

The first section of the questionnaire is filling in the blanks. The purpose of this part is mainly to investigate the learners' ability in discriminating the structural auxiliary words "de" (的) and "de" (地), as well as their ability to identify when to use or to omit "de" (的), "de" (地) and "de" (得). Totally, there are 15 questions in this section. The first 5 items involve the structural auxiliary word "de" (地), the second 5 items the structural auxiliary word "de" (的), and the last 5 the structural auxiliary word "de" (得).

The second section of the questionnaire consists of true or false questions and error correction tasks. There are 10 items adapted from the natural error corpus of South African students' compositions and homework. They are all typical errors made by L2 learners when using structural auxiliary words in Chinese writings. Error correction in this part can reduce the avoidance and guessing of grammatical rules by the learners, which can be more conducive to testing the true acquisition process of the leaners and can help our study to draw objective conclusions.

The third section consists of multiple-choice questions. There are 15 items among which 5 are related to the structural auxiliary words "de" (的) and the rest are related to the structural auxiliary words "de" (地) and "de" (得). In these 10 items, we have selected seven types of attributives to design the questionnaire. They are nouns or noun phrases, personal pronouns, monosyllabic adjectives, bisyllable adjectives, repetitive adjective, verbs or verb phrases, and localizers.

The fourth section consists of sorting or sequencing questions. This section mainly investigates the learners' acquisition of "de" (的) and "de" (地) in two contexts, including sentences with multiple attributives and adverbials, and sentences with mixed attributive adverbials. It also investigates the learners' acquisition of the syntactic position of "de" (的) and "de" (地).

2.3 Statistics and Analysis of Results Obtained from the Questionnaires

From the statistical results shown in Table 1, we can see that the average error rate of using "de" (的) with various types of attributives is 48.5%. When disyllabic adjectives, verbs or verb phrases and localizers are used as attributives, the use of "de" (的) is not skillful or well understood, and the error rate is more than 50%, higher than the average error rate. When monosyllabic adjectives, repetitive adjective, nouns or noun phrases, and personal pronouns as attributives, the command and usage of "de" (的) are relatively

Table 1. The error rate of using "de" (的) in seven types of attributives

Types of attributives	Error rate
Monosyllabic adjectives as attributives	35.4%
Disyllabic adjectives as attributives	56.1%
Repetitive adjective as attributives	39.7%
Nouns or noun phrases as attributives	30.2%
Personal pronouns as attributives	27.6%
Verbs or verb phrases as attributives	57.7%
Localizers as attributives	64.3%
Average error rate	48.5%

better, and the error rate is below 40%, lower than the average error rate. We reckon that the reason is mainly restricted by the following two factors.

First, it is influenced by the frequency of attributive types. One function of personal pronouns is to act as an attributive. As personal pronouns are the first set of attributives L2 learners learnt at the earlier stage, the learners are more familiar with the personal pronouns. Thus, among the uses of "de" with different types of attributives, the uses of "de" (的) behind the pronouns are the most familiar use for the learners. The error rate is the lowest, only 27.6%. On the contrary, the learners are not familiar with attributives like disyllabic adjectives, verbs or verb phrases, and localizers. It is a challenge for the learners to decide whether to add "de" (的) behind these attributives. Since they have relatively poor grasp of the knowledge, the error rate is relatively high.

Second, it is by the clarity of the grammar rules also play a role to explain the error rate mentioned above. Adults' acquisition of the grammar in a second language of is different from children's acquisition. As adults' acquisition focus more on the acquisition of grammar rules, so the clarity of grammar rules directly affects adults' acquisition of the rules. As far as nominal attributives are concerned, the frequency of monosyllabic nouns as attributives is lower than that of disyllabic nouns as attributives. Thus, normally L2 learners should have a poorer grasp of "de" (的) when using monosyllabic nouns as attributives. Yet, the result is the opposite. The reason is that the grammatical rules for the use of "de" (的) after monosyllabic nouns as attributives are clear, while the grammatical rules for the use of "de" (的) after disyllabic nouns as attributives are relatively vague. Sometimes disyllabic nominal attributives can modify monosyllabic nouns. To ensure the harmony of syllables, "de" (的) can be omitted. When no ambiguity occurs in the sentence, "de" (的) will not be used so as to avoid overuse of "de" (的) in a sentence (Huang Borong, Liao Xudong, 2011). For example, "Honghu shui, lang da lang. (Honghu water, waves hitting waves)" and "Yingmian chui guolai de haifeng. (The sea breeze blowing towards us)", where "de" (的) is not added in the phrase of "Honghu water" and the phrase "sea breeze" to form "Honghu de water" (lit.: Honghu's water) and "sea de breeze" (lit.: sea's breeze) respectively. It can be seen that there is a lack of clarity in rules for the attributive of disyllabic nouns, and this explains why there is a low error rate for attributives with low frequency in use, but a high error rate for attributives with high frequency in use.

3 The Discrimination of "de" (的), "de" (地) and "de" (得) by L2 Learners

The pronunciation of the three structural auxiliary words 的, 地 and得 is approximately the same. They all pronounce as unstressed "de". Yet, when it comes to the writing, they must be written into three different characters. In the past, many scholars advocated that there was no distinction among the three structural auxiliary words. Wu Zongyuan (1984) and Jiang Shenli (2008) suggested that "de" (的) took the place of "de" (地) and "de" (得). Sun Rujian (1995), Liu Xiaoyun (2003), Xu Yangchun (2004) and other scholars proposed the similarity between "de" (的) and "de" (地), because of the closeness of the words "de" (的) and "de" (地) used in attributive structures. As for the actual usage of the three words, it is found there is no distinction among the three structural auxiliary words in some newspaper articles and novels. Yet still, most people tend to make distinction among these three structural auxiliary words.

3.1 The L2 Learners' Ability to Differentiate "de" (的) from "de" (地)

3.1.1 Questionnaire Design

In terms of the usage conditions of "de" (的) and "de" (地), the former is generally used before a noun or noun phrase, while the latter is generally used before a verb or verb phrase. Thus, according to the parts of speech of the headwords, it is not so difficult for second language learners to choose whether to use "de" (的) or "de" (地). Relatively speaking, second language learners' confusion about "de" (的) and "de" (地) mainly lies in the structure pattern "adjective (adjectival phrases) + de (的)/de (地) + verb (verb phrase)". The rule is relatively vague for second language learners to follow and hence use the auxiliary word correctly. When the above structure pattern appears in the predicate position, "de" (地) should be used (see sentence (1a) and (2a)) and when it is in the object position, "de" (的) should be used (see sentence (1b) and (2b)). That is, once the verb-preposition structure appears in front of the adverbial phrase with "de" (地), the "de" (地) needs to be converted into "de" (的).

(1a) 我们 热烈 地 讨论 了 这 个 计划。
Women relie de taolun le zhe ge jihua
Women relie de taolun le zhe ge jihua
We enthusiastic DE discuss LE this CL plan
'We have discussed this plan enthusiastically.'.

(1b) 我们 对 这 个 计划 进行 了
Women dui zhe ge jihua jinxing le
We toward this CL plan carry-out LE
热烈 的 讨论。
relie de taolun
enthusiastic DE discussion
'We have made an enthusiastic discussion about the plan'

(2a) 我们 严厉 地 打击 了 这 些 票贩子。
Women yanli de daji le zhe xie piaofanzi
We strict DE crackdown LE this CL ticket-trader
'We have struck a severe blow at these illegal ticket traders.'

(2b) 我们 必须 给 票贩子 以 严厉 的 打击。
Women bixu gei piaofanzi yi yanli de daji
We must give ticket-trader PRT strict DE crackdown
We must give ticket-trader PRT strict DE crackdown
'We must strike a severe blow at these illegal ticket traders.'

Learners need to use appropriate structural auxiliary words at appropriate structures. Therefore, we have designed questionnaires surveying learners' acquisition of different *de*'s in different sentence structures. In the questionnaires, we have focused on the structure pattern "adjectives (adjective phrases) + de (的)/de (地) + verbs (verb phrases)" in both predicate and object positions.

3.1.2 Statistics and Analysis of Questionnaire Results

Table 2. Foreign students' ability to differentiate "de" (的) from "de" (地)

	Error rate
Object position	32.2%
Predicate position	22.0%
Average error rate	27.1%

From the statistics shown in Table 2, it can be seen that the average error rate is 27.1%. The average error rate for the use of *de* in an object position is 32.2%, and the average error rate for that in a predicate position is 22.0%. This shows that the learners' performance in differentiating the use of *de* in the predicate position is relatively poor. Generally speaking, in Chinese classroom teaching, when teachers teach the distinction between "de" (的) and "de" (地), they usually clearly state that "de" (的) is used before nouns while "de" (地) is used before verbs.

Students generally discriminate between "de" (的) and "de" (地) according to this rule. So when the structure "adjective (or adjectival phrase) + de + verb (or verb phrase)" is used in the object position where only "de"(的) can only be used, students' ability to distinguish between "de" (的) and "de" (地) is relatively poor. This also proves that the acquisition of grammar knowledge by adult students is more based on the acquisition of rules. The generalization and applicability of grammatical rules themselves will affect the students' grasp of grammatical features. This reminds Chinese teachers of the fact that when teaching the differences between the structural auxiliary words "de" (的) and "de" (地), teachers should try to avoid to simplify the rules and given out absolute rules. Some subtle differences should also be mentioned to the students promptly so that the students can master their usage quicker and better.

3.2 The L2 Learner's Ability to Differentiate "de" (的) from "de" (得)

In terms of the functional distinction between "de" (的) and "de" (得), "de" (的) is a marker of attributives while "de" (得) is a marker of complement. In terms of the occurrence rules, the structural auxiliary word "de" (的) appears before head words and to modify and restrict the head words; and the structural auxiliary word "de" (得) appears after the predicate for supplementary explanation. For example:

(3) 我把刚写完的一篇文章递给了爸爸。
Wo ba gang xie wan de yi pian wenzhang digei le baba
I BA just write finish DE one CL article hand LE father
'I handed an article I just finished to my father.'

(4) 飞机飞得很低。
Feiji fei de hen di
Airplane fly DE very low
'The plane flies very low.'

In order to investigate the L2 learners' ability to distinguish "de" (的) from "de" (得), we have designed the corresponding questionnaires surveying their acquisition of the use of *de* in the structures of "verb/adjective + de + noun" and "verb/adjective + de + complement".

Table 3. The L2 learners' distinction error rate of the use of "de" (的) and "de" (得)

	Error rate
Attributive position	24.3%
Complement position	20.0%
Average error rate	16.8%

From the statistics shown in Table 3, it can be concluded that the L2 learners' average error rate of distinguishing between "de" (的) and "de" (得) is 16.7%, 10% lower than the error rate of distinguishing between "de" (的) and "de" (地) shown in Table 2 (in Sect. 3.1.2). This shows that the learners' ability to distinguish between "de" (的) and "de" (得) is relatively good. The reason may be that the rule of distinguishing between "de" (的) and "de" (得) is clearer. From Table 3, we can also see that there is no great difference in the ability to use "de" (的) in the attributive position and the ability to use "de" (得) in the complement position. The reason is that no matter which sentence structure is involved, "verb/adjective (predicate)" must be there in the sentence, i.e., "predicate + de (的) + noun "or "predicate + de (得) + complement". The second language learners can evaluate and choose "de" (的) or "de" (得) to use, according to the position of the structural auxiliary words.

4 Pedagogical Suggestions for Structural Auxiliary Word "de" (的)

4.1 Textbook Compilation

As far as the compilation of Chinese textbooks for foreigners is concerned, the grammar teaching content should be detailing the structural auxiliary word "de" (的) appropriately and moderately. Textbook compilers should expose genuine contexts and example sentences to Chinese learners, enabling them to learn the usage of the structural auxiliary word "de" (的) in a natural environment and to use it flexibly and competently. At the elementary and intermediate level, students are expected to understand the basic usage of the structural auxiliary word "de" (的) through texts, dialogues, etc., and lay a good foundation. At the advanced stage, various structural auxiliary words "de" (的) can be incorporated into the text and exercises. In this way, the previous knowledge can be summarized and acquired, and the distinction among the structural auxiliary words "de" (的) and "de" (地) and "de" (得) can be examined and the corresponding teaching activities can be carried on.

The textbooks can also be compiled around various themes. At the intermediate and advanced stages, therefore, the comparisons among the structural auxiliary words "de" (的), "de (地) and de (得)" can be expounded and elaborated as a major topic. Some less common grammatical points can also be supplemented and perfected in teaching so that the foreign students' knowledge structure can be complete, their autonomous error correction ability can be greatly improved, and their learning impression can be intensified accordingly. In this regard, the teaching model of *The Explanation of 201 Grammar Cases in Teaching Chinese as a Foreign Language* by Peng Xiaochuan, etc. serves as a role model in which the grammatical features are presented with common errors made by students and explained through diagrams, text, schematic diagrams. The general grammar rules are deduced and summarized and then are intensified and examined by exercises of blank filling, multiple choice and error correction.

4.2 Exercise Design

The exercise designed in textbooks should connect grammar rules closely into related exercises. A variety of question types should be designed, such as blank filling, multiple-choice, error correction, and replacing. When designing the exercises, we must also pay attention to the background of the learners at different learning stages. For example, due to the limitation of the elementary and intermediate learners' vocabulary and insufficient knowledge on sentence structures and grammatical constituents, we should provide exercises on sentence pattern drilling and sentence completion exercises, enabling learners to master the most basic usage. For example, by presenting picture clues to leaners, teachers can ask to students to construct phrases containing *de*, e.g. "我的书" (wo de shu; my book), "热烈地讨论 (*relie de taolun;* discuss heatedly), "跑得快" (*pao de kuai*; running fast), etc. At the advanced stage, they can fill in the blanks in essays with "de" (的), "de" (地) and "de" (得). By so doing, students' acquisition of them can be comprehensive.

4.3 Classroom Teaching

4.3.1 The Comparison Method

By comparison, we mean the comparison made between Chinese and other language and the comparison made within Chinese languages. For the comparison between Chinese and foreign languages, taking English as an example, it can avoid students' negative transfer due to English as a mother tongue or interlanguage. To do so, teachers can guide students to compare and analyze the similarities and differences between grammatical features of English possessives, verbs + adverbs, and the Chinese usage of "de" (的), "de" (地) and "de" (得). Comparisons among Chinese languages are mainly recommended for the intermediate and advanced learners. Teachers can make complete comparisons among these three structural auxiliary words, and finally conduct an overall analysis and provide practices for students to consolidate the knowledge they obtain at class.

4.3.2 The Replacement Method

The effective use of the replacement teaching method can lead students to focus on the meaning and parts-of-speech of the lexical items. Hence, it can effectively help students recognize the sentence components and to choose the correct structural auxiliary word. When teachers use the replacement method, they need to ask some easy questions and guide students to answer by using complete sentences which contain the structural auxiliary words. For example, as to the sentences "他的笔" (*ta de bi; his pen*), "他们幸福地生活在一起" (*tamen xingfu de shenghuo zai yiqi*; They live happily together) and "我开心得唱起歌来" (*Wo kaixin de chang qi ge lai*; I am so happy that I sing), teachers may ask students such questions as "这是谁的笔?" (*zhe shi shui de bi*; Whose pen is this?), "他们生活在一起有什么感受?" (*tamen shenghuo zai yiqi shi shenme ganshou*; what do they feel when they live together?) and "我因为开心做了什么?" (*Wo yinwei kaixin zuo le shenme*; What did I do because I'm happy?) By asking questions, teachers can guide the students to repeat the original complete sentences. To improve the accuracy, this replacement teaching method can be carried out in writing as well.

4.3.3 The Task-Based Teaching Method

The task-based teaching method should be better applied for the intermediate and advanced students with large numbers of Chinese vocabulary. Taking the word "de" (的) as an example, teachers can divide the whole class into several groups. Some students in each group act as reporters to interview other L2 learners. The content of the interview is not limited, but the usage of "de" (的) should be reflected and presented. Reporters may ask many questions such as "你的国家是?" (Ni de guojia shi; What country are you from?), "你在哪里学的汉语?" (Ni zai nali xue de hanyu; Where did you learn Chinese?) and "你怎么来的中国?" (Ni zenme lai de Zhongguo; How did you come to China?) The students being interviewed should answer all these questions in complete sentences. Finally, each group should make a presentation in front of the class. In this task, they need to sort out and present their questions, and answers in written form to the teacher. Through task-based teaching, not only should the knowledge of related vocabulary and grammar be consolidated, but also the students' ability to communicate in Chinese is cultivated and therefore, integrative ability of listening, speaking, reading, and writing is greatly improved as well.

5 Conclusions

This paper mainly applies the second language acquisition theory, interlanguage theory, and related theories of error analysis to examine and analyze the foreign students' acquisition of structural auxiliary word "de" (的). The subjects under analysis are native speakers of English but Chinese as second language learners. Through the "HSK Dynamic Corpus" and the natural corpus of South African students' compositions and homework collected by the author, the corpus of structural auxiliary words "de" (的) is exhaustively searched, and the data is statistically analyzed. Conclusions can be drawn as follows:

(1) The average error rate made by L2 learners in using "de" (的) with various types of attributives is 48.5%. Among them, when the attributives are disyllabic adjectives, verbs or verb phrases, or localizers, the learners' error rate of use of "de" (的) is higher than the average; when the attributives are monosyllabic adjectives, repetitive adjective, nouns or noun phrases, or personal pronouns, the learners' error rate of use of "de" (的) are below average. It is mainly caused and restricted by the various types of attributives and grammatical rules.
(2) The performance of the L2 learners of distinguishing between "de" (的) and "de" (地) in the object position is poorer than the performance to distinguishing that in the predicate position. The reason is that the headwords are verbs (or verb phrases). This also proves that the grammar acquisition by adult second language learners is more based on the acquisition of rules. The generality and applicability of the rules themselves will affect students' mastery of the grammatical rules.
(3) The performance of L2 learners of distinguishing between "de" (的) and "de" (得) is better than the performance of distinguishing between "de" (的) and "de" (地). The reason is that foreign students are more influenced by the rule to use between "de" (的) and "de" (得). L2 learners can completely judge and determine whether to use "de" (的) or "de" (得) according to the position of the structural auxiliary words before or after the predicate.

When distinguishing among the structural auxiliary words "de" (的), "de" (地) and "de" (得) for the teaching Chinese as a second language, the characteristics of adult L2 learners should be taken into consideration. For grammatical rules, L2 learners should understand the causes of errors, thus it is essential to deepen their memory, and help them focus on the characteristics of the semantic domain, auxiliary meaning, sememe system and syntagmatic relationship in the target language words. It is also important to improve the pedagogy of the structure of auxiliary words by textbook compilation, exercise design, classroom teaching, etc., in the hope to achieve better teaching and learning outcomes.

Acknowledgments. This paper presents the partial outcome of the program "Research on Chinese Grammaticalization Mode" supported by "the Fundamental Research Funds for the Central

Universities WUT (2020VI055)". This paper also presents the partial outcome of the Cooperative Education Program "Online Class Construction of 'Introduction to Teaching Chinese as the Second Language', Based on Numerous Terminals" of the Chinese Education Ministry (The Program Code: 201801049043), and of the Program of OEB teaching: An Investigation of Constructing and Reforming the Class and Teaching Mode of 'Introduction to Teaching Chinese as the Second Language', based on the SPOC platform, (The Program Code: w2020049), which is a key program of WUT's research of teaching method reform. The program is supported by "The Fundamental Research Funds for the Central Universities" (WUT: A study on the semantic evolution of willing verbs from the perspective of Typology: 2021Vl033).

References

Shi Dingxu, J.: "De" and grammatical structure of "de". Contemp. Linguist. **4**, 298–307 (2008)

Zhu Dexi, J.: On "de". China Lang. **12**, 1–15 (1961)

Lv Shuxiang, J.: On "the uniformity of language units" and the like. Chin. Lang. **11**, 483–495 (1962)

Lu Jianming, J.: Issues on the Splitting and Integrating of "de" and Others. Linguistics Issues, vol. 5. Self-collected Works By Lu Jianming Zhengzhou: Henan Education Press 1993: 73–161 (1965)

Ji Yongxin, J.: On "On "de"" Chin. Lang. **5**, 363–364 (1965)

Shi Yuzhi, J.: On the unity of grammar function of "de" Teach. Chin. World **1**, 16–27

Lu Bingfu, J.: The distribution and fundamental and derivative function of "de". Teach. Chin. World, **1**, 14–29 (2003)

Liu Danqing, J.: Syntactic types of Chinese noun phrases. China Lang. **1**, 3–20 (2008)

Zhangbin, Z., Yisheng, M.: Function Words in Modern Chinese. East China Normal University Publishing House, 195–200 (2002)

Tong Huijun, M.: An Analysis of Incorrect Chinese Sentences from Foreign Learners. Beijing Language and Culture University Publishing House, 164–211 (1988)

Yang Qinghui, M.: Analysis of Difficulties in Teaching Chinese as a Foreign Language (1996)

Lv Shuxiang, M.: Eight Hundred Words of Modern Chinese. The Commercial Publishing House, 132–143 (1982)

Chinese Faculty of Peking University, M.: Example Analysis of Function Words in Modern Chinese. The Commercial Publishing House, 151–157 (1996)

Liu Yuehua, M.: Practical Modern Chinese Grammar (2001)

Zhang Guoxian, J.: An Analysis of the Syntax, Semantics and Pragmatics of "de". J. of Huaibei Norm. Univ. **3**(1994), 226–255. Beijing Normal University Publishing House (1996)

Cui Xiliang, M.: The Problem of "de"'s appearance in personal pronouns modifying nouns. Beijing Language and Culture University Publishing House, 90–101 (2001)

Li Xiaoqi, M.: Lecture Notes on Modern Chinese Function Words. Beijing University Press, Beijing (2005)

Cheng Meizhen, M.: Analysis of 900 incorrect Chines sentences. Chinese Education Press, Beijing (1997)

Yang Jizhou, M.: Chinese Coursebook. Beijing Language University Press, Beijing (2016)

Borong, H., Liao Xudong, M.: Modern Chinese. Advanced Education Publishing House, 66–67 (2011)

Wu Zongyuan, J.: Integrate structural auxiliary words "de", "di" and "de"—consult with Niu Chunsheng. J. Ningxia Univ. **3**, 72–75 (1984)

Jiang Shengli, J.: Can structural auxiliary words "de", "di" and "de" be regarded as one word? J. Huangshan Coll. **6** ,101–104 (2008)

Sun Lvjian, J.: The Splitting and integrating issues of structural auxiliary words “de” and “di”. Yangzhou Norm. Coll. **4**, 80–82 (1995)

Liu Xiaoyun, J.: Commonalities of structural auxiliary words “de” and “di”. J. Shangxi Radio Telev. Univ. **2**, 87–88 (2003)

Xu Yangchun, J.: The splitting and integrating issues of structural auxiliary words “de” and “di”. J. Nanchang Univ. **6**, 151–186 (2004)

On the Adequacy of Grammatical Description: The Case of *Ròujiámó* (肉夹馍)

Liyong Wang[1](✉), Xiaoyan Xu[2], and Ren Zhang[1]

[1] School of International Studies, Shaanxi Normal University, Xi'an, China
wang_ly@snnu.edu.cn
[2] School of Foreign Studies, Xi'an Jiaotong University, Xi'an, China
xxiaoyan@mail.xjtu.edu.cn

Abstract. One criterion of linguistic theory is descriptive adequacy, i.e., whether it can capture the speaker's intuition [1]. Traditional structuralist analyses and contemporary generative grammar largely confine themselves to structural description and analyses, which fails to explain the precise use of linguistic units. For example, one would find that for the Chinese compound word *ròujiámó*, 1) the speaker does not have to possess the structural knowledge to know how it is understood and used; and 2) people might have different perceptions towards words with supposedly identical structures, say, *ròujiámó* and *dànchǎofàn*. It is argued in this paper that Chinese compounds in general and *ròujiámó* in particular support two central claims of Cognitive Grammar, i.e., 1) linguistic units are stored in our brain in the form of networks of constructions; 2) the sanction of linguistic expressions is not the simple computation of rules but motivated by the whole network of constructions.

Keywords: Grammatical description · Adequacy · *Ròujiámó* · Network of constructions

1 Introduction

Grammatical analysis in the structuralist tradition is essentially the description of structures on various levels, while structural derivation also plays a central role in contemporary generative grammar. This structure/syntax-centered approach is also applied to lexicon, holding that the internal structure of words is subject to the same set of rules as that of sentences [2, 3]. Chinese scholars, for examples [4–8], have explored the derivational mechanism of Chinese compounds from the perspective of syntactic operations (See the review by [9] for details).

As is pointed out in [1], a desirable linguistic theory should meet three levels of adequacy, i.e., 1) observational adequacy, 2) descriptive adequacy, and 3) explanatory adequacy. By descriptive adequacy, it means the ability to precisely capture the speaker's knowledge of his mother tongue [10]. However, it is found that neither the traditional structuralist analysis nor the generative study of Chinese compounds can reflect the speaker' intuition accurately.

M. Liu et al. (Eds.): CLSW 2020, LNAI 12278, pp. 333–345, 2021.
https://doi.org/10.1007/978-3-030-81197-6_28

Take one particular compound word *ròujiámó* (肉夹馍) 'meat sandwich pancake' as an example. It is the name for the Chinese Hamburger. The correct structural representation of this word has aroused great attention among scholars and people outside the academia as well (See [11] for details). However, two very important properties remain unnoticed concerning its understanding and use. In the first place, it is found that one does not necessarily possess the structural knowledge to know how the word is understood and used. Secondly, speakers might have totally different perceptions towards linguistic units with apparently identical structural representations. A case in point is between *ròujiámó* and another compound *dànchǎofàn* (蛋炒饭) 'egg fry rice', meaning 'egg fired rice', where only the former arouses confusion in terms of structural representation.

These two properties are largely ignored by either the traditional structuralist grammatical description or the generative analysis. Yet, they are nevertheless the important parts in the speaker's linguistic knowledge.

In this paper, we will argue that *ròujiámó* in particular and Chinese compounds in general support a Cognitive Grammar [12–14] account of linguistic knowledge, i.e., 1) linguistic knowledge is stored in our brain in the form of network of constructions; 2) the sanction of linguistic expressions is not the simple computation of rules but involves the motivation of the entire network of constructions.

The remaining part of the paper is organized as follows. Section 2 first reviews the disputes centered around the structural representation of *ròujiámó*, and then, based on the two important observations, raises the theoretical issue of the paper, i.e., how to meet the criterion of descriptive adequacy in grammatical analysis. Section 3 explores the sanctioning mechanism of *ròujiámó* within the theoretical framework of Cognitive Grammar (henceforth CG). Section 4 concludes the paper.

2 A Look At Descriptive Adequacy from *Ròujiámó*

The structural analysis of *ròujiámó* generates a great interest among linguists as well as non-professionals for a long time, with some discussions and debates in the academia and on the Internet. This section first offers a brief introduction to three representative views, i.e., 1) the preposition omission account; 2) the homonymy account; and 3) the compound word account. Then, based on the two important observations concerning this word, we raise the theoretical question this paper tries to answer, i.e., how to achieve descriptive adequacy in grammatical analysis.

2.1 Disputes Concerning the Structural Representation of *Ròujiámó*

Of the three representative views of the structural analysis of *ròujiámó*, the first two folk explanations, i.e., the preposition omission account and the homonymy account, share the basic assumption that the construction of *ròujiámó* is a special grammatical phenomenon (e.g. [15]). What is special about it lies in the fact that what we hear (*ròujiámó* 'meat sandwich pancake') contradicts with what we see (meat sandwiched in pancake). The last view among the three representative ones, on the other hand, treats *ròujiámó* as having the same structural representation as other compound words like *dànchǎofàn*.

2.1.1 The Preposition Omission Account

The basic idea of the preposition omission account is that the word *ròujiámó* is dated back to ancient Chinese with its original full structure as *ròu jiá yú mó* (肉夹于馍) 'meat sandwich Prep. pancake', meaning 'meat is sandwiched in pancake', and the preposition *yú* (于) was dropped over time. The question is, however, what is the mechanism of the dropping of the preposition? According to one explanation, people in Shaanxi, China, the place where the snack *ròujiámó* is originated, are hot-tempered and therefore drop the preposition just to save time. Supporters of this view have argued that preposition omission is quite common in ancient Chinese. For instance, a netizen named Natt Li on Zhihu, a Chinese online Q&A platform, cites the following examples (https://www.zhihu.com/question/19944004?rf=21888079).

(1) a. 是儿少 (于) 秦武阳二岁

shì ér shào (yú) Qín Wǔyáng èr-suì

this child young (Prep.) Qín Wǔyáng two-year

'This child is two years younger than Qin Wuyang.'

(*The Legend of the Child Ou Ji* 童欧记传)

b. 武陵人 (以) 捕鱼为业

wǔlíng-rén (yǐ) bǔyú wéi yè

wǔlíng-people (Prep.) fishing as survival

'People in Wuling take fishing as their way of living.'

(*The Peach Blossom Spring* 桃花源记)

c. 西人长 (于) 火器而短 (于) 技击

xīrén cháng (yú) huǒqì ér duǎn (yú) jìjī

westerner good (Prep.) firearm but bad (Prep.) combat

'Westerners are good at using firearms but not so good at physical combat.'

(*The Story of Feng Wanzhen* 冯婉贞)

However, omission in (1) is possible only when the meaning of the sentence is unaffected. Take (1) as an example. No matter whether the preposition is present or not, the sentence is a comparative construction. Returning to *ròu jiá yú mó* and *ròujiámó*, we would find that while the former is a subject-predicate expression denoting a relation, the latter is a noun phrase denoting an entity. As it is pointed out by another Zhihu netizen called East Street Pancake Boss (东街饼老板), *ròujiámó* is an entity, whereas *ròu jiá yú mó* is a property description, which does not fit in with the naming habits (see the discussions at https://www.zhihu.com/question/23254050).

2.1.2 The Homonymy Account

On this second account, *mó* (馍) is homonymous with *mó* (没), meaning 'no', in Shaanxi dialect, making the supposedly correct structure *mójiáròu* (馍夹肉) 'pancake sandwich meat' sound like *mójiáròu* (没夹肉), meaning 'meat not sandwiched'. As a result, *mójiáròu* (馍夹肉) is changed to *ròujiámó*, so as not to leave customers the false impression that no meat is sandwiched in the pancake (see the discussions at https://www.zhihu.com/question/40075724). Just like the previous account, this one is not a reasonable explanation. In the first place, the snack ròujiámó[1] is typically made in the presence of the customer. It is very easy to decide whether the meat is sandwiched or not. Misunderstanding is therefore unlikely.

Secondly, from the perspective of scientific research, this account fails to capture higher-level regularities. This is because there are many expressions with structures identical to *ròujiámó* which resist the homonymy account (the same criticism also applies to the preposition omission account). According to this account, for instance, the original structure of *yóutiáo juǎnbǐng* (油条卷饼) 'doughnut stick wrap pancake', meaning 'doughnut stick wrapped in pancake', should be *bǐngjuǎn yóutiáo* (饼卷油条) 'pancake wrap doughnut stick'. Homonymy is not supposed to be involved with the latter form, making it unnecessary to adopt the former form for the sake of avoiding misunderstanding. The fact, however, is that such a form as *bǐngjuǎn yóutiáo* does not seem to exist at all.

2.1.3 Ròujiámó as an Endocentric Compound

A more reasonable analysis is to treat *ròujiámó* as an endocentric compound, with its internal structure as *jiámó* modified by *ròu*. In other words, *ròujiámó* is a type of *jiámó*, and *ròu* is a modifier, denoting the ingredients involved. Supporting this analysis are examples of various types of *jiámó* apart from *ròujiámó*, illustrated in (2).

(2) *cài jiámó* (菜夹馍) 'vegetable jiámó', *jīdàn jiámó* (鸡蛋夹馍) 'egg jiámó', *huāgān jiámó* (花干夹馍) 'dried toufu jiámó', *làzi jiámó* (辣子夹馍) 'pepper jiámó', *zīránjiámó* (孜然夹馍) 'cumin jiámó', *liángpí jiámó* (凉皮夹馍) 'cold rice noodle jiámó', *tǔdòu jiámó* (土豆夹馍) 'potato jiámó', *chòudòufǔ jiámó* (臭豆腐夹馍) 'stinky toufu jiámó'

([15:119])

Interestingly, these various types of *jiámó* share some common structural properties with different types of *chǎofàn* (炒饭) 'fried rice' illustrated in (3). To be specific, they have the same constructive pattern, i.e., 'N_1 V N_2', where N_1 can be different nouns denoting the ingredients of *jiámó* or *chǎofàn*. Besides, in the two constructions, verbs serve as the attributive of N_2 [16].

[1] Note that the italicized form indicates the name of the snack, whereas the non-italicized form indicates the snack itself.

(3) *dàn chǎofàn* (蛋炒饭) 'egg *chǎofàn*', *ròu chǎofàn* (肉炒饭) 'meat chǎofàn', *huǒtuǐ chǎofàn* (火腿炒饭) 'ham chǎofàn', *hóngcháng chǎofàn* (红肠炒饭) 'red sausage chǎofàn', *pàocài chǎofàn* (泡菜炒饭) 'pickle chǎofàn', *ròusī chǎofàn* (肉丝炒饭) 'shredded meat chǎofàn', *luóbo xiārén chǎofàn* (萝卜虾仁炒饭) 'carrot shrimp meat chǎofàn'

Moreover, just as a netizen named Mu Daochou (穆纛雔) on Zhihu pointed out, compounds represented by *ròujiámó* and compounds represented by *dànchǎofàn* constitute a dish-denoting construction with the sequence of 'material + action + material' (https://www.zhihu.com/question/23254050). Other examples include *fěnzhēng páigǔ* (粉蒸排骨) 'rice powder steam pork rib', meaning 'steamed pork ribs with rice powder', *jiàngbàoròu sūbing* (酱爆肉酥饼) 'sauce fry meat flaky pastry', meaning 'quick fried meat cake with soy sauce', and *yóulín sǔnjī* (油淋笋鸡) 'oil shower young chicken', meaning 'fried chicken with oil', and etc.

As a matter of fact, apart from dish-denoting compounds, an abundance of '$N_1 VN_2$ (NP)' compounds exist in Chinese, e.g. words like *qìchē xiūlǐgōng* (汽车修理工) 'car repair worker', meaning 'car mechanic'; *dàxióngmāo bǎohù jìhuà* (大熊猫保护计划) 'panda protect plan', meaning 'panda protection plan'; and *zhǐzhāng fěnsuìjī* (纸张粉碎机) 'paper shred machine', meaning 'shredder' and so on (See [5, 7, 8, 17] for detailed discussions).

2.2 The Theoretical Issue Raised by Two Important Observations

As a matter of fact, what arouses our great interest in the word *ròujiámó*, is not merely its structural representation, but also the following two important observations which are largely ignored in previous studies.

Firstly, a closer scrutiny suggests that a great majority of common people have no idea of the structural representation of *ròujiámó*. The interesting fact, however, is that this lack of structural knowledge does not hinder their understanding and use of it.

The second observation is that speakers might have totally different perceptions towards linguistic units with apparently identical structural representations. For instance, we would take *ròujiámó* as the misused form of *mójiáròu* upon hearing it, but such a processing is unlikely to happen with *dànchǎofàn* (cf. ? *fànchǎodàn*).

A natural question that arises is that, how do we offer an adequate description of the mechanism of the understanding and use of *ròujiámó*? If we accept the task of a linguist set by [1] to model the linguistic knowledge in the brain of the speaker, is it enough to only work out the structural representation of this word? Or in other words, can simple structural description really explain what is going on when a sentence is produced and understood?

In fact, a wider range of linguistic facts including Chinese compounds in general lead us to the conclusion that neither simple structural description nor derivational analysis in generative grammar can adequately capture the speaker's intuition. We argue that the two important observations concerning *ròujiámó* lend more support to the CG account of linguistic knowledge, i.e., 1) linguistic knowledge is stored in the brain in the form of constructions, with the different constructions constituting an interconnected network;

2) the sanction of linguistic expressions is not the simple computation of rules but involves the motivation of the entire network of constructions.

3 *Ròujiámó* from a CG Perspective

In the previous sections, the two CG claims about the nature of linguistic knowledge have been mentioned. Guided by the two claims, this section will address the theoretical issue raised by *ròujiámó* and offer an account of the two important observations concerning this word. Our core argument is as follows. First, *ròujiámó* is an entrenched linguistic unit, i.e., construction. Second, the sanction of the word is heavily influenced by related constructions in the Chinese language.

3.1 *Ròujiámó* as an Entrenched Linguistic Unit

As is mentioned above, the generative study of 'N_1VN_2 (NP)' constructions (e.g. [8]) focuses its attention on their derivational process. However, recalling the first important observation we make concerning the disputes on the structural representation of *ròujiámó*, we have observed that one does not have to possess the structural knowledge to know how the word is understood and used. It is difficult for the generative study to acknowledge such a fact.

Yet, it is quite a natural phenomenon, from a CG perspective. According to CG, even expressions completely complying with rules can be stored as a whole in the brain of the speaker if used with sufficiently high frequency, and hence their status as constructions. As soon as a linguistic unit becomes a construction, it can independently serve the function of sanctioning novel linguistic expressions. That is to say, in actual language use, the speaker can directly evoke this entrenched linguistic unit without resorting to abstract constructional schemas.

Figure 1 is our simulation of the sanctioning process of *ròujiámó* from a CG perspective of linguistic knowledge. Firstly, of the three linguistic expressions, two are put in boxes with square corners, i.e., the 'N_1VN_2 (NP)' construction and *ròujiámó*, indicating their status as entrenched linguistic units. There is yet another *ròujiámó*, which is included in a box with rounded corners (at the bottom), indicating its status as an actual language use. Secondly, there exist three categorization relations, i.e. 1) from the 'N_1VN_2 (NP)' construction to *ròujiámó* as an entrenched linguistic unit; 2) from the 'N_1VN_2 (NP)' construction to *ròujiámó* as an actual language use; and 3) from *ròujiámó* as an entrenched linguistic unit to *ròujiámó* as an actual language use.

That is to say, *ròujiámó* as an actual language use is sanctioned by two linguistic units, i.e., the 'N_1VN_2 (NP)' construction and *ròujiámó* as an entrenched linguistic unit. A question that naturally arises is, which is more likely to be taken as the standard in online processing (See [18: 104–107] for the discussion of the standard and the target of categorization relations)?

[14: 230] proposed three principles regarding the selection of the standard of categorization, i.e., 1) degree of entrenchment; 2) contextual priming; and 3) degree of overlap. Obviously, of the two sanctioning units of *ròujiámó*, *ròujiámó* as an entrenched linguistic unit is more concrete than the 'N_1VN_2 (NP)' construction, hence a higher degree

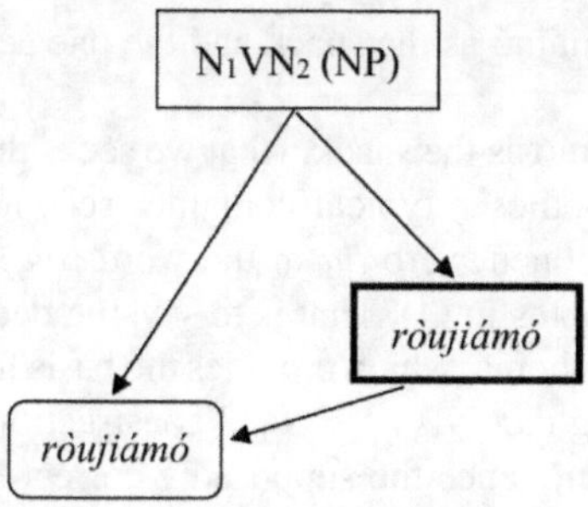

Fig. 1. The sanctioning mechanism of *ròujiám.*

of overlap with the target. Moreover, it is more entrenched than the abstract constructional schema (the 'N_1VN_2 (NP)' construction) because of its higher token frequency, indicated by the thickness of lines. According to usage-based models (e.g. [12: 494, 18: 145]), compared with high-level schemas, low-level schemas play a more important role in sanctioning novel linguistic expressions.

It is also worth noting that because of the loss of analyzability, the speaker cannot necessarily establish the relation between the abstract constructional schema and the concrete linguistic expressions. Just as we may not know that the English word *drawer* is composed of the verb *draw* and the suffix *-er*, we do not necessarily know *ròujiámó* is an endocentric compound comprising *ròu* and *jiámó.*

Lastly, because linguistic knowledge comes from language use according to CG and other usage-based models, and there exist individual differences in terms of linguistic experience, the linguistic knowledge stored in different speakers' brain will not be the same. Specifically, in the case of *ròujiámó*, the perception towards the structural representation of this word is surely different for locals in Shaanxi and non-locals. *Ròujiámó* has been established as an entrenched linguistic unit for the former group because of high frequency in actual language use, but not for the latter one. That explains why non-locals are more puzzled by its structural representation compared with the locals.

3.2 *Ròujiámó* in the Network of Constructions

The idea of network of constructions also explain nicely our second observation that the speaker might have different perceptions towards words with supposedly identical structural representations. The explanation includes two aspects, i.e., 1) why does *ròujiámó* cause puzzles concerning its structural representation? 2) Why do words like *dànchǎofàn* with supposedly an identical structure not cause such puzzles? Let us look at them one by one.

3.2.1 *Ròujiámó* vs. *Mójiáròu*

The fact that *ròujiámó* is perceived to be the misused form of *mójiáròu*, we believe, is a result of conceptual blending [19] between the image schema [20] activated by the structural properties of ròujiámó as the snack and the one activated by the 'N $_{Ag.}$ *jiá* N $_{Pt.}$'[2] construction.

To be specific, in ròujiámó as the snack, what we see is that the pancake sandwiches the meat in it, which embodies a typical container schema. On the other hand, the basic meaning of the component verb *jiá* in the word *ròujiámó* is that of containing, as is illustrated in the examples in (4). That is to say, the two share the common image schema, i.e., the container schema, which provides the basis for the speaker to categorize the snack ròujiámó by using the 'N $_{Ag.}$ *jiá* N $_{Pt.}$' construction or the even more abstract 'N $_{Ag.}$ V N $_{Pt.}$' construction, hence the supposedly correct form *mójiáròu.* Figure 2 is the demonstration of the process of conceptual integration involved in obtaining the supposedly correct form *mójiáròu*, where input space 1 is the structural properties of ròujiámó as the snack (pancake sandwiching meat) and input space 2 is the 'N $_{Ag.}$ *jiá* N $_{Pt.}$' construction, the basis for the integration (the generic space) being the container schema shared by the two input spaces.

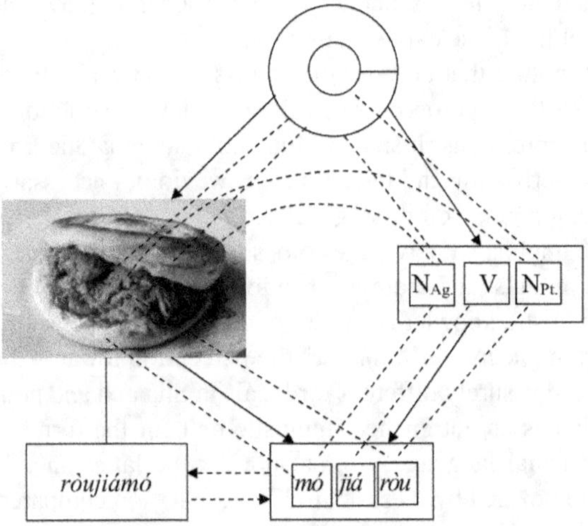

Fig. 2. *Ròujiámó* and *mójiáròu*

(4) a. 史密斯当时左手夹着雪茄烟

shǐmìsī dāngshí zuǒshǒu jiá-zhe xuějiāyān

Smith then left hand sandwich-zhe cigar

'At that moment, Smith held a cigar between his left-hand fingers.'

b. 一阵凉风夹着雨点迎面扑来

[2] Note that the subscript Ag. and Pt. are short for agent and patient respectively. Strictly speaking, *mó* and *ròu* are not agent and patient per se. However, if agent and patient are defined in a broader sense as two asymmetrical entities in in terms of force dynamics, *mó* and *ròu* can be taken as such. See [21] for the discussion of the distinction between proto-agent and proto-patient roles.

yí-zhèn liángfēng jiá-zhe yǔdiǎn yíngmiàn pūlái.
one-Cl cold wind sandwich-zhe rain head-on come
'Breeze and drizzle came head-on.'
c. 两块大饼夹着一根油条
liǎng-kuài dàbǐng jiá-zhe yī-gēn youtiao.
two-Cl pancake sandwich-zhe one-Cl doughnut stick
'One doughnut stick sandwiched in two pancakes'
(All adapted from the SLC corpus)

The reason why we are puzzled by the structural representation of *ròujiámó* is that the form obtained from the process of conceptual integration illustrated in Fig. 2, i.e., *mójiáròu* collides with the actual name of the snack (*ròujiámó*), which is indicated by the dotted-line arrows denoting the relations of extension in CG. Moreover, the box with dotted lines and rounded corners indicates that *mójiáròu* does not exist as a linguistic unit (for the majority of Chinese speakers).

It should be noted that the 'N $_{Ag.}$ *jiá* N $_{Pt.}$' construction or the even more abstract 'N $_{Ag.}$ V N $_{Pt.}$' construction is deeply rooted in our brain because they embody what Langacker (e.g. [14: 357]) calls the canonical event model. In this model, an entity exerts a force on another, causing it to undergo a change of state or location. Because of this deep-rootedness, upon hearing the word *ròujiámó*, we easily take it to be the misused form of *mójiáròu*. Very interestingly, when the first author paid a visit years ago to Shanxi, a neighboring province of Shaanxi, the snack which is almost identical to *ròujiámó* in Shaanxi is named as *mójiáròu*. A conclusion we can get from this is that unlike people in Shaanxi who conceptualize the snack by using the 'modifier + head' compounding pattern or simply store word as a whole in their brain, people in Shanxi are more inclined to employ the force-dynamic image schema.

3.2.2 Ròujiámó vs. Dànchǎofàn

A question left unanswered is, why do words like *dànchǎofàn*, instantiations of the same 'N_1VN_2 (NP)' construction, not generate puzzles concerning their structural representation? We believe there are at least two possible reasons for this.

First, it might be related to the high degree of entrenchment of this class of expressions. Like *ròujiámó*, this class of expressions have been stored as a whole as linguistic units (hence constructions) because of their high frequency, their analyzability is therefore greatly reduced, which results in the low possibility of generating puzzles of structural representation.

However, this cannot be the whole story. The question is, why only expressions like *ròujiámó* generate puzzles of structural representation but not the expressions like *dànchǎofàn,* since they are both entrenched linguistic units? We argue that this difference is directly tied to the semantics of the different cooking verbs (e.g., *jiá* vs. *chǎo*) involved. Corpus investigation finds that the verb *jiá* can either denote a dynamic action, illustrated in (5a), or a static relation, illustrated in (5b). When it denotes a static relation, the entity that carries out the relevant action (the agent) is placed in the background, and our focus becomes the containing relation between the sandwiching entity and the entity being sandwiched, as it is demonstrated in (5b). The asymmetry between the two entities in terms of force dynamics very easily evokes the 'N $_{Ag.}$ *jiá* N $_{Pt.}$' construction in our brain, which collides with its actual name (See the discussion in Sect. 3.2.1).

(5) a. 父亲又夹了块肉送到我的碗里

fùqīn yòu jiá-le kuài ròu song-dào wǒde wǎn-lǐ

father again pick-le Cl meat put-Comp. my bowl-Loc

'My father picked up another piece of meat and put it into my bowl.'

(Adapted from the SLC corpus)

b. 那女孩红红的指头夹着一个便士

nà-nǚhái hónghóngde zhǐtou jiá-zhe yī-gè biànshì

That-girl reddish finger sandwich-zhe one-CL penny

'That girl held a penny between her reddish fingers.'

(Adapted from the BCC corpus)

By contrast, the verb *chǎo* only denotes a dynamic action, since it can only convey an ongoing event rather than an existing state when suffixed with the imperfective marker *-zhe* (着), as is illustrated in (6). Sentence (6a) is an existential construction, signaled by the clause initial locative *guōlǐ* (锅里) 'pot-Loc', the verb plus *-zhe* should convey an existential state. However, the sentence is odd. The dynamic nature of its meaning is further supported by (6b), which describes an ongoing event. The chunk *chǎo-zhe yī-gè cài* (炒着一个菜) 'cook-zhe a dish', meaning 'cooking a dish', is perfectly acceptable.

The reason why *dànchǎofàn* does not evoke the 'N $_{Ag}$. *chǎo* N $_{Pt.}$' construction or the more abstract 'N $_{Ag.}$ V N $_{Pt.}$' construction is due to two possible factors. First, the verb *chǎo* can only take the agent as its subject, which (the agent) is absent in the compound. Second, there does not exist an obvious asymmetry between *dàn* (egg) and *fàn* (rice) like the one between *mó* (pancake) and *ròu* (meat) in terms of force dynamics, making them unable to be conceptualized as the agent/patient relation.

(6) a. ??屋里没人, 锅里炒着一点青菜

wū-lǐ méi-rén, guō-lǐ chǎo-zhe yìdiǎn qīngcài

house-Loc no-people pot-Loc cook-zhe some vegetable

'No one is in the house, and some green vegetables are cooked in the pot.'

b. 康南正在炒着一个菜

kāngnán zhèngzài chǎo-zhe yī-gè cài

kāngnán in progress cook-zhe one-Cl dish

'Kangnan is cooking a dish.'

(Adapted from the BCC corpus)

3.2.3 The Network Representation of *Ròujiámó*

Based on the analysis in Sects. 3.2.1 and 3.2.2, we can now build a network of constructions centered around *ròujiámó*, represented as Fig. 3. In this network, we can first observe that there is an abstract 'N_1VN_2 (NP)' construction, which subsume at least three types of compounds, exemplified by *ròujiámó*, *dànchǎofàn* and *qìchē xiūlǐgōng* respectively (cf. Sect. 2.1). Among these, *ròujiámó* and *dànchǎofàn* constitute the so-called 'material + action + material' construction (cf. Sect. 2.1). Details will not be provided due to space limitations.

Secondly, because of the high degree of entrenchment (indicated by the thickness of lines), these three types of compounds (individual instances) can independently serve to sanction actual language use. In other words, *ròujiámó* in actual language use is more likely to be sanctioned by *ròujiámó* as an entrenched linguistic unit than the more abstract 'N_1VN_2 (NP)' construction (for locals in Shaanxi). Although everyone knows the reference of *ròujiámó*, not everyone knows it is a kind of endocentric compound.

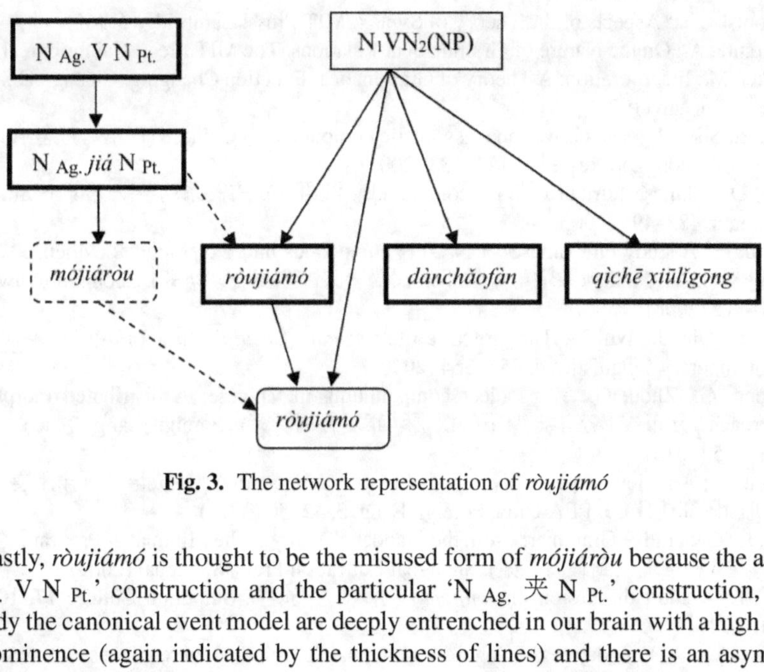

Fig. 3. The network representation of *ròujiámó*

Lastly, *ròujiámó* is thought to be the misused form of *mójiáròu* because the abstract '$N_{Ag.}$ V $N_{Pt.}$' construction and the particular '$N_{Ag.}$ 夹 $N_{Pt.}$' construction, which embody the canonical event model are deeply entrenched in our brain with a high degree of prominence (again indicated by the thickness of lines) and there is an asymmetry between pancake (*mó*) and meat (*ròu*) in terms of force dynamics. Moreover, in compounds like *dànchǎofàn*, the meaning of the verbs as well as the relations between entities are not sufficient to activate the '$N_{Ag.}$ V $N_{Pt.}$' construction, which explains the difference in perception towards different compounds with supposedly identical structural representations.

4 Conclusion

The ultimate purpose of contemporary linguistic study is to model the linguistic knowledge in the brain of the speaker in order to explain how language is understood and used.

From the disputes concerning the structural representation of the compound word *ròu-jiámó*, we can see that the speaker does not necessarily need the structural knowledge to be able to understand and use a certain linguistic expression, and we might have utterly different perceptions towards linguistic units with supposedly identical structures. All these indicate that the traditional structuralist analysis fails to uncover the nature of language. Meanwhile, although generative grammar attaches importance to the relation between language and mind, their specific analysis is featured by structural derivation, which cannot fully capture the native speaker's intuition either. Cognitive linguistics in general and CG in particular, by contrast, emphasize the usage basis of language and the psychological plausibility of linguistic description, thus fostering an organic combination of structural description and online processing and a perfect fusion between language and cognition.

References

1. Chomsky, N.: Aspects of the Theory of Syntax. MIT Press, Cambridge (1965)
2. Marantz, A.: On the Nature of Grammatical Relations. The MIT Press, Cambridge (1984)
3. Baker, M.: Incorporation: A Theory of Grammatical Function Changing. Chicago University Press, Chicago (1988)
4. Gu, Y., Shen, Y.: The derivation of synthetic compounds in Chinese (汉语合成复合词的构造过程). Zhongguo Yuwen **2**, 122–133 (2001)
5. Shi, D.: Chinese attributive V-N compounds (汉语的定中关系动-名复合词). Zhongguo Yuwen **6**, 483–495 (2003)
6. Zhou, R.: A study on Chinese VON/OVN compounds under typological evidence: Universality vs. individuality (共性与个性下的汉语动宾饰名复合词研究). Zhongguo Yuwen (4), 301–312 (2006)
7. Pei, Y., Qiu, J., Wu, Y.: The structure of '*zhizhang fensuiji*' ("纸张粉碎机"的层次结构). Contemporary Linguistics **4**, 356–364 (2010)
8. Cheng, G., Zhou, G.: Verb-object compounding in Chinese: A distributed morphology approach (分布式形态学框架下的汉语动宾复合词研究). Foreign Lang. Teach. Res. **2**, 163–175 (2015)
9. Zhou, G.: A review of generative approaches to compounding (生成语法视角下复合词构词机制研究述评). J. PLA Univ. Foreign Lang. **3**, 42–50 (2018)
10. Xu, L.: Generative Grammar: From the Standard Theory to the Minimalist Program (生成语法理论：标准理论到最简方案). Shanghai Education Press, Shanghai (2009)
11. Xu, Y.: A short discussion on *ròujiámó* (小议"肉夹馍"). Sinogram Culture **17**, 105–107 (2019)
12. Langacker, R.: Foundations of Cognitive Grammar: Theoretical Prerequisites, vol. 1. Stanford University Press, Stanford (1987)
13. Langacker, R.: Foundations of Cognitive Grammar: Descriptive Applications, vol. 2. Stanford University Press, Stanford (1991)
14. Langacker, R.: Cognitive Grammar: A Basic Introduction. Oxford University Press, Oxford (2008)
15. Liu, J.: *Ròujiámó*: a special linguistic phenomenon ("肉夹馍": 一种特殊的语言现象). J. Hum. **2**, 119–121 (1997)
16. Li, J.: Verbs as Attributives in Modern Chinese (现代汉语动词直接做定语研究). Commercial Press, Beijing (2008)
17. Chen, Y.: The relator principle and the N1 VN2 (NP) construction (联系项居中原则与N1 VN2 (NP) 结构). Chinese Teach. World **2**, 51–62 (2006)

18. Langacker, R.: Grammar and Conceptualization. Mouton de Gruyter, Berlin (1999)
19. Fauconnier, G., Turner, M.: The Way We Think: Conceptual Blending and the Mind's Hidden Complexities. Basic Books, New York (2002)
20. Johnson, M.: The Body in the Mind: The Bodily Basis of Meaning, Imagination and Reason. Chicago University Press, Chicago (1987)
21. Dowty, D.: Thematic proto-roles and argument Selection. Language **67**(3), 547–619 (1991)

An Analysis of the Inferential Meaning of Chinese Even-Construction for Semantic Parsing

Long Chen(✉) and Weidong Zhan(✉)

Peking University, Beijing, China
{chenlong,zwd}@pku.edu.cn

Abstract. The even-construction in Chinese, whose form is *lian* + X + *dou* + Y, has been thoroughly investigated in Chinese linguistics. However, in natural language processing, the representation and acquisition of the full meaning of the construction remains to be a tough question. From the perspective of construction grammar, we compare the semantic functions of the even-construction and the corresponding subject-predicate structure. The meanings of the even-construction include a basic proposition and an inferential meaning. The basic proposition of the construction is the occurrence of an unlikely event. There are three kinds of interrelated but distinguishable inferential meanings: (1) the occurrence of an unlikely event means some related and more likely events will happen or have happened; (2) the occurrence of an unlikely event indicates an entity involved in the event has some special attributes; (3) the speaker expresses some subjective attitudes towards the occurrence of an unlikely event. Based on the analysis of the meaning of the even-construction, the ways to compute the meaning are discussed in this paper. It is also pointed out that the semantic computation of the construction depends on much linguistic knowledge, world knowledge and common sense, and temporary contextual information in communication.

Keywords: Even-construction · Semantic computation · Construction grammar · Event extraction

1 Introduction

Many natural language processing tasks intend to acquire the sentence meanings of a language. For example, the task of event extraction (Linguistic Data Consortium 2005) aims at extracting an event from the sentence and so also aims at finding the trigger of an event and event elements from a sentence, such as participants, time, modality, and polarity, thus. However, in language use, the trigger of an event may not appear in the text.

M. Liu et al. (Eds.): CLSW 2020, LNAI 12278, pp. 346–360, 2021.
https://doi.org/10.1007/978-3-030-81197-6_29

(1a)

战士	上	战场，	别说	奖金	了，	可能	连	命	都	搭上
soldier	go	battlefield	let.alone	reward	FP	possible	*lian*	life	*dou*	lose

"When soldiers go to the battlefield, they may not survive, let alone getting rewards."

(1b)

他们	生活	在	那个	连	麻雀	都	吃	不	饱
3PL	live	in	DEM_{DIST}	*lian*	sparrow	*dou*	eat	NEG	full

的	年代。
POSS	age

"They lived in the age when even sparrows did not have enough to eat."

There are three events in the sentence (1a): soldiers go to the battlefield; soldiers may lose their lives; and soldiers may not get rewards. The trigger of the third event, i.e. the verb "get", is not in the text. This kind of event is hard to be detected in natural language processing. In (1b), the event "sparrows did not have enough to eat" can be extracted from the text. However, the sentence intends to express that life was very poor and people lacked food in that age. This property of "the age" is the more important piece of information that needs to be extracted, which is also not presented by a trigger in the sentence.

In (1a-b), even-construction is used and it is such construction that challenges the semantic computation of the sentences. The even-construction in Chinese takes the form of T + *lian* + X + *dou* + Y, where the constant *dou* can also be replaced by *ye* (also) with no constraints. The construction has two common variations: X + *dou* + Y and *lian* + X + T + *dou* + Y. The variable T is the topic of the clause containing the construction, the variable X can be a noun phrase, a verb phrase, a morpheme, etc., and Y is usually an adjective or a verb. Among the two variations, the variation X + *dou* + Y is ambiguous. For example, the sentence "*cai* (dishes) *dou* (*dou*) *liang* (cold) *le* (FP)" can mean "the dishes were all cold", or "even the dishes were cold".

The construction usually functions as a predicate or a clause. It can also be used as a modifier or be used in a complex sentence. When the construction is used in a complex sentence, the other clause of the sentence may contain connectives such as *genghekuang*/*bieshuo* (let alone/not to mention) and *bujin* (not only).

The basic proposition of the even-construction T + *lian* + X + *dou* + Y is its literal meaning, which is the proposition composed of T, X, and Y. The relationships between the variables can be complicated. T and X can be the arguments of Y, and they may also have no thematic relationships. For example:

(2a) 第一 题 连 玛丽 都 不 会 做。
first question *lian* Mary *dou* NEG can do
"Even Mary cannot solve the first question."

(2b) 约翰 连 第一 题 都 不 做。
John *lian* first question *dou* NEG do
"John even did not solve the first question."

(2c) 连 一 分钟 约翰 都 等 不 了。
lian one minute John *dou* wait NEG can
"John cannot wait even for a minute."

(2d) 约翰 一次 都 没 及格。
John once *dou* NEG pass
"John did not pass even once."

X in (2a) and (2b) is "Mary" and "the first question" respectively; the predicates in Y are "cannot do" and "not do". In (2a), X is the agent of the predicate in Y while in (2b), X is the patient. Both of them are the core arguments. In (2c) and (2d), X is "one minute" and "once" respectively, which are the non-core arguments of time and quantities of the predicate in Y (i.e. "cannot wait" in (2c) and "not pass" in (2d)).

(3a) 在 计算机 方面, 他 连 自学 都 不 知
in computer aspect 3SG *lian* self-study *dou* NEG know
如何 下手。
how start
"As for computers, he did not even know how to start self-studying them."

(3b) 玛丽 脸色 苍白, 连 哭 都 不 敢 大声。
Mary look pale *lian* cry *dou* NEG dare loud
"Mary looked pale and did not even dare to cry loudly."

(3c) 这 本 书 他 连 目录 都 没 读。
DEM_{DIST} CL book 3SG *lian* content *dou* NEG read
"He had not even read the contents of this book."

Contrarily in some cases, X may not be a direct argument of the main predicate in Y. For example, in (3a), X is "self-studying" and Y is "not know how to start". In this case, X is the theme of the main predicate in Y. In (3b), the variable X "cry" is not an argument of Y "loudly". Instead, Y is a modifier of X. The variable T is usually an argument of the main predicate in Y, but can also be a non-argument entity related to the event represented by Y, such as the "book" in (3c).

In the CCL corpus (Zhan, et.al. 2019), 22685 documents contain the construction *lian* + X + *dou*/*ye* + Y in which X is a string of no more than 3 characters, and nearly 95% of which are instances of the even-construction. That is to say, in the CCL corpus, there are about 3.2 documents containing the even-construction in every 1000 documents. The proportion will be higher if we also consider the variations of the construction and instances whose variables are longer. This shows that the even-construction is a construction with relatively high frequency in use. So, it should not be neglected in semantic computation.

Besides, the premise of the literal meaning of even-construction is that the literal meaning is related to an unlikely event, of which the unlikeliness is determined usually by world knowledge, such as the knowledge "sparrows feed easily" in (1b). Therefore, the semantic computation of the even-construction can also help computers acquire some world knowledge.

The even-construction has been thoroughly investigated in linguistics, but hardly received any attention in computational linguistics and natural language processing. In this paper, we analyze the meanings of the construction from the perspective of construction grammar and hence discuss the representation and acquisition of the inferential meaning of the construction.

2 A Brief Review of Previous Studies

Based on the semantic functions of the components of the even-construction, there have been many previous studies on the meaning of the construction from different perspectives. Linguists including Lv (1980) and Zhu (1982) concluded the meaning of the construction as "emphasizing"; Paris (1979), Cao (1987), Liu and Xu (1998), and Yuan (2006) discussed the properties of the variables of the construction from the perspective of topic and focus; Ding (1994) and Yuan (2006; 2011) suggested that the construction had the semantic function of counter-expectation; Zhou (1990) and Yuan (2006) also suggested that the construction has the meaning of contrast.

Using sets or scale semantics is a common way to analyze the meaning of even-constructions. Based on the meanings of the constants *lian*, *dou*, and *ye*, Paris (1979), Cui (1984), Cao (1987), Zhou (1990), and Yuan (2011) pointed out that the variable X is at the top or bottom of a scale. For example:

(4) 约翰 连 香蕉 都 不 吃。
John *lian* banana *dou* NEG eat
"John even did not eat bananas."

(5) 约翰 连 澡 都 不 洗。
John *lian* bath *dou* NEG wash
"John even did not take a bath."

The even-construction in (4) can be analyzed as introducing a set of foods or fruits, in which "bananas" are the most unlikely to be not eaten. Zhou (1990) and Liu (2005) also noticed that some instances of *lian* + X + *dou* + Y did not introduce a scale of entities including X, but introduced a series of parallel events that can be linked together including the event composed of X and Y. This kind of even-constructions is called "atypical even-constructions" by Liu (2005). For example, in (5), "taking a bath" may be associated with events such as "sleeping", "eating", "resting", etc., all of which are common activities in daily life.

We believe that the above analyses are untenable to a certain extent. First, the set cannot be determined easily. A speaker knows what is included in a set relatively easily, but the set is hard to be defined for computation. For example, in the set of "foods" mentioned above, it is difficult for us to list the items of the set precisely, such as whether the set contains rice, apples, or meat. Sometimes, it is even uncertain as to whether there is such kind of a set. For example:

(6) 胡适 应酬 太 多， 连 学问 都 退步 了。
Hu.Shi social.activity too much *lian* knowledge *dou* fall.behind FP
"Hu Shi had so many social activities that he even lost his knowledge."

In (6), it is hard to find a set containing "knowledge", in which "knowledge" is the most unlikely information to be lost by the great scholar Hu Shi. The members of the set are unknown. Thus, it is difficult to define a set containing the event of "losing knowledge".

Second, even if we find a set/scale, the even-construction is usually not used to express the existence of the scale or the position of the variable X on the scale. For example, in (6), the construction mainly expresses the huge impact of Hu Shi's social activities on him.

Other than the above analyses, many linguists examined the meanings of the even-construction from different aspects, including its literal meaning/basic proposition, presupposition/premise, and implication/inference of the construction, including Paris (1979), Cui (1984), Tsai (2004), Yuan (2011) and others. Most of them suggested that the presupposition/premise of the construction is that there is an entity of the same kind as X with the property Y, and the implication/inference of the construction is that any entity of the same kind as X has the property Y. As is discussed above, the construction does not necessarily express a set/scale related to X, so these analyses need to be studied further.

Above all, the previous studies were mainly based on the composition of the components of the construction and deduced the meaning of the construction from the meanings and semantic functions of the variables and constants. These analyses differ from each other in the actual meaning and the function of the even-construction in a sense and need to be improved for the semantic computation of the construction.

3 Inferential Meaning of Even-Construction

This paper takes the constructional approach and explores the meaning of the even-construction by comparing the construction with the construction without "even".

Consider a situation in which John and Mary took part in an examination. In the examination, the first question is the easiest, and the rest of the questions are much harder. Based on this situation, we designed two interrogative sentences and two declarative sentences used for their dialogue. (8a) is an instance of even-construction and (8b) is the corresponding subject-predicate structure.

(7a) Mary: 你 第三 题 做 对 了 吗？
Mary 2SG third question do right FP QU
Mary: "Did you do the third question correctly?"

(7b) Mary: 你 第一 题 做 对 了 吗？
Mary: 2SG first question do right FP QU
Mary: "Did you do the first question correctly?"

(8a) John: 我 连 第一 题 都 做 错 了。
John: 1SG *lian* first question *dou* do wrong FP
John: "I even made a mistake in the first question."

(8b) John: 我 第一 题 做 错 了。
John: 1SG first question do wrong FP
John "I made a mistake in the first question."

If the question is (7a), (8a) is a suitable response. Mary knows that John answered the third question wrongly from the response. On the contrary, (8b) is an irrelevant answer. Logically, the first question being easier than the third question does not necessarily mean that one will also answer the third question wrongly if he/she answers the first question wrongly, but if a person hears the dialogue in (7a-8a), he/she will be sure John answered the third question wrongly. This indicates that the even-construction makes an assertion about the third question.

If the question is (7b), (8b) is a suitable response, and (8a) is not, indicating that although literally an assertion about the first question is made, it is not the focal information of the construction. Therefore, it can be concluded that the primary information of the even-construction is the inference of the literal meaning rather than the literal meaning itself.

The even-construction has several types of inferences. For example, when the question is (7a), the response (8a) infers that the speaker answered the third question wrongly; if the question is whether John did well in the exam, (8a) infers that the speaker did not do well; if John says (8a) out of blue, the inference is likely to be that John is angry or believes that it is terrible to answer the first question wrongly. When (8a) is used in other contexts, it can draw different inferences. For example:

(9a) 别说 其他 题， 我 连 第一 题 都 做
let.alone other question 1SG *lian* first question *dou* do
错 了。
wrong FP
"I was not even correct in the first question, let alone the other questions."

(9b) 我 连 第一 题 都 做 错 了， 简直
1SG *lian* first question *dou* do wrong FP absolutely
太 粗心 了。
too careless FP
"It is too careless for me to even make a mistake in the first question."

In (9a), the inference is that the speaker answered the other questions wrongly. In (9b), the inference is that the speaker was very careless.

In a word, there are three types of inferential meanings of the even-construction:

1) IM1: The occurrence of an unlikely event→the occurrence of a series of more likely events. This is the deductive inference of the even-construction. The speaker and hearer deduce the occurrence of other related events from the occurrence of the unlikely event. All of the related events form a scale, with the event denoted by the basic proposition locating at the minimum point, and the event with comparatively high probability at the maximum end. The contrasting event is the inferred event of the construction.
2) IM2: The occurrence of an unlikely event→the special attribute of the entity causing the occurrence of the event. This is the abductive inference of the even-construction. The speaker and hearer abduce the unlikely event from its occurrence.
3) IM3: The occurrence of an unlikely event→the subjective attitude of the speaker. This is the subjective attitudinal inference of the even-construction. The speaker expresses his/her attitudes and/or emotions towards the unlikely event.

We randomly selected 529 instances of the even-construction from the CCL corpus and labeled the types of inferences of each instance. It should be noted that the three types of inferences are not mutually exclusive. A construction instance can draw all the three types of inferences, but usually only one or two types are more prominent. For example, in the sentence (9b), the basic proposition is that the speaker answered the first question wrongly. The deductive inference is that the speaker made mistakes in some other easier questions. The abductive inference is that the speaker was very careless. The subjective attitudinal inference is that the speaker was dissatisfied. The context indicates that the abductive inference is the prominent type of inference of (9b). Table 1 shows the statistics of different types of inferences of the even-construction.

Table 1. Statistics of different types of inferences of the even-construction

Inference	Number	Percentage
IM2	276	52.2%
IM1	98	18.5%
IM2 + IM3	73	13.8%
IM1 + IM3	31	5.9%
IM3	29	5.5%
IM1 + IM2	17	3.2%
IM1 + IM2 + IM3	5	0.9%
	529	100%

According to the Table 1, the most common inference of the even-construction is the abductive/backward inference, and only about 1/5 of the inferences are deductive/forward inferences, indicating that the even-constructions may not necessarily be used in a typical way to draw deductive inferences, as suggested by the frequency shown above.

4 Requirements in Semantic Parsing of Inferential Meaning of Even-Construction

In current NLP practice, deep learning methods are widely used. In these methods, the input is usually the original texts, and the meaning of a language is acquiesced as being acquired from the distribution of the texts. However, as is discussed in the examples above, the inference of even-constructions may not appear in the texts. For example, the inference of the even-construction in sentence (6), i.e., social activities had great impact on Hu Shi, is hidden from the text. The text does not provide words that are directly relevant with "great impact". We acquire the inference from our understanding of the basic proposition of the construction, i.e. Hu Shi lost his knowledge. Therefore, the inferences are related to contexts including the speaker, the hearer, and the world surrounding them, rather than only the words used in the construction. The following sub-sections discuss the knowledge needed in the acquisition of the inferential meanings of the even-constructions.

4.1 Semantic Parsing of Forward Inferences

If an unlikely event has happened or is going to happen, usually some more likely event has also happened or is also going to happen. For example, in (8a) "I even made a mistake in the first question", the more likely event is "I made a mistake in the third question"; in (9a) "I was not even correct in the first question, let alone the other questions", the more likely event is "I made mistakes in the other questions".

To acquire the forward inference of an even-construction, other events should be extracted from the context of the construction. Forward inference appears in the context in different forms. The predicate of the forward inference may appear in the context, being the same as the predicate of the basic proposition, or being hidden from the context.

The forward inference may be expressed on the surface of the context. For example:

(10) 所有 人 都 要 交 人口税， 连 小孩 都 要。
all people all need pay poll.tax *lian* kid *dou* need
"Everyone needed to pay poll taxes, even children had to."

(11) 约翰 不 敢 这么 写， 或许 连 想 都 不
John NEG dare such write maybe *lian* think *dou* NEG
敢 这么 想。
dare such think
"John dared not write about this, maybe even dared not think about this."

In (10) and (11), the forward inferences appear as subject-predicate structures in the contexts. The inference in (10) is "all people needed to pay taxes", and the one in (11) is "John dared not write in such manner". The inferences of the two even-constructions are identical to the proposition expressed by the clause before the even-constructions, and the even-constructions function to emphasize the events in the previous clauses. If the basic proposition and the inferences both involve events that have already happened, such as in (10), the construction emphasizes that the inferred event happened indeed. If the event denoted by the basic proposition and also the inferred event has not happened yet, such as the ones in (11), the construction emphasizes the confidence of the estimations.

It is relatively easier to extract the main predicate and the participants of the inferred event in these even-constructions. However, sometimes the modality and polarity of the inferred event are hard to determine.

(12a)

下雨	天，	别说	打	不	到	车，	连
rain	day	let.alone	call	NEG	can	Taxi	*lian*
地铁	都	会	挤爆。				
subway	*dou*	will	crowded				

"On rainy days, even the subways will be crowded, not to mention that you cannot find a taxi."

(12b)

下雨	天，	别说	打	到	车，	连	地铁
rain	day	let.alone	call	can	taxi	*Lian*	subway
都	会	挤爆					
dou	will	crowded					

"On rainy days, even the subways will be crowded, not to mention if you can find a taxi."

(12c)

下雨	天，	别说	打	车	，	连	地铁
rain	day	let.alone	call	taxi		*Lian*	subway
都	会	挤爆。					
dou	will	crowded					

"On rainy days, even the subways will be crowded, not to mention finding a taxi."

In (12a-c), the basic proposition is "the subways will be crowded on rainy days", and the inference is "you cannot find a taxi on rainy days". What differ among the three sentences is that in (12b) the negator *bu* is omitted which is used in (12a), and in (12c), the bare verb is used without the negator or the modality marker *dao*. In order to correctly understand the polarity and the modality involved in the inference, it is necessary to have some world knowledge to fully understand the events in the basic proposition and those inferred ones. To be specific, one needs to know that the subways being crowded means being unable to get on a subway, and it is usually easier to get on a subway than to find a taxi.

The inferred event may be manifested by the same predicate as that by the basic proposition. In this case, the predicate may not be repeated in the context. For example:

(13) 连 我 都 吓坏 了， 别说 其他 人 了。
lian 1SG *dou* scared FP let.alone other people FP
"Even I was scared, not to mention other people."

(14) 别说 是 春天， 连 寒冬腊月 这里 都 绿 草
let.alone COP spring *lian* winter here *dou* green grass
如 茵
like carpet
"The grass looks like a green carpet here even in winter, not to mention in spring."

The inference in (13) is "other people were scared", and the inference in (14) is "the grass looks like a green carpet here in spring". In current semantic computing tasks, a predicate is usually used in only one proposition, so for this kind of even-constructions, only the basic propositions can be acquired. The events in the inferences will be neglected. However, when the basic proposition and the inference of an even-construction share the same predicate, some participants of the inferred event are not participants of other events. For example, the "other people" in (13) and the "spring" in (14) are not participants of other events. Therefore, these entities can help draw potential inferences.

In more complicated cases, the predicate of the inferred event may not appear in the context. For example:

(15) 在 这 家 百货公司， 有的 货 一 天 顾客
in DEM_{DIST} CL department.store some goods one day customer
就 抢 光 了， 有的 货 顾客 连 看 都 不 看。
Just rob all FP some goods customer *lian* see *dou* NEG see
"In this department store, some goods are sold out in only one day, while some are not even looked at."

(16) 有人 质疑 编辑部， 读者 的 来稿 是否 连
someone question editorial.office reader POSS contribution whether *lian*
信封 都 不 拆 就 扔 进 纸篓。
envelope *dou* NEG open just throw into bin
"Someone suspected that the editorial office might have thrown the contributions from readers into the bin without even opening the envelope."

The inferred event in (15) is "customers do not buy some of the goods", and the inferred event in (16) is "the editorial office did not reply to the contributions from readers". The main predicates "buy" and "reply" do not appear in the context. Readers acquire these inferred events from the understanding of the contexts and the relationships between the events. In (15), the clause before the even-construction expresses that some goods are sold out quickly, in contrast to the event inferred from the even-construction. Thus the even-construction expresses that customers do not buy some of the goods. In (16), the person intended to express his/her dissatisfaction with not getting a reply, because the contribution can only be considered after the editors opened the envelope. To acquire the inferences in these even-constructions, a computerized system thus should be trained to have a thorough understanding of the contexts and the relevant events.

Above all, it is relatively easy to acquire the inference when the predicate of the inferred event is in the context. When the predicate of the inferred event is the same as that of the basic proposition, the participants of the inferred event can help make the inference. When the predicate of the inferred event is not in the text, a thorough understanding of the context is required in the acquisition of the inference.

4.2 Semantic Parsing of Backward Inferences

If an entity causes the occurrence of an unlikely event, the entity is likely to have some special properties. For example:

(17) 在 努力 后， 他 写出 了 连 自己 都 不
at strive after 3SG write FP *lian* REFL *dou* NEG
敢 相信 的 文章
dare believe POSS article
"After working hard, he wrote an article that even he himself dared not believe he could have written."

(18a) 约翰 看 爸爸 妈妈 吵架， 吓 到 连 哭 都 不
John see father mother argue scared to *lian* cry *dou* NEG
敢 哭 。
dare cry
"Witnessing his father and mother argue, John was so scared that he did not even dare to cry."

(18b) 约翰 看 爸爸 妈妈 吵架， 连 哭 都 不 敢 哭
John see father mother argue *lian* cry *dou* NEG dare cry
"Witnessing his father and mother argue, John did not even dare to cry."

The even-construction in (17) indicates that "the article he wrote" is very good. In (18a), the property of John is "scared". The special properties of entities may not be directly expressed in the contexts. For example, the property of "being scared" in (18a) can be omitted and the sentence becomes (18b), which is synonymous with (18a).

It is easy to identify the entity and its attribute if they are expressed directly in the text, such as John's attribute in (18a). When the attributes are not in the contexts, a few syntactic features can indicate that backward inferences are expressed. For example, in (17), the even-construction is a modifier; in (18a), the even-construction is a complement. Their grammatical function is to modify entities, so these even-constructions expresses the properties of entities.

(19) 在 新泽西， 人口 每 22 年 将 增加 一倍；
at New.Jersey population per 22 year FUT increase double
而 在 罗得岛， 连 22 年 都 不 要。
but at Rhode.Island *lian* 22 year *dou* NEG want
"In New Jersey, the population will double every 22 years; but on Rhode Island, it takes lesser time than 22 years for the growth of population."

If the even-construction is used in a complex sentence, and the other clause expresses an event contrary to the basic proposition of the even-construction, the even-construction expresses backward inference. For example, in (19), we infer that "the speed of population growth on Rhode Island is very fast".

In most cases, there are no syntactic features to help determine whether an even-construction expresses a backward inference. Theoretically, the understanding of events and attributes is needed in the discovery of backward inferences. For example:

(20a)

约翰	说好	请	我	吃饭	，但	他	走	的
John	promise	invite	1SG	eat	but	3SG	leave	POSS
时候	连	招呼	都	不	打。			
time	*lian*	hail	*dou*	NEG	say			

"John promised to invite me to dinner, but when he left, he did not even let me know."

(20b)

时代	抛弃	你	的	时候	，连	招呼	都	不
time	throw	2SG	POSS	time	*lian*	hail	*dou*	NEG
打。								
say								

"When the time throws you away, it will not even let you know that."

The even-constructions in (20a-b) are both "even not letting someone know". They have different inferences in the two sentences. The inference in (20a) is "John was impolite", and the inference in (20b) is "the time throws you away suddenly". According to our common knowledge of daily communication, we know that if someone wants to do something concerning another person and to make that person prepared, he/she needs to let that person know about his actions. If a person does not let the other person know what he is going to do, the other person would be unprepared, and his/her action would be unexpected. If the "letting someone know" has a metaphoric meaning, whose agent is not a human, such as in (20b), the phrase mainly expresses the suddenness of the event. The automatic acquisition of these inferences requires plenty of world knowledge.

Contextual information is also needed in the acquisition of backward inferences. For example, in (20b), besides the meaning of "politeness", a computerized system also needs to know that John had invited the speaker before he left, and so it was improper for him to leave without informing the speaker. The invitation John made is a piece of contextual information that appears in the clause before the even-construction.

Above all, linguistic knowledge, world knowledge, and contextual information is needed in the analysis of backward inference of the even-construction.

4.3 Semantic Parsing of Subjective Attitudes

The basic proposition of an even-construction is the occurrence of an unlikely event, and people usually have their subjective attitudes or estimations on an unlikely event. This kind of inferences is common in spoken languages, but relatively rare in formal texts such as news. If the speaker is willing to see the occurrence of the unlikely event, the attitude and emotion of the speaker are usually positive, and vice versa. For example, in (21), the attitude of the speaker can be considered as "appreciative".

(21) 他 连 每个 角落 都 打扫 得 干干净净。
3SG *lian* every corner *dou* wipe COMP clean
"He made even every corner clean."

Compared with the acquisition of forward and backward inferences, the acquisition of subjective attitudes is more difficult. If the even-construction is in a quotation, the subjective attitude of the speaker may appear outside the quotation. For example, in (22), the emotion of the speaker is angry.

(22) 玛丽 生气 了， 说： "我 的 话 你 连 一 句
Mary angry FP say 1SG POSS word 2SG *lian* one CL
都 不 信。
dou NEG believe
"Mary got angry and said, "you did not believe even one word of mine.""

However, in most cases, the subjective attitude of the speaker does not appear in the context. To identify the emotion and evaluation of the speaker, it is necessary to understand the basic proposition of the even-construction and the contextual information, and to know about the relationship between the emotion and the events.

(23) 我 三 岁 时 连 球拍 都 拿 不 稳。
1SG three year time *lian* racquet *dou* hold NEG steady
"I could not even hold the racquet well when I was three years old."

The context of (23) is that the speaker saw a three-year-old boy playing badminton with an adult for several rounds. The basic proposition of the even-construction contrasts with the context, indicating that the speaker believed the boy was exceptional and his attitude was admiration for the boy.

As is shown in the above analysis, similar to the acquisition of the backward inference of the even-construction, the acquisition of the subjective attitude of the even-construction also requires linguistic knowledge, world knowledge, and contextual information.

5 Conclusive Remarks

In Chinese information processing, the representation and acquisition of the meaning of the even-construction remain to be a tough question. This paper discusses the meaning of this commonly-used structure from the perspective of constructions. We believe that previous studies have been affected by the meanings and functions of the constants of the construction and have only been able to account for a part of the construction meaning. By comparing the use of the even-constructions and the corresponding phrase structures, we discover that the even-construction mainly expresses the inference beyond the basic proposition. There are three types of inferences: (1) forward inference: emphasizing the occurrence of a more likely event that is relevant to the basic proposition; (2) backward inference: highlighting a special property of an entity related to the basic proposition; (3) subjective attitude: expressing the emotion or evaluation of the speaker on the basic proposition.

This paper discusses the requirements for the automatic acquisition of the meanings of the even-construction. The acquisition of the basic proposition requires linguistic knowledge, including argument structures and collocations. The acquisition of forward inference also requires world knowledge, which is the constant common knowledge of social communities. The acquisition of the backward inference and subjective attitude also requires contextual information, including the temporary background of the communication. The requirements for the semantic computation of the even-construction are concluded in Fig. 1.

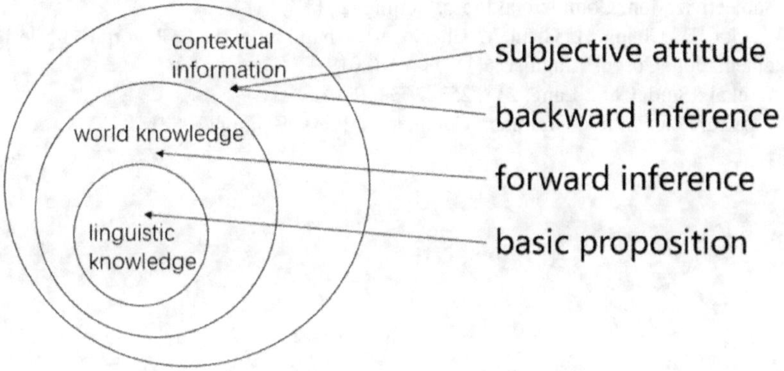

Fig. 1. Requirements for semantic computation of the even-construction

Besides the even-constructions, many other constructions also pose challenges to Chinese information processing, such as constructions expressing subjective quantities, complex constructions, etc. The full meaning of the constructions cannot be acquired directly from the texts. These linguistic phenomena need further research from the perspective of semantic computation.

Acknowledgement. This paper was supported by Major Project of the 'New Generation of Artificial Intelligence' funded by Ministry of Science and Technology (Project NO. 2020AAA0106701) and Major Project of Humanities and Social Sciences of Ministry of Education, P.R. China (Project NO. 15JJD740002).

References

Tsai, W.: On formal semantics of *zhi* and *lian* in Chinese. Stud. Chin. Lang. **299**, 99–111 (2004)

Cao, F.: Further discussion on topic and "Lian…Dou/Ye" structure. In Dai, H., Xue, F. (eds.) Functionalism and Chinese grammar. Beijing Language and Culture University Press, Beijing (1987)

Cui, Y.: An analysis on "Lian…Dou/Ye…" Structure. Lang. Teach. Linguist. Stud. **4**, 30–44 (1984)

Ding, X.: A survey on the contrast of even-constructions. Linguist. Res. **3**, 23–29 (1994)

Linguistic Data Consortium. ACE (Automatic Content Extraction) Chinese Annotation Guidelines for Events (2005). https://www.ldc.upenn.edu/sites/www.ldc.upenn.edu/files/chinese-events-guidelines-v5.5.1.pdf

Liu, D., Xu, L.: Focus, background, topic and Chinese Lian_clause. Stud. Chin. Lang. **265**, 243–252 (1998)
Liu, D.: The atypical Lian_clause as a typical construction. Lang. Teach. Linguist. Stud. **4**, 1–12 (2005)
Lv, S.: Eight Hundred Words in Modern Chinese. Commercial Press, Beijing (1980)
Paris, M.C.: Some aspects of the syntax and semantics of the" lian…. ye/dou" construction in mandarin. Cahiers de linguistique Asie orientale **5**(1), 47–70 (1979)
Yuan Y.: The information structure of lian-construction in Mandarin **21**, 14-28 (2006)
Yuan, Y.: The subjective expressive function of Lian-construction: Also on some relevant counter-expectation and counter-counter-expectation structures. In: Wu, F. Research on Subjectivity and Subjectivization. Commercial Press, Beijing (2011)
Zhan, W., Guo, R., Chang, B., Chen, Y., Chen, L.: The building of the CCL corpus: Its design and implementation. Corpus Linguist. **6**(1), 71–86 (2019)
Zhou, X., et al.: Stud. Chin. Lang. **217**, 258–263 (1990)
Zhu, D.: Lecture on Chinese Grammar. Commercial Press, Beijing (1982)

On the Interaction of Future and Perfective Aspect in Mandarin Chinese

Jiun-Shiung Wu(✉)

Institute of Linguisitcs, National Chung Cheng University, Minhsiung, Chiayi County 62102, Taiwan
Lngwujs@ccu.edu.tw

Abstract. Contrary to the previous literature suggesting incompatibility between future and perfective *le*, this paper presents examples where the three future modals in Mandarin Chinese *jiāng*, *huì* and *yào* are compatible with perfective *le* under appropriate circumstances. I argue that *jiāng*, *huì* and *yào* are rendered compatible with *le* if the following conditions hold. First, a reference time is provided to induce a reasonably close specific time for *le*. Second, the reference time for *le* must be set to a time other than the present time. If these two conditions are satisfied and no temporal clash arises, future is compatible with the perfective aspect. If a reference time is provided for *le* but this reference time does not guarantee a reasonable close specific time, then, the future modals need to provide a specific time, in order to go with the perfective aspect.

Keywords: *Jiang* · *Hui* · *Yao* · Future and perfective aspect · Mandarin Chinese

1 Introduction

In this paper, I examine the interaction between future and the perfective aspect in Mandarin Chinese (hereafter, Chinese). Specifically, I explore the (in)compatibility of three future modals[1] *jiāng*, *huì* and *yào* with perfective *le*.[2] [2] suggests that, in Chinese, epistemic *huì*[3] is not compatible with the perfective aspect. [3] also presents an implication of similar kind because [3] suggests that perfective *le* describes realis whereas *huì* irrealis. However, there are examples to show that the situation is more complicated than what [2, 3] anticipate. Furthermore, there has been little, if any, attention paid to how *jiāng* and *yào* interact with perfective *le* and whether *jiāng* and *yào* behave in a way similar to or different from *huì* in this respect. In this paper, I discuss this issue in detail.

This paper is organized as follows. Section 2 is literature review, where I review literature that addresses or discusses the interaction between epistemic *huì* and perfective

[1] In this paper, I follow [1] and regard *jiāng*, *huì* and *yào* as future modals.

[2] The interaction between future modals and the experiential aspect marker *guò* is a little different from that involving perfective *le*. This issue is left for future studies so that adequate space can be spent on discussion.

[3] Here we follow [4] that *huì* expressing future is an epistemic modal. And hence in this paper *huì* denoting future is referred to as epistemic *huì*.

M. Liu et al. (Eds.): CLSW 2020, LNAI 12278, pp. 361–368, 2021.
https://doi.org/10.1007/978-3-030-81197-6_30

le. In Sect. 3, I present examples where future modals are compatible or incompatible with perfective *le*. I also present generalizations to account for the discrepancy observed. Finally, Sect. 4 summarizes this paper.

2 Literature Review

In this section, I briefly review [2, 3] and show that more should be discussed, based on the generalizations established in these two works.

[2] provides a comprehensive discussion on how Chinese indicates the temporal location in a sentence, given that Chinese is not morphologically marked for tense. On p18, [2] states that future modal *huì* locates, after the evaluation time or the topic time, which is indicated by the functional head Asp. That is, *huì* has a semantics as follows:

(1) $[\![\text{huì}]\!] = \lambda P_{<i,t>}\, \lambda t\, \lambda t_0\, [P(t) \wedge t_0 < t]$

The semantics in (1) says that *huì* locates a situation *P* in the future of another time t_0. (1) produces a relative future reading because t_0 can be instantiated to a past time, the present time or a future time. In terms of its interaction with perfective *le*, [2] suggests that *huì* rules out the possibility to go with the perfective aspect. For example,

(2) *Lǐsì huì líkāi-guò/-le bànggōngshì.
Lisi HUI[4] leave-Exp/Pfv office.

[2] claims that *huì* is incompatible with perfective *le* or experiential *guò* because of type mismatch: *huì* selects a constituent of type $< i, t >$ but perfective *le* and experiential *guo* yield a constituent of type $< i, < i, t > >$, given the semantics of *le* and of *guò* below:

(3) a. $[\![\text{guò}]\!] = \lambda P_{<i,t>}\, \lambda t_{Top}\, \lambda t_0\, t\, [P(t) \wedge \text{IStage}(t, P) \subseteq t_{Top} \wedge t_{Top} < t_0 \wedge$

b. $[\![\text{le}]\!] = \lambda P_{<i,t>}\, \lambda t_{Top}\, \lambda t_0\, t\, [P(t) \wedge \text{IStage}(t, P) \subseteq t_{Top} \wedge t_{Top} < t_0 \wedge$
$t_{ana} \subseteq R_{state}(t, P)]$

As indicated by the two semantics in (3), if the perfective aspect marker and the experiential aspect marker take a VP, a constituent of type $< i, < i, t >>$ is produced. Since *huì* takes a constituent of type $< i, t >$, naturally, perfective *le* and experiential *guò* are not compatible with *huì*.

[2] is significant in that it provides a thorough explanation for how to determine the temporal location of a situation in Chinese. However, with respect to the incompatibility of *huì* and the perfective aspect, I show in Sect. 3 that the picture is more complicated than what is presented in [2]. Moreover, the interaction between perfective *le* and other future modals is not taken into consideration in [2], which leaves some room for discussion in this paper.

[4] The abbreviations used in this paper include: 1st for first-person singular pronoun, 2nd for second-person singular pronoun, 3rd for third-person singular pronoun, ASSO for an associative marker, CL for a classifer/measure word, Exp for an experiential aspect marker, JIANG/HUI/YAO for future modals *jiāng/huì/yào* respectively, Pfv for a perfective marker, PL for a plural marker, Q for an interrogative marker.

[3] utilizes the distinction between realis vs. irealess to conduct a comprehensive discussion on the mood and aspectual systems in Chinese, Taiwan Southern Min and Hakka.

[3] proposes that perfective *le* denotes realis, while *huì* expresses irrealis. Moreover, [3] suggests that *le*/*huì* do not denote absolute (ir)realis; rather, they express relative ones. That is, the reference time[5] they rely on is contextually determined and is by default set to be the speech time.

Although [3] does not specifically discussed the compatibility between perfective *le* and epistemic *huì*, her generalization has an implication on this issue. Since *le* expresses realis whereas *huì* irrealis, they are not compatible since a situation is not likely to be realized or unrealized with respect to the same given reference time.

The implication by [3] concerning perfective *le* and epistemic *hui* may need further attention. Moreover, just like [2, 3] fails to take other modal words into account.

Given the inadequacies of [2, 3] discussed above, in the following section, I will show that epistemic *huì* can and cannot go with perfective *le* under different circumstances and also examine the interaction of three future modals, namely *jiāng*, *huì* and *yào* with perfective *le*, so that a complete picture can be provided.

3 Future and Perfective Aspect

As discussed in the previous section, [2, 3] suggest or, at least, imply incompatibility between perfective *le* and epistemic *huì*. This generalization is on the right track in the sense that, in some cases, perfective *le* cannot go with epistemic *huì*. Yet, the picture is more complicated than what [2, 3] claim, for three reasons: (i) there are examples where perfective *le* is compatible with *huì*, (ii) *jiāng* and *yào* are not taken into account, and (iii) the three future modals *jiāng*, *huì*, and *yào* behave differently with respect to their compatibility with perfective *le* in some other instances. Let's examine examples that include *jiāng*, *huì* and *yào*.

(4) a. *Tā-men míngtiān jiāng/huì/yào kāi-le suǒyǒu-de huì.
3rd-PL tomorrow JIANG/HUI/YAO have-Pfv all-ASSO meeting
b. Tā-men míngtiān jiāng/huì/yào kāi-le suǒyǒu-de huì
3rd-PL tomorrow JIANG/HUI/YAO have-Pfv all-ASSO meeting
zài chīfàn.
and.then eat
'Tomorrow, they will not eat until they finish all meetings.'

[5] The term *reference time* used in this paper simply represents a time point relative to which a situation is located. For perfective *le* and epistemic *huì*, *reference time* is used to identify the time before or after the time which a situation is located. Readers should not confuse the term with R used in [5].

(5) a. *Tā-men míngtiān jiù jiāng/huì/yào xiě-le shí-fēng xìn.
3rd-PL tomorrow JIU[6] JIANG/HUI/YAO write-Pfv ten-CL letter
b. *Tā-men míngtiān xiàwǔ yì-diǎn jiù jiāng/huì/yào
3rd-PL tomorrow afternoon one-o'clock JIU JIANG/HUI/YAO
xiě-le shí-fēng xìng.
write-Pfv ten-CL letter
c. Tā-men míngtiān xiàwǔ yì-diàn jiù ?jiāng/huì/*yào
3rd-PL tomorrow afternoon one-o'clock JIU ?JIANG/HUI/*YAO
yǐjīng xiě-le shí-fēng xìn
already write-Pfv ten-CL letter
'By one o'clock in the afternoon tomorrow, they will have already written ten letters.'

In (4a), *jiāng/huì/yào* are not compatible with perfective *le*, while in (4b) the three future modals are rendered compatible. While in (5a, b) the three future modals cannot go with perfective *le*, yet, in (5c), *huì* is fine, *jiāng* is marginal but *yào* remains bad, to go with perfective *le*.[6]

If we examine these two sets of examples carefully, we can observe two things. First, in (4b), perfective *le* describes an event which occurs first in a series of two consecutive events. Second, *yǐjīng* 'already' appears in (5c), but not in (5a, b).

Given (4) and (5), two questions immediately arise. First, why can a series of consecutive events and *yǐjīng* 'already' render the three future modals compatible with perfective *le*? Second, why can't *yào* become compatible with perfective *le* with the assistance of *yǐjīng* 'already' and why is *jiāng* marginal with respect to *yǐjīng* 'already'?

For the first question, I argue that the reference time with respect to which perfective *le* locates a situation plays a key role. This explains the ill-formedness of (4a) and (5a). As argued in [8], under a neutral circumstance, perfective *le* locates a situation before the speech time. When it relies on the speech time, as suggested in [2, 3], perfective *le* is not compatible with the future modals, which locates a situation in the future of the speech time under the same circumstance.[7] Naturally, a temporal clash arises.

However, if the reference time for perfective *le* can be set to a time rather than the speech time, then the temporal clash can potentially be avoided. [9] proposes a mechanism to contextually determine the reference time on which perfective *le* depends. Describing a series of two situations, perfective *le* which presents the first one shifts

[6] Here, *jiù* expresses the speaker's comment that a situation occurs earlier than expectation. Please refer to [6,7] for a detailed discussion on *jiù*.

[7] Please note that I agree with [3] in that the future modals in Chinese are relative. That is, a future modal locates a situation in the future of a reference time, which is contextually determined. Under unmarked circumstances, the reference time is set to be the speech time, as discussed here.

the reference time it needs to time provided by the second situation.[8] That is, in (4b), perfective *le* locates the situation *they have all the meetings* before the time of *they eat*, instead of the speech time under neutral circumstances.

Yet, the shifting of reference time does not affect *jiāng*, *huì* and *yào* because the three future modals scope over the shifting. As a result, the reference time for the future modals is still the speech time. That is, the future modals locate the series of two consecutive events *they have all the meetings* and *they eat* in a time in the future of the speech time, i.e. *míngtiān* 'tomorrow' in this example. Now that in this example, perfective *le* depends on a reference time provided by the *eat* event, instead of the speech time, and that the future modals rely on the speech time, no temporal clash occurs. This is the reason why (4b) is well-formed.

However simply providing a different reference time is not adequate, and *yǐjīng* 'already' supports this statement. *Yǐjīng* 'already' has a very interesting behavior in that a temporal phrase preceding or following it plays different roles in the sentence. See the example below.

(6) a. Tā zuótiān yǐjīng líkái Táiběi.
3rd yesterday already leave Taipei
'He had left Taipei by tomorrow.'

b. Tā yǐjīng zài zuótiān líkāi Táiběi.
3rd already at yesterday leave Taipei
'He (already) left Taipei yesterday.'

As we can see from (6), a pre-*yǐjīng* temporal phrase functions as a reference time and this is why (6a) gets a *by yesterday* reading. Contrarily, a post-*yǐjīng* temporal adverbial functions as a time for a situation to be realized and this is why (6b) receives a *leave Taipei yesterday* reading. To put it in another way, *yǐjīng* 'already' provides the time indicated by a temporal phrase syntactically preceding it as a reference time.

But, as we can see from (5c), *yǐjīng* cannot render the future modals compatible with perfective *le*, at least not all of the future modals. While the pre-*yǐjīng* temporal phrase, other than the present time, serves as a reference time, why does *yǐjīng* in (5c) fail to improve the sentence? The answer lies in the fact that perfective *le* does not allow indefinite distance from a reference time. To state it differently, an event presented by perfective *le* needs to be 'reasonably' close to its reference time. For example,

(7) Sānmíngzhì, nǐ chī-le ma?
Sandwich 2nd eat-Pfv Q
'Sandwiches, did you eat them?'

Uttering (7), the speaker is not asking any *eat sandwiches* event that occurs in the past. Instead, the speaker is asking an *eat sandwiches* event that is most relevant to

[8] In some cases, perfective *le* has a function very similar to rhetorical relation *Narration* discussed in [10]. Please refer to [11] for an interesting discussion on the role of perfective *le* on determining a rhetorical relation.

the speech time. Most likely, (7) is asking whether the listener eats the sandwiches for today's breakfast (since in Taiwan, people tend to eat sandwiches only for breakfast) while the time for today's breakfast is understood to be close to the speech time. The closeness must be reasonable in the sense that it is close according to normal, unmarked standards.

The reference time induced by *yǐjīng* 'already', however, does not guarantee reasonable closeness. See the example below.

(8) Qíxiàn qián, wǒ huì yǐjīng xiěwān zuòyè.
Deadline before, 1st HUI already write-finish homework
'I will have finished the homework by deadline.'

For (8), the *finish the homework* event does not have to be (reasonably) close to the deadline. Rather, the event can be realized in an indefinite past by the deadline. That is, it is a possible scenario that the speaker has not finished the homework yet, but he will finish it by the deadline. Meanwhile, it is also possible that the speaker has finished the homework by the present time, i.e. he finished the homework at some indefinite past time.

A reference time of this type causes problems for perfective *le* because, as pointed out in [12], followed by [9, 13], perfective *le* presents a specific situation, which occurs at a specific time, although the specific time may or may not be explicitly stated in the sentence. A reference time that allows for an indefinite past cannot provide a specific time for a situation presented by perfective *le*. This is why *yǐjīng* 'already' in (5c) cannot successfully render all of *jiāng*, *huì* and *yào* compatible with perfective *le*.

The above generalization implies that, in (5c), if something can provide a specific time for the situation presented by perfective *le*, then (5c) should become well-formed. Epistemic *huì* salvages this situation because it relies on an epistemic modal base, as suggested in [14]. Using epistemic *huì*, the speaker makes a prediction based on some information (s)he has. The information that allows the speaker to make an affirmative statement of future occurrence of an event provides a specific time for the event.

Jiāng is marginal because of the future readings it expresses. [1] presents (at least) two types of future readings for *jiāng*: a report of future event and a planned future event. A report of future event does not necessarily provide a specific time and hence this reading is not good with perfective *le*. Nevertheless, if *jiāng* describes a planned future, since a plan for an event definitely includes a specific time for that event, therefore, this reading is compatible with perfective *le*.

As for the volitional *yào*, it does not rely on any information, which provides a specific time. Moreover, volition does not necessarily provide a specific time, as well. This is why volitional *yào* is incompatible with perfective *le*.

To sum up, we present two contexts under which perfective *le* is compatible with future modals *jiāng*, *huì* and *yào*: a sequence of (two) consecutive events and *yǐjīng* 'already'. For a sequence of consecutive events, because the second in the sequence provides a new reference time for perfective *le* to avoid temporal clash with future, the perfective *le* is rendered compatible with the future modals. For *yǐjīng* 'already', although it can provide a new reference time, yet it allows for indefinite past from the reference time, which contradicts the specific time requirement for perfective *le*. The modal base for epistemic *huì* provides a specific time for perfective *le*. Describing a planned event, *jiāng* can provide a specific time for the event, but *jiāng* does not have this function. *Yào* does not necessarily provide a specific time. Whether to provide a specific time or not determines the compatibility of perfective *le* with *jiāng*, *huì* and *yào*.

4 Conclusion

In this paper, I argue for the following points, concerning the compatibility of perfective *le* with future modals. First, the incompatibility between perfective *le* with *jiāng*, *huì* and *yào* is due to temporal clash. Second, if a new reference time is provided for perfective *le*, which can be distinguished from the reference time needed for future modals, then no temporal clash arises and perfective *le* can be rendered compatible with the future modals. Third, even though *yǐjīng* can provide a different reference time, it cannot resolve the incompatibility between perfective *le* and the future modals because it allows for indefinite past and hence cannot provide a specific time for perfective *le*. *Jiāng*, *huì* and *yào* behave differently in their compatibility with perfective *le* in a sentence where *yǐjīng* 'already' provides a new reference time, depending on whether they can provide a specific time for a situation presented by perfective *le*.

Acknowledgements. The research reported here is partially funded by National Science Council (now known as Ministry of Science and Technology), Taiwan, under the grant number NSC 97-2410-H-415-015. I hereby acknowledge the financial support of National Science Council. This NSC project originates from a conference course I took with late Professor Carlota Smith. I am deeply grateful for her enlightening and insightful viewpoints on linguistics.

References

1. Wu, J., Kuo, J.: Future and modality: a preliminary study of *jiang*, *hui*, *yao* and *yao… le* in mandarin Chinese. In: Clemens, L., Liu, C. (eds): On-line Proceedings of NACCL 22 & IACL 18. Columbus, Ohi, pp. 54–71 (2010)
2. Lin, J.: Time in a language without tense: the case of Chinese. J. Semantics **23**, 1–53 (2006)
3. Liu, H.: Guó-Mǐn-Kè Yǔ de dòngtài wénfǎ tǐxì jí dòngtàicí de shàngjiā dòngmào yǔyì [The Mood System and Semantic Interaction between Mood and Aspects in Mandarin Chinese, Taiwan Southern Min and Hakka]. Crane Taipei (1997)
4. Wu, J.: Ambiguity of epistemic *Hui* in mandarin Chinese revisited: toward a unified semantics. In: Su, Q., Zhan, W. (eds): From Minimal Contrast to Meaning Construct: Corpus-based Near-synonym-driven Chinese Lexical Semantics. Frontier in Chinese Linguistics 9. Beijing University Press, Beijing, Springer-Verlag, Berlin, Heidenberg, New York, pp. 51–62 (2020)
5. Reichenbach, H.: The tense of verbs. In: Elements of Symbolic Logic. Macmillan, New York (1947)
6. Big, Y.: The Semantics and Pragmatics of *Cai* and *Jiu* in Mandarin Chinese. Crane, Taipei (1987)
7. Lai, H.: Rejected expectations: two scalar particles *Cai* and *Jiu* in mandarin Chinese. Linguistics **37**, 625–661 (1999)
8. Lin, J.: Temporal reference in mandarin Chinese. J. East Asian Ling. **12**, 259–311 (2003)

9. Wu, J.: The semantics of perfective *le* and Its context-dependency: an SDRT approach. J. East Asian Ling. **14**, 299–366 (2005)
10. Asher, N., Lascarides, A.: Logics of Conversation. Cambridge University Press, Cambridge (2003)
11. Wu, J.: Interactions between aspect and temporal relations: a case study of perfective le. Lang. and Ling. **11**, 65–98 (2010)
12. Li, T., Thompson, S.: Mandarin Chinese: A Functional Reference Grammar. University of California Press, Berkeley (1981)
13. Smith, C.: The Parameter of Aspect, 2nd edn. Kluwer, Berlin (1997)
14. Kratzer, A.: Modality. In: von Stechow, W. (ed.) Semantics: An International Handbook of Contemporary Research. de Gruyter, Berlin (1991)

AI, Big Data, and NLP

Comparison and Sense Induction of Temporal Adverbs *Reng* and *Hai*: A Corpus-Based Study

Wenxi Li(✉)

Department of Chinese Language and Literature, Peking University, Beijing, China
liwenxi@pku.edu.cn

Abstract. The identification of near-synonyms is a major concern in semantic research, whose key is listing their senses clearly. Using *People's Daily Corpus* as data support and frame semantics as a research idea, this paper presents the general principles and methods of word sense induction. After applying this method, a comprehensive comparison of *reng* and *hai* is conducted based on the lists of senses: both *reng* and *hai* have the original meaning "continually". However, if the verbs modified by *hai* are action verbs, it may derive the sense of "additionally"; when co-occurring with contrast conjunctions, another sense of *hai* modifying adjectives, "almost" is triggered. These findings provide strong support for the validity of this inductive approach to lexical meaning.

Keywords: Word sense induction · *Reng* · *Hai* · People's daily corpus · Frame semantics

1 Introduction

As members of Chinese temporal adverbs more or less equivalent to *still* in English, *reng* and *hai* both indicate that the states or actions denoted by the verbs continue, and their senses and functions are similar to some extent. It can be observed that they are interchangeable in some contexts:

(1) *Zhe zhong xuexi fangshi* ***reng/hai*** *shi jiaoyu*
this type. Learning method still is education
de weiyi xuanze
DE only choice
'This learning method is still the only choice of education.'

(2) *Sanshi nian lai, zhe zhong bizhong* ***reng/hai*** *zai*
thirty years from this kind share still in-progress
zengjia
increase
'This share is still increasing over the last three decades.'

Nonetheless, they lack interchangeability in some sentences since the substituted results are ungrammatical or their meanings are different from the original sentences, showing subtle nuances in meaning and usage of *reng* and *hai*:

M. Liu et al. (Eds.): CLSW 2020, LNAI 12278, pp. 371–385, 2021.
https://doi.org/10.1007/978-3-030-81197-6_31

(3) *Wo na shihou nianling* **hai/*reng** *hen xiao*
I that time age still very young
'I was very young at that time.'

(4) **Hai/*Reng** *bu dao liming, yuanzi li bian you le*
still not come dawn yard in then have ASP
yueguang
moonlight
'Before dawn, there is moonlight in the yard.'

As a pair of near synonyms, what kind of similarity exists in their meanings, and under what conditions they cannot be substituted by each other? In order to answer these questions, what follows first is a brief introduction to the previous research on the lexicography of *reng* and *hai*.

Many dictionaries use *reng* to interpret the senses of Hai and vice versa. For instance, The Eight Hundred Words of Modern Chinese lists four categories and thirteen subcategories of the senses of *hai*, and when *hai* is used "to indicate the continuing action or state" or "to denote growth of number or expansion of scope", it can be substituted by *reng*. In the case of *reng*, its senses include (i) an unchanged situation, which is less common than *haishi* colloquially and (ii) restoration to the previous state. Another representative example is Modern Chinese Dictionary, where the first sense of *hai* is "to indicate that the state or action continue to exist or to be carried on". However, these criteria are not clear enough since many linguistic phenomena cannot be clarified after applying them. For example, (5) and (6), where *hai* both indicate an increase in the number or scope, are different from each other on whether it can be replaced by *reng*. *Reng* and *hai* in example (7) are interchangeable, although *reng*'s sense is "restoration to the previous state" and not interpreted by *hai*.

(5) *Jiu de maodun jiejue le, xin de maodun*
old DE contradiction resolve ASP, new DE contradiction
hai/reng *hui chansheng*
still can arise
'New contradictions arise when old ones are resolved.'

(6) **Hai/*Reng** dei zai lai ji ge ren.
still need again come a-few M person
'We need a few more people.'

(7) Ta **hai/*reng** *danren chejian zhuren*
he still serve workshop manager
'He still serves as the workshop manager.'

In addition to dictionaries, a number of articles on the semantics of adverbs also study *reng* and *hai*. Ma and Lu (1985) discuss *hai*, identifying its use in three ways: indicating degree, repetition or addition of an action, and time. They also point out that when being used to express temporal concepts, *reng* and *hai* are basically synonymous. Zhu and Wang (2013) specifically compare the two adverbs. *Reng* and *hai* are interchangeable in a context where an action or state may change as time goes by. However, if the destabilizing factors are stated in the forgoing sentences, especially when these factors are initiatives taken by animate subjects, *reng* cannot generally be replaced by *hai*. On the other hand, when expressing some changes are inevitable over time but not at the time point of change, *hai* cannot be substituted by *reng*.

The above review provides us with some useful information on the semantic or functional comparison of these two adverbs, but the accuracy and explanatory power of the existing studies are still insufficient. *Reng* and *hai* are treated independently but not as a part of strings. None of them pays attention to what classes of words (adjectives, adverbs, or verbs) are modified by this pair of synonymous adverbs or what specific types of adjectives/verbs they are most likely to modify respectively. In view of this limitation, this study aims to verify the existing research findings mentioned above and reveal the unexplored semantic information of *reng* and *hai* through a more comprehensive investigation of relevant linguistic phenomena, and to conduct a comparative study of them.

2 Methodology

2.1 Corpus-Based Approach to Word Sense Induction

Research on synonyms begins with a clarification of their senses, a task that is challenged by polysemy. The same is true of *reng* and *hai* and thus the induction of senses is a key to compare their semantic features since only after senses consistent with cognition and logic are established can researchers discuss their relationship. In this process, the most essential question to answer is to what extent can a semantic difference be considered as an independent sense but not a derived meaning.

Cognitive linguists view polysemous words as a web of interrelated meanings, and the criterion for inducting senses is whether the related underlying mental representations are different. An example is Lakoff's (1987) study of the English preposition *over*, arguing that *over* followed by vertical or planar locations reflects different mental processes, and thus can be grouped into two senses. Nonetheless, this "full-specification" approach is controversial because it is not only highly arbitrary in the selection of features and inconsistent for different words, but also leads to too fine-grained senses (Sandra and Rice, 1995). Therefore, Tyler and Evans (2001) provide a new provision for inducting senses that a sense can be separated only in the case where *over* involves a totally different spatial structure and this difference cannot be inferred from encyclopedic knowledge or contextual information.

Although this cognitive-based approach pays some attention to the correlation between meaning and collocations: *over* is dependent on the characteristics of its following nouns, it still relies heavily on the subjective judgments of linguists and lexicographers, and a comprehensive examination of the corpus data is not provided, resulting in many contradictions and dissimilarities. Researchers thus resort to corpus linguistics in the hope that it can provide evidence for semantic delineation. What follows in this section is an introduction to a corpus-based approach to word sense induction.

Distribution is one of the most important methods of corpus analysis that is available to researchers and corpus-based lexical semantics is not an exception to do away with it. The basic assumption is that the distributional features of a word can reveal its syntactic and semantic characteristics. As corpus research progresses, more and more linguists are aware of that the influences on semantics may come from syntactic and contextual levels. Firth (1957) famously summarizes this principle as "you shall know a word by the company it keeps"; Bolinger (1968) states that "a difference in syntactic form always spells a difference in meaning". In other words, the interpretation of a word's meaning depends on the frequencies and features of its co-occurring words. Some scholars even hold the view that conventionally-defined *meaning* of a word does not exist, and instead it is actually derived from a string co-occurring with this word. Moreover, these research theories also make it clear that the use of a corpus is not only an aid to test the results of induction but also an important and necessary tool to facilitate the inducting process.

2.2 Frame-Semantic Approach to Word Sense Induction

There are infinite combinations of words. Without restricting the data in the corpus, meanings of words are considered to be different in different contexts and word sense induction will be meaningless. Therefore, it is necessary to find the most essential collocations that determine the meanings of a word.

Previous researches on lexical semantics taking a corpus-based approach vary in the selection of syntactic constituents according to the features of target words. Atkins (1987) names the constituents that can make a distinction between meanings as "ID tags" when studying the English verb *danger.* These "syntactic or lexical tags in a string" that point to a specific meaning include the roles of phrases containing *danger*, complements of *danger* and seven words that go with *danger*. Levin (1993) argues that for verbs that have different argument structures, they should be treated as two separate words. Certain morphological features can also be used to determine senses. Furthermore, for adjectives, both Justeson and Katz (1995) and Gries and Otani (2010) believe that the best solution for generalizing adjectival meanings is to examine the semantic features of nominal components and the syntactic functions of adjectives themselves. However, research in the past did not paid adequate attention to adverbs. This is not only because of the enormous semantic difference of adverbs, but also because the syntactic positions of adverbs are flexible. In light of this, this paper aims to answer how to define the distributional elements that affect the definition of an adverb so as to better portray its semantics. The studies of lexical semantics of verbs and adjectives described above inspire us to opt for the following elements that influence the semantic generalization:

(1) Syntactic elements. For example, the heads on which the target word depends and the subordinate elements that the target word modifies.
(2) Lexical elements. For example, the part-of-speech and morphological variations of the target word and words in a window of certain sizes.
(3) Semantic elements. For example, the semantic features of the target word and words in a window of certain sizes.

Therefore, for the adverbs to be studied in this paper, the above three categories of elements can be identified as follows: syntactic elements are the types of core components, such as verbs and adjectives modified by adverbs, or the types of modifying components with *reng* and *hai* as core components; lexical elements, such as word class differences, are also an important basis for classifying adverbial senses, but morphological changes are not taken into account because they are not that rich in Chinese; as for semantically related elements, they mainly refer to semantic features related to predicates modified by adverbs.

The above-mentioned research methods highlight the contribution of generalized arguments and semantic roles to lexicon, which is in line with frame semantics. This scheme treats semantics as a frame to hold elements and their relations. Frames are assumed to reflect underlying semantic similarities and differences. All the factors above can be viewed as generalized arguments required by different subcategorized frames, for the purpose of putting senses in separate groups. English lexical semantic repositories, as in VerbNet, FrameNet, PropBank, and etc., though varying in the criteria and granularity of semantic roles, they all provide a description of the semantic roles that words carry. Moreover, extracting the elements that affect the meaning of a word and then generalizing the meaning can also be seen as a precursor to computational work on word sense disambiguation and induction requiring feature extraction and vector representation of contextual information and its basic work is to define features by observing and describing corpus data. Although they may take different approaches to sense induction, a general idea is that certain contextual elements determine the meaning of words, and these elements need to be properly extracted and represented. They are called "semantic roles" in frame semantics and in computational linguistics "features". The generalization of senses is the process to classify these "semantic roles" or "features".

It should be noted that this method is not identical to the co-occurrence method of studying the collocated components of target words, which dismembers the context to some extent in the sense that considering only co-occurring components will ignore the fact that components in different syntactic structures contribute differently to semantics. Therefore, the following discussion will focus on the components that are grammatically and semantically related to *reng* and *hai*, including mainly the lexical and semantic characteristics of heads modified by adverbs, as well as the lexical and semantic characteristics of the components co-occurring with targets words within a certain distance from the adverbs.

2.3 Corpus Data and Sources

Using the raw texts of the People's Daily in 1998, Fujitsu Research & Development Center and the Institute of Computational Linguistics, Peking University, have jointly developed a Chinese corpus, which is named People's Daily Corpus. It provides fundamental linguistic data such as word segmentation and POS tags. One limitation is that the core components and other syntactically related components of *reng* and *hai* cannot be directly examined since the corpus does not include any information related to syntactic structure. Therefore, we applied Stanford Parser, a syntactic analysis tool developed by Stanford University, to obtain the dependency relations between different components based on Dependency Grammar.

3 Results

3.1 Comparison of the Core Elements Modified by *Reng* and *Hai*

Types of Words Modified by *Reng* and *Hai*. This paper first examines the lexical distribution of the components modified by *reng* and *hai* in the corpus. It should be noted that the People's Daily Corpus is based on a more detailed tagging specification than the commonly used POS system and thus the data need to be carefully filtered and analyzed. For example, this specification sets the classes /*i* and /*l* for idioms and proverbs respectively. However, in counting the distribution of target words, we should classify these words according to the syntactic functions in the context. The resulting distribution of the core components modified by *reng* and *hai* is shown in the following table.

Table 1. Distribution of core elements modified by *reng* and *hai*.

	Verb		Adjective	
reng	270	91.44%	22	8.56%
hai	1448	90.33%	155	9.67%

As we can see, *reng* and *hai* mainly modify verbs, with verbs accounting for more than 90% of the total and adjectives for less than 10%. Moreover, *reng* and *hai* are consistent in the distribution of modified word types, which cannot be influenced by the distance between the core components and target words. If the proportions of verbs and adjectives appearing in different positions are counted separately, they still present a situation that the verbs dominate overwhelmingly.

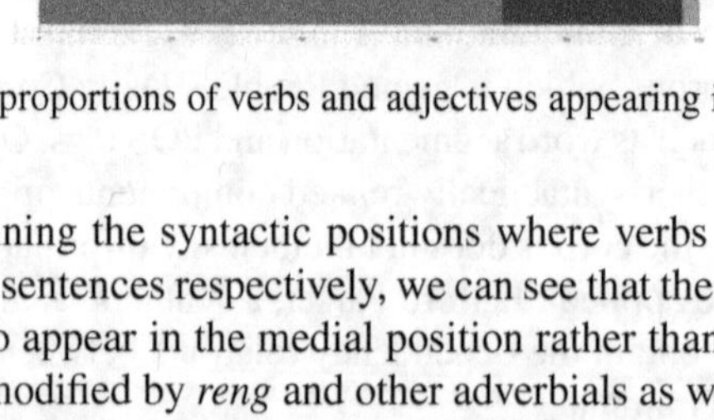

Fig. 1. Visualized proportions of verbs and adjectives appearing in different position

However, if examining the syntactic positions where verbs and adjectives tend to appear in *reng* and *hai* sentences respectively, we can see that the adjectives modified by *reng* are more likely to appear in the medial position rather than in the initial position, mean they tend to be modified by *reng* and other adverbials as well. But for verbs, their tendency to appear in either position is essentially the same. This distinction between verbs and adjectives is illustrated visually in the following table.

Why does the distribution of adjectives differ so much for *reng* and *hai*? This paper examines what happens when adjectives are located in the initial position of *hai* and

Table 2. Proportions of verbs and adjectives appearing in different positions.

		Initial	Medial	Final
reng	Verb	97.22%	78.16%	100%
	Adj	2.78%	21.84%	0%
hai	Verb	92.66%	84.35%	88.46%
	Adj	7.34%	15.65%	11.54%

Table 3. Proportions of verbs and adjectives appearing in different positions.

		Initial	Medial	Final
Adj	*hai*	50.98%	45.10%	3.92%
	reng	5.00%	95.00%	0%
Verb	*hai*	70.20%	26.52%	3.28%
	reng	70.28%	27.31%	2.41%

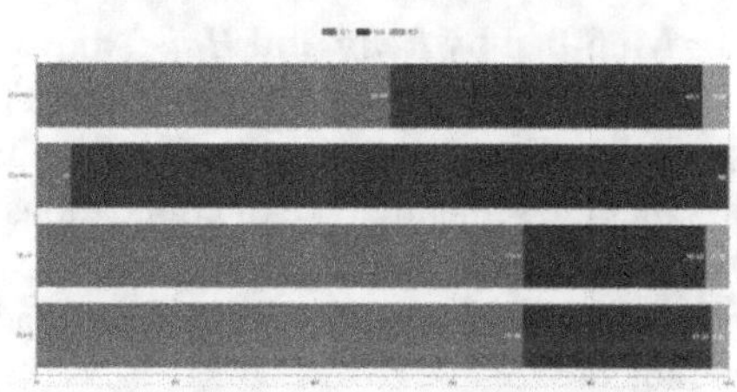

Fig. 2. Visualized respective proportions of verbs and adjectives appearing in different position

finds that its typical context is comparative sentences, for which a number of dictionaries encode the meaning of *hai* as "even more". Nonetheless, the meaning of *hai* in this context is actually a derivation of the original meaning "continually". A usage like example (8) gives a sequence of running speeds, and after comparing the running speeds of "many lines" with others from high to low, the speaker finds that even compared with "standards in the early 1940s" which were already very low, it still maintains a state of "low". This can explain why *hai* in comparative sentences often implies that objects being compared share a similar property.

(8) *Xuduo xianlu de yunxing sudu shenzhi bi ben shiji*
many line DE running speed even than this century
40 niandai de biaozhun ***hai*** *di*
40 decade DE standard still low
'The running speed of many lines is even much lower than the standard in 1940s.'

In addition to comparative expressions, *hai* in the sense of "still" can also directly modify an adjective, subject to certain restrictions: (i) in interrogative sentences; and (ii) one morpheme of the adjective is "not". They are shown in example (9) and (10) respectively.

(9) *Ni* *xianzai* ***hai*** *hao* *ma?*
You now still okay Q
'Are you okay now?'

(10) *Changpian* *xiaoshuo* *de* *chuangzuo* ***hai*** *bugou*
long novel DE writing still not-enough
'Long novel writing is still not enough.'

As shown in example (11), *hai*, to denote a degree of "barely", can also directly modify an adjective. The mechanism to derive this sense is discussed in more detail in Sect. 3.2.

(11) *Zhe* *jia* *fandian* *zhuangshi* *ye* ***hai*** *kaojiu.*
this M hotel decoration also barely exquisite
'The decoration of this restaurant is also barely exquisite.'

Semantic Types of Verbs Modified by *Reng* and *Hai*. Above examination of the frequency and distribution of words modified by *reng* and *hai* provides us with some insights into their syntactic and semantic information. Furthermore, as stated in Sect. 2.1, corpus-based research also shows that the semantic type of collocations is also a very effective element to determine their functions and semantics. Therefore, we queried the top 20 verbs most frequently occurring with and modified by these two adverbs. The proportions of them are presented in Table 4.

Table 4. Proportions of top 20 verbs modified by *reng* and *hai.*

reng									
you	*shi*	*chuyu*	*jixu*	*cheng*	*buneng*	*baochi*	*cunzai*	*weichi*	*chixu*
13.48%	9.57%	3.48%	3.48%	3.04%	2.61%	2.17%	2.17%	2.17%	1.74%
neng	*wu*	*ju*	*wei*	*chi*	*gaoyu*	*da*	*ying*	*yao*	*dadao*
1.74%	1.74%	1.74%	1.74%	1.30%	1.30%	1.30%	1.30%	1.30%	1.30%
hai									
you	*shi*	*keyi*	*shuo*	*meiyou*	*hui*	*neng*	*ying*	*cunzai*	*biaoshi*
6.10%	6.00%	3.05%	2.67%	2.38%	2.29%	2.00%	1.71%	1.43%	1.24%
jueding	*baokuo*	*dei*	*qiangdiao*	*xuyao*	*juban*	*chuyu*	*dao*	*zuo*	*chuxian*
1.24%	1.14%	1.14%	1.14%	1.05%	0.86%	0.76%	0.76%	0.76%	0.67%

Moreover, with regard to *reng* and *hai* indicating the "aspect" of a verb, that is, the way in which an action is started, completed, and continued, it is necessary to examine the temporal properties of the verb. Temporal properties do not simply refer to the time words that are associated with the action. It should also include the intrinsic temporal properties of a verb, i.e., the beginning, continuation, and end point of an event and how these elements are represented. Guo (1993) divides Chinese verbs into five types according to the presence or absence of three elements and the differences in their strengths and weaknesses. We apply them to classify the temporal properties of the above verbs. But a few points need to be noted (i) The classification system is based on the Dictionary of Chinese Verbs, which gives the temporal attributes of each verb. However, a verb may have several senses whose temporal properties are different, so that we need to make judgments according to the context. (ii) Some verbs are not included in the Dictionary of Chinese Verbs and we need to judge their attribution according to the classification criteria. (iii) There are some verbs that cannot find their proper place according to Guo's (1993) classification criteria, and then have to be eliminated. (iv) The People's Daily corpus is biased towards the written style, with some bisyllabic words tending to be used in monosyllabic forms and should be reduced. After these corrections, we obtain the following results.

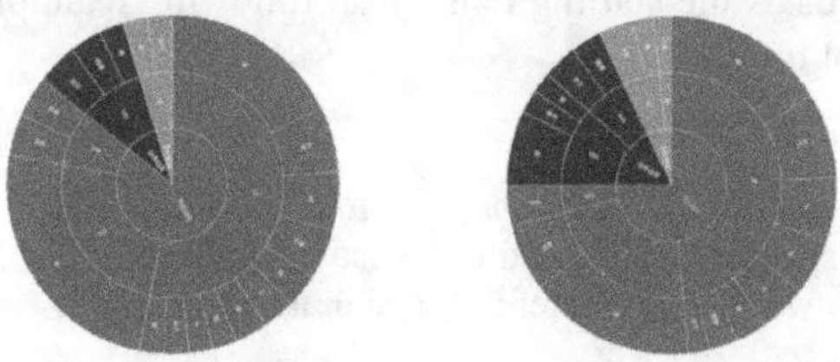

Fig. 3. Visualized semantic types of top 20 verbs according to the criteria of Guo (1993)

From the visualized results above, we can conclude that although the verbs modified by *reng* and *hai* are mainly state verbs, the most obvious difference is that a verb modified by *hai* is more likely to be an action verb than one modified by *reng*. Not only is this reflected in the higher proportion of action verbs in the diagram but also in that *reng* only modifies the action verb meaning "sustain", which is consistent to original sense of *reng*. On the contrary, the most common action verbs modified by Hai are "say" and "do". In addition, the top 20 verbs most frequently modified by *reng* do not include the most typical action verbs which belong to verb type V5.

The difference in the temporal properties of verbs is also the reason why *hai* derives a new sense. State verbs which are static and represent a homogeneous process are semantically compatible with *reng* and *hai*: when the speaker specifies a time point and goes back along the time axis, if the event indicated by the state verb continues to this time point, it can be modified by *reng* and *hai*. This usage further extends to action verbs: if the event denoted by an action verb happens twice or more times, it can also be regarded as a continuation to some extent. Nonetheless, since action verbs present discrete actions in time axis, the semantics of "continually" is progressively weakened and "additionally" is preferred to denote this state. A new sense of *hai* is thus derived.

Of course, not all sentences with action verbs modified by *hai* indicate "additionally", and nor do all sentences with state verbs indicate "continually":

(12) *Ta* **hai** *fendou zai gongzuo gangwei shang*
he still struggle in job post in
'He still works hard at his job.'

However, the meaning of *reng* is relatively fixed. As the Example (14) shows that it can also modify action verb. But this usage is infrequent and does not extend to the meaning of "additionally".

(13) *Zhe zhang heka* **hai** *you yi duan gushi.*
this M card additionally have one M story
'This card even has a little story.'

3.2 Comparison of the Components Modifying by *Reng* and *Hai*

Comparison of Temporal Words. Following the above examination of the intrinsic temporal properties of a verb in *reng* and *hai* sentences, this section moves on to external temporal properties, i.e., where the events denoted by the verbs locate in the time axis, taking temporal words as the starting point. The following examples show how *hai* is expressed at different times.

(14) *Beifang chuntian* **reng** *chi dongtian de baicai*
North spring still eat winter DE cabbage
'In the north, we still eat winter cabbage in the spring.'

(15) *1976 nian, ta* **hai** *shi shate de*
1976 year, he still is Saudi Arabia DE
neizheng buzhang.
interior minister
'He was still Saudi Arabia's Minister of Interior in 1976.'

(16) *Zhe zhong buliang yingxiang* **hai** *wei xiaochu.*
this kind negative effect still not eliminate
'This negative effect has not yet been completely eliminated.'

Do *reng* and *hai* have different tendencies in selecting temporal words? To answer this question, we extract the temporal words from the People's Daily corpus and analyze their frequencies, and obtain a list of tokens with a frequency $\geq$2%. The temporal words in *reng* sentences are: *muqian* (26.39%), *xianzai* (6.94%), *qunian* (6.94%), *1997* (6.94%), *jinhou* (5.56%), *1995* (2.78%) and *1998* (2.78%); the *hai* sentence includes temporal words such as *xianzai* (9.89%), *muqian* (6.59%), *jinnian* (6.59%), *jintian* (4.95%), *dangshi* (3.30%), *qunian* (3.30%), *1998* (3.30%), *1997* (2.75%), *yishi* (2.75%), *dangnian* (2.20%) and chunjie (2.20%). These words are used very frequently with *reng* and *hai*, accounting for 58.33% and 47.82% proportion of sentences with temporal words, which is a relatively comprehensive reflection of the distribution of various temporal words. Using past, present and future as the criteria to make a detailed distinction between these temporal words, we obtain the following results:

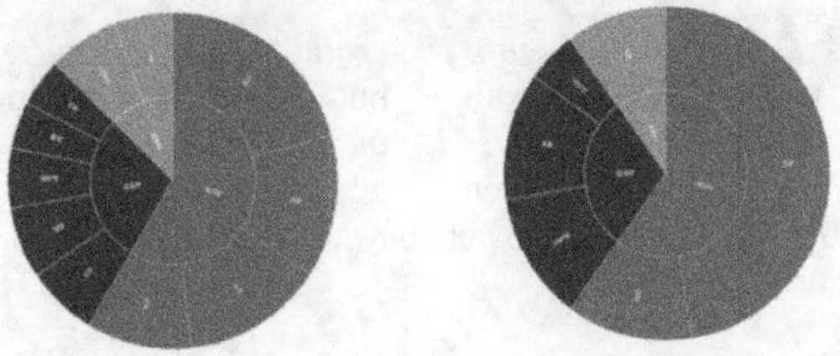

Fig. 4. Visualized distributions of temporal words in *reng* and *hai* sentences

As can be seen, frequencies of *reng* and *hai* involving in past, present and future are 28.56%, 57.14%, 14.30% and 28.75%, 58.60%, 12.65%, respectively, indicating no sharp differences. However, though the examination of corpus data, we find that *hai* is very sensitive to whether a temporal word refers to a time point or a time period. This have a significant impact on the semantic of *hai.* On the contrary, temporal words in *reng* sentences have no such restrictions:

(17) Ni	mingtian	***hai***	zai	zhe	deng	wo.
you	tomorrow	still	in	here	wait	I

'You can still be here to wait for me tomorrow.'

(18) ***Jinnian,***	*guojia*	*jiaowei*		*hai*	*he*	*youguan*
this-year,	state	education-commission		additionally	with	relevant
bumen	*zhaokai*	*le*	*jiaoyu*	*fazhi*	*huiyi*	
departments	hold	ASP	education	legal	conference	

'This year, the State Education Commission, together with other relevant departments, held a national conference on the legal system in education.'

In (18), *hai* means "additionally" while in (19) it is used in the same sense as *reng*, meaning "continually" or "still", i.e., the State Education Commission continues the practice or tradition of holding a national conference on education and legal systems.

This affects sense of *hai* to a greater extent than the semantic type of verbs modified by *hai.* However, this cannot be applied to state verbs: a *hai* sentence still conveys a continuation, regardless of the type of temporal words.

Comparison of Conjunctions. Conjunctions in *hai* sentences may determine the sense of *hai* to some extent, which can be illustrated in the following two sentences.

(19) ***Zhidao***	***jinnian,***	*guojia*	*jiaowei*		*hai*	*he*	*youguan*
Until	this-year	state	education-commission		still	with	relevant
bumen	zhaokai		le	jiaoyu	fazhi	huiyi.	
departments	hold		ASP	education	legal	conference	

'Until this year, the State Education Commission, together with other relevant departments, still held a national conference on the legal system in education.'

(20) ***Budan*** *xi* *le* *zao,* ***erqie*** *ming* *hai/reng* *chi*
not-only take ASP bath, but-also Ming additionally/still eat
le *zaocan* *cai* *qu* *xuexiao*
ASP breakfast then go school
'Not only did Ming take a shower, but Ming also/still ate his breakfast before going to school.'

We can see that the action verb "eat" in (20) modified by *hai* is a supplement. Furthermore, this sense is accentuated by the coordinate conjunctions of (20), while it is suppressed and can only be understood as "continually" or "still" in sentences associated by contrast conjunctions. However, in *reng* sentences, regardless of conjunctions, they can only be interpreted as "still". This difference is also reflected in the use of conjunctions in both cases. The frequencies of two types of conjunctions are presented below:

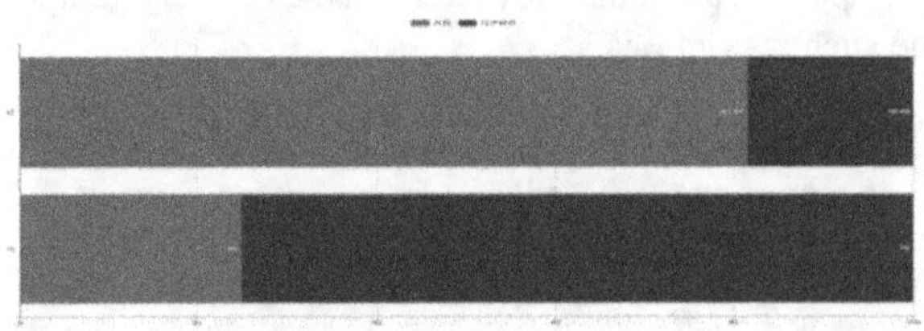

Fig. 5. Visualized frequencies of two types of conjunctions

In other words, in a sentence where two contrast ideas are joined, the former clause always expresses a negative condition while *hai* is used to convey the meaning of keeping doing it or keeping certain properties. However, in such an unsatisfactory situation, the degree of continuation is likely to be weakened. In this way, the sense of barely and almost of *hai* is triggered, as shown in the following example.

(21) ***Suiran*** *jintian* *shijian* *jinji,* ***dan*** *ming* *hai/reng* *chi*
although today time urgent, but Ming additionally/still eat
le *zaocan* *cai* *qu* *xuexiao*
ASP breakfast then go school
'Despite of the limited time today, Ming additionally/still ate his breakfast before going to school.'

Example (22) indicates that even though there is a drought in the autumn, the summer harvest remains in a good state, but this state is obviously affected by the "drought" and thus can only reach a barely satisfactory level.

4 Findings

The Modern Chinese Dictionary lists six senses of *hai*: (1) indicating that a phenomenon continues to exist or an action continues; (2) indicating an increase or addition to a certain extent; (3) meaning that the level is barely adequate and basically meets some requirement; (4) joining with *ne* to provide a comparative standard, equivalent to *shangqie*; (5)

expressing "unreasonably", "unusually" or "unexpectedly" (mostly for tone of admiration, reproach, or sarcasm); (6) indicating that this is already the case, often echoing *jiu*.

Nonetheless, as examined and discussed above, the meanings of *hai* that can be reflected in syntactic form only include the following.

(1) "Still", modifying state verbs in interrogative and comparative sentences, or modifying adjectives with a morpheme *bu* or being modified by other degree adverbs.
(2) "Additionally", modifying action verbs with no temporal words to indicate time period or contrast conjunctions.
(3) "Barely", modifying common adjectives.

The main difference between this system and that of the Modern Chinese Dictionary is that "still" is used to explain *hai*, which leads to a hierarchical and derivable sequence, so that the sense "also" in the comparative sentence is also included (see Sect. 3.1). In this way, we can also explain why some above mentioned senses of the Modern Chinese Dictionary actually have a strong correlation and should be interpreted as the same meaning, and the so-called "unreasonably" or "unexpectedly" are only a difference due to pragmatic factors.

There is also an intrinsic connection between the above three senses. Given the frequency of *reng* modifying action verbs increases, its core meaning "still" contradicts the action verb, which gives a rise to the meaning of "additionally". At the same time, the meaning "still" and sentences with contrast conjunctions are semantically compatible. Nonetheless, the former clause expresses negative conditions and weakens the possibility maintaining the action or state. *Hai* absorbs this contextual meaning to derive the sense "barely".

As for the meaning of the adverb *reng* is concerned, the Modern Chinese Dictionary identifies only one sense, "still". On the other hand, the Eight Hundred Words of Modern Chinese, identifies two: "a situation continues unchanged" and "restitution". In our view, there is no difference between the two apart from contexts where an action or state persist or disrupted. Therefore, for the sake of simplicity, only one definition needs to be used here.

Lin (1998) states that "if a word is used in the same linguistic context, it carries the same meaning. If two different words appear in the same linguistic context, they are likely to have the same meaning". This observation can also be applied to *reng* and *hai*: given the differences and similarities of the contexts in which they are used, they are synonymous to some extent. However, *reng* cannot be used frequently as modifiers of action verbs while *hai* has this function, thus deriving a new sense "additionally".

5 Conclusion

The principles of word sense induction under different linguistic domains have different emphases on the relationship between language and cognition. Theories of cognitivism emphasize the importance of a person's understanding and reasoning about the world, which are reflected by the relationships between senses. Corpus linguistics, on the other hand, examines semantics primarily through the syntactic contextual objects which are observable in the language. The frame semantics combines certain cognitive patterns with lexical-syntactic information. Cognitive linguistics and corpus linguistics follow processes of top-down decomposition or bottom-up induction respectively, the two ways that we study language: starting from linguistic experience already developed in our mind, or examining language-specific phenomena. However, both can lead to the problem of over-classification without restrictions, so it is necessary for frame semantics to standardize the results and achieve a better balance between the scope and depth of classification, thus better facilitating linguistic research.

The results of our study of two near-synonymous adverbs *reng* and *hai* also indicate that corpus-based and frame semantic analysis are viable and effective methods for adverb study. They are applicable not only to verbs and adjectives, but also to adverbs. However, the unique natures of adverbs, e.g., their vacuous meanings and individual variation, also bring us with significant challenges. In order to address these issues, we propose the following research methods and principles.

First of all, it is necessary to adopt diverse semantic frameworks for adverbial lexemes in response to their particularity. Existing research mainly examines the semantic roles that verbs carry or the typical nouns that adjectives modify, but does not take some key elements of adverbs which are not identical to those of adjectives into account. This may be because adverbs can modify verbs, adjectives, and adverbs, and have flexible and versatile usages. This phenomenon requires researchers to examine more than argument structure, so as to obtain comprehensive semantic information. In this paper, for instance, conjunctions and temporal words bring important semantic information to the interpretation of *reng* and *hai* though most modifiers cannot affect the semantics of verbs and adjectives. Therefore, words with vague meanings should not be neglected in studying adverbs.

We also need to identify the difference between adverbial lexemes and find the ones that contribute the most to semantics. In this paper, *reng* and *hai* are temporal adverbs indicating how an action occurs and continues, so the most noteworthy thing is the predicates to be modified and their semantic types. However, for synergistic adverbs like *yiqi* and *gongtong*, research should focus more on the subject of the sentence, namely on the originator of an action. For some adverbs, their positions in a sentence can even affect their function and usage (Liu 2010). Generally speaking, common factors that contribute to an adverb's meaning include (i) the position of the adverb as a modifier; and (ii) the semantic features of words modified by the adverb.

However, there are still limitations of this approach as it cannot be widely used in automatic sense disambiguation and induction. One of the biggest problems is that there is no prescribed procedure for the induction. A researcher needs to look closely at the features of the synonyms under examination to determine which micro-procedure is the best. This brings a great challenge to computational lexical semantics.

Acknowledgement. This work is supported by "the History of Modern Chinese Common Language" (18JJD740001), a major project of the key research base in humanities and social sciences of the Ministry of Education. We also wish to thank reviewers and editors for their helpful comments.

References

Atkins, B.T.: Semantic ID tags: corpus evidence for dictionary senses. In: Proceedings of the Third Annual Conference of the UW Centre for the New Oxford English Dictionary, vol. 1736 (1987)

Biber, D.: Co-occurrence patterns among collocations: a tool for corpus-based lexical knowledge acquisition. Comput. Linguist. **19**(3), 531–538 (1993)

Firth, J.R.: Papers in Linguistic Analysis 1934–1951. Oxford University Press, London (1957)

Gries, S.T.: Corpus-based methods and cognitive semantics: the many senses of to run. In: Corpora in cognitive linguistics: Corpus-based approaches to syntax and lexis, pp. 57–99 (2006)

Gries, S.T., Otani, N.: Behavioral profiles: a corpus-based perspective on synonymy and antonymy. ICAME J. **34**, 121–150 (2010)

Hanks, P.: Contextual dependency and lexical sets. Int. J. Corpus Linguist. **1**(1), 75–98 (1996)

Hanks, P.: Do word meanings exist? Comput. Humanit. **34**(1/2), 205–215 (2000)

Kishner, J.M., Gibbs, R.W., Jr.: How "just" gets its meanings: polysemy and context in psychological semantics. Lang. Speech **39**(1), 19–36 (1996)

Lakoff, G.: Women, Fire, and Dangerous Things. University of Chicago Press, Chicago (2008)

Levin, B.: English Verb Classes and Alternations: A Preliminary Investigation. University of Chicago Press, Chicago (1993)

Liu, D., Espino, M.: Actually, genuinely, really, and truly: a corpus-based behavioral profile study of near-synonymous adverbs. Int. J. Corpus Linguist. **17**(2), 198–228 (2012)

Miller, G.A., Charles, W.G.: Contextual correlates of semantic similarity. Lang. Cogn. Process. **6**(1), 1–28 (1991)

Sandra, D., Rice, S.: Network analyses of prepositional meaning: mirroring whose mind—the linguist's or the language user's? Cogn. Linguist. (includes Cogn. Linguist. Bibliography) **6**(1), 89–130 (1995)

Tyler, A., Evans, V.: Reconsidering prepositional polysemy networks: the case of over. In: Language, pp. 724–765 (2001)

Jianming, L., Zhen, M.: Chinese Function Words: A Prose Introduction. Chinese Language Press, Beijing (1999)

Rui, G.: The process structure of Chinese verbs. Chin. Lang. **6**, 410–419 (1993)

Wang, X., Zhu, D.: A comparative study of the semantics of the temporal adverb "hai" and "reng." Young Lit. Fig. **13**, 174–175 (2013)

Dictionary Office, Institute of Linguistics, Chinese Academy of Social Sciences. The Dictionary of Modern Chinese. Sixth edition. The Commercial Press (2012)

Shuxiang, L.: The Eight Hundred Words of Modern Chinese. The Commercial Press, Beijing (1999)

Citation Network, Knowledge and Evaluation: The Role of Reporting Markers in Citation Content Analysis

Rui Liu[1](✉), Zhihua Chen[2], and Jie Xu[1]

[1] Department of Chinese Language and Literature, University of Macau, Taipa, Macau, China
{liuruioscar,jiexu}@um.edu.mo

[2] Department of General Foundation Requisite Teaching and Research, Guangxi Police College, Nanning, Guangxi, China

Abstract. Linguistic features play an important role in the transition from traditional citation analysis to citation content analysis (CCA). In this paper, a novel language feature, Reporting Marker (RM), is introduced into CCA and acts as a key to open the black box of information structure of citation content. Through discourse analysis performed on a small-scale corpus, we clarify the mapping relationship between information pattern and language structure, and use the Language Technology Platform (LTP) to identify that relationship. A framework is constructed to realize CCA in three dimensions: citation network, knowledge, and evaluation. It is likely to contribute to the improvement of information retrieval and knowledge mining from academic documents.

Keywords: Reporting marker · Citation content analysis · Information structure · Citation network · Knowledge mining · Citer evaluation

1 Introduction

Citation content analysis (CCA) is an interdisciplinary research field involving applied linguistics (especially English for research purposes), sociology of science and information science (White 2004). In contrast to traditional citation analysis, which is based on citation frequency, CCA bases its analysis upon the text surrounding the citation (Small and Klavans 2011). Hence, it covers a wide range of research topics, including studies of citer motivations, citation location, citation classification, citation topic recognition, citation sentiment analysis and so forth (Liu 2014).

Linguistic features have always been an important support for CCA. Almost all aspects of the analysis of citation content can be converted into corresponding language analysis. White (2004) pointed out that analysis of citation content can be achieved by analyzing formulaic words or phrases in the content. Athar (2011) found that 3-grams and dependency relations can be effectively used for citation sentiment analysis. Zhang et al. (2013) attempted to use syntactic and semantic analysis to establish a citation content analysis framework to analyze the social and cultural context of academic

M. Liu et al. (Eds.): CLSW 2020, LNAI 12278, pp. 386–397, 2021.
https://doi.org/10.1007/978-3-030-81197-6_32

articles. Bertin and Atanassova (2014) found that citation act verbs are limited to a small number of verbs, and that citations play different roles according to their position in the rhetorical structure of an academic paper. Liu (2014) comprehensively used sentence segmentation technology, part-of-speech tagging, noun phrase extraction, and semantic analysis technology in citation content analysis. Liao et al. (2018) used feature words and a general sentiment dictionary to identify the sentiment tendency of citation content.

As one of the linguistic characteristics of academic papers (Hyland 2002), Reporting Markers (RMs) are words, phrases, or constructions reporting research actions, research results and findings, or quoted contents in scientific literature, such as, "…研究了… […yanjiu le…] (…studied…)", "根据…的研究 [genju…de yanjiu] (according to the study of…)", "…指出… […zhichu…] (…point out…)". They are mainly used in the citation context (i.e. the text surrounding a reference). Therefore, RMs always co-occur with citation and thus act as **signals** of the latter (Liu and Wang 2019). Citations have multiple motivations, types, and forms, and the RMs also appear accordingly in various vocabulary forms, constructions, and functions. There has been research on reporting verbs and reporting markers in English and Chinese academic papers (such as Swales 1990; Thompson and Ye 1991; Hyland 2002; Liu and Wang 2019; Liu and Huang 2019, etc.). The existing studies mainly focus on the characteristics and usage of the RM itself but pay little attention to the relationship between the RM and the quoted content, as well as the role of the RM in the analysis of citation content.

In this paper, we attempt to introduce a novel linguistic feature, i.e. RM, into CCA. Our hypothesis is that RM plays a key role in the analysis of the information structure of citation content. The structure of the paper is as follows. Section 2 introduces the methodology and materials used in the study. Section 3 generalizes the information structure of citation content and elaborates on its linguistic representation and its identification using the Language Technology Platform (LTP). Section 4 discusses the three dimensions of citation content analysis; Sect. 5 proposes a citation content analysis framework based on RM. In Sect. 6, we draw our conclusion and discuss the future work.

2 Methodology and Materials

Discourse analysis was used in this study to carry out a detailed analysis of two sample scientific articles. They are two highly cited papers in CNKI from the field of linguistics and economics respectively:

- *Foreign Language Learning Anxiety and Its Impact on Foreign Language Learning (外语学习焦虑及其对外语学习的影响), written by Wang Yinquan (王银泉) and Wan Yushu (万玉书) and published in Foreign Language Teaching and Research (《外语教学与研究》), 2001, issue 2.*
- *Estimation of China's Provincial Physical Capital Stocks: 1952–2000 (中国省际物质资本存量估算:1952–2000), written by Zhang Jun (张军), Wu Guiying (吴桂英) and Zhang Jipeng (张吉鹏) and published in Economic Research (《经济研究》), 2004, issue 10.*

The material was prepared through the following steps. Firstly, we identified all RMs and their function categories in these two papers. The criteria come from our previous studies on RMs in Chinese academic papers (Liu and Wang 2019; Liu and Huang 2019). Secondly, contexts containing RMs were extracted and confirmed to be valid citations. Thirdly, citation sentences were parsed manually and by an NLP program (*vide post*) respectively, and the results were cross-validated.

The small-scale material is appropriate for these reasons. The aim of this research is to confirm that it is sound to conduct citation analysis through syntactic and semantic analysis on RM-related context. A detailed discourse analysis of sample articles is proper and effective for this qualitative research. In addition, deficiencies in labor and technology[1] were also a limitation. Nevertheless, once its theoretical correctness is confirmed, the framework in this paper can be used on a large-scale corpus.

In order to explore the possibility of integrating our ideas with automated analysis, an NLP platform was used to assist the analysis. We used the Language Technology Platform (LTP v.4.0.4) developed by the Harbin Institute of Technology Social Computing and Information Retrieval Research Center (Che et al. 2010).

3 Information Structure of Citation Content and Its Linguistic Representation

3.1 Information Structure of Citation Content

In academic papers, a writer (citer) is generally trying to convey the following information by citing the works of others:

i. which paper is cited, which is coded as [CITED] in this paper;
ii. what kind of action and process the researchers do in [CITED], coded as [ACT];
iii. the scope and objects of [ACT], coded as [SCOPE];
iv. results of [ACT], or views given by [CITED], coded as [RESULT];
v. evaluation of the citer, coded as [EVALUATION].

To sum up, the typical information structure of a citation is [CITED] - [ACT] - [SCOPE] - [RESULT] - [EVALUATION], which contains the five most essential elements of a citation. In addition, information such as time and place of research may also be mentioned, but they are not core elements and will not be discussed here.

Taking (1) as an example, [CITED] is "*McCroskey, Fayer & Richmond (1985)*"; [ACT_1] is "研究了 [*yanjiu le*] (*studied*)", of which [SCOPE] is "美国的波多黎各大学生在学习西班牙语和英语过程中的交际畏惧与自我知觉之间的关系 (*the relationship between communication apprehension and self-perceived competence of Puerto Rican college students in the United States who learn Spanish and English*)"; [ACT_2] is "发现 [*faxian*] (*found*)", of which [RESULT] is "对自己英语能力的自我评估低的学生更容易出现较高程度的英语交际畏惧 (*students with low self-assessment of their English proficiency are more likely to have a higher level of fear of English communication*)".

[1] There is no good way to identify RMs in Chinese scientific documents for now. We are yet to explore it in another project.

(1) *McCroskey, Fayer & Richmond (1985)* 研究了美国的波多黎各大学生在学习西班牙语和英语过程中的交际畏惧与自我知觉之间的关系，结果发现对自己英语能力的自我评估低的学生更容易出现较高程度的英语交际畏惧。
McCroskey, Fayer & Richmond (1985) studied the relationship between communication apprehension and self-perceived competence of Puerto Rican college students in the United States who learn Spanish and English, and found that students with low self-assessment of their English proficiency are more likely to have a higher level of fear of English communication.

Most citations do not provide all components for the typical information structure. Some report [CITED], [ACT] and [SCOPE] without [RESULT] and [EVALUATION], like example (2); some, like example (3), provide [CITED], [ACT] and [RESULT] without [SCOPE] and [EVALUATION].

(2) *Spielberger (1972)*对考试焦虑的性质和测定作了评述并且验证了各种心理治疗对焦虑的效果。
Spielberger (1972) made comments on the characteristics and measurement of examination anxiety and verified the effects of various kinds of psychotherapy on anxiety.

(3) Hu and Khan (1997)以及宋海岩等 (2003)估计中国1958年的资本存量为2352亿元(1978年价格)。
Hu and Khan (1997) and Song et al. (2003) estimated that the capital stock of China in 1958 was 235.2 billion yuan (price in 1978).

3.2 Linguistic Representation of Citation Information

The information contained in a citation is organized and expressed with linguistic devices by the citer. That is, citation content is a (written) language representation of the information structure of a citation. Thus, citation content is shaped by language. Some formulaic languages are formed and used by the academic community, resulting in a paradigm of academic writing and its widespread application. Because language structure has an intrinsic influence on the representation of citation content, we can analyze its information structure based on language features.

RM is one of the language features that are able to reveal the information structure of citation content. In our previous study (Liu and Wang 2019), we summarized 4 forms of RMs and 5 constructions of reporting sentences in Chinese academic papers. Based on these findings, we further explored the mapping relationship between reporting sentences and the information structure of citation content.

The mapping mode from information structure to reporting sentence is basically as follows:

[CITED]s are always mapped as the subjects of reporting sentences. They are usually marked in the form of “*Spielberger (1972)*” or “黄永峰等(2002) (*Huang Yongfeng et al. (2002)*)”, which can be recognized by computers (White 2004). Sometimes, they are also marked in the form of discourse anaphora, such as “他/她 [*ta*] (*s/he*)”, “该文献 [*gai wenxian*] (*this paper*)”. [ACT]s are generally mapped as verb phrases containing RM, as well as discontinuous constructions and noun phrases (see Liu and Wang

2019). [SCOPE]s can be either objects of the reporting verb phrases or of prepositions; [RESULT]s are objects of the reporting verb phrases; and [EVALUATION]s are represented as some attached constituencies, including those required by the lexical semantics of RM itself, adverbials, and so forth (see 4.3).

Taking (4) as an example, [CITED], "黄永峰等(2002) (*Huang Yongfeng et al. (2002)*)", is the subject of the root sentence. [ACT] is "估算出 [*gusuan chu*] (*estimates*)". [SCOPE] is the preposition phrase, "在一项对中国制造业资本存量的研究中 (*in a study of China's manufacturing capital stock*)". [RESULT] is a clausal-object of the verb phrase "估算出 [*gusuan chu*] (*estimates*)", that is, "设备的经济折旧率为17%, 建筑为8% (*that the economic depreciation rate of equipment is 17% and construction 8%*).

(4) 黄永峰等(2002)在一项对中国制造业资本存量的研究中， 估算出设备的经济折旧率为17%, 建筑为8%。
In a study of China's manufacturing capital stock, *Huang Yongfeng et al. (2002)* estimates that the economic depreciation rate should be 17% for equipment and 8% for construction.

If we further summarize the above analysis, citation contents can be simplified into hierarchical chunk structures, where [ACT] take [CITED], [SCOPE] and [RESULT] as its associated components in the same way as how a verb takes noun phrases as its arguments.

- [CITED… […SCOPE…] […estimated…[RESULT]…]]

However, the order of information structure changes sometimes. In (5), the order is [RESULT i] - [CITED] - [EVALUATION] - [ACT] - [RESULT i], where [RESULT], "与没有畏惧的人相比,交际畏惧症患者在介入他人的会话以及追求社交方面显得更加勉强 (*Compared with people of less fear, patients of social phobia are more reluctant to join in others' conversations or to pursue social interactions*)", is preposed before [CITED] and an anaphoric constituency "这一点 [*zhe yidian*] (*this*)" is left in situ. There is also an [EVALUATION] clue, "都 [*dou*] (*already*)", an adverbial of the [ACT].

(5) 与没有畏惧的人相比，交际畏惧症患者在介入他人的会话以及追求社交方面显得更加勉强, *Daly & Stafford (1984)*的研究都证实了这一点。
Compared with people of less fear, patients of social phobia are more reluctant to join in others' conversations or to pursue social interactions. The research of *Daly & Stafford (1984)* has already confirmed this.

3.3 Identifying Information Elements by LTP

Semantic Role Labeling (SRL) in LTP is a function to identify the relations between predicates in a sentence and their associated arguments based on syntactic parsing (Che et al. 2010). Since SRL has been proved to be effective in extracting linguistic structure containing citation information in our pilot study, we use this function to assist linguistic analysis.

In a linguistic structure, a verb always plays a central role in organizing and relating all the components in a sentence. According to the mapping relationship between linguistic structure and information structure of citation, it is reasonable to deduce that [ACT] also plays a core role in organizing and associating the other information components. Therefore, we can parse the information in a citation as long as [ACT], where the RM usually resides, is anchored.

In the SRL result of LTP, the verb phrase expressing [ACT] is identified as the head of the semantic argument structure, and [CITED] is usually labeled as A0 (agent argument). The SRL of [SCOPE] or [RESULT] is related to its syntactic function. If it is the object, it will be labeled as A1 (patient argument). If it is the object of a preposition, it may have another semantic role, such as A3 (starting point), LOC (locative), MNR (manner), BNF (beneficiary), PRP (purpose or reason) or ADV (adverbial). If it is an embedded clause, it will be tagged according to its function in the sentence. [EVALUATION] can be implicit in the semantics of RM or realized as ADV (adverbial).

The analysis outputs of (4) and (5) are shown in the following box. In (4), [CITED] "黄永峰等(2002) *(Huang Yongfeng et al. (2002))*" is labeled as A0. [SCOPE] "在一项对中国制造业资本存量的研究中 (*in a study of China's manufacturing capital stock*)" is ARGM-LOC. [RESULT] "设备的经济折旧率为17%, 建筑为 8% (*that the economic depreciation rate should be 17% for equipment and 8% for construction*)" is A1. In (5), [CITED] "*Daly & Stafford (1984)*的研究 (*The research of Daly & Stafford (1984)*)" is tagged as A0. [EVALUATION] "都 [*dou*] (*already*)" is ARGM-ADV. [RESULT] "这一点 [*zhe yidian*] (*that*)" is A1.

```
(4): 估算出  ->  [
                 A0:       黄永峰等(2002),
                 ARGM-LOC:在一项对中国制造业资本存量的研究中
                 A1:       设备的经济折旧率为 17%，建筑为 8%
                 ]

(5): 证实     -> [
                 A0:       Daly & Stafford (1984)的研究
                 ARGM-ADV: 都
                 A1:       这一点
                ]
```

The scheme presented here, however, is a basic and ideal one. Due to the complexity of syntax and semantics, there is a lot of work to do in identifying information elements. How to decide the actual denotation of [RESULT] "这一点 [*zhe yidian*] (*this*)" in (4) is an example.

4 Analysis of Citation Content

Citation content can be analyzed from three perspectives, namely citation network, knowledge, and citer evaluation, after the information structure and elements have been extracted from citation sentences.

4.1 Citation Network

As the most important basis of (traditional) citation analysis, the citation relationship between citing and cited papers is embedded in [CITED]. Each [CITED] represents a link between the citing paper and the cited work. We can get a citation network of the citing paper if all of those links are captured. The citation networks of the two papers discussed in this paper are shown in Fig. 1.

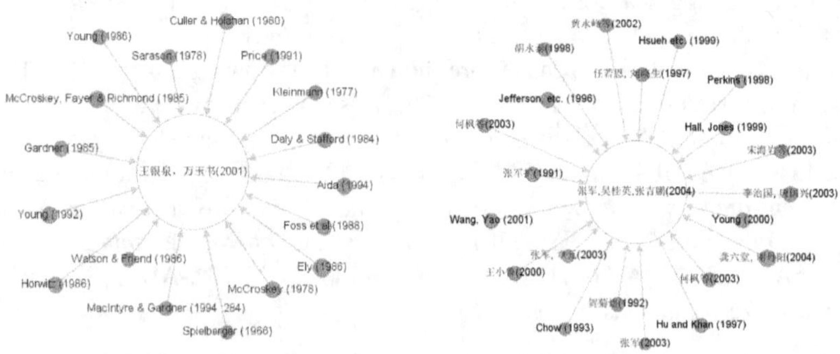

Fig. 1. Citation networks of the two papers

Although gleaning data from bibliographies of scientific documents is the most direct and fastest way to obtain citation networks, this traditional method works at the document-to-document level, which is very insufficient (Liu et al. 2015). Extraction of citation networks based on reporting sentences, however, can filter out **invalid citations** to a certain extent, because the citation links come from actual citation contents in the context, rather than from bibliographies listed at the end of scientific writings.

4.2 Knowledge Contained in [SCOPE] and [RESULT]

[SCOPE] and [RESULT] are connected to RMs with different functions. Liu and Wang (2019) categorized RMs in Chinese academic papers into four functions: 1) **Procedural Markers** introducing research actions, methods and procedures in the research process, 2) **Finding Markers** reporting results and findings, 3) **Discourse Markers** expressing ideas, points of view and discussions, and 4) **Cognition Markers** emphasizing citers' judgments, speculations and appeals. We find that [SCOPE] is obviously connected to Procedural Markers, while [RESULT] is mostly connected with Finding Markers, Discourse Markers and Cognition Markers.

Taking (6) as an example, there are three [ACT]s:

(6) *Young (1986)* 以法语、 德语和西班牙语学习者为对象进行了调查，试图证明口头表达的熟练程度是否会受到焦虑的负影响，结果发现学生的口语考试(Oral Proficiency Interview)成绩与一定的焦虑测量手段有关。
Young (1986) investigated learners of French, German and Spanish, and tried to find out if anxiety has negative impact on foreign language oral performance. He found that the subjects' OPI scores were related to anxiety measures.

[ACT_1] "进行了调查 [*jinxingle diaocha*] (*investigated*)" is a procedural marker, which takes "以法语 、德语和西班牙语学习者为对象 (*learners of French, German and Spanish*)" as its [SCOPE]. [SCOPE] contains these key words: learners, French, German, Spanish.

[ACT_2] "试图证明 [*shitu zhengming*] (*tried to find out*)" is another procedural marker. Its [SCOPE], "口头表达的熟练程度是否会受到焦虑的负影响 (*if anxiety has negative impact on the foreign language oral performance*)" establishes a potential (negative) relationship between two concepts: anxiety and foreign language oral performance.

[ACT_3] "发现 [*faxian*] (*found*)" is a typical finding marker, which takes an assertion "学生的口语考试成绩与一定的焦虑测量手段有关 (*the subjects' OPI scores were related to anxiety measures*)" as its [RESULT].

As we can see, different RMs and their [SCOPE]s and [RESULT]s are associated with various kinds of knowledge. Actually, [SCOPE] and [RESULT] contain a high density of knowledge including but not limited to the following:

- **Proposition.** A proposition, *the subjects' OPI scores were related to anxiety measures*, can be extracted from [RESULT] in (6).
- **Terminology, proper nouns, and subject terms,** such as "制造业资本存量 [*zhizaoye ziben cunliang*] (*manufacturing capital stock*)", "经济折旧率 [*jingji zhejiulv*] (*economic depreciation rate*)", "学习者 [*xuexizhe*] (*learners*)", "心理治疗 [*xinli zhiliao*] (*psychotherapy*)", "口语考试 [*kouyu kaoshi*] (*Oral Proficiency Interview*)", "焦虑测量手段 [*jiaolv celiang shouduan*] (*anxiety measures*)" from (4), (5) and (6).
- **Survey data, statistical data, calculation results,** such as 17% (the economic depreciation rate of equipment), 8% (the economic depreciation rate of construction) in (4).

4.3 Clues for Citer Evaluation

It is necessary to capture the citer's evaluations of the cited works in CCA. The reason is that by identifying whether the citer's attitude towards the citation is supportive, opposed or neutral, it can help to assess whether the citation relationship between the citing paper and the cited work is positive, negative or neutral. The citer's evaluation, however, is generally not presented explicitly because academic papers are required to be as impersonal as possible. Traditional citation analysis does not pay much attention to the citers' evaluation due to the limitation of its methodology. This is likely to cause obscurity and errors in the analysis results. In fact, some highly cited papers may be cited either for positive reasons or for negative reasons, which cannot be distinguished by the analysis of citation frequency. To overcome this shortcoming, many research efforts focus on attitude and sentiment analysis of citation content (Athar 2011; Liao et al. 2018).

RMs can provide clues to clarify a citer's attitude and evaluation about citations. For example, the RM "证实 [*zhengshi*] (*confirmed*)" in (7) shows that the citer agreed with the results of the survey made in *Ely (1986)*, while the RM "推测 [*tuice*] (*speculated*)" in (8) suggests that the citer took the speculation made by *Culler & Holahan (1980)* with a grain of salt.

(7) *Ely (1986)*对一年级西班牙语学生的调查证实了外语熟练程度与外语学习情感变量之间的关系, 这些情感变量包括语言课堂上是否大胆、课堂交际以及课堂不适等。
A survey conducted by *Ely (1986)* of students enrolled in first year university Spanish classes confirmed the relationship between foreign language proficiency and variables related to L2 learning emotion, including confidence, sociability and discomfort in language classes.

(8) *Culler & Holahan (1980)* 推测考试焦虑可能是学生学习技能中的缺陷引发的。
Culler & Holahan (1980) speculated that examination anxiety might be triggered by students' inadequate learning skills.

In our previous work (Liu and Wang 2019), it was discovered that Discourse Markers in Chinese academic papers more often reflect the citer's attitude towards the quoted point of view. Through our investigation of the corpus, we found that RMs, such as "指出 [*zhichu*] (*point out*)", "警告 [*jinggao*] (*warn*)", "呼吁 [*huyu*] (*appeal*)", "主张 [*zhuzhang*] (*advocate*)", "所……指出 [*suo…zhichu*] (*indicated by*)", etc., indicate a citer's affirmative attitude towards the cited content, 'say' verbs like "报道 [*baodao*] (*report*)" and "披露 [*pilu*] (*reveal*)" express a neutral attitude, while "认为 [*renwei*] (*believe*)", "强调 [*qiangdiao*] (*emphasize*)" are used to express a tentative stance. To sum up, an RM itself implies the evaluation of the citer, which can be used as a clue to evaluation analysis. Of course, a more thorough corpus survey is needed to establish an effective evaluation scale of RMs with a wide enough range.

In addition, some adverbial components can also serve as clues for evaluation. For example, "也 [*ye*] (*also*)" in (9) indicates that 'a similar conclusion' was also found in the study of Foss et al. (1988), which signals that the citer holds a positive attitude towards this conclusion.

(9) Foss et al. (1988)的类似研究也得出了相似的结论: 在英语作为第二语言的课堂中存在交际畏惧, 它似乎成了学生掌握英语的绊脚石。
A similar study by Foss et al. (1988) also reached a similar conclusion: there is communicative anxiety in the class of English as a second language, and it seems to have become a stumbling block for students to master English.

Evaluation can be used for weighting calculation in CCA. Taking (7) as an example, since there is a positive clue, i.e. the RM "证实 [*zhengshi*] (*confirmed*)", we can increase the weight of the positive link between the citing paper and the cited paper, i.e. *Wang*

& Wan (2001) and *Ely (1986)*. More specifically, we can increase the credibility of the relationship between "*foreign language proficiency*" and "*variables related to L2 learning emotion, including confidence, sociability and discomfort in language classes*", because the citer positively believed that it was 'confirmed' by *Ely (1986)*. With the weights of the links between papers and the credibility of knowledge, it is possible to go beyond the simple linear relationship (Zhang et al. 2013) represented by traditional citation analysis and to make the analysis of citation content more semantically based and accurate.

5 A Citation Content Analysis Framework Based on RM

CCA can be conducted in three dimensions according to Sect. 4, namely, citation network, knowledge, and evaluation. The *citation network* dimension establishes the connection between the citing document and the cited documents ([CITED]), which is a document-to-document analysis. We can obtain the macroscopic relationship between scientific works. While this analysis is the main part of traditional citation analysis, it only plays the role of a bridge between documents in our framework. The *knowledge* dimension is the extraction of the knowledge information contained in the citation content. The *evaluation* dimension is the citer's attitude, evaluation and weighting of the citation content. The latter two dimensions are not fully addressed in traditional citation analysis. They are where RM-based CCA can contribute more.

The three dimensions are mutually reinforcing. Knowledge analysis can generate a knowledge network using the citation network. In this way, the citation network is not only just a simple document-to-document linear relationship, but a network with full details of knowledge. Evaluation can give weights to the connections (including the citation network and the knowledge network). As a result, the network is no longer homogeneous, static, but dynamic. A detailed and dynamic citation content network can provide a good basis for information retrieval and knowledge mining.

Syntactic analysis oriented towards citation information structure can support a basic framework for CCA. After anchoring the RMs [ACT] in a paper to be analyzed, a syntactic analysis of the reporting sentences is performed, which can be achieved through LTP as discussed in 3.3.

As illustrated in Fig. 2, based on the results of syntactic analysis, a citation relationship between the citing document and the cited document can be established by identifying [CITED]. Then, by extracting [SCOPE] and [RESULT] information, we can capture propositions, terminology, proper nouns, survey data, calculation results and other knowledge contents. Finally, analysis of the RMs and related constituencies can produce evaluation information [EVALUATION], which can be used in weighting calculation. To sum up, the three-dimensional analysis forms three levels to realize in-depth analysis of citation content.

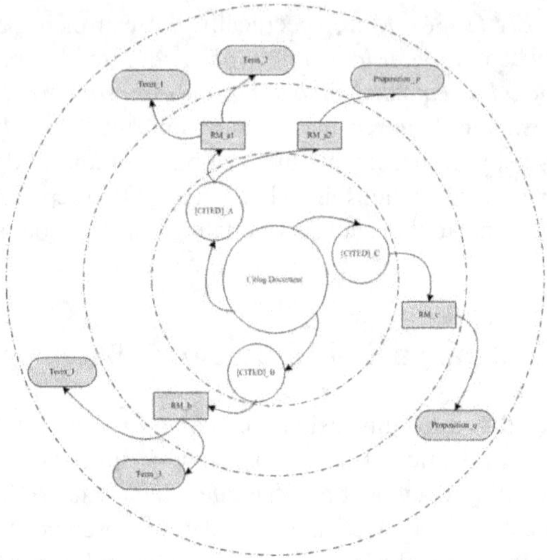

Fig. 2. A citation content analysis framework based on RM

6 Conclusion

This paper takes reporting markers, a language feature of academic articles, as the starting point to analyze the information structure of citation content, and summarizes its pattern, i.e. [CITED]-[ACT]-[SCOPE]-[RESULT]-[EVALUATION]. Elaborating through specific examples, we clarify the mapping relationship between information pattern and language structure, and then discuss the identification of this relationship by LTP in order to achieve automatic analysis of the citation content. A framework is constructed to realize CCA that covers three dimensions, namely, citation network, knowledge and evaluation, achieving more refined, semantically based, and dynamic results than before. This framework is likely to contribute to the improvement of information retrieval and knowledge mining from academic documents. We will make the framework more effective, automatic, and reliable on large-scale corpus in the future.

Acknowledgment. The authors gratefully acknowledge the financial support of the Research Committee of the University of Macau through the research project entitled "Grammatical Violations in Chinese Slogans and Titles and Their Linguistic Implications" (MYRG2018–00201-FAH), which made the research work reported in this article possible.

References

Athar, A.: Sentiment analysis of citations using sentence structure-based features. In: Proceedings of the ACL 2011 Student Session, pp. 81–87 (2011)

Bertin, M., Atanassova, I.: A study of lexical distribution in citation contexts through the IMRaD standard. PloS Negl. Trop. Dis, 1(200,920), 83–91 (2014)

Che, W., Li, Z., Liu, T.: LTP: a Chinese language technology platform. In: Proceedings of the Coling 2010: Demonstrations, pp. 13–16 (2010)

Hyland, K.: Activity and evaluation: reporting practices in academic writing. In: Flowerdew, J. (ed.) Academic Discourse, pp. 115–130. Longman, London (2002)

Liao, J., Liu, Z., Chen, J.: Citation sentiment recognition method based on citation content analysis. Libr. Inf. Serv. **62**(15), 112–121 (2018). (in Chinese) (基于引文内容分析的引用情感识别研究. 图书情报工作 **62**(15), 112–121 (2018))

Liu, R., Wang, S.: How citation is signaled: a corpus-based study on reporting verbs in chinese academic papers. In: Tao, Hongyin, Chen, Howard Hao-Jan. (eds.) Chinese for Specific and Professional Purposes. CLLS, pp. 53–71. Springer, Singapore (2019). https://doi.org/10.1007/978-981-13-9505-5_3

Liu, R., Huang, Q.: The function and distribution of reporting markers in Chinese academic papers. In: The 20th Chinese Lexical Semantics Workshop (CLSW2019), Beijing (2019), (in Chinese) (汉语学术语篇中转述标记的功能与分布, 北京: 第20届汉语词汇语义学研讨会 (CLSW2019) (2019))

Liu, S., Kun, D., Zhang, C.: New stage of citation analysis: from citation description analysis to citation context analysis. Document. Inf. Knowl. (3), 25–34 (2015). (in Chinese) (引文分析的新阶段: 从引文著录分析到引用内容分析. 图书情报知识 (3), 25–34 (2015))

Liu, S.: The citation content analysis of scientific papers and its application. PhD Thesis. Dalian University of Technology (2014). (in Chinese) (科学论文的引用内容分析及其应用, 大连: 大连理工大学博士学位论文 (2014))

Small, H., Klavans, R.: Identifying scientific breakthroughs by combining co-citation analysis and citation context. In: Proceedings of the 13th International Conference of the International Society for Scientometrics & Informetrics, pp. 783–793 (2011)

Swales, J.: Genre Analysis: English in Academic and Research Settings. Cambridge University Press, Cambridge (1990)

Thompson, G., Yiyun, Y.: Evaluation in the reporting verbs used in academic papers. Appl. Linguis. **12**(4), 365–382 (1991)

White, H.D.: Citation analysis and discourse analysis revisited. Appl. Linguis. **25**(1), 89–116 (2004)

Zhang, G., Ding, Y., Milojević, S.: Citation content analysis (CCA): a framework for syntactic and semantic analysis of citation content. J. Am. Soc. Inform. Sci. Technol. **64**(7), 1490–1503 (2013)

Multi-task Learning Neural Networks for Comparative Elements Extraction

Dianqing Liu, Lihui Wang, and Yanqiu Shao(✉)

School of Information Science, Beijing Language and Culture University, Beijing 10083, China
shaoyanqiu@blcu.edu.cn

Abstract. Comparative sentences are common in human languages. In online comments, a comparative sentence usually contains the subjective attitude or emotional tendency of a reviewer. Hence, comparative elements extraction (CEE) is valuable for opinion mining and sentiment analysis. Most of the existing CEE systems use rule-based or machine learning approaches that need to construct a rule base or spend a huge amount of effort on feature engineering. These approaches usually involve multiple steps, and the performance of each step relies on the accuracy of the previous step, risking error cascading oversteps. In this paper, we adopt a neural network approach to CEE, which supports end-to-end training and automatic learning of sentence representation. Furthermore, considering the high relevance of CEE and comparative sentences recognition (CSR), we propose a multi-task learning model to combine the two tasks, which can further improve the performance of CEE. Experiment results show that both our neural network approach and multi-task learning are effective for CEE.

Keywords: Comparative elements extraction · Neural networks · BERT-CRF · Multi-task learning

1 Introduction

Comparative sentences indicate similarities or differences between things. Their predicates contain comparative words or comparative structures. Ding [1] precisely defined comparative sentences as sentences containing comparative items and expressing comparative meanings. From the perspective of language engineering, each comparative sentence can contain several comparative relations, and each comparative relation consists of several elements, which forms a quintuple of comparative subject (also called object of comparison in linguistics literature), comparative object (also called standard of comparison), comparative mark, comparative aspect and comparative result [2]. Such quintuples are essential to the extraction of comparative relations. For example, the sentence "P20在尺寸上比P10大" (P20 is bigger than P10 in size) can be represented as a quintuple <P20, 比, P10, 尺寸, 大> (<P20, than, P10, size, bigger>), in which "P20" is the comparative subject, "P10" is the comparative object, "比" (than) is used as a comparative marker to introduce the comparative object, "尺寸" (size) is the comparative aspect, which can be regarded as a common attribute of the comparative subject

M. Liu et al. (Eds.): CLSW 2020, LNAI 12278, pp. 398–407, 2021.
https://doi.org/10.1007/978-3-030-81197-6_33

and object, and "大" (bigger) is the comparative result indicating their difference in terms of the common attribute.

Most early works of comparative elements extraction used rule-based methods. For example, Jindal and Liu [3] adopted label sequential rules (LSR) to extract comparative relations. However, because of the complexity of language, it was difficult for rules to generalize about all comparative sentences. To address this problem, Song et al. [4] established a Condition Random Field (CRF) model with several types of features (N-gram, Part-of-Speech (POS) tags, position of words, etc.), which led to a significant improvement in accuracy. Wang et al. [5] introduced a variety of linguistic features, such as shallow syntactic information and comparative word candidates, into the CRF model, which effectively improved performance further.

Recently, deep learning technology has been applied to various tasks of natural language processing. Motivated by this, in this paper, we adopt BERT [6] with CRF [12] as a BERT-CRF architecture for comparative elements extraction. This is an end-to-end system without a domain dictionary or a comparative mark dictionary. It can automatically extract features from data, avoiding the shortcoming of Feature Engineering and rule-based systems. Furthermore, because comparative elements extraction is highly related to the task of comparative sentence recognition, we propose a multi-task learning model. All modules for the two tasks are trained jointly by sharing a representation layer of sentences and being optimized alternately. Experimental results show that this BERT-CRF model achieves competitive performance in comparative elements extraction and our multi-task learning model can result in further improvement.

2 Related Work

2.1 Comparative Elements Extraction

Existing methods that have been proposed for CEE can be divided into two categories roughly: rule-based methods and machine learning methods. Jindal and Liu [3] proposed LSR, which extracted comparative elements in the form of quintuples through the rules applied to POS tag sequences. Zhou et al. [7] built a CEE rule base according to the semantic classification of words and comparative sentences, with which they achieved the best performance in the task 2.2 of COAE 2012. Some works used a semantic role labeling (SRL) system to mark and extract comparative elements [8, 9]. Their performance depended on the accuracy of the SRL system. These rule-based methods are easy to implement, but cannot cover all of the myriads of linguistic phenomena. As for the machine learning methods, most of the previous works have utilized a CRF model to extract comparative elements, and different works may have chosen different feature sets. For example, Xing et al. [10] combined keywords, noun phrases and sentence position information into a CRF model to extract comparative elements. And Song et al. [10] used N-grams, POS tags, and syntactic and semantic features of sentences.

2.2 Comparative Sentences Recognition

Rule-based methods [1, 6, 13] and machine learning methods [4, 11, 14] have been adopted as traditional methods of comparative sentences recognition, and CSR can be considered as a text classification task. Recently many neural network methods have been proposed for text classification. TextCNN proposed by Yoon [15], which combined convolutional neural network and fully-connected layers, has achieved good results on multiple datasets. Recently, many works have achieved competitive performance in text classification by fine-tuning a BERT model [6] in a variety of fields, such as medicine [16], sentiment [17], etc.

2.3 Multi-task Learning

Multi-task learning (MTL) can be seen as a way to leverage useful information in related tasks by sharing parameters and introducing inductive bias [18]. Qi and Ji [19] proposed jointly extracting entity mentions and relations. This approach has achieved better performance than pipelined approaches. Yuan and David [20] showed that joint training of dependency parsing (DP) and POS tagging are beneficial for DP. Following the example of [20], the purpose of our work is to improve the performance of CEE by jointly training CEE and CSR.

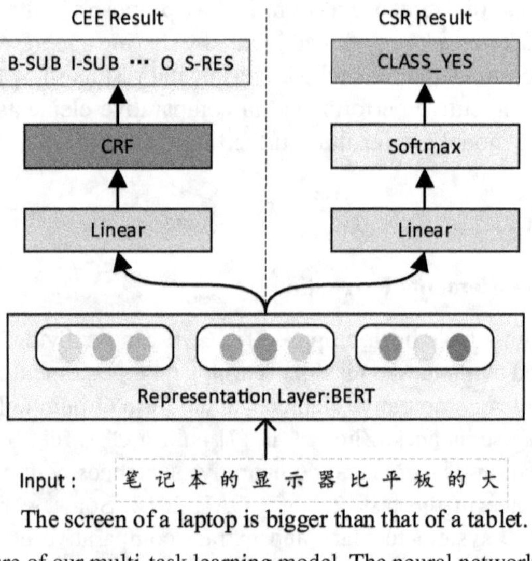

Fig. 1. The architecture of our multi-task learning model. The neural networks on the left are for CEE, and the neural networks on the right are for CSR.

3 Proposed Approach

In this section we describe the details of our model and strategies of training. Given a comparative sentence *S*, comparative elements extraction is to extract quintuples of comparative elements: $Q = \langle Sub, Obj, Mark, Asp, Res \rangle$. For this task, we propose a multi-task learning model that combines comparative elements extraction and comparative sentence recognition. The architecture of our model is shown in Fig. 1.

3.1 BERT-CRF

We treat CEE as a sequence tagging task and port the BERT-CRF architecture to it, which has been applied in many works of sequence tagging.

BERT, which stands for Bidirectional Encoder Representations from Transformers, endows the same word with different representations in different contexts. Unlike word2vec [21], BERT is not simply used to generate word embeddings (feature extraction), but rather to fine tune the whole network so as to finally obtain powerful contextualized word representations. For a given token, its BERT input t_i is constructed as the sum of the corresponding token, position, and segment embeddings [6]. For a sentence of up to n tokens with hidden dimension H, BERT outputs its encoded representation sequence E, where $E \in \mathbb{R}^{\mathrm{n \times H}}$:

$$t_i = emb_i^{tok} + emb_i^{pos} + emb_i^{seg} \tag{1}$$

$$E = BERT(t_0, t_1, \ldots, t_n) \tag{2}$$

Then we use a linear layer to project the H-dimensional representation of each token to dimension K:

$$P = EW_1 + b_1 \tag{3}$$

where $P \in \mathbb{R}^{\mathrm{n \times K}}$, $W_1 \in \mathbb{R}^{\mathrm{H \times K}}$. W_1 is the parameter matrix of projection, K is the number of tags to be predicted, and $b_1 \in \mathbb{R}^{\mathrm{K}}$, which is a bias item.

We utilize a CRF layer to enable our model to learn constraints among tags to be predicted in a sentence. For an input sequence $X = \{x_1, x_2, \ldots, x_n\}$, and a predicted sequence of tags $y = \{y_1, y_2, \ldots, y_n\}$, the score of this predicted sequence is defined as follows:

$$Score(X, y) = \sum_{i=0}^{n} A_{y_{i-1}, y_i} + \sum_{i=0}^{n} P_{i, y_i} \tag{4}$$

where $A \in \mathbb{R}^{\mathrm{(K+2) \times (K+2)}}$ is a parameter matrix of tag transitions. For example, A_{y_{i-1}, y_i} represents the probability of transition from tag y_{i-1} to tag y_i. Two additional tags *start* and *end* are used in the CRF layer to represent the beginning and the end of a sequence, respectively.

The model is optimized by maximizing the log-probability of the label sequence y given the input sequence X:

$$\log(p(y|X)) = log \frac{e^{Score(X,y)}}{\sum_{y' \in Y} e^{Score(X,y')}} \tag{5}$$

where Y represents all the possible label sequences for X.

3.2 Multi-task Learning

Our multi-task learning model combines two tasks: comparative elements extraction and comparative sentences recognition. The module of CSR is similar to that of CEE. Both of them are based on fine tuning BERT. Hence, we have made the two tasks share the same representation layer and fine-tune BERT, so that the CSR module can pass some task-specific information to CEE.

We have defined the loss function of CSR via cross entropy:

$$L_{CSR}(\theta_{\mathrm{R}}) = -\tfrac{1}{M}\sum_{i=1}^{M} c_i \log(p_i) \tag{6}$$

where M is the number of sentence classes, c_i is the true class label, and p_i is the probability of this class. θ_{R} includes parameters of BERT and other networks for CSR.

And the loss function of CEE is:

$$L_{CEE}(\theta_{\mathrm{E}}) = \log(p(y|X)) \tag{7}$$

Similarly, θ_{E} includes all parameters of BERT-CRF networks.

During training, the two tasks are optimized alternately. That is to say, if the CEE module is being trained at step *i*, then the CSR module will be trained at step *i* + *1*, and vice versa. At each step, task-specific parameters will be updated by backpropagation [22]. When extracting comparative elements from the test set, only the trained CEE module is used.

4 Experiments

4.1 Dataset and Evaluation Metrics

Our dataset for CEE consisted of two parts: one part was the dataset of COAE 2013 task 2.2, which could be classified into two fields: auto and digital product. The other part was a dataset of PC comments collected from JD.com. We worked with the annotated triples of COAE 2013, and annotated the data into quintuples. Statistics about the CEE dataset are shown in Table 1. We randomly selected 20% of the sentences as the test set and the rest as the training set. As for CSR, the dataset we used came from COAE2013 task 2.1, which contained 1000 comparative sentences and 1000 non-comparative sentences. We selected 10% of the sentences for CSR randomly as the validation set.

Table 1. Statistics of the dataset

Field	#sent	#subject	#object	#mark	#aspect	#result
Auto	1000	1116	1346	1056	1121	1362
Digital	1000	580	1133	982	897	1156
PC	503	380	476	518	272	510
Total	2503	2076	2955	2556	2290	3028

Precision (P), recall (R) and F-score (F) were used to evaluate the extraction results. Because the boundaries of comparative elements are difficult to determine definitely, the boundary of an element annotated by different people may be different. So, we adopted strict evaluation and lenient evaluation simultaneously. Strict evaluation means that an extracted element is correct only if it matches the answer exactly, and lenient evaluation means the result is correct as long as an extracted element overlaps with the answer.

4.2 Settings

We used the released Chinese BERT-Base model,[1] and fine-tuned the BERT model with its default configuration. The batch size was 8, and the maximum length of sentences was limited to 300. Dropout probability before linear layers was 0.33. We used the AdamW optimizer with a learning rate of $2e^{-5}$, $\beta_1 = 0.9$, $\beta_2 = 0.999$, and $\in = e^{-8}$. For CEE, we used the IOBES tagging scheme.

4.3 Performance Evaluation

To validate the advantage of our proposed model, we used the CRF model with the feature template proposed by Wang [5] as the baseline model for comparison. We applied fivefold cross-validation for each system. The average of the 5 results was used as the final result. Experiment results are shown in Table 2.

Table 2. Experiment results on test set.

Element	Method	Strict			Lenient		
		P	R	F	P	R	F
Subject	CRF	43.21	22.86	29.85	55.14	29.24	38.16
	BERT-CRF	79.13	76.08	77.52	**84.54**	81.31	82.83
	BERT-CRF (mtl)	**80.40**	**82.99**	**81.63**	84.25	**86.99**	**85.55**
Object	CRF	52.02	39.77	45.07	74.55	56.99	64.59
	BERT-CRF	76.74	81.91	79.24	83.07	**88.67**	85.78
	BERT-CRF (mtl)	**80.12**	**83.43**	**81.73**	**84.91**	88.40	**86.61**
Mark	CRF	73.10	63.61	68.02	76.36	66.45	71.06
	BERT-CRF	78.99	88.90	83.58	82.66	93.09	87.49
	BERT-CRF (mtl)	**81.78**	**90.62**	**85.96**	**84.83**	**94.02**	**89.17**
Aspect	CRF	45.71	36.71	40.70	53.83	43.25	47.94
	BERT-CRF	63.22	77.17	69.10	66.86	81.65	73.09
	BERT-CRF (mtl)	**67.22**	**78.31**	**72.25**	**70.11**	**81.69**	**75.36**
Result	CRF	58.29	45.92	51.36	64.94	51.15	57.22
	BERT-CRF	69.86	80.18	74.54	72.62	83.35	77.49
	BERT-CRF (mtl)	**70.88**	**82.35**	**76.05**	**72.91**	**84.66**	**78.21**

[1] https://github.com/google-research/bert.

From the results, we can see that both BERT-CRF and the multi-task learning model outperform the traditional CRF model, which shows the advantages of deep learning. Deep learning can extract features automatically, without the need to manually construct feature sets, and it avoids error cascading among multiple tasks. Furthermore, we can see the MTL model outperforms the single-task BERT-CRF module for CEE in all indicators of strict evaluation and in the vast majority of indicators of lenient evaluation. This indicates that the comparative sentences recognition task can provide extra useful information for comparative elements extraction and boost its performance by joint training. Because the purpose of our work is to improve the performance of CEE through multi-task learning, we will not discuss in detail the influence of MTL on CSR task, but according to the results of the validation set, the accuracy of CSR is also improved.

4.4 Case Study

We have conducted several case studies to further explore the effectiveness of multi-task learning. Some cases are shown in Table 3.

Table 3. Several comparative elements extraction examples in our dataset.

Comparative Sentence	Method	Comparative Elements Extraction Results
对比了多家笔记本，最后还是选择了这款，颜值比想象中的要高。	Answer	1.<这款, 对比, 笔记本, null, null> (<this one, comparing, laptop, null, null>) 2.<颜值, 比, 想象中的, null, 高> (<appearance, than, imaginary, null, better>)
	BERT-CRF	<颜值, 比, 想象中的, null, 高>
	BERT-CRF(mtl)	<(这款，颜值), (对比, 比), (笔记本, 想象中的), null, 高>
随机配的播放器感觉比JAVA里的播放器好一些。	Answer	<播放器, 比, 播放器, null, 好> (<player, than, player, null, better>)
	BERT-CRF	<播放器, 比, JAVA里的播放器, null, 好>
	BERT-CRF(mtl)	<随机配的播放器, 比, JAVA里的播放器, null, 好>
采用散点设计，扬声器效果更好。	Answer	<null, 更, null, 效果, 好> (<<null, more, null, effect, good>)
	BERT-CRF	<null, null, null, 效果, 好>
	BERT-CRF(mtl)	<null, 更, null, 效果, 好>

Experiment results have shown that the MTL model has the best capability to extract comparative relations in a long sentence with complex grammatical structures, multiple comparative relations and flexible word order. For example, "对比了多家笔记本, 最后还是选择了这款, 颜值比想象中的要高" (After comparing many laptops, I have finally chosen this one, which is more beautiful than imagined) is a complex sentence, in which the comparative subject "这款" (this one), the comparative object "笔记本" (laptop) and the comparative marker "对比" (compare) are distributed over the first two clauses, while the last clause contains another comparative relation. From Table 3, we can see that the model based on multi-task learning can understand the context of this sentence better and recognize its comparative elements more accuratelyThe results also indicate that the multi-task learning model helps to recognize phrases that are used as comparative elements. In practical applications, if the comparative subject and the comparative object are both phrases and have the same headword, it would be better to label them in their entirety for the sake of completeness and clarity. For example, in the sentence, "随机配的播放器感觉比JAVA里的播放器好一些" (The randomly installed player is better than the player provided by JAVA) only the multi-tasking model identified "随机配的播放器" (randomly installed player) and "JAVA里的播放器" (player provided by JAVA) as the comparative subject and object respectively.

As for the lexical level, the method based on the single-task BERT-CRF model can recognize prepositions and verbs that express comparative meanings, but it is not ideal for recognizing adverbs, such as "更" (more), "较" (compare), "相对" (comparatively) etc. They are seldom used as comparative markers in comparative sentences. For example, in the sentence "采用散点设计, 扬声器效果更好" (With the scatter point design, the sound effect of the speaker is better), "更" is the comparative marker, but its frequency of usage is much lower than "比" (compare, than) in our dataset. So, the single-task BERT-CRF model failed to recognize it, while the BERT-CRF module of multi-task learning model recognized it accurately. We reckon the reason is because the networks for CSR can learn knowledge about comparative markers more comprehensively, and the knowledge can be disseminated to the networks for CEE during the process of multi-task learning.

5 Conclusion

In this paper, we have proposed using BERT-CRF to extract comparative elements, which has a great capability to extract features and learn representations automatically. Moreover, we have proposed a multi-task learning model that trains a comparative element extraction module and a comparative sentence recognition module jointly, utilizing the strong relatedness between the two tasks. Experiment results have proved that neural networks approaches can significantly improve the accuracy of comparative element extraction. Furthermore, the results have shown that our multi-task learning model is effective in further improvement.

Acknowledgements. This research project is supported by the National Natural Science Foundation of China (61872402), the Humanities and Social Science Project of the Ministry of Education (17YJAZH068), Science Foundation of Beijing Language and Culture University (supported by "the Fundamental Research Funds for the Central Universities") (18ZDJ03), the Open Project Program of the National Laboratory of Pattern Recognition (NLPR), the Fundamental Research Funds for the Central Universities, and Research Funds of Beijing Language and Culture University (20YCX147).

References

1. Ding, C.: A Course for Mandarin Chinese Grammar, pp. 81–95. Peking University Press, Beijing (2009)
2. Wang, S., Zhao, G., Liu, H.: Comparison element ellipsis identification based on rules and sequence patterns. J. Shanxi Univ. (Nat. Sci. Ed.) **38**, 85–92 (2015)

3. Jindal, N., Bing, L.: Mining comparative sentences and relations. In: Proceedings of the 21st National Conference on Artificial Intelligence, pp. 1331–1336. AAAI Press, Menlo Park (2006)
4. Song, R., Lin, H., Chang, F.: Chinese comparative sentences identification and comparative relations extraction. J. Chin. Inf. Process. **23**, 102–107 (2009)
5. Wang, W., Zhao, T., Xin, G., Xu, Y.: Extraction of comparative elements using conditional random fields. Acta Automatica Sinica **41**, 1385–1393 (2015)
6. Jacob, D., Chang, M., Kenton, L., Kristina, T.: BERT: pre-training of deep bidirectional transformers for language understanding. In: Proceedings of the 2019 Conference of the North American Chapter of the Association for Computational Linguistics: Human Language Technologies, vol. 1, pp. 4171–4186. ACL, Stroudsburg (2019)
7. Zhou, H., Hou, M., Hou, M., Teng, Y.: Chinese comparative sentences identification and comparative elements extraction based on semantic classification. J. Chin. Inf. Process. **28**, 136–141 (2014)
8. Kessler, W., Kuhn, J.: Detection of product comparisons-how far does an out-of-the-box semantic role labeling system take you? In: Proceedings of the 2013 Conference on Empirical Methods on Natural Language Processing, pp. 1892–1897. ACL, Stroudsburg (2013)
9. Hou, F., Li, G.: Mining chinese comparative sentences by semantic role labeling. In: Proceedings of the 2008 International Conference on Machine Learning and Cybernetics, pp. 2563–2568. IEEE, Piscataway (2008)
10. Xing, L., Liu, L.: Chinese standard comparative sentence recognition and extraction research. In: Proceedings of the 2013 International Conference on Information Engineering and Applications, pp. 415–422. Springe, London (2013)
11. Huang, G., Yao, T., Liu, Q.: Mining Chinese comparative sentences and relations based on CRF algorithm. Appl. Res. Comput. **27**, 2061–2064 (2010)
12. John, L., Andrew, M., Fernando, P.: Conditional random fields: probabilistic models for segmenting and labeling sequence data. In: Proceedings of the 18th International Conference Machine Learning, pp. 282–289. ACM, New York (2001)
13. Bai, L., Hu, R., Liu, Z.: Recognition of comparative sentences based on syntactic and semantic rules-system. Acta Scientiarum Naturalium Universitatis Pekinensis **51**, 275–281 (2015)
14. Zhang, C., Feng, C., Liu, Q., Shi, C., Huang, H., Zhou, H.: Chinese comparative sentence identification based on multi-feature fusion. J. Chin. Inf. Process. **27**, 110–116 (2013)
15. Yoon, K.: Convolutional neural networks for sentence classification. In: Proceedings of the 2014 Conference on Empirical Methods in Natural Language Processing (EMNLP), pp. 1746–1751. ACL, Stroudsburg (2014)
16. Liadh, K., et al.: Overview of the CLEF eHealth evaluation lab 2019. In: Experimental IR Meets Multilinguality, Multimodality, and Interaction. CLEF 2019. Lecture Notes in Computer Science, vol. 11696, pp. 322–339. Springer, Cham (2019)
17. Yu, J., Jiang, J.: Adapting BERT for target-oriented multimodal sentiment classification. In: Proceedings of the Twenty-Eighth International Joint Conference on Artificial Intelligence, pp. 5408–5414. Morgan Kaufmann, San Francisco (2019)
18. Rich, A.C.: Multitask learning: a knowledge-based source of inductive bias. In: Proceedings of the Tenth International Conference Machine Learning, pp. 41–48. ACM, New York (1993)
19. Li, Q., Ji, H.: Incremental joint extraction of entity mentions and relations. In Proceedings of the 52nd Annual Meeting of the Association for Computational Linguistics, vol. 1, pp. 402–412. ACL, Stroudsburg (2014)
20. Yuan, Z., David, W.: Stack-propagation: improved representation learning for syntax. In Proceedings of the 54th Annual Meeting of the Association for Computational Linguistics, vol. 1, pp. 1557–1566. ACL, Stroudsburg (2016)

21. Tomas, M., Ilya, S., Kai, C., Gregory, S.C., Jeffrey, D.: Distributed representations of words and phrases and their compositionality. In: Advances in Neural Information Processing Systems, NIPS 2013, vol. 26, pp. 3111–3119. MIT Press, Cambridge (2013)
22. Rumelhart, D.E., Hinton, G.E., Williams, R.J.: Learning internal representations by error propagation. In: Parallel distributed processing: explorations in the microstructure of cognition, vol. 1, pp. 318–362. ACM, New York (1986)

Who Killed Sanmao and Virginia Woolf? A Comparative Study of Writers with Suicidal Attempt Based on a Quantitative Linguistic Method

Shucheng Zhu[1], Xi Wang[2], and Pengyuan Liu[1,3](✉)

[1] School of Information Science, Beijing Language and Culture University, 15th Xueyuan Road, Haidian District, Beijing 100083, China
201721198327@stu.blcu.edu.cn, liupengyuan@pku.edu.cn

[2] Faculty of Foreign Studies, Beijing Language and Culture University, 15th Xueyuan Road, Haidian District, Beijing 100083, China
wangxi@xiao100.com

[3] Center of National Language Recourse and Monitoring Research Print Media Branch, Beijing Language and Culture University, 15th Xueyuan Road, Haidian District, Beijing 100083, China

Abstract. Writing style can reflect a writer's change of mood. It is thus important to discover quantitative linguistic markers of writing styles to be used to analyze the works of writers with suicidal attempt. Using five quantitative linguistic markers, this paper investigates the works of Sanmao and Virginia Woolf -- two writers with suicidal attempt -- diachronically. Although they wrote in different languages, the two writers shared some similarities in terms of these quantitative linguistic markers. Via analyzing quantitative linguistic markers of these two writers' works, one can clearly see their change of mood and how their writing styles were influenced. This paper also attempts to find out the reasons why they chose to end their lives with this quantitative linguistic method. These quantitative linguistic markers can be utilized in future studies, such as suicidal ideation detection.

Keywords: Quantitative linguistics · Suicidal attempt · Sanmao · Virginia Woolf

1 Introduction

On 14th October 2019, the famous Korean superstar Sulli committed suicide in her own house. Her premature death at the age of 25 drew public attention back to the topic of suicide. Suicide has become the third chief cause of death in today's world. It is estimated that there are 1.6 million suicidal attempts each year, and approximately 0.8 million people die of suicide [1]. Suicide has always been a hot topic for discussion. In the 1900s, the suicides of two famous female writers -- Sanmao and Virginia Woolf, from China and the U.K., shocked the literary world. Therefore, it is crucial to find methods to detect suicidal ideation or tendency in order to prevent suicide.

Many significant efforts have been made recently. As a first step, several psychological dictionaries have been compiled [2–6]. With the help of these resources and computing technologies, remarkable progress has been made in detecting suicidal ideation

M. Liu et al. (Eds.): CLSW 2020, LNAI 12278, pp. 408–420, 2021.
https://doi.org/10.1007/978-3-030-81197-6_34

and attempts. Most of such research has been conducted on social media materials or has analyzed suicide notes, and has helped to prevent people with suicidal tendencies. Most of the psychological dictionaries have compiled specific words related to different themes, such as death, love, anger, etc. However, language is full of duplicity, especially when it is employed to express negative emotions. Therefore, it is problematic to merely use specific words to detect suicidal ideation. With the help of quantitative linguistics techniques, we can find markers reflecting language phenomena that cannot be consciously hidden by writers. These markers may be used to analyze works in different languages, simultaneously manifesting the general nature of language and the psychology of human beings.

In this paper, we use five quantitative linguistic markers –Richness, Pronoun, Sensory, Lambda and Zipf's Exponent -- to analyze works of two female writers, Virginia Woolf and Sanmao. Virginia Woolf's works are in English whereas Sanmao wrote in Chinese. We can clearly see how big events have influenced their writing styles and their courses of life, respectively. After analysing the values of these markers, we also try to find out different reasons why they committed suicide.

2 Related Works

Literary scholars tend to consider a writer's life as auxiliary information for textual interpretation and analysis. Few of them have attempted to probe into the mental conditions of creators of literary works by analyzing their texts. Most of them have applied Freudian psychoanalysis in their studies. Here we take the works of Virginia Woolf as an example. Virginia Woolf's life and novels are both fascinating and perplexing to literary scholars. In tracing Woolf's protagonists' remarks about death, notably in her later works, Brombert has proclaimed that Woolf considered writing "as a form of salvation", "a struggle against death, an act of defiance, a way of protecting her sanity" [7]. Jean Thomson has focused particularly on *Mrs. Dalloway* and believed that Woolf weaved her own experiences of excitement and depression into this book in order to give an authentic portrayal of the mind of an ex-soldier, Septimus Smith. Thomson has also seen *Mrs. Dalloway* as a kind of "Dance of Death" [8]. Other critical biographers, underlining Woolf's manic-depressive side, have denied her intention of control over her life [9]. C. Caramagno has also expressed his complaints about Woolf's biographies in his contentious essay in the rather subjective nature [10]. Although literary scholars have tended to concentrate on biographical details, especially suicidal attempts, most of their focuses have been put on plot and characterization.

With the help of combining linguistic features and computing technologies, linguists and computing experts are making efforts to study the relation between features of writing style and writers with suicidal ideation. Earlier studies have focused on which professions are more prone to mental disorder [11] or used simple linguistic features to distinguish between ordinary letters and suicidal notes [12]. After psychological dictionaries have been compiled, especially LIWC, the analysis of suicide and language has become a problem of classification by using natural language processing methods. These works include detecting whether the users have suicidal ideation [13–16], or mental disorder [17], and analyzing the emotions and language features of people with suicidal ideation and attempts [18–20]. By using Suicide Probability Scale (SPS) and a classification model, scholars can distinguish people with a high suicidal probability and those with a low probability [21–23]. In distinguishing suicide notes and ordinary notes, computers do a better job than mental health professionals [24]. Technologies like sentiment analysis, opinion mining and depression detection are not perfect for use with suicide [25]. Besides, most studies on detecting suicide in real life have focused on the language used in social-media blogs. The method by using psychological dictionaries is just sentiment analysis. No real analysis of language is performed. In addition, no studies has focused on the chronological changes in the writing styles of the writers with suicidal ideation, as if they somehow all had suicidal ideation throughout their lifetimes.

In suicide detection studies, some scholars have used dictionaries to study the change of words in the blogs of a Chinese adolescent one year before s/he committed suicide [26]. Distinctively, this study concentrates on such change in chronological order, but the time is too short to reflect the subject's lifetime. Some others have mainly used four quantitative linguistic markers: Richness, Pronoun, Activity Power and Sensory (RPAS) [27, 28], as well as Critical Slowing Down [29], to compare two female English writers, Iris Murdoch and P. D. James, to see how big events and the mental disease Alzheimer influence writers' writing style. The quantitative markers RPAS were used in identifying personality in previous studies [30–32] and were proved to have good results. It is also believed that these markers can be used to identify mental diseases like Alzheimer, depression and extreme behaviour like terrorism. But these studies have only focused on the English language. Therefore, it is necessary to prove that these measures are also effective with works in different languages.

3 Methodology

3.1 Materials

In this paper, we have chosen the works of Sanmao and Virginia Woolf to study their change of writing style by using quantitative linguistic markers. Sanmao and Virginia Woolf were both famous female writers who died of suicide. Tables 1 and 2 show the works of Virginia Woolf and Sanmao that we have chosen.

We have only chosen full-length novels of Virginia Woolf to control the variable of genre. All the works are from Project Gutenberg[1] and their contents are in English. The Python package NLTK is used to tag parts-of-speech. We have only selected the novels and proses of Sanmao to control the variable of genre. It should be noticed that the first published novel of Sanmao is *The Story of the Sahara* and the second one is *Rainy Season Won't Come Again*. But *Rainy Season Won't Come Again* is a collection of essays and short stories Sanmao wrote between the age of seventeen and twenty-two. So, we have placed *Rainy Season Won't Come Again* as her first book to analyze. In order to keep the purity of the selected works, everything written by others (including forewords, epilogues, etc.) is removed. The Python package Jieba is used to segment and to tag the parts-of-speech.

[1] http://gutenberg.net.au.

3.2 Markers

Quantitative Index Text Analyzer (QUITA) [33] is used to compute the markers R4, Hapax Percentage and Lambda. Altmann Fitter is used to compute Zipf's Exponent [34].

Richness. Richness is a measure of a person's ability to use vocabulary (with focus on its size). It reflects not only the person's age and education background, but also his/her mood and psychological activities [27, 28, 31, 32]. People in a normal or joyful mood tend to have higher Richness than people in a bad mood, such as people with suicidal ideation [12]. The traditional Richness measure Type-Token Ratio (TTR) is subject to the influence of text size. Therefore, R4 and Hapax Percentage are chosen as the Richness markers as these two measures can eliminate the influence of text size.

Table 1. Works of Virginia Woolf we have chosen to analyze in this paper.

Number	Name	Publication time
V1	*The Voyage Out*	March, 1915
V2	*Night and Day*	October, 1919
V3	*Jacob's Room*	October, 1922
V4	*Mrs. Dalloway*	May, 1925
V5	*To the Lighthouse*	May, 1927
V6	*Orlando: A Biography*	June, 1928
V7	*The Waves*	October, 1931
V8	*The Years*	March, 1937
V9	*Between the Acts*	April, 1941

R4. R4 is one of the many ways of calculating the vocabulary richness of a text. The bigger it is, the higher Richness the text has. Formula (1) shows how R4 is calculated. Here V is the total number of types in a text, N is the total number of tokens in the text, r is the rank and f(r) is the frequency in the rank-frequency distribution of the text.

$$R4 = 1 - \frac{1}{V}(V + 1 - \frac{2}{N}\sum_{r=1}^{V} rf(r)) \tag{1}$$

Hapax. Hapax Percentage is the simple ratio of the number of hapax legomena in a text to the number of tokens. Hapax legomena are words that occur only once in a text. Formula (2) shows how Hapax Percentage is calculated. Here, NH is the number of hapax legomena in a text.

$$Hapax = \frac{NH}{N} \tag{2}$$

Table 2. Works of Sanmao we have chosen to analyze in this paper.

Number	Name in English	Publication time
S1	*Rainy Season Won't Come Again*	July, 1976
S2	*The Story of the Sahara*	May, 1976
S3	*The Scarecrow Notes*	June, 1977
S4	*Crying Camel*	August, 1977
S5	*Tender Night*	February, 1979
S6	*Back*	August, 1981
S7	*Dream Whispering Color*	August, 1981
S8	*Travelling Far to Travel Throughout*	May, 1982
S9	*Give You A Horse*	July, 1983
S10	*Fallen City*	March, 1983
S11	*My Baby*	July, 1987
S12	*Trouble Learning to Remember*	July. 1988

Pronoun. Pronoun is generally used to identify the author's gender [27, 28, 31, 32]. It has been shown that females prefer to use more personal pronouns and that the more pronouns a writer uses, the more emotional s/he is [35, 36]. In addition, people with suicidal ideation use more personal pronouns in their social media [18]. Formula (3) shows how Pronoun Percentage is calculated. Here P is the total number of personal pronouns in a text.

$$Pronoun = \frac{P}{N} \tag{3}$$

Sensory. Sensory words refer to visual, auditory, haptic, olfactory and gustatory adjectives [37]. Sensory words have been proved to have the capacity to stimulate our cerebral cortex [38]. Less sensory adjectives and smaller values of exclusivity can also be a signal of mental disease [28, 32]. In this paper, we use formula (4) to calculate Sensory [32]. Here $\boldsymbol{\varphi_i}$ is the number of the i-th adjective in the text and $\boldsymbol{\vartheta_i}$ is the value of exclusivity of the adjective, NK is the total number of the sensory adjective list. For the English texts, the sensory adjective list with exclusivity is from [39] and for the Chinese texts, the sensory adjective list with exclusivity is from [40]. To eliminate the influence of text size, all of the terms in the summation are normalized.

$$Sensory = \sum_{i=1}^{i=NK} \frac{\varphi_i \vartheta_i}{N} \tag{4}$$

Lambda. Lambda is an indicator that deals with the frequency structure of a text. It reflects both vocabulary richness and the relationship between the neighboring frequencies in a text. This measure also eliminates the influence of text size. Formula (5) shows its calculation.

$$Lambda = \frac{\sum_{r=1}^{V-1} \left[(f(r) - f(r+1))^2 + 1 \right]^{\frac{1}{2}} (log_{10}N)}{N} \tag{5}$$

Zipf's Exponent. Formula (6) shows Zipf's formula. Here P_r is the frequency of the word with rank r. C and b are parameters with b being an exponent. Studies have proved that the exponent b can distinguish languages. The exponent of English is around 1.11 which is the highest among natural languages [41, 42]. Other studies also found that Zipf's exponent might be related to language cognition and other aspects [43–46]. Zipf's exponent is related to language complexity, meaning that children's Zipf's exponents are higher than adults' exponents [47]. We believe that Zipf's exponent is also related to one's mood. If the writer is in the good condition, the Zipf's exponent of his/her language should be low.

$$P_r = Cr^{-b} \tag{6}$$

4 Analysis

4.1 Correlation Analysis

The two matrices in Fig. 1 show the Pearson correlations between the quantitative linguistic markers of Virginia Woolf's works and Sanmao's works respectively. It is clearly seen that the markers Richness, Sensory and Lambda show homogeneity as the higher the markers are, the better mood and mental condition the writers are in. The markers Pronoun and Zipf's Exponent also exhibit homogeneity as the higher the markers are, the worse mental conditions the writers are in. However, it is improper to suggest that the higher the Pearson Correlations are, the better efforts the quantitative linguistic measures give as the different quantitative linguistic markers represent different aspects.

4.2 Markers Visualization

Richness, Sensory and Lambda Visualization. Figure 2 shows how the markers' values change diachronically in the works of Virginia Woolf and Sanmao. R4, Hapax percentage, Sensory and Lambda share a similar tendency in Woolf's works. In the beginning, when she wrote her first two novels (V1 and V2), the values of the markers were relatively low. It was painstaking and difficult for her both physically and mentally. Leonard Woolf, her husband, believed *The Voyage Out* (V1) "to be the root cause of Virginia's distress" [48]. While waiting for the publication of *The Voyage Out*, Woolf was extremely anxious about its critical reception. In September 1913, she attempted to commit suicide [49]. On 23rd February 1915, Woolf became hallucinated, seeing her mother in the room talking wildly to her. It was only until September that she began to lead a normal life. Woolf remained relatively stable in 1916 and 1917, but afterwards, Vanessa's desperate wish to get pregnant and the need to finish *Night and Day* (V2) made her depression deteriorate [48]. The next decade witnessed the prime time of Woolf's literary career. The values of V3 to V7 were at a high level, and Woolf's condition was relatively stable. Though the Sensory values of V4 to V5 and the values of three other markers of V5 were low, they were still relatively high compared to the values of V1, V2 and V8. Then, when she wrote and revised her eighth novel (V8), an anti-war novel, she was horrified and threatened by the imminent war. It made her extremely angst-ridden.

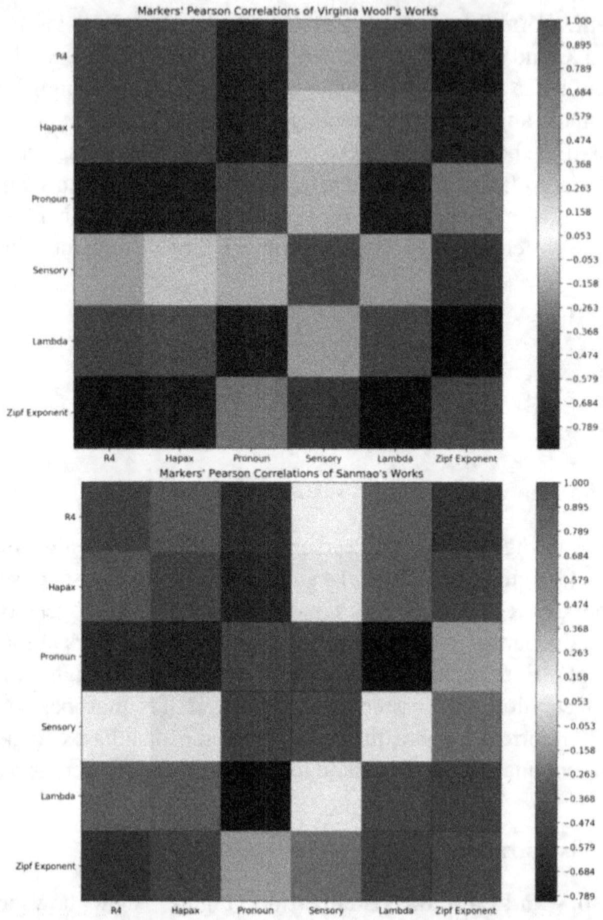

Fig. 1. Pearson Correlations between quantitative linguistic markers: Virginia Woolf's Works (first image) and Sanmao's Works (second image)

The pressure of outside cruelty and the stress of finishing the novel forced Virginia Woolf into insomnia, headaches and a breakdown that followed [48]. For V9, the values of the three markers other than Sensory were higher than the values of the corresponding markers of V8. This is because the mood of Virginia Woolf got better in her last years. It was a sudden event -- the horrific war (the World War II) destroyed her beloved city London -- that accelerated Woolf's suicidal tendency.

With Sanmao, we can see a very low point in *Dream Whispering Color* (S7). In this book, Sanmao wrote down her sad feelings about the loss of her husband Jose (Hexi). She suffered a lot during that time. Before S7, Sanmao wrote about her happy marriage life in the Sahara. After S7, she travelled to Latin America and Europe. This, in a sense, alleviated her pain. So, the Richness values during this period were higher. At the end of her writing career (S10, S11 and S12) the marker values fall as they predict her suicide. This also means that there was no sudden event that led to her death like Woolf.

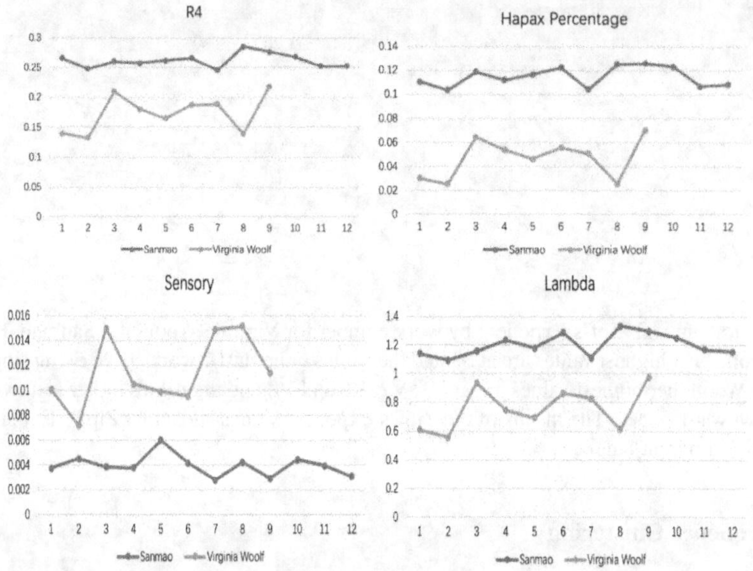

Fig. 2. R4, Hapax, Sensory and Lambda by work number for Virginia Woolf and Sanmao. Sanmao's lowest marker values are at S7, and the marker values of her last several works the go down. Woolf's lowest marker values are at V1, V2 and V8, and the marker values of her last work go up. Sensory is an exception. It reflects her worse mood while the other markers show that her sudden suicide was because of the shock of the war.

Pronoun and Zipf's Exponent Visualization. Figure 3 shows the Pronoun and Zipf's Exponent values changing diachronically in the works of Virginia Woolf and Sanmao. Pronoun and Zipf's Exponent values share a similar tendency in Woolf's works. The value of markers indicates high emotional level at the beginning (V1 & V2), in the middle (V5) and almost at the end (V8). Emotional elation can be caused by personal temperament and high involvement in the process of writing. As we have shown above, when Woolf wrote *The Voyage Out* (V1), *Night and Day* (V2), and *The Years* (V8), she suffered from cyclothymic depression and her mood alternated between extreme depression and euphoria. In these periods, Woolf was highly depressive. As for the case of *To the Lighthouse* (V5), it is Woolf's most "self-revealing" novel [7]. This novel is personal, for it is an elegy for Woolf's parents. As for V4, even though she was in a mentally stable condition, the content of her writing was disturbing: Mrs. Dalloway dealt with an ex-soldier, who suffered from PTSD and committed suicide at the end of the novel. The emotional depth Woolf revealed through these works is profound.

Sanmao's Pronoun Percentage and Zipf's Exponent values reach their highest points in S7. She wrote S7 when she was suffering badly from the loss of her husband. The marker values of her last two books have rising slopes. This shows that Sanmao's mental condition was deteriorating. The Zipf's Exponent graphs also shows that the exponent of English work is higher, confirming the findings of previous study [42].

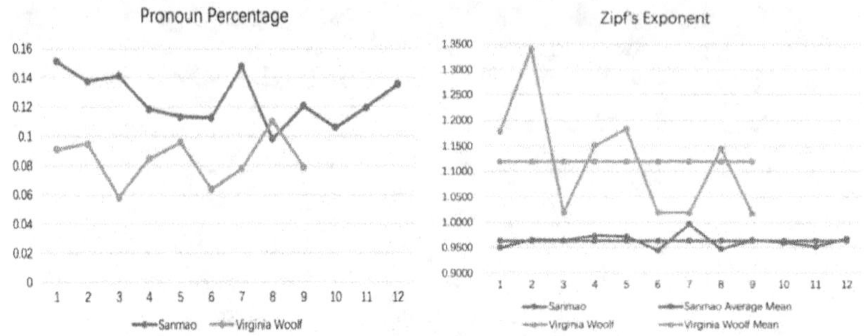

Fig. 3. Pronoun and Zipf's Exponent by work number for Virginia Woolf and Sanmao. For Sanmao's works, her highest values are at S7 and the values of her last 2 works have rising slopes. For Virginia Woolf, her highest values are at V1, V2, V5 and V8 and the values of her last two works have downward slopes. The means of the Zipf's Exponent values show that Zipf's Exponent can distinguish different languages.

4.3 Sequence Clustering

After normalization of all the markers, system clustering is used for sequence clustering [50]. Figure 4 shows the sequence clustering of the works of Sanmao and Woolf.

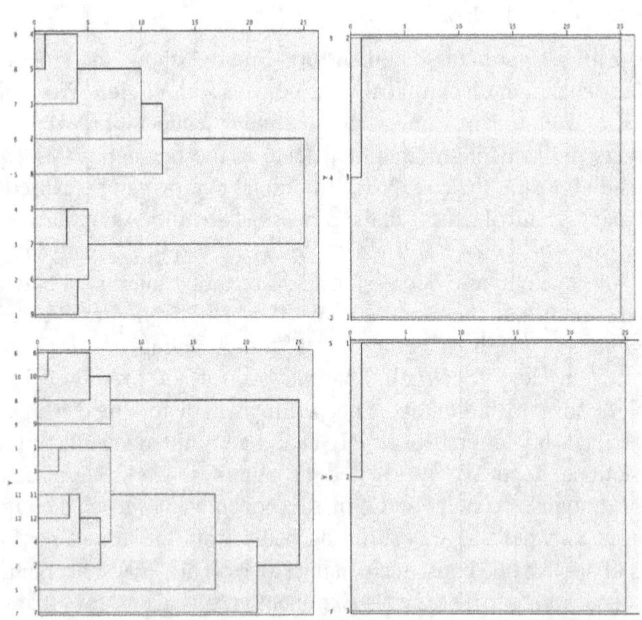

Fig. 4. System clustering of Virginia Woolf's (top 2) and Sanmao's (bottom 2) works

For Virginia Woolf, we understand:
V1, V2: Virginia Woolf's earlier writings. As a novice in literature, Woolf suffered greatly from the pressure to finish the novels. So, her mood alternated between depression and euphoria.
V3, V4, V5: These three books can be seen as Woolf's best literary works. Also, this period was the primetime of her career and personal life.
V6, V7: These two books are generally considered as the best of her later career.
V8: As the outcry of antiwar by Woolf, *The Years* was consuming for Woolf's sensitive nerves, unstable mentality and vulnerable physical condition.
V9: We can see a recovery tendency in the period between Woolf's last two novels. However, the horrific and inexorable war accelerated Woolf's suicidal tendency.
For Sanmao, we understand:
S1, S2, S3, S4, S5, S6: These are her early works and the reflection of her life in the Sahara.
S7: It reveals the darkness of her life because of her beloved husband's death.
S8, S9, S10: She created these works during or after her travels in Latin America and Europe.
S11, S12: Her last 2 works indicate her suicidal tendency because of her negative emotion.

5 Conclusion

We use five quantitative linguistic markers to analyze the English and Chinese works of two female writers. Our results show that all the markers we have chosen can reflect the psychological conditions of the two writers. With the help of these markers, we can find out the psychological changes in the minds of writers who never told the public why they chose to commit suicide to end their lives. We also find that big events, such as mental disorder and the loss of spouse and war, can influence one's life. These influences may be hidden by using emotional words, but cannot be hidden when the language of a writer's works is analyzed using quantitative linguistic markers. However, the markers we use are far from enough. More and more materials should be gathered to analyze the people's psychological activities. We hope that more and more useful features will be found for detecting suicidal ideation. We hope that anyone suffering from mental problems will indeed get the help they need -- indeed and in need.

Acknowledgments. This work is supported by the Fundamental Research Funds for the Central Universities, Research Funds of Beijing Language and Culture University 20YCX152, and Ministry of Education of Humanities and Social Science Project 18YJA740030. Thanks to Prof. Su Qi, Dr. Wan Mingyu and Zhou Jie from Peking University, Zhang Ying, Pan Yue, Du Bingjie and Zhang Sanle from Beijing Language and Culture University for their discussions.

References

1. World Health Organization: Mental Health Action Plan 2013–2020. World Health Organization, Geneva, Switzerland (2013)

2. Coltheart, M.: The MRC psycholinguistic database. Q. J. Exp. Psychol. Sect. A **33**(4), 497–505 (1981)
3. Yao, Z., Wu, J., Zhang, Y., et al.: Norms of valence, arousal, concreteness, familiarity, imageability, and context availability for 1,100 Chinese words. Behav. Res. Methods **49**(4), 1374–1385 (2017)
4. Bucci, W., Kabasakalian, R.: The RA Research Group: Instructions for scoring Referential Activity (RA) in transcripts of spoken narrative texts. Ulm, Germany: Ulmer Textbank (1992)
5. Bucci, W., Maskit, B.: Building a Weighted Dictionary for Referential Activity. Computing Attitude and Affect in Text, pp. 49–60 (2005)
6. Pennebaker, J.W., Boyd, R.L., Jordan, K., et al.: The Development and Psychometric Properties of LIWC2015 (2015)
7. Brombert, V.: Virginia woolf – "death is the enemy." Hudson Rev. **63**(3), 429–444 (2010)
8. Thomson, J.: Virginia Woolf and the case of Septimus Smith. San Francisco Jung Inst. Libr. J. **23**(3), 55–71 (2004)
9. Kenney M.S., Kenney, Jr. J.E.: Virginia woolf and the art of madness. Massachusetts Rev. 23(1), 161–185 (1982)
10. Caramagno, T.C.: Manic-depressive psychosis and critical approaches to Virginia Woolf's life and work. PMLA **103**(1), 10–23 (1988)
11. Post, F.: Verbal creativity, depression and alcoholism: an investigation of one hundred American and British writers. Br. J. Psychiatry **168**(5), 545–555 (1996)
12. Osgood, C.E., Walker, E.G.: Motivation and language behavior: a content analysis of suicide notes. J. Ab. Soci. Psychol. **59**(1), 58 (1959)
13. Xu, L., Song, W.: Suicidal idea detection based on chinese microblog language features. J. North Univ. China (Nat. Sci. Ed.) **40**(4), 350–357 (2019)
14. Coppersmith, G., Leary, R., Crutchley, P., et al.: Natural language processing of social media as screening for suicide risk. Biomed. Inf. Insights **10**, 1178222618792860 (2018)
15. Guan, L., Hao, B., Liu, T., Cheng, Q., et al.: A pilot study of differences in behavioral and linguistic characteristics between Sina suicide microblog users and Sina microblog users without suicide idea. Chin. J. Epidemiol. **36**(5), 421–425 (2015)
16. Huang, X., Zhang, L., Chiu, D., et al.: Detecting suicidal ideation in chinese microblogs with psychological lexicons. In: 2014 IEEE 11th Intl Conf on Ubiquitous Intelligence and Computing and 2014 IEEE 11th Intl Conf on Autonomic and Trusted Computing and 2014 IEEE 14th Intl Conf on Scalable Computing and Communications and Its Associated Workshops, pp. 844-849. IEEE (2014)
17. Wang, T., Zhou, Z., Zhu, T., et al.: Mood disorder patients' language features on their microblogs. Int. J. Embedded Syst. **7**(1), 34–42 (2014)
18. Cheng, Q., Li, T.M.H., Kwok, C.L., et al.: Assessing suicide risk and emotional distress in chinese social media: a text mining and machine learning study. J. Medical Internet Res. **19**(7), e243 (2017)
19. Egnoto, M.J., Griffin, D.J.: Analyzing Language in Suicide Notes and Legacy Tokens. Crisis (2016)
20. Pestian, J., et al.: Sentiment analysis of suicide notes: a shared task. Biomed. Inf. Insights **5s1**, BII.S9042 (2012). https://doi.org/10.4137/BII.S9042
21. Guan, L., Hao, B., Cheng, Q., et al.: Behavioral and linguistic characteristics of microblog users with various suicide ideation level: an explanatory study. Chin. J. Public Health **31**(3), 349–352 (2015)
22. Guan, L., Hao, B., Cheng, Q., et al.: Identifying Chinese microblog users with high suicide probability using internet-based profile and linguistic features: classification model. JMIR Mental Health **2**(2), e17 (2015)

23. Zhang, L., Huang, X., Liu, T., Li, A., Chen, Z., Zhu, T.: Using linguistic features to estimate suicide probability of chinese microblog users. In: Qiaohong, Zu., Bo, Hu., Ning, Gu., Seng, Sopheap (eds.) HCC 2014. LNCS, vol. 8944, pp. 549–559. Springer, Cham (2015). https://doi.org/10.1007/978-3-319-15554-8_45
24. Pestian, J., Nasrallah, H., Matykiewicz, P., Bennett, A., Leenaars, A.: Suicide note classification using natural language processing: a content analysis. Biomed. Inf. Insights **3**, BII.S4706 (2010). https://doi.org/10.4137/BII.S4706
25. Gómez, J.M.: Language technologies for suicide prevention in social media. In: Proceedings of the Workshop on Natural Language Processing in the 5th Information Systems Research Working Days (JISIC), pp. 21–29 (2014)
26. Li, T., Chau, M., Yip, P., Wong, P.: Temporal and computerized psycholinguistic analysis of the blog of a Chinese adolescent suicide. Crisis **35**(3), 168–175 (2014). https://doi.org/10.1027/0227-5910/a000248
27. Kernot, D., Bossomaier, T., Bradbury, R.: The impact of depression and apathy on sensory language. Open J. Modern Linguist. **7**(1), 8–32 (2016)
28. Kernot, D., Bossomaier, T., Bradbury, R.: The stylometric impacts of ageing and life events on identity. J. Quant. Linguist. **26**(1), 1–21 (2019)
29. Dakos, V., Bascompte, J.: Critical slowing down as early warning for the onset of collapse in mutualistic communities. Proc. Natl. Acad. Sci. **111**(49), 17546–17551 (2014)
30. Kernot, D.: The Identification of Authors using Cross Document Co-Referencing. The University of New South Wales Master Degree Thesis (2013)
31. Kernot, D., Bossomaier, T., Bradbury, R.: Novel text analysis for investigating personality: identifying the dark lady in Shakespeare's sonnets. J. Quant. Linguist. **24**(4), 255–272 (2017)
32. Kernot, D., Bossomaier, T., Bradbury, R.: Using Shakespeare's sotto voce to determine true identity from text. Front. Psychol. **9**, 289 (2018)
33. Kubát, M., Matlach, V., Čech, R.: Quantitative Index Text Analyzer. RAM-Verlag, Germany (2014)
34. Altmann-Fitter. Altmann-Fitter User Guide (2013)
35. Rayson, P., Leech, G.N., Hodges, M.: Social differentiation in the use of English vocabulary: some analyses of the conversational component of the British national corpus. Int. J. Corpus Linguist. **2**(1), 133–152 (1997)
36. Schmid, H.J., Fauth, J.: Women's and Men's style: fact or fiction? new grammatical evidence. In: Corpus Linguistics Conference, Lancaster (2003)
37. Zhou, J., Su, Q., Liu, P.: A metaphorical analysis of five senses and emotions in mandarin Chinese. In: Hong, Jia-Fei., Zhang, Yangsen, Liu, Pengyuan (eds.) CLSW 2019. LNCS (LNAI), vol. 11831, pp. 607–617. Springer, Cham (2020). https://doi.org/10.1007/978-3-030-38189-9_62
38. Fernandino, L., Binder, J.R., Desai, R.H., et al.: Concept representation reflects multimodal abstraction: a framework for embodied semantics. Cereb. Cortex **26**(5), 2018–2034 (2015)
39. Lynott, D., Connell, L.: Modality exclusivity norms for 423 object properties. Behav. Res. Methods **41**(2), 558–564 (2009)
40. Chen, I.H., Zhao, Q., Long, Y., et al.: Mandarin Chinese modality exclusivity norms. PloS one **14**(2),(2019)
41. Liu, H., Huang, W.: Quantitative linguistics: state of the art, theories and methods. J. Zhejiang Univ. (Human. Soc. Sci.) **42**(2), 178–192 (2012)
42. Mehri, A., Jamaati, M.: Variation of Zipf's exponent in one hundred live languages: a study of the holy bible translations. Phys. Lett. A **381**(31), 2470–2477 (2017)
43. McCowan, B., Doyle, L.R., Hanser, S.F.: Using information theory to assess the diversity, complexity, and development of communicative repertoires. J. Comp. Psychol. **116**(2), 166 (2002)

44. Ferrer-i-Cancho, R., Elvevåg, B.: Random texts do not exhibit the real Zipf's law-like rank distribution. PLoS One **5**(3) (2010)
45. Kalankesh, L.R., Stevens, R., Brass, A.: The language of gene ontology: a Zipf's law analysis. BMC Bioinf. **13**(1), 127 (2012)
46. Perline, R.: The Random Division of the Unit Interval and the Approximate -1 Exponent in the Monkey-at-the-Typewriter Model of Zipf's Law. arXiv preprint 1512.02177 (2015)
47. Baixeries, J., Elvevåg, B., Ferrer-i-Cancho, R.: The evolution of the exponent of Zipf's law in language ontogeny. PloS one **8**(3) (2013)
48. Dally, P.: The Marriage of Heaven and Hell: Manic Depression and the Life of Virginia Woolf. St Martin's Press, New York (1999)
49. Bishop, E.: A Virginia Woolf Chronology. Macmillan, Basingstoke (1989)
50. Yan, G., Lu, Y.: Study on clustering method of multidimensional sequence samples. Stat. Decis. **256**(4), 29–30 (2008)

A Study on Semantic Features of Antonymous Verbs Based on Lexical Knowledge System

Gaofeng Shi, Caiying Yang, and Hongbing Xing(✉)

Beijing Language and Culture University, Xueyuan Road 15, Beijing 100083, China
xinghb@blcu.edu.cn

Abstract. Extraction of lexical semantic features is a key issue in lexical knowledge research. Based on the connectionism theory, this paper adopts the corpus and statistical approach to analyze the semantic features of antonymous verbs and construct their lexical knowledge systems. We take the antonymous verbs "买(*mai3,buy*)" and "卖(*mai4,sell*)" as examples in our analysis. We also construct lexical knowledge systems of collocation, semantic features of collocation and information about syntactic constructions. On this basis, this paper focuses on the effects of the types and frequencies of word collocations and the syntactic usage of the words, hoping to provide some insights for the studies on lexical semantic features of antonymous verbs.

Keywords: Antonymous verbs · Semantic feature · Lexical knowledge · Corpus

1 Introduction

With the development of corpora and cognitive science, representational research on language knowledge has been paying more attention to the statistical attributes of language. For example, research based on connectionism has emphasized the relationships among words in actual language use, which provides a good research framework for corpus-based analysis on vocabulary knowledge extraction [1]. With the help of the corpus, we can statistically analyze the actual use of words and extract information about the distribution of use, so as to have a deeper understanding of lexical semantics. Wang, Jia and Yu [2] investigated the semantic distribution of Chinese quantifiers and their collocations based on a large-scale corpus. Li [3], utilizing a database, explored the knowledge representation and the acquisition methods of word meaning from multiple perspectives.

In addition, there is a close relationship between the statistical attributes of language and cognitive processing of language. As lexical semantic studies suggest, a complex representational system exists in human brain, where words with a common feature form a clustering relationship. The frequency of combinations of words affects the lexical semantic representation, and it also plays an important role in lexical cognitive processing [4–6]. Scholars are very concerned about the usage of words and the effects of frequencies on lexical semantics and thus multiple perspectives such as language computing and cognition are taken to study the issues. They believe that lexical semantics is

M. Liu et al. (Eds.): CLSW 2020, LNAI 12278, pp. 421–431, 2021.
https://doi.org/10.1007/978-3-030-81197-6_35

influenced by multidimensional knowledge in the process of lexical usage. Therefore, a comprehensive quantitative analysis of usage of words based on a corpus is of great significance to the study of lexical semantic knowledge.

As is known to all, antonymous verbs are verbs with semantic dependency and opposition, and their asymmetry in semantics and syntactic combinations has been drawing scholars' attention [7–9]. However, the complex syntactic and semantic relations of antonymous verbs cannot be simply analyzed from the perspective of asymmetry in the use of words, and there is a need to delve deep into the knowledge system used by antonymous verbs. "买(*mai3, buy*)" and "卖(*mai4, sell*)" are a typical pair of antonymous verbs in Chinese. They denote a kind of transaction behavior involving money and commodities, seemingly without any differences in terms of collocation or syntax. For example, we can say "买/卖车(*mai3/mai4 che1, buy/sell a car*)" and "我买/卖了一辆车(*wo3 mai3/mai4 le yi1 liang4 che1, I bought/sold a car.*)". But by comparing "我把汽车买/卖了(*wo3 ba3 qi4 che1 mai3/mai4 le, I bought/sold a car.)*", we have found that they obviously differ from each other. The sentence "我把汽车卖了(*wo3 ba3 qi4 che1 mai4 le, I sold a car.*)" is more natural while the other sentence is less natural. It is difficult to explain this phenomenon with the notion of asymmetry.

So what is the relationship between the semantics and the usage of antonymous verbs? To study the relationship between them, we will take "买(*mai3, buy*)" and "卖(*mai4, sell*)" as examples and make use of a corpus to provide a comprehensive analysis of the distributional differences between these two words. More importantly, we will focus on the collocational usage, frequencies in collocation and frequencies of the occurrences of the words in different sentence structures; and we will also explore the relationship between lexical knowledge and the semantic features of antonymous verbs.

2 Construction of Lexical Knowledge System of Antonymous Verbs

2.1 The Lexical Knowledge System of Antonymous Verbs

Cronbach [10] first proposed five elements of lexical knowledge, including concepts denoted by the word, appropriateness, meaning items, accuracy and flexibility of the word. Richards [11], Nation [12], Melka [13] and other works studied the framework of lexical knowledge and concluded that lexical knowledge mainly included morphology, meaning, function and usage. With the development of connectionism, researchers have gradually focused on the storage modes of distributional representations of lexical knowledge. Xing [1, 14] believed that lexical knowledge included the pronunciation, form, meaning, mutual connection, frequency information, syntactic function distribution, collocation knowledge, syntactic construction, combination and aggregation relationship from the perspective of corpus. He pointed out that lexical knowledge is knowledge representation formed by abstract cognitive processing of language features in usage. He also emphasized that the distribution of lexical attributes was the source of lexical knowledge acquisition.

Hence, we believe that the collocations of the words are the most direct means to obtain the knowledge of the usage of words which is the key to constructing a lexical knowledge system. Words can express complete semantic information in sentences. Various syntactic constructions in which a word is used and the frequency of use provide important information regarding the semantics of a lexical item. Therefore, we mainly analyze the semantic features of antonymous verbs from the perspectives of collocation knowledge system and the information of syntactic constructions.

2.2 Construction of Lexical Knowledge System of Antonymous Verbs

To obtain the usage of the two words comprehensively and systematically, we have selected 500 examples containing "买(*mai3, buy*)" and "卖(*mai4, sell*)" respectively from the Modern Chinese Corpus of the National Language Commission, and screened the sentences before labeling. The annotated sentences have met the following requirements: (1) the verb should be the main predicate in the sentence; (2) the sentence containing the verb should express the complete meaning; (3) repeated sentences should be eliminated. Finally, we got 326 sentences containing "买(*mai3, buy*)" and 315 containing "卖(*mai4, sell*)". We have annotated the syntactic components (including subjects, objects, adverbials and complements), the semantic types of the components, and the types of syntactic constructions in which the two target words are used.

Collocation Knowledge Systems of Antonymous Verbs. By annotating and analyzing the corpora, we have obtained four collocation relations formed by "买*(mai3, buy)*" and "卖*(mai4, sell)*", namely subject-verb collocation, verb-object collocation, adverbial-verb collocation and verb-complement collocation. We have also obtained the types of the syntactic components of their collocate words and the collocational frequencies. With this information, we have constructed the collocation knowledge systems of the two words, as shown in Fig. 1 and Fig. 2.

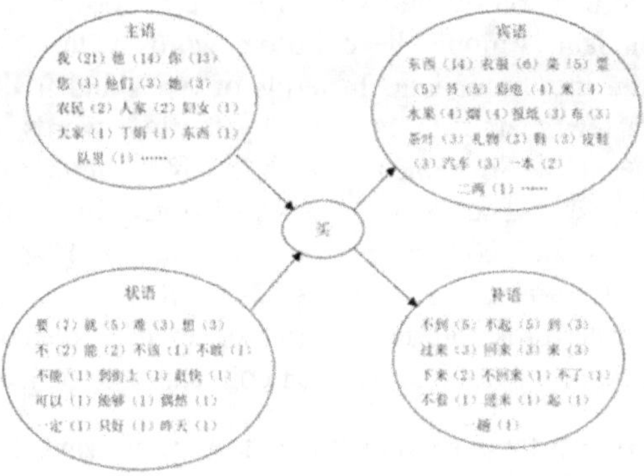

Fig. 1. Collocation knowledge system of "买(*mai3, buy*)"

It is found that 39 words among words that collocate with "买(*mai3, buy*)" are subject collocates. They are mainly personal pronouns and appellation nouns, among which personal pronouns account for 72.5% and appellation nouns account for 18.7%. Besides, there are 21 adverbial collocate words, and the most frequently used collocate word is the auxiliary verb "要(*yao4, want*)". There are a total of 126 object collocate words, which are mainly common nouns. There are 13 complement collocate words with the most frequently used of which are "不到(*bu2dao4, not available*)", "不起(*bu4qi3, can't afford*)" and other words expressing direction and result.

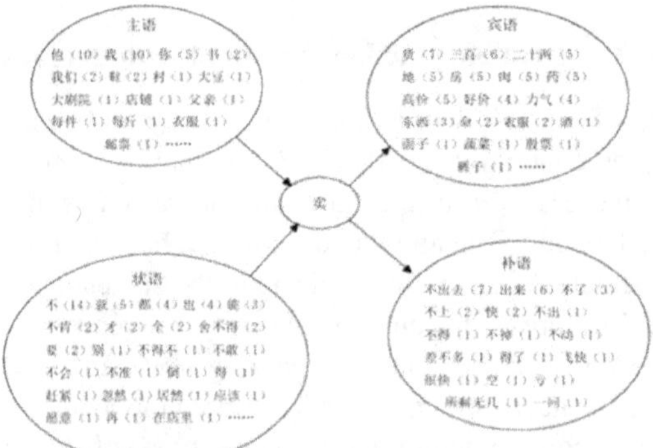

Fig. 2. Collocation knowledge system of "卖(*mai4, sell*)"

Analyzing the collocations of "卖(*mai4, sell*)", we have found that there are 86 subject collocate words, which are mainly common nouns, followed by personal pronouns. 32 words are adverbial collocates. The negative adverb "不(*bu4, no*)" has the highest collocational frequency and mainly negates the objective state of the action; There are 108 object collocate words which are mainly common nouns. 22 words are complement collocates, among which the most frequent words are "不出去(*bu4chu1qu4, cannot be out*)" and "出去(*chu1qu4, out*)".

Comparing the collocation of the two words, the collocate words with "卖(*mai4, sell*)" are more abundant. Among the collocate words with "卖(*mai4, sell*)", subject collocates are the most abundant. In terms of types and frequencies of collocations, there are very few shared collocations by the two words, except for the personal pronouns "我/我们(*wo3/wo3men, I/we*)", "你/你们(*ni3/ni3men, you*)" and "他/他们(*ta1/ta1men, he/they*)". The frequencies are also very different. For example, the negative adverb "不(*bu4, no*)" collocating with "买(*mai3, buy*)" accounts for 5.4%, and when it collocates with "卖(*mai4, sell*)", it is 22.6%. It can be seen that the two verbs have different collocate words and demonstrate different frequencies in collocational relations, indicating that they both have a rich collocation knowledge system.

Syntactic Construction Information of the Antonymous Verbs. The syntactic construction information of words is an important part of the lexical knowledge system [1], which refers to the types of verbs and their frequencies of use in this paper. A verb can be used in many syntactic constructions, and there is a selective relationship between the verb and the syntactic construction in which it appears. For example, the verb "当作(*dang1zuo4, take as*)" being used in "*ba*-sentences" accounts for 71% in terms of proportion, and the verb being used in the "Subject-Verb-Object" construction accounts for only 4% in terms of proportion. In this paper, the annotation principles of the syntactic constructions are mainly guided by the classification criteria of Shi [15]. Table 1 below shows the usage of the two words in different syntactic constructions.

Table 1. Distribution of syntactic constructions

Syntactic constructions	买(*mai3, buy*)			卖(*mai4, sell*)		
	Number	Proportion	Z-score	Number	Proportion	Z-score
(Subject)-Verb	14	4.29	−0.42	16	5.08	−0.46
(Subject)-Verb-Object	130	39.88	2.68	62	19.68	1.21
(Subject)-Verb-Object1-Object2	3	0.92	−0.71	0	0.00	−1.04
(Subject)-Verb-Complement	23	7.06	−0.18	79	25.08	1.83
(Subject)-Verb-Complement-Object	59	18.10	0.78	32	10.16	0.12
Shide-sentence	0	0.00	−0.79	6	1.90	−0.82
Ba-sentence	10	3.07	−0.52	54	17.14	0.92
Bei-sentence	13	3.99	−0.44	0	0.00	−1.04
Existential sentence	0	0.00	−0.79	3	0.95	−0.93
Serial verb constructions(verb after)	60	18.40	0.81	54	17.14	0.92
Pivotal sentences(verb after)	14	4.29	−0.42	9	2.86	−0.71

We have found that totally there are 11 types of syntactic constructions in which "买(*mai3, buy*)" and "卖(*mai4, sell*)" appear, while only 9 of them contain both of the two verbs. The two verbs are quite different in terms of construction types and frequency. Among the 9 types of constructions, it is the "(Subject)-Verb-Object" construction in which "买(*mai3, buy*)" most frequently appears, accounting for the highest proportion of 39.88%. In case of "卖(*mai4, sell*)", it is the "(Subject)-Verb- Complement" construction which accounts for the highest proportion (25.08%), followed by the "(Subject)-Verb- Object" construction. As for the shape of the distribution of the syntactic constructions, "买*(mai3, buy)*" is more concentrated, tending to be used mainly with the "(Subject)-Verb-Object" construction, with a Z-score[1] of 2.68. The shape of the distribution of the constructions in which "卖(*mai4, sell*)" appears is more balanced, as the constructions of "(Subject)-Verb-Complement", "(Subject)-Verb-Object" and "*ba*-sentence" all accounts for a higher proportion.

3 The Lexical Knowledge System and the Semantic Features of Antonymous Verbs

3.1 Collocation Knowledge and Semantic Features

Collocation knowledge of words is the core of lexical knowledge. The meanings of a word will be affected by the semantic features of its collocate words. For example, the word "我们(*wo3men, we*)" is a subject collocate of "买(*mai3, buy*)", which attributes the "subjective" characteristics to the action doer. When "买(*mai3, buy*)" collocates with the adverbial "难(*nan2, difficult*)", the adverbial will provide "买(*mai3, buy*)" with the semantic characteristics of "degree". These collocate words form different collocational

[1] Z-score is a process of dividing the difference between the number and the mean by the standard deviation, and the calculation formula is: $Z = (X-\bar{X})\ /s$.

relations with "买(*mai3, buy*)". We can classify these words according to their semantic types, such as "things", "clothes" and "place".

Since a word and its collocate words can form a system of lexical relationships, we can further explore the semantic features of a word by analyzing the collocate words in different collocational relations and their frequencies in collocation, and the semantic features of the collocate words and their frequencies. Hence, the words in the four collocation relations are classified according to their semantic types, and the semantic forming process can be obtained by examining the differences in the distribution of semantic types in different collocational relations.

Firstly, according to the semantic characteristics of the subject collocates of " 买(*mai3, buy*)" and "卖(*mai4, sell*)" and the classification standard in *同义词词林*(*tong2yi4ci2ci2lin2, Word Similarity*) [16], we have divided the collocate words into four semantic types, namely "appellation", "article", "place" and "quantity", as shown in Table 2.

Table 2. Semantic type distribution of subject collocates

Semantic type	买(*mai3, buy*)		卖(*mai4, sell*)	
	Number	Proportion	Number	Proportion
Appellation	83	91.21	42	35.29
Things	7	7.69	63	52.94
Place	1	1.10	10	8.40
Quantity	0	0	4	3.36

We have found that the semantic types of the subject collocates of " 买(*mai3, buy*)" and "卖(*mai4, sell*)" are different. The semantic types of subject collocates of "买(*mai3, buy*)" are more concentrated, with the proportion of "appellation" reaching 91.21%. The semantic types of subject collocates of "卖(*mai4, sell*)" are more diverse and the collocational concentration is lower, among which the proportion of "things" is the highest (52.5%), followed by "appellation" (35%). It follows that the semantic types of their subjects have a tendency: "买(*mai3, buy*)" collocates frequently with "appellation", while "(*mai4, sell*)" coexists with subjects denoting "things". The differences in the semantic types of subject collocates indicate the differences in the lexical semantics of the two verbs. For example, subjects belonging to the "appellation" category are generally the doers of actions and have strong autonomy, while subjects in the "things" category are generally the recipient of the action and have the character of non-autonomy. As a result, the semantic features of the two verbs are distinct: "买(*mai3, buy*)" has stronger autonomy while the top priority semantic feature of "卖(*mai4, sell*)" is non-autonomy.

To further discuss the semantic features of "买(*mai3, buy*)" and "卖(*mai4, sell*)", we have analyzed the semantic types of the subjects to see whether they are agents or patients (see Table 3). The results have shown that "买(*mai3, buy*)" is more likely to have agentive subjects (80.98%), and "卖(*mai4, sell*)" is more likely to have patient subjects (61.83%), which strongly agrees with the semantic type distribution of subject

Table 3. The distributions of feature of 'autonomy' of the subject collocates

Semantic type	买(*mai3, buy*)		卖(*mai4, sell*)	
	Number	Proportion	Number	Proportion
Agentive	264	80.98	107	33.75
Patient	36	11.04	196	61.83
Neutral	20	6.13	14	4.42

collocates analyzed above. Li [17] believed that there was a continuum from agentive to non-agentive. According to the statistical results, the subject types of "买(*mai3, buy*)" are mainly in the agentive part of the continuum, but the subjects of "卖(*mai4, sell*)" are mainly in the patient end (while the agentive subject collocating with "卖(*mai4, sell*)" accounts for 33.75% in proportion). This shows that the semantic types of the subjects of "卖(*mai4, sell*)" are more balanced on the continuum than that of "买(*mai3, buy*)", indicating that "卖(*mai4, sell*)" mainly expresses the semantic feature of "patient", while its semantic feature of "agent" are also obvious.

Table 4. The distribution of the semantic types of object collocates

Semantic type	买(*mai3, buy*)		卖(*mai4, sell*)	
	Number	Proportion	Number	Proportion
Things	188	94.00	109	62.64
Quantity	9	4.50	3	1.72
Abstraction	3	1.5	14	8.05
Money	0	0	43	24.71

According to Table 4, the majority of the object collocates of "买(*mai3, buy*)" are "things" (94%) which accounts for 62.64% in proportion, indicating that "买(*mai3, buy*)" is more agentive than "卖(*mai4, sell*)". Among them, the object of "money" collocating with "卖(*mai4, sell*)" accounts for 24.71%. Note that "money" is not the direct object of "卖(*mai4, sell*)", but the result obtained by the action instead. It follows that this collocational semantic type gives "卖(*mai4, sell*)" the semantic feature of "obtain". For example, in the sentence "一斤卖了十块钱 (*yi4jin1mai4leshi2kuai4qian2, selling out a catty can earn ten yuan)*", the result of the action of "卖(*mai4, sell*)" is ten dollars. Meanwhile, "买(*mai3, buy*)" also has the semantic feature of "obtain", which is believed to be the result of the strong "agentive" feature of "买(*mai3, buy*)". Be that as it may, the two words interpret the meaning of "obtain" in different ways.

We have divided the semantics of adverbial collocates of "买(*mai3, buy*)" and " 卖(*mai4, sell*)" into eight semantic types (see Table 5). Among the semantic types of adverbial collocates of "买(*mai3, buy*)", the one accounting for the highest proportion (48.65%) is "subjective willingness" which accounts for only 27.59% while collocating

Table 5. Semantic type distribution of adverbial collocates

Semantic type	买(*mai3, buy*)		卖(*mai4, sell*)	
	Number	Proportion	Number	Proportion
Subjective willingness	18	48.65	16	27.59
Time	8	21.62	8	13.79
Extent	3	8.11	2	3.45
Negativity	3	8.11	17	29.31
Place	2	5.41	3	5.17
Scope	1	2.70	7	12.07
Mood	1	2.70	2	3.45
Mode	1	2.70	3	5.17

with "卖(*mai4, sell*)". This indicates that "买(*mai3, buy*)" has a stronger tendency to express "subjectivity". Its semantic distribution is also very concentrated, indicating that it has a stronger semantic feature of "agentive".

However, "卖(*mai4, sell*)" is concentrated in the semantic features of "subjective willingness" and "objective negation", indicating that the subject collocates of "卖(*mai4, sell*)" can be either person with strong "agentive" feature or things with strong feature of "patient". We can see that "买(*mai3, buy*)" has a strong agentive semantic characteristic, and "卖(*mai4, sell*)" has obvious agentive and patient characteristics, which is consistent with the results on the semantic features of "subject" above.

Table 6. Semantic type distribution of complement collocates

Semantic type	买(*mai3, buy*)		卖(*mai4, sell*)	
	Number	Proportion	Number	Proportion
Tendency	14	46.67	1	3.45
Result	8	26.67	18	62.07
Possibility	7	23.33	1	3.45
Quantity	1	3.33	1	3.45
Status	0	0.00	8	27.59

The semantic types of the complements of "买(*mai3, buy*)" and "卖(*mai4, sell*)" are mainly divided into five types (as shown in Table 6), and the top semantic type of the complements of "买(*mai3, buy*)" is "tendency" (46.67%), while for "卖(*mai4, sell*)", the semantic type of "result" takes up the first place (62.07%). From the semantic features of the verb complement, "买(*mai3, buy*)" tends to express the function of "displacement", emphasizing the transfer of things from a buyer to a seller, while "卖(*mai4, sell*)" mainly expresses the sense of "disposal", emphasizing that the "result" is realized through the "disposal".

3.2 Syntactic Construction Knowledge and Semantic Features

Regarding the semantic features of words, a few studies have been carried out from the perspective of syntax. Previous studies have found that there is a dependency relationship between semantic expressions and syntactic structures, and different verbs tend to select specific syntactic structures to express their meanings. According to the statistical analysis of the syntactic construction information of "买(*mai3, buy*)" and "卖(*mai4, sell*)" above, there are great differences between the two verbs. "卖(*mai4, sell*)" tends to be used in "(Subject)-Verb-Complement" construction and "*ba*-sentences", while "买(*mai3, buy*)" has low proportions in these two sentence structures.

Furthermore, the complement also plays an important role in conveying the semantic meanings of sentences. It is generally believed that the more complements are used in a sentence, the richer the meanings of the verb can be, and "disposal" meaning of the verb will be more obvious. As one of the most distinctive syntactic structures in Chinese, the "*ba*-sentence" emphasizes that the action denoted by the verb is "disposition" of the object. It is proved that the "*ba*-sentence" has selectional restriction on verbs. In general, the verb used in "*ba*-sentence" needs to carry the semantic feature of [+motion]. Thus, we can preliminarily conclude that the "disposal" feature of "卖(*mai4, sell*)" is stronger than "买(*mai3, buy*)" according to their different tendency to be used in a "*ba*-sentence"

Based on the above analysis, we have discussed the semantic features of antonymous verbs from the perspective of collocation knowledge and syntactic construction, as is illustrated in Fig. 3. It is found that "买(*mai3, buy*)" and "卖(*mai4, sell*)" have "overlapping", "inclusion" and "separation" relations in collocation knowledge and the information of syntactic construction. They share many collocate types and syntactic

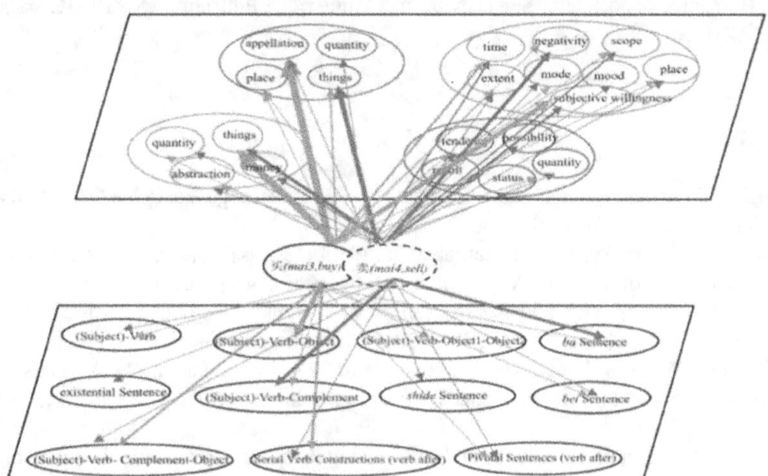

Fig. 3. Semantic feature system of "买*(mai3, buy)*" and "卖*(mai4,sell)*"

structures, but there are significant differences reflected in the frequencies, as shown by the arrows of different widths in Fig. 3. Collocates in different collocation relations are classified according to their lexical meanings. Considering the collocational frequencies between the two words and their semantic types and also the frequencies between the two words and their syntactic constructions, we have discussed the semantic features of the two antonymous verbs and considered that the “agency” and “disposal” are the core features to distinguish them.

4 Conclusion

In this paper, we have taken the antonymous verbs “买(*mai3, buy*)” and “卖(*mai4,sell*)” as examples and constructed the lexical knowledge systems of antonymous verbs. We have then discussed the semantic characteristics of antonymous verbs from perspective of collocation and syntax. We believe that “agentive” and “disposition” are the two important semantic features for “买(*mai3, buy*)” and “卖(*mai4,sell*)”. This paper has also proposed a research approach to the semantic features of antonymous verbs. The semantic knowledge of words is formed based on the knowledge of usage, and the semantic features of words are reflected through the lexical knowledge system. The distribution of semantic features of the collocate words and the tendency of occurrence in different syntactic constructions are the main causes of the formation of lexical semantics. Meanwhile, the knowledge system of antonymous verbs not only includes the lexical knowledge system that regards one word as a core, but also the relationship between the two antonymous verbs. In the future, we will expand the scope of antonymous verbs and further improve the research on the semantic features of antonymous verbs.

Acknowledgments. We thank the anonymous reviewers for their constructive comments, and gratefully acknowledge the support of the Fundamental Research Funds for the Central Universities, and the Research Funds of Beijing Language and Culture University (No.19YCX045) and the Humanities and Social Sciences by the Ministry of Education (No.19XJC740002 and No.20YJAZH110).

References

1. Xing, H.: Vocabulary Acquisition of Chinese as a Second Language. Peking University Press, Beijing (2016). [in Chinese]
2. Wang, M., Jia, Y., Yu, S.: A quantitative study on the classification of classifiers based on classifier-noun collocations. Appl. Linguis. **4**, 61–67 (2014). [in Chinese]
3. Li, B.: The Construction and Application of the Corpus of Cognitive Attributes of Words. World Publishing Corporation, Beijing (2017). [in Chinese]
4. Ellis, N.C., Wulff, S: Usage-based approaches to SLA. In: VanPatten, B., Williams, J. (eds.) Theories in Second Language Acquisition: An Introduction, 2nd edn. Routledge, New York (2014)
5. Xing, H.: Collocation knowledge and second language lexical acquisition. Appl. Linguis. **4**, 117–126 (2013). [in Chinese]
6. Xu, Y., Wang, X., Wang, X., et al.: Doctor, teacher, and stethoscope: neural representation of different types of semantic relations. J. Neurosci. Off. J. Soc. Neurosci., 2562–2517 (2018)

7. Zhang, G.: Marked and unmarked phenomena in language units. Lang. Teach. Linguis. Stud. **4**, 77–87 (1995). [in Chinese]
8. Yang, L.: The Symmetry and Asymmetry of Antonymous verbs. Master Thesis of Qinghai Normal University (2010). [in Chinese]
9. Li, L.: The Statistical Analysis of the Asymmetrical Syntactic Structure and Function of Monosyllabic Dependent Antonymic Verbs. Master Thesis of Beijing Language and Culture University (2017). [in Chinese]
10. Cronbach, L.J.: An analysis of techniques for diagnostic vocabulary testing. J. Educ. Res. **3**, 206–217 (1942)
11. Richards, J.C.: The role of vocabulary teaching. TESOL Q. **1**, 77–89 (1976)
12. Nation, P.: Teaching and Learning Vocabulary. Newbury House, New York (1990)
13. Melka, F.: Receptive vs. productive aspects of vocabulary. In: Schmitt, N., McCarthy, M. (eds.) Vocabulary: Description, pp. 84–102. Cambridge University Press, Cambridge, Acquisition and Pedagogy (1997)
14. Xing, H.: Research on Chinese Vocabulary Knowledge and Acquisition. Jiuzhou Press, Beijing (2019). [in Chinese]
15. Shi, G.: The Mechanism of Syntactic Framework Knowledge of Chinese Verb Processing. Master Thesis of Beijing Language and Culture University (2018). [in Chinese]
16. Mei, J., et al.: Word Similarity. Shanghai Lexicographical Publishing House, Shanghai (1983). [in Chinese]
17. Li, L.: Agent, patient and syntactic analysis. Linguis. Res. **4,** 8-17 (1984). [in Chinese]

The Construction of Knowledge Base on Pre-Qin Chinese Reduplication

Yonghong Ke(✉) and LingJie Leng

Research Center for Folklore, Classics and Chinese Characters, Beijing Normal University, Beijing 100875, China
kyh@bnu.edu.cn

Abstract. Reduplication is an important manifestation of Chinese morphological change. While there are numerous reduplications among polysyllabic words of Pre-Qin Chinese, there is no dedicated knowledge base on reduplication. This project designs a reduplication extraction and annotation framework, extracts reduplications from 29 Pre-Qin texts, and annotates their unique attributes to construct a syntactic and lexical semantic knowledge base of Pre-Qin Chinese reduplication. More specifically, the annotated attributes include base attributes, reduplication information, lexical meaning, grammatical function, additional meanings, collocation information, archaic Chinese phonology, documentary evidence and so on. The proposed knowledge base will play a role in reduplication retrieval, reduplication research, Pre-Qin Chinese information processing, lexicographic compilation, Chinese teaching and learning.

Keywords: Pre-Qin Chinese · Reduplication · Knowledge base

1 Introduction

Reduplication, also called fùdié 复叠, is a linguistic device used to make a linguistic element appear repeatedly. Compared to the Indo-European languages, Chinese lacks morphological changes. Reduplication is not only an important manifestation of Chinese morphological changes but also a major focus of research on Chinese word-formation and grammar. The phenomenon of Chinese reduplication has existed for a long time. For example, in the bronze inscription of Shenziye gui, which was made in the reign of King Kang of the Zhou Dynasty three thousand years ago, yàn 晏 is reduplicated to yàn-yàn 晏晏 "gentle, peaceful", and there are numerous reduplications in *Shijing*. Most of the existing research on reduplication takes modern Chinese as its object of study, ignoring ancient Chinese or just applying certain definitions and methods of modern Chinese reduplication research to ancient Chinese. As a result, some questions about reduplication cannot be answered clearly, limiting exploration of the regularity and direction of reduplication development.

Reduplication in Pre-Qin Chinese is an early form of Chinese reduplication. Although some scholars have conducted research on reduplication in a few special

M. Liu et al. (Eds.): CLSW 2020, LNAI 12278, pp. 432–444, 2021.
https://doi.org/10.1007/978-3-030-81197-6_36

books from the Pre-Qin period, there is currently no knowledge base on Pre-Qin Chinese reduplication. While it is professionally rigorous and time-consuming work to build a knowledge base on Pre-Qin Chinese reduplication, its construction is of great significance as "polysyllabic words in Pre-Qin Chinese account for more than 20% and at the same time, the number of reduplications, as parts of polysyllabic words, is also large, constituting one of the important characteristics of polysyllabic words in Pre-Qin Chinese" [1].

To address the basic problems in the research on Chinese reduplication, (1) this paper identifies the origins, features, reduplicated patterns, composition, phonetic features and type distribution of early reduplication and describes early Chinese reduplication and compares it to modern Chinese reduplication from a diachronic perspective. (2) This paper also examines the semantic features and syntactic functions of special forms of Pre-Qin reduplication systematically and comprehensively by researching the earliest cases, semantics traits, syntactic components, semantic collocation tendencies of all reduplicated patterns to enrich research on special forms of Pre-Qin reduplication. (3) A knowledge base in this area will play a role in research on Pre-Qin Chinese polysyllabic words and in information processing of Pre-Qin documents and can serve as a basic resource for Chinese reduplication research. This paper, based on an accurate and reliable Pre-Qin corpus, designs an extraction and annotation framework for reduplication and annotates its unique attributes to construct a syntactic and lexical semantic knowledge base of Pre-Qin Chinese reduplication. Annotated attributes include base properties, reduplication information, lexical meaning, syntactic function, additional senses, collocation information, archaic Chinese phonology, documentary evidence and so on. We hope to provide a comprehensive knowledge base for research on Chinese reduplication.

2 Existing Research

2.1 Existing Research on Pre-Qin Chinese Reduplication

By comparing the rich and diverse research results on modern Chinese reduplication, it is observed that existing research on Pre-Qin Chinese reduplication is scattered and unsystematic. Scholars often referred reduplication in ancient times as chóngyáncí 重言词 "reduplicated words" and considered it a subcategory of Pre-Qin liánmiáncí连绵词 "alliterative two-syllable terms" due to their similarities in meaning expression; or placed it under the broader category of polysyllabic words, leading to incomprehensive data on Pre-Qin reduplication. Research on Pre-Qin Chinese reduplication can be roughly divided into three groups according to their foci: i) research on reduplication in special books, including Zhang Yinghui (2007) [2] and Fan Hua (2009) [3], who examined reduplication in *Shijing*, *Chuci* and *Lunyu*; ii) research on the reduplication of words in special word classes or reduplicated patterns, including Cui Yingxian (2011) [4], Shi Qin (2010) [5], and Wang Yuan (2003) [6], in which verb, adjective and adverb reduplication were studied; and iii) comprehensive research, including Ma Zhen (1980) [7], Yang Dingfu (1991) [8] and so on. In addition to these works, a small number of studies examined reduplication in newly identified documents such as Ren Haichao's work *Zhanguo chuxi jianbowenxian dieyinci yanjiu.*

Generally speaking, the research on ancient Chinese and Pre-Qin Chinese reduplication in particular is scattered and unsystematic and lacks comprehensive investigation based on a large-scale corpus.

2.2 Current State of Pre-Qin Chinese Reduplication Knowledge Base

A linguistic knowledge base forms the basis of natural language processing, playing an important role in word sense disambiguation, syntactic parsing, semantic understanding, and other tasks. In recent years, databases on modern Chinese such as HowNet [9] and CLKB [10] have been widely used. Given difficulties associated with the automatic segmentation of monosyllabic and polysyllabic words in Old Chinese, the construction of ancient Chinese lexical knowledge bases has been very slow. In spite of the large number of ancient documents available (for example, the Sikuquanshu Wenjin Pavilion collection includes 3462 books and 79338 volumes, containing approximately 800 million words), knowledge bases on ancient Chinese, and Archaic Chinese especially, are scarce. There are only two large-scale word-tagged corpora in the world: an ancient Chinese tagged corpus [11] constructed by the Institute of Linguistics of Academia Sinica, Taiwan and The Sheffield Corpus of Chinese [12]. However, the former cannot be visited from Mainland China, while the latter has many annotation issues, making it unsuitable for high-quality and multi-dimensional research on ancient Chinese.

With the development of Chinese information processing technologies, some scholars have tried to use corpora to conduct research on reduplication. For example, Xing Hongbing (2000) [13] has analyzed types and forms of reduplication constructions in modern Chinese and collected statistics on major reduplication patterns; Shang Ying (2004) [14], based on a large-scale investigation of verb reduplication in modern Chinese, has designed a set of automatic recognition methods for verb reduplication; Zhou Xiaoyun (2012) [15], based on a large-scale corpus, has conducted research on the extraction, inconsistent segmentation comparison, recognition and statistics of reduplication. However, all these studies have focused on the reduplication of modern Chinese. In fact, the phenomenon of Chinese reduplication has existed for a very long time. Roughly 20 reduplicative words have been found in bronze inscriptions of the Western Zhou Dynasty. Later in the Spring and Autumn and Warring States periods, reduplication developed very rapidly. According to Wu Zongwen (2001), more than 650 reduplicative words can be found in Pre-Qin books and frequent reduplication is the major feature of polysyllabic words in Pre-Qin Chinese. Despite this, there is no special lexical knowledge base on Pre-Qin Chinese reduplication.

3 The Construction of a Pre-Qin Chinese Reduplication Knowledge Base

3.1 Definition and Classification

This paper focuses on reduplication in the Pre-Qin period. Reduplication, also called fùdié 复叠, is a linguistic device used to make a linguistic element appear repeatedly, consisting of a 'base' and its' reduplicant'. The "Base" refers to the original form, while "reduplicant" refers to the segment produced by the reduplication process, and the "Base" together with "reduplicant" forms a "reduplication". Sun Jintao (2008) [16], depending on whether the sound of the base changes after being reduplicated, has classified two reduplication types, namely total reduplication and phonologically modified reduplication. The former, which completely reduplicates the base syllable without involving any phonetic changes, is often manifested as the consecutive appearance of two identical Chinese characters. The latter is roughly equivalent to liánmiáncí 连绵词. This paper only discusses total reduplication.

Scholars have different views on the definition and classification of reduplication. We adopt Li Yuming's classification given in *Hanyu fudie leixing zongshu* [17] to classify the reduplication phenomena into three types: diéhé 叠合, chóngdié 狭义的重叠 and chóngfù 重复 as shown below (Table 1):

Table 1. Classification of reduplication phenomena

Overlapping types	Corresponding level	Examples
diéhé 叠合	Word-formation level (lexical reduplication)	mán-mán蛮蛮 "a bird", zhēng-zhēng 丁丁 "the sound of cutting trees", bà-bà爸爸 "dad"
chóngdié 重叠	Syntactic level (morphological reduplication)	rén-rén人人 "everybody", shì-shì世世 "generation after generation", kàn-kan看看 "have a look"
Chóngfù 重复	Pragmatic level (clausal reduplication)	chǎo-chǎo-chǎo, yītiān dào wǎn jiù zhī dào chǎo. 吵吵吵, 一天到晚就知道吵。"Quarrel, quarrel, quarrel all day long"
		zuǒ-zhī-zuǒ-zhī, jūnzǐ yí zhī; yòu-zhī-yòu-zhī, jūn zǐ yǒu zhī. 左之左之, 君子宜之; 右之右之, 君子有之。"There are people around to help, so a gentleman can do things with ease"

This paper only discusses the first two kinds shown above for two reasons. (1) Lexical reduplication and morphological reduplication share the same form, but as lexical reduplication appeared earlier, examining it may reveal the origins of morphological reduplication. In addition, overlapping phenomena are very common in Pre-Qin Chinese, which cannot be ignored. (2) Clausal reduplication, triggered by pragmatic factors, is more relevant to the speech situation rather than to the language.

3.2 Corpus Selection

The corpus used for the construction of the knowledge base is literature from the Pre-Qin period. The Pre-Qin period refers to the period before the Qin Dynasty unified China (from the 21st century BC to 221 BC). Pre-Qin documents include discovered documents and documents that have been passed down. For the former kind of documents, we have not applied bottom-level data preparation or reduplication extraction and instead directly used existing research results. The latter kind of documents have been studied and categorized by scholars during their long histories. Thus, the documents include finalized text and have consistent interpretations, leading to their important role in the study of the history of the Chinese language. According to the statistics of Li Ling [18], 60 Pre-Qin books have been found, which are divided into 7 categories: liùyì (六艺, books on the six arts), zǐshū (子书, sophistic books), shǐshū (史书, history books), bīngshū (兵书, military books), shīfù (诗赋, books of poems and odes), fāngjì (方技, medical books) and shùshù (术数, numerology books).

The six arts can be divided into jīngzhuàn (经传, the classics) and xiǎoxué (小学, Chinese Philology). While it would be ideal to include all of these books in our corpus, this would be difficult to achieve under the existing conditions. Furthermore, because some Pre-Qin documents are quite controversial in terms of year of completion and authorship, we have studied only reliable materials following the "do only what we can do" principle. For these reasons, we have studied the 29 Pre-Qin books shown below (Table 2):

Table 2. Pre-Qin documents studied

Pre-Qin literature	Category	Pre-Qin literature	Category	Pre-Qin literature	Category
Zuozhuan	jīngzhuàn	*Lunyu*	zǐshū	*Yanzi chunqiu*	zǐshū
Guliangzhuan	jīngzhuàn	*Mengzi*	zǐshū	*Guoyu*	shǐshū
Gongyangzhuan	jīngzhuàn	*Mozi*	zǐshū	*Yizhoushu*	shǐshū
Zhouyi	jīngzhuàn	*Xunzi*	zǐshū	*Mutianzi zhuan*	shǐshū
Shangshu	jīngzhuàn	*Zhuangzi*	zǐshū	*Sima fa*	bīngshū
Liji	jīngzhuàn	*Guanzi*	zǐshū	*Sunzi bingfa*	bīngshū
Zhouli	jīngzhuàn	*Lüshi chunqiu*	zǐshū	*Wuzi*	bīngshū
Yili	jīngzhuàn	*Laozi*	zǐshū	*Weiliaozi*	bīngshū
Shijing	jīngzhuàn	*Shangjun shu*	zǐshū	*Chuci*	shīfù
Han Feizi	Zǐshū	*Zhanguo ce*	zǐshū	——	——

3.3 Automatic Extraction and Manual Annotation

We have keyed in regular expressions to extract reduplicated forms from the electronic text library, and then manually proofread, annotated the attributes and designed a framework for reduplication annotation. The attributes and corresponding descriptions of the annotations are shown in the following table (Table 3):

In practice, there are several forms that are seemingly similar to reduplication in terms of form that must be excluded. To be specific:

Table 3. Table of reduplication annotation attributes

	Attribute	Description
Number	ID	Unique number, as the primary key, for indexing and retrieval
Entry information	Reduplication	The reduplicative word extracted according to the judgment standard
	Reduplicated patterns	External forms, such as AA, AAB, ABB, AABB, ABAB, A里AB and so on
	Sentence	The sentence with reduplication
Base information	Archaic Chinese phonology	Adopting Wang Li's ancient sound system to mark the initial consonant and final of the base
	Whether related to reduplication meaning	When content relates to the meaning of reduplication, the base is a word that expresses meaning; when it does not, the base is a syllable
	Lexical meaning	The notion refers to the sense of the base contributing to the construction of reduplication meaning; *Ciyuan*, *Chinese Dictionary* and *Wang Li Ancient Chinese Dictionary* are the references of the lexical meanings; source information for the given dictionary is indicated
	Word class	The word class of the base when it is used as a word
	Example sentence and literature source	The example sentence and Pre-Qin period literature source of the base when it is used as a word
Reduplication information	Property	Word formation duplication or morphological duplication
	Lexical meaning	Lexical meaning of reduplication; *Ciyuan*, *Chinese Dictionary* and *Wang Li Ancient Chinese Dictionary* are the references of the lexical meanings; source information for the given dictionary is indicated
	Word class	The word class of the reduplicated form
	Additional meaning	The grammatical meaning brought about by reduplicating the base. Adjective reduplication: state/emphasis/other; verb reduplication: time duration/momentum repeatability/time duration and momentum repeatability/other; noun reduplication: thorough reference/individual reference/other; quantifier reduplication: overall meaning/higher quantity/individual reference/continuous meaning/other

(*continued*)

Table 3. (*continued*)

	Attribute	Description
	Syntactic function	The grammatical function of the reduplicated form in the sentence, e.g. subject, predicate, object, attribute, adverbial, complement
	Explanatory material	Comments on the reduplicated form written by former scholars
	Inclusion in dictionary	Whether it is included in *Ciyuan* or *Chinese Dictionary*
Information for documentary evidence	Literature source	The literature source of the example sentence containing reduplication
	Literature genre	The literature is a work of verse or prose
	Sentence types	declarative/imperative/interrogative/exclamatory
	Preceding component	The previous collocating component
	Later component	The following collocating component

1) The same continuous form of written symbols without structural relations. For example, for *Shijing* poem T*aishuyutian*: shū yú tián, chéng shèng mǎ. 叔于田, 乘乘马。"Shu comes hunting, riding a horse" and *Maoshi zhengyi*: chéng shèng, shàng rú zì, xià shéng zhèng fǎn. 乘乘, 上如字, 下绳证反。"chéng shèng, the former is pronounced as chéng while the latter is pronounced as shèng." Accordingly, the former 乘 is a verb meaning "to ride" and is pronounced as chéng; the latter 乘 is a quantifier and is pronounced as shèng. The former 乘 is a predicate, while the latter is an attribute. Therefore, the two are not only different words but also do not have a direct syntactic relationship with each other.
2) The same continuous form of written symbols with structural relations. For instance, *Lunyu·Yanyuan*: kǒngzǐ duì yuē: jūn-jūn, chén-chén, fù-fù, zǐ-zǐ. 孔子对曰: "君君, 臣臣, 父父, 子子。" "Confucius answers: Let the king be a king, the minister be a minister, the father be a father and the son be a son." The first uses of 君 "king", 臣 "minister", 父 "father", and 子 "son" respectively, are nouns, while the second uses are verbs, forming four subject-predicate structures. Although they look like noun reduplication, actually noun reduplication is very rare in Chinese.
3) Segmentation error. Because there is currently no high-quality corpus of Pre-Qin with word segmentation, errors are often made when regular expressions are used to extract words automatically. For instance, *Lüshi chunqiu·Tingyan*: bú chá zé shàn bú shàn bú fèn. 不察, 则善, 不善不分。"Without careful examination, one cannot tell good from bad." Using regular expressions may lead to incorrect interpretations (shàn bú shàn bú善不善不). Similar situations must be eliminated through manual proofreading.
4) Pragmatic reduplication. For instance, *Shijin*: fá-kē-fá-kē, qí zé bú yuan. 伐柯伐柯, 其则不远。"Chop handle and chop handle, its rule is near." In *Shijing*, such repeated

chanting is common. In theory, the phrase "伐柯" can be repeated indefinitely, and as components are loosely combined, this cannot be treated as an instance of ABAB reduplication.

During manual annotation, some special cases require special treatment:

1) Reduplication in citations. In the literature of the Pre-Qin period, previous literature is often cited (e.g., *Shijing* or *Shangshu*). Such citations also contain reduplication. For instance, *Lunyu* quotes *Shijing*: xiàng wéi pì gōng, tiān zǐ mù-mù. 相維辟公, 天子穆穆。"All those who facilitated the sacrifice were princes. The emperor in charge of the sacrifice was sincere and respectful." Via comprehensive and complete annotation, this paper also annotates reduplications in citations. As the sources of these reduplications are evident, they can be automatically de-duplicated by the program used or the user can determine whether to accept them or not.
2) Reduplication brought about by different editions of the same book. This can take two forms: (i) the same reduplication using different word forms (e.g., in *Chuci*: shí lài xī qiǎn-qiǎn, fēi lóng xī piān-piān. 石濑兮浅浅, 飞龙兮翩翩。"The water flowed swiftly on the rocky beach, and the dragon boat skimmed lightly and swiftly." 浅浅is written as 濺濺 in another edition), and (ii) some editions using the base and others using the reduplicated form (e.g., in *HanFeizi*: chén wéi xuē zhě yě, zhū wēi wù bì yǐ xuē-xuē zhī, ér suǒ xuē bì dà yú xuē. 臣為削者也, 諸微物必以削削之, 而所削必大於削。"I'm the sharpener. All tiny things must be carved with a paring knife, and carved things will be bigger than the paring knife." Some editions only use 削 to replace 削削). We focus on the word form of the selected version without considering different versions.

4 Preliminary Knowledge Base Construction Results

We have preliminarily searched for reduplications in 29 Pre-Qin literature works and a total of 2057 reduplications were identified.

Regarding reduplication properties, reduplication in the Pre-Qin period is dominated by word formation reduplication, accounting for 86.10%, and morphological reduplication only accounts for 13.90%. The following table presents the statistics of part of speech classification for word formation reduplication (Table 4).

Among the above cases of word formation reduplication, adjective reduplication is the dominant form, accounting for 87.13%, followed by onomatopoeic word and noun reduplication, accounting for 7.45% and 3.78%, respectively.

As for morphological reduplication, the base is mainly used as an adjective. Unlike the base, the reduplication adds grammatical meaning, intensifies the degree of modification, or adds visual features. Few cases of morphological reduplication of verbs are observed. For example, in *Zhuangzi*: fū fán zhī wáng bú zú yǐ sàng wú cún, zé chǔ zhī cún bú zú yǐ cún-cún. 夫凡之亡不足以丧吾存, 則楚之存不足以存存。"As the destruction of the Kingdom of Fan is not enough to destroy my existence, the existence of the state of Chu is not enough to make its king exist for a long time." The base存expresses the meaning of "existence" which means "long term survival." When the verb is reduplicated, it

Table 4. Word class distribution of word formation reduplication

Serial number	Word class	Number	Proportion
1	Adjective	1543	87.13%
2	Onomatopoeic word	132	7.45%
3	Noun	68	3.84%
4	Verb	20	1.13%
5	Adverb	5	0.28%
6	Interjection	3	0.17%

expresses the grammatical meaning of existing for a long time. Similar reduplications include wàng-wàng望望 "looking back again and again to show attachment", liàn-liàn 恋恋 "thinking for a long time, reluctant to give up" and piāo-piāo漂漂 "drifting on the water for a long time". The parts of speech of these bases are the same as those of the reduplicated forms, and their grammatical meanings also seem to be related to the length of time and degree of action. However, there remains a difference between this and the common grammatical meanings of modern Chinese verb reduplication (e.g., transience, attempts and arbitrariness). Furthermore, these cases of "verb reduplication" involve intransitive verbs with limited grammatical functions.

In terms of reduplicated forms, the form of AA is in absolute dominance in Pre-Qin reduplication. It is followed by the form of AABB and only very few cases of ABAB are found (e.g., wēi-yí-wēi-yí 委佗委佗 "graceful and poised"). In terms of reduplication distributions, 82.01% cases are found in verse materials. Such materials in the Pre-Qin literature include mainly *Shijing*, *Chuci* and *Zhouyi*. They account for a small proportion of the annotated corpus but represent 54.30% of reduplications found in Pre-Qin Chinese. It is thus evident that reduplication is mainly found in several books of verse, while only very few reduplications are found in other books. Less than 10 reduplications are found in the studied works (e.g., *Sunzi bingfa*, *Weiliaozi*, *Shangjunshu* and *Yili*).

Parts of speech often change when verbs or nouns are reduplicated, and in most cases the reduplicated form becomes an adjective. Yet, generally speaking, part of speech do not change when adjectives are reduplicated.

5 Main Applications of the Knowledge Base

5.1 Retrieval

Our knowledge base can be used for retrieval purposes. Users can retrieve matching reduplications by searching with keyword, part of speech, time, title, reduplicated pattern and structure combination information. The application is shown in the figure below (Fig. 1):

At present, there is no reduplicated vocabulary database for retrieval purposes, and thus our knowledge base on Pre-Qin Chinese reduplication will be offered as part of an open platform to provide basic resource retrieval services for research and teaching.

Fig. 1. Reduplication retrieval

5.2 Research on Reduplication

There are numerous reduplications of polysyllabic words of Pre-Qin Chinese, and reduplication serves as an important means of word formation. Consequently, research on reduplication also constitutes an important facet of research on Chinese word formation and syntax. However, at present, reduplication research resources are unorganized. The open knowledge base constructed in this project will serve as a reliable and comprehensive corpus for reduplication research. The resource can be used to examine Pre-Qin Chinese reduplication, analyze special forms of reduplication in depth, and conduct diachronic comparative studies of modern Chinese reduplication.

5.3 Pre-Qin Chinese Information Processing

Distinguishing between monosyllabic and polyphonic words of Pre-Qin Chinese is complex, while monosyllabic words being dominant. Reduplication is an important feature of polysyllabic words in Pre-Qin Chinese and poses information processing challenges. The proposed knowledge base can facilitate Pre-Qin Chinese information processing in the following three respects:

1) Word segmentation and part of speech tagging. Rule-based, statistical, annotation-based and dictionary-based methods of automatic word segmentation of ancient Chinese have been developed. Rule-based methods require the use of a large corpus to generate complex rules that are easy to miss; statistical methods cannot be used for reduplication, as many reduplications in Pre-Qin Chinese only appear in verse and are used infrequently (e.g., guān-guān关关 "birdsong", only appears once in *Shijing*, making it difficult to obtain valid statistical information); annotation-based methods involve using a large number of well-annotated documents. Thus, the use of a lexical knowledge base for the information processing of Pre-Qin Chinese, in which monosyllabic words are dominant, is effective. At present, there is no Pre-Qin Chinese lexical knowledge base for information processing. Only the *Chinese Dictionary*, *Ciyuan* and other dictionaries can be referenced, in which many reduplicated forms are not included and those that are included are only explicating the meanings of words without reference to their exclusive attributes and complex features. In determining the boundaries of reduplicated words of Pre-Qin Chinese, one cannot simply apply the methodology which is used to study modern Chinese, as incorrect reduplications (e.g., jūn-jūn君君, chén-chén臣臣, fù-fù父父 and zǐ-zǐ子子) and trans-layer structures (e.g., jūn-jūn军军 in sān jūn jūn yú qián三军军于前 "The third army is stationed in the front") will be misidentified. The proposed corpus can provide a knowledge base for the information processing of Pre-Qin literature.
2) Syntactic parsing. Most previous works argued that reduplicated words of ancient Chinese were essentially adjectives. According to Wang Yun: fán chóngyán jiē xíngróng zhī cí.凡重言皆形容之词 "All reduplicated words are adjectives". However, reduplication in ancient Chinese is not limited to adjectives, as there are also verb reduplication (e.g., juàn-juàn睠睠"looking back", wàng-wàng 望望 "looking back again and again to show attachment" and liàn-liàn 恋恋 "thinking for a long time"), noun reduplication (e.g., rén-rén人人 "everybody", shì-shì 世世 "generation after generation" and shì-shì 事事 "everything"), adverb reduplication (e.g., jǐn-jǐn 仅仅 "only", dú-dú 独独 "just" and jiàn-jiàn 渐渐 "gradually") and so on. The proposed reduplication knowledge base of Pre-Qin Chinese annotates the syntactic functions and pre- and post-collocation components of reduplicated words (e.g., suffixes attached to adjectives include "rán然", "yān焉" and "ěr尔"). These data can be used for automatic syntactic parsing.
3) Semantic analysis. The meaning of reduplication is not compositional. Many adjective reduplications (e.g., yáo-yáo摇摇 "anxious and preoccupied", náo-náo挠挠 "chaotic", tǐng-tǐng挺挺 "tightful") express special meanings. As such words cannot be strictly defined and they refer to impressions, they are defined as cases of impression duplication by Sun Jintao. In some cases, the meaning of reduplication may not

be related to the meaning of the base. For instance, "zhēng-zhēng丁丁" "the sound of logging", yīng-yīng嚶嚶 "birdsong", tián-tián闐闐 "thunder" are onomatopoetic words whose bases serve as phonetic notation. Some reduplications are related to their bases and even generate whole new meaning after the base is reduplicated. For instance, yán-yán延延 "long" is an adjective, while its base is a verb. 延延 is the result of word formation reduplication. The other example is yú-yú 愉愉 "happy and joyful". Its base has the same part of speech and lexical meaning to the reduplicated form. Yet, the reduplicated form does not generate whole new meaning, instead it increases the degree of happiness. As reduplication increases the degree of modification by adding grammatical meaning, 愉愉 demonstrates morphological reduplication.

The above complications of reduplication analysis and processing are difficult to resolve by traditional statistical learning or other means without the support of a syntactic and semantic knowledge base. The use of a Pre-Qin reduplication syntactic semantic knowledge base with statistical learning thus serves as an effective approach to ancient Chinese information processing.

5.4 Dictionary Compilation

The *Chinese Dictionary* serves as an important reference for research on lexical semantics and reduplication. However, it does not include all instances (e.g., cōng-cōng總總 and zǐ-zǐ訿訿).

Some dictionaries of Pre-Qin reduplication have been developed. Ancient dictionaries include *Erya* and *Dieya*. More recent references include the books such as *Lianmian zidian* by Fu Dingyi, *Lianmian zipu* by Wang Guowei, *Maoshi lianmian zipu* by Du Qirong, and *Hanyu chongyanci cidian* by Wang Weimao, most of which adopt annotated material from previous scholars and include simple interpretations while lack analyses of internal components, lexical properties, semantic features and syntactic functions of Pre-Qin reduplication. As many ancient commentaries do not define words but only translate texts, they produce occasional interpretation errors and incorrectly merge dictionary entries.

Therefore, the knowledge base constructed in this paper can be used to provide supplementary information to the dictionaries.

5.5 Chinese Teaching

In the Pre-Qin period, reduplication was relatively developed. For instance, adjective reduplication in the form of AA (e.g., chōng-chōng忡忡 "worried", fēi-fēi霏霏 "big rain and snow", and sù-sù 肅肅 "respectful"), which are widely found in verse, often vividly depict scenes with special semantic functions. *Chuci, Han yuefu,* Tang poetry, Song ci, Yuan qu, Ming and Qing poetry and modern poetry inherit AA reduplication in *shijing*. Some reduplications are even found in prose. With the completion of annotation, the Pre-Qin Chinese reduplication corpus project described in this paper will allow access to all resources and allow one to query word forms so as to retrieve related information. Besides, it can provide teachers with superior reduplication teaching materials and help students understand how the special semantic functions of the reduplication present the artistic features of an article.

6 Conclusion

Despite China's large collections of ancient documents, research on ancient Chinese does not meet information processing needs. The presence of reduplication complicates Pre-Qin Chinese information processing. Based on existing research results, we have designed an annotation framework of knowledge base by annotating of 29 Pre-Qin documents and constructed a Pre-Qin reduplication syntactic and semantic knowledge base to facilitate text analysis and semantic interpretation of Pre-Qin documents.

Acknowledgments. This research is sponsored by the General project of research and application of ancient Chinese characters such as oracle bone inscriptions of the State Language Commission (YWZ-029) and the Major Project of National Social Science Foundation of China (15ZDB096).

References

1. Wu, Z.: Xianqin hanyu fuyinci yanjiu. Bashu Publishing House, Chengdu (2001)
2. Zhang, Y.: Shijing chongyan yanjiu. Lanzhou University, Lanzhou (2007)
3. Fan, H., OuYang, X.: Lunyu chongyan xiaoyi. Anhui Wenxue (05), 302 (2009)
4. Cui, Y.: Hanyu dongci de lishi kaocha. Guangming Daily Press, Shenzhen (2011)
5. Shi, Q.: Hanyu xingrongci chongdiexingshi de lishifazhan. Commercial Press, Beijing (2010)
6. Wang, Y.: Chongdieshi fuci chongdie yanjiu. President's foundation papers of Peking University (2003)
7. Ma, Z.: Xianqin fuyinci chutan. J. Peking Univ. (Phil. Soc. Sci.) (05), 55–64 (1980)
8. Yang, D.: Guhanyu dieyinci de chansheng jiqifazhan. J. Jinan Univ. (Phil. Soc. Sci.) (03), 119–126 (1991)
9. Dong, Z., Dong, Q.: HowNet[EB /OL]. http: //www.keenage.com
10. Yu, S.: Jianshen zonghexing yuyanzhishiku de linianyuchengguo de jiazhi. J. Chin. Inf. Process. **21**(6), 3–12 (2007)
11. Institute of linguistics, Academia Sinica, Taiwan. Ancient Chinese marker corpus [EB/OL]. Accessed 5 Oct 2019, http://ancientchinese.sinica.edu.tw/
12. Xiaoling, H., Jamie, M.: The Sheffield Corpus of Chinese [EB/OL]. Accessed 17 Feb 2014, http://www.Hrionline.ac.uk/scc/
13. Xing, H.: Hanyu ciyu chongdiejiegou tongjifenxi. Lang. Teach. Linguis. Stud. (01), 32-37 (2000)
14. Shang, Y.: Jiyu daguimodiaocha de dongcichongdie xingshi zidongshibiefangfa yu xing-shitezheng yanjiu. Beijing Language and Culture University, Beijing (2004)
15. Zhou, X.: Xiandaihanyu chongdieshi de shibie yu tongji fenxi. Nanjing Normal University, Nanjing (2012)
16. Sun, J.: Guhanyu chongdiegoucifa yanjiu. Shanghai Educational Publishing House, Shanghai (2008)
17. Li, Y.: Hanyu fudie leixing zongshu. Center for language and language Education. Central China Normal University, Beijing (2000)
18. Li, L.: Jianbo gushu yu xueshu yuanliu. Joint Publishing, Beijing (2008)

Exploring Reasoning Schemes: A Dataset for Syllogism Figure Identification

Shiya Peng, Lu Liu, Chang Liu, and Dong Yu(✉)

Beijing Language and Culture University, Beijing, China

Abstract. Argument mining aims at extracting structured arguments from texts. Both argument schemes and their structures are studied. A syllogism is a traditional argument scheme and its figures are important in classical logic. Existing research has not yet investigated syllogism figures. Here, we fill this gap by presenting a study on automatic identification of syllogism figures. We prepared a novel dataset of 8.6k syllogisms. We annotated their figures and carried out identification tasks using both supervised and weakly-supervised approaches. Experimental results show that both approaches are adequate.

Keywords: Syllogism figure · Enthymeme reconstruction · Reasoning scheme

1 Introduction

A syllogism is a traditional argument scheme. With the rapid development of argument mining research [1, 2], syllogisms and other types of argument schemes have gained a lot of attention over the last few years [3–6]. There are now several studies on syllogisms, but the majority have mainly investigated enthymemes [7–9], i.e., the eclipsed forms of syllogisms. The properties of syllogisms in its complete form is still not much considered.

A syllogism in classical logic can appear in the form of one of its four figures. Figure 1 shows a standard example of the first figure of syllogisms.

Major Premise: *Human* is *mortal.*
Minor Premise: *Scorates* is *human.*
Conclusion: Therefore, *Scorates* is *mortal.*

Middle Term: *Human*	
Subject Term: *Scorates*	
Predicate Term: *Mortal*	*Syllogism figure* 1

Fig. 1. An example showing the structure of the first figure of syllogism.

Previous works in formal logic have suggested that the figures of syllogisms are significant [10, 11]. Grasping the properties of syllogisms, such as figures, can contribute

M. Liu et al. (Eds.): CLSW 2020, LNAI 12278, pp. 445–451, 2021.
https://doi.org/10.1007/978-3-030-81197-6_37

to further understanding of the detailed argument process of syllogisms. As less research focuses on the properties of a specific argument scheme, the syllogism figure is a good cut-off point for further studies on the details of an argument scheme. In this paper, we propose a pilot study on syllogism figures and present a task of syllogism figure identification.

To study syllogism figures, the first issue we have to face is the lack of an available syllogism dataset. Since syllogisms often exist in the form of enthymemes, it is difficult to directly extract syllogisms from natural language texts. An accepted way to get syllogism data is to reconstruct enthymeme [3, 12]. Many works have attempted to investigate enthymeme reconstruction. However, most of them have focused on reconstruction methods based on logic-based frameworks [7, 8], or just tried to reconstruct the eclipsed arguments using natural language on a small scale [9].

In this work, first, we have attempted to carry out large-scale enthymeme reconstruction using natural language to solve the lack of syllogism data for our research. We have constructed a sizable syllogism dataset, comprising more than 8.6k syllogisms with annotated figures. Based on the labeled syllogism dataset, we have then performed the syllogism figure identification task with a supervised approach. Furthermore, we have been interested in whether automatically labeling a large set of unlabeled and similar data could assist deep learning models for the syllogism figure identification task. So we have also attempted a weakly-supervised approach in our investigation.

Our contributions are summed up as follows: A dataset for the research on the syllogism and its figures consisting of more than 8.6k syllogisms is established. An automatic identification task of syllogism figures is presented; and the task performed under both supervised and weakly-supervised approaches is attempted.

2 Related Work

Argument Mining is a multi-disciplinary research field focusing on the study of identifying and analyzing arguments in natural language [2, 13, 14]. Existing works in argument mining have mostly concentrated on extracting arguments from texts in certain domains, such as online debates, online product reviews, philosophical essays, and Wikipedia articles [15–17].

Several researchers have also attempted to study argument schemes and structures. [4] has studied the task of classifying arguments by analyzing the structure of different inference schemes, such as exemplification or causality. [6] has studied argument schemes in the "results" and "discussion" sections in biological articles. And [5] has proposed a set of guidelines for the annotation of argument schemes with different structures, aiming to overcome the challenges in annotating a broad range of schemes. Although there are several kinds of scheme studies, the syllogism as a traditional argument scheme is still not much considered.

A syllogism is a kind of logical argumentation method in which a conclusion is derived from two premises. The classical structure of a syllogism is determined by its figure. Several researchers have claimed that the figures of the syllogism have a historical importance that can hardly be over-estimated. To build a syllogism dataset for research, the first issue we have to face is enthymeme reconstruction. Many works have provided shared common knowledge as a solution to the reconstruction. Several studies have attempted to reconstruct enthymeme using logic-based frameworks or natural language techniques. But large-scale enthymeme reconstruction is still lacking. So, we attempt to perform large-scale enthymeme reconstruction in this work.

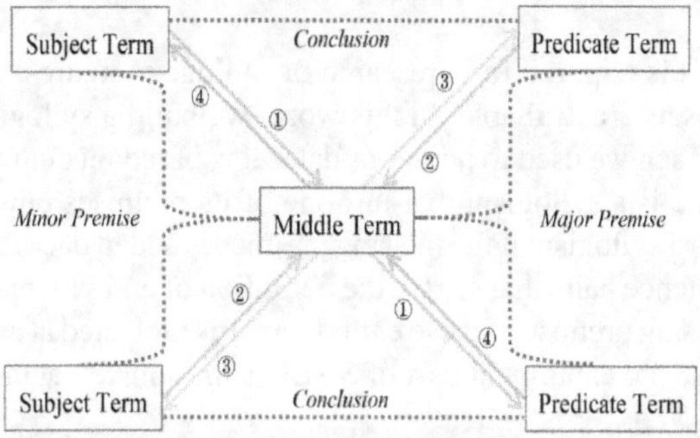

Fig. 2. The relation between the syllogism, terms, and figures. Each dashed line shows the relations between the syllogism and the terms. For example, the minor premise consists of a subject term and a middle term. And each pair of arrows shows the structure of a syllogism figure. For example, cardinal 4 indicates syllogism Fig. 4, the structural orders of which are from predicate to middle term in the major premise and from middle term to subject term in the minor premise.

Furthermore, [18] has used a small annotated dataset and weakly supervised methods to automatically annotate a large dataset. In this paper, we attempt to investigate the task of identifying syllogism figures using supervised and weakly-supervised methods.

3 The Syllogism and Its Figure

The syllogism is a classical argument scheme. A typical syllogism includes a minor premise, a major premise, and a conclusion. It has three terms and four figures. Relations between the syllogism, terms and figures are shown in Fig. 2.

3.1 Relations Between Syllogism and Terms

A syllogism generally has three terms: the predicate term (P), the middle term (M), and the subject term (S). Each component of the syllogism is an assertion about the relationship between the terms. Traditionally, the major premise depicts the relation between the M and P. The minor premise depicts the relation between S and M. And the conclusion, using deductive reasoning, comes up with a relation between S and P derived from the information provided by the two premises.

3.2 Relations Between Terms and Figures

The syllogisms also has four figures that reflect the structure of reasoning. The positions of the middle term in the major and minor premises define the figure of a syllogism. In

Fig. 1, the orders of the terms in the major and minor premises, respectively, are M-P, S-M; in Fig. 2, the orders are P-M, S-M; in Fig. 3, the orders are M-P, M-S; in Fig. 4, the orders are P-M, M-S.

4 Dataset Creation

An adequate dataset is required for a research on syllogism figures. To our knowledge, few syllogism datasets are available. In this work, we build a syllogism dataset for our research. The approach we used to build our dataset is based on enthymeme reconstruction. An enthymeme is a syllogism that has one of its premises omitted, and it can be restored to a standard syllogism if the missing premise is added back. SNLI [19] includes some inference sentence pairs that satisfy the definition of enthymeme, and the difficulty of restoring the missing premise of these enthymemes is indicated at an appropriate level. Therefore, we chose the entailment part of SNLI as the source data for our annotation.

Table 1. Data statistics of each syllogism figure.

Category	Number	Proportion
Figure 1	6428	74.10%
Figure 2	1808	20.80%
Figure 3	363	4.20%
Figure 4	74	0.85%

Our dataset was annotated by ten annotators. All of them are at the graduate level with the linguistics background. This ensures that they have the basic ability to annotate the data for our reconstruction task. To guarantee the annotation quality, the annotators were required to attend a pre-training procedure. We utilized Fleiss' kappa [20] to compute IAA between the annotators. After the training, the Kappa score was raised from 0.469 to 0.746, indicating substantial agreement (within [0.6, 0.8]). The formal annotation included three steps: analyzing the enthymeme, restoring the missing premise, annotating the syllogism figure.

Our data source is the entailment part of SNLI. The inference relations between entailment pairs in the dataset are mostly repetition or an abbreviation. They are not enthymemes. These invalid sentence pairs should be filtered out. The annotators were required to check the entailment pairs to determine whether they were enthymemes in the first step. Then the valid labeled enthymemes would be further analyzed by the annotator. They would decide which one of the premises was missing. Then they would be required to restore the missing premise using the terms of the syllogism. After completing the syllogism, the annotators would annotate the figure of the syllogism. In the annotation process of step 1, totally we annotated 79,700 data, and finally got 8,673 valid annotated enthymemes. The statistics of the four syllogism figure types are shown in Table 1.

5 Experiments

We address the syllogism identification task as multi-class classification. Experiments were performed using two different approaches: supervised and weakly-supervised approaches. The widely used Macro-Average and Micro-Average scores were used as evaluation methods.

Table 2. Performance of different models using two approaches. sp stands for supervised method, wsp stands for weakly-supervised method.

	maP	miP	maF1	miF1	ACC
sp_SVM	0.55	0.74	0.29	0.74	0.74
sp_BiLSTM	0.59	0.85	0.53	0.85	0.86
sp_BERT	0.62	0.91	0.61	0.91	0.92
wsp_BERT	0.62	0.90	0.61	0.90	0.90

5.1 Supervised Approach

Only human-annotated syllogisms data was used in this approach. Our labeled dataset was split into training (6937 components), validation (868 components) and test (868 components) sets, keeping a component proportion similar to that in the dataset. We chose three methods for the supervised approach, including Support Vector Machines (SVM) [21], Bidirectional Long Short-Term (BiLSTM) [22] and Bidirectional Encoder Representations from Transformers (BERT) [23]. These methods were used separately in our identification task.

5.2 Weakly-Supervised Approach

A large scale of weakly labeled data, annotated by classifiers, was used in this approach. E-snli [24] is an entailment explanation dataset. It contains a certain amount of data which has structures similar to the syllogism, and this makes it a suitable source for the weakly-supervised approach. In this experiment, E-snli was selected as our unlabeled data source. BERT was selected as the empirical method in this approach. We first used our labeled data to train a classifier to identify the similar syllogism data in the dataset. Then the human-annotated data were randomly divided into three training sets that were used to train another classifier to get the data with labeled figures for the weakly-supervised approach. These weakly labeled data were used to train BERT for our syllogism figure identification task.

5.3 Results and Analysis

Table 2 contains the experimental results of the supervised and weakly supervised approaches. The accuracies of the models using the supervised approach are 0.74, 0.82

and 0.92. BERT has performed better than the other two methods on the task. This may be because the BERT model, which has been pre-trained on large-scale data of common sense, has a lot of external knowledge, and was able to better enhance the ability to perform the task on a small dataset. The best model achieves 0.92 in accuracy using the supervised approach and 0.90 in accuracy using the weakly-supervised approach. The performance of the model using the weakly supervised approach and trained with large data is same to the model using the supervised approach.

We conjure that both of the approaches have the ability to identify syllogism figures. Among the metrics for the BERT model using a supervised approach, maP is 0.62, and miP is 0.91. The Macro-Average metrics are much lower than the Micro-average metrics. They have a significant difference. This difference indicates that although the average accuracy over the four classes is good, a class with a small size may not be identified effectively. This is because the dataset has an imbalanced class distribution. For this issue, we will attempt other learning methods and focus on the way of how to better improve the model's learning of labeling with a small dataset in the future.

6 Conclusion

This work has investigated a task of syllogism figure identification. We have prepared a syllogism dataset with annotated figures and carried out the task using both supervised and weakly-supervised approaches. Our results have shown that using a weakly-supervised approach is as good as using a supervised approach for the identification task. In future work, we propose to utilize these automatic approaches to assist human annotation for syllogism. And more balanced data will also be used in future syllogism dataset construction.

Acknowledgments. This work is funded by the Humanity and Social Science Youth foundation of Ministry of Education (19YJCZH230) and the Fundamental Research Funds for the Central Universities in BLCU (No.17PT05).

References

1. Stede, M., Schneider, J.: Argumentation mining. Synth. Lect. Hum. Lang. Technol. **11**, 1–191 (2018)
2. Lawrence, J., Reed, C.: Argument mining: a survey. In: Computational Linguistics, pp. 1–55 (2019)
3. Walton, D., Reed, C., Macagno, F.: Argumentation Schemes. Cambridge University Press, Cambridge (2008)
4. Feng, V.W., Hirst, G.: Classifying arguments by scheme. In: Proceedings of the 49th Annual Meeting of the Association for Computational Linguistics: Human Language Technologies, Portland, Oregon, USA, pp. 987–996. Association for Computational Linguistics (2011)
5. Musi, E., Ghosh, D., Muresan, S.: Towards feasible guidelines for the annotation of argument schemes. In: Proceedings of the Third Workshop on Argument Mining, pp. 82–93. Association for Computational Linguistics (2016)

6. Green, N.: Proposed method for annotation of scientific arguments in terms of semantic relations and argument schemes. In: Proceedings of the 5th Workshop on Argument Mining, Brussels, Belgium, pp. 105–110. Association for Computational Linguistics (2018)
7. Black, E., Hunter, A.: A relevance-theoretic framework for constructing and deconstructing enthymemes. J. Log. Comput. **22**, 55–78 (2012)
8. Rajendran, P., Bollegala, D., Parsons, S.: Contextual stance classification of opinions: A step towards enthymeme reconstruction in online reviews. In: Proceedings of the First Workshop on Argumentation Mining, pp. 31–39. Association for Computational Linguistics (2016)
9. Razuvayevskaya, O., Teufel, S.: Finding enthymemes in real-world texts: a feasibility study. Argu. Comput. **8**, 1–17 (2017)
10. Hubien, H.: John buridan on the fourth figure of the syllogism. Revue Internationale de Philoso-phie **113**(113), 271–285 (1975)
11. Dickstein, L.: The effect of figure on syllogistic reasoning. Mem. Cognit. **6**, 76–83 (1978)
12. Hosseini, S.A., Modgil, S., Rodrigues, O.: Enthymeme construction in dialogues using shared knowledge. Front. Artif. Intell. Appl. **266**, 325–332 (2014)
13. Lippi, M., Torroni, P.: Argumentation mining: state of the art and emerging trends. ACM Trans. Internet Technol. **16**(2), 10:1-10:25 (2016)
14. Cabrio, E., Villata, S.: Five years of argument mining: a data-driven analysis. In: International Joint Conferences on Artificial Intelligence Organization, IJCAI-18, pp. 5427–5433 (2018)
15. Lawrence, J., Reed, C., Allen, C., McAlister, S., Ravenscroft, A.: Mining arguments from 19th century philosophical texts using topic based modeling. In: Proceedings of the First Workshop on Argumentation Mining, Baltimore, Maryland, pp. 79–87. Association for Computational Linguistics (2014)
16. Aharoni, E., et al.: A benchmark dataset for automatic detection of claims and evidence in the context of controversial topics. In: Proceedings of the First Workshop on Argumentation Mining, pp. 64–68 (2014)
17. Garca-Villalba, M., Saint-Dizier, P.: A framework to extract arguments in opinion texts. Int. J. Cogn. Inf. Nat. Intell. **6**, 62–87 (2012)
18. Rajendran, P., Bollegala, D., Parsons, S.: Is something better than nothing? automatically predicting stance-based arguments using deep learning and small labelled dataset. In: Proceedings of the 2018 Conference of the North American Chapter of the Association for Computational Linguistics, pp. 28–34 (2018)
19. Bowman, S.R., Angeli, G., Potts, C., Manning, C.D.: A large annotated corpus for learning natural language inference. In: Proceedings of the 2015 Conference on Empirical Methods in Natural Language Processing, pp. 632–642 (2015)
20. Fleiss, J.: Measuring nominal scale agreement among many raters. Psychol. Bull. **76**, 378 (1971)
21. Cortes, C., Vapnik, V.: Support vector network. Mach. Learn. **20**, 273–297 (1995)
22. Graves, A., Schmidhuber, J.: Framewise phoneme classification with bidirectional lstm andother neural network architectures. Neural Netw. Off. J. Int. Neural Netw. Soc. **18**, 602–610 (2005)
23. Devlin, J., Chang, M.W., Lee, K., Toutanova, K.: BERT: pre-training of deep bidirectional transformers for language understanding (2018). CoRR, abs/1810.04805.
24. Camburu, O.-M., Rocktäschel, T., Lukasiewicz, T., Blunsom, P.: e-SNLI: natural language inference with natural language explanations. Adv. Neural. Inf. Process. Syst. **31**, 9539–9549 (2018)

Attentive Intra-modality Fusion for Multimodal Sentiment Analysis

Jianxun Ding[2], Yongqiang Zheng[2], Xia Li[1,2(✉)], Minping Chen[2], and Junshan Huang[2]

[1] Guangzhou Key Laboratory of Multilingual Intelligent Processing, Guangzhou, China
xiali@gdufs.edu.cn

[2] School of Information Science and Technology, Guangdong University of Foreign Studies, Guangzhou, China

Abstract. The growing trend of sharing opinion videos on social media platforms leads to more and more attention to multimodal sentiment analysis research. A number of approaches in multimodal sentiment analysis have been proposed and continual improved results were produced. Most of the existing studies have paid more attention to cross-modalities fusion, ignoring valuable information fusion within each uni-modality. To this end, we propose an AGTL (Attentive Global and Temporal Local information) component to capture the global semantic information and the local semantic representation containing temporal relationships and get the different contributions of global or local information in each uni-modality. After later fusion of representations of different modalities learned by AGTL, we get the final representation of multi modalities. Different comparative and ablation experiments are conducted on a public multimodal sentiment analysis dataset, and the experimental results demonstrate the effectiveness of our model in multimodal sentiment analysis.

Keywords: Sentiment analysis · Multimodal fusion · Intra-modality fusion

1 Introduction

Sentiment analysis aims to automatically extract and classify sentiments in a text [1]. It is a popular research area of Natural Language Processing and is widely used in various fields, such as customer review analysis, product feedback analysis and public opinion analysis. With developments and successes in video, image, audio and language processing, multimodal sentiment analysis has been studied extensively in recent years.

The main challenge in multimodal sentiment analysis is to implement and obtain effective fusion strategies of different modalities. A number of multimodal fusion strategies have been proposed and have achieved impressive results [2–7]. These strategies can be categorized as modal representation learning based methods, memory changing for cross-view fusion methods, late fusion decision based methods and early fusion decision based methods. For example, Pham et al. [5] proposed a strategy based on a machine translation system to learn joint representations of multimodal data. Once the translation model was trained, any unimodal source data can be used for final sentiment prediction

M. Liu et al. (Eds.): CLSW 2020, LNAI 12278, pp. 452–463, 2021.
https://doi.org/10.1007/978-3-030-81197-6_38

in the tests. Zadeh et al. [8] introduced a Delta-memory Attention Network to identify cross-view interactions by associating a relevance score with the memory dimensions of each LSTM and summarize through time with a Multi-view Gated Memory.

Previous works have paid more attention to cross-modality fusion to get better performance, such as the work of Zadeh et al. [8, 9] and Liang et al. [3], ignoring better independent intra-modality fusion strategies. In this paper, we focus on intra-modality independent fusion for multimodal sentiment analysis. First, for each uni-modality, an AGTL (Attentive Global and Temporal Local information) component is proposed in our model to capture the global semantic information and the local semantic representation containing temporal relationships. We use an attention mechanism to capture different contributions of global information or local temporal relationship information in AGTL. The core design idea is that the AGTL component inside a uni-modality can fully represent the hidden semantic representation of each independent mode. After getting the representations of multiple attentive independent uni-modalities learned by AGTL, a simple late fusion method is used to get the multimodal representation. Overall, the contributions of our method are as follows:

1) We propose an AGTL component to independently capture different contributions of global information and temporal local relationship information in each uni-modality and a late cross-modality fusion is used to get the cross-modality representation.
2) Several experiments have been carried out on a public dataset, and the results show that our method achieves better or comparable experimental performance compared with the baseline models. Ablation experiments show that the AGTL component actually improves the performance of our model.

2 Related Work

Sentiment analysis refers to using natural language processing technology to extract people's views, emotions, attitudes about entities and their attributes from texts [1]. From the perspective of analysis granularity, it can be divided into coarse-grained sentiment analysis [10, 11] and fine-grained sentiment analysis [12, 13]. Coarse-grained sentiment analysis includes text level and sentence level sentiment analysis. Fine-grained analysis is based on the analysis of aspects and their attributes. Multimodal sentiment analysis aims to use vision and acoustic features to assist text features to perform sentiment prediction more accurately, which has been studied extensively in recent years.

In this paper we focus on multimodal sentiment analysis at sentence level. Different fusion strategies have been proposed for multimodal sentiment analysis and good performance has been achieved. However, most of the studies have paid more attention to cross-modalities fusion, ignoring the better prospects of intra-modality fusion for multimodal sentiment analysis. In this paper, we will review works from the two sides: modeling cross-modalities fusion and modeling intra-modality fusion.

2.1 Modeling Cross-Modalities Fusion

Existing works have proposed different strategies for modeling cross-modalities or cross-view fusion in order to capture effective interactions of different modalities. For instance,

Rajagopalan et al. [14] extended LSTM into Multi-view LSTM by dividing the memory cells and gates into multiple regions corresponding to different views. Williams et al. [15] constructed a DNN multimodal sentiment classification model, which was different from the feature fusion method mentioned in [14]. Its feature fusion was performed in mid-term. Zadeh et al. [9] introduced a Tensor Fusion Network to learn inter-modality dynamics to integrate multiple uni-modal representations into one compact multimodal representation. Addressing the problem of computation cost of [9], Liu et al. [16] proposed Low-Rank multimodal fusion, leveraging the low-rank weight tensors to make the fusion efficient. Mai et al. [17] designed an ABS-LSTM structure to make sense of the cross-modality dynamics from an integral perspective, in which a bidirectional residual connection of memory cells and hidden states as well as integrate attention mechanisms are introduced to transmit information and learn overall representations more effectively.

2.2 Modeling Intra-modality Fusion

Less previous studies thoroughly focused on intra-modality fusion in multimodal sentiment analysis. Most of the above studies which have just been reviewed in the previous sub-section used simple intra-modality fusion strategies. For example, Zadeh et al. [9] used LSTM to get a matrix of language, vision, acoustic representations and then used it as an input to a fully-connected network that generated language, vision and acoustic embeddings respectively. Verma et al. [18] utilized a factorization machine to help estimate the target and use a single feed-forward linear layer to get the latent embeddings of vision and acoustic. It used LSTM followed by a single feed-forward layer to get the latent embedding of language.

In this paper, we focus on intra-modality fusion and try to investigate whether sufficient intra-modality fusion and simple inter-modality fusion can also achieve better results in multimodal sentiment analysis tasks. We propose an AGTL component to capture the global semantic information and the local semantic information containing temporal relationships. We use an attention mechanism to capture different contributions of global information or local temporal information in each uni-modality. After getting the representations of multiple attentive independent uni-modality learned by AGTL, a simple concatenation of a later fusion method is used to get the multimodal representation.

3 Our Model

Similar to most of multimodal sentiment analysis methods, our model consists of two components aiming to capture intra-modality fusion and inter-modality fusion. However, unlike previous methods, our model focuses on intra-modality fusion. We propose an attentive fusion of intra-modality component for capturing more about global information and local temporal relationship information and obtaining the contributions of different local or global information. And we just use a simple concatenation and fully connection operation for later inter-modality fusion. The whole architecture of our model is shown in Fig. 1.

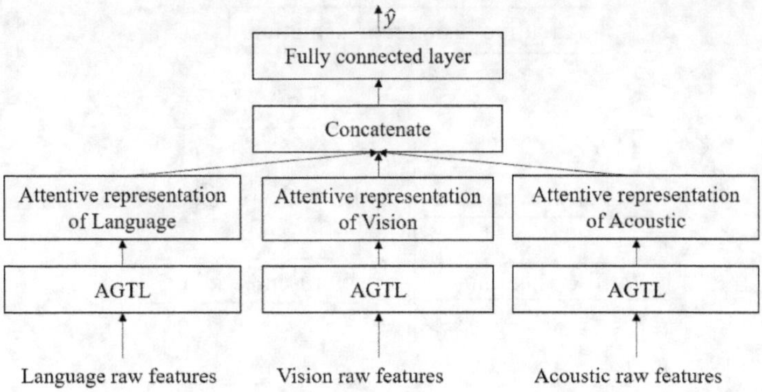

Fig. 1. Overall architecture of our model.

In the following sections, we will describe our model in more detail. Section 3.1 introduces the attentive fusion of intra-modality. This includes the AGTL component, the representation of independent uni-modality and attentive intra-modality fusion. Section 3.2 introduces later inter-modality fusion and Sect. 3.3 describes the loss function of our model.

3.1 Attentive Intra-modality Fusion

AGTL Component. As described earlier, an attentive intra-modality fusion component (AGTL, Attentive Global and Temporal Local information) is proposed in our model to get the attentive global semantic information and local temporal relationship semantic information. The component of AGTL for the language modality is shown in Fig. 2, which is the same as the components of the vision modality and the acoustic modality. As shown in Fig. 2, AGTL consists of two parts. The first part is a fully connected layer used for uni-modality raw features. The global information of the modality, termed as hglobal, is also obtained from this part. The second part is a Bi-LSTM [19] encoding layer with the temporal features of different modalities as inputs and the output of temporal information of the modality is obtained from this part, denoted as $\{h1,h2,...,hT\}$.

Independent Uni-Modality Representation. Firstly, we use the Bi-LSTM system to model the raw data of each modality based on time series, which can be used to model independent modalities and obtain local semantic representations according to time series relationships. The raw features of acoustic, vision and language modalities are denoted as $\{Z_L, Z_A, Z_V\}$. For language modality Z_L, we use Bi-LSTM to encode it and obtain its temporal local semantic information, which is represented as $L_{local} = \{h_1, h_2, \cdots, h_T\}$, where T is the temporal information of the video in the multimodal dataset. Simultaneously, we input the initial representation of the language modality into a fully connected layer to obtain its global semantic representation L_{global}. Then we can obtain an internal fusion representation of the language modality L by concatenating L_{local} and L_{global}, as shown in Eqs. (1)–(3). Analogous to the above language modality, we can obtain the fusion representation of the acoustic modality and the vision modality

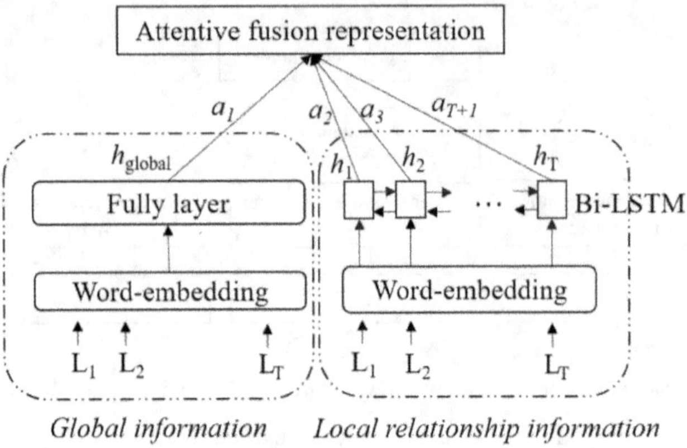

Fig. 2. The architecture of AGTL components of our model.

and combine temporal local semantics and global semantics, as shown in Eqs. (4)–(6) and Eqs. (7)–(9) respectively. Where $W_{L_{\text{global}}}$, $b_{L_{\text{global}}}$, $W_{A_{\text{global}}}$, $b_{A_{\text{global}}}$, $W_{V_{\text{global}}}$, $b_{V_{\text{global}}}$ are the weight and bias parameters.

$$\mathrm{L}_{\text{local}} = BiLSTM\,(Z_L) \tag{1}$$

$$L_{\text{global}} = W_{L_{\text{global}}} \cdot Z_L + b_{L_{\text{global}}} \tag{2}$$

$$\mathrm{L} = \left[L_{\text{global}},\ \mathrm{L}_{\text{local}}\right] \tag{3}$$

$$\mathrm{A}_{\text{local}} = BiLSTM\,(Z_A) \tag{4}$$

$$A_{\text{global}} = W_{A_{\text{global}}} \cdot Z_A + b_{A_{\text{global}}} \tag{5}$$

$$\mathrm{A} = \left[A_{\text{global}},\ \mathrm{A}_{\text{local}}\right] \tag{6}$$

$$\mathrm{V}_{\text{local}} = BiLSTM\,(Z_V) \tag{7}$$

$$V_{\text{global}} = W_{V_{\text{global}}} \cdot Z_V + b_{V_{\text{global}}} \tag{8}$$

$$\mathrm{V} = \left[V_{\text{global}},\ \mathrm{V}_{\text{local}}\right] \tag{9}$$

Attentive Intra-modality Fusion. Attention mechanisms are widely used in many tasks such as machine translation and image classification [20]. Attention mechanisms are inspired by human attention. For example, humans' visual processing system tends to focus selectively on parts of the image, ignoring other irrelevant information. In order to selectively obtain more effective local or global semantic information within a single modality, we use the attention mechanism for the corresponding representations of each uni-modality *L*, *A*, and *V* described in the above.

As shown in Fig. 2, the language representation L is input to an attention component to calculate the attention coefficient of the global and local temporal information, as shown in Eqs. (10–12). The final attentive representation of the language inter-modality fusion, termed as a_L, is calculated as in Eq. (12). Similar to the above language modality, we can obtain the final attentive representations of the vision inter-modality fusion and acoustic inter-modality fusion, denoted as a_A and a_V, as shown in Eqs. (13–15) and (16–18) respectively.

$$o_{L_i} = tanh(W_L \cdot L_i + b_L) \tag{10}$$

$$w_{L_i} = softmax\left(u_{L_i} \cdot o_{L_i}\right) \tag{11}$$

$$a_L = \sum\nolimits_{i=1}^{T} w_{L_i} L_i \tag{12}$$

$$o_{A_i} = tanh(W_A \cdot A_i + b_A) \tag{13}$$

$$w_{A_i} = softmax\left(u_{A_i} \cdot o_{A_i}\right) \tag{14}$$

$$a_A = \sum\nolimits_{i=1}^{T} w_{A_i} A_i \tag{15}$$

$$o_{V_i} = tanh(W_V \cdot V_i + b_V) \tag{16}$$

$$w_{V_i} = softmax\left(u_{V_i} \cdot o_{V_i}\right) \tag{17}$$

$$a_V = \sum\nolimits_{i=1}^{T} w_{V_i} V_i \tag{18}$$

3.2 Later Inter-modality Fusion

As mentioned earlier, we use a simple later inter-modality fusion operation to get the final multiple modalities fusion. As shown in Fig. 1, we use a concatenation operation for three modalities to get the final inter-modality fusion. The final representation of the multimodal data, termed as $a = [a_L, a_A, a_V]$, is concatenated from each attentive intra-modality fusion representations a_L, a_A and a_V. In order to better obtain the encoding of the representation of multimodal fusion, as shown in Fig. 1, we input the final multimodal representation obtained to a fully connected layer, and then input it to the classification agent to obtain the predicted sentiment value $\hat{y}$.

3.3 Loss Function

We estimate the parameters of our model via minimizing the L1 loss function in Eq. (19). Where X denotes the training set of multimodal training data instances, $y(x)$ denotes the correct label of instance x, and $\hat{y}(x)$ denotes the prediction obtained from our model.

$$loss = \frac{1}{n} \sum_{\forall x \in X} \left| y(x) - \hat{y}(x) \right| \tag{19}$$

4 Experiments

4.1 Dataset

We carried out several experiments on the public multimodal sentiment analysis dataset CMU-MOSI[1] (Multimodal Corpus of Sentiment Intensity and Subjectivity Analysis) [21, 22]. The CMU-MOSI dataset contains 93 videos from YouTube, each of which involved a different speaker, who was expressing his or her opinions towards a movie. The videos in the CMU-MOSI dataset were split into 2199 video clips, and each clip has a sentiment label. Similar to the work of [8], we have trained our model on 52 videos (1284 clips), validated with 10 videos (229 clips) and conducted tests on 31 videos (686 clips). The statistical information of the CMU-MOSI dataset is shown in Table 1.

Table 1. Statistical information of the experimental dataset.

Dataset	CMU-MOSI
#Train	1284
#Valid	229
#Test	686

4.2 Evaluation Metrics

We have conducted several experiments on the CMU-MOSI dataset. Following the ACL2018 Human Multimodal Language Challenge[2], we have evaluated the performances of the baseline models and our model on binary classification tasks and regression tasks. For regression tasks, we have reported Mean Absolute Error (MAE) and Pearson's Correlation (Correlation). For binary classification, we have reported accuracy and F1 score. For all the metrics, higher values denote better performance, except for MAE, where lower values denote better performance.

4.3 Experimental Setting

The parameters of our model are as follows. For Bi-LSTM on language modality, the hidden size is set to 140. For Bi-LSTM on acoustic modality, the hidden size is set to 40. For Bi-LSTM on vision modality, the hidden size is set to 50. For Bi-LSTM on multi-modalities representation, we set its hidden size to 100. In order to reduce the risk of over-fitting, several dropout layers are added to the network. The initial learning rate is 0.0015 and the batch size is 70. The model has been trained for 170 epochs.

1 http://immortal.multicomp.cs.cmu.edu/raw_datasets/processed_data/cmu-mosi/
2 http://multicomp.cs.cmu.edu/acl2018multimodalchallenge/

4.4 Baseline Models

We have used several methods for comparison. These baseline models are as follows.

We first chose the traditional machine learning methods as the baseline models for comparison. They are: SVM [23] and RF (Random forest classifier) [24]. The two models used support vector machines and conditional random fields to implement sentiment classification and regression, respectively. We also used THMM[25] and MV-HCRF [26] as other baseline models. THMM (Tri-modal Hidden Markov Model) [25] used the hidden Markov model to realize classification of multimodal features. MV-HCRF [26] is an extension of HCRF for Multi-view data, explicitly capturing view-shared and view specific sub-structures.

We also chose neural networks based models as our baseline models. MV-LSTM [14], BC-LSTM [2], and GME-LSTM [27] are the models based on the improved LSTM structure to fuse multiple modal features from different perspectives. TFN[9] proposes a Tensor Fusion model to fuse the language, vision and acoustic modalities by taking outer products of the augmented unimodal representations, which can obtain raw unimodal representations, bimodal (or trimodal) interactions in the form of 2D-tensor (3D-tensor) simultaneously. LMF [16] is similar to TFN but alleviates the drawback of a large number of parameters through low-rank factor. MFN [8] uses attention and memory mechanisms to learn multimodal representations.

4.5 Experimental Results on CMU-MOSI.

We carried out several experiments on the CMU-MOSI dataset. Results using our model and the baseline models are shown in Table 2 for comparison.

As shown in Table 2, compared with the traditional non-neural network models, our model achieves the best results in accuracy and F1 score for the binary classification task. For the regression task, our model also gets the best results in MAE and Pearson correlation coefficient evaluation. Compared with MV-LSTM, BC-LSTM, TFN, LMF, GME-LSTM and MFN, our model also achieves the best results in accuracy and F1 score for the binary classification task. For the regression task, our model gets the second best result in correlation compared with all the baseline models. But as we can see, for the MAE metric in the task of regression, our model is better than MV-LSTM, BC-LSTM and TFN, but performs not as good as the LMF, GME-LSTM and MFN models scoring 0.096, 0.053 and 0.043. Overall, the results on the CMU-MOSI dataset show the effectiveness of our model.

4.6 Ablation Studies

In order to verify the effectives of different components in our model, we performed ablation experiments on the CMU-MOSI dataset.

Firstly, we removed the attention mechanism in the AGTL component and other parts remained the same, which was labeled as Ours(Δ). And then, we used Bi-LSTM to capture local temporal information of single modalities and also used the attentive mechanism to capture different contributions of local information of uni-modality, but did not use the global part, which was labeled as Ours(β). Finally, we used the raw

Table 2. Experimental results of different models on test set.

Model	Binary		Regression	
	Acc.	F1	MAE	Corr
SVM	47.7	36.6	1.426	0.129
RF	53.9	52.7	1.409	0.204
THMM	50.7	45.4	–	–
MV-HCRF	65.6	65.7	–	–
MV-LSTM	73.9	74.0	1.019	0.601
BC-LSTM	75.2	75.3	1.079	0.614
TFN	74.6	74.5	1.040	0.587
LMF	76.4	75.7	0.912	0.668
GME-LSTM(A)	76.5	73.4	0.955	-
MFN	77.4	77.3	0.965	0.632
Ours	**77.7**	**77.7**	**1.008**	**0.653**

semantic information to replace the global part and also used the temporal local semantic information captured by Bi-LSTM, and other components remained the same. We use Ours(γ) to represent it. The ablation experimental results are shown in Table 3.

Table 3. Ablation experimental results of our models on CMU-MOSI dataset.

Model	Binary		Regression	
	Acc	F1	MAE	Corr.
Ours(Δ)	76.9	76.9	0.969	0.651
Ours(β)	77.1	77.1	1.031	0.635
Ours(γ)	75.4	75.3	1.013	0.617
Ours	77.7	77.7	1.008	0.653

We first compare the complete model of Ours with Ours (Δ). As shown in Table 3, Ours (Δ) has lower accuracy and F1 score than that of Ours, decreasing by 0.7% points on F1 score. This indicates that the attention mechanism does effectively improve the performance of overall multimodal sentiment analysis. The experimental results show that Ours(γ) has significantly lower accuracy and F1 values than that of Ours. For instance, Ours(γ) decreases by 2.4% in F1 score. The performance on the regression task is also reduced at the same time. This indicates that obtaining global semantic information as proposed in this paper has indeed played an important role in improving the performance of the overall model. The comparison of the experimental results of Ours and Ours(β) indicates that the global semantic information of the single modality obtained through the fully-connected layer is more effective. Overall, experiments show the effectiveness of fusing global semantic information with local temporal semantic information in multimodal sentiment analysis.

5 Conclusions

In this paper, we propose an attentive intra-modality fusion component for multimodal sentiment analysis. The core of our model uses the AGTL component to capture the attentive representation of uni-modality, which contains the global semantic information and local temporal information with contributions of different local or global information for multimodal sentiment analysis prediction. As for inter-modality fusion, we use a simple concatenating operation on the attentive representations of three modalities. Several comparison experiments have been carried out on a public dataset with baseline models. The experimental results indicate that the model proposed in this paper does effectively improve the overall sentiment analysis performance. At the same time, the results of ablation experiments also indicate that fusing global semantic information within a modality and using an attention mechanism for global and local information in uni-modality does effectively improve the overall model performance proposed in this paper.

Acknowledgements. This work is supported by Science and Technology Program of Guangzhou(201904010303),National Natural Science Foundation of China (61976062), Special Funds for the Cultivation of Guangdong College Students' Scientific and Technological Innovation (pdjh2019b0173) and the Ministry of Education of Humanities and Social Science project in China (No. 20YJCZH124 and No. 20YJA740031).

References

1. Bo, P., Lillian, L.: Opinion mining and sentiment analysis. Found. Trends® Inf. Retrie. **2**(1–2), 1–135 (2008)
2. Poria, S., Cambria, E., Hazarika, D.: Context-dependent sentiment analysis in user-generated videos. In: Proceedings ofthe 55th Annual Meeting ofthe Association for Computational Linguistics, pp. 873–883 (2017)
3. Liang, P.P., Liu, Z., Zadeh, A., Morency, L.-P.: Multimodal language analysis with recurrent multistage fusion. In: Proceedings of the 2018 Conference on Empirical Methods in Natural Language Processing, pp. 150–161 (2018)
4. Xu, N., Mao, W., Chen, G.: A co-memory network for multimodal sentiment analysis. In: 41st International ACM SIGIR Conference on Research and Development in Information Retrieval, SIGIR 2018, pp. 929–932 (2018)
5. Pham, H., Liang, P.P., Manzini, T., Morency, L., Póczos, B.: Found in translation: learning robust joint representations by cyclic translations between modalities. Proc. AAAI Conf. Artif. Intell. **33**, 6892–6899 (2019)
6. Tsai, Y.H., Bai, S., Liang, P.P., Kolter, J.Z., Morency, L., Salakhutdinov, R.: Multimodal transformer for unaligned multimodal language sequences. In: Proceedings of the 57th Annual Meeting of the Association for Computational Linguistics, pp. 6558–6569 (2019)

7. Wang, Y., Shen, Y., Liu, Z., Liang, P.P., Zadeh, A., Morency, L.: Words can shift: dynamically adjusting word representations using nonverbal behaviors. Proc. AAAI Conf. Artif. Intell. **33**, 7216–7223 (2019)
8. Zadeh, A., Liang, P.P., Mazumder, N., Poria, S., Cambria, E., Morency, L.-P.: Memory fusion network for multi-view sequential learning. In: Thirty-Second AAAI Conference on Artificial Intelligence (2018)
9. Zadeh, A., Chen, M., Cambria, E., Poria, S., Morency, L.P.: Tensor fusion network for multimodal sentiment analysis. In: Conference on Empirical Methods in Natural Language Processing, Proceedings, pp. 1103–1114 (2017)
10. Saif, H., Fernandez, M.: Contextual semantics for sentiment analysis of twitter. Inf. Process. Manag **52**(1), 5–19 (2015)
11. Yang, B., Cardie, C.: Context-aware learning for sentence-level sentiment analysis with posterior regularization. In: Proceedings ofthe 52nd Annual Meeting ofthe Association for Computational Linguistics, pp. 325–335 (2014)
12. Nguyen, T.H.: PhraseRNN: phrase recursive neural network for aspect-based sentiment analysis. In: Conference on Empirical Methods in Natural Language Processing, Proceedings, pp. 2509–2514 (2015)
13. Zeng, Z., Ma, J., Chen, M., Li, X.: Joint learning for aspect category detection and sentiment analysis in chinese reviews. In: Zhang, Qi., Liao, X., Ren, Z. (eds.) CCIR 2019. LNCS, vol. 11772, pp. 108–120. Springer, Cham (2019). https://doi.org/10.1007/978-3-030-31624-2_9
14. Rajagopalan, S.S., Morency, L.-P., Baltrušaitis, T., Goecke, R.: Extending long short-term memory for multi-view structured learning. In: Leibe, B., Matas, J., Sebe, N., Welling, M. (eds.) ECCV 2016. LNCS, vol. 9911, pp. 338–353. Springer, Cham (2016). https://doi.org/10.1007/978-3-319-46478-7_21
15. Williams, J., Comanescu, R., Radu, O., Tian, L.: DNN multimodal fusion techniques for predicting video sentiment. In: In Proceedings of Grand Challenge and Workshop on Human Multimodal Language (Challenge-HML), pp. 64–72 (2018)
16. Liu, Z., Shen, Y., Lakshminarasimhan, V.B., Liang, P.P., Zadeh, A., Morency, L.-P.: Efficient low-rank multimodal fusion with modality-specific factors. In: Proceedings of the 56th Annual Meeting of the Association for Computational Linguistics, vol. 1: Long Papers, pp. 2247–2256 (2018)
17. Mai, S., Hu, H., Xiong, S.: Divide , Conquer and combine: hierarchical feature fusion network with local and global perspectives for multimodal affective computing. In: Proceedings of the 57th Annual Meeting ofthe Association for Computational Linguistics, pp. 481–492 (2019)
18. Verma, S., Wang, C., Zhu, L., Liu, W.: DeepCU: integrating both common and unique latent information for multimodal sentiment analysis. In: Proceedings of the 28th International Joint Conference on Artificial Intelligence, pp. 3627–3634 (2019)
19. Hochreiter, S., Schmidhuber, J.: Long short-term memory. Neural Comput. **9**(8), 1735–1780 (1997)
20. Chaudhari, S., Polatkan, G., Ramanath, R., Mithal, V.: An Attentive Survey of Attention Models (2019). arXiv preprint arXiv:1904.02874
21. Zadeh, A., Zellers, R., Pincus, E., Morency, L.-P.: MOSI: multimodal corpus of sentiment intensity and subjectivity analysis in online opinion videos (2016)
22. Zadeh, A., Zellers, R., Pincus, E., Morency, L.P.: Multimodal sentiment intensity analysis in videos: facial gestures and verbal messages. IEEE Intell. Syst. **31**(6), 82–88 (2016)
23. Cortes, C., Vapnik, V.N.: Support vector networks. Mach. Learn. **20**(3), 273–297 (1995)
24. Breiman, L.: Random forests. Mach. Learn. **45**(1), 5–32 (2001)
25. Morency, L.P., Mihalcea, R., Doshi, P.: Towards multimodal sentiment analysis: Harvesting opinions from the web. In: Proceedings of the 2011 ACM International Conference on Multimodal Interaction, pp. 169–176 (2011)

26. Song, Y., Morency, L., Davis, R.: Multi-view latent variable discriminative models for action recognition. In: IEEE Conference on Computer Vision and Pattern Recognition (CVPR) (2012)
27. Chen, M.: Multimodal sentiment analysis with word-level fusion and reinforcement learning. In: ICMI2017, pp. 163–171 (2017)

Construction of a General Lexical-Semantic Knowledge Graph

Yi Li, Yanqiu Shao(✉), and Yuhang Zhao

School of Information Science, Beijing Language and Culture University, Beijing, China

Abstract. One of the main factors that restricts the performance of deep learning models is the lack of exogenous knowledge, which makes the model fail to understand the lexical semantic relationships in real situations. For this reason, building a general lexical-semantic knowledge graph to give the model external knowledge to improve the performance in generalization and robustness of the model has become a hot research topic. This article introduces a method of constructing a general lexical-semantic knowledge graph (GLSKG). GLSKG stores useful knowledge such as definitions, parts of speech, and examples of vocabularies. Besides, it contains detailed knowledge such as lexical semantic relationships and semantic dependency relationships among words. The construction of GLSKG introduced in this paper is carried out in two steps: i) redesign of the architecture of the knowledge graph and ii) integrate existing knowledge graphs. The scale of GLSKG is considerable, including 295,734 entities, 170 semantic dependency relations, and eight lexical-semantic relations. GLSKG can provide not only external entity information for the semantic dependency parser (SDP) but also provide semantic dependency information for the SDP to improve its performance.

Keywords: General lexical-semantic knowledge graph · Knowledge graph integration · Semantic dependency relations · Lexical-semantic knowledge

1 Introduction

Two kinds of knowledge are essential for constructing knowledge graphs, namely semantic dependency knowledge and lexical semantic knowledge. Semantic dependencies can very well describe the lexical relationships in natural languages in actual use and therefore it can provide necessary data support for Semantic Dependency Parsing tasks. This kind of knowledge is essential for constructing a knowledge graph due to the fact that the information held by nodes (representing words) in the knowledge graph and the relationships between adjacent nodes can determine the design of a knowledge graph. Lexical semantic knowledge describes the essential attributes of vocabularies for interpretation, for example, pronunciation, parts of speech, etc., which can be used as external data to provide models with accurate lexical-semantic knowledge as a support.

However, in practical applications, these two kinds of knowledge are often used jointly and need to be searched in different knowledge graphs. On the one hand, this increases the complexity of retrieval and poses significant challenges to the performance

M. Liu et al. (Eds.): CLSW 2020, LNAI 12278, pp. 464–472, 2021.
https://doi.org/10.1007/978-3-030-81197-6_39

of the model. On the other hand, combining diverse sources of knowledge is still an enormous challenge. To remedy these defects, this article aims to i) explore the design and construction of a unified knowledge graph; ii) highly integrate the knowledge contents and functions of existing lexical-semantic knowledge graphs and semantic dependency knowledge graphs; and, iii) represent lexical knowledge in a unified form using the same graph.

1.1 Related Work

The construction of GLSKG in this paper relies on existing research [1, 2]. The following is an introduction to the related works.

The Semantic Dependency Bank (SDB) of BLCU-HIT [3]: The BLCU-HIT semantic dependency bank includes 61 coarse-grained dependency labels, 170 fine-grained dependency labels, and 45000 annotated sentences. The BLCU-HIT semantic dependency bank covers semantic dependency phenomena in modern Chinese.

HowNet [4, 5]: As a large-scale language knowledge base maintained by Dong Zhendong and Dong Qiang for 30 years, HowNet contains rich semantic information, which are mainly concepts represented by Chinese and English words. It aims to reveal the relationships between concepts and the attributes of concepts. HowNet uses the most basic and indivisible semantic units, i.e., sememes, to mark the semantic information of the words. In the structure of HowNet, words are composed of meaning items, and meaning items are defined by sememes and relationships. In detail, dynamic information of characters or semantic information is marked by sememes, which can accurately portray the meaning of words. At present, HowNet contains more than 800 sememes, 212541 meanings, and 118347 words.

Semantic Dependency Knowledge Graph (SDKG) [6]: Based on BLCU-HIT and HowNet, SDKG has been constructed. It is a knowledge graph with rich dependency knowledge and also dependency labels between sememes. According to its documentation, SDKG ultimately inherits the features of HowNet and describes each concept in detail. It contains about 500000 nodes and 88 relationships.

Multi-source Lexical Semantic Knowledge Base (MLSKB) [7]. MLSKB is a large-scale knowledge base which is easy to expand and contains general lexical-semantic knowledge and semantic annotations. Also, it possesses dynamic semantic characteristics. This knowledge base functionally inherits from HowNet, Tongyici-Cilin, and Semantic Knowledge Base of Contemporary Chinese, and can interpret words in many aspects. It can provide accurate knowledge and convenient services for users. At the same time, it uses a large amount of semantic knowledge resources to provide support for downstream research such as semantic information retrieval, question answering system, and relationship extraction. The MLSKB contains 153754 words, 1598356 triples, and 137 relationships.

1.2 A Brief Description of the GLSKG

In this experiment, the general lexical-semantic knowledge graph (GLSKG) we have built contains two kinds of knowledge, namely lexical-semantic knowledge and semantic dependency knowledge. GLSKG provides a comprehensive introduction to the definition of words, parts of speech, and examples of words. At the same time, the knowledge of semantic dependency relations, hyponymy relationships, synonymy relationships among words are described in detail.

For the construction of the knowledge graph, we have reused the data from SDKG and MLSKB and stored these data in a unified frame. To represent different kinds of knowledge, we have redesigned the structure of the knowledge graph and implemented entity disambiguation through some strategies. Besides, GLSKG includes the corpus of the SDB, so the semantic dependencies are further supplemented. The overall framework is shown in Fig. 1. The overall construction process consists of three steps: (1) the structural design of knowledge graph, including ontology design and data structure design, (2) the integration of process-based knowledge graphs, (3) the further expansion of GLSKG.

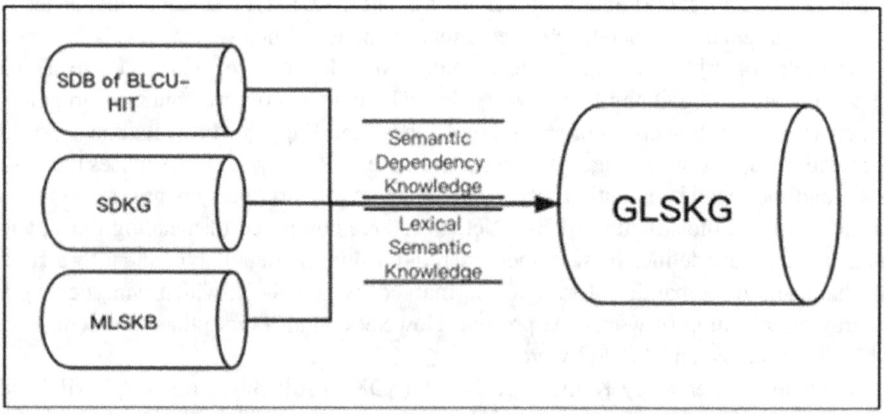

Fig. 1. Overall architecture: Semantic dependency knowledge and lexical-semantic knowledge is included in a unified framework – our GLSKG.

2 Structural Design of the GLSKG

In this section, we introduce the specific design method of knowledge graph structure which consists of two layers, namely the ontology layer and the data layer. For the ontology layer, we have restructured the ontology of words. For the data layer, we have designed its storage structure. It is believed that the redesigned structure is more suitable for the storage and for displaying existing words and meanwhile helps to expand the vocabulary further.

2.1 The Ontology Layer of the GLSKG

The ontology layer of GLSKG describes the essential attributes and classification of words. In the design of the ontology layer, we have made reference to the existing research and have chosen to construct the ontology of nouns and verbs.

Our first step of design is to inherit and revise the noun ontology of MLSKB. Noun ontology allows nouns existing in the knowledge graph and being connected with ontological elements to obtain a precise classification. An example of the noun ontology is shown in Fig. 2.

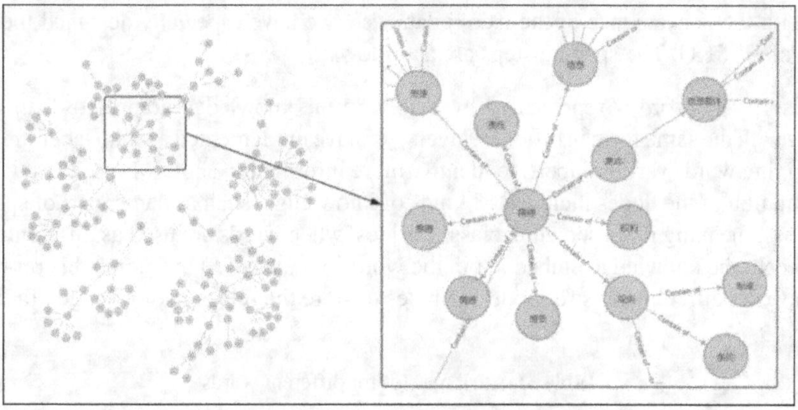

Fig. 2. This illustration shows an example of the Noun ontology

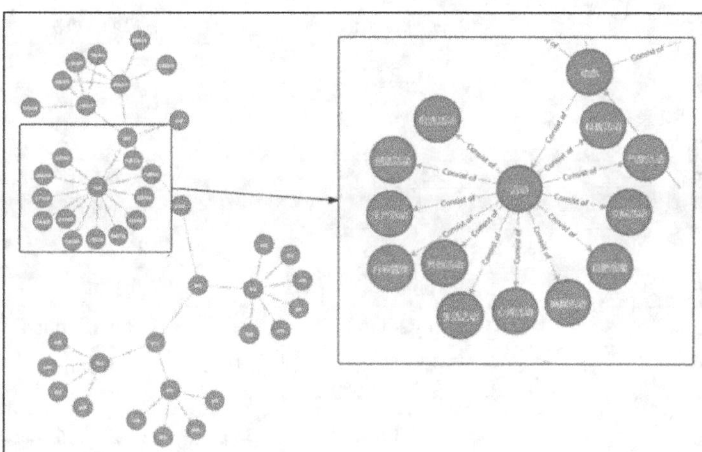

Fig. 3. This illustration shows an example of the Verb ontology

Next, we have redesigned the ontology of verbs in an expert-driven manner by which a new classification system for verb meaning is constructed. It follows that the verb ontology in our study is significantly different from those in the previous works. Due to the discrepancies between our new design and the original verb system, we have built a table mapping elements from the new verb ontology to the original verb system, aiming to integrate all the elements into one. In this way, the task of incorporating the data could be done later. An example of the verb ontology is shown in Fig. 3.

2.2 The Data Layer of the GLSKG

For data storage of GLSKG, we have used the popular graph database framework-Neo4j [8]. Neo4j is a native graph database. Its data storage method is customized and optimized for graph data, so it can better display data of the graph structure. It has better storage and retrieval performance, which help to construct and maintain GLSKG.

With the convenience of the Neo4j database, we have especially designed the data layer of GLSKG. The specific steps are as follows:

Nodes: Every word is represented by a node in the knowledge graph. To distinguish words with the same form in different layers, we have implemented a multi-label strategy, i.e., giving words various labels, to identify more information about words. The attribute information of the nodes includes 13 kinds of knowledge, such as name, part of speech, subclass, meaning item, semantic class. Besides, when words are used as different parts of speech, the knowledge embedded in the words may vary. To overcome this problem, we have set different slots for node attributes to store the relevant knowledge (Table 1).

Table 1. Attribute slot for different words

Attribute slot	Node category	Example - "苹果 Apple"	Example-"传授 Teach"
Name	All words	苹果 Apple	传授 Teach
POS		名词 Noun	动词 Verb
SubClass			
Meaning		4	1
Homomorphic			
SemanticClass		植物 Plant	领属转移 subordinative transfer
Valence			3
Eng		Apple	Teach
Pronunciation		Ping2Guo3	Chuan3shou4
Definitions		电脑 Computer; 水果 Fruit; 用具 Tool; 树 Tree	教 Teach
Hypernym		水果 Fruit	科教活动 Science and education activities
Ground	Noun		/
Target			/
Subject	Verb	/	人类 Mankind
Object		/	抽象事物 Abstract things
DativeRole		/	人类 Mankind

Relationships: The semantic dependency relationship between two nodes may vary and there may also be multiple relations among lexical items, such as synonymy, hypernym, and instance. Therefore, when establishing a node, different relationships should be distinguished. We have added the characteristics of multiple relationships in this knowledge graph and have identified the relationships by various labels, names, and attributes. At the same time, in the storage of knowledge graph relationship, the relationship labels and attributes are made available for modification in future.

3 Knowledge Graph Construction Based on Data Fusion

The redesigned architecture of the knowledge graph supports the integration of previously constructed knowledge graphs. This section introduces the method of fusing two knowledge graphs and explains how to use this method to obtain a more practical and complete knowledge graph - GLSKG. GLSKG is larger in scale, more abundant in knowledge, and more powerful. The redesigned architecture of the knowledge graph will also streamline redundant nodes, bringing great convenience for retrieval and usage. The steps for knowledge fusion are as follows:

1) The first step is to store all the words as nodes in the data layer. Word attributes stored in the original knowledge graphs will be assigned to the attribute information slots of each word. The previous work established sememes as nodes, which seems to be inappropriate. So, the newly constructed knowledge graph removes the sememe nodes and only retains word nodes.
2) The second step is to import the ontology layer of GLSKG, mark all the ontology elements at the ontology layer, create the ontology elements as nodes, and then copy the category, vocabulary name, and definition to the preset attribute slots of the nodes. Relationships between the ontology elements are directly inherited from the original knowledge graphs.
3) The third step is to modify the ontology and add a new branch to mark the ontology information which is extracted from other resources and contains the category information of the newly added nodes.
4) The fourth step is to use the relationships stored in the original data to connect the ontology layer and the data layer. Meanwhile, the similarity between newly added words and ontology items is calculated, which helps us to obtain a precise classification for new words.
5) In the fifth step, we preserved the relationships in the original data when constructing GLSKG. The similarity among the nodes is obtained by comparison among the elements in the first-order neighbor node set of the nodes in the previous knowledge graphs. Nodes with high similarity are connected by a Synonym relationship.
6) The sixth step is to process the node attribute information in the original knowledge graphs. According to the attribute information such as Hypernym, Valence number, Subject, Object, etc., the relevant nodes are linked in the data layer according to the above information.

7) The seventh step is to process the Semantic Dependency Bank (SDB) of BLCU-HIT corpus. We examined whether the words in the sentence existed in the knowledge graph and added the nodes for the words if they were not there. At the same time, the dependency relationship is established among the words according to the dependency label, and each sentence is stored in the Example Sentence slot of the dependency relationship attributes. Semantic dependency knowledge is shown in Fig. 4.

After the above steps, we have completed the construction of GLSKG. GLSKG stores knowledge in a more reasonable manner, and its structure is also easy to modify in the future. On the one hand, it can be used as a knowledge base to supplement external knowledge for deep learning models. On the other hand, it can be used as a dictionary independently, which brings convenience to retrieval of words.

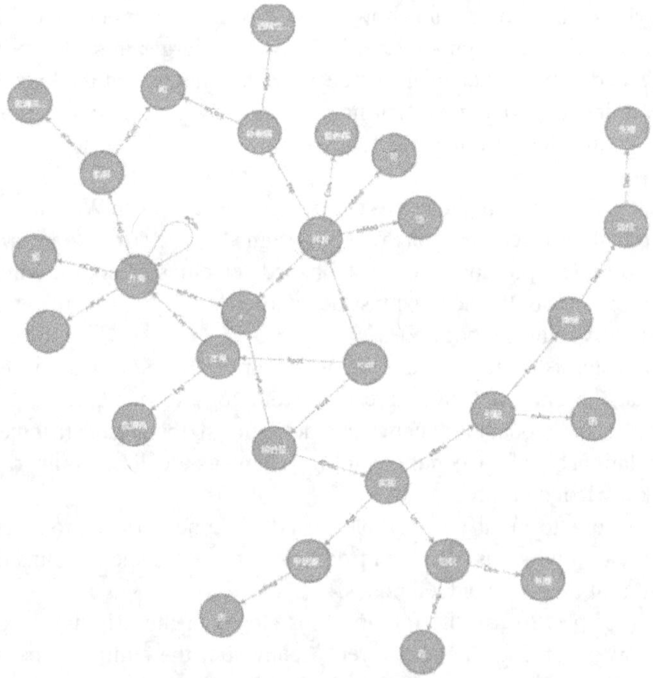

Fig. 4. This illustration shows a sample of semantic dependency knowledge in the data layer of GLSKG

4 Experimental Results and Analysis

Through the above process, we have built a GLSKG with rich lexical-semantic knowledge and semantic dependency knowledge. Regarding the amount of data, GLSKG contains 293,676 words, which are stored together with word-related knowledge in terms of pronunciation, definitions, examples, usage, example sentences, and other. And the knowledge graph contains 170 fine-grained dependency labels, which describe the semantic dependency information at the lexical level in detail. It also saves sentence information for the dependency relationships among words.

As for functions, the knowledge graph inherits the functions of multiple knowledge bases and can accurately describe words in various aspects. The lexical-semantic knowledge and semantic dependency knowledge contained in it can meet the needs of most downstream tasks and different application scenarios. Figure 5 shows some examples in the ontology and data layers, which contain lexical-semantic knowledge and semantic dependency knowledge.

But there are still many shortcomings that need to be improved. First, the attribute knowledge of many word nodes is incomplete and needs further supplementation. Second, the relationships between many words have not been completed. We can use knowledge extraction methods to obtain them in future. Third, Named Entity is a vital lexical resource. The knowledge graph has not yet had a reasonable way to deal with such vocabulary items, and further research is needed.

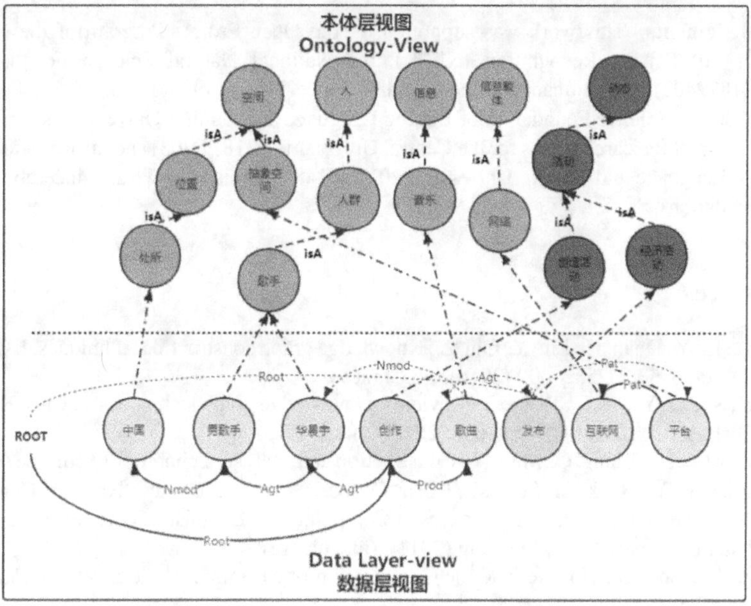

Fig. 5. The illustration shows a sample of GLSKG

5 Conclusion and Future Work

This paper has studied how to construct a general lexical-semantic knowledge graph (GLSKG). Methods used in the construction process include the design of the structure of the knowledge graph and the construction of a lexical-semantic knowledge graph based on existing data. Our knowledge graph stores information in the graph database

Neo4j. In detail, we have stored words in the form of nodes and stored detailed knowledge as node attributes and relationship attributes. We have also connected words through the relationships among words. Reusing of existing data enables the work to be carried out quickly, and the overall quality of the knowledge graph can be guaranteed. The GLSKG is constructed by this method and it contains not only lexical-semantic knowledge but also semantic dependency knowledge. It can be used for related tasks such as Semantic Dependency Parsing to provide sufficient external knowledge for deep learning models.

Our initial exploration still leaves many problems to be solved. For example, how to improve the quality of knowledge and the scale of the knowledge graph, and its application value need further research. In future, we will continue to carry out two aspects of research on this basis. On the one hand, we will continue to build this knowledge graph, supplement it with accurate and reliable knowledge, and expand the vocabulary and coverage of the knowledge graph. It can be enhanced in overall quality to meet the needs of higher-level research. On the other hand, we will explore the application of knowledge graphs by applying them to practical tasks to confirm the value of knowledge graphs in applications.

Acknowledgments. This work was supported by the Open Project Program of the National Laboratory of Pattern Recognition (NLPR), the National Natural Science Foundation of China (61872402), the Humanities and Social Science Project of the Ministry of Education (17YJAZH068), Science Foundation of Beijing Language and Culture University (supported by the Fundamental Research Funds for the Central Universities) (18ZDJ03) and the Research Funds of Beijing Language and Culture University (20YCX148). In this way, we are sincerely grateful to the foundations.

References

1. Liu, Q., Li, Y., Duan, H., Liu, Y., Qin, Z.: Knowledge graph construction techniques. J. Comput. Res. Develop. **53**(3), 582–600 (2016). (in Chinese)
2. Xu, Z., Sheng, Y., He, L., Wang, Y.: Review on knowledge graph techniques. J. Univ. Electron. Sci. Technol. China **45**(04), 589–606 (2016). (in Chinese)
3. Shao, Y., Qiu, L., Liang, C., Mao, N.: Construction and analysis technology of chinese semantic dependency TreeBank. In: Sun, M., Chen, Q. (eds.) The 11th Chinese National Conference on Computational Linguistics. Advances of Computational Linguistics in China, pp. 236–241. Tsinghua University Press, Luoyang (2011). (in Chinese)
4. Dong, Z., Dong, Q.: HowNet and the Computation of Meaning. World Scientific Publishing Co. Inc. (2006)
5. HowNet Homepage. http://www.keenage.com. Accessed 20 Jan 2020
6. Zhu, S., Li, Y., Shao, Y., Wang, L.: Building semantic dependency knowledge graph-based on HowNet. In: Hong, J.F., Zhang, Y., Liu, P. (eds.) Chinese Lexical Semantics, CLSW 2019. Lecture Notes in Computer Science, vol. 11831. Springer, Cham (2020).https://doi.org/10.1007/978-3-030-38189-9_54
7. Zhu, S., Li, Y., Shao, Y.: Research on construction and automatic expansion of multi-source lexical semantic knowledge base. In: Zhu, X., Qin, B., Zhu, X., Liu, M., Qian, L. (eds.) Knowledge Graph and Semantic Computing: Knowledge Computing and Language Understanding, CCKS 2019. Communications in Computer and Information Science, vol. 1134. Springer, Singapore (2019). https://doi.org/10.1007/978-981-15-1956-7_7
8. Neo4j Homepage. https://neo4j.com. Accessed 20 Jan 2020

A Study of the Polysemy Distribution of Mongolian

Dulan(✉) and Dabhvrbayar(✉)

School of Mongolian Studies, Inner Mongolia University, College Road 235, Hohhot 010021, China

Abstract. Previous studies of Mongolian paraphrasing were mainly based on qualitative methods, but few on its quantitative characteristics. Based on *Mongolian Dictionary*, this paper uses the research method of quantitative linguistics to make a statistical analysis of Mongolian polysemy, in order to explore not only its distribution but also its commonality with its counterparts in other languages. The fitting of polysemy distribution of Mongolian indicates that the right truncated modified Zipf-Alekseev distribution gives the best result, and there are common characteristics between Mongolian and other languages in terms of polysemy.

Keywords: Mongolian · Polysemy distribution · Right truncated modified Zipf-Alekseev distribution · Waring distribution · Quantitative linguistics

1 Introduction

Polysemy refers to the number of meanings of a word in the dictionary (Liu 2017). For example, the Mongolian verb " ᠲᠠᠲᠠᠬᠤ (pull)" has 24 meanings, and thus its polysemy is 24. By reviewing the former research, we find that the polysemy distribution of German (Köhler 2005), French (Nemcová 2010), English (Liu 2011) and Chinese (Wang 2014; Gao 2014; Liu and Liu 2017) has already been studied.

Specifically, according to the previous studies, we have known that the distribution of polysemy of verbs fits to the positive negative binomial distribution. Moreover, it is confirmed that the distribution of polysemy of verbs fits the positive negative binomial distribution in German (Köhler 2005), English (Liu 2011) and Chinese (Gao 2014). In addition, in Chinese (Liu and Liu 2017), based on HowNet, it is confirmed that the polysemy distribution and mixed geometric-logarithmic distribution fit best, and the polysemy distribution of Chinese noun abides by the right truncated modified Zipf-Alekseev distribution. Based on the studies of the polysemy distribution of the whole dictionary, the polysemy distribution of French (Nemcová 2010) and Chinese (Wang 2014) fits Waring distribution. So, what is the polysemy distribution of Mongolian as a typical adhesive language? Does it share a common feature with other languages?

Most of the researches on Mongolian polysemy are qualitative researches. In recent years, there have been studies conducted to find out the statistics of the part-of-speech of Mongolian polysemy (Sarana 2012), the research on the meaning differentiation rules

M. Liu et al. (Eds.): CLSW 2020, LNAI 12278, pp. 473–481, 2021.
https://doi.org/10.1007/978-3-030-81197-6_40

of Mongolian polysemous verbs (Hureltux 2016) and the research on the construction of Mongolian polysemy database (Qi 2009). However, from the perspective of quantitative linguistics, there were not any in-depth and systematic studies. Therefore, it is meaningful and is worth studying of the distribution of Mongolian polysemy with the method of quantitative linguistics.

2 Research Materials

The materials used in this study are all sourced from *Mongolian Dictionary*. *Mongolian Dictionary* is a modern Mongolian dictionary compiled by the compilation group of *Mongolian Dictionary*, which was published by Inner Mongolia People's Publishing House in November 1997. After more than 40 years of work on dictionary compilation, the dictionary has contained more than 69,000 entries, including the written pronunciation, the spoken pronunciation of Cyrillic transcription, parts of speech, the category of grammatical inflection, paraphrases, examples and other information of the entries. *Mongolian Dictionary* is the result of painstaking efforts of many senior experts and scholars in the Mongolian academic circle, and as one of the most popular authoritative dictionaries, it is commonly used in the life of Mongolian language both at home and abroad. We have electronized *Mongolian Dictionary* and made it as a database to provide data support for subsequent processing such as comparison and statistical analysis. The database of *Mongolian Dictionary* made use of Microsoft Office Access 2007 to store relevant information such as entry, written pronunciation, paraphrase, etc. The dictionary database contains 69,302 entries, including 41,129 main entries and 28,173 sub-entries. The main entries are single words, and the secondary entries are phrases. This paper takes the main entries as the research object. Among them, 40,349 main entries are used for research while 780 unexplained main entries (mostly onomatopoeic words) are deleted (leaving paraphrases under the secondary entry). Before analysis, the corpus of *Mongolian Dictionary* has been proofread. Errors such as duplication, missing, misspelling and wrong matches of the main entries have been corrected. And it can assist in further statistical analysis as a good corpus.

3 Research Results

3.1 Polysemy Distribution of Mongolian

Based on the *Mongolian Dictionary*, we have classified and counted the number of words in terms of polysemy to obtain the polysemy distribution data. In order to investigate the polysemy distribution, we have used the distribution fitting software Altmann fitter 3.1 to fit the data of polysemy distribution. The results show that Mongolian polysemy distribution fits to 47 distribution models of Altmann fitter 3.1, and the specific models and their parameter values are shown in Table 1. It can be seen from Table 1 that the fitting results of these 47 models are very good, with 0.9976 as the lowest value of the determination coefficient R^2 and 1.0000 the highest ($R^2 > 0.90$, very good; $R^2 > 0.80$, good; $R^2 > 0.75$, acceptable; $R^2 < 0.75$, unacceptable;) and 0.0007 as the lowest value of C and 0.0093 the highest ($C < 0.01$, very good; $C < 0.02$, good). Among the 47

Table 1. Altmann fitter fitting results of Mongolian polysemy distribution.

Distribution model	X^2	$P(X^2)$	C	DF	R^2
Right truncated modified Zipf-Alekseev (a, b; n = x-max, α fixed)	29.12	0.0022	0.0007	11	1.0000
Waring (b, n)	46.30	0.0000	0.0011	11	1.0000
Mixed negative binomial (k, p1, p2, α)	48.14	0.0000	0.0012	14	0.9999
Right truncated Waring (b, n)	47.16	0.0000	0.0012	11	1.0000
Inverse Polya (a, k, p)	50.50	0.0000	0.0013	11	1.0000
Positive Yule (b)	68.83	0.0000	0.0017	14	0.9996
Extended logarithmic (θ, α)	74.95	0.0000	0.0019	9	1.0000
Zipf-Mandelbrot (a, b; n = x-max)	81.76	0.0000	0.0020	13	0.9996
Beta-Pascal (m, n, k)	99.24	0.0000	0.0025	8	1.0000
Mixed logarithmic (q, θ, α)	130.25	0.0000	0.0032	7	0.9999
Jain-Consul-logarithmic (α, β)	130.23	0.0000	0.0032	9	0.9999
Mixed geometric (q1, q2, α)	133.08	0.0000	0.0033	7	0.9999
Consul-Jain-Poisson (a, b)	131.67	0.0000	0.0033	9	0.9998
Jain-Consul-negative binomial (n, α, β)	131.82	0.0000	0.0033	7	0.9999
Shenton-Skees-logarithmic (a, b, θ)	136.37	0.0000	0.0034	8	0.9998
Bissinger-negative binomial (k, p)	144.34	0.0000	0.0036	9	0.9998
Consul (n, p)	146.88	0.0000	0.0036	9	0.9995
Borel (a)	165.34	0.0000	0.0041	10	0.9992
Borel-Tanner (a; r = x-min)	165.34	0.0000	0.0041	9	0.9992
Mixed geometric-logarithmic (q, θ, α)	168.69	0.0000	0.0042	7	0.9999
Left truncated negative binomial (k, p; L = x-min)	175.84	0.0000	0.0044	7	0.9998
Dacey-negative binomial (k, p, α)	175.92	0.0000	0.0044	7	0.9998
Right truncated logarithmic (q; R = x-max)	187.07	0.0000	0.0046	8	0.9998
Logarithmic (q)	187.07	0.0000	0.0046	9	0.9998
Doubly truncated logarithmic (q; L = x-min, R = x-max)	187.07	0.0000	0.0046	7	0.9998
Stirling Type 1 (θ; n = x-min)	187.07	0.0000	0.0046	8	0.9998
Good-Engen (p)	189.34	0.0000	0.0047	11	0.9995
Doubly truncated negative binomial (k, p; L = x-min, R = x-max)	191.41	0.0000	0.0047	6	0.9998
Right truncated negative binomial (k, p; R = x-max)	215.01	0.0000	0.0053	7	0.9996
Gegenbauer (a, b, k)	214.33	0.0000	0.0053	7	0.9997

(*continued*)

Table 1. (*continued*)

Distribution model	X^2	$P(X^2)$	C	DF	R^2
Consul-Mittal-binomial with 3 parameters (n, p, θ)	217.55	0.0000	0.0054	6	0.9999
Positive negative binomial (k, p)	222.87	0.0000	0.0055	7	0.9999
Bissinger-geometric (p)	220.01	0.0000	0.0055	8	0.9999
Lingappaiah-Poisson (a, b, α)	232.94	0.0000	0.0058	6	0.9999
Feller-Arley (k, p1, p2)	233.73	0.0000	0.0058	6	0.9999
Negative binomial (k, p)	292.20	0.0000	0.0072	8	0.9984
Polya (s, p, n)	291.44	0.0000	0.0072	6	0.9996
Negative binomial-logarithmic (k, p, θ)	293.55	0.0000	0.0073	6	0.9997
Good (a, p)	301.62	0.0000	0.0075	7	0.9990
Right truncated Good (a, p)	301.62	0.0000	0.0075	7	0.9990
Geometric-negative binomial (k, p, θ)	313.69	0.0000	0.0078	6	0.9996
Mixed Poisson (a, b, α)	312.81	0.0000	0.0078	5	0.9997
Negative hypergeometric (K, M, n)	319.30	0.0000	0.0079	6	0.9992
Darwin (B; M = x-max)	358.82	0.0000	0.0089	7	0.9992
Non-central negative binomial (a, k, p)	359.75	0.0000	0.0089	5	0.9999
Geometric-geometric (p, θ)	372.82	0.0000	0.0092	7	0.9994
Digamma (a, c)	375.30	0.0000	0.0093	18	0.9976

fitting distribution models, the best one is the right truncated modified Zipf-Alekseev distribution model. We show the fitting results of the polysemy distribution and this model in Fig. 1, and Table 2 shows word number and theoretical value of polysemy distribution as well as the specific values of each parameter.

From Fig. 1 and Table 2, we can see that the polysemy of Mongolian is ranged from 1 to 24, and the average polysemy is 1.36 per word. In polysemy distribution, words with one and two polysemy account for 92.84% of the dictionary. More than half of the words involving one polysemy account for 75.67%, while words involving two polysemy account for 17.17%, and words involving more than two polysemy account for 7.16%. This means that the lexicon is mainly composed of words with fewer meanings. It can be seen that with the increase of polysemy, the number of words decreases rapidly.

3.2 The Commonality Between Mongolian and Other Languages in Terms of Polysemy Distribution

After analyzing the overall characteristics of polysemy distribution, we have conducted a study on whether there is a commonality between Mongolian and other languages in terms of polysemy. According to previous studies, we know that the French (Nemcová

Table 2. Fitting right truncated modified Zipf-Alekseev to Mongolian polysemy.

Polysemy	Number of words	Right truncated modified Zipf-Alekseev
1	30532	30532.00
2	6926	6924.73
3	1878	1833.07
4	614	609.97
5	207	237.46
6	100	103.63
7	38	49.35
8	21	25.17
9	9	13.58
10	7	7.68
11	4	4.51
12	1	2.75
13	2	1.72
14	2	1.10
15	0	0.73
16	3	0.49
17	3	0.33
18	0	0.23
19	1	0.16
20	0	0.12
21	0	0.08
22	0	0.06
23	0	0.05
24	1	0.03
$a = 1.8643$, $b = 0.7890$, $\alpha = 0.7567$, $X^2 = 29.1201$, $P(X^2) = 0.0022$, $DF = 11$, $C = 0.0007$, $R^2 = 1.0000$		

2010) and Chinese (Wang 2014) polysemy distribution fits with the Waring distribution. Therefore, in order to study the commonalities between Mongolian and other languages, we use distribution fitting software Altmann-fitter 3.1 to fit Mongolian polysemy distribution into Waring distribution (see Table 3 and Fig. 2).

It can be seen from the values of the coefficient of determination R^2 ($1.00 > 0.90$) and C value ($0.0011 < 0.01$) that the Waring distribution fits Mongolian polysemy distribution very well, which is only the second in the right truncated modified Zipf-Alekseev distribution. The results demonstrate that the distribution of Mongolian polysemy accord with the Waring distribution, and there are commonalities between Mongolian and other

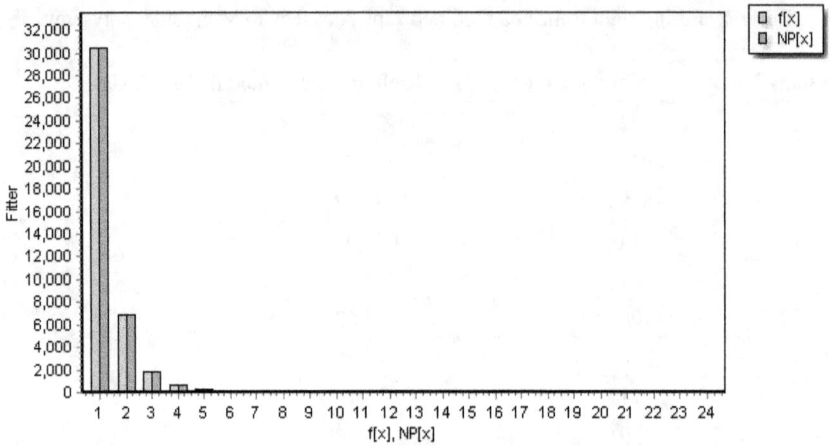

Fig. 1. Fitting right truncated modified Zipf-Alekseev to Mongolian polysemy.

languages such as French and Chinese. According to the parameters $X^2 = 336.1304$, $P(X^2) = 0.0000$, $DF = 14$, $C = 0.0066$, $R^2 = 0.9995$ from Chinese polysemy distribution fitting, we know that Mongolian polysemy distribution fits even better, compared with the fitting result of Chinese polysemy distribution (Wang 2014). Since only $X^2{}_{12} = 16.72$, $P = 0.16$ were provided in the parameters from French polysemy distribution fitting, no comparison can be made here.

Table 3. Fitting Waring distribution to Mongolian polysemy.

Polysemy	Number of words	Waring
1	30532	30532.00
2	6926	6792.09
3	1878	1927.70
4	614	648.08
5	207	247.20
6	100	104.05
7	38	47.41
8	21	23.06
9	9	11.85
10	7	6.38
11	4	3.58
12	1	2.08
13	2	1.25

(continued)

Table 3. (*continued*)

Polysemy	Number of words	Waring
14	2	0.77
15	0	0.49
16	3	0.31
17	3	0.21
18	0	0.14
19	1	0.10
20	0	0.07
21	0	0.05
22	0	0.03
23	0	0.02
24	1	0.07
$b = 8.0758$, $n = 2.5966$, $X^2 = 46.3031$, $P(X^2) = 0.0000$, $DF = 11$, $C = 0.0011$, $R^2 = 1.0000$		

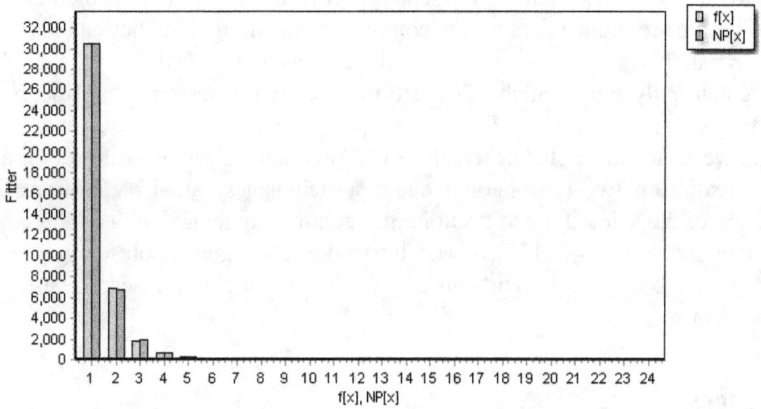

Fig. 2. Fitting Waring distribution to Mongolian polysemy.

4 Conclusion

In this paper, 40,349 main entries in *Mongolian Dictionary* are categorized and counted by its polysemy, the Mongolian polysemy distribution is analyzed. The commonality between Mongolian polysemy and polysemy in other languages is discussed. Firstly, the distribution fitting software Altmann-fitter 3.1 is used to fit the Mongolian polysemy distribution. It is found that the distribution of Mongolian polysemy fits into 47 kinds of distributions in Altmann-fitter 3.1. Among them, the right truncated modified Zipf-Alekseev distribution is fitted best. Secondly, in order to explore the commonality between Mongolian polysemy distribution and the related distribution in other languages, we used distribution fitting software Altmann-fitter 3.1 to fit Mongolian polysemy distribution into Waring distribution. The results show that the distribution of Mongolian polysemy also accords with Waring distribution, and the commonalities with other languages are observed.

The current study of Mongolian polysemy distribution can be applied to natural language processing and lexical semantic research. In terms of natural language processing, Mongolian polysemy distribution can provide more precise linguistic data for syntactic and semantic analysis, that is, polysemy information for each word. Polysemy of Mongolian helps to observe whether words, phrases and sentences are ambiguous, thereby improving the efficiency of corpus processing, text semantic understanding and computer automatic recognition of polysemous words in machine translation. In terms of lexical semantic research, one can compare the existing Mongolian polysemy distribution with the historical distribution of Mongolian polysemy at different times in history and observe the historical changes of polysemy of a specific word or the entire Mongolian vocabulary polysemy distribution. The Mongolian polysemy distribution can be applied to analyze the diachronic study of the semantic changes of Mongolian vocabulary. Similarly, based on the existing Mongolian polysemy and its distribution, one can compare polysemy and its distribution of the vocabulary contained in different dictionaries in the same period to observe the collection of vocabulary polysemy. Such comparison can be applied to the synchronic study of the semantic coverage of Mongolian dictionaries and can also provide reference data for the compilation of Mongolian dictionaries.

In general, this paper uses quantitative linguistics method to discuss the distribution of Mongolian polysemy, which is new to the previous research on polysemy which adopted the qualitative method. From a new perspective of quantitative linguistics, not only can we better understand the rules of Mongolian polysemy distribution and the common features between Mongolian and other languages, but also can we realize the importance of the scientific and accurate research on language studies. Therefore, we believe that this research will help enrich the studies of language polysemy distribution, and it will also be of great significance to promote Mongolian research based on the philosophy of science.

References

Compilation Group of Mongolian Dictionary: Mongolian Dictionary. Inner Mongolia People's Publishing House, Hohhot (1997)

Gao, S., Zhang, H., Liu, H.: Synergetic properties of Chinese verb valency. J. Quant. Linguist. **21**(1), 1–21 (2014)

Köhler, R.: Quantitative Untersuchungen zur Valenz deutscher verbena. Glottometrics **9**, 13–20 (2005)

Liu, H.: Quantitative properties of English verb valency. J. Quant. Linguist. **18**(3), 207–233 (2011)

Liu, H.: An Introduction to Quantitative Linguistics, 1st edn. The Commercial Press, Beijing (2017)

Liu, Y., Liu, P.: Quantitative analysis of synergetic properties in Chinese nouns. In: Wu, Y., Hong, J., Su, Q. (eds.) Chinese Lexical Semantics, pp. 341–348 (2017)

Nemcová, E., Rensinghoff, S.: On word length and polysemy in French. Glottotheory **1**, 83–88 (2010)

Qi, H.: The Construction of Mongolian Polysemy Database. Inner Mongolia University Master Thesis (2009)

Sarana, D.: A quantitative study on the part-of-speech distribution of Mongolian polysemy. Mongolian Stud. China **40**(6), 157–162 (2012)

Sarana, D., Hureltux: An econometric analysis of meaning differentiation rule of polysemous verbs in modern Mongolian Language. J. Minzu Univ. China (Philos. Soc. Sci. Edn.) **43**(4), 130–136 (2016)

Wang, L.: Quantitative and Synergetic Studies on Lexical Units in Chinese. der Universität Trier Doctoral Thesis (2014)

Generating Dialogue Responses with Latent Lexical Meaning

Xikai Liu, Hongfei Lin(✉), Yumeng Li, Yuqi Ren, and Bo Xu

Dalian University of Technology, Dalian, China
{ws_lxk,lym199286,ryq13}@mail.dlut.edu.cn, {hflin, xubo}@dlut.edu.cn

Abstract. Response generation is an important component of a dialogue system. One major drawback of the existing models is that they tend to generate generic responses, and specific and latent lexical information is ignored. However, in the real world, latent lexical meaning may better reflect the user's background, which may be used to make our responses more coherent and interesting. To solve this problem, we propose a response generation model with latent lexical meaning. The model generates responses based on latent lexical meaning, which is identified by a multi-head attention mechanism, and is aligned with the output responses by lexical representation to guarantee the diversity and coherence of responses. Extensive experiments are conducted on the Sina Weibo Corpus. Experimental results show that the generated responses are evaluated as more interactive and diverse in both automatic and human evaluations.

Keywords: Latent lexical meaning · Response generation · Multi-head attention

1 Introduction

Open domain dialogue systems aim to allow machines to communicate with people in daily life. This means they need to understand and generate meaningful and coherent dialogue responses like humans do. Many research projects have applied the generation model to open domain dialogue systems. However, the generation model tends to generate "safe responses" [1]. The model ignores the latent lexical meaning of the query. This is one of the burning issues of this model. Thus, the model is difficult to model the user and give a specific reply according to the user's attributes. For example, if we input "I'm going to college" and "I'm going to retire", the model will respond to both of them with "I'm glad to hear that." But obviously, there are better ways to respond manually, such as "you have to study hard" for the first sentence and "you don't look that old!" for the second. For the two examples, the second type of response pays more attention to the latent lexical meaning and uses it to model the user. Therefore, how to use latent lexical meaning for user modeling deserves a deeper exploration and a study.

The latent meaning of words can often correspond to the user's profiles of the query. In the existing research, using latent lexical meaning to enhance the quality of response generation is not yet explored. It is complex to use user's attributes because the model is

M. Liu et al. (Eds.): CLSW 2020, LNAI 12278, pp. 482–490, 2021.
https://doi.org/10.1007/978-3-030-81197-6_41

unable to know the user's attributes in advance and to generate corresponding responses accordingly. For instance, the first example contains the age attribute embedded in the lexical item ‘college’ and ‘retire’. It is therefore necessary for the model to analyze such latent lexical meaning contained in the query. At the same time, the model has to align the latent attributes with the output responses, which need to obtain the related words corresponding to different attributes.

To solve this problem, in this paper, we propose a response generation model with latent lexical meaning (LLM) by using latent lexical meaning to model users' attributes and improve the quality of responses. Gender, age, region of origin are the user attributes which are representatives and well-studied. More importantly, these attributes can often indicate the basic information about a person. We first propose a multi-head attention framework to predict the above user attributes. These attributes can be regarded as the latent lexical meaning of the query. To align the attributes with the output responses, we integrate the attributes into the generation model using a lexical representation matrix. The matrix obtains the related words corresponding to different attributes and aligns the information of the attributes with the output responses through model training.

Our contributions are summarized as below:

1. We develop a novel response generation model LLM, which uses latent lexical meaning to model users' profiles and improve the quality of responses.
2. Experimental results on automatic evaluation and human evaluation show that LLM significantly outperforms other models. It generates more diverse responses.

2 Related Works

Making a chatbot that can naturally converse with humans has been attracting more and more research interests recently. The chatbot should be able to converse with human-beings consistently and naturally on open-domain topics. Early research on chatbots used manual features and rules to generate responses. But this is time-consuming and ineffective [2]. With the development of machine learning and deep learning, many works [2] tried to use data-driven methods to make a more effective chatbot. Recently, a common approach to making a chatbot is to adopt an encoder-decoder framework [3]. On this foundation, the attention mechanism has been proved successful in dialogue generation tasks due to the enhancement of the accuracy of the generated responses. But the model still encounters the problem of “safe response” [1]. Many works tried to improve model construction to solve this problem. [4] proposed a two-step method that generated keywords and the final response in two steps. [5] extracted multiple references for response generation. Such approaches improved the model structure but were still limited by the training dataset. [6] introduced topic knowledge into the task to generate topic-aware responses.

All these works have greatly improved the quality of response generation in dialogues. However, none of these works has considered the analysis of latent lexical meaning in the query. We try to use latent lexical meaning to improve the capability of the chatbot.

3 Method

In this work, we propose a response generation model with latent lexical meaning (LLM) that models the users' profiles and then generates the response based on the information of the profiles. In this way, our model can analyze the query and the latent lexical meaning which can be used to model the user. The step that's follow such analysis is that the model will align the users' profiles with words for response to enhance the quality of response generation. The framework of the proposed model is illustrated in Fig. 1. We propose a neural-based model for predicting users' profiles. To align the attributes in the profiles with the responses, we indicate the probability of the attributes in a trainable lexical matrix to get the final latent representation. In the process of response generation, we add the attributes to each step of the generation to determine whether the users' attributes need to be used in each step.

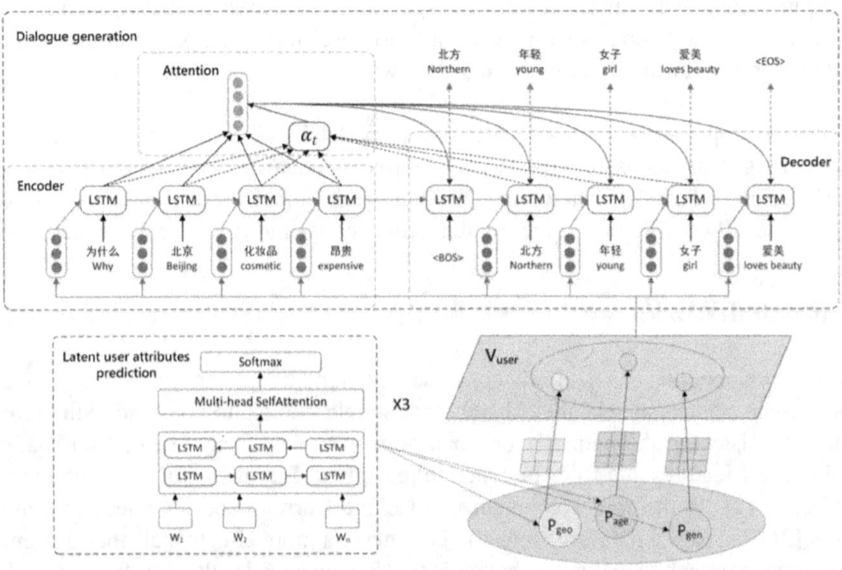

Fig. 1. The structure of the response generative model with latent lexical meaning (LLM). A multi-head attention-based model is used for predicting users' latent attributes. This generation model is for generating responses with attribute information. The latent lexical representation aligns latent attributes with the output responses.

3.1 Prediction of Users' Profiles by Latent Lexical Meaning

The latent meaning of words usually represents the users' profiles, which can help the model analyze specific information in the users' query. Meanwhile, this information is the fundamental difference among different sentences. Because our dialogue dataset cannot provide enough relevant information for predication of users' profiles, we need to use a pre-trained model to strengthen our model. We use the SMP CUP 2016 competition dataset to pre-train our model. The SMP CUP 2016 competition dataset is from Sina Weibo. The users' personal information and publicized texts are available. Three representative attributes are selected as training labels: Region (South China, North China, Northeast China, East China, Central China, Northwest China, Southwest China, unrecognized), Age (0–20, 20–30, 30–40, 40+, unrecognized) and Gender (Female or Male). Certainly, any other dataset built for prediction of users' profiles could also be used in our task.

For each query q = $\{w_1, w_2, \ldots, w_n\}$, where w_i denotes the vector for the i-th word, we use bi-directional Long Short-Term Memory (Bi-LSTM) to encode the vector representation of each word. For this task, different words in the sentence have different contributions, so different weights need to be set. However, most traditional attention mechanisms only focused on a specific representation, which only reflected single sentence semantic information of the input sequence. In fact, understanding a sentence requires learning of semantic information from multiple sources, especially when it comes to a long sentence. To learn semantic information from different sources, we get different sentence representations by multiple attentions. In this paper, we propose a multi-head self-attention mechanism for predicting users' attributes to obtain multiple sources of semantic information in a sentence. The representation vector $m_{i,j}$ of the j-th word learned by the i-th attention is computed by a weighted summation of all word representation vectors in H, which is formulated as follows:

$$\hat{\beta}^i_{j,k} = w_j^T U_i w_k \tag{1}$$

$$\beta^i_{j,k} = \frac{\exp\left(\hat{\beta}^i_{j,k}\right)}{\sum_{m=1}^{M} \hat{\beta}^i_{j,m}} \tag{2}$$

$$m_{i,j} = W_i\left(\sum_{m=1}^{M} \beta^i_{j,m} w_m\right) \tag{3}$$

where $\hat{\beta}^i_{j,k}$ denotes the relative importance of the chunk between the j-th and k-th words, w_k denotes the k-th word.

The SoftMax function is then used for normalization. U_i and W_i are the parameters of the i-th self-attention. In this way, various sentence representations are obtained by a multi-head self-attention mechanism which employs the attention mechanism to learn different semantic information on the same input for many times. Then, the multi-head representations of the j-th word are concatenated as the final sentence representation. In this way, the users' attributes of the query are obtained.

3.2 Response Generation with Predicted Attributes

Seq2Seq is a conditional language model that operates on query-response pairs (Q, R), where Q = $\{q_1, q_2, \ldots, q_n\}$ and R = $\{r_1, r_2, \ldots, r_n\}$ are sentences consisting of sequences of words. Although there are the variants in Seq2Seq models, the models usually consist of two major components which are encoder and decoder. We adopt this setup and use an encoder to embed a query message.

On the basis of the traditional model, we hope to add information of users attributes. So, the first step of the model is to determine how to express the attribute information needed. In the step of predicting attributes, we train a classification model to classify the attributes of the users. After the model has used SoftMax to predict the labels, the model gets a probability vector which includes the probability that each tag is assigned. But the probability cannot be aligned with the output responses. So, we use an attribute matrix to represent the users' attributes. The matrix is randomly initialized and updated by backpropagation. We use probability to weigh the attributes matrix so as to get the vector of users' attributes. The formula is as follows:

$$v_{user} = \left[P_{reg} \cdot M_{reg} \cdot P_{gen} \cdot M_{gen}, P_{age} \cdot M_{age}\right] \tag{4}$$

where M_{reg} is a matrix denoting the attribute of the users' region of origin and P_{reg} is the probability of the user's region of origin, which is predicted by the classification model. M_{gen}, P_{gen}, M_{age}, P_{age} are also defined similarly. When we use the model in actual runs, we may obtain users' attributes sometimes, or we may hope to specify the information of users' attributes. To do these, we can generate directional responses related to an attribute. In order to deal with such particularities in some tasks, we set the values of P_{reg}, P_{gen}, P_{age} manually to forcibly output relevant attribute information in model testing.

We integrate vectors of users' attribute into the encoder and apply an LSTM cell to process at each time step. The formulas are as follows:

$$\begin{bmatrix} i_t \\ f_t \\ o_t \\ l_t \end{bmatrix} = \begin{bmatrix} sigmoid \\ sigmoid \\ sigmoid \\ tanh \end{bmatrix} W \cdot \begin{bmatrix} x_t \\ h_{t-1} \\ v_{user} \end{bmatrix} \tag{5}$$

$$c_t = f_t \cdot c_{t-1} + i_t \cdot l_t \tag{6}$$

$$h_t = o_t \cdot \tanh(c_t) \tag{7}$$

where i_t, f_t, o_t denote the input gate, forget gate and output gate for the LSTM, c_t denotes the memory cell and h_t denotes the hidden state as well as the output. We integrate v_{user} into every hidden layer to help the model get the information of the attribute. The advantage of this is that the model can learn at every step whether the information of the attributes should be used.

Similar to the encoder, the decoder network uses an LSTM layer with the same structure but different parameters as the encoder network. An attention mechanism is used to capture word-level dependency.

4 Experiments

We have conducted experiments to prove that our response generation model could generate quality responses. In the remaining sub-sections, we will show our verification of whether our method can improve the effectiveness of the dialogue task. Since classification model is not the focus in our research and many works have proved that this type of structure can achieve good results in text categorization, we have not carried out experiments to verify the effectiveness of the classification model alone. In a competition (SMP2016, the competition where the data comes from), the final comprehensive score of our classification model was 0.700, which ranked fifth and was an acceptable result.

4.1 Dataset and Experimental Details

For the response generation task, we have conducted experiments on the Sina Weibo Corpus, a large dataset built by [7]. The corpus has contained about three million dialog data from Weibo queries and responses. We have obtained the dataset and split the dataset into three parts: 3M for training, 50K for validation and 10K for testing. On one hand, this dataset has contained Weibo texts produced by people of different ages, genders and from different regions. The diversity of texts was suitable for our task. Moreover, the large scale of the dataset could better reflect the effectiveness of the response generation model.

In order to keep our model comparable with the state of the art, we have set most parameter values of our model to the same values as in [8]. We have built our vocabulary dictionary based on the most frequent 40000 words. Beam search has been used as the sampling strategy. Based on the performance on the validation set, we have set the learning rate at 0.001 and set the beam search size to 10. We have used the Adam algorithm to update the parameters and trained the model in batches with a batch size of 128. The dimension of the attribute matrix is 30. The hyper-parameter λ for information control has been set at 1.5 based on the performance on the validation set.

4.2 Evaluation Metrics and Baselines

In this work, we have used five kinds of automatic metrics for quantitative evaluation and human metrics for judging the quality of the responses.

BLEU: BLEU has been shown to correlate reasonably well with human judgment on the response generation task. We have used BLEU-1 to reflect the adequacy of dialogue generation and BLEU-2 to reflect the fluency of dialogue generation.

Word embedding score (wemb): Word embedding score [9] has been used to evaluate semantic similarities between responses generated by the model and real responses.

Distinct-1(dis-1)&Distinct-2(dis-2): Dis-1 and dis-2 [1] have been applied to count numbers of distinct unigrams and bigrams in the generated responses.

Human evaluation: Besides the automatic evaluation methods, we have made use of human evaluation to evaluate the validity of the model. We have followed protocols defined by [1], employing crowed-sourced judges to evaluate a random sample of 100 items. For measuring the quality of responses, we have presented 2 metrics: fluency and coherence. Fluency indicates whether the sentence is without grammatical mistakes. Coherence indicates whether the generated response is relevant to the query. We have used the geometric mean (G-score) of these metrics as the final human evaluation metric.

To evaluate our model, we have compared our model with the following baseline systems.

S2S: S2S is a seq2seq model following the same implementation as in [3]. The seq2seq model serves as a baseline. We have used beam search with beam size of 15 to decode when testing.

S2SATT: S2SATT is a seq2seq model with an attention mechanism. Attention mechanism has been proved to be effective in language models. We have used an attention mechanism with LuongAttention [10] for decoding. And we have used Long short-term memory units for the model.

MMI: MMI is a bidirectional-MMI decoder based on the Seq2Seq model as in [1], which has shown better performance over the anti-LM model. We have followed the suggestion and set the parameter λ at 0.5/150 candidates per context sampled for re-ranking.

TA-S2S: TA-S2S is the topic-aware sequence-to-sequence model proposed in [6]. The model is also a knowledge enhanced method that introduces topic information into the task.

Table 1. Automatic evaluation results and human evaluation results of the LLM model and other baseline models.

Model	BLEU-1	BLEU-2	wemb	dis-1	dis-2	Fluency	Coherence	G-score
S2S	70.32	33.26	84.04	0.102	0.191	6.38	4.62	5.43
S2SATT	75.26	41.66	85.10	0.142	0.274	6.20	4.18	5.09
MMI	75.73	44.01	85.23	0.158	0.301	5.87	**5.38**	5.62
TA-S2S	76.01	45.78	85.43	0.226	0.426	6.61	5.12	5.82
LLM	**76.05**	**46.35**	**85.92**	**0.238**	**0.466**	**6.84**	5.15	**5.93**

4.3 Results

The results of automatic evaluation are shown in Table 1. Our model has achieved higher BLEU-1, BLEU-2 and wemb scores than other models. This shows our model can generate more fluent and accurate sentences. This is encouraging because keeping the coherence of the sentence while adding more information to the model is important. If the sentence cannot maintain coherence, it is useless to generate more diverse sentences. To be specific, if information added to sentences fails to correspond to the input sentences reasonably, diversity is useless no matter how. With the encouraging results mentioned above, we can carry out our proposed work deeper and further as our model can generate more fluent and accurate sentences. According to the distinct-1 and distinct-2 metrics, our model has achieved significantly better diversity compared with other models.

The results have shown that our approach can be more aware of the users by integrating users' attributes into the model. Since our approach can add the user attribute information to the model, we can generate better sentences and more diverse sentences.

This is proved by the distinct-1 and distinct-2 results. To guarantee the coherence of the generated results, these two results can be significantly improved, which in return can prove that our model can play a good role in dialogue generation.

In the task of dialogue generation, it is sometimes not enough to judge a model only by automatic evaluation. So, we have used human evaluation to analyze the model. Results of human evaluation in Table 1 have shown that our model has a high score in the human evaluation. LLM is the best performer in Fluency score. This shows that our model can generate the most fluent response. This is really important because this proves that our model does not lose fluency after adding users' attributes. Note that our model is worse than MMI on coherence. This means that the generated results of our model, compared to that of MMI, are less consistent with the input texts. The reason is that MMI adds mutual information to the model, which increases the correlation between responses and input texts. However, our model uses additional information, and so there may be situations where the generated results are not consistent with the input texts.

5 Conclusion and Future Work

In this work, we have proposed a response generation model with latent lexical meaning (LLM) to generate more diverse responses. The model analyzes the latent lexical meaning of the query to predict users' attributes and accordingly apply a generation model to generate responses with representation of the attributes. More importantly, our model is extensible, which means that users can tailor their specific attributes to their needs. Results have shown that our model can generate more appropriate responses, with users' attributes, which are thereby more likely to be specific and diverse.

The model can generate responses using users' attributes trained in the training process. Due to the personal privacy policy, obtaining users' attributes are constrained by social media. To get around this, the model can improve its effectiveness by integrating multimodal methods to construct users' attributes, such as photos, videos, etc.

Acknowledgments. This research is supported by the National Natural Science Foundation of China (No: 61632011, 61772103, 61702080).

References

1. Li, J., Galley, M., Brockett, C., et al.: A diversity-promoting objective function for neural conversation models. arXiv preprint arXiv:1510.03055 (2015)
2. Ritter, A., Cherry, C., Dolan, W.B.: Data-driven response generation in social media. In: Proceedings of the Conference on Empirical Methods in Natural Language Processing, pp. 583–593. Association for Computational Linguistics (2011)
3. Sutskever, I., Vinyals, O., Le, Q.V.: Sequence to sequence learning with neural networks. In: Advances in Neural Information Processing Systems, pp. 3104–3112 (2014)
4. Mou, L., Song, Y., Yan, R., et al.: Sequence to backward and forward sequences: a content-introducing approach to generative short-text conversation. arXiv preprint arXiv:1607.00970 (2016)

5. Qiu, L., Li, J., Bi, W., et al.: Are training samples correlated? Learning to generate dialogue responses with multiple references. In: Proceedings of the 57th Annual Meeting of the Association for Computational Linguistics, pp. 3826–3835 (2019)
6. Xing, C., Wu, W., Wu, Y., et al.: Topic aware neural response generation. In: 31st AAAI Conference on Artificial Intelligence (2017)
7. Shang, L., Lu, Z., Li, H.: Neural responding machine for short-text conversation. arXiv preprint arXiv:1503.02364 (2015)
8. Serban, I.V., Klinger, T., Tesauro, G., et al.: Multiresolution recurrent neural networks: an application to dialogue response generation. In: 31st AAAI Conference on Artificial Intelligence (2017)
9. Serban, I.V., Sordoni, A., Bengio, Y., et al.: Building end-to-end dialogue systems using generative hierarchical neural network models. In: 30th AAAI Conference on Artificial Intelligence (2016)
10. Luong, M.T., Pham, H., Manning, C.D.: Effective approaches to attention-based neural machine translation. arXiv preprint arXiv:1508.04025 (2015)

The Effect of Word Frequency and Position-in-Utterance in Mandarin Speech Errors: A Connectionist Model of Speech Production

I.-Ping Wan[1](✉) and Marc Allassonnière-Tang[2]

[1] Research Center for Mind, Brain and Learning, Graduate Institute of Linguistics, National Chengchi University, Taipei, Republic of China
ipwan@g.nccu.edu.tw

[2] Lab Dynamics of Language UMR 5596, CNRS/University of Lyon, Lyon, France
marc.tang@univ-lyon2.fr

Abstract. The connectionist model of speech processing infers that word frequency and position-in-utterance play a major role in the occurrence of speech errors. First, words that are not frequently used are more likely to result in speech errors since they generally receive less activation than frequently occurring words and require more activation to be chosen. Second, speech errors are more likely to occur near the end of utterances since, according to the given-before-new-principle, utterance-final words convey new information that has not yet been activated in the preceding context. The information of word frequency and position-in-utterance is extracted automatically from 382 utterances of a Mandarin speech error corpus and fed to generalized linear mixed models and a decision-tree based classifier. The results show that word frequency and position-in-utterance can predict the occurrence of speech errors with a performance over (but close to) the majority baseline. Therefore, additional information is required to improve the accuracy in the predictions.

Keywords: Speech errors · Mandarin · Frequency · Position-in-utterance

1 Introduction

This paper focuses on lexical errors, i.e., erroneous selections of lexical items that involve a meaningful morpheme or word. They typically occur when the 'lemmas' of semantically or phonologically appropriate candidates for lexical items are activated. For instance, when a speaker says *glass* instead of *cup*, or *book* instead of *cook*. The other types of errors that result in meaningless strings of phoneme (e.g., when a speaker says *perple* instead of *person* or *people*) and errors that originate from the surrounding context (e.g., when a speaker says *the glass is in the glass* instead of *the glass is in the fridge*) are excluded from the current study for theoretical and practical reasons. First, this study investigates the frequency of the target (i.e., the intended word) and the error

M. Liu et al. (Eds.): CLSW 2020, LNAI 12278, pp. 491–500, 2021.
https://doi.org/10.1007/978-3-030-81197-6_42

word in utterances. Thus, meaningless words must be excluded since they cannot be assigned a frequency in corpora. Second, context-induced errors and purely phonological errors are less relevant to the cognitive representation of language since the cause of context-induced errors is due to interference of the surface structure of language rather than its deep inner processing.

The main contributions of this paper are as follows. First, most of the literature focused on speech errors resulting in meaningless strings of phonemes due to their higher occurring frequency and relevance to phonology [1–4]. By focusing on lexical errors, this study provides another type of data that can verify the predictions made by language processing models at the semantic level. Second, previous studies found several tendencies predicted by language-processing models within corpora of lexical errors [5–10]. However, these results were mostly obtained from Germanic and Romance languages and may be subject to Galton's problem [11], i.e., the tendencies observed in speech errors may be language-family-specific rather than universal. This paper thus provides data on speech errors in Mandarin, which enhances linguistic diversity in the results. Last but not least, previous studies investigating speech processing models with speech errors seldom provide quantitative analysis on the predictive power of word frequency and position-in-utterance with regard to speech errors. A few studies known to the authors investigated quantitatively the negative correlation between word frequency and the probability of occurrence of speech errors [12]. Nevertheless, the effect of position-in-utterance was not considered simultaneously with word frequency in these studies. This paper also aims at filling this gap. The detailed research questions and hypotheses are developed in Sect. 2, while the data and analysis are explained in Sect. 3 and 4. The summary and conclusion are provided in Sect. 5.

2 Hypotheses and Research Questions

In parallel models of speech production, multiple levels of processing take place simultaneously [13–15]. This approach assumes that speech processing is not a serial motion but a simultaneous activation of distinct units of speech (e.g., phonemes, morphemes) represented as nodes that interact with each other in parallel across the semantic, word, and sound levels. An example is demonstrated in Fig. 1.

Assuming that a speaker intends to retrieve and convey the concept of *cat*, the semantic features related to the target word *cat* are activated and spread to neighbor nodes. For instance, the semantic feature *four-legs* is activated and the activation spreads to *cat* at the word level. However, the semantic feature *four-legs* also activates the words that share this feature, e.g., *dog* and *rat*. The activation of the word *cat* spreads at the same time to the phoneme level, e.g., to the phoneme [t]. The same process applies for other nodes that received activation, e.g., *rat* and *dog*. The activation of nodes is simultaneous and multidirectional. By way of illustration, the activation of *rat* by the feature *four-legs* can also spread back to the semantic level and activate other semantic features that were not activated by *cat,* e.g., *is grey.* The same logic applies between the word level and the sound level. Finally, the most highly activated word of the appropriate grammatical category is chosen.

Several predictions can be made only under models of the connectionist approach. First, word frequency is expected to play a significant role in predicting the occurrence

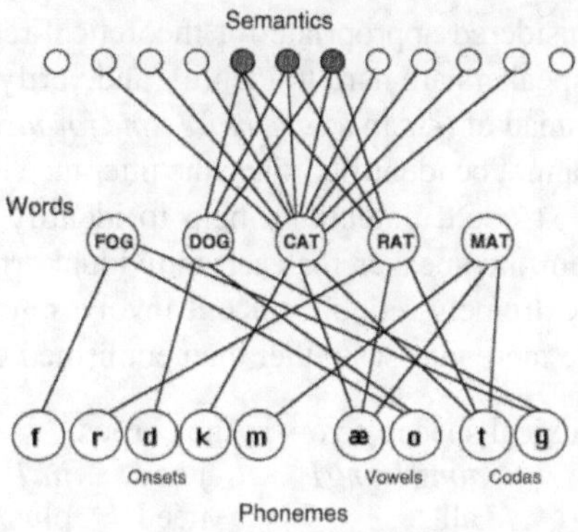

Fig. 1. A simplified overview of the connectionist approach [16].

of speech errors. Words that are frequently used are activated more frequently and are thus more likely to reach the required level of activation before potential erroneous candidates. Few previous studies investigated the effect of word frequency on speech errors in Mandarin [12]. However, the results were not analyzed in connection with the predictions of speech processing models. Second, the position-in-utterance of the target word is expected to play a significant role in predicting the occurrence of speech errors. This prediction relates to the given-before-new-principle in discourse analysis [17–22], which states that old information comes first in utterances, while new information comes later. Under such premise, speech errors are less expected at the beginning of an utterance since old information would already have received activation from the preceding context and would thus be less likely to result in erroneous activation. The opposite statement would be assumed for utterance-final words conveying new information that has not yet been activated. No previous studies known to the authors have investigated the predictive effect of position-in-utterance on speech errors.

3 Data

The 382 speech errors investigated in this study are from the same source as [12]. The speech errors are retrieved from a conversational corpus produced in a naturalistic setting by approximately 100 native speakers of Taiwan Mandarin between 1995 and 2009 [3, 23, 24]. Each audio file of conversational speech is automatically transcribed based on a Speech-to-Text software modeled by Taiwan AI Lab with an average accuracy rate of 70%. The entire transcript is then automatically segmented by the Academia Sinica word segmenter [25, 26]. Further editing and correction is made by two research assistants. Speech errors are identified based on repair (self-correction) initiated by the speakers. For example in (1a), the target *dong4ci2* 'verb' is erroneously replaced by *ming2ci2* 'noun'. However, the speaker immediately initiates repair with the intended target. In (1b), the target *na4* 'that' is erroneously replaced by *zhe4* 'this', and the speaker also initiates repair directly after the speech error. Separate analyses are made by research assistants working on the corpus and inconsistencies are resolved by analysis of the context and the phonetic realization of the words [7]. This method restricts the sample size of speech errors. However, it is considered appropriate for theoretical reasons. Errors that occur without repair from the speakers are hard to identify and verify. By way of illustration, if a speaker says *more* instead of *less* in *I have more time for myself* but does not initiate repair, the speech error cannot be identified since the utterance is perfectly grammatical. An analysis of the context could potentially help to identify errors of this type, but without confirmation from the speaker for each individual error (which is practically impossible and still not extremely reliable since it involves an off-line judgment from the speakers), it would be mere guesses rather than confirmed errors.

(1) Examples of lexical speech errors in the corpus

a. *zhe4xie1* *tong1tong1* *yao4* *jia1* *dan1shu4*
these all need plus singular
ming2ci2 ... *dong4ci2*
noun verb
'These must be used with singular nouns … singular verbs.'

b. *zhe4* ... *na4* *tian1* *zai4* *bian4lun4* *de0*
this … that day at debate DE
shi2hou4 […]
time
'This … that day during the debate […]'

Each speech error is annotated with the following information: target, error, preceding context, and following context. The corpus has 382 speech errors and includes 3022 words when considering the preceding and following words of each speech error. Due to the small size of the entire corpus (382 sentences in total), information about word frequency is added based on word frequency from the Academia Sinica Corpus [25], which has 11,245,330 words and is the first fully POS-tagged balanced Chinese Corpus [27]. Then, to facilitate comparison between high and low frequencies of words, the logarithm of the raw frequency is used. Information about position-in-utterance is also added by counting how many words separate the speech error from the beginning of the utterance in which it occurred. Each error is assigned a number based on the numbers of words which separate the error from the beginning of the utterance. By way of illustration, in (1a), the error *ming2ci2* 'noun' occurs at the sixth word of the utterance, its position-in-utterance is thus annotated as 5. This value is then normalized by dividing the position-in-utterance by the total number of words in the utterance (not including the error nor the repair of the target), i.e., 5/5 = 1. A value of 1 indicates that the error is found at the end of the utterance. A value of 0 indicates that the error is at the beginning of the utterance.

4 Analysis

In terms of word frequency, previous studies [12] have already pointed out that (i) the targets and the errors have higher frequency than most of the other words in the lexicon, and (ii) the word frequency of the targets and the errors is commonly lower than the frequency of words in the surrounding context. We thus only provide an overview of the interaction between word frequency and position-in-utterance. In Fig. 2, a PCA analysis of the two variables is shown. Since we only have two variables, the two components refer to word frequency (PC1 on the x axis) and position in utterance (PC2 on the y axis, coded as dist_to_start). First, we can visualize the effect of word frequency attested in previous studies. The errors are more likely to occur on the left side of the plot, in the opposite direction of the loading word_freq, which indicates that errors are found more frequently with words of small frequency. Second, a similar effect seems to be found with position-in-utterance, as more errors are located at the bottom of the plot. However, the effect is less obvious than with frequency. As a reminder, this is only a visualization of the data. The following paragraphs provide the quantitative analysis for the interaction between word frequency and position-in-utterance.

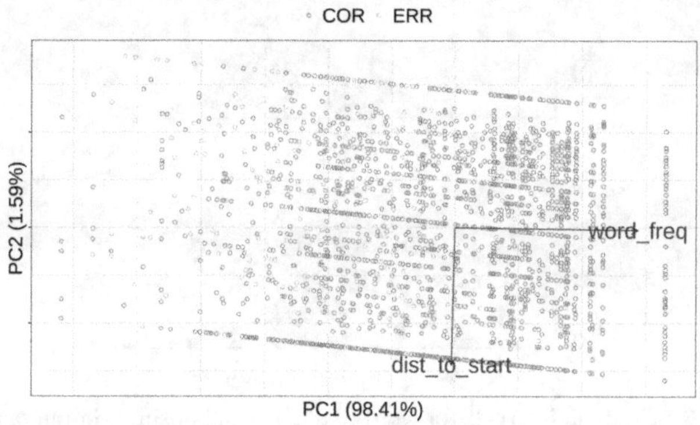

Fig. 2. PCA visualization of the interaction between word frequency and position-in-utterance with regard to speech errors.

We then feed the information of word frequency and position-in-utterance to two different machine learning methods. First, we use generalized linear mixed models (GLMMs) [28, 29] to interpret the effect of the two variables on the occurrence speech errors. Second, we try to reproduce the results from previous studies [12] by using a decision tree based classifier to predict the occurrence of speech errors based on the two variables.

GLMMs are a further developed model of generalized linear model (GLM), as GLMMs include random effects in addition to the usual fixed effects. That is to say, the model also evaluates the possible variation of effect for individual entries in the data. Moreover, it provides a more general view of the effect of each variable by assessing the credible intervals of the effects. The parameters were set as follows: 4 chains with 500 of iterations each, including 200 iterations as warm-up. The results reported are from the model that does not consider interaction for the two variables. This choice is made on the theoretical premise that we want to directly assess the effects of the two variables on speech errors. We did test random effects and interaction in other models and compared their fit with leave-one-out cross validation. The model considering word frequency as a random effect gives the best fit, but the divergence across the models is not big. Thus, we consider that the results reported in the current paper are sufficient for identifying the effect of word frequency and position-in-utterance on speech errors. The credible intervals of the output are shown in Fig. 3. First of all, neither of the two 'humps' cross 0 (Rhat = 1, ESS = 763, 690), which indicates that the two variables have a clear positive/negative effect. In our case, word frequency has a negative effect (all the values are negative) while position-in-utterance (represented by dist_to_start, i.e., distance to the beginning of a sentence), has a positive effect as all the values are positive. In other words, the model indicates that i) the higher the frequency of a word, the less likely it is to be a speech error (est = −0.32) ii) the bigger distance to the beginning of sentence, i.e., the closer a word is to the end of a sentence, the more likely it is to be a speech error (est = 0.59). Therefore, the output of the model supports the two predictions based on the connectionist model.

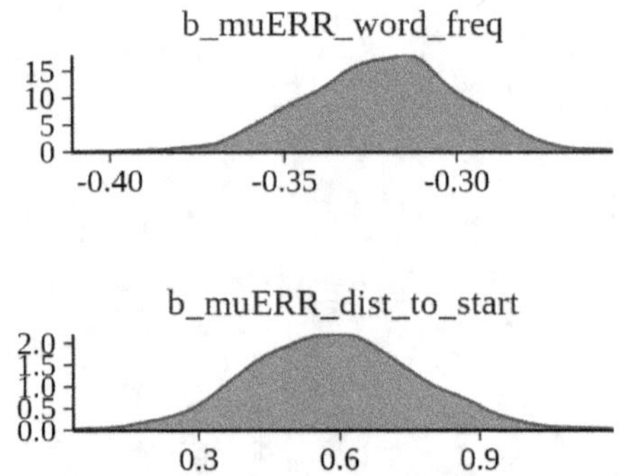

Fig. 3. Credible intervals for word frequency and position-in-utterance.

Then, the computational classifier based on decision trees is used to predict the presence of speech error for each word of each utterance in the speech error corpus. These decision trees are based on binary recursive partitioning [30, 31]. The main functioning of the classifier is explained as follows. Binary splits recursively divide the data into homogeneous or near-homogeneous buckets. The split is considered ideal if the homogeneity of the buckets is improved after the split. To assure a low variation in the output, the model does not use all the variables and observations. Instead, the model uses a bootstrap sample of the original data and also selects a random subset of variables for each split. That is to say, first, the algorithm scans through the variables and selects the strongest association with the response. Then, the data set is divided into two subsets based on the chosen variable. These two steps are repeated for every subset until no variables may split the data with statistical significance. The main advantage of decision trees is their visualization of the interaction between the variables. That is to say, a decision tree is generated and can be read to make a prediction on a specific data point.

To train the classifiers, the data is split into two sets, one for training, the other for testing. The training set contains 70% of the data while the test set has 30%. To avoid bias from a specific combination of tokens between the training and test sets, the same process was conducted ten times with different training and test sets. Since the results did not vary across the sets, we report the output of the tenth set in the current paper. Moreover, to replicate previous studies [12], we also assess the performance of the classifier in five different window sizes. That is to say, we first ask the classifier to identify the speech error from the target and its preceding and following word. Then, we expand the window to the two preceding and following words, and so on, until the maximum size of five preceding and following words, which capture the full length of each sentence in the corpus.

The performance of the decision tree is evaluated based on its accuracy, precision, and recall. The accuracy provides an overview of the performance by dividing the correctly predicted tokens with the total of the tokens. The precision evaluates how many tokens are correct among all the output of the classifier and the recall quantifies how many tokens are correctly retrieved among all the expected correct output [32]. Finally, since the quantity of correct words and speech errors is unbalanced within the data set, we use the majority rule as a benchmark of accuracy. Taking window size five as an example, i.e., when we consider the entire corpus: the corpus of 3022 words only contains 382 speech errors, and the computational classifier may reach an accuracy of 87.4% simply by guessing that all the words do not undergo speech errors. In such case, the computational classifier should have an accuracy higher than 81.8% to be considered as having a good performance. The same logic is applied for the other window sizes. The performance on the tenth test set is reported for each window size in Table 1.

Table 1. The performance of the decision-tree based classifier.

Window	1		2		3		4		5	
Type	cor	err	cor	err	cor	err	cor	err	cor	err
Precision	0.80	0.62	0.94	0.40	0.95	0.38	1	0	1	0
Recall	0.77	0.66	0.82	0.64	0.87	0.61	0.84	0	0.88	0
Accuracy	0.73		0.80		0.84		0.84		0.87	
Baseline	0.62		0.76		0.81		0.84		0.87	

As observed in previous studies [12], the classifier does not perform well with large window sizes. That is to say, speech errors are hard to detect in large corpora due to their scarcity. On the other hand, for window sizes smaller than four, the performance of the classifier increases and exceeds the majority baselines. The performance on individual categories is also analyzed. First, the model performs well at detecting the absence of speech errors (cor). This is not surprising since the majority of the data points are without speech errors. The model therefore has less difficulty to identify this category. Second, starting from window size three, the recall on detecting errors does not increase by much, however, its precision increases. An interesting fact is the increase of accuracy in comparison with the model used in [12], which only considered word frequency. In the current model, the accuracy of window size 1–3 is 3–4% higher (similar results are found when our classifier is fed with information on word frequency and position-in-utterance individually), which suggests that adding the information of position-in-utterance does result in better predictions.

The regularities found by the classifier in window size one are visualized in Fig. 4. The upper nodes refer to the decision path and the buckets at the bottom indicate the ratio of speech errors (ERR) and correctly uttered words (COR). As an example, if the logarithm of word frequency is smaller than 9.2 (node 1 → node 3) and the distance to the beginning of the sentence is larger than or equal to 9.6, i.e., if the word is either the final word or the penultimate word of a sentence (node 3 → node 7), the model predicts that a speech error occurs. Within the entire data set, 77 tokens are categorized under such pathway and 91% (70/77) of them are classified correctly.

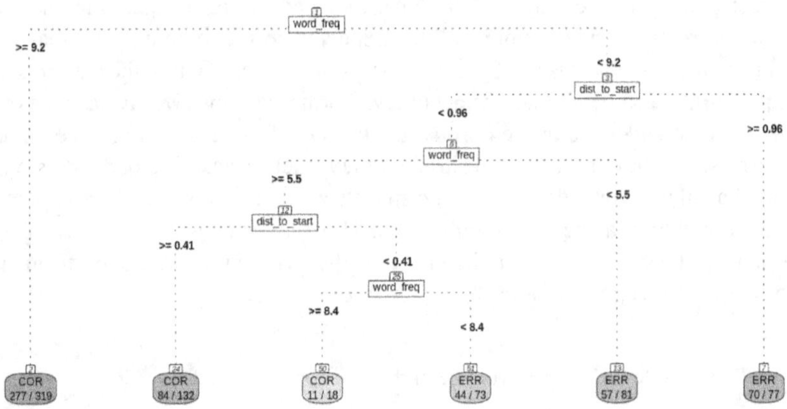

Fig. 4. Decision tree for the occurrence of speech errors in the corpus of 1888 words.

The tree shows that position-in-utterance and word frequency both play a role in predicting speech errors. On the one hand, words with high frequency are less likely to result in speech errors (e.g., node 1 → node 2). On the other hand, words located at the end of an utterance are extremely likely to result in speech error (node 1 → 3 → 7). Similar tendencies are found further away from these two extremes. As an example, words closer to the beginning of a sentence are more likely to result in speech errors if they have a low frequency (e.g., node 1 → 3 → 6 → 13).

5 Conclusion

The connectionist model predicts that i) frequently used words are less likely to result in speech errors since they are easily activated and ii) utterance-initial words are less likely to result in speech errors since they convey old information that already has been activated in the preceding context. The results of the current analysis support these two hypotheses, but also indicate that other variables should be added in the model to result in an accurate prediction of speech errors. Potential candidates for additional variables are part-of-speech tags and semantic/phonological distance with the preceding and following contexts.

Acknowledgements. The first author is thankful for the support of grants from the University of Lyon (ANR-10-LABX-0081, NSCO ED 476), the IDEXLYON Fellowship (2018–2021, 16-IDEX-0005), and the French National Research Agency (ANR-11-IDEX-0007). The corresponding author would like to thank Dr. Chain-wu Lee for his continuous cutting-edge high-tech programming support in constructing all the corpora in Phonetics and Psycholinguistics lab at National

Chengchi University. All remaining errors are our own. The research reported in this paper was funded to the corresponding author by MOST three-year grant, MOST 98-2410-H-004-103-MY2, in Taiwan.

References

1. Alderete, J., Davies, M.: Investigating perceptual biases, data reliability, and data discovery in a methodology for collecting speech errors from audio recordings. Lang Speech. **62**, 281–317 (2019). https://doi.org/10.1177/0023830918765012
2. Alderete, J., Tupper, P.: Connectionist approaches to generative phonology. In: Hannahs, S.J., Bosch, A.R.K. (eds.) The Routledge Handbook of Phonological Theory, pp. 360–390. Routledge (2017). https://doi.org/10.4324/9781315675428-13
3. Wan, I.-P.: Mandarin speech errors into phonological patterns. J. Chin. Linguist. **35**, 185–224 (2007)
4. Wan, I.-P.: Consonant features in Mandarin speech errors. Concentric Stud. Linguist. **42**, 1–39 (2016). https://doi.org/10.6241/concentric.ling.42.2.01
5. Fay, D., Cutler, A.: Malapropisms and the structure of the mental lexicon. Linguist Inq. **8**, 505–520 (1977)
6. Fromkin, V.: Speech Errors as Linguistic Evidence. Mouton de Gruyter, The Hague (1973)
7. Harley, T.A., MacAndrew, S.B.G.: Constraints upon word substitution speech errors. J. Psycholinguist. Res. **30**, 395–418 (2001). https://doi.org/10.1023/A:1010421724343
8. Jaeger, J.J.: Kids' Slips: What Young Children's Slips of the Tongue Reveal About Language Development. Lawrence Erlbaum Associates, Mahwah (2004)
9. Jaeger, J.J., Wilkins, D.: Semantic relationships in lexical errors. In: Jaeger, J.J. (ed.) Kids' Slips: What Young Children's Slips of the Tongue Reveal About Language Development, pp. 311–384. Lawrence Erlbaum Associates, Mahwah (1993)
10. Nooteboom, S.G.: The tongue slips into patterns. In: Fromkin, V. (ed.) Speech Errors as Linguistic Evidence, pp. 87–95. Mouton de Gruyter, The Hague (1973)
11. Naroll, R.: Galton's problem. In: Naroll, R., Cohen, R. (eds.) A Handbook of Method in Cultural Anthropology, pp. 973–989. Columbia University Press, New York (1973)
12. Tang, M., Wan, I.-P.: Predicting speech errors in Mandarin based on word frequency. In: Su, Qi., Zhan, W. (eds.) From Minimal Contrast to Meaning Construct. FCL, vol. 9, pp. 289–303. Springer, Singapore (2020). https://doi.org/10.1007/978-981-32-9240-6_20
13. Dell, G.S.: Representation of serial order in speech: evidence from the repeated phoneme effect in speech errors. J. Exp. Psychol. Learn. Mem. Cogn. **10**, 222–233 (1984). https://doi.org/10.1037/0278-7393.10.2.222
14. Dell, G.S.: A spreading-activation theory of retrieval in sentence production. Psychol. Rev. **93**, 283–321 (1986). https://doi.org/10.1037/0033-295X.93.3.283
15. Dell, G.S.: The retrieval of phonological forms in production: tests of predictions from a connectionist model. J. Mem. Lang. **27**, 124–142 (1988). https://doi.org/10.1016/0749-596X(88)90070-8
16. Dell, G.S., Chang, F., Griffin, Z.M.: Connectionist models of language production: lexical access and grammatical encoding. Cogn. Sci. **23**, 517–542 (1999). https://doi.org/10.1207/s15516709cog2304_6
17. Clifton, C., Frazier, L.: Should given information come before new? Yes and no. Mem. Cogn. **32**, 886–895 (2004). https://doi.org/10.3758/BF03196867
18. Gundel, J.K.: 'Shared knowledge' and topicality. J. Pragmat. **9**, 83–107 (1985). https://doi.org/10.1016/0378-2166(85)90049-9

19. Gundel, J.K.: Universals of topic-comment structure. In: Hammond, M., Moravcsik, E.A., Wirth, J. (eds.) Typological Studies in Language, p. 209. John Benjamins Publishing Company, Amsterdam (1988). https://doi.org/10.1075/tsl.17.16gun
20. Hu, C.: Information structure in English, Mandarin Chinese and Taiwanese southern min: argument realization of ditransitive objects (2015)
21. Ibrahim, S.: A corpus-based investigation of the given before new principle in Tanzanian English. Papers from the 9th Lancaster University Postgraduate Conference in Linguistics & Language Teaching 2014, pp. 70–70 (2014)
22. Junge, B., Theakston, A.L., Lieven, E.V.M.: Given–new/new–given? Children's sensitivity to the ordering of information in complex sentences. Appl. Psycholinguist. **36**, 589–612 (2015). https://doi.org/10.1017/S0142716413000350
23. Wan, I.-P.: Mandarin phonology: evidence from speech errors. Ph.D. dissertation. State University of New York at Buffalo, N.Y., U.S.A. (1999)
24. Wan, I.-P.: On the phonological organization of Mandarin tones. Lingua **117**, 1715–1738 (2007). https://doi.org/10.1016/j.lingua.2006.10.002
25. CKIP (Chinese Knowledge and Information Processing): Part-of-speech analysis of Academia Sinica Balanced Corpus of Modern Chinese, Technical Report (no. 93-05), version 3. Academia Sinica (2004)
26. Ma, W.-Y., Chen, K.-J.: Introduction to CKIP Chinese word segmentation system for the first international Chinese word segmentation bakeoff. In: Proceedings of ACL, 2nd SIGHAN Workshop on Chinese Language Processing, pp. 168–171 (2003)
27. Huang, C.-R., Lee, L.-H., Hong, J.-F., Yu, S.: Quality assurance of automatic annotation of very large corpora: a study based on heterogeneous tagging systems. LREC **2008**, 2725–2729 (2008)
28. Bürkner, P.-C.: brms: an R package for Bayesian multilevel models using Stan. J. Stat. Softw. **80**(1) (2017). https://doi.org/10.18637/jss.v080.i01
29. Bürkner, P.-C.: Advanced Bayesian multilevel modeling with the R package brms. R J. **10**(1), 395 (2018). https://doi.org/10.32614/RJ-2018-017
30. Breiman, L., Friedman, J., Stone, C.J., Olshen, R.: Classification and Regression Trees. Taylor & Francis, New York (1984)
31. Breiman, L.: Random forests. Mach. Learn. **45**, 5–32 (2001)
32. Zeugmann, T., et al.: Precision and recall. In: Sammut, C., Webb, G.I. (eds.) Encyclopedia of Machine Learning, pp. 781–781. Springer US, Boston, MA (2010). https://doi.org/10.1007/978-0-387-30164-8_652

Morpheme-Based Chemical Term Analysis and Recognition

Wang Guirong[1,2], Xiao Ye[2], Qian Qingqing[2], Xia Bo[2], Rao Gaoqi[1(✉)], and Xun Endong[2(✉)]

[1] Research Institute of International Chinese Language Education, Beijing Language and Culture University, Beijing 100083, China
raogaoqi@blcu.edu.cn

[2] Institute for Language Intelligence, Beijing Language and Culture University, Beijing 100083, China

Abstract. Term recognition is the foundation of natural language processing, especially in chemical domain. The most important feature of domain text is terminology. Because of the variety and the complexity of chemical terms, chemical term NER is too difficult to be used in a production environment such as patents search. Based on the theory of morphemes and data of Chinese patents, we construct a chemical morpheme classification according to the composition characteristics and formation rules of chemical terms. A simple and stable HMM (The Hidden Markov Model) model is used to model the word formation relationship among different morphemes. The improved forward algorithm is applied to estimate the word formation probability in terminology recognition. It solves the problem of poor recognition of the long terms in low frequency in traditional statistical methods. It also solves the problem of subjectivity and completeness in establishing rules to some extent. Finally, the F1 reaches 91.58% in chemical patents.

Keywords: Term recognition · Chemical term · Morpheme · Word-formation rules · Morpheme classification

1 Introduction

Term recognition, especially domain term recognition, is an important part of NLP tasks in practical scenarios. Mainstream features such as boundary words, contextual conjunctions are not sufficient to solve the current problem caused by diverse word formation and large word lists in chemical term recognition.

By observing a large number of scientific and technical terms, especially the names of chemical substances, we have found that chemical terms have a stable composition pattern and a limited set of chemical morphemes. From the linguistic point of view, morpheme is the smallest unit of sound and meaning, and the main way to construct a word is to combine morphemes. For example, in Chinese chemical term, 氢氧化钠 (sodium hydroxide) consist of morphemes expressing “hydrogen”, “oxygen”, “compound”, “sodium”, i.e. 氢, 氧, 化 and 钠 respectively.

M. Liu et al. (Eds.): CLSW 2020, LNAI 12278, pp. 501–514, 2021.
https://doi.org/10.1007/978-3-030-81197-6_43

Inspired by term morpheme characteristics, this paper introduces the notion of morpheme and morpheme classification into chemical term recognition and models the word formation rules of morpheme-constituted terms to term recognition. Specifically, through the statistical analysis of chemical compound names and the use of corpora, this paper has classified chemical morphemes into four categories, namely i) core morpheme, ii) definite morpheme, iii) auxiliary morpheme and iv) general morpheme. Then, the word formation rules of morpheme contained by a large term list are fitted by the parameters of the Hidden Markov Model (HMM). Hence, the recognition of chemical terms is transformed into estimating the effective word formation probability of chemical morphemes, and the remaining problems of heavy labeling task, low-frequency terms and over-long terms in the existing work are solved.

2 Related Work

At present, the mainstream methods of named entity and term recognition are combining rules and using statistics. Liang L. [6] identified drug names and other terms in commodity texts based on constructive features and word frequency distribution; Song D. et al. [10] classified chemical-specific words into three categories and proposed a rule-based approach for identifying chemical substance names; Li N. et al. [4] used rule-based methods and CRF models to translate term recognition into a process of valid word sequence recognition; Ma J. [7] initially identified the use of statistical methods based on the linguistic patterns and features of the chemical text, and then corrected the method with the use of lexicographic and rule-based methods.

The use of statistical methods for term recognition requires a large amount of specific annotation resources and corresponding algorithms, which makes information integration difficult. Although some scholars used the context information, they still could not solve the long-tail problem of many low-frequency terms in chemistry. When it comes to knowledge rules, the word formation rules of morphemes are various and not strict in the field of chemistry, and the word list is huge while chemical words cannot be exhausted. Long terms with low frequency in use and generic lexicons also bring some interference, so it is difficult to build a complete database of chemical proper names.

Yet, analyzing the features of morphology from the linguistic point of view can bring some help in term recognition. According to Yu et al. [14], the word formation of complex terms in a specific discipline can be classified as basic morphemes, inherent morpheme groups and complex terms. Li N. [5] et al. categorized morphemes in the names of chemical substances. Wang Q. et al. [12] defined and classified chemical morphemes and established the knowledge calculation model of chemical terms by a rule-based method.

The existing method of chemical term recognition from the perspective of morphemes is mostly based on rules, with a certain degree of accuracy, but fail to solve the problems of flexible combination and long words. In this paper, morphemes are regarded as the basic unit in the process of chemical term recognition, the relationship between different chemical morphemes is predicted by HMM model. The problem of chemical term recognition is transformed into the calculation of the probability of chemical morphemes forming words, which solves the remaining problems in the existing work to a certain extent.

3 Chemical Terms and Their Morphemes

3.1 Chemical Terms

The chemical terms can be defined broadly by including basic concept terms, chemical elements, chemical formulae, chemical instrument names, chemical reaction principle names, and physical chemical terms and biochemical terms formed by intersection with other disciplines, etc. The chemical terms in this paper refer to the chemical substance terms represented by chemical formulae. These words are often long, large in number, easy to expand, and cannot be fully included in the chemical vocabulary, causing difficulty for automatic identification of chemical terms.

At the same time, these chemical terms also have some regular characteristics. By observing a large number of chemical terms, it is found that there is a robust regularity in the word formation of chemical terms. Here are some rules for naming organic compounds [9]:

(1) Heavenly Stems + substituent group/substance generic words: e.g., "methyl" (*jia3ji1*, 甲基), "butane" (*yi3wan2*, 乙烷), "decane" (*gui1wan3*, 硅烷), etc.;
(2) Chemical element + substance generic words: e.g., "boric acid" (*peng2suan1*, 硼酸) and "bromic acid" (*xiu4suan1*, 溴酸);
(3) Chinese numbers + Heavenly Stems + substituent + Heavenly Stems + substance generic words: e.g., "trimethylbutane" (*san1jia3ji1ding1wan2*, 三甲基丁烷);
(4) Chinese numbers + chemical element + "化 (*hua4*)" + Chinese numbers + chemical element name: e.g., "carbon dioxide" (*er4yang3hua3tan4*, 二氧化碳), "vanadium trioxide" (*san1yang3hua4er4fan2*, 三氧化二钒);
(5) Structure words + Chinese numbers / Heavenly Stems + chemical element name: e.g., "polytetrafluoroethylene" (*ju4si4fu2yi3xi1*, 聚四氟乙烯), "p-toluene" (*dui4jia3ben3*, 对甲苯), "m-bromophenylacetone" (*jian4xiu4ben3bing3tong2*, 间溴苯丙酮), etc.;
(6) Arabic numeral + '–' + Chinese numbers + Heavenly Stems + chemical elements/substituent + Heavenly Stems + chemical elements, e.g., "2,7,8-trimethyldecane" (*2,7,8-san1jia3ji1gui1wan2*, 2,7,8-三甲基硅烷).

In addition, from the above rules, we can find that the characters used in chemical terms are limited and exhaustible, including chemical elements, generic characters of chemical substances (e.g., 酸 (*suan1, acid*), 醇 (*chun2, alcohol*)), Arabic numerals, Chinese numerals, Heavenly Stems, structure characters (e.g., 正 (*zheng4*), 偏 (*pian1*), 交 (*jiao1*)) and connective symbols. Moreover, morphemes and the smallest unit of chemical terms show a one-to-one correspondence. Therefore, this paper will establish HMM model to learn the combination relationship between different chemical morphemes and evaluate the probability of candidate string sequence becoming chemical terms.

3.2 Chemical Morphemes and the Classification

3.2.1 Chemical Morphemes

In modern Chinese grammar, the term "语素 (*morpheme, yu3su4*)" is the Chinese translation of the English word "morpheme". This concept was proposed by the American structural linguist Bloomfield. In Zhang Bin's [18] *New Modern Chinese*, morpheme is defined as that "morpheme is the smallest combination of sound and meaning in language and the smallest linguistic unit capable of distinguishing meaning. The role and function of a morpheme is primarily to construct words [18]." In terms of the relation between content and meaning, a morpheme can no longer be decomposed, otherwise it leads to meaning change or the loss of meaning. In Chinese, a character is usually used as a morpheme. Some morphemes are consisted by the words on their own, and morphemes can also be made into words by combining them with one another. In fact, the combination of monophonic morphemes does produce a lot of new words, and even becomes one of the most important ways of generating new words in Chinese.

The basic semantic functions of morphemes in chemical terms are consistent with those of general morphemes, both referring to the basic unit of meaning, and are the smallest word formation unit. Most morphemes in the field of chemistry are monosyllabic morphemes with clear and professional meanings. For example, "alkane (烷, *wan2*)", "hydrocarbon (烃, *ting1*)" and "acid (酸, *suan1*)" indicate the species of chemical substances. In addition, there are many transliterated words in chemical terms. Once being subdivided into single words, they no longer have practical significance. We regard these transliterated words as a morpheme, such as "pyridine (吡啶, *bi3ding4*)" and "furan (呋喃, *fu2nan2*)", etc. Therefore, the chemical morphemes discussed in this paper refer to the smallest word formation units with the characteristics of chemical terms or auxiliary naming of chemical substances, which can be a number, a letter, a Chinese character, or a word.

3.2.2 Chemical Morphemes Classification

According to the position and function of a chemical morpheme in chemical terms, chemical morphemes are divided into four categories: core morpheme, definite morpheme, auxiliary morpheme and general morpheme.

Core morpheme is a kind of morpheme which can clearly indicate the nature or category of substances in chemical terms, so as to be distinguished from other types of substances and play the most important role of identification. These morphemes only appear in the texts related to chemistry. The core morphemes consist of two subcategories, that is, chemical elements and chemical proper noun, which are labeled as category A and category B, respectively. Among them, chemical elements are a fixed set of elements as there are 118 chemical elements published at present. Chemical proper nouns refer to morphemes that indicate the category of chemical substances and such nouns are some polysyllabic morphemes, such as "acid (酸, *suan1*)", "fat (脂, *zhi3*)", "salt (盐, *yan2*)" and "pyridine (吡啶, *bi3ding4*)" and so on.

Definite morpheme refers to the morpheme that restricts the nature, type or structure of the substance based on the core morpheme. It can further limit the type or scope of the chemical substance, which is conducive to clarifying the concept of the substance. It is labeled as category C. It mainly includes the restrictive morphemes that often appear at the beginning of words in chemical terms, such as "正 (*zheng4*), 原 (*yuan2*), 偏 (*pian1*), 交 (*jiao1*), 重 (*zhong4*), 亚 (*ya4*), 次 (*ci4*), 高 (*gao1*), 联 (*lian2*)" and so on, and the morphemes that indicate the method of chemical substance formation can be exemplified by "化 (*hua4*), 合 (*he2*), 聚 (*ju4*), 缩 (*suo1*), 代 (*dai4*)" and so on.

Auxiliary morpheme refers to the capital numbers, Roman numerals, Arabic numerals, Heavenly stems, English letters, Latin letters, and some chemical connection symbols in chemical terms. It is used with core morphemes and definite morphemes to improve the specificity of chemical terms. It is labeled as category D.

General morpheme refers to a morpheme that can appear in terms of chemistry and other fields. It has a meaning which has no relevance to the field and no effect on the specificity of chemical terms. This category of morphemes mainly includes the morphemes that constitute the basic terms of chemistry, such as "电 (*dian4*), 解 (*jie3*), 溶 (*rong2*), 液 (*ye4*)", etc., which are labeled as category E; and morphemes in the general domain, such as "水 (*shui3*), 子 (*zi3*), 白 (*bai2*)", which are labeled as category F (Table 1).

Table 1. Classification of chemical morphemes

Categories		Examples of chemical morpheme
Core morpheme	Chemical elements (A)	Hydrogen (氢), helium (氦), lithium (锂), beryllium (铍), boron (硼), nickel (镍)……
	Chemical proper noun (B)	Acid (酸), salt (盐), alkali (碱), fat (脂), alcohol (醇), ether (醚)……
		Pyridine (吡啶), furan (呋喃), pyrimidine (嘧啶), indole (吲哚)……
Definite morpheme	Chemical preposition (C)	化 (*hua4*), 合 (*he2*), 聚 (*ju4*), 缩 (*suo1*), 代 (*dai4*)……
	Specific prefix (C)	原 (*yuan2*), 偏 (*pian1*), 交 (*jiao1*), 亚 (*ya4*), 次 (*ci4*)……
Auxiliary morpheme (D)		Capital numbers, Roman numerals, Arabic numerals, Heavenly stems, English letters, Latin letters and chemical connection symbol
General morpheme	Chemical basic morpheme (E)	电 (*dian4*), 解 (*jie3*), 溶 (*rong2*), 液 (*ye4*)……
	Universal basic morpheme (F)	水 (*shui3*), 子 (*zi3*), 白 (*bai2*)……

3.2.3 Construction of Chemical Morpheme List

In order to train the chemical term recognition model, a chemical term lexicon is constructed in this paper. The lexicon mainly includes Sogou's chemical vocabulary[1], Wikipedia's organic[2] and inorganic chemistry terms[3], with a total of more than 22,000 chemical term words. After that, the chemical morpheme list has been constructed by combining statistics with artificial method. First, 2045 monosyllabic morphemes were obtained according to their frequency in the chemical term list. Then, 105 polysyllabic morphemes were collected by manual annotation. Because the number of terms collected in the chemical lexicon is limited, it cannot cover all chemical morphemes, and the set of core and auxiliary morphemes is relatively closed, so morphemes were artificially supplemented, and the monosyllabic morphemes were selected according to the morpheme frequency. Finally, a total of 718 morphemes were obtained. According to the statistics, the number of chemical morphemes is much smaller than the number of chemical terms. It can thus be seen that the combination of chemical morphemes is flexible, and the word formation ability is strong. It is simple and feasible to recognize chemical terms by using the combination rules of chemical morphemes.

Table 2. The top 20 in the chemical morpheme list

M	F	M	F	M	F	M	F
酸 (*suan1, acid*)	6707	基 (*ji1, base*)	3421	二 (*er4, two*)	2721	化 (*hua4, transform*)	2510
甲 (*jia3, A*)	2368	苯 (*ben3, benzene*)	2176	乙 (*yi3, B*)	1805	氯 (*lv4, chlorine*)	1649
氧 (*yang3, oxygen*)	1312	胺 (*an1, amines*)	1246	醇 (*chun2, alcohol*)	1181	氨 (*an,1 ammonia*)	1120
硫 (*liu2, sulfur*)	1077	酰 (*xian3, acyl*)	1046	酯 (*zhi3, ester*)	1019	丙 (*bing3, C*)	921
三 (*san1, three*)	913	盐 (*yan2,salt*)	871	酮 (*tong2, ketone*)	866	烷 (*wan2, alkane*)	841

Table 2 shows the top 20 chemical morphemes in the chemical morpheme list sorted by frequency. In the table, "M" stands for the morpheme and "F" the frequency. There are 16 core morphemes, the frequency of morpheme "acid" is the highest, of which the frequency is 6707. Figure 1 shows the frequency distribution of all the morphemes in the chemical term lexicon, which also shows the same trend. That is, the top of the distribution is basically the core morphemes, followed by the definite morphemes.

After obtaining the chemical morpheme list, the chemical morphemes were manually classified according to the above-mentioned chemical morpheme classification, and the

[1] 搜狗细胞词库-化工(The Sogou Vocabulary List of Cell – Chemical Engineering): https://pinyin.sogou.com/dict/cate/index/109.

[2] https://zh.wikipedia.org/wiki/有机化合物列表.

[3] https://zh.wikipedia.org/wiki/无机化合物列表.

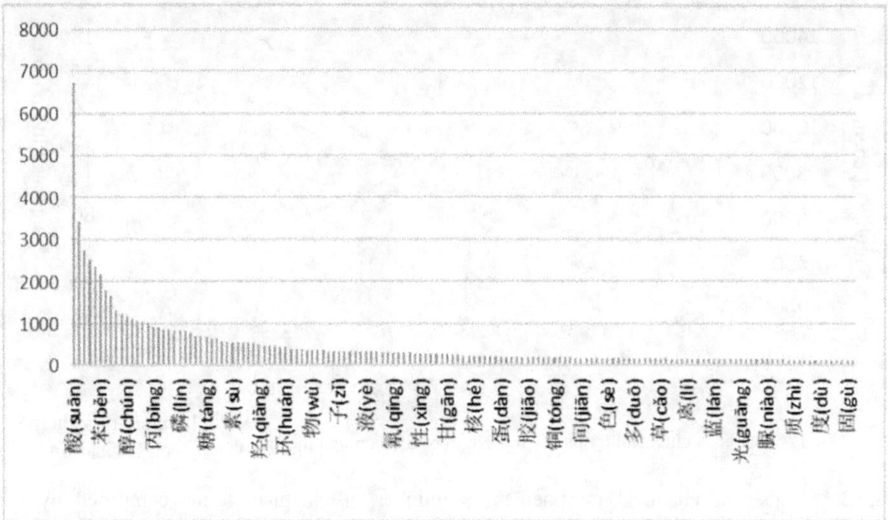

Fig. 1. Frequency distribution of chemical morphemes (partial)

number of chemical terms formed by different types of chemical morpheme was counted. According to Table 3, the number of chemical proper noun (B) and definite morphemes (C) are the largest, accounting for 22.98% and 20.75% of all morphemes respectively, and the number of general morphemes (F) is the least, accounting for 10.58% of all morphemes, and the numbers of other morphemes are close. However, categories A, B, C and D are relatively closed sets with limited number of morphemes. Therefore, the use of morphemes for chemical term recognition can achieve a multiplier effect, while ensuring the recognition effect and reducing manpower investment.

Table 3. Statistics on the classification of chemical morphemes

	Chemical element (A)	Chemical proper noun (B)	Definite morpheme (C)	Auxiliary morpheme (D)	Chemistry basic morpheme (E)	Universal basic morpheme (F)
Number	109	165	149	101	118	76
Percent	15.18%	22.98%	20.75%	14.08%	16.43%	10.58%

Figure 2 shows the number of chemical morphemes in each category and the number of chemical terms formed by the chemical morphemes in that category correspondingly. It is observed that the number of chemical morphemes in categories C, E, and F is relatively large, but the number of chemical terms it constitutes is relatively small. Category C contains chemical definite morpheme, and there are many different definite morphemes among chemical terms in the chemical term lexicon, such as "亚 (*ya4*)/C 铁 (*tie3*)/A 氰 (*jing1*)/B 化 (*hua4*)/C 钙 (*gai4*)/A (Calcium ferrocyanide)"; category E and

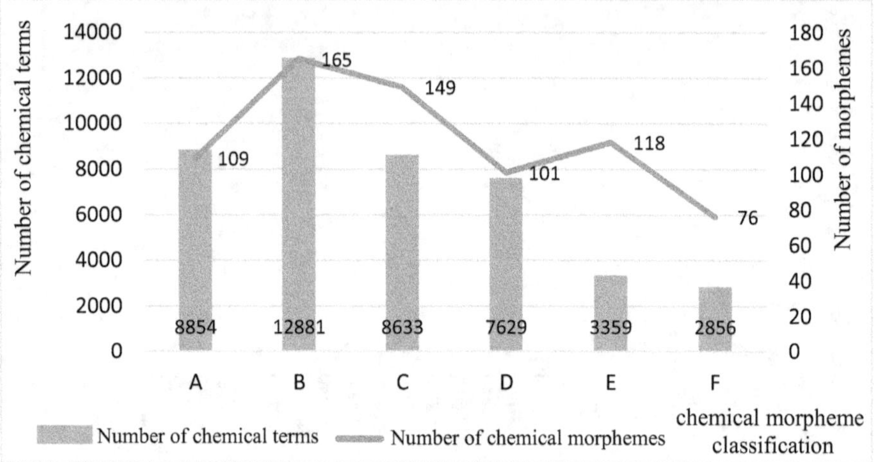

Fig. 2. Number of chemical morphemes vs number of chemical terms composed by the morphemes

F are relatively common morphemes with various types. They only play an auxiliary role in chemical terms, and their occurrence is irregular. Therefore, the number of morphemes is large, yet the number of chemical terms formed by these morphemes is small.

4 Approaches

In this paper, the chemical morphemes are used as the basis to model formation rules of morphemes in the chemical lexicon. In practice, we calculate the probability of a term consisting of continuous morphemes in a sentence, so as to identify chemical terms. Our approaches take the following features of morphemes as the starting point for term recognition: fixed number, less semantic ambiguity, stable formation rules, and high degree of morphological adhesion.

Specifically, *HMM* is used to model the word formation rules of morphemes. As shown in Fig. 3, the *improved forward algorithm* is applied to calculate the probability of forming chemical terms by the morphemes. The probability is used to determine whether to merge the morphemes or not, and finally the merging result was the recognition result. At last, threshold binarization is used to distinguish the probability of a number of morphemes forming a term and to determine whether a number of morpheme combinations can constitute a term.

4.1 Modeling the Word Formation Rules of Morpheme

The Hidden Markov Model (HMM) is used to model the word formation rules of morpheme. HMM is a formal foundation for making probabilistic models of linear sequence 'labeling' problems. HMM is applied to the recognition of chemical terms in Chinese. Formally, the morphological pattern model λ and λ is specified by the following components:

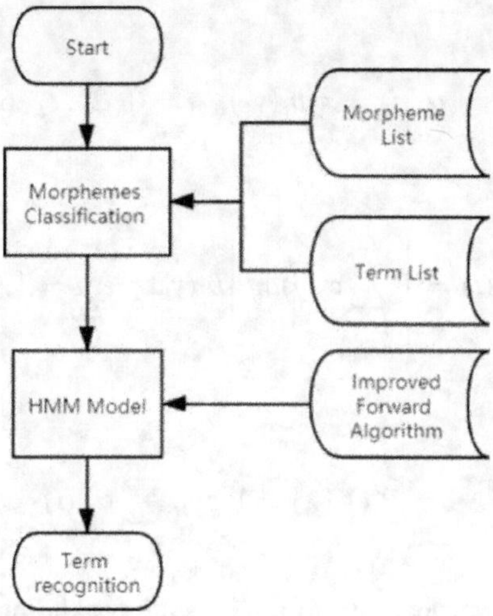

Fig. 3. The workflow of chemical term recognition from patents

$$\lambda = (\Omega_x, \Omega_o, A, B, \pi) \tag{1}$$

Here, morpheme classification is the hidden state, $\Omega_x = \{q_1, q_2, \ldots, q_6\}$, including six tags; a morpheme is the observation state, $\Omega_o = \{y_1, y_2, \ldots, y_{718}\}$, including 718 morphemes; an initial probability distribution π over morphemes; a transition probability matrix A, in which each a-i-j representing the probability of moving from morpheme tag i to tag j; a sequence of observation likelihoods B, with each expressing the probability of a morpheme being generated from a morpheme tag as following:

$$\frac{\text{frequency of } y_j}{\text{frequency of all morphemes in the category of } y_j} \tag{2}$$

We simply use add-one smoothing for 0 item in B.

4.2 The Improved Forward Algorithm

Given the model λ, forward algorithm can be applied to compute the probability of the word formation rules of morphemes. However, in the process of iterative computation, the forward probability tends to be zero as the length of sequence increases.

To ensure that the final probability does not arithmetically underflow, we introduce the improved forward algorithm, which is a treatment of observation sequence length and can be used to compute the probability safety. The formal definition of the algorithm is as follows.

Given model $\lambda = (A, B, \pi)$, the observation sequence $\Omega_o = \{y_1, y_2, \ldots, y_N\}$ is computed as:

1) Initialization:

$$\alpha_1(i) = \pi_i b_i(y_1), \quad i = 1, 2, \ldots, 6 \tag{3}$$

2) Recursion:

$$\alpha_{t+1}(i) = \left[\sum_{j=1}^{6} \alpha_t(j) a_{ij}\right] b_i(y_{t+1}), \quad i = 1, 2, \ldots, 6 \tag{4}$$

3) Termination:

$$P'(Y \mid \lambda) = \log_{1/t} \sum_{i=1}^{6} \alpha_T(i) \tag{5}$$

As we compute each probability to a new state for the next observation, the final probability $P'(\mathrm{O} \mid \lambda)$ leads us to a chemical term.

5 Experiments

5.1 Datasets

The datasets in our work include a chemical morpheme list, a lexicon of chemical terms, and a recognition test set. The chemical morpheme list contains 718 morphemes, and the lexicon of chemical terms contains 22,356 words and the average word length is 4.3, as detailed in Sect. 3.2.3.

A test set is randomly selected from Chinese patent documents under C01 (inorganic chemistry) and C07 (organic chemistry). Patent terminological identification is one of the most important scenarios for NLP. And the International Patent Classification (IPC classification) is the basis for patent classification. The test set contains 531 sentences with an average term length of 6.998, maximum term length of 48, average sentence length of 24.537, maximum sentence length 172.

Data and Code are available on Github[4].

5.2 Implementation and Metric

In our experiments, there are two parts of implementation: preparation and recognition. In preparation, HMM model parameters are calculated and probability threshold is determined, followed by merging morphemes according to the probability to complete recognition. The pseudo code flow is as follows:

[4] https://github.com/xiabo0816/ChemNer.

```
// Define return values and temporary arrays of morphemes
results = []
term    = []
// Fitting the parameters of the training model and determine probability thresholds
Pi, A, B, threshold = TrainHMM (morpheme, wordlist)
// Traversing morphemes in sentences
for morpheme in sentence:
        if prob(term + morpheme) > threshold:
            //to be a term
              term += morpheme
        else:
            //or not to be
              results.append(term)
              term.clear()
return results
```

The threshold for terms in length 2 is 15, for length 3 is 15.8, for term length 4 is 18.7. Reported metrics are precision, recall, F1-score, which evaluate the performance of the recognition.

5.3 Results and Analysis

As shown in Fig. 4 and Fig. 5, our approaches have achieved impressive performance of long and complex chemical term recognition with modeling the word formation rules of morphemes.

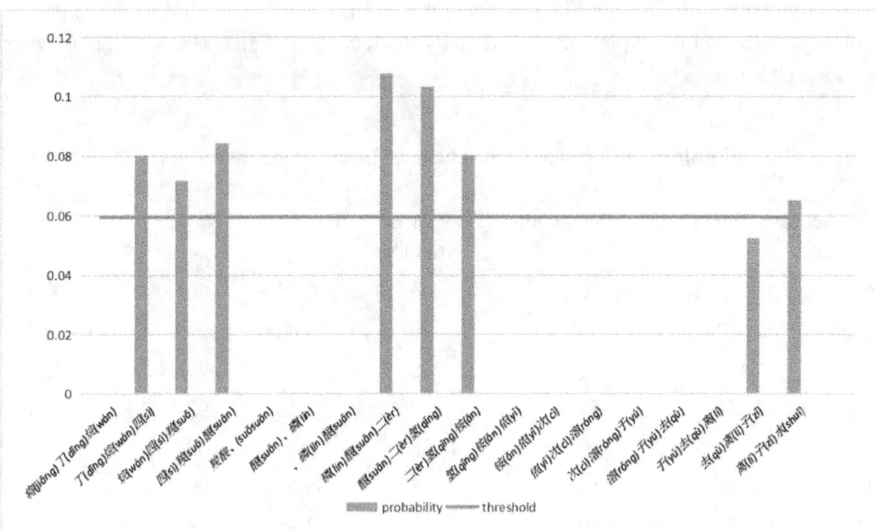

Fig. 4. Word formation threshold and term recognition results in sentence 1

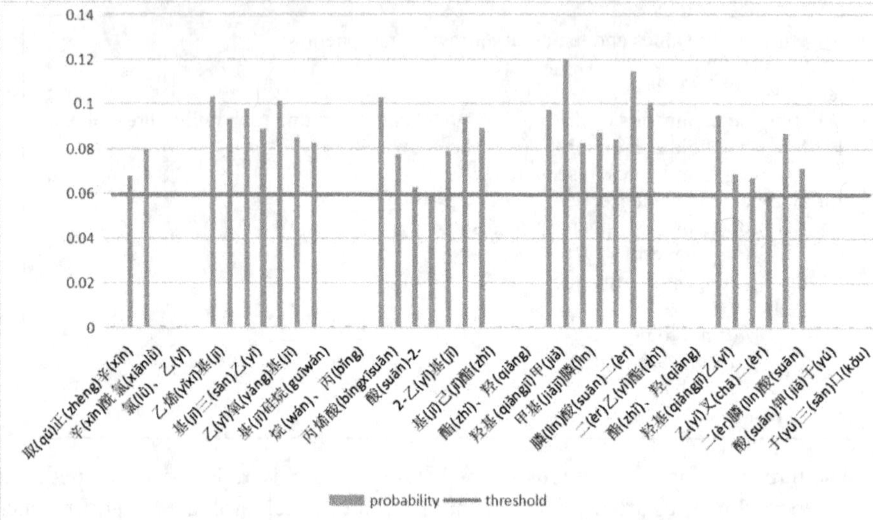

Fig. 5. Word formation threshold and term recognition results in sentence 2

Take sentence 1, of which the analysis is shown in Fig. 4. Sentence 1 is "将丁烷四羧酸、磷酸二氢铵依次溶于去离子水中" (Dissolve butane tetracarboxylic acid and ammonium dihydrogen phosphate in deionized water in proper order). The linked morphemes such as "丁烷四" "烷四羧" "四羧酸" can merge to form terms like "丁烷四羧酸" (Butane tetracarboxylic acid), "磷酸二氢铵" (ammonium dihydrogen phosphate) filtered by probability thresholds. Furthermore, in Fig. 5, it shows that we have got the result correct even in complex terms with a hyphen, e.g., "丙烯酸-2-乙基己酯"(2-ethyl hexyl acrylate).

Table 4. Test results of the chemical patent literature test set

Morpheme number	Total terms	Recall number	Correct number	P	R	F1
≤3	6	6	6	100.00%	100.00%	100.00%
4	10	10	9	90.00%	100.00%	94.74%
5	288	286	275	96.15%	99.31%	98.39%
6	54	53	47	88.68%	98.15%	93.17%
8	30	30	28	93.33%	100.00%	96.55%
10	12	12	8	66.67%	100.00%	79.80%
≥12	131	130	75	57.69%	99.24%	72.97%
All	531	527	448	**85.01%**	99.25%	**91.58%**

From the experimental results summarized in Table 4, we see that:

1. In practice, the term length is often higher than 5. As the term length increases, the degree of the term complexity increases even rapidly.
2. The term morphemes are kept stable despite of various chemical term length.
3. The terms have high degree of adhesion and low morpheme ambiguity.

In conclusion, we can clearly see the benefit of modeling the word formation rules of morphemes, which make chemical term recognition more reliable, efficient and interpretable.

In terms of running efficiency, the test set takes 3500 ms for a single thread on experimental machine[5], with an average of 7 ms per sentence. Roughly estimated, it can support at least 30+ connections with response time less than 300 ms.

However, a bad case is observed. There are several reasons for poor recognition: i) Among the categories of chemical morphemes, there are few terms including morphemes such as numbers, "–", and brackets ("2-氯-6-三氯甲基吡啶" "四溴邻苯二甲酸双 (2-乙基己基) 酯"), which require the recognition model to be optimized; ii) when it comes to the length of the chemical terms, it appears that our approach can compensate for conditioning on term length, especially when the term length is within 10. However, considering the diversity of chemical terms, we still need to expand the morpheme lists.

6 Conclusions

In this paper, we present a method for chemical term recognition by modeling the word formation rules of morphemes. This approach can compensate for the lack of term structural features and solve the problems caused by flexible combination and long words. In conclusion, the main advantage of introducing the notion of morphemes into a recognition system is that morphemes come with their ability to capture not only syntactic but also semantic features.

For future direction, we are interested in exploring the combination of morphemes and Conditional Random Field and anticipate to how they can be better incorporated location and boundary information of terms within our method. Furthermore, morphemes used in other domains require more experiments, such as news and sports terms.

Acknowledgments. This paper is supported by Research Funds of National Language Committee (YB13590).

References

1. Frederick, J.: Statistical Methods for Speech Recognition. The MIT Press (1997)
2. Huang, C.: Reflection on automatic word segmentation. In: Proceedings of the 7th National Joint Conference on Computational Linguistics, pp. 36–48. Tsinghua University (2003)

[5] Experimental machine configuration: CPU i7-8568u, memory 16.0 GB.

3. Li, Z.: Research on Chinese word segmentation based on context information. Beijing Jiaotong University (2018)
4. Li, N., Zheng, R., Ji, J., Teng, Q.: Research on Chinese chemical substance naming and recognition based on heuristic rules. Mod. Libr. Inf. Technol. **05**, 13–17 (2010)
5. Li, N., Sun, J., Ji, J.: A preliminary study of Chinese morpheme semantics and knowledge discovery. J. Libr. Inf. Work **57**(17), 109–113 (2013)
6. Liang, L., Li, Y.: A study on recognition of drug names and chemical names in commodity texts. J. Yantai Univ. (Nat. Sci. Eng.) **04**, 280–285 (2002)
7. Ma, J., Wang, L., Yao, S.: Named entity recognition for chemical resource text. J. Zhengzhou Univ. (Sci. Edn.) **50**(04), 14–20 (2018)
8. Ma, W., Chen, K.: A Bottom-up Merging Algorithm for Chinese Unknown Word Extraction (2003)
9. Principles for Naming Organic Compounds 2017. Chinese Chemical Society (2017)
10. Song, D., Sun, J., Song, D., et al.: Automatic index of chemical feature words based on rules. J. Intell. **28**(5), 689–692 (2009)
11. Terminology online. http://www.termonline.cn/index.htm
12. Wang, Q., Chen, R., Li, N., Sun, J.: Research on calculation model of term knowledge for chemical name. Libr. J. **34**(10), 59-62+98 (2015)
13. Xia, B., Xun, E.: Distributed representation of Chinese collocation. In: Proceedings of the 32nd Pacific Asia Conference on Language, Information and Computation (2018)
14. Yu, D., Rao, G., Tang, G., Xun, E.: Morphemization in the formation of complex scientific and technical terms. Chin. Sci. Tech. Terminol. **17**(02), 15–20 (2015)
15. Yue, J.: Research on Chinese word segmentation technology for patent documents. Beijing Jiaotong University (2013)
16. Zeng, H., Zhan, E., Zheng, J., Wang, Y.: Recognition of unknown words based on extended rules and statistical features. Appl. Res. Comput. **36**(09), 2704–2707+2711 (2019)
17. Zhang, J.: Review and prospect of automatic indexing technology. Mod. Intell. **29**(04), 221–225 (2009)
18. Zhang, B.: Modern Chinese. Fudan University Press (2017)
19. Zhai, Z., et al.: Improving Chemical Named Entity Recognition in Patents with Contextualized Word Embeddings (2019)

Revisiting Tibetan Word Segmentation with Neural Networks

Sangjie Duanzhu[1,2,3,4], Cizhen Jiacuo[1,2,3,4], and Cairang Jia[1,2,3,4](✉)

[1] The State Key Laboratory of Tibetan Intelligent Information Processing and Application, Qinghai Normal University, Xining, Qinghai, China

[2] Tibetan Information Processing Engineering Technology and Research Center, Qinghai Normal University, Xining, Qinghai, China

[3] Key Laboratory of Tibetan Information Processing, Ministry of Education, Xining, China

[4] Tibetan Information Processing and Machine Translation Key Laboratory, Xining, Qinghai, China

Abstract. Tibetan Word Segmentation is a basic and essential task in Tibetan Natural Language Processing workflow. Performance of TWS can directly affect many other downstream Tibetan NLP tasks since errors propagate in a multi-stage NLP pipeline. Traditionally the majority of researchers leverage linear statistical approaches to tackle Tibetan Word Segmentation, which often requires hand-crafted linguistic feature engineering with great care. In this work, we propose a neural network architecture for Tibetan Word Segmentation, which is a stacked combination of CNN, Bi-LSTM and CRF. By using tagged data for supervised learning and unlabeled data for representation learning, with no involvement in feature engineering, our model can produce promising performance on the test set, surpassing our baseline models by a large margin, and indicating the effectiveness of the proposed neural model.

Keywords: Tibetan word segmentation · Neural networks · Tibetan Natural Language Processing

1 Introduction

Unlike European languages such as English, there is no explicit delimiter between words in Tibetan. Therefore, Tibetan Word Segmentation (TWS) is usually the first and essential subtask to tackle with in Tibetan NLP workflow. The performance of TWS would have a crucial impact on many downstream tasks including Tibetan semantic analysis, reading comprehension, machine translation, etc., since all these systems directly consume the output of a TWS system as their input, leading to the issue of error propagation in a multi-stage NLP pipeline.

Given the significance of TWS in Tibetan NLP, researchers began to address it using the maximum match method [1] back in 1999. Later on, rule-based methods incorporating Tibetan case auxiliary words and continuation features have been proposed. Qi et al. presented a context-aware multi-stage TWS method [2]. The ambiguity issue in TWS

M. Liu et al. (Eds.): CLSW 2020, LNAI 12278, pp. 515–524, 2021.
https://doi.org/10.1007/978-3-030-81197-6_44

was effectively alleviated by detecting overlapping language units and using large-scale word frequency statistics support [3]. Li et al. proposed a Conditional Random Fields (CRF) based model for TWS and a Maximum Entropy (ME) based model for Tibetan part-of-speech (POS) tagging [4], and open-sourced the implementation[1]. Li et al. also proposed to leverage unsupervised features learned from unlabeled text data, gaining a notable boost in system performance [5].

Among previous works, dictionary-based methods [3] relied heavily on dictionaries, linguistic rules and other forms of knowledge hand-crafted with great care by linguistic experts, and thus revealed shortcomings in terms of disambiguation and OOV recall. Statistics-based methods, on the other hand, such as Hidden Markov Model (HMM), hold very strong assumptions on conditional independence [6], limiting the capacity for feature selection, and hence failing to model the contextual signals. Even though the EM model [4] can select any arbitrary form of features, due to the normalization at each node, the model is prone to be stuck in local minimum. The CRF model [7] applies normalization on global context, and accommodates arbitrary contextual features, making it very suitable for conditional sequence modeling. The CRF models are widely used in many NLP tasks such as tokenization, POS-tagging, etc. Like other traditional statistical models, the CRF model requires discrete representations of basic units of the language, which leads to a moderate amount of semantic information loss and imposes constraints on the modeling capacity.

In recent years, with the widespread adoption of deep learning methods in the field of NLP, Tibetan NLP communities have started to embrace the new paradigm shift. To the best of our knowledge, [8] was the first work to explore the neural network approach to TWS. In their work, the authors conducted a series of experiments with different RNN (Recurrent Neural Network) variations and performed a concrete analysis on model selection and system performance.

In this work, we use pre-trained models for both character-level and syllable-level contextual representations [9]. And then by combining the CNN (Convolutional neural network) and Bi-LSTM (Bi-directional Long Short-Term Memory) networks, we build a neural architecture to fully capture Tibetan sentence-level representation. And finally, a subsequent CRF layer is appended to serve as the inference component of our model. In experiments, the accuracy, recall rate and F_1 score of the proposed model reach *93.4%*, *94.2%* and *94.1%*, respectively, surpassing our baseline models by a large margin, indicating the effectiveness of the proposed neural model.

2 Methods

Word segmentation is a process to decompose sentences and paragraphs into word units to facilitate the analysis of the subsequent processing. The TWS task can be formalized similarly to what was described in [10] for Chinese Word Segmentation (CWS). The fundamental cause of ambiguity in TWS is not the absence of natural delimiters between words, but inexplicitness of word boundaries. Therefore, the key to TWS is the correct resolution of this ambiguity. When TWS is treated as a sequence labeling task, it becomes a process of inferring correct positional information on each Tibetan syllable.

[1] https://github.com/liyc7711/tip-las.

2.1 Tagging Schemes

The *BMES* (Beginning, Middle, End, Singleton) tagging scheme is commonly used in both TWS and CWS tasks [4]. However, in Tibetan many functional suffixes such as འིའང ས ནོ ར can be agglutinated with certain syllables, and the Tibetan syllable is thus not necessarily the smallest language unit that requires tagging. As shown in Table 1, the *BMES* scheme will potentially produce a large number of invalid Tibetan character combinations such as ནམ་མཁ. Furthermore, these combinations will also challenge the representation learning if neural network based methods are used to model the TWS tagging task. To eliminate this problem, an extra tag can be introduced to label agglutinated Tibetan syllables (as shown in Table 1, མཁར is tagged as *N*, because the character ར is actually a functional suffix). Introducing the extra tag *N* not only simplifies syllable tagging for TWS, but also eases the process of lemmatization, which is very important in many Tibetan NLP tasks such as lexical analysis and machine translation. Although there is no explicit delimiter to serve as word boundary in Tibetan as aforementioned, there is a dedicated character, ་*wylie: tsheg*), to delimit syllables. If the syllable delimiter is also tagged then there will be redundant tagging information. During inference, when a syllable is tagged as *N*, then it is easy to identify and remove its suffix by applying simple rules. Based on the "*don't guess if you know*" principle [11], we dispense with all *tsheg*'s in the experiments.

Table 1. An example of Tibetan sentence with different tagging.

Original sentence	Tagging schemes	Tokenized sentence	Tagged sentence
ནམ་མཁར་འཕུར།	BMES	ནམ་མཁ ར་ འཕུར།	ནམ[B]་[M]མཁ[E]ར[B]་[E]འཕུར[S]།[S]
(*Fly into the sky*)	BMESN with tsheg		ནམ[B]་[M]མཁར་[N]འཕུར[S]།[S]
	BMESN without tsheg		ནམ[B]མཁར[N]འཕུར[S]།[S]

2.2 The CRF Model

Word segmentation could be formalized as a sequence labeling task, which not only requires the modeling of input tokens to the corresponding labels, but also the dependencies between predicted labels. Chained CRF among the family of probabilistic graph models [7] implements sequential dependencies in the predictions, which allow unconstrained features to represent the conditional probability of the output y for a given input x, as shown in the Fig. 1.

Let X be the sequence of random variables that need to be labeled, that is, the character sequence of the input sentence. Y is the corresponding label sequence, and $\forall$ i, $y_i \in \{B, E, M, S, N\}$. Although the input and output variables are jointly distributed, discriminant-based modeling frameworks flavor conditional probability distribution P(Y | X) estimation with known input and output sequence pairs as shown in Eq. 1.

$$p_\theta(y \mid x) \propto \exp\left(\sum_{e \in E, k} \lambda_k f_k\left(e, y|_e, x\right) + \sum_{v \in V, k} \mu_k g_k\left(v, y|_t, x\right) \right) \quad (1)$$

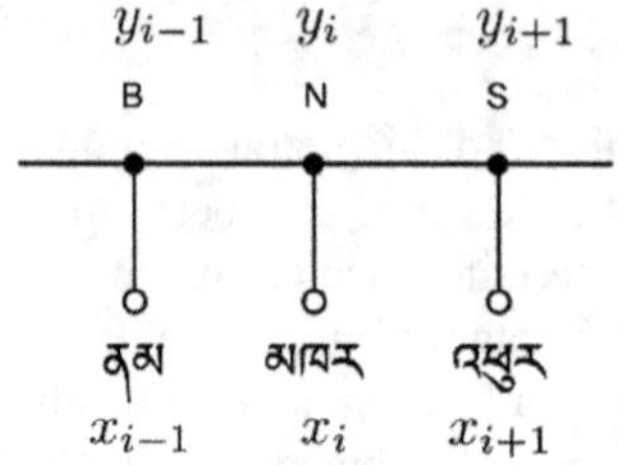

Fig. 1. An illustration for CRF probabilistic graph

where x is the input sequence, y is the corresponding tag sequence, f_k and g_k are the feature functions, and $\theta = \{\lambda_1, \lambda_2, \cdots, \lambda_k; \mu_1, \mu_2, \cdots, \mu_k\}$ are the parameters that need to be optimized iteratively by maximizing the log-likelihood estimation as shown in Eq. 2.

$$\begin{aligned}\mathcal{L}(\theta) &= \sum_{i}^{N} log\, p_\theta\left(y^{(i)}|x^i\right) \\ &\propto \sum_{x,y} \tilde{p}(x,y) log\, p_\theta(y\,|\,x)\end{aligned} \tag{2}$$

where $\tilde{p}(x, y)$ is the empirical distribution attained from training data $D = \left(x^i, y^i\right)_{i=1}^N$.

2.3 Character-Level Representation Learning with CNNs

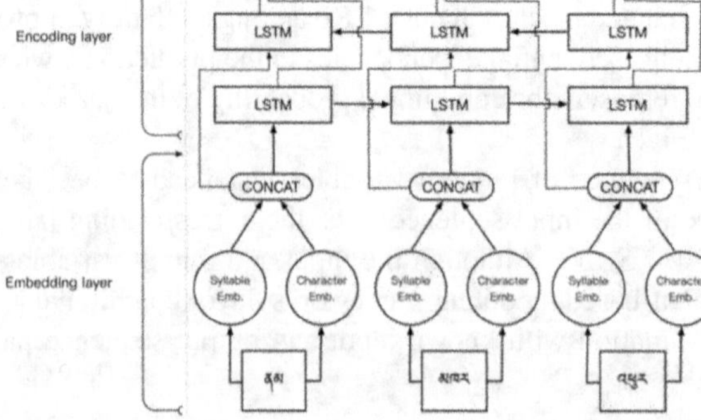

Fig. 2. TWS neural network architecture

The convolutional neural network (CNN) [12] is currently the primary modeling architecture dominating in computer vision [13]. It is also widely used in NLP [14]. Unlike RNN, which is a time-series network, CNN is computationally much lightweight and more suitable for parallelism [15]. In this work, we adopt a framework similar to that proposed in [16] to capture the character-level representation of a given syllable (such as suffixes like ཤཞེཞུའོའང) and use the output of the Tibetan syllable representation as input for subsequent Bi-LSTM networks. For a syllable w consisting of M characters $\{c_1, \cdots, c_M\}$, when CNN is used for modeling, the input of the network is the set represented by each character vector $\varsigma(\{r_1, \cdots, r_M\})$, and a series of convolution operations are performed on ς with a window size of k. Then representation of the mth character in syllable w (denoted as z_m) can be computed as in Eq. 3 (Fig. 2).

$$z_m = \left(r_{m-\frac{(k-1)}{2}}, \ \ldots, \ r_{m+\frac{(k-1)}{2}} \right)^T \tag{3}$$

2.4 Syllable-Level Representation Learning with Bi-LSTM

Long Short-term Memory (LSTM) Networks [17] as a variant of RNN, through some special gate structures of RNN, enable information to selectively affect the state of each time in memory, thus effectively prevent the gradient exploding or vanishing problem and can learn the long-distance dependencies in sequence [18]. To capture both forward and backward contextual information of the given syllable sequence, we use Bi-LSTM to encode Tibetan sentences. A forward LSTM starts to encode the sentence from its first syllable to the end of the sentence (*<eos>*), and learns a forward hidden representation $\vec{h}$. Similarly, a backward LSTM encodes a sentence from *<eos>* and learns a backward hidden representation $\overleftarrow{h}$. These two hidden states are combined in a simple concatenation to serve as the final syllable-level representation h ($h = \left[\vec{h}, \overleftarrow{h} \right]$).

2.5 Inference

An inference layer is the last component in the network and is used to predict label tags for the hidden states of the syllable sequences obtained by CNN and Bi-LSTM. In this work, we experiment with two inference frameworks, namely softmax and CRF. One of our neural network baseline [8] uses a softmax layer for label inference.

Softmax Inference. Suppose x is the syllable sequence of the input Tibetan sentence ($x = \{x_1, \cdots, x_n\}$), and the corresponding hidden state of Bi-LSTM layer is $h = \{h_1, \cdots, h_n\}$, and then softmax normalization operates on h, we get x_1 probability distribution s_i on the label set. See Eq. 4. The loss function is defined by cross entropy. See Eq. 5, where y_i is the ground-truth label for x_i. During decoding, the label $\hat{y}$ for x_t is selected from s_i according to its probabilistic values. See Eq. 6.

$$s_i = \frac{e^{h_i}}{\sum_{i=1}^{n} e^{h_i}} \tag{4}$$

$$\mathcal{L} = -\sum_{i=1}^{n} y_i \log s_j \tag{5}$$

$$\hat{y_i} = \max\left(s_{i,B}, s_{i,M}, s_{i,E}, s_{i,S}, s_{i,N}\right) \tag{6}$$

What is worth noting is that the softmax inference layer does not explicitly model dependencies between labels, and its output is just the probability distribution of the label set. So softmax inference suffers from the label bias problem [19]. For example, this architecture cannot prevent invalid tag sequences, such as *BB, BS, EN, and SN,* from appearing in the predictions.

CRF Inference. Unlike the independent label inference in softmax, as described in Sect. 2.2, CRF can leverage contextual information and features for global optimization objectives. Its input is a label probability matrix P with dimension n × k where n is the number of syllables in the sentence and k is the size of label set. $P_{(i,j)}$ denotes the probability of ith syllable in the sentence being labeled as j. Let the input be $X = (x_1, x_2, \cdots, x_n)$ and the corresponding tag output be $Y = \left(y_1, y_2, \cdots, y_n\right)$, then the tag probability score can be defined as shown in Eq. 7. Where A is the transition matrix of label scores, A_{ij} denotes the transition probability from label i to j, and y_0 and y_1 represent *<bos>* and *<eos>* symbols, respectively. Obviously, the size of the matrix A is $(k+2)^2$. The model can be optimized by maximizing the probability of correctly labeled tag sequences, as shown in Eq. 9, where Y_X is the set of all possible tags of the input sequence X. In the decoding process, the inference of sequence tags is completed by finding the sequence with the highest probability score, as shown in Eq. 10.

$$s(X, y) = \sum_{i=1}^{n} A_{y^i, y^{i+1}} + \sum_{i=1}^{n} P_{i, y_i} \tag{7}$$

$$p(y \mid X) = \frac{e^{s(X,y)}}{\sum_{\tilde{y} \in Y_X} e^{s(X,\tilde{y})}} \tag{8}$$

$$log(p(y \mid X)) = s(X, y) - log\left(\sum_{\tilde{y} \in Y_X} e^{s(X,\tilde{y})}\right) \tag{9}$$

$$y^* = \arg\max_{\tilde{y} \in Y_X} s(X, \tilde{y}) \tag{10}$$

3 Experiments

3.1 Dataset Settings

In the experiments, the training dataset consists of *150K* sentences, and another *10K* sentences serve as validation and testing dataset. The tag distribution in all these datasets is fairly closed as shown in Table 2. All sentences are limited to a maximum length of *125* syllables and a minimum length of *40* syllables. The data scale for training Tibetan syllable vectors is *500K* sentences, and the data scale for training Tibetan character vectors is *100K* sentences. There is no data sample overlapping in the three sets of data.

Table 2. Syllable tag distribution in training, validation and test datasets.

Dataset	B (%)	M (%)	E (%)	S (%)	N (%)
Training	27.60	7.35	26.79	32.29	5.98
Validation	27.66	7.29	26.83	32.33	5.89
Testing	27.71	7.26	26.77	32.33	6.03
Overall	27.61	7.34	26.79	32.29	5.97

3.2 Experiment Settings

CRF Model Settings. The training of the traditional CRF TWS model is based on the Stanford Word Segmenter[2]. In the course of the experiment, a combination of unigram and bigram syllable-level n-gram features extracted from text data is employed.

Embeddings. When pretraining Tibetan syllable and character continuous vector representations, we used Facebook's *fasttext*[3].

Table 3. Syllable and character representation learning algorithms, training parameters, and number of tokens.

Embedding type	Training algorithm	Parameter settings	Total token	Unique token
Syllable embedding	cbow	Iteration number: 10	14M	10206
		Window size: 3		
		Learning ratio: 0.005		
		Minimum frequency:10		
		Dimension: 50		
Character embedding	cbow	Iteration number:10	60M	306
		Window size: 7		
		Learning ratio: 0.005		
		Minimum frequency: 1000		
		Dimension: 30		

As shown in Table 3, we use a large minimum word frequency threshold (1000) when training character representations to filter out uncommon non-Tibetan characters.

2 https://nlp.stanford.edu/software/segmenter.shtml.
3 https://fasttext.cc.

Neural Network Models. The neural network model is implemented using *Tensorflow* as ML framework, GPU is Nvidia *Tesla P4*, and primary parameters are shown in Table 4. A Tibetan syllable contains a maximum of 7 Tibetan characters (such as བསྒྲིགས), so the window size of the convolutional layer is set to 7 in the experiment to obtain complete features of all Tibetan syllable and character levels. In addition, in the experiment, there are two schemes for the initialization of the character vector and the syllable vector. One is the pre-training vector representation, and the other is the random initialization. The effects of the two initialization schemes are as shown in Table 5.

Table 4. Hyper-parameters selection

Parameter name	Parameter value
Character embedding dimension	30
Syllable embedding dimension	50
Kernel size in convolutional layers	7
Number of Convolutional layers	4
Batch size	50 (sentences)
Iteration numbers	25
Dropout	0.5
Optimizer	Adam
Initial learning ratio	0.005

3.3 Results

The experimental results of different models are as shown in Table 5. The model performance parameters are all word-based automatic evaluation results. We choose to use the *perl* script from the *backoff2005* dataset in the *Second International Chinese Word Segmentation (SIGHAN)*[4] as evaluation tool for all our experiments.

Table 5. Experimental results of four types of models

Models	Embedding	Accuracy (%)	Recall (%)	F_1
CRF	-	89.0	86.6	89.5
LSTM+SOFTMAX	Random	89.7	87.9	88.6
	Pretrained	90.1	88.6	87.5
LSTM+CRF	Random	90.9	89.6	89.7
	Pretrained	92.0	90.4	90.5
CNN+LSTM+CRF	Random	92.5	91.3	90.0
	Pretrained	**93.4**	**94.2**	**94.1**

4 http://sighan.cs.uchicago.edu/bakeoff2005.

4 Conclusion

In this work, we explore a Tibetan Word Segmentation model with multiple neural network architectures, and compare it with traditional statistical methods, and finally verify that the CNN + LSTM + CRF neural architecture preforms best on the test data set. Due to limited labeled data, the model can only be used for the Tibetan word segmentation task. We plan to use this work as a basis in the future to study and implement a general Tibetan sequence labeling framework for Tibetan word segmentation, part-of-speech tagging, and Named Entity Recognition (NER). Recently, Transformer-based models in NLP have truly changed the way researchers work with text data. It is potential to further improve our model by using Transformer models to encode Tibetan syllable or even character sequences, but we will leave this for future work.

Acknowledgments. This work was supported by Science and Technology Department of Qinghai Province (grant numbers: 2020-ZJ-Y05, 2020-ZJ-704) and The National Key Research and Development Program of China (grant number: 2017YFB1402200).

References

1. Tsering, T.: Design of an interative Tibetan word segmentation and word registering system (1999)
2. Qi, K.: Tibetan word segmentation designed for information processing. J. Northwest Univ. Nationalities **4**, 92–97 (2006)
3. Liu, H., et al.: Segt: a practical Tibetan word segmentation tool. J. Chin. Inf. Process. **26**(1), 97–104 (2012)
4. Li, Y., et al.: Tip-las: an opensource Tibetan tokenization and pos-tagging system. J. Chin. Inf. Process. **29**(6), 203–207 (2015). (in Chinese)
5. Li, Y., et al.: An hybrid Tibetan word segmentation with unsupervised features. J. Chin. Inf. Process. **31**(2), 71–75 (2017)
6. Rabiner, L.R., Juang, B.: An introduction to hidden Markov models. IEEE ASSP Mag. **3**(1), 4–16 (1986)
7. Lafferty, J.D., McCallum, A., Pereira, F.C.N.: Conditional Random Fields: probabilistic models for segmenting and labeling sequence data. In: Proceedings of the 18th International Conference on Machine Learning (2001)
8. Li, B., et al.: Deep learning based Tibetan word segmentation methods. Comput. Eng. Des. **1**, 194–198 (2018)
9. Bengio, Y., et al.: A neural probabilistic language model. J. Mach. Learn. Res. **3,** 1137–1155 (2003)
10. Xue, N.: Chinese word segmentation as character tagging. Int. J. Comput. Linguist. Chin. Lang. Process. **8**, 29–48 (2003)
11. Tapanainen, P., Voutilainen, A.: Tagging accurately: don't guess if you know. In: ANLP (1994)
12. Cun, Y.L., et al.: Handwritten digit recognition with a back-propagation network. Adv. Neural. Inf. Process. Syst. **2**(2), 396–404 (1990)
13. Krizhevsky, A., Sutskever, I., Hinton, G.E.: ImageNet classification with deep convolutional neural networks. In: International Conference on Neural Information Processing Systems (2012)
14. Collobert, R., et al.: Natural language processing (almost) from scratch. J. Mach. Learn. Res. **12**, 2493–2537 (2011)

15. Lample, G., et al.: Neural Architectures for Named Entity Recognition (2016)
16. Kim, Y., et al.: Character-aware neural language models. In: 30th AAAI Conference on Artificial Intelligence (2016)
17. Hochreiter, S., Schmidhuber, J.: Long short-term memory. Neural Comput. **9**(8), 1735–1780 (1997)
18. Yao, K., et al.: Spoken language understanding using long short-term memory neural networks. In: 2014 IEEE Spoken Language Technology Workshop (SLT) (2014)
19. Sutton, C., McCallum, A., et al.: An introduction to conditional random fields. Found. Trends Mach. Learn. **4**(4), 267–373 (2012)

A Study of *KIND* Metaphors and Simile Annotation Using Dependency Parsing and ConceptNet

Menghsien Shih[1(✉)], Siawfong Chung[2], Yuhsiang Shen[3], and Hengchia Liao[3]

[1] Center for General Education, National Chung Cheng University, No. 168, Section 1, University Road, Min-hsiung, Chia-yi 621301, Taiwan (R.O.C.)
cgesmh@ccu.edu.tw

[2] Department of English, National Chengchi University, No. 64, Section 2, ZhiNan Road, Taipei City 11605, Taiwan (R.O.C.)
sfchung@nccu.edu.tw

[3] Master's Program in Teaching Chinese as a Second Language, National Chengchi University, No. 64, Section 2, ZhiNan Road, Taipei City 11605, Taiwan (R.O.C.)
{108161001,106161011}@nccu.edu.tw

Abstract. Metaphors are pervasive in human language, but the understanding of metaphors still poses a challenge for artificial intelligence (AI), cognitive science, and linguistics. To understand metaphorical expressions, a metaphor resource is important. Among different metaphorical expressions, the mapped concepts of KIND metaphors ('X IS A KIND OF Y') provide important knowledge resources for metaphor understanding. In this study, we examined the correlation between metaphoricity and concept distance in KIND metaphors. We first automatically captured the mapped nominal concepts in 144 KIND metaphor sentence candidates using parsing, and then estimated the concept distance using ConceptNet. Three annotators then rated the degree of metaphoricity of these sentences using a 5-point Likert scale. The results indicated a significant negative correlation between concept distance and metaphoricity, implying that the shorter the concept distance, the higher the metaphoricity of the expression. Our findings showed that KIND metaphors form a different type of metaphorical mapping.

Keywords: Conceptual metaphor · Corpus annotation · Metaphor identification

M. Liu et al. (Eds.): CLSW 2020, LNAI 12278, pp. 525–535, 2021.
https://doi.org/10.1007/978-3-030-81197-6_45

1 Introduction

It has been claimed [1] that "the essence of metaphor is understanding and experiencing one kind of thing in terms of another". It was further elaborated that conventional metaphors are metaphors that are conceptualized in a conceptual system and that language contains a rigorous use of conceptual metaphors. The pervasiveness of metaphors is evident in studies on metaphors in various fields, ranging from politics, economics, and psychology to linguistic and discourse studies. One scholar [2] commented that with "language being a major component of culture, metaphor is no doubt one of the most important linguistic devices that reflect cognitive vision and epitomize cultural context." We found abundant research that was based on the Conceptual Metaphor Theory and its applications in various languages (e.g., [3, 4, 5], to name a few).

Despite these many studies, we claim that there is still a need to observe how concepts are mapped in different languages for the following reasons. First, not all metaphors can be captured using the Conceptual Metaphor Theory. Second, many previous studies only focused on several target or source domains at the same time due to the high cost of manual annotation. Third, because computer programming has become more accessible, the possibility of creating a metaphor resource requires a continual investment of energy and time.

In this paper, we will focus on two types of metaphors less studied by others, namely KIND metaphors and similes. Unlike metaphors that have designated source and target domains, KIND metaphors set no restrictions on target and source domains. KIND metaphors are often intentionally created to explain one thing in terms of another. Some examples are shown in (1) below, taken from the Academia Sinica Balanced Corpus (Sinica) and Corpus of Contemporary Taiwanese Mandarin (COCT). The possible mapped concepts are presented using small capitals (although they may not be the same as the domains used in the Conceptual Metaphor Theory). Other useful clues are underlined.

(1) (a) 因為 換 工作 對 個人 來 CHANGE IS EXHAUSTION
yīnwèi huàn gōngzuò duì gèrén lái
because change job for individual come
說 也 是 一 種 耗損 (Sinica)
shuō yě shì yì zhǒng hàosǔn
speak also SHI one kind exhaustion
'Because changing job is a kind of exhaustion for individuals.'

(b) 全民健保 其實 只 是 SYSTEM IS A DESIGN
quánmínjiànbǎo qíshí zhǐ shì
National Health Insurance actually only SHI
一 種 「財務 設計」(Sinica)
yì zhǒng cáiwù shèji
one kind financial design
'Actually National Health Insurance is just a kind of financial design'

(c) 宗教 是 一 種 定形 的 而 精密 RELIGION IS ANSWER
zōngjiào shì yì zhǒng dìngxíng de ér jīngmì
religion SHI one kind fixed DE and precise
的 答案 (COCT)
de dáàn
DE answer
'Religion is a kind of fixed and precise answers.'

(d) 其實 被 關心 也 是 一 種 EMOTION IS A PRESSURE
qíshí bèi guānxīn yě shì yì zhǒng
actually BEI care also SHI one kind
壓力 (COCT)
yālì
pressure
'Actually to be cared about is also a kind of pressure.'

Like most metaphors, KIND metaphors consist of two concepts that are related by a language user to achieve a certain discourse effect. Since KIND metaphors have predictable patterns, we hypothesized that we could semi-automatically capture the mapped concepts using computational methods. After that, we intended to observe how the two nominal concepts were related using a lexical knowledgebase system called ConceptNet [6]. This paper will address the following two research questions:

1. How can the two mapped concepts of KIND metaphors and similes be captured semi-automatically?
2. Can the similarity distance between two concepts facilitate KIND metaphor and simile annotation/identification?

We will answer both questions using computational methods.

2 Previous Metaphor Identification Procedure

To better understand how metaphors are identified, we first introduced a model that was based on the lexical identification of metaphors (i.e., the Metaphor Identification Procedure Vrije Universiteit [MIPVU]) [7][1]. Although this is not the only model for metaphor identification, it is the only model that states explicitly how a lexical item can be considered an instance of metaphors. Unlike the Conceptual Metaphor Theory [1], MIPVU pays no attention to identifying the source or target domains. Instead, it aims to recognize words that carry metaphorical meanings when they are used in sentences. In MIPVU, metaphors are identified based on the basic sense and the metaphorical sense in the dictionary, and even the most conventional metaphors need to refer to the dictionary to determine whether their senses are metaphorical. For example, the English adjective *valuable* in the expression "to do valuable work" is identified as a metaphorical use [8] because, in addition to the sense of "very useful and important," the word *valuable* also has a more basic sense of "worth a lot of money." In other words, one of MIPVU's major steps is to find a more basic meaning other than the metaphorical meaning already identified by the annotator. The presence of a more basic meaning other than the metaphorical one is an indicator that a lexical item is being used metaphorically.

Although both MIPVU and our investigation of KIND metaphors and similes are based on the lexical level, they are different in the following way. MIPVU is based on the identification of metaphorical senses and adopts a bottom-up approach that does not assume that "related conceptual metaphors guide linguistic metaphor identification" [8]. In the current study, however, we examined the central role of concepts in metaphor identification. In other words, although we took a lexical approach in identifying KIND metaphors and similes (using keywords such as 'a kind of' and 'as…as' in Chinese), we also wanted to know whether the two nominals we identified were close or distant in meaning. The similarity of our approach and that of MIPVU is that neither claims whether a particular nominal belongs to a certain higher and conceptual target or source domain.

Another advantage of MIPVU is that it developed a corpus interface based on the VU Amsterdam Metaphor Corpus [7, 8], which consists of 200,000 words of manually annotated English metaphorical expressions from the British National Corpus in four registers: news, fiction, academic, and spoken. For the Chinese language, such a metaphor resource is still under development. Using a semi-automatic approach to identify Chinese metaphorical expressions with the help of computational linguistic tools, the current study will contribute to the building of a Chinese metaphor resource in the future.

In addition to MIPVU, there have been other studies that have used different techniques to identify metaphors. In a metaphor labeling task, one study [9] recruited 100 Mechanical Turks to label the metaphoricity of 60 sentences (with the labels "Not Metaphoric," "Slightly Metaphoric," and "Very Metaphoric") from the Corpus of Contemporary American English. The results showed that "the metaphoricity of a sentence was computed by taking the percentage of participants who identified it as metaphoric" [9]. That is, if all participants labeled a sentence as "Very Metaphoric," the metaphoricity

[1] The MIPVU is modified from the MIP proposed by [12].

of the sentence would be computed as 1.0, whereas if half of the participants labeled the sentence as "Not metaphoric," it would be 0.5.

Although the study [9] introduced a gradient measurement of metaphoricity in the labeling task, the results were not evaluated by inter-annotator agreement. In this study, we annotated the metaphoricity of a sentence with a more fine-grained 5-point Likert scale and evaluated the results using Krippendorff's α agreement, a method that we hoped would provide a better evaluation of the nominal concepts we retrieved semi-automatically from the corpus.

3 Methodology

The first step of this work was to collect potentialKIND metaphor and simile sentences from the corpora. This was done by searching for sentences containing the keywords 是一種 *shì yì zhǒng* 'is a kind of' and [X 一樣的 Y] X *yīyàng de* Y 'Y is like an X'. A total of 144 metaphor sentence candidates of 'is a kind of' were collected from the Academia Sinica Balanced Corpus 4.0,[2] of which 19 out of the 144 sentences were first annotated as metaphorical in a binary metaphor identification task by one annotator. This step was taken to ensure that KIND metaphor identification was possible by analyzing these 'is a kind of' sentences. Table 1 below shows a summary of the number of KIND metaphor and simile sentence candidates:

Table 1. Total number of KIND metaphor and simile sentence candidates and number of positively annotated sentences (annotated as KIND metaphors or similes).

	KIND	Simile
Positive	19	293
Negative	125	107
Total	144	400

Similarly, a total of 400 [X 一樣的 Y] X *yīyàng de* Y 'Y is like an X' sentences were collected from the Corpus of Contemporary Taiwanese Mandarin (the written corpus of version 2017),[3] of which 293 out of the 400 sentences were annotated as similes. Of the [X 一樣的 Y] X *yīyàng de* Y 'Y is like an X' sentences, 107 were deleted because they were identified either as repeating expressions, incorrect sentences or POS tagging errors. For example, the adjective 溫馴 *wēnxùn* 'tame' in the sentence 像綿羊一樣的溫馴 *xiàng miányáng yīyàng de wēnxùn* 'as tame as a sheep' was incorrectly tagged as a noun in the corpus and was therefore deleted in our study of nominal concepts in simile sentences.

Once we were certain that the sentences we collected could be used to identify KIND metaphors and similes, we used a dependency parser to identify the nominal concepts

[2] http://asbc.iis.sinica.edu.tw.

[3] http://coct.naer.edu.tw.

we needed for the study. To locate the two mapped noun concepts in each KIND metaphor sentence, we searched for a dependency pattern between the two nouns in the sentence. For the 'is a kind of' sentences, the 144 copular sentences were first parsed by the SyntaxNet dependency parser [10]. In the dependency grammar, the two nouns in each copular sentence were analyzed as a nonverbal predicate and its nominal subject, as shown below in Fig. 1.

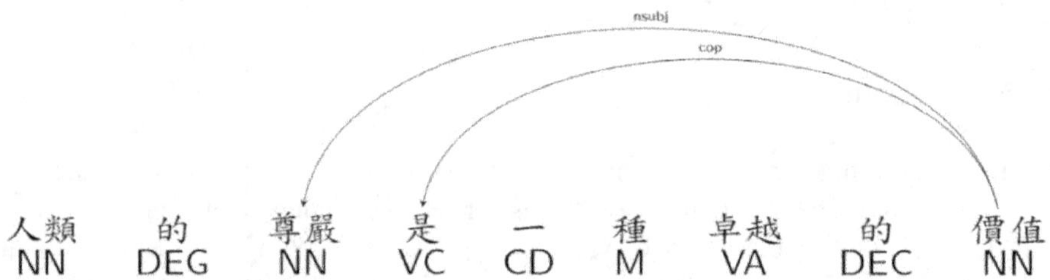

rénlèi *de* *zūnyán* *shì* *yī* *zhǒng* *zhuóyuè* *de* *jiàzhí*

human.being DE dignity SHI one kind excellent DE value

'Human beings' dignity is a kind of excellent value'

Fig. 1. Nominal concept identification through dependency parsing

For the example shown in Fig. 1, we located the two nominal concepts in the parsed sentence (e.g., the nouns 尊嚴 *zūnyán* 'dignity' and 價值 *jiàzhí* 'value' in the above sentence) connected by the dependency relation of *nsubj* (the nominal subject). All 144 sentences were parsed in the same way so that all nominal concepts were captured. Then, the captured nominal concepts, along with the sentences, as illustrated in Table 2 below, were given to three annotators to rate (using the 5-point Likert scale) so that the metaphoricity of the usage of the two nominal concepts in the sentences could be determined.

Three skilled annotators were required to rate the 144 KIND metaphor sentence candidates (collected from the Academia Sinica Balanced Corpus 4.0) in the copular 是一種 *shì yì zhǒng* 'is a kind of' construction, and the two nominal concepts in the sentences were first captured using a dependency parser. If an annotator strongly agreed that a sentence was metaphoric, the metaphoricity of the sentence was rated 5 on the Likert scale; if the annotator strongly disagreed that the sentence was metaphoric, the metaphoricity was rated 1. The complete set of 5-point Likert items and their corresponding metaphoricity statements are listed in Table 3.

After the annotation was completed, we measured the annotator agreement scores. Then, we investigated the correlation between the concept distance and the average metaphoricity scores given by the two annotators.

Concept distance was estimated by the shortest path between the two concepts in question in ConceptNet 5.5 [6]. For example, the concept 尊嚴 *zūnyán* 'dignity' was directly related to the concept 人生 *rénshēng* 'life' (i.e., 人生 *rénshēng* 'life' was linked by 尊嚴 *zūnyán* 'dignity' in ConceptNet with a 'motivated by' relation), and 人

Table 2. Example data for annotation.

Concept 1	Concept 2	Sentence
市場 *shìchǎng* 'market'	神話 *shénhuà* 'myth'	完全 自由的 市場 只 是 一 種 神話 *wánquán zìyóu de shìchǎng zhǐ shì yì zhǒng shénhuà* absolute free DE market just SHI one kind myth 'absolutely free market is just a kind of myth'
想像 *xiǎngxiàng* 'imagination'	遊戲 *yóuxì* 'game'	所以 我們 這 一 種 想像 是 一 種 *Suǒyǐ wǒmen zhè yì zhǒng xiǎngxiàng shì yì zhǒng* therefore we this one kind imagination SHI one kind 不 容易 玩 的 遊戲 *bù róngyì wán de yóuxì* not easy play DE game 'therefore, this kind of imagination for us is a kind of game which is not easy to play'
聲音 *shēngyīn* 'sound'	染料 *rǎnliào* 'dye'	原來， 聲音 就 是 一 種 染料 *yuánlái shēngyīn jiù shì yì zhǒng rǎnliào* it.turns.out.that sound JIU SHI one kind dye 'it turns out that sound is indeed a kind of dye'

Table 3. The 5-point Likert items and corresponding metaphoricity statements.

Likert item	Statement that the sentence is metaphoric
1	Strongly disagree
2	Disagree
3	Neither agree nor disagree
4	Agree
5	Strongly agree

生 *rénshēng* 'life' was further related to 價值 *jiàzhí* 'value' in a 'located at' relation. Therefore, the shortest path between 尊嚴 *zūnyán* 'dignity' and 價值*jiàzhí* 'value' in ConceptNet was 尊嚴 *zūnyán* 'dignity' → 人生 *rénshēng* 'life' → 價值 *jiàzhí* 'value'. We then computed the concept distance as the length of the path. In this case, the distance between 尊嚴 *zūnyán* 'dignity' and 價值 *jiàzhí* 'value' was 3. If we could not find any connection between two concepts in ConceptNet, we assigned such a sentence with an extremely long distance of 99 to create a distinctive distance between the not-found distances and the ones that we detected. For similes, the shortest path between the two mapped nominal concepts was also computed, but the metaphoricity of the [X 一樣的 Y] X *yīyàng de* Y 'Y is like an X' sentences was not rated because it is a relatively robust pattern in simile identification (as three-quarters of [X 一樣的 Y] X *yīyàng de* Y 'Y is like an X' sentences are the intended simile expressions). In other words, almost all instances of [X 一樣的 Y] X *yīyàng de* Y 'Y is like an X' were similes and there was no need to annotate which ones were more metaphorical than the others, although such a step could be added in a future study to see whether there are different types of similes with different levels of figurative meanings.

4 Results and Discussion

After removing 33 KIND metaphor data entries with concept parsing errors or with annotation differences among three annotated metaphoricity points greater than or equal to 3 points, we evaluated the inter-annotator agreement (IAA) of the annotated metaphoricity points using Krippendorff's α. The value of α among the 111 annotated sentences of 是一種 *shì yì zhǒng* 'is a kind of' was 0.645, near a tentative agreement among the three annotators.

The correlation between concept distance and metaphoricity was also examined by the Pearson's R test. We first computed the mean of the metaphoricity scores given by the two annotators. For example, if one annotator rated the metaphoricity of a sentence 5 points ('Strongly agree') and the other annotator rated it 4 points ('Agree'), the mean of the metaphoricity of the sentence was computed as 4.5.

The Pearson's R test was then performed to examine the correlation between the mean metaphoricity and the distance[4] of the two mapped nominal concepts in the sentence. The statistical test indicated a significant negative correlation (Pearson's R = −0.203, $p < 0.05$), implying that the longer the concept distance, the lower the metaphoricity of a sentence (see Table 4).

Table 4. The results of the Pearson's correlation test.

Correlations

		Metaphoricity	ConceptDist
Metaphoricity	Pearson Correlation	1	-.203*
	Sig. (2-tailed)		.033
	N	111	111
ConceptDist	Pearson Correlation	-.203*	1
	Sig. (2-tailed)	.033	
	N	111	111

*. Correlation is significant at the 0.05 level (2-tailed).

We also examined the concept distances in the simile sentences of [X 一樣的 Y] X *yīyàng de* Y 'Y is like an X' and compared the KIND metaphor sentences 是一種 *shì yì*

4 The average concept distance of the 111 concept pairs is 21.78.

zhǒng 'is a kind of'. The two barplots in Fig. 2 below (the concept distances of 99 were omitted in the barplots but are displayed as outliers in the boxplots in Fig. 3) and the two boxplots in Fig. 3 below show the distribution of the concept distances in the 111 KIND metaphor sentences and the 293 simile sentences.

The boxplots in Fig. 3 show the interquartile range from the first to the third quartile of concept distances. We observed that the range of concept distances in the simile expressions (the right panel, mean = 32.92, and the median of 5.0 is represented by the bold horizontal line in the box) was enormously wider than the small range box (mean = 21.64, median = 4.0) in the left panel of the KIND metaphor expressions. These results indicate that for the KIND metaphor expressions of 是一種 *shì yì zhǒng* 'is a kind of', there were only 23 out of the 111 sentences (20.53%) where the mapped concepts did not have any connection in ConceptNet and were thus assigned a distance of 99 between the nominal concepts. These 23 KIND metaphor sentences were treated as outliers, as plotted on the top of the left panel.

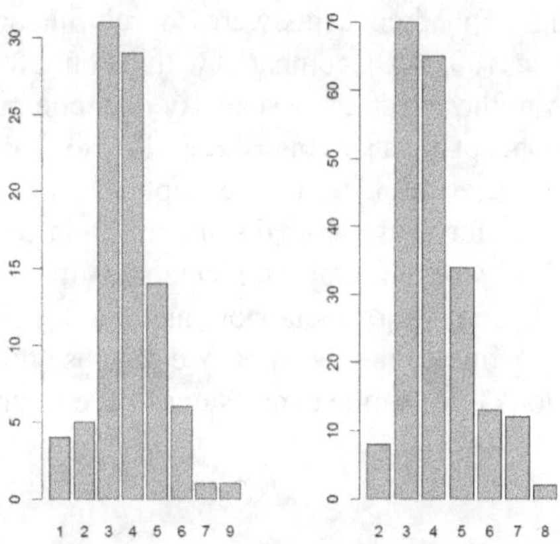

Fig. 2. Two barplots of the raw counts of the concept distances in the 111 KIND metaphor sentences (left panel) and the 293 simile sentences (right panel), respectively.

For the simile expressions of [X一樣的 Y] X *yīyàng de* Y 'Y is like an X', there were up to 90 out of the 293 sentences (30.71%) in which there was not any connection (in ConceptNet) between the mapped concepts. These 90 sentences broadened the interquartile range of the concept distances in the simile sentences of [X 一樣的 Y] X *yīyàng de* Y 'Y is like an X'.

As for interpretation, on average, the mapped concepts in the KIND metaphor expressions of 是一種 *shì yì zhǒng* 'is a kind of' were closer to each other (mean = 21.64) compared with the concepts in the simile expressions (mean = 32.92). It was easier to find a closer connection between the mapped concepts in the KIND metaphor expressions based on ConceptNet (only 23 out of the 111 mapped concepts without any connection).

As for the simile expressions of [X 一樣的 Y] X *yīyàng de* Y 'Y is like an X', they demonstrated a behavior different from the KIND metaphor expressions. We observed in the barplot in the right panel of Fig. 2 that more mapped concepts in the simile expressions were connected with longer distances (compared with concept distances in the

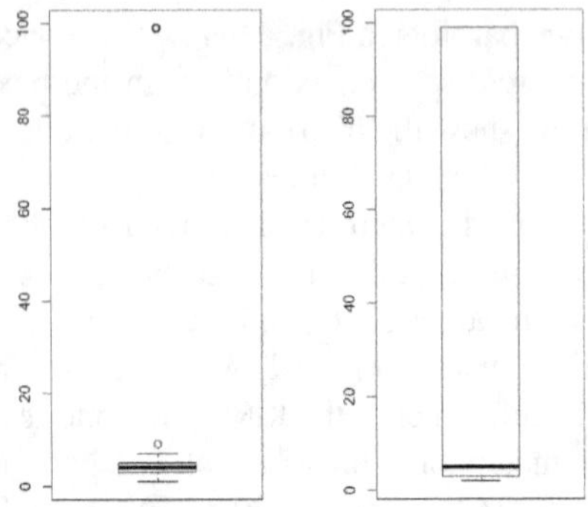

Fig. 3. Two boxplots of the concept distances in the KIND metaphor sentences and the simile sentences, respectively.

KIND metaphor expressions in the left panel). Meanwhile, there were many more simile expressions in which the mapped concepts were entirely unconnected in ConceptNet (90 out of the 293 sentences [30.71%], compared with 23 out of the 111 KIND metaphor sentences [20.53%]). From these results, we can say that, comparing the two, the KIND metaphors have shorter concept distances than the similes do. This means that in creating KIND metaphors, language users prefer to use concepts that are closer and more related, whereas for similes, they prefer to use more distant or unrelated concepts. This distinction was not mentioned in any of the studies on metaphors in the past. Furthermore, not many studies have investigated KIND metaphors and similes in the metaphor models and in the way that was conducted in this work. We may use this feature to distinguish KIND metaphor expressions from simile expressions in the future.

5 Conclusion

This paper has addressed the importance of KIND metaphors and similes in the understanding of metaphorical expressions. For KIND metaphors, we automatically captured the mapped nominal concepts in the KIND metaphor sentence candidates by dependency parsing and estimated the concept distance using ConceptNet. Then, two annotators rated the metaphoricity (using a 5-point Likert scale) of the 144 KIND metaphor sentence candidates. A statistical test was performed to examine the correlation between metaphoricity and concept distance in the KIND metaphor sentence candidates. The results indicated a significant negative correlation (Pearson's $R = -0.213$, $p < 0.05$) between metaphoricity and concept distance, implying the shorter the concept distance, the higher the metaphoricity of a KIND metaphor sentence candidate. This finding is new, for we previously assumed that metaphorical concepts are more distant. Our findings showed that the KIND metaphors formed a different type of metaphorical mapping, which was [11] claimed to be almost literal. Compared with the KIND metaphors, the similes showed a mapping of more distant concepts, which was different from the KIND metaphors. The differences between KIND metaphors, similes, and literal sentences is a good direction for further studies. In the future, a larger-scale experiment is necessary to automatically identify KIND metaphor expressions using dependency parsing and ConceptNet.

This study is the first of its kind to analyze data that were not "Conceptual Metaphors" but metaphors with markers [11], such as 'is a kind of' and 'as ... as'. This is a significant contribution to metaphor research and to the lexical-based analysis of metaphors. In the future, it should be possible to work toward a metaphor resource in Chinese that contains metaphors with such marking.

Acknowledgments. We appreciate the assistance from Chi-Ling Lee, Man-Hua Huang, and Wei-Ting Tseng (from the Master's Program in Teaching Chinese as a Second Language, National Chengchi University). This research was supported by the Ministry of Science and Technology (MOST 108-2410-H-004-095- and 109-2811-H-004-503), Taiwan.

References

1. Lakoff, G., Johnson, M.: Metaphors We Live By. University of Chicago Press, Chicago (1980)
2. Su, L.I.-W.: What can metaphors tell us about culture? Lang. Linguist. **3**, 589–613 (2002)
3. Yu, N.: The Contemporary Theory of Metaphor : A Perspective from Chinese. John Benjamins Publishing Company (1998)
4. Ibarretxe-Antuñano, I.: The relationship between conceptual metaphor and culture. Intercult. Pragmat. **10**, 315 (2013)
5. Charteris-Black, J., Ennis, T.: A comparative study of metaphor in Spanish and English financial reporting. English Specif. Purp. **20**, 249–266 (2001)
6. Speer, R., Chin, J., Havasi, C.: ConceptNet 5.5: An open multilingual graph of general knowledge. In: Proceedings of the 31st AAAI Conference on Artificial Intelligence. pp. 4444–4451. Association for the Advancement of Artificial Intelligence (2017)
7. Steen, G.J., Dorst, A.G., Herrmann, J.B., Kaal, A., Krennmayr, T., Pasma, T.: A Method for Linguistic Metaphor Identification. John Benjamins Publishing Company, Amsterdam (2010)
8. Krennmayr, T., Steen, G.: VU Amsterdam Metaphor Corpus. In: Ide, N., Pustejovsky, J. (eds.) Handbook of Linguistic Annotation, pp. 1053–1071. Springer, Dordrecht (2017). https://doi.org/10.1007/978-94-024-0881-2_39
9. Dunn, J.: Measuring metaphoricity. In: Proceedings of the 52nd Annual Meeting of the Association for Computational Linguistics, pp. 745–751. Association for Computational Linguistics (2014)
10. Alberti, C., et al.: SyntaxNet Models for the CoNLL 2017 Shared Task (2017)
11. Goatly, A.: The Language of Metaphors. Routledge (1997)
12. Pragglejaz Group: MIP: A method for identifying metaphorically used words in discourse. Metaphor Symb. **22**, 1–39 (2007)

A Research on the Generation Model and Evaluation Model of Chinese Wu-Qing Couplets

Yuyao Zhang, Yuru Jiang(✉), Yu Wu, and Jing Su

Beijing Information Science and Technology University, Beijing, China
jiangyuru@bistu.edu.cn

Abstract. The Wu-Qing couplet is a unique form of expressions in couplets and an important part of Chinese traditional culture. It shows the profoundness and interesting features of Chinese. However, there is still no research on the automatic generation of Wu-Qing couplets because the corpus of Wu-Qing couplets is scarce and it cannot support the training process of deep learning. This paper proposes a sequence-to-sequence Wu-Qing couplet generation model based on the idea of transfer learning. At the same time, in order to further improve the effectiveness of the model, based on the characteristics of Wu-Qing couplets, such as coherence, rhythm change, and semantic separation, this paper proposes an evaluation model, which can reorder the output of the generation model for better results. Finally, a complete Chinese Wu-Qing couplet automatic generation system is constructed based on the generation model and the evaluation model.

Keywords: Wu-Qing couplet · Transfer learning · Sequence to sequence

1 Introduction

Chinese couplets, called "duìlián", is a form of Chinese traditional culture. The couplet was born from poetry. It can be traced back to the Three Kingdoms period, which was about two thousand years ago. It is inseparable from people's lives, as it shows people's living circumstances and spiritual wishes in different periods. The Spring Festival couplets are best known to people. In addition, couplets have different categories according to their applications, meanings, techniques, etc. Wu-Qing couplets is a special kind among Chinese couplets.

Wu-Qing couplets originated in the late Qing dynasty and was first developed as a popular word game among scholars and bureaucrats. It is generally said that it was created by Zhang Zhidong. At a banquet, Zhang Zhidong and his friends drank and wrote couplets. One person said "树 (tree) 已 (already) 半 (half) 寻 (an ancient measure of length equal to about eight feet) 休 (not) 纵 (though) 斧 (axe)" as the first sentence, and Zhang Zhidong replied "果 (fruit) 然 (function word) 一 (one) 点 (little) 不 (no) 相 (mutually) 干 (shield)". This couplet is seemingly not neat, but in fact, the words correspond one by one. "树" and "果" are plants. "已" and "然" are both function words.

M. Liu et al. (Eds.): CLSW 2020, LNAI 12278, pp. 536–548, 2021.
https://doi.org/10.1007/978-3-030-81197-6_46

"半" and "一" are both numerals. As eight feet equals one 'xun (寻)' in ancient times, "寻" is a unit of length, which is similar to the quantifier "点". "休" and "不" are both function words expressing negative meaning and "纵" and "相" are also function words. "斧" and "干" are both ancient cold weapons [1]. It is hard to come by such writing of the ingenious couplet without roughness.

Wu-Qing couplet shows broadness, profoundness and an interesting side of Chinese, which is a very important art form. In spite of the lack of inheritance and development, Wu-Qing couplets have a relatively complete system. Wu-Qing couplets are written with the methods such as metonymy, conversion, interpreting words as other meanings, changing the part of speech of words in Chinese homophones, polysemy, transliteration, and temporary cancellation of the original meaning in proper nouns. In these ways that emphasize the neatness of a single word, especially the subclass of nouns (although the semantics are irrelevant [2]), the couplets are created. Therefore, Wu-Qing couplets are opposite to the general couplets in structure and are committed to breaking the consistency of rhythm between the first and second sentences, to reach a strong contrast [3].

On the basis of all the above, it is necessary to ensure that the meaning of each sentence is natural and coherent. Hence, the following characteristics of Wu-Qing couplets are being put forward: (1) Matching between single characters: using antonyms or similar words in the same category would create better couplets. It can be appropriately relaxed in practical application. (2) Connotation: using different meanings in polysemy, changing the category, part of speech and the structure of the word. (3) Semantic separation: from matching characters to non-matching words, from non-matching words to non-matching sentences, it shows the irrelevance of sentence meaning. (4) Fluency and coherence: each sentence is natural and coherent, and the meaning of the second sentence is often common, clear, and easy to understand.

Some researchers have completed the automatic generation of couplets with the natural language processing technology based on the sequence-to-sequence model and have got excellent outcomes. However, these works are all based on a large-scale corpus. Wu-Qing couplets, as a unique branch of the couplets, have a late origin, and the magnitude of the collected corpus cannot support the training process of deep learning. In order to solve this problem, in this paper, a two-stage sequence-to-sequence Wu-Qing couplet automatic generation model based on the idea of transfer learning is proposed. In the first stage, a large-scale general couplet corpus is being used to pretrain. In the second stage, the Wu-Qing couplet corpus is being used to fine-tune the generation model. To further ensure the generation effect of the second stage model, this paper also proposes an evaluation model that can reorder the output of the generation model for results with higher quality. Finally, by combining the generation model and evaluation model, a complete Chinese Wu-Qing couplet automatic generation system is constructed.

The main contributions of this paper are as follows: (1) Construct a two-stage Wu-Qing couplet automatic generation model based on the transfer learning to solve the problem that Wu-Qing couplet corpus is deficient; (2) For ensuring the effect of Wu-Qing couplet automatic generation model, an evaluation model with the characteristics of Wu-Qing couplet is proposed; (3) Construct the first Wu-Qing couplet automatic generation system. It is expected to help beginners and enthusiasts to break through the shackles of thinking and provide some help and possibilities for the creation of Wu-Qing couplets.

2 Related Work

With the development of artificial intelligence and natural language processing, there has been extensive research in the field of couplet generation, and many researchers have proposed different research methods. Dr. Yue [4] used neural network to study the semantics of the image-thought level. He used vocabularies in the Spring Festival couplets to conduct experiments, and constructed the Spring Festival couplet generation system, in which the Case Grammar is adopted to represent semantics. In 2005, researchers from Shandong University of Technology proposed a couplet generation system based on the grammar development platform using Lexical-Functional Grammar. The system obtained parallel words or phrases by describing the constraints of the first sentence [5]. In 2008, Microsoft Research Asia proposed a couplet system based on multi-stage statistical machine translation [6]. In 2009, researchers from Tsinghua University proposed a method of generating couplets based on the combination of statistics and rules which were distinguished into hard rules and soft rules, and established a directed probability graph model of couplet response under the guidance of soft rules, in which Expectation-Maximization algorithm is used to estimate parameters [7]. In 2019, Fan et al. [8] adopted the sequence-to-sequence neural network method based on attention mechanism to generate couplets, and reordered the candidates of the generated couplets.

3 Method

3.1 System Overview

The automatic generation system of Wu-Qing couplets consists of two parts: a generation model and an evaluation model. The flow of the generation of Wu-Qing couplets is shown in Fig. 1. First, input the first sentence and prediction is made through the generation model to obtain a series of candidates of the second sentence. Next, pair all the candidates of the second sentence with the first sentence, and send them to the evaluation model for evaluation, and then reorder the candidates of the second sentence according to the evaluation results to obtain the final output result.

3.2 Generation Model

The generation model of Wu-Qing couplets uses a sequence-to-sequence model based on the attention mechanism. The model structure is shown in Fig. 2, where the encoder and decoder are in both Long Short-Term Memory. The sequence-to-sequence model was first proposed by Sutskever et al. [9]. In the standard sequence-to-sequence model, the network structure is mainly divided into encoder and decoder. The encoder converts a variable-length sequence into a fixed-length vector, and the decoder converts the fixed-length vector into a variable-length target sequence. In order to further improve

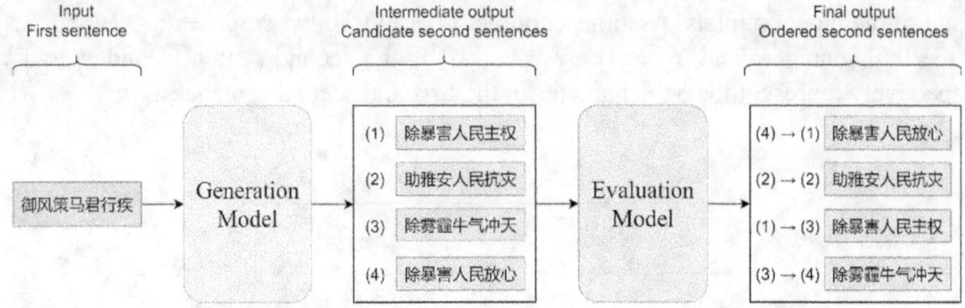

Fig. 1. The complete system structure and generation process

the syntactic and semantic correlation between the first sentence and the second sentence, attention mechanism [10] is often introduced when decoding. With the attention mechanism, the decoder no longer simply depends on the final encoding result of the encoder, but dynamically obtains the encoder result as a context at each generation time to decode and generate better.

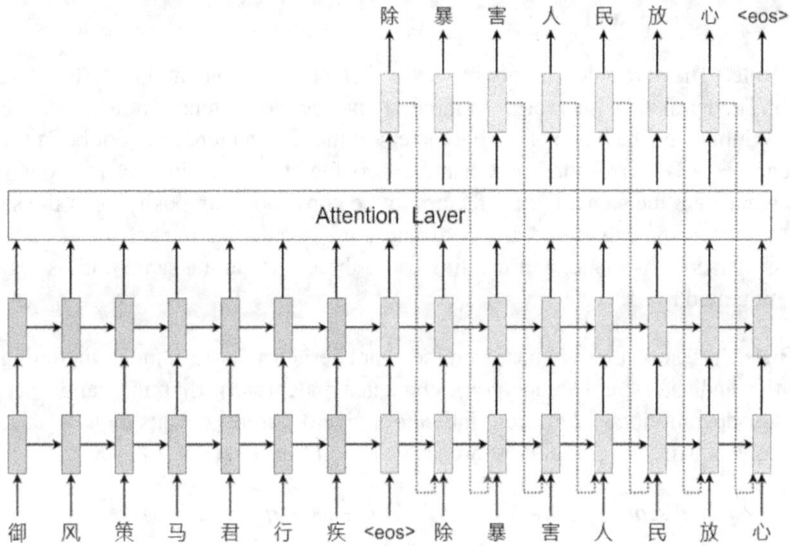

Fig. 2. A sequence-to-sequence model based on attention mechanism

3.3 Evaluation Model

Based on the characteristics of Wu-Qing couplets, we have built a rule-based evaluation model, as shown in Fig. 3. The input of the evaluation model is the first sentence and the second sentence of the couplets. First, the hard rules are used to make a basic judgment on the format, and the couplets that do not meet the hard rules are filtered. For the couplets that meet the hard rules, the content is scored by soft rules, and the final output is the score result of the input couplets. Assume a couplet $D = \{X, Y\}$, where $X = \{x_1, x_2, \ldots, x_m\}$ is the first sentence, and $Y = \{y_1, y_2, \ldots, y_n\}$ is the second sentence, and x_i and y_i respectively represent the i-th character in the first and second sentences.

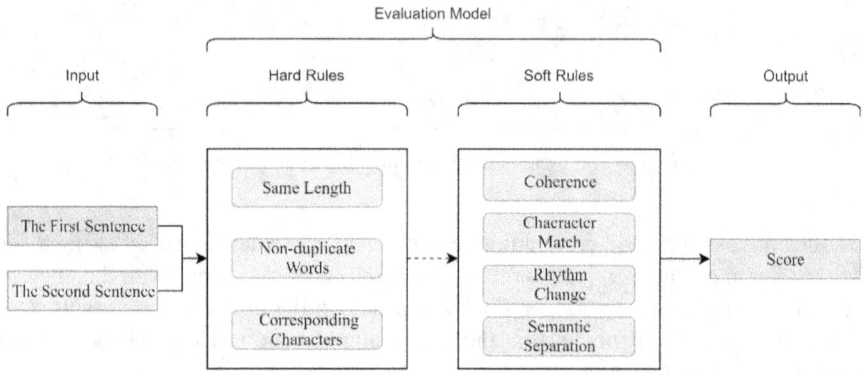

Fig. 3. The evaluation model structure

Hard Rule. The hard rules mainly judge the format of the content, including: (1) Same Length: The number of characters in the first and second sentences must be consistent; (2) Non-duplicate Characters: The characters in the first sentence cannot be in the second sentence; (3) Corresponding Characters: If the characters in two positions in the first sentence are the same, the characters in the corresponding positions in the second sentence must also be the same. For hard rules, only the distinction between satisfaction and dissatisfaction is made, and meeting those three rules at the same time is regarded as meeting hard rules.

Soft Rule. The soft rules are based on the characteristics of the content, including four evaluation indicators: coherence(*con*), character match(*ctm*), rhythm change(*rhy*) and semantic separation(*ses*). Different indicators give different weights $\omega_i, i = 1, 2, 3, 4$, and finally constitute a complete score, as shown in formula (1).

$$score = \omega_1 * con + \omega_2 * ctm + \omega_3 * rhy + \omega_4 * ses \quad (1)$$

Coherence. This article uses statistical language models to calculate coherence. The core idea of the statistical language model is to judge the probability of a sentence appearing in the text. Due to the lack of Wu-Qing couplets, this paper has chosen couplets and poetry in similar fields as the training corpus. However, compared to poetry and general mood couplets, the language of Wu-Qing couplets is often more popular and interesting. Therefore, we have also added some proverbs to the training corpus of the statistical language model to better match the characteristics of Wu-Qing couplets. As shown in formula (2), the coherence score *con* can be obtained after passing the language model.

$$con = SLM(Y) \quad (2)$$

Character Match. Parts of speech (POS) and character meaning determine the result of the character match. In terms of POS, Wu-Qing couplets involve a large number of POS changes during the escaping of the sense of meaning, so the consistency of part-of-speech here refers to "any POS of a character in the second sentence is consistent with any POS of the corresponding character in the first sentence". We have checked the part-of-speech of 20,948 Chinese characters up on the online Chinese dictionary[1], and judge whether the POS is consistent based on what we obtained from the dictionary. In terms of character meaning, what is being evaluated is the degree of semantic relevance between two characters. In this paper, we have chosen to use vectors to measure, and the result is obtained by calculating the cosine similarity among character vectors. The calculation method of POS and character meaning is shown in formula (3), (4), where pos_i indicates whether the POS of the i-th pair is consistent, sim_i indicates the degree of relevance of the meaning of the i-th pair, and a_i and b_i represent the i-th character vector in the context respectively.

$$pos_i = \begin{cases} 0 \; POS \; is \; not \; consistent \\ 1 \quad POS \; is \; consistent \end{cases} \tag{3}$$

$$sim_i = \frac{a_i \cdot b_i}{\|a_i\| * \|b_i\|} \tag{4}$$

After obtaining the results of POS and meaning, respectively, a complete character match is calculated, as shown in formulas (5), (6). ctm_i represents the character match score of the i-th pair of characters, α and β represent the weights of the two parts of POS consistency and semantic correlation, and *ctm* represents the character match score of the entire couplet.

$$ctm_i = \alpha * pos_i + \beta * sim_i \tag{5}$$

$$ctm = \frac{1}{n} \sum_{i=1}^{n} ctm_i \tag{6}$$

Rhythm Change. We use the word segmentation results to evaluate the value of rhythm change. A single word segmentation tool has accuracy problems. In order to achieve better word segmentation results, we use Hanlp[2], Jieba[3], PyNLPIR[4], THULAC[5], pkuseg[6], FoolNTLK[7], SnowNlp[8], totally seven word segmentation tools, according to the voting method to get the final word segmentation results. For the word segmentation results of the first and the second sentences, the rhythm breakpoints at the same position are deleted, and the maximum number of remaining breakpoints are taken instead, which is the number of rhythm change of this pair of couplets. As shown in the case of Fig. 4, the value of the rhythm change is 5.

[1] http://xh.5156edu.com.
[2] https://github.com/hankcs/HanLP.
[3] https://github.com/fxsjy/jieba.
[4] https://github.com/tsroten/pynlpir.
[5] https://github.com/thunlp/THULAC-Python.
[6] https://github.com/lancopku/pkuseg-python.
[7] https://github.com/rockyzhengwu/FoolNLTK.
[8] https://github.com/isnowfy/snownlp.

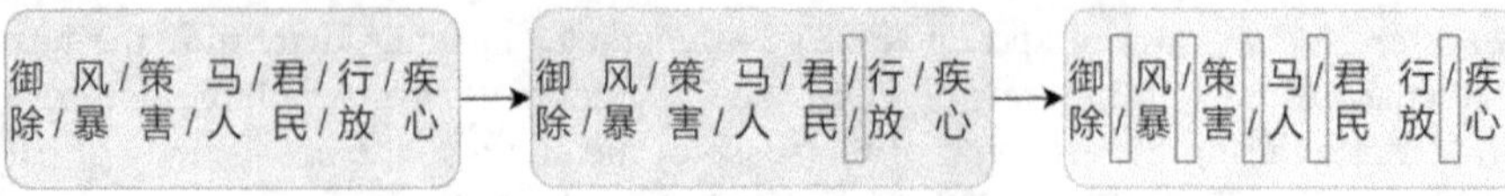

Fig. 4. An example of the rhythm change calculation method

Semantic Separation. We consider the semantic information of the entire sentence to calculate the semantic separation. Therefore, it is necessary to obtain the sentence vectors of the first sentence and the second sentence to represent, and the calculation method is to average the word vectors, as shown in formulas (7), (8). In addition, due to the calculation of the degree of semantic separation, it is necessary to make some adjustments to the cosine similarity, as shown in formula (9).

$$A = \frac{1}{n}\sum_{i=1}^{n} a_i \tag{7}$$

$$B = \frac{1}{n}\sum_{i=1}^{n} b_i \tag{8}$$

$$ses = 1 - \frac{A \cdot B}{\|A\| * \|B\|} \tag{9}$$

4 Experiment Results and Analysis

4.1 Experiments of the Generation Module

In the experiment of the generation model, this article uses 770,000 general couplets[9] collected by Wang Bin and 1,815 Wu-Qing couplets collected on the Internet. According to the ratio of 8:2, the two parts of the dataset are divided. The data distribution is shown in Table 1.

Table 1. Couplet and Wu-Qing couplet scale.

Dataset	Scale	Average length
General couplet—Training set	616392	9.35
General couplet—Validation set	154099	9.32
Wu-Qing couplet—Training set	1452	6.6
Wu-Qing couplet—Validation set	363	6.62

We have used Fairseq [11], a sequence modeling toolkit for machine translation proposed by the Facebook Artificial Intelligence Research Institute, to build and train the automatic generation model of Wu-Qing couplets. The training process is divided

[9] https://github.com/wb14123/couplet-dataset.

into two stages: pre-training and fine-tuning. In the pre-training phase, ordinary couplet data is used. When the loss on the verification set no longer drops, the pre-training stops, and the fine-tuning starts. The Wu-Qing couplets are used to fine-tune the generation model. Similarly, the fine-tuning stops until the loss on the verification set no longer drops, and the training is completed.

In the pre-training stage, the learning rate is 0.001. The optimizer is Adam, and the batch size is 128. In addition, the gradient clipping value is 0.1, and the dropout rate is 0.3. In the fine-tuning stage, the learning rate is adjusted to 0.0005, and the batch size is set to 16. Other parameters are consistent with those in the pre-training stage. The changes in loss and perplexity during the two phases of training are shown in Fig. 5 and Fig. 6.

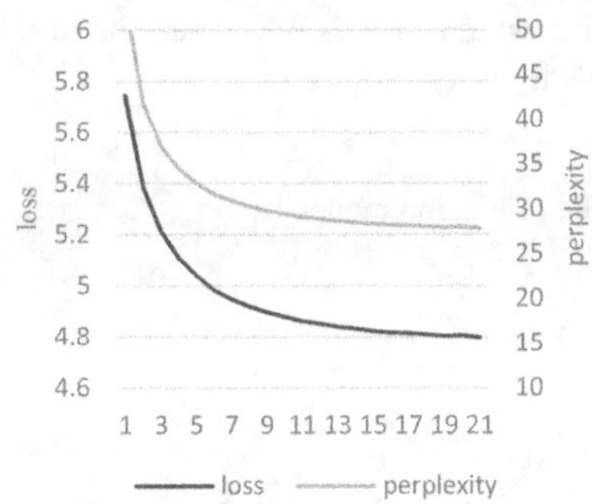

Fig. 5. Pre-training loss and perplexity

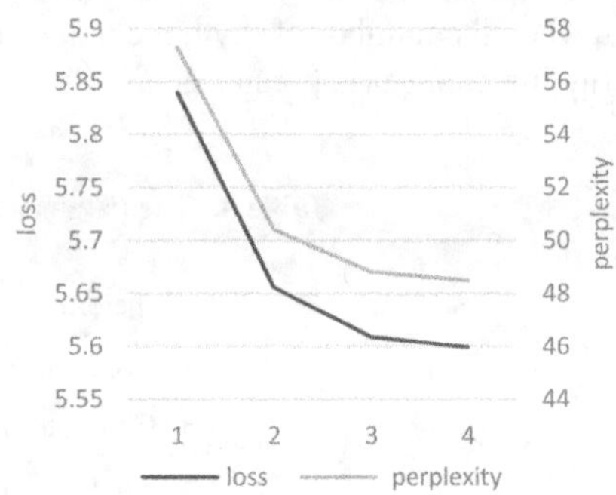

Fig. 6. Fine-tuning loss and perplexity

At the same time, in order to prove the effectiveness of this method, this paper has also set up a group of comparative experiments, which only uses Wu-Qing couplets to train the corpus, and the setting of training parameters is consistent with that in the pre-training stage. The results of different training methods are shown in Table 2. It can be seen that the model cannot be optimized basically by using only single-stage training, with a loss value of 8.892 and a perplexity of 475.14. After using the method in this paper for training, the performance has been significantly improved, with a loss of 5.599 and a perplexity of 48.48, almost one-tenth of the perplexity in direct training, which shows the validity of the method proposed in this paper.

4.2 Experiments on the Evaluation Module

According to the characteristics of Wu-Qing couplets, we have constructed the evaluation model of Wu-Qing couplets. In the soft rule part of the evaluation model, the parameters involved are shown in Table 3. In addition to the parameters listed in Table 3, in the coherence section, the statistical language model is a bi-gram language model, and the training corpus includes the 770,000 couplets (which are mentioned in the previous subsection), 62,000 poems and 14,000 proverbs. In addition, because the number of rhythm changes is large, a mapping is made here. When the number of rhythm changes is 1, the value of rhy is 0.5; when the number of rhythm changes is 2, the value of rhy is

Table 2. The results of different training strategies.

Training strategy	Training detail	Loss	Perplexity
Single-stage training	Training Wu-Qing couplets	8.892	475.14
Two-stage training	Pre-training general couplets	4.799	27.84
	Fine-tuning Wu-Qing couplets	5.599	48.48

0.6; when the number of rhythm changes is 3, the value of rhy is 0.8; when the number of rhythm changes is greater than 3, the value of rhy is 1.

Table 3. The parameters in the evaluation model.

Parameters	Values
ω_1 (Coherence)	0.25
ω_2 (Character match)	0.35
ω_3 (Rhythm change)	0.26
ω_4 (Semantic separation)	0.14
α (Part of speech)	0.40
β (Character meaning)	0.60

The role of the evaluation model in this paper is to score the candidates from the generation model and to select better Wu-Qing couplets. To some extent, it is actually required that the evaluation model has the ability to distinguish general couplets from Wu-Qing couplets. By ignoring the data that do not satisfy the hard rules, this paper has randomly selected 100 general couplets and 100 Wu-Qing couplets, and scored them using the evaluation model. The final results are shown in Fig. 7. It can be clearly seen that in the test data, the evaluation system scores the Wu-Qing couplets more clearly than the general couplets. The mean value of the 100 Wu-Qing couplets is around 0.45, while the mean value of the general couplets is around 0.32. This shows that the evaluation model designed in this paper has the ability to distinguish general couplets from Wu-Qing couplets.

4.3 The Experiment Result of the Complete System

To evaluate the quality of couplets, we need to judge from multiple angles and the evaluator needs to have certain professional knowledge. Therefore, it is difficult to evaluate. In order to ensure the objectivity of the evaluation results, researchers often use the

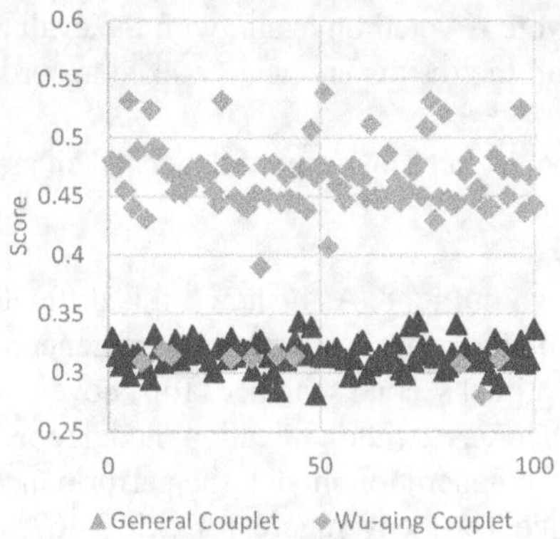

Fig. 7. Scoring results

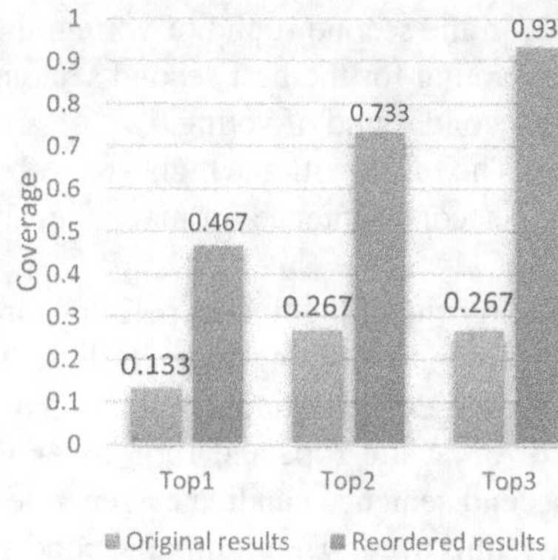

Fig. 8. TopN coverage results

method of BLEU automatic evaluation combined with the manual evaluation to evaluate the quality of the generated couplets. However, the BLEU does not apply in the assessment of Wu-Qing couplets. The main reason lies in the unique characteristics of Wu-Qing couplets. It is committed to seeking changes in content, structure and other aspects between the first and the second sentences. The use of BLEU for evaluation is rigid and restrictive, which limits the possibility of Wu-Qing couplets. Therefore, this paper has only used manual evaluation to screen the better results in candidate couplets and used TopN coverage to measure the system performance.

TopN coverage refers to the manual evaluation of whether the best second sentence appears in the top N of the final candidates of the second sentences. For example, if the first sentence is '花(flower)明(bright)柳(willow)媚(charming)", the top three candidates for the second sentence generated by the model are "李(plum)白(white)桃(peach)夭(lush)", "草(grass)莽(reckless)杨(poplar)雄(powerful)" and "草(grass)莽(reckless)桃(peach)英(fine)", and the best second sentence for manual evaluation is "草莽杨雄", then Top1 is not covered, while Top2 and Top3 are covered. The calculation of TopN coverage is shown in the formula (10), where p represents the total number of couplets and c_k indicates whether the k-th pair of couplets have coverage.

$$TopN_{Coverage} = \frac{1}{p}\sum_{k=1}^{p} c_k \tag{10}$$

The experiment uses the Wu-Qing couplets obtained in the generative experiment as the test corpus. We randomly selected 15 groups of Wu-Qing couplets. Each group of Wu-Qing couplet data includes 20 different second sentences corresponding to the same first sentence, and 20 different second sentences are the results of Beam Search. The manual evaluation judged the quality of the Wu-Qing couplets from four perspectives: Coherent, character match, connotation, semantic separation. There are seven participants in the evaluation, and the evaluation process includes three steps:

Step1: In the first round of voting, for each group of Wu-Qing couplets, the evaluator selects the best three sentences and ranks them according to the vote. It would be moved to Step 2 if there is a tie vote, otherwise skipping to Step 3.

Step2: In the second round of voting, the tie vote is voted on again, with the evaluator simply voting for the best second sentence, and the results are again ranked according to the second round of voting.
Step3: The result with the highest number of votes in each group of data is taken as the optimal second sentence for manual evaluation.

The results of the final experiment are shown in Fig. 8. As it shows, it is difficult to get the optimal Wu-Qing couplets directly if only the results generated by the generation model are used. For manual evaluation of the optimal second sentence, Top1 coverage is only 0.133, while Top2 and Top3 coverage is 0.267. After the evaluation model reorders the second sentence candidates generated by the generation model, the performance is greatly improved. The optimal second sentence Top1 coverage can reach 0.467, and Top2 coverage reaches 0.733, and Top3 coverage reaches 0.933, which means that the final Top 3 candidate second sentences generated by the system in this paper will essentially contain the optimal second sentence. Therefore, the effectiveness of this system is illustrated.

Table 4 shows the top 3 results of some Wu-Qing couplets automatically generated by the system.

Table 4. Cases of system generated Wu-Qing couplets.

No	Wu-Qing couplets
1	斜 (oblique) 阳 (sun) 古 (old) 道 (say) 中 (middle) 日 (sun) 新 (new) 闻 (hear)
2	内 (inner) 蒙 (ignorant) 走 (go) 亲 (intimate) 戚 (sad) 中 (middle) 庸 (mediocre) 行 (do) 善 (friendly) 良 (fine)
3	听 (hear) 泉 (spring) 石 (stone) 上 (up) 花 (flower) 无 (empty) 语(word) 踏 (tread) 雪 (snow) 梅 (plum) 前 (front) 李 (plum) 白 (plain) 诗 (poem)
4	白 (white) 日 (day) 梦 (dream) 游 (vagrant) 活 (life) 见 (see) 鬼 (ghost) 黄 (yellow) 昏 (dusk) 思 (think) 故 (old) 土 (land) 伤 (hurt) 人 (human)
5	雁 (wild goose) 排 (row) 一 (single) 字 (character) 南 (south) 方 (region) 去(go) 马 (horse) 列 (column) 全 (full) 书 (book) 北 (north) 国 (country) 来 (come)

Take no. 5 as an example for illustration, which is "雁排一字南方去,马列全书北国来". In the character-to-character analysis, "雁" and "马" are animals, belonging to the same subcategory. "排" corresponds with "列". "一" and "全" are quantity. "字" and "书" are for character interpretation. "南" and "北" both indicate direction. "方" indicates region corresponding with "国". "去" corresponds with "来". In the meaning of the whole couplet, "马列" is the abbreviation of Marxism-Leninism, and the result of the segmentation is "雁/排/一字/南方/去, 马列/全书/北国/来". The rhythm changes by borrowing and escaping meanings. The meanings of the first and second sentences have nothing to do with it, and the effect of Wu-Qing couplet is obvious.

It can be seen that the Wu-Qing couplet in this case is naturally smooth, and basically achieves trimness. Moreover, there are a lot of borrowings and transferring of meanings, showing that semantics disconnect between the first and the second sentences. Apart from the couplets shown above, there are yet many other ingenious and intriguing Wu-Qing couplets.

5 Conclusions and Future Work

This paper has launched a research from the point of view of the automatic generation and evaluation of Wu-Qing couplets, and proposed a two-stage sequence-to-sequence automatic generation model of Wu-Qing couplet based on transfer learning. The first stage is to pretrain on a large-scale general couplet corpus. Then, the second stage is to use the Wu-Qing couplet corpus to fine-tune the generation model. Such model has solved the problem that of the lack of Wu-Qing couplet corpus which prevents the training from being carried out directly. Based on the characteristics of Wu-Qing couplets, this paper has also proposed an evaluation model, which can score the results of the generation model and reorder the output according to the scores for better results. Finally, based on the research in the two parts, the first complete automatic generation system of Chinese Wu-Qing couplets is constructed; and the results of the experiment have proved the effectiveness of the system proposed in this paper.

There are certainly still some deficiencies in the work of this paper, including: (1) the setting of some parameters in soft rules is the result of a manual attempt; (2) the current mainstream pre-trained models such as GPT are not used; (3) the evaluation rules can continue to be improved and detailed; (4) the test corpus size is small and needs to be increased to further ensure the effectiveness of the system. The above four shortcomings are the focus of our future work.

Acknowledgments. This work is supported by the National Natural Science Foundation of China (Grant No. 61602044) and the funds for improving the quality of personnel training in 2020 of Beijing Information Science and Technology University (Grant No. 5102010805).

References

1. Wu, X.: On "the WuQing couplet". Chin. Cult. (02), 28–29 (2008). (in Chinese)
2. Tian, Z.: Generation, development and creation of The WuQing Couplet. Zhong Hua Ying Lian, (3) 15 April 2012. (in Chinese)
3. Zou, Z.: The several peculiar antithesis forms of couplet. Chongqing Soc. Sci. (08), 105–111 (2008). (in Chinese)
4. Fei, Y.: Research on multi-level integration of Chinese semantics and system design of spring festival couplets [Ph.D. thesis]. Institute of Automation Chinese Academy of Science, Beijing (1999). (in Chinese with English abstract)
5. Hua, B.: Conduct couplet matching system based on grammar development platform. In: Xiamen University: Proceedings of the CLSW6. Xiamen University: Fujian Language Association, pp. 90–95 (2005). (in Chinese with an English abstract)

6. Jiang, L., Zhou, M.: Generating Chinese couplets using a statistical MT approach. In: Proceeding of the Conference on COLING 2008, International Conference on Computation Linguistics, 18–22 August 2008, pp. 377–384. DBLP, Manchester (2008)
7. Zhang, K., Sun, M.: A Chinese couplet generation model based on statistics and rules. J. Chin. Inf. Process. **23**(1), 101–105 (2009). (in Chinese)
8. Fan, H., Wang, J., Zhuang, B., Wang, S., Xiao, J.: Automatic acrostic couplet generation with three-stage neural network pipelines. In: Nayak, A.C., Sharma, A. (eds.) PRICAI 2019. LNCS (LNAI), vol. 11670, pp. 314–324. Springer, Cham (2019). https://doi.org/10.1007/978-3-030-29908-8_25
9. Sutskever, I., Vinyals, O., Le, Q.V.: Sequence to sequence learning with neural networks. In: Advances in NIPS (2014)
10. Bahdanau, D., Cho, K., Bengio, Y.: Neural machine translation by jointly learning to align and translate. arXiv preprint arXiv:1409.0473 (2014)
11. Ott, M., et al.: FairSeq: a fast, extensible toolkit for sequence modeling. In: Proceedings of NAACL-HLT: Demonstrations (2019)

The Development of the Semantic Network of Chinese as Second Language Learners' Production: From the Perspective of Collocations

Yali Mu, Joerg Heseler, and Hongbing Xing(✉)

Beijing Language and Culture University, Xueyuan Road 15, Beijing 100083, China
{myl,xinghb}@blcu.edu.cn

Abstract. Making use of the data from narrative writings of the same topic, this paper has used the common high-frequency adjective "大" (big) and the common high-frequency noun "城市" (city) as the main vocabularies to examine the frequency of all adjective-noun collocations produced by Chinese as Second Language (CSL) learners. From the perspective of lexical semantics and the methodology of quantitative linguistics, the development of the semantic network of the CSL production has been investigated.

Keywords: CSL production · Semantic network · Adjective-noun collocation · Development

1 Introduction

It is generally believed that vocabulary does not exist in isolation in the user's mind but is interconnected according to meaning so as to form a network with certain structural characteristics. The position of each word in the network is different, and the number of connected words and the ways of connection are also different. According to the principles of semantics, the semantic relations among lexical items are usually divided into the following types: synonymous, antonymous, hypernyms and hyponyms, whole-part relationship, individual-collection, and component-total. However, Feng [1] pointed out that such classification excludes the collocation relationship which is closely related to the semantic relationships. He believed that the collocation characteristics are effective components of the lexical meaning, and that the collocation relationship is an indispensable and important relationship in the lexical semantic network. Previous studies basically took the same view as such, when adopting the test method of vocabulary association to study the representation of mental lexicon, regardless of whether it is the phonetic-horizontal vertical-aggregation triad, or the phonetic-semantic-syntactic triad, or the form-based response meaning-based response position-based response triad. The studies emphasized equal importance between collocations and semantics, or they have placed horizontal combination and vertical aggregation in semantic responses [2–4]. Based on the theory of connectionism, Xing [5] concluded that Chinese as Second

M. Liu et al. (Eds.): CLSW 2020, LNAI 12278, pp. 549–561, 2021.
https://doi.org/10.1007/978-3-030-81197-6_47

Language (CSL) learners' vocabulary knowledge includes pronunciation, word form and meaning, and their mutual connections, among which meaning knowledge is the core. The collocation knowledge is the core content of the word relationship knowledge in meaning knowledge. Based on the above viewpoints, the collocation knowledge of words is an important part of the lexical semantic network. Therefore, we explore the development of the CSL production semantic network at the lexical collocation level.

Regarding adjective-noun collocations, research on language ontology are quite abundant. There are, for examples, the studies on the classification and grammatical function of adjectives in adjective-noun collocations [6, 7]; studies on the concealed and restrictive factors of "的" in adjective-noun collocations [8–10]; the studies on the convertibility of the adjective-noun collocations in attributive headword structures and subject-predicate structures [11–14]; the studies on the rules of mutual selection of adjectives and nouns in adjective-noun collocations [15–17]; and the studies on adjective-noun collocations from the perspective of computational linguistics [18–20] and so on. The use of adjectives and nouns as well as the acquisition of adjective-noun collocation knowledge are the difficult but key aspect in CSL learning. However, in the field of CSL teaching and acquisition, the research on adjective-noun collocations started relatively late. There are, for instances, studies adopting a corpus-based approach to discuss the acquisition of "的" in adjective-noun collocations of international students [10, 21, 22]; an experimental research discussing the acquisition of international students' awareness of the adjective "同" [23]; and also comparative studies that simply studied the acquisition situation of a certain type of adjectives and nouns by international students [24, 25].

Both the methods and the results of the above research have given us great inspiration. At the same time, we find that there are still some problems worth further discussions. For example, will collocations of adjectives and nouns develop with the improvement of Chinese proficiency? If so, what are the levels of the development involved?

Under the task of writing an essay titled "我的家乡" (My Hometown), we compared and analyzed the written output of international students of different levels with the written output of native speakers. The collected composition data was segmented, tagged, and all adjective-noun collocations were counted. The high-frequency adjective is "大" (big), the high-frequency noun is "城市" (city), and "大城市" (big city) is a high-frequency collocation shared by CSL learners with "大" and "城市" as core words. Based on this observation, we investigated whether the selection and collocation of adjectives and nouns of CSL learners of different levels will develop with their improvement in Chinese language proficiency.

2 Research Design

2.1 Research Questions

In the composition "我的家乡" (my hometown), do CSL learners share collocations of the high-frequency adjective "大" and the high-frequency noun "城市"? If not, what will happen as the level goes up? What is the difference between CSL learners and native speakers in terms of adjective-noun collocations ?

2.2 Research Methods

We took the common high-frequency adjective "大" and the common high-frequency noun "城市" in the composition "我的家乡" as the core vocabulary and listed all adjective-noun collocations related to "大" and "城市" in an exhaustive manner. In terms of the adjective "大", we looked at the set of nouns with which it matched, and in terms of the noun "城市", we looked at the set of adjectives with which it matched. At the same time, the similarities and differences between the two sets of CSL and native speakers were compared. Through the semantic selection and application of nouns and adjectives as well as the characteristics of adjective-noun collocations in the output, the development of the semantic collocation knowledge of adjective-noun collocations of international students was observed.

2.3 Research Objects

In the experimental group, students at intermediate and advanced[1] levels were all from universities in Beijing. Among them, there were 22 students at intermediate level, of which 13 students whose mother tongue was from Asia (namely Korea, Japan, Thailand, Vietnam and Mongolia), 9 students were from other countries (namely France, Italy, Chile, Russia and Pakistan). There were 26 students at advanced level, of which 14 students were from Asia (namely Korea, Japan, Indonesia, Thailand and Mongolia) and 12 students were from other countries (namely Germany, France, Poland, Belgium, Romania, Tunisia and Algeria). Moreover, there were 20 Chinese students in the reference group, all of whom were from universities in Beijing with a bachelor's degree or a postgraduate degree.

2.4 Implementation Process

Due to the differences in Chinese proficiency and writing ability between the two groups of subjects, the writing test was designed to be a narrative writing expressing a common emotion and common experiences – "我的家乡" (My hometown). The form was a propositional composition and the test time was 50 min. Participants were required to write about 400 words, and no auxiliary materials such as dictionaries were allowed. The test requirements of the reference group were the same as those of the experimental group.

2.5 Data Collection

A total of 68 answers were collected. Among them, 22 intermediate student produced a total of 7,066 characters, 26 advanced students a total of 11,206 characters and 20 native speakers a total of 9,058 characters. The original corpus collected in this study was handwritten material written on square paper. In order to facilitate the analysis, the

[1] Due to the limited number of sentences produced by CSL learners at elementary level, this paper considers the development of CSL learners at intermediate and advanced levels for the time being.

author has transferred these materials into electronic documents, and then they were annotated. The number of characters and words as well as the frequency of characters and words etc. were then examined. At first, the word segmentation and part-of-speech tagging tool "Corpus Word Parser 3.0"[2] performed a preliminary segmentation and statistics, while adjective-noun collocations were manually tagged and counted. On this basis, a manual inspection, re-classification and statistics calculation were carried out. The tagging was done manually by two authors and cross-checked.

2.6 Measurement Indicators and Statistical Tools

(1) Type: refers to the number of different words in the corpus.
(2) Token: refers to the number of times a certain type of word is used in the corpus.
(3) Lexical diversity (U): refers to the diversity of a certain type of word in the corpus, which can be calculated by the following formula:

$$U = \frac{(\log Tokens)^2}{\log Tokens - \log Types} \tag{1}$$

(4) Semantic similarity (S): refers to the semantic similarity between two words. This paper uses a programming method to calculate semantic similarity based on the "同义词词林" ('The dictionary of synonyms') (Second Edition) [26]. The specific principle is as follows:
The "同义词词林" ('The dictionary of synonyms') contains more than 640,000 entries. Each entry points to a semantic code consisting of five levels and one mark-bit. For example, "城市" points to "Cb25A01 =", where "C" is the first level, "b" is the second level, "25" is the third level, "A" is the fourth level, "01" is the fifth level, and "=" is the mark-bit. Mark-bits are mainly used to distinguish regular synonyms, related words and independent words with neither synonyms nor related words, which are represented by the characters "=", "#" and "@". We assign a certain weight to each level of the "同义词词林" ('The dictionary of synonyms'), namely 1.2, 1.2, 1.0, 1.0, 0.8 and 0.4, whose sum is 5.6. Then, when calculating the semantic similarity, we first obtain the corresponding codes of the two words, and then we compare the codes one by one to see whether each level is equal. The weight of all equal levels is added up and divided by the sum to get the semantic similarity.
(5) Collocation compactness (MI): refers to the degree of closeness of two words. In this paper, the mutual information value is used to represent the collocation compactness.

Mutual information is a common method of model analysis in computational linguistics, which measures the mutuality between two objects. Mutual information is a concept in information theory, which is used to represent the relationship between information. It is a measure of the statistical correlation between two random variables. In

[2] Copyrighted by the Research Department of Computational Linguistics, Institute of Language and Character Applications, Ministry of Education, produced by Xiao Hang, download address: www.cncorpus.org.

this paper we use it to measure the collocation strength of two words, the formula is as follows:

$$MI(X, Y) = \log_2 \frac{P(X, Y)}{P(X)P(Y)} \quad (2)$$

P(X,Y) is the frequency of the co-occurrence words X and Y, P(X) is the frequency of word X, and P(Y) is the frequency of word Y.

3 Statistical Results

3.1 Output Diversity of Adjectives and Nouns

First, we investigated the overall distribution of adjectives and nouns in the corpus. The statistical results showed that with the improvement of Chinese proficiency, the diversity of adjectives produced by CSL learners has increased. Although the diversity of the adjectives produced by CSL learners was still rather different from the diversity of the adjectives produced by native speakers, the former was gradually approaching the target language level. As for the diversity of nouns, with the improvement of Chinese proficiency, the output of CSL learners has not improved significantly, and there was still a big gap between them and the native speakers. See Table 1 and Fig. 1.

Table 1. Summary table of types, tokens and diversity of adjectives and nouns.

Proficiency	Adjective types	Adjective tokens	Adjective diversity U	Noun types	Noun tokens
Intermediate	108	234	16.716	520	1,222
Advanced	186	381	21.390	709	1,887
Native	176	293	27.492	920	1,545

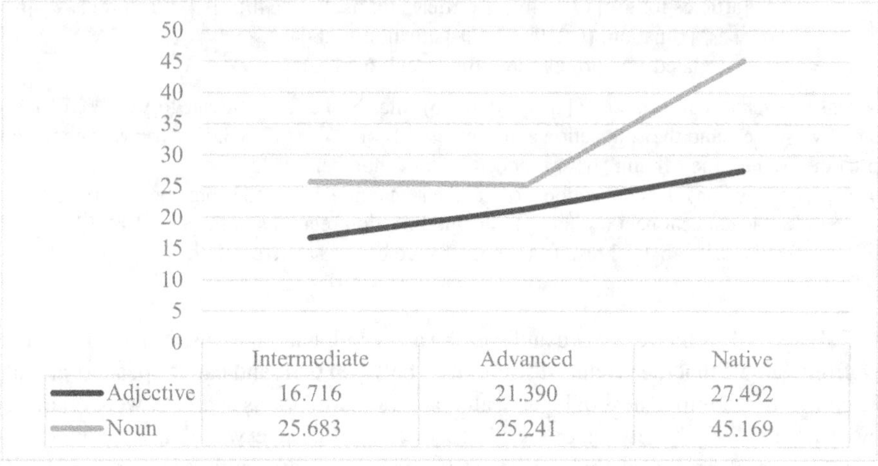

Fig. 1. Diversity line chart of productive adjectives and nouns.

3.2 The Lexical Semantic Adjective-Noun Set Composed of "大" and "城市"

Starting from the two high-frequency words "大" and "城市", we found the adjective-noun collocations containing these two core words. For example, "(很/特/最/不) 大 (的) + N" found the set of N, and "(很/特/最) A (的) + 城市" found the set of A. According to the semantic classification of "同义词词林" ('The dictionary of synonyms') (Second Edition), the words in the set were labeled with semantic categories. Then we used the MI formula to calculate the corresponding collocation compactness. See Table 2 and Table 3.

Table 2. The set of nouns N modified by "大" (For the semantic categories in the noun set, please refer to the word classification of "同义词词林" ('The dictionary of synonyms') (Second Edition) edited by Mei Jiaju et al. We made some adjustments for the words not included in the book).

Proficiency	(很/最/不) 大 (的) + N[a]
Intermediate	城市 [space] (5.167b), 国家 [community] (4.752b), 游乐园 [institution] (6.752b), 公园 [architecture] (5.945b), 医院[institution] (7.752b), 百货商店 [institution] (7.752b), 超市 [institution] (7.167b), 珍珠 [material] (5.945b), 淡水湖[landform] (7.752b), 风 [meteorology] (6.167b), 面积 [object part] (7.752b), 特色 [performance] (6.167b), 意思 [situation] (6.167b)
Advanced	城市 [space] (4.453b), 都市[space] (7.754b), 面积[object part] (6.17b), 波兰 [community] (6.754b), 群岛 [landform] (7.754b), 海 [landform] (6.754b), 草原 [landform] (7.754b), 热带雨林 [landform] (5.170b), 物种 [biological unit] (7.754b), 声音 [natural thing] (6.754b), 亮光 [natural thing] (6.754b), 星星 [celestial body] (5.170b), 花园 [architecture] (7.754b), 公园 [architecture] (5.754b), 特点 [performance] (5.170b), 差异 [performance] (6.170b), 区别[performance] (6.754b), 幸福 [circumstance] (6.170b), 影响[nature] (7.754b)
Native	县城[space] (5.057b), 村庄[space](6.864b), 庄 [space] (6.864b), 院子 [architecture] (5.864b), 人 [human generic term] (4.057b), 雪[meteorology] (6.864b), 淡水湖 [landform] (7.864b), 梨[plant] (7.864b), 汤锅 [articles for use] (7.864b), 桌[articles for use] (7.864b), 个头 [posture] (7.864b), 事情 [situation] (6.279b), 乐事 [situation] (7.864b), 不同[related similarities and differences] (6.542b)

(Note: In the set of N above, "城市 [space] (5.055b)" means: the semantic category of "城市" is the category "space", and the collocation compactness MI of "大城市" is 5.055. To avoid confusion with other values, the MI unit "b" was added after the number.)

[a]This paper only considers the collocations at the semantic level of the core vocabulary and does not consider the concealment of "的" and the modification relations of the adverbs "很", "最", "不", etc., that is, the statistical results are not affected by the above conditions.

Taking "大" as the core word and looking at its collocation nouns, it is not difficult to see from Table 2 that the produced similarities between CSL and native speakers are that the collocation compactness of [space] and [architecture] nouns and "大" were relatively low, such as "城市" (city) and "公园" (park). A difference was that the collocation compactness of produced [human general term] nouns by native speakers with "大" was also low, such as "人" (person). Taking "城市" as the core word and looking at

Table 3. The set of adjectives A that qualifies "城市" (The adjectives in this paper are counted according to their broad meanings, including distinguishing words that express the attributes or distinguishing features of things).

Proficiency	(很/不/最) A (的) + 城市
Intermediate	大 [shape] (5.167b), 美丽 [representation] (6.305b), 热闹 [circumstance] (4.720b), 发展 [state of affairs] (4.720b), 热 [nature] (4.583b), 民主化 [performance] (5.820b), 一样 [similarities and differences] (5.820b), 现代 [performance] (5.457b)
Advanced	大 [shape] (4.453b), 小 [shape] (6.355b), 小小的 [shape] (5.733b), 美丽 [representation] (3.733b), 秀丽 [representation] (5.733b), 干净 [representation] (6.733b), 古老 [representation] (5.733b), 很多 [sparse] (4.412b), 有名 [nature] (4.139b), 最好 [nature] (4.587b), 热闹 [circumstance] (6.148b), 国际化 [performance] (6.733b), 熟悉 [psychological activity] (4.733b)
Native	美丽 [representation] (7.414b), 典型 [nature] (8.414b), 喧嚣 [circumstance] (8.414b), 繁华 [circumstance] (8.414b)

(Note: The labeling of set A is the same as that of set N above.)

the adjectives with which it was matched, it is not difficult to see from Table 3 that intermediate CSL learners produced [nature] adjectives with "城市" that had a low degree of collocation compactness, such as "热" (hot). Advanced CSL learners produced [representation], [shape], and [nature] adjectives with "城市" that had a low degree of collocation compactness, such as "美丽" (beautiful), "大" (big) and "热" (hot). Native speakers produced [human general term] nouns with "大" that also had a low degree of collocation compactness, such as "人" (person).

3.3 Adjective-Noun Collocations Formed by "大" and "城市"

From the above, it is not difficult to see that in the two sets of A and N, and there were many common semantic categories in the output by intermediate and advanced CSL learners and native speakers. We used the semantic approximation to roughly estimate

Table 4. Summary table of types, tokens and diversity of adjectives and nouns.

Class	Semantic similarity S of N	Standard deviation SD	Semantic similarity S of A	Standard deviation SD
Intermediate – Native	0.127	0.183	0.208	0.276
Advanced – Native	0.118	0.171	0.271	0.219
Intermediate – Advanced	0.121	0.035	0.175	0.236

the semantic differences among the three. It was found that in the production of nouns, the average semantic similarity between intermediate level CSL learners and native speakers was higher. In the output of adjectives, the average semantic similarity between advanced level CSL learners and native speakers was higher, as shown in Table 4:

Due to the limited space, we only drew the semantic collocation diagram of adjectives and nouns of "城市" based collocations listed in Table 3 and hyponyms related to "城市", as shown in Figs. 2, 3 and 4:

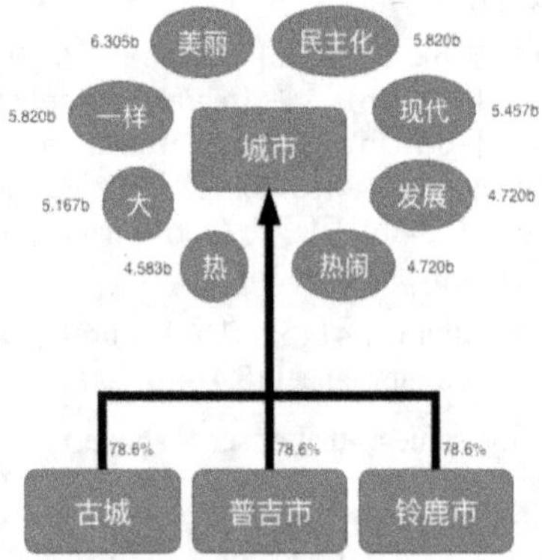

Fig. 2. Adjective-noun collocation network diagram of "城市" produced by intermediate level CSL learners.

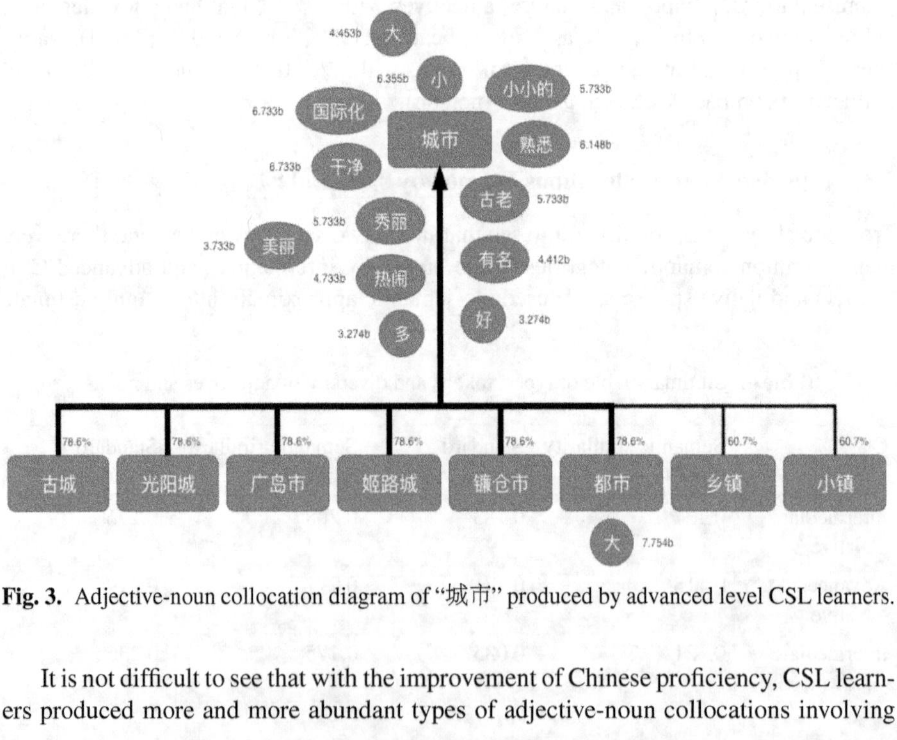

Fig. 3. Adjective-noun collocation diagram of "城市" produced by advanced level CSL learners.

It is not difficult to see that with the improvement of Chinese proficiency, CSL learners produced more and more abundant types of adjective-noun collocations involving

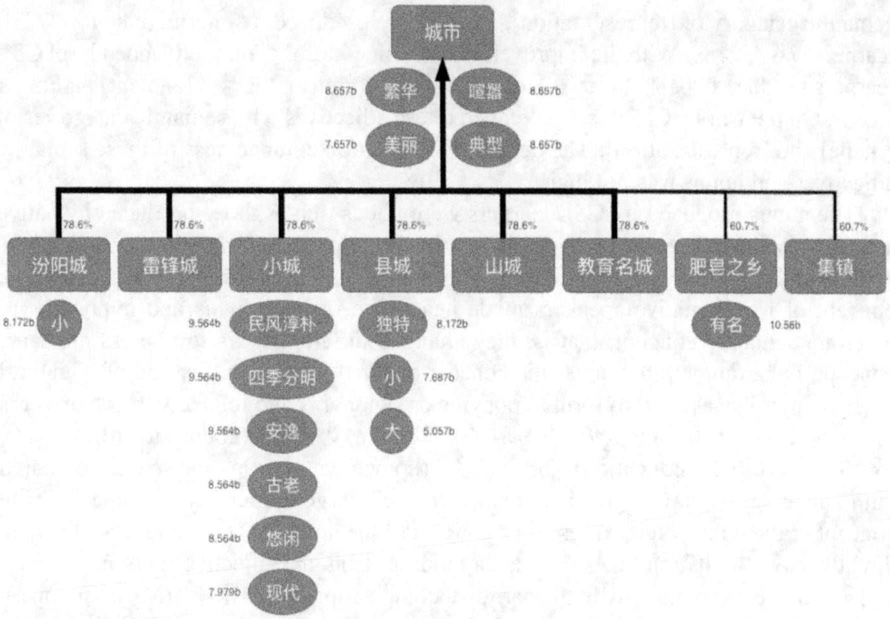

Fig. 4. Adjective-noun collocation diagram of "城市" produced by native speakers.

"城市". Not only the semantic categories of adjectives have been more diverse, but also the total number of words within the same semantic category have also increased, such as in the category [shape] and the way the advanced level CSL learners producing "大", "小" (small), "小小的" (little). The use of hyponyms of "城市" was increasing, but collocations of hyponyms and adjectives were still relatively few. This was still a difference from how native speakers produced adjective-noun collocations involving "城市".

4 General Discussion

4.1 Developmental Characteristics of CSL Learners Regarding Productive Adjectives and Nouns

Generally speaking, productive adjectives of CSL learners were more diversified and had more semantic types. With the improvement of Chinese proficiency, the richness of adjectives produced by CSL learners has increased. It included not only the increase of semantic categories, but also the number of semantic sub-categories. For example, there were specific adjectives such as those in the categories [shape] and [representation], as well as abstract adjectives, e.g., those in the categories [circumstance], [nature], [state of affairs], [performance] and [psychological activity] that modify and define "城市". At the intermediate Chinese level, CSL learners mastered a lot of modifying semantic categories of adjectives that define a type of "城市". As they continued to learn, the number of vocabularies under each semantic category increased. For example, under the semantic category of [representation], an adjective produced by intermediate level CSL learners was "美丽". With the improvement of Chinese proficiency, advanced level CSL learners produced "美丽", "干净" (clean), "秀丽" (pretty), "古老" (ancient) and other words. On the whole, CSL learners tended to use adjectives with semantic categories of [shape] and [representation]. The degree of collocation compactness of these semantic adjectives and nouns was not high.

The nouns produced by CSL learners were not as rich as those produced by native speakers. With the improvement of Chinese language proficiency, there was no obvious change in the diversity of noun output. This was mainly because, when expressing the concept of nouns, native speakers made heavy use of hypernyms and hyponyms of a certain semantic concept, that is, they changed different ways to express the same concept. For example, to express the concept of "城市", the combination of "[feature] + [hypernym]" was used to form hyponyms or synonyms and related words to express the same concept; for example, "小城" (small town), "县城" (county town), "教育名城" (famous city for education), "肥皂之乡" (hometown of soap) and so on. Compared with native speakers, CSL learners tended to use a large amount of the superordinate concept of the noun when expressing a concept of the noun. For example, CSL learners directly used "城市", and then selected a modifier from the adjective set to match with it. This can also explain why in the narrative composition "我的家乡" (My hometown), although the word "城市" was a high-frequency noun common to CSL learners, it was not a high-frequency noun for native speakers[3]. With the improvement of Chinese proficiency, the hyponyms of nouns that can be used by CSL learners have increased by a certain amount, mainly in the form of "[description] + [hypernym]". For example, in addition to the category [space] representing the noun "城市", CSL learners of the intermediate level also used its hyponyms "古城" (ancient city), "普吉市" (Phuket City), and "铃鹿市" (Suzuka City). CSL learners at advanced and intermediate level wrote in similar manner in the fact that in addition to using "城市", they also used the hyponyms "古城", "广岛市" (Hiroshima City), "光阳市" (Gwangyang City), "姬路城" (Himeji Castle) or "镰仓市" (Kamakura City). In addition, they also used hyponyms such as "都市" (metropolis), "乡镇" (township), and "小镇" (small town). But in short, with the improvement of Chinese proficiency, CSL learners were limited to producing more noun hyponyms. The collocations of hyponyms and adjectives were still limited.

4.2 The Development of the Semantic Network of CSL Learners When Using Adjective-Noun Collocations – from "Many to One", "One to Many" to "Many to Many"

The output of adjective-noun collocations can be said to be a two-way selection process. It is necessary to select the accurate noun form from the noun concept set and also to select the accurate adjective form from the adjective set that represent such characteristics according to the semantic characteristics of the noun. For example, when choosing noun concepts, CSL learners tended to use the hypernym N_1 and to choose the suitable

[3] The output usage rate of "城市" was 74.2% at intermediate level, 93.6% at advanced level, and 25.1% for native speakers. Usage rate = the number of times a word is used/the total number of words in the corpus.

adjective A_1 in the adjective set. When expressing other similar noun concepts, CSL learners continued to choose the corresponding form from the noun set. Due to the limited number of vocabularies they learned, they tended to continue to select the hypernym N_1, and then selected the adjective A_2 from the adjective set, and so on. Noun N_1 may have many adjectives A_1, A_2, A_3 … A_n which are compatible with it, forming "many to one" "A (的) + N" adjective-noun semantic collocations. Just like "城市" in this paper, there were "many-to-one" adjective-noun semantic collocations such as "大城市" (big city), "美丽的城市" (beautiful city), "热闹的城市" (busy city) and "干净的城市" (clean city). There was also a situation, where CSL learners tended to use adjectives of [shape] or [representation]. This led to the selection of the adjective A_1 with [shape] or [representation] meaning in the adjective set after N_1 was selected. When they wanted to express the second noun concept N_2, the selected adjective was still A_1, and so on. Maybe N_1, N_2, N_3 … N_n selected the adjective A_1, forming a "one-to-many" "A (的) + N" semantic collocation of adjective-nouns. For example, in this paper, "one-to-many" adjective-noun collocations such as "大城市", "大公园" (big park), "大医院" (big hospital), "大超市" (big supermarket) and "大珍珠" (big pearl) were formed. In the output of native speakers, the choice of nouns included not only the hypernym N_1 but also the hyponyms or related words N_2, N_3 … N_n. The adjectives corresponding to the semantic features included A_1, and A_2, A_3 … A_n representing other semantic features. Therefore, "many to many" adjective-noun collocations such as "喧嚣的城市" (noisy city), "繁华的城市" (prosperous city), "安逸的小城" (easy town), "独特的县城" (unique county town) and so on were formed. We believe that the semantic network of adjective-noun collocations in CSL production is a gradual development process from "many to one" and "one to many" to "many to many", as shown in Fig. 5:

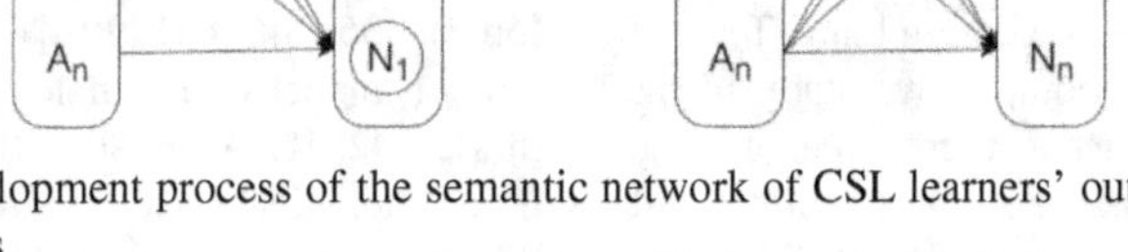

Fig. 5. The development process of the semantic network of CSL learners' output of adjective-noun collocations.

5 Conclusion

Word collocation knowledge is an important part of vocabulary knowledge, and it is also important for CSL learners when they are obtaining the vocabulary knowledge of their target language. Moreover, it is the difficulty that CSL learners encounter when learning target language vocabulary [5]. Based on the written output regarding the narrative of "我的家乡" (My hometown), this paper compared the use of the high-frequency adjective "大" and the high-frequency noun "城市" and the use of related collocations by international students of different Chines proficiency levels. We found that with the improvement of Chinese language level, the richness of adjectives produced by CSL learners has increased, including not only the increase of semantic types, but also the number of vocabularies of each semantic type has increased. However, there was no obvious change trend in the richness of nouns, which may be related to the fact that CSL learners used more noun hypernyms. The collocations of adjectives and nouns of CSL learners mainly focused on collocations of noun hypernyms and adjectives, while the collocation compactness of the adjectives of [shape], [representation], [nature] was relatively weak. With the improvement of the Chinese proficiency, the number of noun hyponyms used by CSL learners has increased, but the collocations of hyponyms and adjectives were still limited. In short, the development of the semantic network of adjective-nouns produced by the CSL learners was a slow development process from "many to one" and "one to many" to "many to many".

The research in this paper also has certain insights for CSL teaching. If teachers use collocation semantic network teaching, it will be an effective way to expand the vocabulary of students; for example, while teaching the word "城市" (city), its hyponyms "小城" (small city), "古城" (ancient city) and "县城" (county city) could be taught together to students. From the perspective of collocations, teachers can choose adjectives that have a high degree of semantic collocation compactness with the noun, such as "古老" (ancient), "熟悉" (familiar) and so on, and teach adjective-noun collocations according to the student's proficiency level. This is undoubtedly an effective way for CSL learners to use the target language accurately and authentically.

Acknowledgement. This research project is supported by the Fundamental Research Funds for the Central Universities, and the Research Funds of Beijing Language and Culture University (No. 20YCX110) and the Humanities and Social Sciences by the Ministry of Education (No. 19XJC740002 and No. 20YJAZH110).

References

1. Feng, X.: Research on the semantic network in the psychological dictionary of Chinese English learners. Foreign Lang. Teach. Res. **46**(03), 435–445+481 (2014)
2. Zhang, P.: A comparative study of the association model of psychological Lexicons for Chinese English learners. Foreign Lang. Teach. Res. **42**(01), 9–16+80 (2010)
3. Liu, S., Fu, B., Hu, A.: An empirical study of the psychological vocabulary representation network of people with different CSL levels. J. PLA Univ. Foreign Lang. **35**(02), 57–60+70+128 (2012)
4. Wu, J.: An Empirical Study of Psychological Lexical and Semantic Networks of People with Different CSL Levels. Journal of Zhaotong University, pp. 101–104 (2018)
5. Xing, H.: Research on Vocabulary Acquisition of Chinese as a CSL. Peking University Press (2016)
6. Lü, Sh.: A Preliminary Study of the Problems of Single and Double Syllables in Modern Chinese. Chinese Language (1963)
7. Lin, X.: On Word Collocation and Its Research. Language Teaching and Research, pp. 18–25 (1994)

8. Shen, J.: The Marking Mode of the Syntactic Function of Adjectives. Chinese Language, pp. 242–250 (1997)
9. Xu, Y.: Restrictive Factors for the Emergence of the Word "的". Rhetoric Study, pp. 33–34 (2003)
10. Wang, L., Xue, X.: A Study on the Implicit Acquisition of the "的" after the Attributive Adjective. Chinese Learning, pp. 83–90 (2007)
11. Liu, D.: Co-Occurrence of Adjective-Nouns and Direction of Adjectives. Nanjing Normal University Journal (Social Science Edition), pp. 56–61 (1987)
12. Zhang, G.: On Qualitative Adjectives. World Chinese Teaching, pp. 5–17+2 (2006)
13. Li, J.: Syntactic Limitation and Expression Function of Adjective-Noun Collocations. Journal of Shanghai Normal University (Philosophy and Social Sciences Edition), pp. 128–132 (2007)
14. Zhao, Ch.: The Principle of the Co-Occurrence of Emotion Adjectives and Nouns. Chinese Language and Literature, pp. 125–132+191 (2007)
15. Qi, F.: A Selective Study of "Adjective-Noun Collocations" in the Attributive Head Word Form [$A_{单}$ + N]. Shanghai Normal University (2003)
16. Zhao, Ch.: Static and Dynamic Research on Adjective-Noun Collocations. Jinan University (2006)
17. Liu, Z.: Analysis of Semantic Class Combination of Compound Words of Adjective Names and Morphemes. Nanjing Normal University (2008)
18. Wen, Y., Yuan, Ch., Huang, Ch.: Chinese Adjective-Noun Clustering Based on Collocation Pairs. Journal of Chinese Information Processing, pp. 45–50 (2000)
19. Qi, F., Duanmu, S.: A Quantitative Study on Attributive Headword Adjective-Noun Collocations of Word Lengths. Language Teaching and Research, pp. 83–91 (2015)
20. Hu, R., Xiao, H.: Research on the Construction and Application of Chinese Collocation Knowledge Base for CSL Teaching. Application of Language and Characters, pp. 135–144 (2019)
21. Xie, Ch.: An Investigation of Occurrence Rules of the "的" Character in Multiple Attributive Head Word Structures. Beijing Language and Culture University (2008)
22. Lei, Y.: A Quantitative Study on the Attribution of Multiple Attributives and the Word "的". Peking University (2012)
23. Zhang, Y.: A Study on the Acquisition of the Syntactic and Semantic Relations of Noun Collocation of Adjectives by International Students. Beijing Language and Culture University (2009)
24. Wu, K.: A contrastive analysis of interlanguage in the acquisition of adjective-nouns in polish rod Adjective predicate sentences for international students. J. Huainan Teach. Coll. **19**(02), 58–61 (2017)
25. Zhang, Y.: An Analysis of the Errors of Adjective-Nouns and Similar Words of International Students. Ludong University (2018)
26. Mei, J,. et al.: Synonym Cilin. Shanghai Dictionary Publishing House (1983)

Classifying Wu-Qing Couplets and General Couplets with Structural and Semantic Features

Yang Xu, Yuru Jiang([✉]), Yu Wu, and Yuyao Zhang

Beijing Information Science and Technology University, Beijing, China
jiangyuru@bistu.edu.cn

Abstract. The couplet custom is one of China's intangible cultural heritage, and it occupies a pivotal position in Chinese traditional culture. The general couplet requires that the first sentence and the second sentence be opposite and semantically coherent. As a particular type of couplets, Wu-Qing couplets are related on the character level but irrelevant at the sentence level. Such characteristics make the Wu-Qing couplet significantly different in structure and semantics compared to the general couplets. In this paper, Convolutional Neural Network is used to extract structural features. Long Short-Term Memory network and vector operations are used to extract semantic features, and the Attention mechanism is used to strengthen structural and semantic information. Finally, we propose a model with structural and semantic features between general couplets and Wu-Qing couplets. The final F1 score reached 82.6, which was an increase of 5.4 compared to the baseline model.

Keywords: Text classification · Wu-Qing couplet · Convolutional neural network · Long short-term memory network · Attention mechanism

1 Introduction

Chinese couplets are the treasure of Chinese traditional culture and have a long history. Before the Qin and Han dynasties, people wrote the legendary names of the gods "Shen Cha" and "Yu Lei" on Taomu (Peachwood) and hung them on both sides of the door so as to drive away ghosts and ward off evil spirits. In the Three Kingdoms period, people began to write couplets on Taomu, which was the prototype of couplets.

According to the logical relationship between the first and the second sentence of the couplets, couplets can be divided into two categories [1], namely general couplets and Wu-Qing couplets. Nowadays, there is a lot of couplet data on the Internet, but there is no classification. It is very unfavorable for the couplet generation task.

The couplets generally known to people in daily life is called general couplets, like the Spring Festival couplets. Wu-Qing couplets, as a particular type of couplet, was created by Zhang Zhidong in the late Qing dynasty. The general couplet requires that the first sentence and the second sentence be literally opposite but semantically coherent. Yet, for the Wu-Qing couplets, the characters are related, while the sentences are irrelevant. In other words, if the semantics of the first and second sentences in a couplet is not coherent, it is likely to be the Wu-Qing couplet [2, 3].

M. Liu et al. (Eds.): CLSW 2020, LNAI 12278, pp. 562–575, 2021.
https://doi.org/10.1007/978-3-030-81197-6_48

Example 1. 公门桃李争荣日，法国荷兰比利时。
The lexical items involved in the couplets are interpreted as: "公门(Ancient government, we call it "yamen" in Chinese) 桃 (peach) 李 (plum) 争 (vie for) 荣 (glory) 日 (day), 法国(France) 荷 (lotus) 兰 (orchid) 比 (vie for) 利 (benefit) 时 (hour)". Literally, each character of the first sentence and the second sentence corresponds to each other. To be specific, "公门" corresponds to "法国" as they both denote places. "桃" corresponds to "荷" while "李" corresponds to "兰", and they are all plant names. "争" corresponds to "比" as they have similar meanings to each other, while "荣" corresponds to "利" with the same reason. "日" correspond to "时", and they are both time-measure words. As for the sentence level, the first sentence is from the ancient story, and the second sentence contains three country names, from which we can see the semantics of these two sentences are unrelated. The characters in each sentence of the couplet correspond to those in the other sentence, but the contents of each sentence are different, which is intriguing and shows that Wu-Qing couplets unveil the charm of the Chinese Language [4]. Apart from the uniqueness at the lexical level and the sentence level, the conversion of parts of speech and the change of rhythm also make Wu-Qing couplet rarer and interesting. Few Wu-Qing couplets are straightforward and obvious in meaning while most of them are gentle, requiring a certain amount of knowledge accumulation.

At present, no one has carried out the classification work between Wu-Qing couplets and general couplets, and thus there is still many issues left for studies. This paper has adopted *the General rules of couplet rhyme(revision)* published by the Yinglian Society of China [5] to provide the formation rules of couplets. We have then systematically compared Wu-Qing couplets and general couplets in terms of the formation rules. Hence, we have found that the differences between the two are mainly reflected in structures and semantics (The details are listed in the third section). We then carried out the classification work based on the structural and semantic features.

Our contributions in this paper are as follows: 1. the Wu-Qing couplet dataset is constructed in this paper; 2. A classification model that combines structural and semantic features of Wu-Qing couplets and general couplets is proposed to lay the foundation for further researches; 3. Our study is the first study using natural language processing technology to distinguish Wu-Qing couplets from general couplets. We hope this work will attract more attention to the couplet culture.

2 Related Work

Text classification has always been important in the research in natural language processing. How to extract text features more effectively has become the focus of text classification research. With the development of deep learning technology, Long Short-Term Memory Network (LSTM) and Convolutional Neural Network (CNN) have been well applied. In the research process of text classification, Peng [6] mentioned that BiLSTM could effectively express the text's semantic features. The model they built has shown certain advantages in topic semantic coherence and text classification accuracy. Ma [7] found that in short text classification, using CNN alone can only extract local features. Using LSTM alone will cause the problem that the contextual connection will be weakened as the text length increases. So, they proposed a hybrid model based on the two and achieved a good effect. In recent years, the Attention Mechanism [8] has begun to be applied to the field of natural language processing. Its primary role is to pay full attention to and to learn the critical features of input data, extract essential information of input data, and improve output data quality. Zhao [9] has done practice and proposed that the CNN model with the Attention Mechanism significantly affects extracting the non-local importance of features.

3 Differences Between Wu-Qing Couplets and General Couplets

Yinglian Society of China published *the General rules of couplet rhyme (revision)* [5] in 2008. For most couplet lovers, a series of specifications have been formulated for couplet creation, evaluation, appreciation, etc. The list of rules is shown in Table1.

Table 1. Six rules for completing a couplet.

	Description
Rule 1	Corresponding characters and sentences
Rule 2	Corresponding part of speech
Rule 3	Corresponding structures
Rule 4	Corresponding rhythm
Rule 5	Corresponding level and oblique tones
Rule 6	Corresponding meanings

A couplet that meets the above six rules is complete. Based on the six rules above, a comparison between Wu-Qing couplets and general couplets in terms of their characteristics is made, as shown in Table 2.

From the comparisons shown in Table 2, we have obtained the following observations:

(1) Wu-Qing couplets and general couplets both are the same in terms of Rule 1 and Rule 5.
(2) Rule 2 and Rule 6 are related to the semantic features of the couplets. There exists a difference between the two kinds of couplets.
(3) Rule3 and Rule4 are related to the structural features of the couplets. There also exists a difference between the two.

There are apparent differences between Wu-Qing couplets and general couplets in terms of semantic and structural features, as shown in Table 3. This paper takes these two as the main features and adds them to the classification model to enhance the discrimination.

Table 2. Comparisons between Wu-Qing couplet and general couplets based on the six rules

	Wu-Qing couplets	General couplet
Rule 1	Equal number of characters; equal number of sentences	Equal number of characters; equal number of sentences
Rule 2	Try to subvert the part of speech in the same position	Keep the same part of speech in the same position
Rule 3	The sentence patterns are not necessarily the same literally	The context has the same sentence structure or phrase structure
Rule 4	The rhythm of the sentences is as asynchronous as possible	Try to keep the rhythm of the sentences consistent
Rule 5	Opposite tones	Opposite tones
Rule 6	Semantically and contextually irrelevant	Semantically and contextually related

Table 3. Differences in structure and semantics. Between Wu-Qing couplets and general couplets.

	Wu-Qing couplets	General couplets
Structure	Inconsistent structure and subversive rhythm	Consistent structure and consistent rhythm
Semantics	Irrelevant content, disjointed semantics, subversive parts of speech	Relevant content, connective sentences, same parts of speech

4 Model Architecture

4.1 Baseline Model

The overall structure of the baseline model is shown in Fig. 1. The input of the model is pre-processed couplet data. The first step is encoding the first and the second sentences of the couplet separately by the embedding layer (*EMBEDDING*). We used the randomly initialized word embedding to obtain matrix W_u of the first sentence and matrix W_l of the second sentence. Next, we encoded matrix W_u and matrix W_l, respectively, with the BiLSTM encoding layer (*BiLSTM*) to capture the couplet's semantic dependency information. Then, we obtained the initial coding results L_u and L_l containing the couplet's information, as shown in formula (1) (2).

$$L_u = \overleftrightarrow{BiLSTM}(W_u) . \quad (1)$$

$$L_l = \overleftrightarrow{BiLSTM}(W_l) . \quad (2)$$

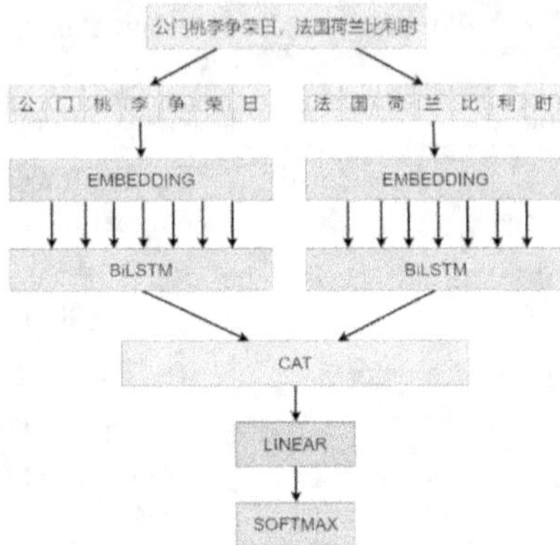

Fig. 1. Baseline model structure

What come next are to connect (*CAT*) the encoding information L_u and L_l then put them into the linear layer (*LINEAR*) to obtain the final coding information $Final_{enc}$, and to use the activation function (*SOFTMAX*) to get the final classification prediction result *Predict_classify*, as shown in formula (3) (4).

$$Final_{enc} = LINER([L_u; L_l]) . \tag{3}$$

$$Predict_{classify} = SOFTMAX(Final_{enc}) . \tag{4}$$

4.2 The Semantic and Structural Feature Combination Model

The semantic and structural feature combination model proposed in this paper adds semantic features and structural features based on the baseline model. The complete model structure is shown in Fig. 2. The top part of the model is the input layer, and the bottom part is the output layer.

The main structure is divided into three parts: the first part is the BiLSTM encoding module, which encodes text information; the second part is the semantic feature combination module, which extracts and combines the semantic information of text; the third part is the structural feature combination module, which extracts and combines the structural information of the text.

4.2.1 BiLSTM Encoding

The structure of the BiLSTM encoding module is the same as that of the baseline model. The difference is that the pre-trained word embedding Word2Vec is used to embed each word in the input part. Compared with the stochastically initialized word embedding in the baseline model, the pre-trained word embedding contains more comprehensive information. The formula is the same as that of the baseline model.

After putting the input information into the BiLSTM encoding layer, we added semantic and structural features to the model.

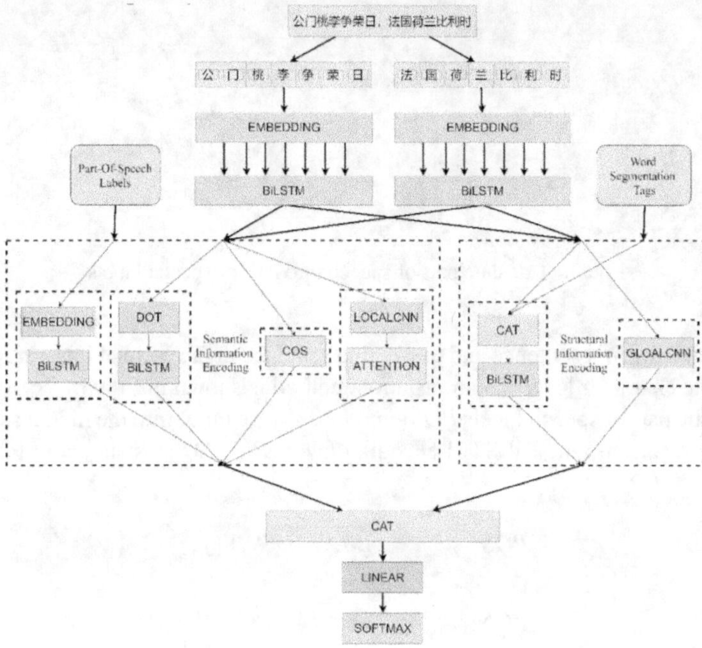

Fig. 2. The semantic and structural feature combination model

4.2.2 Semantic Information Encoding

For semantic information encoding module, this part contains four different semantic feature representation methods:

(1) Part of speech feature (*POS*)
There are cases of Wu-Qing couplets where a word in a sentence changes its part of speech so as to be compatible with the part of speech of the word corresponding it in the other sentence. The semantics of the word will also change accordingly when the part of speech of the word changes.

Example 2. 陆地, 十天。
In the above example, the first sentence contains "陆地 (land)", and the second sentence contains "十天 (ten days)". The part of speech of the character "陆 (land or six)" in Chinese is often noun or numeral. When understood at the character level, here "陆(six)" is a numeral corresponding to "十(ten)", but when apprehended at the sentence level, "陆(land)" becomes a noun. Overall, this Wu-Qing couplet has related word meaning and unrelated semantic content.

We collected the parts of speech of 20948 Chinese characters from the Online Chinese Dictionary[1], including 12 categories. According to statistics, each word has at most seven parts of speech. In the pre-processing stage, the part of speech dictionary is constructed.

[1] http://xh.5156edu.com.

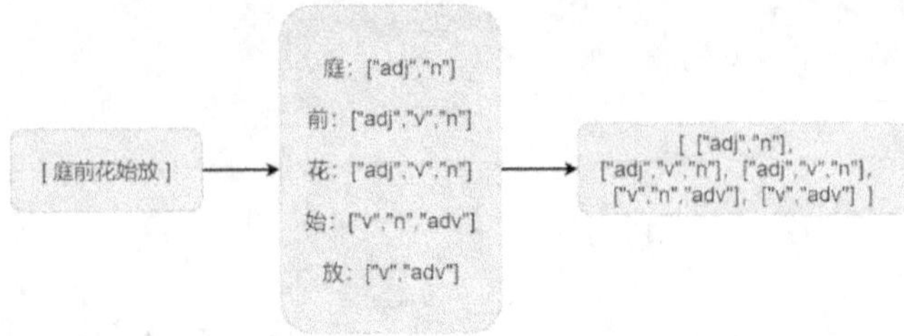

Fig. 3. Translate part of speech into part of speech labels

We translated the first sentence and the second sentence in the couplet into part of speech labels at the character level. The translation method is shown in Fig. 3.

After the part of speech labels POS_1 and POS_2 were input into the model as features, we put the result into the BiLSTM encoding layer ($BiLSTM$), as shown in formula (5) (6).

$$POS_{1\_enc} = \overleftrightarrow{BiLSTM}(POS_1)\ . \tag{5}$$

$$POS_{2\_enc} = \overleftrightarrow{BiLSTM}(POS_2)\ . \tag{6}$$

(2) Local semantic feature (*Local CNN + Attention*)

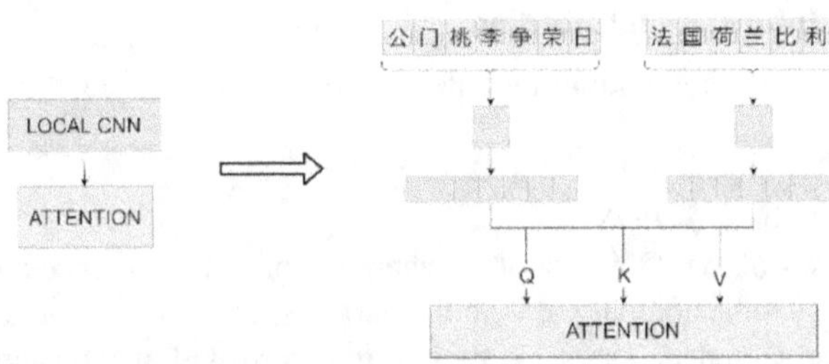

Fig. 4. Local CNN + Attention Model

Wu-Qing couplets and general couplets have a big difference in the character level's correspondence. To strengthen local semantic information, we used convolutional neural networks and attention mechanism methods. The structure of this part is shown in Fig. 4. We adopted the convolution neural networks of convolution, *Kernel* was 1 and step length *Stride* was 1. Then, we put the initial encoding results L_u and L_l into the one-dimensional convolution, and the step was expressed as $Conv1d$. Next, we put the convolution result into the activation function $f(= ReLU)$ to get Cnn_{word_1} and Cnn_{word_2}. The result of the addition was recorded as Cnn_{word}. Finally, we utilized the attention mechanism to extract information of both sentences of the couplet in the same position, getting Att_{enc}, which captured the semantic features of the couplet, as shown in formula (7)–(10).

$$Cnn_{word_1} = f(Conv1d(L_u, Kernel(1), Stride(1)))\ . \tag{7}$$

$$Cnn_{word_2} = f(Conv1d(L_l, Kernel(1), Stride(1))) \,. \tag{8}$$

$$Cnn_{word} = Cnn_{word_1} + Cnn_{word_2} \,. \tag{9}$$

$$Att_{enc} = Attention(Cnn_{word1}, Cnn_{word2}). \tag{10}$$

(3) Semantic similarity feature1 (*Dot*)

The vector operation can measure the similarity of the texts. First, we took the dot product operation of vectors L_u and L_l. Then, we put the operation result *Dot* into the BiLSTM encoding layer (*BiLSTM*). It is shown in formula (11) (12).

$$Dot = L_u \cdot L_l. \tag{11}$$

$$Dot_{enc} = \overleftrightarrow{BiLSTM}(Dot) \,. \tag{12}$$

(4) Semantic similarity feature2 (*Cosine*)

The most common method used to calculate the text similarity is the cosine similarity. We took the vectors L_u and L_l to calculate cosine similarity directly, as shown in Formula (13).

$$Cosine_{enc} = \frac{L_u \cdot L_l}{\|L_u\| * \|L_l\|} \,. \tag{13}$$

4.2.3 Structural Information Encoding

For structural information encoding module, the part contains two different structural feature representation methods:

(1) Global structural feature (*Global CNN*)

Its rhythmic division can show the structure of a couplet. To strengthen global structural information, we used convolutional neural networks. This part is shown in Fig. 5. In a couplet, most of the rhythm is usually divided in the unit of one character, two characters or three characters, while less cases divide the rhythm in the unit of more than three characters. To extract the structural features of the information of the couplet more accurately and comprehensively, we did three convolution operations. Respectively, convolution kernels and stride were set to 1, 2, 3. Three times of convolution of different parameters are used to extract the structural information corresponding to one character, two characters, three characters.

We connected the couplet encoding information L_u and L_l to obtain the couplet encoding information *Couplet*, as shown in formula (14).

$$Couplet = [L_u; L_l] \,. \tag{14}$$

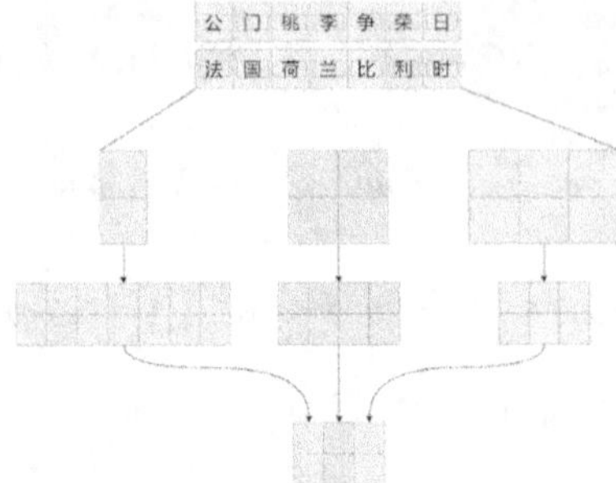

Fig. 5. Global CNN model

Then, the convolution kernels *Kernel* were set respectively with the three window sizes of 1, 2, 3, and the step length *Stride* were set as 1, 2, 3. Then, we carried out three one-dimensional convolution operations on the encoding information of both sentences of *Couplet*, expressed as $Conv1d$. After three convolution operations, we put the results of three convolutions into the activation function $f(=ReLU)$ to obtain the final result Cnn_i, as shown in formula (15).

$$Cnn_i = f(Conv1d(Couplet, Kernel(i), Stride(i)))\ (i = 1, 2, 3)\ . \tag{15}$$

In order to avoid overfitting, we put the result Cnn_i into *Dropout* and *MaxPooling* layer to obtain Cnn_{enc_i}. Finally, the three results were added to obtain the encoding result Cnn_{final}, as shown in formula (16) (17).

$$Cnn_{enc_i} = MaxPooling(Dropout(Cnn_i))\ (i = 1, 2, 3)\ . \tag{16}$$

$$Cnn_{final} = Cnn_{enc_1} + Cnn_{enc_2} + Cnn_{enc_3}. \tag{17}$$

(2) Word segmentation feature (WS)

The rhythm of the couplet can also be shown by word segmentation. Wput both sentences in the couplet into the word segmentation tool. The result was the sentence division of a couplet. It is also the structural feature of the couplet.

To make up for the lack of accuracy of a single word segmentation tool, we chose five different tools HanLP[2], jieba[3], pynlpir[4], THULAC[5] and pkuseg[6] in this paper, and then selected the most appropriate word segmentation results according to the voting method. Many traditional expressions in the general couplets, such as "玉 (jade) 楼 (tower)" and "金 (gold) 乌 (crow)", are difficult to be distinguished by the existing word segmentation tools. People also rarely express in this way in modern Chinese. Therefore, we collated

[2] https://github.com/hankcs/HanLP.
[3] https://github.com/fxsjy/jieba.
[4] https://github.com/tsroten/pynlpir.
[5] https://github.com/thunlp/THULAC-Python.
[6] https://github.com/lancopku/pkuseg-python.

these words in the corpus and added them to the word segmentation dictionary for further word segmentation.

Fig. 6. Translate word segmentation results into word segmentation labels

After the word segmentation operation, the word segmentation results were translated into word segmentation labels, which were to be input into the model together as the features of both sentences of the couplet. This part is shown in Fig. 6.

Because the word segmentation result was related to the couplet's information, we connected the word segmentation labels and the initial encoding results, and then sent them into the BiLSTM layer (*BiLSTM*). At last we did subtraction. The difference in structural features was obtained. It is shown in formula (18)–(20).

$$WS_{x_1} = \overleftrightarrow{BiLSTM}\left(\left[L_u, Tag_1\right]\right). \quad (18)$$

$$WS_{x_2} = \overleftrightarrow{BiLSTM}\left(\left[L_l, Tag_2\right]\right). \quad (19)$$

$$Sub_{enc} = WS_{x_1} - WS_{x_2}. \quad (20)$$

At last, we connected all the results of the obtained semantic encoding layer and the structural encoding layer. Then, we put them into a linear layer (*LINEAR*) to obtain the final coding information $Final_{enc}$ and used the activation function (*SOFTMAX*) to get the last classification prediction result *Predict_classify*, as shown in formula (21) (22).

$$Final_{enc} = LINEAR\left(\begin{bmatrix} L_u;\, L_l;\, POS1_{enc};\, POS2_{enc};\, Cnn_{word};\, Att_{enc}; \\ Dot_{enc};\, Cosine_{enc};\, Cnn_{final};\, WS_{x1};\, WS_{x2};\, Sub_{enc} \end{bmatrix}\right). \quad (21)$$

$$Predict_{classify} = SOFTMAX\left(Final_{enc}\right). \quad (22)$$

5 Experiment Settings

5.1 Dataset

The Wu-Qing couplet data[7] and the general couplet data[8] used in the experiment are collected online. The number of the existing Wu-Qing couplet is relatively rare and has no official available dataset. After collecting and collating, we have obtained 2,354 Wu-Qing couplets in total. To evaluate the performance of the model better, we have used two ways to balance the dataset. One is setting the Wu-Qing couplet to a maximum length of 10 characters, and the other is selecting 3,595 general couplets from the general couplets we collected.

[7] http://www.zhgc.com.

[8] http://tool.xdf.cn/chunlian.

Table 4. Dataset statistics

Word count	The Wu-Qing couplet	The general couplet
2	20	33
3	98	14
4	321	328
5	636	1061
6	88	162
7	1083	1763
8	59	119
9	26	56
10	23	59
Total	2354	3595

5.2 Evaluation Metrics

The model that combines the semantic and structural features is to solve the classification problem, and the evaluation metrics usually used are *Accuracy*, *Precision*, *Recall* and F_1 *score*. The values of them are used to measure the prediction of the Wu-Qing couplets, as shown in formula (23)–(26).

$$Accuracy = \frac{\text{Total number of the couplets correctly classified into the corresponding categories}}{\text{Total number of the couplets of all categories}}. \tag{23}$$

$$Precision = \frac{\text{Total number of couplets correctly classified into the Wu-Qing couplet}}{\text{Total number of the predicted Wu-Qing couplets}}. \tag{24}$$

$$Recall = \frac{\text{Total number of couplets correctly classified into the Wu-Qing couplet}}{\text{Total number of the Wu-Qing couplets in dataset}}. \tag{25}$$

$$F_1 = \frac{2\text{*Recall}}{\text{Precision} + \text{Recall}}. \tag{26}$$

5.3 Experiment Parameters

The experiment adopts the method of five-fold cross-validation for the dataset. The PyTorch version is 1.3.0. In the process of model training, word embedding dimension is set as 100, learning rate 0.001, the epoch of training rounds 20, the dropout rate in word embedding layer and convolutional layer 0.3, and prediction layer 0.5. The optimizer is Adam. The dataset used in Word2Vec word embedding training include 2,354 Wu-Qing couplets and 3,595 general couplets we collected, and 770,000 general couplets collected by Wang Bin[9].

[9] https://github.com/wb14123/couplet-dataset.

6 Experiment Results

6.1 Baseline Model

Figure 7 shows the results of the baseline model by using five-fold cross-validation method, with the abscissa representing different folds and the ordinate representing scores. The F1 score and accuracy of each fold in the baseline model are shown in the table below. After calculation, the final average of the F1 score is 77.2 and the accuracy 81.1.

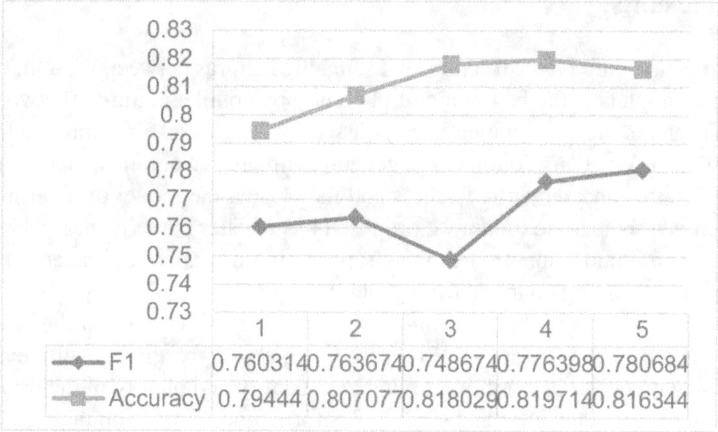

Fig. 7. The Results of F1 score and accuracy on the baseline model

6.2 The Semantic and Structural Feature Combination Model

As can be seen from Table 5, the F1 score of the baseline model is 77.2. After adding the pre-trained word embedding, the F1 score is improved by 3.2, increasing to 80.4. Then, by adding the semantic and structural features, the F1 score is further enhanced by 2.2, rising to 82.6. It is shown that the semantic and structural features proposed in this paper can promote the performance of the classification between Wu-Qing couplets and general couplets. However, the enhancement effect is weaker than that of the pre-trained word embedding for the baseline model. The effect of semantic and structural features on the baseline model was also tested. Compared with the baseline model, the F1 score increased by about 0.6, proving the validity of the semantic and structural features proposed in this paper again.

Table 5. The ablation experiment results of the semantic and structural features combination model

System	F1	Accuracy	Precision	Recall
All	82.6	86.0	81.5	83.9
Semantic and structural features	80.4	83.8	77.0	84.2
Pre-trained word embedding	77.2	81.1	74.0	80.7

7 Conclusions

The classification method of structural and semantic features between Wu-Qing couplets and general couplets is the beginning of our study on couplet culture. The work of this paper can promote the development of couplet generation tasks. First, through the analysis of features of Wu-Qing couplets and general couplets, we obtained their quantifiable structural features and semantic features and then tested the model on a certain corpus scale. The result is that the model we proposed has a better performance, which means that the structural and semantic features proposed in this paper have a certain effect in distinguishing Wu-Qing couplets from general couplets.

However, considering the experimental corpus, it is observed that the scale of the corpus is limited by the few available resources, which may result in insufficiency in the deep learning model. It is worth improving the quantity and quality of the training data to learn and express semantic and structural information better. Simultaneously, using the phenomenon that the meanings of words are mostly different under different parts of speech has obtained a certain effect. Yet, how to better recognize and represent the meanings of words can also be explored in future studies. Moreover, to better represent word meaning in semanteme is also one of our future works.

Acknowledgments. This work is supported by the National Natural Science Foundation of China (Grant No.61602044) and the funds for improving the quality of personnel training in 2020 of Beijing Information Science and Technology University (Grant No. 5102010805).

References

1. Luo, W.: The types and characteristics of couplets. J. Beijing Inst. Graph. Commun. **02**, 38–43 (2003). (in Chinese)
2. Li, X.: How to recognize the Wu-Qing couplet. Couplets (Folk Couplets Stories). 2006(12). (in Chinese)
3. XiaoHua, W.: Talk about "the Wu-Qing couplet." Chinese Culture **02**, 28–29 (2008). (in Chinese)
4. Yan, S.: Chinese character is a special phenomenon of Chinese culture. Chinese Character Culture **2018**(16), 152–154+158 (2018). (in Chinese)
5. General rules of couplet rhyme (revision). Couplets (Folk Couplets Stories) 0(12), 8–9 (2008). (in Chinese)

6. Peng, M., Yang, S., Zhu, J.: Topic model based on bidirectional LSTM semantic enhancement . J. Chinese Inf. Process. **32**(04), 40–49 (2018). (in Chinese)
7. Ma, Z., Hu, J., Long, M.: Short text classification using CNN-LSTM mixed model. J. Air Force Early Warning Acad. **33**(04), 295–297+302 (2019). (in Chinese)
8. Vaswani, A., et al.: Attention is all you need. Advances in Neural Information Processing Systems (2017)
9. Zhao, Y., Duan, Y.: Convolutional neural network text classification model based on attention mechanism . J. Appl. Sci. **37**(04), 541–550 (2019). (in Chinese)

The Construction of a Collocation List Based on Academic Papers of Teaching Chinese to Speakers of Other Languages

Qihong Zhou[1,2(✉)]

[1] School of Chinese as a Second Language, Peking University, Beijing 100871, China
zhouqihong@pku.edu.cn

[2] Department of Chinese Language and Literature, Sichuan International Studies University, Chongqing 400031, China

Abstract. This paper builds a collocation list in academic papers of teaching Chinese to speakers of other languages, based on the dependency parsing analysis. It adopts python to extract the Chinese academic collocations, analyzes the accuracy of collocation extraction, and compares with the collocation distribution of Wang Shuo's works. After setting the frequency and automatic threshold of mutual information, 12,357 pieces of collocations were obtained from academic papers of teaching Chinese to speakers of other languages, and there are 5,298 collocations from Wang Shuo's works. It is found that in the academic Chinese collocation list, the collocation frequency of attributive is the highest, followed by adverbials. The most overlapping part between academic Chinese and Wang Shuo's works is the adverbial structure. This study attempts to construct the collocation list in academic papers of teaching Chinese to speakers of other languages, which bears certain insights to academic Chinese teaching.

Keywords: Collocation extraction · Dependency parsing · Academic Chinese · Wang Shuo's works

1 Introduction

The increasingly in-depth studies on international Chinese education boost the development of Chinese learning, and accordingly, learners' demand for academic Chinese has further been raised. According to the statistics on foreign students which was published by the Ministry of Education in China in 2018, a total of 258,122 foreign students received academic education, accounting for 52.44% of the total number of foreign students in China and representing an increase of 16,579 over 2017, or a 6.86% rise over the same period last year. There are 85,062 master and doctoral students, representing a growth of 12.28% over 2017, of which 25,618 are doctoral students and 59,444 are master students. On the one hand, it is shown that the quantity of students with academic qualifications is going up. On the other hand, foreign students in China are involved in reading Chinese academic papers and conducting academic exchanges, showing a

M. Liu et al. (Eds.): CLSW 2020, LNAI 12278, pp. 576–592, 2021.
https://doi.org/10.1007/978-3-030-81197-6_49

higher demand for academic Chinese. Compared with the booming demand, the attention put to academic Chinese is insufficient for teaching Chinese to speakers of other languages. For instance, in CNKI database of academic papers, when the phrase "academic Chinese" is used as the key word for searching, there are only 9 journal papers and conference papers, and 1 master's thesis shown up, indicating that the research results are limited and the research scope is scattered. Overall, the studies on the relevant topics on academic Chinese are relatively weak, and as Liu Lening [1] proposed, strengthening the research and teaching of academic Chinese has become a top priority in advanced Chinese teaching. In this context, this paper mainly discusses the collocations used in academic papers in the field of teaching Chinese to speakers of other languages.

In the course of foreign language learning, the common words in the target language as well as their major usage patterns and typical collocations are critical [2]. Collocations are crucial for language learning because the ability to use idiomatically constitutes is a key indicator of language proficiency [3–5].

In every language, vocabulary collocation is a significant topic, especially in Chinese [6]. Lin Xingguang [7] pointed out that the practice of teaching Chinese as a foreign language proves that it is often difficult for foreigners to master the collocation patterns of Chinese vocabularies, and hence, the problem of vocabulary collocation is vital in this subject.

2 Literature Review

Howard Ho-Jan Chen and Tao Hongyin [8] reviewed the construction of written academic Chinese corpora and discussed the theoretical and practical issues in academic Chinese. Liu Yu [9] compiled a list of Chinese Academic Formulas for Presentation based on research articles in liberal arts and science, oral presentation recording of academic conferences and a list of Chinese parentheses. Liu, C., et al. [10] extracted and classified 3-g, 4-g academic lexical bundles based on 9 million academic journal corpora of liberal arts. Compared with the research of academic English, academic Chinese is still in its infancy. There are few publicly available academic Chinese corpus resources, and there is a lack of academic collocation lists, and academic Chinese dictionaries, and a lack of research on how to effectively integrate the corpus with academic Chinese teaching.

From the view of computational linguistics, Sinclair [11] put forward the concepts of span, node, casual collocation, significant collocation, collocate. Sun Maosong et al. [12] proposed three statistical indexes to determine collocation, including intensity, dispersion and tip, based on the results of a computational linguistics research on collocations. Subsequently, on the basis of Sun Maosong's proposal, some scholars developed the automatic identification and extraction system of collocation from different perspectives [13–23].

Hu Renfen and Xiao Hang [24] extracted 22,298 pieces of collocations from 2.4 million Chinese words as a second language textbook corpus, and extracted 219,451 collocations from the web corpus of 138 million words. Hence they constructed a large-scale collocation list in two different fields, and also proposed four properties and nine types of Chinese collocations. Hu Renfen and Xiao Hang [25] built the collocation list of Chinese vocabulary, which put imputes into the research of Chinese collocation as a second language. Nevertheless, the collocation list was derived from the existing Chinese teaching materials that were suitable for general Chinese teaching, while academic Chinese belongs to specialized Chinese, featuring strong specialty yet scare collocation research results. This paper attempts to employ computational linguistics in the research of academic Chinese collocation and to construct the collocation list of academic Chinese.

3 Methodology

3.1 Corpus Design

The corpus used in this present study consisted of two sub-corpora, namely the academic corpora and fiction. The academic corpora were retrieved from leading journals of teaching Chinese to speakers of other languages, including *Chinese Teaching in the World, Language Teaching and Linguistic Studies, Applied Linguistics* published from 2016 to 2018. The selected papers involve second language acquisition research, Chinese pronunciation research, grammar teaching research, Chinese language research, Chinese teaching research, Chinese textbook compilation, and test research of Chinese as a second language. After normalization, the Chinese classification number, document identification code, article number, DOI, English abstract, figures, tables, formulas, data, footnote, endnote, reference, appendix, author profile, and etc. were removed. As a result, 168 journal papers were selected, with a total of 1,892,794 words (Table 1).

Table 1. Quantity of the selected academic papers

Journal name	Quantity (pieces)
Chinese Teaching In The World	39
Language Teaching and Linguistic Studies	84
Applied Linguistics	45
Total	168

For fiction, Wang Shuo's works were chosen as a collocation list for reference. A complete collection of 27 works (see Appendix) in 1,520,692 words, reflecting the contemporary language of urban life in northern China, especially in Beijing, are within the range.

3.2 Dependency Parsing

The concept of dependency parsing (DP) was put forward by L. Tesniere in 1959. The study focused on the syntactic structure and analyzed the interdependencies among linguistic units [22]. The dependency parsing defines a core word (usually a verb) in a sentence, which is not subject to other words but can dominate other words by certain dependency.

The Language Technology Platform (LTP) was developed by the Research Center for Social Computing and Information Retrieval of Harbin Institute of Technology. The LTP provides five models, including word segmentation model *cws.model*, part-of-speech annotation model *pos.model*, named entity recognition model *ner.model*, dependency parsing syntactic analysis model *parser.model*, and semantic role annotation model *pisrl.model*. This paper takes the word segmentation model *cws.model*, part-of-speech annotation model *pos.model* and dependency parsing syntactic analysis model *parser.model* [26] as our basis. Python 3.6 is applied as the programming tool for this research.

4 Results

This part is largely the discussion of the collocation list compiled based on academic papers of teaching Chinese to speakers of other languages and the collocations of Wang Shuo's works, covering the collocations, the frequency distribution of collocations, the correct rate, and the contrast between the two texts.

4.1 Academic Chinese Collocation Results

4.1.1 The Results of Academic Chinese Collocation List

A total of 51,327 sentences were extracted from the self-built corpus of academic papers of teaching Chinese to speakers of other languages (hereinafter referred to as "the academic Chinese corpus"), and 1,148,912 sets of dependency pairs are automatically extracted. According to the analysis of dependency parsing, 14 kinds of dependencies were extracted, namely, ATT (attribute), ADV (adverbial), RAD(right adjunct), SBV(subject-verb), HED (head), LAD (left adjunct), COO (coordinate), VOB(verb-subject), CMP (complement), POB (preposition-object), WP (punctuation), FOB (fronting-object), DBL (double), IOB (indirect-object).

Table 2. Dependency parsing types, quantity and percentage of collocations

Type	Quantity	Percentage	Rank
ATT	262686	22.86%	1
ADV	150777	13.12%	3
RAD	81711	7.11%	7
SBV	86255	7.50%	6
HED	51327	4.46%	8
LAD	20282	1.76%	10
COO	108577	9.45%	4
VOB	107486	9.35%	5

(*continued*)

Table 2. (*continued*)

Type	Quantity	Percentage	Rank
CMP	13004	1.13%	11
POB	39962	3.47%	9
WP	222258	19.34%	2
FOB	2295	0.19%	12
DBL	2196	0.19%	13
IOB	96	0.008%	14
Total	1148912	100%	

From the quantitative point of view (Table 2, the proportion of ATT in academic Chinese collocation list occupies the highest, in a total of 262,686, making up 22.86%. It is followed by WP of 222,258, accounting for 19.34%. The ADV takes the third place with a total of 150,777, i.e., 13.12%. 107,486 VOB ranks fifth, constituting 9.35% and 86,255 SBV ranks sixth for 7.5%. CMP stands the 11th of 13,004, with the proportion being 1.13%.

4.1.2 Frequency Distribution of the Chinese Collocation List

Amidst the academic Chinese collocation list, the 14 relationships extracted automatically are relatively macroscopic; hence, we selected the top ten collocation in terms of frequency among the five collocations, namely SBV, VOB, CMP, ATT and ADV, with the purpose to observe their specific usage distribution, as shown in Table 3.

As shown in Table 3, the collocation with the highest frequency in VOB is "认为是" (to think to be) which occurs for 382 times, followed by "进行分析" (to conduct analyses) which occurs for 238 times. The collocation with the highest frequency in ATT is "两个" (two) with 1,153 times of occurrences, followed by "一种" (a kind) with 928 times of occurrences. In ADV, the collocation with the highest frequency is "不是" (to be not), with a total of 1,286 occurrences, and the one with the second high frequency is "不能" (cannot) which occurs for 774 times. When it comes to SBV, "这是" (this is) takes the highest by appearing 628 times, which is followed by "我们认为" (we think) that occurs for 269 times in total.

4.1.3 Accuracy Rate

In order to testify the effect of the automatic extraction of collocations, the accuracy rate of the first 1,000 items of VOB extracted from the self-built academic Chinese collocation list were manually evaluated. It is found that 776 were reasonable while 224 were unreasonable, and the accuracy rate of collocation extraction reached 77.6%. The unreasonable collocation includes incomplete extraction, e.g., "需要处" (need to place), "具有语法学" (have syntax), "得研究" (to study); verb recognition error, e.g., "附缀化方面" (root-fix aspects), "由于知识" (due to knowledge); and collocation error, e.g., "过时间" (pass the time), "经历还是" (the experience still), "强调主" (stress the major).

Table 3. Top 10 frequency of five collocations in academic Chinese corpus

VOB		ATT		ADV		CMP		SBV	
Collocation	Frequency	Collocation	Frequency	Collocation	Frequency	Collocation	Frequency	Collocation	Frequency
认为是 To think to be	382	两个 Two	1153	不是 To be not	1286	表现出To eflect that	149	这是 This is	628
进行分析 To conduct analyses	238	一种 A/one kind	928	不能 Cannot	774	出现在 To appear in	130	我们认为 We think	269
是有 To have	234	母语者Native speakers	615	以为 To hold that	636	倾向于 To be prone to	123	结果显示 the results show	247
是指 To refer to	216	两种 Two kinds	558	都是 both	630	用来 To be used to	121	差异显著 The Difference(s) is/are notable	193
说是 To be	185	第二语言 The second language	527	也是 Also to be	618	注意到 Notice that	108	我们发现 We find that	186
存在差异 To present differences	129	三个 Three	471	就是 To be precisely	580	表现为 To be shown as	105	一是 The first is	169
进行研究 To carry out research	127	过程中 In the process	429	本研究 This study	366	呈现出 To show that	104	这说明 It means that	153

(*continued*)

Table 3. (*continued*)

VOB		ATT		ADV		CMP		SBV	
Collocation	Frequency	Collocation	Frequency	Collocation	Frequency	Collocation	Frequency	Collocation	Frequency
是时候 To be the time	126	汉语教学 Chinese teaching	418	也可以 Can also	316	打死 To spare no effort	94	二是 The second is	147
学习汉语 To learn Chinese	120	华文教育 Chinese education	358	也有 And also	313	相当于 To be Equal to	85	我们进行 We carry out	123
有差异 To have differences	118	这类 This kind	348	更多 More	290	给出 To give	85	研究是 The study is	119

4.2 Collocation Results in Wang Shuo's Works

4.2.1 Collocation Relationships in Wang Shuo's Works

In Wang Shuo's works, there are 60,769 sentences, among which 1,010,912 collocations were automatically extracted. Based on the analysis of dependency parsing, 14 kinds of dependencies were selected, which is consistent with the results of academic Chinese collocation list.

Table 4. Dependency parsing types, quantity and percentage of Wang Shuo's works

	Quantity	Percentage	Rank
WP	192502	19.04%	1
ADV	164732	16.29%	2
ATT	132324	13.08%	3
VOB	107121	10.59%	4
SBV	99800	9.87%	5
COO	94964	9.39%	6
RAD	87895	8.69%	7
HED	60769	6.01%	8
POB	31355	3.10%	9
CMP	28098	2.77%	10
DBL	5055	0.50%	11
LAD	4188	0.41%	12
FOB	1346	0.13%	13
IOB	763	0.07%	14
Total	1010912	100%	

From the quantitative perspective (Table 4), the appearance frequency of WP ranks the top, with a total of 192,502, taking up 19.04%, followed by ADV, appearing for 164,732 times, i.e., 16.29%. The third place is ATT, occupying 13.08% with 132,324 times of appearances. VOB ranks fourth with total of 107121 items, accounting for 10.59%. Collocation frequency of SBV ranks fifth with 99,800 in total, or 9.87%. CMP takes the tenth place with overall 8,098 items, with the proportion of 2.77%.

4.2.2 Frequency Distribution of Collocation in Wang Shuo's Works

In Wang Shuo's works, the top ten among the five collocations involving VOB, ATT, ADV, CMP and SBV were extracted referring to the collocation extraction from academic Chinese collocation list, and the analysis of their frequency and usage.

According to Table 5, the collocation with the highest frequency in VOB is "说是" (to say to be), appearing totally 785 times, followed by "是人" (to be a human) appearing

Table 5. Top 10 frequency of five collocations in Wang Shuo's works

VOB		ATT		ADV		CMP		SBV	
Collocation	Frequency	Collocation	Frequency	Collocation	Frequency	Collocation	Frequency	Collocation	Frequency
说是 To say to be	785	两个 Two	500	不是 To be not	2422	坐在 To sit at	529	我是 I am/was	1239
是人 To be a human	320	一个人 A/one person	482	就是 To be precisely	1008	站在 To stand at	455	我说 I'd say	1179
知道是 To know to be	256	几个 Some	403	也是 Also to be	872	站起来 To stand up	293	这是 This is	1018
是说 To say	252	一眼 A/one glimpse	376	不能 Cannot	824	走到 To walk to	280	你是 You are	882
是有 To have	232	一种 A/one kind	366	都是 All to be	806	躺在 To lie on	137	我想 I consider	689
说什么 To say what	207	一片 A/one piece	277	不知道 To have no idea	504	走去 On foot	111	你说 You say	576
说有 To say there to be	200	一声 A/one sound	270	不会 Will not	474	走进 To walk in	111	那是 That is	559
看我 To look at me	176	一条 A/one strip	269	不想 To be reluctant to	311	走过来 To walk towards	97	我看 I see	475
是想 To be intend to	147	一次 A/one time	267	不了 Would not	218	走来 To come	91	我知道 I know	421
问我 To ask me	141	一句 A/one sentence	257	不出 Not out	212	出现在 To appear in	89	我觉得 I think	370

for a total of 320 times. In ATT, the collocation with the highest frequency is "两个" (two) with the frequency of 500 times, followed by "一个人" (a/one person) appearing for 482 times. As for ADV, the collocation with the highest frequency is "不是" (to be not) with 2,422 times of occurrences and the second highest is "就是" (to be precisely) which appears for 1,008 times. In SBV, "我是" (I am) ranks the first with 1,239 times, followed by "我说"(I'd say) which is the second highest, appearing for 1018 times.

4.2.3 Accuracy Rate

The first 1,000 items of VOB in Wang Shuo's works were extracted automatically. After checking, 879 items were extracted correctly with 121 errors, showing that the accuracy rate of collocation extraction is 87.9%. Among unreasonable collocation in evaluation are extraction errors, e.g., "是度" (is the degree), "载进行" (carry it on), "告诉粗" (tell the coarse), "是干" (is dry), "漫公路" (roam the road), "敢K" (dare to K), "搞错" (mess up), "停飞" (stop the flight), "别生气" (don't be angry); and incomplete extractions, e.g., "打量舰" (observe at vessel), "戴徽" (wear the emblem), "进坞" (enter the dock), "派用场" (play the role), "吃日子" (ear the days), "帮佬" (helpful man), "学会左耳进" (learn to hear with left ear) and "打算死皮赖" (pan to be shameless).

4.3 Comparison

4.3.1 Comparisons Between Two Types of Text Sentences

Given the length of the sentence (Table 6), the average sentence length of the academic text is 31.14, the average sentence length of the novels is 29.62, and hence, the sentence of the academic text is longer than that of the novel text.

Table 6. Average sentence length

	Total words	Sentence number	Average length
Academic corpus	1892794	60769	31.14
Wang Shuo's works	1520692	51327	29.62

4.3.2 Contrasts in Collocation Frequency

From the usage frequency, the ranking of the five relationships in the academic Chinese collocation list is as follows: ATT > ADV > VOB > SBV > CMP (see Table 2). It can be seen that the usage frequency of ATT is the highest, followed by ADV. The ranking in Wang Shuo's works is: ADV > ATT > VOB > SBV > CMP (see Table 4), which indicates that it contrasts with academic Chinese, as ADV ranks the top while ATT is the second.

It is interesting to find that the collocation with the highest frequency in ATT is "两个" (two + measure word), the one in ADV is "不是" (are not). ATT is the category with the highest frequency, in consistence with the extracted results from two collocation lists. From the stylistic point of view, the two collocations are universal application. In VOB, the top ten of the frequency of the VOB structure in the novels are all action verbs + prepositions, such as, "坐在" (to sit at), "站在" (to stand at), and "站起来" (to stand up). The top ten of the VOB structure in the academic Chinese collocation list are not limited to action verbs, like "表现出" (to show that), and "体现在" (to reflect in).

Table 7. Overlapping collocation comparison in academic Chinese collocation list and Wang Shuo's works

Collocation	Academic Chinese	Wang Shuo's works	Overlapping	Overlapping ratio	Non-overlapping ratio
(ATT)	262686	132324	30503	23.05%	76.95%
(VOB)	107486	107121	24005	22.40%	77.60%
(ADV)	150777	164732	60603	36.78%	63.22%
(CMP)	13004	28098	9051	32.21%	67.79%
(SBV)	86255	99800	28774	28.83%	71.17%

It can be seen from Table 7 that through the contrast between the collocations in academic Chinese collocation list and the collocations in Wang Shuo's works, the categories with the highest frequency in the two lists are identical, i.e., ADV structure, and the overlapping reaches 36.78%. What comes next is CMP, with the overlapping rate of 32.21%. The overlapping of SBV structure is 28.83% while that of ATT is 23.05% and that of VOB is 22.40%.

5 Optimized Collocation Automatic Extraction

With the aim to improve the accuracy rate of collocation automatic extraction, we have made improvements in two aspects. One is to build a user-defined dictionary. The purpose is to enhance the accuracy of word segmentation in professional fields. The other is to introduce mutual information, to set the threshold, and to eliminate the collocation of low frequency.

5.1 Constructing a User-Defined Dictionary

In the field of natural language processing, it is common to improve the accuracy rate of word segmentation in specialized fields by constructing a user-defined dictionary.

The applied collocation list is mainly from papers teaching Chinese to speakers of other languages. The major problems in word segmentation is to separate the technical terms, such as "疑/问句" (Interrogative sentence), "存现/句" (existential sentence), " 句/法" (syntax), "印/欧语" (Indo European languages), "反/义词 " (antonym). In terms of such problems, there are three sources in our user-defined dictionary. The first one is derived from the extraction process found in the word segmentation errors, such as "指/称" (Reference), "句/法学" (syntax), "习/得" (acquisition). The second is the key words published in papers in the past decade (2010–2019) by the three journals of *Language Teaching and Linguistic Studies, Chinese Teaching In The World,* and *Applied Linguistics*, with topics ranging from output-oriented method, vocabulary incidental acquisition, to speech distribution. The third is the technical terms used in the textbook *Introduction to Linguistics*, such as "subject-predicate statement", "expiratory sound", and "aggregate relationship". The user-defined dictionary consists of 6,358 terms. This dictionary employing specific operations in programming requires to call the function *segmentor.load_with_lexicon* provided by LTP, and simultaneously imports the word segmentation model *cws.model* and custom dictionaries (terminology). As shown in Fig. 1:

```
segmentor = Segmentor() #initialization
segmentor.load_with_lexicon('D:/NLP/ltp_data_v3.4.0/cws.model','D:/NLP/terminology.txt') #load model and user-defined dictionary
```

Fig. 1. The method for user-defined dictionary import

The commonly applied approaches to evaluate collocation intensity are mutual information (MI), T-score, Z-score, Chi-squared, and log-likehood. Based on the method of Hu Renfen and Xiao Hang [21] (referred to as "Hu's method"), we set the frequency threshold of collocation occurrence of academic Chinese collocation list to 3, and the threshold of mutual information of two words in collocation to 0. The calculation formula is as follows:

$$I(W_1, W_2, R) = \log \frac{(P(W_1, W_2, R))}{P(R) \times P(W_1|R) \times P(W_2|R)}$$

W1 and W2 represent the words in the combination, R is the type of collocation, and P is probability. Through statistical feature filtering, the low frequency collocation and partial extraction of wrong collocation can be eliminated, thus improving the accuracy rate.

Referring to the threshold set by Hu Renfen and Xiao Hang (2019), we got a total of 43,493 collocations in the academic Chinese collocation list, and 31,264 collocations in Wang Shuo's works, as shown in Table 8.

Table 8. The Filtered collocations

Type	Wang Shuo's works		Academic Chinese collocation list	
	Hu's method	After modifications	Hu's method	After modifications
ATT	3574	1493	11216	5167
WP	6031	133	7591	643
HED	1723	0	1513	0
COO	1620	231	2933	1569
VOB	3240	1152	4307	1580
SBV	2871	347	3149	1154
RAD	3939	324	2963	262
ADV	5663	1045	6334	1453
POB	940	129	1568	307
LAD	129	15	1158	17
CMP	1329	420	578	166
DBL	159	3	86	11
IOB	27	0	8	1
FOB	19	6	89	27
Total	31264	**5298**	43493	**12357**

After several experiments, the mutual information threshold was set to 3.8, and the frequency threshold to 4 for the highest extraction accuracy. 12,357 collocations were obtained in the academic collocation list, and 5,298 collocations were obtained in Wang Shuo's works, as shown in Table 8. In Table 9, taking VOB as an example, "受影响" (be affected), "存在差异" (have differences), and "受到影响" (be influenced) are the highly-frequent collocations in academic papers of teaching Chinese to speakers of other languages.

To verify the accuracy rate of extraction collocation after adjusting the parameters, 1,580 collocations in the extracted VOB were analyzed in the results of the academic collocation list. After manual check, there were 104 mismatched collocations, 1,476 correct collocations, and the accuracy rate reached 93.42%. Based on the study, the combination of user-defined dictionary and mutual information can improve the accuracy of collocation extraction in academic Chinese.

Table 9. Top 5 of the five combinations after modifications

NNO	SBV	Frequency (Mutual information values)	VOB	Frequency (Mutual information values)	ATT	Frequency (Mutual information values)	ADV	Frequency (Mutual nformation values)	CMP	Frequency (Mutual information values)
11	结果显示 Results are	249 (5.13)	受影响 Be affected	99 (4.36)	两个 Two	1153 (3.86)	更多More	290 (4.07)	听懂 Understand	51 (5.79)
22	差异显著 Distinct differences	193 (5.38)	存在差异 Have differences	97 (4.43)	这一 This one	964 (4.16)	较高 Higher	218 (4.52)	用来 Used for	43 (4.19)
33	这说明 It show	153 (4.31)	受到影响 Be influenced	92 (4.54)	名词短语Noun phrases	215 (5.46)	较大Larger	178 (4.17)	洗干净Clean up	37 (6.58)
44	结果表明 Results show	101 (4.59)	起作用 Take effect	86 (6.09)	第二语言学习者 Second language learners	153 (4.86)	很大 Enormous	168 (4.15)	结合起来 Combine together	36 (4.87)
55	效应显著 Notabal effect	69 (5.54)	见表 See table	85 (5.54)	一定程度Certain degree	135 (5.19)	最大 Largest	158 (4.19)	打死 Beat to death	33 (6.08)

6 Discussion

Lei [27] pointed out that the collocations in English academic papers are distinct from those in general English. For instance, "The article says/said" is used in general English while "The article describes/explores" is adopted in English academic papers. Inspired by such difference, it is believed that there is a need to also distinguish Academic Chinese and general Chinese. In regard to our research results, academic Chinese and general Chinese collocations are not the same. In academic style, when it comes to the SBV structure in which "认为" (think) is the predicate, the subjects which highly frequently appear with it include "我们" (we) (267 times), followed by "笔者" (the author) (39 times), and "本文" (this article) (27 times), while "我认为" (I consider) appears for only 14 times. Authors tend to use "我们认为" (we think) when expressing views with the aim of enhancing persuasion, interacting with readers and reflecting interactive subjectivity, according to our collocation list of academic papers of teaching Chinese to speakers of other languages.

The evaluation of the accuracy rate of the collocation extraction is 93.42%. The results were indispensable to further improve the accuracy of automatic extraction according to the study of ontology and computational linguistics.

The follow-up research can be conducted with the following regards:

(1) The boundaries of words and phrases to be determined. Whether "建设/银行" (construction / bank) is a proper noun (i.e., China Construction Bank) or is in VOB structure (i.e., to construct a bank) in the text requires further clarification.
(2) Specific sub-style. From a stylistic point of view, the collocation list should be subdivided. For example, in the study of spoken language, there are a lot of examples of oral language. Given that academic language belongs to the language in study, oral materials should not be included. Even in the academic papers of teaching Chinese to speakers of other languages, there are differences in the expressions of each sub-field, such as acquisition, grammar, vocabulary, pronunciation, etc.
(3) The extracted collocations call for expert intervention. The collocations with high frequency need to be further screened so as for us to select the commonly used collocations in the field of teaching Chinese to speakers of other languages. In the meantime, it is essential to further explore the classroom teaching of academic Chinese, so as to comprehend the demands of teachers and students and improve the collocation list commonly applied in teaching Chinese to speakers of other languages.
(4) It is required to identify the node words, enhance the algorithm, expand the scale of the collocation list, and construct an online platform to share the research results.

7 Conclusion

This paper has constructed a collocation list involving two different fields. The collocation list consists of a total of 12,357 collocations of academic collocations in Chinese international teaching, and 5,298 collocations of Wang Shuo's collocation list. With it, the high-frequency collocations of five commonly used structures are found out. Vocabulary collocation teaching is an indispensable part in Chinese teaching as well as in academic Chinese writing teaching. The academic collocation list can provide a reference for foreign students learning academic writings and also for teaching Chinese as a foreign or second language.

Acknowledgments. This research project was supported by the 2016 scientific research project of Sichuan International Studies University "A Comparative Study of Vocabulary Class Annotations in Two Dictionaries" (sisu201638) and Sichuan International Studies University Department of Chinese language and literature "13th Five-Year Plan" key discipline construction basic scientific research project "A Study HSK New Syllabus Vocabulary in Dictionary" (135 ZDXKF001).We are grateful to the editors and anonymous referees for the constructive comments and suggestions on an earlier version of the paper. Any remaining shortcomings are entirely our own.

Appendix

The specific language sources are as follows: *Fly Girls, Farewell My Love, Half Flame Half Brine, The Blue Ghost, Mortal Transfer, Wild Beasts, Rubber Man, Mr. Xu, I am the Wolf, Skip the Bill, Hold on, Nobody Applauds, Liu Huifang, I'm Your Daddy, No One Dares Harm Me, Ignorant, The Troubleshooters, Play the Game, Not to be a Common Man, The Fools, Please Don't Call Me Human, Publish after Amendments, All Clever Man, In Vain, Little Red Flowers Preface, Little Red Flowers, Wang Shuo's Selected Works.*

References

1. Liu, L.: Linguistic characteristics and realization of Academic Chinese. In: Proceeding of The 13th International conference on Chinese Language Teaching and Research, Secretariat of International Society for Chinese Language Teaching/Beijing University School of Chinese as a Foreign Language Education (2018)
2. Sinclair, J.: Corpus, Concordance, Collocation. Oxford University Press, Oxford (1991)
3. Crossley, S., Salsbury, T., McNamara, D.: Assessing lexical proficiency using analytic ratings: a case for collocation accuracy. Appl. Linguis. **36**(5), 570–590 (2015)
4. Howarth, P.: Phraseology and second language proficiency. Appl. Linguis. **19**(1), 24–44 (1998)
5. Pawley, A., Syder, F. H.: Two puzzles for linguistic theory: Nativelike selection and nativelike fluency. In: Richards, J.C., Schmidt, R.W.: (eds.) Language and Communication, pp. 191–226. Longman, London (1983)
6. Zhang, Z.: Letter from Mr. Zhang Zhigong to All Editors [C], Zhang Shoukang, Lin Xingguang, Modern Chinese Notional Collocation Dictionary, Beijing: Commercial Press (1992)
7. Lin, X.: On Vocabulary Collocation and Its Research, Language Teaching and Linguistic Studies, vol. 4 (1994)
8. Chen, H.H., Tao, H.: Academic Chinese: From corpora to language teaching language teaching. In: Lu X., Chen B. (eds.) Computational and Corpus Approaches to Chinese Language Learning. Chinese Language Learning Sciences. Springer, Singapore pp. 57–79 (2019)
9. Liu, Y.: Disfluency in L2 Chinese Academic Oral Presentations and Formulaic Language Instruction. In: Tao, H., Chen, H.-J. (eds.) Chinese for Specific and Professional Purposes. CLLS, pp. 95–130. Springer, Singapore (2019). https://doi.org/10.1007/978-981-13-9505-5_5

10. Liu, C., Chen, H., Yang, H.: Study on the lexical bundles in Chinese academic writing. J. Chinese Lang. Teach. **14**(1), 119–152 (2017)
11. Sinclair, J.: Beginning the study of lexis. In: Bazell, C.E., Catford, J.C., Halliday, M.A.K., Robins, R.H. (eds.) In Memory of J. R. Firth. Longman, London (1966)
12. Sun, M., Huang, C., Fang, J.: A preliminary study on quantitative analysis of Chinese Collocation. Chinese language 1997 (1), pp. 29–38
13. Jian, S., Wei, W., Yixin, Z.: Statistics-based Methods of Finding Common Collocation, Journal of The China Society For Scientific and Technical Information, vol.1 (2002)
14. Deng, Y.: Statistical in Collocation Study, Journal of Dalian Maritime University (Social Sciences Edition) vol. 4 (2003)
15. Qu, W., Chen, X., Ji, G.: A frame-based Approach to Chinese Collocation Automatic Extracting. Computer Engineering, vol. 4 (2004)
16. Deng, Y., Wang, T.: Statistical in Collocation Extraction and Computer Implementation. Technology Enhanced Foreign Language Education, vol. 5 (2005)
17. Quan, C., Liu, H., He, T.: Analysis and Comparison about the Approaches of Collocation Retrieval Based on Statistical Model, Application Research of Computers, vol. 9 (2005)
18. Wang, S., Yang, J., Zhang, W.: Automatic Acquisition of Chinese Collocation. Journal of Chinese Information Processing, vol. 20, No. 6 (2006)
19. Zhang, C., Qi, K.: Collocation Research Method Based on Mutual Information. Journal of Northwest University for Nationalities(Natural Science), vol. 3 (2009)
20. Bai, M., Zheng, J.: Autolabeling of Chinese Verb-Verb Collocation Based on Maximum Entropy Principle, Computer Engineering and Applications, vol. 3 (2009)
21. Wang, S., Yang, A.: A Method of Collocation Orientation Identification Based on Hybrid Language Information, Journal of Chinese Information Processing, vol. 3(2010)
22. Wang, L., Zhang, Y.: Quantitative Analysis and Extracting Arithmetic of Collocations Basic on Typical Patterns, Computer Science, vol. S1 (2012)
23. Sun, T.: Automatic Collocation Extraction Based on Quintuple, Electronic Design Engineering, vol.19 (2015)
24. Hu, R., Xiao, H.: The Construction of Chinese Collocation Knowledge Bases and Their Application in Second Language Acquisition. Applied Linguistics, vol. 1 (2019)
25. Liu, H.: Dependency Parsing and Machine Translation, Applied Linguistics, vol. 3 (1997)
26. Che, W., Li, Z., Liu, T.: LTP: A Chinese language technology platform. In: Proceedings of the Coling: Demonstrations. 08, pp. 13–16, Beijing, China (2010)
27. Lei, L., Liu, D.: The academic English collocation list: a corpus-driven study. Int. J. Corpus Ling. **23**(2), 216–243 (2018)

Militarization of Newspaper Language – Diachronic Study of War Metaphor in *People's Daily*

Qi Li and Gaoqi Rao(✉)

Research Institute of International Chinese Language Education, Beijing Language and Culture University, Beijing 100083, China
raogaoqi@blcu.edu.cn

Abstract. Taking the corpus of *People's Daily* from 1946 to 2015 as the research object, this paper discusses the use of war metaphors with the development of society and the influence of major events by comparing news reports made in different period of time and in different themes. It is found that war metaphors combining with military terms are used frequently and widely in newspaper language. Meanwhile, the texts are classified using the metaphorical data of war over the years. By comparing the resonance of source domain of war metaphors in different fields, it is found that metaphorical keywords are more widely used in the fields of economy. By conducting research on the metaphors of war in newspapers for many years, a new method is found for the study of metaphors in newspapers.

Keywords: *People's Daily* · Metaphor of war · Diachronic evolution

1 Introduction

People's Daily has witnessed the development of various fields as well as the change of social ideological and cultural consciousness in China throughout the entire diachronic process since its publication. Its reports not only present the lexical features of newspaper languages in different social and historical periods, but also reflect the dynamic evolution of the modern Chinese vocabulary system along with social and historical changes, all of which provide a good perspective for studying of the diachronic evolution of war metaphors and meanwhile are of high value in lexical research.

American cognitive linguists George Lakoff and Mark Johnson put forward the theory of conceptual metaphor from the perspective of cognition in Metaphors We Live By [1], which clearly pointed out that metaphor is the way of cognizing the world, and that in our daily life metaphor exists everywhere.

Metaphor is widely used in news reports. As an authoritative newspaper, *People's Daily* uses metaphors to better explain the current national policies and major decisions for the country, and also outline real-time hot events, making the language of newspapers clear and concise. With the methods in corpus linguistics and the metaphor theory, it is believed that a diachronic and quantitative research targeted at the corpus of newspapers can contribute insights of the diachronic development of metaphors.

M. Liu et al. (Eds.): CLSW 2020, LNAI 12278, pp. 593–602, 2021.
https://doi.org/10.1007/978-3-030-81197-6_50

This paper takes the 70-year newspaper of *People's Daily* as the research object and builds a diachronic corpus of it. It aims to conduct a comparative study of war metaphors in various fields in the newspaper from a diachronic point of view and under the framework of metaphor theory, the related theories of corpus linguistics as well as the quantitative analysis methods.

2 Research Status

As the theory of conceptual metaphor was put forward, research on metaphors in political discourse has gradually become popular, mostly taking national policies, or leaders' speeches, or the decree issued as research objects. Using the relevant analytical methods of corpus linguistics, researchers used the theory of conceptual metaphor to make qualitative or quantitative analysis, and to explore the functions of metaphors, cognitive mechanism and ideological issues.

Shu Dingfang [2] believed that metaphors have a function of naming, evaluation, and persuasion in political discourse. Charteris-Black [3] discussed the important role of metaphors in the political issue of immigration legalization proposed by the right wing at the 2005 British election. Wu Enfeng [4] extracted news headlines from economic reports to build a corpus. The study used three types of conceptual metaphors to quantify the metaphorical phenomena that appeared in headlines and explained the types of metaphors from the cognitive and cultural perspective.

Charteris-Black [5] also proposed the concept of "resonance of source domain", which measured the extent to which the metaphorical source domains involved in a particular corpus. Some scholars studied metaphors based on this theory. Wu Danping and Pang Jixian [6] took the content of the US president Bush's weekly radio speech in 8 years as the research object and used the resonance of source domain to determine the most frequently used conceptual metaphor type in the corpus.

Some scholars also conducted synchronic research based on specific types of conceptual metaphors. Luo Lingyun [7] took the linguistic strategies by American politicians as an example to illustrate the social factors highlighted and covered by different war metaphors. Liu Xiangyu [8] took the report on the theme of "Sino-US Trade Disputes" as the research object. It was explained in the study that the contexts of the metaphors used and the ideological issues are involved in three aspects, namely metaphor identification, interpretation, and explanation.

Compared to synchronic research on metaphors, there are few studies using corpus taking a diachronic perspective. Huang Qiulin and Wu Benhu [9] tried to explore the relationship between the development of society and the composition and use of various conceptual metaphors in the editorials of the *People's Daily* in 30 years.

3 Construction of Corpus and Identification of War Metaphors

3.1 Construction of Corpus

Construction of corpus is the foundation of text analysis. Based on the relevant theories and methods of corpus linguistics and with the help of the modern Chinese vocabulary diachronic retrieval system [10], the diachronic corpus of *People's Daily* from 1946 to 2015 is established in this project. Before conducting the comparative analysis of newspaper corpus quantitatively, we must process the corpus, of which the procedures mainly include corpus collation, corpus proofreading, word segmentation and part-of-speech tagging.

The report format collected in the corpus of *People's Daily* from 1946 to 2015 is not the same. Yet, it is convenient for a computerized system to recognize and calculate the text after making the date and title format consistent. The corpus processing carried out in this study mainly includes deduplication, stop-words deletion, annotation, and proofreading. Among them, Jieba[1] (a tool for word segmentation) was used to annotate text segmentation, adopting Peking University's word segmentation and part-of-speech tagging standards. During the manual labeling and proofreading process, we mainly focused on correcting the phrases bracketed in quotation marks.

3.2 Identification of War Metaphors

The Latent Dirichlet Allocation (LDA) is a generative model for text topic extraction, involving knowledge such as probability distribution and Bayesian prior theory. At the same time, the LDA theme model consists of a document level, a theme level, and a vocabulary level, with all of which form a clear hierarchical structure. Assuming there are D documents, the formal expression of its potential LDA model is:

$$\theta_i \sim Dir(\alpha), \phi_{z_{ij}} \sim Dir(\beta), z_{ij} = j \sim \theta_i, x_{ij} \sim \phi_{z_{ij}} \quad (1)$$

where *Dir ()* is Dirichlet distribution, and α and β are the parameters of *Dir ()*. That is, a document is generated by sampling from Dirichlet distribution α to generate the theme distribution θ_i of document I, which is a polynomial distribution. In the polynomial distribution θ_i, the theme of the jth word of document I is z_{ij}. Samples from Dirichlet distribution β generate the word distribution corresponding to theme z_{ij} as $\phi_{z_{ij}}$ and finally generate word x_{ij} in polynomial distribution of words $\phi_{z_{ij}}$.

We adopted the metaphor identification standard proposed by Charteris-Black [5] used by Wu Danping and Pang Jixian [6] to identify war metaphors. Firstly, we used the military terminology in the Sogou lexicon as the specific source domain of war metaphors, that is, the war metaphor keywords. At the same time, through the observation of the corpus in the early stage, we have found that there are many phrases in different fields that are combined with "战 (war)" and "仗 (battle)". Based on the Sogou lexicon, a military glossary of 487 words are formed. As shown in Table 2, "减价战 (price reduction war)", "秋收仗 (autumn harvest battle)", "货币战 (currency war)" etc. are the phrases formed by the "field nouns" and "战 (war)" or "仗 (battle)". After that, the candidates of the war metaphor keywords are used to match with the newspaper corpus one by one. Finally, we segmented the matching candidates in the newspaper corpus into clauses, and used LDA in the scikit-learn toolkit[2] to extract the topic words in the news and classified the reports according to their content.

[1] https://pypi.org/project/jieba/.
[2] https://scikit-learn.org/stable/.

4 Diachronic Distribution of War Metaphors

Because of the identification of war metaphors mentioned above, we cannot use machines to determine whether military terms in the newspaper corpora are war metaphors. Instead, to sort out candidates from different fields in the newspaper corpus, we randomly selected 100 sentences in 34 fields in every year's publication and hence identified the metaphors manually, thus obtaining the use of war metaphors in different fields over the years.

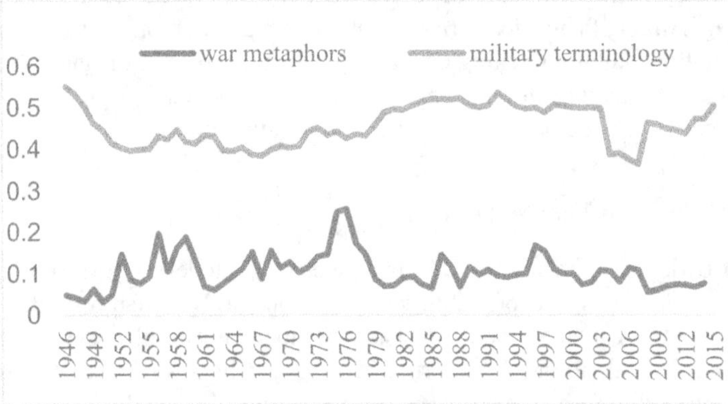

Fig. 1. Proportion chart of military terms and war metaphors

Figure 1 shows the comparison of the proportion of military terminology and war metaphors from 1946 to 2015. It can be found from Fig. 1 that the proportions of military terminology and war metaphors both show ups and downs. From 1946 to 1980, great changes have taken place in these two kinds of linguistic units. During this period, the use of military terms and the use of war metaphors are inconsistent, while in many cases the changing trends are even opposite. When the proportion of military terms shows a downward trend, the proportion of war metaphors shows an upward trend, which is more obvious from 1946 to 1959 in the figure. On the contrary, when the proportion of military terminology is on the rise, the proportion of war metaphors is on the decline, which is even obvious from 1976 to 1988. From 1980 to 2015, the changes of military terminology and war metaphors are relatively small.

"The essence of metaphor is to understand and experience the present through another thing." This shows that the use of metaphor and the frequency of use are closely related to our current understanding and experience of our social life. Therefore, it can be considered that the changes in the use of military terms and war metaphors are also related to the era we are living in. In times of social unrest or frequent wars, military terms account for a large proportion, while war metaphors account for a small proportion. For example, when the Liberation War broke out in 1946, military terms accounted for a large proportion, while war metaphors accounted for a small proportion. The chapters that directly reported and commented on military operations occupied a large part in the newspaper, and these parts were directly related to war. With the end of the civil war, society became stable and the country was eager for development. On the one hand, the national leaders began to invest in national construction after the war. They used military terms to describe things, such as "劳动生力军 (labor forces)", "石油大会战 (oil battles)", and "工业战线 (industrial fronts)". On the other hand, national leaders emphasized the mass production movement in the early days of the founding of the People's Republic of China, and soldiers also actively engaged in mass production, which is directly related to the increase in the use of war-related words.

This is the mapping of some characteristics of war (such as urgency and importance), from the source domain to target domain, (such as society, politics, economy, and culture), so as to bring tension to production, to stimulate the public to shape a national consciousness, and to encourage the public to actively participate in social construction. At the same time, the public can have a new understanding of the construction of society, and know the importance of politics, economy and culture.

5 Diachronic Filed Distribution of War Metaphor

5.1 Resonance of War Metaphor in Different Fields

Through LDA topic words, we simply classified the candidate corpus and manually sorted out the war metaphor corpus. In order to facilitate subsequent data processing, we re-integrated the classified texts in the selected war metaphor field, and obtained war metaphor texts in 12 fields, including economy, poverty alleviation, science and technology, and ecological environment. Charteris-Black [5] put forward the concept of "resonance of source domain", which means the discovery degree of metaphorical source domain in a particular corpus, and it is an indicator to measure the productivity of metaphor keywords. The specific algorithm is as follows:

$$\text{resonance of source domain} = \sum \text{keyword type} \times \sum \text{frequency} \tag{2}$$

Among them, "$\sum$ keyword type" refers to the number of war metaphor keywords appearing in this field, and "$\sum$ frequency" refers to the number of times military terms appearring in this field. We used this formula to calculate the war metaphor corpus in 12 fields.

Table 1 is arranged in descending order according to the resonance of source domain. It can be found that in the war corpus of People's Daily, the war metaphorical keywords related to the five fields, namely economy, science and technology, social security, ecological environment and disaster, are more productive. Therefore, these five fields will be seen as the focus of our diachronic research.

5.2 Diachronic Analysis of War Metaphor in Different Fields

Based on the proportion of war metaphors in different fields over the years, and with the use of N-gram to analyze metaphorical keywords in different fields, we tried to explain the close relationship between the use of war metaphors and social development. We sorted out the war metaphor keywords in top ten high-frequency in five fields, and used bigram to tabulate the war metaphor keywords in five fields.

Table 1. Metaphorical resonance value of war in in different fields of People's Daily's 70-year

Fields	∑ Keyword type	∑ Frequency	Resonance of source domain
Economy	298	22943	6837014
Science and technology	271	16351	4431121
Social security	269	15436	4152284
Ecological environment	258	14003	3612774
Disaster	256	12945	3313920
News media	243	7800	1895400
Education	229	7293	1670097
Policy	220	6701	1474220

Table 2. The top five war metaphorical keywords in different fields

Economy	Science and technology	Social security	Ecological environment	Disaster
战线 (battlefront)	第一线 (the first line)	战线 (front line)	斗争 (combat)	斗争 (combat)
斗争 (combat)	战线 (front line)	武器 (weapon)	战胜 (defeat)	宣战 (declaration of war)
夺取 (capture)	攻坚战 (tough battle)	占领 (occupation)	搏斗 (fighting)	第一线 (the first line)
第一线 (the first line)	战略 (strategy)	斗争 (combat)	开战 (starting wars)	战斗 (battle)
贸易战 (trade war)	阵地 (position)	前线 (front line)	战斗 (battle)	争夺 (scramble)

'Indirect-discriminatory metaphor' refers to a metaphor that does not directly map the source domain to the target domain, but uses a segmented framework through analogy of specific elements to imply the overall connection between the two domains." [10] For example, in Table 2, the top ten high-frequency keywords in five fields have been sorted out, including "争夺 (scramble)", "占领 (occupation)" and "宣战 (declaration of war)" that are used to describe the course of the war; "战胜 (defeat)", "夺取 (capture)" that are used to describe the results of war; "战线 (battlefront)", "前线 (front line)" and "第一线 (the first line)" that are used to describe the location of war, as well as the "贸易战 (trade war)" and "攻坚战 (tough battle)" that are used to describe the form of war.

Combining Table 3 it can be found that the economic, technological, and social security fields place more emphasis on both parties, locations, and forms of warfare, such as "价格战 (price war)", "贸易战 (trade war)", "勘探攻坚战 (prospecting battle)", "科研战线 (scientific front)", "科学堡垒 (science bastion)" and "收购大军 (acquisition army)", "科学后备军 (science reserve army)" and so on.

Table 3. Metaphorical collocation of war in different fields

Fields	Collocation of metaphorical keywords
Economy	价格战 (price war) 贸易战 (trade war) 收购堡垒 (acquisition bastion) 收购大军 (acquisition army)
Science and technology	勘探攻坚战(prospecting battle) 科学后备军 (science reserve army) 科学堡垒 (science bastion) 科学大军 (scientific army) 科研战线 (scientific front)
Disaster	减灾第一线 (the first line of disaster reduction) 鏖战洪水(fight against the flood) 迎战大暴雨 (fight against heavy rain) 战胜灾难 (defeat disaster)
Ecological environment	征战荒漠 (conquer the desert) 战胜环境污染 (conquer environmental pollution) 治污硬仗 (hard fight against pollution)
Social security	交通战线 (traffic battlefront) 法律宣战 (declare war on the law) 法律战线 (legal battlefront)

The use of the above phrases is to explain the characteristics of economic and scientific development, as well as the difficulties and the long duration of social security construction. The words "宣战 (declaration of war)", "搏斗 (fighting)" and "开战(starting wars)" used to describe the course of the war and the words "战胜 (defeat)" and "夺取 (capture)" used to describe the result of the war are more used in the fields of disasters and ecological environment, such as, "鏖战洪水 (fight against the flood)", "迎战大暴雨 (fight against heavy rain)", "征战荒漠 (conquer the desert)", "治污硬仗 (hard fight against pollution)", "战胜环境污染 (conquer environmental pollution)", etc. In the face of disasters and pollution of the ecological environment, it is difficult to put aside the military terminology and use concise language to describe the event clearly. The military terms are conducive to visual and auditory associations, and they increase tension, suffocation, and tragic feelings. Therefore, when mobilizing the public to deal with disasters and carry out environmental governance, the use of war metaphors can reflect the urgency of the incident and increase the public's awareness of self-protection.

We reckon that this type of metaphors is more clearly reflected in the metaphorical proportion of the diachronic figure. In Fig. 2 and Fig. 3, it can be seen that the changing trend in the economic field, the scientific and technological field, the social security field, and the ecological environment field from 1946 to 1980 tends are consistent.

The metaphor of "indirect-discriminatory metaphor" highlights some of the characteristics of the source domain and maps it to the target domain. The war metaphors used in the fields of economy, science and technology, and social security highlight the importance and the urgency of wars, or the duration of the war and the difficulty of winning in it. For example, from 1953 to 1957, China promulgated the first five-year plan. In order to make the public better understand national policies and encourage everyone to actively respond to national calls, the proportion of war metaphors is on the rise. With the rapid economic development, corresponding technology and social security have further developed. For example, from 1964 to 1976, while the "大跃进 (Great Leap Forward)" and "人民公社化运动 (People's Commune Movement)" were

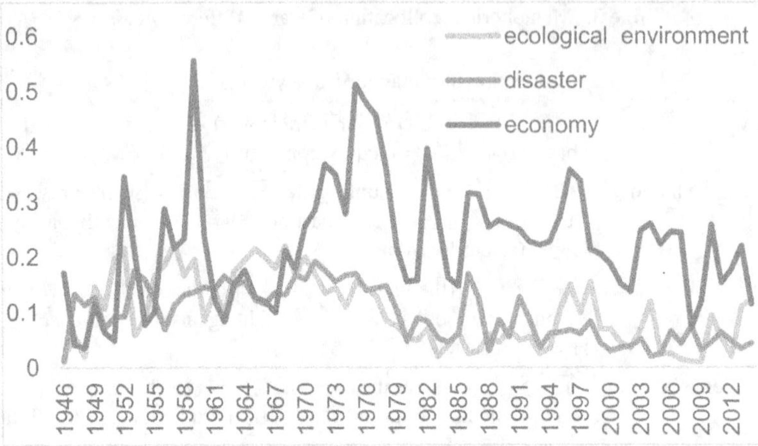

Fig. 2. The proportion of war metaphors in different fields of *People's Daily*

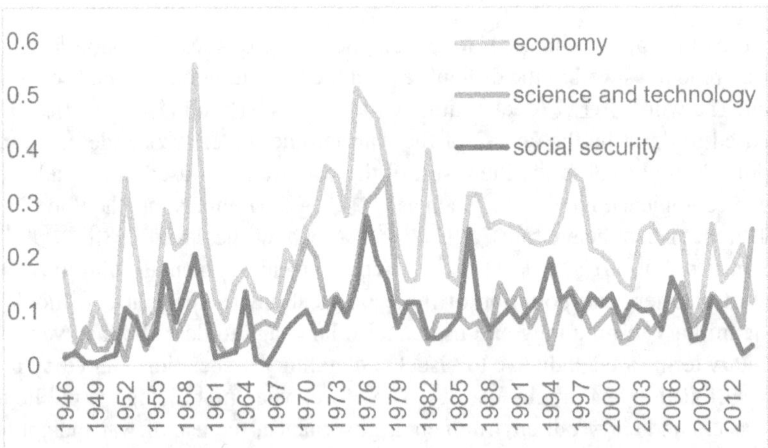

Fig. 3. The proportion of war metaphors in different fields of *People's Daily*

launched nationwide, the project of "two bombs and one satellite" in the field of science and technology had achieved significant results, and the poverty alleviation policy in the field of social security had also been further improved. All these can reflect the importance and timeliness of war metaphors. Therefore, we suggest that the war metaphors in the fields of economy, technology, and social security are the "indirect-discriminatory war metaphors."

In Fig. 2, it can be seen that from 1980 to 2015 the proportion of war metaphors in ecological field is smaller than that of war metaphor in economic field, which tends to be flat. Compared with the war metaphors in the economic field, the war metaphors in the disaster field have changed gradually during the 70 years, from 1946 to 2015. This is different from "indirect-discriminatory metaphor of war". "Indirect-discriminatory metaphor" appears more often in the time when new things and major events occur. However, the metaphors of war in the field of disaster do not. For instance, in the metaphor "conquering disaster is a war.", the words of this kind do not reflect the characteristics of war from the side to be mapped to the target domain, but directly use the words that are related to war, such as "战胜灾难 (overcoming disaster)", "鏖战洪水 (fighting the flood)" and "迎战大暴雨 (facing the heavy rain)" in the disaster field from Table 3. The changes in the ecological environment from 1980 to 2015 also made people realize the importance of ecological environmental protection, so there was no need to use too many metaphors to warn everyone about the importance of environmental protection. On the contrary, it is more important to directly tell people that it is particularly urgent to manage the damaged ecological environment, so the "indirect-discriminatory war metaphor" was no longer used. The war metaphor key words in the field of ecological environment from 1980 to 2015 are consistent with the war metaphor keywords in the field of disasters, such as "减排攻坚战 (reducing the tough battle against emissions)", "战胜荒漠 (conquering the desert)", and "治理攻坚战 (controlling the tough battle)", as shown in Table 3.

6 Conclusions

In this paper, through topic modeling and metaphor selection, it is observed that war metaphors in the language of newspapers are frequently and widely used in the 70-year time span. The use of military terms and war metaphors are used in different frequency. The frequency of the latter is closely related to major events and social ideologies. In the non-military field, the economic and technological fields have the highest resonance. For most non-military events and fields, the collocation of war metaphor keywords and the proportion of war metaphors in the five fields collated by N-gram are compared diachronically. It can be found that "indirect narrative war metaphor" is more prominent in diachronic analysis. At the same time, the metaphors of war in the fields of disasters and those in the field of ecological environment after 1980 are different. Through the changes of the proportion of military terms and war metaphors, combined with the specific social changes over the years, it can be seen that the use of war metaphors is closely related to the development of society and the occurrences of major events.

The language of newspaper uses a lot of metaphors to describe and comment on national policies and major events, which is conducive to people's comprehensive understanding of current affairs. Through the diachronic study of war metaphors in newspapers, a method for the diachronic study of metaphors is provided. In addition, by using the diachronic corpus of war metaphors, a series of war metaphors has been studied, such as the change of metaphor keywords in different fields over the years and the derivative metaphors of war metaphors including defensive metaphors, attack metaphors, and struggle metaphors.

Acknowledgments. This paper is supported by Graduate Innovation Foundation in 2019 (19YCX151).

References

1. Lakoff, G., Johnson, M.: Metaphors We Live by. The University of Chicago Press, Chicago, Chicago and London (1980)
2. Shu, D.: Metaphorical studies. Shanghai Foreign Language Education Press, Shanghai (2000).(in Chinese)
3. Charteris-Black, J.: Britain as a container: Immigration metaphors in the 2005 election campaign. Discourse&Society **17**(5), 563–581 (2006)
4. Wu, E.: Research on Conceptual Metaphor Based on Economic Report Title Corpus. Zhejiang University, Zhejiang (2008).(in Chinese)
5. Charteris-Black, J.: Corpus Approaches to Critical Mrtaphor Analysis. Palgrave Macmillan, New York (2004)
6. Wu, D., Pang, J.: Persuasiveness of Metaphors in Political Discourse and Discourse Strategies: A Corpus-based Study. Foreign Languages and Foreign Language Teaching, (04), 38–42+47(2011). (in Chinese)
7. Luo, L.: A Brief Analysis of War Metaphor and Ideology. J. Xidian Univ. (Soc. Sci. Edition) **02**, 137–140 (2005). (in Chinese)
8. Liu, X.: War Without Seeing Smoke. Zhejiang University, Zhejiang (2019).(in Chinese)
9. Huang, Q., Wu, B.: A diachronic analysis of political metaphors based on editorials on "two sessions" in people's daily (1978–2007). Lang. Teach. Linguistic Stud. **05**, 91–96 (2009). (in Chinese)
10. Xun, E., Rao, G., Xie, J., et al.: Diachronic retrieval for modern chinese word: system construction and its application. J. Chin. Inf. Process. **29**(03), 169–176 (2015). (in Chinese)

The Construction and Application of an Emotion Knowledge Base for Prose Reading Comprehension

Dian Wang[1], Suge Wang[1,2](✉), Xin Chen[1], Jing Zhang[3], Jing Li[1], and Ru Li[1,2]

[1] School of Computer and Information Technology, Shanxi University, Taiyuan 030006, Shanxi, China
{wsg,liru}@sxu.edu.cn

[2] Key Laboratory of Computational Intelligence and Chinese Information Processing of Ministry of Education, Shanxi University, Taiyuan 030006, Shanxi, China

[3] School of Foreign Languages, Shanxi University, Taiyuan 030006, Shanxi, China
jennyzhang@sxu.edu.cn

Abstract. In natural language processing, emotion analysis research in prose are relatively few for the lack of relevant knowledge base. This study establishes an category system of emotions in prose and builds up an emotion knowledge base for prose reading comprehension. In addition, because the proposed emotion categories are associated with topics, we introduce topic information into emotion knowledge base, label emotion types and topic elements. The constructed knowledge base will help to provide data resources for the solution of emotion type problems in prose reading comprehension.

Keywords: Prose · Emotion knowledge base · Emotion classification system · Topic · Reading comprehension

1 Introduction

As a the natural language processing task, machine reading comprehension aims to allow the computer to obtain knowledge from text data to reason and answer questions. Nowadays, most of the tasks of machine reading comprehension are about facts and opinions, and their background materials are often objective texts. True artificial intelligence does not only have the ability to think and inference like humans do, but also needs perception and emotion. Prose integrates rich emotional factors into the lines. It is a genre that expresses one's own emotions by describing people, events, and objects [1]. Moreover, questions asking about emotions are mostly in the form of open questions in reading comprehension of prose as the background material.

For example: 作者写岸边石缝间的几朵勿忘我表达什么情感? (What emotion does the author express with several forget-me-not flowers between the stone crevices on the shore?).

M. Liu et al. (Eds.): CLSW 2020, LNAI 12278, pp. 603–617, 2021.
https://doi.org/10.1007/978-3-030-81197-6_51

The reference answer given to the above question is as follows:

表达了对荷花的喜爱。(It expresses the author's love for lotus.)

It is obvious that the above question and answer are not closely connected. However, through the analysis and understanding of the background materials, it is found that "forget-me-not" is a foil for "lotus" and contains positive emotions, expressing the author's love for lotus. Through a large number of studies, it is found that the emotions expressed in prose are more durable, profound and implicit than ordinary emotions. It is composed of ordinary emotions. For example, in "An Alternative Memoir", the author demonstrated the process of generating "Eternal Regret" as a series of emotional changes in the following ways: out of ignorance, the author neglected his mother's existence and left her, and then lost her forever; the author's sadness was conveyed with "heart-broken and liver-torn, tears welled up in the eyes", which shows a condensed deep regret and feeling of missing his mother. According to this analysis, it is reckoned that the general emotion of this prose is the basic emotion expressed by the author. Thus, based on the general emotion category system, we build the emotion category system and knowledge base according to the needs of emotion analysis in prose reading comprehension, which is related to the author's emotion orientation, emotion attitude, thought emotion, thought viewpoint and so on.

2 Related Work

There is little work on the emotional analysis of prose. Thus, this paper will make a review on the establishment of emotion category system and the construction of an emotion corpus.

2.1 Emotion Classification System

There are three concepts related to emotion, namely, affect, emotion and mood [2], all of which have been studied in different fields. In psychology, Ekmanhe et al. proposed the six basic emotions of humans, namely happiness, sadness, fear, surprise, anger and jealousy [3]. These basic emotions, after being combined with each other, generate more complex emotions, such as "depression, tension, anxiety" etc. Plutchik's et al. proposed an evolutionary theory of psychological emotions and pointed out that there are eight basic emotions in human beings, namely, acceptance, happiness, expectation, sadness, fear, surprise, anger and hatred, and that other emotions can be derived from these emotions [4]. Lin et al. divided emotions into 18 categories including joy, silence, grief, compassion, anger, sorrow, boredom, anger, fear, horror, awe, respect, caress, hatred, pride, greed, jealousy, guilt and shame [5]. Qiu et al. divided emotions into 4 categories according to their objectives, namely, emotions for objects, emotions for people, emotions for oneself and emotions for society [6]. In computational linguistics, Xu et al. believed that language is the basis of thinking and divided Chinese emotion into three systems, namely basic emotions, psychological feeling based emotions and expressiveness based emotions [7]. The basic emotion system is formulated by referring to the basic emotion types of Chinese tradition, which is divided into seven categories: goodness (love, respect), evil, joy (happiness), anger, sorrow, fear and desire. In the psychological feeling-based emotion system, "joy" and "happiness", "sorrow" and "depression", "shame" and "guilt" are incorporated into one category respectively. In view of this, those psychological feeling-based emotions can be divided into "joy, happiness; love; worry, boredom; sadness; panic; respect; excitement; shame, guilt; annoyance; urgency; pride; surprise; anger; disappointment; reassurance; hatred; jealousy; contempt; regret; grievance; shock; faith; doubt" and "others". The expressiveness-based emotion system includes attitude words, character words, voice words and other adjectives. Lee et al. classified Chinese emotions into five categories in their research, namely happiness, sadness, anger, fear and surprise [8]. Lin et al. classified emotions into seven categories according to the intensity of emotion words in the process of building emotion vocabulary ontology: "joy, goodness, anger, sadness, fear, evil and surprise" [9]. Although the above category systems were elaborated on emotions, it may be helpful for the category of prose genre.

2.2 The Construction of Emotional Vocabulary and Corpus Resources

In the construction of emotional dictionary, Philip et al. have established general inquirer lexicon, which provides polarity, intensity and other information for each word [10]. In the Opinion Finder system, Subjective lexicon[1] contains 8221 subjective words. The HowNet[2] based emotion dictionary contains 9193 Chinese emotion analysis word sets and 9142 English emotion analysis word sets. The emotion dictionary established by Taiwan University contains 8325 negative words and 2846 positive words. Lin et al. divided the emotional vocabulary into seven categories and 21 sub-categories, and 27466 emotional words have been included in the emotional vocabulary [9]. Yang et al. extracted emotional knowledge from an emotional corpus to build emotional knowledge database and used HowNet and synonym Cilin[3] to expand the scale and annotation manually [11]. In recent years, some local and foreign research institutions have released related emotion analyses and evaluation tasks in corpus construction, and provided a certain scale of an emotion corpus. For example, a series of organizations, such as Semval[4], TREC[5], NTCIR[6] and so on, regularly have organized relevant reviews. At the same time, there are text orientation analyses and evaluation COAE [12–14] in China. Lee et al. constructed an emotional reason corpus in the Chinese emotional corpus by matching and using emotional keywords [8]. Pak et al. constructed a subjective corpus by using Twitter Weibo, which contains positive and negative emotions [15]. Toprak et al. released a corpus DSRC for emotional information extraction, which contains 234 articles in the service field and 256 articles for universities [16]. By establishing a tagging system for a Chinese corpus of information extraction, Dai et al. annotated the documents in notebook, hotel and cosmetics fields, and constructed a corpus of emotional

1 http://mpqa.cs.pitt.edu/lexicons/subj_lexicon/.
2 http://www.keenage.com/html/e_index.html.
3 http://www.ltp-cloud.com/download.
4 http://alt.qcri.org/semeval2020/.
5 https://trec.nist.gov/pubs/call2020.html.
6 http://research.nii.ac.jp/ntcir/index-en.html.

information extraction for Chinese texts [17]. Xu et al. established an emotion corpus tagging system and annotated primary school textbooks, movie scripts and other collected corpora [18]. Li et al. constructed a large-scale Chinese movie emotion corpus, which can provide data resources for emotion analysis research [19]. Focusing on the answers for the appreciation-type questions of literary reading comprehension, Chen et al. established an emotional knowledge base and a rhetorical emotional knowledge base but did not consider the emotions implied by the topic information in the text [20]. From the above works, we can find that the existing emotion classification systems are not unified. Although researchers have tried to build emotional knowledge bases, the constructed emotional knowledge bases for prose reading comprehension task are still in its preliminary stages.

3 Emotion Classification System in Prose Genre

In this study, a category system of the more commonly used emotion in Chinese emotion analysis is proposed [9]. The Chinese emotion vocabulary is divided into seven categories, which are "happy, good, anger, sadness, fear, hatred and surprise". The proposed system is based on Ekman's six categories of emotion category system, including "happiness, sadness, fear, surprise, anger, jealousy", with a newly added category "preference". Among them, positive affective lexical ontology is divided into two categories: joy and liking. However, this system is far from enough to solve the problem of prose reading comprehension. Therefore, in order to improve the performance of reading and understanding emotional questions, this paper specially divides "happiness" into "love", "respect", "praise" and "wish" according to the questions and answers in the constructed database of understanding emotional questions in prose reading. Moreover, this paper has also developed 10 types of emotions and has given each type a corresponding description. The specific emotion categories and descriptions are shown in Table 1.

Table 1. Emotion categories and their descriptions related to prose reading comprehension

Type	Description
Happiness	It refers to the emotion generated when one has realized his dream and pursuit
Love	It refers to the emotion generated when one adores somebody or something deeply, as well as the satisfaction of someone's initiative or conscious expectation
Respect	It refers to dedicating politely, devoutly and respectfully, and it also refers to seriousness
Praise	It refers to high approval or admiration for somebody or something, which is thus followed by the advocacy and promotion of it
Wish	It refers to the hope and expectation that the development of something is consistent with one's own idea

(continued)

Table 1. (*continued*)

Type	Description
Anger	It refers to a certain emotion when the goal pursued is hindered, or when the development of something cannot reach an agreement with one's idea
Sad	It refers to someone's emotional experience in the face of losing something they love, or in the context of disillusionment. It also refers to someone's helpless emotion in the face of something
Fear	It refers to the feeling that people face some dangerous situation and try to get rid of it but cannot do anything about it
Hatred	It refers to the feeling of being disgusted or very unhappy with someone or something
Surprised	It refers to the sudden emotional experience caused by unexpected changes, as well as a passion for the unknown, and driving one to find the meaning behind it

4 The Construction of an Emotional Knowledge Base of Prose

In order to facilitate the answers to the emotion questions in reading comprehension, the data format of the knowledge base of emotion analysis of prose is set to a 6-tuple: EmotionModel = (Topic, ETopic_aspect, E_aspect, E_word, E_category, E_holder), where Topic is the topic involved in prose, E_aspect includes the author's emotional orientation, emotional attitude, thoughts-emotions, ideas and so on. ETopic_aspect is the carrier or object assigned to E_aspect. E_word is the emotion words in prose. E_Categoy is the type of emotion, and E_Holder is the emotion holder (e.g., the author). A prose usually has the expression way of "put emotion in emotion", so the author may use the emotions of others to express his emotions.

The construction process of the knowledge base mainly includes the following steps, namely data collection, manual annotation, and data expansion. The specific construction process is shown in Fig. 1.

(1) Data collection

Our goal of constructing the emotional knowledge base of prose is to answer the emotional questions in prose reading comprehension. It is difficult to annotate the prose knowledge base because annotators need to have high literary competence. The knowledge base established mainly relies on the answers in the emotional question bank. Therefore, we collected 7878 questions of prose reading comprehension in the real and simulated questions of Beijing college entrance examination over the years as the original data, and selected 577 questions, answers and corresponding prose texts related to emotional analysis as the basic data.

(2) Manual annotation

The 6-tuple needs to be annotated according to the data format in the knowledge base.

EmotionModel = (Topic, ETopic_aspect, E_aspect, E_word, E_category, E_holder). Examples of reading materials, emotional questions, and answers are as follows:

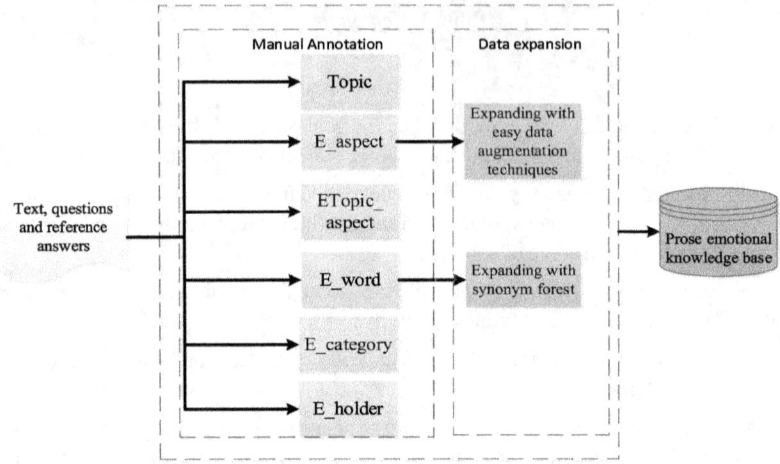

Fig. 1. Construction processing of the emotional knowledge base of prose

Reading materials: 还是这方荷塘, 却已不是以往模样……爱这荷的高洁, 爱这荷的清香, 更爱这荷出淤泥而不染的品行, 她时刻净化着我的思想。(It's still this lotus pond, but it's no longer what it used to be …… I love the nobleness of this lotus, love the fragrance of this lotus, love the conduct of this lotus of keeping clean in the dirty soil. She purifies my mind all the time.)

Emotional question: 作者写岸边石缝间的几朵勿忘我表达什么情感? (What kind of emotion does the author express in the several forget-me-not flowers between the stone crevices on the shore?).

Answer: 表达了对荷花的喜爱。(It expresses the author's love for lotus.)

The annotation of the above example is as follows:

{荷塘, 荷花, 表达了作者对荷花的喜爱, 爱, 爱, Author} ({lotus pond, lotus, it express the author's love for lotus, love, love, Author}).

(3) Quality monitoring

In the part of emotion annotation, three annotators were selected, and a related description of emotion types is provided as the guidance of standard annotation. However, since different annotators have different understanding and thus annotate different results, the annotation process needs to be discussed repeatedly. In the process, we have chosen the voting method for the result of annotation. If the opinions are still inconsistent, they need to be discussed again.

In order to monitor the quality of the annotation results of the three annotators, we specially calculated the Kappa value for the annotation results of the emotion types. Usually, the Kappa value is between 0–1, where 0.0–0.20 represents that the annotation result has very low consistency, 0.21–0.40 represents that the annotation result has general consistency, 0.41–0.60 represents that the annotation result has medium consistency, 0.61–0.80 represents that the annotation result has high consistency, and 0.81–1 represents that the annotation results are completely consistent. The mean value of this annotation is 0.618, which shows high consistency.

(4)Knowledge base expansion

In view of the relatively small scale of the existing questions and answers, and a limited scale of the knowledge base, we will expand E_aspect and E_word in the knowledge base.

E_aspect expansion: This paper uses EDA [21] data enhancement technology. The types and examples that can be expanded are as follows:

a) Replacement of "Topic" in E_aspect

Original sentence: 表达了对梅花精神的赞美. (It expresses the author's praise for the spirit of plum blossom.)

Data enhancement: It expresses the author's praise for spirit of lotus. (表达了对荷花精神的赞美.)

b) Replacement of "E_word" in E_aspect

Original sentence: 表达了作者对黑妮的仰慕与称赞. (It expresses the author's admiration and praise for Heini.)

Data enhancement: 表达了对作者对黑妮的称赞与敬仰. (It expresses the author's praise and reverence for Heini.)

c) Replacement of "core verb" in E_aspect

Example: Original sentence: 表现了作者对炼油工人幸福心灵的赞颂. (It expresses the author's admiration for the joyful spirit of the refiner.)

Data enhancement: 传达了作者对炼油工人幸福心灵的赞颂. (It conveys the author's admiration for the joyful spirit of the refiner.)

After using EDA, it is possible to enhance Emotion_aspect of emotion as well as introducing noise to the data. In order to eliminate the noisy data, we designed the following three exclusion rules, with the effect of removing 23% of the total instances expanded by E_aspect and retaining 1513 instances to be used.

Rule 1: Mistakes in collocation.

Original sentence: 形象生动地表达了对美丽家乡的赞美和思念之情. (It vividly expresses the author's admiration and yearning for his/her hometown.)

Data enhancement: 样貌动人地表达了对美丽于家乡的赞美和思念之情. (*It expresses the author's admiration and yearning for hometown, in the manner of beautiful appearance.)

Rule 2: Confusion of expression.

Original sentence: 主张人类应与自然和谐相处, 启示人类应亲近自然, 善待生灵. (It advocates that human beings should live in harmony with nature, and enlightens that human beings should be close to nature and be kind to living beings.)

Data enhancement: 主张交往人类人的关系应与自然和谐称许称道见解相处启示. (*It advocates that the relationship between human beings and nature should be in harmony with nature. Opinions get along and inspire.)

Rule 3: Semantic errors.

Original sentence: 对昌都的热爱, 对寻梦者的称许. (It expresses the enthusiasm for Changdu and the praise for dreamers.)

Data enhancement: 对热爱的称许, 对昌都的寻梦者. (It expresses the praise for enthusiasm and expresses the dreamers of Changdu.)

By using the above three exclusion rules, the final expanded E_aspect (E_asp) data distribution in E_category (E_cate) is shown in Table 2.

Table 2. Instance distribution after the expansion of the knowledge base

E_category	E_aspect	E_category	E_aspect
Happiness	228	Suprised	24
Respect	148	Hatred	346
Sad	468	Wish	282
Fear	24	Love	266
Anger	18	Praise	401

The expansion of E_word is with the aid of the synonym Cilin. The specific process is to find the synonym of E_word that has been marked in the "Synonym Cilin Extended Edition (See footnote 3)", and then removed the duplicates and deleted the unwritten expressions. The expanded E_word is shown in Table 3.

When expanding E_word, it is necessary to remove some synonyms with improper collocation. For example, the synonym of "欣喜" (happiness) is "喜滋滋" (joy), but it is not appropriate to express the author's "喜滋滋" (joy).

Table 3. Distribution of emotional words after the expansion

E_category	E_word	E_category	E_word
Happiness	44	Suprised	7
Respect	28	Hatred	46
Sad	90	Wish	25
Fear	14	Love	71
Anger	15	Praise	20

5 The Analysis of E_category in the Knowledge Base

In order to analyze the E_category data in the knowledge base, it is necessary to calculate the E_word and E_aspect distribution under E_category and the relationship between E_category and Topic_aspect.

(1) E_word and Emotion_aspect distribution

This paper used the material of the real questions in the college entrance examination and simulated questions as the basic corpus. Through statistical analysis of 577 test questions related to emotions, the E_word and E_aspect statistical results of 10 types of E_category in the knowledge base were obtained, as shown in Table 4.

Table 4. Statistical results of E_word and E_aspect in 10 types of E_category

E_category	E_word	E_aspect	E_category	E_word	E_aspect
Happiness	17	64	Suprised	5	8
Respect	12	56	Hatred	1	1
Sad	15	47	Wish	101	101
Fear	39	138	Love	18	6
Anger	11	101	Praise	65	159

From Table 4, we can see that there are more "sad" data. the main reason is that since all the data come from the college entrance examination questions, the data sources are limited. Due to such limitation, the distribution of data is unbalanced. In addition, in Table 4, there are more pieces of prose expressing the emotion of "love", and less pieces of prose expressing "fear" and "surprise", which has a great relationship with prose genre. Next, we will consider introducing resources from Chinese textbooks or other sources.

(2) ETopic_ Aspect and E_ Category relationship

The annotation result indicates that when the text is oriented to a topic ETopic_aspect, it may have certain relevance to certain types of emotion E_category. To this end, we counted topics that have appeared more than 5 times, and then gave the ETopic_aspect and calculated the probability of the corresponding E_category by using formula (1). We hence used this probability as the correlation between ETopic_aspect and E_category.

$$P(\text{E\_ category}|\text{ETopic\_ aspect}) = \frac{P(\text{E\_ category,ETopic\_ aspect})}{\text{P(ETopic\_ aspect)}} \tag{1}$$

See Table 5 for the details of emotion categories related to some topics.

According to Table 5, two different topic information, ETopic_aspect and E_category, have different connections with emotions. For example, the topics of "hometown" and "mother" have the "sorrow" E_category, but the topics of "Beijing" and "nature' have no "sorrow" E_category.

Table 5. The correlations between ETopic_aspect and E_category

Topic	ETopic_aspect	E_category	Correlation degree%
Beijing	Hutong in Beijing, traditional life in Beijing, etc.	Love	60.00
	The future development of Beijing	Wish	40.00
Hometown	The snows in winter, the downfall of hometown, etc.	Grief	73.08
	Home, homeland, etc.	Love	26.92
Nature	The primitive beauty of nature, the landscape of nature, etc.	Praise	66.67
	The life of nature, etc.	Wish	33.33
Mother	Mother's quality, maternal love, etc.	Praise	18.18
	Mother's departure, etc.	Grief	50.00
	Mother	Respect	13.63
		Love	18.18

6 The Application of Emotional Knowledge Base

In order to apply the established emotional knowledge base to the emotional questions and answers in prose reading comprehension, it is necessary to establish relevant answer templates. On this basis, we need to use the emotional knowledge base to answer.

6.1 Establishment of Answer Templates

The answers to the emotional questions of prose reading comprehension include the following categories, Topic, ETopic_aspect, and E_aspect. The background material includes Topic, ETopic_aspect and E_word. Therefore, with the aid of ETopic_aspect, answer templates are generated. On the other hand, ETopic_aspect and E_category are related in the knowledge base. And E_category also contains a certain E_word. According to the above characteristics, the answer templates are constructed with E_category as the core, as shown in Table 6.

Table 6. The answer templates for answering emotional questions

E_category	ETopic_aspect	E_word	Answer templates	Examples
Happiness	{平静生活\黄山的雨…} ({peaceful life\rain of Huangshan Mountain…})	{陶醉\惬意…} ({intoxicated\comfor- -table\…})	表达了/表现了/…对 ETopic_aspect的 E_word (It expresses/shows/…for E_word for ETopic _aspect)	表现了作者为黄山的雨所陶醉 (It shows author's intoxicated feeling for Huangshan Mountain)
Love	{窗外绿色世界\故乡气候…} ({green world outside the window\ hometown climate…})	{喜爱\挚爱…} ({like\love…})	表现了/表达了/.. 对ETopic_aspect的E_word (It expresses/ shows /…for E_word for ETopic_ aspect)	表现了作者对窗外绿色世界的喜爱之情 (It shows author's love for green world outside the window)
Respect	{父亲\大自然…} ({father\nature…})	{敬重\敬畏…} ({respect\awe…})	表达了/抒发了/..对 ETopic_aspect的 E_word (It expresses/ shows /…for E_word for ETopic_ aspect)	表达了作者对父亲的敬重之情 (It expresses author's respect for his/her father)
Praise	{梅花精神\共产党人…} ({spirit of plum\Communists…})	{赞美\赞颂…}({praise\extol….})	表达了/抒发了/..对 ETopic_aspect 的 E_word (It expresses/voices /…for E_word for ETopic_ aspect)	表达了作者对梅花的赞美(It expresses author's respect for plum blossom.)
Wish	{自由\美好生活…} ({freedom\good life…})	{向往\期盼…}({longing\expect…})	表达了/展现了/..对 E_aspect 的 E_word (It expresses/emerges/…for E_word for ETopic_ aspect)	表达了作者对自由的向往 (It expresses author's longing for freedom.)

(continued)

Table 6. (*continued*)

E_category	ETopic_aspect	E_word	Answer templates	Examples
Anger	{故乡风貌被毁\ 世间百态…} ({hometown destroyed\ all kinds of things in the world…})	{愤怒\激愤…} ({anger\worry…})	表现了/表达了/..对 E_aspect 的E_word (It shows/ expresses /…for E_word for ETopic_ aspect)	表现了作者对世间百态而忧愤 (It shows author's anger for all the events in the world.)
Sad	{逝去时光\底层人民…} ({lost time\people at the bottom…})	{怀念\同情……} ({yearn\sympathy……})	表现了/表达了/..对 E_aspect 的E_word (It shows/expresses /…for E_word for ETopic_ aspect)	表达了作者对逝去时光的怀念 (It expresses author's yearning for the past.)
Fear	{死亡\病痛…} (Death\sickness…})	{恐惧\恐慌…} ({fear\panic…})	表现了/突显出/..对 E_aspec t的 E_word (It shows/emerges /… for E_word for ETopic_ aspect)	表达了作者对死亡的恐惧 (It expresses author's panic for death)
Hatred	{名贵花木\专制社会…} ({precious flowers and trees\autocratic society…})	{厌恶\批判…} ({disgust/criticize…})	表达了/引发了/..对E_aspect 的 E_word (It expresses/initiates/… for E_word for ETopic_ aspect)	表达了作者对那个名贵花木的厌恶 (It expresses author's disgust for precious flowers and trees.)
Surprised	{初生生命\浩大的瀑布…} ({newborn life\the great waterfall…})	{惊喜\惊讶…} ({surprise/amaze…})	表达了/表现了/…对 E_aspect 的 E_word (It expresses/shows/… for E_word for ETopic_aspect)	表达了作者对初生生命的惊喜 (It expresses author's pleasant astonishment for new life.)

6.2 Process of Answer

In the questions and answers of the emotion comprehension, the system needs to obtain Topic/ETopic_aspect and E_word from background materials. Then, it combines E_word and E_category to obtain the connection between Topic/ETopic_aspect and E_category. On this basis, it needs to use the answer templates to get the final answers. The specific process is shown in Fig. 2:

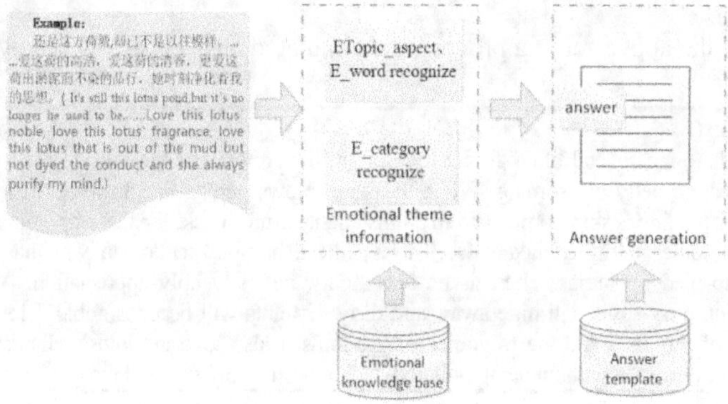

Fig. 2. The flow chart of answering emotional question

According to the process shown in Fig. 2, the question answering process mainly includes the acquisition of emotional subject information and the generation of answers.

(1)Information on emotional topic.

Identifying of E_category: At present, only the existing text classification method is used in the experiment to finish E_category identifying. DPCNN [22] and RCNN [23] are the best methods to achieve the experimental results. The specific experimental results are shown in Table 7.

Table 7. The experimental results of emotion recognition

Experimental method	Precision	Recall	F1-score
DPCNN	0.2679	0.2783	0.2583
RCNN	0.3286	0.2654	0.2490

Identifying Topic/ETopic_aspect and E_word: In order to obtain "Topic/ETopic_aspect", we need to combine the topic with the location of the prose topic, and the association relationship between "E_category" and "ETopic_aspect" mentioned in Sect. 5.2. E_word can be obtained using the emotion knowledge base. At the same time, the recognized E_word can support the subsequent answers.

(2) Generation of answers

In order to make the answer more fluent, we use Eq. (1) to obtain emotional category information, and then combine it with the constructed answer templates to generate the answers.

By using the above solving process, the simulated answers are as follows:

Text:

There are several blue "forget me not" in the crevices of the rocks on the bank, which are quietly blossoming. When the wind blows gently, the flower occasionally touches the fairy's skirt corner, and in an instant, it shrinks back, for fear of polluting the fairy's clothes. How excited it is? This is like a humble gentleman who has been accompanied by fairies. There is no blasphemy, but only holy appreciation. At this moment, dirty souls will hide away, and dirty thoughts will become noble. I love the purity of the lotus and the fragrance of the lotus, I also love the lotus's character of coming from mud without being polluted, which purifies my spirit all the time.

Emotional topic information:

E_category: love
Topic: Lotus pond
ETopic_aspect: Lotus
E_word: love

Generated answers:

(1) Using the answer template of "love"
(2) Match ETopic_aspect: lotus; match E_word: love
(3) Answer generation: this expresses the author's love for lotus

7 Conclusion

Using the constructed prose emotion category system, the collected data in this paper are annotated to establish an emotional knowledge base and answer templates for prose reading comprehension. The data enhancement technology is used to expand annotated resources. At the same time, the expansion of knowledge base is still in progress. For the next step, we will extract resources from Chinese teaching materials to enrich the construction of knowledge base and will explore the semantic reasoning technology and the method of emotion recognition in complex prose.

Acknowledgments. The authors would like to thank all anonymous reviewers for their valuable comments and suggestions which have significantly improved the quality of this work. The work is supported by the National Key Research and Development Project of China (No. 2018YFB1005103) and the National Natural Science Foundation of China (62076158).

References

1. Zeng, X.: Research on affection education and its implementation in the high school prose teaching. Kashi Normal University (2013)
2. Liu, B.: Sentiment Analysis: Mining Opinions, Sentiments, and Emotions. Jun Zhao translate. China Machine Press, Kang Liu (2017)
3. Ekman, P.: An argument for basic emotions. Cogn. Emot. **6**(3), 169–200 (1992)
4. Plutchik, R.: The multifactor-analytic theory of emotion. J. Psychol. **50**(1), 153–171 (1960)
5. Lin, C.: Emotional problems in socialist psychology. Soc. Psychol. Sci. **1**, 37 (2006)
6. Qiu, D.: Unified value theory. China Science and Technology Press (1998)
7. Xu, X., Tao, H.: A study on the division of emotion in Chinese emotional system. In: Proceedings of the First Chinese Conference on Affective Computing and Intelligent Interactio, pp. 199–205 (2003)
8. Lee, S., Mei, Y.: Emotion and Cause. Springer (2019). https://doi.org/10.1007/978-981-10-6194-3
9. Xu, L., Lin, H., Pan, Y, et al.: Constructing the affective lexicon ontology. J. China Soc. Sci. Tech. Inf. **27**(2), 180–185 (2008)
10. Stone, P.J., Dunphy, D.C., Smith, M.S.: The General Inquirer: A Computer Approach to Content Analysis. MIT Press, Cambridge (1996)
11. Liang, Y., Feng, Z., Lin, H., Wang, J., Zhang, S.: Chinese emotion commonsense knowledge base construction and its application. In: Chinese Lexical Semantics 19th Workshop, pp. 136–146 (2018)
12. Liao Xiangwen, X., Hongbo, T.Y.: Construction and analysis of the third Chinese opinion analysis evaluztion (COAE2011) corpus. J. Chin. Inf. Process. **27**(1), 56–63 (2013)
13. Tan, S., Liu, K., Wang, S., Liu, K.: Overview of Chinese opinion analysis evaluation 2013. In: Proceeding of the COAE2013, pp. 5–33 (2013)
14. Jing, Y., Weiran, X., et al.: Task and data designing of sentiment sentence analysis evaluation in COAE2014. In: Proceedings of the COAE2014, pp. 5–25 (2014)
15. Pak, A., Paroubek, P.: Twitter as a corpus for sentiment analysis and opinion mining. In: Proceedings of the International Conference on Language Resources and Evaluation, pp. 1320–1326 (2010)
16. Toprak, C., Jakob, N., Gurevych, I.: Sentence and expression level annotation of opinions in user-generated discourse. In: Proceedings of the 48th Annual Meeting of the Association for Computational Linguistics, pp. 575–584 (2010)
17. Min, D., Zhu, Z., Shousha, L., et al.: Corpus construction on opinion information extraction in Chinese. J. Chin. Inf. Process. **29**(4), 67–73 (2015)
18. Lin-hong, X., Hong-fei, L., Jing, Z.: Construction and analysis of emotional corpus. J. Chinese Inf. Process. **22**(1), 116–122 (2008)
19. Li, C., Wang, Y., Li, C., Liji, C., Liu, Q.: Building corpus with emoticons for sentiment analysis. In: Proceedings of the 7th CCF International Conference on Natural Language Processing and Chinese Computing, pp. 309–318 (2018)
20. Chen, X., Li, Y., Wang, S., Li, D., Mu, W.: Emotional knowledge corpus construction for deep understanding of text. In: Proceedings of Chinese Lexical Semantics 19th Workshop, pp. 655–667 (2018)
21. Wei, J., Zou, K.: EDA: easy data augmentation techniques for boosting performance on text classification tasks. Computation and Language (2019). arXiv preprint arXiv: 2019: 1901.11196
22. Johnson, R., Zhang, T.: Deep pyramid convolutional neural networks for text categorization. In: Proceedings of the the Association for Computational Linguistics, pp. 562–570 (2017)
23. Lai, S., Xu, L., Liu, K., et al.: Recurrent convolutional neural networks for text classification. In: Proceedings of the National Conference on Artificial Intelligence, pp. 2267–2273 (2015)

Diachronic Changes and Multi-dimensional Quantitative Analysis of Sentence Complexity in Contemporary Chinese Written Language

Enshang Xia and Gaoqi Rao(✉)

Research Institute of International Chinese Language Education,
Beijing Language and Culture University, Beijing 100083, China
raogaoqi@blcu.edu.cn

Abstract. Using the corpus of *People's Daily* from 1946 to 2015 as our data, this paper conducts a quantitative study on sentence complexity based on average length, dispersion, fragmentation and lexical density to investigate its historical changes. This paper also aims to conduct a multi-dimensional analysis on the correlation between sentence complexity and structure or function. Our results reveal that the trend of sentence complexity changes from stable to a rise. The amount of verbs in a sentence has a strong positive correlation with sentence length and fragmentation. Emotional value has a positive correlation with average sentence length, as well as dispersion of sentence lengths. The proportion of interrogative sentences is negatively correlated with average sentence length. Furthermore, the proportion of exclamatory sentences is negatively correlated with lexical density, while the proportion of declarative sentences positively correlates with lexical density.

Keywords: Sentence complexity · *People's Daily* · Quantitative analysis · Correlation · Diachronic change

1 Introduction

As the most influential newspaper in China, *People's Daily* has recorded the linguistic and cultural characteristics of different historical periods. Its development reflects the changes in the contemporary Chinese written language to a considerable extent. Meanwhile, it has attracted wide attention in academic circles. Previous scholars conducted vocabulary research on new loan words from Japanese, high-frequency nouns, and subject words over the years, such as Huang [1]. Gong [2] and Zhong [3] analyzed news headlines and editorials from the aspects of language memes, words usage, rhetoric, etc. Research of newspapers represented by *People's Daily* mostly focused on the study of language features and structural elements such as headlines and editorials, and seldom involved large-scale diachronic investigation of sentences. The study of sentences has a history of nearly a hundred years, mainly focusing on sentence category, sentence pattern and sentence structure. There is also a certain amount of research on sentence length.

M. Liu et al. (Eds.): CLSW 2020, LNAI 12278, pp. 618–630, 2021.
https://doi.org/10.1007/978-3-030-81197-6_52

Zhang [4] conducted a sample survey on sentence length, and calculated the average sentence length of political papers and novels. Qiu [5] and Huang [6] statistically analyzed the average sentence length and sentence length distribution to lay the foundation for the analysis of the factors restricting sentence length. Shi [7] used the basic principles and methods of the multivariate normal population mean to compare the characteristics of long and short sentences in literary works and revealed significant differences between the two. To sum up, most of the research efforts on sentences focused on the synchronic level, while diachronic studies on sentences and their complexity are scarce.

It is generally believed that the complexity of sentences comes from the complexity of cognitive processing, which is highly related to the complexity of the symbols expressed by the sentence in the carrier. This paper selects about 1.17 billion Chinese characters in *People's Daily* from 1946 to 2015 as data and conducts statistical research on the average sentence length, dispersion of sentence lengths, sentence fragmentation, and lexical density in terms of sequence length and structure depth. The paper also analyzes the changes in the complexity of language in the past 70 years and performs relevant analysis from the perspectives of sentence complexity and emotion, sentence types, the amount of verbs, in an attempt to explore the relationship between sentence complexity and its structure or function.

2 Complexity of Sentences and Changes Over Time

The complexity of sentences is mainly affected by the length of symbol sequence, the depth of a symbol structure and the difficulty of understanding the symbols. Among these three aspects, the length of the symbol sequence is the sentence length, which is generally calculated from the number of words or characters, and its variation can be reflected by the dispersion of sentence lengths. The complexity of the symbol structure and the difficulty in understanding symbols can be measured by sentence fragmentation and lexical density.

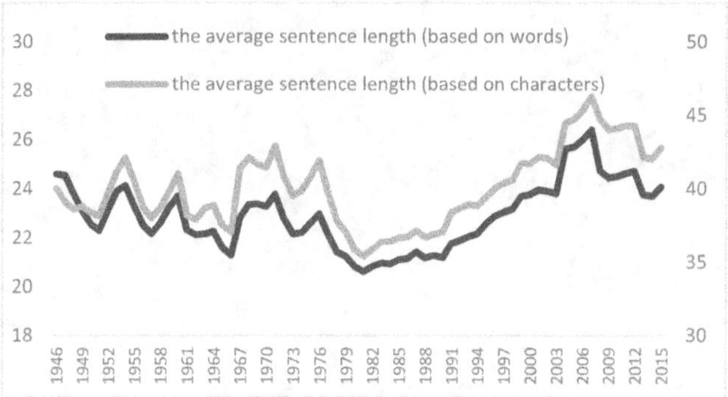

Fig. 1. The average sentence lengths (based on characters/words)

The British statistician Ure put forward a formula to calculate the average sentence length: *average sentence length = total number of words or characters/total number of sentences*. Based on this, the average sentence length of *People's Daily* in the unit of 'year' was calculated. From Fig. 1, it can be seen that the average sentence length shows a trend from fluctuating to steadily rising and then gradually decreasing. Specifically, average sentence length fluctuated greatly before the 1980s, with several lows around 1951, 1957, and 1966, and several peaks around 1954, 1960, 1971, and 1976. After the ups and downs, average sentence length began to rise steadily in the early 1980s and reached the peak around 2008 before falling.

The dispersion of sentence lengths (D) is the standard deviation of the sentence length in a text, which is the degree of the sentence length deviating from the average sentence length of the text. It can measure the distribution of sentence length and reflect the diversity and change of sentence length, and thus reflect the rhythm of a language. If the degree of dispersion is low, the distribution of sentence length will be relatively concentrated, and so the language is characterized by fluency and repetition. However, if the degree of dispersion is high, the distribution of sentence length will be relatively scattered, and the sentence length is characterized by diversity and variability. In this case, the language will also be rich and colorful. The formula for calculating the dispersion of sentence lengths [8] is as follows:

$$D = \sqrt{\frac{1}{N} \sum (Li - Lo)^2} \tag{1}$$

It can be seen from Fig. 2 that the dispersion of sentence lengths shows a downward trend prior to the 1960s. It means that, during this period, sentences gradually changed from 'varying length' to 'almost the same length', while the language also changed from 'fluctuation' to 'stabilization'. From 1960s to 1970s, the dispersion fluctuated haphazardly, and the language rhythm maintained its momentum. Around 1976, the dispersion dropped sharply with an overall increase in the stability of language. After the reform and opening-up in the Chinese history, the dispersion increased slightly. In 2000, it began to increase obviously, which is explained by the use of both long and short sentences, as well as the richness in change.

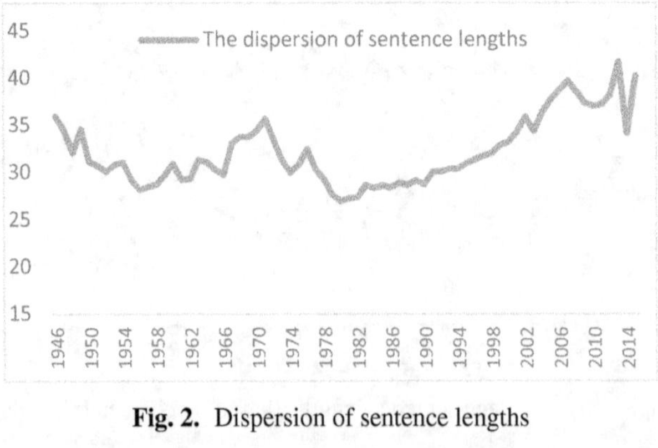

Fig. 2. Dispersion of sentence lengths

Sentence fragmentation refers to the number of pauses in a sentence [9]. The degree of sentence fragmentation is calculated based on three punctuation marks at the end of the sentence (i.e., the full-stop, the question mark and the exclamation mark) and four punctuation marks within the sentence (i.e., comma, colon, slight-pause mark and semicolon). Punctuation is an important manifestation of the sentence structure. It can reflect the complexity of sentence structure to a certain extent. Generally speaking, the more punctuation marks in a sentence, the more complex the sentence structure is. The formula of sentence fragmentation is as follows:

$$\text{Sentence fragmentation} = \text{total number of pauses in the sentence}/\text{total number of sentences} \tag{2}$$

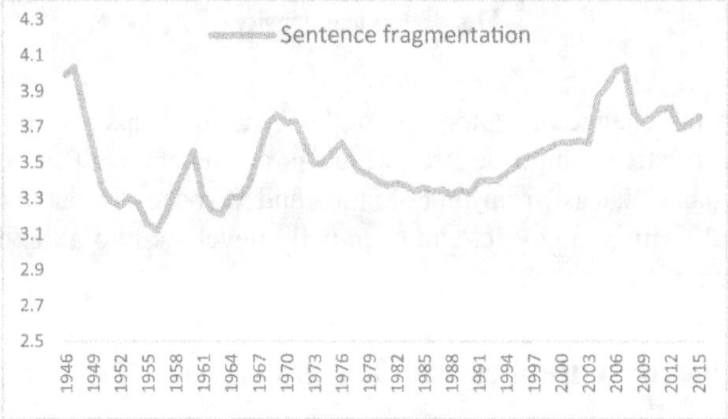

Fig. 3. Sentence fragmentation

In Fig. 3, the graph of sentence fragmentation shows the following trend: it declines suddenly first, then fluctuates unsteadily, and finally rises to a gentle up and down. Therefore, it can be inferred from the curve that the spoken language was greatly weakened after the founding of the People's Republic of China. In the stage of frequent political and cultural movements, language exhibited instability, but eventually, it turned to the integration of fragmentation and standardization gradually. Meanwhile, the sentence structure underwent ups and downs and evolved from stability and regularity to various changes.

The calculation method of lexical density was proposed by Ure [10]. In his opinion, lexical density is the ratio of the number of content words to the total number of words. Halliday [11] believed that the carrier of information is not only words, but also sentence structure, and thus he proposed that the sentence lexical density should be measured by the average number of the content words in a sentence. Above all, we believe that lexical density refers to the ratio of the number of content words to the total number of words in the text or sentence. It can reflect the amount of information in a text or a sentence and is also one of the important indicators to measure the level of difficulty in understanding. It can be seen in Fig. 4 that the lexical density of sentences in People's Daily was always between 80.5% and 86.7% over the past 70 years, while the overall trend is a decrease at first and then an increase steadily. From the 1950s to the early 1970s, the lexical density was relatively small but fluctuated greatly. Around 1978, the lexical density generally continued to rise, and the amount of information in sentences increased steadily.

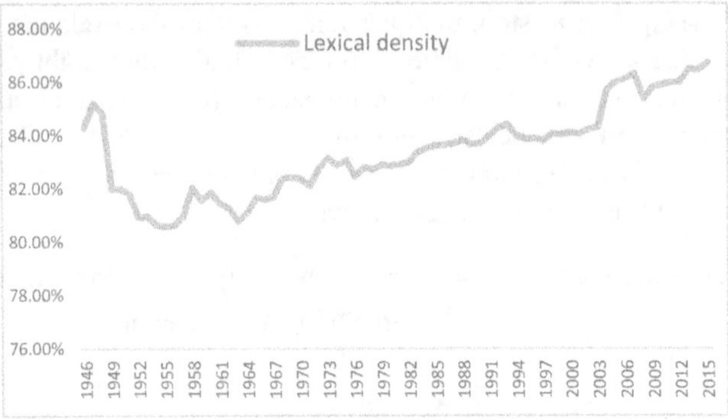

Fig. 4. Lexical density

In summary, the change of sentence complexity is from ups and downs to stability and finally to a relatively high degree, which is consistent with the changing trend of popular language, that is, from monotonous and verbose discourse to an objective and standardized written language, and eventually develops into a flexible but diverse language for the public.

3 Complexity of Sentences and Verbs

The verb is the core of a sentence, with other elements being associated with it and drawn by it [12]. The "Verb Core" is a widely accepted grammatical concept. Therefore, we counted the number, proportion and types of verbs in the sentences of *People's Daily*, and also compared them with the indicators affecting the complexity of sentences (length, length dispersion, sentence fragmentation, and lexical density) to perform correlation analysis. The "year" served as the unit in our diachronic survey for the number, proportion, and types of verbs in a sentence.

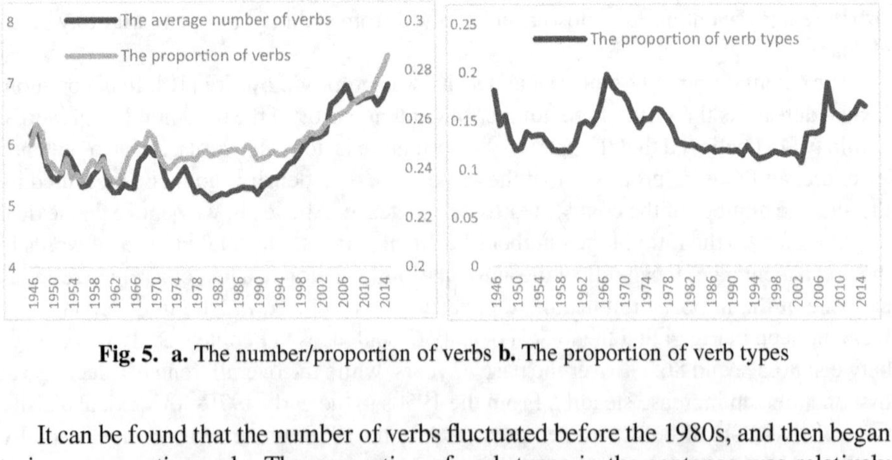

Fig. 5. a. The number/proportion of verbs **b.** The proportion of verb types

It can be found that the number of verbs fluctuated before the 1980s, and then began to increase continuously. The proportion of verb types in the sentence was relatively high in the late 1960s and early 1970s. It declined in the 1980s-1990s and began to rise around 2000s.

The length, fragmentation, dispersion, and lexical density of sentences containing different number of verbs in the *People's Daily* corpus were calculated, in order to find out the relationship between verbs and sentence complexity. The classification standard of the correlation coefficient 'r' is as follows (Table 1):

Table 1. The correlation coefficient |r|

Correlation coefficient	Highly strong correlation	Strong correlation	Moderate correlation	Weak correlation	Extremely weak or unrelated
|r|	0.8–1.0	0.6–0.8	0.4–0.6	0.2–0.4	0–0.2

The Pearson correlation coefficient and the Spearman correlation coefficient are two commonly used correlation coefficients at present. Since the former requires conformation of data to a normal distribution, the Spearman correlation coefficient is utilized in this paper as it boasts wider applications for correlation tests.

Table 2. The correlation between verbs and sentence complexity

y	x				
	Average sentence length (based on characters)	Average sentence length (based on words)	Sentence fragmentation	Sentence length dispersion	Lexical density
The number of verbs	0.9917	0.9890	0.9908	0.7902	0.6833
The proportion of verb types	0.9974	0.9979	0.9966	0.8569	0.5621

In Table 2, it can be found that the number of verbs and the proportion of verb types in the sentence have a strong positive correlation with the average sentence length, the sentence fragmentation and the dispersion of lengths.

In addition, we performed a diachronic correlation test on the indicators of verbs shown in Fig. 5 a, b, and on the relevant indicators of sentence complexity.

It can be seen in Table 3 that the average number of verbs in a sentence has a strong positive correlation with the average sentence length and dispersion of sentence lengths over the years. Sentence fragmentation is highly correlated with the number of verbs and the proportion of verb types, which is consistent with the so-called "verb core" concept, that is, the verbs at the core of a sentence always dominate or connect with other components. Thus, the number of verbs in a sentence is positively correlated with

Table 3. The correlation between verbs and sentence complexity over years

y	x				
	Average sentence length (based on characters)	Average sentence length (based on words)	Sentence fragmentation	Sentence length dispersion	Lexical density
The average number of verbs	0.910594	0.964238	0.781716	0.899571	0.554475
The proportion of verbs	0.512204	0.524031	0.7666	0.611058	0.830041
The proportion of verb types	0.589222	0.588452	0.797043	0.667781	0.764185

the sentence length. The lexical density shows a strong correlation with the proportion of verbs in a sentence, which may be interpreted in the way that the verb is loaded with information and serves as an important indicator for calculating lexical density. The proportion of verb types in sentences are closely correlated with sentence dispersion but are moderately related to average sentence length. In other words, the proportion of verb types in sentences has little effect on diachronic change of sentence lengths.

4 Complexity of Sentences and Semantic Relationships

Sentences can be divided into simple sentences and complex sentences according to their structure. In the *People's Daily* corpus over the past 70 years, the proportion of complex sentences in contemporary Chinese is more than 70%. The classification of complex sentences is based on the semantic relationships between clauses. Therefore, we explored the relationship between semantics and sentence complexity based on the types of complex sentences in an effort to better describe the development of sentences.

According to the 'trichotomy' of Chinese complex sentences, complex sentences can be divided into three types: compound complex sentences, causal complex sentences, and transitional complex sentences, with connectives in the sentences serving as a useful tool to classify the sentences. On this basis, we identified and extracted complex sentences in the corpus, and, at the same time, calculated the proportion of complex sentence types. However, for those complex sentences without obvious signs, statistical calculations were skipped. In Fig. 6, the left coordinate axis is the proportion of compound complex sentences, and the right coordinate axis is the proportion of transitional and causal complex sentences. On the whole, the proportion of complex sentences shows a downward trend, and the proportion of compound complex sentences is much higher than that of transitional complex sentences and causal complex sentences.

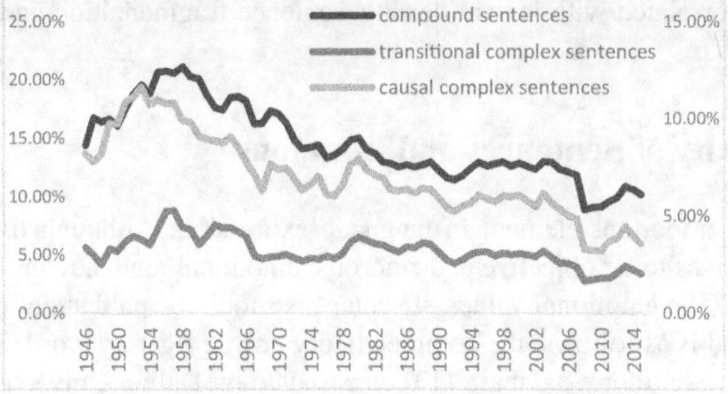

Fig. 6. The proportions of complex sentence types

The correlation between the proportion of each type of complex sentences and the sentence complexity over the years has been tested. The final results are as follows (Table 4):

Table 4. The correlation between the proportion of complex sentence types and sentence complexity

y	x				
	Average sentence length (based on characters)	Average sentence length (based on words)	Sentence fragmentation	Sentence length dispersion	Lexical density
Compound complex sentences	−0.3373	−0.2204	−0.4923	−0.4022	−0.8832
Transitional complex sentences	−0.6647	−0.5344	−0.7926	−0.7052	−0.7614
Causal complex sentences	−0.5101	−0.3218	−0.5981	−0.5057	−0.8642

The results reveal a weak correlation between the average sentence length and the proportion of various types of complex sentences. Except for transitional complex sentences, most of them are moderately or weakly related to sentence length, while lexical density of the sentences has a strong negative correlation with the proportion of various types of complex sentences. Sentence fragmentation and dispersion of lengths have a strong or moderate negative correlation with various types of complex sentences. That is to say, semantics has little effect on sentence length, but has a role in information density, colloquialism and rhythm. To a certain extent, the proportion of complex sentences is negatively correlated with lexical density, sentence fragmentation and dispersion of sentence lengths.

5 Complexity of Sentences and Emotions

Emotion is an important element in language expression. Although the language of newspapers strives to be objective and rigorous, emotional tendency in the language is still inevitable. The emotional values of sentences should be paid attention to. The calculation method is as follows: the commendatory and derogatory words in the Chinese affective lexicon ontology database [13] constructed by Dalian University of Technology are selected as the basic emotional words, and then their emotional intensity (1–9) is extracted to assign values to the emotional words, and meanwhile, positive and negative values of emotional tendencies should be distinguished. Specifically, commendatory words have positive emotional values (1–9), while derogatory words have negative emotional values (−9–1). For those sentences with negative words in front of emotional words, we should reverse the emotional state (from positive to negative or from negative to positive) according to the number of negations, and then calculate the emotional values of the sentence.

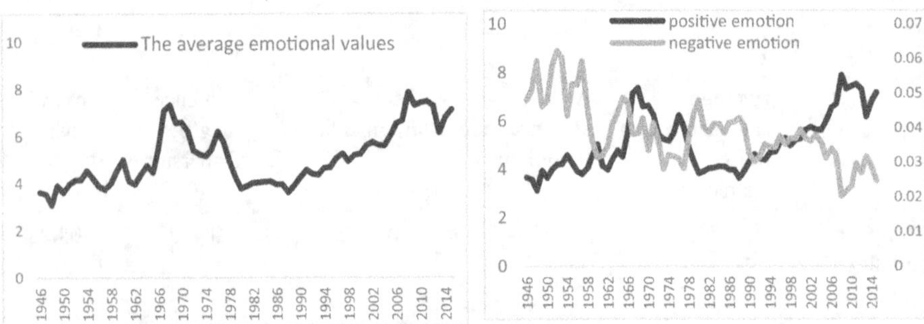

Fig. 7.a. The average emotional values **b**. Negative and positive emotion

Based on the above method, we counted the average emotional values of sentences over the past 70 years and calculated the absolute values of negative and positive emotions in the sentences. As shown in the Fig. 7a and Fig. 7b, the left axis represents the average emotional values and positive emotional values, and the right axis represents the absolute values of negative emotional values. It can be seen that, on the whole, the language of newspapers in the 1960s and 1970s was relatively rich in emotion. After experiencing the low levels in the 1980s and 1990s, the emotional value began to rise sharply in the early 2000s. Combined with the emotional tendency, the negative emotions in sentences before 1950s and 1960s were relatively obvious and reached the peak around 1966. Since the 1980s, the emotional value rose steadily, with negative emotions remaining low, and positive emotions increasing.

Similarly, we also calculated the correlation between the emotional value of sentences and the diachronic data of sentence complexity. It can be seen in Table 5 that there is a

Table 5. The correlation between emotions and sentence complexity

y	x				
	Average sentence length (based on characters)	Average sentence length (based on words)	Sentence fragmentation	Sentence length dispersion	Lexical density
The average affective values	0.8116	0.6413	0.6292	0.7014	0.3830
Positive emotion	0.8133	0.6586	0.6285	0.7025	0.3831
Negative emotion	0.5471	0.4809	0.7418	0.6218	0.7943

strong positive correlation between emotional value and sentence length dispersion. The average emotional value and positive emotional value have a strong positive correlation with average sentence length. However, negative emotional value is weakly related to this. This shows that the emotional components are positively correlated with dispersion of sentence lengths. With the increase of sentence length and sentence fragmentation, there are more emotional components in the sentence, especially positive emotions, while negative emotions have a low correlation with them.

6 Complexity of Sentences and Sentence Functions

Sentences can be classified into declarative sentences, interrogative sentences, exclamatory sentences and imperative sentences according to their functions. Declarative sentences, interrogative sentences and exclamatory sentences all have obvious features in the written language. They are identified by full-stops, question marks, and exclamation marks. Therefore, we classified sentence types according to punctuation marks and calculated their proportions to study the relationship between sentence functions and sentence complexity.

Based on the punctuation marks, the proportions of declarative sentences, interrogative sentences and exclamatory sentences over the years are calculated. As shown in Fig. 8, the left coordinate axis represents the proportion of declarative sentences, while the right coordinate axis represents the proportions of interrogative sentences and exclamatory sentences. It can be found that the proportion of declarative sentences is complementary to the proportions of exclamatory sentences and interrogative sentences in terms of their changing trend. After the reform and opening up, the proportion of declarative sentences has basically increased, while the proportions of exclamatory sentences and interrogative sentences have declined since 1980. Exclamatory sentences had more changes in their proportions than interrogative sentences did, especially in 1950s-1970s, when the proportion of exclamatory sentences exceeded the proportion of interrogative sentences. After the 1980s, the proportions of exclamatory sentences and interrogative sentences have changed relatively little, and the proportion of interrogative sentences is mostly higher than that of interrogative sentences.

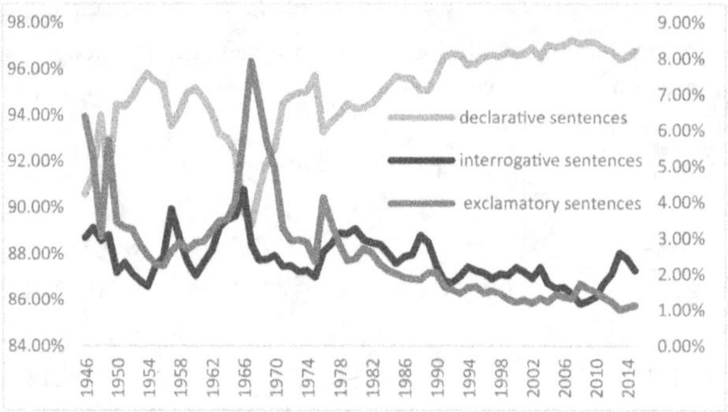

Fig. 8. The proportion of sentence types

The correlation between different sentence types and their complexity has been examined, and the final results are shown in Table 6:

Table 6. The correlation between sentence type and sentence complexity

y	x				
	Average sentence length (based on characters)	Average sentence length (based on words)	Sentence fragmentation	Sentence length dispersion	Lexical density
Sentence types	0.887309	0.897603	0.920981	0.758681	0.269954

It can be seen that sentence function is strongly correlated with average sentence length and fragmentation, and it has a positive correlation with dispersion of sentence length. However, there is a weak positive correlation between sentence function and lexical density. That is to say, these three types of sentences have great differences in average length and fragmentation, certain differences in length dispersion and almost no difference in lexical density.

In addition, we also examined the correlation between proportion of sentence types and complexity of sentences over the years. The results are shown in Table 7. There is a strong negative correlation between proportion of interrogative sentences and average sentence length (based on characters). The proportion of exclamatory sentences and declarative sentences is strongly correlated with the lexical density of sentences, while others are moderately or weakly correlated. This shows that, to a certain extent, the length of the sentence is inversely proportional to the proportion of interrogative sentences. Lexical density is negatively related to the proportion of exclamatory sentences and positively connected with the proportion of declarative sentences.

Table 7. The correlation between the proportion of sentence types and sentence complexity over years

y	x				
	Average sentence length (based on characters)	Average sentence length (based on words)	Sentence fragmentation	Sentence length dispersion	Lexical density
Interrogative sentences	−0.6607	−0.5695	−0.3661	−0.4788	−0.4376
Exclamatory sentences	−0.3843	−0.3098	−0.2774	−0.2582	−0.7236
Declarative sentences	0.5223	0.4375	0.3245	0.3664	0.6838

7 Conclusions

Language is closely connected with society. The main reason for language change is the development of society, but the development of society is inconsistent, so we should pay attention to a longer time span in order to precisely find out the changes in language [14].

This paper summarizes the diachronic trend of sentence complexity in the newspaper language in the past 70 years: After the turmoils in the early days of the founding of the People's Republic of China, the complexity of sentences became stable, while the language tended to be gentle. During this period, the average sentence length was short but there was a tendency to rise slowly. The lexical density of sentences also showed an upward trend and the amount of information in sentences was increasing steadily. The dispersion of sentence length and degree of fragmentation showed a downward trend. The sentence structure and text rhythm remained stable. However, since the beginning of the new century, the complexity of sentences has increased rapidly, and the language has become popular and dynamic. The average sentence length has gradually decreased after reaching the peak, but the lexical density of sentences has still been increasing, and the amount of information has risen sharply. The dispersion of sentence lengths shows a steady increase, and the rhythm of text is bright and varied. The degree of fragmentation is larger, and the sentence structure tends to be diversified. In summary, the changing trend of sentence complexity is from ups and downs to stability, and finally to a rise, which is consistent with the changing trend of popular language. In other words, Chinese written language has changed from monotonous and verbose discourse to an objective and standardized written language, and eventually developed into a flexible and diverse language for the public.

However, there are many other factors that can affect the complexity of sentences. After exploring the relationship between verbs, semantics, sentence functions, emotions and sentence complexity, it can be found that there is a strong positive correlation between number of verbs and sentence length. Semantics has no obvious relationship with sentence length, but it is related to information density and rhythm. As for emotion, the affective value has a very strong positive correlation with average sentence length and has a relatively strong positive correlation with dispersion of sentence lengths. That is, the dispersion decreases with the decrease in emotional components, while sentence length increases with the emotional components in sentences, especially positive emotions. Regarding sentence functions, the proportion of interrogative sentences decreases with sentence length. The proportion of exclamatory sentences is inversely proportional to the lexical density of sentences, and the proportion of declarative sentences is proportional to lexical density.

Acknowledgments. This paper is supported by Graduate Innovation Foundation in 2019 (19YCX148).

References

1. Huang, G.: Research on the Meanings of High-frequency Nouns in the Complete Works of the Editorials in People's Daily. Doctor, Nanning Normal University, Nanning (2016). (in Chinese)
2. Gong, S.: The Study on the Linguistic Variation of the News Headlines in the People's Daily. Master thesis, Jiangxi Normal University, Nanchang (2018). (in Chinese)
3. Zhong, R.: Rhetoric Identification: An analysis of the discourse of the new year editorial in People's Daily (1979-2018). J. Fujian Normal Univ. (Philosophy and Social Sciences Edition). **1**, 48–56 (2019). (in Chinese)
4. Zhang, S., Li, M.: Computer statistics of the similarities and differences of classical Chinese style between novels and political commentaries (experimental report). J. Tianjin Normal Univ. **04**, 82–86 (1986). (in Chinese)
5. Qiu, X.: The distribution of sentence length and its restrictive factors. Doctor, Anhui University, Hefei (2017). (in Chinese)
6. Huang, Z.: The mean length and distribution of sentence in Chinese characters. J. Qiqihar Univ. **01**, 133–138 (2018). (in Chinese)
7. Shi, Z., Xia, X.: Multivariate statistical analysis of long and short sentences in the study of Redology. J. Mianyang Normal Univ. **30**(11), 16–19 (2011). (in Chinese)
8. Zhang, J.: Text style analysis based on statistical method. Doctor, Shandong University, Shandong (2012). (in Chinese)
9. Kan, M.: A quantitative comparative study of several stylistic parameters: a case study of news reports and interviews. J. Lang. Lit. Stud. **17**, 46–48 (2011). (in Chinese)
10. Ure, J.: Lexical density and register differentiation. Appl. Linguist. 443–452 (1971)
11. Halliday, M.A.K.: Spoken and written language. Geelong Vict. Deakin University. pp. 61–67(1985)
12. Chinese Language Research Institute. Sentence patterns and verbs. Language and Culture Press, Beijing (1987). (in Chinese)
13. Xu, L.H., Lin, H.F., Pan, Y.: Constructing the affective lexicon ontology. J. Chin. Soc. Sci Tech. Inf. **27**(2), 180–185 (2008). (in Chinese)
14. Diao, Y.: History of Modern Chinese. Fujian People's Publishing House, Fuzhou (2006). (in Chinese)

ARSCC: An Assisted Reading System for Contemporary Chinese Based on Multi-aspect Analysis

Yingjie Han, Yingjie Yan, Lingling Mu, Kunli Zhang, and Hongying Zan(✉)

School of Information Engineering, Zhengzhou University, No.100 Science Avenue, Zhengzhou 450000, Henan, China
{ieyjhan,iellmu,ieklzhang,iehyzan}@zzu.edu.cn,
yjyan@gs.zzu.edu.cn

Abstract. With the on-going Chinese fever all over the world, there is an increasing number of scholars studying methods and tools for assisting Chinese learning. This paper discusses the construction of an **A**ssisted **R**eading **S**ystem for **C**ontemporary **C**hinese based on multi-aspect analysis (ARSCC). First of all, short essays about Chinese culture are selected as reading materials according to preliminary statistics. Secondly, a multi-aspect analysis of these essays is conducted in three stages: word segmentation and POS tagging, sense of content words and usage of function words annotation, and syntactic analysis. Finally, the above analysis results are displayed hierarchically and systematically. The system provides second-language learners with multi-aspect assisting information to improve their analysis-oriented reading comprehension ability. It also provides a platform for teaching Chinese as a foreign language.

Keywords: Word segmentation · Sense of content words · Usage of function words · Syntactic analysis

1 Introduction

With the enhancement of China's comprehensive strength and the improvement of its international status, a Chinese fever has emerged throughout the world. The Silk Road and 21st Century Maritime Silk Road have been attracting many foreigners. In addition to this, traditional Chinese culture has also been attracting many foreigners who want to learn Chinese. Up to December 2019, 550 Confucius Institutes [1] and 1,172 Confucius Classrooms in 162 countries and regions have been established. Some countries have incorporated Chinese into their national education systems. Therefore, the importance of teaching Chinese as a foreign language is self-evident.

Reading is an essential skill for study, work and social life. The development of reading comprehension ability is critical in Chinese learning, which is long-term, challenging, and requires lots of trainings. Professor Bai's research results showed that inserting a space between two adjacent words can help learners to understand reading texts, and correct word segmentation can improve reading efficiency [2].

M. Liu et al. (Eds.): CLSW 2020, LNAI 12278, pp. 631–639, 2021.
https://doi.org/10.1007/978-3-030-81197-6_53

In Chinese, words can be divided into content words and function words [3] according to whether they can form sentences or can be syntactic components independently. Content words are essential for reading comprehension. Not knowing the senses of content words would lead to difficulties in reading. Function words also play a vital semantic role in Chinese. Different function words may have similar usage. Incorrect analyses of the usage of function words can cause confusion. However, many function words do not correspond to specific words in foreign languages, which increases the difficulty of understanding [4]. Knowledge of syntax is the basis of language learning, and syntactic analysis can solve the problems of having difficulties with meaning and the vocabulary in reading.

This paper presents an **A**ssisted **R**eading **S**ystem of **C**ontemporary **C**hinese based on multi-aspect analysis (ARSCC). First of all, to solve the problem that there is no clear boundary between two Chinese words, we conduct word segmentation and POS tagging after selecting reading materials. Secondly, to avoid ambiguity and confusion, we analyze and annotate the senses of content words and the usage of function words in the reading materials. Because of negative transfer from the mother tongue and over-generalization in learning a second language, sentences with typical grammatical errors concerning function words are also analyzed. Thirdly, to promote understanding, we use syntactic trees to represent dependency relationships between words. Finally, we display the above analysis results in a multi-level and multi-aspect manner.

2 Background

At present, there are many websites for Chinese as a Foreign Language (CFL) learners to learn Chinese, which can be divided into four categories: tool websites, specialized skill training websites, online video teaching websites, and comprehensive websites. Tool websites mainly refer to the websites of Chinese-English dictionaries [5]. Specialized skills training websites include pinyin practice websites [6], vocabulary practice websites [7], Chinese character practice websites [6], grammar practice websites [8], listening and reading practice websites [9], etc. Online video teaching websites include Panda Chinese [10] and others. Comprehensive learning websites mainly include the website of Confucius Institute [11]. These Chinese learning websites have some limitations. For example, tools websites are used for looking up new words, which are only suitable for CFL learners with a solid foundation. For online video teaching websites, second language learners are in a passive learning mode, resulting in undesirable acquisition effects.

In this paper, the construction of ARSCC is described. Natural language processing techniques are applied to provide CFL learners with assisted reading information from multi-aspects, including Pinyin, parts of speech (POS), senses of content words, usages of function words, and syntactic analysis.

3 The Construction of ARSCC

During the construction of ARSCC, we have selected reading materials carefully and produced statistics. And then, we have conducted a comprehensive analysis in three phases. Firstly, we have done word segmentation, POS tagging, and phonetic tagging. Secondly, we have analyzed and annotated the senses of content words and the usages of function words. Besides, we have compared some examples of correct and incorrect usages of function words. Thirdly, we have conducted syntactic analyses and constructed a dependency tree for each sentence in the reading materials. A diagram showing the construction of ARSSC is shown in Fig. 1.

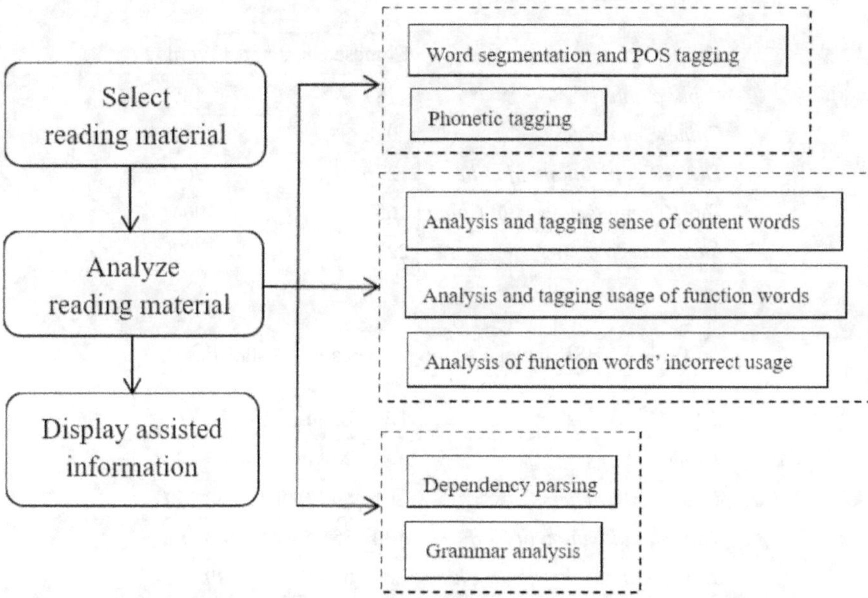

Fig. 1. There are three stages in the construction of ARSSC: selecting reading materials, analyzing reading materials, and displaying assisting information. Analyzing reading materials is the most important stage. It includes three steps: preliminary processing, adding auxiliary information for content words and function words, and adding auxiliary information for grammar.

3.1 Statistics

The reading materials analyzed in ARSSC were mainly from CCTV websites [12], which were about traditional Chinese culture. We selected five essays by considering the number of Chinese characters and words involved in the essays. They were *Paper Cuttings, Chinese Characters, The Self-Cultivation of Foodies, Top 10 Famous Flowers in China*, and *Auspicious Patterns.*

We compiled statistics on these essays from two aspects: Chinese characters and words, which are shown in Table 1. The number of Chinese characters ranged from 700 to 2,300, and the number of words ranged from 400 to 1,100.

The Hanyu Shuiping Kaoshi (HSK) is a test for international Chinese language proficiency for non-native Chinese speakers (including foreigners, Overseas Chinese, and ethnic-minority people). It focuses on non-native speakers' ability to communicate in daily life and to study and work in Chinese. The test is divided into HSK Level 1 (150 words), HSK Level 2 (300 words), HSK Level 3 (600 words), HSK Level 4 (1,200 words), HSK Level 5 (2,500 words) and HSK Level 6 (5,000 words). We compiled statistics on the HSK vocabulary entries used in the five essays, which are shown in Table 2. L1, L2, L3, L4, and L5 are short forms for HSK Level 1 to level 5, respectively.

Table 1. The statistics of Chinese characters and words in five essays

Title	Chinese characters	Words
Paper cuttings	787	485
Chinese characters	978	644
The self-cultivation of foodies	1230	963
Top 10 famous flowers in China	1653	1064
Auspicious patterns	2213	787

Table 2. HSK Level 1 to Level 6 vocabulary statistics

Title	L1	L2	L3	L4	L5	L6
Paper cuttings	123	46	46	33	47	33
Chinese characters	201	59	61	49	30	27
The self-cultivation of foodies	130	35	21	40	43	8
Top 10 famous flowers in China	264	64	89	43	79	37
Auspicious patterns	180	48	49	61	71	69

3.2 Multi-aspect Analysis

We analyzed the five essays and added assisted reading information on the words and sentences to help CFL learners. This analysis consists of three steps.

In the first step, PKUSeg[1] was used to do word segmentation [13, 14] and POS tagging according to reference [15]. The accuracy of automatic word segmentation was between 90% and 95%. To ensure the correctness of the results, we adopted multi-person, multi-round, and back-to-back proofreading.

In the second step, we analyzed and annotated the senses of content words and the usages of function words in the essays. The semantic features of content words play an essential role in reading comprehension. To help learners understand the senses of content words, we annotated the senses of nouns, verbs, and adjectives according to the Chinese Semantic Dictionary (CSD) [16] and the Chinese Large-scale Knowledge Base (CLSKB) [17]. They were developed by the Institute of Computational Linguistics

[1] https://github.com/lancopku/pkuseg-python.

of Peking University and the Natural Language Processing Laboratory of Zhengzhou University, respectively. We used CLSKB to do semantic tagging when we could not find the content word in CSD. We described content words according to the Chinese Function Word Usage Knowledge Base (CFKB) [18]. CFKB described content words with the following attribute information: word, word category, source, Pinyin, frequency, meaning, relevant word (synonym, hypernym, associative word and antonym), genus and remarks. There are six types of function words: adverb, preposition, conjunction, auxiliary word, modal word and positional word. We put our focus on examples, usages, and correct and incorrect sentence pairs. The frequencies of content words and function words in the five essays were counted. They are shown in Table 3.

Table 3. Statistics of content words and function words in the five essays

Title	Content words	Function words
Paper cuttings	346	139
Chinese characters	452	192
The self-cultivation of foodies	856	107
Top 10 famous flowers in China	790	274
Auspicious patterns	582	205

In the third step, with the help of LTP (Language Technology Platform) [19], the syntactic structure of sentences in the five essays were analyzed. Specifically, dependency parsing marks the grammatical elements in the sentences as subject, predicate, object, attribute, adverbial or complement. This is helpful to CFL learners with knowledge of parsing and semantics. According to LTP, there are 14 relationships between sentence components, such as subject-verb relationship (SBV), verb-object relationship (VOB), and indirect-object relationship (IOB). We marked the main components in sentences by combining Chinese with English.

4 System Display

ARSCC is built for CFL learners. Its menu is in English, including submenu items: Home, Preparation, Text and Follow up. A screenshot of Home is shown in Fig. 2, displaying the opening guideline of ARSCC. Under the submenu 'Preparation', the backgrounds of the essays are introduced. Under the submenu 'Text', the essays and assisting information are displayed. Under 'Follow up', in-depth discussion of function words is displayed. We take *Auspicious Patterns* as an example to introduce the contents under 'Preparation', 'Text' and 'Flow up' in detail.

With a click on 'Preparation', the introductions of the backgrounds of the essays and reading skills involved will be displayed. The system proposes preview questions, expands the contents of the essays in the form of pictures and texts, and contains information on easily confusable words and proper nouns.

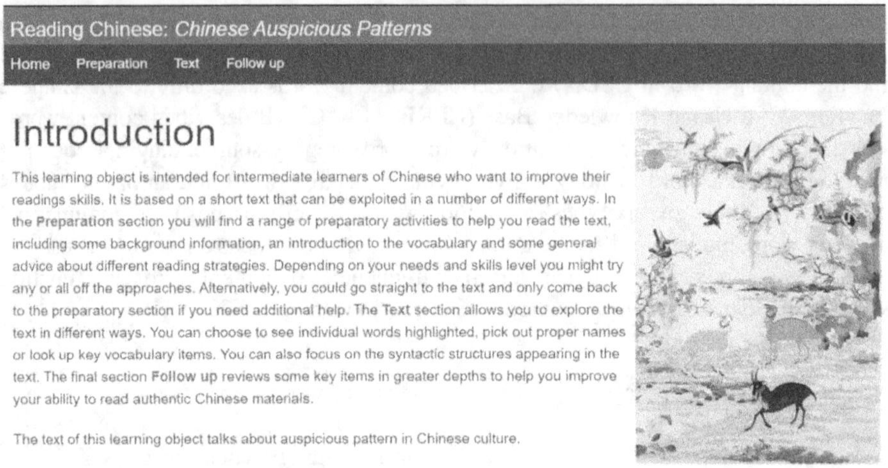

Fig. 2. Home page of ARSCC shows the main functions of the system and contains links to three other submenus: Preparation, Text and Follow up.

Table 4. Functions provided under Text (the submenu)

No.	Title	The detail
(1)	Show proper words	Display proper words in essays
(2)	Show words on hover	Display word segmentation
(3)	Show function words	Display usages of function words in context
(4)	Show CSD words	Display paraphrases of content words in context
(5)	Activate grammar	Display syntax of sentences
(6)	Activate translations	Display corresponding English sentences
(7)	Show dependency tree	Displays syntactic dependency trees of sentences
(8)	Show HSK1–6	Display HSK level

There are eight functions under 'Text'. Details are shown in Table 4.

Show proper words displays proper words in the essays due to their particularity. This can be used with other functions, such as Activate translations.

Show words on hover which displays word segmentation delimits the words with their context.

Show function words highlights all the function words in the essay. This information comes from CFKB. Figure 3 shows the information about the auxiliary word '之[*zhi*](*of*)' as an example.

Show CSD words focuses on content words. Content words are different components of a sentence according to context and sense, which is difficult for learners. Here, POS in context, pinyin, paraphrase, and usage examples of the content words are displayed.

word	之
单义	no
quanpinyin	zhi1
interpretation	同"的"。
example	光荣~家 公民有受教育和参加政治活动~权利 中国~最
usage	用在修饰语和中心语之间。用在修饰语和中心语之间，表示偏正关系，作用与"的"相当。

Fig. 3. Click the word '之[*zhi*](*of*)' to display information about it. '之[*zhi*](*of*)' is a function word. ARSCC displays its Pinyin, interpretation, examples and usage.

Information about '应用[*yingyong*](*apply*)' is displayed by clicking the word which is demonstrated in Fig. 4 as an example.

word	应用
wcat	v
pinyin	ying4yong4
interpretation	使用。
example	~新技术。 这种方法~得最为普遍。

Fig. 4. Click the word '应用[*yingyong*](*apply*)' to display information about it. '应用[*yingyong*](*apply*)' is a content word. ARSCC displays its POS, Pinyin, paraphrase and examples.

Activate grammar displays the most important elements in a sentence in different colors to reveal relationships between words.

Activate translations displays the English translation of the Chinese sentences.

Show dependency tree displays syntactic structures. The syntax analyzer displays the syntactic dependencies of sentences in the form of syntactic trees as shown in Fig. 5.

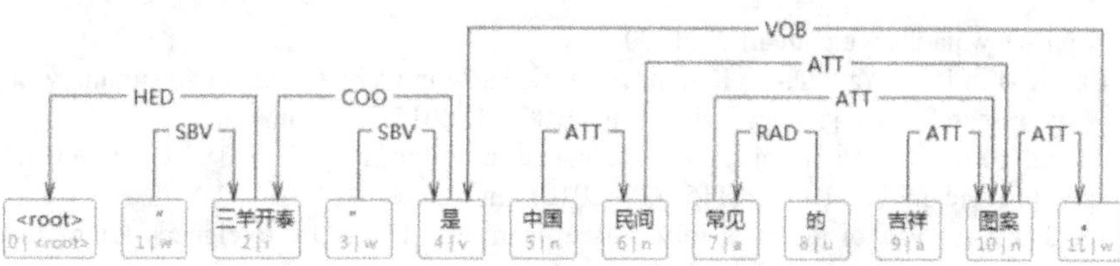

Fig. 5. The syntactic tree of ' "三羊开泰"是中国民间常见的吉祥图案。'. SBV, HED, COO, etc. are from LTP [19].

Show HSK1–6 serves the CFL learners with words and corresponding HSK levels.

The Follow up submenu gives an in-depth discussion about function words. On the one hand, we rely on the function word dictionary to provide examples and usages of function words. On the other hand, we display typical wrong usage of function words and give modification suggestions.

5 Conclusions

We have described the construction of ARSCC, an Assisted Reading System for Contemporary Chinese based on analyses from multi-aspects. We have selected five essays as reading materials. Besides, we have analyzed and displayed the five essays in multi-aspects. Mainly,

1) word segmentation and POS tagging are displayed;
2) senses of words and usages of function words are analyzed and presented, respectively; and
3) syntactic analyses are displayed in the form of syntactic trees.

Moreover, the analysis results of each stage have been manually proofread to provide correct information. ARSCC displays the assisting detailed information about the essays in multi-aspects to help CFL learners to learn Chinese vocabulary and improve their reading comprehension ability. It also provides an online learning tool for CFL learners, as well as a high-quality assisted teaching and learning platform for teaching Chinese as a foreign language.

In our future work, we will enrich the reading materials and recommend reading materials that are suitable for CFL learners according to the HSK vocabulary sets. Furthermore, we will make full use of natural language processing technology to reduce manual annotation and proofreading.

Acknowledgments. We thank the anonymous reviewers for their constructive comments, and gratefully acknowledge the support of the National Key Research and Development Project under Grant No. 2017YFB1002101; and the Henan Social Science Planning Project under Grant No. 2019BYY016.

References

1. http://www.jiaohanyu.com/article/1500
2. Bai, X.-J., Li, X., Yan, G.-L.: Eye movement control in Chinese reading: a summary of 20 years' research. Psychol. Dev. Educ. **31**(01), 85–91 (2015). (in Chinese)
3. Xiong, Y.: Discussion on some controversial issues in the study of modern Chinese parts of speech. J. Huangshan Univ. **1**, 115–117 (2010). (in Chinese)
4. Lu, J.-J.: Analysis of grammatical errors in learning Chinese by foreigners. Lang. Teach. Linguist. Stud. **1**, 49–64 (1994). (in Chinese)
5. https://dict.naver.com/linedict/zhendict/#/cnen/home
6. http://remembr.it/chinese-Chinese characters

7. http://www.nciku.cn/
8. http://resources.allsetlearning.com/chinese/grammar/
9. http://www.slow-chinese.com/
10. http://www.prcba.com/
11. http://wz.chinesecio.com/
12. http://english.cntv.cn/program/learnchinese/chineseculture/
13. Huang, C.-N., Zhao, H.: A ten-year Review of Chinese participles. J. Chin. Inf. Process. **03**, 8–19 (2007). (in Chinese)
14. Jiang, M., Huang, T., Zhang, M., et al.: An improved neural network for domain adaptive Chinese word segmentation. J. Comput. Methods Sci. Eng. **20**(4), 1–11 (2020)
15. Yu, S.-W., Duan, H.-M., Zhu, X.-F., et al.: Specification of corpus processing in Peking University: segmentation, part-of-speech tagging, phonetic transcription. J. Chin. Lang. Comput. **13**(2), 121–158 (2003). (in Chinese)
16. Hui, W., Zhan, W.-D., Yu, S.-W.: Structure and application of modern Chinese semantic dictionary. Appl. Lingust. **1**, 134–141 (2006). (in Chinese)
17. Shi, J.-M., Zan, H.-Y., Han, Y.-J.: Specification of the large-scale Chinese lexical semantic knowledge base building. J. Shanxi Univ. Nat. Sci. Ed. **38**(4), 581–587 (2015). (in Chinese)
18. Zhang, K.-L., Zan, H.-Y., Chai, Y.-M., et al.: Survey of the Chinese function word usage knowledge base. J. Chin. Inf. Process. **29**(3), 1–8 (2015). (in Chinese)
19. https://www.ltp-cloud.com/

BIPRE: Bi-directional Inter-Personal Relationship Extraction Task

YongSheng Tian[1], Pengyuan Liu[1,2(✉)], and Bingjie Du[1]

[1] School of Information Science, Beijing Language and Culture University, 15th Xueyuan Road, Haidian District, Beijing 100083, China
liupengyuan@pku.edu.cn

[2] Center of National Language Recourse and Monitoring Research Print Media Branch, Beijing Language and Culture University, 15th Xueyuan Road, Haidian District, Beijing 100083, China

Abstract. The existing inter-personal relationship extraction task (the IPRE task) is a one-way task, and it does not take relationship in the opposite direction into consideration. For example, the IPRE task can extract the relationship in "A is the father of B", but in such a circumstance, one still does not know whether the relationship between B to A is a son-father relationship or a daughter-father relationship. For many natural language processing tasks, it is important to identify specific and clear bi-directional relationships, especially in dialogue and question answering systems. Hence, this paper proposes a Bi-directional InterPersonal Relationship Extraction Task (BIPRE), establishes a corresponding dataset, and implements a Bert-based relationship extraction baseline model. Also, we propose a new model which combines the position and gender information of the persons. The experimental results show that our new model is effective and there is still much room for improvement in this task.

Keywords: Bi-directional · Inter-personal relationship extraction · Gender information · Task

1 Introduction

Relationship extraction is an important subtask of information extraction, which aims to extract the relationship between entity pairs when given the entity pairs and sentences containing entity pairs. The character relationship extraction task is a subtask of relation extraction, in which entity pairs are limited to the names of characters (i.e., persons), and the purpose is to extract the relationship between the characters in the entity pair when given the characters' name pair and the sentence containing them. The current character relationship extraction task extends the setting of previous relationship extraction tasks, for which the relationship is unidirectional. To be specific: There is an entity pair (E1, E2). The relationship label R represents the relationship relating character entity E2 to character entity E1. In the sentences *E2 is the father of E1* and *E4 is the sister of E3*, E1 ~ E4 correspond to four character-entities. The relationship between the two members of the character-entity pair (E1, E2) is generally set as father, while the relationship

M. Liu et al. (Eds.): CLSW 2020, LNAI 12278, pp. 640–651, 2021.
https://doi.org/10.1007/978-3-030-81197-6_54

between the members of the character-entity pair (E3, E4) is generally set as sister. This relationship is set in one direction, which only represents the relationship relating one character to another. This unidirectional relationship is not complete, because we do not know if E1 is a son or daughter of E2, or whether E3 is a brother or a sister of E4. Clarifying the complete relationship between characters and entities is very important for many natural language processing tasks, especially in Q&A and dialogue systems.

Moreover, from the perspective of the model, there are often a lot of non-target characters in the sentence, which could bring lots of noise to the model. For example, in the sentence: *Cao Zhi, a character of The Three Kingdoms period, whose father is Cao Cao and whose brother is Cao Pi*, if we only extract the relationship between *Cao Zhi* and *Cao Cao*, the relationship between *Cao Pi* and the other characters would be missed. Furthermore, gender information is significant in the determination of bidirectional relations. For example, in the sentence: *Cao Chong is the son of Cao Cao and Madam Huan*, without knowing the genders of the characters *Cao Cao* or *Madam Huan*, it would be difficult to get a correct bidirectional relationship between *Cao Chong* and them. If gender information is available, the range of the correct relationships can be narrowed down.

This paper proposes a Bi-directional Relationship Extraction task: BIPRE (Bidirectional Inter-personal Relationship Extraction), in which the original unidirectional relationship is accurately accessed through Bi-directional Relationship Extraction. As shown in Fig. 1. The father-son relationship indicates that character 1 is the father of character 2, while character 2 is the son of character 1. In accordance with this model, the corresponding data sets are constructed manually, and the baseline model based on Bert Bi-directional relationship extraction is realized. Based on the baseline model, the gender and location information of the characters are introduced simultaneously, and a new model is established, which is specially used for the task of bidirectional character relationship extraction.

The contributions of this paper are as follows:

1) Proposing a new task, namely a bi-directional character relation extraction task, and establishing the corresponding data set;
2) establishing a bi-directional relationship baseline model based on Bert and proposing a new model to introduce gender and location information.
3) The F1 values of the model under the conditions of inclusion and non-inclusion are 16.3% and 22.5%, which are higher than those of the baseline model.

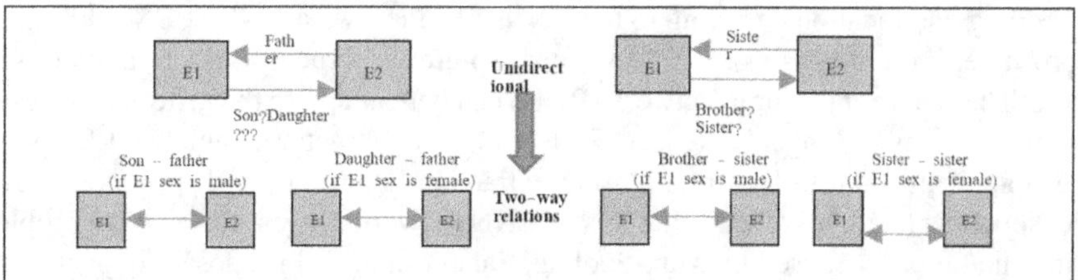

Fig. 1. An example of bi-directional character relationship. In the figure, E1 represents person 1 and E2 represents person 2.

2 Related Works

2.1 Tasks and Data Sets

The most commonly used data sets in Relationship extraction are NYT and ACE-2005. The NYT (New York Times) dataset is a widely used dataset for remote supervised relationship extraction tasks. This data set is generated by aligning the relationships in the Freebase with the New York Times (NYT) corpus. Ace2005 datasets are various types of data published by the Language Data Federation (LDC) consisting of the entities, relationships, and event annotations. There are 7 relationship types in AC-2005, and the data sets include Chinese, Arabic, and English. The main data set of character relationship extraction is the CCKS2019-IPRE data set. The training data are obtained through remote supervision, and the verification and test data are obtained through manual annotation. Table 1 shows the basic statistics of these datasets.

Table 1. ACE-2005 (ACE-2005 dataset can be downloaded at: http://curtis.ml.cmu.edu/w/courses/index.php/ACE_2005_Dataset), NYT, CCKS2019-IPRE data sets [1]. The units of the ACE-2015 dataset are words, and the units of the remaining datasets are sentences.

Data sets	Language	Training sets	Test set	Relationship
ACE-2005	English	300,000	50,000	7
	Arabic	150,000	50,000	7
	Chinese	300,000	50,000	7
NYT	English	522,611	172,448	53
CCK S2019-IPRE	Chinese	287,400	38,417	35

2.2 Models and Methods

The Neural Network: In recent years, the neural network model has shown excellent performance in the relation extraction task. The Convolutional Neural Network (CNN) is the first neural network model applied to relational extraction. To introduce the location information of entities, Zeng et al. applied the combined representation of word embedding and location encoding to the CNNs [2]. Piecewise-CNN (PCNN) [3] made improvements on the basis of CNN to better capture the structural information of sentences. The Recurrent Neural Network (RNN) can better capture the historical information of sentences in longer sentences. RNN with CNN structure, namely CLDNN [4], is also applied to the task of relationship extraction. Hinton et al. [5] proposed Capsule Networks to address the limitations of CNN networks. Then, to solve the problem of information loss caused by Max-Pooling, Sabour et al. [6] proposed the method of replacing Max-Pooling with Dynamic Routing on the capsule network. Capsule networks have been used in NLP fields in recent years, such as in the text classification tasks [7]. Based on the capsule network, Xinsong Zhang [8] combined an attention mechanism with the Capsule Network, which significantly improved the effectiveness of the multi-tag relationship extraction task.

Remote Supervision: Most Neural Network relationship extraction tasks require a lot of labeled data for training, but obtaining such a large amount of annotated data is very costly in manpower and material resources. To solve this problem, a method of aligning plain texts based on knowledge graph is proposed to obtain a large amount of training data [9] However, the data obtained by remote supervision contains a large amount of noise. To solve the noise problem caused by remote supervision, a method is proposed to select the most representative sentence among all sentences extracted by remote supervision for an entity pair. Although this method effectively reduces the noise caused by remote supervision, it wastes a lot of useful information. Therefore Yankai Lin et al. [10] proposed a method of giving attention mechanisms with different weights to different sentences in the package to solve the problem of information waste caused by selecting only one sentence in the package. In 2018 Qin P. et al. [11] proposed to use reinforcement learning to de-noise the data generated by remote supervision and improve the classification performance of the model.

The Pre-training Model: The pre-training model has a long history. In recent years, some pre-training models such as CoVe [12], ELMo [13], Bert [14] and so on have shown good performance in various tasks of natural language processing. Bert, a pretraining model based entirely on the attention mechanism, is a very popular pretraining language model and has become the best baseline model for relationship extraction tasks. However, using Bert directly on the native corpus for relationship extraction cannot achieve good results, because relationship extraction is different from other natural language processing tasks, and entity words and the position of entity words are of great importance in relationship extraction. So, the method of entity position information enhancement is applied in relation classification based on Bert [15]. This method effectively avoids the influence of relationship information between other entity words and non-target entity words in the sentence and significantly improves the performance of relationship extraction.

3 Data Set Building

For the sake of the direct comparison with the unidirectional relationship extraction task, the bi-directional character relationship data set used in this paper is based on the CCKS2019IPRE data set. The construction steps are as follows:

1) *Relationship and Instance Selection.* There are 35 kinds of relationships in the data set of CCKS2019-IPRE. This paper selects 8 kinds of relationships and examples that can be easily extended into bi-directional relationships.
2) *Expanding into bi-directional relationships.* A Knowledge Graph and a Search Engine are used to mark the gender information of the characters. On this basis, the character relationships are extended to 17 kinds. The relationship label is composed of two words. The first word represents the relationship that relates the first character to the second character, and the second word represents the relationship

that relates the second character to the first character. For example, there is a sentence: 'Cao Chong, son of Cao Cao and Lady Huan, courtesy name Cangshu, was born at the end of the Eastern Han Dynasty at Qiao, administration centre of the province of Yuzhou in the Eastern Han Dynasty (now Bozhou)'. In this sentence, the first character is *Cao Chong* and the second character is *Lady Huan.* The original unidirectional relationship is son. The gender of Lady Huan obtained from the Knowledge Graph or with the Search Engine is female. So we can set the relationship tag of the two characters as son-mother.

3) *Manual tagging.* Each piece of data was marked by three people, and two rounds of marking were carried out. The average consistency rate of first-round of labeling was 96.1%. Then, all the data with inconsistent labels were annotated in the second round, mainly by querying the search engine. The second annotation had a 100% consistency rate.

As shown in Table 2, there are a total of 5,347 training data items which include 3,889 irrelevant data items (NA), and 3,053 test data items which include 2,500 irrelevant data items. We also manually constructed a name-gender dictionary with 2,604 entries for all the names (all celebrities) in the data set through consulting encyclopedia and querying the search engine.

Table 2. BIPRE dataset relationship statistics. Original unidirectional relationships are the original character relationships extracted from CCKS. Bi-directional relationships are the corresponding bidirectional relationships in the data set established in this paper, that is, the extensions of the original relationships.

Original one-way relationship	Unidirectional relationship	Number of training sets	Number of test sets and validation sets
NA	NA	3889	2500
Father	Son - father	544	202
	Daughter - father	82	57
Mother	Son - Mother	82	31
	Daughter - Mother	87	57
Son	Father - Son	60	21
	Mother - Son	157	67
Daughter	Father - Daughter	57	33
	Mother - Daughter	48	10
Younger Brother	Elder Sister - Younger Brother	16	4
	Elder Brother - Younger Brother	61	17

(continued)

Table 2. (*continued*)

Original one-way relationship	Unidirectional relationship	Number of training sets	Number of test sets and validation sets
Younger Sister	Elder Brother - Younger Sister	32	7
	Elder Sister - Younger Sister	37	10
Elder Sister	Younger Brother - Elder Sister	18	7
	Younger Sister - Elder Sister	35	12
Elder Brother	Younger Brother - Elder Brother	100	24
	Younger sister - Elder Brother	42	11
Total	/	5347	3053

4 Method

4.1 Overview

To facilitate processing, this paper combined the package-level data obtained by remote supervision into long sentences and introduced additional information on the genders and positions of the characters. Finally, we classified the long sentences that carried additional information with pre-training model Bert and FNN. The model overview is shown in Fig. 2.

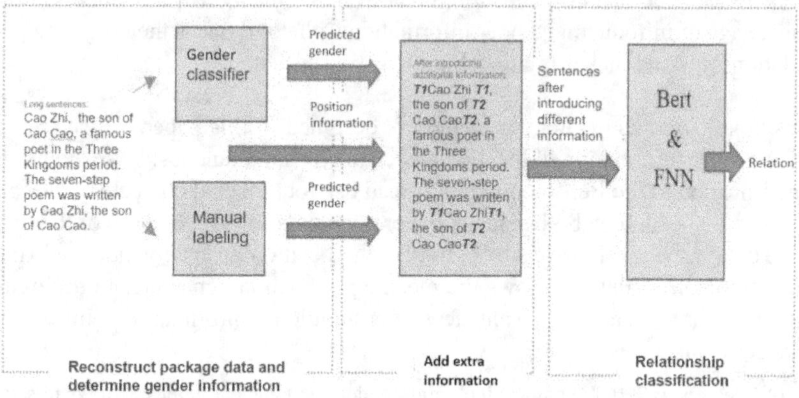

Fig. 2. Overview of the bi-directional relationship extraction process. T1 represents the information introduced by character 1 and T2 represents the information introduced by character 2. Specific details can be seen in Table 3 and 4.

4.2 Re-package Data

After examining the *CCKS2019-IPRE* data set, we found that the quality of the package-level data is higher than that of the sentence-level data, and most of the relationships in the sentence-level data do not correspond to their labels, which was not obvious in the package-level data. So, we experimented with package-level data and spliced all the sentences in each package into one long sentence for ease of processing. Since Bert allows a maximum length of input of 512, we only kept the first 500 characters of the spliced sentence.

4.3 Gender Information Confirmation

As described in Sect. 3 of this paper, we manually annotated gender information. At the same time, we also trained a gender classifier[1] with a large corpus of names with gender information. It could then predict the genders of characters when given their names. We then compared the predicted sex and the actual sex.

4.4 Information Introduction

There are two main challenges in the bi-directional character relationship extraction task:

1) The location of the target character is uncertain: There are a large number of entity words and relationships of non-target characters in the data set, which will bring great noise to relationship extraction. Therefore, it is necessary to introduce the location information of the target.
2) The lack of relationship information. Many sentences in the original one-way relationship data sets often lack information on bi-directional relationships. To make up for this defect and extract complete bidirectional character relationships, we introduced easily accessible gender information through manual tagging, and pre-trained a gender classifier with a gender-name corpus. Predicted genders and the real genders in our relational classification model were compared experimentally. Another advantage of introducing gender information is that it reduces the range of optional relationships and makes relationship extraction easier.

A simple automatic matching method is designed in this paper, which can introduce the gender information of location and character simultaneously. We did not make any modification to the Bert model, but instead use four labeled characters: #, ￥, &,* , which were embedded on both sides of the entity words in the sentences of the training and test data respectively, so as to introduce the position and the gender information of the entity words. Table 3 shows the meaning of each tag character in different circumstances. Table 4 shows the sentences after additional information is introduced in different circumstances.

[1] We collected about 300,000 names, including modern and ancient names, from data sets containing gender information, and used a classifier that could predict the gender of characters when given their names through FNN + CNN training. The F1 performance of the classifier was 85.6%.

Table 3. Marked characters on both sides of a character entity and their meanings in different situations.

Experimental group	Marked character	Meaning
Introduction of gender location information	￥	The character is the first one, and the sex is male.
	#	The character is the first one, and the gender is female.
	*	The character is the second one, and the gender is male.
	&	The character is the second one, and the gender is female.
Position information only	#	The character is the first character.
	&	The character is the second character.
Gender information only	#	The gender is male.
	&	The gender is female.

4.5 Relationship Classification

The pre-training model Bert gives very good performance on the task of relationship extraction. In this paper, the reconstructed additional information data introduced were encoded directly using Bert, and the coding representing the whole sentence was taken out and classified using the FNN network.

5 Experiments

To verify the effectiveness of the proposed method, a number of controlled ablation experiments were designed to verify the influence of the introduction of gender and location information on the bidirectional relationship extraction, as well as the gap between the prediction and the real gender information. Not only did we take Bert implemented in this paper as the baseline model to conduct multiple ablation experiments on the BIPRE bidirectional character relationship extraction data set, but we also compared it with the baseline model officially provided by CCKS2019 for unidirectional character relationship extraction evaluation. CNN+ATT is the baseline model provided by the authorities, based on the paper *Neural relation extraction with selective attention over Instances* published by *Yankai Lin et al.* in 2016 [10].

Table 4. Raw sentences and sentences with additional information. The markers on the two sides of the entities represent different additional information. Refer to Table 3 for specific meanings.

Original sentence	Example 1: Cao Chong, courtesy name Cang Shu, eastern han dynasty last years, the son of cao cao and Madam Huan. Example 2: Huang Zhiqing, female, is Huang Rihua and his wife Liang Jiehua's daughter.
Introduction of gender location information	Example 1: ¥Cao Chong¥, courtesy name Cang Shu, eastern han dynasty last years , the son of * Cao Cao* and Madam Huan. Example 2:# Huang Zhiqing #, female, is Huang Rihua and his wife & Liang Jiehua &'s daughter.
Position information only	Example 1:# Cao Chong # , courtesy name Cang Shu, eastern han dynasty last years , the son of & Cao Cao & and Madam Huan . Example 2:# Huang Zhiqing #, female, is Huang Rihua and his wife & Liang Jiehua &'s daughter.
Gender information only	Example 1:# Cao Chong # , courtesy name Cang Shu, eastern han dynasty last years , the son of # Cao Cao # and Madam Huan. Example 2:& Huang Zhiqing &, female, is Huang Rihua and his wife & Liang Jiehua&'s daughter.
No additional information	Example 1: Cao Chong , courtesy name Cang Shu, eastern han dynasty last years, the son of Cao Cao and Madam Huan . Example 2: Huang Zhiqing, female, is Huang Rihua and his wife Liang Jiehua's daughter.

5.1 Parameter Setting

Experimental hyper-parameters are shown in Table 5.

Table 5. Hyper-parameters of bidirectional character relationship extraction experiment.

Parameters	Value	Note
Learn rate	$5*10^{-5}$	Learning rate
Epoch num	5	Number of iterations
Batch size	8	Batch size
Sentence max length	256	The maximum length of input sentence
Hidden state dimension	768	Bert hidden state dimension

5.2 Experimental Results

The experimental results are shown in Table 6. It can be seen that in the unidirectional relationship extraction task, Bert-based models give excellent performance, but in the bi-directional relationship extraction task, the performance of the model drops dramatically. When location and gender information are not introduced into the model, the experimental results are very poor in the bi-directional character relationship extraction task. The F1-MICRO value is only 16.9%, almost half of the previous unidirectional performance. Even with the introduction of position information, performance degradation is also significant (from 47.8% to 32.7%). This is consistent with our guess. Bi-directional relationship extraction is more difficult than unidirectional. With location and gender information added, the performance of the model was greatly improved under various conditions. In the case of NA/NA inclusion, F1 values were improved by 16.3% and 22.5%, respectively. This shows that location and gender information of the characters are very important for bi-directional character relationship extraction. Location information is more important for the results of ablation experiments that introduced location and gender information (+13.4/+11 vs. +7.4/+5.9, 11.3/+8.2). Model performance was optimal after simultaneously introducing gender and location information of the characters. Moreover, the model performed better with real gender information marked manually than with automatic gender information (+11.3/+8.2 vs. +7.4/+5.9).

Table 6. Bi-directional Inter-Personal Relationship Extraction Task experimental results. PG represents predicted gender information and TG represents manually labeled true gender information. CCKS2019-IPER experimental results are directly cited in the existing experimental results.

Data sets	Model	F1-micro (including NA)%	F1-micro (excluding NA)%
CCKS2019-IPRE (35 types)	CNN+ATT	/	32.1
	Bert[3] (ours)	80.6	30.4
	Bert+pos (ours)	84.3	47.8
BIPRE (17 types)	Bert (ours)	40.4	16.9
	Bert+Pos (ours)	53.8 (+13.4)	27.9 (+11)
	Bert+PG (ours)	47.8 (+7.4)	22.8 + (5.9)
	Bert+TG (ours)	51.7 (+11.3)	25.1 (+8.2)
	Bert+Pos+PG (ours)	54.3 (+13.9)	32.7 (+15.8)
	Bert+Pos+TG (ours)	56.9 (+16.3)	39.4 (+22.5)

6 Conclusions and Future Work

This paper has proposed a bi-directional character relationship extraction task and established the corresponding data set. For this task, based on Bert, this paper has introduced gender and location information to assist the extraction of bidirectional character relationships and conducted experiments and analysis. The experimental results show that compared with the baseline model, the performance of this model has been improved significantly and reached the current optimal level. Besides, the location information and gender information of character entities can effectively improve the performance of character relationship extraction. At the same time, the best F1 values of this task are 56.7% and 39.4% respectively, which indicates that this task still has a lot of room for improvement. On this basis, the next step is to sort out and investigate the relationship extraction task, to find out whether there is a similar bidirectional relationship among relationships other than those between characters, and to propose corresponding tasks and establish an open data set for academic research.

Acknowledgments. This work is supported by the Research Funds of Beijing Language and Culture University 20YCX151(Fundamental Research Funds for the Central Universities) and Beijing Natural Science Foundation(4192057).

References

1. Wang, H., He, Z., Ma, J., Chen, W., Zhang, M.: Ipre: a dataset for inter-personal relationship extraction. In: Tang, J., Kan, M.-Y., Zhao, D., Li, S., Zan, H. (eds.) NLPCC 2019. LNCS (LNAI), vol. 11839, pp. 103–115. Springer, Cham (2019). https://doi.org/10.1007/978-3-030-32236-6_9
2. Zeng, D., Liu, K., Lai, S., et al.: Relation classification via convolutional deep neural network. In: Proceedings of COLING 2014, the 25th International Conference on Computational Linguistics: Technical Papers, pp. 2335–2344 (2014)
3. Zeng, D., Liu, K., Chen, Y., et al.: Distant supervision for relation extraction via piecewise convolutional neural networks. In: Proceedings of the 2015 Conference on Empirical Methods in Natural Language Processing, pp. 1753–1762 (2015)
4. Zhu, Q., Zhou, X., Tan, J., et al.: Knowledge base reasoning with convolutional-based recurrent neural networks. IEEE Trans. Knowl. Data Eng. (2019)
5. Hinton, G.E., Krizhevsky, A., Wang, S.D.: Transforming auto-encoders. In: Honkela, T., Duch, W., Girolami, M., Kaski, S. (eds.) ICANN 2011. LNCS, vol. 6791, pp. 44–51. Springer, Heidelberg (2011). https://doi.org/10.1007/978-3-642-21735-7_6
6. Sabour, S., Frosst, N., Hinton, G.E.: Dynamic routing between capsules. Adv. Neural Inf. Process. Syst. 3856–3866 (2017)
7. Zhao, W., Ye, J., Yang, M., et al.: Investigating capsule networks with dynamic routing for text classification. arXiv preprint arXiv:1804.00538 (2018)
8. Zhang, X., Li, P., Jia, W., Zhao, H.: Multi-labeled relation extraction with attentive capsule network. Proce. AAAI Conf. Artif. Intell. **33**, 7484–7491 (2019)
9. Mintz, M., Bills, S., Snow, R., et al.: Distant supervision for relation extraction without labeled data. In: Proceedings of the Joint Conference of the 47th Annual Meeting of the ACL and the 4th International Joint Conference on Natural Language Processing of the AFNLP, pp. 1003–1011 (2009)
10. Lin, Y., Shen, S., Liu, Z., et al.: Neural relation extraction with selective attention over instances. In: Proceedings of the 54th Annual Meeting of the Association for Computational Linguistics (vol. 1: Long Papers), pp. 2124–2133 (2016)
11. Qin, P., Xu, W., Wang, W.Y.: Robust distant supervision relation extraction via deep reinforcement learning. arXiv preprint arXiv:1805.09927 (2018)

12. McCann, B., Bradbury, J., Xiong, C., et al.: Learned in translation: contextualized word vectors. Adv. Neural Inf. Proc. Syst. 6294–6305 (2017)
13. Peters, M.E., Neumann, M., Iyyer, M., et al.: Deep contextualized word representations. arXiv preprint arXiv:1802.05365 (2018)
14. Devlin, J., Chang, M.W., Lee, K., et al.: Bert: Pre-training of deep bidirectional transformers for language understanding. arXiv preprint arXiv:1810.04805 (2018)
15. Peng, C.: A Bert Based Relation Classification Network for Inter-Personal Relationship Extraction. CCKS2019-shared task (2019)

Building a Pediatric Medical Corpus: Word Segmentation and Named Entity Annotation

Zan Hongying[1,3](✉), Li Wenxin[1](✉), Zhang Kunli[1,3], Ye Yajuan[1], Chang Baobao[2,3], and Sui Zhifang[2,3]

[1] School of Information Engineering, Zhengzhou University, Zhengzhou 450001, Henan, China
{iehyzan,ieklzhang}@zzu.edu.cn, wencyli@gs.zzu.edu.cn

[2] Key Laboratory of Computational Linguistics of Ministry of Education, Peking University, Beijing 100871, China
{chbb,szf}@pku.edu.cn

[3] The Peng Cheng Laboratory, Shenzhen 518052, Guangdong, China

Abstract. Word segmentation and named entity annotation are essential foundations for medical text information extraction. This paper focuses on clinical pediatric diseases and takes the existing medical named entities and entity-relationship labeling systems as references. Under the guidance of the Chinese word segmentation and named entity labeling, the specifications for pediatric medical texts have been constructed in this paper. This paper also applies a self-developed distributed annotation platform to pre-annotate and manually proofread the named entities for many times. The corpus consists of 38,805 medical entries which can be divided into nine categories. Among the medical entries, there are 504 entries of common pediatric diseases, 7,085 entries of body parts, 12,907 entries of clinical manifestations, and 4,354 entries of medical procedures. This paper constructs the largest corpus with pediatric medical word segmentation and named entity annotation, which provides a data basis for related research.

Keywords: Medical text · Medical word segmentation · Named entities · Annotation norms · Corpus construction

1 Introduction

Chinese word segmentation and named entity recognition are essential tasks in information extraction. In the medical field, medical text information processing technology has become a new research field. Medical text contains a lot of medical-related information such as case summary, admission records, inspection and test reports, and drug use. Most of this information is written in narrative text summaries, which are essential resources for disease prediction, drug recommendation, and auxiliary diagnosis and treatment. Most of these texts are written by medical professionals and contain a lot of professional medical knowledge. Through analysis, medical knowledge closely related to patients can be mined. For example, in the medical phrase "nitrogen qualitative thematic disease after gastrointestinal massive hemorrhage, blood urea nitrogen often rise", "nitrogen qualitative thematic disease" is a disease, "gastrointestinal massive hemorrhage", "blood urea

M. Liu et al. (Eds.): CLSW 2020, LNAI 12278, pp. 652–664, 2021.
https://doi.org/10.1007/978-3-030-81197-6_55

nitrogen often rise" are symptoms, "gastrointestinal" describes a body part, "blood urea nitrogen" is a substance measured in a medical test. These are called named entities in medical information extraction research, which are to be automatically mined from medical texts. We need to automatically identify all kinds of named entities in medical texts which are closely related to patients' health. In recent years, the application of natural language processing, information extraction and other technologies to medical text for clinical research has attracted a lot of attention. This process is mainly divided into two stages, namely natural language processing and information extraction. Natural language processing mainly focuses on the preprocessing of medical texts, including word segmentation, sentence boundary recognition, part-of-speech tagging, syntactic analysis and so on. Based on natural language processing, information extraction mainly focuses on the identification of various named entities or medical concepts, which are the medical knowledge in medical texts.

At present, many research institutions have compiled knowledge bases of medical resources. For example, the medical term system UMLS [1], which has been constantly under construction at the National Medical Library of the United States for 20 years, contains about 2 million medical concepts and more than 5 million medical words. Neveol et al. constructed an annotated corpus of named entities using machine annotation and manual proofreading. The corpus contains 15 types of objects [2]. East China University of Science and Technology and Shanghai Shuguang Hospital constructed the TCM knowledge map [3]. The above resources have provided formal descriptions of medical knowledge, which are widely used in therapeutic areas such as health services, medical literature classification, and clinical work. Furthermore, they have a milestone significance in the development of the Chinese medical field.

Word segmentation processing and named entity recognition are the foundation of in-depth semantic analysis or mining. Chinese word segmentation is a classical problem of Chinese information processing [4]. Most of the early Chinese word segmentation methods are based on dictionaries, and these methods rely on the dictionary size. Methods proposed later are mainly rule-based and statistics-based [5]. Xu Dong has built a cross-domain core medical dictionary, designed a data-driven framework using unsupervised algorithms and SVM, and processed Chinese writing conventions to convert free texts from electronic records into structured texts for describing the events [6]. There are many popular Chinese word segmentation technologies, such as the N-shortest path method, the word-based n-gram syntax model method, and neural network technology [7]. Annotated Chinese corpora usually need manual proofreading. Errors are inevitable, and even annotation and word segmentation results obtained in the same context can be inconsistent. Consistency testing is also an important step in constructing corpus resources. In recent years, consistency testing has mainly focused on the research of word segmentation, part-of-speech, and part-of-speech tagging. Han Dongxu used chi-square statistics and boundary entropy, which are combined with bootstrap learning technology, to improve the ability to process unregistered words [8]. This method has effectively improved the effect of cross-domain Chinese word segmentation and increased the recall rate of registered words slightly. Based on the Gaussian distribution characteristics of word length noise, Zhang Yi proposed to use the word length feature of context as the feature for word segmentation, and used the CRF model to improve the accuracy of word segmentation [9]. Peng Pai proposed a probabilistic language model for Chinese word cohesion [10]. Based on word order modeling analysis, Chinese words are converted into digital vectors, and the similarity of Chinese synonyms in digital space features is ensured. Tu Wenbo used the PCNN model to classify words based on the upper and lower windows of word vectors [11]. They have verified that the network model without a pooling layer is superior to the model with a pooling layer.

Medical texts contain medical knowledge, and how to effectively use this information is the main subject of current research. This study will provide support for clinical decisions and drug recommendation, so as to improve the quality of medical services. The construction of a medical named entity corpus will lay a solid foundation for this study. In this paper, a labeling system suitable for naming entities in pediatric medical texts is proposed, and the corpus is from Clinical Pediatrics [12]. We adopt an annotation standard suitable for pediatrics and formulate detailed labeling norms. Having been constantly optimizing labeling norms under the guidance of professional doctors, the largest named entity corpus for pediatric medicine has been constructed. The purpose of the annotation standard is to construct a high-quality medical corpus knowledge map. It mainly aims at the two tasks of named entity recognition and textual segmentation of medical texts. Currently, word segmentation proofreading has been completed for the corpus and nine types of medical entities have been annotated. According to our statistics, the corpus contains nearly 2.2 million words, 47,194 sentences and 38,805 named entities.

The structure of this paper is as follows. Section 2 introduces the specifications of the medical text word segmentation and named entity annotation system. Section 3 introduces the proofreading process and a summary of the problems during such process. Section 4 conducts data statistics and analysis of the labeled corpus. The final section concludes this paper.

2 Word Segmentation and Named Entity Annotation System for Medical Texts

2.1 Medical Text Word Segmentation

The quality of medical text segmentation directly affects whether the text content can be understood correctly [13], and it also has great significance for further data mining research. The specific process of word segmentation is to let the computer add a word breaker between words in Chinese text [14]. This paper uses spaces as word breakers.

In previous works on word segmentation, the standard commonly used by researchers is "Peking University Modern Chinese Corpus Basic Processing Specification (2002 Edition) [15]", which uses 26 part-of-speech codes and adds classification marks for proper nouns, morpheme subclasses, verbs, and adjective special usage. Combining the existing medical named entities and entity-relationship labeling standards, this paper develops a detailed labeling standard for naming entities and word segmentation labels under the guidance of medical experts. In such standard, which is suitable for clinical pediatrics, named entities in medical text are divided into nine categories. The entity division criteria are introduced in Sect. 2.2.

The basic principles of word segmentation in Chinese medical text are shown as follows:

(1) Use a half-width space " " as a separator between words.
(2) There are nine types of entities. Except for the "clinical manifestation" entity, all medical entities are treated as a word segmentation unit. And they are required to be separated from other words by a word segmentation symbol.
(3) Segmentation may be allowed in the entity of "clinical manifestation", but not in other entities.

An annotated sample is presented below:
Original text: In an emergency, tracheal foreign body leads to severe dyspnea and asphyxia, or children dental tightly closed cannot open, can consider tracheal incision Open and Cylar puncture to improve ventilation.

Text after labeling: < In an emergency, [tracheal foreign body]dis leads to [severe dyspnea]sym and [asphyxia]sym, or children [[dental]bod tightly closed cannot open]sym, can consider [tracheal incision Open]pro and [Cylar puncture]pro to improve ventilation. >

2.2 Classification of Medical Named Entities

The construction of the entity resource database is based on Medical Subject Headings (MeSH) [16]. MeSH is an authoritative medical subject vocabulary compiled by the National Library of Medicine. It is a standardized and extensible dynamic thesaurus.

Named entity recognition is the first step in information extraction [17]. This study divides the appointed entities of medical text into nine categories, including "disease", "clinical manifestations", "drugs", "medical equipment", "medical procedures", "body", "medical test items", "microorganisms", "Department", and uses different reference standards to define the scope of each entity.

Disease (dis): the cause of the patient's unhealthy symptoms or the doctor's diagnosis, mainly including illnesses or syndromes, poisoning or injuries, and organ cell damage.

Clinical manifestation (sym): the patient's unhealthy performance due to a particular disease, including the patient's abnormal sensations or the results of examinations as measured with medical instruments. This mainly includes symptoms and signs. For example, a patient says that he is dizzy. This is a symptom or clinical manifestation, and the doctor uses a sphygmomanometer and finds that the patient's blood pressure increases which is also the clinical manifestation.

Medical procedure (pro): measures, methods, and systems for treatment or diagnosis. This mainly includes inspection procedures, surgery, and prevention procedures. For example, "auscultation of both lungs" belongs to examination, and "intravenous injection" belongs to treatment.

Medical equipment (equ): medical tools and instruments used by doctors during diagnosis and treatment. This mainly includes inspection equipment and treatment equipment.

Drugs (dru): substances used by doctors to diagnose, treat and prevent diseases.

Physical medical examination item (ite): refers to the physiological measurement and humoral examination items, such as blood routine examinations, leukocyte examinations.

Body (bod): human body tissues, cells, organs, limbs, including substances produced by or dissected from the human body, such as human milk and human feces. This mainly includes body parts and body substances.

Department (dep): departments established by medical institutions, such as pediatrics, ophthalmology.

Microorganism (mic): small protists such as bacteria, viruses, fungi, and also toxins, enzymes, and other substances produced by organisms.

Regarding the labeling of named entities in medical texts, the annotators strictly follow the "Medical NER Standard [18]" and use "[named-entity]tag" to mark them tightly. There is no space between the left or right bracket, and the first or last character of the marked medical entity. And the brackets must appear in pairs. In Table 1, the subcategories and formal labels of the nine medical named entities are shown.

Table 1. Medical named entity classification.

Named entity	Subclass	Label
Disease	Disease or syndrome	Dis
	Poisoned or injured	
	Organ or cell damage	
Symptom	Symptom	sym
	Signs	
Procedure	Inspection procedures	pro
	Treatment or prevention procedures	
Equipment	Inspection equipment	equ
	Treatment equipment	
Drug	Drug	dru
Medical examination items	Medical test items	ite
Body	Body parts	bod
	Body substances	
Department	Department	dep
Microorganism	Microorganism	mic

The basic principles for labeling medical named entities include:

(1) "Clinical manifestation" entities are allowed to contain other entities.
(2) Medical entities other than "clinical manifestation" entities follow the "label-the-biggest-unit method" when labeling, that is, if an entity contains other entities, only the biggest entity needs to be labeled.

(3) Necessary punctuation marks can be included inside the entity, and the entity can be a word, a phrase or a sentence.

An annotated sample is given below:

Original text: Respiratory Muscle Paralysis and Respiratory Center Involved patients can be complicated by pneumonia symptom pneumonia, pulmonary atelectasis and so on.

Text after labeling: < [[Respiratory Muscle]bod Paralysis]sym and [[Respiratory Center]bod Involved]sym patients can be complicated by [pneumonia]sym symptom [pneumonia]dis, [pulmonary atelectasis]dis and so on. >

3 Proofreading Process and Problem Summary

3.1 Labeling Process

The development of medical text word segmentation and named entity annotation norms requires professional knowledge. The standards referenced in this work are from the Institute of Computational Linguistics of Peking University, which uses multiple rounds of annotation mode to provide feedback and annotation work on annotation norms [19]. The flow chart of the labeling process is shown in Fig. 1.

In the first stage, tagging personnel is organized to learn medical text word segmentation and named entity tagging specifications, and an entity resource library is built. At the same time, labeling personnel is organized to carry out pre-labeling, the purpose is to get familiar with the medical text word segmentation and entity labeling specifications and to collect problems that occur in the actual labeling of medical texts. After two rounds of pre-labeling, the labeling standard is further improved to make the labeling standard more adequate for the research task.

In the second stage, the medical entity resource library built in the first stage is used to label named entities in medical text on the annotation platform. The labeling process adopts a multi-round iteration mode. That is, each medical text is handled by two labeling assistants. After the first assistant completes the labeling task, the places with doubts are recorded, and then the second assistant is responsible for checking and documenting the inconsistencies and uncertainties. A unified solution is obtained by consulting with medical experts. After the discussion, the first assistant is responsible for revising the corpus to form the final documents. Then, the labeling specifications will be modified according to the feedback from the labeling staff, so that the labeling specifications will be more applicable to medical texts.

The third stage is to proofread the word segmentation. Labeling tools are developed which is used to modify word segmentation. Annotation personnel correct word segmentation errors on the new labeling platform and check whether the entity labeling conforms to the currently updated specifications. At the same time, problems such as missing or misaligned entities are also checked to improve the quality of labeling.

3.2 The Annotation Tool

To improve the labeling efficiency, a medical text entity labeling platform is developed [20]. On the labeling platform, different medical entities can be selected for labeling

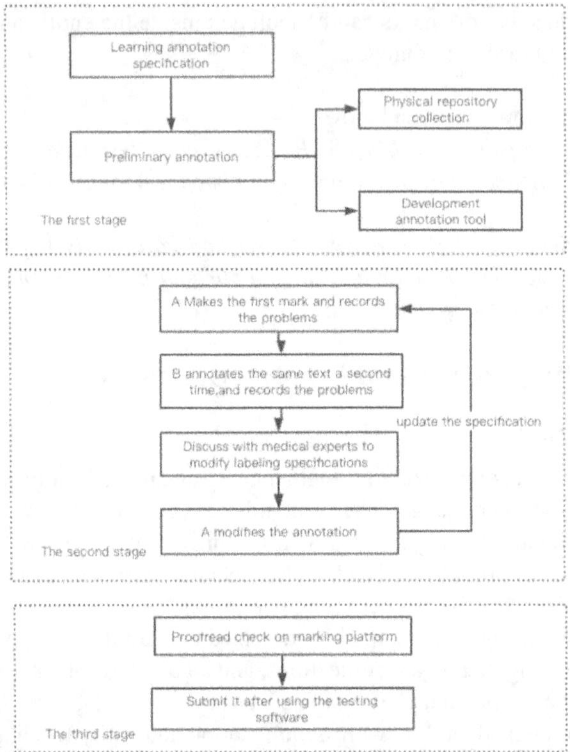

Fig. 1. Process of annotation in three stages: preparation of annotation tool, annotation of named entities, annotation inspection and proofreading

entities in the texts of clinical pediatrics. Among them, different types of medical objects are represented by different color blocks to facilitate distinction. The platform also provides file management functions that can view the progress of the first and second texts and facilitate multiple labeling rounds. The labeling platform is shown in Fig. 2.

3.3 Problems and Solutions

Four problems were found during the labelling task and the corresponding solutions are provided.

1) Entity types were confusing. The same entity could be labeled as a disease or clinical manifestation, and it was often impossible to determine which category the entity belongs to. The solution of this paper is to emphasize the boundary range of each entity by elaborating on the specification. The category of an entity can be determined according to its part of the speech. An annotated sample is provided below:
The text after labeling: *<clinical can be divided into typical and atypical. [[DOC]bod increase]sym due to [[11β-OH]bod deficiency]sym can cause [high blood sodium]dis, [low blood potassium]dis, [alkaline poisoning]dis and [high Blood volume]dis causes [[hypertension]dis symptom]sym.>*

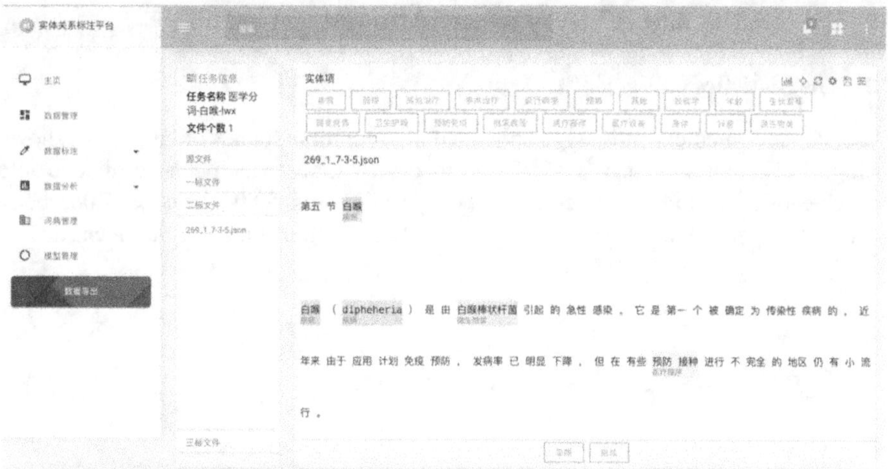

Fig. 2. In the second stage of labeling, we use the medical Entity Identification platform to complete the labeling of named entities.

In this example, "11β-OH deficiency" and "increased DOC" are the physical condition of the patient through a test because these symptoms lead to the four diseases ("high blood sodium", "low blood potassium", "alkaline poisoning", "high blood volume").

2) Whether punctuation marks could be included in the entity was a problem: in certain labeling process, it was found that if there was a lack of punctuation, the integrity of the entity meaning would be lost. An annotated sample is presented below:
The original text: Corrosive poisons can cause skin, facial features, digestive tract and respiratory mucosal damage.
The text after labeling: <Corrosive poison can cause [skin]bod, [facial features]bod, [digestive tract and respiratory mucosal damage]dis.>
The revised text after the specification: <Corrosive poison can cause [skin, facial features, digestive tract and respiratory mucosal damage]dis.>
The solution of this paper is to allow punctuation in the middle of all entity classes while ensuring the integrity of the meaning of the entity.
3) When the physical entity was a single word, such as when "back pain" and "headache" were labeled as clinical manifestations, whether "back" and "head" should be nested and labeled as body was a problem. After discussion, we have decided that it depends on whether the body part is part of the word. No nested tags are required as long as body parts can be combined with the headwords.
The revised text after the specification: <[brain atrophy]sym, [pulmonary edema]sym, [proteinuria]sym>
4) No processing of English data. Since there are no original documents available for our references, and medical-specific English vocabulary is usually more complicated, this paper only corrects the Chinese word segmentation and entity names.

4 The Results and Analysis of the Annotated Corpus

4.1 The Statistics on the Entity Types

The number of medical entity types obtained after multiple rounds of labeling is shown in Fig. 3. The corpus consists of 38,805 medical entities which can be divided into nine categories, including 504 common pediatric diseases, 8,802 diseases, 7,085 body entities, 4,354 medical procedure entities, and 12,907 clinical manifestation entities.

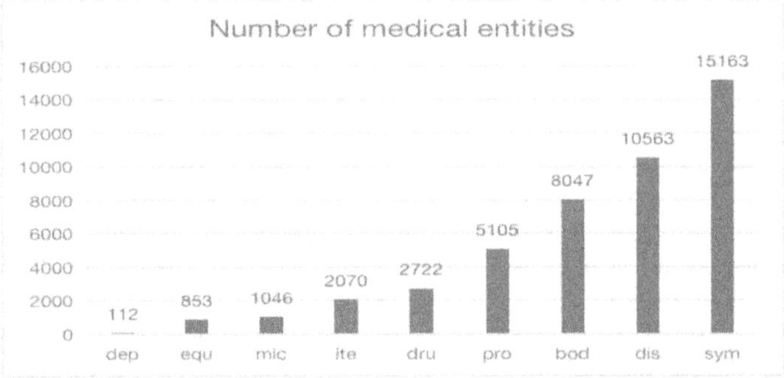

Fig. 3. Statistical information on the number of de-weighted medical entity types.

4.2 The Analysis of Word Segmentation Proofreading

Generally speaking, the accuracy rate is used to measure the accuracy of a system. Accuracy means that when the number of predictions and answers are consistent, the number of correct judgments made by the system is divided by the total number of tests. Generally, the word-level accuracy, recall rate, and F value on the standard data set are used to measure the performance of the word classifier in the Chinese word segmentation task.

When determining the consistency of word segmentation, words can be considered to be consistent only when they have the same starting and ending positions in the text. To test the medical word segmentation after our proofreading, the entity labels are removed, and then the final segmented text is compared with the original medical file obtained initially. According to the statistics, for most of the word segmentation, the F values of the proofreading are lower than 0.6. After analysis, it is found that most of the entity parts involved in the modifications are "disease," "body," and "virus".

During the proofreading process, it is found that a large part of the disease entities consists of "a certain virus + infection," such as "cytomegalovirus infection," "HCMV infection." In the initial segmentation, "mega cell", "virus," "infection," are considered as three words. When "cytomegalovirus infection" is used as a medical vocabulary during proofreading, many word boundaries will be modified. Besides, since a disease often has many pathological types according to different causes, different degrees of infection, and different infection times, there are some disease entities composed of "disease + type," "onset period/onset location + disease," such as < *RF negative* >, < *acquired immunodeficiency syndrome in childhood*>, < *intrauterine HIV infection* >.

When dealing with body entities, we classify the body parts and the substances produced by people as body entities, so the proofreading errors mainly come from two aspects. First, the description of specific parts of the human body is often composed of more than one word. The common ones include "body organs + tissue cells", "hormones/a certain disease + antibodies" such as < *middle cerebral artery supply area* >, < *cerebrospinal fluid protein* >, < *Fibrous or bone connective tissue cells* >, < *Anti-insulin receptor antibody* >, etc. Second, in medical texts, the substances produced by humans are often composed of "body parts + substances", such as < rhinitis secretions >. The first part is "rhinitis," and the second part is "secretions", which also together form a word. According to our labeling norms, we treat "nasopharyngeal secretions" as a medical entity.

In addition to the two types of entities analyzed above, the other seven types of objects also have similar word segmentation processing, so there are many texts with low consistency in word segmentation. The results of word segmentation proofreading in this paper are quite different from the ordinary word segmentation, but they are relevant in the medical field. The study of Chinese text word segmentation is essential.

4.3 The Statistics on the Performance of Entity Labeling

Annotation consistency is used to describe the degree of agreement between two annotation results. This is generally indicated by Kappa value [21] and F value [22]. In the study of the corpus of manual annotation of entities, F value is usually used to calculate annotation consistency [23]. The specific method is to compare the first and the third annotation texts, it takes the result of the third bid as the answer A, and records the result of the opening bid as A. The calculation formula is (1)-(3):

$$P = \frac{A \cap B}{B_{total}} \tag{1}$$

$$R = \frac{A \cap B}{A_{total}} \tag{2}$$

$$F = \frac{2 \times P \times R}{P + R} \tag{3}$$

When determining entity consistency, it is only consistent when the entity type labels, and the starting and ending positions are identical. When calculating this consistency, the number of all medical entities are firstly counted at the time of the first annotation and such number is called A. In the same way, the number of various medical objects in the three criteria is recorded as B. Then, the final F is obtained. After the statistics of the ultimate entity labels are obtained, the entity recognition consistency rate of the corpus reaches 0.869. This result indicates that the corpus built in this study is reliable [24].

5 Conclusion

In this paper, clinical pediatrics are used as the fundamental corpus for labeling, and the main tasks include word segmentation of medically named entities and labeling proofreading. We analyzed the characteristics of clinical pediatrics, and developed labeling norms for the pediatric corpus. Medical texts were labeled through multiple rounds of iteration, and the labeling norms were improved continuously during the labeling process. This paper describes the labeling system and the corpus-based approach adopted in this paper. We have proofread the word segmentation after the labeling of a medical entity. In January 2020, nine categories of entities had been constructed, including 504 entities of pediatric diseases, 8,802 entities of diseases, 12,907 clinical entities, 7,085 physical entities, and 4,354 entities of medical procedures, which total 38,805 medical entity concepts in all. Based on word segmentation and named entity recognition by manually labeling medical text, this study deals with the medical entity concepts with fine granularity, and lays a foundation for subsequent research of relationship extraction.

Acknowledgement. We gratefully acknowledge the grant supports of the National Key Research and Development Project (Grant No. 2017YFB1002101), Major Program of National Social Science Foundation of China (Grant No. 17ZDA138), Science and Technique Program of Henan Province (Grant No. 192102210260), Medical Science and Technique Program Co-sponsored by Henan Province and Ministry (Grant No. SB201901021), Key Scientific Research Program of Higher Education of Henan Province (Grant No. 19A520003, 20A520038).

References

1. Nadkarni, P., Chen, R., Brandt, C.: UMLS concept indexing for production databases. J. Am. Med. Inf. Assoc. **8**, 80–91 (2001)
2. Névéol, A., Grouin, C., Leixa, J., Rosset, S., Zweigenbaum, P.: The QUAERO French medical corpus: a ressource for medical entity recognition and normalization. In: Fourth Workshop on Building and Evaluating Ressources for Health and Biomedical Text Processing - BioTxtM2014, pp. 24–30 (2014)
3. 阮彤,孙程琳,王昊奋,方之家,殷亦超.中医药知识图谱构建与应用. 医学信息学杂志 **37**(04), 8–13 (2016). Ruan, T., Sun, C.-L., Wang, H.-F., et al.: Construction and Application of TCM Knowledge Graph. J. Med. Inf. **37**, 8–13 (2016) (in Chinese)

4. 孙茂松,左正平,黄昌宁.汉语自动分词词典机制的实验研究.中文信息学报 **01**, 1–6 (2000). Sun, M.-S., Zuo, Z.-P., Huang, C.-N.: Experimental research on the dictionary mechanism of Chinese automatic segmentation. J. Chin. Inf. Process. **1**, 1–6 (2000) (in Chinese)
5. 袁向铎. 基于统计和规则的中文地址分词系统设计与实现.东南大学 (2018). Yuan, X.-D.: Design and Implementation of Chinese Address Word Segmentation System Based on Statistics and Rules. Southeast University, Nanjing (2018) (in Chinese)
6. Dong, X., et al.: Data-driven information extraction from Chinese electronic medical records. PLoS ONE **10**(8), e0136270 (2015)
7. 周寅. 融合深度学习特征与浅层机器学习特征的中文分词关键技术研究. 华中师范大学 (2017). Zhou, Y.: Research on Key Technologies of Chinese Word Segmentation Combining Deep Learning Features and Shallow Machine Learning Features. Central China Normal University, Wuhan (2017) (in Chinese)
8. 韩冬煦,常宝宝.中文分词模型的领域适应性方法.计算机学报 **38**(02), 272–281 (2015). Han, D.-X., Chang, B.-B.: Domain adaptation method of Chinese word segmentation model. Chin. J. Comput. **38**, 272–281 (2015) (in Chinese)
9. 张义,李治江.基于高斯词长特征的中文分词方法.中文信息学报 **30**(05), 89–93 (2016). Zhang, Y., Li, Z.-J.: Chinese word segmentation method based on Gaussian word length feature. J. Chin. Inf. Process. **30**, 89–93 (2016) (in Chinese)
10. 彭湃. 自然语言处理—中文词和短文本向量化的研究.华中师范大学 (2019). Peng, P.: Natural Language Processing—Research on Vectorization of Chinese Words and Short Texts. Central China Normal University, Wuhan (2019) (in Chinese)
11. 涂文博,袁贞明,俞凯.无池化层卷积神经网络的中文分词方法. 计算机工程与应用 **56**(02), 120–126 (2020). Tu, W.-B., Yuan, Z.-M., Yu, K., et al.: Chinese word segmentation method without pooling layer convolutional neural network. Comput. Eng. Appl. **56**, 120–126 (2020) (in Chinese)
12. 沈晓明, 桂永浩. 临床儿科学.第2版. 人民卫生出版社 (2013). Shen, X.-M,. Gui, Y.-H.: Clinial Pediatrics. 2nd edu. People's Medical Publishing House, Beijing (2013) (in Chinese)
13. 冯俐.中文分词技术综述.现代计算机(专业版) **34**, 17–20 (2018). Feng, L.: Overview of Chinese word segmentation technology. Modern Comput. **34**, 17–20 (2018) (in Chinese)
14. 李原. 中文文本分类中分词和特征选择方法研究.吉林大学 (2011). Li, Y.: Research on word segmentation and feature selection methods in Chinese text classification. Jilin University, Jilin (2011) (in Chinese)
15. 俞士汶,段慧明,朱学锋,孙斌.北京大学现代汉语语料库基本加工规范.中文信息学报 **05**, 49–64 (2002). Yu, S.-W., Duan, H.-M., Zhu, X.-F., et al.: Basic processing standards of Peking university modern Chinese Corpus. J. Chin. Inf. Process. **5**, 49–64 (2002) (in Chinese)
16. Lipscomb, C.: Medical subject headings (MeSH). Bull. Med. Libr. Assoc. **88**, 265-266 (2000)
17. 张明淘,韩普. 医疗实体识别研究进展. 计算机技术与发展 **04**, 1–10 (2020). Zhang, M.-T., Han, P.: Research progress of medical entity recognition. Comput. Technol. Dev. **4**, 1–10 (2020) (in Chinese)
18. 黎绍武.基于文本挖掘的胶质瘤蛋白质相互作用抽取方法的研究.华南理工大学 (2018). Li, S.-W.: Research on Extraction Method of Glioma Protein Interaction Based on Text Mining. South China University of Technology, Guangzhou (2018) (in Chinese)
19. 昝红英,窦华溢,贾玉祥,关同峰,奥德玛,张坤丽,穗志方. 基于多来源文本的中文医学知识图谱的构建. 郑州大学学报(理学版) **52**(02), 48–54 (2020). Zan, H.-Y., Dou, H.-Y., Jia, Y.-X., et al.: Construction of Chinese medical knowledge graph based on multi-source Text. J. Zhengzhou Univ. Nat. Sci. Ed. **52**, 45–51 (2020) (in Chinese)

20. 张坤丽,赵旭,关同峰,等. 面向医疗文本的实体及关系标注平台的构建及应用. 中文信息学报 **34**(6), 117–125 (2020). Zhang, K.-L. Zhao, X., Guan, T.-F., et al.: Construction and application of entity and relationship annotation platform for medical text. J. Chin. Inf. Process. **34**, 117–125 (2020) (in Chinese)
21. Carletta, J.: Assessing agreement on classification tasks: The kappa statistic. Comput. Linguist. **22**, 249–254 (1996)
22. Hripcsak, G., Rothschild, A.-S.: Agreement, the f-measure, and reliability in information retrieval. J. Am. Med. Inf. Assoc. JAMIA. **12**, 296–298 (2005)
23. Ogren, P., Savova, G., Chute, C.: Constructing evaluation corpora for automated clinical named entity recognition. In: Proceedings of the 12th World Congress on Health (Medical) Informatics. Marrakech, Morocco: European Language Resources Association (ELRA), pp. 2325–2330 (2008)
24. Artstein, R., Poesio. M.: Inter-Coder Agreement for Computational Linguistics. Comput. Linguist. **34**, 555–596 (2008)

Construction of Chinese Medicine Knowledge Base

Kunli Zhang[1,3], Xiaohui Ren[1(✉)], Lei Zhuang[1], Hongying Zan[1,3], Weicong Zhang[1], and Zhifang Sui[2,3]

[1] School of Information Engineering, Zhengzhou University, Beijing, China
{ieklzhang,ielzhuang,iehyzan}@zzu.edu.cn

[2] Key Laboratory of Computational Linguistics, Ministry of Education, Peking University, Beijing, China
szf@pku.edu.cn

[3] Peng Cheng Laboratory, Beijing, China

Abstract. Drug information includes indications, contraindications, adverse reactions and other specific descriptions, which can provide a basis and support for rational clinical rational drug use and clinical decision-making. A Chinese medicine knowledge base with interrelated disease and drug information interrelated can effectively monitor the therapeutic effects of drugs on diseases and provide knowledge resources for drug research. In this paper, the multi-source Chinese Medicine Knowledge Base (CMKB) is constructed in view of the important role of drug information in clinical medication, auxiliary diagnosis, disease risk prediction and other practical applications. CMKB takes the mainstream medical resources such as the China Disease Knowledge Total Database and Dingxiangyuan as data sources. On the basis of the knowledge description system and classification system of the medicine database, CMKB standardizes the classification of drugs from multiple sources and forms corresponding detailed knowledge descriptions. CMKB constructed contains 14,141 drugs, which are described in detail from 7 aspects such as indications, contraindications and classification. Knowledge association between drug entities and disease entities is established through named entity recognition. The accuracy of named entity recognition reaches 90.6% and 91.4% as assessed by different models. CMKB constructed in this paper can be a useful supplement for the Chinese Medical Knowledge Graph (CMeKG) and will provide support for intelligent diagnosis and clinical decision-making.

Keywords: Chinese Medicine Knowledge Base · CMeKG · Knowledge description system · Classification system · Knowledge association

1 Introduction

With the rapid development of medical technology, all kinds of new drugs are continuously listed and enter the hospital pharmacies, making it difficult for clinicians to grasp the information of all drugs such as the usage, dosage, adverse reactions and contraindications [1]. In addition, the variety of names and complicated usages of new drugs may

M. Liu et al. (Eds.): CLSW 2020, LNAI 12278, pp. 665–675, 2021.
https://doi.org/10.1007/978-3-030-81197-6_56

also affect the safety and efficacy of clinical drugs [2]. At the same time, medical data is known for its difficulty in quality control and its complicated sources, which makes it difficult to tap the potential high value of big data in medicine and pharmacy. Therefore, faced with the huge amount of medical data, it is urgent to construct a medical knowledge base to retrieve knowledge efficiently and accurately. It can provide a sufficient and reliable basis for medical research and clinical decision-making, and finally achieve the goals of precise prevention, precise diagnosis and precise treatment [3].

In recent years, many scholars have conducted research on the construction of medical knowledge base. Chongqing VIP Information Co., Ltd et al. have jointly developed the China disease knowledge total database (CDD), which has a disease knowledge base, a medicine knowledge base, an auxiliary examination knowledge base and an evidence-based medicine knowledge base [4]. Hou et al. [5] constructed an ontology-based clinical medical knowledge base under the guidance of ontology theory comprising three medical knowledge bases: a disease knowledge base, a medicine knowledge base, and an examination knowledge base. In addition to the comprehensive knowledge base, there are also some medicine knowledge bases. For example, Ge et al. [6] established a knowledge base linking drugs with adverse reactions. Wu et al. [7] constructed a medicine knowledge base system for the clinical pathway support system. Xu et al. [8] constructed a knowledge base of drug attribute classification. Zhang et al. [9] constructed a digestive system medication knowledge base.

Although many medical knowledge bases have been constructed in China, it is found that different medicine knowledge bases have their own unique knowledge representation. For example, the classification standards of medicines are different between the CDD and other medicine knowledge bases. And the medicine knowledge contained in different medicine knowledge bases also has certain limitations, such as different quantities of drug knowledge, differences in drug knowledge description about drug attributes, level of detail and so on.

The Chinese Medicine Knowledge Base (CMKB) in this paper will eventually become an organic part of the Chinese Medical Knowledge Graph (CMeKG). CMeKG [10, 11] is a Chinese medical knowledge graph developed by using natural language processing and text mining technology, based on large-scale medical text data. The construction of CMeKG refers to authoritative international medical standards such as ICD [12], ATC, SNOMED-CT [13, 14], MeSH [15] and medical text information such as large-scale, multi-source and heterogeneous clinical guidelines, industry standards [10]. Many drugs are included in CMeKG at present, but drug descriptions are not perfect. Detailed information on drugs is very important to assist clinical doctors in the rational use of drugs, the intelligent use of drugs and improvement of medical quality. Therefore, we construct a multi-source Chinese Medicine Knowledge Base (CMKB) in this paper for the purpose of providing CMeKG with comprehensive drug descriptions and establishing the relationship between drugs and diseases.

The construction process of CMKB can be divided into two parts: the schema layer construction and the data layer construction, as shown in Fig. 1. The construction of schema layer is mainly divided into two steps. The first step is to construct the knowledge description system, and the second step is to construct the classification system. Finally, the data layer is constructed by adding specific drug information on the basis of the schema layer. In what follows, Sect. 2 and 3 will discuss the construction process of the schema layer. Section 4 will introduce the construction process of the data layer. The last section will conclude the paper.

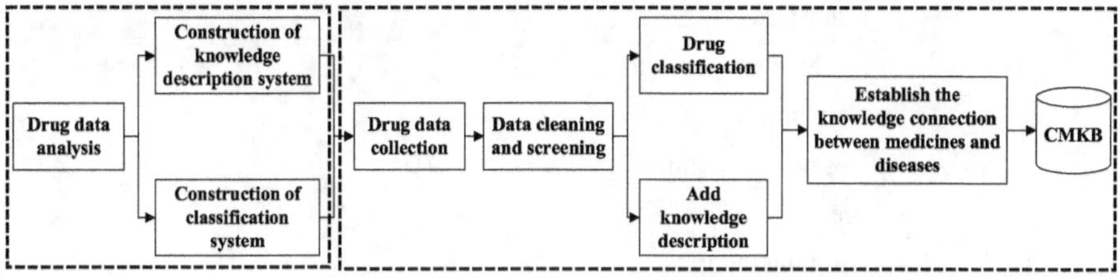

Fig. 1. Schematic diagram of CMKB construction process

2 Construction of Knowledge Description System

The medicine knowledge base with complete information can provide timely services for clinicians and nurses in handling medical orders, rationally selecting and using drugs [16]. Therefore, in order to ensure the safety and effectiveness of clinical medicine and make CMKB more valuable for clinical application, the medicine knowledge base constructed in this paper needs to be added with detailed drug knowledge descriptions.

Before adding detailed knowledge descriptions of the drugs into the medicine knowledge base, a knowledge description system needs to be established firstly. After analyzing multiple domestic drug resource databases, we select the drug data in CDD, Dingxiangyuan[1] (DXY, for short) and MCDEX China physician pharmacist clinical medication guide (MCDEX, for short) as the sources of drug knowledge descriptions and establish the CMKB knowledge description system. Among them, CDD is a knowledge base for clinical medicine professionals. DXY is the largest professional social network in China for doctors, medical institutions, and medical practitioners, which contains thousands of drug package inserts [17]. As for MCDEX, it is a clinical medication reference book compiled by the Ministry of Health's Rational Medication Expert Committee [18].

The knowledge description system of drugs can provide multiple drug attributes, including indications, precautions and contraindications, etc. [8] CDD, DXY and MCDEX all describe the knowledge about drugs in detail according to the attributes of the drugs. Table 1 shows statistics of drug attributes from the three sources. (Where "√" means a certain source has a certain attribute).

As can be seen from Table 1, different sources describe different attributes of the drugs. But in reality, there are many kinds of diseases, and a disease may have many different symptoms and different medication schemes may be used for different symptoms of diseases. So the constructed medicine knowledge base needs to cover as many drug attributes as possible, and include all the basic information that must be available for the sake of rational use of drugs. Therefore, we select 7 drug attributes for the drug

[1] http://www.dxy.cn/.

Table 1. Attributes of drugs in different sources

Drug attribute	CDD	DXY	MCDEX
Ingredient		√	
Indication	√	√	√
Adverse reaction	√	√	√
Contraindication	√	√	√
Dosage and administration	√	√	√
Precaution	√	√	√
Drug interaction	√	√	√
Pharmacological action	√		√
Pharmacokinetic	√		√
Dosage form	√		√
Classification	√	√	√

description system established in CMKB, including indication, contraindication, adverse reaction, dosage and administration, precaution, drug interaction and classification.

3 Construction of Classification System

After completing the construction of the knowledge description system, a standard classification system needs to be constructed to classify the drugs in CMeKG, so as to realize rapid drug inquiry, improve the efficiency of clinical application, and effectively reduce the occurrence of unreasonable drug use and improve the medical quality.

When constructing a classification system, it is necessary to specifically analyze the classification of drugs from multiple sources. Therefore, in addition to CDD, DXY and MCDEX, another mainstream pharmaceutical website, MENET[2], is also analyzed. MENET is a leading platform for medical and health information, including medical insurance drugs, reference drugs and listed drugs. Through the analysis, we find that the drug classifications in different drug resources are different. Statistics on comparisons among the first-level categories from different sources are shown in Table 2.

It can be seen from Table 2 that the number of different first-level categories in the drug classifications of different sources is much more than the same number of first-level categories, and there is no first-level category in DXY. CMKB, which is constructed in this paper, hopes to realize the rapid inquiry of drugs and effectively avoid the situation of irrational drug use. Therefore, it is necessary to establish a classification system that is not only universally recognized, but also can cover a variety of drug classifications.

We compare the classification of drugs from the four sources of CDD, DXY, MENET, and MCDEX with the classification standard of the National Essential Medicines List (2018 version), which is the reference for the use of drugs at all levels of medical and

[2] http://www.menet.com.cn/.

Table 2. Statistics on comparisons among the first-level categories from different sources

Comparison of different sources	Num of same first-level categories	Num of first-level categories
CDD and MENET	3	19 (CDD)
MENET and MCDEX	4	25 (MENET)
CDD and MCDEX	8	34 (MCDEX)

health institutions. Based on clinical pharmacology, the drug list is mainly divided into 26 first-level categories and 102 s-level categories. After analysis, we find that there are still some classifications of drugs that do not meet the classification standard, although the classification standards in the National Essential Medicines List are highly recognized and can best cover the classification of drugs from multiple sources.

Therefore, the CMKB constructed in this paper uses the classification system of the National Essential Medicines List as the basic classification framework, and uses the classification standards of the four sources as references to make certain adjustments. Finally, a drug classification system that can adapt to multi-source drug classification is formed. The classification system finally adopted contains 27 first-level categories and 119 s-level categories in total. Table 3 lists the six first-level categories and the number of second-level categories and examples of second-level categories.

Table 3. Classification system of Chinese medicine knowledge base after adjustment

First-level category name	Num of secondary category	Examples of secondary category
Antimicrobial drug	15	青霉素类 (Penicillin)
Antiparasitic drug	6	抗疟药 (Antimalarial drug)
Anesthetic	4	局部麻醉药 (Local anesthetic)
Nervous system drug	7	抗癫痫药 (Anticonvulsant)
Mental disorders drug	6	抗抑郁药 (Antidepressant)
Cardiovascular drug	7	抗心绞痛药 (Antianginal drug)

In addition to the six listed above, the first-level categories in the constructed classification system also include analgesic, antipyretic, anti-inflammatory drug, anti-rheumatic, anti-gout drug, respiratory system drug, digestive system agent, urinary system drug, hematologic system drug, hormone and endocrine-disrupting drug, antineoplastic drug, vitamins, minerals drug, drug for regulating water, electrolyte and acid-base balance, antidote, diagnostic drug, dermatologic agent, ophthalmic drug, obstetrics and gynecology drug, family planning drug, pediatric drug, antianaphylaxis drug, immune system drug, biological product, otorhinolaryngology drug and other drugs.

4 Construction of the Data Layer

After analyzing the data to construct the CMKB schema layer, the next step is to construct the CMKB data layer. This section will build on the schema layer constructed in Sect. 2 and 3 by introducing specific drug data. The construction of the data layer is mainly divided into four steps, which are data collection, data cleaning and screening, drug classification and improvement of drug information, and establishment of knowledge associations between drugs and diseases. We will describe the details below.

4.1 Drug Data Collection

The detailed drug information in the CMKB constructed in this paper comes from the CDD, DXY and MCDEX. According to the established drug knowledge description system, information of seven attributes of all drugs are extracted, including the indications, contraindications, adverse reactions and so on.

4.2 Drug Data Cleaning and Screening

Because drug data are multi-sourced, it is necessary to process the knowledge about the drugs before constructing the data layer. However, the following problems are found when collating the data:

(1) The CMKB constructed in this paper targets at chemical drugs, but there are mixed Chinese patent medicines or folk medicines in the drug data collected.
(2) For the same drugs, there are some symbolic differences in the attribute data, resulting in data content duplication, slight differences in the descriptions of attribute data, resulting in high content similarity or large differences in the descriptions of attributes.
(3) There are different forms of drug names in different sources. Some are chemical ingredients, such as Erythromycin, while others are specific drug names, namely chemical ingredients and dosage form, such as Roxithromycin Dispersible Tablets.

To address the above problems, different measures are adopted for data cleaning and screening. Firstly, we clean the collected data according to the "Chinese patent medicines" and "ethnic medicines" in different sources to ensure the accuracy of the medicines classified as "chemicals". Then all the duplicate attribute data generated due to the differences in punctuation marks or special symbols are cleared.

Next, data screening is performed on the remaining drug data. According to the similarity of the data, data items with a similarity rate of more than 80% are selected for each drug. Then, we select the data item with the longest string length from among the data items with a high similarity rate and the data items with big differences. Such items are then retained as the finalized drug knowledge descriptions. And for the differences in the forms of drug names, we temporarily ignore the subordinate relation between the names and retain the corresponding drug data in this paper.

4.3 Classifying and Perfecting Drug Information

After the data has been processed, we then need to start building a specific data layer, in which two main steps are involved. The first step is to classify the drug knowledge from multiple sources according to the classification system, in which different measures are adopted to deal with different category conflicts, as demonstrated below:

(1) Because the National Essential Medicine List is the basis for the allocation and use of drugs in medical and health institutions at all levels, drugs from three sources should first be organized according to its drug classification. Because the names of medicines in the National Essential Medicines List are listed in the form of chemical ingredients, for medicines with chemical ingredients and dosage forms, the drug names are matched with the drug names in the National Essential Medicines List. For example, "Amoxicillin" matches "Amoxicillin tablets" and "Amoxicillin capsules". Finally, the classification of the National Essential Medicines List shall prevail.
(2) For the classifications of the drugs listed in different sources, some of the classification names are similar, and some are very different. Therefore, the similarity of the classification names of the same drugs from multiple sources is matched. The classification name with high similarity shall be based on the sorted drug classification. If the similarity is low, considering that the drug itself may have multiple curative effects, that is, it may have multiple classifications, the classification with larger differences is also retained.
(3) When the drug data cannot determine the specific classification, the first concern is whether there is a correct secondary classification. If not, we will adjust the classification according to the relevant first-level classification (except for pediatric medicine, all of which are first in the pediatric first-level category)
(4) If there are still some drug data that are not assigned appropriate categories at the end, these drug data will be classified into "other drugs" under the first-level category. For example, "nutritional drugs", "muscle relaxants" and so on.

After that, the collected and processed knowledge descriptions are added to the classified drug list to form the preliminary version of CMKB.

4.4 Establish Knowledge Association Between Disease and Drug Entities

In view of the initially formed drug knowledge base, the next step is to consider how to use the CMKB constructed to serve CMeKG. CMeKG takes the disease entities as the center, annotates the relation types between the disease entities and other entities, and forms a broad knowledge association between the entities [11, 19]. Therefore, the CMKB in this paper needs to extract the disease information from the detailed knowledge descriptions of the drugs and establishes the knowledge relations between the disease entities and the drug entities.

After the analysis of the knowledge of multiple attributes of drugs, it is found that the drug knowledge descriptions of indications, contraindications and adverse reactions are mostly closely related to diseases. Therefore, we use knowledge obtained from these three attributes to establish knowledge associations in this paper.

To begin with, for the unlabeled knowledge descriptions, the disease information in the MeSH vocabulary, ICD-10 and CMeKG is used for disease labeling. Then, in order to ensure the accuracy in the disease labeling, manual proofreading is carried out to clean up some wrong information.

Based on the characteristics of the text, methods based on deep learning are used to realize the automatic extraction of disease information from the multi-sourced drug knowledge descriptions. The knowledge descriptions of three attributes of medicine is taken as the original corpus, and the corpus is used for named entity recognition of disease entities.

In this paper, we use nearly 30,000 drug knowledge descriptions of indications, contraindications and adverse reactions as the original corpus, and the BiLSTM-CRF and T-BiLSTM-CRF are also used for named entity recognition of disease entities. The T-BiLSTM-CRF model is trained on the BiLSTM model using data sets from non-medical fields, and the model parameters that contain the key knowledge of non-medical fields are taken as the initial data. The next step is to improve the performance of the BiLSTM-CRF by incorporating key features of external resources. In the experiment, the annotated corpus after manual proofreading is taken as the standard result of entity recognition, and the final experimental results are shown in Table 4.

Table 4. Experimental results

Model	Precision	Recall	F1-score
BiLSTM-CRF	91.7%	89.5%	90.6%
T-BiLSTM-CRF	90.7%	92.2%	91.4%

The literature [20] indicates that the accuracy of entity annotation can be considered reliable when the value of F1 is above 80%. The experimental results also show that the disease entities extracted by the neural network model are reliable.

Moreover, we also construct the triplet data of inter-entity association among the drug entities, the automatically extracted disease entities and the corresponding attribute knowledge descriptions of drugs. Since the relations between the drug entities and the disease entities are known, the final drug knowledge base can form the knowledge association form of <drug - disease - knowledge description - attribute relation>. The value of F1 also shows that the knowledge association between disease entities and drug entities in CMKB is reliable. Examples are shown in Table 5.

Table 5. Examples of drug and disease knowledge correlation analysis

Drug-disease association	Description of drug knowledge	Drug attributes
头孢克洛分散片(Cefaclor Dispersible Tablets)-肺炎(**Pneumonia)**	It is mainly used for respiratory tract infection caused by sensitive bacteria such as **pneumonia**	Indication
氨苄西林胶囊(Ampicillin capsules)-痛风(**Gout)**	It is contraindicated for patients with acute attack of **gout**	Contraindication
环酯红霉素片(Erythromycin cyclic ester tablets)-肝功能损伤(**Liver function injury)**	**Liver function injury** may occur in a very small number of patients	Adverse reaction

4.5 CMKB Classification Statistics

According to the multi-source drug data, the schema layer and the data layer are constructed, and finally the Chinese Medicine Knowledge Base CMKB is formed, CMKB contains a total of 27 first-level categories and 119 secondary categories, and covers 14,141 chemicals. Figure 2 shows the distribution of the number of drugs in each category.

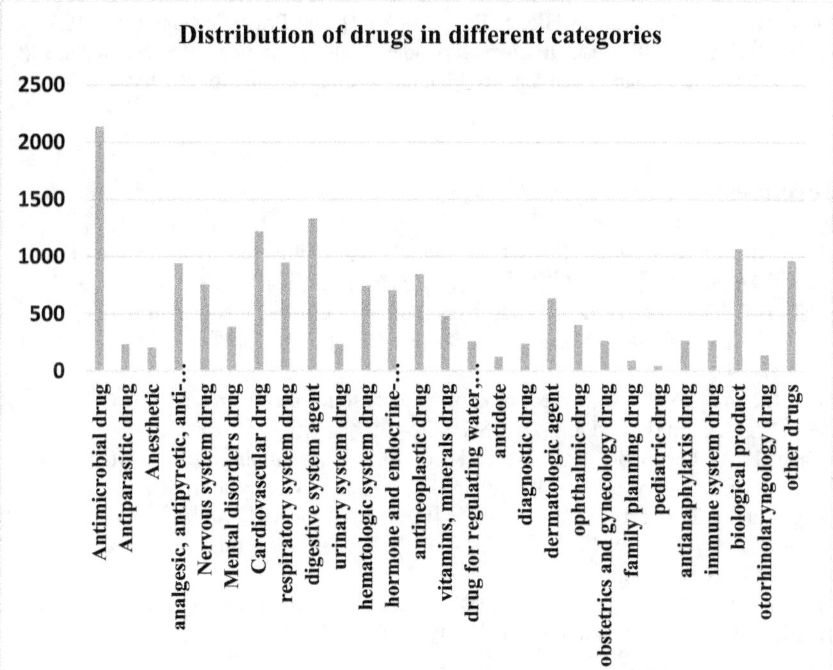

Fig. 2. Distribution of drugs in different categories

5 Conclusions

In order to perfect the drug description in CMeKG, we construct the multi-source Chinese Medicine Knowledge Base CMKB in this paper. Firstly, a knowledge description system and a classification system are constructed based on multi-source drug data. Then, according to the established architecture and the drug information after data processing, we construct the data layer, and finally CMKB is constructed. The knowledge of drugs in the CMKB constructed in this paper can improve the effectiveness and rationality of clinical medication, and can be used to add new features in natural language processing applications, including medical Q & A. Knowledge association established in CMKB will help to complete the clinical medical decision support system of multi-disciplinary cooperation and the whole process of diagnosis and treatment. It will also help to assist diagnosis and intelligent guidance. In the next step, we will try to use semi-supervised information extraction and other new technologies to automatically extract a wider range of medicine knowledge and further expand the scale of CMKB.

Acknowledgment. We thank the anonymous reviewers for their constructive comments, and gratefully acknowledge the support of National Key Research and Development Project (Grant No. 2017YFB1002101), Major Program of National Social Science Foundation of China (Grant No. 17ZDA138), China Postdoctoral Science Foundation (Grant No. 2019TQ0286), Science and Technique Program of Henan Province (Grant No. 192102210260), Medical Science and Technique Program Co-sponsored by Henan Province and Ministry (Grant No. SB201901021), Key Scientific Research Program of Higher Education of Henan Province (Grant No. 19A520003, 20A520038, 20A320070), the MOE Layout Foundation of Humanities and Social Sciences (Grant No. 20YJA740033), Henan Social Science Planning Project (Grant No. 2019BYY016).

References

1. Lv, W.-H., Jiang, X., Wang, S.-H.: Establishment and application of drug instruction database (药品说明书数据库的建立与应用). Chin. Pharm. (中国药房). **18**, 316–318 (2007)
2. Cai, Y.-P.: Guidance of drug product instruction in clinical medication (药品说明书在临床用药中的指导作用). Nursing Res. Second Half Month (护理研究:下半月). **20**, 1850–1851(2004).
3. Liu, L., Wang, X.: Development of knowledge base for precision medicine (精准医学知识库的构建). Chin. J. Med. Libr. Inf. Sci. (中华医学图书情报杂志). **27**, 1–9 (2018)
4. Yang, Z.: A knowledge database based on knowledge mining and knowledge organization——Introduction to clinical medicine knowledge service system of China disease knowledge total database (一种基于知识挖掘与知识组织的知识型数据库——中国疾病知识总库之临床医药学知识服务系统介绍). Chin. J. Med. Libr. Inf. Sci. (中华医学图书情报杂志) **17**, 63–65 (2008)
5. Hou, L., Qian, Q., Huang, L.-H., et al.: Discussion on clinical medicine knowledge base system construction based on ontology (基于本体的临床医学知识库系统构建探讨). J. Med. Inf. (医学信息学杂志). **32**, 42–47 (2011)
6. Ge, C.-X., Zhang, Y.-S., Chen, W.-H., Duan, H.-L., Li, H.-Y.: Construction and statistical analysis of an adverse drug reaction knowledgebase (药物不良反应知识库的建设以及统计分析). Chin. J. Hosp. Pharm. (中国医院药学杂志). **35**, 765–769 (2015)

7. Wu, J., Zhang, Q., Qi, D.-X., Lin, W.-W.: Medicine knowledge base system for clinical path support system (面向临床路径支撑系统的药品知识库系统). Herald Med. (医药导报). **29**, 1106–1110 (2010)
8. Xu, F., Zeng, S.: Construction of drug attribute classification knowledge base and its application in data analysis of pharmaceutical administration in our hospital (药品属性分类知识库的构建及其在我院药事管理数据分析中的应用). Chin. Pharm. (中国药房). **30**, 2737–2741 (2019)
9. Zhang, L., Yang, Y.-F., Zhu, J.-P., Hu, J,-Q.: Application of intelligent management system of medicine knowledge base for digestive system (消化系统用药知识库智能化管理系统的应用). China Pharmacist (中国药师). **22**, 378–380 (2019)
10. Zan, H.-Y., Dou, H.-Y., Jia, Y.-X., et al.: Construction of Chinese medical knowledge graph based on multi-source corpus (基于多来源文本的中文医学知识图谱的构建). J. Zhengzhou Univ. (Natural Science Edition) (郑州大学学报(理学版)). **52**, 45–51 (2020)
11. Ao, D.-M., Yang, Y.-F., Sui, Z.-F., et al.: Preliminary study on the construction of Chinese medical knowledge graph (中文医学知识图谱CMeKG构建初探). J. Chin. Inf. Process. (中文信息学报). **33**, 1–9 (2019)
12. World Health Organization. International Statistical Classification of Diseases and Related Health Problems. (10th Revision) [OL] https://icd.who.int/browsel0/2016/en
13. Mou, D.-M., Zhang, Y.-X., Huang, L.-L., et al.: Constructing medical ontology based on SNOMED CT and FCA (基于SNOMEDCT和FCA的医学领域本体构建研究). J. Chin. Soc. Sci. Tech. Inf. (情报学报). **32**, 653–662 (2013)
14. Amarilli, A., Galárraga, L., Preda, N., Suchanek, F.M.: Recent topics of research around the YAGO knowledge base. In: Chen, L., Jia, Y., Sellis, T., Liu, G. (eds.) APWeb 2014. LNCS, vol. 8709, pp. 1–12. Springer, Cham (2014). https://doi.org/10.1007/978-3-319-11116-2_1
15. Nelson, S.J., Douglas Johnston, W., Humphreys, B.L.: Relationships in medical subject headings (MeSH). In: Bean, C.A., Green, R. (eds.) Relationships in the Organization of Knowledge, pp. 171–184. Springer, Dordrecht (2001). https://doi.org/10.1007/978-94-015-9696-1_11
16. Meng, L.-Y..: To establish a complete and standardized drug information database to improve the level of drug management and pharmacy services (建立完整、规范的药品信息数据库，提高药品管理和药学服务水平). Med. Inf. (医学信息:医学与计算机应用). **14**, 672–673 (2001)
17. Zhu, X.-D.: Sharing in the mobile medical industry: Taking Dingxiangyuan and Weiyi as examples (移动医疗行业的共享——以丁香园、微医为例). Hangzhou Weekly (杭州(我们)), pp. 49–52 (2017)
18. Organized and compiled by the Expert Committee on rational drug use of the Ministry of Health: MCDEX Clinical Medication Guidelines for Chinese Physicians and Pharmacists (MCDEX中国医师药师临床用药指南). Chongqing Publishing Group. **6**, 2737–2741 (2009)
19. Yue, D., Zhang, K., Zhuang, L., Zhao, X., Byambasuren, O., Zan, H.: Annotation scheme and specification for named entities and relations on Chinese medical knowledge graph. In: Hong, J.-F., Zhang, Y., Liu, P. (eds.) CLSW 2019. LNCS (LNAI), vol. 11831, pp. 563–574. Springer, Cham (2020). https://doi.org/10.1007/978-3-030-38189-9_58
20. Artstein, R., Poesio, M.: Inter-coder agreement for computational linguistics. Comput. Linguist. **34**, 555–596 (2008)

MDPeQA: A Pediatric Question Answering System Based on Multiple Data Sources

Junzhuo Li[1,2], Hongying Zan[1,2(✉)], Yingjie Yan[1,2], and Kunli Zhang[1,2]

[1] School of Information Engineering, Zhengzhou University, No.100 Science Avenue, Zhengzhou 450000, Henan, China
jzlee@stu.zzu.edu.cn, {iehyzan,ieklzhang}@zzu.edu.cn, yjyan@gs.zzu.edu.cn

[2] Peng Cheng Laboratory, No.2 Xingke 1st Street, Shenzhen 518052, Guangdong, China

Abstract. Child healthcare and disease diagnosis and treatment are a focus of most families in our society. The goal of our research is to provide accurate and useful pediatric information consultation. In this paper, we report our development of a pediatric question answering system, which generates answers from multiple data sources including knowledge graphs, medical encyclopedia, medical textbooks and medical records. When a user inputs a question, the system first uses AC automata to map the question into a logical form and then retrieves information accordingly from knowledge graphs and medical documents. We formulate a scoring mechanism to decide which answer to be returned for a question. One month after the system was put into trial use in a Grade III Level A hospital in China, its satisfaction rate reached 86.7% by user rating.

Keywords: Medical question answering system · Multiple data sources · Pediatric diseases

1 Introduction

In recent years, with the development of Internet technology, information explosion takes place on the Internet. A huge amount of complicated information increases the possibility for users to obtain the knowledge they need from the Internet.

The traditional search engine allows users to do keyword searches, using a logical combination of keywords and involving little at the semantic level. Search results are mostly returned in the form of web links, and a user needs to screen the results [1]. With such a massive amount of information on the Internet, search engines often cannot effectively improve the efficiency of users' information access, and may even provide biased or misleading information. Therefore, it is necessary to develop a question and answer system that can accurately understand user questions and quickly give accurate answers.

In the 1960s and 1970s, the early NLIDB (Natural Language Interface to Data Base) began to be developed [2], and its two representatives were the BASEBALL system in 1961 [3] and the LUNAR system in 1972 [4]. This kind of question answering systems

M. Liu et al. (Eds.): CLSW 2020, LNAI 12278, pp. 676–685, 2021.
https://doi.org/10.1007/978-3-030-81197-6_57

were generally limited to specific fields, allowing users to query a structured knowledge base using natural language. They were generally oriented to a specific domain, using special parsing rules, and it was difficult to transfer them to other fields of application.

The Information Retrieval based Question-Answering System (IRQA) [5] extracted answers from a large number of texts or related Internet pages based on keywords and semantic relationships in given questions. This approach solved the problem of narrow coverage to a certain extent, but because of the quality of the texts and the quality of the information on the web, the quality of the extracted answers was difficult to reach a satisfactory level.

In 2012, Google proposed the concept of Knowledge Graph (KG), which provided a model for knowledge base construction [6]. The graph structure database provides a new direction for the development of question answering systems, namely KBQA (Knowledge-Based Question Answering). KBQA can make full use of the structured data in the knowledge graph to provide users with concise and accurate answers. It has become an important research direction in the field of automatic question answering.

In this paper, we report our development of a multiple data sources based pediatric disease oriented medical question answering system MDPeQA (**M**ultiple **D**ata Sources Based **Pe**diatric **Q**uestion-**A**nswering System), which fully utilizes the advantages of IRQA and KBQA. It can provide comprehensive and accurate answers to more than ten kinds of questions related to common pediatric diseases, covering symptoms, examination, treatment, prognosis, and other related knowledge. At present, the system has been put into trial use in Grade III Level A hospital in Henan Province.

2 Related Work

Research on automatic question-answering in the medical field started early abroad and produced question answering systems for medical professionals, which are usually not friendly enough to the public.

Lee et al. (2006) completed the MedQA system [9], a genuine online medical question answering system. This system is oriented to doctors, answering definition-type questions in the medical field. The results that it returns are problem-related paragraphs produced by means of information retrieval techniques. AskHERMES [10] is also a question-and-answer system for doctors, which can answer complicated clinical practice questions. Answer extraction of MEANS [11] is performed in response to a semantic query. A shortcoming of this system is that questions are divided into ten categories, but only those in four of them can be handled. EnquireMe [12] is a question-and-answer system for non-medical professionals. Its data come from question-and-answer pairs provided by online communities.

There are many medical websites in China, e.g., 39 Health Net[1] and Dr. Chun Yu.[2] The main purpose of these websites is to promote daily health encyclopedias and patient messages by way of providing answers and explanations that are offered by relevant doctors. However, only manual answers are available. Due to the limitation of the number

[1] http://www.39.net/.

[2] https://www.chunyuyisheng.com/.

of participating doctors and their working hours, it is often difficult to give users timely answers.

In summary, most existing automatic question-answering systems in the medical field have the following limitations. (1) The source of answers is often limited to knowledge maps or medical document data, and the advantages of the two are not fully utilized. (2) Most question answering systems are aimed at light-weight questions and answers of widespread diseases and rarely provide in-depth professional knowledge through question-answering services that consider the peculiarities of pediatric diseases.

MDPeQA is aimed at solving these problems. It combines the characteristic strengths of IRQA and KBQA and also provides effective consultation services for pediatric diseases. It is promoted by means of WeChat mini-programs. It is characterized by free download and open use by the public, providing greater convenience than Web programs.

3 MDPeQA's Structure

The MDPeQA system is divided into three modules: a user interaction module, a question matching module, and an answer retrieval module. The system flow is shown in Fig. 1. The user interaction module is mainly responsible for receiving questions input by a user and returning answers to the user. In the testing phase, the statistical analysis of users' evaluation scores on the answers is also completed in this module. The sematic parsing module is responsible for identifying the key words in the questions, matching them against questions and templates, and generating query statements. The answer retrieval module is mainly for generating answers from both the medical knowledge graph and the medical documents in use. Considering that the authority and correctness of answers from different sources are different, we provide an answer scoring mechanism in this module.

After the users enter their questions, the MDPeQA system first uses AC automata [7] to identify keywords in order to map the questions into logical forms. Then it will retrieve the answers from the Chinese medical knowledge graph [8] and various medical documents and generate candidate answers. Each candidate answer will get a score and all of them will be sorted by their scores. The one with the highest score will be presented to the user first.

4 Key Technologies

4.1 Mapping Questions into Logical Forms

We consider keyword identification and template matching as part of the problem of mapping natural language utterances into logical forms. They all together carry out the task of semantic parsing.

We first use a simple deterministic procedure to construct a manageable set of candidate logical forms. Ideally, we would generate all logical forms for all possible questions that the users may ask, but this is intractable. So, we opt to construct the logical forms mainly from the types of the nodes and edges in the Chinese medical knowledge graph.

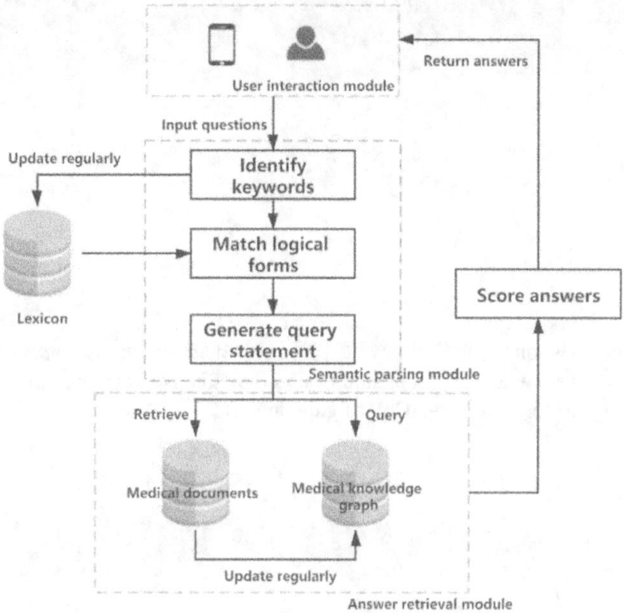

Fig. 1. The MDPeQA system consists of three modules: a user interaction module, a semantic parsing module, and an answer retrieval module, each of which is responsible for a unique task. The function of answer scoring is grouped into the answer retrieval module.

In the same time, we create a series of lexicons. For example, the disease lexicon includes the word "*allergic purpura*".

The next step is to map an input sentence (question) into a logical form. Our method is to identify its keywords, and construct the corresponding logical form through the category of the identified keywords. We consider the keywords in the lexicons we have constructed as effective keywords, because they can show the real intention of the person who asks the question. To make this procedure more efficient, we use an AC automaton to identify these effective words in an input question. Figure 2 shows an illustration of this process.

4.2 Generation of Query Statement

Cypher Statements for Knowledge Graph. The knowledge graph of our system is stored in the graph database Neo4j, and the standard Cypher statement is used to query the results. When the logical form of a question is available, we can map it into a Cypher statement. Figure 3 gives an example for this mapping.

Unstructured Medical Document Retrieval Based on Text Hierarchy. Medical documents are not purely unstructured texts. Most of them carry certain semi-structured information.

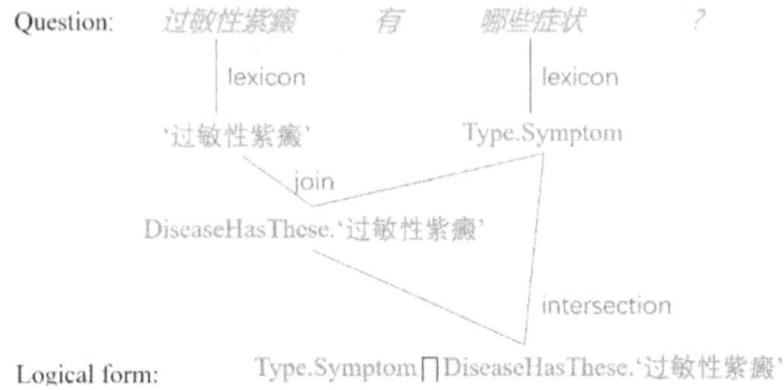

Fig. 2. Given the question "过敏性紫癜有哪些症状?" (*What are the symptoms of allergic purpura?*) in Chinese, we use an AC automaton to find the effective keywords in order to map the question into a logical form (*in red*). (Color figure online)

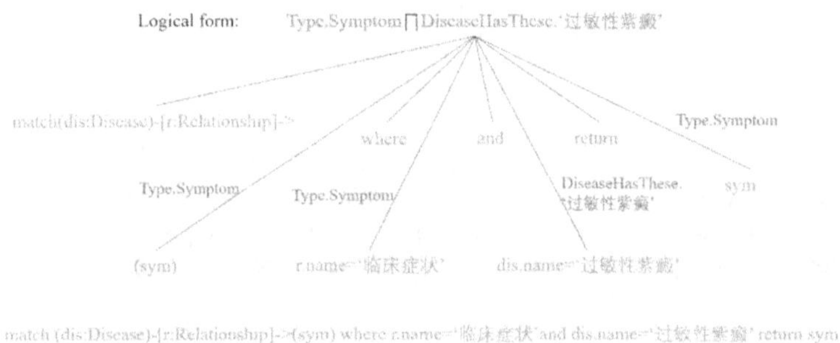

Fig. 3. Given the logical form "*Type.Symptom⊓DiseaseHasThese.'过敏性紫癜'*" (*in red*), we consider the elements of the logical form as conditions (*in blue*) to generate the Cypher statement (*in green*). (Color figure online)

Taking the textbook "Pediatrics" [14] as an example, we can find that the subsection titles are disease names, and the text of a subsection is usually divided into "Causes", "Pathology", "Clinical Manifestations", "Laboratory Examinations", "Diagnosis", "Treatment" and "Prognosis and Prevention". A semi-structured retrieval system based on the hierarchical structure of headings, subheadings and paragraphs can not only effectively find the information that is required by users, but also save the time that would otherwise be wasted by blind search through massive data. This method reduces system complexity and significantly improves system performance.

4.3 Data Sources and Processing

Since a question answering system in the medical field requires high accuracy of answers, it is necessary to select high authority data. MDPeQA combines structured data (such as a knowledge graph) and unstructured text data into its database. It uses is the Chinese Medical Knowledge Graph jointly developed by Peking University, Zhengzhou University and Peng Cheng Laboratory [13]. Its unstructured data come from a medical encyclopedia, medical textbooks, clinical pathways, clinical practice literature, and a set of pediatric medical records provided by a Grade III Level A hospital.

Chinese Medical Knowledge Graph (CMeKG). CMeKG involves a total of 3671 diseases (including more than 500 common diseases in pediatrics), 11 entities, and 54 relationships, which can cover most patients' consultation needs. Figure 4 shows a visual display of CMeKG.

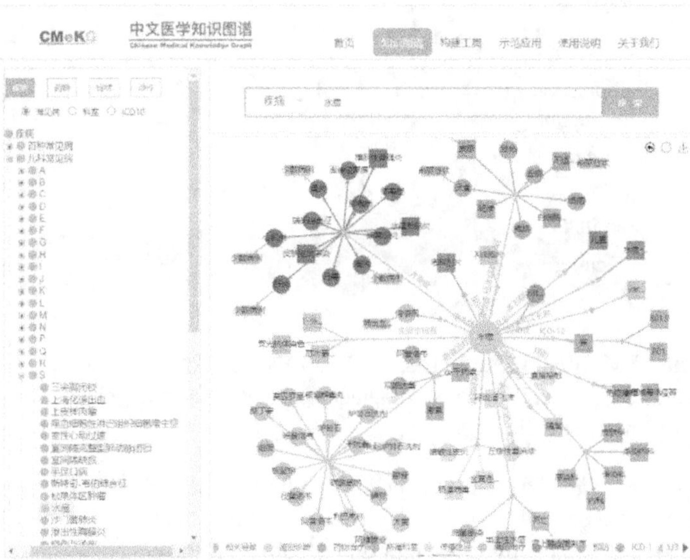

Fig. 4. An illustration of CMeKG visual display

The coverage of CMeKG is wide, and the entity relationships that it uses are also devided very carefully. Considering that the users of our question and answer system are patients or the general public, rather than medical experts, there is no need to scrutinize the relationships in full detail. This is because (1) the users do not have a professional medical education, usually knowing not much about the meaning of medical terms and the basis for classification, and (2) over-refinement of the relationships may instead lead to insufficient answers to users questions.

Therefore, we merge some of the relationships of CMeKG into more general ones. Using the original relationships as an attribute of the merged relationship (as shown in Table 1), our system can find answers that meet the needs of users better, and the complexity of the system can be reduced too. This merging of relationships can ensure the comprehensiveness of answers, in the sense of not losing the professionalism of the knowledge graph. The users are provided with a choice whether to use this feature: for a common question, a user may press a button to choose merged relationship search; when asking about more specific content, a user may select the other attributes as search conditions.

Table 1. Partially merged CMeKG relationship table.

Original relationships			Merged relationship
Entity 1	Relationship	Entity 2	
Disease	Clinical signs	Symptom	< Disease-Clinical symptoms-Symptom >
	Clinical symptoms		
	Auxiliary examination	Check	< Disease-Related Check-Check >
	Endoscopy		
	Pathological examination		
	Film degree exam		
	Laboratory examination		
	Screening		

Text-Structured Medical Documents. MDPeQA's medical documents can be divided into five categories: medical encyclopedia, medical textbooks, clinical pathways, clinical practice, and pediatric medical records.

Medical Encyclopedia. The medical encyclopedia is under Wikipedia, acquired by crawling for relevant entry information with the international/national standard terminology. Considering that the authority, rigor and accuracy of the encyclopedia documents cannot be guaranteed, answer retrieval can only be enabled for documents after manual proofreading.

Medical Textbooks. The medical textbooks selected are Pediatrics [14] and Clinical Pediatrics [15]. Both are classic textbooks and reference books for Chinese pediatric clinical medicine majors. The main contents of former include the development and prospects of pediatrics, principles of child health, principles of diagnosis and treatment of pediatric diseases, adolescent health and disease, newborn and neonatal diseases, infectious diseases, respiratory diseases, cardiovascular diseases, hematopoietic system Diseases, children's first aid, etc. The latter is closely related to the clinic, using Chinese and international epidemiological data and focusing on dynamics of diagnosis, treatment, prevention, and prognosis. The book is divided into 17 sections, covering various diseases of the pediatric system with a focus on growth and development (especially neuropsychological development), nutrition, first aid and monitoring, and tumors that reflect the characteristics of pediatrics.

Clinical Pathways. The clinical pathways are to establish a set of standardized treatment models and treatment procedures for particular diseases, involving comprehensive models of clinical treatment that follow medical evidence and guidelines to promote treatment organizations and disease management. A clinical pathway is usually issued by a national authority to ensure its high reliability.

Clinical Practice. The clinical practice integrates the latest research results, diagnostic steps, treatment steps, evidence, etc., to provide clinicians and patients with specific clinical conditions and support tools for guiding clinical decisions.

Pediatric Medical Records. The pediatric medical records come from real pediatric medical record data provided by a Grade III Level A hospital. The advantage of these original medical records is that they are authentic and reliable, but their disadvantage is that the cases are highly personalized and can only be referenced with great caution.

4.4 Answer Scoring

Since MDPeQA retrieves multiple answers, it is necessary to set up a method to choose the best one. For this purpose, we formulate an answer scoring mechanism $S_{priority}$ based on the authority of data source. Given the necessity to prioritize the data sources according to their authority to exclude unreliable answers and avoid misleading users, the idea behind this scoring is that the higher the priority, the higher the authority.

Medical knowledge graphs often require professionals to participate in their construction. In this paper, CMeKG is selected as the knowledge base with the highest authority. The medical textbooks are selected from *Pediatrics* and *Clinical Pediatrics*, both of which are highly recognized by the industry and hence should also have the highest priority. The clinical pathway and clinical practice literature are usually formulated and published by international and national institutions. They are characterized by high authority and openness and reliability of data, but their updating is not always timely. So the priority should be placed second. The actual pediatric medical records are about different individual cases, so their priority should not be too high. As for the medical encyclopedia (Wikipedia), due to the unknown nature of the editor's occupation and education experience, the correctness of its answers is difficult to guarantee. Its priority should be set to the lowest. The authority of various data sources-priority assignments is shown in Table 2.

Table 2. The scores for the authority of various data sources.

Data sources	$S_{priority}$
Medical knowledge graph	5
Medical encyclopedia	1
Medical textbook	5
Clinical pathway	4
Clinical practice	4
Pediatric medical records	3

5 Summary

The evaluation indicators of a question answering system usually include correctness, precision, interpretability, and user-friendliness [16]. One month after MDPeQA was released for the public use, we compiled its statistics on these five indicators. Its correctness rate of question answering for 100 kinds of common pediatric diseases reached 86.7%. The overall precision of the system was maintained at 89%. Since the system's data come from reliable information sources, MDPeQA's answers are highly interpretable. Its user-friendliness, by users' ratings of the quality of answers when using the system, received an average score 2.104 out of 3 over all questions (13,284 total).

Figure 5 shows real examples of questions and answers about common diseases and new diseases. When the user consults "*what tests need to be done for hydrocephalus?*", the system finds the answer with the highest score from CMeKG, and then marks its source and returns it to the user. As for the user input "*What are the symptoms of COVID-19?*", because the current CMeKG system and its medical documents in use have not yet had a record of this disease, MDPeQA returns the answer from Wikipedia. At the same time, the question is recorded for later use.

5.1 Future Work

The size of the lexicons appears to limit MDPeQA's ability to answer the questions. Due to a lack of sufficient text corpus, the lexicons are constructed based on common

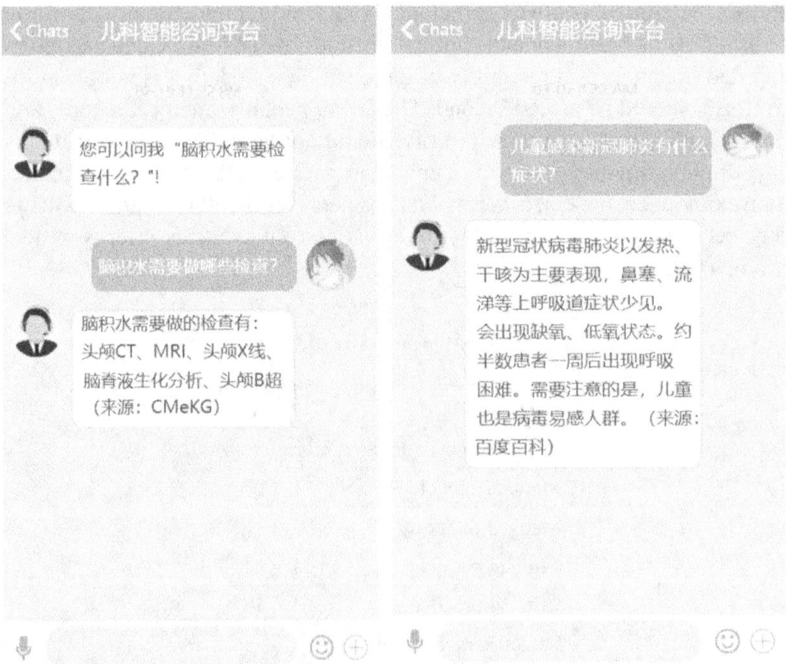

Fig. 5. Examples of questions and answers about common diseases and new diseases.

sense and knowledge graph. For problems that are presented in a colloquial manner, this system has a poor performance of recognizing them. In the process of interaction with users, the system records all questions that result in a weak effect, for the purpose of collecting corpus data to enable more advanced novel methods like deep learning.

Next, we will further enrich MDPeQA's data sources from two aspects of pediatric health care and childcare knowledge, make full use of suitable natural language processing technologies to enhance its performance, provide users with more comprehensive pediatric knowledge, and further improve user experience.

Acknowledgments. We thank the anonymous reviewers for their constructive comments, and gratefully acknowledge the support of the National Key Research and Development Project under Grant No. 2017YFB1002101, the Science and Technique Program of Henan Province under Grant No. 192102210260, the Medical Science and Technique Program Co-sponsored by Henan Province and Ministry under Grant No. SB201901021.

References

1. Zheng, S.-F., Liu, T., Qin, B., Li, S.: Overview of question-answering. J. Chin. Inf. Process. **16**(6), 47–53 (2002). (in Chinese)
2. Androutsopoulos, I., Ritchie, G.D., Thanisch, P.: Natural language interface to databases – an introduction. Nat. Lang. Eng. **1**(1), 29–81 (1995)
3. Green, B., Jr., Wolf, A., Laughery, K., Chomsky, C.: Basebal: an automatic question-answerer. In: Western Joint IRE-AIEE-ACM Computer Conference, pp. 219–224. ACM (1961)
4. Woods, W.: Progress in natural language understanding: an application to lunar geology. In: Proceedings of the National Computer Conference and Exposition, pp. 441–450. ACM (1973)
5. Kolomiyets, O., Moens, M.-F.: A survey on question answering technology from an information retrieval perspective. Inf. Sci. **181**(24), 5412–5434 (2011)
6. Singhal, A.: Introducing the Knowledge Graph: Things, Not String. Official Google Blog (2012). https://googleblog.blogspot.com/2015/05/introducing-knowledeg-graph-things-not.html
7. Aho, A.-V., Corasick, J.-M.: Efficient string matching: an aid to bibliographic search. Commun. ACM. **18**(6), 333–340 (1975)
8. Zan, H.-Y., Liu, T., Niu, C.-Y., et al.: Construction and application of named entity and entity relations corpus for pediatric diseases. J. Chin. Inf. Process. **34**(5), 19–26 (2020). (in Chinese)
9. Lee, M., Cimino, J., Zhu, H.-R., et al.: Beyond information retrieval – medical question answering. AMIA Annu. Symp. Proc. **06**(13), 469–473 (2006)
10. Cao, Y.-G., Liu, F.-F., Simpson, P., et al.: AskHERMES: an online question answering system for complex clinical questions. J. Biomed. Inform. **44**(2), 277–288 (2011)
11. Abacha, A.-B., Zweigenbaum, P.: MEANS: a medical question-answering system combining NLP techniques and semantic web technologies. Inf. Process. Manag. **51**(5), 570–594 (2015)
12. Wong, W., Thangarajah, J., et al.: Contextual question answering for the health domain. J. Assoc. Inf. Sci. Technol. **63**(11), 2313–2327 (2012)
13. Byambasuren, O., Yang, Y.-F., Sui, Z.-F., et al.: Preliminary study on the construction of chinese medical knowledge graph. J. Chin. Inf. Process. **33**(10), 1–7 (2019). (in Chinese)
14. Wang, W.-P., Sun, K., Chang, L.-W.: Pediatrics. 9th edn. People's Medical Publishing House, Beijing (2018). (in Chinese)
15. Shen, X.-M., Gui, Y.-H.: Clinical Pediatrics, 2nd edn. People's Medical Publishing House, Beijing (2013). (in Chinese)
16. Wang, H.-F., Qi, G.-L., Chen, H.-J.: Knowledge Graph: Method. Practice and Application. Publishing House of Electronics Industry, Beijing (2019). (in Chinese)

Comparison of Changes Between Mainland China and Taiwan

Shan Wang[1,2](✉) and Lei Tang[1]

[1] Faculty of Arts and Humanities, University of Macau, Taipa, Macau
shanwang@um.edu.mo
[2] Institute of Collaborative Innovation, University of Macau, Taipa, Macau

Abstract. The Chinese language is the common language for ethnic Chinese all over the world. The study of variants of the Chinese language in different regions not only lays the foundation for understanding its situation, but is also an important way for knowing about the social changes and development of the Chinese communities. Focusing on the commonly used word 改變 *gǎibiàn* "change", this article compared the similarities and differences of its collocates in the subject-predicate relation and the verb-object relation based on the large-scale data of Mainland China and Taiwan from The Tagged Chinese Gigaword Corpus. It also explored the common and specific concerns of the changes between the two sides from the beginning of the 1990s to the early 21st century. This study provides an important window for understanding the changes in Mainland China and Taiwan, which is conducive to further enriching the research achievements on the Chinese language and strengthening communication and exchange between Chinese communities.

Keywords: The Chinese language · Vocabulary · Change · Mainland China · Taiwan

1 Introduction

Global Chinese is the lingua franca of all Chinese people in the world based on Putonghua. It has a certain range of flexibility and tolerance in variants in phonetics, vocabulary, grammar and pragmatics [1]. Based on this perspective, the Chinese language in various regions with certain differences are called Chinese variants. The study of Chinese variants not only lays the foundation for a comprehensive understanding of the status of the Chinese language, but is also an important way to understand the development of the Chinese communities. It also provides a reference for language planning. Guo and Zhu [2] proposed that Chinese language is of great significance in terms of resources, culture and strategy. Li [3] pointed out that research on Chinese language is associated with the language identity of Chinese people all over the world. Recently, the depth and breadth of Chinese language research have greatly increased. However, Li [4] concluded that Chinese language research is still at an initial stage, which needs to be further extended through various studies. Specifically, in order to facilitate the

M. Liu et al. (Eds.): CLSW 2020, LNAI 12278, pp. 686–710, 2021.
https://doi.org/10.1007/978-3-030-81197-6_58

convergence of Chinese, it is necessary to strengthen the coordination of language use in Mainland China and Taiwan.

This article selected the commonly used word 改變 *gǎibiàn* 'change' as the research object to compare the similarities and differences of its collocates in the subject-predicate relation and the verb-object relation based on the large-scale data in Mainland China and Taiwan from the Chinese Gigaword Corpus. It further explored the common and specific concerns of changes on the two sides from the beginning of the 1990s to the early 21st century.

2 Related Work

2.1 Research on the Chinese Language

Diverse studies have been conducted on the Chinese language. First, the theoretical exploration of Chinese language issues has been conducted, such as its definition, academic value, and research scope. For example, Guo [5] pointed out that global Chinese is an integral part of the study about the Chinese language, which extends the existing research. Besides, the Chinese language in overseas Chinese communities is a standard language rather than a dialect of Chinese. Li [4] proposed that Mandarin in Mainland China, Taiwan, Hong Kong, Macau, and Southeast Asia are the most important variants, all of which are important topics in Chinese studies. He also mentioned that Chinese language research is still at an initial stage, so it is necessary to deeply study it from a global perspective. Lu [1] pointed out that the study of the Chinese language is beneficial for strengthening the sense of belonging and cohesion of the Chinese communities.

Second, studies on Chinese phonetics have been conducted. The phonetic features of Chinese variants and their differences from Mandarin are mainly analyzed. For example, Shang and Chew [6] investigated the phonetics of the Chinese language in Singapore and analyzed the phonological variations of initials, vowels, and polyphonic words. Their analysis showed that emergence of these variations is mainly caused by the influences of southern Chinese dialects, Taiwan Mandarin, and English. You [7] analyzed the phonetics of Mandarin in Taiwan, focusing on three distinctive phonetic phenomena including symmetrical tonal change, continuous reading, and Chinese-English reading. He also analyzed the impact of these phonetic phenomena on TV programs and online neologisms of Mainland China.

Third, research on Chinese grammar has been conducted. For example, Lu [8, 9] analyzed the syntactic features of the Chinese language in Singapore from three aspects: sentence structure, syntactic structure, and the function of reduplicative words, and explained the reasons for the differences. Diao [10] compared the differences between the 遭*zāo* 'suffer' and 被*bèi* 'passive bei' sentences as used in Mainland China, Taiwan, Hong Kong and Macau in terms of semantic features, the scope of use, the style and the complexity. He found that the use of 遭 *zāo* 'suffer' in Taiwan, Hong Kong and Macau is more frequent because they retain a lot of classical Chinese usages.

Fourth, research on the Chinese vocabulary has been conducted. For example, Tin [11] investigated the characteristics of the Hong Kong Chinese vocabulary from the aspects of social sources and word formation characteristics, and proposed the concept of community vocabulary. Jia and Xu [12] used the Singapore Newspaper *Lianhe Zaobao* as a corpus to collect 116 Singaporean Chinese-specific words and divided them into five subcategories to discuss the lexical differences in Singapore and Mainland China in terms of structure and sources. Tang [13] chose the common Chinese vocabulary in Hong Kong and Macau as the research object and divided them into three types: regional words, foreign loan words, and Cantonese words. He also emphasized that there are special regional words in both places. Shi [14] discussed the contact between the Thai vocabulary and the Mandarin vocabulary. He pointed out that with further contact, the vocabulary systems in the two places will continue to affect each other, and thus more community words will enter into the communication field of each other. Wang and Luo [15] compared the vocabulary of Macau's tourism Chinese in *The Syllabus of Chinese Vocabulary and Characters' Levels* [16], *The Tourism Chinese Vocabulary Handbook* [17] with Mandarin Chinese of Mainland China in order to provide data and theoretical references for Macau tourism practitioners and international students studying or traveling in Macau. Research on Chinese vocabulary also touches upon dictionary compilation. There are works like *Global Chinese Dictionary* [18] and *Comprehensive Global Chinese Dictionary* [19].

In short, research on the Chinese language has been extended to global Chinese and global Chinese language has become an integral part of the national language strategy [20]. Based on this new research direction, the depth and breadth of the studies of Chinese phonetics, grammar, and vocabulary have been continuously increased.

2.2 Research on Vocabulary of Mainland China and Taiwan

Vocabulary is the most active factor in a language. Differences in economics, politics, and culture between Mainland China and Taiwan inevitably affect their vocabulary. Therefore, the main difference of language use between the two sides lies in the vocabulary. This section focuses on the research on Chinese vocabulary on both sides, which germinated in the 1980s. First, patterns of cross-strait vocabulary are different. Existing research focuses on three aspects, namely special regional words, synonyms and homonyms, and new words. To be specific, Huang [21], Zhu [22], Tang [23], and Liu [24] studied the unique regional words in Taiwan. Su [25] compared 200 homonyms from dictionaries of Taiwan and Mainland China, and found that these differences mainly come from the evolution of lexical meanings. Hou [26] collected 290 different words from modern Taiwan novels and classified them according to different internal structures to investigate the synonymous vocabulary. Research on new words is also conducted. Su [27] pointed out that the new words on both sides of the Taiwan Strait are basically the same in structure and with the same word-making motivation, but their meanings and distributions are different. Zhao [28] investigated new words in the fields of politics, society, and technology in Mainland China and Taiwan, and analyzed the causes of their differences from a cognitive perspective.

Second, previous studies have observed the differences in certain words. For example, through investigating the word 而已 *éryǐ* 'only', Diao [29] analyzed language contact between Mainland China and Taiwan through the two patterns of "differences tending to integration" and "only differences without integration". Chu and Zhang [30] compared the verbal and prepositional usage of the word 透過 *tòuguò* 'through' in Mainland China and Taiwan, and found that it showed an integrating tendency.

Third, previous studies have adopted different research methods to examine cross-strait vocabulary. Mainly, methods based on dictionaries and literary works, and methods based on large-scale corpora were adopted. As for the former, Hou [26] and Diao [29], for instance, used novels to describe the differences in the vocabulary across the Taiwan Strait; and Dang, Wang and Su [31] analyzed the trends and reasons for vocabulary fusion in Mainland China and Taiwan by comparing vocabulary items in Mainland that were absorbed by Taiwan Mandarin and were collected in three dictionaries. As for the corpus-based method, for example, Li and Shi [32] used the CCL corpus of Mainland China and the Sinica Corpus of Taiwan to study vocabulary contact between the two sides. Wang and Wang [33] selected the word 安全 *ānquán* 'safe; safety' to explore the safety issues on the two sides based on the data of Xinhua News Agency (XIN) and Central News Agency (CNA) in the Tagged Chinese Gigaword Corpus from the beginning of the 1990s to the early 21st century.

In sum, existing research showed that scholars have carried out analyses on the vocabulary systems of the two sides of the Taiwan Strait from different perspectives. However, there has been no research on cross-strait changes based on large-scale corpora yet. Therefore, this article selected the commonly used word 改變 *gǎibiàn* 'change' and analyzed the similarities and differences of its collocates in the subject-predicate relation and the verb-object relation based on the large-scale data of Mainland China and Taiwan from the Tagged Chinese Gigaword Corpus. It explored the common concerns and unique concerns of the changes on the two sides from the beginning of the 1990s to the early 21st century. This study provides an important window for understanding the changes in Mainland China and Taiwan, which is conducive to further enriching the research achievements of the Chinese language and strengthening the communication and exchange between Chinese communities.

3 Research Methods

3.1 Corpus Selection

To conduct a corpus-based research, this paper selected the Chinese Gigaword Corpus (Second Edition), which was established by David Graff, Ke Chen, Junbo Kong, and Kazuaki Maeda [34] in 2005. In the corpus, a large number of texts from Mainland China and Taiwan were collected and processed by segmentation and part-of-speech tagging [35]. The items used in this article are from two subcorpora, namely XIN and CNA. The former includes data from the Xinhua News Agency of Mainland China from 1991 to December 2004 and the latter includes data from the Central News Agency of Taiwan from 1990 to December 2004.

3.2 Chinese Word Sketch Engine

Word Sketch is a corpus analysis tool that can automatically analyze grammatical collocation relationships on the basis of a large corpus [34]. Previously, it was used to analyze English; and later it was developed to be applicable to the analysis of other languages, such as Chinese. Quite a number of studies have adopted this tool [33, 36–45]. The

Chinese Word Sketch Engine generates a large amount of grammatical knowledge by combining with a Chinese billion-word corpus [34, 46]. Hence, not only can it search Chinese keywords and context, but it can also show grammatical collocational relations and synonyms.

3.3 Research Procedures

This paper focuses on the study of the word 改變 *gǎibiàn* 'change'. To begin with, the co-occurring words with 改變 *gǎibiàn* were extracted from CNA and XIN with a minimum frequency of 3, a maximum grammatical range of 999, and a minimum saliency of 0.00. Collocations in a subject-predicate relation or a verb-object relation were extracted. Then, among the extracted data, the correct collocations were manually selected and the wrong ones were deleted, as shown in Table 1.

For subject-predicate collocations, in the form of "XX *gǎibiàn*" (where XX stands for *subject*), there are 395 different collocations (types, not tokens) and 361 of them are correct in the XIN corpus; in the CNA corpus, there are 497 different collocations and 438 of them are correct. For verb-object collocations, in the form of "*gǎibiàn* XX" (where XX stands for *object*), the XIN corpus provides 894 different collocations, among which 393 of them are correct; the CNA corpus provides 926 different collocations among which 539 of them are correct. After manually filtering out the correct data, we can see that CNA has more collocations than XIN: CNA provides 77 more subject-predicate collocations and 146 more verb-object collocations.

After the selection of data, this study compared the common collocates and unique collocates in XIN and CNA. We have annotated their semantic categories by referring to *A Thesaurus of Modern Chinese* [47], and then further analyzed the causes of similarities and differences. This thesaurus contains more than 80,000 modern Chinese words. It is an important dictionary for classifying words due to its clear classification system. Through using the semantic categories of this thesaurus, all the collocates were classified according to their meanings. In addition to the collocates of 改變 *gǎibiàn* 'change' that are directly collected in this dictionary, we also manually classified those that were not in this dictionary by referring to the classification of this thesaurus.

4 Common Words Collocating with 改變 *gǎibiàn* 'change' in XIN and CNA

Common words refer to words used in both XIN and CNA as collocates of 改變 *gǎibiàn* 'change'. Analyzing common words can help us understand the common issues that concern people on the two sides of the strait.

4.1 Semantic Analysis of Common Words in the *Subject* Position

The collocated words in the subject position were classified into the first-level semantic categories and the second-level semantic categories according to *A Thesaurus of Modern Chinese* [47]. The common words in the subject position both in XIN and CAN belong to six first-level semantic categories, as shown in Table 2. They are mainly Abstract

Table 1. The number of different collocations of two grammatical relations in XIN and CNA

Classification	Subject-predicate collocations in XIN		Verb-object collocations in XIN		Subject-predicate collocations in CNA		Verb-object collocations in CNA	
No. of collocations	Original	Correct	Original	Correct	Original	Correct	Original	Correct
	395	361	894	393	497	438	926	539

Things, accounting for 72.4%, followed by words expressing Living Things, accounting for 15.7%; and very few words belong to Concrete Things, Time and Space, Social Activities, and Biological Activities. They indicate that the subjects that changed something or changed on both sides of the strait were very widely distributed in six semantic categories. Moreover, both Mainland China and Taiwan were concerned about abstract things. Among the second-level semantic categories of Abstract Things, as shown in Table 3, the distribution is also very wide. The subjects are distributed in 11 categories, such as Matters, Attributes, Politics, and Consciousness. Among the second-level semantic categories of Abstract Things, the top categories are Matters and Attributes, such as 立場 *lìchǎng* 'standpoint', 原則 *yuánzé* 'principle', 方針 *fāngzhēn* 'policy', 本質 *běnzhì* 'essence', and 單方面 *dānfāngmiàn* 'one side', reflecting the fact that both Mainland China and Taiwan were concerned about cross-strait political relations. Specifically, they both focused on the political standpoint, principles, and policies.

However, though there are many common subjects, the two sides have very different degrees of attention on specific issues. We further compared the words based on saliency difference. Table 4 and Table 5 list the top 10 words with highest saliency difference in XIN and CNA in the *subject* position. They show the following characteristics through comparison: (1) 狀況 *zhuàngkuàng* 'situation' and 局面 *júmiàn* 'situation' are the top two words with the highest saliency difference in Mainland China compared to Taiwan, reflecting that Mainland China attached importance to economic system reforms, focusing on developing economy, changing the situation of poverty and backwardness, and reforming the industrial structure. (2) 氣候 *qìhòu* 'climate' and 地貌 *dìmào* 'landform' are among the most saliently different words in Taiwan compared to Mainland China, indicating that Taiwan is much more concerned about issues of environmental change. Taiwan is an island and is often affected by natural disasters such as typhoons and earthquakes, which have a major impact on people's livelihood and the economy. For example, the earthquake in 1999, which was the biggest earthquake in Taiwan at the end of the 20th century, killed more than 2,000 people and caused economic losses of up to US$9.2 billion. It also changed the topography of Taichung and Nantou.

4.2 Semantic Analysis of Common Words in the *Object* Position

The distribution of common words in the *object* position is shown in Table 6. The major semantic category is Abstract Things, accounting for a high ratio of 81.4%, followed by words that belong to Time and Space, Living Things, and Concrete Things, accounting for 6.5%, 4.5%, and 4.5% respectively. Words of Social Activities, Biological Activities, and Movement and Changes only account for 2%, 0.8%, and 0.4% respectively.

The distribution shows the following characteristics: (1) Objects that have undergone changes on both sides of the strait were also very extensive, which are distributed in six semantic categories. (2) Both sides paid attention to Abstract Things. The second-level semantic distribution of words in this category is shown in Table 7. They are distributed in ten categories such as Matters, Attributes, Awareness, and Politics. (3) Words of Matters and Attributes account for a relatively high proportion in the second-level semantic categories of Abstract Things. Among Matters and Attributes, it can be observed that both Mainland China and Taiwan were concerned about health and diet, such as 生活 *shēnghuó* 'life', 飲食 *yǐnshí* 'diet', 成份 *chéngfèn* 'ingredient', 品質 *pǐnzhì* 'quality'.

Table 2. Distribution of *common words* functioning as *subjects* among the first-level semantic categories

First-level semantic category	Number	Percentage	Examples
抽象事物*chōuxiàng shìwù* 'Abstract Things'	134	72.4%	國家 *guójiā* 'country', 戰略 *zhànluè* 'strategy', 因素 *yīnsù* 'factor', 行爲 *xíngwéi* 'behavior', 社會 *shèhuì* 'society'
生物 *shēngwù* 'Living Things'	29	15.7%	人 *rén* 'humans being', 人們 *rénmen* 'people', 人類 *rénlèi* 'human beings', 個人 *gèrén* 'individual', 大家 *dàjiā* 'everyone'
具體物 *jùtǐwù* 'Concrete Things'	8	4.3%	衛星 *wèixīng* 'satellite', 飛機 *fēijī* 'airplane', 土地 *tǔdì* 'land', 地貌 *dìmào* 'landform', 顔色 *yánsè* 'color'
時空 *shíkōng* 'Time and Space'	7	3.8%	方向 *fāngxiàng* 'direction', 日期 *rìqī* 'date', 時間 *shíjiān* 'time', 時代 *shídà*i 'era', 任期 *rènqī* 'term of office'
社會活動 *shèhuì huódòng* 'Social Activities'	6	3.2%	發明 *fāmíng* 'invention', 威脅*wēixié* 'threat', 協議 *xiéyì* 'agreement', 承諾*chéngnuò* 'commitment', 規定 *guīdìng* 'regulation'
生物活動*shēngwù huódòng* 'Biological Activities'	1	0.5%	習慣 *xíguàn* 'habit'
Total	185	100.0%	/

Note: 習慣 *xíguàn* 'habit', 決定 *juédìn* 'decision', 位置*wèizhì* 'positions', 規定*guīdìng* 'regulation', and 路綫*lùxiàn* 'route' are polysemous, and thus belong to two semantic categories

In reality, people's concerns can be reflected by several issues. For instances, doctors in both Mainland China and Taiwan suggested that people change their diet and develop exercise habits to prevent diseases such as high blood pressure and hypercholesterolemia. Health service and products consumption have become a new type of consumption in Mainland people's life, and more and more people do exercise outdoors or in fitness clubs. Many people put health first in their lives, especially after the outbreak of SARS.

This part compares the saliency difference of the common words functioning as objects in Mainland China and Taiwan. The top 10 words with the highest saliency difference are listed in Table 8 and Table 9. They show the following characteristics. (1) The top two words with the highest saliency difference in Mainland China are 面貌 *miànmào* 'appearance' and 狀况 *zhuàngkuàng* 'situation'. This may indicate that Mainland China was highly concerned about getting rid of economic and social backwardness, such as

Table 3. Distribution of *common words of Abstract Things* functioning as subjects among the second-level semantic categories

Second-level semantic category	Number	Percentage	Examples
事情 *shìqíng* 'Matters'	37	27.6%	立場 *lìchǎng* 'standpoint', 原則 *yuánzé* 'principle', 方針 *fāngzhēn* 'policy', 秩序 *zhìxù* 'order', 局勢 *júshì* 'situation'
屬性 *shǔxìng* 'Attributes'	34	25.4%	本質 *běnzhì* 'essence', 單方面 *dān fāngmiàn* 'one side', 格局 *géjú* 'pattern', 態度 *tàidù* 'attitude', 力量 *lìliàng* 'strength'
政治 *zhèngzhì* 'Politics'	21	15.7%	國家 *guójiā* 'country', 機構 *jīgòu* 'agency', 部門 *bùmén* 'department', 當局 *dāngjú* 'authority', 政府 *zhèngfǔ* 'government'
意識 *yìshí* 'Awareness'	17	12.7%	價值觀 *jiàzhíguān* 'value', 理念 *lǐniàn* 'idea', 看法 *kànfǎ* 'opinion', 願望 *yuànwàng* 'wish', 需求 *xūqiú* 'need'
社會 *shèhuì* 'Society'	12	9.0%	企業 *qǐyè* 'enterprise', 單位 *dānwèi* 'unit', 集團 *jítuán* 'group', 地位 *dìwèi* 'status', 關係 *guānxì* 'relationship'
科教 *kējiào* 'Science and Education'	6	4.5%	方案 *fāng'àn* 'proposal', 決議 *juéyì* 'resolution', 格局 *géjú* 'pattern', 基調 *jīdiào* 'tone', 風格 *fēnggé* 'style'
數量 *shùliàng* 'Quantity'	3	2.2%	比例 *bǐlì* 'proportion', 速度 *sùdù* 'speed', 温度 *wēndù* 'temperature'
經濟 *jīngjì* 'Economy'	1	0.7%	歐元 *ōuyuán* 'Euro'
文體 *wéntǐ* 'Culture and Sports'	1	0.7%	運動 *yùndòng* 'sport'
生活 工作*shēnghuó gōngzuò* 'Life and Work'	1	0.7%	生活 *shēnghuó* 'life'
軍事 *jūnshì* 'Military'	1	0.7%	戰術 *zhànshù* 'tactics'
Total	134	100.0%	/

changing the situation of the economy, transportation, health, and other aspects. (2) The top two words with the most saliency difference in Taiwan are 戲路 *xìlù* 'acting style' and 現狀 *xiànzhuàng* 'current situation', indicating that Taiwan paid close attention to entertainment and the situation across the strait.

Table 4. Words with *subject* function that have higher saliency in XIN than CNA

Word	XIN subject.frequency	*XIN subject.saliency*	CNA subject.frequency	*CNA subject.saliency*	Saliency difference
狀況*zhuàngkuàng* 'situation'	260	*35.81*	55	*14.31*	21.5
局面 *júmiàn* 'situation'	121	*29.96*	18	*11.07*	18.89
面貌 *miànmào* 'appearance'	59	*26.4*	7	*10.15*	16.25
秩序 *zhìxù* 'order'	49	*17.82*	6	*3.14*	14.68
開放政策*kāifàng zhèngcè* 'open policy'	19	*19.69*	4	*7.6*	12.09
它 *tā* 'it'	98	*24.95*	50	*15.79*	9.16
人們 *rénmen* 'people'	38	*14.76*	4	*5.83*	8.93
形象 *xíngxiàng* 'image'	18	*10.88*	5	*2.06*	8.82
北約 *běiyuē* 'NATO'	26	*11.78*	5	*3.6*	8.18
現象 *xiànxiàng* 'phenomenon'	32	*13.53*	15	*6.1*	7.43

Table 5. Words as *subject* that have higher saliency in CNA than XIN

Word	XIN subject.frequency	*XIN subject.saliency*	CNA subject.frequency	*CNA subject.saliency*	Saliency difference
氣候 *qìhòu* 'climate'	8	*6.67*	92	*32.3*	25.63
結構 *jiégòu* 'structure'	53	*12.47*	250	*34.83*	22.36
地貌 *dìmào* 'landform'	3	*6.56*	20	*27.02*	20.46
體系 *tǐxì* 'system'	4	*0.19*	4	*19.9*	19.71
政策 *zhèngcè* 'policy'	184	*21.7*	1104	*40.61*	18.91
環境*huánjìng* 'structure'	48	*9.23*	301	*27.84*	18.61
方向 *fāngxiàng* 'direction'	16	*8.92*	119	*24.47*	15.55
時代 *shídài* 'era'	6	*3.75*	65	*18.8*	15.05
制度 *zhìdù* 'system'	22	*5.26*	131	*19.99*	14.73
原則 *yuánzé* 'principle'	16	*6.11*	104	*20.62*	14.51

Table 6. Distribution of *common words* functioning as objects among the first-level semantic categories

First level semantic category	Number	Percentage	Examples
抽象事物*chōuxiàng shìwù* 'Abstract Things'	201	81.4%	世界 *shìjiè* 'world', 機制 *jīzhì* 'mechanism', 觀念 *guānniàn* 'concept', 體制 *tǐzhì* 'system', 思維 *sīwéi* 'thinking'
時空 *shíkōng* 'Time and Space'	16	6.5%	日期 *rìqī* 'date', 時間 *shíjiān* 'time', 地點 *dìdiǎn* 'location', 軌迹 *guǐjì* 'track', 航道 *hángdào* 'waterway'
生物 *shēngwù* 'Living Things'	11	4.5%	人 *rén* 'human', 人類 *rénlèi* 'human beings', 我們 *wǒmen* 'us', 細胞 *xìbāo* 'cells', 基因 *jīyīn* 'gene'
具體物 *jùtǐ wù* 'Concrete Things'	11	4.5%	河道 *hédào* 'river course', 地貌 *dìmào* 'landform', 時間表 *shíjiānbiǎo* 'timetable', 顏色 *yánsè* 'color', 裝備 *zhuāngbèi* 'equipment'
社會活動*shèhuì huódòng* 'Social Activities'	5	2.0%	用藥 *yòngyào* 'medication', 决定 *juédìng* 'decision', 承諾*chéngnuò* 'commitment', 待遇 *dàiyù* 'treatment', 裝備*zhuāngbèi* 'equipment'
生物活動*shēngwù huódòng* 'Biological Activities'	2	0.8%	設計 *shèjì* 'design', 過去 *guòqù* 'past'
運動與變化*yùndòng yǔ biànhuà* 'Movement and Changes'	1	0.4%	循環 *xúnhuán* 'cycle'
Total	247	100.0%	/

From the above analysis, we can see that Mainland China and Taiwan share many common concerns about changes, but they pay different amount of attention to the issues of changes. This was mainly related to their respective social and cultural backgrounds.

5 Unique Words Collocating with 改變 *gǎibiàn* 'change' in XIN and CNA

Unique words refer to words that only appear in either the XIN corpus or the CNA corpus. These words can directly reflect differences between the two sides of the strait on issues of changes in economy, politics, and culture. It can help to reduce barriers against communication across the Taiwan Strait, and hence improve mutual understanding. In the XIN corpus, there are 181 word types functioning as subjects and 153 word types

Table 7. Distribution of *common words of Abstract Things* functioning as objects among the second-level semantic categories

Second-level semantic category	Number	Percentage	Examples
事情 *shìqíng* 'Matters'	62	30.8%	生活 *shēnghuó* 'life', 飲食 *yǐnshí* 'diet', 環境 *huánjìng* 'environment', 人生 *rénshēng* 'life', 生涯*shēngyá* 'career'
屬性 *shǔxìng* 'Attributes'	49	24.4%	成份 *chéngfèn* 'ingredient', 品質 *pǐnzhì* 'quality', 品種 *pǐnzhǒng* 'variety', 種類 *zhǒnglèi* 'type', 形態 *xíngtài* 'form'
意識 *yìshí* 'Awareness'	26	12.9%	主意 *zhǔyì* 'idea', 信仰 *xìnyǎng* 'belief', 觀點 *guāndiǎn* 'opinion', 理念 *lǐniàn* 'idea', 意願 *yìyuàn* 'wish'
政治*zhèngzhì* 'Politics'	18	9.0%	體制 *tǐzhì* 'system', 制度 *zhìdù* 'system', 所有權 *suǒyǒu quán* 'ownership', 憲法 *xiànfǎ* 'constitution', 政權 *zhèngquán* 'regime'
科教 *kējiào* 'Science and Education'	17	8.5%	內容 *nèiróng* 'content', 文字 *wénzì* 'text', 方案 *fāng'àn* 'proposal', 地圖 *dìtú* 'map', 課程 *kèchéng* 'course'
社會 *shèhuì* 'Society'	14	7.0%	習俗 *xísú* 'customs', 習慣 *xíguàn* 'habit', 風氣 *fēngqì* 'atmosphere', 地位 *dìwèi* 'status', 社會 *shèhuì* 'society'
數量*shùliàng* 'Quantity'	11	5.5%	產量 *chǎnliàng* 'volume of production', 含量 *hánliàng* 'content', 配額 *pèi'é* 'quota', 匯率 *huìlǜ* 'exchange rate', 稅率 *shuìlǜ* 'tax rate'
經濟 *jīngjì* 'Economy'	2	1.0%	經濟 *jīngjì* 'economy', 待遇 *dàiyù* 'treatment'
文體 *wéntǐ* 'Culture and Sports'	1	0.5%	節奏 *jiézòu* 'rhythm'
軍事 *jūnshì* 'Military'	1	0.5%	戰術 *zhànshù* 'tactics'
Total	201	100.0%	/

functioning as objects. In CNA corpus, there are 273 word types functioning as subjects and 305 word types functioning as objects, as shown in Table 10.

The semantic categories of all unique words in the XIN corpus and the CNA corpus were annotated. Their distribution is shown in Table 11 and Table 12. In the first semantic

Table 8. Words functioning as *objects* that have higher saliency in XIN than CNA

Word	XIN object.frequency	*XIN object.saliency*	CNA object.frequency	*CNA object.saliency*	Saliency difference
面貌*miànmào* 'appearance'	1363	*65*	106	*34.43*	30.57
狀況*zhuàngkuàng* 'situation'	469	*32.7*	69	*10.41*	22.29
原狀 *yuánzhuàng* 'original state'	38	*33.01*	7	*11.78*	21.23
過去*guòqù* 'past'	1213	*53.36*	517	*35.25*	18.11
局面 *júmiàn* 'situation'	337	*33.44*	53	*15.5*	17.94
用途 *yòngtú* 'use'	158	*40.63*	83	*24.53*	16.1
邊界 *biānjiè* 'boundary'	71	*17.95*	7	*2.43*	15.52
格局 *géjú* 'pattern'	187	*29.84*	33	*15.17*	14.67
處境 *chǔjìng* 'situation'	27	*19.36*	7	*5.38*	13.98
生態環境 *shēngtài huánjìng* 'ecological environment'	111	*22.99*	15	*11.74*	11.25

Table 9. Words functioning as *objects* that have higher saliency in CNA than XIN

Word	XIN object.frequency	*XIN object.saliency*	CNA object.frequency	*CNA object.saliency*	Saliency difference
戲路 *xìlù* 'acting style'	3	*12.74*	40	*40.51*	27.77
現狀*xiànzhuàng* 'current situation'	426	*44.35*	1247	*67.1*	22.75
地貌 *dìmào* 'landform'	4	*5.11*	26	*25.78*	20.67
生態 *shēngtài* 'ecology'	13	*2.96*	133	*22.22*	19.26
本質 *běnzhì* 'essence'	9	*7.8*	57	*25.31*	17.51
航道 *hángdào* 'sailing track'	5	*4.45*	35	*21.49*	17.04
心態 *xīntài* 'mentality'	17	*12.97*	136	*29.96*	16.99
印象 *yìnxiàng* 'impression'	22	*10.07*	112	*25.87*	15.8
政策 *zhèngcè* 'policy'	517	*22.97*	1847	*38.4*	15.43
一生 *yīshēng* 'a life time'	25	*14.47*	88	*29.46*	14.99

Table 10. Number of unique words in two grammatical relations in XIN and CNA respectively

Collocations	Unique words in *subject-predicate collocations* in XIN	Unique words in *verb-object collocations* in XIN	Unique words in *subject-predicate collocations* in CNA	Unique words in *verb-object collocations* in CNA
No. of collocations	181	153	273	305
Total	283		420	

Words that appear as subject as well as object are not repeatedly counted, so the total number is not the sum of the two kinds of relations.

level, Abstract Things is the largest category in both Mainland China and Taiwan. In the Mainland, the unique words are distributed in eight categories at the first semantic level, with Abstract Things accounting for 67.1%, Living Things accounting for 16.8%, Concrete Things accounting for 8.7%. Time and Space, Social Activities, Biological Activities, Movement and Changes, and Auxiliary Words only account for 3.8%, 1.4%, 1.0%, 0.7%, and 0.3% respectively. In Taiwan, unique words are also distributed in eight categories, with Abstract Things accounting for 69.8%, followed by Living Things 14.1%. Time and Space 5.6%, Concrete Things 5.4%, Social Activities 2.8%, Biological Activities 1.2%, Nature and State 0.7%, and Movement and Changes 0.5%.

Table 11. Distribution of *unique words* in XIN among the first-level semantic categories

First-level semantic category	Number	Percentage	Examples
抽象事物*chōuxiàng shìwù* 'Abstract Things'	192	67.1%	傳統農業 *chuántǒng nóngyè* 'traditional agriculture', 鄉鎮企業 *xiāngzhèn qǐyè* 'township enterprise', 村 *cūn* 'village', 知識 *zhīshí* 'knowledge', 支出 *zhīchū* 'expenditure'
生物*shēngwù* 'Living Things'	48	16.8%	村民 *cūnmín* 'villager', 鄉親們 *xiāngqīn men* 'folks', 職工 *zhígōng* 'staff', 群衆 *qúnzhōng* 'the masses', 基因組 *jīyīn zǔ* 'genome'
具體物 *jùtǐ wù* 'Concrete Things'	25	8.7%	自然 *zìrán* 'nature', 土壤 *tǔrǎng* 'soil', 建築 *jiànzhù* 'building', 道路 *dàolù* 'road', 導彈 *dǎodàn* 'missile'
時空 *shíkōng* 'Time and Space'	11	3.8%	農村 *nóngcūn* 'rural area', 山河 *shānhé* 'mountains and rivers', 家鄉 *jiāxiāng* 'hometown', 領土 *lǐngtǔ* 'territory', 大地 *dàdì* 'land', 保護期 *bǎohùqī* 'the protection period'
社會活動*shèhuì huódòng* 'Social Activities'	4	1.4%	選舉 *xuǎnjǔ* 'election', 高考 *gāokǎo* 'college entrance examination', 分類 *fènlèi* 'classification', 出臺 *chūtái* 'appear publicly'
生物活動*shēngwù huódòng* 'Biological Activities'	3	1.0%	預算 *yùsuàn* 'budget', 期望 *qīwàng* 'expectations', 抉擇 *juézé* 'choice'
運動與變化*yùndòng yǔ biànhuà* 'Movement and Changes'	2	0.7%	終場 *zhōngchǎng* 'end of event', 建成 *jiànchéng* 'built'
輔助詞 *fǔzhù cí* 'Auxiliary Words'	1	0.3%	自然 *zìrán* 'naturally'
Total	286	100.0%	/

Note: 自然 *zìrán* 'nature/ naturally', 道路 *dàolù* 'the way/ road', and 出路 *chūlù* 'way out' and polysemous, and thus each belong to two semantic categories

Since Abstract Things is the largest semantic category of these words, we further examined the second-level semantic categories in XIN and CNA. The results are shown

Table 12. Distribution of *unique words* in CNA among the first-level semantic categories

First-level semantic category	Number	Percentage	Examples
抽象事物*chōuxiàng shìwù* 'Abstract Things'	298	69.8%	選制 *xuǎnzhì* 'election system', 選情 *xuǎnqíng* 'the situation of election', 唱法 *chàngfǎ* 'singing style', 潮流 *cháoliú* 'trend', 聯准會 *liánzhǔn huì* 'The Federal Researve System'
生物*shēngwù* 'Living Things'	60	14.1%	選民 *xuǎnmín* 'voter', 導演 *dǎoyǎn* 'director', 駕駛人*jiàshǐ rén* 'driver', 心 *xīn* 'heart', 植被 *zhíbèi* 'vegetation'
時空 *shíkōng* 'Time and Space'	24	5.6%	選區 *xuǎnqū* 'electoral district', 天空 綫*tiānkōngxiàn* 'skyline', 栖地*qīdì* 'habitat', 未來 *wèilái* 'future', 周期 *zhōuqī* 'cycle period'
具體物 *jùtǐ wù* 'Concrete Things'	23	5.4%	選票 *xuǎnpiào* 'ballot', 食譜*shípǔ* 'recipe', 食物 *shíwù* 'food', 船隻*chuánzhī* 'boat', 空調 *kōngdiào* 'air conditioning'
社會活動*shèhuì huódòng* 'Social Activities'	12	2.8%	公投 *gōngtóu* 'referendum', 研究 *yánjiū* 'research', 決策 *juécè* 'decision-making', 軍售 *jūnshòu* 'arms sales', 訴求 *sùqiú* 'demand'
生物活動*shēngwù huódòng* 'Biological Activities'	5	1.2%	認知 *rèzhī* 'cognition', 品味 *pǐnwèi* 'taste', 意識 *yìshí* 'awareness', 預期 *yùqī* 'expectation', 作息 *zuòxī* 'work and rest'
性質與狀態 *xìngzhì yǔ zhuàngtài* 'Nature and State'	3	0.7%	權威 *quánwēi* 'authority', 重要性 *zhòngyào xìng* 'importance', 高層 *gāocéng* 'senior management'
運動與變化 *yùndòng yǔ biànhuà* 'Movement and Changes'	2	0.5%	動脉 *dòngmài* 'artery', 組合 *zǔhé* 'combination'
Total	427	100.0%	/

Note: 心 *xīn* 'heart', 決策 *juécè* 'decision-making', 標志 *biāozhì* 'sign', 動向 *dòngxiàng* 'trend', 意識 *yìshí* 'awareness', 未來 *wèilái* 'future', and 出路 *chūlù* 'way out' belong to two semantic categories each.

in Table 13 and Table 14, and both of them include 10 s-level semantic categories. The largest category in Mainland China is Society, accounting for 21.9%, and the others are Politics, Matters, Attributes, Science and Education, Quantity, Economic, Stylistic, Military, and Awareness. In comparison, the largest category is Attributes in Taiwan, followed by Politics, Matters, Science and Education, Awareness, Society, Quantity, Economic, Military, and Culture and Sports.

As shown in Table 11 and Table 13, the unique words in Mainland China show the following characteristics. (1) Among the categories of Living Things, Space and Time,

Table 13. The second-level semantic category distribution of *unique words of Abstract Things* in XIN

Second-level semantic category	Number	Percentage	Examples
社會 *shèhuì* 'Society'	42	21.90%	傳統農業 *chuántǒng nóngyè* 'traditional agriculture', 鄉鎮企業 *xiāngzhèn qǐyè* 'township enterprise', 歐佩克 *ōupèikè* 'OPEC', 歐共體 *ōugòngtǐ* 'European Community', 國際排聯 *guójì páilián* 'FIVB', 國際足聯 *guójì zúlián* 'FIFA', 國家隊 *guójiā duì* 'national team', 奧委會 *àowěihuì* 'Olympic Committee'
政治*zhèngzhì* 'Politics'	34	17.70%	村 *cūn* 'village', 準則 *zhǔnzé* 'guideline', 責任 *zérèn* 'responsibility', 使命 *shǐmìng* 'mission', 任命制 *rènmìng zhì* 'appointment system'
事情 *shìqíng* 'Matters'	33	17.20%	電子商務 *diànzǐ shāngwù* 'E-commerce', 信息 *xìnxī* 'information', 戰局 *zhànjú* 'war situation', 治安 *zhìān* 'public security', 渠道 *qúdào* 'channel', 成果 *chéngguǒ* 'achievement'
屬性 *shǔxìng* 'Attributes'	33	17.20%	主場 *zhǔchǎng* 'home field', 名詞 *míngcí* 'noun', 次序 *cìxù* 'order', 素質 *sùzhì* 'quality', 技術 *jìshù* 'technology'
科教 *kējiào* 'Science and Education'	25	13.00%	信息技術 *xìnxī jìshù* 'information technology', 網絡 *wǎngluò* 'network', 互聯網 *hùlián wǎng* 'internet', 因特網 *yīntè wǎng* 'internet', 教育 *jiàoyù* 'education'
數量 *shù liàng* 'Quantity'	9	4.70%	數據 *shùjù* 'data', 深度 *shēndù* 'depth', 密度 *mìdù* 'density', 尺寸 *chǐcùn* 'size', 參數 *cānshù* 'parameter'
經濟 *jīngjì* 'Economy'	7	3.60%	支出 *zhīchū* 'expenditure', 價格 *jiàgé* 'price', 成本 *chéngběn* 'cost', 油價 *yóujià* 'petroleum price', 製造商 *zhìzàoshāng* 'manufacturer'

(*continued*)

Table 13. (*continued*)

Second-level semantic category	Number	Percentage	Examples
文體 *wéntǐ* 'Culture and Sports'	5	2.60%	足球 *zúqiú* 'football', 世界杯 *shìjièbēi* 'World Cup', 國安隊 *guóāndui* 'Guo'an Football Club', 非典 *fēidiǎn* 'SARS', 畫面 *huàmiàn* 'picture'
軍事 *jūnshì* 'Military'	2	1.00%	戰爭 *zhànzhēng* 'war', 海灣戰爭 *hǎiwān zhànzhēng* 'the Gulf War'
意識 *yìshí* 'Awareness'	2	1.00%	偏見 *piānjiàn* 'prejudice', 愛心 *àixīn* 'the sense of love and compassion'
Total	192	100.00%	/

and Abstract Things, there are a large number of words related to agriculture, farmers, and rural areas, such as 村民 *cūnmín* 'villager', 農村 *nóngcūn* 'rural area', 傳統農業 *chuántǒng nóngyè* 'traditional agriculture', 鄉鎮企業 *xiāngzhèn qǐyè* 'township enterprises', and 村 *cūn* 'village'. The issues of agriculture, rural areas, and farmers have always been a major concern in Mainland China. Like other developing countries, Mainland China is also facing the challenge of sustainable agricultural development, such as how to properly plan and use natural resources like land and water, how to improve agricultural production conditions, and how to increase farmers' income. (2) In the second-level semantic categories of Abstract Things, there are many words related to the internet, such as 電子商務 *diànzǐ shāngwù* 'E-commerce', 信息 *xìnxī* 'information', 信息技術 *xìnxī jìshù* 'information technology', 網絡 *wǎngluò* 'network', 互聯網 *hùliánwǎng* 'internet', and 因特網 *yīntèwǎng* 'internet'. In the 20th century, with the rapid rise of the internet economy, e-commerce has profoundly and comprehensively changed the economy and life in Mainland China. For example, hotels have been transformed from traditional management methods to online approaches, because the majority of travelers tend to search accommodation and book hotels on the internet. (3) In the second-level categories of Abstract Things, there are many words related to the international situations and organizations, such as 歐佩克 *ōupèikè* 'OPEC', 歐共體 *ōugòngtǐ* 'European Community', 戰局 *zhànjú* 'war situation', 戰爭 *zhànzhēng* 'war', and 海灣戰爭 *hǎiwān zhànzhēng* 'the Gulf War'. These words reflect that people in Mainland China pay close attention to issues of changes in the international, political and economic situations. (4) In the second-level categories of Abstract Things, there are words related to *Society* and *Culture and Sports*, such as 國際排聯 *guójì páilián* 'FIVB', 國際足聯 *guójì zúlián* 'FIFA', 國家隊 *guójiā duì* 'national team', 奥委會 *àowěihuì* 'Olympic Committee', and足球 *zúqiú* 'football', revealing that Mainland China paid attention to promoting sports and reforming the sports system.

In comparison, the unique words in Taiwan have different characteristics, as shown in Tables 12 and 14. (1) There are a lot of election-related words among the first-level categories of Abstract Things, Living Things, Space and Time, and Concrete Things, such as 選制 *xuǎnzhì* 'election system', 選情 *xuǎnqíng* 'situation of election', 選區 *xuǎnqū* 'electoral district', 選民 *xuǎnmín* 'voter', and 選票 *xuǎnpiào* 'ballot', which show that Taiwan was very concerned about changes of the election. It may arise from

Table 14. Distribution of *unique words of Abstract Things* in can among the second-level semantic categories

Second-level semantic category	Number	Percentage	Examples
屬性 *shǔxìng* 'Attributes'	67	22.5%	黨名 *dǎngmíng* 'the name of the political party', 積弊 *jībì* 'accumulated disadvantage', 席次 *xícì* 'seats (of political groups in Parliament / Legislative Council)', 型貌 *xíngmào* 'appearance', 特質 *tèzhì* 'characteristics'
政治 *zhèngzhì* 'Politics'	53	17.8%	南向政策 *nánxiàng zhèngcè* 'Southern Policy', 稅制 *shuìzhì* 'tax system', 稅法 *shuìfǎ* 'tax law', 選制 *xuǎnzhì* 'elective system', 新黨 *xīndǎng* 'New Party'
事情 *shìqíng* 'Matters'	51	17.1%	選情 *xuǎnqíng* 'situation of election', 病程 *bìngchéng* 'th course of illness', 制程 *zhìchéng* 'the process of production', 現勢 *xiànshì* 'current situation', 世局 *shìjú* 'world situation'
科教 *kē jiào* 'Science and Education'	38	12.8%	報紙 *bàozhǐ* 'newspaper', 科目 *kēmù* 'subject', 文化 *wénhuà* 'culture', 議題 *yìtí* 'issue', 輿論 *yúlùn* 'public opinion'
意識 *yìshí* 'Awareness'	37	12.4%	民意 *mínyì* 'public opinion', 共識 *gòngshí* 'consensus', 人心 *rénxīn* 'popular support', 野心 *yěxīn* 'ambition (negative)', 成見 *chéngjiàn* 'prejudice'
社會 *shè huì* 'Society'	20	6.7%	選風 *xuǎnfēng* 'election style', 陣營 *zhènyíng* 'camp', 組織 *zǔzhī* 'organization', 邦交 *bāngjiāo* 'diplomatic relation', 台塑 *táisù* 'Formosa Plastic Corporation'
數量 *shù liàng* 'Quantity'	13	4.4%	比率 *bǐlǜ* 'ratio', 費率 *fèilǜ* 'rate', 酸碱值 *suānjiǎn zhí* 'PH value', 人數 *rénshù* 'number of people', 數字 *shùzì* 'number'

(*continued*)

Table 14. (*continued*)

Second-level semantic category	Number	Percentage	Examples
經濟 *jīngjì* 'Economy'	8	2.7%	稅收 *shuìshōu* 'tax', 外資 *wàizī* 'foreign capital', 幣值 *bìzhí* 'currency value', 價值 *jiàzhí* 'value', 獎金 *jiǎngjīn* 'bonus'
軍事 *jūnshì* 'Military'	6	2.0%	戰法 *zhànfǎ* 'tactics', 戰事 *zhànshì* 'warfare', 軍隊 *jūnduì* 'military', 美軍 *měijūn* 'U.S. Military'
文體 *wéntǐ* 'Culture and Sports'	5	1.7%	電影 *diànyǐng* 'movie', 藝術 *yìshù* 'art', 曲目 *qǔmù* ' repertoires of songs/music, etc.', 手術 *shǒushù* 'surgery'
Total	298	100.0%	/

Taiwan's electoral system. The political party and politician elections were fiercely contested and their political opinions were often promoted on various platforms to win over voters. This is an important part of the political life in Taiwan. (2) In the second-level category of Abstract Things, especially in Politics, there are many words related to economic policies, such as 南向政策 *nánxiàng zhèngcè* 'Southern Policy', 稅制 *shuìzhì* 'tax system', and 稅法 *shuìfǎ* 'tax law'. For example, Taiwan has been implementing the southbound policy successively since the 1990s, seeking a way out for its labor-intensive industry and promoting relocation of investment to Southeast Asian countries. Taiwan made some changes in the tax system and laws during 1990s.

6 Conclusion

This paper selected the commonly used word 改變 *gǎibiàn* 'change' and extracted its grammatical collocates from the Tagged Chinese Gigaword Corpus using Chinese Word Sketch Engine. We then compared the similarities and differences of the collocates, including subject collocates in subject-predicate relations and object collocates in the verb-object relations. It further explored the common and specific concerns of changes between the two sides of the strait from the beginning of the 1990s to the early 21st century.

Common words in XIN and CNA help us to understand the common concerns of the changes on both sides of the Taiwan Strait. The study has the following findings. (1) For the common words in *subject* positions, it is found that these words distributed in six categories of the first level semantic categories, with Abstract Things accounting for the most. The second-level semantic categories of Abstract Things are distributed in 11 categories. Both indicate that the subjects that have changed on both sides of the strait were very widely distributed. Moreover, among the second-level semantic categories of Abstract Things, the larger categories are Matters and Attributes, reflecting the fact that both Mainland China and Taiwan are very concerned about the changes of political relations. Further comparisons of the saliency difference of common words in *subject* positions reveals that some of the commonly concerned issues in Mainland China and Taiwan differ in terms of attention: Mainland China is highly concerned about the the changes in economy and development situation, while Taiwan pays more attention to climate and landform issues. (2) For the common words in the *object* position, it is found that these words are distributed in seven categories of the first semantic level, with Abstract Things also accounting for the most. The second-level semantic categories of Abstract Things are distributed in 10 categories. This indicates that the objects that changed on both sides of the strait were also very widely distributed, the same as the denotations of the subject collocates. Secondly, among the second-level semantic categories of Abstract Things, the largest categories are Matters and Attributes, showing that both Mainland China and Taiwan were concerned about health and diet, issues of people's livelihood. But the saliency of common words in *object* positions indicates that Mainland China was highly concerned about its changing economic and social backwardness, while Taiwan closely focused on the film and television culture, and the status quo across the Strait.

Unique words can directly reflect differences between the two sides through the issues of changes in economy, politics, and culture. Firstly, the first-level semantic categories of these words in Mainland China and Taiwan are both distributed in eight categories, seven of which are the same, with the Abstract Things being the largest category. But there are some differences in the second-level semantic category analysis of Abstract Things. The largest proportion of Abstract Things is Society in Mainland China, whereas Attributes has the largest proportion in Taiwan. Secondly, a further comparative analysis of the semantic characteristics of unique words, especially in the category of Abstract Things, found that Mainland China is more concerned with the issues of agriculture, farmers and rural areas, e-commerce economy, the international situation, and sports, while Taiwan is more concerned about changes in the election situation and economic policies that can affect the investments of Taiwan's businessmen.

With the increasingly frequent economic and cultural exchanges between the Chinese-speaking communities and other areas of the world, global Chinese vocabulary will inevitably develop and change. Vocabulary can also provide a perspective to observe society. This fertile ground for academic research is worth further exploring in the future. Investigating the characteristics of the global Chinese vocabulary and conducting comparative studies can provide references for a comprehensive understanding of the situation of the Chinese language and promote the development of Chinese studies. It can also facilitate communication and exchanges among different Chinese speaking regions.

Acknowledgement. This research is supported by University of Macau (MYRG2019–00013-FAH).

References

1. Lu, J.: Establish and confirm the concept of "greater Chinese" (shùlì bìng quèrèn "dà huáyǔ" gàiniàn). Chin. Cult. Teach. World (shìjiè huáwén jiàoxué), 3–5 (2019)

2. Guo, X., Zhu, X.: Theories and practices of "language life" in China (yǔyán shēnghuó yánjiū shínián). Chin. J. Lang. Policy Plan. (yǔyán zhànlüè yánjiū), 24–33 (2016)
3. Li, Y.: A pioneer of global Chinese: a preface to language selection and language education by Zhou Qinghai ("dà huáyǔ" de yīmiàn qízhì——xù Zhōu Qīnghǎi xiānshēng《yǔyán xuǎnzé yǔ yǔwén jiàoyù》). TCSOL Stud. (huáwén jiàoxué yǔ yánjiū), 1–4+23 (2020)
4. Li, Y.: The greater Chinese: lingua franca of the global Chinese (dà huáyǔ: quánqiú huárén de gòngtóngyǔ). Appl. Linguis. (yǔyán wénzì yìngyòng), 2–13 (2017)
5. Guo, X.: On the research of Huayu (lùn huáyǔ yánjiū). Appl. Linguis. (yǔyán wénzì yìngyòng), 22–28 (2006)
6. Shang, G., Chew, C.: Singapore Mandarin: sound and pronunciation variations (Xīnjiāpō huáyǔ de yǔyīn yǔ liúbiàn). J. Int. Chin. Stud. (guójì hànyǔ xuébào), 205–216 (2016)
7. You, Y.: The phonetic features of "Mandarin" in Taiwan and its impact on the Mainland (Táiwān "guóyǔ" de yǔyīn tèsè jí qí duì Dàlù dìqū de yǐngxiǎng). Mod. Chin. (Lang. Res. Ed.) (xiàndài yǔwén yǔyán yánjiū bǎn), 34–36 (2014)
8. Lu, J.: Syntactic features and standardization of Singapore Chinese (I) (Xīnjiāpō huáyǔ jùfǎ tèdiǎn jí qí guīfàn wèntí- shàng). Overs. Chin. Educ. (hǎiwài huáwén jiàoyù), 1–11 (2001)
9. Lu, J.: Syntactic features and standardization of Singapore Chinese (II) (Xīnjiāpō huáyǔ jùfǎ tèdiǎn jí qí guīfàn wèntí- xià). Overs. Chin. Educ. (hǎiwài huáwén jiàoyù), 1–10 (2002)
10. Diao, Y.: The contrast study about "zao (遭)" sentenses in the four places across the Taiwan Strait (liǎng'àn sìdì de "zāo" zìjù jí qí yǔ "bèi" zìjù de chàyì). Lang. Teach. Linguis. Stud. (yǔyán jiàoxué yǔ yánjiū), 70–77 (2012)
11. Tin, S.: Research on Hong Kong community words (Xiānggǎng shèqūcí yánjiū). Linguis. Sci. (yǔyán kēxué), 29–37 (2004)
12. Jia, Y., Xu, Y.: The supplementary examples of Singapore Chinese and its variance analysis with Mandarin (Xīnjiāpō huáyǔ tèyǒu cíyǔ bǔlì jí qí yǔ Pǔtōnghuà cíyǔ chāyì fēnxī). J. Coll. Chin. Lang. Cult. Jinan Univ. (Jìnán dàxué huáwén xuéyuàn xuébào), 32–38+49 (2005)
13. Tang, Z.: On "HongKong-Macau words" and "Macau-specific words" (lùn "Gǎng-Ào cíyǔ" yǐjí "Àomén tèyǒu cíyǔ"). J. Jiangsu Univ. (Soc. Sci. Ed.) (Jiāngsū dàxué xuébào shèhuì kēxué bǎn), 24–29 (2008)
14. Shi, C.: On the problem of Chinese community words from the features of Thai Chinese words (cóng tàishì huáwén de yòngcí tèzhēng kàn huáwén shèqūcí wèntí). Linguis. Res. (yǔwén yánjiū), 26–34 (2015)
15. Wang, S., Luo, H.: A corpus-based study of the vocabulary of Macau tourism Chinese. In: Tao, H., Chen, H.-J.H. (eds.) Chin Lang Learn Sci, pp. 373–391. Springer, Singapore (2019). https://doi.org/10.1007/978-981-13-9505-5_16
16. Examination Center of the National Chinese Proficiency Test Committee. The Syllabus of Chinese Vocabulary and Characters' Level (hànyǔ shuǐpíng cíhuì yǔ hànzì děngjí dàgāng). Economic Science Press, Beijing (2001)
17. Tourism Chinese Vocabulary Project Group, College of Chinese as a Foreign Language, Shanghai Normal University: The Tourism Chinese Vocabulary Handbook (lǚyóu hànyǔ cíhuì shǒucè). Shanghai World Book Publishing Company, Shanghai (2008)
18. Li, Y.: Global Chinese Dictionary (quánqiú huáyǔ cídiǎn). The Commercial Press, Beijing (2010)
19. Li, Y.: Comprehensive Global Chinese Dictionary (quánqiú huáyǔ dà cídiǎn). The Commercial Press, Beijing (2016)
20. Zhu, X., Zhou, T.: A review of the domestic research of global Chinese (quánqiú huáyǔ guónèi yánjiū zōngshù). Chin. J. Lang. Policy Plan. (yǔyán zhànlüè yánjiū), 49–59 (2017)
21. Huang, G.: Overseas Chinese language and Mandarin (hǎiwài hànyǔ yǔ Pǔtōnghuà). Chin. Lang. Learn. (hànyǔ xuéxí), 49–51 (1984)
22. Zhu, Y.: Analysis of language differences in Taiwan's novels (Táiwān xiǎoshuō yǔyán xīyì). Chin. Literat. (huáwén wénxué), 57+61–63 (1988)

23. Tang, Z.: Examples of the lexical variants among Mainland China, Taiwan, Hong Kong, and Singapore (Zhōngguó Dàlù、Táiwān、Xiānggǎng、Xīnjiāpō hànyǔ cíhuì fāngmiàn ruògān chāyì jǔlì). J. Xuzhou Normal Univ. (Xúzhōu shīfàn xuéyuàn xuébào), 103–109 (1995)
24. Liu, X.: The Contemporary Chinese dictionary Across the Straits Investigation into the Special Meaning of the Unique Terms in Taiwanese Chinese (《liǎng'àn xiàndài hànyǔ chángyòng cídiǎn》Táiwān hànyǔ tèyǒu cíyǔ jí cíyǔ tèyǒu yìxiàng kǎochá). vol. Master. Beijing Language and Culture University, Beijing (2006)
25. Su, J.: Research on homographs on both sides of the Taiwan Strait (hǎixiá liǎng'àn tóngxíng-yìyì-cí yánjiū). Stud. Chin. Lang. (Zhōngguó yǔwén), 107–117 (1995)
26. Hou, C.: On the synonymous heteromorphic words on both sides of the Taiwan Strait (shìtán hǎixiá liǎng'àn de tóngyì yìxíng cíyǔ). J. Zhanjiang Teach. Coll. (philosophy & social sciences) (Zhànjiāng shīfàn xuéyuàn xuébào zhéxué shèhuì kēxué bǎn), 52–55 (1999)
27. Su, X.: Studies of neologisms in Taiwan (Táiwān xīncíyǔ jí qí yánjiū tèdiǎn). J. Xiamen Univ. (Arts Soc. Sci.) (Xiàmén dàxué xuébào zhéxué shèhuì kēxué bǎn), 23–29 (2003)
28. Zhao, C.: An investigation into the new word cognitive differences between Mainland China and Taiwan (Dàlù hé Táiwān xīncíyǔ chāyì de rènzhī kǎochá). J. Huzhou Teach. Coll. (Húzhōu shīfàn xuéyuàn xuébào), 105–108 (2012)
29. Diao, Y.: A microscopic comparison of cross-strait varieties: the case of Eryi (而已) (shìlùn hǎixiá liǎng'àn yǔyán de wēiguān duìbǐ yánjiū——yǐ "éryǐ" yīcí de kǎochá fènxī wèilì). J. Beijing Normal Univ. (Soc. Sci.) (Běijīng shīfàn dàxué xuébào shèhuì kēxué bǎn), 44–51 (2012)
30. Chu, Z., Zhang, Q.: The diversity and tendentiousness investigation of "Tou Guo" in both sides of the Taiwan Strait (hǎixiá liǎng'àn "tòuguò" yòngfǎ de duōyàngxìng yǔ qīngxiàngxìng kǎochá). Appl. Linguis. (yǔyán wénzì yìngyòng), 70–79 (2013)
31. Dang, J., Wang, L., Su, J.: On the developing tendency of Taiwan Mandarin absorbing vocabularies from Putonghua (Táiwān guóyǔ xīshōu Dàlù Pǔtōnghuà cíyǔ qūshì yánjiū). Appl. Linguis. (yǔyán wénzì yìngyòng), 113–121 (2017)
32. Li, Y., Shi, C.: A study on the interaction between the words on both sides of the Taiwan Strait (hǎixiá liǎng'àn cíyǔ hùdòng guānxì yánjiū). Contemp. Rhetoric (dāngdài xiūcí xué), 64–73 (2011)
33. Wang, S., Wang, X.: The attention to safety issues from mainland China and Taiwan. In: Hong, Jia-Fei., Su, Qi., Wu, Jiun-Shiung. (eds.) CLSW 2018. LNCS (LNAI), vol. 11173, pp. 801–818. Springer, Cham (2018). https://doi.org/10.1007/978-3-030-04015-4_69
34. Kilgarriff, A., Huang, C.-R., Rychlý, P., Smith, S., Tugwell, D.: Chinese word sketches. In: ASIALEX 2005: Words in Asian Cultural Context, Singapore (2005)
35. Chinese Gigaword version 2.0. (https://catalog.ldc.upenn.edu/LDC2005T14), http://www.ldc.upenn.edu/Catalog/catalogEntry.jsp?catalogId=LDC2009T14
36. Wang, S., Huang, C.-R.: Adjectival modification to nouns in Mandarin Chinese: Case Studies on "cháng+noun" and "adjective+túshūguǎn. In: Otoguro, R., Ishikawa, K., Umemoto, H., Yoshimoto, K., Harada, Y. (eds.) Proceedings of The 24th Pacific Asia Conference on Language, Information and Computation (PACLIC-24), pp. 701–705. Tohoku University, Sendai (2010)
37. Wang, S., Huang, C.-R.: Type construction of event nouns in mandarin Chinese. In: Proceedings of The First Workshop on Generative Lexicon for Asian Languages (GLAL-1), Workshop of The 26th Pacific Asia Conference on Language, Information and Computation (PACLIC-26), pp. 582–591. Faculty of Computer Science, Universitas Indonesia, Bali (2012)
38. Wang, S., Huang, C.-R.: The semantic type system of event nouns. In: Jing, S.Z. (ed.) Increased Empiricism: Recent Advances in Chinese Linguistics, vol. 2,pp. 205–221. John Benjamins Publishing Company, Amsterdam/Philadelphia (2013)
39. Wang, S., Huang, C.-R.: Word sketch lexicography: new perspectives on lexicographic studies of chinese near synonyms. Lingua Sinica **3**, 1–22 (2017)

40. Wang, S., Huang, C.-R., Xu, H.: Compositionality of NN compounds: a case study on [N_1+Artifactual-Type Event Nouns]. in: Proceedings of The 26th Pacific Asia Conference on Language, Information and Computation (PACLIC-26), pp. 70–79. Faculty of Computer Science, Universitas Indonesia, Bali (2012)
41. Wu Y., Wang, S.: Applying Chinese word sketch engine to distinguish commonly confused words. In: Dong, M., Lin, J., Tang, X. (eds.) Chinese Lexical Semantics. CLSW 2016. Lecture Notes in Computer Science, vol. 10085, pp. 600–619 (2016). https://doi.org/10.1007/978-3-319-49508-8_55
42. Wang, S.: Semantics of Event Nouns. vol. Ph.D. The Hong Kong Polytechnic University, Hong Kong (2013)
43. Wang, S., Yin, J.: A Corpus-based Study of Keywords in Legislative Chinese and General Chinese. In: Hong, J.-F., Zhang, Y., Liu, P. (eds.) Chinese Lexical Semantics, pp. 639–653. Springer, Cham, Switzerland (2020)
44. Wang, S., Yin, J.: A comparative study of the collocations in legislative Chinese and general Chinese. In: Hong, J.-F., Zhang, Y., Liu, P. (eds.) CLSW 2019. LNCS (LNAI), vol. 11831, pp. 710–724. Springer, Cham (2020). https://doi.org/10.1007/978-3-030-38189-9_72
45. Wang, S.: Chinese Multiword Expressions: Theoretical and Practical Perspectives. Springer Singapore, Singapore (2020). https://doi.org/10.1007/978-981-13-8510-0
46. Huang, C.-R., et al.: Chinese Sketch Engine and the extraction of grammatical collocations. In: Proceedings of the Fourth SIGHAN Workshop on Chinese Language Processing, Jeju Island, Korea, pp. 48–55 (2005)
47. Su, X.: A Thesaurus of Modern Chinese (xiàndài hànyǔ fēnlèi cídiǎn). The Commercial Press, Beijing (2013)

The Collocations of Chinese Tactile Adjectives

Shan Wang[1,2(✉)], Le Wu[1], and Qiaomin Gong[3]

[1] Faculty of Arts and Humanities, University of Macau, Taipa, Macau, China
shanwang@um.edu.mo
[2] Institute of Collaborative Innovation, University of Macau, Taipa, Macau, China
[3] School of Art Design, Hankou University, Wuhan, China

Abstract. Embodied cognition is a basic way for people to understand the world. As the basic sense of people, the tactile sense plays an important role in the process of understanding the world. Existing research focused on the definition, preliminary comparison between Chinese and foreign tactile adjectives, and the extension of word meanings of tactile adjectives, but there is rarely any systematic study on the distribution of the collocations of Chinese tactile adjectives. In this paper, the typical adjectives 軟 *ruǎn* 'soft' and 硬 *yìng* 'hard' of the tactile domain were selected as representatives. We used a large-scale corpus to extract the collocational data of the two adjectives and annotated the word meaning of 軟 *ruǎn* 'soft' and 硬 *yìng* 'hard' in each sentence, as well as the semantic categories of their collocates. We then compared the characteristics of their collocates in the experiential domain and the abstract domain by exploring the distribution of their collocates. This study is helpful in deepening understanding of lexicon expansion in different cognitive domains and better understanding the cognitive process of the tactile sense.

Keywords: Tactile adjectives · Collocations · Semantic categories · The experiential domain · The abstract domain

1 Introduction

Embodied cognition is a basic way for people to understand the world. The tactile sense, as the basic sense of human beings, plays an important role in the process of understanding the world. Tactile adjectives have attracted a great amount of attention in linguistic research. Existing research mainly focused on the definition of such adjectives, preliminary comparison of Chinese and foreign tactile adjectives, and the extension of word meaning of tactile adjectives. However, previous research is unclear about how to judge experiential and abstract domains of the collocates of Chinese tactile adjectives and there is a lack of systematic research on the semantic distribution of the collocates. This study used a large-scale corpus to explore the collocations of the two selected tactile adjectives 軟 *ruǎn* 'soft' and 硬 *yìng* 'hard'. Specifically, collocates of the two tactile adjectives were first automatically extracted from subject-predicate relations and attributive-head relations. Then the collocates were assigned to the experiential domain and the abstract domain one by one through sense annotation. The semantic categories

M. Liu et al. (Eds.): CLSW 2020, LNAI 12278, pp. 711–733, 2021.
https://doi.org/10.1007/978-3-030-81197-6_59

of these collocates were also manually annotated. The study of the semantic distribution rules in the experiential and abstract domains provide a new perspective for studying tactile adjectives and deepen the understanding of the role of the tactile sense in people's experience of the world.

2 Related Research

The existing research on adjectives in the tactile domain mainly focused on three aspects. First, it is the definition of tactile adjectives. Wang [1] compared the definitions of 'touch' in various dictionaries and pointed out that the unified understanding is that the meaning of touch is the sense caused by external stimuli to the skin, but it is difficult to describe the degree of stimulation accurately. Therefore, she defined tactile adjectives as feelings caused by external stimuli touching the skin surface, which is a general description for mechanical stimuli, such as contact, sliding, and pressure perception. Wen [2] divided tactile adjectives into three semantic fields: the semantic field of the sense of touch and pressure (軟 *ruǎn* 'soft', 硬 *yìng* 'hard', 松 *sōng* 'loose', 緊 *jǐn* 'tight'), the semantic field of the sense of temperature (冷 *lěng* 'cold', 熱 *rè* 'hot', 温 *wēn* 'warm', 暖 *nuǎn* 'warm'), and the semantic field of the sense of pain (痛 *tòng* 'pain', 癢 *yǎng* 'itchy'). He [3] integrated the above two views and defined tactile sense as the feeling experienced by the human skin due to contact with external objects through contacting, sliding, or squeezing. It mainly includes three categories: the sense of touch and pressure, the sense of temperature, and the sense of pain. She also pointed out that humans use the sense of touch to identify the physical characteristics of an object.

Second, studies have been conducted to explore the similarities and differences of various cognitions through the comparative study of tactile adjectives in two languages. Park [4] made a comparison between Chinese and Korean on the syntactic and semantic levels. In terms of the semantic level, they can both express the senses of hearing, taste and smell, but for tactile adjectives with the same basic meaning, their extended meanings are not necessarily the same. Chu [5] found that synesthesia exists in both Chinese and Japanese tactile adjectives; that is, adjectives in the tactile domain can be extended to the domains of sight, hearing, and smell. Lu [6] summarized the extension of Chinese and English adjectives to six categories, including from solid to liquid, from concrete to abstract, from objective to subjective, from natural objects to social objects, from space to time, and from non-entities to entities. These extensions conform to the development of society and the laws of human cognition.

Third, studies have also been conducted to explore the extension of the meanings of tactile adjectives from the perspective of semantics. Wang [1] explained the extension and construction of the meanings of tactile adjectives from a cognitive perspective. She found that the extension rules are projected from concrete feelings to abstract psychology and emotion. Wen [2] explored the semantic characteristics of English and Chinese adjectives and found that the characteristics of the derivative meanings of English and Chinese tactile adjectives are basically the same; if their basic meanings are opposite, then their derived meanings are also opposite in general.

In addition, some scholars have explored adjectives from the perspective of constructions and metaphors. Tang [7] compared the similarities and differences of the subjects, constructions and configurations of 12 groups of adjectives in Chinese and English, including the taste domain, the temperature domain, the tactile domain and the visual domain. He proposed an internal consistency hypothesis: the higher is the internal consistency in Chinese-English adjective, the higher their metaphorical extension will be; otherwise, the extension will be lower. Based on corpus research, Tang, Dong and Li [8] found that besides differences in cognition and script, the conventional system based on usage also has a certain influence on the generation of metaphors. Tang and Tian [9] proposed a prioritized sequence of internal consistency elements of physical adjectives: experience > cognitive framework > adjective nature > cultural experience. Li and Zhan [10] described the subjects and constructions of English taste adjectives based on BNC and further explained the factors that affect the metaphorical extension of adjectives in the taste domain from the perspective of prototypical experience and from the construction of sentences with adjectives as predicates.

In sum, the research on adjectives in the tactile domain has laid a foundation for further exploration of their characteristics, but there is still a lack of analysis on the collocations of Chinese tactile adjectives in the experiential and abstract domains based on a large-scale corpus. Referring to existing research, this article defines tactile adjectives as follows: they are adjectives that are used to describe the feeling of the human skin when it is exposed to the stimulation of external things. This kind of feeling includes the sense of touch and pressure, the sense of temperature, and the sense of pain. This article selected the typical tactile adjectives 軟 *ruǎn* 'soft' and 硬 *yìng* 'hard' and analyzed their collocations in subject-predicate relations and attributive-head relations to discover the rules of extension of their meanings. These findings further reveal the role of touch in people's understanding of the world.

3 Research Methods

3.1 Research Tools and Corpus Selection

Sketch Engine is a grammatical knowledge generation system on the basis of a large corpus [11–13]. In addition to providing general keywords and contextual queries, it also provides a quick description of vocabulary features, which can be applied to dictionary compilation, language teaching, linguistic research, and natural language processing. In addition to be used to analyze English, it was developed to be applicable to the analysis of other languages, such as Chinese. The Chinese Word Sketch Engine (CWS) is a word sketching system based on Chinese corpora [14, 15]. There have been many studies using CWS [16–26]. Collocations of tactile adjectives studied in this paper are based on the grammatical knowledge list displayed as quick descriptions of lexical features in CWS.

3.2 Corpus Selection and Annotation

After selecting CWS as the search tool, this article set the same search conditions for 軟 *ruǎn* 'soft' and 硬 *yìng* 'hard': minimum frequency was set to '1', maximum range of its grammatical relation was set to '999', and minimum saliency was set to '0.00'. Then the tagged Chinese Gigaword Corpus was searched. The displayed grammatical lists of the two adjectives have only two grammatical relations: the subject-predicate relation (軟 *ruǎn* 'soft' and 硬 *yìng* 'hard' are used as predicates, while the collocated words function as subjects) and the attributive-head relation (軟 *ruǎn* 'soft' and 硬 *yìng* 'hard' are the attributives and the collocated words are the heads). Then we manually screened the collocations of 軟 *ruǎn* 'soft' and 硬 *yìng* 'hard' one by one and deleted wrongly extracted collocations. The results are shown in Table 1. For example, the collocation of 硬隊伍 *yìngduìwǔ* 'hard team' appears to have the attributive-head relation. However, the original sentence is 造就出了一支素質過硬的隊伍 *zàojiù chū le yī zhī sùzhì guòyìng de duìwǔ* '(It) created a team of excellent quality'. Obviously, 硬 *yìng* 'hard' in the sentence does not modify 隊伍 *duìwǔ* 'team', but is the predicate of the previous noun 素質 *sùzhì* 'quality'. Therefore, such collocations were removed during manual screening. The number of different types (not tokens) of collocations are shown in Table 1[1]. The number of original collocations in Table 1 are those automatically extracted from the corpus and the correct collocations are those after manual screening. It shows that there are many wrongly extracted collocations. Out of 182 different collocations of 軟 *ruǎn* 'soft', there are 61 correct ones; and out of 445 different collocations of 硬 *yìng* 'hard', there are 202 correct ones. Table 2 shows the number of correct collocations in different grammatical relations. Regarding the subject-predicate relation and the attributive-head relation, the collocations of 軟 *ruǎn* 'soft' account for 39.3% and 60.7% respectively, the collocations of 硬 *yìng* 'hard' account for 25.7% and 74.3% respectively.

Table 1. Number of different collocations for 軟 *ruǎn* 'soft' and 硬 *yìng* 'hard'

Tactile adjectives	軟 *ruǎn* 'soft'		硬 *yìng* 'hard'	
Collocations	Original collocations	Correct collocations	Original collocations	Correct collocations
Number	182	61	445	202

Table 2. Number of different grammatical relations

Relation	軟 *ruǎn* 'soft'		硬 *yìng* 'hard'	
	Number	Percentage	Number	Percentage
The Subject-Predicate Relation	24	39.3%	52	25.7%
The Attributive-Head Relation	37	60.7%	150	74.3%
Total	61	100.0%	202	100.0%

In this study, we further manually annotated the meaning of 軟 *ruǎn* 'soft' and 硬 *yìng* 'hard' in each collocation and divide them into the experiential and abstract domains

[1] The number refers to types (different collocates), not tokens (all collocates, including repeated ones) in the whole paper.

according to the meaning. We also further annotated the semantic categories of the nouns collocated with 軟 *ruǎn* 'soft' and 硬 *yìng* 'hard' in the two types of grammatical relations. Then we explored the semantic distribution and metaphorical extension of the collocations.

4 The Experiential and Abstract Domain of the Tactile Adjectives

According to the senses in *The Contemporary Chinese Dictionary* (the seventh edition) [27], we manually annotated the senses of 軟 *ruǎn* 'soft' and 硬 *yìng* 'hard' in each collocation. The distribution of different senses is shown in Table 3 and Table 4. Sense ⑩ of 軟 *ruǎn* 'soft' is a newly added sense by us. We judged the domain of the subjects in the subject-predicate relation and the heads in the attributive-head relation according to their senses (this article uses "collocates" to refer to both the subjects and the heads) and decided that if the original meaning is used, the collocate belongs to the experiential domain; otherwise it belongs to the abstract domain. The original meanings of 軟 *ruǎn* 'soft' and 硬 *yìng* 'hard', ① for both adjectives, reflect the characteristics of their experiential domains, while the extended meanings reflect the use of their abstract domains.

Table 3. Sense distribution of 軟 *ruǎn* 'soft'

Senses	The subject-predicate relation		The attributive-head relation	
	Number	Percentage	Number	Percentage
① Adj. (an object) loose-textured and easy to change shape under an outside force (as opposed to 硬 *yìng* 'hard', the same with ②④⑤⑥⑧)	10	41.7%	29	78.4%
② Adj. mild; gentle	3	12.5%	2	5.4%
③ Adj. physical weakness	2	8.3%	0	0.0%
④ Adj. weak	2	8.3%	2	5.4%
⑤ Adj. poor in ability, quality, etc	1	4.2%	2	5.4%
⑥ Adj. soft-hearted; easy to change one's mind	2	8.3%	0	0.0%
⑦ V. not firm; soften	0	0.0%	0	0.0%

(*continued*)

Table 3. (*continued*)

Senses	The subject-predicate relation		The attributive-head relation	
	Number	Percentage	Number	Percentage
⑧ Adj. no rigid requirement; there is room for expansion and contraction	1	4.2%	2	5.4%
⑨ N. a surname	0	0.0%	0	0.0%
⑩ Adj. (economy) weak; (price) cheap [a newly added sense by the researchers]	3	12.5%	0	0.0%
Total	24	100.0%	37	100.0%

Table 3 shows the sense distribution of 軟 *ruăn* 'soft'. In *The Contemporary Chinese Dictionary* [27], 軟 *ruăn* 'soft' has 9 senses. Sense ① reflects the use of 軟 *ruăn* 'soft' in the experiential domain and the other senses reflect its use in the abstract domain. In the subject-predicate relation, the experiential domain of 軟 *ruăn* 'soft' accounts for 41.7% and its abstract domain accounts for 58.3%. In the attributive-head relation, the experiential domain accounts for 78.4% and the abstract domain accounts for 21.6%. It can be seen that 軟 *ruăn* 'soft' tends to be used with metaphorical extension in the subject-predicate relation, while it is mostly used to express the original meaning in the attributive-head relation. In addition, in the experiential domain, 軟 *ruăn* 'soft' of the attributive-head relation is 36.7% higher than that of the subject-predicate relation, indicating that it is more inclined to be used in an attributive-head relation to express the physical property of the collocates.

Table 4 shows the sense distribution of 硬 *yìng* 'hard'. In *The Contemporary Chinese Dictionary* [27], it has 6 senses. Sense ① reflects the use of 硬 *yìng* 'hard' in the experiential domain and the other senses reflect its use in the abstract domain. In the subject-predicate relation, senses of 硬 *yìng* 'hard' from the experiential domain account for 34.6% and senses from the abstract domain accounts for 65.4%; in the attributive-head relation, senses from the experiential domain account for 44.7% and sense from the abstract domain accounts for 55.3%. It can be seen that 硬 *yìng* 'hard' tends to be used through metaphorical extensions in both relations in the abstract domain.

5 Semantic Analysis of Collocates of the Tactile Adjectives

After dividing the use of 軟 *ruăn* 'soft' and 硬 *yìng* 'hard' into the experiential domain and the abstract domain by using *A Thesaurus of Modern Chinese* [28], we annotated the semantic categories of the collocates of 軟 *ruăn* 'soft' and 硬 *yìng* 'hard'. In this section, through a comparative analysis, we will examine the semantic distribution of the collocates of 軟 *ruăn* 'soft' and 硬 *yìng* 'hard' in different grammatical relations.

Table 4. Sense distribution of 硬 *yìng* 'hard'

Senses	The subject-predicate relation		The attributive-head relation	
	Number	Percentage	Number	Percentage
① Adj. stiff to the touch; (an object) not easily yielding to the pressure of an external force because of its tight texture (as opposed to 軟 *ruǎn* 'soft', the same with ②④⑥)	18	34.6%	67	44.7%
② Adj. (character) strong; (will) firm	9	17.3%	7	4.7%
③ Adv. resolutely or stubbornly (to do something)	0	0.0%	0	0.0%
④ Adv. reluctantly (to do something)	0	0.0%	0	0.0%
⑤ Adj. (ability) strong; (quality) good	22	42.3%	48	32.0%
⑥ Adj. rigid	3	5.8%	28	18.7%
Total	52	100.0%	150	100.0%

The distribution of first-level semantic categories of the collocates is shown in Table 5. Most collocates belong to one semantic category and only some collocates of 硬 *yìng* 'hard' in the attributive-head relation belong to two semantic categories. The semantic categories can comprehensively reflect the preference for semantic choice.

Table 5. The number of the first-level semantic categories of 軟 *ruǎn* 'soft' and 硬 *yìng* 'hard'

Relation	軟 *ruǎn* 'soft'		硬 *yìng* 'hard'	
	Number	Percentage	Number	Percentage
Subject-predicate relation	24	39.3%	52	25.1%
Attributive-head relation	37	60.7%	155	74.9%
Total	61	100.0%	207	100.0%

(Note: The words 思想 *sīxiǎng* 'thought', 東西 *dōngxi* 'thing', 皮 *pí* 'skin', 班子 *bānzi* 'team', and 队伍 duìwǔ 'team' in the attributive-head relation of 硬 *yìng* 'hard' belong to two semantic categories. Thus the total number of senses of 硬 *yìng* 'hard' is larger than in Table 2.)

5.1 Comparison Under the Subject-Predicate Relation

This part compares the semantic categories of the collocates of 軟 *ruǎn* 'soft' and 硬 *yìng* 'hard' under the subject-predicate relation. The number of first-level semantic categories of the collocates of 軟 *ruǎn* 'soft' and 硬 *yìng* 'hard' in the experiential domain and the abstract domain is shown in Table 6. The total number of the collocates of 軟 *ruǎn* 'soft' and 硬 *yìng* 'hard' is 24 and 52 respectively, with the latter about twice in number of the former. Collocates of the two words fall into the same four first-level semantic categories and also share the same frequency order: Abstract Things > Living Things > Concrete Things > Social Activities. Abstract Things is the largest category for the collocates of both軟 *ruǎn* 'soft' and 硬 *yìng* 'hard', reaching a high percentage of 41.7% and 51.9% respectively. The second largest category is Living Things, accounting for 37.5% and 25.0% respectively. In the following, we will make a more detailed comparison of the collocates in the experiential domain and the abstract domain in Table 6 to explore the rules of their distribution.

(1) Comparison of collocates of 軟 *ruǎn* 'soft' in the experiential domain and the abstract domain.

As can be seen in Table 6, in the subject-predicate relation, when 軟 *ruǎn* 'soft' is used as a predicate, it has 10 collocates in the experiential domain and 14 in the abstract domain, indicating that 軟 *ruǎn* 'soft' tends to be more often used with metaphorical extension. The experiential domain reflects the original meaning and the abstract domain reflects the extended meaning. In the process of meaning extension, the original meaning of 軟 *ruǎn* 'soft', that is, '(an object) loose-textured and easy to change shape under an outside force', is more than the foundation for embodied cognition. In the following paragraphs, we will compare the semantic categories of the collocates of 軟 *ruǎn* 'soft' in the experiential domain and in the abstract domain to explore the semantic distribution.

The collocates of 軟 *ruǎn* 'soft' in the experiential domain are mainly concentrated in the semantic category of Living Things, accounting for a high proportion of 40.0%, indicating that the prototypical experience of 軟 *ruǎn* 'soft' is strong, because Living Things are all entities (呼吸道 *hūxīdào* 'respiratory tract', 麥粒 *màilì* 'wheat grain', 繭子 *jiǎnzi* 'cocoon', 顱骨 *lúgǔ* 'skull'). In addition to Living Things, the collocates of 軟 *ruǎn* 'soft' in the experiential domain belong to the two semantic categories of Abstract Things (質地 *zhìdì* 'texture', 土地 *tǔdì* 'land', 訓練場*xùnliànchǎng* 'training ground') and Concrete Things (路面 *lùmiàn* 'road surface', 床鋪 *chuángpù* 'sleeping bunk', 安全帽 *ānquánmào* 'safety hat'), both accounting for a percentage of 30.0%. Among them, the second-level semantic categories of the collocates belonging to Abstract Things include Attributes (質地 *zhìdì* 'texture', 土地 *tǔdì* 'land') and *Culture and Sports* (訓練場 *xùnliànchǎng* 'training field'); the second-level semantic categories of the collocates belonging to Concrete Things include General Names (路面 *lùmiàn* 'road surface') and Daily Necessities (床鋪 *chuángpù* 'sleeping bunk', 安全帽 *ānquánmào* 'safety hat').

Up to 50% of the collocates of 軟 *ruǎn* 'soft' in the abstract domain are concentrated in Abstract Things (態度 *tàidù* 'attitude', 方式 *fāngshì* 'manner', 精神 *jīngshén* 'spirit', 地產股 *dìchǎngǔ* 'real estate stock', 澳幣 *àobì* 'Australian dollar', 新元 *xīnyuán* 'SGD', 售價 *shòujià* 'selling price'), followed by Living Things (人 *rén* 'person', 对手 *duìshǒu* 'opponent', 腿 *tuǐ* 'leg', 心 *xīn* 'heart', 手 *shǒu* 'hand'), accounting for 35.7%. We further annotated the second-level semantic categories of the collocates belonging to Abstract

Table 6. The semantic category distribution of the collocates of 軟 *ruǎn* 'soft' and 硬 *yìng* 'hard' in the subject-predicate relation

First-level semantic category	軟 *ruǎn* 'soft'						硬 *yìng* 'hard'					
	The experiential domain		The abstract domain		Total		The experiential domain		The abstract domain		Total	
	Number	Percentage	Number	Percentage	Number	Percentage	Number	Percentage	Number	Percentage	Number	Percentage
生物 *shēngwù* 'Living things'	4	40.0%	5	35.7%	9	37.5%	6	33.3%	7	20.6%	13	25.0%
抽象事物 *Chōuxiàng shìwù* 'Abstract things'	3	30.0%	7	50.0%	10	41.7%	5	27.8%	22	64.7%	27	51.9%
具體物 *jùtǐ wù* 'Concrete things'	3	30.0%	1	7.1%	4	16.7%	7	38.9%	3	8.8%	10	19.2%
社會活動 *shèhuì huódòng* 'Social activities'	0	0.0%	1	7.1%	1	4.2%	0	0.0%	2	5.9%	2	3.8%
Total	10	100.0%	14	100.0%	24	100.0%	18	100.0%	34	100.0%	52	100.0%

Things and found that they are from either economic or non-economic categories. Economic collocates are words related to economy, such as 地產股 *dìchǎngǔ* ‘real estate stock’, 澳幣 *àobì* ‘Australian dollar’, 新元 *xīnyuán* ‘SGD’, and 售價 *shòujià* ‘selling price’. Non-economic collocates are mainly words that describe things, attributes, and consciousness, such as 方式 *fāngshì* ‘manner’, 態度 *tàidù* ‘attitude’, and 精神 *jīngshén* ‘spirit’. An example of the former is 西瓜因天雨銷售大減, 售價又趨軟 *xīguā yīn tiānyǔ xiāoshòu dàjiǎn, shòujià yòu qū ruǎn* ‘Watermelon sales are greatly reduced due to the rain and the sales price is softening’. Actually the sales price itself is neither hard nor soft. Instead, people use a metaphoric mechanism to extend the original meaning of 軟 *ruǎn* ‘soft’, indicating the price is lower. *The Contemporary Chinese Dictionary* of 軟 *ruǎn* ‘soft’ has nine senses but none is related to economics, showing that there is no suitable sense for such sentences. Thus, we added the sense “(economy) weak; (price) cheap” in the process of manual annotation. The non-economic use of 軟 *ruǎn* ‘soft’ can refer to Matters, Attributes, and Consciousness. The original meaning of 軟 *ruǎn* ‘soft’ is loose and easy to change shape. On this basis, people use metaphorical mechanisms to further express its reference to modes, attitudes, or spirits that are not firm. The second-level semantic categories of the collocates that belong to the biological category include Human (人 *rén* ‘person’, 对手 *duìshǒu* ‘opponent’) and Biological Parts (腿 *tuǐ* ‘leg’, 心 *xīn* ‘heart’, 手 *shǒu* ‘hand’). Although they are also entities, what is different from the experiential domain is that 軟 *ruǎn* ‘soft’ does not mean that these entities themselves are soft. They express abstract concepts. For example, 人软 *rénruǎn* (human-soft) means that the human’s character is weak; 对手软 *duìshǒu ruǎn* (opponent-soft) means that the opponent is not strong; 心软 *xīnruǎn* (heart-soft) means soft hearted; 腿软 *tuǐruǎn* (leg-soft) expresses fear.

To sum up, the collocates of 軟 *ruǎn* ‘soft’ in the experiential domain and the abstract domain are concentrated in the two first-level semantic categories of Living Things and Abstract Things, but they belong to different second-level semantic categories. In the experiential domain of 軟 *ruǎn* ‘soft’, the collocates are mainly distributed in Living Things. In the abstract domain, the collocates belonging to Living Things are distributed in two second-level semantic categories: Biological Parts and Human. In the experiential domain, the collocates belonging to Abstract Things are distributed in two second-level categories: Attributes, Culture and Sports, while in the abstract domain, the collocates belonging to Abstract Things are mainly concentrated in the vocabulary related to Economics. The characteristics of 軟 *ruǎn* ‘soft’ in the abstract domain and the experiential domain show that the use of 軟 *ruǎn* ‘soft’ in the experiential domain is often based on physical experience, while collocates in the abstract domain are mainly used through metaphorical extension.

(2) Comparison of collocates of 硬 *yìng* ‘hard’ between the experiential domain and the abstract domain.

It can be seen from Table 6 that when 硬 *yìng* ‘hard’ is used as a predicate, there are 18 subjects in the experiential domain and 34 subjects in the abstract domain. The latter is about twice as many as the former. It can be seen that 硬 *yìng* ‘hard’ has been extended from the basic usage of expressing human tactile experience to more abstract fields. In the following, we will compare the semantic categories of the collocates of 硬

yìng 'hard' in the experiential domain and the abstract domain to explore the semantic distribution of the subjects.

In the experiential domain, the collocates of 硬 *yìng* 'hard' are mainly distributed in three first-level semantic categories: Concrete Things (外殼 *wàiké* 'shell', 它們 *tāmen* 'they', 地板 *dìbǎn* 'floor', 污垢 *wūgòu* 'dirt', 地層 *dìcéng* 'stratum', 土壤 *tǔrǎng* 'soil', 地質 *dìzhì* 'geology'), Living Things (花柄 *huābǐng* 'flower stalk', 麥粒 *màilì* 'wheat grain', 雙臂 *shuāngbì* 'arms', 翼 *yì* 'wings', 骨頭 *gǔtóu* 'bone', 翅膀*chìbǎng* 'wing'), and Abstract Things (立場 *lìchǎng* 'position', 質地 *zhìdì* 'texture', 鞋質 *xiézhì* 'shoe quality', 肉質*ròuzhì* 'meat quality', 腫瘤 *zhǒngliú* 'tumor'), accounting for 38.9%, 33.3% and 27.8% respectively. These percentages do not differ significantly. Then we further analyzed the second-level semantic categories of the three major first-level semantic categories and found that they are quite different. The collocates of 硬 *yìng* 'hard' from Concrete Things belong to these second-level categories: Natural Objects (污垢 *wūgòu* 'dirt', 地層 *dìcéng* 'stratum', 土壤 *tǔrǎng* 'soil', 地質 *dìzhì* 'geology'), General Names (外殼 *wàiké* 'shells', 它們 *tāmen* 'they'), and Buildings (地板 *dìbǎn* 'floor'), of which the category of Natural Objects contains the most collocates. The collocates from Living Things belong to the second-level category of Biological Parts. The collocates from Abstract Things belong to these second-level categories: Attributes (地質 *dìzhì* 'geology', 鞋質 *xiézhì* 'shoe quality', 肉質 *ròuzhì* 'meat quality'), Matters (立場 *lìchǎng* 'position'), and Culture and Sports (腫瘤 *zhǒngliú* 'tumor'), of which the category of Attributes contains the most collocates.

In the abstract domain, 64.7% of the collocates of 硬 *yìng* 'hard' are concentrated in the category of Abstract Things; 20.6% are in the category of Living Things; 8.8% in Concrete Things; and only 5.9% are in Social Activities. We further analyze their second-level semantic categories. The collocates under Abstract Things have the following second-level categories: the Attributes category (e.g., 態度 *tàidù* 'attitude', 水準 *shuǐzhǔn* 'level', 條件 *tiáojiàn* 'condition', 作風 *zuòfēng* 'style', 素質 *sùzhì* 'quality' etc.) contains the most collocates, followed by categories of Consciousness (思想 *sīxiǎng* 'ideology', 心理 *xīnlǐ* 'Psychology'), Society (業務 *yèwù* 'business', 黨組織 *dǎngzǔzhī* 'party organization'), Matters (措施 *cuòshī* 'measures', 軍事 *jūnshì* 'military', 指標 *zhǐbiāo* 'indicators') and so on. Among the second-level semantic categories of Living Things, the Human category (他 *tā* 'he', 法官 *fǎguān* 'judge', 自身 *zìshēn* 'self', 职工 *zhígōng* 'employee') have the most collocates, followed by the category of Biological Parts (肩膀 *jiānbǎng* 'shoulder', 嘴 *zuǐ* 'mouth', 身子骨 *shēnzǐgǔ* 'body bone').

To sum up, the Concrete Things category contains the most collocates of硬 *yìng* 'hard' in the experiential domain, indicating the prototypical experience of natural objects under the category Concrete Things is strong. That is to say, people often use 硬 *yìng* 'hard' to describe the tactile perception of natural objects in embodied experience. This is different from the collocates of 硬 *yìng* 'hard' referring to Concrete Things in the abstract domain, which are mainly General Names and Cultural Items. The collocates of 硬 *yìng* 'hard' in the abstract domain are mainly Abstract Things, most of which belong to the subcategory of Attributes. The same is true in the word 軟 *ruǎn* 'soft'.

(3) Comparison of collocates of 軟 *ruǎn* 'soft' and 硬 *yìng* 'hard' in the experiential domain.

In the experiential domain, the collocates of 軟 *ruǎn* 'soft' and 硬 *yìng* 'hard' are distributed in three primary semantic categories, Concrete Things, Abstract Things, and Living Things. In the experiential domain, the collocates of 軟 *ruǎn* 'soft' are mostly Living Things, accounting for 40.0%, while the collocates of 硬 *yìng* 'hard' are mostly Concrete Things, accounting for 38.9%. This indicates that Living Things often have soft characteristics, while Concrete Things often have hard characteristics. Although 軟 *ruǎn* 'soft' and 硬 *yìng* 'hard' are mainly distributed in these three semantic categories, the second-level semantic categories of each first-level category are different, reflecting in: (1) both 軟 *ruǎn* 'soft' and 硬 *yìng* 'hard' can describe Concrete Things, but 軟 *ruǎn* 'soft' always describes Daily Necessities (床鋪 *chuángpù* 'sleeping bunk', 安全帽 *ānquánmào* 'hard hat'), while 硬 *yìng* 'hard' always describes Natural Objects (污垢 *wūgòu* 'dirt', 地層 *dìcéng* 'stratum', 土壤 *tǔrǎng* 'soil', 地質 *dìzhì* 'geology'); (2) for Abstract Things, the collocates of both 軟 *ruǎn* 'soft' and 硬 *yìng* 'hard' tend to express tactile perception of the attributes of things, and they all focus on the physical sense of things. For example, the collocates of 軟 *ruǎn* 'soft' have texture like 質地 *zhìdì* 'texture' and the collocates of 硬 *yìng* 'hard' include 地質 *dìzhì* 'geology', 鞋質 *xiézhì* 'shoe quality', and 肉質 *ròuzhì* 'meat quality'; (3) in the category of Living Things, the collocates of 軟 *ruǎn* 'soft' and 硬 *yìng* 'hard' are all Biological Parts; the collocates of 硬 *yìng* 'hard' that belong to Living Things are 花柄 *huābǐng* 'flower stalk', 麥粒 *màilì* 'wheat grain', 雙臂 *shuāngbì* 'arms', 翼 *yì* 'wings', 骨頭 *gǔtóu* 'bone', and 翅膀 *chìbǎng* 'wing'; and the collocates of 軟 *ruǎn* 'soft' belong to Living Things are 呼吸道 *hūxīdào* 'respiratory tract', 麥粒 *màilì* 'wheat grains', 繭子 *jiǎnzi* 'cocoons', and 顱骨 *lúgǔ* 'skulls'.

(4) Comparison of collocates of 軟 *ruǎn* 'soft' and 硬 *yìng* 'hard' in the abstract domain.

The semantic category distribution of the collocates of 軟 *ruǎn* 'soft' and 硬 *yìng* 'hard' is consistent in the abstract domain. Abstract Things from both of them account for 50.0% and 64.7% respectively, followed by Living Things, accounting for 35.7% and 20.6% respectively. However, their distributions are different in the second-level semantic categories. The collocates of 軟 *ruǎn* 'soft' focus on the description of Economics (地產股 *dìchǎngǔ* 'real estate stock', 澳幣 *àobì* 'Australian dollar', 新元 *xīnyuán* 'SGD', 售價 *shòujià* 'selling price') under Abstract Things, while the collocates of 硬 *yìng* 'hard' focus on Attributes (e.g., 態度 *tàidù* 'attitude', 作風 *zuòfēng* 'style', 水準 *shuǐzhǔn* 'level', 品質 *pǐnzhì* 'quality', etc.) under Abstract Things. The difference not only reflects the asymmetry of the semantics of the collocates, but also reflects the difference in cognition and metaphorical extension. In the category of Living Things, the collocates of both 軟 *ruǎn* 'soft' and 硬 *yìng* 'hard' describe Biological Parts and Humans, but the difference lies in that the collocates of 硬 *yìng* 'hard' focus on the description of persons (法官 *fǎguān* 'judge', 职工 *zhígōng* 'employee', 他 *tā* 'he', 自身 *zìshēn* 'self'), while the collocates of 軟 *ruǎn* 'soft' focus on the description of Biological Parts (腿 *tuǐ* 'leg', 心 *xīn* 'heart', 手 *shǒu* 'hand').

In addition, through statistical analysis of the data, we found that the distribution of the collocates of 軟 *ruǎn* 'soft' and 硬 *yìng* 'hard' in the subject-predicate relation also has the following characteristics. First, when 軟 *ruǎn* 'soft' and 硬 *yìng* 'hard' are used as predicates, they need to be modified by degree adverbs, such as 很 *hěn* 'very'. Thus the collocates often occur in this construction: ___很软 or ___很硬. A construction like this is static. That is, 軟 *ruǎn* 'soft' and 硬 *yìng* 'hard' are used to describe the nature of the subject and are rarely used to describe the dynamic changes of the subject. Second, some collocates often appear together with軟 *ruǎn* 'soft' and 硬 *yìng* 'hard' to show a strong tendency of lexicalization, such as 手软 *shǒuruǎn* (soft-hand) 'be soft-hearted; be irresolute when firmness is needed' and 嘴硬 *zuǐyìng* (mouth-hard) 'stubborn and reluctant to admit mistakes or defeats'. Such expressions are used to express metaphorical concepts. Finally, comparing the experiential domain and the abstract domain of 軟 *ruǎn* 'soft', we found that the prototypical experience and metaphorical extension are different as shown by the collocates' semantic categories.

5.2 Comparison Under the Attributive-Head Relation

The number of collocates of 軟 *ruǎn* 'soft' and 硬 *yìng* 'hard' in the experiential domain and the abstract domain under the attributive-head relation is shown in Table 7, in which 37 different collocates of 軟 *ruǎn* 'soft' and 155 collocates of 硬 *yìng* 'hard' appear as heads. The number of the latter is much larger than the former; that is, 4.2 times of the former. Comparing the six semantic categories, it can be seen that the collocates of both 軟 *ruǎn* 'soft' and 硬 *yìng* 'hard' belong to Concrete Things, Living Things, Abstract Things, and Social Activities. No collocates of 軟 *ruǎn* 'soft' are from the category of 運動與變化 *yùndòng yǔ biànhuà* 'Movement and Changes' and no collocates of 硬 *yìng* 'hard' are from the category of Time and Space. As for the number of collocates in different semantic categories, the number of the collocates of 軟 *ruǎn* 'soft' in Concrete Things is the largest, accounting for 43.2%, followed by Abstract Things, accounting for 29.7%; most of the collocates of 硬 *yìng* 'hard' are concentrated in Abstract Things, accounting for 47.7%, followed by 31.6% from Concrete Things. The semantic category that ranked third in both of the collocates of 軟 *ruǎn* 'soft' and 硬 *yìng* 'hard' is Organisms, accounting for 21.6% and 16.8% respectively. This following section will further explore the semantic distribution of these collocates.

(1) Comparison of collocates of 軟 *ruǎn* 'soft' in the experiential domain and the abstract domain.

As shown in Table 7, under the attributive-head relation, there are 29 collocates in the experiential domain and 8 in the abstract domain. The former is 3.6 times of the latter. This indicates that people usually use its original meaning; that is, 軟 *ruǎn* 'soft' provides a stronger sense of prototypical experience and its metaphorical extension meaning is rarely used when 軟 *ruǎn* 'soft' is used as an attributive modifier to modify the head.

The collocates of 軟 *ruǎn* 'soft' in the experiential domain are mainly distributed in Concrete Things, with a proportion of up to 55.2%, followed by Living Things and Abstract Things, with a proportion of 24.1% and 17.2% respectively. This study further annotated the second-level semantic categories of the collocates. (1) The collocates belong to these second-level categories under Concrete Things include: Daily Necessities (椅子 *yǐzi* 'chair', 鞋 *xié* 'shoe', 塑膠帽 *sùjiāomào* 'plastic cap', 内視鏡 *nèishìjìng* 'endoscope'), Materials (橡皮管 *xiàngpíguǎn* 'rubber tube', 葉岩 *yèyán* 'leaf rock', 沙粒 *shālì* 'sand grain'), General Names (東西 *dōngxi* 'thing', 物質 *wùzhì* 'material'), Food Items (漢堡 *hànbǎo* 'hamburger', 食物 *shíwù* 'food', 飯糰 *fàntuán* 'rice ball'),

Table 7. Semantic distribution of collocates of 軟 *ruǎn* ‘soft’ and 硬 *yìng* ‘hard’ in the attributive-head relation

First-level semantic category	軟 *ruǎn* ‘soft’						硬 *yìng* ‘hard’					
	The experiential domain		The abstract domain		Total		The experiential domain		The abstract domain		Total	
	Number	Percentage	Number	Percentage	Number	Percentage	Number	Percentage	Number	Percentage	Number	Percentage
具體物 *jùtǐ wù* ‘Concrete things’	16	55.2%	0	0.0%	16	43.2%	47	68.1%	2	2.3%	49	31.6%
生物 *shēngwù* ‘Living things’	7	24.1%	1	12.5%	8	21.6%	17	24.6%	9	10.5%	26	16.8%
抽象事物 *chōuxiàng shìwù* ‘Abstract things’	5	17.2%	6	75.0%	11	29.7%	4	5.8%	70	81.4%	74	47.7%
時空 *shíkōng* ‘Time and space’	1	3.4%	0	0.0%	1	2.7%	0	0.0%	0	0.0%	0	0.0%
社會活動 *shèhuì huódòng* ‘Social activities’	0	0.0%	1	12.5%	1	2.7%	0	0.0%	4	4.7%	4	2.6%
運動與變化 *yùndòng yǔ biànhuà* ‘Movement and changes’	0	0.0%	0	0.0%	0	0.0%	1	1.4%	1	1.2%	2	1.3%
Total	29	100.0%	8	100.0%	37	100.0%	69	100.0%	86	100.0%	155	100.0%

Utensils (輪胎 *lúntāi* 'tire'), Stationery (球 *qiú* 'ball'), and Natural Objects (黑土 *hēitǔ* 'black soil'). (2) The collocates belong to these second-level categories under Living Things include: Plants (柿子 *shìzi* 'persimmon', 竹子 *zhúzi* 'bamboo', 水果 *shuǐguǒ* 'fruits', 花 *huā* 'flowers'), Biological Parts (鼻梁 *bíliáng* 'nose bridge', 脂肪 *zhīfáng* 'fat', 胸部 *xiōngbù* 'chest'. (3) The collocates belong to these second-level categories under Abstract Things include: Attributes (材質 *cáizhì* 'material', 組織 *zǔzhī* 'organization'), Culture and Sports (跑道 *pǎodào* 'runway', 腫瘤 *zhǒngliú* 'tumor'), and Matters (伙食 *huǒshí* 'food').

The collocates of 軟 *ruǎn* 'soft' in the abstract domain are mainly distributed in Abstract Things, accounting for 75.0%. Under Abstract Things, their distribution in second-level semantic categories is more scattered, including Matters (方式 *fāngshì* 'method'), Attributes (身段 *shēnduàn* 'figure'), Politics (黨支部 *dǎngzhībù* 'party branch'), Economy (經濟 *jīngjì* 'economy'), Quantity (數據 *shùjù* 'data'), and Awareness (策略 *cèlüè* 'strategy').

(2) Comparison of collocates of 硬 *yìng* 'hard' in the experiential domain and the abstract domain.

Unlike 軟 *ruǎn* 'soft', there are more collocates of 硬 *yìng* 'hard' in the abstract domain (86 types) than in the experiential domain (69 types). This indicates that there are more metaphorical extensions when 硬 *yìng* 'hard' is used as an attributive. Most collocates of 硬 *yìng* 'hard' in the experiential domain are Concrete Things, accounting for 68.1%, followed by Living Things, accounting for 24.6%. The collocates belong to the following second-level semantic categories under Concrete Things: Materials (e.g., 磚 *zhuān* 'brick', 鋼筋 *gāngjīn* 'steel bar', 漆板 *qībǎn* 'lacquered board', etc.), General Names (e.g., 東西 *dōngxi* 'thing', 金屬物 *jīnshǔwù* 'metal objects', etc.), Daily Necessities (e.g., 風箏線 *fēngzhēngxiàn* 'kite line', 枕頭 *zhěntóu* 'pillows', etc.), Food Items (e.g., 麵包 *miànbāo* 'bread', 饅頭 *mántóu* 'bun', etc.), Natural Objects (果嶺 *guǒlǐng* 'putting green', 沾塗層 *zhāntúcéng* 'clay layer', 岩層 *yáncéng* 'rock layer', 冰 *bīng* 'ice'), Stationery (撐竿 *chēnggǎn* 'pillar', 皮球 *píqiú* 'ball', 球 *qiú* 'ball') and Utensils (紙版 *zhǐbǎn* 'paper edition'). The collocates belong to the following second-level categories under Living Things include: Biological Parts (e.g., 樹根 *shùgēn* 'tree root', 脂肪 *zhīfáng* 'fat', etc.), Plants (e.g., 水果 *shuǐguǒ* 'fruit', 樹 *shù* 'tree', etc.), and Humans (傢伙 *jiāhuǒ* 'guy').

The collocates of 硬 *yìng* 'hard' in the abstract domain are mainly distributed in Abstract Things, accounting for 81.4%, followed by Living Things, accounting for 10.5%. Under Abstract Things, the collocates are widely distributed in second-level semantic categories, including Attributes (e.g., 本領 *běnlǐng* 'skill', 態度 *tàidù* 'attitude', etc.), Matters (e.g., 做法 *zuòfǎ* 'practices', 證據 *zhèngjù* 'evidence', etc.), Military (e.g., 軍隊 *jūnduì* 'army', 部隊 *bùduì* 'troops', etc.), Science and Education (e.g., 思想 *sīxiǎng* 'idea', 知識 *zhīshí* 'knowledge', etc.), Awareness (e.g.,手段 *shǒuduàn* 'means', 觀點 *guāndiǎn* 'opinion', etc.), Politics (e.g.,黨支部 *dǎngzhībù* 'party branch', 紀律 *jìlǜ* 'discipline', etc.), Economy (開支 *kāizhī* 'expense'), Society (企業 *qǐyè* 'enterprise'), and Culture and Sports (班子 *bānzi* 'team', 車技 *chējì* 'car skills', 推拿術 *tuīnáshù* 'the technique of massage'), with Attributes being the largest category. Under Living Things, except for a small number of modifications to Biological Parts (面孔 *miànkǒng* 'face', 骨頭 *gǔtóu* 'bone'), the rest are mostly descriptions of People (e.g., 演員 *yǎnyuán* 'actor', 精兵 *jīngbīng* 'elite soldier', etc.). Although some collocates of 硬 *yìng* 'hard' overlapped in the experiential domain as well as the abstract domain and all belong to Living Things, they are used differently. In the experiential domain, they are mainly Biological Parts, which illustrate the characteristics of things, whereas in the abstract domain, they are mainly descriptions of humans, indicating the qualities or abilities of people through metaphorical extension.

(3) Comparison of collocates of 軟 *ruǎn* 'soft' and 硬 *yìng* 'hard' in the experiential domain.

In the attributive-head relation, most of the collocates of both 軟 *ruǎn* 'soft' and 硬 *yìng* 'hard' belong to Concrete Things, accounting for 55.2% and 68.1%, respectively, followed by Living Things, accounting for 24.6% and 10.5%, respectively. Under the semantic category of Concrete Things, the collocates' distribution of 軟 *ruǎn* 'soft' and 硬 *yìng* 'hard' in the second-level semantic categories are mostly consistent: they are both distributed in seven categories, including Materials, General Names, Daily Necessities, Food Items, Cultural Items, Natural Objects, and Utensils. The difference is that the collocates of 軟 *ruǎn* 'soft' cannot be found in Buildings. Under the semantic category of Living Things, the collocates of both 軟 *ruǎn* 'soft' and 硬 *yìng* 'hard' are further distributed into the second-level categories of Plants and Biological Parts. But the collocates of 軟 *ruǎn* 'soft' describe Plants (柿子 *shìzi* 'persimmon', 竹子 *zhúzi* 'bamboo', 水果 *shuǐguǒ* 'fruit', 花 *huā* 'flower'), while the collocates of 硬 *yìng* hard' mainly describe Biological Parts (e.g., 樹根 *shùgēn* 'tree root', 脂肪 *zhīfáng* 'fat' etc.).

(4) Comparison of collocates of 軟 *ruǎn* 'soft' and 硬 *yìng* 'hard' in the abstract domain.

In the abstract domain, the collocates of 軟 *ruǎn* 'soft' are distributed in three semantic categories: Living Things, Abstract Things, and Social Activities, while the collocates of 硬 *yìng* 'hard' are distributed in five semantic categories: Concrete Things, Living Things, Abstract Things, Social Activities, and *Movement and Changes*. The collocates of 軟 *ruǎn* 'soft' and 硬 *yìng* 'hard' are similar in that they are both concentrated in Abstract Things, with a percentage of 75.0% and 81.4%, followed by Living Things, with a percentage of 12.5% and 10.5%, respectively. We further analyzed the second-level semantic categories of Abstract Things, which is the largest first-level semantic category. We found that the collocates of 硬 *yìng* 'hard' have richer semantic categories than those of 軟 *ruǎn* 'soft'. While they both belong to Attributes, Matters, Politics, Economy, and *Culture and Sports*, the collocates of 軟 *ruǎn* 'soft' specifically belong to Number and *Culture and Sports*, and the collocates of 硬 *yìng* 'hard' specifically belong to Military, Science and Education, Consciousness, and Society. Unlike 軟 *ruǎn* 'soft', 硬 *yìng* 'hard' is often used to modify words of Military and *Science and Education*, which reflects their deep differences in extended metaphorical use. In the category of Living Things, 軟 *ruǎn* 'soft' is only used to describe people, while 硬 *yìng* 'hard' is mainly used to describes people. In addition, 硬 *yìng* 'hard' is also used to describe a small number of Biological Parts.

5.3 Similarities and Differences of Collocates' Distributions Under the Subject-Predicate Relationship and the Attributive-Head Relationship

The semantic category distribution of the collocates of 軟 *ruǎn* 'soft' and 硬 *yìng* 'hard' under the subject-predicate relation and the attributive-head relation is shown in Table 8. In general, in the subject-predicate relation, the collocates belong to four semantic categories: Living Things, Abstract Things, Concrete Things, and Social Activities. In addition to these four categories, in the attributive-head relation, the collocates belong to two more categories: Time and Space, Movement and Changes.

Specifically, there are differences in the semantic categories of the collocates of 軟 *ruǎn* 'soft' under the subject-predicate relation and under the attributive-head relation. Under the subject-predicate relation, the collocates belong to four semantic categories, while under the latter, the collocates belong to five categories. Under the subject-predicate relation, the collocates are mainly distributed in the two categories of Abstract Things and Living Things, with a ratio of up to 41.7% and 37.5% respectively. Under the attributive-head relation, the collocates mainly distribute in Concrete Things and Abstract Things, accounting for 43.2% and 29.7% respectively. Though the collocates in both of the two relations belong to Abstract Things, the second-level categories show that the collocates of 軟 *ruǎn* 'soft' under the subject-predicate relation focus on the description of Attributes and *Culture and Sports*, while they are mostly used to describe Matters, Attributes, Politic, Economy, Quantity, and Awareness under the attributive-head relation.

Under the subject-predicate relation, the collocates of 硬 *yìng* 'hard' are mainly distributed in Abstract Things and Living Things, accounting for 51.9% and 25.0%, respectively. Under the attributive-head relation, the collocates mainly are distributed in Abstract Things and Concrete Things, accounting for 47.7% and 31.6%. Whether in a subject-predicate relation or attributive-head relation, the collocates of 硬 *yìng* 'hard' has the highest proportion in the first-level semantic category of Abstract Things and the Attributes category contains the most collocates in the second-level categories of Abstract Things.

Table 8. Semantic category distribution of collocates of 軟 *ruǎn* 'soft' and 硬 *yìng* 'hard'

First-level semantic category	The subject-predicate relaTION						The attributive-head relation					
	軟 *ruǎn* 'soft'		硬*yìng* 'hard'		Total		軟 *ruǎn* 'soft'		硬 *yìng* 'hard'		Total	
	Number	Percentage	Number	Percentage	Number	Percentage	Number	Percentage	Number	Percentage	Number	Percentage
生物 *shēngwù* 'Living things'	9	37.5%	13	25.0%	22	28.9%	8	21.6%	26	16.8%	34	17.7%
抽象事物 *chōuxiàng shìwù* 'Abstract things'	10	41.7%	27	51.9%	37	48.7%	11	29.7%	74	47.7%	85	44.3%
具體物 *jùtǐ wù* 'Concrete things'	4	16.7%	10	19.2%	14	18.4%	16	43.2%	49	31.6%	65	33.9%
社會活動 *shèhuì huódòng* 'Social activities'	1	4.2%	2	3.8%	3	3.9%	1	2.7%	4	2.6%	5	2.6%
時空 *shíkōng* 'Time and space'	0	0.0%	0	0.0%	0	0.0%	1	2.7%	0	0.0%	1	0.5%
運動與變化 *yùndòng yǔ biànhuà* 'Movement and changes'	0	0.0%	0	0.0%	0	0.0%	0	0.0%	2	1.3%	2	1.0%
Total	24	100.0%	52	100.0%	76	100.0%	37	100.0%	155	100.0%	192	100.0%

6 Conclusion

Embodied cognition is a basic way of people's understanding of the world. The tactile sense plays an important role in the process of knowing the world. But there has been little systematic analysis of the collocations of tactile adjectives in Chinese. This paper used a large-scale corpus to extract and manually screen collocations of the representative tactile adjectives 軟 *ruǎn* 'soft' and 硬 *yìng* 'hard' in the subject-predicate relation and the attributive-head relation. We annotated the sense of 軟 *ruǎn* 'soft' and 硬 *yìng* 'hard' in each sentence. Based on whether the original meaning or the extended meaning is used, the collocations were divided into the experiential domain and the abstract domain. Then we carried out semantic category annotation and statistical analysis on the distribution of collocates of the two adjectives in the two domains. The results are summarized in Table 9 and Table 10.

In sum, this paper systematically analyzed the rules of the distribution of the collocates of Chinese tactile adjectives through using a large-scale corpus, with the aim to help deepen the understanding of the prototypical experience and the metaphorical extension of tactile adjectives in different cognitive domains and to gain a deeper understanding of the characteristics of touching in the cognitive world of the Chinese people.

Table 9. First-level semantic categories of the collocates of 軟 *ruǎn* 'soft' and 硬 *yìng* 'hard'

Collocates	Domain	First-level semantic categories of the subject-predicate relation	First-level semantic categories of the attributive-head relation
Collocatesof 軟 *ruǎn* 'soft'	The experiential domain	Living things (40.0%) Abstract things (30.0%) Concrete things (30.0%)	Concrete things (55.2%) Living things (24.1%) Abstract things (17.2%) Time and space (3.4%)
	The abstract domain	Living things (35.7%) Abstract things (50.0%) Concrete things (7.1%) Social activities (7.1%)	Living things (12.5%) Abstract things (75.0%) Social activities (12.5%)
Collocatesof 硬 *yìng* 'hard'	The experiential domain	Living things (33.3%) Abstract things (27.8%) Concrete things (38.9%)	Concrete things (68.1%) Living things (24.6%) Abstract things (5.8%) Movement and changes (1.4%)

(*continued*)

Table 9. (*continued*)

Collocates	Domain	First-level semantic categories of the subject-predicate relation	First-level semantic categories of the attributive-head relation
	The abstract domain	Living things (20.6%) Abstract things (64.7%) Concrete things (8.8%) Social activities (5.9%)	Concrete things (2.3%) Living things (10.5%) Abstract things (81.4%) Social activities (4.7%) Movement and changes (1.2%)

Table 10. Second-level semantic categories of the collocates from the largest first-level semantic category

		The subject-predicate relation		The attributive-head relation	
Collocates	Domain	Largest first-level semantic category of the subject-predicate relation	Second-level semantic categories of the collocates from the largest first-level semantic category	Largest first-level semantic category of the attributive-head relation	Second-level semantic categories of the collocates from the largest first-level semantic category
Collocates of 軟 *ruǎn* 'soft'	The experiential domain	Living things (40.0%)	Entities	Concrete things (55.2%)	Daily Necessities Materials General Names Food Items Utensils Stationery Natural Objects

(*continued*)

Table 10. (*continued*)

		The subject-predicate relation		The attributive-head relation	
	The abstract domain	Abstract things (50.0%)	Economy Matters Attributes Consciousness	Abstract things (75.0%)	Matters Attributes Politics Economy Quantity Awareness
Collocates of 硬 *yìng* 'hard'	The experiential domain	Concrete things (38.9%)	Natural Objects General Names Buildings	Concrete things (68.1%)	Materials General Names Daily Necessities Food Items Natural Objects Stationery Utensils
	The abstract domain	Abstract things (64.7%)	Attributes Consciousness Society Matters	Abstract things (81.4%)	Attributes Matters Military Science and Education Awareness Politics Economy Society Culture and Sports

Acknowledgement. This research is supported by University of Macau (MYRG2019–00013-FAH).

References

1. Wang, X.: The Study on Tactile Adjective in Modern Chinese(yīnghàn chùjué xíngróngcí yǔyì duìbǐ yánjiū), vol. Master. Guangxi Normal University, Guilin (2010)
2. Wen, Y.: A Semantic Analysis of Tactile Adjective in English and Chinese (yīnghàn chùjué xíngróngcí yǔyì duìbǐ yánjiū), vol. Master. Yanbian University, Yanbian (2013)
3. He, M.: A Comparative Analysis of Metaphors of Tactile Adjectives in Russian and Chinese (éhàn chùjué xíngróngcí yǐnyù duìbǐ fēnxī), vol. Master. Jilin University, Changchun (2018)
4. Park: A Comparative of Korean and Chinese tactile Adjective (hànhányǔ chùjué xíngróngcí duìbǐ), vol. Master. Jilin University, Changchun (2018)
5. Chu, H.: A comparative Study of Synesthesea Matphor in Chinese and Japanese Tactile Adjective (guānyú hànrì chùjué xíngróngcí gònggǎnjué bǐyù xiànxiàng de duìzhào yánjiū), vol. Master. Dalian Maritime University, Dalian (2010)

6. Lu, J.: A Comparative Study on the Ways and Mechanisms of Semantic Expansion of Chinese and English Tactile Adjectives (hànyīng chùjué xíngróngcí cíyì kuòzhăn tújīng yŭ jīzhì duìbĭ yánjiū), vol. Master. Shanghai Normal University, Shanghai (2019)
7. Tang, S.: Why are some metaphors impossible (yǒuxiē yǐnyù wèishénme búkěnéng), vol. PH.D. Shanghai International Studies University, Shanghai (2010)
8. Tang, S., Dong, Y., Li, F.: Constructions and metaphorical extensions——a comparative study of Chinese and English temperature domain predicative adjectives and their metaphorical extension constraints (hànyīng wēndùyù wèiyŭjù xíngróngcí yǐnyùtuòzhăn chāyì jí chéngyīn tànxī). J. Fore. Lang. (wàiguó yŭ) **34**, 50–57 (2011)
9. Tang, S., Tian, Z.: The overlapping predictability of semantic extensions of physical quality adjectives in taste domain (jīyú wèijuéyù de hànyīng wùxìngxíngróngcí yŭyì chónghédù yánjiū). Fore. Lang. Teach. Res. (Wàiyŭ jiàoxué yŭ yánjiū) **46**, 666–677 (2014)
10. Li, Y., Zhan, Q.: A corpus-based study of English taste domain predictive adjectives and their metaphorical extension (jīyú yŭliàokù de yīngyŭ wèijuéyù xíngróngcí wèiyŭgòushì jí yǐnyù tuòzhăn). J. Hunan Inst. Eng. (Soc. Sci. Ed.) (Húnán gōngchéng xuéyuàn xuébào) **24**, 52–57 (2014)
11. Kilgarriff, A., et al.: The sketch engine: ten years on. Lexicography **1**, 7–36 (2014)
12. Kilgarriff, A., Rychly, P., Smrz, P., Tugwell, D.: The sketch engine. In: Information Technology, pp. 105–116 (2004)
13. Kilgarriff, A., Tugwell, D.: Sketching words. In: Corréard, M.-H. (ed.) Lexicography and Natural Language: A Festschrift in Honour of B.T.S.Atkins (Euralex 2002), pp. 125–137 (2002)
14. Huang, C.-R., et al.: Chinese Sketch Engine and the extraction of grammatical collocations. In: Proceedings of the Fourth SighanI Workshop on Chinese Language Processing, Jeju, Island, Korea, pp. 48–55 (2005)
15. Kilgarriff, A., Huang, C.-R., Rychlý, P., Smith, S., Tugwell, D.: Chinese word sketches. In: Asialex 2005: Words in Asian Cultural Context Singapore, Singapore (2005)
16. Wang, S., Huang, C.-R.: Adjectival modification to nouns in mandarin chinese: case studies on "cháng+noun" and "adjective+túshūguăn. In: Otoguro, R., Ishikawa, K., Umemoto, H., Yoshimoto, K., Harada, Y. (eds.) Proceedings of The 24th Pacific Asia Conference on Language, Information and Computation (PACLIC-24), pp. 701–705. Tohoku University, Sendai (2010)
17. Wu, Y., Wang, S.: Applying Chinese word sketch engine to distinguish commonly confused words. In: Dong, M., Lin, J., Tang, X. (eds.) Chinese Lexical Semantics, pp. 600–619. Springer, Switzerland (2016)
18. Wang, S., Wang, X.: The attention to safety issues from mainland China and Taiwan. In: Hong, J.-F., Su, Q., Wu, J.-S. (eds.) Chinese Lexical Semantics, pp. 801–818. Springer Nature, Switzerland (2018)
19. Wang, S., Huang, C.-R., Xu, H.: Compositionality of NN compounds: a case study on [N1+Artifactual-Type Event Nouns]. In: Proceedings of The 26th Pacific Asia Conference on Language, Information and Computation (PACLIC-26), pp. 70–79. Faculty of Computer Science, Universitas Indonesia, Bali (2012)
20. Wang, S., Huang, C.-R.: The semantic type system of event nouns. In: Jing, S.Z. (ed.) Increased Empiricism: Recent Advances in Chinese Linguistics, vol. 2, pp. 205–221. John Benjamins Publishing Company, Amsterdam / Philadelphia (2013)
21. Wang, S.: Semantics of Event Nouns, vol. Ph.D. The Hong Kong Polytechnic University, Hong Kong (2013)
22. Wang, S., Huang, C.-R.: Type construction of event nouns in mandarin Chinese. In: Proceedings of The First Workshop on Generative Lexicon for Asian Languages (GLAL-1), Workshop of The 26th Pacific Asia Conference on Language, Information and Computation (PACLIC-26), pp. 582–591. Faculty of Computer Science, Universitas Indonesia, Bali (2012)

23. Wang, S., Huang, C.-R.: Word sketch lexicography: new perspectives on lexicographic studies of chinese near synonyms. Lingua Sinica **3**, 1–22 (2017)
24. Wang, S., Yin, J.: A comparative study of the collocations in legislative Chinese and general Chinese. In: Hong, J.-F., Zhang, Y., Liu, P. (eds.) CLSW 2019. LNCS (LNAI), vol. 11831, pp. 710–724. Springer, Cham (2020). https://doi.org/10.1007/978-3-030-38189-9_72
25. Wang, S., Yin, J.: A corpus-based study of keywords in legislative Chinese and general Chinese. In: Hong, J.-F., Zhang, Y., Liu, P. (eds.) Chinese Lexical Semantics, pp. 639–653. Springer, Cham (2020)
26. Wang, S.: Chinese Multiword Expressions: Theoretical and Practical Perspectives. Springer Singapore, Singapore (2020). https://doi.org/10.1007/978-981-13-8510-0
27. Lv, S., Ding, S.: The contemporary Chinese dictionary (xiàndài hànyǔ cídiǎn). In: Dictionary Editing Office, I.o.L.S., Chinese Academy of Social Sciences (ed.). The Commercial Press, Beijing (2016)
28. Su, X.: A Thesaurus of Modern Chinese (xiàndài hànyǔ fēnlèi cídiǎn). The Commercial Press, Beijing (2013)

A Study on the Classification of Chinese Interrogatives Based on Pragmatic Functions

Li Jiangtao(✉) and Rao Gaoqi

Research Institute of International Chinese Language Education, Beijing Language and Culture University, Beijing, China

raogaoqi@blcu.edu.cn

Abstract. This paper emphasizes the importance of regular questions in question classification, analyzes and summarizes four types of interrogative sentences and their corresponding pragmatic functions. On this basis, we have manually annotated interrogative sentences in selected novels and BaiduQA and have created an interrogative sentences corpus. A formal feature set for question classification is constructed based on corpus, and classification experiments based on rules and statistics are conducted respectively to check the classification effect of each feature. The set of feature rules is formed by optimizing feature combinations iteratively through multiple rounds of experiments, which provides a formal classification basis for the existing Question-answering systems. Through the comparison of various classification models, it is found that the method of finite state automaton can achieve a macro average F1-score of 0.94, whereas the random forest model achieving a macro average F1-score of 0.98, has better performance in classification.

1 Literature Review

Question-answering systems are generally composed of modules performing the following tasks: question classification, query expansion, search engine, answer extraction and answer sorting and selection [1]. Question classification is an important sub-module of a question answering system. Its main task is to classify questions into corresponding semantic categories according to the types of answers, producing a good guiding effect on subsequent answer extraction and answer selection modules of the question answering system [2]. The effectiveness of question classification directly affects the understanding of questions.

There is no unified standard for the current question classification systems. The most authoritative question classification system so far is UIUC, which is a hierarchical classification system based on answer types. In this system, questions are divided into 6 major categories and 50 minor categories, and each major category contains non-repeating minor categories [3, 4]. Previously, many Chinese question classification systems were developed using relevant methods adopted for English question classification. However, the process of Chinese question classification is more complicated than that of English question classification. Because of such discrepancy, Chinese question classification has developed on its own path.

M. Liu et al. (Eds.): CLSW 2020, LNAI 12278, pp. 734–746, 2021.
https://doi.org/10.1007/978-3-030-81197-6_60

Traditional Chinese question classification is developed mainly based on the types of things asked and answered about, such as people, places, time, numbers, etc. Cao [5] added classification of question phrases, standard question types and feature word segmentation to enhance the computer capability to identify questions. Liu [6] further mapped interrogative word patterns to question types and conducted a question understanding study based on interrogative sentence pattern recognition. In those tasks, the forms of the questions are only used as an auxiliary feature for classification. Yet, in fact, the categories of certain question forms can correspond to certain question functions. Such relationship between forms and function has not been paid much attention to in the studies on question understanding practices.

With the increase of datasets and the more extensive coverage of questions, features of complex question forms have been observed as patches to solve new problems. Yet, in this way, question classification standards become more and more complex. If questions can be formally classified first and then further classified according to the corresponding question functions expressed by different question forms, all questions can be formally covered, and at the same time the specific function information of questions can be obtained in the classification process. Therefore, it is of profound significance to advocate the formal classification of questions on the basis of the existing research (Fig. 1).

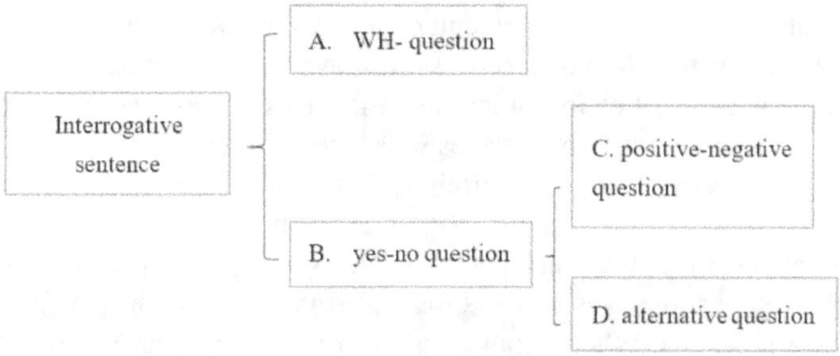

Fig. 1. Lv's question classification system [7]

Lv [7] put forward a "derivation system-based" classification method according to the derivation relation of subclasses of interrogative sentences. He suggested that WH-questions and yes-no questions were two basic types, while positive-negative questions and alternative questions were derived from yes-no questions, and both positive-negative and alternative questions could construct yes-no questions. For example:

你去? (positive) 你不去? (negative) → 你去不去? (yes-no question).

[Ni qu? ni bu qu? → ni qu bu qu?].

'Are you going? You're not going? → Will you go?'

In fact, Lv's system merely reflected the current state of question comprehension. So far, analyses of questions mostly focused on WH-questions. However, yes-no questions have more complex and various formal characteristics compared with WH-questions. The language information to answer yes-no questions thus involves more details, which are rarely mentioned in question comprehension research. If we can distinguish the four different types of questions and classify them formally at the initial stage of question recognition, then research on question recognition and understanding will be more on target, and the subsequent semantic classification can also be carried out under such formal classification system.

Section 2 will start from identifying words that related to interrogatives and the specific sentence patterns of questions, without involving the semantic understanding of questions. On this basis, this article emphasizes that forms should have a higher priority in question classification. Section 3 introduces the construction of corpus of interrogative and the research on question classification based on formal features is to be carried out in Sect. 4. Section 5 concludes this paper.

2 Classification Features Based on Language Forms

At present, there is a question classification system for academic use, which is established based on answer type and semantic information of questions. This classification is oriented towards problem solving and is beneficial to understanding and answering questions. While the system classifies questions according to their semantics, it also uses formal features as reference for classification. In the existing classification methods, formal features are used as supplementary features. However, this kind of processing methods often makes the classification system imperfect since it cannot give full play to the screening and classification functions of formal features, and it cannot cover all questions. Besides, different forms of questions have different pragmatic functions, and the corresponding answers are also different. For example, the four questions of "Is the weather hot or not?", "Is it hot?", "Is the weather hot or not hot?" and "How is the weather?" all belong to the weather category in terms of answer types, but they belong to positive-negative questions, yes-no questions, alternative questions and WH-questions in linguistic classification. When a conversation participant is to answer the above questions, s/he needs to pay attention to the differences in the focus of the questions and also to the amount of information needed. If we only look at the above four questions, in Chinese, the answers could be the same, i.e. 'hot', despite the degree of naturalness in the answers. This drives us to give the formal classification of questions a higher priority before considering other classifications of questions.

In question-and-answer systems developed in the past, the questions that can usually be identified and answered mostly are WH-questions. However, there are three more types of questions if classified in terms of form, namely yes-no questions, alternative questions and positive-negative questions. If the coverage of question recognition is to be expanded, these three types of questions must also be included in the research objective [6]. The categories of questions are defined as follows:

Yes-no questions. The structure of this kind of questions is similar to that of declarative sentences, in the sense that the question is formed with a declarative sentence which sometimes ends with an interrogative particle "吗[*ma*]", "啊[*a*] or 哇[*wa*]" (but not the particle "呢[*ne*]"). Even if sometimes the sentence does not end with these particles, the sentence is uttered with a rising intonation (while declarative sentences are not uttered with such intonation).. There is usually a "?" at the end of the sentence. For example:

21 世纪人类将要开发月球吗?

[er shi yi shi ji ren lei jiang yao kai fa yue qiu ma?].

Will mankind explore the moon in the 21st century?

WH-questions. Interrogative pronouns are used in-situ to replace the questioned element. Commonly used interrogative pronouns are "谁[*shui*](*who*), 什么[*shen me*](*what*), 哪儿[*na er*](where), 怎么[*zen me*](*how*), 多少[*duo shao*](*how many*)" and so on. The question is ended with "呢[*ne*]" or "啊[*a*]" (but not "吗[*ma*]"). For example:

你有什么要求呢?

[ni you shen me yao qiu ne?].

What request do you have?

Alternative questions. There are several parallel clauses in alternative questions. Two clauses are often connected with "是[*shi*]" and "还是[*hai shi*]". Sometimes the mood particle "呢[*ne*]" or "啊[*a*]" is used, but "吗[*ma*]" is not used. In addition, alternative questions contain both the mood particles and the conjunctions. For example:

他是去了北京还是去了天津啊?

[ta shi qu le bei jing hai shi qu le tian jin a?].

Did he go to Beijing or Tianjin?

Positive-negative questions. It usually contains a negative word, e.g. "不[*bu*]" or "没有[*mei you*]", and does not take the form of complex sentences. The affirmation and negation of the verb or adjective are coordinated in the question and the focus of the question can be the verb/adjective or the complement [8]. The patterns and the examples are shown in the following table (Table 1):

Table 1. Patterns and examples of positive-negative questions

Pattern	Example
V/Adj + 不[*bu*] + V/Adj	你饿不饿? (*Are you hungry?*)
V + 不[*bu*] + V + X	你吃不吃饭? (*Do you eat or not?*)
V + X + 不 + V	你吃饭不吃? (*Do you eat or not?*)
V + X + 不[*bu*] + V + X	你吃饭不吃饭? (*Do you eat or not?*)
V + 不[*bu*]/没有[*mei you*]/否[*fou*]	你吃饭不? (*Do you eat?*)
V + C + V + Neg + C	这饭你吃得了吃不了? (*Can you eat this meal?*)

The choice of question features depends on the form of question classification. Lin [9] once put forward five formal markers of questions: interrogative pronouns, "是[*shi*]…还是[*hai shi*]…" (the form of alternative questions) and "X不[*bu*]X" (the form of positive-negative questions), modal particles and sentence intonation. However, Huang and Liao

[10] believed that questions can take the form of intonation, interrogative words, modal adverbs and interrogative forms. Li [11] suggested that interrogative markers include interrogative intonation, interrogative modal particles, interrogative pronouns of WH-questions and interrogative formats. He also divided interrogative markers into upper, middle and lower levels according to the distribution of interrogative markers. It was explicated that the upper level included interrogative intonation, the middle level modal particles e.g. "吗[*ma*], 吧[*ba*]" and "呢[*ne*]", and the lower level interrogative pronouns of WH-questions and interrogative formats.

Sentence intonation is a phonological issue. The intonation of questions is mostly marked by a question at the end of the interrogative. The data used in this article all contain the question mark "?", and so sentence intonation is not taken as an interrogative feature.

Instead, what could be taken as significant interrogative feature include i) mood particles, ii) interrogative pronouns, iii) interrogative format and iv) modal adverbs. To be specific: a particle "呢[*ne*]" can distinguish yes-no questions from non-yes-no questions and questions ending with particles such as "吧[*ba*], 吗[*ma*], 么[*me*]" must be yes-no questions. So mood particles can be listed as a feature of questions. Interrogative pronouns are the most prominent markers of WH-questions. All WH-questions contain interrogative pronouns. So interrogative pronouns can also be listed as an interrogative feature. In addition, each question category also has its own interrogative formats, which belong to the lower level in the classification system of Li [11], and happen to be the feature type that carries the most question information. Questions containing specific interrogative formats can often directly and accurately locate their question classification. Therefore, we regard the interrogative formats of yes-no questions, WH-questions, alternative questions, and positive-negative questions as a question feature. As for modal adverbs, they do not have obvious correlation with question classification. Yet, considering that redundant features can be filtered in the later stage with it, modal adverbs are also taken as a question feature here. These four features will be adopted in this paper for question classification.

3 Construction of Interrogative Sentence Corpus

3.1 Corpus Construction

In this paper, question marks are used as markers to search for questions in corpora. As a result, 4300 questions were randomly selected from a batch of modern novel corpora and a BaiduZhidao's question data set. The datasets are then used to constitute the corpus of interrogative sentences, which will be open to academia. The questions were annotated with question categories manually. The annotating standard of questions mainly follows Huang and Liao's definitions of questions.

'YN' refers to yes-no questions; 'WH' refers to WH- questions; 'AL' refers to alternative questions; 'PN' refers to positive-negative questions; 'BD-data' refers to the dataset from the website "BaiduZhidao".

As can be seen in Table 2, the amount and proportions of WH questions are the highest across different datasets, followed by yes-no questions, positive-negative questions and alternative questions, which to some extent reflects the natural distribution of these

Table 2. Datasets of the interrogative sentences

	Novels		BD-data		Total	
	Count	Percent	Count	Percent	Sum	Percent
YN	651	38.8%	527	20.1%	1178	27.4%
WH	749	44.6%	1857	70.9%	2606	60.6%
AL	23	1.4%	40	1.5%	63	1.5%
PN	256	15.2%	197	7.5%	453	10.5%
Sum	1679		2621		4300	

four types of questions in the language. The proportion of alternative questions and positive-negative questions is not high both in novels and in Baidu. This also shows that WH-questions and yes-no questions are more likely to be selected when people ask questions.

Despite the similarities mentioned above, the distributions of the four types of questions are slightly different across the two data sets. In the novels, the proportion of WH-questions is slightly higher than that of yes-no questions. However, in BD-data, WH-questions account for more than 70%, far more than the yes-no questions which account for 20.1%. To some extent, this shows that the distributional characteristics of questions in novels and in BD-data are different despite a small degree of similarity. BD-data is an encyclopedic question dataset, with a large proportion of questions about concepts, causes of events, etc. Therefore, there are more questions containing interrogative pronouns, which explains why more WH-questions are distributed in BD-data. However, there is no such obvious tendency in novels, which makes the distribution of yes-no questions and WH-questions more balanced.

3.2 Feature Selection and Question Feature Set Construction

Syntactic formats and interrogative markers play a major role in question classification. According to the linguistic definitions of yes-no questions, WH-questions, alternative questions and positive- negative questions, the syntactic formats and interrogative markers can be further divided into four categories: interrogative formats, modal particles, modal adverbs and interrogative pronouns. These four categories can be further divided into seven sub-categories according to the actual corpus, namely, i) modal particle "呢[*ne*]", ii) modal particle "吗[*ma*]" and other particles, iii) interrogative pronouns, iv) modal adverbs, v) yes-no question format, vi) positive-negative question format and vii) alternative question format.

One thing worth noting regarding yes-no questions is that some sentences have too few explicit question markers and do not contain any of the features of the seven sub-categories, such as "她走了?(*Has she gone?*)" So, in order to avoid the situations where there is no feature matching for a yes-no question, we need to add a supplementary feature. To elaborate, when there is no interrogative pronoun and the question is not in the positive-negative question format or the alternative question format, the sentence has the supplementary feature by default, and otherwise there is no such feature.

We use F1-F8 to stand for the above eight types of features. In the corpus, the quantitative statistics of these formal features are as follows:

Table 3. Distribution of question features

Feature	Category	Count/Percent	Explanation
F1	Modal particle	253/5.88%	"呢[*ne*]"
F2	Modal particles	802/18.65%	"吗[*ma*], 么[*me*], 吧[*ba*]", etc
F3	Interrogative pronouns	2790/64.88%	什么(*what*), 啥(*what*), 如何(*how*), 哪(where), 几(*how* many), 谁(*who*), 为啥(why), 为何(*why*), 为什么(*why*), 咋(*how/why*), 干吗(*why*), 多X(*how* + *x*), 怎么(*how*), etc
F4	Interrogative format	1292/30.05%	Yes-no questions: modal verb + modal particle
F5	Interrogative format	66/1.53%	Alternative questions: X还是[*hai shi*]X
F6	Interrogative formats	479/11.14%	Positive-negative questions: X不[*bu*]X, X不[*bu*], X没有[*mei you*], X不成[*bu cheng*], etc
F7	Modal adverbs	96/2.23%	莫非[*mo fei*], 莫不是[*mo bu shi*], 难道[*nan dao*], 难不成[*nan bu cheng*], 到底[*dao di*], 何必[*he bi*], 究竟[*jiu jing*], 岂[*qi*], etc
F8	Supplementary feature	1053/24.49%	supplementary features of F3, F5, F6

It can be seen from the above table that the distribution of features is related to the distribution of different types of questions, and some feature distributions can even directly reflect the overall distribution of questions. For example, the proportions of F3, F4, F5 and F6 are the distribution of four types of questions in the dataset, reflecting that WH-questions and yes-no questions account for a larger proportion of questions, and the number of alternative questions is less than that of positive-negative questions. On the other hand, the total sum of the proportions of these interrogative forms and interrogative pronouns is greater than 100%, which indicates that the result of question classification is not determined only by interrogative forms. That means some questions contain multiple interrogative forms or interrogative pronouns. The complexity of question classification is also reflected by this.

4 Automatic Classification Based on Question Forms

4.1 Finite State Automaton Based on Formal Feature Sets

The contributions of different question features to question classification are different. According to Table 3, we can regard the coverage rate of features as the degree of contribution of question features to question classification. They are in the following order: modal words (吗[*ma*], 么[*me*], 吧[*ba*]) = interrogative formats > interrogative pronouns > others (question adverbs, etc.). Then, based on the contribution ranking of these question features, we can let question features with large contribution values participate in question judgment preferentially, and questions that cannot be covered by question features can be classified as yes-no questions. In this way, the classification of question is carried out within such limited rule. As long as a question is input, the category to which the question belongs can definitely be output. Then the preparation for the construction of a finite state automaton based on formal features is complete.

4.2 Multi-feature Classification Based on Statistical Machine Learning

Based on the question features in Table 3, we carry out feature vectorization on the questions in the corpus. The vector coordinate of the dimension containing the specified feature is marked as 1. Otherwise the vector coordinate is marked as 0. We use 1, 2, 3 and 4 to mark the classification of questions. An example is shown in Table 4.

Table 4. An example of feature transformation

例:《亮剑》中李云龙求婚对白是第几集啊? e.g. In *which episode of DRAWING SWORD did Li Yunlong propose?*								
F1	F2	F3	F4	F5	F6	F7	F8	CLASS
0	0	1	0	0	0	0	0	2

After obtaining the multi-dimensional vector and its corresponding classification label, the task of classifying questions according to the feature distribution has essentially started. We propose to use six machine learning methods, namely support vector machine, linear classifier, Bayesian classifier, K nearest neighbor, decision tree and random forest to verify the classification effectiveness of question features according to the experience of previous classification tasks.

In addition, the number of selected features will also affect the result of question classification. Features such as F1 to F8 are similar to listing question forms from a linguistic perspective. However, further experiments are needed to prove which combination of features can achieve the best question classification results. Therefore, this paper arranges and combines F1 to F8, totaling 225 combination results.

We used 1679 manually annotated questions in novels as a training corpus, and 2621 subsequently annotated BD-data as a test corpus. After combining machine learning methods with feature combination results, we analyze the classification performance of the model from multiple perspectives in following sub-section.

4.3 Experiments and Analysis

4.3.1 Analysis of Experimental Results and Features

Since there are four types of questions, we mainly analyze the changes of models with the number of features from a macro perspective. That is, we analyze the overall advantages and disadvantages of question classification through the macro-average and micro-average of F1-score of different model classifications. While a certain amount of features involved, different combinations of the features can affect the accuracy of classification results. Consider this, we only select the best result for the certain number of features for comparison.

	1	2	3	4	5	6	7	8
BYS	0.21	0.72	0.72	0.97	0.98	0.98	0.98	0.98
DTC	0.47	0.72	0.96	0.98	0.98	0.98	0.98	0.97
LNC	0.47	0.72	0.95	0.97	0.97	0.97	0.97	0.97
RFR	0.47	0.72	0.96	0.98	0.98	0.98	0.98	0.98
KMS	0.47	0.72	0.95	0.97	0.97	0.97	0.97	0.95
SVM	0.47	0.72	0.96	0.98	0.98	0.98	0.97	0.97

BYS DTC LNC RFR KMS SVM

Fig. 2. Macro average of F1-score of each model and the number of features used

	1	2	3	4	5	6	7	8
BYS	0.7081	0.9599	0.9611	0.9760	0.9912	0.9916	0.9916	0.9912
DTC	0.8909	0.9611	0.9748	0.9912	0.9916	0.9916	0.9916	0.9908
LNC	0.8909	0.9611	0.9741	0.9901	0.9908	0.9908	0.9908	0.9905
RFR	0.8909	0.9611	0.9748	0.9912	0.9928	0.9928	0.9920	0.9916
KMS	0.8909	0.9611	0.9741	0.9905	0.9908	0.9908	0.9908	0.9657
SVM	0.8909	0.9611	0.9748	0.9908	0.9908	0.9908	0.9908	0.9908

BYS DTC LNC RFR KMS SVM

Fig. 3. Micro average of F1-score of each model and the number of features used

In Fig. 2 and 3, the classification effects of all models other than the Bayesian model are similar, and the graphs coincide in the figure. On the whole, the classification performance of a model becomes better with the increase of features in the early stage, and the classification performance reaches the best when there are 5–7 features. But after that, for most models, the increase of features will make the classification performance worse. This shows that for classification of question forms, the increase in the amount of features has only a certain effect on the classification performance of the model. If new question features are to be added, their contribution to classification must be checked.

Examining macro-averages and micro-averages of F1-scores of the random forest model, it is concluded that the macro-averages and micro-averages of F1-scores of the classification model reach the highest scores of 0.99 and 0.98 respectively when the number of features used with the random forest model is 5.

Table 5. Random forest: classification effect and selected features

Number of features	Micro-average of F1-score	Macro average of F1-score	Features
1	0.8909	0.47	F2
2	0.9611	0.72	F2, F6
3	0.9748	0.96	F2, F6, F5
4	0.9912	0.98	F2, F6, F5, F3
5	0.9928	0.98	F2, F6, F5, F3, F4
6	0.9928	0.98	F2, F6, F5, F3, F4, F8
7	0.9920	0.98	F2, F6, F5, F3, F4, F8, F1
8	0.9916	0.98	F2, F6, F5, F3, F4, F8, F1, F7

Table 6. Random forest: the gain effect of the formal features of questions

Feature	Micro-average gain of F1-score	Macro average gain of F1-score	Intensity of features
F6	+ 0.0702	+ 0.25	Strong
F5	+ 0.0137	+ 0.24	Strong
F3	+ 0.0164	+ 0.02	Sub-strong
F4	+ 0.0016	0	Sub-strong
F8	0	0	Weak
F7	−0.0008	0	Weak
F1	−0.0004	0	Weak

According to Table 5, we can sort the features according to the gain generated by the classification model using the newly added different features. Firstly, the question forms are divided into strong formal features and weak formal features. Strong formal features are those formal features of questions that are beneficial to the classification of questions, otherwise they are weak formal features. Results are shown in Table 6.

Therefore, we can further divide the eight features of F1–F8 into strong formal features, less-strong formal features and weak formal features. Strong formal features include modal particles (F2), positive-negative question format (F6), and alternative question format (F5); less-strong formal features include interrogative pronouns (F3) and yes-no question format (F4); weak formal features include supplementary features (F8), modal adverbs (F7) and the modal particle "呢[*ne*]" (F1). The greater the strength of the feature, the greater its contribution to question classification. At the same time, this result can be compared with Li's question hierarchy theory mentioned in Sect. 2. Li's interrogative markers correspond to the formal features of questions mentioned in this paper. Li's hierarchy division is based on the distribution range of interrogative markers, so it can be concluded that there is no absolute correlation between the strength of formal features and their distribution range.

4.3.2 A Comparison Between Random Forest Model and Finite State Automaton

With BD-data as the test set, the results of classification performance using finite state automaton and random forest model are obtained, as shown in Table 7:

Table 7. Classification results of random forest model and finite state automaton

Method		Precision	Recall	F1-score
Finite state automaton	Yes-no questions	0.96	0.87	0.91
	WH-questions	0.95	0.99	0.97
	Alternative questions	0.93	0.95	0.94
	Positive-negative questions	0.99	0.88	0.93
	Mac-avg of F1-score		0.94	
	Mic-avg of F1-score		0.96	
Random forest model	Yes-no questions	0.99	0.99	0.99
	WH-questions	1.00	0.99	1.00
	Alternative questions	0.97	0.95	0.96
	Positive-negative questions	0.96	0.99	0.98
	Mac-avg of F1-score		0.98	
	Mic-avg of F1-score		0.99	

Examining the overall performance of the model, it is observed that the macro-average and micro-average of F1-score of random forest are 0.04 and 0.03, which are higher than those of finite state automaton. This shows that the method of finite state automaton classification also has good performance in question classification, and most questions can be effectively covered by specific question rules. However, this method often has a low recall rate and cannot deal with sentences containing some combinations of features. Therefore, this also suggests that the random forest model actually has better classification performance in question classification.

Among the classification results of individual question forms, the F1-score of WH-questions is the best. Yet, the F1-score of finite state automaton is much lower than that of the random forest for yes-no questions and positive-negative questions. This reflects the obvious diversity of formal features for identifying yes-no questions and positive-negative questions, and a single formal feature is inadequate to cover most of these questions. As for positive-negative questions, the method of finite state automaton has higher accuracy than the random forest model, which reveals that the formal features play a strong role in identifying positive-negative questions. Yet, the recall rate is lower than that of the random forest model, which reflects the diversity of formal features of positive-negative questions.

5 Conclusions and Prospects

In this paper, the role of question forms in question classification is analyzed in detail. A question corpus has been then constructed, and the distribution of the corpus has been counted. It is noted that the number and proportion of WH-questions are the highest in the data set, followed by yes-no questions, positive-negative questions and alternative questions. This reflects the natural distribution of the four types of questions to a certain extent. Finally, according to the gain generated by the question features in classification performance, the question formal features are divided into strong formal features, less-strong formal features and weak formal features. It is concluded that the question classification can achieve the best results without the weak formal features. The classification results will provide an analytical basis for semantic classification and semantic understanding of interrogatives. In addition, a question classification model is constructed using a machine learning method based on the formal features of questions. Automatic classification experiments show that when the formal feature set is modal particles "吗[*ma*], 么[*me*], 吧[*ba*]", yes-no question format, interrogative pronouns, alternative question format, and positive-negative question format, question classification has high accuracy. This indicates that the modal particles at the end of a sentence not only distinguishes questions, but also is a powerful feature for question classification.

Classification of question forms is a problem with clear characteristics and strong rules. Using a rule-based system can also achieve good results in question classification. Therefore, we believe that a question form classification interface can be added when classifying questions. On the one hand, the accuracy of automatic classification based on question forms is guaranteed; on the other hand, all questions can uniformly be dealt with through the question form classification interface, which provides a basis for further classification of questions.

References

1. Wen, X., Zhang, Y., Liu, T., Ma, J.S.: Syntactic structure parsing based Chinese question classification. Chinese Inf. Process. (02) (2006). (in Chinese)

2. Liu, C.T.: Study on Sentence Type Theory and Its Applications in Chinese Question Answer System. Chongqing University (2010). (in Chinese)
3. Zhen, L.H., Wang, X.L., Yang, S.C.: Overview on question classification in question-answering system. Anhui University of Technology (Nature. Sci) (01) (2015). (in Chinese)
4. Niu, Y.Q., Chen, J.J., Duan, L.G., Zhang, W.: Study on classification features of Chinese interrogatives. Comput. Appl. Softw. (03) (2012). (in Chinese)
5. Cao, Z.J., Li, Z.S., Liu, C.T.: Study of question analysis in question-answering system. Comput. Sci. (11) (2005). (in Chinese)
6. Liu, C.T., Li, Z.S.: Research on question analysis based on sentence type recognition of interrogative sentences. Comput. Sci. (12) (2008). (in Chinese)
7. Lv, S.X.: Question, negation and affirmation. Zhongguo Yuwen (04) (1985). (in Chinese)
8. Yuan, Y.L.: Positive and Negative Questions and Related Typological References. Beijing Language and Culture University Press, Beijing (1994). (in Chinese)
9. Lin, Y.W.: Vocabulary, Grammar and Rhetoric. Shanghai Education Press, Shanghai (1985). (in Chinese)
10. Huang, B.R., Liao, X.D.: Modern Chinese. Higher education Press, Beijing (2017). (in Chinese)
11. Li, Y.M.: Reuse of question markers and decay of marker function. Zhongguo Yuwen (02) (1997). (in Chinese)

A Research on the Evolution of Chinese Semantics Based on Distributed Representation

Huiping Wang(✉) and Luyao Wang

Institute of Chinese Information Processing, Beijing Normal University, Beijing, China
201821090023@bnu.edu

Abstract. The diachronic evolution of word meaning has always been an important topic in linguistics, with new achievements in traditional linguistics. However, traditional "field work" can only be analyzed qualitatively, which needs accurate data collection and consumes a lot of manpower and material resources. In recent years, with the gradual rise of deep learning technology, more and more researchers have used distributed semantic representation to address semantic related problems. This new method not only opens up a new way of studying the semantic evolution of ancient Chinese, but also breaks the limitations of traditional linguistics. Under this background, this paper uses the BERT model, which is trained with the "Siku Quanshu" (*Imperial Collection of Four*) to get the distributed representation, (1) to practise the automatic annotation in ancient Chinese, with an accuracy of 92%; (2) to practise the automatic detection of semantic changes, including extension of meaning, synchronic distribution of meaning and diachronic evolution of meaning.

Keywords: Distributed representation · Ancient Chinese · Diachronic evolution

1 Introduction

Language is in a dynamic process of development, and vocabulary is the best embodiment of this development and change. Only by making clear the main line of the development and changes of the Chinese vocabulary from ancient to modern times can a history of vocabulary be established. Many achievements have been made in the study of the semantic evolution of ancient Chinese. Traditional "field work" requires a large amount of data, and it is time-consuming and laborious to rely on manual annotation to obtain data. Therefore, traditional linguistic studies often focused on individual cases. If we can automatically label the meaning of ancient Chinese words, it will be helpful to study the historical evolution of ancient Chinese words.

In recent years, with the rise of deep learning technology, a growing number of studies have used the distributed semantic representation approach to give rise to related semantic issues. Distributed semantic representation, which represents text as a low-dimensional, dense continuous vector, has achieved relatively good results in many natural language processing tasks. This new approach breaks the research limitation of traditional linguistics and provides a new methodologyto diachronic semantic evolution.

M. Liu et al. (Eds.): CLSW 2020, LNAI 12278, pp. 747–755, 2021.
https://doi.org/10.1007/978-3-030-81197-6_61

Based on the distributed representation, we attempt to apply the automatic annotation on polysemous words in ancient Chinese and to study the synchronic distribution and diachronic evolution of polysemous words in ancient Chinese.

2 Related Work

Studies on diachronic evolution in Chinese semantics have a long history and a lot of research results have been achieved. Traditional linguistic research focused on qualitative research. In terms of research objects, some studies have comprehensively analyzed the usage of some common words (Xu Shiyi 2004, Lina Zhang 2004). Some studies have focused on the development of a common word (Lu Chuanfeng 2006, Yang Fengxian 2006, Li Huixian 2008, Xu Shiyi 2010, Dacheng 2012). In regard of research theory, some studies have focused on traditional exegesis (Ma Jinglun 1997, Song Yongpei 2001, He Shu 2006, Peng Hui, 2007, Sheng Lin 2008), some on traditional semantics (Wei Huiping 2006, Wang Weihui and Gu Jun 2012), some on structuralist semantics (Wang Weihui 2003, Li Juan 2006, Wang Feng 2008), and other on cognition and pragmatics (Jiang Shaoyu 2007, Li Zongjiang 2009, Xiao Zhiye and Shen Jianxuan 2009, Li Ziling 2014). While many studies share the same concern on the evolution of Chinese vocabulary from ancient times to the present, there are also dynastic evolution studies concerning the lexical evolution in a certain period.

In addition to qualitative research, there are many linguistic studies taking quantitative approach to study the diachronic evolution of word meaning. With the establishment of large-scale corpora, scholars have adopted the "data-driven" quantitative research method (Traugott 2017). Foreign scholars have summasrized the semantic evolution of words by searching in corpora and analyzing the frequency of words and specific use of words in context. (Gries 2008, Hilpert et al. 2008, Kerremans et al. 2010, Gulordava et al. 2011). By analyzing the distribution of lexical items and parts of speech in different dynasties, Chinese scholars have studied the motivation and mechanism of lexical evolution (Li Jiequn 1986, Wang Lei 2003, Yu Jinhai 2005, Zhang Juan 2009, Yang Fengxian 2012, Yin Si 2015, Yu Xiao 2015 Li Juan 2016). All these studies can describe the evolution of word meaning from a diachronic perspective and scientifically explain the context of word meaning. But the diachronic evolution of word meaning needs a lot of data to justify. Traditional linguistic and statistical methods have tended to focus on individual cases and lacked an overall vision. They have not taken full advantage of the large amount and randomness of natural language information.

In recent years, more and more researchers have used distributed semantic representation to explore the semantic relationship among words, focusing on the following two tasks: detecting semantic changes of words (Kim et al. 2014, Kulkam et al. 2014, Liu Zhiyuan 2014, Li Peng 2017, Zheng Yi 2018, Liang Shichen 2019), tracking diachronic changes in relationships among words (Dubossarsky et al. 2015, Zhang et al. 2015, Eger and Mehler 2016, Hamilton 2016, Dubosssarsky et al. 2017, Kutuzov et al. 2017, Rosin et al. 2017, Szymanski 2017). At present, there are still some limitations and challenges in the study of evolution of word meaning based on distributed representation. Most of the current research efforts have only focused on establishing technical means to analyse semantic changes, but more detailed and in-depth analyses on, for instance, time, reasons and degrees of semantic changes should also be called for. Meanwhile, due to the limitation that a word can only be represented by one vector in Word2vec, Glove and other word vector models, evolution of the meaning of a word can only be examined from the overall level. Different meanings of polysemous words can be missed out.

3 Methods

Based on the distributed representation, we attempt to apply the automatic annotation on polysemous words in ancient Chinese and to study the synchronic distribution and diachronic evolution of polysemous words in ancient Chinese.

3.1 Training the BERT Classical Chinese Model

BERT, proposed by Google in 2018, provides approaches to polysemy. Compared to conventional fixed word embeddings, BERT can construct dynamic word embeddings according to context, which helps to model the phenomenon of polysemy. BERT predicts a word from its context. This learning process is called self-surpervised learning. Self-surpervised learning enables the model to train on large-scale corpora. With pretraining and finetuning, BERT achieves state-of-the-art performance in various natural language processing tasks.

Since our data is ancient Chinese, we have chosen to train the BERT classical Chinese model with the corpus of "Siku Quanshu" (Imperial Collection of Four) which is the largest collection of books in Chinese history. The parameters are shown in Table 1.

Table 1. BERT classical Chinese model parameters.

Number of layers	Size of hidden layers	Number of heads	Total parameters
12	768	12	110 M

3.2 Preparing a Polysemy Dictionary

There are great differences between ancient Chinese and modern Chinese. Compared with modern Chinese, ancient Chinese involves mainly monosyllabic morphemes, with most of which are polysemous words. Due to limited time and energy, we chose 10 typical polysemous words in ancient Chinese according to the following five principles: (1) a large number of senses; (2) high frequency of use; (3) special usages such as *Tongjia* (通假) and *Jiajie* (假借); (4) substantial changes in meaning from ancient to modern time. We synthesized the words from The Chinese Dictionary, Etymology, The Ancient Chinese Dictionary, Kangxi Dictionary and so on, and artificially merged similar meanings to construct a polysemy dictionary. The dictionary contains polysemous words, meanings and corresponding examples. The 10 polysemous words are "要 (yao)", "遗 (yi)", "除 (chu)", "书 (shu)", "病 (bing)", "信 (xin)", "知 (zhi)", "间 (jian)", "见 (jian)" and "从 (cong)", as shown in Table 2.

Table 2. Ten typical polysemous words in ancient Chinese.

Words	Multiple senses	High frequency	Special usages	Great changes
要 (yao)	+	+	+	+
遗 (yi)	+	+	+	+
除 (chu)	+	+	+	+
书 (shu)	+	+	+	–
病 (bing)	+	+	+	–
信 (xin)	+	+	+	+
知 (zhi)	+	+	+	–
间 (jian)	+	+	+	+
见 (jian)	+	+	+	–
从 (cong)	+	+	+	+

3.3 Getting the Sense Vector

Each sense of the polysemous words is formalized as a vector in n-dimensional space. We used BERT classical Chinese model to obtain the vector representation of each sense of the polysemous words. Each meaning of the polysemous words has some corresponding examples. First, we input all the sample sentences with each of which contains only one sense of one polysemous word into the BERT classical Chinese model. Then, we sumed up and averaged the vector representations of the sense in each sentence. The final vector is the distributed representation of the polysemy. The formula for calculating the sense vector is as follows, where c_m is the meaning vector, c_i is the vector representation of the polysemous word in each sample sentence, and n is the number of sample sentences of each sense.

$$c_m = \frac{\sum_{i=1}^{n} c_i}{n}. \tag{1}$$

Let's take the polysemous word "信 Xin" as an example to show how to get each sense vector of the polysemous word. There are 12 meanings of "Xin", which are "Xin 1 - Honesty", "Xin 2 - Honest", "Xin 3 - Trust", "Xin 4 - True", "Xin 5 - Truth", "Xin 6 - Certificate", "Xin 7 - Messenger", "Xin 8 - Message", "Xin 9 - Letter", "Xin 10 - Random", "Xin 11 - Uphold", and "Xin 12 - Two nights accommodation". Some meanings and the corresponding examples are shown in Table 3.

3.4 Polyseme Tagging

With the representation of the sense of a polyseme, we used the following method to tag the sense of a polyseme in context. First, we used BERT to get the representation of a word in context. Second, the representation of the word was compared with each representation of the sense of the word with respect to cosine similarity. The sense with

Table 3. Some senses and example sentences of Xin.

Senses	Examples Sentences
1-Honesty	**信**者，诚也。专一不移也。
	《国风》曰：“言笑晏晏，**信**誓旦旦。”
	牺牲玉帛，弗敢加也，必以**信**。
	期果言当谓之**信**。
	信者，言之端也。
2- Random	要不拿出纲纪来，**信**着他胡行乱做
	低眉**信**手续续弹，说尽心中无限事。
	贾蓉只管**信**口开河，胡言乱道。
	杜陵布衣老且愚，**信**口自比契与稷。
	自山中归来，灯下裁答，**信**笔而书。

highest similarity is the sense of the word in the context. Cosine similarity is computed as:

$$S = \frac{C_m \times C_n}{\sqrt{{C_m}^2 + {C_n}^2}}. \tag{2}$$

Where S is the similarity of M and N, c_m is the representation of sense M, c_n is the representation of sense N.

For example, “朋党比周以弊主, 言曲以使私者, 必信于重人矣”. To get the sense of “信 (xin)” in this sentence, we fed the sentence into the BERT model to get the context representation of “信 (xin)”. Then the consine similarities of representation of “信 (xin)” with all senses of “信 (xin)” was computed. Finally, the sense with the highest similarity was considered as the sense of “信 (xin)”.

3.5 Synchronic Distribution and Diachronic Evolution of Sense

The State Language Commission Corpora has collected 70 millions characters of classical Chinese, with the source of literature works from the Zhou Dynasty to the Qing Dynasty and most of the “Sikuquanshu”. We studied the synchronic distribution and diachronic evolution of sense based on the classical Chinese corpora. We used the above methods to tag the senses of polysemous words in various contexts, and then calculated the frequencies of all the senses, including the representative sense of a word used in each of the dynasties, frequencies of all the senses of a word in each of the dynasties, the frequencies of a sense in different dynasties, which showed the synchronic distribution and diachronic evolution of senses. The frequency of a sense is computed as:

$$P_m = \frac{C_m}{\sum_{i=1}^{n} c_i}. \tag{3}$$

Where P_m is the frequency of sense M in a dynasty, C_m is the number of instances of sense M, n is the number of senses.

4 Experiment Analysis

4.1 Accuracy of Polyseme Tagging

We constructed a test dataset by manually labeling some randomly sampled instances. As shown in Table 4, the accuracy of polyseme tagging for "信 (xin)" is 92.06%, which is highly satisfactory.

Table 4. The accuracy of polyseme tagging for "信 (xin)".

Words	Polyseme senses	Context sentences	Number of correct	Accuracy
"信 (xin)"	1-Honesty	432	415	96.06%
	2-Honest	429	385	89.74%
	3-Trust	332	315	94.88%
	4-True	451	416	92.24%
	5-Truth	392	369	94.13%
	6-Certificate	378	361	95.50%
	7-Meaaenger	512	482	94.14%
	8-Messege	395	332	94.94%
	9-Letter	415	394	91.18%
	10-Random	363	331	84.05%
	11-Uphold	318	280	88.05%
	12-Accommodation	295	265	89.83%

The word with the highest accuracy is "Honesty"(96.06%). Words with relatively low accuracy are "Uphold" and "Random". Their accuracy rates are 88.05% and 84.05% respectively.

The average accuracy rate is 92.06%. When we added more training data with instances of "Uphold" and "Random", the accuracy rate increased to 94.23%. The complete results are shown in Table 5.

Table 5. The accucacy of polysemy tagging for all words

Words	Accuracy	Words	Accuracy
要 (yao)	193/200	信 (xin)	183/200
遗 (yi)	196/200	知 (zhi)	194/200
除 (chu)	189/200	间 (jian)	189/200
书 (shu)	182/200	见 (jian)	179/200
病 (bing)	191/200	从 (cong)	184/200

4.2 Diachronic Evolution of Sense

We selected three typical polysemes, “看 (kan)”, “危 (wei)”,”算 (suan)”, to study the synchronic distribution and diachronic evolution of senses (Fig. 1).

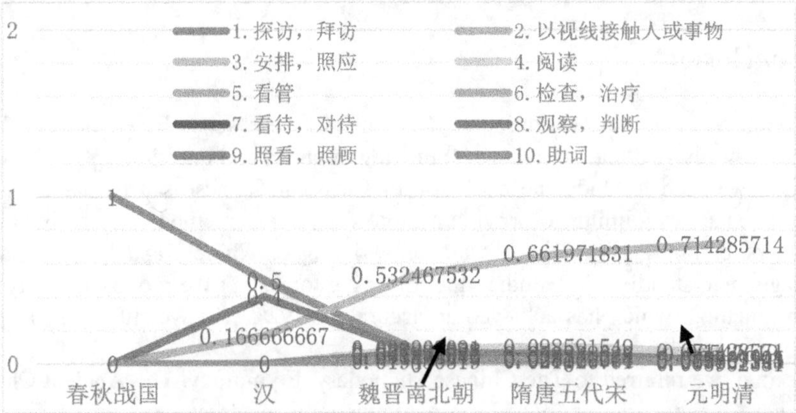

Fig. 1. The synchronic distribution and diachronic evolution of the meaning of “看 (kan)”

We found that: First, in the pre-Qin period, there was only one instance of “看 (kan)”: “梁车新为邺令, 其姐往看之, 暮而后, 门闭, 因逾郭而入”. Its sense was ‘visit’. The sense of “see” was denoted by “视 (shi)”. Second, in the Wei Jin and the Northern and Southern Dynasties, most senses of “看 (kan)” became complete. Third, in the Yuan-Ming-Qing period, “看 (kan)” obtained a new sense, namely “assign”., as in “看茶 (arrange the tea)”, “看座 (arrange the seat)”, “看饭 (arrange the food)”. The emergence of these expressions can be given rise by the flourishing of poetry and the prevalence of vernacular writing at that period of time (Fig. 2).

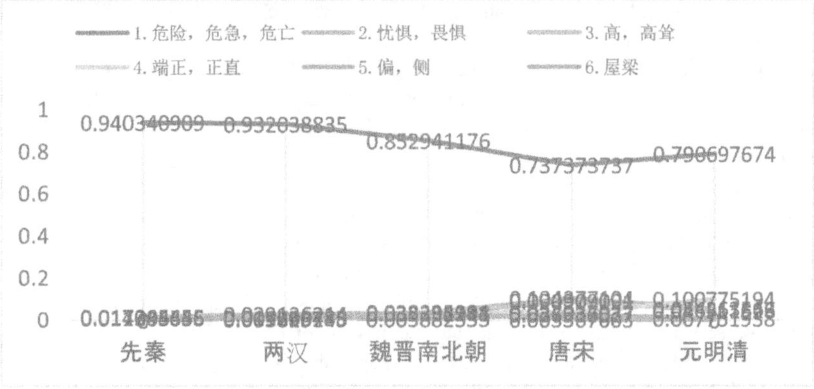

Fig. 2. The synchronic distribution and diachronic evolution of the meaning of “危 (wei)”

In general, "危 (wei)" has evolved from a word with single sense to a polyseme, and the degree of polysemy has decreased. In the Han Dynasty, "危 (wei)" evolved and obtained the new sense of "房梁 (beams)". "危急, 危险, 危亡 (danger)" have remained the most commonly used senses in "危". The use of the other senses has decreased. The senses of "偏; 侧 (lateral)" and "屋梁 (beams)" have even vanished.

5 Conclusion

During our study of the evolution of Classical Chinese sense based on distributed semantics, we have come up with the following thoughts and questions.

First, we tried to study the evolution of Classical Chinese senses based on distributed semantics, aiming to provide a more convenient methodology to linguistics reserchers, computing lingutics researchers working on Chinese and Chinese natural language understanding. We studied the following topics:(1) We proposed a polyseme tagging method, which has achieved an accuracy of 92%. (2) We studied synchronic distribution and diachronic evolution of senses.

Second, we referred to The Chinese Dictionary, Etymology, The Ancient Chinese Dictionary, Kangxi Dictionary and so on, to manually construct a dictionary of senses. Regardless of the discussion and modification during our studies, the senses may still have flaws. This needs to be improved in future studies.

Third, most words in classical Chinese are monosyllabic morphemes. Thus, the method proposed in this paper only focuses on single character words. Further studies on multi-character polysemes are therefore called for.

Acknowledgments. National Language Commission Scientific Research Plan 2017 Key Project (ZDI135–42).

References

Baldonado, M., Chang, C.-C.K., Gravano, L., Paepcke, A.: The stanford digital library metadata architecture. Int. J. Digit. Libr. **1**, 108–121 (1997)

Baugh, A.C., Cable, T.: A History of the English Language. Routledge, London (1993)

Bruce, K.B., Cardelli, L., Pierce, B.C.: Comparing object encodings. In: Abadi, M., Ito, T. (eds.) TACS 1997. LNCS, vol. 1281, pp. 415–438. Springer, Heidelberg (1997). https://doi.org/10.1007/BFb0014561

Kerremans, D., Stegmayr, S., Schmid, H.-J.: The NeoCrawler: identifying and retrieving neologisms from the internet and monitoring ongoing change. In: Allan, K., Robinson, J.A. (eds.) Current Methods in Historical Semantics, pp. 130–160. De Gruyter Mouton (2010)

Traugott, E.: Semantic Change. Oxford Research Encyclopedias: Linguistics (2017)

Azarbonyad, H., Dehghani, M., Beelen, K., Arkut, A., Marx, M., Kamps, J.: Words are malleable: computing semantic shifts in political and media discourse. In: Proceedings of the ACM on Conference on Information and Knowledge Management, Singapore, pp. 1509–1518 (2017)

Gulordava, K., Baroni, M.: A distributional similarity approach to the detection of semantic change in the Google Books Ngram corpus. In: Proceedings of the GEMS 2011 Workshop on Geometrical Models of Natural Language Semantics, Edinburgh, UK, pp. 67–71 (2011)

Hilpert, M.: Germanic Future Constructions: A Usage-based Approach to Language Change. Benjamins, Amsterdam (2008)

Michalewicz, Z.: Genetic Algorithms + Data Structures = Evolution Programs, 3rd edn. Springer, Heidelberg (1996)

Juola, P.: The time course of language change. Comput. Humanit. **37**(1), 77–96 (2003)

Turney, P., Pantel, P., et al.: From frequency to meaning: vector space models of semantics. J. Artif. Intell. Res. **37**(1), 141–188 (2010)

Gries, S.T.: Particle movement: a cognitive and functional approach. Cogn. Linguist. **10**(2), 105–145 (1999)

Eger, S., Mehler, A.: On the linearity of semantic change: investigating meaning variation via dynamic graph models. In: Proceedings of the 54th Annual Meeting of the Association for Computational Linguistics, Berlin, Germany, pp. 52–58 (2016)

Mikolov, T., Sutskever, I., Chen, K., Corrado, G.S., Dean, J.: Distributed representations of words and phrases and their compositionality. Adv. Neural Inf. Process. Syst. **26**, 3111–3119 (2013)

Leeuwen, J. (ed.): Computer Science Today. LNCS, vol. 1000. Springer, Heidelberg (1995). https://doi.org/10.1007/BFb0015232

Kulkarni, V., Al-Rfou, R., Perozzi, B., Skiena, S.: Statistically significant detection of linguistic change. In: Proceedings of the 24th International Conference on World Wide Web, Florence, Italy, pp. 625–635 (2015)

Zhang, Y., Jatowt, A., Bhowmick, S., Tanaka, K.: Omnia mutantur, nihil interit: connecting past with present by finding corresponding terms across time. In: Proceedings of the 53rd Annual Meeting of the Association for Computational Linguistics and the 7th International Joint Conference on Natural Language Processing, Beijing, China, pp. 645–655 (2015)

Kim, Y., Chiu, Y.-I., Hanaki, K., Hegde, D., Petrov, S.: Temporal analysis of language through neural language models. In: Proceedings of the 52nd Annual Meeting of the Association for Computational Linguistics, Baltimore, USA, pp. 61–65 (2014)

Lexical Semantic Evolution of Polysemous Words in Ancient Chinese—Taking *Zǒu* as an Example

Tun-Yu Hsu and Jia-Fei Hong(✉)

National Taiwan Normal University, Taipei, Taiwan
{60884002i,jiafeihong}@ntnu.edu.tw

Abstract. This study will explore the contexts of the lexical semantic change of the word *zǒu* "走, to walk" through the concepts of metonymy and metaphor. By examining the documents from ancient times, middle times to early modern times, this study has found that *zǒu* "走, to walk" in ancient times meant "to escape". During the Wei Jin Southern and Northern Dynasties, the meaning became "to run". In early modern times, the major meaning of "to walk" appeared. The semantics of the word *zǒu* "走, to walk" in ancient times, middle times, and early modern times has not changed much in basic concepts, following the general concept of "starting point → path → ending point".

Keywords: Polysemy · Ancient Chinese · Metaphor · Metonymy

1 Introduction

Meanings have been changing in language. The verb *zǒu* "走, to walk", which is closely related to our daily life, appeared as early as in the Oracle era, which the semantics has changed a lot since then.

This study explores the lexical relevance of the Chinese polysemous word *zǒu* "走, to walk" by examining its semantics, its metonymies and metaphorical uses. While some previous research have explored the cognitive meaning of *zǒu* "走, to walk" in modern Chinese, few studies have examined the meanings of this word in the ancient times. This study has taken ancient Chinese as a starting point to explore the polysemy of *zǒu* "走" in Chinese from a diachronic and statistical perspective.

2 Previous Studies

In "Introduction to Linguistics" [1], the author mentioned that *zǒu* "走" in ancient Chinese meant "striding away with two arms" (which differed from *xíng* "行" which expressed the sense of "walking step by step on two feet alternately".) Until now, the meaning of *zǒu* "走" has already evolved into the meaning of "to walk".

With the development of culture and our cognition, *zǒu* "走" not only expresses its original meaning of "to run" but is also constantly differentiated into other meanings

M. Liu et al. (Eds.): CLSW 2020, LNAI 12278, pp. 756–763, 2021.
https://doi.org/10.1007/978-3-030-81197-6_62

related to the original meaning. Previous studies [2–7] mentioned that after the Wei Jin Southern and Northern Dynasties, the meaning of "to run" of *zǒu* "走" gradually shrank. Until the Ming Dynasty, the meaning of "to walk" became the basic meaning of *zǒu* "走". In the process of semantic change, *zǒu* "走" has developed into many other meanings, according to [8], for example, *zǒu* "走" expresses the sense of "leaving a place for one destination". The process of the event can be regarded as "passing" of the entire journey, and so "to walk" can extend the meaning of "to pass". Such word extension is observed within the metaphorical cognitive model. [9] suggested that in addition to the meaning of "to pass", the meaning of *zǒu* "走" now has developed into "experience", "implementation", "visiting", "leaking", "leaving", etc., with all of which are related to the cognitive model of metaphor and metonymy.

3 Data Collection

In this study, the data were collected from the Academia Sinica Tagged Corpus of Old Chinese [10], Academia Sinica Tagged Corpus of Middle Chinese [11] and Academia Sinica Tagged Corpus of Early Mandarin Chinese [12]. The corpus for Old Chinese includes Chinese from the pre-Qin to Western Han. The Middle Chinese covers the Eastern Han to the Wei Jin Southern and Northern Dynasties. The early modern Chinese includes the important corpora after the Tang Dynasty and the Five Dynasties and Ten Kingdoms Periods.

In this study, "*Shǐjì*" (Records of the Grand Historian) has been selected as the corpus of Old Chinese. "*Sōu Shén Jì*" (Anecdotes about Spirits and Immortals) and "*Shì Shuō Xīnyǔ*" (A New Account of Tales of the World) were used as the corpus of middle Chinese. "*Xīyóu Jì*" (Journey to the West) has been chosen as the corpus of Early Modern Chinese corpus. The selected corpora are relatively well-known classic literary works. Based on the corpora of old, middle and early modern times, we have observed the diachronic semantic evolution of *zǒu* "走", and the metaphorical use of *zǒu* "走, to walk" in ancient Chinese.

219 records of *zǒu* "走" have been obtained from the ancient Chinese corpus. The mid-ancient Chinese corpus which consists of "*Sōu Shén Jì*" (Anecdotes about Spirits and Immortals) and "*Shì Shuō Xīnyǔ*" (A New Account of Tales of the World), has provided 51 and 7 records respectively. The early modern Chinese corpus "*Xīyóu Jì*" (Journey to the West) has provided 1171 instances.

4 Data Analysis

4.1 The Lexical Semantic Evolution of Zǒu in Ancient Chinese

After analyzing the corpora, this study has found that the lexical semantics of *zǒu* "走" in the old Chinese corpus can be roughly divided into eight categories, which are ranked in the following order based on frequency of use: to escape, to leave, to go, to seek refuge, to move, to run, to walk, and to attack. In the middle Chinese corpus, the lexical semantics of *zǒu* "走" can be roughly divided into five categories, which are sorted in order based on frequency of use: to run, to escape, to walk, to move, and to leave. In

Table 1. Lexical semantic distribution of *zǒu* "走" in ancient, middle and modern times.

Corpus / The number of occurrences / Semantic	Old Chinese corpus	Middle Chinese corpus	Early Modern Chinese corpus
bùxíng (步行, to walk)	4	5	**464**
líkāi (離開, to leave)	22	1	311
yídòng (移動, to move)	8	1	208
táo (逃, to escape)	**147**	15	105
xièlòu (洩漏, to leak)	0	0	24
qiánwǎng (前往, to go to)	15	0	18
pǎo (跑, to run)	12	**36**	16
jīngguò (經過, to go through)	0	0	10
diūshī (丟失, to lose)	0	0	6
chuándì (傳遞, to pass)	0	0	3
gōngjí (攻擊, to attack)	4	0	3
sànshī (散失, to dissipate)	0	0	2
zhuǎnbiàn (轉變, to change)	0	0	1
tóubēn (投奔, to seek refuge)	7	0	0
Total	219	58	1171

the early modern Chinese corpus *zǒu* "走" is found to express twelve kinds of senses including: to walk, to leave, to move, to escape, to leak, to go, to run, to pass, to lose, to go through, to dissipate and to change. The results of the above semantic analysis are shown in Table 1.

It can be seen from the table that *zǒu* "走" in the old Chinese corpus mostly expresses the meaning of "to escape", accounting for 147 out of the 219 corpora, which takes up 70%. This is related to the choice of corpus. "*Shǐjì*" (Records of the Grand Historian) has recorded more than 2,000 years of historical deeds. It is the first general history of Chronicles. A considerable number of writings about dynasties and victorious battles between countries are recorded in the compilation. It follows that the meaning of "to escape" will be substantially used. In the middle Chinese corpus, it can be found that the main sense expressed by *zǒu* "走" is "to run". Based on the early modern corpus, the statistics shows that the semantics of *zǒu* "走" is remarkably diverse. Since the middle of the Ming Dynasty, people's preferences on literature had begun to change. Short articles in the past were preferred to long and popular pieces written in various themes. Therefore, the semantics of *zǒu* "走" was also changing.

The diachronic change in lexical semantics is thus observed from the above statistics that the lexical meaning of *zǒu* "走" has changed from the original meaning of "to escape" and "to run" into "to walk".

4.2 The Metaphors and Metonymies of Ancient *zǒu*

As for metonymy and metaphorical meanings of *zǒu* "走", the amount of metonymies is more than metaphorical meanings in the old Chinese corpus. In the middle Chinese corpus, only metonymies are found and there is no metaphorical use of *zǒu* "走". In early modern Chinese corpus, there are many metaphorical uses of the word. The number of metaphors is almost the same as the number of metonymies.

Table 2. Metaphors and metonymies of zǒu "走" in ancient, middle and modern times

	Core-meanings	Metaphorical meaning	Metonymies
Old Chinese corpus	to escape	to attack, to seek refuge	to leave, to go to, to run, to move, to walk
Middle Chinese corpus	to run		to escape, to move, to walk, to leave
Early modern Chinese corpus	to walk	to leak, to lose, to pass, to dissipate, to change	to leave, to move, to escape, to go to, to run, to go through

From the Table 2, it is found that there is no significant change in metonymies from ancient times to early modern times. In ancient times, there were meanings such as "to leave, to go to, to run, to move, to walk". In modern times, due to human creation and cognition, the ways to achieve semantic changes in vocabulary have become more diverse.

4.2.1 The Metaphors of Ancient *Zǒu* "走, to Walk"

In our corpora, it is observed that several metaphorical uses of *zǒu* "走" have been demonstrated. They include the sense of "to attack", "to seek refuge", "to leak", "to dissipate", "to lose", "to change" and "to pass", exemplified and explicated below:

(1) **燕王北定代、雲中，摶胡眾入蕭關,**走長安。(史記/列傳-吳王濞列傳) "*Yàn wángběidìng dài, yún zhōng, tuán hú zhòng rù xiāo guān,* ***zǒu cháng ān***. *(Shǐjì/lièzhuàn-wú wáng bì lièzhuàn)*" (King Yan put down Dai County and Yunzhong County in the north. He led the Hu's army into Xiaoguan and took Chang'an directly. (Shiji/biography-Biography of Wú Wáng Bì))

(2) **楚雖彊,**後必屬漢。故耳走漢。(史記/列傳-張耳陳餘列傳) "*Chǔ suī qiáng, hòu bì shǔ hàn. Gù ěr* ***zǒu hàn***. *(Shǐjì/lièzhuàn-zhāng ěr chén yú lièzhuàn)*" (Even though Chu is strong, it will certainly belong to Han. So seek refuge in Han. (Shiji/biography-Biography of Zhāng Ěr Chén Yú))

(3) 只說是他自不小心,走了火,將我禪堂都燒了。(西遊記/第十六回) "*Zhǐ shuō shì tā zì bù xiǎoxīn,* ***zǒule huǒ***, *jiāng wǒ chántáng dōu shāole. (Xīyóu jì/dì shíliù huí)*" (All he said was that he was careless, leaking a fire and burning my Zen Court. (The Journey to The West/The 16th chapter))

(4) 行者道: 「老官, 走了牛也!走了牛也!」 (西遊記/第五十二回) "*Xíngzhě dào: Lǎo guān,* ***zǒule niú yě! Zǒule niú yě!****(Xīyóu jì/dì wǔshí'èr huí)*" (The pedestrian said, "Old man, you have lost the cow! You have lost the cow!" (The Journey to The West/The 52th chapter))

(5) 這是你坐下的一個青毛獅子, 卻怎麼走將來成精, 你就不收服他?」 (西遊記/第三十九回)"*Zhè shì nǐ zuò xià de yīgè qīngmáo shīzi, què zěnme* ***zǒu jiānglái chéng jīng***, *nǐ jiù bù shōufú tā?'(Xīyóu jì/dì sānshíjiǔ huí)*" (This is the one blue-haired lion that you sit on. Why didn't you subdue it, and let it become the monster? (The Journey to The West/The 39th chapter))

(6) **言訖，各坐座位，走斝傳觴,**簪花鼓瑟,果好會也。(西遊記/第七回) "*Yán qì, gè zuò zuòwèi,* ***zǒu jiǎ chuán shāng***, *zān huāgǔ sè, guǒ hǎo huì yě. (Xīyóu jì/dì qī huí)*" (After finishing the speech, everyone took a seat, and passed the wine glasses and played the zither. It was really a good party. (The Journey to The West/The 7th chapter))

In the old Chinese corpus, it can be found that the object after *zǒu* "走" is often added with the location. Interesting, it also takes an object which is the target to be attacked. In this case, *zǒu* "走 expresses the sense of "to attack". *Zǒu cháng ān* in (1) means to attack the people living in Chang'an, emphasizing the consequences of the attack. When *zǒu* "走" means "to seek refuge", usually the object following it is the refuge, e.g., "han" in (2). *Zǒu* "走," also expresses the meanings of "to leak", "to dissipate", "to lose". In (3), "*zǒu le huǒ*" can be understood as that the fire has left the place where it should be and moved to a place where it should not exist, which caused the accident. *"Zǒu le niú"* in (4) means that the cow was lost. The cow left from where it was supposed to be and located at some unknown place afterwards. These two examples, i.e. (3) and (4) are related to the meaning of "to leave". Yet, the difference is that the meaning of "to leak" in (3) has a more stable path than the meaning "to lose" in (4). To elaborate, there is no exact end point for the action of "losing", i.e., it may be lost to any place, but "leaking" has a clear destination. The sense of "leak" is more in line with the concept of "starting point → path → ending point". *Zǒu* "走" also means "to change", as in (5). In this case, the blue-haired lion having transformed into a lion monster also shows a clear path. Thus, the action of *zǒu* here conforms to the process of "starting point → ending point". Another semantic meaning that matches "starting point → path → ending point" is "to pass" as in (6). In the expression of *"zǒu jiǎ chuán shāng"*, *"jiǎ"* and *"shāng"* refer to wine containers. The whole expression thus helps visualize how the wine containers are passed from one guest to the other, which moves similarly to how a man walks from one place to another.

4.2.2 The Metonymies of Ancient *Zǒu*

Zǒu "走, to walk" had different basic meanings in ancient, middle and early modern times. In ancient times, the basic meaning was "to escape". In the middle times, it was "to run". In early modern times, it has become "to walk". Regardless of which meaning it expresses, there are some rules to use the word. For example, the subject must fulfil the premise of the ability of moving feet, and the subject must pass somewhere to finally reach his/her destination. When *zǒu* "走" is used as "to leave", the subject must be animate, such as a person or an animal. Consider the example:

(7) 有二赤蛇, 可長二尺, 分南北走。(新校搜神記/卷九) "*Yǒu èr chì shé, kě cháng èr chǐ, **fēn nánběi zǒu**. (Xīn xiào sōu shén jì/juǎn jiǔ)*" (Two red snakes, about two feet long, crawled south and north respectively. (Anecdotes About Spirits and Immortals/Vol. 9th))

The semantics of *zǒu* "走" extends to "to leave", emphasizing that the subject leaves the original place. When *zǒu* "走" is used as "**to go to**", the subjects are also animate. Consider:

(8) 著他代管蟠桃園，他即偷桃；又走至瑤池，偷殽，偷酒，攪亂大會(西遊記/第七回) "*Zhe tā dài guǎn pán táoyuán, tā jí tōu táo; yòu **zǒu zhì yáochí**, tōu xiáo, tōu jiǔ, jiǎoluàn dàhuì (xīyóu jì/dì qī huí)*" (He took over Peach Garden, and he stole peaches; he went to Yaochi again, stolefruits and alcohol, and disturbed the party. (The Journey to The West/The 7th chapter))

The emphasis of "going to" is the process of movement rather than the whole process of *zǒu* "走", and so it can be regarded as metonymy, which replaces the word or phrase entirely with a related concept. *Zǒu* "走" is also used to express "**to move**" in ancient, middle times, and early modern times. Its subject can be animate or inanimate, such as a vehicle. The example is below:

(9) 乃兩個狼頭妖怪,朗朗的說著話, 向西北上走。(西遊記/第八十九回) "*Nǎi liǎng gè láng tou yāoguài, lǎnglǎng de shuōzhe huà, **xiàng xīběi shàng zǒu**.(Xīyóu jì/dì bāshíjiǔ huí)*" (There are two wolf-headed monsters spoke loudly and headed northwest. (The Journey to The West/The 89th chapter))

When *zǒu* "走", is used to convey the sense of "to move", the subject is very diverse. It emphasizes the process and direction of movement rather than the complete process of "starting point → path → ending point". When *zǒu* "走" expresses the sense of "**to go through**", it selects an animate subject, exemplified:

(10) 我也與你走過好幾處松林，不似這林深遠。(西遊記/第八十回) "*Wǒ yě yǔ nǐ* ***zǒuguò****hǎojǐ chù sōnglín, bù shì zhè lín shēnyuǎn.(Xīyóu jì/dì bāshí huí)*" (I've walked through several pine forests with you, not as far as this forest. (The Journey to The West/The 80th chapter))

When it comes to the meaning of "to go through", a place is usually added in the sentence, emphasizing the places and routes involved, rather than the whole process of *zǒu* "走".

5 Conclusion and Future Work

From the data obtained from the ancient Chinese corpus, this study has found that the basic meaning of *zǒu* "走, to walk" in ancient times was "to escape", and its metonymies included "to leave, to go, to run, to walk, to move, etc.", and the related metaphorical meaning included "to attack" and "to seek refuge". The basic meanings of the middle Chinese "to run", "to escape, to move, to walk, and to leave". These meanings were all extended meanings of "to run". In early modern times, the basic meaning of *zǒu* "走" was "to walk", and the conceptual metaphorical use of the word expressed the sense of "to leak, to lose, to pass, to dissipate, to change", and its conceptual metonymies included the meanings of "to leave, to move, to escape, to go to, to run and to go through".

Based on the abovementioned examples, this study has found that the metaphors of *zǒu* "走" throughout the history included the meanings of "to leave, to go to, to walk, to move, and to escape", and all of them conformed to a certain part of "starting point → path → ending point". Some meanings emphasized the starting point, and some emphasized the path of the process, and yet others emphasized the ending point. The metaphor of *zǒu* "走" was a process of "starting point → path → ending point".

With the results presented above, we suggest that more data from corpora of Modern Chinese should be included in the future research for in depth analyses and for comparison with the results provided in this project, so as to to better understand the semantic changes of *zǒu* "走".

Acknowledgement. This work was supported by National Taiwan Normal University's Chinese Language and Technology Center. The center is funded by Taiwan's Ministry of Education (MOE), as part of the Featured Areas Research Center Program, under the Higher Education Sprout Project.

References

1. Ben-Yi, G.: Introduction to Linguistics. Wu-Nan Book Co., Lid, Taipei (2017). (in Chinese)
2. Yi-San, Z.: Evolution of the grammatical function of the verb "*Zǒu*" from ancient times to middle age. J. Hubei Normal Univ. **18**(1) (2005). (in Chinese)

3. Yun, B.: A comparative study of the diachronic and synchronic system of the word meaning system of "*Zǒu*". J. Shanxi Univ. **30**(2) (2019). (in Chinese)
4. Jing-Min, Z.: Diachronic investigation of "*Xíng, Zǒu*". J. Guangxi Sci. Technol. Normal Univ. **30**(1) (2016). (in Chinese)
5. Shi-Juan, S.: On the evolution of the meanings of "*Zǒu, Xíng, Bēn*" in ancient Chinese. Chin. Culture 17 (2019). (in Chinese)
6. Xiao-Yan, M.: Discrimination of synonyms of "*Zǒu Qū*" in Pre-Qin dynasty. Guizhou Soc. Sci. 3 (2008). (in Chinese)
7. Jin-Ming, Y.: Diachronic replacement investigation of "*Zǒu, Pǎo*". J. Hubei Normal Univ. **39**(3) (2019). (in Chinese)
8. Wen-Wen, C.: The new meaning of "*Zǒu xīn*", Language Construction, 7 (2015). (in Chinese)
9. Lee, S.-C.: A Study of the Polysemous Verb "*Zǒu*" in Modern Chinese: A Cognitive Approach. Master's thesis, National Tsing Hua University, Hsinchu (2016). (in Chinese)
10. Academia Sinica Tagged Corpus of Old Chinese. http://lingcorpus.iis.sinica.edu.tw/ancient/
11. Academia Sinica Tagged Corpus of Middle Chinese. http://lingcorpus.iis.sinica.edu.tw/middle/#
12. Academia Sinica Tagged Corpus of Early Mandarin Chinese. http://lingcorpus.iis.sinica.edu.tw/early/

A DM-Based Approach to the Lack of Monomorphemic Causative Verbs in Mandarin Chinese

Daran Yang[1] and Xiao Chang[2(✉)]

[1] School of International Studies, Hangzhou Normal University, Hangzhou 311121, China
[2] School of Graduate Studies, Hangzhou Normal University, Hangzhou 311121, China

Abstract. In this paper, we attempt to explain the paucity of monomorphemic causative verbs in Mandarin under the framework of Distributed Morphology. We argue that there is no difference between English and Mandarin with respect to the syntactic configuration encoding causative accomplishments, but the two languages may interpret the syntactic representations by imposing different insertion frames on the functional node *Path*, resulting in different morpho-syntactic properties of their causative constructions.

Keywords: Distributed morphology · Monomorphemic causative verb · Causative constructions

1 Introduction

One of the noticeable differences between English and Mandarin in causative constructions is that while simplex predicates in English can express achievements and accomplishments with the same form (as in (1)), monomorphemic change-of-state verbs in Mandarin are primarily unaccusatives, only expressing transition from one state to another and hence are excluded from causative constructions, as shown in (2).

(1) The window broke ⇄ John broke the window.
(2) a. Huaping po-le.
vase break-Asp
'The vase broke.'
b. *Zhangsan po-le huaping.
Zhangsan break-Perf vase
'Zhangsan broke the vase.'

This contrast was first pointed out by Tai (1984), who argues that Mandarin monomorphemic verbs encode only three types of events in Vendler's (1967) classification, i.e. states, activities, and results (roughly corresponding to achievements), but not accomplishments. In order to denote causative meaning, Mandarin usually resorts to resultative verb compounds (or RVCs), which can be rendered as compensation for its

M. Liu et al. (Eds.): CLSW 2020, LNAI 12278, pp. 764–771, 2021.
https://doi.org/10.1007/978-3-030-81197-6_63

lack of simplex causatives (Huang 2015: 23). This finding was advocated and elaborated on by many subsequent studies (Lin 2001, Lin and Liu 2005, Huang Li and Li 2009, Huang 2015, etc.), but none of them have come up with a satisfactory explanation of why Mandarin has few monomorphemic verbs encoding causative meaning.

This paper will approach the contrast between (1) and (2) from the perspective of the Distributed Morphology (hereinafter DM), aiming at a reasonable account for the paucity of monomorphemic causative verbs in Mandarin. In Sect. 2, a brief review will be made on the relevant studies, with some comments on their pros and cons. Section 3 will introduce the main tenets of DM with a focus on the hypothesis of Late Insertion. A detailed analysis based on DM will be carried out in Sect. 4. Section 5 is our conclusion.

2 Literature Review

2.1 Lin (2001)

Lin (2001) attributes different patterns of argument realization between English and Mandarin to the extent to which event structures undergo lexicalization into individual lexical items, which he labels as Lexicalization Parameter.

(3) Lexicalization Parameter: Languages may vary on the phrase-structural heights which L-Syntax reaches (Lin 2001: 114).

Based on this parameter, Lin claims that what syntax has been doing in Mandarin is very much parallel to what lexicon has been doing in English. Specifically speaking, the lexicalization process in English is applied at the highest point of light verbs, i.e. English verbs have full-fledged light verb bundles in L-Syntax. By contrast, the lexicalization process in Mandarin is applied at the lowest point of light verbs. That is to say, Mandarin verbs are bare roots in L-Syntax, which have the possibility of incorporating different light verbs in S-Syntax, resulting in its flexibility in argument realization.

Lin's proposal has been advocated by many subsequent studies (c.f., Lin and Liu 2005, Huang et al. 2009, etc.), but it has also left unexplained the contrast between (1) and (2). As for (1), following Lin's analysis, *break* can be equipped with two light verbs, v_{Cause} and v_{Bec} in L-syntax, resulting in a causative verb, or with only v_{Bec}, resulting in an inchoative verb. By the same token, it is not unreasonable to assume that Mandarin verb root *po*'break' can also incorporate both v_{Cause} and v_{Bec}, yielding a causative verb (as in (2b)), although this process occurs in S-syntax rather than L-syntax. The result, however, is ungrammatical, which shows that Lin's proposal cannot rule out the overgeneration of (2b).

2.2 Huang (2015)

In view of a number of systematic differences between English and modern Chinese, Huang (2015) proposes a macro-parameter of analyticity/syntheticity in an effort to capture the large-scale correlations of parametric differences in a unified way. Huang maintains that modern Chinese is a highly analytic language while English is more synthetic. One of strong evidence for that is that Mandarin has few simplex predicates expressing accomplishments, which, however, abounds in English.

Following the feature checking mechanism (Chomsky 2001), Huang (2015) holds that verb movement in resultative constructions is driven by the need to check the affixal feature of the light verb CAUSE. As for the contrast between (1) and (2), Huang assumes that CAUSE in English has a [+strong] affixal feature, triggering overt verb movement and resulting in a synthetic causative verb like *break*. By contrast, CAUSE in Mandarin can be realized as overt light verbs like *da* 'hit'/*nong* 'make', or action verbs like *ti* 'kick', blocking the movement of lower verbs and resulting in analytic verb compounds like *da-po* 'hit-broken' or *ti-po* 'kick-broken'.

(3) Zhangsan	da/ti-CAUSE	le	chuangzi	po.
Zhangsan	*hit/kick*	*Perf*	*window*	*break*

This proposal may explain why Mandarin lacks monomorphemic causative verbs since no strong affixal feature is to be checked in English, but the problem is that if the movement of *po*'break' is blocked due to the phonetic realization of CAUSE, the surface order obtained for (3) should be "Zhangsan da/ti le chuangzi po", which is, however, not licit in modern Chinese. In other words, Huang (2015)'s proposal has failed to explain why the action verb (V) and the result verb (R)in an RVC have to be adjacent rather than separated from each other.

3 Theoretical Framework: Distributed Morphology (DM)

Differing from the previous studies, in this paper we approach the contrast between (1) and (2) from a morpho-syntactic perspective by adopting the main tenets of Distributed Morphology (Halle and Marantz 1993, Harley and Noyer 1999, Siddiqi 2009, Embick 2015, etc.). DM makes a radical departure from the traditional lexicalist models of syntax by assuming that a single generative system is responsible for both word structure and phrase structure. Accordingly, there would be no operations within the lexicon, and the traditional lexicon is distributed across three different components (or lists), as shown below (Marantz 1995, 1997):

(4) a. List 1 (Narrow Lexicon), containing bundles of purely morphosyntactic features called *morphemes*.
b. List 2 (Vocabulary), containing Vocabulary Items (VI), which are rules of correspondence between a phonological exponent and an underspecified set of morphosyntactic features.
c. List 3 (Encyclopedia), containing encyclopedia entries, which are rules of correspondence between a phonological exponent and a set of world knowledge properties.

According to DM, there is no such a process as lexicalization going on in Lexicon, since the primitive elements available to narrow syntax are solely morphemes from List 1, including f-morphemes and l-morphemes (or roots). Roots cannot appear "bare" and must be categorized by combining with category-defining functional heads (Harley and Noyer 1999). This idea has much in common with the Neo-constructionist approach advocated by Borer (2005, 2013) in her exo-skeleton model. If this idea is on the right track, then the previous studies reviewed earlier resorting to the so-called Lexicalization Parameter cannot work out under the framework of DM.

Another enlightening idea of DM we will adopt is Late Insertion, which assumes that the morphemes manipulated by syntactic computation are devoid of any phonological content and that information is inserted post-syntactically when the representation is shipped to the PF interface. In other words, there are morphological processes that alter phonological forms at PF, which may give rise to cross-linguistic variations (Siddiqi 2009). In this vein, our general idea is that the syntactic representations encoding different types of events are language universal, but different languages may vary in the means to interpret these representations at PF. This idea is very much in line with the latest biolinguistic perspective of linguistic universals and variations, which maintains that parameterization and diversity would be mostly or possibly entirely restricted to an externalization process (Berwick and Chomsky 2011: 37).

With the above tenets of DM in mind, what we will explore in the following is two-fold: first, we are going to work out the syntactic configuration which is universal to the causative construction in different languages; second, we are going to capture the differences between English and Mandarin causative constructions by investigating their language-specific morphological operations of externalization that apply at the PF branch.

4 A New Analysis Based on DM

4.1 Syntactic Configuration of the Causative Construction

Most of the previous studies on causative sentences within the generative framework propose that causative accomplishment is underlyingly a complex structure consisting of an upper projection headed by a causative light verb and a lower projection headed by a light verb denoting change of state. Although a consensus has been reached that the upper head is *Cause*, opinions may vary about the nature and denomination of the lower head such as v_{Bec} (Xiong 2004), v_{Go} (McIntyre 2004), Res (Ramchand 2008). In this paper, we follow Talmy (2000) and Wechsler (2005) by assuming that resultative verbs denoting change of states essentially involve an abstract "path" argument, which corresponds to a degree along the scale denoted by the resultative. In this sense, we designate the lower head as *Path.* Accordingly, the syntactic configuration encoding causative accomplishments involve two layers of light verbs, i.e., *Cause*P and *Path*P. This is shown in (5), where DP1 and DP2 represent external and internal arguments respectively.

(5) $[_{CauseP}$ DP1 $[_{Cause'}$ *Cause* $[_{PathP}$ DP2 $[_{Path'}$ *Path* $\sqrt{}$ROOT]]]]

Based on the idea of syntax-semantics isomorphism held by DM, we assume that the above configuration is universal to causative accomplishments in all languages. In other words, it is applicable to both English and Mandarin causative constructions, regardless of whether the predicates are monomorphemic or compound verbs. Next, we will see how the differences between English and Mandarin in causative constructions are defended to rely on different morphological operations the two languages employ to meet morpho-phonological requirements of the node *Path.*

4.2 Different Insertion Frames for *Path*

According to DM, cross-linguistic variations may result from PF operations, that is, different languages may establish different insertion frames for a given functional node. Acedo-Matellán (2016) assumes that the functional node *Path* is subject to different insertion frames in different languages, which results in typological differences in directed motion constructions. Specifically, *Path* in some languages is exponent-defective, which requires that *Path* be strictly adjacent to the upper light verb, usually *Cause*, to achieve proper exponence at the time of Vocabulary Insertion. This results in the univerbation of *Path* and *Cause*, a typical property of verb-framed languages such as Romance languages. In other languages, however, the adjacency condition between *Path* and *Cause* is not so strict that *Path* should be adjacent to a span of *Cause*, i.e. a stretch of contiguous nodes containing *Cause*, rather than *Cause* itself. This allows a verb root to intervene between *Path* and *Cause*, a typical feature of weak satellite-framed languages. There is still a third type of language in which *Path* is not exponent-defective, which means *Path* can acquire exponence independent of *Cause*, a typical feature of strong satellite-framed languages.

Based on Acedo-Matellán (2016)'s proposal, we assume that *Path* in Mandarin Chinese is exponent-defective, which requires that it undergo morphological merger of some kind with the upper functional head to acquire exponence. Moreover, we further assume that the insertion frame of *Path* in Mandarin is non-strict adjacency, i.e. not strictly adjacent to *Cause* but adjacent to a span of *Cause.* This assumption is in accordance with Acedo-Matellán (2016)'s classification of Mandarin as a weak s-framed language. Thus, the functional node *Path* in Mandarin, with exponent *e*, will have the following insertion frame.

(6) Mandarin: *Path* ⟷ *e* / [...*Cause*...] -_.

Having established the above insertion frame, let's see why monomorphemic verbs are excluded from Mandarin causative constructions, like in (2b). According to DM, the verb *po* "break" enters the derivation as a bare root, which is base-generated in the complement of *Path.* In the Narrow Syntax, the root will move to *Path* in order to be categorized. Since the affixal feature of *Cause* is weak (Huang 2015), the complex object *Path-*$\sqrt{po}$ need not move further to *Cause.* After Spell-out, this structure will be transferred to the PF component, where *Path-*$\sqrt{po}$ will move to *Cause* by means of PF raising to check the weak affix feature of the latter. The movement process can be shown as below.

(7) [$_{CauseP}$ Zhangsan [$_{Cause'}$ *Cause* [$_{PathP}$ huaping [$_{Path'}$ *Path* $\sqrt{po}$]]]]

after Spell-out before Spell-out

After feature checking, linearization will come into play to convert the hierarchical structure into a linear order so that Vocabulary Insertion may apply. After linearization, the linear order of *Path* with respect to other terminal nodes is shown as follows.

(8) Linear order 1: *Cause- Path-√po*

In this linear order, *Path* is strictly adjacent to *Cause*, which violates the insertion frame of *Path* in Mandarin, which requires *Path* to be adjacent to a span of *Cause* rather than to *Cause* itself. As a result, *Path* will not receive any exponence during vocabulary insertion, leading to a failure in PF interpretation of the structure.

The way employed by Mandarin to rescue this structure is to conflate another verb root with *Cause*, creating a span of *Cause*. According to Haugen (2009) and Mateu and Acedo-Matellán (2012), manner roots can be adjoined to a functional head by direct merge through an operation called "Manner Conflation", yielding a complex head as in compounding. This operation is by nature a morphological one, which occurs after spell-out but before linearization. Then after manner roots like *da* "hit" are introduced in the PF component, the linear order we will obtain is as follows.

(9) Linear order 2: *Cause-√da -Path-√po*

In this linear order, *Path* is adjacent to the complex [*Cause-√da*], a span of *Cause*. In other words, *Path* is adjacent, but not strictly adjacent to *Cause*, which conforms to the insertion frame of *Path* in Mandarin. In this way, we can achieve an adequate explanation of why Mandarin causative constructions only allow resultative verb compounds rather than monomorphemic verbs to serve as predicates.

Now let us turn to English. Unlike Mandarin, English is commonly characterized as a strong satellite-framed language (Talmy 2000, etc.). Therefore, *Path* in English is not exponent-defective and does not require adjacency with any functional node. We agree with Huang (2015) that *Cause* in English has strong affixal feature, which must be checked via movement of a lower head before Spell-out. That means the final landing site of the root *√break* before Spell-out will be *Cause*, resulting in a complex object *Cause- Path-√break.* That's why English allows simplex predicates to appear in causative constructions expressing completed events or accomplishments.

In addition to simplex predicates, English has another type of causative construction which roughly corresponds to Mandarin RVCs, as in (10). However, the difference lies in that in English causative constructions, the action verb and the result are separated from each other, while in Mandarin causative constructions they are agglutinated as a compound.

(10) a. John hammered the metal flat.
b. Zhangsan zaping le tiekuai.
Zhangsan hammer-flat Perf metal
'Zhangsan hammered the metal flat.'

The above difference can also be captured in our proposal. In (10a), the verb root *√hammer* is introduced in narrow syntax via manner conflation to check the strong feature of *Cause*; therefore, the lower complex *Path-√flat* has no motivation to move and will remain in-situ. Since *Path* in English is not exponent-defective, the linear order *Path-√flat* can acquire proper exponence independently at PF and everything goes fine with (10a). But for (10b), as we mentioned earlier, *Path* in Mandarin is exponent-defective, which means even if the affixal feature of *Cause* can be checked by directly merging the verb root√*za*"hammer", the complex object *Path-√ping* will still have to move to *Cause* under the adjacency requirement of *Path*, resulting in the combination of the action verb and the result verb in linear order.

5 Conclusion

In this paper, we have made an attempt to explain the lack of Chinese monomorphemic causative verbs from the perspective of morphology-syntax interface under the DM framework. Our conclusion is that there is no variation across languages with respect to the syntactic configuration encoding a given type of events. Accordingly, English and Mandarin causative constructions share the same syntactic representation, which consists of two functional projections headed by *Cause* and *Path* respectively. However, English and Mandarin may interpret the syntactic representation by imposing different insertion frames on the functional node *Path*. Mandarin imposes a non-strict adjacency condition on *Path*, which requires the movement of *Path* to *Cause* and the introduction of an action verb root to intervene between *Path* and *Cause*. That's why Mandarin causative accomplishments are invariably expressed by complex forms like RVCs instead of simplex verbs monomorphemic verbs. By contrast, *Path* in English is not exponent-defective and the affixal feature of *Cause* is strong. English has two options to check that strong feature, i.e. either by moving the lower complex head *Path-√ROOT* or by directing merging a manner verb root. In the former we have monomorphemic causative verbs like *break*, and in the latter, we have separated causative constructions like *John hammered the metal flat*. Our proposal is line with the latest biolinguistic perspective on language diversity, which attributes parametric variations to post-syntactic morphological processes that convert syntactic representations to entities accessible to the sensorimotor system.

References

Acedo-Matellán, V.: The Morphosyntax of Transition. Oxford University Press, Oxford (2016)

Berwick, R.C., Chomsky, N.: The biolinguistic program: the current state of its development. In: Di Sciullo, A.M., Boeckx, C. (eds.) The Biolinguistic Enterprise, pp. 19–41. Oxford University Press, Oxford (2011)

Borer, H.: The Normal Course of Events (Structuring Sense vol. 2). Oxford University Press, Oxford (2005)

Borer, H.: Structuring Sense, vol.3: Taking Form. Oxford University Press, Oxford (2013)

Chomsky, N.: Derivation by phase. In: Kenstowicz, M. (ed.) Ken Hale. A Life in Language, pp. 1–52. MIT Press, Cambridge (2001)

Embick, D.: The Morpheme: A Theoretical Introduction. De Gruyter Mouton, Berlin (2015)

Halle, M., Marantz, A.: Distributed morphology and the pieces of inflection. In: Hale, K., Keyser, S.J. (eds.) The View from Building 20: Essays in Linguistics in Honor of Sylvain Bromberger, pp. 111–176. The MIT Press, Cambridge (1993)

Harley, H., Noyer, R.: State-of-the-article: distributed morphology. Glot Int. **4**, 3–9 (1999)
Haugen, D.: Hyponymous objects and late insertion. Lingua **119**, 242–262 (2009)
Huang, C.-T.J.: On syntactic analyticity and parametric theory. In: Li, A., Simpson, A., Dylan Tsai, W.-T. (eds.) Chinese Syntax in a Cross-Linguistic Perspective, pp. 1–48. Oxford University Press (2015)
Huang, C.-T., James, Y.H., Li., Y.F.: The Syntax of Chinese. Cambridge University Press, Cambridge (2009)
Lin, T.-H.J.: Light Verb Syntax and the Theory of Phrase Structure. Ph.D Dissertation, University of California, Irvine (2001)
Lin, T.-H.J., Liu, C.-Y.: Coercion, event structure and syntax. Nanzan Linguist. **2**, 9–31 (2005)
Marantz, A.: 'Cat' as a Phrasal Idiom: Consequences of Late Insertion in Distributed Morphology, ms. MIT(1995)
Marantz, A.: No escape from syntax: don't try morphological analysis in the privacy of your own lexicon. In: Dimitriadis, A., Siegel, L., Surek-Clark, C., Williams, A. (eds.) Proceedings of the 21st Penn Linguistics Colloquium, UPenn Working Papers in Linguistics, Philadelphia, pp. 201–225 (1997)
Mateu, J., Acedo-Matellán, V.: The manner/result complementarity revisited: a syntactic approach. In: Cuervo, C., Roberge, Y. (eds.) The End of Argument Structure, pp. 209–228. Bingley, Emerald Group Publishing Limited (2012)
McIntyre, A.: Event paths, conflation, argument structure and VP shells. Linguistics **3**, 523–571 (2004)
Ramchand, C.G.: Verb Meaning and the Lexicon: A First-Phase Syntax. Cambridge University Press, Cambridge (2008)
Siddiqi, D.: Syntax Within the Word: Economy, Allomorphy, and Argument Selection in Distributed Morphology. John Benjamins North America Publishing Co., Philadelphia (2009)
Tai, J.H.-Y.: Verbs and times in Chinese: Vendler's four categories. In: Testen, D., Mishra, V., Drogo, J. (eds.) Papers from the Parasession on Lexical Semantics. Chicago Linguistic Society, Chicago (1984)
Talmy, L.: Toward a Cognitive Semantics, vol.2: Typology and Process in Concept Structuring. The MIT Press, Cambridge (2000)
Vendler, Z.: Linguistics in Philosophy. Cornell University Press, Ithaca (1967)
Wechsler, S.: Resultatives under the 'event-argument homomorphism' model of telicity. In: Erteschik-Shir, N., Rapoport, T. (eds.) The Syntax of Aspect. Deriving Thematic and Aspectual Interpretation, pp. 255–273. Oxford University Press, Oxford (2005)
Xiong, Z.R.: Causative Constructions in Modern Chinese. Anhui University Press, Hefei (2004)

Cognitive Science and Experimental Studies

A Study of Complement Coercion in Mandarin Chinese: Evidence from an Acceptability Judgment Task

Wenting Xue[1(✉)], Meichun Liu[1], and Stephen Politzer-Ahles[2]

[1] Department of Linguistics and Translation, City University of Hong Kong, 83, Tat Chee Avenue, Kowloon, Hong Kong SAR, China
wtxue3-c@my.cityu.edu.hk

[2] Department of Chinese and Bilingual Studies, The Hong Kong Polytechnic University, Hung Hom, Kowloon, Hong Kong SAR, China

Abstract. This study explores whether or not Mandarin sentences with complement coercion are acceptable to native speakers. Complement coercion is a phenomenon in which some event-selecting verbs (e.g., *begin*), which semantically select an event-denoting complement, can occur with an entity-denoting complement (e.g., *Mary began the book*). While the phenomenon has been extensively studied in English, it is debatable whether complement coercion sentences are also available in Mandarin. To provide empirical evidence, Mandarin speakers were invited to rate the acceptability of coercion sentences composed of aspectual verbs with entity-denoting complements (e.g., 完成这篇文章 'finished (writing/reading) the article'), and non-coercion counterparts (e.g., 翻译这篇文章 'translate the article'). Results show that the coercion sentences are generally acceptable to native speakers of Mandarin. The results support previous findings in relevant studies on Mandarin, and further provide cross-linguistic evidence for this universal phenomenon.

Keywords: Complement coercion · Mandarin Chinese · Acceptability ratings

1 Introduction

Complement coercion is an intriguing phenomenon in sentence comprehension. This phenomenon is found in sentences like *Mary began the book.* In this sentence, the verb *begin* appears to be an event-selecting verb, which normally takes an eventive complement [9], as in *begin reading/to read the book*. But this verb can also co-occur with a nominal complement denoting an entity/object [4], as in *begin the book*. What is noticeable in *begin the book* is that there seems to be a type mismatch between the verb and its nominal complement since the involved activity is not overtly indicated. To remedy the mismatch and to satisfy the semantic selection of the verb, the complement noun is supposed to undergo a type shift from an entity to an event [17, 18], which triggers a semantic enrichment [7].

M. Liu et al. (Eds.): CLSW 2020, LNAI 12278, pp. 775–784, 2021.
https://doi.org/10.1007/978-3-030-81197-6_64

Complement coercion has been widely examined in the past two decades with both linguistic analysis [e.g., 7, 15, 17, 18] and empirical evidence [e.g., 1, 6, 8, 14, 24]. However, most studies are done on Germanic languages, such as English and German. Languages outside the Indo-European family, such as Mandarin Chinese, are relatively under-researched. The lack of research in Mandarin might be due to a common belief, as implied by Lin and Liu [12], that the linguistic phenomenon does not exist in Mandarin. This, however, has been challenged by subsequent studies through both linguistic analyses and corpus-based observations [e.g., 11, 13, 22, 23].

The purpose of the current study is to provide empirical evidence for Mandarin complement coercion through an acceptability judgment task with this research question: Are Mandarin sentences with complement coercion structure (i.e., event-selecting verb + entity-denoting complement) acceptable to Mandarin speakers? Based on previous relevant linguistic analyses and corpus-based observations, we predict that this type of sentences could be accepted to some degree by native speakers of Mandarin.

1.1 Complement Coercion Phenomenon

Verbs differ in their complement types. Certain English verbs, such as *begin, finish,* appear to be event-selecting verbs, which semantically select a complement expressing an event or activity. The eventive complement can be represented with two syntactic forms, as either a gerund (e.g., *Mary began reading the book*) or an infinitival (e.g., *Mary began to read the book*) [9]. In addition, these verbs can also be combined with a simple noun phrase (NP) [4] (e.g., *the lecture/the book*). The NPs differ in their semantic denotations. When *began* is followed by an event-denoting NP as in *Mary began the lecture,* the semantic selection of an eventive complement is still satisfied. However, when *began* is followed by an entity-denoting NP as in *Mary began the book*, a type mismatch seems to occur between the verb and its complement. To resolve the type mismatch, the verb is assumed to coerce the entity NP into an event reading.

The coerced entity NP involves an enriched form of semantic composition [7]. The enriched composition contains certain event information associated with the complement, which is expressed neither by the lexical items nor by the surface structure of the coercion sentence. According to Pustejovsky [17], the underspecified event can be recovered by resorting to *qualia roles* lexically associated with the complement NP, and in particular, the *agentive* role (i.e., 'factors involved in the origin' of an object) and the *telic* role (i.e., 'purpose and function' of an object). Consider the following example:

1. (a) Jane Austen began the novel.
   (b) Jane Austen began *writing* the novel.
   (c) Jane Austen began *reading* the novel.

In (1a), the entity NP *the novel* is lexically associated with an agentive role (event of 'writing') and a telic role (event of 'reading'). Thus, its full interpretation could be (1b) or (1c). Besides the two qualia roles, the interpretation of a coercion sentence can also be affected by other factors, such as contextual information [3] and world knowledge [1]. In the case of (1), among the plausible readings of (1a), the agentive activity 'writing the novel' is preferred to the telic activity 'reading the novel', due to the fact that the subject

Jane Austen refers to a famous novelist and hence a writing activity is more relevant, as in (1b).

1.2 Complement Coercion in Mandarin

As mentioned above, complement coercion has been studied extensively in English and other Germanic languages, but not much is known about the relevant linguistic phenomenon in Mandarin Chinese. The reason why little attention has been paid to Mandarin Chinese is probably due to the presumption that complement coercion may not exist in Mandarin. According to Lin and Liu [12], the Mandarin sentence 张三开始一本书 *Zhāng-sān kāi-shǐ yī-běn shū*, a direct translation of the English sentence *John began a book*, does not sound good to native Mandarin speakers, and it has to be modified by inserting a verb (e.g., 读 *dú* 'to read' or 写 *xiě* 'to write') between the matrix verb 开始 *kāi-shǐ* 'begin/start' and the NP一本书 *yī-běn shū* 'a book'.

However, the assumption has been challenged by subsequent studies [e.g., 11, 13, 22, 23, 25, 26]. These studies provided some evidence to confirm the existence of complement coercion in Mandarin. Liu [13] presented a case study of the transitive use of the manner-denoting verb 赶 *gǎn* 'rush' in the construction [赶 *Gan* + NP] from the perspective of Construction Grammar [5]. It was found that the meaning of the construction cannot be directly retrieved from either the verb or the NP but has to resort to a constructionally coerced interpretation. For instance, 赶报告 *gǎn bào-gào* 'to rush (to finish writing) a report' requires a semantic inference from the NP's agentive role ('writing a report'). Subsequently, a corpus-based study [11] provided evidence that these types of expressions are used in Mandarin. Similar arguments were elaborated by Song [23] from the perspective of Generative Lexicon [18]. Nevertheless, Song [23] also claimed that although event coercion is a universal linguistic mechanism, it is less pervasive in Mandarin compared with in English, because some Mandarin equivalents of the English coercion verbs (e.g., 企图/试图 *qǐ-tú/shì-tú* 'attempt') are not available in the coercion construction in Mandarin. Song's claim may need to be verified since certain Mandarin verbs can be used in the coercion construction, which may not have a corresponding counterpart in English (e.g., the verb 赶 *gǎn* 'rush (to finish)'). Although complement coercion in Mandarin has drawn attention of some linguists, little experimental evidence has been offered. Therefore, the current study aimed at verifying previous claims about the existence of complement coercion in Mandarin through an acceptability judgment task. This kind of task has been acknowledged as one of the most widely used methods in linguistic research to provide information and evidence for the acceptability of expressions, for instance, complement coercion expressions, which are not readily available, not naturally produced, or fail to appear in a corpus [21].

2 The Current Study

The current study was to test the acceptability of Mandarin sentences with complement coercion structure. The acceptability rating of the coercion sentences was compared with that of non-coercion equivalents. Detailed information is described as follows.

Participants
60 native Mandarin speakers (28 women; age range = 19–35) were recruited to rate sentences. They all had college/university as their highest level of education. Participants received a monetary reward after completing the rating task.

Materials
22 triplets of experimental sentences were generated in total, all of which contain 14 to 16 Chinese characters. One example set is shown in Table 1. In these stimuli, three verb types were manipulated to co-occur with an entity-denoting complement: a) Event-selecting verbs (i.e., EventVs), which semantically select an event-denoting complement; b) Preferred verbs (i.e., PreferredVs), which denote a preferred interpretation of the expressions with the EventVs; c) Non-preferred verbs (i.e., Non-preferredVs), which denote a non-preferred but plausible interpretation of the expressions with the EventVs. The event-selecting verbs were five aspectual verbs: 开始 *kāi-shǐ* 'begin/start', 继续 *jì-xù* 'continue', 完成 *wán-chéng* 'finish', 结束 *jié-shù* 'end', 停止 *tíng-zhǐ* 'stop'. The preferred verbs and the non-preferred verbs were determined with a *preference norming* test, as described below.

As for the aspectual verbs, they are considered as raising predicates [2, 20]. The raising predicate does not impose selectional restrictions on their surface subject; but rather, the subject is restricted by the embedded predicate [20]. Furthermore, aspectual verbs do not introduce an independent event; rather, they function as 'aspectual modifiers' which modify the event expressed by the embedded predicate and its argument (ibid.). For instance, in a sentence like 顾客开始这份问卷之前看过身份证 *gù-kè kāi-shǐ zhè-fèn wèn-juàn zhī-qián kàn-guò shēn-fèn-zhèng* 'The customer took a look at the ID card before starting the questionnaire', the surface subject 顾客 *gù-kè* 'customer' can be also interpreted as the subject of the implicit predicate, e.g., 填写 *tián-xiě* 'fill in', such that the sentence has the meaning of 'it is the customer who filled in the questionnaire', but not 'it is someone else who filled in the questionnaire'. Also, the sentence does not involve two events described by *start* and *filling in the questionnaire* but rather introduces only one single event of 'start to fill in'. Note, nevertheless, that although the aspectual verbs have similar usages, subtle differences may also exist. For example, the meaning of完成[1] *wán-chéng* 'finish' indicates not only 'the completion of an event', but also 'the completion of the event as expected' [23]. The aspectual verb 结束 *jié-shù* 'end', differing from the other aspectual verbs, is more likely to pair with an eventive NP as its object rather than a verb phrase, such as 结束战争 *jié-shù zhàn-zhēng* 'end the war', 结束参观 *jié-shù cān-guān* 'end the visiting', as observed in the BCC Corpus (Beijing Language and Culture University Corpus Center, BLCU Corpus Center, BCC) [27]. The subtle differences among the aspectual verbs, however, did not have much influence on the present study since this study was carried out to test the acceptability when the verbs precede an entity complement rather than an eventive complement relevant to the potential differences.

The preferred and non-preferred verbs were derived from a preference norming test. 25 native Mandarin speakers were invited to complete 22 coercion sentences by providing two-character verbs like 翻译员完成 () 这篇文章 *fān-yì-yuán wán-chéng* ()

[1] We thank one of the reviewers' comments for this.

zhè-piān wén-zhāng 'The translator finished () the article'. The preferred verbs (e.g., 翻译 *fān-yì* 'to translate') were those occurred most frequently, 11.5 times (out of 25) on average, ranging from 6 to 23 times; the non-preferred verbs (e.g., 浏览 *liú-lǎn* 'to skim') were those occurred least frequently, 1.73 times on average, ranging from 1 to 5 times. Finally, 22 triplets of sentences were determined.

The 22 triplets of experimental sentences were randomly distributed to three lists with Latin Square design. This was to make sure that none of the three conditions of each item occurred in the same list. Note that since the 22 items could not be divided into three equal parts for putting into the three lists, each list contained two conditions with 7 items and the third condition with 8 items. Besides the experimental sentences, fillers were also included in each of the lists so that no two experimental sentences were adjacent.

Table 1. One example set of experimental sentences.

Condition	Sample sentence (CL = Classifier)
EventVs (Coercion)	翻译员完成这篇文章之后就去喝咖啡。 *fān-yì-yuán wán-chéng zhè-piān wén-zhāng zhī-hòu jiù qù hē kā-fēi* translator finish this-CL article after then go drink coffee The translator went for a coffee after finishing the article
PreferredVs (Non-coercion)	翻译员翻译这篇文章之后就去喝咖啡。 *fān-yì-yuán fān-yì zhè-piān wén-zhāng zhī-hòu jiù qù hē kā-fēi* translator translate this-CL article after then go drink coffee The translator went for a coffee after translating the article
Non-preferredVs (Non-coercion)	翻译员浏览这篇文章之后就去喝咖啡。 *fān-yì-yuán liú-lǎn zhè-piān wén-zhāng zhī-hòu jiù qù hē kā-fēi* translator skim this-CL article after then go drink coffee The translator went for a coffee after skimming over the article

Procedures

The acceptability judgement task was conducted using the City University of Hong Kong Qualtrics survey tool. 60 native Mandarin speakers, who did not participate in the preference norming test, were asked to rate the acceptability on a scale of 6, with each level having a label (1 = 完全不可接受 'completely unacceptable', 2 = 基本不可接受 'almost unacceptable', 3 = 有点不可接受 'marginally unacceptable', 4 = 有点可接受 'marginally acceptable', 5 = 基本可接受 'almost acceptable', 6 = 完全可接受 'completely acceptable'). They were randomly assigned a list, 20 participants for each. After the rating task, all participants were also required to complete a five-question questionnaire about biographical information, such as age, gender, highest level of education, mother tongue, dialect.

Data Analysis

No participant was excluded. Descriptive analysis and Analysis of Variance (ANOVA) were performed in the R statistical environment [19]. The overall mean acceptability scores for the three conditions were obtained through averaging across participants within items first and then averaging across items. ANOVA was performed to examine the differences among the conditions.

3 Results

The mean acceptability ratings and standard deviations (SD) are listed in Table 2, which shows a relatively high acceptance rate (4.4–4.7 out of 6), and indicates the acceptability of the three types of experimental stimuli. Individual data for participants and items are shown respectively in Fig. 1, which further illustrates the extent to which the rating differences are or are not robust across participants and items [16].

Table 2. Mean acceptability scores and SDs of the experimental sentences.

Condition	Mean	SD
EventVs	4.45	0.59
Non-preferredVs	4.65	0.40
PreferredVs	4.68	0.47

All mean scores presented in Table 2 are higher than 3.5 (midpoint on the scale of 6), which indicates that all experimental sentences, particularly complement coercion sentences, were generally acceptable to native Mandarin speakers. Among the three conditions, complement coercion sentences show a relatively lower mean score than the other two non-coercion conditions, which can also be observed in Fig. 1, in which most of the mean ratings (either by participants or by items) were lower for EventV sentences than for non-preferredV and preferredV sentences.

Results of One-Way ANOVA showed that there was no statistically significant difference among the mean ratings of the three conditions (F (2,63) = 1.392, $p > .05$), although the mean score for EventVs was lower numerically than PreferredVs and Non-preferredVs. Post hoc analysis (with correction of Tukey test) showed that no significant difference was observed in each of the three varied pairs: EventVs vs. PreferredVs ($t = 1.545$, $p > .05$), EventVs vs. Non-preferredVs ($t = 1.318$, $p > .05$), PreferredVs vs. Non-preferredVs ($t = 0.227$, $p > .05$).

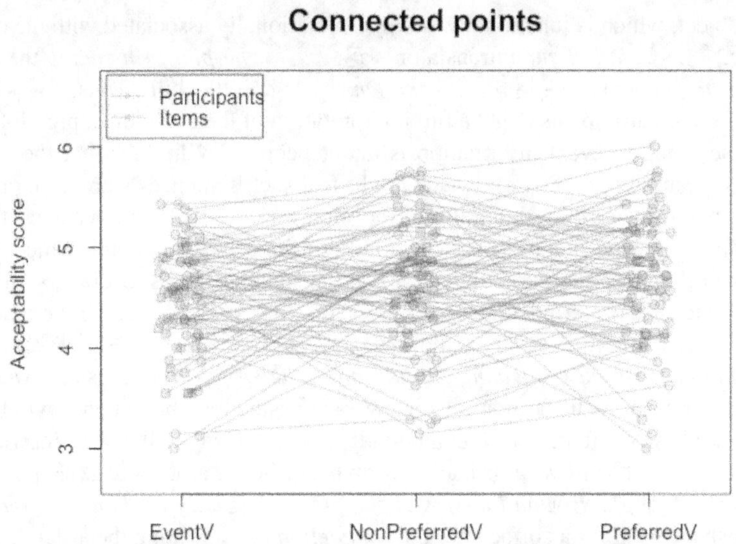

Fig. 1. Individual connected data. Each red circle represents one participant's mean acceptability score for sentences in that condition; the circles connected by red lines show one participant's means for different conditions. Each blue square represents one item's mean acceptability score in that condition; the squares connected by blue lines show one item's means in different conditions.

4 Discussion

The current study addresses the critical issue of whether Mandarin sentences with complement coercion configuration can be accepted by Mandarin native speakers. Based on acceptability judgment ratings, three main findings were obtained: 1) Complement coercion sentences were rated with relatively higher acceptability scores (which were numerically higher than the midpoint of the scale); 2) The complement coercion expressions were less acceptable than their non-coercion counterparts, i.e., preferred and non-preferred ones; 3) No significant difference was obtained among the mean ratings of the three types of sentences.

The first finding indicates that the complement coercion sentences in the current study were generally acceptable to Mandarin speakers, which supports previous relevant linguistic analysis [13, 22, 23] and corpus-based studies [11] in Mandarin complement coercion.

The result that the complement coercion sentences obtained lower acceptability ratings than their non-coercion counterparts is compatible with the semantic constraint of the event-selecting verbs; that is, they are commonly combined with a complement describing an event/activity [9]. When these verbs are paired with a complement describing an entity/object, although the sentences are still acceptable, they are just not as common as their counterparts in which no eventive complement is required, i.e., sentences in PreferredVs and Non-preferredVs conditions.

The lack of the significant difference between the coercion and non-coercion sentences may be owing to two possibilities. First, most of the stimuli have a semantically biased subject, which is functionally and interpretationally associated with the complement, e.g., 翻译员 *fān-yì-yuán* 'translator' – 这篇文章 *zhè-piān wén-zhāng* 'the article', 会计 *kuài-jì* 'accountant' – 这份账单 *zhè-fèn zhàng-dān* 'the bill'. Thus, it was not that difficult for the participants to get a full interpretation of the coercion expressions, even though the required event information is underspecified. When reading the coercion sentences, even though the participants might feel such sentences were little odd, they were not paying very close attention to the mild awkwardness but accepted the sentences. This may lead to a higher acceptability of the coercion stimuli, which resulted in a smaller difference in the ratings between the sentence types. Second, some of the preferred and non-preferred sentences may sound not very natural[2]. Take a sentence in the preferred condition as an example, 翻译员翻译这篇文章之后就去喝咖啡 *fān-yì-yuán fān-yì zhè-piān wén-zhāng zhī-hòu jiù qù hē kā-fēi* 'The translator went for a coffee after translating the article'. A reviewer suggested that the sentence will be more natural if a phase resultative verb compound 完 *wán* 'finish' [10] can be inserted after the matrix verb 翻译 *fān-yì* 'translate' to form the sentence as 翻译员翻译完这篇文章之后就去喝咖啡 *fān-yì-yuán fān-yì wan zhè-piān wén-zhāng zhī-hòu jiù qù hē kā-fēi* 'The translator went for a coffee after the *completion* of translating the article'. A small number of such non-coercion sentences may influence their acceptability scores more or less, which consequently decreased the mean ratings' difference between the coercion and non-coercion conditions. But since all mean acceptability scores of our experimental sentences were higher than the midpoint of the scale, it indicates that, overall, all the sentences were acceptable to the participants. To verify whether the complement coercion sentences are significantly more acceptable, further research can include a condition with genuine unacceptable sentences as baseline to make comparisons between the coercion sentences and the unacceptable sentences. The inclusion of such a condition can also be a manipulation check to see whether participants pay attention to the experimental sentences.

5 Conclusion

The current study aims to address the controversial issue about the grammaticality and acceptability of complement coercion in Mandarin. By conducting an acceptability judgment task, this study provides preliminary but significant empirical evidence to show that Mandarin sentences with complement coercion are acceptable and interpretable to native speakers in general, which supports the theoretical analyses and corpus-based observations in previous works about the existence of Mandarin complement coercion. This study further supports the argument that complement coercion is a universal linguistic phenomenon and it is at work cross-linguistically.

Appendix: Selected Stimuli

Five triplets of stimuli samples are presented. Each sentence of a triplet varies in the matrix verb only (separated by slashes). The first verb is an aspectual verb (i.e., coercing

[2] We express our appreciation to one of the reviewers for such comments.

verb); the second is a preferred verb; and the third is a non-preferred verb. The abbreviations involved were citied from Li and Thompson [10] (CL = Classifier; CRS = Currently relevant state *le*).

1) 作家开始/创作/编写这本小说之前已经构思很久。
zuò-jiā kāi-shǐ/chuàng-zuò/biān-xiě zhè-běn xiǎo-shuō zhī-qián yǐ-jīng gòu-sī hěn jiǔ
author start/write/edit this-CL novel before already design very long
'The author has spent a long time to design the plots before starting/writing/editing the novel.'
2) 助理继续/整理/打印这份材料之前需要做好记录。
zhù-lǐ jì-xù/zhěng-lǐ/dǎ-yìn zhè-fèn cái-liào zhī-qián xū-yào zuò-hǎo jì-lù
assistant continue/sort/print this-CL material before need make-well note
'The assistant needs to take note before continuing/sorting/printing the materials.'
3) 翻译员完成/翻译/浏览这篇文章之后就去喝咖啡。
fān-yì-yuán wán-chéng/fān-yì/liú-lǎn zhè-piān wén-zhāng zhī-hòu jiù qù hē kā-fēi
translator finish/translate/skim this-CL article after then go drink coffee
'The translator went to get some coffee after finishing/translating/skimming through the article.'
4) 患者停止/服用/注射这种药品大约已经 5 年了。
huàn-zhě tíng-zhǐ/fú-yòng/zhù-cè zhè-zhǒng yào-pǐn dà-yuē yǐ-jīng wǔ nián le
patient stop/take/inject this-CL medicine around already five year CRS
'The patient has stopped/taken/injected this kind of medicine for around five years.'
5) 董事长结束/阅读/浏览这份报纸是在接到电话前。
dǒng-shì-zhǎng jié-shù/yuè-dú/liú-lǎn zhè-fèn bào-zhǐ shì zài jiē-dào diàn-huà qián
chairman end/read/skim this-CL newspaper be at answer phone before
'The chairman ended/read/skimmed through the newspaper before answering the phone.'

References

1. Baggio, G., Choma, T., Van Lambalgen, M., Hagoort, P.: Coercion and compositionality. J. Cogn. Neurosci. **22**, 2131–2140 (2010)
2. Cao, F.: Raising verbs in Chinese. Zhongguo yuwen. 汉语的提升动词. 中国语文 **3**, 172–182 (1996)
3. Culicover, P.W., Jackendoff, R.: Simpler syntax. Oxford University Press on Demand (2005)
4. Egg, M.: Beginning novels and finishing hamburgers: remarks on the semantics of to begin. J. Semant. **20**, 163–191 (2003)
5. Goldberg, A.E.: Constructions: A Construction Grammar Approach to Argument Structure. University of Chicago Press, Chicago (1995)
6. Husband, E.M., Politzer-Ahles, S.: Coercing events or inserting structure? Eye-tracking evidence for a distinction between semantic and syntactic enrichment. In: Poster presented at the 22nd Architectures and Mechanisms in Language Processing (AMLaP) Conference, Bilbao Spain (2016)
7. Jackendoff, R.: The Architecture of the Language Faculty. MIT Press, Cambridge, MA (1997)
8. Lai, Y.Y.: The complement coercion phenomenon: Implications for models of sentence processing (Doctoral dissertation). Yale University (2017)
9. Levin, B.: English Verb Classes and Alternations: A Preliminary Investigation. University of Chicago press, Chicago (1993)
10. Li, C.N., Thompson, S.A.: Mandarin Chinese: A Functional Reference Grammar. University of California Press, Berkeley (1981)
11. Lin, S.Y., Hsieh, S.K., Huang, Y.J.: Exploring Chinese type coercion: a web-as-corpus study. In: Proceedings of the 5th International Conference on Generative Approaches to the Lexicon. Pisa Italy, pp. 43–50 (2009)

12. Lin, T.H.J., Liu, C.Y.C.: Coercion, event structure, and syntax. Nanzan Linguist. **2**, 9–31 (2005)
13. Liu, M.C.: Lexical information and beyond: Meaning coercion and constructional inference of the Mandarin verb gan. J. Chin. Linguist. **33**, 310–332 (2005)
14. McElree, B., Traxler, M.J., Pickering, M.J., Seely, R.E., Jackendoff, R.: Reading time evidence for enriched composition. Cognition **78**, B17–B25 (2001)
15. Piñango, M.M., Deo, A.: Reanalyzing the complement coercion effect through a generalized lexical semantics for aspectual verbs. J. Semant. **33**, 359–408 (2015)
16. Politzer-Ahles, S., Piccinini, P.: On visualizing phonetic data from repeated measures experiments with multiple random effects. J. Phonet. **70**, 56–69 (2018)
17. Pustejovsky, J.: The generative lexicon. Computat. Linguist. **17**, 409–441 (1991)
18. Pustejovsky, J.: The Generative Lexicon. MIT Press, Cambridge, MA (1995)
19. R Core Team: R: A language and environment for statistical computing. R Foundation for Statistical Computing, Vienna Austria (2018)
20. Rochette, A.: The selection properties of aspectual verbs. In: Johnson, K., Roberts, I.G. (eds.) Beyond principles and parameters, pp. 145–165. Springer, Dordrecht (1999). https://doi.org/10.1007/978-94-011-4822-1_6
21. Schütze, C.T., Sprouse, J.: Judgment data. In: Podesva, R.J., Sharma, D. (eds.) Research Methods in Linguistics, pp. 27–50. Cambridge University Press, Cambridge (2013)
22. Song, Z.: Light verb event and complement coercion in Mandarin Chinese. Studies of the Chinese Language. 轻动词、事件与汉语中的宾语强迫. 中国语文, **3**, 205–217 (2011)
23. Song, Z.: The Generative Lexicon and Event Coercion in Mandarin Chinese. Peking University Press, Beijing. 生成词库理论与汉语事件强迫现象研究. 北京大学出版社 (2015)
24. Traxler, M.J., Pickering, M.J., McElree, B.: Coercion in sentence processing: evidence from eye-movements and self-paced reading. J. Mem. Lang. **47**, 530–547 (2002)
25. Xue, W., Liu, M.: Chinese complement coercion: Evidence from real-time sentence processing. In: Presented at the 26th Annual Conference of International Association of Chinese Linguistics (IACL-26) & The 20th International Conference on Chinese Language and Culture (ICCLC-20). Madison USA (2018)
26. Xue, W., Liu, M., Politzer-Ahles, S.: Processing cost of complement coercion in Mandarin Chinese: Evidence from a self-paced reading study. In: Poster presented at the 25th Architecture and Mechanisms of Language Processing (AMLaP) Conference, Moscow Russia (2019)
27. Xun, E.D., Rao, G.Q., Xiao, X.Y., Zang, J.J.: The construction of the BCC Corpus in the age of Big Data. Yuliaoku Yuyanxue. 大数据背景下BCC语料库的研制. 语料库语言学, 93–118 (2016)

Significance and Application of Time Parameters in Intelligibility Test

Han Wen(✉)

Zhejiang International Studies University, No. 299 Liuhe Road, Xihu District, Hangzhou 310023, China

Abstract. The intelligibility test is widely used in the field of second language acquisition and language distance measuring. However, the current intelligibility test does not take the effect of response time into account, which causes deviations in comprehension results. This paper attempts for the first time to demonstrate the necessity of adding time parameters to the comprehension test. On the basis of the previous calculation formula, it is found that reaction time should be converted into time parameters through normalization, classification and reference calculation, so that a more complete and more unified paradigm can be established for the intelligibility test among different language variants. The addition of the time parameter extends the interpretation of the test results from one dimension to two dimensions, which not only makes the intelligibility test process more consistent with the actual communicative state, but also makes the intelligibility test more psychologically realistic, thereby further improving the theory and practice of comprehension.

Keywords: Intelligibility · Intelligibility test · Time parameters

1 Intelligibility and Intelligibility Test

Intelligibility is regarded as the behavioral standard of communication (Kent 1992: 1). Intelligibility may be defined as the match between the intention of the speaker and the response of the listener to the speech passed through the transmission system. When all the words in the listener's response list match all the words intended to be produced by the speaker, speech intelligibility is perfect. When none of the words in the listener's response list matches the words intended to be produced by the speaker, speech intelligibility is zero. In between these extremes of perfect and zero speech intelligibility lies a continuum on which we may quantify the degree to which the response list of the listener matches the intended productions of the speaker (Schiavetti 1992: 13).

In the field of linguistics, the earliest studies on intelligibility were those of Bansal (1969) and Tiffen (1974), who measured the intelligibility of Indian and Nigerian English accents respectively. At present, international research mainly aims at the intelligibility test of English with accent, and the research objects are mostly English as a second language. The subjects are usually native speakers of German, Chinese, Japanese and other languages, as in Brennan and Brennan (1981), Wardrip-Fruin and Constantinou

M. Liu et al. (Eds.): CLSW 2020, LNAI 12278, pp. 785–795, 2021.
https://doi.org/10.1007/978-3-030-81197-6_65

(1993), Munro and Derwing (1995), Munro et al. (2006), Hardman (2010), Hayes-Harb and Watzinger-Tharp (2012), Nejjari et al. (2012). In addition, the intelligibility theory is also applied to measure language distance to determine the affinity between languages, such as Voegelin and Harris (1951), Comrie (2009), Melinger (2018). In the field of Chinese linguistics, the genealogical relationship between dialects is mainly measured by degree of intelligibility, such as Tang and Heuven (2008), Tang and Heuven (2009), Tang (2009).

Intelligibility test, as the name implies, is a test of intelligibility. The test of intelligibility presupposes that intelligibility is an objective value that can be measured by specific methods. In the present literature on the intelligibility test, researchers have adopted different methodologies without a fixed research paradigm. This makes it difficult to compare experimental results quantitatively, while comparisons are made only at the conclusion level. In addition, this paper finds that at present all intelligibility tests do not measure the reaction times of the subjects, nor include reaction times in the test results. The lack of the time variable to reflect the degree of understanding will cause errors in the test results. So, how to incorporate response times into intelligibility tests is a problem for intelligibility tests. This article will illustrate the importance of response time for intelligibility tests, the necessity and the methodology to add time parameters, thus further improving the paradigm of the intelligibility test.

2 The Necessity of Adding Time Parameters

In previous intelligibility tests, only the accuracy of the subjects in completing the corresponding task after hearing the speech was considered, and reaction time (RT) was not involved in the trial as a variable. In an actual test, however, accuracy alone cannot reflect the real level of intelligibility. At present, there are two commonly used intelligibility testing methods: one uses a transliteration task; the other uses a selection task. In a selection task, the process of answering is simple and easy to learn. After hearing the recording, the subjects can make a quick response without transcribing. Therefore, the subjects can focus their attention on understanding and analyzing. The advantage of this method is its simplicity and convenience, while its disadvantage is that even if the subjects do not understand the input very well, they can choose an option through analysis or association, which is commonly called "Guess". In this way, the subjects may get the questions right by means of association, recalling and other methods instead of showing their complete ignorance. Apparently, such a way of understanding does not accord with the actual understanding of the target language. It is only the result of processing. It takes more time and attention for brain and attention to understand complex tasks. Attention is like a resource and its capacity is limited (Gui 2011). For a complex problem that cannot be solved in a unit of time, one has to spend more time to deal with it. If the subject is in a real communication circumstance with the speed of natural language flow, the brain of the subject does not have enough time to process, associate and process the language of the listener. To be specific, usually the second or third word is spoken before the first word is understood. As a result, every word is processed too late and therefore fails to be interpreted, and the communicative purpose is not achieved. Results obtained in this way do not really reflect intelligibility. In addition, when dealing with different branches of the same language or different variations in the same dialect, the differences of the phonetic systems between the two sides are subtle, so that the final scores of the traditional means evaluated on the basis of only accuracy rates may be inaccurate. This cannot lead to objective conclusions.

Tang (2009:76–78) measured the intelligibility of Chinese dialects compared with the intelligibility test without time parameter. She conducted intelligibility experiments on sentences and vocabularies respectively. In the sentence test, the subjects were required to translate the last word of the sentence into their own dialect. In the lexical test, the subjects had to classify the words they heard into certain categories. According to the results of this mutual intelligibility experiment, each dialect has 100 questions. The author used accuracy rate directly as intelligibility. The lexical intelligibility of Cantonese subjects to other dialects are as below: 23% for Xiang dialect; 31% for Min dialect; 19% for Xiang dialect. As for the lexical intelligibility of speakers of Xiang dialect, the results are: 19%and 12% for Cantonese and Min dialect respectively. The lexical intelligibility of Min dialect-speaking subjects to Cantonese and Xiang dialect is 25% and 24% respectively.

The above figures seem to reflect certain objective phenomena. However, the accuracy rate of native speakers understanding their own languages is rather low, e.g. the Xiamen natives is only 39%, the Guangzhou natives 55% and the Changsha natives 48%. Theoretically, the correct rate of the tested dialect should be close to 100%, but the results are just about 50%. Moreover, the accuracy rate is directly converted into intelligibility, ignoring the important influence of time parameters, which results in a large error in the results.[1]

Therefore, this paper suggests that RT should be added to the intelligibility test being calculated as a time parameter in the calculation of results. The reasons for using reaction time instead of processing time in this paper are as follows:

1. Intelligibility experiments belong to the level and category of behavioral experiments. In behavioral experiments, reaction time is a common practice. At the neurological level, more precise instruments are needed. Reaction time is usually around 1,000 ms, but understanding time in neural experiments may be 10–100 ms. They are not on the same scale.
2. Reaction time can be divided into two parts. One is the comprehension time used by the brain to process information, and the other is the behavioral time of the brain to direct the body to react after receiving information. The amount of time

[1] With reference to Wen (2019), the mutual understanding among the dialects of Yue (Cantonese), Xiang and Min is calculated. The one-way intelligibility of Cantonese and Xiang dialects is 10.16%; the one-way intelligibility of Xiang and Cantonese dialects is 19.20%; the one-way intelligibility of Cantonese and Min dialects is 5.03%; the one-way intelligibility of Xiang and Cantonese dialects is 14.06%; the one-way intelligibility of Xiang and Min dialects is 7.64%. The one-way intelligibility of Cantonese dialect is 0.01%. The mutual intelligibility rate between Cantonese and Xiang dialects is 14.68%; the mutual intelligibility between Cantonese and Min dialects is 9.54%; and the mutual understanding between Xiang and Min dialects is 3.82%. Among them, the internal comprehension of Cantonese is 100%; the internal comprehension of Xiang dialect is 95.65%; the internal comprehension of Min dialect is 92.87%. Compared with Tang (2009:76–78), its data are more consistent with the observed language phenomenon.

every time the body takes to respond to an information is usually similar and so the behavioral time will not be prolonged due to the difficulty in understanding. Yet, the comprehension time can change according to the level of difficulties in understanding and the reaction time will thus also change with it. Thus, it is the comprehension time that makes a difference.

3 The Methodology of Adding Time Parameters

3.1 Improve the Formula

The relevant literature that tests intelligibility does not provide a good calculation method, because the main calculation method is to calculate the average after counting the accuracy of the subjects:

$$I = \frac{A1 + A2 + \ldots\ldots + An}{n} \quad (1)$$

In a Chinese linguistic study, Zheng (1994) proposed to calculate the degree of intelligibility among Chinese dialects by using phonological materials. His methodology was to list the types of initials, vowels and tones into different groups according to their features, and calculate a parameter depending on the number of elements in that group, then add it into his equation as below:

$$D = \frac{F1 \times W1 + F2 \times W2 + \ldots\ldots + Fn \times Wn}{F1 + F2 + \ldots\ldots + Fn} \quad (2)$$

D is the language distance, F represents frequency, and W represents weight (c.f. Chen (1996:367)[2]). According to Zheng (1994), in the process of calculating the degree of intelligibility among dialects, the calculated weights set for various situations were 1, −1, 0.5, −0.5, 0.25 and −0.25, which were too subjective and lacked the support of objective basis. In addition, Zheng only obtained such an algorithm through the induction and arrangement of phonological materials, but did not perform a practical experimental test of intelligibility. So, the formula needs improvement. Regardless of this, the concept of weights proposed by Zheng (1994) is consistent with the actual situation in the communication process. Therefore, it is believed that the concept of weights plays an important role in the intelligibility test.

This paper proposes that, first, calculation of the average of the correct values can no longer meet the needs of the development of intelligibility theory and practice. Secondly,

[2] Chen Hailun (1996: 361) proposed that the one-way intelligibility of the dialect can be obtained by Y(I, J) = ΣX(I,j) × P(I,j) × R(I,j), where Y(I, J) is the one-way intelligibility of the dialect A to B dialect, X(I, j) is the similarity of a corresponding type of speech between the dialect A and B dialect, P(I, j) is the proportion of the type in the whole, R(I, j) is the influence of this type on the corresponding rule. The unidirectional communicability of the dialect vocabulary can be obtained by C(I, J) = ΣKv/(ΣKv + Σjv).Where C(I, J) is the one-way communicability of the dialect vocabulary, Kv is the occurrence rate of words shared by both dialects, ΣKv is the sum of the occurrence rates of all words shared by both dialects, and jv is unique to a dialect The occurrence rate of a word of Σjv is the sum of the occurrence rates of all words unique to the dialect.

the use of weights in previous literature should be inherited. Finally, a time parameter should be incorporated into the calculation formula by adding it into the calculation as a weight. Thus, the following formula is obtained:

$$I = \sum_{i=1}^{n} \frac{An \times T1}{n} = \frac{An \times T1 + An \times T2 + \ldots\ldots + An \times Tn}{n} \quad (3)$$

A is the score of the subject who answers the question, one standing for a correct answer and zero for a wrong answer. T is the time parameter, which adds a weight to the score. The time parameter is not a fixed coefficient, which varies from person to person and from question to question in order to ensure objectivity to the greatest extent, but calculated by the reaction time of each subject in each question. It should be noted that I is not the final intelligibility score of the languages, but the intelligibility score of one person. The final intelligibility between the two languages is determined by a series of Is. There is a relatively simple and easy way to calculate the average of I. One should keep in mind that the final inter-language intelligibility should be a range of intelligibility, which can be obtained only by establishing a norm.

3.2 Time Parameter Operation Method

In this paper, RT is converted into a time parameter through three steps of normalization, classification and reference operation. The time parameter here is not a fixed constant but will change with the subject and his reaction time. The time parameter is multiplied by the accuracy score to obtain the subject's intelligibility score for that question. In this paper, in order to demonstrate the process of how the time parameter is calculated, the raw data are cited from the third Chapter of Wen (2019:102) (Table 1)[3].

Table 1. Table of raw data of reaction times (ms)

Question No.	Question 1	Question 2	Question 3	Question 4		Question n
Subject 1	1143	3085	971	2552	……	t_{1-n}
Subject 2	776	1268	1265	3114	……	t_{2-n}

3.2.1 Normalization

The average RT for the same sentence varies from person to person. Some people have a shorter RT, while others have a longer one, e.g. 200 ms is a relatively normal response time for a sprinter, but a fast response time for people at large. Such difference in reaction

[3] The data in Chapter 3 of Wen (2019) contains 122 subjects who are mostly from universities. Each subject in that experiment needs to finish a test including 30 questions using different types of experiment material by the software of E-Prime, and the computer collects accuracy and reaction time for each question. Therefore, there are 30 × 2 × 144 = 8640 data items.

time is systematic but not independent. Therefore, taking into account the individual differences of the subjects, the reaction time is first normalized. The normalized coefficient is abbreviated to N. The coefficient should have the following characteristics: (1) the value range of N is the closed interval from zero to one ([0,1]). (2) There is a negative correlation between the N value and reaction time. The longer the reaction time is, the smaller the N value is; on the contrary, the shorter the reaction time is, the larger the N value is. In this way, the time parameters calculated in the next step can show that the longer the reaction time is, the lower the score will be. (3) The N value varies from person to person, and so the N value of each subject will be calculated separately.

In this paper, a linear normalization method is adopted to normalize RT. The final *t value* is 1 minus the normalized value, and the effect of negative correlation with RT can be achieved. The formula is:

$$t = 1 - \frac{RT - MinRT}{MaxRT - MinRT} \tag{4}$$

In this formula, RT is the reaction time of a certain subject, and MaxRT and MinRT respectively correspond to the longest and the shortest reaction time of the subject in the whole test. Then the results are as shown below (Table 2):

Table 2. Table showing results of the normalization process

Question No.	Question 1	Question 2	Question 3	Question 4		Question n
Subject 1	0.934	0.189	1	0.394	......	t_{1-n}
Subject 2	0.962	0.759	0.761	0	......	t_{2-n}

The advantage of normalization is to convert absolute reaction times in the data into relative normalized data, so that the possibility of comparison among data can be obtained. At the same time, it is also a simplified calculation method, which converts dimensional expressions into dimensionless expressions. It gives scalar results. The numerical value representing the length of time is converted into a parameter for calculation, comparison and analysis. The original data are mapped into the interval of [0, 1], which can make data processing more convenient and faster.

3.2.2 Classification

In intelligibility tests, different kinds of experimental materials are often used, such as different phonetic features, syntactic structures, and questions types. Different properties of experimental materials will certainly have a corresponding impact on the RT. Normalization can only eliminate the influence of individual factors, but not the effect of the properties of experimental materials on the reaction time. Therefore, it is necessary at this step to process the normalized data according to the properties of experimental materials.

According to the properties of the experimental materials, they are divided into several groups, and the average reaction time data of different types of experimental materials are obtained. For example, if there are experimental materials in the subjects' native language and in a foreign language, they can be divided into two categories. If the materials contain lexical and sentential materials, they need to be divided into four categories (Table 3).

Table 3. Table showing the classification process

Category	Category A		Category B		Category n	
Question number	Question 1	Question 2	Question 3	Question 4	……	Question n
Subject 1	0.934	0.189	1	0.394	……	t_{1-n}
Subject 2	0.962	0.759	0.761	0	……	t_{2-n}

3.2.3 Reference Operation

The parameter criteria are selected according to the nature of the research. For example, in a study of the intelligibility of a non-native language, the native intelligibility of the language should be taken as the reference standard, and the average value $\overline{t}_A$ of this set of data should be taken as the reference value R. Finally, the normalized value of each subject is divided by the reference value of that category. The theory is based on the relative position of the test value on the same question as the reference value. This method essentially looks at each item individually and vertically with other items and it also calculates the relative relationship of each data to the reference value. Doing so can better show the differences between real intelligibility and post-processing on a level of time (Table 4).

$$T = \frac{t}{R} \tag{5}$$

Table 4. Table showing the reference operation process

	$R_A = \bar{t}_A = 0.825$		$R_B = \bar{t}_B = 0.779$			
Category	Category A		Category B		Category n	
Question number	Question 1	Question 2	Question 3	Question 4	……	Question n
Subject 1	0.847	0.172	1.284	0.506	……	t_{1-n}
Subject 2	0.962	0.759	0.977	0	……	t_{2-n}

So, the time parameter, which converts the original reaction time into a parameter, can participate in the final calculation of intelligibility. If the subject answers the question correctly and the response time is faster than $\bar{t}_A$ ($t > \bar{t}_A$), the time parameter T will be greater than 1, and then the score of the question will be increased. The reaction time of the subject can also be slower than $\bar{t}_A$ ($t < \bar{t}_A$). There is a circumstance when the time parameter is 0, which is the slowest reaction time to that question by the subject. It thus shows that even though the subject has given the correct answer, he could fail to understand the sentence or word in the real speech flow. Anastasi and Urbina (1997) has argued that the design for the derived scores should serve dual purposes. First, the derived scores should indicate the relative positions of individuals in the group, so their scores can be evaluated by referring to other individuals. Second, the derived scores should provide a measure so that individuals' scores in different tests can be compared.

In this process, the three steps of normalization, classification and reference operation cannot be swapped, and all of them are necessary. If there were not normalization, there would be no way to eliminate differences between individuals. If the classification step is removed, the normalized data will be directly referenced. It follows that all subjects are measured using only one set of criteria, which will cause a huge error in the result.

Finally, it needs to be explained again that calculating the time parameters in this way is dynamic but can finally fall within a certain interval. As the number of samples increases, the R value will fluctuate within a certain range, so that the overall understanding parameter will fluctuate upwards or downwards slightly.

4 Meaning and Function of the Time Parameter

This paper first proposes that a time parameter should be added in intelligibility tests. In previous intelligibility researches and practices, time parameters have not been really added in the calculation of intelligibility formula. Yorkston and Beukelman (1981) first put forward the concept of communication efficiency ratio, where the time concept was mentioned. They contemplated that different speakers, even when saying the same intelligible words, also showed differences. The more intelligible words a speaker can say in "unit time", the higher degree of effective communication the person can achieve. This leads to a similar conclusion to the present study. The more words the listener understands per unit time, the more familiar the listener is with the language.

Early intelligibility tests belonged to the field of pathology and primarily focused on the speech disorder patients. Therefore, the priority concern was right or wrong, not fast or slow. In addition, when testing mutual understanding between different languages, if the language differences were considered obvious, time was not considered as an essential factor in the effectiveness of communication, because no matter how much time the subjects were given, it was believed that they could not get the answers right. In that case, it was not necessary to introduce the time parameter. Finally, the traditional intelligibility testing methods, whether it was the transliteration method or the selection method, did not have resources to record the time of each choice made by the subjects. That is to say, it was not technically adequate to record the time.

The significance of the time parameters for intelligibility test firstly concerns the data. The introduction of the time parameters can reduce the errors found in the original intelligibility tests. A new dimension is added to the original data on the basis of right and wrong. Through the reaction time, one can test whether the subjects really understand or not. And the time consumed is different according to the comprehension ability of the subjects. The so called "understanding" with a longer RT is not true understanding, but understanding with a discount. If the subject is a native speaker, he/she will "understand instantly" with very short RT.

Additionally, the significance of the time parameter for intelligibility tests is to introduce intelligibility tests into the field of psychology. This is because reaction time is an important value that must be collected in psychological testing, which can bring many issues into deep discussions. Adding a time parameter can also make intelligibility tests move further towards formalization and scaling up.

5 Conclusion

Intelligibility tests have been widely used in pathology, second language acquisition and language kinship studies. Due to the influence of history of theoretical development and research problems, the testing methodology of intelligibility test has not been theoretically discussed. Therefore, materials selection, testing methods and result statistics are all not uniform, which results in the failure of intelligibility tests to play a bigger role in practicability and credibility.

This paper discusses the concept of the time parameter, which is calculated based on the RT of the subjects and then used in the calculation of intelligibility. In normal language flow, only "understand instantly" can ensure normal communication. Therefore, in order to distinguish the above cases, it is necessary to introduce a time parameter to the operation. Especially when dealing with the situation where there is little linguistic difference between two varieties of languages, or two different dialects, it is highly necessary to introduce the time parameter. Moreover, it is the intelligibility between two similar language variants within a language that is often the key point of debate. Therefore, it is more necessary to introduce time parameters to intelligibility tests to make detailed distinctions and investigations.

Then, the calculation formula of communication effectiveness and intelligibility in the previous literature is improved, and the time parameter is added into the calculation as weights, reflecting the influence of the time parameter on intelligibility. It is also clear that reaction times need to be converted into time parameters through three steps of normalization, classification and reference operation, before finally participating in intelligibility operations.

Finally, this paper discusses the significance of adding time parameters. It is believed that the introduction of time parameters can provide more refined analysis results for language research. By adding a dimension of data, the right and wrong opposition can be extended into a two-dimensional matrix, and the space for interpretation of intelligibility tests can be expanded. Simultaneously, using the time parameter may decrease errors in the actual tests, so that more accurate results will be obtained. At last, the purpose is to incorporate intelligibility tests into the scientific tests, fully drawing on the usual paradigm of psychological testing, so as to make the intelligibility test itself more scientific and to establish a better model and standardized intelligibility test. The addition of time parameters greatly enriches the theory and practice of intelligibility, plays a significant role in paving the way for the scaling up and normalizing of intelligibility tests, further improves the quantitative research paradigm of intelligibility, and helps to build a platform for intelligibility analysis between any two languages or language variants.

References

Anastasi, A., Urbina, S.: Psychological testing. Prentice Hall/Pearson Education (1997)
Bansal, R.K.: The Intelligibility of Indian English: Measurements of the Intelligibility of Connected Speech, and Sentence and Word Material, Presented to Listeners of Different Nationalities. Ph.D. dissertation. Hyderabad: Central Institute of English (1969)
Brennan, E.M., Brennan, J.S.: Measurements of accent and attitude toward Mexican-American speech. J. Psycholinguist. Res. **10**(5), 487–501 (1981)
Chen, H.: Studies on dialect relevance, similarity, and communication indicators. Stud. Chin. Lang. (5), 361–368 (1996)
Comrie, B. (ed.): The World;s Major Languages. Routledge, Milton Park (2009)
Gui, S.: What is Psycholinguistic. Shanghai Foreign Language Education Press, Shangai (1996)
Hardman, J.: The Intelligibility of Chinese-accented English to International and American Students at a U.S. University. (Electronic Thesis or Dissertation) (2010)
Hayes-Harb, R., Watzinger-Tharp, J.: Accent, intelligibility, and the role of the listener: perceptions of English-Accented German by native German speakers. Foreign Lang. Ann. **45**(2), 260–282 (2012)
Kent, R.D. (ed.): Intelligibility in Speech Disorders: Theory, Measurement, and Management, vol. 1. John Benjamins Publishing, Amsterdam (1992)
Melinger, A.: Distinguishing languages from dialects: a litmus test using the picture-word interference task. Cognition **172**, 73–88 (2018)
Munro, M.J., Derwing, T.M.: Foreign accent, comprehensibility, and intelligibility in the speech of second language learners. Lang. Learn. **45**(1), 73–97 (1995)
Munro, M.J., Derwing, T.M., Morton, S.L.: The mutual intelligibility of L2 speech. Stud. Second Lang. Acquisition **28**(1), 111–131 (2006)
Nejjari, W., Gerritsen, M., Van der Haagen, M., Korzilius, H.: Responses to Dutch-accented English. World Engl. **31**(2), 248–267 (2012)
Schiavetti, N., Kent, R.D.: Scaling procedures for the measurement of speech intelligibility. Intelligibility in speech disorders, Intelligibility in Speech Disorders: Theory, Measurement, and Management, vol. 1, pp. 11–34. John Benjamins Publishing, Amsterdam (1992)
Tiffen, B.W.: The Intelligibility of Nigerian English. (Doctoral dissertation, University of London) (1974)
Tang, C., van Heuven, V.J.J.P.: Mutual intelligibility of Chinese dialects tested functionally. Linguist. Netherlands **25**(1), 145–156 (2008)
Tang, C., Van Heuven, V.J.: Mutual intelligibility of Chinese dialects experimentally tested. Lingua **119**(5), 709–732 (2009)
Tang, C.: Mutual Intelligibility of Chinese Dialects: An Experimental Approach. LOT, Utrecht (2009)
Van Kuyk, S., Kleijn, W.B., Hendriks, R.C.: An instrumental intelligibility metric based on information theory. IEEE Sign. Proces. Lett. **25**, 115–119 (2017)
Wardrip-Fruin, C., Constantinou, A.: Intelligibility of Mandarin speakers of English: Correlation of acoustic and perceptual measures. J. Acoust. Soc. Am. **94**(3), 1866 (1993)
Wen, H.: Theory and Practice of Mutual Intelligibility Measurement between Languages. University of Macau, Ph.D desertation (2019)
Voegelin, C.F., Harris, Z.S.: Methods for determining intelligibility among dialects of natural languages. Proc. Am. Philos. Soc. **95**(3), 322–329 (1951)

Yorkston, K.M., Beukelman, D.R.: Communication efficiency of dysarthric speakers as measured by sentence intelligibility and speaking rate. J. Speech Hearing Disord. **46**(3), 296–301 (1981)
Zheng, J.: Calculation of the communication degree of Chinese dialects. Stud. Chinese Lang. **1**, 35–43 (1994)

A Preliminary Study of Confusable Words in L2 Chinese Textbooks

Chun-Ting Chou and Li-ping Chang(✉)

Graduate Program of Teaching Chinese as a Second Language, National Taiwan University, Taipei 106319, Taiwan
lchang@ntu.edu.tw

Abstract. The present study adopts a corpus-based method to investigate the effectiveness and reliability of the confusable word sets listed in the L2 Chinese textbooks. Through the quantitative analysis based on the interlanguage corpus, we have found that, among 134 sets of words listed in the textbooks, only 50 sets have met the criterion we had set for "confusable words". We hence provide a suggested word list based on the results of the analysis, a list of the sets that do not meet the criteria but are found in the corpus with more obvious mistaken usage, as well as a list of the sets that meet the criteria but are not the most typical ones. Through the preliminary research, it is hoped to provide teachers and textbooks editors with a more scientific method of selecting confusable words to improve the efficiency of teaching and learning.

Keywords: Confusable words · Chinese teaching · Corpus · Misuse rate

1 Introduction

Along with the expansion of L2 learners' lexicons while learning a second language, more and more words sharing semantic similarities to some extent (i.e., the so-called "near synonyms (*jìnyìcí*, 近义词)") may cause confusion to the learners, and subsequently can be misused by learners. Therefore, the ability of discriminating near synonyms is important for advanced learners. In order to raise the learners' awareness of near synonyms, language teachers or editors of teaching materials have designed materials to remind the learners of the subtle differences among near synonyms, which should not be neglected.

A corpus-based approach can offer data and results for analysis based on massive and real language use. Some previous studies on differentiating synonyms have adopted the corpus-based approach to provide evidence for their analyses. For instance, Tsai et al. conducted a series of research since 1999 to analyze the lexical semantics and syntactic features of synonyms (e.g. discriminating *gāoxìng* 高兴 and *kuàilè* 快乐, both of which are often translated into "happy" in English). Xiao and McEnery (2006) analyzed collocations and semantic prosody of synonyms and they also set up criteria for discriminating synonyms, both in Chinese and English. They cited multiple sets of examples of synonyms to illustrate that both Chinese and English have semantic prosody, which

M. Liu et al. (Eds.): CLSW 2020, LNAI 12278, pp. 796–809, 2021.
https://doi.org/10.1007/978-3-030-81197-6_66

can be identified by native speakers' intuitions. L2 learners, however, lack such kind of intuitions. Thus, it is necessary to teach the learners to understand the differences among synonyms, in order to assist them to learn the language more effectively.

Near synonyms can be misused due to the semantic similarities among each other. Nevertheless, not every set of words misused by learners were caused by semantic similarities. For examples, (1)–(2) are the error sentences made by learners of Chinese as Second Language (henceforth CSL), which were found in the interlanguage (L2) corpus[1]:

(1) 那个 时候 世界 上 刚 受到 战争 的
Nàge shíhòu shìjiè shàng gāng shòudào zhànzhēng-de
That time world up just receive war-PARTICLE
*印象， 经济 起居 是 很 困难 的。
**yìnxiàng, jīngjì qǐjū shì hěn kùnnán-de.*
*impression economic living be very difficult-PARTICLE
"At that time, the world had just undergone the *impression of wars, and the financial situations were very poor."

(2) 我 来到 中国 以后， 对 北京 的
Wǒ láidào Zhōngguó yǐhòu, duì Běijīng-de
I arrive China after towards Beijing-PARTICLE
*影响 非常 好。
**yǐngxiǎng fēicháng hǎo.*
impact very good
"After I came to China, my *impact toward Beijing is very good."

In the examples above, the misused words, *yìnxiàng* (印象, impression) and *yǐngxiǎng* (影响, impact) were not synonyms or near synonyms. They had no similarities in meaning. They were often similar in Chinese L2 learners' auditory sense instead. These examples reflect that language teachers and editors of teaching materials should not only put their attention to teaching how to discriminate near synonyms, but also the words that had the similar pronunciation, forms, etc.

Laufer noticed in 1988 that not only semantic features would cause confusion. In her studies, the concept of "synforms" was firstly mentioned. "Synforms" was defined as sets of words which have similar forms. The similarity on sound, script or morphology are what cause confusion. With similar concept of synforms, Carpenter (1993) collected 400 sets of "confusable words" from the perspectives of EFL learners. He indicated that the confusable words may "look or sound similar". Taking the sentences (3)–(4) below for examples, which are found in the L2 corpus, *lǚxíng* (旅行, to travel) and *yóuxíng* (

[1] The examples in this paper are taken from HSK Dynamic Composition Corpus 2.0, http://hsk.blcu.edu.cn/. In the corpus, we have found 25 tokens of 印象 ("impression") that should be replaced with 影响 ("impact"), which account for 80.6% of the 31 misuse instances of 印象 ("impression") in total. Meanwhile, 13 tokens of 影响 ("impact") have been found to be misused and should be replaced with印象 ("impression"), accounting for 36.1% of the 36 misuse instances of影响 ("impact") in total.

游行, to parade) contain not only the same character "*xíng* (行)", but also the characters *lǚ* (旅) and *yóu* (游) which share similar shape[2].

(3) 每个人 来 中国 的 目的 不同，
Měi-ge-rén lái Zhōngguó-de mùdì bùtóng,
Everyone come China-PARTICLE purpose different
有的 来 *游行， 有的 来 谈 生意，
*yǒude lái *yóuxíng, yǒude lái tán shēngyì,*
some come parade some come talk business
还有 不少 外国人 来 学习。
háiyǒu bù-shǎo wàiguórén lái xuéxí.
also many foreigner come study
"Everyone came to China for different purposes. Some came for *parade and some for business. And yet, many foreigners came to study."

(4) 如果 自己的 国家的 人 去
Rúguǒ zìjǐ-de guójiā-de rén qù
If self-PARTICLE country-PARTICLE person go
外国 *游行 时 随地 扔掉 烟，
*wàiguó *yóuxíng shí suídì rēngdiào yān,*
foreign-country parade time random throw cigarette
这 不 是 丢人 的 事 吗？
zhè bù shì diūrén-de shì ma?
this not be shame-PARTICLE matter Q-PARTICLE
"If people from one's own country litter cigarettes during their *parade in foreign countries, it is a shame, isn't it?"

Both *yóuxíng* (游行, to parade) in the sentences above should be replaced with *lǚxíng* (旅行, to travel). *Lǚxíng* and *yóuxíng* do not share semantic features, but they can still be misused. The sentences (1)–(4) illustrate that the errors made by the learners are not only caused by the near synonyms but also by the "synforms".

To solve the problem in learning, the concept of "confusable words" (*yìhǔnyáocí*, 易混淆词) has been formed and adopted into Chinese teaching materials and lessons in recent years. Nevertheless, what are the criterion for identifying confusable words so that a more accurate list of confusable words can be included in teaching materials? Are the confusable words in the teaching materials worth being taught? These issues have not been discussed before.

In order to address these issues, the present study investigates the effectiveness and reliability of confusable words listed in L2 Chinese textbooks. By adopting a corpus-based approach, this paper finds out a set the criterion for the selection of confusable words for teaching materials.

---

[2] 5 tokens of 游行 ("to parade") have been found in the corpus which should be replaced with 旅行 ("to travel"), which account for 62.5% of the 8 instances of misuse of 游行 ("to parade") in total. No case has been observed that旅行 ("to travel") is misused to express 游行 ("to parade").

2 Confusable Words and the Previous Studies (*yìhǔnyáocí*, 易混淆词)

The term "*yìhǔnyáocí* (confusable words)" was proposed by Zhang (2007). In her schema, confusable words should not be analyzed within the frameworks of synonyms or near synonyms. Zhang believed the research on near synonyms in Chinese had contributed to the theoretical aspect but had little help in Chinese teaching. For learners, not only near synonyms cause errors, some words which have no semantic relations also cause confusion. Therefore, the studies of words discrimination for learners should not be limited to near synonyms. They should turn the perspectives to learners' interlanguage instead. Furthermore, the selection of confusable words should be based on the frequency of misusing, and the words that are frequently misused should be discriminated in priority. She also classified seven different confusable word categories according to the factors of learners' misuse and misunderstanding, which are listed below:

Category A. Influenced by the forms, sounds and meanings of the words:

1. Words with basically the same meaning. E.g. *jiěshì* (解释, to explain), *shuōmíng* (说明, to explain).
2. Words with the same morpheme. E.g. *yǒu diǎnr* (有点儿, a little bit), *yī diǎnr* (一点儿, little bit); *kàn* (看, to see), *kànjiàn* (看见, have seen).
3. Words with the same or similar pronunciation. E.g. *wénmíng* (文明, civilization), *wénmíng* (闻名, famous); *yǒu* (有, there be), *yòu* (又, again).
4. Words with the similar script. E.g. *dà* (大, big), *tài* (太, too); *tíshì* (提示, hint), *jiēshì* (揭示, unfold).

Category B. Influenced by learners' first languages (L1):

5. The word in learners' first language while translated to Chinese is polysemy. E.g. "by" in English corresponds to *zuò* (坐, to sit) or *qí* (骑, to ride) in Chinese.
6. Characters in the learner's first language consist of Chinese characters, e.g., "经验" in Japanese corresponds to the meaning of *jīngyàn* (经验, experience) or *jīnglì* (经历, background) in Chinese.
7. Two or more dialect words become one word while translated to Mandarin. E.g. Indonesian students with Hakka dialect background fuse *chī* (吃, to eat) and *shí* (食, to have food).

The classification demonstrates that the scope of confusable words is wider than near synonyms. Zhang pointed out that those words that have no similar semantic features, but cannot be classified as synonyms or near synonyms, were easily misused by learners. Thus, it can be helpful if such kinds of words can also be classified as confusable words.

In recent years, some textbooks have been designed to contain confusable words as teaching items to raise learners' consciousness. However, confusable words can be caused by many factors, as mentioned, and the scope of confusable words is relatively wider. Learners can often spend only a few hours per week or per month in second language learning. How to maximize the effectiveness of the teaching materials is the key to success. If no scientific method is dedicated to determining which words should be included in teaching, the selection might be arbitrarily decided by the teachers or textbook editors. The bias of asymmetry selection might happen. It may be counterproductive for L2 learning and the words that are not confusing in actual use might become problematic as a result.

As for the selection of confusable words, some studies have adopted the quantitative method. For instance, Zhang (2013) reckoned that one of the purposes of discriminating confusable words was to correct the learners' habitual misusing. For this reason, the selection of confusable words should be "misuse-driven". That is, the words with high error rate should be the confusable words to be discriminated. So, she proposed that the degree of confusion of the words should be determined by two indices. One was the frequency that the words were confused with others in the L2 corpus; and the other one was the ratio of the times that the words were correctly used to the times the words were used. Confusable words were ranked and given with scores based on these two indices respectively. The sum of these two scores represented the degree of confusion. She deemed that the words with a higher score should be included in the dictionary of confusable words.

Moreover, Xiao (2008) collected corpus data of Indonesian learners and conducted sample analysis according to their language background (e.g., whether they had ever studied in Chinese schools). The words which appeared at least 3 times and were misused for more than 2 times were defined as "quasi-confusable words". She then examined the "quasi-confusable words" in the whole corpus. If the words were misused for more than 15% in frequency, the words were considered as confusable words. By the quantitative method, 346 sets of confusable words for Indonesian learners were selected.

Conducting a contrastive analysis, Soontornthamniti (2018) made a prediction on confusable words based on the vocabulary in *New HSK Syllabus Level 6*. Through the results of the language tests and questionnaires for the Thai-native learners that surveyed whether they found the certain set of confusable words difficult or not, 96 sets of words with high degree of confusion were selected for "the table of Chinese confusable words for Thai-native speakers".

The previous studies mentioned above selected confusable words by quantitative methods. They analyzed the learners' corpus data, conducting tests, questionnaires, etc. The results suggest the word lists or dictionaries of confusable words. However, the time spent in learning is limited, the teaching materials may not be able to include an exhaustive list of all confusable words. Therefore, the effectiveness of the confusable-word list is more important than the comprehensiveness. The confusable words included in textbooks should be the words that the learners would misuse in the real situations. The frequency of misusing should also be taken into consideration.

3 The Research Method and Analysis

The present study examined the confusable words listed in *A Course in Contemporary Chinese* (Teng 2018), which is a series of textbooks being often used in Chinese language training centers in Taiwan. In addition, the fifth and sixth volumes of this series are currently the only two teaching materials that include the section for confusable words (*yìhǔnyáocí*, 易混淆词) in Taiwan. In such sections, the differences between each set of words are interpreted with illustrative sentences and supplemented by word meanings, usages, collocations, etc. Besides, exercises are provided to help learners to be familiar with the main usages of the words. In total, there are 134 sets of confusable words collected in the two volumes[3]. Our study has examined the actual use of those words according to the L2 corpus, HSK Dynamic Composition Corpus 2.0. Based on the corpus, we aim to find out whether the word selection is appropriate.

For example, "*táotuō* (逃脱, to outrun), and *táolí* (逃离, to escape)" are listed as a set of confusable words in the textbook. In the L2 corpus, there are 66 tokens of *táolí* (逃离, to escape), among which 6 tokens are the cases of misuse. Yet, none of the six tokens include *táotuō* (逃脱, to outrun). Instead, *jiějué* (解决, to solve), *líkāi* (离开, to leave), *táobì* (逃避, to avoid), etc., are used. In other words, even though "*táotuō* (逃脱, to outrun), and *táolí* (逃离, to escape)" are listed as confusable words in the textbook, they are not confusing with each other in actual use. Similarly, "*chéngnuò* (承诺, commitment), *dāyìng* (答应, promise)" and "*fúchí* (扶持, to help to sustain), *zhīchí* (支持, to support)" were two sets of confusable words that are not really confusing with each other.

On one hand, based on these findings, we have found that the criterion for picking confusable words are mostly based on the judgement of the textbook editors. The asymmetric bias is obvious. On the other hand, we have found that some sets of confusable words listed in the textbooks indeed caused great confusion to the learners. Take "*dàodá* (到达, to arrive), *dádào* (达到, to reach)" for an example, of which the data of the usage is listed in Table 1 below. The total amount of tokens of *dàodá* (到达, to arrive) in the corpus is 52, of which 7 are the cases of misuse. The misuse rate is 13.5%. There are 6 other instances in which the learners used other incorrect words to substitute the *dàodá* (到达, to arrive).

As for *dádào* (达到, to reach), its frequency in the corpus is much higher than the frequency of *dàodá* (到达, to reach). 540 tokens of it are identified in the corpus. Among them, 38 tokens are found to be the cases of misuse, with a misuse rate of 7%; and 39 tokens are found to the cases of substitution with incorrect words by the learners.

In Table 1, the phenomenon of the learner often misusing the original word to express meanings of certain sets of words can been found. Furthermore, most of the time, the misuse situations are diverse in the sense that the substituting words are rather various. For instance, we found some cases in the corpus misusing *dádào* (达到) as *jiějué* (解决, to solve), *dádào* (达到) as *duōdá* (多达, much as), *dádào* (达到) as *huòdé* (获得, to gain) etc., each appeared only once. Since such unsystematic errors are not the focus of our study, we examine the systematic ones.

Back to the issue we have discussed: whether "*dàodá* (到达), *dádào* (达到)" listed in textbook are really confusable. We try to set a standard to solve this problem. As long as the misuse rate exceeds 10%, words are considered as confusable ones. For example, among the 7 error-usage in *dàodá* (到达), 4 of them should be replaced by *dádào* (达到). So, the misuse rate is 57% (4/7). Then, the pair "*dàodá* (到达), *dádào* (达到)" meets the criterion and is a set of confusable words. Besides, among 38 misuse cases of *dádào*

[3] There are 141 sets in total, with 7 sets are repeated which are not counted as a result.

Table 1. The usage of *dàodá* (到达, to arrive) and *dádào* (达到, to reach) in the L2 corpus

***dàodá* 到达, to arrive (52 tokens)**	
Situation A (incorrect uses)	4 times of misusing *dàodá* (到达) as *dádào* (达到) 3 times of misusing *dàodá* (到达) as the other words, once each
Situation B (using incorrect words as substitutes)	2 times of misusing *dádào* (达到) as *dàodá* (到达) 2 times of misusing *dào* (到) as *dàodá* (到达) 2 times of misusing the other words as *dàodá* (到达), once each
***dádào* 达到, to reach (540 tokens)**	
Situation A (incorrect uses)	5 times of misusing *dádào* (达到) as *qǔdé* (取得, to get) 4 times of misusing *dádào* (达到) as *shíxiàn* (实现, to achieve) 3 times of misusing *dádào* (达到) as *dédào* (得到, to obtain) 5 times of misusing *dádào* (达到) as *dàodá* (到达) 5 times of misusing *dádào* (达到) as *dào* (到, to arrive) 22 times of misusing *dádào* (达到) as other words, once each
Situation B (using incorrect words as substitutes)	14 times of misusing *dédào* (得到) as *dádào* (达到) 4 times of misusing *dàodá* (到达) as *dádào* (达到) 3 times of misusing *dáchéng* (达成, to achieve) as *dádào* (达到) 2 times of misusing *chéngjiù* (成就, achievement) as *dádào* (达到) 16 times of misusing other words as *dádào* (达到), once each

(达到), only 2 should be replaced by *dàodá* (到达). The misuse rate does not exceed 10%, as a result, and this situation is not considered as a notable misused one[4].

So far, we have examined 134 sets of confusable words in the textbooks in the method described above. Only 50 sets have met the criterion while the learners do not "obviously" misuse the words in other 84 cases. Appendix 1 displays more details of it.

4 Results and Discussion

Our research findings are summarized in the following, categorized into four aspects:

[4] A set of confusable words could be divided into one-way misusing and two-way misusing, according to the directionality of misusing. The one-way misusing refers to the situations that only A is often misused as B, but B is not often misused as A. The two-way misusing refers to the situations that both A and B are often misused as each other.

A. 50 sets of confusable words listed in the textbooks have met our criterion. However, not all of them are the most typical sets.

As mentioned in session 3, we have found only 6 times that "*dàodá* (到达, to reach) and *dádào* (达到, to arrive)" were misused, whereas "*dádào* (达到, to arrive) and *dédào* (得到, to obtain)" were misused 14 times. The set of "*dádào* (达到, to arrive) and *dédào* (得到, to obtain)" is therefore more typical. Furthermore, the misuse rate of "*sīkǎo* (思考, thought), *kǎolǜ* (考虑, consideration)" met our criterion, while only 9 cases of misuse are found. However, as many as 22 cases of misuse of "*sīxiǎng* (思想, thought) and *sīkǎo* (思考, thought)" are found. For effectiveness, the set of "*sīxiǎng* (思想, thought) and *sīkǎo* (思考, thought)" is better than the set of "*sīkǎo* (思考, thought) and *kǎolǜ* (考虑, consideration)". Therefore, in addition to investigating the practical usage of confusable words in the corpus, we list out the words that were more likely to confuse the learners based on the quantitative analysis as well. 15 new sets of words are recommended: "*biǎodá* (表达, to express), *biǎoxiàn* (表现, to display)"; "*dádào* (达到, to reach), *dédào* (得到, to get)"; "*guīzé* (规则, rule), *guīdìng* (规定, stipulation), *guīlǜ* (规律, regular pattern)"; "*yìnxiàng* (印象, impression), *yǐngxiǎng* (影响, impact)"; "*sīxiǎng* (思想, thought), *sīkǎo* (思考, thought)"; "*shíhòu* (时候, times), *shíjiān* (时间, time), *shí* (时, when)"; "*jiāowǎng* (交往, to associate), *jiāoliú* (交流, to contact)"; "*quèshí* (确实, indeed), *shízài* (实在, certainly)"; "*quēdiǎn* (缺点, drawback), *huàichù* (坏处, disadvantage)"; "*cānguān* (参观, sight-seeing), *fǎngwèn* (访问, visit)"; "*yǒu* (有, there be), *jùyǒu* (具有, have), *yǒngyǒu* (拥有, possess)"; "*tǎolùn* (讨论, to discuss), *yìlùn* (议论, to speak at)"; "*jiàoyù* (教育, to educate), *jiàoxùn* (教训, to teach someone a lesson)"; "*fāshēng* (发生, to happen), *fāxiàn* (发现, to discover)"; "*gòngtóng* (共同, in common), *hùxiāng* (互相, mutual)".

B. 84 sets that do not meet our criterion. However, we found other candidates.

Take "*yǒuguān* (有关, have to do with) and *guānyú* (关于, about)" for examples. There are 714 tokens and 712 tokens respectively in the L2 corpus, without any case being misused as the other. On the contrary, the misusing of "*guānyú* (关于, about), *duìyú* (对于, for)" is serious. 712 and 861 tokens are found for "*guānyú* (关于, about) and *duìyú* (对于, for)" respectively in the corpus, among which 56 tokens and 64 tokens respectively are found misused. 13 examples of "*guānyú* (关于, about) being misused as *duìyú* (对于, for)" and the misuse rate was as high as 53.6%. 18 cases are found to be the misuse of "*duìyú* (对于, for) as *guānyú* (关于, about)", and the misuse rate is 42.9%. The examples are listed as (5)-(6) below. It is obvious that "*guānyú* (关于, about), *duìyú* (对于, for)" have made a greater confusion to leaners than "*yǒuguān* (有关, have to do with) and *guānyú* (关于, about)". Clearer explanations on their subtle differences are more in need for L2 learners. In column B of Appendix 1, we list out the additional sets of words of which the misuse rate exceeds 10%, for reference.

(5) 从 这 来 看， 父母 *关于
*Cóng zhè lái kàn, fùmǔ *guānyú*
From this come see parent about
孩子的 未来 发展的
háizi-de wèilái fāzhǎn-de
children-PARTICLE future development-PARTICLE
责任 是 很 明显的。
zérèn shì hěn míngxiǎn-de.
responsibility be very obvious-PARTICLE
"From this point of view, the parents' responsibility *concerning their children's future development is obvious."

(6) 我 以前 看过 *对于 "安乐死"的
*Wǒ yǐqián kàn-guò *duìyú "ānlèsǐ"-de*
I before see-EXPERIENTIAL for euthanasia-PARTICLE
电视剧。
diànshìjù.
TV-show
"I have seen TV shows *for euthanasia before."

C. Instead of the misuse rate, the error rate of words is taken into consideration for the selection of sets of confusable words.

As described in Sect. 3, we have suggested that high-frequently misused words should be given priority in language teaching. However, we have also found that in some sets of confusable words, the frequency of each in the set may differ tremendously. Take "*gòngtóng* (共同, in common) and *gòngtōng* (共通, applicable to both/all)" as an example. 515 tokens of *gòngtóng* (共同, in common) are found in the corpus, but only 9 tokens of *gòngtōng* (共通, applicable to both/all) are found. The error rate of *gòngtōng* (共通, applicable to both/all) exceeds 10%, but the misuse rate of *gòngtóng* (共同, in common) is only 2%. It seems the learners have acquired the word *gòngtóng* (共同, in common). Nevertheless, of the 5 misuses of *gòngtōng* (共通, applicable to both/all), 4 should be replaced with *gòngtóng* (共同, in common). The misuse rate is actually very high. Although it seems to be a one-way misuse, the similarities on the form and sound still make them candidates of confusable words. While the error rate of words has not been set into the criterion for the selection, more further studies are needed.

D. According to our statistical data and results (refer to Appendix 1), the set of confusable words are mostly those with the same morpheme.

These confusable words are not influenced by L1 transferring. As most of the impacts of L1 transfer usually occurs in the early stages of learning and the writers of the compositions in the HSK (L2) corpus are advanced learners, the result is not too surprising. Our finding is actually consistent to the L2 language acquisition theory.

5 Conclusion

This study examines the confusable words used in textbooks through the learners' authentic usages in an interlanguage corpus. We have found that some sets of confusable words are misused by learners, but only 50 of the 134 sets got misuse rate higher than 10%, which is the criterion we set for confusable words. The confusable words selected by teachers or textbook editors might only be judged by their subjective experiences. The prediction is not really successful as the accuracy is less than 50%.

We are still working on the improvements of this study. As mentioned above, the directionality of misuse can definitely affect the teaching strategies. How confusable words should be treated in L2 Chinese teaching also requires further discussion. For example, should confusable words be taught or just listed as supplementary teaching materials? If confusable words should be included in teaching, how should they be taught? These questions still need further studying. After all, words that in similar forms without semantic relations in Mandarin Chinese are numerous. The misuse of words is inevitable in communication and writing.

The present study has provided a preliminary analysis and examination on the criterion for identifying confusable words used in textbooks, despite the unsolved problems. We have done the quantitative analysis. The qualitative analysis is still in progress.

Acknowledgments. The work was supported by the Ministry of Science and Technology, grant no. MOST 109-2410-H-002-189.

Appendix 1: Empirical Results of Confusable Words in Textbooks

A. Sets with misuse (50 sets)
表达 ‘to express’, 表示 ‘to show’
到达 ‘to arrive’, 达到 ‘to reach’
具有 ‘to have’, 具备 ‘to be equipped with’
权利 ‘right’, 权力 ‘authority’
缺陷 ‘defect’, 缺点 ‘drawback’
满足 ‘to satisfy’, 满意 ‘satisfied’
现实 ‘reality’, 实际 ‘practical’
发明 ‘to invent’, 发现 ‘to discover’
相关 ‘be related to’, 有关 ‘have to do with’
命运 ‘fate’, 运气 ‘luck’
具有 ‘to have’, 拥有 ‘to possess’
目标 ‘goal’, 目的 ‘purpose’
真诚 ‘sincerity’, 忠诚 ‘loyalty’
共通 ‘applicable to both/ all’, 共同 ‘in common’
达成 ‘to achieve’, 达到 ‘to reach’
面前 ‘before the eyes of’, 前面 ‘in front of’
采访 ‘interview’, 访问 ‘visit/ interview’
辽阔 ‘vast’, 开阔 ‘broad’
交流 ‘to contact’, 交换 ‘to exchange’
的确 ‘certainly’, 确实 ‘indeed’
出现 ‘to appear’, 发现 ‘to discover’
激发 ‘to motivate’, 刺激 ‘to stimulate’
成立 ‘to set up’, 建立 ‘to establish’
尽情 ‘to one’s heart content’, 尽量 ‘as far as possible’
既然 ‘since’, 竟然 ‘unexpectedly’
培养 ‘to cultivate’, 培训, ‘to train’
兴趣 ‘interest’ 有趣 ‘interesting’
谈论 ‘to talk about’, 讨论 ‘to discuss’
证据 ‘evidence’, 证明 ‘proof’
领悟 ‘to comprehend/ to grasp’, 醒悟 ‘to come to realize’
温和 ‘gentle’, 温柔 ‘tender’
思考 ‘to think’, 考虑 ‘to consider’
意义 ‘significance’, 意思 ‘meaning’
发展 ‘to develop’, 发达 ‘developed’
平静 ‘unruffled’, 安静 ‘quiet’
印象 ‘impression’, 形象 ‘image’
接近 ‘to approach’, 拉近 ‘to bring sth./ sb. closer’
回答 ‘to answer/ respond’, 答案 ‘answer’

时刻 ‘moment’, 时候 ‘times’
个人 ‘individual’, 各人 ‘each one’
本性 ‘nature’, 本能 ‘instinct’
教养 ‘upbringing’, 教育 ‘education’
适合 ‘suitable for’, 适当 ‘appropriate, 合适 ‘suitable’
推广 ‘to popularize, 推行 ‘to promote’, 执行 ‘to carry out’
规则 ‘rule’, 规定 ‘stipulation’, 规范 ‘norm’
产业 ‘industry’, 行业 ‘business’, 职业 ‘profession’
通常 ‘usually’, 经常 ‘often’, 平常 ‘normally’
念头 ‘idea’, 观念 ‘notion’, 概念 ‘concept’
后果 ‘consequence’, 结果 ‘result’, 效果 ‘effect’
渴望 ‘desire’, 希望 ‘hope’, 愿望 ‘wish’, 心愿 ‘aspiration’
B. Sets without significant misuse (84 sets) (Words in brackets are additional sets suggested by this study)
缺乏 ‘be short of’, 缺少 ‘lack of’ (减少 ‘to reduce’, 缺少 ‘lack of’)
明白 ‘to understand’, 清楚 ‘clear’ (知道 ‘to know’, 明白 ‘to understand’)
理念 ‘concept’, 理想 ‘ideal’ (希望 ‘hope’, 理想 ‘ideal’)
承受 ‘to bear with’, 忍受 ‘to endure’ (承受 ‘to bear with’, 受 ‘to receive’)
此外 ‘in addition’, 另外 ‘besides’ (另外 ‘besides’, 还有 ‘also’)
逃离 ‘to escape’, 脱离 ‘to get out of’ (脱离 ‘to get out of’, 离开 ‘to leave’)
庇护 ‘to asylum’, 保护 ‘to protect’ (保护 ‘to protect’, 保持 ‘to maintain’)
连结 ‘to connect’, 连络 ‘to contact’ (连络 ‘to contact’, 连系 ‘to communicate’)
温馨 ‘warmth’, 温暖 ‘warm’ (温暖 ‘warm’, 温柔 ‘tender’)
原本 ‘originally’, 原来 ‘it turns out that’ (本来 ‘at first’, 原来 ‘it turns out that’)
信任 ‘to trust’, 相信 ‘to believe’ (信任 ‘to trust’, 信用 ‘credit’)
有关 ‘have to do with’, 关于 ‘about’ (关于 ‘about’, 对于 ‘for’)
引发 ‘to evoke’, 引起 ‘to lead to’ (引起 ‘to lead to’, 造成 ‘to bring about’)
地点 ‘place’, 地方 ‘local’ (方面 ‘area’, 地方 ‘local’)
热忱 ‘earnest’, 热心 ‘be enthusiastic about’ (热情 ‘enthusiastic’, 热忱 ‘earnest’)
阶段 ‘stage’, 阶层 ‘level’ (阶段 ‘stage’, 段落 ‘paragraph’)
周遭 ‘around’, 周边 ‘surrounding’ (周围 ‘round’, 周边 ‘surrounding’)
事业 ‘enterprise’, 职业 ‘profession’ (职业 ‘profession’, 行业 ‘business’)
景象 ‘scene’, 现象 ‘phenomenon’ (现象 ‘phenomenon’, 影响 ‘impact’)
远行 ‘go on long journey’, 旅行 ‘to travel’ (游行 ‘to parade’, 旅行 ‘to travel’)
愕然 ‘stunned’, 已然 ‘be already so’, 偶然 ‘by accident’, 自然 ‘natural’ (自然 ‘natural’, 当然 ‘certain’)
风险 ‘risk’, 危险 ‘danger’, 冒险 ‘adventure’ (危害 ‘harm’, 危险 ‘danger’)
身为 ‘in the capacity of’, 成为 ‘to become’ (成为 ‘to become’, 当成 ‘to consider as’, 成 ‘have become’)
应对 ‘to reply’, 应变 ‘to meet contingency’
诠释 ‘to interpret’, 解释‘to explain’

创业 'to start up business', 创新 'to innovate'
灾害 'disaster', 灾难 'misfortune'
承诺 'commitment', 答应 'promise'
扶持 'to help to sustain', 支持 'to support'
安置 'to place', 设置 'to set up'
争相 'to vie with each other', 竞争 'to compete'
点出 'to point out', 指出 'to indicate'
制造 'to manufacture', 建造 'to build'
训练 'to train', 练习 'to practice'
根本 'at all', 原本 'at first'
品牌 'brand', 招牌 'sign(broad)'
洞察 'to have insight into', 观察 'to observe'
场所 'place', 场合 'occasion'
舞动 'to dance/ to brandish', 舞蹈 'dance'
特地 'specially', 特别 'special'
传承 'to pass down through generations', 传人 'successor'
预期 'to expect', 期待 'to look forward to'
传授 'to pass on', 传承 'to pass down through generations'
技巧 'technique', 技术 'technology'
力道 'physical strength', 力量 'power'
抱持 'to hold ', 保持 'to maintain'
借着 'by means of', 借由 'through'
回覆 'to reply', 回答 'to answer/ respond'
紊乱 'confusion', 错乱 'confused'
焦急 'vexed', 焦虑 'anxious'
天分 'natural gift', 天才 'genius'
热衷 'fall over oneself for', 热心 'be enthusiastic about'
宣布 'to announce', 宣告 'to declare'
收益 'profit', 收入 'income'
名义 '(in the) name (of)', 意义 'significance'
盲点 'blind spot', 盲目 'blindly'
破绽 'weak point', 破洞 'hole'
开明 'enlightened', 开放 'open'
纪录 'record/ documentary', 记载 'to put down in writing'
展现 'to show', 表现 'to display'
宣称 'to assert', 宣布 'to declare'
庞大 'huge', 巨大 'giant'
预估 'to estimate', 预算 'budget'
亏本 'to lose one's capital', 回本 'to earn its cost'
决选 'final election', 竞选 'to run for'
建设 'to construct', 兴建 'to build'
随着 'in pace with', 接着 'after that'

支撑 'to prop up', 支持 'to support'
维护 'to maintain', 保护 'to protect'
梦想 'dream', 理想 'ideal'
多重 'multiple', 多元 'multiplex'
景致 'view', 景观 'landscape'
家当 'family belongings', 家用 'family expense'
脱逃 'to run away', 脱离 'to get out of'
学术 'academics', 学问 'learning'
调控 'to regulate', 调整 'to adjust'
忧郁 'depressed', 焦虑 'anxious'
区域 'region', 领域 'field'
促使 'to spur', 促发 'to trigger'
野生 'wild', 野放 'to free captive animals', 野外 'open country'
居留 'residency', 居住 'to reside', 居所 'residence'
华裔 'ethnic Chinese', 华侨 'overseas Chinese', 华人 'Chinese people'
行销 'to market', 销售 'to sell', 畅销 'to sell well'
专精 'specialized', 专业 'professional', 专家 'specialist'

References

Carpenter E.: Confusable Words. HarperCollins (1993)

Laufer, B.: The concept of 'Synforms' (similar lexical forms) in vocabulary acquisition. Lang. Educ. **2**(2), 113–132 (1988)

Soontornthamniti, N.: Tàiguó Xuéxízhě Hànyǔ Yì Hǔnyiáo Cí Yánjiù (泰国学习者汉语易混淆词研究), Minzu University of China (2018)

Teng, S.H. (ed.): A Course in Contemporary Chinese Volume 5, 6, Taipei. Linking Publishing, Taiwan (2018)

Tsai, M.C., Huang, C.R., Chen, K.J.: From near-synonyms to the interaction between syntax and semantics. In: Yin, Y.M., Yang, I.L., Chan, H.C. (eds.): Proceedings of the Fifth International Symposium on Chinese Languages and Linguistics. Academia Sinica, Taipei, pp. 439–459 (1999)

Xiao, R., McEnery, T.: Collocation, semantic prosody, and near synonymy: a cross-linguistic perspective. Appl. Linguist. **27**(1), 103–129 (2006)

Xiao, P.: The Research on Confusable Words of Indonesian Students in Chinese Interlanguage, Beijing Language and Culture University (2008)

Zhang, B.: Synonymy, Near-Synonymy and Confusable Word: A Perspective Transformation from Chinese to Interlanguage, Chin. Teach. World (3), 98–107 (2007)

Zhang, B.: Pertinence: a fundamental principle for compiling a dictionary of confused words. Chin. Teach. World **27**(2), 214–231 (2013)

Speaker Commitment in Mandarin Conditional Connectives: Distributional and Experimental Evidence

Mingya Liu[1(✉)] and Yuting Wang[2]

[1] Department of English and American Studies, Humboldt University of Berlin, Unter den Linden 6, Berlin, Germany
mingya.liu@hu-berlin.de

[2] School of Linguistic Sciences and Arts, Jiangsu Normal University, Xuzhou, China
wyt0205@jsnu.edu.cn

Abstract. The interpretation of conditionals is a long but ongoing debate in linguistics (cf. von Fintel 2007, 2011, 2012). In this paper, we focus on the meaning contribution of conditional connectives (CCs). According to Kratzer (1986, 1991), the English CC *if* has no semantics on its own and *if*-clauses are used to restrict modal or generic frequency operators. The restrictor analysis of conditionals and CCs has inspired many insightful follow-up studies through which it becomes clear that the interpretation of conditional sentences is subject to a process of semantic and pragmatic modulation. However, the role of CCs in the modulation process remains far from clear in the current literature, although CCs can influence the interpretation of conditional sentences in various ways. In this paper, we will argue with distributional observations and experimental evidence that Mandarin CCs can express different degrees of speaker commitment (Giannakidou and Mari 2015) towards the antecedent proposition, as non-at-issue meaning.

Keywords: Mandarin · Conditional connective · *Wanyi* · Speaker commitment · Experiment

1 Introduction

Conditionals are a vital research topic in cognitive science attracting great attention from linguistics, logic, psychology, and philosophy. In linguistics, it is a long but ongoing debate concerning the compositional meanings of conditional sentences (cf. [2–4]). In this paper, we focus on the meaning contribution of conditional connectives (CCs). According to [10] and [11], the English CC *if* does not have a distinctive conditional meaning and *if*-clauses are used to restrict modal operators or generic frequency operators. The restrictor analysis of conditionals and CCs has inspired many insightful follow-up studies through which it becomes clearer that the interpretation of conditionals is subject to semantic and pragmatic modulations. However, the role of CCs in the modulation process remains far from clear in the current literature, although CCs can influence the interpretation of conditionals in various ways. In this paper, we will

M. Liu et al. (Eds.): CLSW 2020, LNAI 12278, pp. 810–823, 2021.
https://doi.org/10.1007/978-3-030-81197-6_67

argue with distributional observations and experimental evidence that Mandarin CCs can express different degrees of speaker commitment (see [5]) towards the antecedent proposition, as non-at-issue meanings.

The paper is structured as follows: Sect. 2 provides a general description of Mandarin CCs. Section 3 presents the distribution of some commonly-used Mandarin CCs in different kinds of conditionals. For these, Sect. 4 proposes an analysis and reports on an experiment in support of it. Section 5 concludes the paper.

2 General Notes on Mandarin Conditional Connectives

Mandarin possesses a variety of conditional markers. The generally-recognized CCs are the ones occurring at the beginning of the antecedent, and they differ in style and frequency. For example, the CCs *ru, ruo, ruruo, tang, tanghuo, tangshi, sheruo, tangruo* are not used much in modern Mandarin, and even less so in colloquial speeches. In contrast, *ruguo, yaoshi, jiaru, jiashe* and *wanyi* are used more often, as demonstrated in (1), but even among these, there are non-trivial differences. They are all used in hypothetical conditionals; therefore, some studies consider them hypothetical markers.

(1) *Ruguo/yaoshi/jiaru/jiashe/wanyi mingtian xiayu, shiwai huodong hui quxiao.*
if tomorrow rain outdoor activity will cancel
'If it rains tomorrow, the outdoor activities will be canceled.'

In addition to these CCs, Mandarin also has conditional markers at the end of the antecedent, e.g., namely *de-hua* 'speaking of/hypothetically speaking'. *De-hua* is not as independent as other CCs. According to a survey on CCs [15], it tends to co-occur with a CC in the antecedent (e.g., *ruguo...dehua*) and the frequency of the co-occurrence in written materials (79%) is higher than in colloquial speeches (50.9%).

(2) *Mingtian xiayu de-hua, shiwai huodong hui quxiao.*
tomorrow rain hypothetically-speaking outdoor activity will cancel
'If it rains tomorrow, the outdoor activities will be canceled.'

Furthermore, Chinese also features conditional markers in the consequent. While English *then* can occur optionally in the consequent in combination with a CC in the antecedent, Mandarin *name, jiu, ze* (which only occurs in formal style) 'then', and *hai, ye, you, hui* 'still/also/again/will', can occur with or without any CCs in the antecedent.

(3) *(Ruguo) Mei ren lai, women jiu hui jia.*
if no person come we then go home
'If no one comes, we will go home.'

(4) (*Ruguo*) *Yi ge ren haipa shibai,*
if one CL person fears failure
(name) ta yongyuan bu hui chenggong.
then he forever not will succeed
'If a man is afraid of failure, he will never succeed.'

(5) *Mei ren qu, women ye bu qu.*
no person go we also not go
'If no one goes, we also won't go.'

Name 'then' and *jiu* 'then' are used with CCs to construct conditionals, while *ye* often co-occurs with *jishi* 'even if' in concessional conditionals, as in (6). Interestingly, CCs such as *ruguo, yaoshi, jiaru, wanyi* can occur together with *ye* 'also' to construct concessive conditional readings [17] (see (7)).

(6) *Xianzai cun dian qian, jishi diu le gongzuo,*
now save some money even-if lose LE job
**name / *jiu / ye bu hui lu-su-jie-tou.*
then / then / also not will live-on-street
'Save some money now. Even if you lose the job, you won't be homeless.'

(7) *Xianzai cun dian qian, ruguo/yaoshi/jiaru/wanyi diu le gongzuo,*
now save some money if lose LE job
ye bu hui lu-su-jie-tou.
also not will live-on-street
'Save some money now. Even if you lose the job, you won't be homeless.'

Besides the aforementioned CCs, Mandarin uses specific CCs (e.g. *yaobushi 'if not'*) or adverbs (e.g. *zao 'early'*, usually appearing in *zao zhidao 'had I known'*) to express counterfactual conditionals (see [7–9] and [6]).

(8) *Yaobushi Luomiou ai Zhuliye, ta bu hui he xia duyao.*
if-not Romeo love Juliet he not will drink down poison
'If Romeo hadn't loved Juliet, he would not have drunk the poison.'

The counterfactual CC *yaobushi* 'if-not' contains *yao(shi)* 'if' and *bu* 'not'. The antecedent *Luomiou ai Zhuliye* 'Romeo loves Juliet' is true, which means the negation of it is false, and the consequent *ta bu hui he xia duyao* 'he would not drink the poison' is false. *Yaobushi* is considered to be a counterfactual CC instead of a simple combination of a CC and a negation, and it is in contrast with the example below: Sentences such as (9) are ambiguous between indicative and counterfactual readings, the latter of which can only arise through contextual support.

(9) *Yaoshi Luomiou bu ai Zhuliye, ta bu hui he xia duyao.*
if Romeo not love Juliet he not will drink down poison
'If Romeo didn't love Juliet, he would not drink the poison.'

Furthermore, some Mandarin CCs can co-occur with each other such as *ruguo-yaoshi, ruguo-wanyi, wanyi-yaoshi* and *jiaru-yaoshi*, in contrast to languages such as English (**if-when*) or German (**wenn-falls* 'when if') (see [15] and [18]). In this paper, we will leave the issues of co-occurrence to future studies (concerning their syntactic and interpretive effects) and discuss only the cases with just one CC.

3 Distribution of Mandarin Conditional Connectives in Different Kinds of Conditionals

In this section, we will focus on CCs that occur at the antecedent-initial position and discuss their distribution in different kinds of conditionals.

First, most of Mandarin CCs can occur in indicative conditionals, but it is odd to use CCs with *jia-* 'false/hypothetically', e.g., *jiaru, jiashe, jiaruo,* in contexts where there is supporting evidence for the truth of the antecedent. Imagine the following scenario: Teacher A and teacher B are monitoring an exam, and the student Zhangsan is taking the exam. Students are not allowed to use mobile phones during the exam.

(10) A: *Zhangsan gangcai zai kan shouji.*
Zhangsan just-now be look mobile-phone
'Zhangsan was looking at the mobile phone'
B: *Ruguo/yaoshi/wanyi/#jiaru Zhangsan zuobi, women yao zuzhi ta.*
if Zhangsan cheat we will stop him
'If Zhangsan is cheating, we will stop him.'

Second, *wanyi* and *jiaru, jiashe, jiaruo* are odd in premise conditionals.

(11) A: *Zhangsan zai zuobi!*
Zhangsan be cheat
'Zhangsan is cheating'
B: *Ruguo/yaoshi/#wanyi/#jiaru Zhangsan zuobi, women yao zuzhi ta.*
if Zhangsan cheat we will stop him
'If Zhangsan is cheating, we will stop him.'

Third, *wanyi* cannot be used in counterfactual conditionals.

(12) *Ruguo/yaoshi/jiaru/#wanyi wo shi nan-de,*
if I be male-DE
wo hui liu huzi.
I will grow beard
'If I were a man, I would grow a beard.'

Fourth, not only CCs in the antecedent, but conditional markers in the consequent also have limited distributions in counterfactual conditionals. As shown in (13), *jiu* cannot occur in the counterfactual conditionals with *yaobushi* in the antecedent but *ye* can.

(13) a. *Yaobushi zuotian ni qu, wo ye/*jiu bu hui qu de.*
if-not yesterday you go I also/*then not will go DE
'If you hadn't gone yesterday, I wouldn't have gone either.'
b. *Yaobushi Luomiou ai Zhuliye, ta ye/??jiu bu hui*
if-not Romeo love Juliet he also/??then not will
he xia duyao.
drink down poison
'If Romeo had not loved Juliet, he wouldn't have drunk the poison.'

Fifth, w*anyi* is the only CC that cannot be used in optative conditionals, in which the consequent and *jiu* are both obligatory, unlike English where a consequent or 'then' is not necessary. Following Grosz (2012), we assume that (14) has no descriptive but presuppositional content (i.e. Xiaosan is not alive) and expressive content (i.e. the speaker desires that Xiaosan be alive).

(14) *Ruguo/yaoshi/jiaru/*wanyi Xiaosan hai huozhe jiu hao le*!
if Xiaosan still alive then good LE
'If only Xiaosan were alive!'

Sixth, most CCs can be used in biscuit conditionals, except for *jiaru* and CCs with *jia.* In this context, conditional markers e.g., *jiu* or *ye* cannot be used in the consequent.

(15) *Ruguo/yaoshi/wanyi/?jiaru ni e le,*
if you hungry LE
*zhuozi shang (*jiu/*ye) you binggan.*
table on *then/*also have biscuit
'If you are hungry, there are biscuits on the table.'

Seventh, [16][1] pointed out that German CC *falls* cannot co-occur with quantifying adverbs, see his example in (16). The same goes for *wanyi,* see (17) in contrast with other CCs.

1 '…that explicitly quantified c-constrcutures are plural forms, bare c-constructures with particels like *if* are transnumeral forms (unspecified with respect to number), and that bare c-constructures with *falls* are singular forms.' ([16]: 216).

(16) *Immer wenn/*falls Steffi gewinnt, wird gefeiert.*
always if Steffi wins get celebrated.
'Always, wenn Steffi wins, it will be celebrated.'

(17) *Meici ruguo/yaoshi/jiaru/*wanyi shangdian dazhe,*
every-time if store discount
ta jiu qu mai dongxi.
he then go buy thing
'Every time the store is on sale, he goes shopping.'

Last but not least, most CCs license negative polarity items (NPIs) such as *shenme* 'what/any', except for *yaobushi* (cf. [6], in which they analyze *bu* as a negative clitic and different from normal negation).

(18) a. *Ruguo/yaoshi/jiaru/wanyi ni kanjian shenme ren lai*
if you see what person come
gaosu wo.
tell me
'If you see anyone, come and tell me.'
b. **Yaobushi ni kanjian shenme ren...*
if-not you see what person...

4 Analysis of Mandarin Conditional Connectives with a Focus on *Wanyi* with Experimental Evidence

What do the data above reveal about Mandarin CCs? There is not much written on these in general, with a few exceptions: [1] has claimed that these words differ in terms of hypothetical probability about the condition, whereas [14] has argued that these words do not differ in this aspect. The distributional observations of these CCs in Sect. 3, however, suggest that they might differ in this aspect. In this section, we will propose an analysis for *wanyi* in comparison and report on an experiment.

4.1 *Wanyi* in Comparison

According to [13], *wanyi* is derived from the Num-phrase *wanfenzhiyi* (lit. 'ten thousand-portion-of one', i.e. 'one of ten thousand'), which first turned into a NP meaning 'unexpected change with extremely low possibility' (e.g. *haipa wanyi* 'worry about unexpected change', *yifang wanyi* 'to avoid unexpected change') and later became an adverb 'possibly/probably', and then was used as a conditional marker 'if'. In modern Mandarin, *wanyi* is polysemous, which reflects the process of grammaticalization.

(19) a. *Che li yao fang miehuoqi, yifang wanyi.*
car in need put extinguisher avoid unexpected-change
'Put an extinguisher in the car to avoid unexpected change.'

b. *Bu pa yi wan, jiu pa wanyi.*
not worry ten thousand just worry unexpected-change
'Not worry about ten thousand (things), just worry about one unexpected change.'

(20) a. *Xiao tuzi duo zai dong li bu chu lai,*
Little rabbit hide in cave inside not out come
pa wanyi yudao huli.
worry possibly meet fox
'The little rabbit hid in the cave, worrying about it might possibly bump into a fox.'

b. *Jiagu diba, fangzhi wanyi fasheng yiwai.*
reinforce dam prevent possibly happen accident
'Reinforce the dam to prevent accidents from happening possibly.'

Wanyi is an NP in (19) and is an adverb in (20), and we have provided many examples of *wanyi* as a CC in former sections. We have conducted a corpus study with the BCC corpus (http://bcc.blcu.edu.cn/lang/zh): the conditional antecedents modified by *wanyi* typically have negative connotations (see (21a)). When the antecedent is desirable, it adds a meaning to the sentence that the speaker takes the antecedent as highly unlikely, as shown in (21b). It is odd to use *wanyi* for a conditional antecedent that is both desirable for the speaker and likely (based on e.g., speaker's attitude or world knowledge), as shown in (22).

(21) a. *Wanyi duizhang buxing shoushang,*
if captain unfortunately hurt
women zhi neng quxiao bisai.
we only can cancel game
'If the captain, unfortunately, get injured, we can only cancel the game.'

b. *Wanyi dedao le zhen-ai, yiding yao haohao zhenxi.*
if get LE true-love must will greatly cherish
'If you got true love, you should cherish it greatly.'

(22) a. #*Wanyi wo zhaodao yi jia haochi de canting,*
if I find a CL tasty DE restaurant
wo jiu qing pengyou-men chifan.
I then invite friends eat
'If I find a good restaurant, I will then invite our friends to go eat there.'

b. #*Wanyi yingbi zhengmian chao shang,*
if coin front-side towards up
jiu you hong dui faqiu.
then by red team serve
'If it's the coin's front side up, the red team gets to serve.'

In (21a), the speaker holds a negative attitude towards the proposition in the antecedent and *buxing* 'unfortunately' is often found in conditionals with *wanyi*. In

(21b), the proposition in the antecedent is not considered to be negative by the speaker but the event has low possibility of happening. In contrast, (22a) and (22b) are neither negative nor unlikely to happen according to our world knowledge. Based on the distributional observations, we argue that *wanyi* has an attitudinal meaning at the doxastic or bouletic dimension, as illustrated in (23).

(23) *Wanyi mingtian xiayu, women jiu dai zai jiali.*
if tomorrow rains we then stay at home
'If it rains tomorrow, we will stay at home.'
  a. **at-issue meaning: p→q**
  b. **non-at-issue meaning: λp.¬likely/desirable(p,x)**

The non-at-issue meaning is not cancellable, as shown in (24).

(24) a. *Wo xiwang mingtian tian hao.*
I hope tomorrow weather good
Wanyi tian hao, #wo juede hen keneng,
if weather good I think very likely
women keyi chumen qu hubian paobu.
we can out go lake-side jog
'I hope the weather will be good tomorrow. If the weather is good, which I believe to be very likely, we can go jogging at the lakeside.'
b. *Wo renwei mingtian hui xiayu. Wanyi xiayu,*
I think tomorrow will rain if rain
#wo juede ting hao-de, women jiu dai zai jia-li.
I think very good-DE we then stay at home-in
'I think it will rain tomorrow. If it rains, which I find pretty good, we will stay at home.'

Putting the factor of desirability aside, we will first discuss the speaker's epistemic commitment. *Wanyi* is bad in premise conditional due to the clash between its non-at-issue meaning vs. the presupposition triggered by the premise conditionals. It is bad in counterfactual conditionals or counterfactual optatives due to a redundancy between the non-at-issue unlikelihood meaning and presupposition. It is bad with quantifying adverbs due to a clash between the unlikelihood meaning and the presupposition by the quantifying adverbs that the given event at least takes place once.

In addition, we tentatively propose that *jiaru* (literally 'hypothetically if') presupposes the lack of objective evidence for the truth of the antecedent. It is odd in contexts where there is positive or certain evidence for the antecedent, due to presupposition failure. It is fine in counterfactual conditionals/optatives as counterfactuality (i.e. by the counterevidence for the antecedent) is compatible with its presupposition. It is odd in biscuit conditionals (or, speech act conditionals), as the speech act presupposes its relevance. However, this awaits future investigations.

4.2 Experiment

In order to test the analysis of *wanyi* above, we have conducted a psycholinguistic study using the same method of [12] as described below, finding that certain CCs indicate a lower degree of speaker commitment, such as German *falls* in comparison to *wenn*, and so did some NPIs such as *ever/at all.* Accordingly, we have come up with the following scales as in (27). We have tested the validity of these scales using a 4 × 2 factorial design with the factor 'CC' (i.e. *ruguo, yaoshi, jiaru* and *wanyi*) and 'NPI' (i.e. with/out *renhe* 'any'). The NPI factor was included to test the cross-linguistic validity of the NPI effect found for NPIs such as *ever/at all* in English and German ([12]). The reason we used *any* is because Mandarin lacks counterparts like English *ever/at all.* As we are aware that Mandarin *renhe* is multiply ambiguous, we do not have strong predictions on it. We have used 32 items (see Appendix for the complete list of items in Condition 1) such as in (28), and 48 additional fillers.

(27) Scales to be validated:

a. CC: *more committed* <*yaoshi/ruguo, jiaru, wanyi*> *less committed*

b. NPI: *more committed* <~~*renhe*~~, *renhe*> *less committed*

(28) S1: *Xiaotang zheng zai chi zaocan.* ('Xiaotang is having breakfast.')

S2: *Ta xiang: "ruguo/yaoshi/jiaru/wanyi wo xiawu you (renhe) shijian, wo jiu he pengyou qu he kafei."*

('He thinks, "If I have (any) time in the afternoon, I will go have a coffee with my friends."')

S3: *Xiaotang renwei ta xiawu you shijian ma?*

('Does Xiaotang think that he has time in the afternoon?')

S4: *Xiaotang zheng zai chi wucan ma*?

('Is Xiaotang having lunch?')

Subjects read scenarios presented sentence by sentence: S1 sets the context; S2 contains a conditional sentence with 8 combinations; S3 asks the subjects to rate the degree of the protagonist's commitment to the antecedent on a 5-point Likert scale (1 = 'certainly not', 5 = 'certainly yes'). S4 is a comprehension question (with a yes/no answer) for attention check. The dependent variable is the ratings for S3. The predictions concerning S3 are that the rating would be lower with *wanyi* in S2 than with the others and that the rating would possibly be lower with *renhe* in S2 than without. The experiment was set up with Ibex Farm (spellout.net/ibexfarm/). 56 undergraduate students at Nanjing University participated in the study for small payments.

Table 1. Descriptive statistics of the experiment, with standard deviations/errors (SD/SE)

Condition	CC	Rating	SD	SE
1	*ruguo*	3.32	0.81	0.06
2	*jiaru*	3.29	0.87	0.06
3	*yaoshi*	3.28	0.79	0.05
4	*wanyi*	3.10	0.92	0.06

We have used data of 54 subjects whose accuracy rate on the comprehension question was above 80%. We have performed a linear mixed effects analysis of the data using R. The model used CC and NPI (with interaction term) as fixed effects. The random effects are intercepts for subjects and items, as well as by-subject random slopes for the effect of CC * NPI. As the result reveals neither an interaction nor an effect of NPI, we will only discuss the first four conditions without NPIs (see Table 1 for the descriptive statistics). The t-tests show that 1) neither *jiaru* nor *yaoshi* differ from *ruguo*, 2) *wanyi* and *yaoshi* do not differ, but 3) *wanyi* differ significantly from *ruguo* ($p = 0.001$) and from *jiaru* ($p = 0.04$). That is, we are not able to confirm the scale in (27a), but the results provide evidence for the difference between *wanyi* and *jiaru* and *ruguo*. The lack of differences between *yaoshi* and *wanyi* remains to be investigated in future studies.

5 Conclusion

In this paper, we provide an overview of Mandarin CCs. Based on their distribution, we argue that they differ in terms of speaker commitment at the non-at-issue dimension. We report on a rating study, which shows that *wanyi* expresses a lower degree of speaker commitment than *ruguo* and *jiaoru*, but not from *yaoshi*. While these results are inconclusive, they show that CCs can differ in lexical pragmatics.

Acknowledgments. This work was supported by a DFG (German Research Foundation) grant to Mingya Liu's project "The Semantics and Pragmatics of Conditional Connectives: Crosslinguistic and experimental Perspectives" (Project number: 367088975) within the Priority-Program XPrag.de. We thank the anonymous reviewers for their helpful comments.

Appendix: Experimental Materials (in Condition 1)

Item	S1 / S2 / S3 / S4
1	莎莎是小学老师。/ 她想："如果有学生会使用乐器，我会带着他们一起玩。" / 莎莎认为有学生会使用乐器吗？/ 莎莎在幼儿园工作吗？
2	小萍在生日时收到了许多书。/ 她想："如果我有机会 去度假 的话，我就带一些书去读。" / 小萍认为自己有机会去度假吗？/ 小萍在生日收到了许多书吗？
3	张先生和妻子一起去电影院看电影。/ 他想："如果 有电影适合青少年看的话，下次我就带孩子们一起来。" / 张先生认为有电影适合青少年看吗？/ 张先生一个人去看电影吗？
4	娜娜有一台旧电脑。/ 她想："如果它有硬件坏了， 我就送去维修。" / 娜娜认为她电脑有硬件坏了吗？/ 娜娜有一台新电脑吗？
5	小刘感冒了打算吃两粒药。/ 他想："如果药里有抗 生素，我就向医生咨询。"/ 小刘认为他吃的药里有抗生素吗？/ 小刘感冒了吗？
6	小杨最近在家养病。/ 他想："如果我明天会有好 转，我就去训练。" / 小杨认为他明天会有好转吗？ / 小杨最近在训练吗？
7	小李有朋友来家里做客。/ 他想："如果客人对我的书有 兴趣，我就再订几本送他们。" / 小李认为客人对他的书有兴趣吗？/ 小李家里有客人要来吗？
8	小柯今天想出去活动一下。/ 她想："如果外头有雾霾，我 就不去散步了。" / 小柯认为外头有雾霾吗？/ 小柯今天想呆在家里吗？
9	小孙晚上经常加班。/ 她想："如果我这周晚上有时间，我就和朋友去看最新的电影。" / 小孙认为她这周晚上有时间吗？/ 小孙晚上经常加班吗？

10 小陈喜欢园艺。/ 她想："如果超市有玫瑰花种子，我就买些回家。" / 小陈认为超市有玫瑰花种子吗？/ 小陈喜欢园艺吗？

11 冰冰打算早上在家练琴。/ 她想："如果有邻居在家，可能就被我的琴声吵醒了。"/ 冰冰认为有邻居在家吗？/ 冰冰早上要去上班吗？

12 王老师想去看医生。/ 他想："如果下周我有空闲时间，我就去医院。" / 王老师认为他下周有空闲时间吗？/ 王老师已经见过医生了吗？

13 孙博士在空间站维修设备。/ 她想："如果有安全扣脱扣了，我们就回不到地面了。" / 孙博士认为有安全扣会脱扣了吗？/ 孙博士在空间站工作吗？

14 阿云下载了最新的消防 APP。/ 她想："如果会有火灾发生，APP 就会告诉我着火点在哪里。"/ 阿云认为会有火灾发生吗？ / 阿云下载了最新的购物 APP 吗？

15 暴雨中小林坐的公交车突然停了。/ 他想："如果公交车有故障，就只能原地等待救援。" / 小林认为公交车有故障吗？/ 小林在暴雨中自己开车吗？

16 天天喜欢做蛋糕。/ 他想："如果有新同事喜欢吃蛋糕，我就做一个给他们吃。"/ 天天认为有新同事喜欢吃蛋糕吗？/ 天天喜欢做蛋糕吗？

17 小安的照相机被偷了。/ 他想："如果商店里有打折款相机，我就买一台。" / 小安认为商店里有打折款相机吗？/ 小安买了新相机吗？

18 小刘喜欢诗歌。/ 他想："如果最近剧院有诗歌分享会，我就去参加。" / 小刘认为最近剧院有诗歌分享会吗？/ 小刘喜欢诗歌吗？

19 小黄正在旅行。/ 他想："如果明天酒店有空房，我就去续住。" / 小黄认为明天酒天有空房吗？/ 小黄正在家里吗？

20 老张喜欢听歌剧。/ 他想："如果最近歌剧院有演出，我就和妻子一起去参加。"/ 老张认为最近歌剧院有演出吗？/ 老张喜欢听歌剧吗？

21 小李正在挤公交上班。/ 他想："如果明天我有精力早起，我会骑单车上班。"/ 小李认为他明天有精力早起吗？/ 小李正在骑单车吗吗？

22 老李喜欢吃烧烤。/ 他想："如果今晚家门口有烧烤铺开门，我就会去吃。"/ 老李认为今晚家门口有烧烤铺开门吗？/ 老李喜欢吃烧烤吗？

23 老刘连续工作两周了。/ 他想："如果下周有休假机会，我就休息两天。"/ 老刘认为他下周有休假机会吗？/ 老刘连续工作两周了吗？

24 小顺正在写年终总结。/ 他想："如果下个月部门有新员工入职，我就多陪陪家人。" / 小顺认为下个月部门有新员工入职吗？/ 小顺正在写年终总结吗？

25 小唐正在吃早餐。/ 他想："如果我下午有时间，我就和朋友去喝咖啡。"/ 小唐认为他下午有时间吗？/ 小唐正在吃午餐吗？

26 小代睡眠质量不好。/ 他想："如果今晚外头有噪音，我就会睡不着。"/ 小代认为今晚外头有噪音吗？/ 小代睡眠质量好吗？

27 小程马上要毕业了。/ 他想："如果学院还有保研名额，我就去申请。" / 小程认为学院还有保研名额吗？/ 小程已经毕业了吗？

28 老迟爱好摄影。/ 他想："如果今天天气有好转，我就去街上拍照。" / 老迟认为今天天气有好转吗？/ 老迟爱好摄影吗？

29 老许最近心情不好。/ 他想："如果明天外头有阳光，我就去跑步。" / 老许认为明天外头有阳光吗？/ 老徐最近心情不好吗？

30 小陈最近谈恋爱了。/ 他想："如果女朋友这周末有时间，我们就去泡温泉。" / 小陈认为他女朋友这周末有时间吗？/小陈最近谈恋爱了吗？

31 老陈喜欢自己在晚上听音乐。/ 他想："如果家里还有红酒，今晚我就听听音乐放松放松。" / 老陈认为家里还有红酒吗？/ 老陈喜欢在晚上听音乐吗？

32 老黄抽烟二十年了。/ 他想："如果体检报告有发现问题，我就戒烟。" / 老黄认为体检报告有发现问题吗？/ 老黄抽烟三十年了吗？

References

1. Chao, Y.R.: A Grammar of Spoken Chinese. University of California Press, Berkeley/Los Angeles (1968)
2. von Fintel, K.: If: The biggest little word (2007). http://mit.edu/fintel/gurt-slides.pdf
3. von Fintel, K.: Subjunctive conditionals. In: Fara, D.G., Russell, G. (eds.) Routledge Companion to the Philosophy of Language to the Philosophy of Language. Routledge (2011)
4. von Fintel, K.: Conditionals. In: Maienborn, C., von Heusinger, K., Portner, P. (eds.) Semantics: An International Handbook of Natural Language and Meaning, pp. 1515–1538. de Gruyter, Berlin (2012)
5. Giannakidou, A., Mari, A.: Future and universal epistemic modals: reasoning with nonveridicality and partial knowledge. In: Tense, Mood, and Modality: New Answers to Old Questions. Cambridge University Press, Cambridge, New York (2015)

6. Grosz, P.: On the Grammar of Optative Constructions. John Benjamins, Amsterdam (2012)
7. Ippolito, M., Su, J.: Counterfactuals, negation and polarity. In: Crnĭc, L., Sauerland, U. (eds.) The Art and Craft of Semantics: A Festschrift for Irene Heim, pp. 225–243. MITWPL, Cambridge (2014)
8. Jiang, Y.: Counterfactual interpretations of Chinese conditionals. Stud. Explor. Syntax (Chin.) **10**, 257–279 (2000)
9. Jiang, Y.: On the lexical meaning of conditional connectives in Chinese. In: Su, X., He, T. (eds.) CLSW 2014. LNCS (LNAI), vol. 8922, pp. 43–54. Springer, Cham (2014). https://doi.org/10.1007/978-3-319-14331-6_5
10. Jiang, Y.: Ways of expressing countnerfactual meaning in context. Linguist. Vanguard **5**(s3), 1–7 (2019). https://doi.org/10.1515/lingvan-2019-0009
11. Kratzer, A.: Conditionals . Chicago Linguist. Soc. **22**(2), 1–15 (1986)
12. Kratzer, A.: Conditionals. In: von Stechow, A., Wunderlich, D. (eds.) Semantics: An International Handbook of Contemporary Research, pp. 651–656. De Gruyter, Berlin (1991)
13. Liu, M.: The elastic nonveridicality property of conditionals. Linguist. Vanguard **5**(s3), 1–10 (2019). https://doi.org/10.1515/lingvan-2019-0007
14. Luo, Y., Xu, H.: On the grammaticalization of "wanyi" and related issues. Central China Acad. **2**, 273–285 (2012)
15. Lü, S.: The Complete Works of Lü Shuxiang. Shangwu Yingshuguan, Beijing (2004)
16. Xie, X.: A research on continuous use of hypothetical conjunctions. Chin. Linguist. **2**, 29–36 (2010)
17. Zaefferer, D.: Conditionals and unconditionals: cross-linguistic and logical aspects. In: Semantic Universals and Universal Semantics, pp. 210–236. Foris Publications (1991)
18. Zhang, X.: A textual analysis of Wanyi. Chin. Teach. World **23**(1), 49–64 (2009)
19. Zhang, X.: On continuous use patterns and causes of the conjunctive hypothetical words of "ruguo(shuo)". Chin. Lang. Learn. **2**, 24–31 (2019)

A Corpus-Based Study on the Classification and Processing Mechanism of English-Chinese One-To-Many Translation-Equivalent Word Pairs

Quanbei Zhao[1], Jia Li[2], and Hongbing Xing[1](✉)

[1] Beijing Language and Culture University, Xueyuan Road 15, Beijing 100083, China
xinghb@blcu.edu.cn

[2] Harbin Normal University, Shida Road 1, Harbin 150025, China

Abstract. To study the correspondence relation between one English word and its Chinese equivalent words as well as their processing mechanism in bilinguals' mental lexicons, we make a corpus-based calculation to acquire frequencies of some English-Chinese one-to-many translation-equivalent word pairs (i.e., the respective usage frequency of multiple Chinese words corresponding to an English word), according to which we introduce a classification criterion to divide these word pairs into two groups with the lower and higher maximum values of corresponding frequency. To verify whether this criterion reflects the activation in the actual processing, a semantic consistency decision task under the priming paradigm is carried out. Its results demonstrate that the actual activation of the English-Chinese translation-equivalent word pairs can be predicted by their corresponding frequencies obtained from statistics drawn based on a corpus. Additionally, the research further indicates that current English-Chinese dictionaries are far from sufficient coverage of all the translation-equivalent word pairs, and excluding a large number of words in the actual use of language.

Keywords: One-to-many translation-equivalent word pair · Corresponding frequency · Classification criterion · Processing mechanism · Non-dictionary words

1 Introduction

Lexical connection and relevant processing mechanism are continuing concerns within the study of language processing. Bilingual words are mutually connected in the mental lexicon and interactive in language processing [1]. One of the most key aspects of bilingual acquisition is the processing of "English-Chinese" (L1 to L2) translation-equivalent word pairs. Understanding the relevant processing mechanism will help in predicting the acquisition patterns of L2 learners and hence help the compilation of L2 textbooks and bilingual dictionaries.

A clear classification of English-Chinese equivalent words is of great importance for investigating the processing mechanism. Zgusta [2] used the term "lexical equivalent"

M. Liu et al. (Eds.): CLSW 2020, LNAI 12278, pp. 824–836, 2021.
https://doi.org/10.1007/978-3-030-81197-6_68

to refer to a lexical unit of the target language which has the same lexical meaning as the respective lexical unit of the source language, and further distinguished all equivalents into "absolute equivalents" and "partial equivalents". Most of the "absolute equivalents" are proper nouns and technical terms, for instance, Alzheimer = 阿尔茨海默氏病, nanometer = 纳米. When it comes to "partial equivalents", "translation ambiguity" which suggests the one-to-many semantic mapping between two languages, commonly [3–5] occurs. The very different conceptual systems of English and Chinese engender a high "asymmetry" between English and Chinese words, which differ from each other in a number of respects, including meaning, parts of speech, and usage [6]. The "lexical asymmetry" is widely reflected in English-Chinese bilingual dictionaries, not least by the number of headwords and lexical entries. As we know, "认识", "了解", and "叫做" are all Chinese translation-equivalent words of "know".

Some researchers have attempted to classify English-Chinese one-to-many translation-equivalent word pairs [6–10]. According to Ding [6], English-Chinese translation-equivalent word pairs should be categorized into three groups, namely "full equivalence", "partial equivalence", and "zero equivalence" from the respects of semantics and pragmatics. Yang [9] probed into the non-correspondence between English and Chinese semantics, and divided such non-correspondence into "partial non-correspondence" and "complete non-correspondence". Moreover, some corpus-based research on English-Chinese translation-equivalent word pairs correspondence relation compared the relevant equivalent words in a corpus with those in the dictionary. Based on the corpus statistics, Li [11] investigated the "complete equivalence" and "zero equivalence" and pointed out that the corpus statistical results covered some word meanings excluded from the dictionary. However, the existing body of research may not be reasonable enough because of a paucity of valid supporting data and clear classification criteria. Besides, most studies on the differences between corpora and dictionaries were conducted through presenting examples, whose conclusions did not have any support from specific quantitative indicators and necessary statistical tests.

In terms of the Second Language Acquisition Theory, the Usage-Based Theory of Language Acquisition and Connectionist Models provide a useful account of how important frequency for knowledge learning is [12]. Frequency is a key determinant of acquisition because "rules" of language are structural regularities which emerge from learners' lifetime analysis of the distributional characteristics of the language input [13]. Furthermore, frequency does affect the process of language internalization and the proficiency and fluency of language processing, as well as vocabulary recognition and learning. Even though several studies related to L2 acquisition have confirmed that high-frequency words are more readily judged and extracted than low-frequency words according to learners' performance in the lexical decision tasks [13–15], most of them seem to go into issues about frequency effect merely in the L2 system, instead of the corresponding frequency effect of translation equivalents in two language systems.

Given the above consideration, this study takes the Usage-Based Theory of Language Acquisition [16] and Connectionist Models [17, 18] as the theoretical bases. The corpus-based statistical method of language property is adopted to calculate corresponding frequencies, according to which we put forward a classification criterion.

Furthermore, a semantic consistency decision task under the priming paradigm is conducted to verify whether the classification criterion is in line with the actual activation of learners' processing. Assuming that the activation probability based on corpus statistics reflects the actual activation, the probability that an English word activates each Chinese word during processing can be inferred by the corresponding frequencies of the English-Chinese translation-equivalent word pairs, which can further be taken as a reference for constructing the processing models of English-Chinese translation-equivalent word pairs to simulate the processing of mental lexicon. Meanwhile, the present study compares the equivalent words in a dictionary and a corpus as well, in the hope of providing beneficial data support and a reference for the bilingual dictionary compilation.

2 Classification Study

English-Chinese one-to-many translation-equivalent word pairs can be elucidated as one lexical form in English and its multiple Chinese counterparts which share the same meanings with it, which foregrounds the one-to-many form and meaning mapping between English and Chinese. For example, "like" and "喜欢", "希望", "认为", "赞成", and etc. In order to meet the requirements of choosing experimental materials (i.e., matching the number of strokes, word length, and word frequency), this study has randomly selected 54 common Chinese verbs from the *Chinese Proficiency Test Syllabus* [19], whose English translation-equivalent words were taken from some textbooks teaching Chinese as a foreign language, such as, *Kuaile Hanyu (Volume 1)* [20], *Road to Success (Upper Elementary)* [21]. We have hence calculated the corresponding frequencies of the Chinese translation-equivalent words corresponding to the 54 English words.

2.1 The Experimental Corpus and Corpus Preprocessing

The study used the English-Chinese Bilingual Parallel Corpus of Xiamen University including 278,201 English-Chinese parallel sentence pairs, of which there were about 4.13 million English words and 7.18 million Chinese characters. Concentrating on verbs chiefly and taking some verb and preposition collocations into account, we have made use of the RAND function to select a random sample containing 15,981 corpus instances from the 81,006 corpus instances including the 54 English words. With the instances extracted, we have conducted statistical analysis on those words with no more than 400 corpus instances in an exhaustive way, and drawn corpus instances from the others in the average proportion of 20% of the total number of corpus instances.

2.2 The Calculation of Corresponding Frequency

Corresponding frequency is defined as the respective usage frequency of multiple Chinese words corresponding to an English word. Taking "jump" as an example, we collated all instances in the corpus and obtained 20 Chinese translation-equivalent words, whose corresponding frequencies equal their occurrence numbers divided by the total number of all the instances, as shown in Table 1.

Table 1. The respective number of occurrences and the corresponding frequency of 20 Chinese translation-equivalent words to the English word "jump".

Word	Occurrences number	Corresponding frequency	Word	Occurrence number	Corresponding frequency
跳跃	74	0.379	欢呼	4	0.021
跨越	23	0.118	绷	2	0.010
跳上	22	0.113	匹配	2	0.010
抓住	12	0.062	改变	2	0.010
增加	11	0.056	略去	1	0.005
得出	10	0.051	跳出	1	0.005
移动	9	0.046	支付	1	0.005
袭击	8	0.042	爬	1	0.005
工作	5	0.026	参加	1	0.005
批评	5	0.026	跟风	1	0.005

2.3 Data Transformation and Classification Criteria

For comparison, we have transformed the corresponding frequencies into Z-scores. Then, we have classified these English-Chinese translation-equivalent word pairs according to the Z-scores corresponding to the maximum values of their corresponding frequencies among multiple Chinese translation-equivalent words. The mean value of Z-scores was 3.3, which was set as a benchmark to categorize all of these English-Chinese one-to-many translation-equivalent word pairs into two groups, namely the group with the lower maximum value of corresponding frequency ($0 \le Z \le 3.3$) and the group with the higher maximum value of corresponding frequency ($Z > 3.3$), as shown in Table 2.

Table 2. The classification of the 54 word pairs according to the values of the Z-scores corresponding to the maximum values of the corresponding frequencies.

Group with lower maximum value of corresponding frequency				Group with higher maximum value of corresponding frequency			
English word	Chinese word	Corresponding Frequency	Z-score	English word	Chinese word	Corresponding frequency	Z-score
Tell	告诉	70.1	3.216	Lift	举起	47.8	5.606
Look	看着	40.7	3.178	Take	采取	37.3	5.391
Wash	清洗	81.5	3.170	Work	工作	59.1	5.009
Feel	感到	71.7	3.093	Leave	离开	52.7	4.652

(*continued*)

Table 2. (*continued*)

Group with lower maximum value of corresponding frequency				Group with higher maximum value of corresponding frequency			
English word	Chinese word	Corresponding Frequency	Z-score	English word	Chinese word	Corresponding frequency	Z-score
Wait	等待	87.5	3.015	Run	奔跑	38.5	4.614
Try	尝试	75.0	2.982	Drop	下降	26.7	4.158
Find	发现	39.0	2.931	Call	命名	38.6	4.139
Spend	花费	68.2	2.854	Jump	跳跃	37.9	4.073
Buy	购买	94.8	2.844	Blow	吹动	55.6	4.053
Kick	踢踹	72.1	2.835	Move	移动	44.4	4.024
Swim	游泳	82.4	2.828	Accept	接受	71.9	3.967
Wish	希望	80.0	2.660	Pass	通过	17.6	3.925
Know	知道	77.8	2.647	Pull	拉扯	51.5	3.921
Save	救助	41.5	2.571	Ask	询问	65.0	3.909
Speak	讲话	91.0	2.476	Lead	导致	47.4	3.766
Finish	完成	79.8	2.457	Walk	步行	82.0	3.727
Help	帮助	79.5	2.436	Serve	服务	74.5	3.725
Stop	停止	26.8	2.376	Hold	拥有	29.2	3.721
Learn	学习	69.5	2.374	Meet	会面	38.6	3.714
Exercise	行使	93.8	2.235	Send	发送	52.6	3.677
Talk	交谈	63.8	2.202	Rise	增加	26.8	3.667
Play	玩球	21.8	1.919	Open	打开	53.9	3.661
Listen	听着	88.5	1.780	Read	阅读	50.7	3.584
Drink	喝水	42.9	1.707	Visit	访问	73.2	3.427
Use	使用	83.0	1.476	Join	参加	70.8	3.419
Like	喜欢	63.8	1.272	Lose	失去	58.9	3.408
Tidy	收拾	63.6	1.156	Live	居住	51.2	3.360

3 The Experimental Research

To further verify the classification criterion and explore the influence of the frequency factor on the processing of English-Chinese one-to-many translation-equivalent word pairs for learners, a semantic consistency decision task under the priming paradigm was performed, in which the independent variable was the corresponding frequency and the dependent variables were the reaction time and error rate of the participants' decision on English-Chinese translation-equivalent word pairs.

3.1 Participants

27 graduate students (10 males and 17 females) from Beijing Language and Culture University and China University of Mining and Technology were paid to participate

in the experiment. The average age of these participants was 24.2 years old (the ages ranging from 22 to 30). The participants were all native speakers of Chinese and had English as their L2. Their Chinese and English proficiency levels of listening, speaking, reading, and writing were evaluated through a language history questionnaire, on a 7-point scale, where 1 means 'very poor' and 7 mean 'native-like'. The participant group in this study was tested as an unbalanced bilingual group with a significant difference between its Chinese and English proficiency scores ($M_{L1} = 6.42$ and $M_{L2} = 4.38$).

3.2 Research Materials

All Chinese words were two-character words, and the length of each English word was measured by the number of letters included. After matching the number of strokes, word length, and the frequency of Chinese and English words, 40 pairs of English-Chinese translation-equivalent words (half from lower corresponding frequency group and half from the higher corresponding frequency group) were selected as experimental materials from the 54 pairs in the self-built corpus. Besides, 56 words were selected as fillers.

Table 3. The averages of data related to experimental materials.

Type of group	Z-score	Number of stroke	Word length	Frequency of Chinese word	Frequency of English word
Group of lower corresponding frequency	2.577 (.516)	17.100 (4.051)	4.300 (.733)	.159 (.017)	189680.450 (226798.568)
Group of higher corresponding frequency	3.863 (.442)	15.000 (3.920)	4.150 (.671)	.185 (.027)	146835.800 (85667.079)

The results of the paired-sample t-test of experimental materials in Table 3 showed that there was a significant difference between the two groups in Z-scores (t (19) = $7.417, p < .001$). However, the differences in the number of strokes (t (19) = $-1.418, p = .172$), word length (t (19) = $-.679, p = .505$), the frequency of Chinese words (t (19) = $.411, p = .686$) and English words (t (19) = $.722, p = .479$) were not significant. The statistical results showed that the values of the relevant variables of the material met the grouping requirements, and the control variables were effectively controlled.

3.3 Experimental Procedure

We applied E-Prime 2.0 to present primes and targets on screen in turn. Participants needed to judge whether the meanings of English words and Chinese words presented in sequence were consistent (answering Yes/No) by pressing buttons (the "J" key/the "F" key). The experimental basis is the Spreading-Activation Theory of Semantic Processing [22]. The stimulus words first seen by the participants will activate other related words in the mental lexicon at the lemma level, thus generating relevant expectations for the forthcoming target words. If the target word meets the expectations, the activated residual information will prompt the participants to press the response button quickly [23].

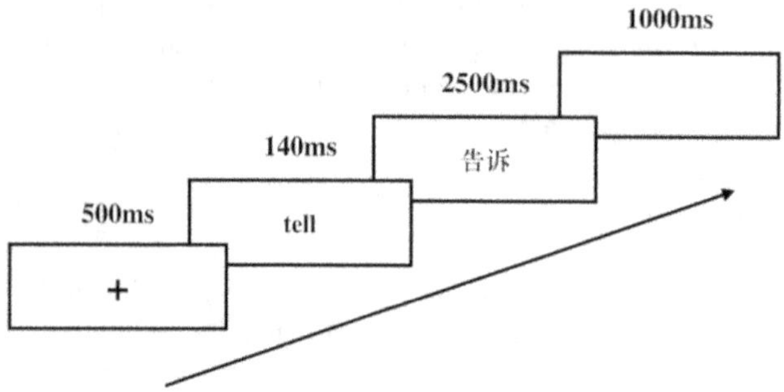

Fig. 1. The presentation sequence of items in the semantic consistency decision task under the priming paradigm

As shown in Fig. 1, in each experimental trial, the fixation sign "+" first appeared for 500ms in the center of the screen (the computers' screen resolution was set to be 1024 × 768 with 32-bit color depth and a refresh rate of 60 Hz), which reminded the participants that a trial was about to start. Then, an English prime word (presented in lowercase Courier New font, size 34, and black) appeared for 140 ms and was then immediately replaced by the Chinese target (font, size, and color were same with the primes). The target remained on the screen for 2500 ms, or until the participants responded. Then a 1000 ms blank screen signaled the end of the trial, followed by a new trial. Experiments were conducted in a one-on-one manner in a quiet psycholinguistics laboratory. In addition to the 96 experiment trials, there were 7 practice trials at the beginning of the experiment for each participant to get familiar with the procedure. The experiment lasted for about 7 min, after which participants were also asked to fill in the questionnaire about their L2 background [24].

3.4 Data Analysis and Research Results

The response data of only 40 critical word pairs (excluding fillers and practice word pairs) entered into our statistical analysis, 10.1% of which were discarded because they were stemmed from incorrect responses. In addition, outliers were trimmed from the data by removing responses with more than 2.5 standard deviations, below or above each participant's mean (which accounts for 2.6% of the total responses). Such a procedure of trimming has been a common practice in the priming literature [25, 26].

The results of the paired-sample t-test in Table 4 showed that there was a significant difference in semantic consistency decision response time between the groups of lower and higher corresponding frequency (t (19) = −2.042, p = .027). Compared with those words with higher corresponding frequency (M = 721.914 ms), participants judged the words with lower corresponding frequency (M = 793.835 ms) more slowly. However, there was no significant difference in the semantic consistency decision error rate between the groups of lower and higher corresponding frequency (t (19) = −1.29,

Table 4. The response time (ms) and error rate (%) of participants in semantic consistency decision for groups of lower and higher corresponding frequency.

	Response time	Error rate
Group of lower corresponding frequency	793.835 (122.496)	0.133 (0.213)
Group of higher corresponding frequency	721.914 (42.954)	0.069 (0.079)
Frequency effect	+71.921*	+0.064
Note: *p < .05		

$p = .213$). The average error rates were 6.9% and 13.3% for the groups with higher and lower corresponding frequency respectively.

4 The Correspondence Between English-Chinese One-To-Many Translation-Equivalent Words Based on Corpus Statistics and Dictionary Words

Taking *Oxford Advanced Learner's English-Chinese Dictionary (9th Edition)* [27] as a reference, we noticed that many English-Chinese one-to-many translation-equivalent words were not included in the dictionary. In other words, many English-Chinese one-to-many translation-equivalent words are "non-dictionary words". For example, 10 Chinese translation-equivalent words ("跳跃", "跨越", "增加", "略去", "移动", etc.) of the English word "jump" were embodied in the dictionary, while the other 10 Chinese translation-equivalent words ("参加", "爬", "批评", "跟风", "匹配", etc.) cannot be seen in the dictionary yet. Both dictionary words and non-dictionary words make up 50% of the Chinese translation-equivalent words of "jump" as shown in Table 5.

The results of the paired-sample t-test showed that the main effect of translation-equivalent word types was significant ($t\,(53) = -4.917, p < .001$), and the proportion of non-dictionary words in English-Chinese one-to-many translation-equivalent words ($M = .594$) was much higher than that of dictionary words ($M = .406$). This result revealed that numerous non-dictionary words were applied together with dictionary words in the actual language use.

5 Discussion

First of all, this study has identified the existence of the significant corresponding frequency effect in bilingual word recognition. Ellis and Schmidt [28] studied the process of L2 learners' learning knowledge about English plural nouns. The result of their study gave a positive response to the presupposition that learners acquired the language projects with higher frequency faster. Considerable work has been done in the frequency effect of words merely in the L2 system [28–31]. These findings are best explicated

Table 5. The proportion (%) of dictionary words and non-dictionary words of the Chinese translation-equivalent words to the 54 selected English words.

English word	Dic word (%)	Non-dic word (%)	English word	Dic word (%)	Non-dic word (%)	English word	Dic word (%)	Non-dic word (%)
Tell	53.8	46.2	Learn	50.0	50.0	Move	30.8	69.2
Look	23.8	76.2	Exercise	33.3	66.7	Accept	38.9	61.1
Wash	25.0	75.0	Talk	62.5	37.5	Pass	55.6	44.4
Feel	50.0	50.0	Play	41.2	58.8	Pull	50.0	50.0
Wait	36.4	63.6	Listen	80.0	20.0	Ask	33.3	66.7
Try	27.3	72.7	Drink	40.0	60.0	Lead	44.4	55.6
Find	40.9	59.1	Use	50.0	50.0	Walk	31.2	68.8
Spend	18.2	81.8	Like	50.0	50.0	Serve	50.0	50.0
Buy	40.0	60.0	Tidy	33.3	66.7	Hold	50.0	50.0
Kick	18.2	81.8	Lift	29.0	71.0	Meet	31.8	68.2
Swim	40.0	60.0	Take	50.0	50.0	Send	23.5	76.5
Wish	33.3	66.7	Work	20.7	79.3	Rise	44.4	55.6
Know	55.6	44.4	Leave	19.2	80.8	Open	31.2	68.8
Save	66.7	33.3	Run	28.1	71.9	Read	58.8	41.2
Speak	25.0	75.0	Drop	36.4	63.6	Visit	28.6	71.4
Finish	62.5	37.5	Call	26.9	73.1	Join	28.6	71.4
Help	50.0	50.0	Jump	50.0	50.0	Lose	50.0	50.0
Stop	53.3	46.7	Blow	63.2	36.8	Live	27.8	72.2

by Morton's [32] logogen model. Morton stated that those words (or logogens) with a higher frequency (or are more common) had a lower threshold. This means that they require less perceptual power in the brain to be recognized and decoded from the lexicon and are recognized faster than those words that are less common. This study, beyond that, has arrived at the conclusion that the corresponding frequencies of translation-equivalent word pairs between two language systems adhere to the same processing mechanism. Specifically, equivalent word pairs with higher corresponding frequency are more readily activated. This finding is best explained by Seidenberg and McClelland's [33] distributed and developmental model and Xing' [34] connectionist models of second language vocabulary acquisition. Seidenberg and McClelland proposed the model to bespeak that the more frequently a word is encountered, the more associated are the systems (e.g., orthography, phonetics, semantics) representing the word. Xing expounded that frequency factors were closely related to associative strength, suggesting that L2 learners learn equivalent word pairs with higher corresponding frequencies more frequently than those with lower corresponding frequencies. This explains why the associative strength of the form, sound, and meaning information for those translation-equivalent word pairs with higher corresponding frequencies is stronger and their form, sound, and meaning are bound up more directly. Such words are more readily activated with higher speed and accuracy by L2 learners.

Moreover, the present study has taken the corresponding frequency as the standard to achieve a quantitative classification on English-Chinese one-to-many translation-equivalent word pairs. Meanwhile, the study has found the obvious difference between the corpus-based statistical result and the actual dictionary compilation, which has been highlighted by some of the existing studies through describing typical instances [35–39]. Grounded on an objective quantitative analysis on the difference and by means of statistical approaches, this study has further revealed the fact that 40.6% of the English-Chinese translation-equivalent words can be found in English-Chinese bilingual dictionaries, while the other 59.4% are still not taken as entries in dictionaries. In accordance with Xing's [40] L2 vocabulary acquisition model, L2 lexical knowledge consists of static lexical knowledge and dynamic lexical knowledge, which explicates that lexical knowledge is primarily obtained from the mother tongue directly and gradually acquired in the process of using language respectively. The present study has made a further discussion, unraveling that lexical knowledge should include not only the static lexical knowledge composed of dictionary words but also the dynamic lexical knowledge made up of non-dictionary words. Mostly originated from the deviation and extension of word meanings due to specific contexts, (e.g., "move" in "After I came out of hospital, I had to have therapy once a week on my arm to help it move more easily" means "康复"), non-dictionary words are indispensable components of learners' vocabulary system, which are deserved to be considered as the constituents of English-Chinese bilingual dictionaries. Huang and Chen [41] concluded three principles for dictionary compilation: (1) historical development; (2) logical connection; (3) usage frequency. However, for English-Chinese bilingual dictionaries, the applicability of the first two principles is questionable. According to Li [42], most of the modern English-Chinese bilingual dictionaries for learning purposes are synchronic descriptive dictionaries, reflecting modern language in use. Apparently, it is not appropriate to take the principle of historical development to arrange lexical entries, which will cause great trouble in word searching for L2 learners. On the other hand, the definition and standard of logical connection are too vague to provide guidance for practice. In addition, we have summarized other objections to current English-Chinese dictionaries, for instance, leaving out some non-dictionary words, giving too many descriptions to those equivalent words that are not commonly used, etc. There is, therefore, a definite need for promoting the principle of usage frequency in the compilation of English-Chinese bilingual dictionaries and textbooks, which conforms to users' practical requirements and cognitive habits. The corpus-based statistical result rationalizes the categorization of non-dictionary words into "non-dictionary words with low frequency" (occurrence number ≤ 1) and "non-dictionary words with high frequency" (occurrence number > 1). Dictionary compilers are supposed to balance many factors, ranging from the usage of every single word, the richness of the entire lexicon, to learners' cognitive rules [43–45]. Thus, it is an inevitable trend and apparently beneficial to take non-dictionary words with high frequency (45.3%

of non-dictionary words) in English-Chinese bilingual dictionaries tentatively. Meanwhile, it is also reasonable to have non-dictionary words with low frequency (54.7% of non-dictionary words) set aside for the present, seeing that most of them are rarely used and unrepresentative, including which may lead to heavy cognitive load on L2 learners.

6 Conclusion

The major findings are as follows:

1. There is a high "asymmetry" between the English lemmas and the Chinese translation-equivalent words. English-Chinese one-to-many word pairs can be categorized into two groups, namely the group with the lower maximum value of corresponding frequency ($0 \leq Z \leq 3.3$) and the group with the higher maximum value of corresponding frequency ($Z > 3.3$).
2. The corresponding frequencies of these English-Chinese one-to-many translation-equivalent word pairs obtained through corpus statistics reflect the actual activation. Specifically, equivalent word pairs with higher corresponding frequency are more readily activated, representing the significant corresponding frequency effect.
3. Insofar as actual language use, 40.6% of the English-Chinese translation equivalents can be found in English-Chinese bilingual dictionaries, while the other 59.4% are still not taken as entries in dictionaries. Hence, it is rational to involve both dictionary words and non-dictionary words with high frequency (occurrence number > 1) in English-Chinese bilingual dictionaries and exclude non-dictionary words with low frequency (occurrence number ≤ 1).

Acknowledgments. We thank the anonymous reviewers for their constructive comments. This research was supported by the Humanities and Social Sciences Planning Fund under the Ministry of Education (20YJAZH110); the Humanities and Social Sciences Research Project in Western and Frontier Areas from Ministry of Education of China (19XJC740002); the Fundamental Research Funds for the Central Universities, and the Research Funds of Beijing Language and Culture University (19YCX059).

References

1. Zhao, X., Li, P.: Simulating cross-language priming with a dynamic computational model of the lexicon. Bilingualism Lang. Cogn. **16**, 288–303 (2013)
2. Zgusta, L.: Manual of Lexicography. Mouton, Hague (1971)
3. Tokowicz, N., Kroll, J.F., de Groot, A.M.B., van Hell, J.G.: Number-of-translation norms for Dutch-English translation pairs: a new tool for examining language production. Behav. Res. Methods Instrum. Comput. **34**, 435–451 (2002)
4. Prior, A., MacWhinney, B., Kroll, J.F.: Translation norms for English and Spanish: the role of lexical variables, word class, and L2 proficiency in negotiating translation ambiguity. Behav. Res. Methods **39**, 1029–1038 (2007)
5. Degani, T., Prior, A., Tokowicz, N.: Bidirectional transfer: the effect of sharing a translation. J. Cogn. Psychol. **23**, 18–28 (2011)

6. Ding, J.: An analysis of translation-equivalent words in English-Chinese dictionaries. Fudan Forum Foreign Lang. Lit. **1**, 53–58 (2013). (in Chinese)
7. Qiu, S.D.: Incomplete meaning equivalency of English-Chinese analogues. Foreign Lang. Teach. Res. **1**, 40–47+80 (1989). (in Chinese)
8. Li, T.S.: A research on English-Chinese corresponding words and terms. Master thesis, Beijing Language and Culture University, Beijing (2001). (in Chinese)
9. Yang, X.X.: Semantic non-correspondence and cultural differences between English and Chinese. Acad. J. Zhongzhou **5**, 297–298 (2006). (in Chinese)
10. Qu, Y.H., Xu, J.X., Wang, X.Y.: Syntactically and semantically unified classification system of English and Chinese verbs. Foreign Lang. Res. **2**, 37–44 (2016). (in Chinese)
11. Li, D.J.: The investigation of full equivalence and zero equivalence and the research on the compilation of Chinese-English bilingual dictionaries—a parallel corpus-based research. Lexicographical Stud. **2**, 55–66 (2009). (in Chinese)
12. Ellis, N., Wulff, S.: Usage-based approaches to SLA. In: VanPatten, B., Williams, J. (eds.) Theories in Second Language Acquisition: An Introduction, 2nd edn., pp. 75–93. Routledge, NY (2015)
13. Ellis, N.: Frequency effects in language processing: a review with implications for theories of implicit and explicit language acquisition. Stud. Second Lang. Acquis. **24**, 143–188 (2002)
14. Yang, J.: A study on frequency-based college english vocabulary teaching model. Doctoral thesis, Shanghai International Studies University, Shanghai (2013). (in Chinese)
15. Wolter, B., Yamashita, J.: Word frequency, collocational frequency, L1 congruency, and proficiency in L2 collocational processing: what accounts for L2 performance? Stud. Second Lang. Acquis. **40**, 1–22 (2017)
16. Bybee, J.L.: The emergent lexicon . Chicago Linguist. Soc. **34**, 421–435 (1998)
17. MacWhinney, B.: Second language acquisition and the competition model. In: MacWhinney, B. (ed.) Tutorials in Bilingualism: Psycholinguistic Perspectives, pp. 113–142. Lawrence Erlbaum Associates, Hillsdale (1997)
18. MacWhinney, B.: A unified model of language acquisition. In: Kroll, J.F., de Groot, A.M.B. (eds.) Handbook of Bilingualism: Psycholinguistic Approaches, pp. 49–67. Oxford University Press, Oxford (2005)
19. Confucius Institute Headquarters (Hanban): Chinese Proficiency Test Syllabus. The Commercial Press, Beijing (2006). (in Chinese)
20. Li, X.Q., Luo, Q.S., Liu, X.Y.: Kuaile Hanyu. People's Education Press, Beijing (2014). (in Chinese)
21. Zhang, L., Qiu, J.: Road to Success. Beijing Language and Culture University Press, Beijing (2009).(in Chinese)
22. Collins, A., Loftus, E.: A spreading-activation theory of semantic processing. Psychol. Rev. **82**, 407–428 (1975)
23. Zhang, P., Wei, J.S., Hu, B., Yu, J.Y., Chen, Y.Y.: A study on priming effect of Chinese native language knowledge in English collocation processing. Foreign Lang. Teach. **1**, 60–72+147–148 (2017). (in Chinese)
24. Li, P., Sepanski, S., Zhao, X.: Language history questionnaires: a web-based interface for bilingual research. Behav. Res. Methods **38**, 202–210 (2006)
25. Zeelenberg, R., Pecher, D.: Evidence for long-term language repetition priming in conceptual implicit memory tasks. J. Mem. Lang. **49**, 80–94 (2003)
26. Zhao, X.W., Li, P., Liu, Y.Y., Fang, X.P., Shu, H.: Cross-language priming in Chinese-English bilinguals with different second language proficiency levels. Proc. Annu. Meeting Cogn. Sci. Soc. **33**, 801–806 (2011)
27. Hornby, A.S.: Oxford Advanced Learner's English-Chinese Dictionary. The Commercial Press, Beijing (2018)

28. Ellis, N., Schmidt, R.: Rules or associations in the acquisition of morphology? The frequency by regularity interaction in human and PDP learning of morphosyntax. Lang. Cogn. Process. **13**, 307–336 (1998)
29. Duyck, W., Vanderelst, D., Desmet, T., Hartsuiker, R.J.: The frequency effect in second-language visual word recognition. Psychon. Bull. Rev. **15**, 850–855 (2008)
30. Wolter, B., Gyllstad, H.: Frequency of input and L2 collocational processing: a comparison of congruent and incongruent collocations. Stud. Second Lang. Acquis. **35**, 451–482 (2013)
31. Fang, N., Zhang, P.: Effects of translational congruency and L2 lexical frequency on English collocational process. Mod. Foreign Lang. **42**, 85–97 (2019). (in Chinese)
32. Morton, J.: Interaction of information in word recognition. Psychol. Rev. **76**, 165–178 (1969)
33. Seidenberg, M.S., McClelland, J.L.: A distributed, developmental model of word recognition and naming. Psychol. Rev. **96**, 523–568 (1989)
34. Xing, H.B.: Study on Vocabulary Acquisition of Chinese as A Second Language. Peking University Press, Beijing (2016).(in Chinese)
35. Wang, R.Q.: On principle of "translating by POS" in bilingual lexicography. J. Sichuan Int. Stud. Univ. 41–46 (2007). (in Chinese)
36. Xie, J.H.: Corpora and bilingual dictionaries . Lexicographical Stud. **5**, 57–64 (1997). (in Chinese)
37. Liu, Z.Q., Zhang, D.D.: The parallel corpus-based research on Chinese-English literary translation and lexicography—taking the idiom related to "eat" in hongloumeng and their English translation as an example. Chin. Translators J. 18–22 (2012). (in Chinese)
38. Wang, W.: Research of speech act verb interpretations about dictionaries of learning Chinese as a foreign language from the perspective of frame semantics. In: Hong, J.-F., Zhang, Y., Liu, P. (eds.) CLSW 2019. LNCS (LNAI), vol. 11831, pp. 84–93. Springer, Cham (2020). https://doi.org/10.1007/978-3-030-38189-9_8
39. Chen, Q.X., Huang, C.N., Cheng, H.: Preliminary study on machine readable dictionary of contemporary Chinese predicate verbs. Comput. Linguist. Res. Appl. 231–236 (1993). (in Chinese)
40. Xing, H.B.: A connectionist research framework of second language vocabulary acquisition. Lang. Teach. Linguist. Stud. **5**, 66–73 (2009). (in Chinese)
41. Huang, J.H., Chen, C.X.: A Dictionary of English Allusions. The Commercial Press, Beijing (2001).(in Chinese)
42. Li, D.G.: A new thought of polysemies' senses order arrangement in English-Chinese dictionary: the order arrangement on the basis of cognition. Mod. Linguist. **6**, 772–780 (2018). (in Chinese)
43. Zhang, Y.H.: Computational Lexicography and New Types of Dictionaries. Shanghai Lexicographical Publishing House, Shanghai (2004).(in Chinese)
44. Li, D.J.: The role of English-Chinese parallel corpus in bilingual dictionary compilation. J. PLA Univ. Foreign Lang. **3**, 41–44+64 (2006). (in Chinese)
45. Hong, J.-F., Zhang, Y., Liu, P. (eds.): CLSW 2019. LNCS (LNAI), vol. 11831. Springer, Cham (2020). https://doi.org/10.1007/978-3-030-38189-9

Cognitive Styles Underlying the Contradiction of Temporal Direction Between Chinese Qián/Hòu and Their English Counterparts *Forward/Back*

Zhong Chen(✉)

E21-2071, Department of Chinese Language and Literature, Faculty of Arts and Humanities, University of Macau, Avenida da Universidade, Taipa, Macau, China
zhongchen@um.edu.mo

Abstract. Drawing on a ground-preference cognitive style on the basis of object-centered value orientation, Chinese prefers to take the structure of temporal sequence as a frame of reference to conceptualize *qián* (preceding) and *hòu* (following) in the horizontally temporal relationship. By contrast, in light of a figure-preference cognitive style on the basis of ego-centered value orientation, English tends to conceptualize *forward/back* in a journey-metaphor frame of reference. In the case of English, a traveler's face toward the destination in future is regarded as *forward*, unless the traveler/observer's direction is eclipsed by the temporal sequence signified in context. The difference between the two cognitive styles leads to a directional discrepancy between *qián* (preceding)/*hòu* (following) and their English counterparts *forward/back*. This rationale clarifies the controversy over whether Chinese takes the past as its normal temporal direction. Further understanding of this issue can help reduce machine mistranslation regarding temporal direction.

Keywords: Object-centered value orientation · Ego-centered value orientation · Ground-preference cognitive style · Figure-preference cognitive style

1 Introduction: Contradictions Between Counterparts of Chinese and English Temporal Directions

Cross-linguistic comparisons can provide an important lens to the understanding of differences in cognitive styles across cultures. The present study focuses on the expressions indicating temporal directions in Chinese and English. Drawing on the contrasts between Chinese 前/*qián* (preceding) and 后/*hòu* (following) and their English counterparts forward/back, this study aims at shedding light upon the different cognitive styles underlying Chinese/English time perception. To begin with, the disagreement between Chinese temporal directions 前/后 and their English counterparts *forward/back*

I'd like to extend my sincere gratitude to reviewers' comments on this paper. This research is funded by University of Macau with the file number: SRG2019-00149-FAH.

M. Liu et al. (Eds.): CLSW 2020, LNAI 12278, pp. 837–849, 2021.
https://doi.org/10.1007/978-3-030-81197-6_69

can be revealed in the following cases. (Sentences (a) demonstrate the English sentences containing *forward* or *back* while sentences (b) show the Chinese counterparts).

(1) a. That was a few years *back*. (Oxford Advanced Learner's English-Chinese Dictionary, fourth edition, 88) [1]

b. 那 是 几 年 *前* 的 事。

Nà shì jǐ nián qián de shì

That be few year *qián* of matter

"*That was a few years back.*"

(2) a. This new discovery pushed *back* the date of the historical event to about 1,100 years ago.

b. 这 项 新 发现 将 历史 事件 的 年代 *前推* 至 大约 1100 年前。

Zhè xiàng xīn fāxiàn jiāng lìshǐ shìjiàn de niándài *qiántuī* zhì dàyuē 1100 nián qián

This CL new discovery make history event DE year *qiántuī* to about 1100 year *qián*

(*Identical meaning with the English sentence above.*)

(3) a. We've put *forward* the date of our wedding by one week. (Oxford Advanced Learner's English-Chinese Dictionary, fourth edition, 1209)

b. 我们 将 婚礼 日期 *提前* 了 一周。

Wǒmen jiāng hūnlǐ rìqī *tíqián* le yīzhōu.

We make wedding date *tíqián* LE one week

(*Identical meaning with the English sentence above.*)

(4) a. From this time *forward* (Oxford Advanced Learner's English-Chinese Dictionary, fourth edition, 581)

b. 从 今 *以后*

Cóng jīn *yǐhòu*

From now *yǐhòu*

"*from now on.*"

In (1) and (2), the English adverb *back* refers to a temporal direction, which appears to be opposite to its Chinese counterpart *后*. In (3), The English adverb *forward* indicates the future, which is consistent with its Chinese counterpart *前* as in (3), yet inconsistent with its Chinese counterpart *后* as in (4).

The discrepancy of temporal direction between Chinese *前/后* and their English counterparts *forward/back* demonstrates a paradox of temporal cognition across cultures. In particular, the same temporal relationship can be manifested in a contradictorily reverse form of direction in Chinese and English.

Temporal directional discrepancy is not limited to the abovementioned cases. Instead, this disagreement pattern widely exists between Chinese *前/后* and their English counterparts *forward/back*. It is therefore interesting to explore the cognitive mechanisms underlying this disagreement. Several related questions warrant discussion here: How can we account for the systematic discrepancy of temporal direction between Chinese and English? What motivates this paradox of cross-linguistic temporal direction? Is it true that language users of Chinese always face the past and those of English tend to face the future?

2 Literature Review

Traditionally, it has been argued that Chinese language views the past as its normal temporal direction, while English takes future as the typical direction [2]. Yu Ning [3] attempted to rebut such claim, arguing that Chinese also views future as forward in line with English. However, they have failed to answer the question why *forward* in (3) and (4) is either in the same direction or in the opposite direction with its Chinese counterpart *前*, while *back* in (1) and (2) corresponds to its reverse direction *前* rather than its literal Chinese counterpart *后*. Liu Ningsheng [4] ascribed this discrepancy of temporal direction to different reference frames, yet he has not pointed out the condition of the disparity in temporal directions. Furthermore, He Liang [5] has distinguished spatial direction (e.g., *前程 future, 前途 prospect, 面前 in face of*) based on an ego-based frame of reference from temporal direction based on a sequence-based frame of reference. His claim is in line with our argument in the current paper that Chinese is a language defining temporal direction canonically based on a sequence-based frame of reference, while English tends to employ an ego-centric reference.

There seems to be a paradox of temporal directions based on two types of frame of reference in English, which are characterized by different metaphorical approaches to conceptualization. For instance, Charles Fillmore has claimed that a contradiction is found in the metaphorical organization of time in English, one of which seems that the future lies in front and the past is behind [6]. Consider the following sentences. (Sentences (a) in (6–9) are extracted from Lakoff and Johnson 1980: 41–43 and sentences (b) are the Chinese counterparts provided by the author of this paper.)

(5) a. In the weeks *ahead of* us… [future]
    b. 数 周 *之后*
    few week back
    "a few weeks later."

(6) a. That 's all *behind* us now… [past]
    b. 那 是 *以前* 的 往 事。
    that be front DE past event
    (*Identical meaning with the English sentence above.*)

(7) a. In the *following* weeks… [future]
    b. *后面* 的 数 周。
    back DE few week
    (*Identical meaning with the English sentence above.*)

(8) a. In the *preceding* weeks… [past]
    b. *前* 几 周。
    front few week
    (*Identical meaning with the English sentence above.*)

The contradictory metaphors can even work well in one sentence, e.g.,

(9) a. We are looking *ahead* to the *following* weeks.
b. *我们 展望 未来 几 周。*
We anticipate future few week
(*Identical meaning with the English sentence above.*)

In fact, the type of temporal direction indicated in (7a) and (8a) differs from that in (5a) and (6a) in terms of the frame of reference. Sentences are referenced to an ego-centered reference in (5a) and (6a), where *us* functions as the ego, whereas (7a) and (8a) employ a temporal sequence-based frame of reference. However, if the ego-reference in (5a) is replaced by a sequence-reference, *ahead of* will indicate the sense of **precedence** rather than the **future**, e.g.,

(10) The work can be completed *ahead of* schedule. [precedence/提前tíqián]

To clarify the disagreement between Chinese and English with respect to temporal directions as shown above, this paper aims to explore the cognitive mechanisms that give rise to this discrepancy. Specifically, we seek to discuss the rationale of ground-preference and figure-preference cognitive styles, which are based on object-centered and ego-centered value orientations respectively. As such, we hope to rectify the misunderstanding of Chinese temporal cognition in the previous studies.

3 Object-Centered and Ego-Centered Cognitive Styles and Their Respective Canonical Frame of Reference of Temporal Direction

There are at least two types of frame of reference in temporal direction cognition in horizontal dimension, namely Temporal Sequence-based Frame of Reference (TSFR) built on Object-centered basis of value orientation [7], and Ego-based Frame of Reference (EFR) based on Ego-centric basis of value orientation [7]. In most cases, Chinese language tends to select object's own structure as its canonical temporal frame of reference. As such, Chinese is thought to be characteristic of conceptualizing the world more from an object-centered perspective than from an ego-centric aspect. As is known, the direction of temporal sequence is marked by its internal direction formed by its own sequential structure. In this manner, Chinese takes temporal sequence rather than ego's orientation as its canonical frame of reference. It denotes the start-point or earlier point as *前* (*preceding, ahead of, former, previous, ex-*) and the endpoint or latter point as *后* (*following, latter, post*). In the cases where the observer is absent or just partly implied in context, the observer himself is either eclipsed by the structure of temporal sequence, or excluded from the reference frame, regardless of the observer's orientation of whether facing the past or the future. The ego-based orientation will not be taken as frame of reference unless under condition that the observer is evidently highlighted in the context by such viewing activity as *展望* (look forward), *回顾* (retrospect), *回首* (look back). These viewing activities implicate observer's semi-temporalized spatial orientation (e.g., *前景 prospect, 前途 prospect, 前头 front*).

By contrast, as a language that is in favor of conceptualizing the world from an ego-centered perspective, English prefers to define temporal relationship with a journey-metaphorical frame of reference. In this system, the observer faces either the future or the upcoming event that can be moved closer to or away from the observer, and the past event can be pushed back. In this manner, the future event being brought forward are regarded as *forward* (corresponding to Chinese *提前 bring forward*); while a future event being pushed away from the observer is viewed as *put off* or *postpone* (c.f. Chinese *推后, 推迟, 延期 put off*). Similarly, a past event being pushed away from the observer toward a further time-point is regarded as being *pushed back* (c.f. Chinese *向前推, 前推 push back*), e.g., (7) and (8). Only in the cases that a temporal sequence is evidently denoted by a sequence-based expression (e.g., *preceding, former, previous, ex-, following, post-* etc.), can the ego-based temporal direction be replaced by sequence-based frame of reference in English. Therefore, we claim that English is characterized by an ego-centered temporal cognition.

Figure 1 illustrates the reverse temporal directions of EFR and TSFR marked by two reverse arrows respectively based on ego-centered and object-centered cognitive styles. It leads to the discrepancy between *前/后* and *forward/back,* as cases in (1)(2)(4)(5).

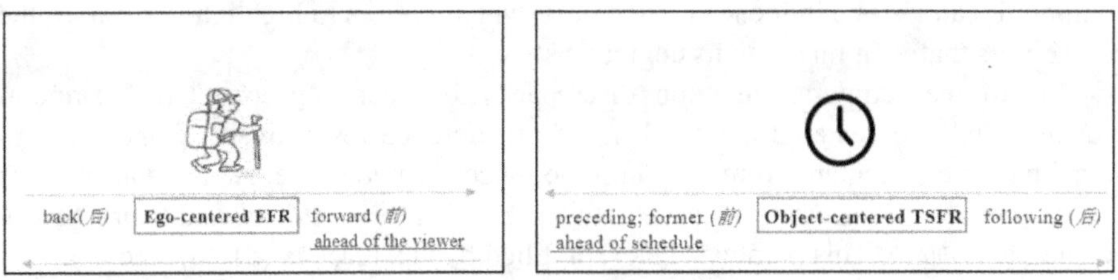

Fig. 1. The discrepancy between *前/后* and *forward/back/ahead of* based on EFR and TSFR.

We found in sentence (5) and (10), the ego-based reference *us* is replaced by the sequence-reference *schedule,* where *ahead of* no longer refers to future, instead, it denotes *preceding* in time sequence in (10). This reverse directions of *ahead of* are led by the alternation from the ego-reference *us* to the temporal sequence-reference *schedule.* Interestingly, this collocation is irreversible. This direction reversal between an ego-reference and a temporal sequence-reference exists in Chinese as well, such as (12) and (13) with *前面.*

Such discrepancy is also evident in cases where a past event is pushed away to a further time-point from the original one. For example, based on an ego-based frame of reference, *push back* in (2a) corresponds to its Chinese counterpart *向前推* (*push earlier*) in (2b), which is based on a temporal sequence frame of reference drawn on the ground-preference cognitive style from an object-centered perceptive. The discrepancy between *向前推* (*push earlier*) *and push back* reveals the fact that under the same condition of temporal relationship, Chinese canonically views and interprets the temporal relationship from an object-centered perspective, whereas English from an ego-centered perspective. The cognitive perspectives underlying the discrepancy reflects the opposition between Chinese ground-preference and English figure-preference cognitive style, in support of the figure-ground opposition in language-specific cognitive modes.

It might be quite common to experience an unconscious attention alternation on a moving train. One's attention can switch the focus between the inside scene of the compartment and the outside scene. This experience is based on the figure/ground reversal effect that might bring about a bewildering perception of mover alternation between the train and the outside scene. Likewise, TSFR based on ground-preference cognitive style from object-centered perspective and EFR based on figure-preference cognitive style from ego-centered perspective can also switch between the dual frames of reference. This shift could be realized by highlighting the internal structure of the temporal sequence as a frame of reference, in which the earlier point is viewed as *preceding* the later one (e.g., *following, next or latter*). Alternatively, this switch could also be completed by highlighting the frame of reference, which is based on the observer in a journey-metaphor. In the latter scenario, the observer is taken as the reference for temporal directions, including facing or moving towards future; bringing *forward* the scheduled event closer, pushing *back* a past or future event-time away further from the original point.

However, this alternation between TSFR and EEFR is on a language-specific basis, which is determined by the ground-preference cognitive style from object-centered perspective and the figure-preference cognitive style from the ego-centered perspective. In addition, it can also be influenced by context, which prefers to highlight the one of the competitors that is in line with its cognitive style.

Due to the temporal direction discrepancy between object-centered temporal sequence and ego-centered temporal frame of reference respectively, Chinese has to alter English ego-centered direction into the object-centered one. As a result, it leads to a reverse-direction correspondence between English *forward/back* and their Chinese counterparts *前/后*. This point can be exemplified by (1), (2), (4), (5), (6), and (9).

In light of ground-preference cognitive style from an object-centered perspective, not only does Chinese convert English ego-centered EFR direction into TSFR one, but also converts Chinese ego-centered spatial direction into object-centered TSFR temporal direction. In the latter case, canonical temporal noun *前* or *后* is added to specify a temporal duration that is measured in a temporal sequence. e.g.,

(11) a.展望 10年 （*后*） 的 前景。
Zhǎnwàng shínián hòu de *qián*jǐng
anticipate ten year (later) DE prospect
"look forward to the prospect ten years later"
a'.*展望 10年 *前* 的 前景。
Zhǎnwàng shínián *qián* de *qián*jǐng
anticipate ten year ago DE prospect
b.回顾 10年 *前* 的 往事。
Huígù shínián *qián* de wǎngshì
look back ten year ago DE past event
"look back on the past events ten years ago"
b'.*回顾 10年*后*的发展前景。
look back ten year ago DE past event
c. Look *forward* to the future in ten years.
d. Look *back* over ten years of experience.

The verb *展望 (look ahead to)* can only be followed by *后 (following)* based on TSFR rather than *前 (forward)* based on EFR. This phenomenon is due to the fact that Chinese is an object-centered language, which takes temporal sequence as its canonical temporal direction rather than ego-centered reference. As such, Chinese regards *前 (preceding)* as the canonical temporal direction, and *展望 (look ahead)*, *前景 (prospect)* as semi-temporalized spatial direction denoting the viewer's orientation, which is excluded from Chinese canonical temporal direction. Yet, *forward* as the English counterpart of *前*, as shown in (11c) is regarded as canonical temporal direction based on ego's orientation, because English is an ego-centered language.

Interestingly, (11a) is more acceptable than (11a'), because both *展望* (look forward) and *前景* (prospect) take the front side of the viewer, namely ego-centered EFR as a space-temporal direction, which is in line with the TSFR temporal direction *后 (following)*, and yet contradicts with TSFR temporal direction of *前*. In other words, in sentences (11a), the space-temporal direction based on viewer's perspective *展望 (look forward)* and *前景 (prospect)* have to be switched into canonical TSFR temporal direction of *后 (following)* in case of describing a sequential span of *10年 (ten years)* in temporal sequence, based on the ground-preference cognitive style from object-centered perspective in Chinese.

The above-mentioned cases demonstrated that in Chinese, ego-centered frame of reference is more limited within the space-temporal directions such as *展望 (look forward)*, *前景 (prospect)*, *回顾 (look back)* than the temporal ones. They are usually switched into canonical temporal direction *前/后 (preceding, previous/following)* in context of specifying a temporal duration that is measured in a temporal sequence, for instance, (11a).

It's obvious that ego-based perception usually involves a traveler's experience of participating in the metaphorical journey rather than a detached description of the sequence from a spectator's perspective. By contrast, a temporal sequence-based perspective as non-human observation addresses the objectively detached description. Therefore, the direction based on the ego-centered frame of reference usually foresees more prospects of participant-involved experiences than the detached ones based on sequence reference. Thus the ego-centered temporal direction is more compatible with initiative words such as *迎来/迎接 (welcome), and 走进 (enter)*, and yet less compatible with the detached prospect such as *出现 (occur) and 发生 (happen)*. On the contrary, the temporal sequence-based direction is more compatible with the detached prospect, e.g., *出现 (occur)*, *发生 (happen*), and yet less compatible with participant-related words (e.g., *迎来/迎接*, *走进*). This pragmatic discrepancy accounts for the grammatical disagreement between sentences (a) and sentences (b) in (12–13).

Specifically, (12a) is more acceptable than (12b), because *前面* in (12a) indicates that the observer is a participant who is experiencing the metaphorical journey to welcome the upcoming dawn. As thus, *前面* based on the participant/observer's orientation highlighting participant's experience of welcoming is more compatible than *后面* based on the temporal sequence from a bystander perspective.

(13a) is not as acceptable as (13b), because *出现* describes a detached non-human action from a bystander perspective, which is more compatible with the temporal sequence direction *后面* rather than *前面* that highlights participant's experience. Additionally, the first *前面* in (13a) has already set a temporal direction in the sequential frame of reference, which sets the restriction on the second *前面* that is unable to reverse it into ego-based frame of reference. The case in (13a), therefore, can only define the direction based on the temporal sequence. (12b) is more in accordance with this condition than (13a).

(12) a. 我们　已经　跨过　了　最　艰难　的　时刻，*前面* 即将 迎来 曙光。 [perspective from a participant/experiencer]

Wǒmen yǐjīng kuàguò le zuì jiānnán de shíkè, qiánmiàn jíjiāng yínglái shǔguāng.

We already step over LE most difficult moment, forthcoming soon embrace dawn

(*see meaning from* a'.)

a'. We are going to embrace the forthcoming dawn after having passed the most difficult moment)

b. 我们已经跨过了最艰难的时刻，**后面*即将迎来曙光。 [conflict between ego perspective and sequential perspective]

Wǒ men yǐ jīng kuà guò le zuì jiān nán de shí kè, hòu miàn jí jiāng yíng lái shǔ guāng

We already step over LE most difficult moment, backward soon embrace dawn

(13) a. 虽然前面的问题已得到解决，**前面*还会出现新的困难。 [conflict between ego perspective and sequential perspective]

b. 虽然　前面　的　问题　已　得到　解决，后面　还会　出现　新　的　困难。 [spectator perspective]

Suīrán qiánmiàn de wèntí yǐ dédào jiějué, hòumiàn háihuì chūxiàn xīn de kùnnán.

Although forthcoming DE problem already get resolved, following still will appear new DE difficulty

(*see meaning from b'.*)

b'. Though the previous problem has been resolved, there will be new difficulties lying ahead.

Interestingly, *前面* in (12a) differs from the one in (13a). First, *前面* in (12a) involves the participant/observer's experience of participation in the metaphorical journey rather than a detached description of an objective sequence. It can be verified by the fact that once a temporal measurement *几天 (a couple of days)* is added to the sentence, the direction *前面* is no longer compatible with *几天* with the denotation of a duration in the temporal sequence.

(14) 我们 已经 跨过 了 最 艰难 的 时刻。
We already step-over LE most difficult DE time
**前面* 几 天 即将 迎来 曙光。 [spatial EFR]
Front few day immediately welcome dawn
(*See pinyin, glossing, and translation from (12).*)

This proves that *前面* in (12a) addresses the participant/observer's experience of participation in the journey rather than refers to the temporal sequence, therefore is incompatible with any kind of duration in the temporal sequence. In summary, the alternative directions of *前面* are usually defined by the following conditions:

First, once followed by words indicating a duration such as *几天 (a couple of days)*, *前面* will be assimilated and switched into a frame of temporal sequence reference referring to *preceding*. As such, it no longer refers to the viewer's orientation of facing or moving towards future (*forward*) based on the ego's frame of reference.

Second, in the context that highlights the detached sequence-based temporal direction from a bystander's perspective, *前面 (forward/ahead)* as a viewer-based temporalized spatial direction have to be switched into sequence-based direction *后(面) (following/posterior)*, e.g., (13b). This reveals the perplexity that *后(面)* in Chinese can only refer to posterior time point, while *前(面)* can indicate either *anterior/preceding* based on temporal sequence or *ahead* based on participant/observer's envisagement, which is characteristic of temporalized spatial meaning. These cases collectively demonstrate that Chinese EFR lacks the dependency of temporal frame of reference. Instead, it tends to be converted into the canonical object/sequence-centered frame of reference, which significantly contrasts with the dominant status of English ego-centered frame of reference.

To sum up, in light of a sequence-preference cognitive style from an object-centered perspective, Chinese prefers to select TSFR as the canonical temporal frame of reference given the context of coexistence of both EFR and TSFR. As a result, 前 alternates its meanings between the EFR temporalized spatial direction (*forward*) based on journey-metaphor and the canonical temporal direction (*preceding/former/-ex*) based on temporal sequence. In the situation where EFR eclipses TSFR by highlighting observer's participatory action such as *embrace* in (13), *前* will be characterized with a temporalized-spatial direction (*forward*), highlighting observer's experience in collocation with such kind of verbs denoting the observer's initial behavior such as *迎来* (embrace) in (12). Yet, it is incompatible with verbs such as *出现* (*lie/appear/occur*) that are based on objective description from bystander's perspective, e.g., (13a). In context of TSFR eclipsing EFR, *前* will resume its canonical temporal direction (*preceding/former/-ex*) based on temporal sequence. In particular, the first *前* in (13a) and (13b) on the one hand tends to show up with verbs such as *出现* (*lie/appear/occur*) that purely highlight objectivity, on the other hand, they are incompatible with words indicating any kind of observer's information such as observation etc.

In the context that refers to a posterior time point, *后* (*following/post-/step-*) will be used to denote the detached perception of temporal direction.

In a nutshell, the sequence/object-centered and ego-centered cognitive styles lead to different approaches for Chinese and English to conceptualize the sequence-based and ego-based temporal directions respectively. This is evident in the directional discrepancy between *前/后* and *forward/back*. Chinese uses ego-centered EFR in the context of highlighting the observer's experience of participation in journey-metaphor, which is assimilated or switched into temporal direction based on TSFR. In other words, Chinese takes TSFR as canonical temporal cognition with EFR as the temporalized spatial direction. This rationale could accommodate the paradox of directions in (11) and (12).

Importantly, it sheds light on the fact that *前(面) (preceding/forward)* and *后(面) (following)* as a pair of contradictive directions refer to reverse directions in an irreversible way.

4 Reverse Correspondence Between Chinese and English Temporal Directions

There are at least two ways of conceptualizing a temporal or temporalized-spatial relationship. The first one would be a sequence-based canonical temporal direction, which is characteristic of the cognitive style based on object-centered perspective in Chinese. The other type is an observer-based direction that is taken as the cognitive style based on ego-centered perspective, such as in English. In the context where both Chinese and English refer to a temporal sequential relationship, the two languages tend to select a consistent temporal direction, for instance, *former president (前总统)*, *ex-girlfriend (前女友)*, *preceding years* (前几年), and cases in (7) and (8).

In context of signifying ego such as *us* in (5) and (6), or the observer's experience of participation in the metaphorical journey such as (11) and (12), English differs from Chinese. In particular, English regards the observer's ego-based direction as the temporal direction, while Chinese tends to regards it as a spatial one, which is converted into temporal sequential direction under condition of time duration occurring in context, or in a context without observer-related information. This pattern is shown in the comparison of cross-linguistic translation e.g. (1), (2), (4), (5), (6), (11), and otherwise, the failure of conversion results in unacceptable collocation, e.g. (11a'), (11b'), (13a) and (14).

Moreover, ego-based perception usually involves the observer's on-site experience of participation in the metaphorical journey, whereas sequence-based perception describes time-point distribution in temporal sequence from a detached perspective of the bystander. This subjective and objective contrast leads to the collocation of *前面/后面* with their respectively compatible constituents regarding subjectivity and objectivity.

To summarize the facts above, the discrepancy between *前/后* and *forward/back* reveals the different ways we perceive and encode the spatial and temporal relationships. Moreover, this encoding process might be a reflection of the object-based TSFR and ego-based EFR cognitive styles.

However, the affection of object-centered and ego-centered cognitive styles on Chinese and English temporal conceptualization might be concealed by the following context. First, when the ego-based orientation is absent or eclipsed by temporal-sequence indication in the context, both Chinese and English show the same direction. It could be seen as a result of conceptualizing the sequence-based direction in the same manners, where Chinese *前* and *后* respectively corresponds to their English counterparts *former, previous, -ex, ahead of and latter, post-, following, post* etc. Additionally, in case that the ego-based orientation appears in the context, Chinese will convert this ego-based direction into the temporal sequence direction as long as such kind of words indicating duration show up in the context. Accordingly, this leads to *前/后* appearing in a discrepant direction with their English counterparts *forward/back.*

The abovementioned rationale clarifies the controversy on whether Chinese takes the past as its normal temporal direction. As a language regarding the temporal-sequence as canonical reference for its temporal direction, Chinese conceptualizes temporal relationship by denoting time points in the anterior or posterior position *前/后* (*preceding/following*) in the sequence. An ego-based orientation is seen as a temporalized-spatial direction, let alone the so-called *facing the past or the future*. Thus, the hypothesis that Chinese takes the past as its normal temporal direction is ill grounded in that besides defining a temporal direction on the basis of a past-future dualistic frame, there could also be a third way, that is, to define it in light of a preceding-following sequence-based frame of reference, as the case of point in Chinese.

Table 1. Discrepancies between *前/后* and *forward/back*

Chinese	English	Direction
(Object-centered TSFR)	(Ego-centered EFR)	
今后	From this time forward	Reverse
提前 (future event)	Bring forward	Same
提前	In advance	Same
推后, 推迟	Put off, postpone	Same
向前推 (past event)	Push back	Reverse
之前, 过期	Back issues of magazines	Reverse

5 Barrier in Object-Centered and Ego-Centered Temporal Direction Conversion in Machine Translation

Based on the discussion above, we can see that the Chinese object-centered and English ego-centered cognitive styles navigate temporal conceptualization in different ways. This disagreement might lead to a discrepancy in temporal conceptualization with ego-based contexts, which entails a conversion between the two cognitive styles. To verify this issue, we have tested whether current machine translation is smart enough to accurately convert the English ego-based temporal direction into its Chinese sequence-based counterpart.

(15) a. The archaeological research pushes back the date of China's earliest known large-scale water engineering project to approximately 5,100 years ago.

Translation by Google:

b. 考古研究将中国最早的大型水利工程工程*推迟*到大约 5100 年前。

Translation by Baidu:

c. 考古研究将中国已知最早的大型水利工程工程的时间*推后*至大约 5100 年前。

Obviously, the directions in Chinese translation in (15b) and (15c) appear to be discrepant with the English one in (15a). This discrepancy shows a barrier in the conversion between the object-centered and ego-centered cognitive styles.

The cases of translations quoted from local and overseas machine translation have verified the fact that Chinese and English temporal direction cognition differ from each other, thus forming a barrier in cross-linguistic translation.

6 Conclusions

Navigated by the object-centered value orientation, Chinese language conceptualizes temporal direction on the basis of temporal sequence frame of reference, which is in line with ground-preference cognitive style. As a consequence on the language system, Chinese prefers to view time as a temporal sequence with the earlier point being defined as *前 (preceding)* and the latter point *as 后 (following)*, even in the case that temporal measurement appears in an ego-based context, e.g., (12). Even though there is the same type of TSFR illustrating same direction as Chinese, English language takes ego's orientation towards future as *forward*, and views past and the time-point moving away from the observer as *back*. This is due to the English figure-preference cognitive style, which is navigated by the ego-centered value orientation. The differing cognitive styles in Chinese and English lead to discrepant approaches of temporal conceptualization between *前/后* and their English counterparts *forward/back.*

There is yet still another question warranting discussion: What underlies the differences between Chinese object-centered cognitive style and English ego-centered cognitive style?

Levinson [8] argued for the centrality of the egocentric and anthropomorphic nature of spatial apprehension. In European value orientation derived from ancient Greek philosophy, human is the measure of all things. This tradition has, at least since Aristotle's six directions, generally placed the human body at the center of European spatial notions.

However, there are languages and cultures where these generalizations seem out of place. For instance, Chinese philosophy of the Oneness of Man and Nature highlights the object-centered notion, underlying sequence-preference frame of reference, which is in contrast to English philosophy of man as the measure of all things, underlying ego-preference frame of reference. The different cognitive styles cause the same ego-based relationship being conceptualized in reverse directions in Chinese and English. The culture-specific temporal cognition discrepancy reveals that what we see from the world might rely on the way we see it. Likewise, the way we see the world depends on the cognitive styles based on object-centered or ego-centered value orientations that navigate the conceptualization process.

As known to linguists, phoneme is the socialized conception of a set of phones with a culture-specific value. Similarly, cognitive styles are culture-based which navigate the approach to the conceptualization in terms of language-specific option of alternative metaphor, schema, and frame of reference.

To conclude, the above-mentioned analyses between Chinese and English temporal directions reveal the different cognitive styles that navigate the culture-specific approaches to conceptualizing temporal relationships. Importantly, the culture-specific affection of cognitive styles on temporal conceptualization is significant not only for machine translation but also for cross-cultural language teaching and learning. Yet, most of the previous research mainly focused on the psychological rationale, while little attention has been paid to the culture-cognition-language interaction mechanism in terms of the affection of cognitive styles on the temporal conceptualization and the culture-specific basis. There is no doubt that this bottleneck is an unavoidable challenge to machine translation and artificial intelligence, which requires nuanced explorations in future work.

References

1. Hornby, A.S.: Oxford Advanced Learner's Dictionary of Current English, 4th edn. The Commercial Press (1997)
2. Alverson, H.: Semantics and Experience: Universal Metaphors of Time in English, Mandarin Hindi and Sesotho. John Hopkins University Press , Baltimore (1994)
3. Yu, N.: The Contemporary Theory of Metaphor–A Perspective from Chinese. John Benjamins Publishing Company, Amsterdam (1998)
4. Liu, N.: The time in language and language acquisition. Chin. Lang. Stud. **5**, 39 (1993)
5. He, L.: The differences of "前/后" in ancient and modern Chinese on the temporal indication and cognitive modes. J. Chongqing Normal Univ. (Soc. Sci.) **4**, 52–59 (2018)
6. Lakoff, G., Johnson, M.: Metaphors We Live By. The University of Chicago Press, Chicago (1980)
7. Chen, Z.: The Chinese cognitive style underlying cultural representation and cross-cultural Chinese teaching. J. Int. Chin. Teach. **3**, 4–10 (2020)
8. Levinson, S.C.: Language and space. Annu. Rev. **25**, 353–382 (1996)

Exploring Chinese Lexicon Knowledge in Biliteracy Development of International Students in Hong Kong

I.-Chiao Hung and Jia-Fei Hong(✉)

National Taiwan Normal University, Taipei, Taiwan
{ichiao.hung,jiafeihong}@ntnu.edu.tw

Abstract. This study explores the Chinese Lexicon knowledge in biliteracy development of teenagers at an international school in Hong Kong. It presents how Diagnostic Adaptive Chinese Competence (DACC) is applied to a total of 137 Grade 9 international students and their reading performance. The lexicon knowledge is one critical subskill [1] contributing to the overall reading performance [2] based on the component-skills approach. This study shows that 1) DACC is justified as a reliable and valid assessment test; 2) in the five dimensions of DACC, the Grade 9 students' performance on literal comprehension, contextual integration, inferential comprehension, and analysis and evaluation are higher than the average of the norm group but lower when it comes to the Chinese lexicon knowledge. More studies focusing on biliteracy reading development with the learning of Chinese as a foreign language are needed in the near future.

Keywords: Lexicon knowledge · Vocabulary · Biliteracy · Chinese reading · Component-skills approach · DACC

1 Introduction

Literacy refers to the ability to read and write [3]. The concept of "reading literacy" means the ability to read, the correlated knowledge to obtain and comprehend through the reading behavior. Reading literacy is an important skill to obtain knowledge and develop correlated intellection among other subjects [4]. The concept of "reading literacy" in the "Programme for International Student Assessment (PISA)" is further expanded to that the learners can engage with reading to identify, analyze, solve and interpret problems in societies [5]. Researchers, educators and policymakers have devoted enormous efforts to providing the research-based assessment system and proven practices to promoting reading for the young generations. How one can implement the appropriate programs and follow the instruction to improving students' reading literacy has been a critical concern among different countries [4, 5].

Various proficiency tests for school-age language learners have been developed with the aim of providing assessments of students' language proficiency in different domains. PISA is one example. In 2018, 600,000 students across 79 countries participated in the

M. Liu et al. (Eds.): CLSW 2020, LNAI 12278, pp. 850–858, 2021.
https://doi.org/10.1007/978-3-030-81197-6_70

English reading assessment of PISA [6]. "Diagnostic Assessment of Chinese Competence (DACC)" is the other example which is a computerized adaptive testing system providing Chinese reading assessment across countries.

As there are only a few studies on adaptive Chinese reading assessment with a primary focus on the younger generations' learning in a biliteracy environment [2] and Hong Kong is an international city that provides a biliteracy learning environment for younger generations, the current study targets at the students from an international school in Hong Kong. We will discuss the students' performance on English reading of PISA 2018 and their performance on Chinese reading of DACC 2019 and compare their performances.

Moreover, the previous studies are inconsistent in the contribution of lexicon knowledge to reading comprehension and its complexity and multidimensional nature, in all the different domains. This study explores students' performances in DACC, in the aspects of Chinese lexicon knowledge (vocabulary), literal comprehension, contextual integration, inferential comprehension, and analysis and evaluation. To justify the validation of DACC, we also compare the reading literary involved DACC with that in PISA. We suggest the Chinese lexicon knowledge is one of the most critical and effective subskills contributing to the overall reading performance [2].

2 Related Backgrounds

2.1 Diagnostic Assessment of Chinese Competence (DACC)

"Diagnostic Assessment of Chinese Competence (DACC)" is a technique included on the online platform "SmartReading"[1], designed by scholars in Taiwan. DACC is a computerized adaptive testing system and has the test norm of more than 40,000 students from Taiwan and Hong Kong which can help access learners' reading ability accurately.

After taking the DACC test, students will get the scores from SR1 to SR1000 (SR stands for SmartReading; the unit of reading value) reflecting their reading literary. The whole reading score is calculated by adding up scores obtained from the five dimensions. The system has been well applied to schools and has showed the positive results of students' reading performance. Five reading dimensions are evaluated in DACC assessment (Table 1).

2.2 Reading Performance of Hong Kong Students in PISA

In this study, we make a reference to PISA to justify the reliability and validity of DACC. PISA is organized by the Organization for Economic Co-operation and Development (OECD) since 1997. It has been first implemented across countries since 2000 and repeated every three years to evaluate 15-year-old students' scholastic performance in

[1] The other technique available on SmartReading is "Chinese Readability Index Explorer (CRIE)" which is a text readability analytic system to evaluate the level of readability of books, based on words, syntax, semantics and cohesion levels involved in the books. It can analyze the SR reading scores and hence automatically generate a list of recommended books to match students' reading level.

Table 1. Five reading dimensions targeted in the test development of DACC

Reading dimension	Description
Vocabulary	Readers can understand the meaning of the word or phrase in a text
Literal comprehension	Readers can understand the meaning of a sentence in a text
Contextual integration	Readers can conclude the content in a brief text
Inferential comprehension	Readers can use their existing experience or knowledge to interpret the messages behind a text
Analysis and evaluation	Readers can construct a deep understanding of a text and judge a text objectively

Mathematics, Science, and reading literacy. Only one of these areas is the primary focus every time the assessment holds. It provides valuable and comparable data for countries to improve their education policies and outcomes. In 2018, the focus aspect was reading in which the PISA sought ways to prepare students to conquer the challenges of today's knowledge societies [5, 7]. Five reading processes have been defined by PISA (since 2009) to guide the reading literacy assessment tasks (Table 2).

In addition, as an international practice, the scores in the evaluation were scaled at a mean score. In 2001, the mean score was 500 and the average score of Hong Kong was 528, which ranked Hong Kong 14th amongst the 35 countries. In 2003, Hong Kong was ranked first in Mathematics, third in Science, and 10th in reading. The measurement of Problem Solving Skills was a new domain in PISA 2003 in which Hong Kong ranked second. Overall, the mean performance of Hong Kong students was well above the international mean (Education Statics, 2006).

As for the PISA 2018 results, compared to the average mean score of 487 points, the mean score of Hong Kong 15-year-old students in reading was 524 points. From 2001 to 2018, the mean scores of Hong Kong in reading were all higher than the average. In 2018, though 38% of students in Hong Kong had an immigrant background, there was no statistically significant difference in reading performance between immigrant students and non-immigrant students [8].

These five aspects of PISA in reading performance are organized into three broad categories, and are considered as reading subscales, namely access and retrieve, integrate and interpret, and reflect and evaluate.

2.3 The Education Reform About Literacy in Hong Kong

The Education Commission (EC) of Hong Kong submitted the "Reformed Proposal for the Education System in Hong Kong" to the government in September 2000. The reform included the curriculum, the assessment mechanisms, and the admission systems for different levels of education. The government hence published a timetable for implementing the reform measures. After three reports made in January 2002, June 2003 and December 2004, the EC issued the fourth "Progress Report on the Education Reform" providing a detailed account of the progress in the seven critical areas of the education

Table 2. Five reading aspects targeted in the test development of PISA

Reading aspect	Description
Retrieving information	Readers can navigate a text to locate a particular piece of explicitly stare message
Forming a broad general understanding	Readers can form a text in a whole and brief perspective
Developing an interpretation	Readers can understand the content and make internal sense of a text
Reflecting on and evaluating the content of a text	Readers can find the information, link to one's knowledge, and assess the text with the difference of other knowledge
Reflecting on and evaluating the form of a text	Readers can draw upon the attitudes beyond the text to one's own conceptual and experiential frames of reference

reform in December 2006 (Education Bureau, 2013; Education Commission, 2017). The educational innovation included canceling the assigned reading texts in class, applying the qualified books or contents into the classroom, providing systematic training courses for teachers, and flexible curriculum frameworks, and incorporating with the family and society.

Besides, the availability of computers at schools and homes for 15-year-old full-time students was 100.0% at schools and 93.2% at homes (PISA 2003, Education Statics 2006). On Hong Kong students' reading habits, the Education Statics (2006) showed that students were more inclined to choose the reading materials by themselves. Regarding the types of books students read, junior primary students who participated in the Online Reading Contract preferred to read picture books and comics books whereas senior primary and secondary liked reading novels, proses, and essays. To conclude, not only did Hong Kong state the importance of reading but the education reform of Hong Kong and the change of the learning environment were all striving to enhance students' reading literacy. By far, there was a lack of studies addressing the importance of learners' Chinese reading literacy and the issue of bilingualism in Hong Kong.

2.4 Component Approach to Reading

The component-skills approach viewed reading as a process development as a complex system that encompassed different subskills or components which contribute to the overall reading performance [1]. With this approach to reading, orthographic, phonological, vocabulary, and morphological knowledge are considered critical component skills in learning to read [2, 7].

For beginning readers, the first step is to decode the printed texts with symbols and interpret the text. To become an excellent decoder, a student needs to be aware when reading the text, sensing the systematic relationship between the symbols, written or spoken message, and the words composed of letters in the case of English, and from the radical to singular or a unit of characters in the case of Chinese [8]. In the case of Chinese, readers had to develop orthographic knowledge of the sound-symbol relationship of the printed text and pronunciations first then interpret the Chinese text. On the other side, in the case of English, learners need to segment the syllable into constituent phonemes and can associate with the letters for the meanings.

3 Methods

3.1 Participants

A total of 137 Grade 9 students in an international school in Hong Kong, aged 14–15, participated in this study. As international students in Hong Kong, taking the curriculum of the secondary education based on the framework of the National Curriculum for England, their linguistic skills in both English and Chinese are equally sharpened to adapt to a bilingual learning environment. There were 116 participants (85%) from local Hong Kong Chinese, learning Chinese as language arts, and 21 participants (15%) from other regions using Chinese as their foreign language.

In this study, the students were divided into different classes according to their Chinese proficiency, including three classes of Chinese as an Additional Language (CAL) and eight classes of Chinese as a First Language (CFL). All the students attended their Chinese classes five times a week, each of which lasted for fifty-five minutes.

3.2 Data Collection

The study uses the online platform "SmartReading" to access students' Chinese reading ability. The DACC is a computerized and adaptive testing system that measures students reading comprehension on five dimensions: vocabulary, literal comprehension, contextual integration, inferential comprehension, analysis and evaluation. The students completed their tests with computers or tablets in fifty-five minutes. Students and teachers could access to the assessment report on the system.

The participants completed the DACC assessment at the beginning of the semester (August 2019). They started the computerized DACC with randomly selected questions that tested their grade level. Then the system estimated their comprehension abilities based on their responses and hence provided adaptive questions to measure their reading level. By this, the accurate measurement was achieved. The reports of the assessment performance showed immediately after the test. The whole process of assessment is visualized in Fig. 1.

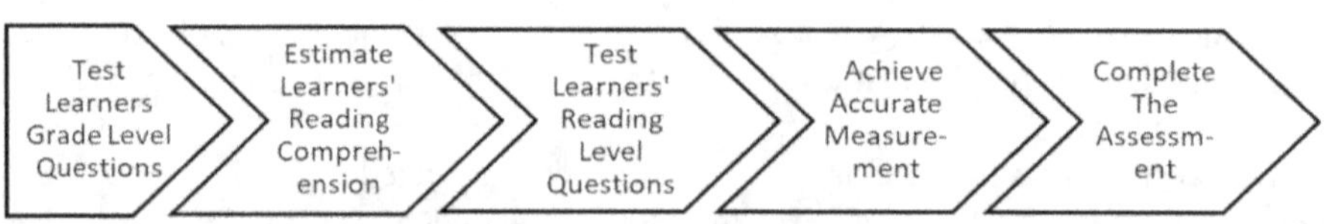

Fig. 1. The process of taking DACC

On the assessment report, the percentile rank of the grade level, the overall reading scores of the student and the norm-referenced, the scores in the five dimensions and the norm-referenced with explanation and comments were presented. The unit of the reading level is "SR" as SmartReading, and the ability score level ranges from SR1 to SR1000. The sum of the scores in the five dimensions is the whole composite reading score. For example, student A in Grade 5 obtained SR472 on the whole reading scores, and the norm-referenced equivalent score for Grade 5 was SR477. The report presented the comparison with a norm-referenced score, and showed SR472, being greater than 46 percentile of the scores of the Grade 5 students. In the five dimensions, the comparisons of scores/norm-referenced scores are as follows: SR92/SR91 in vocabulary (Chinese lexicon knowledge), SR91/SR97 in literal comprehension, SR95/SR97 in contextual integration, SR97/SR95 in inferential comprehension, and SR97/SR97 in analysis and evaluation.

3.3 Justifying DACC

To justify if DACC is a reliable and valid assessment test, we refer to the five reading aspects involved in PISA and compare it with the five reading dimensions of DACC. The comparison shows that the two dimensions ("vocabulary" and "literal comprehension") of DACC correspond to "retrieving information" in PISA, as both of them require the readers show their ability to understand the meaning of the word or phrase in a text [10, 11]. The aspects of "developing an interpretation" and "reflecting on and evaluating the form of a text" in PISA correspond to "inferential comprehension" in DACC. Table 3 shows that the standards and concepts of PISA in reading and DACC are related. It follows that the DACC is a reliable and valid measurement.

Table 3. Comparisons between PISA and DACC

PISA in reading	DACC	Descriptions in common
Retrieving information	Vocabulary	Readers can comprehend a word or phrase in a text
	Literal comprehension	
Forming a broad general understanding	Contextual integration	Readers can conclude the content in a brief text
Developing an interpretation	Inferential comprehension	Readers can apply one's existing experience or knowledge to interpret the messages behind a text
Reflecting on and evaluating the form of a text		
Reflecting on and evaluating the content of a text	Analysis and evaluation	Readers can assess the text with the difference of other knowledge and express own opinions objectively

4 Data Analysis and Discussion

A comparison between general Hong Kong international students and those from the norms of Grade 9 Chinese learning students who took DACC are discussed as follows. The normative sample consists of 137 Chinese learning students in Grade 9. The overall reading ability score, dimensions of vocabulary (Chinese lexicon knowledge), literal comprehension, contextual integration, inferential comprehension, and analysis and evaluation are compared to the average of the norm group (Table 4).

Table 4. The average reading scores of the Grade 9 students

	Overall reading ability	Vocabulary	Literal comprehension	Contextual integration	Inferential comprehension	Analysis and evaluation
Y9-1	SR600	SR116	SR120	SR120	SR123	SR119
Y9-2	SR560	SR109	SR111	SR113	SR115	SR113
Y9-3	SR602	SR118	SR122	SR122	SR124	SR117
Y9-4	SR627	SR121	SR126	SR126	SR129	SR126
Y9-5	SR607	SR118	SR122	SR122	SR124	SR121
Y9-6	SR643	SR125	SR131	SR129	SR132	SR125
Y9-7	SR643	SR125	SR130	SR130	SR133	SR124
Y9-8	SR648	SR126	SR132	SR130	SR133	SR127
Y9CAL1	SR356	SR66	SR62	SR73	SR70	SR85
Y9CAL2	SR433	SR82	SR81	SR88	SR87	SR95
Y9CAL3	SR373	SR70	SR65	SR76	SR72	SR89
Average score of class	SR575	SR111	SR114	SR116	SR118	SR116
Average score of norm group	SR573	SR112	SR114	SR116	SR117	SR114

As Table 4 shows, the overall reading ability (SR575), inferential comprehension (SR118), and analysis and evaluation (SR116) of the Hong Kong international students are above the average of the norm group (SR573). The students' average score on vocabulary (SR111) is lower, and their average score on literal comprehension (SR114) and contextual integration (SR114) are the same as the average of the norm group. Secondly, 21 students learning Chinese as an additional or foreign language from the three CAL classes (Y9CAL1, Y9CAL2, and Y9CAL3) are counted into the average score of the Grade 9, which is one significant reason lowering down the average score. If the 21 students from the three CAL classes were not added into the account, then the 116 students' average score of vocabulary (SR119.75) in the Grade 9 would be above the average score of the norm group (SR112).

To explore how the Grade 9 students' performance on Chinese lexicon knowledge contributes to the overall reading performance of DACC in teenagers biliteracy development, the results show that the overall reading ability of the Grade 9 international students in the Hong Kong international school is generally well above the average. This study has revealed that the students tend to be good at comprehending short texts (with a reference to the dimension of literal comprehension) and summarizing the main idea in a sentence (with a reference to the dimension of a summary of the main ideas). However, the performance on demonstrating Chinese lexicon knowledge (i.e. the dimension of vocabulary) is slightly lower than the average of the norm group, when the average score of CAL students (SR66, SR82, and SR70) from the three classes (Y9CAL1, Y9CAL2, and Y9CAL3) are taken into the account. The average score (SR72.67) of the Chinese lexicon knowledge from the three Grade 9 CAL classes has contrasted to the average score (SR69) of Grade 3 students from the norm group. While all the students have the same amount of time in Chinese classes, they differ in their linguistic background that the CAL students are students learning Chinese as a foreign language. Besides, the norm group includes local schools, international schools, and experimental schools, in which the factors of differences from urban and rural areas and wealth gaps are not discussed in this study.

The results above indicate that the biliteracy teenagers especially those learning Chinese as an additional language require teachers' assistance to immerse them more in Chinese lexicon knowledge learning, since vocabulary was a critical component skill in learning to read [1].

The component-skill approach can be helpful for both teachers and students. The approach views reading as a process of development and as a complex system that encompasses different subskills or components which contribute to the overall reading performance [1]. With this approach to reading, orthographic, phonological, vocabulary, and morphological knowledge are considered critical component skills in learning to read [2, 8]. To be specific, readers need to decode the symbols (or words) in printed texts and so as to interpret the entire piece of text. To become an excellent decoder, a student needs to pay attention when reading the text, sensing the systematic relationship among the symbols, differentiating written from spoken messages, understanding the words composed of letters in case of English, and understand from radicals to a unit of characters in case of Chinese [9]. To understand Chinese texts, readers have to develop orthographic knowledge of the sound-symbol relationship. That is to say, they need to figure out the pronunciation first and hence interpret the Chinese text. In the case of English, learners need to segment the syllable into constituent phonemes and associate among the letters for the meanings.

5 Conclusions

This study has examined the reading abilities of the Grade 9 international students in Hong Kong using the DACC assessment in Chinese as well as the PISA assessment in English. The reading abilities, in general, are above the average. In the Chinese reading assessment, overall, students also performed well, despite the inclusion of students learning Chinese as a foreign language to the study. Moreover, the domain of Chinese lexicon knowledge, refers to students' knowledge of lexicon and their understanding of the depth and breadth of vocabulary, which is crucial for students to develop their reading literacy [10].

Above all, in order to develop the reading ability of younger generations, it is worth arousing the awareness of Chinese lexicon knowledge in biliteracy reading for administrators, teachers, and students [12, 13]. Moreover, the impacts of other domains, including literal comprehension, contextual integration, inferential comprehension, and analysis and evaluation should be further investigated in future studies.

Acknowledgements. This work was supported by National Taiwan Normal University's Chinese Language and Technology Center. The center is funded by Taiwan's Ministry of Education (MOE), as part of the Featured Areas Research Center Program, under the Higher Education Sprout Project.

References

1. Carr, T.H., Brown, T.L., Vavrus, L.G., Evans, M.A.: Cognitive skill maps and cognitive skill profiles: componential analysis of individual differences in children's reading efficiency. In: Carr, T.H., Levy, B.A. (eds.) Reading and Its Development: Component Skills Approaches, pp. 1–55. Academic Press (1990)
2. Lü, C.: Chinese Literacy Learning in an Immersion Program. Springer, Cham (2019). https://doi.org/10.1007/978-3-030-04987-4
3. Oxford University Press. Oxford advanced learners' dictionary (2011). https://www.oxfordlearnersdictionaries.com/definition/english/literacy?q=literacy
4. Chen, M.C., Hsu, W.S.: The direction of reading literacy education: From the comparative perspective of PISA reading assessment. New Horizon Bimonthly for Teachers in Taipei, vol. 181, pp. 4–15 (2012). (in Chinese). http://www3.nccu.edu.tw/~mujinc/journal92/5-60.pdf
5. Brozo, W.G., Shiel, G., Topping, K.: Engagement in reading: lessons learned from three PISA countries. J. Adolesc. Adult Literacy **51**(4), 304–315 (2007)
6. OECD. PISA 2019 results (Volume I-III), PISA, OECD Publishing, Paris (2019). https://www.oecd.org/pisa/publications/PISA2018_CN_HKG.pdf
7. Kuo, L.J., Anderson, R.C.: Morphological awareness and learning to read: a cross-language perspective. Educ. Psychol. **41**(3), 161–180 (2006)
8. Lu, C.: English-Chinese biliteracy acquisition among Chinese heritage language learners. Carnegie Mellon University, Pennsylvania, U.S.A (2009)
9. Pasquarella, A., Chen, X., Lam, K., Luo, Y.C., Ramirez, G.: Cross-language transfer of morphological awareness in Chinese-English bilinguals. J. Res. Reading **34**(1), 23–42 (2011)
10. Shao-long, L.I.U.: Dimension development and acquisition patterns of L2 word knowledge. J. Pla Univ. Foreign Lang. **2** (2002)
11. Stoller, F., Grabe, W.: Implications for L2 vocabulary acquisition and instruction from L1 vocabulary research. In: Second Language Reading and Vocabulary Learning, pp. 24–45 (1993)
12. Wang, M., Cheng, C., Chen, S.W.: Contribution of morphological awareness to Chinese-English biliteracy acquisition. J. Educ. Psychol. **98**(3), 542 (2006)
13. Zhang, D., Yang, X.: Chinese L2 learners' depth of vocabulary knowledge and its role in reading comprehension. Foreign Lang. Ann. **49**(4), 699–715 (2016)

Semantic Classification and Cognitive Mechanism of Chinese "Body-Part" Compounds with Semantic Exocentricity

Qi Li(✉)

School of Chinese as a Second Language, Peking University, No. 5 Yiheyuan Road, Haidian District, Beijing 100871, China
1901111161@pku.edu.cn

Abstract. This study starts from the semantic eccentricity of compound words and analyzed 95 eccentric semantic compound words composed of Chinese "body part" nouns. We then divided them into two classes in terms of their meanings, namely compounds referring to "people" and compounds referring to "thing". The second class can be further divided into two subtypes, i.e., "personal characteristics" and "diseases/symptom". The number of the compounds shows that the morphological pattern [X + body part] is still productive in modern times, which is now mainly reflected in the production of new words. In addition, some compounds have been reanalyzed semantically in the two major classes of compound words as mentioned, and the two word-base subtypes of "tool" and "behavior" have been derived from the compound words of "disease/symptom" type. Such derivation also proves the productivity of [X + body part] with semantic eccentricity. Finally, this study has illustrated the cognitive mechanism of the compounds based on Conceptual Metonymy and Conceptual Integration theory and has found their cognitive path: the word base and the "body part" nouns activate metonymy operation, and then conceptual blending takes place between the two targets via certain semantic relations between them. In the process of metonymy operation, the salience of the word base and body part nouns play important roles at both generic level and basic level of metonymy.

Keywords: Semantic exocentricity · "Body part" nouns · Semantic classification · Conceptual metonymy · Conceptual combination

1 Introduction

"Body-part" nouns are one of the basic concepts in human cognition. There is a class of compound words composed of "body part" nouns whose meanings are special, as shown in the following examples:

(1) *jiàngǔtóu* (贱骨头) ❶ Refers to people who have low self-esteem or cannot tell what is right and what is wrong (swear words). ❷ Refers to people who are willing to suffer what they do not deserve.

M. Liu et al. (Eds.): CLSW 2020, LNAI 12278, pp. 859–870, 2021.
https://doi.org/10.1007/978-3-030-81197-6_71

(2) *míngzuǐ* (名嘴) Refers to a program host, speaker, lawyer, etc. who are known for his or her eloquence.

(3) *shǒunǎo* (首脑) refers to leaders.

Jiàngǔtóu (贱骨头) contains the word 'bone' (i.e., 骨头) but it does not refer to a type of bone (*gǔtóu*). *Míngzuǐ* contains the word 'mouth' (i.e., 嘴) yet it does not refer to a type of mouth (*zuǐ*). Similarly, *shǒunǎo* (首脑) contains the word 'brain' (脑) but it does not refer to a type of brain (*nǎo*). Compound words like these have one feature in common: they are all composed of nouns which denote body-parts as its head element, while the denotation is not a certain body part, but a specific type of people. Some new words of this type come into being in recent years, such as *xiǎoxiānròu* (小鲜肉; lit. 'small fresh meet'), *liànàinǎo* (恋爱脑; lit. 'love-affair brain'), and *shǔbiāoshǒu* (鼠标手; lit. 'mouse hand'). The first two words refer to "young and good-looking males", and "people who regard romantic love as their life priority thus lost their self-value" respectively. While the last one is different, it does not refer to a class of people, but a type of disease, which is "wrist symptoms caused by the long-time use of mouse".

It is demonstrated that this type of "body-part" compounds are compounds with semantic exocentricity, that is, the denotation type of the word is inconsistent with that of the head element (Bauer 2010; Bauer et al. 2013). The semantic complexity caused by the semantic exocentricity may result in understanding difficulties for Chinese as Second Language (CSL) learners. If the compound words with semantic exocentricity are productive, then exploring their semantic classification and cognitive mechanism is necessary. There is not so much research on semantic exocentricity in Chinese. Most research having studied exocentricity tend to illustrate it morphologically, not semantically (Shi 1988; Shi 2007). Carmen (2015) conducted a research on the semantic exocentricity of English "body-part" compounds and analyzed the productivity of them. There is no research of Chinese "body-part" compounds with semantic exocentricity until now.

This paper will take 95 "body-part" compounds as examples, aiming to analyze their semantic categories and discuss the productivity of each category, then hence exploring the cognitive mechanism of them.

2 Semantic Exocentricity of Chinese "Body-Parts" Compounds

Semantic exocentricity refers to the feature that the denotation of the exocentric construction is inconsistent with that of its head element (Bauer 2010; Bauer et al. 2013). Exocentric construction, which is also called "Bèixīn Jiégòu (Wang 1944)" or "Wàixiàng Jiégòu" in Chinese, and along with edocentric construction, are the concepts in Constructionism Linguistics. The concepts were firstly put forward by Bloomfield (1933) and were applied in syntax only at the very beginning. When the concepts are being applied in lexicology, there comes the concept "exocentric compounds". Recent studies held different opinions about the productivity of exocentric compounds: some researchers have suggested that exocentrics are non-productive, atypical, opaque and non-analyzable (Benczes 2013); others have claimed that they are quite common in a large number of languages in the world and they are even the most common pattern in some languages. (Guevara and Scalise 2009).

Exocentricity is often discussed "morphologically" in the research mentioned above. Our study adopts the definition of exocentricity presented by Bauer (2010), which demonstrated exocentricity semantically. Bauer (2010) has noted that there are two classes of exocentric compounds: 1) compounds that are clearly morphologically and semantically exocentric, as they function as a number of word classes which are not the word classes of their head elements. This class of compounds are non-productive; 2) compounds that have a head element of the correct word class, but with apparently the wrong denotation, in which case exocentricity is only semantic.

The exocentric compounds which Carmen (2015) discussed, which are also to be discussed in our study, belong to the second type. This type of compounds is common in both English and Chinese and some of the "body part" nouns are very productive morphemes. Taking "*gǔtóu* (骨头, bone)" as an example:

(4) *ruǎngǔtóu* (软骨头) (lit. soft bone) refers to people who are easy to surrender.

(5) *yìnggǔtóu* (硬骨头) (lit. hard bone) refers to people who are strong and robust.

(6) *lǎngǔtóu* (懒骨头) (lit. lazy bone) refers to lazy people (this phrase belongs to swear words).

(4) (5) and (6) have the head element "*gǔtóu* (骨头)" in common. However, they do not refer to different types of bones. Instead, they refer to different types of people. Morphologically, the word-class of their head elements and compounds are nouns in common. Semantically, the denotations are different. Thus, compounds like these are exocentric semantically. Besides "*gǔtóu* (骨头)", there are other "body part" nouns like "*tóu* (头, head)" "*nǎo* (脑, brain)" and "*liǎn* (脸, face)" which can form compounds with semantic exocentricity to be discussed in the following sections.

3 Semantic Classification and Productivity of Chinese "Body-Part" Compounds

Our study analyzes 95 Chinese exocentric compounds formed by "body-part" nouns. These compounds were selected from dictionaries (i.e. *Modern Chinese Dictionary,* 7TH edition, *MCD*; *New Chinese Words 2008–2017, NCW*) and the Internet as well. Using the Internet as one of our corpus sources, have we considered two purposes: first, to expand the scale of selected words; second, to make full use of the academic value of web language. Shi (2010) illustrated that the morphological patterns of web words (or new words) are consistent with that of conventionalized words. By analyzing new web compounds, we can discuss whether the exocentric words are productive or not in modern times as well.

After analyzing the semantic features of the selected compounds, we have divided the compounds into two main classes semantically, namely (i) the compounds referring to "people" and (ii) compounds referring to "things". We will discuss the classification in detail.

3.1 Compounds Referring to "People"

3.1.1 The Case of "-*shǒu* (手, Hand)" Formations

Compounds formed by the morpheme "*shǒu* (手)" form the most representative subtype in compounds referring to "people" as the morphological pattern [X + *shǒu*] is the most productive, which has produced 40 compounds in total. The word base X covers a wide range of word classes. It can be a noun, forming words like "*gŭshŏu* (鼓手)" (lit., drum hand, i.e., a drum player) and "*qíshŏu* (棋手)" (lit., chess hand, i.e., a chess player); or adjectives, forming words like "*gāo shŏu* (高手)" (lit., tall hand, i.e. an expert) and "*xīnshŏu* (新手)" (lit., new hand, i.e. a novice); or even verbs or VPs, forming words like "*măishŏu* (买手)" (lit., buy hand, i.e. a buyer) and "*qíshŏu* (骑手)" (lit., ride hand, i.e. a rider).

Due to its high productivity, the meaning of 'a certain type of people' has become one sense listed in the dictionary (MCD) under the entry of "*shŏu* (手)", explained as it "refers to people who are good at one skill". In English, "-head" compounds are also the most typical type of exocentric compound as "*-shŏu*" compounds in Chinese (Bauer et al. 2013: 478), which has formed a large number of compounds so far. There is also a sense of "person; individual" of *head* presented in Oxford English Dictionary.

The morphological pattern with "*–shŏu*" still has relatively high productivity in recent years, for the number of "new words"[1] it produced being 10 out of 40. Thanks to the development of time and the emergence of new concepts, compound words such as "*măishŏu* (买手)" (lit., buy hand, i.e. a buyer), "*dàigòushŏu* (代购手)" (lit., shopping hand, i.e., a shopping agent) and "*fābāoshŏu* (发包手)" (lit., distribute red packet hand, i.e. a person who distributes red packets) are constantly generated based on this morphological pattern.

It is worth noting that the meaning of some conventionalized words formed in the early days had been reanalyzed semantically, and then the newly generated words share the homophonic relation with the conventionalized words, such as "*qíshŏu* (骑手)". The definition in MCD is "people who are good at riding, especially [it] refers to athletes in equestrian competitions", while the most widely used meaning is "the take-away delivery man". The semantic reanalysis proves the productivity of *–shŏu*. (See more examples of conventional and new compounds in the Appendix.)

3.1.2 The Cases of Formation By Other "Body Part" Nouns

Apart from–*shŏu*, exocentric compounds are also formed with other "body part" nouns which is a common phenomenon in Chinese language.

As is shown in Table 1, "body part" nouns like "*tóu* (头)" "*liăn* (脸)" "*zuĭ* (嘴)" etc. can form exocentric compounds referring to "people", such as "*cìtóu* (刺头)" "*shēngliăn* (生脸)" and "*míngzuĭ* (名嘴)", but their word formation capability varies. Among them, the capability of "*tóu* (头)" is the highest while "*xīnyăn* (心眼)" is the lowest. The majority of the "body part" nouns are still productive morphemes in recent years. Only "*gŭtóu* (骨头)" "*kŏu* (口)" and "*yăn* (眼)" have not produced any new words. Moreover, some of the "body part" nouns which did not produce any words in the old days are productive in recent years, such as "*wàn* (腕)", forming words like "*dàwàn* (大腕)".

In general, although the way the "body part" nouns (except for "*shŏu* (手)") denote "people" is not conventionalized, these nouns are still productive in forming new words to some extent. Therefore, the morphological pattern of [X + body part] referring to "people" still has relatively high productivity in modern Chinese language.

---

[1] New words refer to words that are selected from NWD and the Internet.

Table 1. Numbers of compounds formed by other "body part" nouns

Body part nouns	MCD	NWD	Total
Tóu (头) 'head'	9	1	10
Liǎn (脸) 'face'	4	3	7
Nǎo (脑) 'brain'	1	4	5
Zuǐ (嘴) 'mouth'	4	1	5
Yǎn (眼) 'eye'	3	1	4
Gǔtóu (骨头) 'bone'	4	0	4
Wàn (腕) 'wrist'	0	2	2
Kǒu (口) 'mouth'	2	0	2
xīnyǎn (心眼) 'heart'	1	0	1

3.2 Compounds Referring to "Things"

Exocentric compounds generated by [X + body parts] do not only refer to "people", but also to "things". These compounds can be further divided into two subtypes, namely (i) compounds referring to "personal characteristics" and (ii) compounds referring to "diseases/symptoms".

3.2.1 Compounds Referring to "Personal Characteristics"

Table 2. Examples of compounds referring to "personal characteristics"

Example	Meaning	Source
Ruìyǎn (锐眼)	'A keen eye (mostly refers to the ability to discern things)'	MCD
Tiěwàn (铁腕)	'Powerful and useful approaches implemented by people'	MCD
Píjiǔyǎn (啤酒眼)	'A phenomenon which people feel that others are more attractive after drinking than before'	NCW
Kōng'ěr (空耳)	'A phenomenon which people deliberately listen to or say another sentence with a similar pronunciation'	NCW
Quánqiúnǎo (全球脑)	'Global awareness'	Internet

The compounds given in Table 2 are generated by the pattern [X + body part] and refer to some personal characteristics. "*Tiěwàn* (铁腕)" as an example of conventionalized words (see Appendix for more examples), refers to "powerful and useful approaches implemented by people"; "píji*ǔyǎn* (啤酒眼)" as an example of new words, refers to "a phenomenon which people feel that others are more attractive after drinking than before".

3.2.2 Compounds Refer to "Diseases/Symptoms"

[X + body part] referring to a certain kind of disease or symptom is not a new morphological pattern in modern Chinese. There are compound words like "*qīngguāngyǎn* (青光眼)", "*báihóu* (白喉)", "*píngzú* (平足)" and "*hántuǐ* (寒腿)" that are conventionalized. The word base X mainly describes two types of issues, namely the symptoms or causes of the disease. The bases of the first 3 words describe the symptoms of 3 diseases respectively while the base "*hán* (寒)" in the last word describes the cause of the disease, that is, working in cold and humid environment for a long time. All the newly generated words belong to the second type of issue, i.e., X describes causes, except for "*jīpífū* (鸡皮肤)".

Besides, under the category of 'diseases/symptoms', it is found that there are two subtypes which can form newly generated words. The subtypes are "behaviors" and "instruments", with both of which describe causes of the disease. Examples are given in Table 3.

Table 3. Examples of compounds referring to diseases/symptoms

Example	Base	Meaning	Source
Shǔbiāoshǒu (鼠标手)	Instrument	'Carpal tunnel syndrome'	NCW
Jiànpánshǒu (键盘手)	Instrument	'Tenosynovitis, a modern disease, often caused by long time use of computer or phone'	NCW
Yóuxìshǒu (游戏手)	Instrument	same as *jiànpánshǒu*	NCW
Jiànpánzhǐ (键盘指)	Instrument	'the wrist pain and finger numbness caused by long-time typing on the keyboard'	Internet
Yóuxìnǎo (游戏脑)	Instrument	'the phenomenon of delayed brain development due to video games'	NCW
Pǎobùxī (跑步膝)	Behavior	'the knee injury caused by running and other sports'	NCW
jiàshǐjiān (驾驶肩)	Behavior	'injury of anterior and posterior labrum of shoulder joint'	NCW
jipífū (鸡皮肤)	Description	'Keratosis pilaris'	NCW

In Table 3, the morphological pattern of "*jiàshǐjiān* (驾驶肩)" and "*pǎobùxī* (跑步膝)" is [behavior + body part], referring to diseases in a certain body part which is caused by the behavior, e.g., driving and running. The pattern of "*shǔbiāoshǒu* (鼠标手)" and "*yóuxìnǎo* (游戏脑)" is [instrument + body part], referring to diseases caused by the use of a certain instrument. For example, "*shǔbiāoshǒu* (鼠标手)" is the wrist disease caused by long-time use of mouse and the instrument involved is the computer mouse.

Both subtypes of base represent causes of diseases. The difference between them is that the "instrument" base will experience construction coercion after it enters the [cause + body part] pattern. Construction coercion requires the constructions coerce lexical items into having systematically related meanings (Goldberg 1995: 238).

When encountered with "*jiàshǐjiān* (驾驶肩)" and "*pǎobùxī* (跑步膝)", it's easy for the hearers to understand that the disease is caused by the behaviors represented by the word base. While "*shǔbiāoshǒu* (鼠标手)" and "*yóuxìnǎo* (游戏脑)" are not that transparent semantically, after the word base of "instrument" enters the [cause + body part] pattern. The pattern which serves as a construction will coerce the instrument base "*shǔbiāo* (鼠标)" and "*yóuxì* (游戏)" into having the related meanings with the construction, and thus activate the encyclopedia knowledge of the two words, say, the function of these instruments. A predicate "use" or "play" will then be added for understanding the words.

Apart from the two subtypes of word bases mentioned above, it is also observed that there are some words that have been reanalyzed semantically, such as "*jiànpánshǒu* (键盘手)", which was once used to refer to a piano player in a band, but now it has been reanalyzed as a disease caused by frequent use of keyboard.

In conclusion, the number of compound words referring to things is smaller than that of compound words referring to person. Yet, when comparing it in terms of the number of newly generated words, it shows the opposite trend. Moreover, the appearance of new word-base subtypes and the semantic reanalysis have proved the productivity of compound words generated by [X + body part] referring to things.

4 Cognitive Mechanisms of Chinese "Body-Part" Compounds

From the view of Embodied Cognition, humans obtain basic concepts through their interactions with the world. These concepts are formed by the embodied experiences of the body, perception, physical space, physical movement, etc. On the basis of the basic concepts, concepts that are more abstract and complex can be formed through cognitive operations like metaphor, metonymy, conceptual integration, etc. (Lakoff and Johnson 1980; Fauconnier and Turner 1998, 2002). "Body part" nouns are basic concepts, which means concepts that are more complex than them can be formed on top of them. The exocentric compounds formed by "body part" nouns we have discussed in this study have experienced cognitive operation of conceptual metonymy and conceptual integration.

Barcelona (2018) was committed to building a cross-linguistic metonymy database and establishing metonymy typology. The field in each metonymy entry is "hierarchical level" which classifies all metonymies into four levels, namely generic level, high level, basic level and low level. There are three types at the generic level which are PART FOR WHOLE, WHOLE FOR PART and PART FOR PART metonymies. There are more types at the basic level, for example, LOCATION FOR LOCATED, INSTRUMENT FOR ACTION, etc. Although the metonymy database is yet to be improved, the idea of classifying the metonymic levels is practical. Our study will analyze the metonymy operation of exocentric compounds at the generic level and the basic level.

4.1 Operations of Metonymy and Conceptual Integration of [X + Body Part] Compounds

Before discussing the cognitive operation of [X + body part] compounds, a question need to be raised: Does the metonymy operate on the level of the compound itself, or does it operate separately on the word base X and the body parts noun at first, and then integrate two target concepts? Our study agrees with the second hypothesis. Suppose the first one is the proper cognitive operation path, the word-forming motivation of the target compounds should be clear. In other words, the compounds should be transparent semantically before the metonymy operation. Yet "*lǎngǔtóu* (懒骨头)" and "*shǔbiāoshǒu* (鼠标手)" do not seem transparent semantically in fact. Therefore, we hold the opinion that metonymy operates separately on the word base X and the body part nouns.

Due to the restriction of semantic exocentricity, metonymy operates on the body part nouns for sure, while it only operates on some of the word bases.

4.1.1 Compounds of Metonymy Operation on Word Base

The relations that most pre-modifiers and head elements of this kind of compounds hold is cause-effect relation are functional relation. Such relation is mainly applied in compounds of two types: (i) compounds with word base of "instrument" and referring to "disease"; (ii) compounds referring to "people" with the semantic feature of a certain profession.

For the first type, taking "*shǔbiāoshǒu* (鼠标手)" as an example, "*shǒu* (手)" activates the basic-level metonymy *THING PERCEIVED FOR PERCEPTION* under the generic-level metonymy *PART FOR PART*, and then maps into the target domain of "disease/symptoms on hand". At the same time, base "*shǔbiāo* (鼠标)" activates the basic-level metonymy *INSTRUMENT FOR ACTION* under the generic-level metonymy *PART FOR WHOLE*, and maps the target domain of "actions that using mouse so often". At last, according to the Conceptual Integration Theory, the two target domains integrate via the cause-effect relation in the blending space, and thus the proper meaning is generated.

The second type is the compounds referring to person with the feature of a certain profession. These compounds have one feature in common, that is the second morpheme of them is "*shǒu* (手)", such as "*bèisīshǒu* (贝斯手)". The definitions of these compounds listed in a dictionary are provided in the same pattern (i.e., … refers to people who …) which is also used the explain other compounds referring people in general. Yet, what is different in the compounds related to profession is that the referents of the compounds have a feature of a certain profession. The pre-modifier shows the professional function of people. In "*bèisishǒu* (贝斯手)", "*shǒu* (手)" activates the basic-level metonymy *PART OF A FORM FOR THE FULL FORM* under the generic-level metonymy *PART FOR WHOLE*, and then maps into the target domain of "a certain kind of people". The base "*bèisi* (贝斯)" activates the basic-level metonymy *INSTRUMENT FOR ACTION* under the generic-level metonymy *PART FOR WHOLE*, and maps into the target domain of "actions of playing bass". Then, the two target domains integrate in the blending space via functional relation, and thus the proper meaning is generated.

4.1.2 Compounds of No Metonymy Operation on Word Base

The relation that most pre-modifiers and head elements of this kind of compounds shares is descriptive relation. Such relation is especially found in two types of compounds, namely the compounds referring to "person" and the compounds referring to "personal characteristics".

Taking "*lǎngǔtóu* (懒骨头)" as an example from the first type, "*gǔtóu* (骨头)" activates the basic-level metonymy *PART OF A FORM FOR THE FULL FORM* under the generic-level metonymy *PART FOR WHOLE*, then maps into the target domain of "a certain kind of people". Then, the metonymy operation on the word base is skipped (given there is none). Finally, it goes directly to the blending process to generate the proper meaning.

As for the second type, taking "*ruìyǎn* (锐眼)" as an example, "*yǎn*" activates the basic-level metonymy *ENTITY FOR CHARACTERISTIC* under the generic-level metonymy *PART FOR PART*, then maps into the target domain of "abilities that eyes possess". Then, it is blended with the base "*ruì* (锐)" and generates the proper meaning at last.

4.2 Salience in Metonymy Operations

Salience is one of the basic concepts of Perceptual Psychology. Things that are salient are easy to identify, process, and remember. The salience effect of metonymy is very common in daily life. As illustrated by Lakoff (1987: 89), a DC-10 aircraft was widely publicized by the media after the crash, so people did not take this type of aircraft, even though the safety record of this aircraft was better than other models. The DC-10 here has a higher degree of salience than other models in the "aircraft crash" event framework.

Shen (1999) noted that in metonymy operation, the whole often has higher degree of salience than the part, just as when one says, "I'm holding the hammer", actually s/he is only holding the handle of the hammer. Sometimes the part may have higher degree of salience than the whole, like the metonymy operation of exocentric compounds this study discusses. Most metonymies belong to *PART FOR WHOLE* or *PART FOR PART* at the generic level, which means the part has higher degree of salience. The reason for it is that the body parts which metonymy takes place are cognitively salient (Carmen 2015). When metonymy operates on body part nouns and blends the target zone with a certain word base to denote a certain type of person, personal characteristics or disease, the body parts are more salient than the whole body. For example, to denote a person with high professional skills, it is better to use *gāoshǒu* (高手) instead of *gāorén* (高人), for the *hand* in Chinese possessing the semantic feature of a certain profession, which makes it more salient than *rén* (人). This example demonstrates that the salience of body part nouns is justified.

Nevertheless, the salience that Carmen (2015) noted can only explain why and what body part nouns can activate semantic transfer via metonymy operations. He did not further explain whether the process of semantic transfer via metonymy is selective on different body part nouns. In other words, would different word base select different body parts nouns to form compounds of certain meaning? Can a certain type of semantic transfer only be activated by one certain body part noun? Our study agrees that the process is selective, and the salience represented by the body part nouns which can activate semantic transfer is determined by both the word base and the body part nouns. For example, to denote a lazy person, we say "*lǎngǔtóu* (懒骨头)" instead of "*lǎntóu* (懒头)" or "*lǎnnǎo* (懒脑)", because "*gǔtóu* (骨头)" is the support of the body, which is often associated with personality or integrity. There are also phrases like "*gǔqì* (骨气)" "*fēnggǔ* (风骨)" in Chinese and *lazy bone* in English. Therefore, when we evaluate a person, we often use "*gǔtóu* (骨头)" to denote, because at this time it is the cognitively salient part of the human body. As for other denotations, such as *liànàinǎo* (恋爱脑), which denotes people who regard love affairs as the priority in their life thus lose their self-value, i.e. a mode of thinking about love affairs. *Nǎo* (脑) "brain" is the cognitively salient part, because it is the only organ of consciousness in human body, responsible for mental activities such as prioritizing love affairs in one's life.

In terms of metonymy at the basic level, the most frequent target domain involves is the domain of "instrument". It covers compounds in different semantic categories, including compounds referring to people, e.g., "*bèisishǒu* (贝斯手)", compounds referring to diseases, e.g., "*shǔbiāoshǒu* (鼠标手)", and compounds referring to personal characteristics, e.g., "*píjiǔyǎn* (啤酒眼)". Instruments are the salient argument in the event frame. There are many arguments in an event frame, such as action, object, subject, instrument etc. Among them, arguments like subject or object can activate several possible action arguments, while instrument is the one which can activate the central argument, say, the action by default.

5 Conclusion

The present study has analyzed 95 eccentric semantic compound words formed by Chinese "body part" nouns and divided their semantic categories into two classes: "person" and "things". The "things" class can be further divided into "personal characteristics" and "diseases/symptoms".

The number of conventional compounds and unconventional new compounds shows the productivity of the morphological pattern [X + body part]. After the analysis of the semantic features of the compounds, it is found in every semantic category that there are compounds that are reanalyzed semantically. Among these compounds, two new subtypes of morphological pattern have emerged, namely [behavior + body part] and [instrument + body parts]. These two pieces of evidence show the productivity of [X + body part], and also the productivity of compounds with semantic exocentricity.

The cognitive mechanisms operated on the target compounds are conceptual metonymy and conceptual blending. More specifically, metonymies operate separately on the word base and the body part noun first. Then, two target concepts will be integrated in the blending space to yield the proper meaning. The salience of metonymy plays an important role in the cognitive process.

It is shown by many examples that exocentric compounds formed by body part nouns reflect some cross-linguistic similarities. Since the examples are limited, the present study has only mentioned some comparisons between Chinese and English. What about the productivity of these compounds in other languages? Are the cognitive mechanisms also the same in other languages? These questions are worth being discussed in future studies.

Appendix: 95 Exocentric Compounds Formed by [X + Body Part]

Body parts	Conventionalized compounds	New compounds
shǒu 手 'hand'	贝斯手, 弓箭手, 提琴手, 炮手, 火枪手, 键盘手1, 打手, 辩手, 乐手, 高手,副手, 舵手, 新手, 老手, 快手1, 国手, 钓手, 铁手, 枪手1, 骑手1, 强手, 拳手, 扒手, 名手, 写手	键盘手2, 快手2, 段子手, 游戏手, 手机手, 买手, 枪手2, 骑手2, 代购手, 发包手, 鼠标手
tóu头 'head'	滑头, 刺头, 冤头, 冤大头, 猎头, 空头, 戆头, 对头, 孱头, 无厘头	N/A
jiǎo 脚 'feet'	国脚	乌龙脚
nǎo 脑 'brain'	首脑	恋爱脑, 事业脑, 游戏脑, 城市脑, 全球脑, 咸鱼脑, 方脑壳
y*ǎn* 眼 'eye'	千里眼, 法眼, 青眼, 势利眼, 二五眼	四眼, 啤酒眼
ěr 耳 'ear'	顺风耳	空耳
ròu 肉 'flesh'	N/A	小鲜肉, 老鲜肉, 老腊肉
zuǐ 嘴 'mouth'	快嘴, 铁嘴, 名嘴	乌鸦嘴, 大嘴巴
bí 鼻 'nose'	酒渣(糟) 鼻	草莓鼻
li*ǎn* 脸 'face'	黑脸, 红脸, 花脸	异国脸, 大众脸, 生脸, 熟脸
zh*ǐ* 指 'finger'	N/A	乌龙指, 键盘指
zhǒu 肘 'elbow'	N/A	网球肘
xī 膝 'knee'	N/A	跑步膝
wàn 腕 'wrist'	N/A	大腕, 小腕
jiān 肩 'shoulder'	N/A	驾驶肩
xīnyàn 心眼	死心眼	N/A
gǔtóu 骨头 'bone'	贱骨头, 软骨头, 硬骨头, 懒骨头	N/A

References

Barcelona, A.: General description of the metonymy database in the Córdoba project, with particular attention to the issues of hierarchy, prototypicality, and taxonomic domain. In: Blanco-Carrión, O., Barcelona, A., Pannain, R. (eds.) Conceptual Metonymy, Methodological, Theoretical, and Descriptive Issues, pp. 26–54. John Benjamin Publishing Company, Amsterdam/Philadelphia (2018)

Bauer, L.: The typology of exocentric compounding. In: Scalise, S., Vogel, I. (eds.) Cross-disciplinary Issues in Compounding, pp. 167–176. John Benjamins Publishing Company, Amsterdam & Philadelphia (2010)

Bauer, L., Lieber, R., Plag, I.: The Oxford Reference Guide to English Morphology. OUP, Oxford (2013)

Benczes, R.: On the non-voability of the endocentric distinction: evidence from linguistic creativity. Explorations English Lang. Linguist. **1**(1), 3–18 (2013)

Bloomfield, Language. Holt, New York (1933)

Carmen, P.-M.: Are smartphone face and Googleheads a real or a fake phenomenon? In: Blanco-Carrión, O., Barcelona, A., Pannain, R. (eds.) Conceptual Metonymy, Methodological, Theoretical, and Descriptive Issues, pp. 261–286. John Benjamin Publishing Company, Amsterdam/Philadelphia (2015)

Dictionary Editing Office of the Institute of Linguistics, Chinese Academy of Social Sciences: Modern Chinese Dictionary (seventh edition). The Commercial Press, Beijing (2016)

Fauconnier, G., Turner, M.: Conceptual integration networks. Cogn. Sci. **22**(2), 133–187 (1998)

Fauconnier, G., Turner, M., Think, T.W.W.: Conceptual Blending and The Mind's Hidden Complexities. Basic Books, New York (1998)

Guevara, E., Scalise, S.: Searching for universals in compounding. In: Scalise, S., Magni, E., Bisetto, A. (eds.) Universals in Language today, pp. 101–128. Springer, Dordrecht (2009). https://doi.org/10.1007/978-1-4020-8825-4_6

Lakoff, G., Johnson, M.: Metaphors We Live By. Chicago University Press, Chicago (1980)

Shen, J.: A metonymic model of transferred designation of de-constructions in Mandarin. Contemporary Linguistics (2) (1997)

Shi, D.: Endocentric structures and exocentric structures. Foreign Lang. Teach. Res. (4) (2007)

Shi, C.: Web language as a language variety and a linguistic issue. Appl. Linguist. (3) (2010)

Shi, G.: Endocentric structures and exocentric structures in modern Chinese. Stud. Chin. Lang. (4) (1988)

Zhou, J.: New Chinese Words 2008. The Commercial Press, Beijing (2009)

Zhou, J.: New Chinese Words 2009. The Commercial Press, Beijing (2010)

Zhou, J.: New Chinese Words 2012. The Commercial Press, Beijing (2013)

Zhou, J.: New Chinese Words 2014. The Commercial Press, Beijing (2015)

Zhou, J.: New Chinese Words 2015. The Commercial Press, Beijing (2016)

Zhou, J.: New Chinese Words 2016. The Commercial Press, Beijing (2017)

Zhou, J.: New Chinese Words 2017. The Commercial Press, Beijing (2018)

Mandarin Children's Acquisition of Numeral Container Classifier Phrases

Xiaohe Yuan(✉)

College of Foreign Languages, Capital Normal University, Beijing, China

Abstract. This study aims to discover native Mandarin-speaking children's knowledge and acquisition of the numeral container classifier phrases in Mandarin Chinese. Such phrases are ambiguous and often have two readings. *san bei shui* 'three cups of water' does not only have a counting reading, referring to three cups, but also has a measure reading. Our measure *de* comprehension experiments showed that both adults and children tended to have a strong bias for counting: in the recipe context, the majority of adults would accept *san bei (de) shui* 'one cup of water' even if the cup was halfway filled with water; 3-to-5-year-olds would have a much stronger bias for counting, as they would accept it regardless of the little amount of water inside.

Keywords: Numeral container classifier phrases · Counting · Measure · Acquisition · Measure *de*

1 Introduction

Many scholars (Doetjes 1997; Partee and Borschev 2012; Rothstein 2009; 2010; 2017) have, in fact, noticed that container classifier phrases are ambiguous with either a counting or a measure interpretation. For instance, the container classifier phrase *three cups of water* does not only refer to three cups, which is a counting interpretation, but also has a measure interpretation. *There are three cups of water on the table* is counting the number of cups. On the other hand, *He poured three cups of water into the soup* is measuring the total volume of water.

The two functions, counting and measure, have their own distinct syntactical representations. Based on Rothstein (2009), counting is putting atomic entities in a one-to-one correspondence with the natural numbers while measuring is assigning a quantity a value on a scale. Li (2013) further delivered the two syntactical representations of Chinese container classifier phases with the two readings:

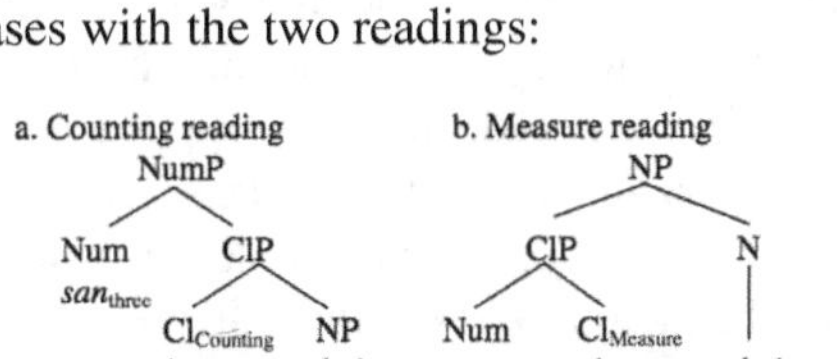

Chinese numeral container classifier phrases have either a counting or a measure interpretation. For instance, *san wan jiu* 'three bowls of wine' can refer to three concrete,

M. Liu et al. (Eds.): CLSW 2020, LNAI 12278, pp. 871–878, 2021.
https://doi.org/10.1007/978-3-030-81197-6_72

individual bowls of wine, or the amount of wine analogous to milliliters/ounces that equal three bowls. The former must have three concrete bowls of wine in reality while the latter does not have to have real bowls.

Li and Rothstein (2012) also pointed out that the particle *de* always induces a measure reading when it follows a classifier head. Unlike Cheng and Sybesma (1998), who argued that *de* only follows count classifiers, they discover that *de* appears in measure contexts, whether the preceding classifier is count or mass. (We shall support this with short examples. Based on Rothstein (2017) and (Li and Rothstein 2012; Li 2013), the Chinese numeral container classifier *de* phrase is a measure term and should induce a full size meaning.

2 Previous Studies on Acquisition

Previous studies have focused more on the acquisition of number words (Carey 2004; Gelman and Gallistel 1978; Wynn 1990; 1992) in contrast to measuring expressions. Some studies revealed that 2-to-6-year-olds are bad at distinguishing amounts, and yet they tend to favor counting individual pieces (Piaget and Szeminska 1952). There is actually little data pertaining to Mandarin-speaking children's acquisition and understanding of numeral container classifier phrases in terms of their counting and measure interpretations.

Only a few studies have explored child language acquisition and knowledge of container phrases in terms of counting and measuring contrast. Lima (2014) conducted an acquisition study on container phrases in Yudja and found that container phrases only indicated the location of a substance in a morphosyntactic aspect, while other factors like the size of the container, the amount of the substance in the container should not be connected with the container phrase itself. Lima (2016) then further took the context and verb categories into consideration. The result showed that both adults and children would more likely interpret container phrases in Yudja by counting the numbers. Lima and Snedeker (2015) also investigated the acquisition of container phrases in English from 37 English-speaking adults and 3–6-year-olds asking the same questions used in the above Yudja experiments. The results showed that 3–5-year-olds answered "yes" in both individuation and measure contexts while 6-year-olds behaved similarly as adults in distinguishing the individuating contexts from measure contexts, who only answered "yes" when the containers were completely full in the measure scenarios.

To summarize, previous studies have demonstrated that Children during the age 2 to 5 show weak abilities in acquiring container phrases when distinguishing counting and measure having a bias towards counting. It was not until the age of 6 that children could behave similarly as adults when distinguishing the two. We aim to test Mandarin children's understanding and knowledge of container classifiers to see whether Mandarin-speaking children could distinguish counting from measure interpretation generated by those phrases. We would like them to answer the following questions. Would the participants react the same towards Chinese numeral container classifier phrases with *de* (a forced measure interpretation) and those without *de* (the ambiguous reading)? Do those phrases with *de* induce a full size measurement reading? Would Chinese 3–5-year-olds also have a strong bias towards counting like those children in other languages?

3 Measure *De* Experiments 1&2 in Comprehension

Our experiments 1 & 2 aimed to test what participants' responses would be when a certain measure context (the recipe context) was given. We used the picture identification task and adopted the "Covered Box" paradigm used in Huang et al. (2013) to investigate the meaning of numbers. Before the real test began, the participants were told that they would see pictures of two cartoon boys—one dressed in red and another dressed in yellow. Each had some materials in front of them. One boy's materials would be completely visible while the materials of others would be partly hidden except the bottom. Participants would then be given a description of one of the characters. Their task was to decide which cartoon boy this description was about.

The participants would be told that there was only one correct answer from two choices, and they were required to choose the answer that better described the sentence. If they thought the description would fit the visible picture, they should select it. If not, they should select the covered picture. The visible pictures and covered ones were arranged in a randomized order. Before the real experiment began,there was a training helping participants to get to know what the recipe context was and children were asked to play a game making the tomato soup with toys based on the recipe.

3.1 Participants, Procedure and Materials

We have 20 adults (all over 18 years old) from Ocean University of China for each experiment, who were tested via qualtrics online survey platform (2 failed in experiment 3 and 5 failed in 4). We used 2 training items to have the participants familiar with the task. Altogether we have 6 controls with 3 empty ones/controls and 3 full ones/controls, 9 targets with 6 half ones/targets and 3 little ones, in addition to 8 fillers (See Appendix 1)

Take *san bei de guozhi* 'three cups of juice' as an example. The description went as follows: "Here are two boys. Both of them want to make a special dessert/dish, but only one boy is able to make such a dessert/dish because according to the recipe, *san bei de guozhi* 'three cups of juice' are needed and only one boy has three cups of juice. Tell me which boy it is." What followed were two pictures with one boy having three visible containers with different levels of liquid on one side, and one boy and three covered containers on the other side.

3.2 Adults' Result

We noticed that the recipe context did not help. For the half-filled targets, less than 25% selected the covered ones while more than 50% selected the covered ones instead of the targets with small amounts of liquid for the little targets. We see that for the recipe *de* context and recipe no *de* context, there is no group effect ($F = 0.12$ (1, 31), $p = .912$); See Fig. 1.

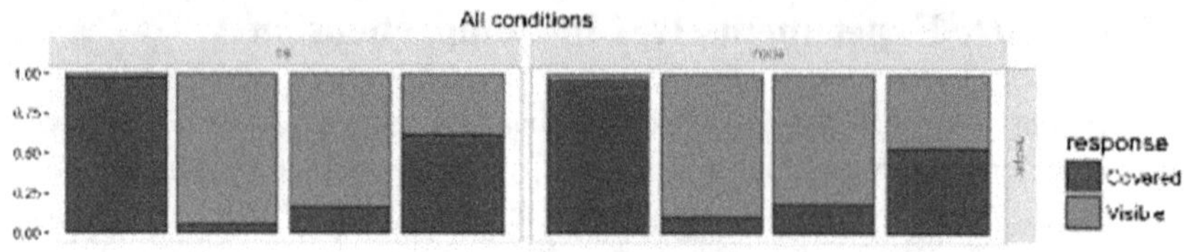

Fig. 1. Results

3.3 Children's Result

We tested 20 younger children (4;07–5;07, m = 4;10) and 18 elder children (6;00–7;06, m = 6;08) adopting the above recipe *de* experiment (4 younger children and 2 elder children failed in 5/6 of the 6 controls). The Mann-Whitney U test showed that younger children behaved similarly as adults in both controls and half-filled targets ($p = 0.129$, $z = 1.673$) but differed greatly from adults in targets with little amounts of liquid ($p = 0.004$, $z = 2.972$). On the contrary, the 6-year-olds behaved similarly to adults when the targets had little amount of liquid ($p = 0.401$, $z = 0.908$).

See Fig. 2.

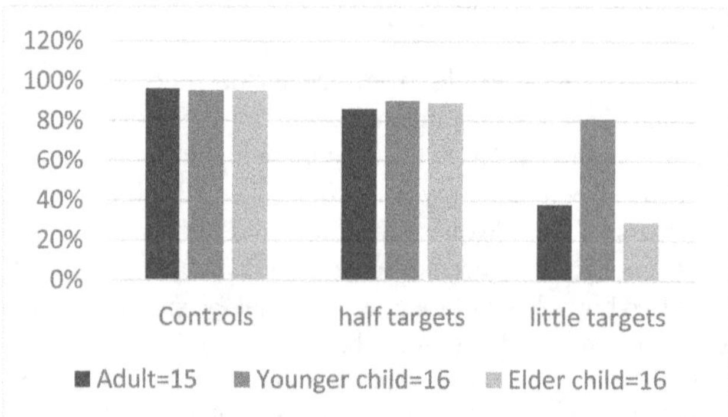

Fig. 2. Acceptance in Recipe *de* condition among three groups

4 Discussion

We altogether had two sub-experiments of the measure *de* comprehension experiment (*de* in recipe context; no *de* in recipe context). Our results can be summarized as follows:

Firstly, the experiments with *de* and without *de* did not have a big difference supporting Rothstein (2010, 2017) and Li (2013), who suggested that Chinese numeral container classifier phrases without *de* have both a counting reading and a measure reading.

Secondly, participants' performance the two experiments suggests that the [Num-Cl-*de*-N] structure does not necessarily entail a full-size interpretation, even in contexts biased towards this meaning (the "recipe context" in our experiment). It seems that the semantic threshold is half-full, rather than almost-full or completely-full, as the adults and the older children in the experiment tended to accept this structure in the completely-full condition, as well as the half-full condition, but not in the almost-empty condition.

Thirdly, the result tells us that the recipe context does not help, which means that participants prefer the counting reading of container classifier *de* phrases even in the strong measure context. In other words, people have a strong bias towards counting.

5 Conclusion

In this study, we get to know 3-to-6-year-olds' knowledge and acquisition of Chinese numeral container classifier phrases. Our experiments are in consistency with previous English studies, both of which have proved that it is not until 6 that children can better distinguish counting and measuring. It also seems to prove that both Chinese adults and children prefer counting instead of measuring or have a very strong bias for counting in comprehension, since adults and elder children totally accept halfway- filled cups as *san bei shui* 'three cups of water'.

Our experiments have further proved that half-full sizes of the contents inside are the minimum amount for Chinese adults' acceptance of the individuation reading, while 3-to-5-year-olds have a much stronger bias for counting and would accept containers with little amounts on the other hand. Thus, we are against Rothstein & Li's argument that Chinese container phrases with *de* would elicit the full size meaning. It seems as if container phrases in Mandarin Chinese share similar functions with those in Yudja. It is more likely that such phrases in Mandarin Chinese belong to the default dotted-type; namely, when speaking of numeral container classifier phrases, we tend to only refer to those containers having the contents of a given kind.

Acknowledgments. This research is funded by the 65th Chinese Postdoctoral Science Foundation (Project No: 2019M650955) and the Project of Dengfeng Plan of Chinese Academy of Social Sciences-Psycholinguistics, Language Acquisition and Development. For insightful comments and suggestions, the author greatly acknowledges Lyn Tieu and Stephen Crain.

Appendix 1: 4 Figures in Experiment 1 & 2

Here are two boys. They both want to make a special dessert, but only one of them will be able to. This is because the recipe requires *san guan (de) guozhi* '3 jars of juice', and only one boy has them. Tell me which boy it is.

Fig. 3. The empty control

Here are two boys. They both want to cook a special dish, but only one of them will be able to. This is because the recipe requires *san bei (de) cha* '3 mugs of tea', and only one boy has them. Tell me which boy it is.

Fig. 4. The full control

Here are two boys. They both want to bake a cake, but only one of them will be able to. This is because the recipe requires *san bei (de) guozhi* '3 glasses of juice', and only one boy has them. Tell me which boy it is.

Fig. 5. The half-full target

Here are two boys. They both want to make a special dish, but only one of them will be able to. This is because the recipe requires *san ping (de) tang* '3 jars of soup', and only one has them. Tell me which boy it is.

Fig. 6. The little amount target

Appendix 2: Contexts in Experiments 1 & 2

Three empty controls

1. Here are two boys. They both want to make a special dish, but only one of them will be able to. This is because the recipe requires *san bei (de) cha* '3 glasses of tea', and only one boy has them. Tell me which boy it is.
2. Here are two boys. They both want to make a special dessert, but only one of them will be able to. This is because the recipe requires *san guan (de) guozhi* '3 jars of juice', and only one boy has them. Tell me which boy it is.

3. Here are two boys. They both want to make a special dessert, but only one of them will be able to. This is because the recipe requires *san ping (de) niunai* '3 jars of milk', and only one boy has them. Tell me which boy it is.

Three full controls

1. Here are two boys. They both want to cook a special dish, but only one of them will be able to. This is because the recipe requires *san bei (de) cha* '3 mugs of tea', and only one boy has them. Tell me which boy it is.
2. Here are two boys. They both want to cook a special dish, but only one of them will be able to. This is because the recipe requires *san tong (de) shui* '3 jugs of water', and only one boy has them. Tell me which boy it is.
3. Here are two boys. They both want to cook a special dessert, but only one of them will be able to. This is because the recipe requires *san ping (de) niunai* '3 bottles of milk', and only one boy has them. Tell me which boy it is.

Six half-full targets

1. Here are two boys. They both want to bake a cake, but only one of them will be able to. This is because the recipe requires *san bei (de) guozhi* '3 glasses of juice', and only one boy has them. Tell me which boy it is.
2. Here are two boys. They both want to bake a cake, but only one of them will be able to. This is because the recipe requires *san bei (de) niunai* '3 glasses of milk', and only one boy has them. Tell me which boy it is.
3. Here are two boys. They both want to cook a special dish, but only one of them will be able to. This is because the recipe requires *san hu (de) shui* '3 pitchers of water', and only one boy has them. Tell me which boy it is.
4. Here are two boys. They both want to cook a special dish, but only one of them will be able to. This is because the recipe requires *san hu (de) cha* '3 pots of tea', and only one boy has them. Tell me which boy it is.
5. Here are two boys. They both want to cook a special dessert, but only one of them will be able to. This is because the recipe requires *san ping (de) suannai* '3 bottles of yogurt', and only one boy has them. Tell me which boy it is.
6. Here are two boys. They both want to cook a special dish, but only one of them will be able to. This is because the recipe requires *san wan (de) tang* '3 bowls of soup', and only one boy has them. Tell me which boy it is.

Three little amount targets

1. Here are two boys. They both want to make a special dish, but only one of them will be able to. This is because the recipe requires *san ping (de) tang* '3 jars of soup', and only one has them. Tell me which boy it is.
2. Here are two boys. They both want to make a special dessert, but only one of them will be able to. This is because the recipe requires *san guan (de) suannai* '3 jars of yogurt', and only one boy has them. Tell me which boy it is.

3. Here are two boys. They both want to make a special dessert, but only one of them will be able to. This is because the recipe requires *san bei (de) guozhi* '3 glasses of juice', and only one boy has them. Tell me which boy it is.

References

Carey, S.: Bootstrapping and the origin of concepts. Daedalus **133**, 59–68 (2004)

Cheng, L.L.S., Sybesma, R.: Yi-wantang, yi-ge tang: classifiers and massifiers. Tsing-Hua J. Chin. Stud. New Ser. XXVIII **3** 385–412 (1998)

Doetjes, J.: Quantifiers and Selection. Ph.D. dissertation, Leiden University (1997)

Gelman, R., Gallistel, C.R.: The Child's Understanding of Number. Harvard University Press, Cambridge (1978)

Huang, Y., Spelke, E., Snedeker, J.: What exactly do numbers mean? Lang. Learn. Dev. **9**, 105–129 (2013)

Li, X., Rothstein, S.: Measure readings of mandarin classifier phrases and the particle *de*. Lang. Linguist. **13** 693–741 (2012)

Li, X.: Numeral Classifiers in Chinese: The Syntax-Semantics Interface. Mouton de Gruyter, Berlin/New York (2013)

Lima, S.: The grammar of individuation and counting. Ph.D. dissertation, University of Massachusetts (2014)

Lima, S., Snedeker, J.: On the acquisition and interpretation of container phrases in English. In: 28th Annual CUNY Conference on Human Sentence Processing (2015)

Lima, S.: Container constructions in Yudja: locatives, individuation and measure. Baltic Int. Yearbook Cognit. Logic Commun. **11**, 1 (2016)

Partee, B.H., Borschev, V.: Sortal, relational, and functional interpretations of nouns and Russian container constructions. J. Semant. **29**, 445–486 (2012)

Piaget, J., Szeminska, A.: The Child's Conception of Number. Routledge & Kegan Paul, London (1952)

Wynn, K.: Children's understanding of counting. Cognition **36**, 155–193 (1990)

Wynn, K.: Children's acquisition of the number words and the counting system. Cogn. Psychol. **24**, 220–251 (1992)

Rothstein, S.: Individuating and measure readings of classifier constructions: evidence from modern Hebrew. Brill Ann. Afroasiatic Lang. Linguist. **1**, 106–145 (2009)

Rothstein, S.: Counting and the mass-count distinction. J. Semant. **27**, 343–397 (2010)

Rothstein, S.: Semantics for Counting and Measuring. Cambridge University Press, Cambridge (2017)

Effects of Lexical Spatial-Temporal Metaphors on Mandarin and Cantonese Speakers' Temporal Conceptualizations

Pan Zhu[1] and Yu-Yin Hsu[2(✉)]

[1] Faculty of Humanities, The Hong Kong Polytechnic University, Kowloon, Hong Kong SAR, China
xiaopan.zhu@connect.polyu.hk

[2] Department of Chinese and Bilingual Studies, The Hong Kong Polytechnic University, Kowloon, Hong Kong SAR, China
yyhsu@polyu.edu.hk

Abstract. This study investigated whether and how different types of temporal metaphors in the lexicon and individual reading habits influence native Chinese speakers' conceptualizations of time. The results indicate that the Cantonese- and Mandarin-speaking participants constructed time expressions differently to some extent. Both groups responded faster on the transverse axis than on the vertical one, which was in accordance with the reading habits produced by the major writing/printing directions in both Mandarin and Cantonese. However, the Cantonese participants made judgments significantly faster than the Mandarin participants did in non-canonical conditions on the vertical axis. This finding, though surprising, is in line with a finding of our linguistic survey that Cantonese speakers use linguistic terms on the vertical axis to express time concepts much more often than Mandarin speakers do. This suggests that, even in the case of Chinese languages, speakers' space and time associations can to some extend be influenced by the use of different temporal metaphors in their lexicon.

Keywords: Temporal compound · Spatial-temporal metaphor · Cantonese · Mandarin · Temporal conceptualization

1 Introduction

People's conceptualization of time has been of great interest in language studies. How do people conceptualize time and express temporal concepts? And how do their language systems and cultures influence those processes [1–5]? Two potentially important factors are spatial-temporal metaphorical expressions adopted in a language system, and reading-writing habits of individual speakers of that language. Accordingly, this paper reports the results of two studies, a survey, and an implicit-task experiment, which were aimed at ascertaining how native speakers of Beijing Mandarin and Hong Kong Cantonese express concepts related to time.

M. Liu et al. (Eds.): CLSW 2020, LNAI 12278, pp. 879–889, 2021.
https://doi.org/10.1007/978-3-030-81197-6_73

Some studies have reported that people who often use vertical metaphors to express time prefer to think about time vertically, whereas those who tend to use horizontal expressions tend to think about time horizontally [1, 6–9]. However, the results of some other studies that used the same type of spatial-priming paradigm did not support this view [6]. Sometimes, Chinese morphemes can exhibit divergence in the representation of spatial axes. Yet, different temporal terms were used in different studies. For instance, some used only *yue* 'month' [10]; some used *xingqi* 'week', *yue* 'month', and *jidu* 'quarter' [6]; some used *tian* 'day' [11]. Moreover, these studies often indulged in direct comparisons of fundamentally dissimilar temporal expressions, for example, words in phrasal expressions of time in English (such as 'ahead' in 'the good times ahead of us') and morphemes in temporal compounds in Mandarin (such as *qian* 'front' in *qian-nian* 'last year'). While speakers of different languages are reasonably assumed to use different linguistic devices in their temporal expressions, it remains unclear what inferences can be directly drawn from such comparisons, given that phrasal processing and lexical processing are usually considered to be distinct, and that the effects of L1-L2 transfer may not be easily identified. Thus, ideally, the cognitive association between linguistic forms and temporal conceptualization may be more effectively observed based on comparisons of languages with discrepancies in their temporal expressions at the same linguistic level. Chinese languages provide us with such an ideal environment of comparisons. In this study, we investigated Hong Kong Cantonese and Beijing Mandarin, which are culturally closely related but linguistically distinct.

It is well-known that Cantonese differs sharply from Mandarin on some linguistic levels, such as pronunciation and the lexicon, as well as in temporal expressions. For example, the concept of 'last year' is expressed in Hong Kong Cantonese as *soeng-nin* 'up-year', and in Beijing Mandarin as *qu-nian* 'backward-year'; while the concept of 'next year' in Cantonese is *haa-nin* 'down-year', and in Mandarin is *ming-nian* 'tomorrow-year'. Crucially, these examples show that Mandarin and Cantonese use morphemes that express categorically different spatial metaphors to form compounds with the same temporal meaning. To the best of our knowledge, however, no prior study has focused on such differences in temporal metaphors across two or more Chinese languages. Besides the collocations with 'year' noted above, other differences between Cantonese and Mandarin temporal expressions exist, and it is worth asking whether and how they affect the construction of temporal concepts by speakers of these two languages.

Individuals' reading and writing habits, meanwhile, have been considered a major factor in temporal conceptualization on the transverse axis (e.g., left/right in English). As opposed to sagittal (backward/forward) and vertical spatial terms, transverse spatial terms are extremely uncommon in temporal expressions cross-linguistically. However, some studies have reported that people's mental representations of time are consistent with the writing direction of their native languages [12–15]. These studies compared pairs of languages with opposite writing directions, including English/Arabic and Spanish/Hebrew. This factor is also potentially relevant to the current study, insofar as (apart from some recent official documents) Hong Kong Chinese tends to be printed vertically – whether on shop signs, in newspapers, or in books – and the lines are read transversely from right to left; whereas most Mandarin speakers only read horizontal and transverse left-to-right texts. Thus, if human constructs such as writing and reading directions can influence readers' mental representations of time, one can reasonably expect that such representations will differ across Cantonese and Mandarin speakers.

In the following section, we report the results from a survey of temporal expressions in Hong Kong Cantonese and Beijing Mandarin, focusing on the lexical similarities and differences on the use of the sagittal and vertical axes in temporal compounds. Then, in Sect. 3, we report the results from an implicit-processing experiment, examining how two groups of speakers associate time concepts with the transverse and vertical axes. In Sect. 4, we briefly conclude our study.

2 Experiment 1: Survey on Spatial Collocations

To facilitate our comparison of the use of temporal compounds which consist of sagittal vs. vertical terms for various sizes of time units, we collected the spatial collocations in temporal compounds in Hong Kong Cantonese and Beijing Mandarin through a questionnaire designed by the research team. This approach was adopted for the following reasons. First, not many Cantonese corpora are available, and the two existing corpora – *the Hong Kong Cantonese adult language corpus* of spontaneous speech recordings [16] and *the Hong Kong Cantonese corpus* of conversations [17] –contain just four temporal expressions (26 tokens in total) that are of interest to the present study. This number is so small that we needed to conduct our own language survey as a supplement. In the case of Beijing Mandarin, although there are some temporal expressions with an extremely high frequency in the corpora created by the Center for Chinese Linguistics at PKU, they are used rarely in spoken Mandarin and are mostly found in journalism. Therefore, we used one parallel questionnaire for Cantonese and Mandarin, which were more likely to capture the current colloquial uses of temporal expressions that would be best suited to our research aims.

2.1 Participants

All participants were asked to fill out a questionnaire about their language background prior to receiving the experiment invitation. We used a self-reported 7-point Likert Scale (1 = "do not know the language", and 7 = "native fluency") for the participants to rate their language fluency in Mandarin and Cantonese. There were 90 Beijing Mandarin speakers (58 females and 32 males) aged 18–30 and living in Beijing; and none had a good command of Cantonese (average fluency rating of Cantonese: 1.08). In another group, 50 Hong Kong Cantonese speakers (28 females and 22 males) aged 18–30 and living in Hong Kong were invited; none of them had learnt Mandarin before the age of six or had received formal Mandarin education (average fluency rating of Mandarin: 4.32); and the language that they feel most natural and comfortable to use is self-reported as only Cantonese.

2.2 Survey Materials and Procedure

Across Chinese languages, the concepts of 'month' and 'week' are often expressed by vertical spatial metaphors such as *shang* 'up' and *xia* 'down', whereas the concept of 'day' is more commonly associated with *qian* 'front' and *hou* 'back'. To avoid system-internal bias being introduced through the stimuli, our questionnaire explored the differences in core/common space-time metaphors used by speakers of Beijing Mandarin and Hong Kong Cantonese involving the four spatial-metaphorical terms (i.e., *qian* 'front', *hou* 'back', *shang* 'up', and *xia* 'down') and 11 time units: *shiji* 'century', *nian* 'year', *jidu* 'quarter', *yue* 'month', *xingqi* 'week', *zhou* 'week', *tian* 'day', *ri* 'day', *xiaoshi* 'hour', *fenzhong* 'minute', and *miao* 'second'. For each time unit, six spatial-metaphoric combinations were created by adding a measure word 'ge', resulting in 66 sets of target items in the survey. For example, a time unit like *shiji* 'century' had six combinations with each of the spatial-metaphorical terms. The six combinations formed by *shiji* with *qian* 'front' are shown in (1). Six similar expressions were created for *shiji* and each of the other three spatial-metaphorical terms, and we created items for the other 10 time units listed above in the same manner.

(1) *shiji* 'century' with the spatial-metaphorical term *qian* 'front':

a. *qian* 'front' + *shiji*
b. *qian* 'front' + *ge* 'MW' + *shiji*
c. *qian* 'front' + *ban* 'half' + *shiji*
d. *qian* 'front' + *yi* 'one' + *shiji*
e. *qian* 'front' + *ban.ge* 'half.MW' + *shiji*
f. *qian* 'front' + *yi.ge* 'one.MW' + *shiji*

All participants were given the task of judging which of the spatial-metaphorical terms was appropriate to each of the six combinations mentioned above in (1). Participants were allowed to accept more than one combination for each time unit. They provided their answers on a hardcopy printout of the survey form.

2.3 Survey Analysis and Results

The questionnaires completed by 21 Mandarin-speaking and 18 Cantonese-speaking subjects were deemed invalid due to incompleteness or the same response being given to all questions. 69 Mandarin and 32 Cantonese results were valid for analysis.

Among the 66 sets of items, 12 ungrammatical sets were included as control items. These were *ge nian / yi ge nian* 'one year', *ban ge nian* 'half year', *ge tian / yi ge tian* 'one day', *ban ge tian* 'half day', *ge fen zhong / yi ge fen zhong* 'one minute', *ban ge fen zhong* 'half minute', *ge miao / yi ge miao* 'one second', and *ban ge miao* 'half a quarter'. The participants' acceptability ratings for these items were extremely low, suggesting that they were concentrating adequately when making their acceptability judgments.

The ratings assigned to these 12 sets were excluded from further data analysis. Although the combinations of "spatial metaphors + *xiao shi* 'hour' / *fen zhong* 'minute' / *miao* 'second'" are not considered grammatical in Mandarin, these three sets were analyzed because more than 50% of the Cantonese participants rated them as grammatical. Among the remaining sets, those containing *tian* 'day' and *ri* 'day' were treated as one category for analytical purposes; and the same was done for *xingqi* 'week' and *zhou* 'week'. Finally, a total of 42 sets were analyzed.

We began by calculating the number and proportion of participants who selected sagittal metaphors and vertical metaphors in each group. We took preference for vertical metaphors as the baseline, and obtained a value for it by subtracting the percentage of sagittal metaphors from that of the vertical metaphors selected by the participants on each item. Interestingly, both groups demonstrated a strong preference for vertical metaphors when it came to longer time units (ranging from *xingqi* 'week' to *shiji* 'century'), as shown in Fig. 1, except *nian* 'year' for which the Mandarin group preferred sagittal metaphors and the Cantonese group had no clear preference. This discrepancy could be related to lexical differences between the groups. That is, Mandarin only uses *qian* 'front' and *hou* 'back' to form compounds with *nian* 'year' and *tian* 'day' directly. Likewise, Cantonese *cin* 'front' and *hau* 'back' can form compounds with *nin* 'year' and *tin* 'day' (e.g., *cin nin* 'last year'). Unlike Mandarin, Cantonese speakers also use very commonly the vertical metaphors *soeng* 'up' and *haa* 'down' when forming compounds with *nin* 'year' (i.e., *soeng nin* 'last year'). As for the shorter time units (ranging from *miao* 'second' to *xiaoshi* 'hour'), as shown in Fig. 1, the Cantonese participants tended to prefer vertical metaphorical expressions, while the Mandarin counterparts exhibited no clear preference. The only exception to this among time units shorter than one week was *tian* 'day', for which both groups preferred sagittal metaphors.

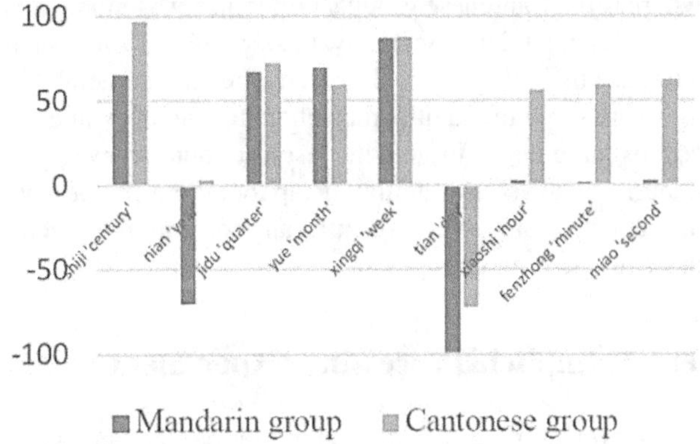

Fig. 1. The overall preference of Mandarin and Cantonese groups for vertical metaphors on time units (x-axis); the y-axis presents the percentage.

We further analyzed the similarities and differences between the two participant groups in their preferences for temporal expressions by items, and found some interesting mixture of temporal expressions. Among the 42 sets of survey items (Table 1), the 20 sets of temporal expressions for which the Mandarin group had a sagittal preference contained all the 8 sets for which the Cantonese group also had a sagittal preference, as well as all the 9 sets for which the Cantonese group expressed unclear preferences. The 8 sets for which both groups had a clear sagittal preference were (7 of the terms in which 'half' modified a shorter time period): *fenzhong* 'minute', *xiaoshi* 'hour', *xingqi* 'week', and *yue* 'month', and *(yi) tian* '(one) day.' Conversely, the 25 sets of temporal expressions for which the Cantonese group had a clear vertical preference contained all the 16 sets for which the Mandarin group had a vertical preference, and all the 6 sets for which the Mandarin group had unclear preferences. The 16 sets for which both groups had a clear vertical preference were the terms of a full longer time period (i.e., *shiji* 'centry', *jidu* 'quarter', *yue* 'month', *xinqi* 'week') and *ge xiaoshi* 'one hour'.

Table 1. Spatial-metaphor preferences, by group.

	Sagittal preference	Vertical preference	Unclear preference
Mandarin	20	16	6
Cantonese	8	25	9

We examined the differences between the two language groups within each type of preferences. We found that the choices of Mandarin speakers seem to be more linguistically compositional, while that of Cantonese speakers' were based more on the temporal concepts expressed by the terms. Firstly, when the concept 'half' was used to modify a temporal term, the Mandarin group preferred sagittal metaphors for most of the terms, even for that of longer time periods (e.g., *shiji* 'centry', *jidu* 'quarter', and *xinqi* 'week'), whereas the Cantonese group's preference was more vague but they still tended to prefer vertical metaphors. Secondly, the term *ge xiaoshi* 'one hour' was rated with vertical preference by both groups. However, when the numeral *yi* 'one' was used, the Mandarin group preferred the sagittal metaphors for 'hour', while this did not affect the Cantonese groups' judgments. In sum, the results of our survey suggest that, of these two groups, Mandarin speakers use sagittal metaphors more frequently and Cantonese speakers use vertical metaphors more frequently, and some more influences of linguistic forms were found with the Mandarin group.

3 Experiment 2: Implicit-Processing Experiment

The results of our survey showed that the Cantonese participants were more inclined to use vertical terms to express time, so we adopted an implicit-processing experiment – the key-pressing-based judgment tasks devised by Fuhrman [4]– as a further test of possible differences between the two groups' temporal conceptualizations.

The participants were instructed to press the keys on a nine-button USB keypad, which was connected to a computer and attached to a tripod mount with a rotating ball head. This enabled the keypad to be placed perpendicular to the table, facing to the participant. The middle key was masked with a blue sticker. In the canonical condition, the left key of the centre row was masked with a black sticker and labelled "earlier", and the right key was masked with a pale grey sticker and labelled "later". In the non-canonical condition, the right key was labelled "earlier" and was masked with a black sticker, and the left key meant "later" and was masked with a pale grey one. The remaining buttons were not labelled. The keys were not labelled by using verbal forms so that the participants would not be disturbed by the linguistic expressions. A picture presenting a middle time point (e.g., an apple being eaten: Fig. 2) was shown on the screen, followed by a picture showing either an earlier time point (e.g., a whole apple) or a later time point (e.g., an apple core). The participants needed to press the key corresponding to their judgments regarding whether the second picture shows a conceptual time point earlier or later than the one expressed by the first picture.

In theory, people who automatically access the canonical key mapping should respond faster when the keys are in a canonical condition; and their performance will be slower or otherwise hindered when the keys are in a non-canonical condition. Based on our survey results, we expected that Cantonese speakers would make judgments faster than Mandarin speakers in the vertical timeline, and that members of the former group would respond faster than the latter group when the keypad was presented in a top-earlier/bottom-later format. Also, because right-to-left reading order is common in printed materials in Hong Kong, we expected that on the transverse axis Cantonese speakers would make judgments faster when given the right-earlier/left-later keypad format than Mandarin speakers, who normally read from left to right.

Fig. 2. An example set.

3.1 Experiment Participants

The participants were asked to fill out a questionnaire covering their language background and reading habits prior to receiving the invitation to the experiment. We used a 7-point Likert Scale to rate how often the speakers of the two groups read vertical text, with 1 = "never" and 7 = "all the time". For the Mandarin group, only those who selected 1 or 2 were invited; for the Cantonese group, only those who selected 6 or 7 were invited. The participants in the Mandarin experiment included 30 native speakers of Beijing Mandarin, who were aged 18–30 and university students in Beijing; none of them had participated in our survey; none had a good command of Cantonese (the mean of their self-reported fluency in Cantonese was 1.03, using a 7-point scale: 1 = "do not know the language" and 7 = "native"); and all read vertical texts rarely. The Cantonese experiment included 30 native speakers of Hong Kong Cantonese (the mean of their self-reported fluency in Mandarin was 4.3, using a 7-point scale: 1 = "do not know the language" and 7 = "native"), who were aged 18–30 and university students in Hong Kong; none of them had participated in our survey; none of them had learnt Mandarin before the age of six; and all of them did not have language-understanding and expression barriers.

3.2 Experimental Materials and Procedure

The materials included 40 sets of pictures created by the research team, all illustrating temporal progressions similar to the set shown in Fig. 2. The participants were tested individually in a quiet lab with a Thinkpad laptop computer and keypad. Taking the transverse axis as an example, at the beginning of each trial, the participant was instructed via a message on the laptop screen to press the blue button to start the trial. Then, this message was replaced by a picture presenting a middle time point for 2000 ms; and that picture, in turn, was replaced by another one showing either an earlier or a later time point. The participant was instructed to press the key corresponding to his/her judgment of whether the second picture showed a time-point earlier or later than the first one. The second picture remained on the screen until the participant made this judgment or – if s/he did not press any key – for 4000 ms. All participants completed four test blocks, each consisting of 40 trials, with two blocks covering the transverse axis (left-earlier/right-later key mapping and its opposite) and two covering the vertical axis (top-earlier/bottom-later key mapping and its opposite). The order of the four blocks was counterbalanced with each item appearing once in each axis. Before each test block, the participants were instructed to perform 10 practice trials to familiarize themselves with the task. The items used in the practice trials were not repeated in the subsequent test blocks.

3.3 Data Analysis and Results

The accuracy of the 60 participants' responses was 99.96%. Linear mixed effects models were created using log-transformed reaction times (log RT) with the lme4 package (Bates et al., [18]) in R (R Team, [19]). The models first included random intercepts for an item and a subject, and then three fixed effects – i.e., Language, SpatialAxis (Vertical vs. Transverse), and Canonicality – and their interactions were then added (powers = .99). Language refers to Mandarin or Cantonese. Canonicality refers to the conditions in which a left or top key was designated as "earlier" and a right or bottom key as "later", while in non-canonical conditions, a left or top key was designated as "later" and a right or bottom key as "earlier". The significance of the main effect was evaluated by likelihood-ratio testing, and post-hoc Tukey's comparisons were conducted using the multcomp package (Hothorn et al., [20]) in R.

The results revealed that SpatialAxis with vertical conditions seemed to be processed more slowly than with transverse conditions, but this effect was not statistically significant (B = .022, SE = .012, p = .078). Canonical key-mapping was processed faster than its non-canonical counterpart (B = .023, SE = .012, p = .064), an effect that was marginally significant. Our analysis revealed significant interaction effects of SpatialAxis, Canonicality, and Language. While both Cantonese and Mandarin speakers processed transverse-axis items with canonical key-mapping significantly faster than items on the same axis with non-canonical key-mapping (ps < .001), differences were found with respect to vertical-axis conditions. Post hoc tests verified that Mandarin speakers processed vertical canonical conditions faster than non-canonical ones, as expected (although such differences were not statistically significant, p = .5), but surprisingly, Cantonese speakers processed non-canonical vertical mapping conditions significantly faster than canonical ones (p = .016).

We also observed significant differences between the Mandarin and Cantonese groups' processing of items with non-canonical key-mapping on the vertical axis (p = .042), with the Cantonese group being faster. No other inter-group comparisons revealed significant differences in reaction time (ps > .168). Analysis of within-group SpatialAxis differences showed that speakers of both languages processed transverse-axis items in the canonical condition significantly faster than vertical-axis items in the same condition (*ps* < .001). While the same SpatialAxis tendency was found within each language group in the non-canonical condition, those differences were not significant (Cantonese, $p = 1$; Mandarin, $p = .57$). Figure 3 illustrates the overall results.

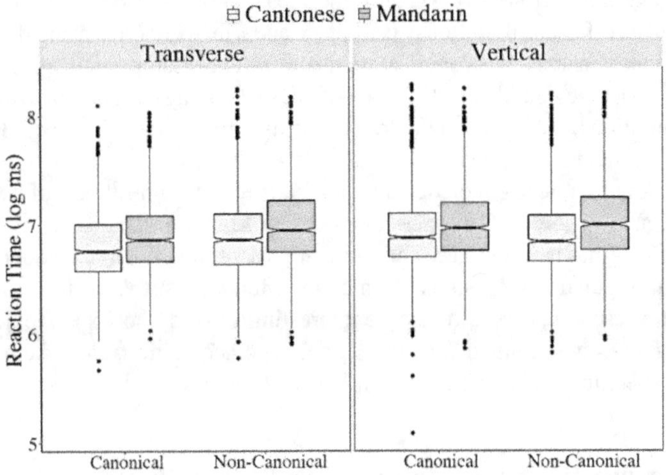

Fig. 3. Boxplots of log-transformed reaction times, by Language, Canonicality, SpatialAxis.

4 Discussion and Conclusion

The current study has resulted in interesting findings about the relation between speakers' conceptualization of time and spatial dimensions in temporal compounds in Hong Kong Cantonese and Beijing Mandarin and in individuals' reading and writing habits.

First, we found some language-specific tendencies. Concerning the vertical and the sagittal axes, both groups demonstrated strong vertical preferences for longer time units (such as *xingqi* 'week' and *shiji* 'century'), and sagittal preferences for shorter time units (such as *miao* 'second' and *xiaoshi* 'hour'). However, while the Cantonese group tended to prefer vertical-spatial metaphorical expressions, the Mandarin group's judgments could be influenced by the linguistic forms of time.

Concerning the time association on the transverse and the vertical axes, the Cantonese-speaking participants processed items on the vertical axis in the condition of non-canonical key-mapping significantly faster than those on the same axis in canonical key-mapping. Mandarin speakers, on the other hand, tended to disregard whether key-mapping was canonical or not, but tended to process items on the transverse axis faster than those on the vertical axis. We speculate that these differences between the groups were related to the fact (as reported in our survey) that Cantonese speakers are more likely to use vertical terms to express time concepts than Mandarin speakers do. Therefore, the Cantonese speakers in our experiment exhibited more flexibility than their

Mandarin-speaking counterparts when it came to expressing time concepts on a vertical axis, and they reflected shorter reaction times for the task.

Second, if the reading habit created by texts' printing orientation is indeed one of the main factors influencing space and time associations, then one would reasonably expect that individuals will process an item faster if it is in accordance with their own language's major writing and printing conventions. Our results support this idea. Both Beijing Mandarin and Hong Kong Cantonese speakers often read texts in the transverse axis (though in different directions: Beijing Mandarin words are printed from left to right horizontally, and Hong Kong Cantonese lines are often printed from right to left vertically). Also, we found that for both of these language groups and in both key-mapping conditions, the items on the transverse axis were mostly processed faster than those on the vertical axis.

We therefore incline to conclude that our results reflect some effects of lexical differences between Hong Kong Cantonese and Beijing Mandarin on speakers' conceptualization of time – that is, Cantonese speakers employ relatively more frequently vertical metaphors in linguistic expressions of time than Mandarin speakers do.

For future research, it would also be interesting to study how such lexical-internal spatial differences may influence learners' language acquisition, and whether and how training of specific spatial metaphorical association has effects on L2 word/phrase acquisition.

Acknowledgments. This study was supported by the Faculty of Humanity in Hong Kong Polytechnic University and the HK PolyU-PKU Research Centre on Chinese Linguistics (RP2U2). We thank Stephen Politzer-Ahles and Chu-Ren Huang for their comments and suggestions on the earlier stages of this study. We also thank Xia Wang for the technical support. Remaining mistakes are exclusively our own.

References

1. Boroditsky, L.: Does language shape thought? English and Mandarin speakers' conceptions of time. Cogn. Psychol. **43**(1), 1–22 (2001)
2. Boroditsky, L., Fuhrman, O., McCormick, K.: Do English and Mandarin speakers think about time differently? Cognition **118**, 123–129 (2010)
3. Lai, V.T., Boroditsky, L.: The immediate and chronic influence of spatio-temporal metaphors on the mental representations of time in English, Mandarin, and Mandarin-English speakers. Front. Psychol. **4**, 142–151 (2013)
4. Fuhrman, O., et al.: How linguistic and cultural forces shape conceptions of time: English and Mandarin time in 3D. Cogn. Sci. **35**, 1305–1328 (2011)
5. Liu, L., Zhang, J.: A study about the psychological mechanism of space-time metaphor. Exploration of Psychology, pp. 332–336 (2009)
6. Chen, J.Y.: Do Chinese and English speakers think about time differently? failure of replicating Boroditsky (2001). Cognition **104**(2), 427–436 (2007)
7. Lan, C.: A cognitive approach to spatial metaphors in English and Chinese. Doctoral The-sis of Hong Kong Polytechnic University (1999)
8. Scott, A.: The vertical dimension and time in Mandarin. Australian J. Ling. **9**, 295–314 (1989)
9. Tai, J.H.Y.: Translation by Ye Feisheng, Meager opinion on Chinese functional grammar based on cognition (2nd half). Linguistics Abroad **1**, 25–33 (1991)

10. Wu, N., Xu, N., Zhang, Y.: The facilitation of spatial schema processing on directional temporal expression comprehension. Psychol. Sci. **4**, 853–856 (2007)
11. Han, C.: Mandarin speakers' preferences in their spatial representation of time (Dissertation). Beijing Foreign Studies University (2014)
12. Chen, J.Y., Friedrich, M., Shu, H.: The effect of immediate and lifetime experience of reading horizontal and vertical texts on Chinese speakers' temporal orientation. J. Cogn. Culture, **15**, 1–12 (2015)
13. Fuhrman, O., Boroditsky, L.: Cross-cultural differences in mental representations of time: evidence from an implicit nonlinguistic task. Cogn. Sci. **34**, 1430–1451 (2010)
14. Tversky, B., Kugelmass, S., Winter, A.: Crosscultural and developmental-trends in graphic productions. Cogn. Psychol. **23**, 515–557 (1991)
15. Ye, H.: Embodied cognition: a new approach in cognitive psychology. Adv. Psychol. Science **5**, 705–710 (2010)
16. Leung, M.T., Law, S.P.: HKCAC: The Hong Kong Cantonese adult language corpus. Int. J. Corpus Linguistics **6**(2), 305–325 (2001)
17. Luke, K.K., Wong, M.L.: The Hong Kong Cantonese corpus: design and uses. J. Chinese Linguistics **25**, 309–330 (2015)
18. Bates, D., Mächler, M., Bolker, B., Walker, S.: Fitting linear mixed-effects models using lme4. J. Stat. Softw. **67**(1), 1–48 (2015)
19. R Core Team: R: A language and environment for statistical computing. R Foundation for Statistical Computing, Vienna, Austria. https://www.R-project.org (2016)
20. Hothorn, T., Bretz, F., Westfall, P.: Simultaneous inference in general parametric models. Biom. J. **50**(3), 346–363 (2008)

Occasion-Level Classifiers vs. Event-Level Classifiers in Child Mandarin -The First Experiment

Xiaohe Yuan[1(✉)], Xiaoqian Zhang[2], Yingyi Luo[2], and Jianhua Hu[2]

[1] College of Foreign Languages, Capital Normal University, Beijing, China
[2] Institute of Linguistics, Chinese Academy of Social Sciences, Beijing, China

Abstract. *Time*-adverbials in English are ambiguous between an occasion-counting reading and an event-counting reading. On the contrary, Mandarin has a rich system of event classifiers that can lexically disambiguate the two readings. The classifier *hui* counts the number of occasions on which the event of *ringing the doorbell* occurs and we call it occasion-level classifier. The classifier *xia* counts the number of the event of *ringing the doorbell* that takes place in one single occasion and thus is called event-level classifier. We adopted the Truth Value Judgment task and our experiment examined whether Mandarin-speaking children had mastered the basic meaning of verbal classifier *hui* and also whether they could distinguish between the two types of verbal classifiers. The result revealed that Mandarin-speaking 4–5-year-old children behaved in the same way as adults and could successfully distinguish between the two types of event classifiers.

Keywords: Acquisition · Occasion-level verbal classifiers · Event-level verbal classifiers

1 Event Classifiers

Event classifier is a semantic category similar to Time-adverbials in many languages. For instance:

(1) a. John went to the movies three times. [English] (Doetjes 2008)
b. Jean est allé trois fois au cinéma. [French] (Doetjes 2008).
c. Jan is drie keer naar de bioscoop gegaan. [Dutch] (Doetjes 2008).

Mandarin has a rich system of event classifiers (Donazzan 2012; Deng 2013; Donazzan and Tovena 2013; Zhang 2017a and 2017b) and can be categorized according to different criteria. In Mandarin, event classifiers or *dongliangci* are used to quantify the verbs and events. It is widely acknowledged that there are four types of event classifiers or verbal classifiers (Huang and Liao 2009: 16), including specialized verbal classifiers, verbal classifiers of nominal borrowings, verbal classifiers of verbal borrowings, and duplication of verbs and verbal classifiers of temporal measure words.

M. Liu et al. (Eds.): CLSW 2020, LNAI 12278, pp. 890–898, 2021.
https://doi.org/10.1007/978-3-030-81197-6_74

1.1 The Hierarchy of Event Units

The distinction between occasion and event comes from Cusic (1981). He proposes a multi-layered division of the event ontology. Specifically, there are three types of event individuals, i.e. phase, event and occasion. The unit in the superior level is composed of a single unit or several units in the lower level. The linguistic evidence can be found in many languages in terms of the three-level contrast.

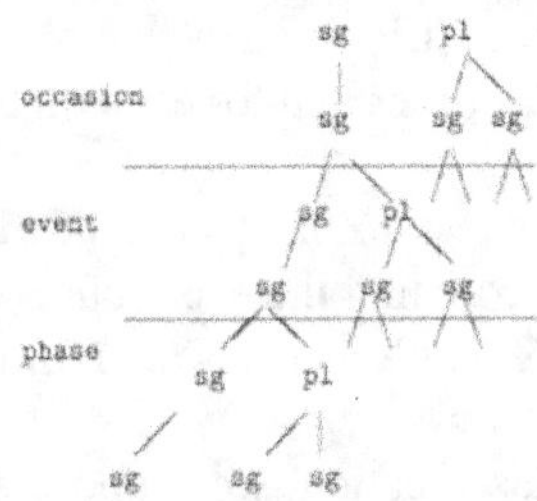

Fig. 1. Cusic's hierarchy of event plurality

Time-adverbials in English are ambiguous between an occasion-counting reading and an event-counting reading, e.g. (2). *Twice* can specify either the number of the occasions on which the event of *ringing the doorbell* occurs or the number of the events of *ringing the doorbell.* The same goes to French *fois*, e.g. (3a), whereby *deux fois* 'twice-occasion' in (3a) is also ambiguous between the event-level interpretation and the occasion-level reading. In addition to *fois*, French has another event modifier *reprise* that only triggers an occasion-counting reading, cf. (3b) in which the ringing-bell-event must take place on two different times.

(2) The salesman rang the doorbell twice. (Cusic 1981)

a. On two separate occasions he rang the bell twice. (occasion-counting reading)
b. On one occasion he rang the bell two times. (event-counting reading).

(3) a. Daniel a sonné deux fois. (Tovena 2012).

'Daniel rang the bell two times.'

b. Daniel a sonné à deux reprises. (Tovena 2012).

'Daniel rang the bell (on) two (different) times.'

1.2 Occasion-Level and Event-Level Classifiers in Mandarin

Based on different types of units the classifiers are counting in the verbal domain, Zhang (2017a) classifies Chinese verbal classifiers into event-level classifiers and occasion-level classifiers.

As in (4), *sanci* 'three CLF[1]$_{ci}$' is a an event-external classifier while *san quan* 'three CLF$_{quan}$' counting the actions internal the event belongs to an event-internal classifier. *san ci* 'three CLF$_{ci}$' specifies the number of the events on which occurs the beating-Yuru-event.

(4) Dalin da-le Yuru san ci.
Dalin beat-PFV Yuru three CLF
Mei-ci dou da-le san quan.
everytime all beat-PFV three CLF
'Dalin has beaten Yuru three times and for each time he punched him three times.'

However, there are problems[2] with Zhang's classification, and we propose that it is more proper to assume *quan* is counting the event but not the phase internal the event. At the same time, in contrast to Zhang's analysis of *ci* 'time' as counting events, we treat *ci* as the occasion counter.

Based on Cusic (1981)'s classification, we propose that Chinese verbal classifiers can be divided into occasion-level classifiers and event-level classifiers. For instance, the classifier *hui* 'CLF$_{time-o}$', which counts the number of occasions on which the event, e.g. *ringing the doorbell* occurs is termed as the occasion-level classifier. We also group the traditional generalized verbal classifiers into occasion-level classifiers represented by *hui* 'CLF$_{time-o}$', *ci* 'CLF$_{ci}$', *bian* 'CLF$_{bian}$', *tang* 'CLF$_{tang}$', *dun* 'CLF$_{dun}$', etc. On the other hand, the classifier *xia* 'CLF$_{time-e}$' counts the number of the event that takes place within one single occasion and we call it event-level classifier. Event-level classifiers consist of *xia* 'CLF$_{time-e}$' as a traditional generalized verbal classifier and some verbal classifiers of nominal borrowings such as *qiang* 'CLF$_{gun}$', *dao* 'CLF$_{knife}$', *shui* 'CLF$_{water}$', *jiao* 'CLF$_{foot}$', *quan* 'CLF$_{fist}$', etc. A crucial difference between the two types of classifiers is that for occasion-classifiers, a clear boundary or interval between the two events should be involved. Accomplishment and achievement predicates are those predicates which denote eventuality consisting of natural final endpoint, and the occasion individuals could be temporally continous. And we propose that the natural endpoint denoted by the predicate plays the role of identifying the occasion as an individual occasion. In our present experiment, we only investigated the activity predicates that have the semelfactive use and the occasion individuals selected by the occasion-level classifier which collocates with activity predicates that refer to atelic or temporally unbounded eventualities.

Previous studies focus more on the acquisition of nominal classifiers (Chien et al. 2003; Li et al. 2008; Lee and Huang 2007; Huang 2009). To our knowledge, there is relatively no or less research on child language acquisition of Chinese verbal classifiers. There is little data pertaining to Mandarin children's acquisition and understanding of verbal classifiers. So our study aims to test whether Mandarin children had mastered the basic meaning of a very typical Chinese verbal classifier *hui* 'CLF$_{time-o}$' as an occasion-level classifier and whether they could distinguish occasion-level classifiers from event-level classifiers.

[1] Abbreviations used in this paper: CLF$_{time-o}$ = occasion-level classifier; CLF$_{time-e}$ = event-level classifier; CLF = classifier; PFV = perfective.

[2] Yuan et al. (2019) states there are all three problems existing in Zhang's (2017) proposal.

2 Experiment

Two targets are involved in the experiment, i.e. target 1 and target 2. Target 1 includes three phrases containing the occasion-level classifier hui '$CLF_{time\text{-}o}$' which are used to form true statements and to contrast with event-level classifier xia '$CLF_{time\text{-}e}$'. Target 2 includes three phrases containing the occasion-level classifier hui '$CLF_{time\text{-}o}$' which are used to form false statements[3] and to contrast with event-level classifiers like kou 'CLF_{mouth}', jiao 'CLF_{foot}' and bazhang 'CLF_{palm}'. Apart from these, 9 controls and 8 fillers are also used in the experiment. (See the appendix.)

2.1 Methods

Participants, Procedure and Materials. We altogether had 20 adult participants from Ocean University of China and 22 children (4;06–5;06; M = 4;11) from Beijing Changqing Kindergarten.

We adopted the Truth Value and Judgment Task (TVJT) (Crain &Thornton, 1998). Each trial was presented as a story with 3 pictures followed by a testing sentence uttered by a puppet Peppa Pig in the end.

In both targets 1 and targets 2, the event-level classifiers were involved in the story, while the testing sentence contained the occasion-level classifier *hui* '$CLF_{time\text{-}o}$'. Targets 1 were testing whether children could distinguish occasion-level classifier *hui* '$CLF_{time\text{-}o}$' from event-level classifier *xia* '$CLF_{time\text{-}e}$' while targets 2 were concerned with the contrast between occasion-level classifier *hui* '$CLF_{time\text{-}o}$' and three Chinese verbal classifiers of nominal borrowings *kou* 'CLF_{mouth}', *jiao* 'CLF_{foot}' and *bazhang* 'CLF_{palm}'. To make sure participants would not be affected by the numeral used in the story, we used a different numeral in the testing sentence. To illustrate, while the numeral 3 and 1 were used in the story, e.g. 'san jiao' (3 CLF_{foot}) and 'ji jiao' (1 CL_{foot}), the numeral 4 was used in the testing sentence, e.g. 'si hui' (4 $CLF_{time\text{-}o}$).

When it comes to false targets, *ti san jiao* 'kick three CLF_{foot}' was involved in the first occasion and *ti yi jiao* 'kick one CLF_{foot}' was in the second occasion, but the final testing numeral phrase was *ti si hui* 'kick four $CLF_{time\text{-}o}$', which was in fact *ti liang hui* 'kick two $CLF_{time\text{-}o}$'. So if participants copy the number of each of the two occasions, they may tend to think it was *ti san hui* 'kick three $CLF_{time\text{-}o}$' or *ti yi hui* 'kick one $CLF_{time\text{-}o}$'. This is the reason why the numeral appearing in the testing sentence had to be different. See the following context as an example for the false target.

[3] In accordance with the Truth and Value Judgment Task (1998), all the targets are usually made false. We could make all the targets false here, but then we could not make further comparisons in terms of children's responses towards true targets and false ones. The true targets should increase the acceptance rate than the false targets if children failed in understanding the difference between occasion-level classifiers and event-level classifiers. We made three targets true and three targets false here to see whether there was a difference in reactions towards event-level classifiers *xia* '$CLF_{time\text{-}e}$' and those of nominal borrowings. Moreover, if there was no significant difference, we may infer that they could make correct judgments based on different situations accordingly regarding both of the two targets as event-level classifiers and have totally mastered the basic meaning of occasions and occasion-level classifiers.

Context: The little rabbit came across a monster in the woods in the morning and she kicked him three times (*san jiao* 'three CLF$_{foot}$') (here is a prerecorded voice). Then the little rabbit came back home, but she still wanted to have a look at the monster in the woods. In the late evening, the little rabbit went to see the monster again and she kicked him once (*yi jiao* 'one CLF$_{foot}$') (here is a prerecorded voice), and she quickly left.

Peppa Pig: The little rabbit kicked the monster four times (*si hui* 'four CLF$_{time\text{-}o}$').

Fig. 2. A typical trial in experiment 2-false target 2

The three true targets went with the same format as the three false targets. The only difference is that the testing sentences in the three true targets are true. Here is an example:

Context: The big wolf went to knock at the rabbit's door. In the early morning, he knocked four times (*si xia* 'four CLF$_{time\text{-}e}$') (here is a prerecorded voice), but the door was not open. Then the big wolf went back home. In the late evening, the big wolf went to knock at the door again once (*yi xia* 'one CLF$_{time\text{-}e}$') (here is a prerecorded voice), and the door was still not open.

Peppa Pig: The big wolf knocked at the rabbit's door twice (*liang hui* 'two CLFtime-o').

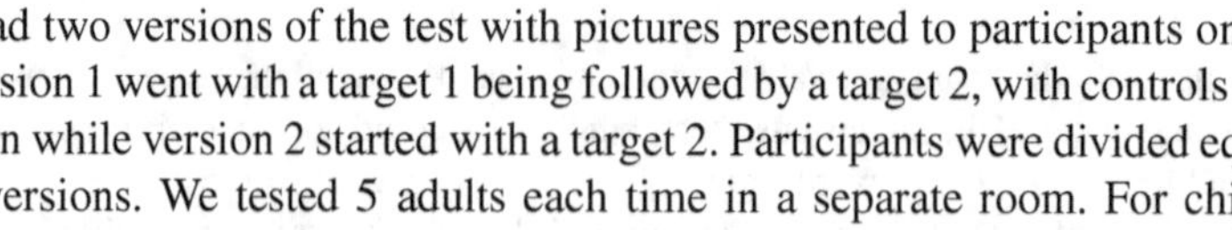

Fig. 3. A typical trial in experiment 2-true target 1

We had two versions of the test with pictures presented to participants on the computer. Version 1 went with a target 1 being followed by a target 2, with controls and fillers in between while version 2 started with a target 2. Participants were divided equally into the two versions. We tested 5 adults each time in a separate room. For children, we invited each of them into a separate room and the testing procedure lasted for 10 to 15 min.

In addition to 3 targets 1 and 3 targets 2, we had 9 controls with 3 on occasion-level classifiers *hui* 'CLF$_{time\text{-}o}$', 3 on event-level classifiers *xia* 'CLF$_{time\text{-}e}$', and 3 on event-level classifiers in terms of borrowing including *dao* 'CLF$_{knife\text{-}e}$', *bang* 'CLF$_{stick\text{-}e}$', and *qianzi* 'CLF$_{pincer\text{-}e}$'. All the three occasion-level classifiers *hui* 'CLF$_{time\text{-}o}$' were made false as the first control group with the format '1 + 1' (one time in the each of the two occasions) while the testing numeral phrase was *san hui* 'three $CLF_{time\text{-}o}$'. The second control group of three event-level classifiers *xia* '$CLF_{time\text{-}e}$' were all made true with the format 4 + 1, 4 + 1, 4 + 1 while the third control group of three event-level classifiers of nominal borrowings were all made false with the format 3 + 1, 3 + 1, 3 + 1. In addition, we had 8 fillers, half true and half false, to balance the total true and false testing statements.

2.2 Results

Immediately after the test, we carried out a post-test in order to make sure that participants had no problems in arithmetic, which involved the arithmetic addition of numbers within 5. Three children failed in the post-test and another two failed in 7/9 controls. So, we finally took the data of 20 adults and 17 children into the analysis. Children made correct choices of 89%, 70% and 69% on the above 9 controls, 3 targets 1 and 3 targets 2 respectively in contrast to 100%, 90% and 79% correct responses made by adults (See Fig. 4). There was no significant difference between groups in target 1 ($z = -1.398$, $p = 0.257$) and target 2 ($z = -0.869$, $p = 0.424$) as per Mann-whitney U test, confirming that children and adult participants behaved in the same way in distinguishing occasion-level classifier *hui* '$CLF_{time\text{-}o}$' from either of the two groups of event-level classifiers. Wilcoxon rank-signed test further showed that there was no significant difference between target 1 and target 2 in terms of adults' responses ($z = -1.941b$, $p = 0.052$) and also no significant difference between two groups of event-level classifiers in terms of children's responses ($z = -.284b$, $p = 0.776$).

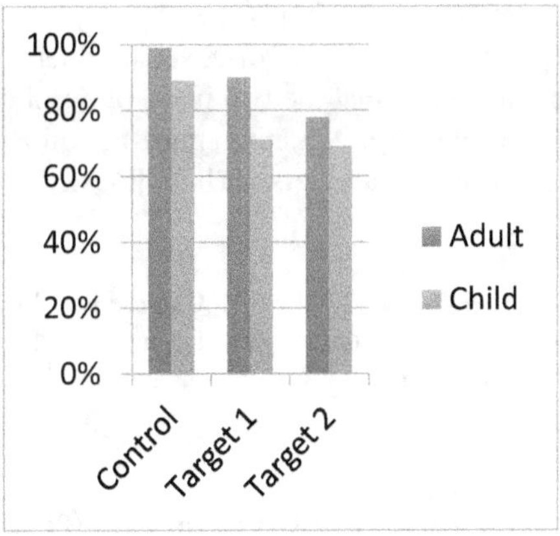

Fig. 4. Experiment result

Out of 20 children, three children failed in passing 7/9 controls. Out of the other 17 children, 3 children did not show an identical performance in the two groups of event-level classifiers; two accepted all and one rejected all[4]. Probably they did not seem to

[4] If participants accepted or rejected 2 out of the 3 in each level, we consider that they accepted or rejected this group of classifiers.

understand the task; 3 other children all rejected targets 1 but accepted targets 2, which was totally contrary to the correct responses. The other 11 children, who made the correct choices, accepted targets 1 and rejected targets 2 at the same time. So, among the children who fully understood the task, there were altogether 11 out of 14 children (79%) who successfully distinguished event-level classifiers from occasion-level classifiers.

3 Discussion and Conclusion

Based on the theoretical division between occasion-level classifiers and event-level classifiers, this study investigated whether Mandarin-speaking-4–5-year-old children had successfully understood the basic meaning of occasion-level classifier *hui* 'CLF$_{time\text{-}o}$' and if they were capable of differentiating between occasion-level classifiers and event-level classifiers.

Our experiments proved that younger participants, that is, 4–5-year-olds, behaved in the same way as adults as they successfully mastered such contrast. They made their judgments based on the sound of knocking the door several times continuously so that such an event or events should be regarded as one occasion. In this way, events and occasions were distinguished and the use of the traditional generalized verbal classifier xia 'CLF$_{time\text{-}e}$' was also differentiated from the use of the occasion-level classifier hui 'CLF$_{time\text{-}o}$'. In addition, Mandarin-speaking children were also sensitive to distinguish between traditional Chinese borrowing verbal classifiers like *jiao* 'CLF$_{foot}$', *quan* 'CLF$_{fist}$', *dao* 'CLF$_{knife}$' and occasion-level classifier *hui* 'CLF$_{time\text{-}o}$'. In both cases, they regarded the former as event-level verbal classifiers. Furthermore, there was no significant difference between the event-level classifier xia 'CLF$_{time\text{-}e}$' and the nominal borrowing classifers in terms of the children's reactions.

In our following experiment, we aim to discover whether Mandarin pre-schoolers (4–5-year-olds) had the knowledge that the two types of event classifiers have distinct aspectual constraints, i.e., event-level classifiers must be followed with homogeneous predicates while the occasion-level classifiers can be collocated with either homogeneous predicates or telic ones.

Acknowledgments. This study is funded by the 65th Chinese Postdoctoral Science Foundation Project (Project No: 2019M650955) and the Project of Dengfeng Plan of Chinese Academy of Social Sciences-Psycholinguistics, Language Acquisition and Development.

Appendix Testing Sentences in Experiment - Version 1

1	target1–1	*qiao lianghui* 'knock two CLF$_{time\text{-}o}$' (T)
2	filler 1	*yaoqing pengyou* 'invite friends' (F)
3	control1–1	*pa sanhui* 'climb three CLF$_{time\text{-}o}$' (F)
4	target 2–1	*ti sihui* 'kick four CLF$_{time\text{-}o}$' (F)

(*continued*)

(*continued*)

5	filler 2	*mai guozhi* 'buy orange juice' (T)
6	control 2–1	*pai si xia* 'pat $CLF_{time\text{-}e}$' (T)
7	filler 3	*he guozhi* 'drink orange juice' (F)
8	control 3–1	*qie liangdao* 'cut two CLF_{knife}' (F)
9	target 1–2	*zhuang lianghui* 'bump two $CLF_{time\text{-}o}$' (T)
10	control 1–2	*shua sanhui* 'brush three $CLF_{time\text{-}o}$' (F)
11	filler 4	*zhuangshi shengdanshu* 'decorate the Christmas tree' (T)
12	target 2–2	*yao sihui* 'bite four $CLF_{time\text{-}o}$' (F)
13	control 2–2	*ke sixia* 'knock against sth. four $CLF_{time\text{-}e}$' (T)
14	filler 5	*zuo binggan* 'make biscuits' (F)
15	control 3–2	*za liangbang* 'smash two CLF_{stick}' (F)
16	target 1–3	*shuai lianghui* 'throw two $CLF_{time\text{-}o}$' (T)
17	filler 6	*zhunbei shuiguo* 'prepare the fruits' (T)
18	control 1–3	*diao sanhui* 'fish three $CLF_{time\text{-}o}$' (F)
19	filler 7	*chi rou* 'eat meat' (F)
20	target 2–3	*da sihui* 'beat four $CLF_{time\text{-}o}$' (F)
21	control 2–3	*nao sixia* 'scratch four $CLF_{time\text{-}e}$' (T)
22	filler 8	*mai xihongshi* 'buy tomatoes' (F)
23	control 3–3	*jia liangqianzi* 'pinch two CLF_{pincer}' (F)

References

Chien, Y.C., Lust, B., Chiang, C.P.: Chinese children's comprehension of count-classifiers and mass-classifiers. J. East Asian Linguis. **12**, 91–120 (2003)

Crain, S., Thornton, R.: Investigations in Universal Grammar: A Guide to Experiments on the Acquisition of Syntax and Semantics. The MIT Press, Cambridge (1998)

Cusic, D.D.: Verbal plurality and aspect. Ph.D. dissertation, Stanford University (1981)

Deng, D.: The syntax and semantics of event quantifiers in Mandarin Chinese. Ph.D. dissertation, University of Wisconsin-Madison (2013)

Doetjes, J.: Counting and degree modification. Recherches linguistiques de Vincennes **37**, 139–161 (2008)

Donazzan, M.: On counting and measuring events. In: Chemla, E., Homer, V., Winterstein, G. (eds.) Proceedings of Sinn und Bedeutung, vol. 17, pp. 219–236 (2012)

Donazzan, M., Tovena, L.: Classifiers and events. Présentation à la 8th Conference of the European Association of Chinese Linguistics, Paris (2013)

Huang, A.: Count-mass Distinction and the Acquisition of Classifiers in Mandarin-speaking Children. M.A. thesis. The Chinese University of Hong Kong (2009)

Huang, B., Liao, X.: Modern Chinese Language (Volume II). Higher Education Press, Beijing (2009)

Lee, T. H.-T., Huang, A.: Classifiers and the count-mass distinction in Chinese revisited. LSHK Annual Research Forum, Hong Kong Baptist University, December 8–9 (2007)

Li, P., Barner, D., Huang, B.: Classifiers as count syntax: individuation and measurement in the acquisition of Mandarin Chinese. Lang. Learn. Dev. **4**(4), 1–42 (2008)

Tovena, L.: Elements for a linguistic ontology in the verbal domain. In Guarino, N. (ed.): Formal Ontology in Information Systems, vol. 239, pp. 235–248. Amsterdam: IOS Press (2012)

Yuan, X., Zhang, X., Luo, Y., Hu, J.: Acquisition of verbal classifiers. Frontiers in Contemporary Linguistics-Towards New Descriptivism (2019)

Zhang, N.N.: The syntax of event-internal and event-external verbal classifiers. Studia Linguistica, pp. 1–35 (2017a)

Zhang, N. N.: Verbal classifiers. In: Sybesma, R., Behr, W., Gu, Y., Handel, Z., Huang, C.-T.J., Myers, J., (eds.) Encyclopedia of Chinese language and Linguistics, Volume 1. Netherland: Brill, pp. 627–631 (2017b)

Author Index